APPLIED DIFFERENTIAL EQUATIONS

with Boundary Value Problems

TEXTBOOKS in MATHEMATICS

Series Editors: Al Boggess and Ken Rosen

PUBLISHED TITLES

ABSTRACT ALGEBRA: A GENTLE INTRODUCTION
Gary L. Mullen and James A. Sellers

ABSTRACT ALGEBRA: AN INTERACTIVE APPROACH, SECOND EDITION
William Paulsen

ABSTRACT ALGEBRA: AN INQUIRY-BASED APPROACH
Jonathan K. Hodge, Steven Schlicker, and Ted Sundstrom

ADVANCED LINEAR ALGEBRA
Hugo Woerdeman

ADVANCED LINEAR ALGEBRA
Nicholas Loehr

ADVANCED LINEAR ALGEBRA, SECOND EDITION
Bruce Cooperstein

APPLIED ABSTRACT ALGEBRA WITH MAPLE™ AND MATLAB®, THIRD EDITION
Richard Klima, Neil Sigmon, and Ernest Stitzinger

APPLIED DIFFERENTIAL EQUATIONS: THE PRIMARY COURSE
Vladimir Dobrushkin

A BRIDGE TO HIGHER MATHEMATICS
Valentin Deaconu and Donald C. Pfaff

COMPUTATIONAL MATHEMATICS: MODELS, METHODS, AND ANALYSIS WITH MATLAB® AND MPI, SECOND EDITION
Robert E. White

A COURSE IN DIFFERENTIAL EQUATIONS WITH BOUNDARY VALUE PROBLEMS, SECOND EDITION
Stephen A. Wirkus, Randall J. Swift, and Ryan Szypowski

A COURSE IN ORDINARY DIFFERENTIAL EQUATIONS, SECOND EDITION
Stephen A. Wirkus and Randall J. Swift

DIFFERENTIAL EQUATIONS: THEORY, TECHNIQUE, AND PRACTICE, SECOND EDITION
Steven G. Krantz

PUBLISHED TITLES CONTINUED

DIFFERENTIAL EQUATIONS: THEORY, TECHNIQUE, AND PRACTICE WITH BOUNDARY VALUE PROBLEMS
Steven G. Krantz

DIFFERENTIAL EQUATIONS WITH APPLICATIONS AND HISTORICAL NOTES, THIRD EDITION
George F. Simmons

DIFFERENTIAL EQUATIONS WITH MATLAB®: EXPLORATION, APPLICATIONS, AND THEORY
Mark A. McKibben and Micah D. Webster

DISCOVERING GROUP THEORY: A TRANSITION TO ADVANCED MATHEMATICS
Tony Barnard and Hugh Neill

DISCRETE MATHEMATICS, SECOND EDITION
Kevin Ferland

ELEMENTARY NUMBER THEORY
James S. Kraft and Lawrence C. Washington

EXPLORING CALCULUS: LABS AND PROJECTS WITH MATHEMATICA®
Crista Arangala and Karen A. Yokley

EXPLORING GEOMETRY, SECOND EDITION
Michael Hvidsten

EXPLORING LINEAR ALGEBRA: LABS AND PROJECTS WITH MATHEMATICA®
Crista Arangala

EXPLORING THE INFINITE: AN INTRODUCTION TO PROOF AND ANALYSIS
Jennifer Brooks

GRAPHS & DIGRAPHS, SIXTH EDITION
Gary Chartrand, Linda Lesniak, and Ping Zhang

INTRODUCTION TO ABSTRACT ALGEBRA, SECOND EDITION
Jonathan D. H. Smith

INTRODUCTION TO ANALYSIS
Corey M. Dunn

INTRODUCTION TO MATHEMATICAL PROOFS: A TRANSITION TO ADVANCED MATHEMATICS, SECOND EDITION
Charles E. Roberts, Jr.

INTRODUCTION TO NUMBER THEORY, SECOND EDITION
Marty Erickson, Anthony Vazzana, and David Garth

INVITATION TO LINEAR ALGEBRA
David C. Mello

PUBLISHED TITLES CONTINUED

LINEAR ALGEBRA, GEOMETRY AND TRANSFORMATION
Bruce Solomon

MATHEMATICAL MODELLING WITH CASE STUDIES: USING MAPLE™ AND MATLAB®, THIRD EDITION
B. Barnes and G. R. Fulford

MATHEMATICS IN GAMES, SPORTS, AND GAMBLING–THE GAMES PEOPLE PLAY, SECOND EDITION
Ronald J. Gould

THE MATHEMATICS OF GAMES: AN INTRODUCTION TO PROBABILITY
David G. Taylor

A MATLAB® COMPANION TO COMPLEX VARIABLES
A. David Wunsch

MEASURE AND INTEGRAL: AN INTRODUCTION TO REAL ANALYSIS, SECOND EDITION
Richard L. Wheeden

MEASURE THEORY AND FINE PROPERTIES OF FUNCTIONS, REVISED EDITION
Lawrence C. Evans and Ronald F. Gariepy

NUMERICAL ANALYSIS FOR ENGINEERS: METHODS AND APPLICATIONS, SECOND EDITION
Bilal Ayyub and Richard H. McCuen

ORDINARY DIFFERENTIAL EQUATIONS: AN INTRODUCTION TO THE FUNDAMENTALS
Kenneth B. Howell

PRINCIPLES OF FOURIER ANALYSIS, SECOND EDITION
Kenneth B. Howell

REAL ANALYSIS AND FOUNDATIONS, FOURTH EDITION
Steven G. Krantz

RISK ANALYSIS IN ENGINEERING AND ECONOMICS, SECOND EDITION
Bilal M. Ayyub

SPORTS MATH: AN INTRODUCTORY COURSE IN THE MATHEMATICS OF SPORTS SCIENCE AND SPORTS ANALYTICS
Roland B. Minton

TRANSFORMATIONAL PLANE GEOMETRY
Ronald N. Umble and Zhigang Han

TEXTBOOKS in MATHEMATICS

APPLIED DIFFERENTIAL EQUATIONS

with Boundary Value Problems

Vladimir A. Dobrushkin

CRC Press is an imprint of the
Taylor & Francis Group, an informa business
A CHAPMAN & HALL BOOK

Chapman & Hall/CRC Press
Taylor & Francis Group
6000 Broken Sound Parkway NW, Suite 300
Boca Raton, FL 33487-2742

First issued in paperback 2022

Version Date: 20170901

ISBN 13: 978-1-4987-3365-6 (hbk)
ISBN 13: 978-1-03-247657-5 (pbk)
ISBN 13: 978-1-315-36978-5 (ebk)

DOI: 10.1201/9781315369785

Publisher's Note
The publisher has gone to great lengths to ensure the quality of this reprint but points out that some imperfections in the original copies may be apparent.

Library of Congress Cataloging-in-Publication Data

Names: Dobrushkin, V. A. (Vladimir Andreevich)
Title: Applied differential equations with boundary value problems / Vladimir Dobrushkin.
Other titles: Differential equations with boundary value problems
Description: Boca Raton : CRC Press, 2017. | Includes bibliographical references and index.
Identifiers: LCCN 2017015454 | ISBN 9781498733656
Subjects: LCSH: Differential equations--Textbooks. | Boundary value problems--Textbooks. | Boundary value problems--Numerical solutions.
Classification: LCC QA372 .D6325 2017 | DDC 515/.35--dc23
LC record available at https://lccn.loc.gov/2017015454

Visit the Taylor & Francis Web site at
http://www.taylorandfrancis.com

and the CRC Press Web site at
http://www.crcpress.com

Contents

List of Symbols

$\lvert a, b\rvert$	any interval (closed, open, semi-open) with end points a and b.
$n!$	factorial, $n! = 1 \cdot 2 \cdot 3 \cdot \ldots \cdot n$.
$\ln x$	$= \log_e x$, natural logarithm, that is, the logarithm with base e.
$n^{\underline{k}}$	$= n(n-1)\ldots,(n-k+1)$ (falling factorial).
$\binom{n}{k}$	$= \dfrac{n^{\underline{k}}}{k!} = \dfrac{n!}{k!\,(n-x)!}$ binomial coefficient.
$\mathtt{D}$	$= \mathrm{d}/\mathrm{d}x$ or $\mathrm{d}/\mathrm{d}t$, the derivative operator.
$\mathtt{D}^n(fg)$	$= \sum_{r=0}^{n} \binom{n}{r} (\mathtt{D}^{n-r} f)\,(\mathtt{D}^r g)$, Leibniz formula.
$\dot{y}$	$= \mathrm{d}y/\mathrm{d}t$, derivative with respect to time variable t.
$\mathrm{H}(t)$	the Heaviside function, Definition 5.3, page 274.
$\mathrm{Si}(x)$	sine integral: $\int_0^x \frac{\sin t}{t}\,\mathrm{d}t$.
$\mathrm{Ci}(x)$	cosine integral: $-\int_x^\infty \frac{\cos t}{t}\,\mathrm{d}t$.
$\mathrm{sinc}(x)$	$= \dfrac{\sin(x\pi)}{x\pi}$, normalized cardinal sine function.
$\frac{1}{2}\ln\left\lvert\frac{1+x}{1-x}\right\rvert$	$= \begin{cases} \mathrm{arctanh}(x) & \text{for } \lvert x\rvert < 1, \\ \mathrm{arccoth}(x) & \text{for } \lvert x\rvert > 1. \end{cases}$
$\int \frac{v'(x)}{v(x)}\,\mathrm{d}x$	$= \ln\lvert v(x)\rvert + C = \ln Cv(x), \quad v(x) \neq 0.$
$\mathbf{I}$	the identity matrix, Definition 6.6, page 359.
$\mathrm{tr}\,(\mathbf{A})$	trace of a matrix $\mathbf{A}$, Definition 6.8, page 360.
$\det\,(\mathbf{A})$	determinant of a matrix $\mathbf{A}$, §7.2.
$\mathbf{A}^T$	transpose of a matrix $\mathbf{A}$ (also denoted as $\mathbf{A}'$), Definition 6.3, page 358.
$\mathbf{A}^*$	or $\mathbf{A}^H$, adjoint of a matrix $\mathbf{A}$, Definition 6.4, page 358.
ODE	ordinary differential equation.
PDE	partial differential equation.
CAS	computer algebra system.
$\mathbf{j}$	unit pure imaginary vector on the complex plane $\mathbb{C}$: $\mathbf{j}^2 = -1$.
$x + y\mathbf{j}$	complex number, where $x = \Re\,(x + y\mathbf{j})$, $y = \Im\,(x + y\mathbf{j})$.

Preface

Applied Differential Equations with Boundary Value Problems is a comprehensive exposition of ordinary differential equations and an introduction to partial differential equations (due to space constraint, there is only one chapter devoted directly to PDEs) including their applications in engineering and the sciences. This text is designed for a two-semester sophomore or junior level course in differential equations and assumes previous exposure to calculus. It covers traditional material, along with novel approaches in presentation and utilization of computer capabilities, with a focus on various applications. This text intends to provide a solid background in differential equations for students majoring in a breadth of fields.

This book started as a collection of lecture notes for an undergraduate course in differential equations taught by the Division of Applied Mathematics at Brown University, Providence, RI. To some extent, it is a result of collective insights given by almost every instructor who taught such a course over the last 15 years. Therefore, the material and its presentation covered in this book were practically tested for many years.

There is no need to demonstrate the importance of ordinary and partial differential equations (ODE and PDE, for short) in science, engineering, and education—this subject has been included in the curriculum of universities around the world for almost two hundred years. Their utilization in industry and engineering is so widespread that, without a doubt, differential equations have become the most successful mathematical tool in modeling. Perhaps the most germane point for the student reader is that many curricula recommend or require a course in ordinary differential equations for graduation. The beauty and utility of differential equations and their application in mathematics, biology, chemistry, computer science, economics, engineering, geology, neuroscience, physics, the life sciences, and other fields reaffirm their inclusion in myriad curricula.

In this text, differential equations are described in the context of applications. A more comprehensive treatment of their applications is given in [14]. It is important for students to grasp how to formulate a mathematical model, how to solve differential equations (analytically or numerically), how to analyze them qualitatively, and how to interpret the results. This sequence of steps is perhaps the hardest part for students to learn and appreciate, yet it is an essential skill to acquire. This book provides the common language of the subject and teaches the main techniques needed for modeling and systems analysis.

The **goals** in writing this textbook:

- To show that a course in differential equations is essential for modeling real-life phenomena. This textbook lays down a bridge between calculus, modeling, and advanced topics. It provides a basis for further serious study of differential equations and their applications. We stress the mastery of traditional solution techniques and present effective methods, including reliable numerical approximations.

- To provide qualitative analysis of ordinary differential equations. Hence, the reader should get an idea of how all solutions to the given problem behave, what are their validity intervals, whether there are oscillations, vertical or horizontal asymptotes, and what is their long term behavior. So the reader will learn various methods of solving, analysis, visualization, and approximation. This goal is hard to achieve without exploiting the capabilities of computers.

- To give an introduction to four of the most pervasive computer software[1] packages: *Maple*™, *Mathematica*®, MATLAB®, and *Maxima*—the first computer algebra system in the world. A few other such solvers are available: *Sage*, *R*, and SymPy, but we cannot afford to present them in the text and refer the reader to the accompanied

[1]The owner of *Maple* is Maplesoft (`http://www.maplesoft.com/`), a subsidiary of Cybernet Systems Co. Ltd. in Japan, which is the leading provider of high-performance software tools for engineering, science, and mathematics. *Mathematica* is the product of Wolfram Research company of Champaign, Illinois, USA founded by Stephen Wolfram in 1987; its URL is `http://www.wolfram.com`. MATLAB® is the product of the MathWorks, Inc., 3 Apple Hill Drive, Natick, MA, 01760-2098 USA, Tel: 508-647-7000, Fax: 508-647-7001, E-mail: info@mathworks.com, URL: `www.mathworks.com`.

web site. Some popular software packages have either similar syntax (such as Octave or GiNaC) or include engines of known solvers (such as MathCad and MuPad—integrated part of MATLAB). Others should be accessible with the recent development of cloud technology such as *Sage*. Also, simple numerical algorithms can be handled with a calculator or a spreadsheet program.

- To give the lecturer a flexible textbook within which he or she can easily organize a curriculum matched to their specific goals. This textbook presents a large number of examples from different subjects, which facilitate the development of the student's skills to model real-world problems. Staying within a traditional context, the book contains some advanced material on differential equations.

- To give students a thorough understanding of the subject of differential equations as a whole. This book provides detailed solutions of all the basic examples, and students can learn from it without any extra help. It may be considered as a self-study text for students as well. This book recalls the basic formulas and techniques from calculus, which makes it easy to understand all derivations. It also includes advanced material in each chapter for inquisitive students who seek a deeper knowledge of this subject.

Philosophy of the Text

We share our pedagogical approach with famous mathematician Paul Halmos [19, pp. 61–62], who recommended the study of mathematics by examples. He goes on to say:

> ...it's examples, examples, examples that, for me, all mathematics is based on, and I always look for them. I look for them first, when I begin to study. I keep looking for them, and I cherish them all.

Pedagogy and Structure of the Book

Ordinary and partial differential equations is a classical subject that has been studied for about 300 years. However, education has changed with omnipresent mathematical modeling technology available to all. This textbook stresses that differential equations constitute an essential part of modeling by showing their applications, including numerical algorithms and syntax of the four most popular software packages.

It is essential to introduce information technologies early in the class. Students should be encouraged to use numerical solvers in their work because they help to illustrate and illuminate concepts and insights. It should be noted that computers cannot be used blindly because they are as smart as the programmers allow them to be—every problem requires careful examination.

This textbook stays within traditional coverage of basic topics in differential equations. It contains practical techniques for solving differential equations, some of which are not widely used in undergraduate study. Not every statement or theorem is followed by rigorous verification. Proofs are included only when they enhance the reader's understanding and challenge the student's intellectual curiosity.

Our pedagogical approach is based on the following principle: follow the author. Every section has many examples with detailed exposition focused on how to choose an appropriate technique and then how to solve the problem. There are hundreds of problems solved in detail, so a reader can master the techniques used to solve and analyze differential equations.

Notation

This text uses numbers enclosed with brackets to indicate references in the bibliography, which is located at the end of the book, starting on page 669. The text uses only standard notations and abbreviations [*et al.* (et alii from Latin) means "and others," or "and co-workers;" *i.e.* from Latin "id est" meaning that is, that is to say, or in other words; *e.g.* stands for the Latin phrase "exempli gratia," which means for example; and *etc.* means "and the others," "and other things," "and the rest"]. However, we find it convenient to type ■ at the end of proofs or at the end of a topic presented; we also use □ at the end of examples (unless a new one serves as a delimiter). We hope that the reader understands the difference between = (equal) and ≡ (equivalence relation). Also $\stackrel{\text{def}}{=}$ is used for short to signal that the expression follows by definition. There is no common notation for complex numbers. Since a complex number (let us denote it by z) is a vector on the plane, it is a custom to denote it by $z = x + y\mathbf{j}$ rather than $z = x\mathbf{i} + y\mathbf{j}$, where the unit vector $\mathbf{i}$ is dropped and $\mathbf{j}$ is the unit vector in the positive vertical direction. In mathematics, this vector $\mathbf{j}$ is denoted by i. For convenience, we present the list of symbols and abbreviations at the beginning of the text.

For students

This text has been written with the student in mind to make the book very friendly. There are a lot of illustrations accompanied by corresponding codes for appropriate solvers. Therefore, the reader can follow examples and learn how

to use these software packages to analyze and verify obtained results, but not to replace mastering of mathematical techniques. Analytical methods constitute a crucial part of modeling with differential equations, including numerical and graphical applications. Since the text is written from the viewpoint of the applied mathematician, its presentation may sometimes be quite theoretical, sometimes intensely practical, and often somewhere in between. In addition to the examples provided in the text, students can find additional resources, including problems and tutorials on using software, at the website that accompanies this book:
`http://www.cfm.brown.edu/people/dobrush/am33/computing33.html`

The focus of the book is upon applications and methods of solutions because most practical problems need mathematical and numerical approximations to gain insight into their behavior. For adequate preparation of a student for study in her or his respective fields, it is imperative to muster in computer applications, in particular, being familiar with numerical solvers and computer algebra systems (CAS for short). In engineering or other application oriented courses, CAS and numerical solvers become a part of education in mathematics because of following reasons:

- They are a part of solving tools.
- CASs allow investigation of algorithms; in particular, they can be helpful to analyze algorithms, their complexity, dependency on input data, and performance.
- They help to understand mathematics, illustrate concepts, and boosts the learning process.
- CAS and numerical solvers usually reveal the underlying mathematics; in particular, its open code could be a part of a mathematical proof.

For instructors

Universities usually offer two courses on differential equations of different levels; one is the basic first course required by curriculum, and the other covers the same material, but is more advanced and attracts students who find the basic course trivial. This text can be used for both courses, and curious students have an option to increase their understanding and obtain deeper knowledge in any topic of interest. A great number of examples and exercises make this text well suited for self-study or for traditional use by a lecturer in class. Therefore this textbook addresses the needs of two levels of audience, the beginning and the advanced.

Acknowledgments

This book would not have been written if students had not complained about the other texts unleashed on them. In addition, I have gained much from their comments and suggestions about various components of the book, and for this I would like to thank the students at Brown University.

The development of this text depended on the efforts of many people. I am very grateful to the reviewers who made many insightful suggestions that improved the text.

I am also thankful to Professors Raymond Beauregard, Constantine Dafermos, Philip Davis, Alexander Demenchuk, Yan Guo, Jeffrey Hoag, Gerasimos Ladas, Anatoly Levakov, Martin Maxey, Douglas Meade, Orlando Merino, Igor Najfeld, Lewis Pakula, Eduard Polityko, Alexander Rozenblyum, Bjorn Sandstede, and Chau-Hsing Su, who generously contributed their time to provide detailed and thoughtful reviews of the manuscript; their helpful suggestions led to numerous improvements. This book would not have been written without the encouragement of Professor Donald McClure, who felt that the division needed a textbook with practical examples, basic numerical scripts, and applications of differential equations to real-world problems.

Noah Donoghue, George Potter, Neil Singh, and Mark Weaver made great contributions by carefully reading the text and helping me with problems and graphs. Their suggestions improved the exposition of the material substantially.

Additional impetus and help has been provided by the professional staff of our publisher, Taylor & Francis Group, particularly Robert Ross, Karen Simon, Shashi Kumar, and Kevin Craig.

Finally, I thank my family for putting up with me while I was engaged in the writing of this book.

Vladimir Dobrushkin,
Providence, RI

Chapter 1

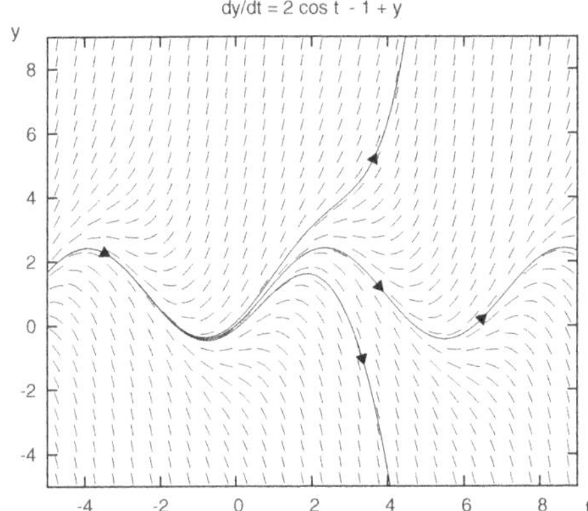

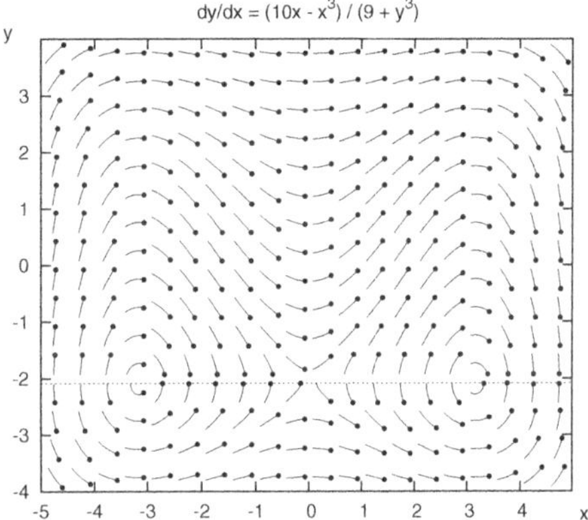

Introduction

The independent discovery of the calculus by I. Newton and G. Leibniz was immediately followed by its intensive application in mathematics, physics, and engineering. Since the late seventeenth century, differential equations have been of fundamental importance in the study, development, and application of mathematical analysis. Differential equations and their solutions play one of the central roles in the modeling of real-life phenomena.

In this chapter, we begin our study with the first order differential equations in normal form

$$\frac{\mathrm{d}y}{\mathrm{d}x} = f(x, y),$$

where $f(x, y)$ is a given single-valued function of two variables, called a slope or rate function. For an arbitrary function $f(x, y)$, there does not necessarily exist a function $y = \phi(x)$ that satisfies the differential equation. In fact, a differential equation usually has more than one solution. We classify first order differential equations and formulate several analytic methods that are applicable to each subclass.

One of the most intriguing things about differential equations is that for an arbitrary function f, there is no general method for finding an exact formula for the solution. For many differential equations that are encountered in real-world applications, it is impossible to express their solutions via known functions. Generally speaking, every differential equation defines its solution (if it exists) as a special function not necessarily expressible by elementary functions (such as polynomial, exponential, or trigonometric functions). Only exceptional differential equations can be explicitly or implicitly integrated. For instance, such "simple" differential equations as $y' = y^2 - x$ or $y' = e^{xy}$ cannot be solved by available methods.

1.1 Motivation

In applied mathematics, a **model** is a set of equations describing the relationships between numerical values of interest in a system. **Mathematical modeling** is the process of developing a model pertaining to physics or other sciences. Since differential equations are our main objects of interest, we consider only models that involve these equations. For example, Newton's second law, $\mathbf{F} = m\mathbf{a}$, relates the force $\mathbf{F}$ acting on a particle of mass m with the resulting acceleration $\mathbf{a} = \ddot{\mathbf{x}} \stackrel{\text{def}}{=} \mathrm{d}^2\mathbf{x}/\mathrm{d}t^2$. The transition from a physical problem to a corresponding mathematical model is not easy. It often happens that, for a particular problem, physical laws are hard or impossible to derive, though a relation between physical values can be obtained. Such a relation is usually used in the derivation of a mathematical model, which may be incomplete or somewhat inaccurate. Any such model may be subject to refining, making its predictions agree more closely with experimental results.

Many problems in the physical sciences, social sciences, biology, geology, economics, and engineering are posed mathematically in terms of an equation involving derivatives (or differentials) of the unknown function. Such an equation is called a **differential equation**, and their study was initiated by Leibniz[2] in 1676. It is customary to use his notation for derivatives: $\mathrm{d}y/\mathrm{d}x$, $\mathrm{d}^2y/\mathrm{d}x^2$, ..., or the prime notation: y', y'', ... For higher derivatives, we use the notation $y^{(n)}$ to denote the derivative of the order n. When a function depends on time, it is common to denote its first two derivatives with respect to time with dots: $\dot{y}$, $\ddot{y}$.

The next step in mathematical modeling is to determine the unknown, or unknowns, involved. Such a procedure is called solving the differential equation. The techniques used may yield solutions in analytic forms or approximations. Many software packages allow solutions to be visualized graphically. In this book, we focus on four popular packages: MATLAB®, *Maple*™, *Mathematica*®, and *Maxima*. Some attention will be given to (free) computer algebra systems *Sage* and *SymPy*. To motivate the reader, we begin with two well-known examples.

Example 1.1.1: (Carbon dating) The procedure for determining the age of archaeological remains was developed by the 1960 Nobel prize winner in chemistry, Willard Libby[3]. Cosmic radiation entering the Earth's atmosphere is constantly producing carbon-14 (${}_6\mathrm{C}^{14}$), an unstable radioactive isotope of ordinary carbon-12 (${}_6\mathrm{C}^{12}$). Both isotopes of carbon appear in carbon dioxide, which is incorporated into the tissues of all plants and animals, including human beings. In the atmosphere, as well as in all living organisms, the proportion of radioactive carbon-14 to ordinary (stable) carbon-12 is constant. When an organism dies, the absorption of carbon-14 by respiration and ingestion terminates. Experiments indicate that radioactive substances, such as uranium or carbon-14, decay by a certain percentage of their mass in a given unit of time. In other words, radioactive elements decay at a rate proportional to the mass present. Let $c(t)$ be the concentration of carbon-14 in dead organic material at time t, counted since the time of death. Then $c(t)$ obeys the following differential equation subject to the initial condition:

$$\frac{\mathrm{d}c(t)}{\mathrm{d}t} = -\lambda\, c(t), \qquad t > 0, \qquad c(0) = c_0, \tag{1.1.1}$$

where at the time of death $t = 0$, c_0 is the concentration of the isotope that a living organism maintains, and λ is the characteristic constant ($\lambda \approx 1.24 \times 10^{-4}$ per year for carbon-14). The technique to solve this type of differential equation will be explained later, in §2.1. We guess a solution: $c(t) = K\, e^{-\lambda t}$, with constant K, using the derivative property of the exponential function $\left(e^{kt}\right)' = k\, e^{kt}$. Since $c(0) = K$, it follows from the initial condition, $c(0) = c_0$, that

$$c(t) = c_0\, e^{-\lambda t}.$$

Suppose we know this formula to be true for $c(t)$. We determine the time of death of organic material from an examination of the concentration $c(t)$ of carbon-14 at the time t. The following relationship holds:

$$\frac{c(t)}{c_0} = e^{-\lambda t}.$$

Applying a logarithm to both sides, we obtain

$$-\lambda t = \ln[c(t)/c_0] = -\ln[c_0/c(t)],$$

from which we can find the time t of death of the organism to be

$$t = \frac{1}{\lambda} \ln\left(\frac{c_0}{c(t)}\right).$$

Recall that the half-life of a radioactive nucleus is defined as the time t_h during which the number of nuclei reduces to one-half of the original value. If the half-life of a radioactive element is known to be t_h, then the radioactive nuclei decay according to the law

$$N(t) = N(0)\, 2^{-t/t_h} = \frac{N(0)}{2^{t/t_h}}, \tag{1.1.2}$$

[2] Gottfried Wilhelm Leibniz (1646–1716) was a German scientist who first solved separable, homogeneous, and linear differential equations. He co-discovered calculus with Isaac Newton.

[3] American chemist Willard Libby (1908–1980).

where $N(t)$ is the amount of radioactive substance at time t and $t_h = (\ln 2)/\lambda$. Since the half-life of carbon-14 is approximately 5,730 years, present measurement techniques utilize this method for carbonaceous materials up to about 50,000 years old.

Example 1.1.2: (RC-series circuit) The most common applications of differential equations occur in the theory of electric circuits because of its importance and the pervasiveness of these equations in network theory. Figure 1.1 shows an electric circuit consisting of a resistor R and a capacitor C in series. A differential equation relating current $I(t)$ in the circuit, charge $q(t)$ on the capacitor, and voltage $V(t)$ measured at the points shown can be derived by applying Kirchhoff's voltage law, which states that the voltage $V(t)$ must equal the sum of the voltage drops across the resistor and the capacitor (see [14]).

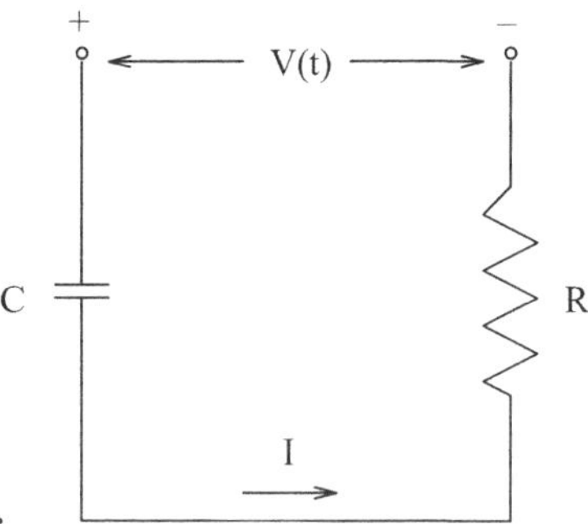

Figure 1.1: RC-Circuit.

It is known that the voltage changes across the passive elements are approximately as follows.

$$\Delta V_R = RI \quad \text{for the resistor,}$$
$$\Delta V_C = q/C \quad \text{for the capacitor.}$$

Furthermore, the current is defined to be the rate of flow of charge: $I(t) = \mathrm{d}q/\mathrm{d}t$. By combining these expressions using Kirchhoff's voltage law we obtain a differential equation relating $q(t)$ and $V(t)$:

$$R\frac{\mathrm{d}q}{\mathrm{d}t} + \frac{1}{C}q(t) = V(t).$$

1.2 Classification of Differential Equations

To study differential equations, we need some common terminology and basic classification of equations. If an equation involves the derivative of one variable with respect to another, then the former is called a *dependent variable* and the latter is an *independent variable.* For instance, in the equation from Example 1.1.2, the charge q is a dependent variable and the time t is an independent variable.

Ordinary and Partial Differential Equations. We start classification of differential equations with the number of independent variables: whether there is a single independent variable or several independent variables. The first case is an ODE (acronym for **O**rdinary **D**ifferential **E**quation), and the second is a PDE (**P**artial **D**ifferential **E**quation).

Systems of Differential Equations. Another classification is based on the number of unknown dependent variables to be found. If two or more unknown variables are to be determined, then a system of equations is required.

Example 1.2.1: We will derive a simple model of an arms race between two countries. Let $x_i(t)$ represent the size (or cost) of the arms stocks of country i $(i = 1, 2)$. Due to the cost of maintenance, we assume that an isolated country will diminish its arms stocks at a rate proportional to its size. We express this mathematically as $\dot{x}_i \overset{\text{def}}{=} \mathrm{d}x_i/\mathrm{d}t = -c_i x_i$, $c_i \geqslant 0$. The competition between countries, however, causes each one to increase its supply of arms at a rate proportional to the other country's arms supplies. The English meteorologist Lewis F. Richardson [43, 44] proposed a model to describe the evolution of both countries' arsenals as the solution of the following system of differential equations:

$$\dot{x}_1 = -c_1 x_1 + d_1 x_2 + g_1(x_1, x_2, t), \qquad c_1, d_1 \geqslant 0,$$
$$\dot{x}_2 = -c_2 x_2 + d_2 x_1 + g_2(x_1, x_2, t), \qquad c_2, d_2 \geqslant 0,$$

where the c's are called *cost* factors, the d's are *defense* factors, and the g's are *grievance* terms that account for other factors. □

The **order of a differential equation** is the order of the highest derivative that appears in the equation. More generally, an ordinary differential equation of the n-th order is an equation of the following form:

$$F\left(x, y(x), y'(x), \ldots, y^{(n)}(x)\right) = 0. \tag{1.2.1}$$

Here $y(x)$ is an unspecified function having n derivatives and depending on $x \in (a, b)$, $a < b$; $F(x, y, p_1, \ldots, p_n)$ is a given function of $n + 2$ variables. Some of the arguments, x, y, ..., $y^{(n-1)}$ (or even all of them) may not be present in Eq. (1.2.1). However, the n-th derivative, $y^{(n)}$, must be present in the ordinary differential equation, or else its order would be less than n. If this equation can be solved for $y^{(n)}(x)$, then we obtain the differential equation in the **normal** form:

$$y^{(n)}(x) = f\left(x, y, y', \ldots, y^{(n-1)}\right), \quad x \in (a, b). \tag{1.2.2}$$

A *first order differential equation* is of the form

$$F(x, y, y') = 0. \tag{1.2.3}$$

If we can solve it with respect to y', then we obtain its normal form:

$$\frac{\mathrm{d}y}{\mathrm{d}x} = f(x, y) \qquad \text{or} \qquad \mathrm{d}y = f(x, y)\,\mathrm{d}x, \tag{1.2.4}$$

where $\mathrm{d}x$ and $\mathrm{d}y$ are differentials in variables x and y, respectively.

Linear and Nonlinear Equations. The ordinary differential equation (1.2.1) is said to be *linear* if F is a linear function of the variables $y(x)$, $y'(x)$, ..., $y^{(n)}(x)$. Thus, the general linear ordinary differential equation of order n is

$$a_n(x)y^{(n)} + a_{n-1}(x)y^{(n-1)} + \cdots + a_0(x)y = g(x). \tag{1.2.5}$$

An equation (1.2.1) is said to be *nonlinear* if it is not of this form.

For example, the van der Pol equation, $\ddot{y} - \epsilon(1 - y^2)\,\dot{y} + \delta y = 0$, is a nonlinear equation because of the presence of the term y^2. On the other hand, $y'(x) + (\sin x)\,y(x) = x^2$ is a linear differential equation of the first order because it is of the form (1.2.5). In this case, $a_1 = 1$, $a_0(x) = \sin x$, and $g(x) = x^2$.

The general forms for the first and second order linear differential equations are:

$$a_1(x)\,y'(x) + a_0(x)\,y(x) = f(x) \quad \text{and} \quad a_2(x)\,y''(x) + a_1(x)\,y'(x) + a_0(x)\,y(x) = f(x).$$

1.3 Solutions to Differential Equations

Since the unknown quantity in a differential equation is a function, it should be defined in some domain. The differential equation (1.2.1) is usually considered on some open interval $(a, b) = \{x : a < x < b\}$, where its solution along with the function $F(x, y, p_1, \ldots, p_n)$ should be defined. However, it may happen that we look for a solution on a closed interval $[a, b]$ or a semi-open interval, $(a, b]$ or $[a, b)$. For instance, the bending of a plane's wing is modeled by an equation on a semi-closed interval $[0, \ell)$, where the point $x = 0$ corresponds to the connection of the wing to the body of the plane, and ℓ is the length of the wing. At $x = \ell$, the equation is not valid and its solution is not defined at this point. To embrace all possible cases, we introduce the notation $|a, b|$, which denotes an interval (a, b) possibly including the end points; hence, $|a, b|$ can denote the open interval (a, b), the closed interval $[a, b]$, or the semi-open intervals $(a, b]$ or $[a, b)$.

Definition 1.1: A **solution** or **integral** of the ordinary differential equation

$$F\left(x, y(x), y'(x), \ldots, y^{(n)}(x)\right) = 0$$

on an interval $|a, b|$ $(a < b)$ is a continuous function $y(x)$ such that y, y', y'', ..., $y^{(n)}$ exist and satisfy the equation for all values of the independent variable on the interval, $x \in |a, b|$. The graphs of the solutions of a differential equation are called their **integral curves** or **streamlines**.

This means that a solution $y(x)$ has derivatives up to the order n in the interval $|a, b|$, and for every $x \in |a, b|$, the point $(x, y(x), y'(x), \ldots, y^{(n)}(x))$ should be in the domain of F. We can show that a solution satisfies a given differential equation in various ways. The general method consists of calculating the expressions of the dependent variable and its derivatives, and substituting all of these in the given equation. The result of such a substitution should lead to the identity.

From calculus, it is known that a differential equation $y' = f(x)$ has infinitely many solutions for a smooth function $f(x)$ defined in some domain. These solutions are expressed either via an indefinite integral, $y = \int f(x)\,dx + C$, or a definite integral with a variable boundary, $y = \int_{x_0}^{x} f(x)\,dx - C$ or $y = -\int_{x}^{x_0} f(x)\,dx + C$, where x_0 is some fixed value. The *constant of integration*, C, is assumed arbitrary in the sense that it can be given any value within a certain range. However, C actually depends on the domain where the function $y(x)$ is considered and the form in which the integral is expressed. For instance, a simple differential equation $y' = (1+x^2)^{-1}$ has infinitely many solutions presented in three different forms:

$$\int \frac{dx}{1+x^2} = \arctan x + C = \arctan\left(\frac{1+x}{1-x}\right) + C_1 = \frac{1}{2}\arccos\left(\frac{1-x^2}{1+x^2}\right) + C_2,$$

where arbitrary constants C, C_1, and C_2 can be expressed in terms of each other, but their relations depend on the domain of x. For example, $C_1 = \frac{\pi}{4} + C$ when $x < 1$, but $C_1 = C - \frac{3\pi}{4}$ for $x > 1$. Also, the antiderivative of $1/x$ (for $x \ne 0$) will usually be written as $\ln Cx$ instead of $\ln |Cx|$ because it would be assumed that $C > 0$ for a positive x and $C < 0$ for a negative x. In general, a function of an arbitrary constant is itself an arbitrary constant.

Given the above observation, one might expect that a differential equation $y' = f(x,y)$ has infinitely many solutions (if any). For instance, the function $y = x + 1$ is a solution to the differential equation

$$y' + y = x + 2.$$

To verify this, we substitute $y = x + 1$ and $y' = 1$ into the equation. Indeed, $y' + y = 1 + (x+1) = x + 2$. It is not difficult to verify that another function $g(x) = x + 1 + e^{-x}$ is also a solution of the given differential equation, demonstrating that a differential equation may have many solutions.

To solve a differential equation means to make its solutions known (in the sense explained later). A solution in which the dependent variable is expressed in terms of the independent variable is said to be in **explicit form**. A function is *known* if it can be expressed by a formula in terms of standard and/or familiar functions (polynomial functions, exponentials, trigonometric functions, and their inverse functions). For example, we consider functions given by a convergent series as known if the terms of the series can be expressed via familiar functions. Also, quadrature (expression via integral) of a given function $f(x)$ is regarded as known.

However, we shall see in this book that functions studied in calculus are not enough to describe solutions of all differential equations. In general, a differential equation defines a function as its solution (if one exists), even if it cannot be expressed in terms of familiar functions. Such a solution is usually referred to as a special function. Thus, we use the word "solution" in a broader sense by including less convenient forms of solutions.

Any relation, free of derivatives, that involves two variables x and y and that is consistent with the differential equation (1.2.1) is said to be a solution of the equation in **implicit form**. Although we may not be able to solve the relation for y, thus obtaining a formula in x, any change in x still results in a corresponding change in y. Hence, on some interval, this could define locally a solution $y = \phi(x)$ even if we fail to find an explicit formula for it or even if the global function does not exist. In fact, we can obtain numerical values for $y = \phi(x)$ to any desired precision.

In this text, you will learn how to determine solutions explicitly or implicitly, how to approximate them numerically, how to visualize and plot solutions, and much more.

Let us consider for simplicity the first order differential equation in normal form: $y' = f(x,y)$. One can rarely find its solution in explicit form, namely, as $y = \phi(x)$. We will say that the equation

$$\Phi(x,y) = 0$$

defines a solution in *implicit form* if $\Phi(x,y)$ is a known function. How would you know that the equation $\Phi(x,y) = 0$ defines a solution $y = \phi(x)$ to the equation $y' = f(x,y)$? Assuming that the conditions of the implicit function theorem hold, we differentiate both sides of the equation $\Phi(x,y) = 0$ with respect to x:

$$\Phi_x(x,y) + \Phi_y(x,y)\,y' = 0, \qquad \text{where } \Phi_x = \partial\Phi/\partial x, \quad \Phi_y = \partial\Phi/\partial y.$$

From the equation $y' = f(x,y)$, we obtain

$$\Phi_x(x,y) + \Phi_y(x,y)\,f(x,y) = 0. \tag{1.3.1}$$

Therefore, if the function $y = \phi(x)$ is a solution to $y' = f(x,y)$, then the function $\Phi(x,y) = y - \phi(x)$ must satisfy Eq. (1.3.1). Indeed, in this case, we have $\Phi_x = -\phi'$ and $\Phi_y = 1$.

An ordinary differential equation may be given either for a restricted set of values of the independent variable or for all real values. Restrictions, if any, may be imposed arbitrarily or due to constraints relating to the equation. Such constraints can be caused by conditions imposed on the equation or by the fact that the functions involved in the equation have limited domains. Furthermore, if an ordinary differential equation is stated without explicit restrictions on the independent variable, it is assumed that all values of the independent variable are permitted with the exception of any values for which the equation is meaningless.

Example 1.3.1: The relation

$$\ln y + y^2 - \int_0^x e^{-x^2}\,\mathrm{d}x = 0 \quad (y > 0)$$

is considered to be a solution in implicit form of the differential equation

$$(1+2y^2)\,y' - y\,e^{-x^2} = 0 \qquad \text{or} \qquad y' = \frac{y}{1+2y^2}\,e^{-x^2}.$$

This can be seen by differentiating the given relationship implicitly with respect to x. This leads to

$$\frac{\mathrm{d}}{\mathrm{d}x}\left[\ln y + y^2 - \int_0^x e^{-x^2}\,\mathrm{d}x\right] = \frac{1}{y}\frac{\mathrm{d}y}{\mathrm{d}x} + 2y\,\frac{\mathrm{d}y}{\mathrm{d}x} - e^{-x^2} = 0.$$

Therefore,

$$\frac{\mathrm{d}y}{\mathrm{d}x}\left(\frac{1+2y^2}{y}\right) = e^{-x^2} \qquad \Longrightarrow \qquad \frac{\mathrm{d}y}{\mathrm{d}x} = \frac{y}{1+2y^2}\,e^{-x^2}.$$

Example 1.3.2: The function $y(x)$ that is defined implicitly from the equation

$$x^2 + 2y^2 = 4$$

is a solution of the differential equation $x + 2y\,y' = 0$ on the interval $(-2,2)$ subject to $g(\pm 2) = 0$. To plot its solution in *Maple*, use the `implicitplot` command:

```
implicitplot(x^2+2*y^2=4, x=-2..2, y=-1.5..1.5);
```

The same ellipse can be plotted with the aid of *Mathematica:*

```
ContourPlot[x^2 + 2 y^2 == 4, {x, -2, 2}, {y, -2, 2},
 PlotRange -> {{-2.1, 2.1}, {-1.5, 1.5}}, AspectRatio -> 1.5/2.1,
 ContourStyle -> Thickness[0.005], FrameLabel -> {"x", "y"},
 RotateLabel -> False]        (* Thickness is .5% of the figure's length *)
```

The same plot can be drawn in *Maxima* with the following commands:

```
load(draw);
draw2d(ip_grid=[100,100],  /* optional, makes a smoother plot */
      implicit(x^2 + 2*y^2 = 4, x,-2.1,2.1, y,-1.5,1.5));
```

MATLAB is capable to perform the same job:

```
[x,y] = meshgrid(-2:.1:2,-2:.1:2); contour(x.^2 + 2*y.^2)
```

The implicit relation $x^2 + 2\,y^2 = 4$ contains the *two* explicit solutions

$$y(x) = \sqrt{2 - 0.5x^2} \qquad \text{and} \qquad y(x) = -\sqrt{2-0.5x^2} \quad (-2 < x < 2),$$

which correspond graphically to the two semi-ellipses. Indeed, if we rewrite the given differential equation $x + 2y\,y' = 0$ in the normal form $y' = -x/(2y)$, then we should exclude $y = 0$ from consideration. Since $x = \pm 2$ correspond to $y = 0$ in both of these solutions, we must exclude these points from the domains of the explicit solutions. Note that the differential equation $x + 2y\,y' = 0$ has infinitely many solutions: $x^2 + 2y^2 = C$ $(|x| \leqslant C)$, where C is an arbitrary positive constant. □

Next, we observe that a differential equation may (and usually will) have an infinite number of solutions. A set of solutions of $y' = f(x,y)$ that depends on one arbitrary constant C deserves a special name.

Definition 1.2: A function $y = \phi(x, C)$ is called the **general solution** to the differential equation $y' = f(x, y)$ in some two-dimensional domain Ω if for every point $(x, y) \in \Omega$ there exists a value of constant C such that the function $y = \phi(x, C)$ satisfies the equation $y' = f(x, y)$. A solution of this differential equation can be defined implicitly:

$$\Phi(x, y, C) = 0 \qquad \text{or} \qquad \psi(x, y) = C. \tag{1.3.2}$$

In this case, $\Phi(x, y, C)$ is called the **general integral**, and $\psi(x, y)$ is referred to as the **potential function** of the given equation $y' = f(x, y)$.

A constant C may be given any value in a suitable range. Since C can vary from problem to problem, it is often called a parameter to distinguish it from the main variables x and y. Therefore, the equation $\Phi(x, y, C) = 0$ defines a one-parameter family of curves with no intersections. Graphically, it represents a family of solution curves in the xy-plane, each element of which is associated with a particular value of C. The general solution corresponds to the entire family of curves that the equation defines.

As might be expected, the inverse statement is true: the curves of a one-parameter family are integrals of some differential equation of the first order. Indeed, let the family of curves be defined by the equation $\Phi(x, y, C) = 0$, with a smooth function Φ. Differentiating with respect to x yields a relation of the form $F(x, y, y', C) = 0$. By eliminating C from these two equations, we obtain the corresponding differential equation.

Example 1.3.3: For an arbitrary constant C, show that the function $y = C\,x + \frac{C}{\sqrt{1+C^2}}$ is the solution of the nonlinear differential equation $y - x\,y' = \dfrac{y'}{\sqrt{1 + (y')^2}}$.

Solution. Taking the derivative of y shows that $y' = C$. Substitution $y = C\,x + \frac{C}{\sqrt{1+C^2}}$ and $y' = C$ into the differential equation yields

$$C\,x + \frac{C}{\sqrt{1 + C^2}} - x\,C = \frac{C}{\sqrt{1 + C^2}}.$$

This identity proves that the function is a solution of the given differential equation. Setting C to some value, for instance, $C = 1$, we obtain a particular solution $y = x + 1/\sqrt{2}$.

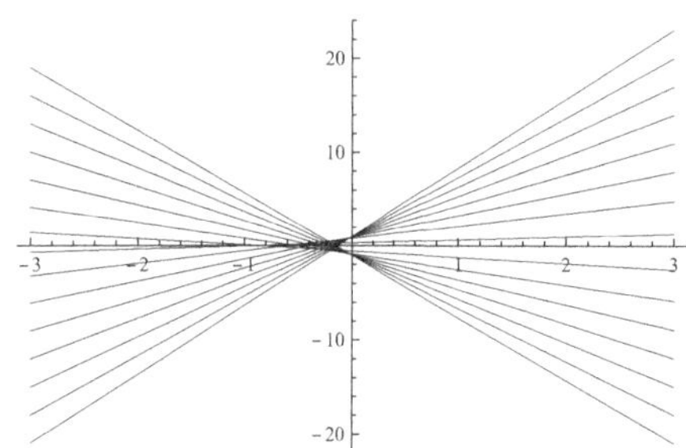

Figure 1.2: Example 1.3.3. A one-parameter family of solutions, plotted with *Mathematica*.

Example 1.3.4: Show that the function $y = \phi(x)$ in parametric form, $y(t) = t\,e^{-t}$, $x(t) = e^t$, is a solution to the differential equation $x^2\,y' = 1 - xy$.

Solution. The derivatives of x and y with respect to t are

$$\frac{\mathrm{d}x}{\mathrm{d}t} = e^t \quad \text{and} \quad \frac{\mathrm{d}y}{\mathrm{d}t} = e^{-t}\,(1 - t),$$

respectively. Hence,

$$\begin{aligned} \frac{\mathrm{d}y}{\mathrm{d}x} &= \frac{\mathrm{d}y/\mathrm{d}t}{\mathrm{d}x/\mathrm{d}t} = \frac{e^{-t}\,(1 - t)}{e^t} = e^{-2t}\,(1 - t) \\ &= e^{-2t} - t\,e^{-2t} = \frac{1}{x^2} - \frac{y}{x} = \frac{1 - yx}{x^2}, \end{aligned}$$

because $x^{-2} = e^{-2t}$ and $y/x = t\,e^{-2t}$.

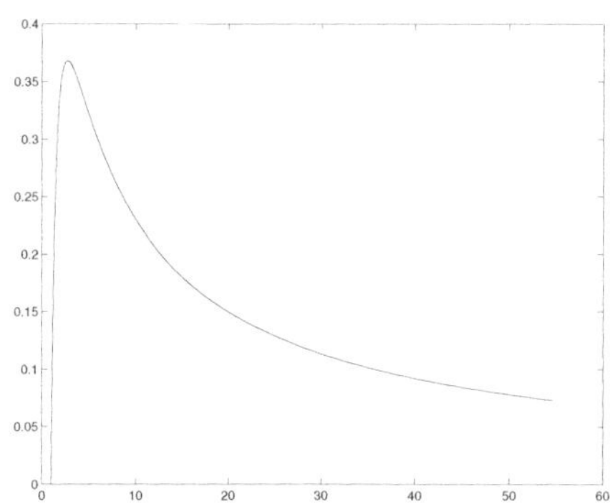

Figure 1.3: Example 1.3.4. A solution of the differential equation $x^2y' = 1 - xy$ (plotted with MATLAB).

Example 1.3.5: Consider the one-parameter (depending on C) family of curves

$$x^2 + y^2 + Cy = 0 \qquad \text{or} \qquad C = -\frac{x^2 + y^2}{y} \quad (y \neq 0).$$

On differentiating, we get $2x + 2y\,y' + Cy' = 0$. Setting $C = -(x^2 + y^2)/y$ in the latter, we obtain the differential equation

$$2x + 2y\,y' - y'\left(\frac{x^2 + y^2}{y}\right) = 0 \qquad \text{or} \qquad y' = 2xy/(x^2 - y^2).$$

This job can be done by *Maxima* in the following steps:

```
depends(y,x);                  /* declare that y depends on x */
soln: x^2 + y^2 + C*y = 0;     /* soln is now a label for the equation */
diff(soln,x);                  /* differentiate the equation */
eliminate([%,soln], [C]);      /* eliminate C from these two equations */
solve(%, 'diff(y,x));          /* solve for y' */
```

Sometimes the integration of $y' = f(x,y)$ leads to a family of integral curves that depend on an arbitrary constant C in parametric form, namely,

$$x = \mu(t, C), \quad y = \nu(t, C).$$

This family of integral curves is called the **general solution in parametric form**.

In many cases, it is more convenient to seek a solution in parametric form, especially when the slope function is a ratio of two functions: $y' = P(x,y)/Q(x,y)$. Then introducing a new independent variable t, we can rewrite this single equation as a system of two equations:

$$\dot{x} \stackrel{\text{def}}{=} \mathrm{d}x/\mathrm{d}t = Q(x,y), \qquad \dot{y} \stackrel{\text{def}}{=} \mathrm{d}y/\mathrm{d}t = P(x,y). \tag{1.3.3}$$

1.4 Particular and Singular Solutions

A solution to a differential equation is called a **particular** (or specific) **solution** if it does not contain any arbitrary constant. By setting C to a certain value, we obtain a particular solution of the differential equation. So every specific value of C in the general solution identifies a particular solution or curve. Another way to specify a particular solution of $y' = f(x,y)$ is to impose an **initial condition**:

$$y(x_0) = y_0, \tag{1.4.1}$$

which specifies a solution curve that goes through the point (x_0, y_0) on the plane. Substituting the general solution into Eq. (1.4.1) will allow you to determine the value of the arbitrary constant. Sometimes, of course, no value of the constant will satisfy the given condition (1.4.1), which indicates that there is no particular solution with the required property among the entire family of integral curves from the general solution.

Definition 1.3: A differential equation $y' = f(x,y)$ (*or, in general,* $F(x,y,y') = 0$) subject to the initial condition $y(x_0) = y_0$, where x_0 and y_0 are specified values, is called an **initial value problem** (*IVP*) or a Cauchy problem.

Example 1.4.1: Show that the function $y(x) = x\left[1 + \int_1^x \frac{\cos x}{x}\,\mathrm{d}x\right]$ is a solution of the following initial value problem:

$$x\,y' - y = x\,\cos x, \quad y(1) = 1.$$

Solution. The derivative of $y(x)$ is

$$y'(x) = 1 + \int_1^x \frac{\cos x}{x}\,\mathrm{d}x + x\,\frac{\cos x}{x} = 1 + \cos x + \int_1^x \frac{\cos x}{x}\,\mathrm{d}x.$$

Hence,

$$x\,y' - y = x + x\,\cos x + x\int_1^x \frac{\cos x}{x}\,\mathrm{d}x - x\left[1 + \int_1^x \frac{\cos x}{x}\,\mathrm{d}x\right] = x\,\cos x.$$

The initial condition is also satisfied since

$$y(1) = 1 \cdot \left[1 + \int_1^1 \frac{\cos x}{x}\,\mathrm{d}x\right] = 1.$$

We can verify that $y(x)$ is the solution of the given initial value problem using the following steps in *Mathematica:*

```
y[x_]=x + x*Integrate[Cos[t]/t, {t, 1, x}]
x*D[y[x], x] - y[x]
Simplify[%]
y[1]                (* to verify the initial value at x=1 *)
```

Definition 1.4: A **singular solution** of $y' = f(x, y)$ is a function that is not a special case of the general solution and for which the uniqueness of the initial value problem has failed.

Not every differential equation has a singular solution, but if it does, its singular solution cannot be determined from the general solution by setting a particular value of C, including $\pm\infty$, because integral curves of the general solution have no common points. A differential equation may have a solution that is neither singular nor a member of the family of one-parameter curves from the general solution. According to the definition, a singular solution always has a point on the plane where it meets with another solution. Such a point is usually referred to as a **branch point**. At that point, two integral curves touch because they share the same slope, $y' = f(x, y)$, but they cannot cross each other. For instance, functions $y = x^2$ and $y = x^4$ have the same slope at $x = 0$; they touch but do not cross.

A singular solution of special interest is one that consists entirely of branch points—at every point it is tangent to another integral curve. An **envelope** of the one-parameter family of integral curves is a curve in the xy-plane such that at each point it is tangent to one of the integral curves. Since there is no universally accepted definition of a singular solution, some authors define a singular solution as an envelope of the family of integral curves obtained from the general solution. Our definition of a singular solution includes not only the envelopes, but *all* solutions that have branch points. This broader definition is motivated by practical applications of differential equations in modeling real-world problems. The existence of a singular solution gives a warning signal in using the differential equation as a reliable model.

A necessary condition for the existence of an envelope is that x, y, C satisfy the equations:

$$\Phi(x, y, C) = 0 \qquad \text{and} \qquad \frac{\partial \Phi}{\partial C} = 0, \tag{1.4.2}$$

where $\Phi(x, y, C) = 0$ is the equation of the general solution. Eliminating C may introduce a function that is not a solution of the given differential equation. Therefore, any curve found from the system (1.4.2) should be checked on whether it is a solution of the given differential equation or not.

Example 1.4.2: Let us consider the equation

$$y' = 2\sqrt{y} \quad (y \geqslant 0), \tag{1.4.3}$$

where the radical takes positive sign. Suppose $y > 0$, we divide both sides of Eq. (1.4.3) by $2\sqrt{y}$, which leads to a separable equation (see §2.1 for detail)

$$\frac{y'}{2\sqrt{y}} = 1 \qquad \text{or} \qquad \frac{\mathrm{d}\sqrt{y}}{\mathrm{d}x} = 1.$$

From chain rule, it follows that

$$\frac{\mathrm{d}\sqrt{y}}{\mathrm{d}x} = \frac{\mathrm{d}}{\mathrm{d}x}(y)^{1/2} = \frac{1}{2}\,y^{-1/2}\,y'.$$

Hence $\sqrt{y} = x + C$, where $x > -C$. The general solution of Eq. (1.4.3) is formed by the one-parametric family of semi-parabolas

$$y(x) = (x + C)^2, \quad \text{or} \quad C = \sqrt{y} - x, \quad x \geqslant -C.$$

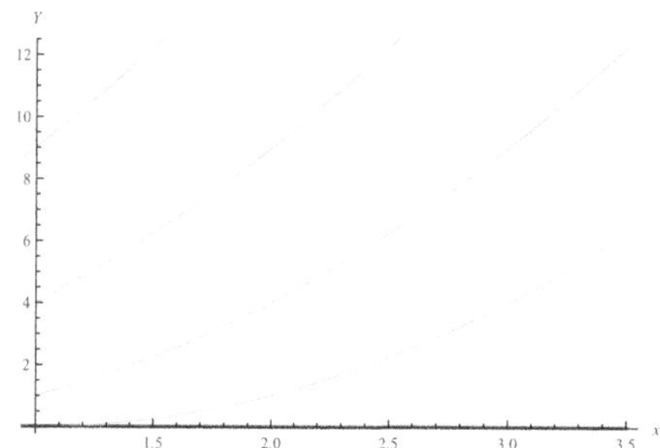

Figure 1.4: Example 1.4.2: some solutions to $y' = 2\sqrt{y}$ along with the singular solution $y \equiv 0$, plotted with *Mathematica*.

The potential function for the given differential equation is $\psi(x, y) = \sqrt{y} - x$. Eq. (1.4.3) has also a trivial (identically zero) solution $y \equiv 0$ that consists of branch points—it is the envelope. This function is a singular solution since $y \equiv 0$ is not a member of the family of solutions $y(x) = (x + C)^2$ for **any** choice of the constant C. The envelope of the family of curves can also be found from the system (1.4.2) by solving simultaneous equations: $(x + C)^2 - y = 0$ and $\partial\Phi/\partial C = 2(x + C) = 0$, where $\Phi(x, y, C) = (x + C)^2 - y$. We can plot some solutions together with the singular solution $y = 0$ using the following *Mathematica* commands:

```
q1 = Plot[Evaluate[(x + C[1])^2 /. C[1] -> {0, 1, -1}], {x, -3.5, 3.5},
  AxesLabel -> {x, Y}]
q2 = Plot[y = 0, {x, -3.5, 3.5}, PlotStyle -> Thick]
Show[q1, q2]
```

Actually, the given equation (1.4.3) has infinitely many singular solutions that could be constructed from the singular envelope $y = 0$ and the general solution by piecing together parts of solutions. An envelope does not necessarily bound the integral curves from one side. For instance, the general solution of the differential equation $y' = 3\,y^{2/3}$ consists of $y = (x+C)^3$ that fill the entire xy-plane. Its envelope is $y \equiv 0$.

Example 1.4.3: The differential equation $5y' = 2y^{-3/2}$, $y \neq 0$, has the one-parameter family of solutions $y = (x-C)^{2/5}$, which can be written in an implicit form (1.4.2) with $\Phi(x,y,C) = y^5 - (x-C)^2$. Differentiating with respect to C and equating to zero, we obtain $y \equiv 0$, which is not a solution. This example shows that conditions (1.4.2) are only necessary for the envelope's existence.

Example 1.4.4: Prove that the function $y(x)$ defined implicitly from the equation $y = \arctan(x+y) + C$, where C is a constant, is the general solution of the differential equation $(x+y)^2\,y' = 1$.

Solution. The chain rule shows that

$$\begin{aligned}\frac{\mathrm{d}y}{\mathrm{d}x} &= \frac{\mathrm{d}}{\mathrm{d}(x+y)}\left[\arctan(x+y)+C\right]\frac{\mathrm{d}}{\mathrm{d}x}\left[x+y\right] \\ &= \frac{1}{1+(x+y)^2}\left[1+\frac{\mathrm{d}y}{\mathrm{d}x}\right] \quad \Longrightarrow \quad \frac{\mathrm{d}y}{\mathrm{d}x} + \frac{\mathrm{d}y}{\mathrm{d}x}(x+y)^2 = 1 + \frac{\mathrm{d}y}{\mathrm{d}x}.\end{aligned}$$

From the latter, it follows that $y'(x+y)^2 = 1$. □

The next example demonstrates how for a function that contains an arbitrary constant as a parameter we can find the relevant differential equation for which the given function is its general solution.

Example 1.4.5: For an arbitrary constant C, show that the function $y = \frac{C-x}{1+x^2}$ is the solution of the differential equation

$$(1+2xy)\,\mathrm{d}x + (1+x^2)\,\mathrm{d}y = 0. \tag{1.4.4}$$

Prove that this equation has no other solutions.

Solution. The differential of this function is

$$\mathrm{d}y = y'\,\mathrm{d}x = \frac{-(1+x^2)-(C-x)2x}{(1+x^2)^2}\,\mathrm{d}x = \frac{x^2-1-2Cx}{(1+x^2)^2}\,\mathrm{d}x.$$

Multiplying both sides by $1+x^2$, we have

$$(1+x^2)\,\mathrm{d}y = \frac{x^2-1-2Cx}{1+x^2}\,\mathrm{d}x = \frac{-x^2-1+2x^2-2Cx}{1+x^2}\,\mathrm{d}x = -\left(1+2x\frac{C-x}{1+x^2}\right)\mathrm{d}x$$

and, since $y = (C-x)/(1+x^2)$, we get

$$-(1+x^2)\,\mathrm{d}y = \frac{x^2-1-2Cx}{1+x^2}\,\mathrm{d}x = (1+2xy)\,\mathrm{d}x.$$

We are going to prove now that there is no solution other than $y = (C-x)/(1+x^2)$. Solving for C, we find the potential function $\psi(x,y) = (1+x^2)y + x$. Suppose the opposite, that other solutions exist; let $y = \phi(x)$ be a solution. Substituting $y = \phi(x)$ into the potential function $\psi(x,y)$, we obtain a function that we denote by $F(x)$, that is, $F(x) = (1+x^2)\,\phi(x) + x$. Differentiation yields

$$F'(x) = 2x\phi(x) + (1+x^2)\,\phi'(x) + 1.$$

Since $\phi'(x) = -\dfrac{1+2x\phi}{1+x^2}$, and we get

$$F'(x) = 2x\,\phi(x) - (1+2x\,\phi(x)) + 1 \equiv 0.$$

Therefore, $F(x)$ is a constant, which we denote by C. That is, $\phi(x) = \dfrac{C-x}{1+x^2}$.

1.5 Direction Fields

A geometrical viewpoint is particularly helpful for the first order equation $y' = f(x, y)$. The solutions of this equation form a family of curves in the xy-plane. At any point (x, y), the slope dy/dx of the solution $y(x)$ at that point is given by $f(x, y)$. We can indicate this by drawing a short line segment (or arrow) through the point (x, y) with the slope $f(x, y)$. The collection of all such line segments at each point (x, y) of a rectangular grid of points is called a **direction field** or a **slope field** of the differential equation $y' = f(x, y)$.

By increasing the density of arrows, it would be possible, in theory at least, to approach a limiting curve, the coordinates and slope of which would satisfy the differential equation at every point. This limiting curve—or rather the relation between x and y that defines a function $y(x)$—is a solution of $y' = f(x, y)$. Therefore the direction field gives the "flow of solutions." Integral curves obtained from the general solution are all different: there is precisely one solution curve that passes through each point (x, y) in the domain of $f(x, y)$. They might be touched by the singular solutions (if any) forming the envelope of a family of integral curves. At each of its points, the envelope is tangent to one of integral curves because they share the same slope.

Direction fields can be plotted for differential equations even if they are not necessarily written in the normal form. If the derivative y' is determined uniquely from the general equation $F(x, y, y') = 0$, the direction field can be obtained for such an equation. However, if the equation $F(x, y, y') = 0$ defines multiple values for y', then at every such point we would have at least two integral curves with distinct slopes.

Example 1.5.1: Let us consider the differential equation not in the normal form:

$$x(y')^2 - 2yy' + x = 0. \tag{1.5.1}$$

At every point (x, y) such that $y^2 \geqslant x^2$ we can assign to y' two distinct values

$$y' = \frac{y \pm \sqrt{y^2 - x^2}}{x} \qquad (y^2 - x^2 \geqslant 0).$$

When $y^2 \leqslant x^2$, Eq. (1.5.1) does not define y' since the root becomes imaginary. Therefore, we cannot draw a direction field for the differential equation (1.5.1) because its slope function is not a single-value function. Nevertheless, we may try to find its general solution by making a guess that it is a polynomial of the second degree: $y = Cx^2 + Bx + A$, where coefficients A, B, and C are to be determined. Substituting y and its derivative, $y' = 2Cx + B$, into Eq. (1.5.1), we get $B = 0$ and $A = 1/(4C)$. Hence, Eq. (1.5.1) has a one-parametric family of solutions

$$y = Cx^2 + \frac{1}{4C}. \tag{1.5.2}$$

For any value of C, $C \neq 0$, y^2 is greater than or equal to x^2. To check our conclusion, we use *Maple:*

```
dsolve(x*(diff(y(x),x))^2-2*y(x)*diff(y(x),x)+x=0,y(x));
phi:=(x,C)->C*x*x+0.25/C;                    # the general solution
plot({subs(C=.5,phi(x,C)),phi(x,-1),x},x=-1..1,y=-1..1,color=blue);
```

Let us consider any region R of the xy-plane in which $f(x, y)$ is a real, single-valued, continuous function. Then the differential equation $y' = f(x, y)$ defines a direction field in the region R. A solution $y = \phi(x)$ of the given differential equation has the property that at every point its graph is tangent to the direction element at that point. The slope field provides useful qualitative information about the behavior of the solution even when you cannot solve it. Direction fields are common in physical applications, which we discuss in [14]. While slope fields prove their usefulness in qualitative analysis, they are open to several criticisms. The integral curves, being graphically obtained, are only approximations to the solutions without any knowledge of their accuracy and formulas.

If we change for a moment the notation of the independent variable x to t, for time, then we can associate the solution of the differential equation with the trajectory of a particle starting from any one of its points and then moving in the direction of the field. The path of such a particle is called a **streamline** of the field. Thus, the function defined by a streamline is an integral of the differential equation to which the field applies. A point through which just one single integral curve passes is called an *ordinary point.*

When high precision is required, a suitably dense set of line segments on the plane region must be made. The labor involved may then be substantial. Fortunately, available software packages are very helpful for practical drawings of direction fields instead of hand sketching.

There is a friendly graphical program, **Winplot**, written by Richard Parris, a teacher at Phillips Exeter Academy in Exeter, New Hampshire. Mr. Parris generously allows free copying and distribution of the software and provides

As we see from Figure 1.5, integral curves intersect each other, which would be impossible for solutions of a differential equation in the normal form. Indeed, solving Eq. (1.5.2) with respect to C, we observe that for every point (x, y), with $y^2 \geqslant x^2$, there are two distinct values of $C = \left(y \pm \sqrt{y^2 - x^2}\right)/x^2$. For instance, Eq. (1.5.1) defines two slopes at the point $(2, 1)$: $2 \pm \sqrt{3}$.
Let us find an envelope of singular solutions. According to Eq. (1.4.2), we differentiate the general solution (1.5.2) with respect to C, which gives $x^2 = -1/(4C^2)$. Eliminating C from these two equations, we obtain $x^2 - y^2 = 0$ or $y = \pm x$. Substitution into Eq. (1.5.1) yields that these two functions are its solution.
Hence, the given differential equation has two singular solutions $y = \pm x$ that form the envelope of integral curves corresponding to the general solution. □

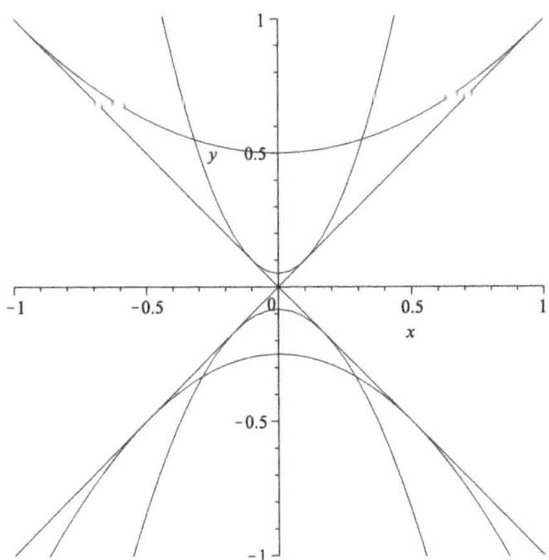

Figure 1.5: Example 1.5.1: some solutions along with two singular solutions, plotted in *Maple*.

frequent updates. The latest version can be downloaded from the website:
`http://math.exeter.edu/rparris/winplot.html`. The program is of top quality and is easy to use.
You can find many online applications for plotting direction fields by entering "desmos direction fields" into any search engine. These include `https://bluffton.edu/homepages/facstaff/nesterd/java/slopefields.html`, `http://www.geogebra.org`, `http://www.mathscoop.com`, and `http://slopefield.nathangrigg.net`.

Maple

It is recommended to clear the memory before starting a session by invocation of either `restart` or `gc( )` for garbage collection. *Maple* is particularly useful for producing graphical output. It has two dedicated commands for plotting flow fields associated with first order differential equations—`DEplot` and `dfieldplot`. For example, the commands

```
restart;  with(DEtools):  with(plots):
dfieldplot(diff(y(x),x)=y(x)+x, y(x), x=-1..1, y=-2..2, arrows=medium);
```

allow you to plot the direction field for the differential equation $y' = y + x$. To include graphs of some solutions into the direction field, we define the initial conditions first:

```
inc:=[y(0)=0.5,y(0)=-1];
```

Then we type

```
DEplot(diff(y(x),x)=y(x)+x, y(x), x=-1..1, y=-2..2, inc, arrows=medium,
linecolor=black,color=blue,title='Direction field for y'=y+x');
```

There are many options in representing a slope field, which we demonstrate in the text. A special option, `dirgrid`, specifies the number of arrows in the direction field. For instance, if we replace *Maple*'s option `arrows=medium` with `dirgrid=[16,25]`, we will get the output presented in Figure 1.10.

The computer algebra system (CAS for short) *Maple* also has an option to plot the direction fields without arrows or with comets, as can be seen in Figures 1.8 and 1.9, plotted with the following script:

```
dfieldplot(x*diff(y(x),x)=3*y(x)+2*x,y(x),
x=-1..1,y=-2..2,arrows=line,title='Direction field for xy'=3y+2x');
DEplot(x*diff(y(x),x)=3*y(x)+x^3,y(x),
x=-1..1,y=-2..2,arrows=comet,title='Direction field for xy'=3y+x*x*x');
```

You may draw a particular solution that goes through the point $x = \pi/2$, $y = 1$ in the same picture by typing `DEplot(equation, y(x), x-range, y-range, [y(Pi/2)=1], linecolor=blue)`.
Maple also can plot direction fields with different colors:

```
dfieldplot(diff(y(x),x)=f(x,y), y(x), x-range, y-range, color=f(x,y)).
```

Mathematica

It is always a good idea to start *Mathematica*'s session with clearing variables or the kernel. With *Mathematica*, only one command is needed to draw the direction field corresponding to the differential equation $y' = f(t, y)$. By choosing, for instance, $f(t, y) = 1 - t^2 - y$, we type:

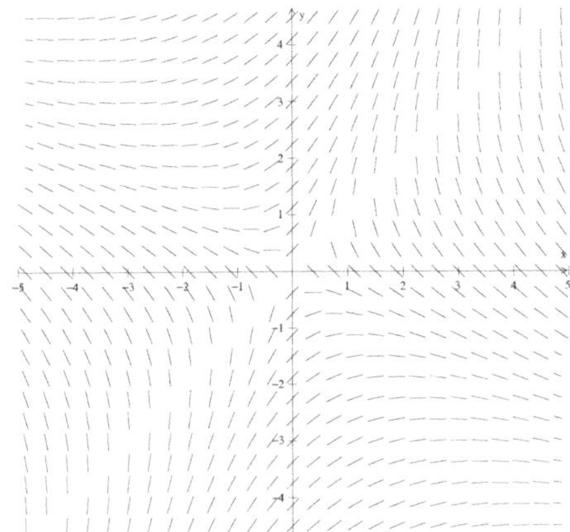

Figure 1.6: Direction field for the equation $y' = (y+x)/(y-x)$, plotted using Winplot. There is no information in a neighborhood of the singular line $y = x$.

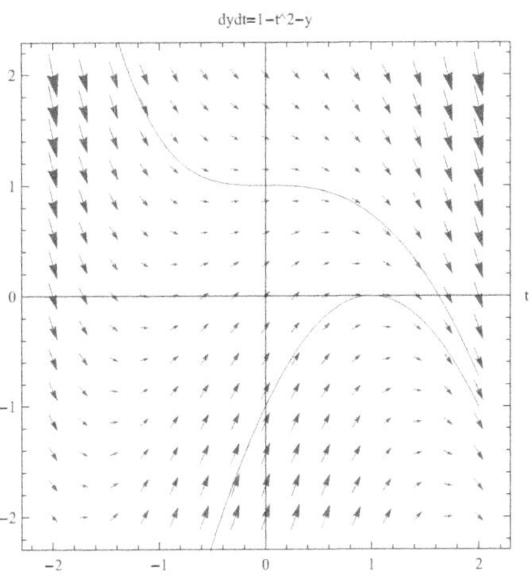

Figure 1.7: Direction field along with two solutions for the equation $y'(t) = 1 - t^2 - y(t)$, plotted with *Mathematica*.

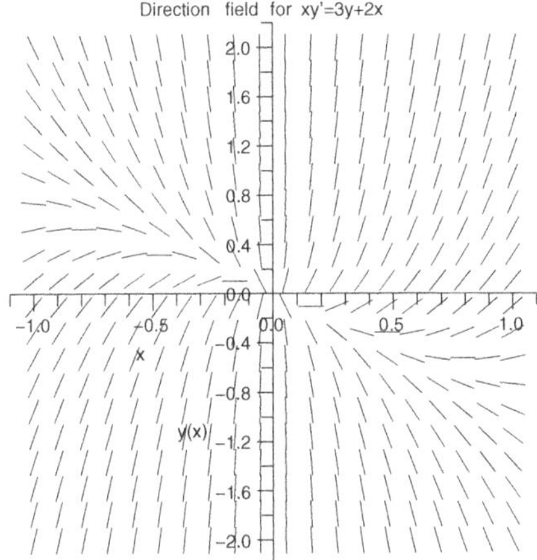

Figure 1.8: Direction field for $xy' = 3y + 2x$, plotted with *Maple*.

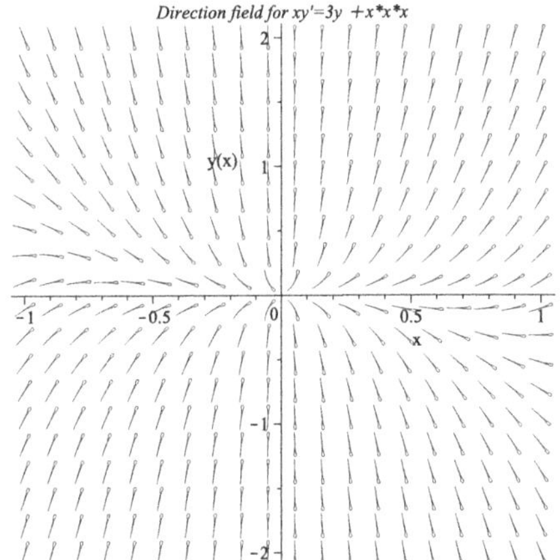

Figure 1.9: Direction field for $xy' = 3y + x^3$, plotted with *Maple*.

```
dfield = VectorPlot[{1,1-t^2-y}, {t, -2, 2}, {y, -2, 2}, Axes -> True,
VectorScale -> {Small,Automatic,None}, AxesLabel -> {"t", "dydt=1-t^2-y"}]
```

The option `VectorScale` allows one to fix the arrows' sizes and suboption `Scaled[1]` specifies arrowhead size relative to the length of the arrow. To plot the direction field along with, for example, two solutions, we use the following commands:

```
sol1 = DSolve[{y'[t] == 1 - y[t] - t^2, y[0] == 1}, y[t], t]
sol2 = DSolve[{y'[t] == 1 - y[t] - t^2, y[0] == -1}, y[t], t]
pp1 = Plot[y[t] /. sol1, {t, -2, 2}]
pp2 = Plot[y[t] /. sol2, {t, -2, 2}]
Show[dfield, pp1, pp2]
```

For plotting streamlines/solutions, CAS *Mathematica* has a dedicated command: `StreamPlot`. If you need to plot a sequence of solutions under different initial conditions, use the following script:

```
myODE = t^2*y'[t] == (y[t])^3 - 2*t*y[t]
IC = {{0.5, 0.7}, {0.5, 4}, {0.5, 1}};
Do[ansODE[i] =
  Flatten[DSolve[{myODE, y[IC[[i, 1]]] == IC[[i, 2]]}, y[t], t]];
```

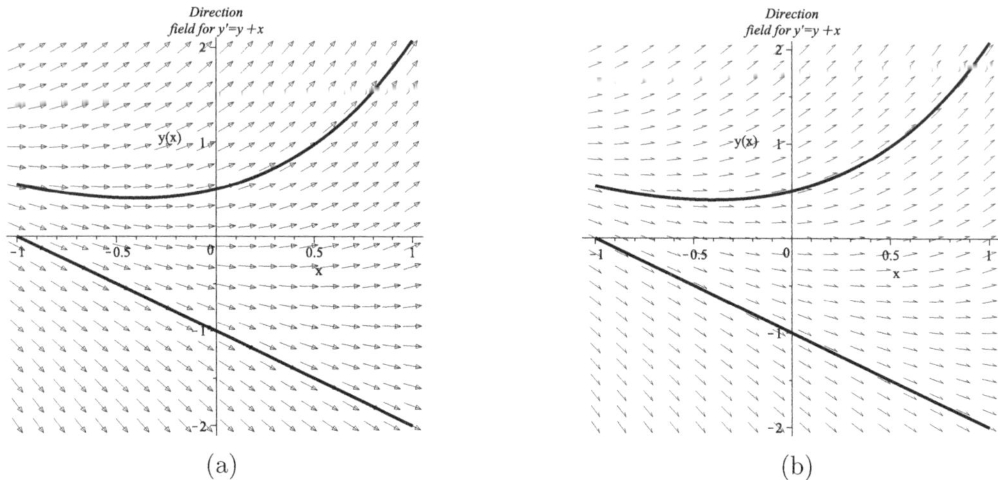

Figure 1.10: Direction field for $y' = y + x$ using (a) `arrows=medium` and (b) `dirgrid`, plotted with *Maple*.

```
myplot[i] = Plot[Evaluate[y[t] /. ansODE[i]], {t, 0.02, 5}];
Print[myplot[i]];  , {i, 1, Length[IC]}]
```

Note that *Mathematica* uses three different notations associated with the symbol "=." Double equating "==" is used for defining an equation or for testing an equality; regular "=" is used to define instant assignments, while ":=" is used to represent the left-hand side in the form of the right-hand side afresh, that is, unevaluated.

MATLAB®

MATLAB is a numerical computing environment that also includes CAS subroutines: MuPad (based on *Maple*) and Live Editor. Before beginning a new session, it is recommended that you execute the `clc` command to clear the command window and the `clear` command to remove all variables from memory. In order to plot a direction field with MATLAB, you have several options. One of them includes creation of an intermediate file[4], say `function1.m`, yielding the slope function $f(x, y)$. Let us take a simple example $f(x, y) = xy^2$. This file will contain the following three lines (excluding comments followed after %):

```
% Function for direction field:          (function1.m)
  function F=function1(x,y);
  F=x*y*y;
F=vectorize(F); % to get a vectorized version of the function, which is optional
```

If a function $f(x, y)$ is not complicated, it could be defined directly within MATLAB code. We demonstrate it in the case of the rational slope function $f(x, y) = (x + 2y - 5)/(2x + 4y - 7)$ that is used in Example 2.2.3 on page 56:

```
der=@(x,y) (x+2*y-5)./(2*x+4*y+7); % define slope function
equ=@(x) (3-4*x)/8;                % define equilibrium
sig=@(x) -(7+2*x)/4;               % define singular
xmin=-5; xmax=5; ymin=-5; ymax=5;  % set the frame
dx=(xmax-xmin)/20;                 % set spatial steps
dy=(ymax-ymin)/20;
[X,Y]=meshgrid(xmin:dx:xmax, ymin:dy:ymax); % generate mesh
Dx=ones(size(X));       % Unit x-components of arrows
Dy=der(X,Y);            % Computed y-components
L=sqrt(Dx.^2 + Dy.^2);  % Initial lengths
Dx1=Dx./L; Dy1=Dy./L;   % Unit lengths for all arrows
quiver(X,Y, Dx1,Dy1, 'b-'); % draw the direction field
axis tight;             % sets the axis limits
% to the range of the data
```

[4] All script files in MATLAB must have extension `m`.

```
xlabel('x','FontSize',16); % set labels, fontsize for them
ylabel('y','FontSize',16,'rotation',0); % and direction for letter y
set(gca, 'FontSize', 12); % set fontsize for axis
hold on
xx=xmin:dx/10:xmax;
plot(xx, equ(xx), 'k', 'LineWidth', 3);
plot(xx, sig(xx), 'k--', 'LineWidth', 3);
% below are several solutions based on different initial conditions
[x,y1]=ode45(der, [-5 5], 3.1); % IC: y(-5)=3.1
plot(x,y1,'r', 'LineWidth', 2);
[x,y2]=ode45(der, [-5 5], 4.0); % IC: y(-5)=4.0
plot(x,y2,'r', 'LineWidth', 2);
  print -deps direction_field.eps;  % or print -deps2 direction_field.eps;
  print -depsc direction_field.eps; % for color image
```

In the above code, the subroutine quiver(x, y, u, v) displays velocity vectors as arrows with components (u, v) at the points (x, y). To draw a slope field without arrows, quiver is not needed as the following code shows. If a graph of a function needs not to be plotted along with the slope field, comment out the last line.

```
func = @(x) 3*x.^2 - 8*x;    % slope function
ifunc = @(x) x.^3 - 4*x.^2; % integral from 0 to x
xmin = -2; xmax = 5;        % set limits for x
dx = (xmax - xmin)/100;     % step for computing ifunc
xx = xmin:dx:xmax;          % arguments for computing ifunc
sing = ifunc(xx);           % computing ifunc
dy = (max(sing)- min(sing))/100; %step along y-axis for slope field
figure; axis tight;   % set the axis limits to the range of the data
for y = min(sing):6*dy:max(sing) % loop for plotting a slope field
  for x = xmin:6*dx:xmax
    c=2/sqrt((1/dx)^2 + (func(x)/dy).^2 ); d=func(x).*c;
    tpt = [x - c, x + c]; ypt=[y - d, y + d];
    line(tpt,ypt); % plot an element of slope field
  end;
end
hold on; plot(xx, sing, 'k-', 'LineWidth', 3); hold off;
```

As you have seen from the above two scripts, although resourceful, MATLAB does not have the ability to plot direction fields naturally. However, many universities have developed some software packages that facilitate drawing direction fields for differential equations. For example, John Polking at Rice University has produced dfield and pplane programs for MATLAB. The MATLAB versions of dfield and pplane are copyrighted in the name of John Polking [38, 39]. While they are not in the public domain, these subroutines are being made available free of charge to educational institutions. Another possibility to plot slope fields provides MuPad, an integrated CAS in MATLAB.

Maxima

Maxima and its popular graphical interface *wxMaxima*[5] are free software[6] projects. This means that you have the freedom to use them without restriction, to give copies to others, to study their internal workings and adapt them to your needs, and even to distribute modified versions. *Maxima* is a descendant of Macsyma, the first comprehensive computer algebra system, developed in the late 1960s at the Massachusetts Institute of Technology.

Maxima provides two packages for plotting direction fields, each with different strengths: plotdf supports interactive exploration of solutions and variation of parameters, whereas drawdf is non-interactive and instead emphasizes the flexible creation of high-quality graphics in a variety of formats. Let us first use drawdf to plot the direction field for the differential equation $y' = e^{-t} + y$:

```
load(drawdf);
drawdf(exp(-t)+y, [t,y], [t,-5,10], [y,-10,10]);
```

[5]See http://maxima.sourceforge.net/ and http://wxmaxima.sourceforge.net/

[6]See http://www.fsf.org/ for more information about free software.

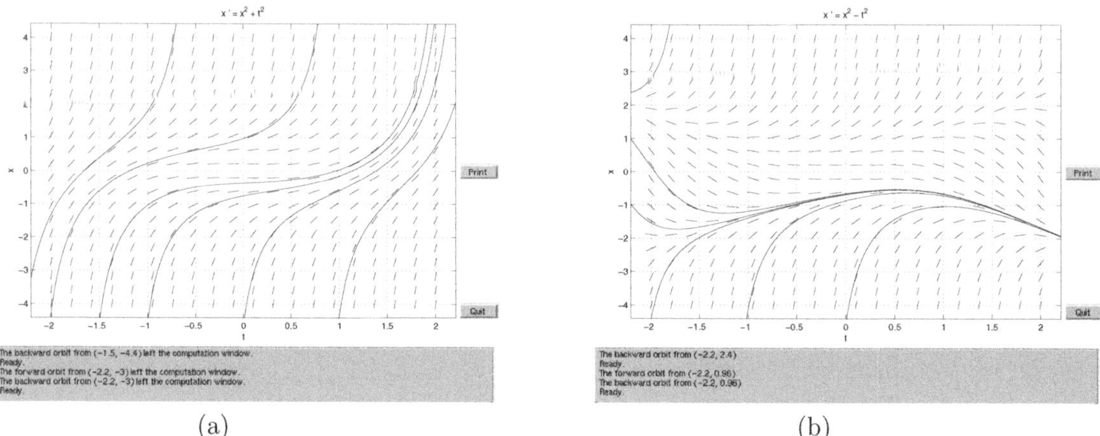

(a) (b)

Figure 1.11: Direction fields and streamlines for the equations (a) $x'(t) = x^2(t) + t^2$ and (b) $x'(t) = x^2(t) - t^2$ using `dfield` in MATLAB.

Note that `drawdf` normally displays graphs in a separate window. If you are using *wxMaxima* (recommended for new users) and would prefer to place your graphs within your notebook, use the `wxdrawdf` command instead. The `load` command stays the same, however.

Solution curves passing through points $y(0) = 0$, $y(0) = -0.5$, and $y(0) = -1$ can be included in the graph as follows:

```
drawdf(exp(-t)+y, [t,y], [t,-5,10], [y,-10,10],
  solns_at([0,0], [0,-0.5], [0,-1]));
```

By adding `field_degree=2`, we can draw a field of quadratic splines (similar to *Maple*'s comets) which show both slope and curvature at each grid point. Here we also specify a grid of 20 columns by 16 rows, and draw the middle solution thicker and in black.

```
drawdf(exp(-t)+y, [t,y], [t,-5,10], [y,-10,10],
  field_degree=2, field_grid=[20,16], solns_at([0,0], [0,-1]),
  color=black, line_width=2, soln_at(0,-0.5));
```

We can add arrows to the solution curves by specifying `soln_arrows=true`. This option removes arrows from the field by default and also changes the default color scheme to emphasize the solution curves.

```
drawdf(exp(-t)+y, [t,y], [t,-5,10], [y,-10,10], field_degree=2,
  soln_arrows=true, solns_at([0,0], [0,-1], [0,-0.5]),
  title="Direction field for dy/dt = exp(-t) + y",
  xlabel="t", ylabel="y");
```

Actual examples of direction fields plotted with *Maxima* are presented on the front page of Chapter 1, and scattered in the text. The following command will save the most recent plot to an encapsulated Postscript file named "plot1.eps" with dimensions 12 `cm` by 8 `cm`. Several other formats are supported as well, including PNG and PDF.

```
draw_file(terminal=eps, file_name="plot1",
  eps_width=12, eps_height=8);
```

Since `drawdf` is built upon *Maxima*'s powerful `draw` package, it accepts all of the options and graphical objects supported by `draw2d`, allowing the inclusion of additional graphics and diagrams. To investigate differential equations by varying parameters, `plotdf` is sometimes preferable. Let us explore the family of differential equations of the form $y' = y - a + b\cos t$, with a solution passing through $y(0) = 0$:

```
plotdf(y-a+b*cos(t), [t,y], [t,-5,9], [y,-5,9],
  [sliders,"a=0:2,b=0:2"], [trajectory_at,0,0]);
```

You may now adjust the values of a and b using the sliders in the plot window and immediately see how they affect the direction field and solution curves. You can also click in the field to plot new solutions through any desired point. Make sure to close the plot window before returning to your *Maxima* session.

Sage

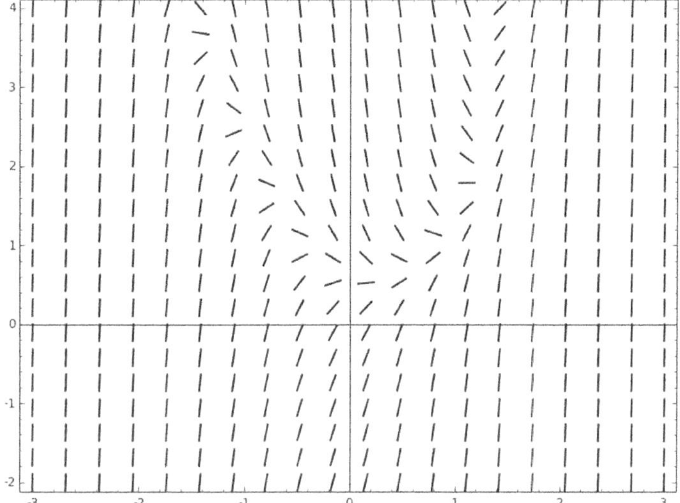

Figure 1.12: Direction field for the equation $y'(x) = 6\,x^2 - 3y + 2\,\cos(x+y)$, plotted with *Sage*.

SageMath is a free open-source mathematics software system licensed under the GPL. It builds on top of many existing open-source packages Python, *R*, Julia (computational geometry), GAP (discrete algebra), Octave, including computer algebra systems *Maxima* and SymPy, and much more. *Sage* comes with two options: one can download it[7] (for free) or use it interactively through cloud version. *SageMath* is available for every platform, and is ubiquitous throughout industry. SageMathCloud supports authoring documents written in LaTeX, Markdown, or HTML. SageMathCloud also allows you to publish documents online.

To plot a direction field for first order differential equation $y' = f(x, y)$ using *Sage* we first declare variables and then we use a standard command, which we demonstrate in the following example: $y' = 6\,x^2 - 3y + 2\,\cos(x+y)$.

```
x,y = var('x,y')
plot_slope_field(6*x^2 -3*y+2*cos(x+y), (x,-3,3), (y,-2,4), xmax=10)
```

Python

Python[8] is a high-level and general-purpose programming language (free of charge). Part of the reason that it is a popular choice for scientists and engineers is the language versatility, online community of users, and powerful analysis packages such as NumPy, SciPy, and, of course, SymPy, a CAS written completely in Python. Anaconda is a free Python distribution from Continuum Analytics that includes many useful packages for scientific computing.

The function **odeint** is available in SciPy for integrating first order vector differential equations. A higher order ordinary differential equation can always be reduced to a differential equation of this type by introducing intermediate derivatives into the vector (see §6.3). There are many optional inputs and outputs available when using *odeint* that can help tune the solver.

Problems

1. For each equation below, determine its order, and name the independent variable, the dependent variable, and any parameters.

   (a) $y' = y^2 + x^2$; **(b)** $\dot{P} = rP(1 - P/N)$; **(c)** $m\ddot{x} + r\dot{x} + kx = \sin t$;

   (d) $L\,\ddot{\theta} + g\,\sin\theta = 0$; **(e)** $(x\,y'(x))^2 = x^2\,y(x)$; **(f)** $2y'' + 3y' + 5y = e^{-x}$.

2. Determine a solution to the following differential equation:

$$(1 - 2x^2)\,y'' - x\,y' + 6y = 0$$

   of the form $y(x) = a + bx + cx^2$ satisfying the normalization condition $y(1) = -1$.

3. Differentiate both sides of the given equation to eliminate the arbitrary constant (denoted by C) and to obtain the associated differential equation.

[7]See `http://www.sagemath.org/`.

[8]`https://www.python.org/`

(a) $x^2 + 4xy^2 = C$; (b) $x^3 + 2x^2y = C$; (c) $y\cos x - xy^2 = C$;
(d) $Cx^2 = y^2 + 2y$; (e) $e^x - \ln y = C$; (f) $3\tan x + \cos^2 y = C$;
(g) $Cx = \ln(xy)$; (h) $Cy + \ln x = 0$; (i) $y - 1 + \frac{x}{1+Cx} = 0$;
(j) $y = \frac{x}{x+C}$; (k) $y + Cx = x^4$; (l) $\sin^2 y - \tan x = C$.

4. Find the differential equation of the family of curves $y_n(x) = \left(1 + \frac{x}{n}\right)^n$, where $n \ne 0$ is a parameter. Show that $y = e^x$ is a solution of the equation found when $n \to \infty$.

5. Find a differential equation of fourth order having a solution $y = Ax\cos x + Bx\sin x$, where A and B are arbitrary constants.

6. Which of the following equations for the unknown function $y(x)$ are *not* ordinary differential equations? Why?
   (a) $\frac{\mathrm{d}}{\mathrm{d}x}\frac{y''(x)}{(1+(y'(x))^2)^{3/2}} = (1 + (y'(x))^2)^{1/2}$; (b) $y'(x) = y(x-1) - y(x)/3$;
   (c) $\int_{-\infty}^{\infty} \cos(kx)y'(x)\,\mathrm{d}x = \sqrt{2\pi}y'(k)$; (d) $y''(x) = g - y'(x)|y'(x)|$;
   (e) $\int_0^{\infty} e^{-xt}y(x)\,\mathrm{d}x = e^{-t}/t$; (f) $y(x) = 1 + \int_0^x y(t)\,\mathrm{d}t$.

7. Determine the order of the following ordinary differential equations:
   (a) $(y')^2 + x^2 y = 0$; (b) $\frac{\mathrm{d}}{\mathrm{d}x}(y\,y') = \sin x$; (c) $x^2 + y^2 + (y')^2 = 1$;
   (d) $x^2 y'' + \frac{\mathrm{d}}{\mathrm{d}x}(xy') + y = 0$; (e) $\frac{\mathrm{d}}{\mathrm{d}x}(y' \sin x) = 0$; (f) $t^3 u''(t) + t^2 u'(t) + t\,u(t) = 0$;
   (g) $\frac{\mathrm{d}}{\mathrm{d}t}\left[t^2 u''(t)\right] = 0$; (h) $(u'(t))' = t^3$; (i) $y'' = \sqrt{4 + (y')^2}$;
   (j) $xy'' + (y')^3 + e^x = 0$; (k) $xy''' + \mathrm{sign}(x) = 0$; (l) $(y'')^2 + y' \sin x = \cos x$.

8. Which of the following equations are linear?
   (a) $y^{(4)} + x^3 y = 0$; (b) $\frac{\mathrm{d}}{\mathrm{d}x}(y\,y') = \sin x$; (c) $\frac{\mathrm{d}}{\mathrm{d}x}[xy] = 0$;
   (d) $y' + y\sin x = 1$; (e) $(y')^2 - x^2 = 0$; (f) $y' - x^2 = 0$,
   (g) $\frac{\mathrm{d}}{\mathrm{d}x}[x^2 + y^2] = 1$; (h) $y'''(x) + x\,y''(x) + y^2(x) = 0$; (i) $y' = \sqrt{xy}$;
   (j) $y'(x) + x^2 y(x) = \cos x$; (k) $y' = x^2 + y^2$; (l) $y'' + y' \sin x = \cos x$.

9. Let $y(x)$ be a solution for the ODE $y''(x) = x\,y(x)$ that satisfies the initial conditions $y(0) = 1$, $y'(0) = 0$. (You will learn in this course that exactly one such solution exists.) Calculate $y''(0)$. The ODE is, by its nature, an equation that is meant to hold for *all* values of x. Therefore, you can take the derivative of the equation. With this in mind, calculate $y'''(0)$ and $y^{(4)}(0)$.

10. In each of the following problems, verify whether or not the given function is a solution of the given differential equation and specify the interval or intervals in which it is a solution; C always denotes a constant.
   (a) $y'' + 4y = 0$, $y = \sin 2x + C$. (b) $y' - y^2(x)\sin x = 0$, $y(x) = 1/\cos x$.
   (c) $2yy' = 1$, $y(x) = \sqrt{x+1}$. (d) $x\,y'' - y' + 4x^3 y = 0$, $y = \sin(x^2 + 1)$.
   (e) $y' = ky$, $y(x) = C\,e^{kx}$. (f) $y' = 1 - ky$, $k\,y(x) = 1 + C\,e^{-kx}$.
   (g) $y''' = 0$, $y(x) = C\,x^2$. (h) $y'' + 2y = 2\cos^2 x$, $y(x) = \sin^2 x$.
   (i) $y'' - 5y' + 6y = 0$, $y = C\,e^{3x}$. (j) $y'' + 2y' + y = 0$, $y(x) = C\,x\,e^{-x}$.
   (k) $y' - 2xy = 1$, $y = e^{x^2}\int_0^x e^{-t^2}\,\mathrm{d}t + C\,e^{x^2}$. (l) $x\,y' - y = x\sin x$, $y(x) = x\int_0^x \frac{\sin t}{t}\,\mathrm{d}t + Cx$.

11. Solutions to most differential equations cannot be expressed in finite terms using elementary functions. Some solutions of differential equations, due to their importance in applications, were given special labels (usually named after an early investigator of its properties) and therefore are referred to as *special functions*. For example, there are two known *sine integrals*: $\mathrm{Si}(x) = \int_0^x \frac{\sin t}{t}\,\mathrm{d}t$, $\mathrm{si}(x) = -\int_x^{\infty} \frac{\sin t}{t}\,\mathrm{d}t = \mathrm{Si}(x) - \frac{\pi}{2}$ and three cosine integrals: $\mathrm{Cin}(x) = \int_0^x \frac{1-\cos t}{t}\,\mathrm{d}t$, $\mathrm{ci}(x) = -\int_x^{\infty} \frac{\cos t}{t}\,\mathrm{d}t$, and $\mathrm{Ci}(x) = \gamma + \ln x + \int_0^x \frac{\cos t}{t}\,\mathrm{d}t$, where $\gamma \approx 0.5772$ is Euler's constant. Use definite integration to find an explicit solution to the initial value problem $x\,y' = \sin x$, subject to $y(1) = 1$.

12. Verify that the indicated function is an implicit solution of the given differential equation.
   (a) $(y+1)y' + x + 2 = 0$, $(x+2)^2 + (y+1)^2 = C^2$;
   (b) $x\,y' = (1 - y^2)/y$, $yx - \ln y = C$;
   (c) $y' = (\sqrt{(1+y)} + 1 + y)/(1+x)$, $C\sqrt{1+x} - \sqrt{1+y} = 1$;
   (d) $y' = (y-2)^2/x^2$, $(y-2)(1+Cx) = x$.

13. Show that the functions in parametric form satisfy the given differential equation.
   (a) $3x\,\mathrm{d}x + 2\sqrt{4-x^2}\,\mathrm{d}y = 0$, $x = 2\cos t$, $y = 3\sin t$; (b) $(x+3y)\mathrm{d}y = (5x+3y)\mathrm{d}x$, $x = e^{-2t} + 3\,e^{6t}$, $y = -e^{-2t} + 5\,e^{6t}$;
   (c) $2x\,y' = y - 1$, $x = t^2$, $y = t + 1$; (d) $y\,y' = x$, $x = \tan t$, $y = \sec t$;
   (e) $2x\,y' = 3y$, $x = t^2$, $y = t^3$; (f) $ax\,y' = by$, $x = t^a$, $y = t^b$;
   (g) $y' = -4x$, $x = \sin t$, $y = \cos 2t$; (h) $9y\,y' = 4x$, $x = 3\cosh t$, $y = 2\sinh t$;
   (i) $2(y+1)\,y' = 1$, $x = t^2 + 1$, $y = t - 1$; (j) $y' = 2x + 10$, $x = t - 5$, $y = t^2 + 1$;
   (k) $(x+1)\,y' = 1$, $x = t - 1$, $y = \ln t$, $t > 0$; (l) $a^2 y\,y' = b^2 x$, $x = a\cosh t$, $y = b\sinh t$.

14. Determine the value of λ for which the given differential equation has a solution of the form $y = e^{\lambda t}$.
    (a) $y' - 3y = 0$; (b) $y'' - 4y' + 3y = 0$; (c) $y'' - y' - 2y = 0$;
    (d) $y'' - 2y' - 3y = 0$; (e) $y''' + 3y'' + 2y' = 0$; (f) $y''' - 3y'' + 3y' - y = 0$.

15. Determine the value of λ for which the given differential equation has a solution of the form $y = x^{\lambda}$.
    (a) $x^2y'' + 2xy' - 2y = 0$; (b) $xy' - 2y = 0$; (c) $x^2y'' - 3xy' + 3y = 0$;
    (d) $x^2y'' + 2xy' - 6y = 0$; (e) $x^2y'' - 6y = 0$; (f) $x^2y'' - 5xy' + 5y = 0$.

16. Show that (a) the first order differential equation $|y'| + 4 = 0$ has no solution; (b) $|y'| + y^2 + 1 = 0$ has no real solutions, but a complex one; (c) $|y'| + 4|y| = 0$ has a solution but not one involving an arbitrary constant.

17. Show that the first order differential equation $y' = 4\sqrt{y}$ has a one-parameter family of solutions of the form $y(x) = (2x + C)^2$, $2x + C \geqslant 0$, where C is an arbitrary constant, and a singular solution $y(x) \equiv 0$ which is not a member of the family $(2x + C)^2$ for any choice of C.

18. Find a differential equation for the family of lines $y = Cx - C^2$.

19. For each of the following differential equations, find a singular solution.
    (a) $y' = 3x^2 - (y - x^3)^{2/3}$; (b) $y' = \sqrt{(x + 1)(y - 1)}$;
    (c) $y' = \sqrt{x^2 - y} + 2x$; (d) $y' = (2y)^{-1} + (y^2 - x)^{1/3}$;
    (e) $y' = (x^2 + 3y - 2)^{2/3} + 2x$; (f) $y' = 2(x + 1)(x^2 + 2x - 3)^{2/3}$.

20. Show that $y = \pm a$ are singular solutions to the differential equation $y\,y' = \sqrt{a^2 - y^2}$.

21. Verify that the function $y = x + 4\sqrt{x + 1}$ is a solution of the differential equation $(y - x)\,y' = y - x + 8$ on some interval.

22. The position of a particle on the x-axis at time t is $x(t) = t^{(t^t)}$ for $t > 0$. Let $v(t)$ be the velocity of the particle at time t. Find $\lim_{t\to 0} v(t)$.

23. An airplane takes off at a speed of 225 km/hour. A landing strip has a runway of 1.8 km. If the plane starts from rest and moves with a constant acceleration, what is this acceleration?

24. Let $m(t)$ be the investment resulting from a deposit m_0 after t years at the interest rate r compounded daily. Show that
$$m(t) = m_0 \left[1 + \frac{r}{365}\right]^{365t}.$$
From calculus we know that $\left[1 + \frac{r}{n}\right]^{nt} \to e^{rt}$ as $n \to \infty$. Hence, $m(t) \to m_0 \exp\{rt\}$. What differential equation does the function $m(t)$ satisfy?

25. A particle moves along the abscissa so that its instantaneous acceleration is given as a function of time t by $a(t) = 2 - 3t^2$. At times $t = 1$ and $t = 4$, the particle is located at $x = 5$ and $x = -10$, respectively. Set up a differential equation and associated conditions describing the motion.

26. A particle moves along the abscissa in such a way that its instantaneous velocity is given as a function of time t by $v(t) = 6 - 3\,t^2$. At time $t = 0$, it is located at $x = 1$. Set up an initial value problem describing the motion of the particle and determine its position at any time $t > 0$.

27. A particle moves along the abscissa so that its velocity at any time $t \geqslant 0$ is given by $v(t) = 4/(t^2 + 1)$. Assuming that it is initially at π, show that it will never pass $x = 2$.

28. The slope of a family of curves at any point (x, y) of the plane is given by $1 + 2x$. Derive a differential equation of the family and solve it.

29. The graph of a nonnegative function has the property that the length of the arc between any two points on the graph is equal to the area of the region under the arc. Find a differential equation for the curve.

30. Geological dating of rocks is done using potassium-40 rather than carbon-14 because potassium has a longer half-life, 1.28×10^9 years (the half-life is the time required for the quantity to be reduced by one half). The potassium decays to argon, which remains trapped in the rocks and can be measured. Derive the differential equation that the amount of potassium obeys.

31. Prove that the equation $y' = (ay + b)/(cy + d)$ has at least one solution of the form $y = kx$ if either $b = 0$ or $ad = bc$.

32. Which straight lines through the origin are solutions of the following differential equations?
    (a) $y' = \dfrac{4x + 3y}{3x + y}$; (b) $y' = \dfrac{x + 3y}{y - x}$; (c) $y' = \dfrac{x + 3y}{x - y}$;
    (d) $y' = \dfrac{3y - 2x}{x - 3y}$; (e) $y' = \dfrac{x}{x + 2y}$; (f) $y' = \dfrac{2x + 3y}{2x + y}$.

33. Phosphorus (^{31}P) has multiple isotopes, two of which are used routinely in life-science laboratories dealing with DNA production. They are both beta-emitters, but differ by the energy of emissions—^{32}P has 1.71 MeV and ^{33}P has 0.25 MeV. Suppose that a sample of 32-isotope disintegrates to 71.2 mg in 7 days, and 33-isotope disintegrates to 82.6 mg during the same time period. If initially both samples were 100 mg, what are their half-life periods?

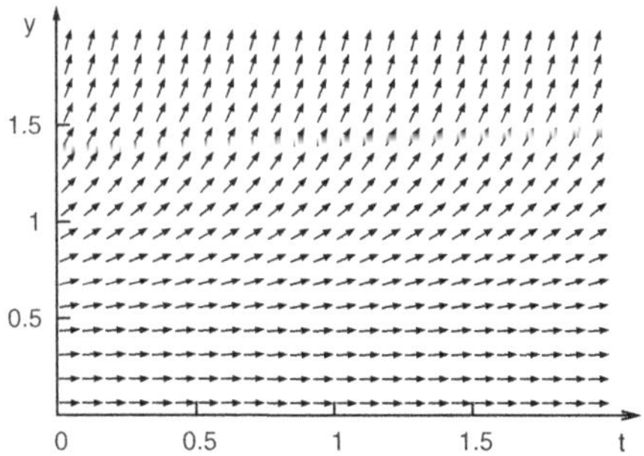

Figure 1.13: Direction field for Problem 36.

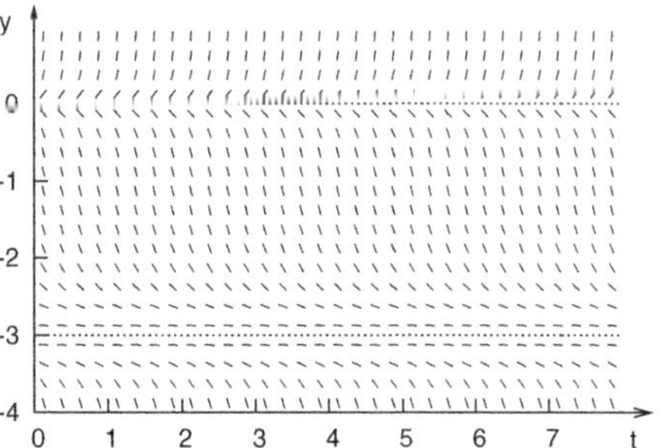

Figure 1.14: Direction field for Problem 37.

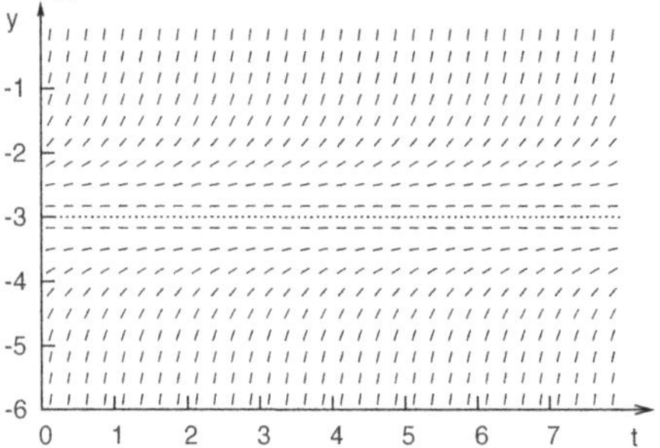

Figure 1.15: Direction field for Problem 38.

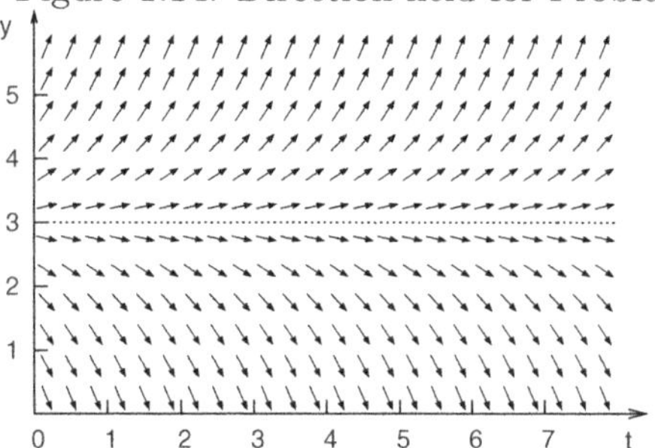

Figure 1.16: Direction field for Problem 39.

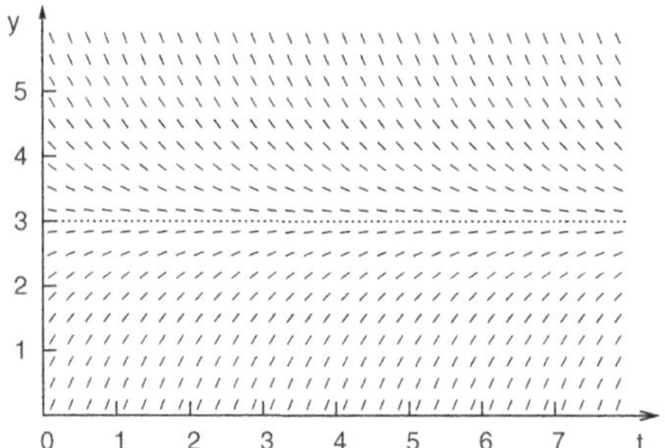

Figure 1.17: Direction field for Problem 40.

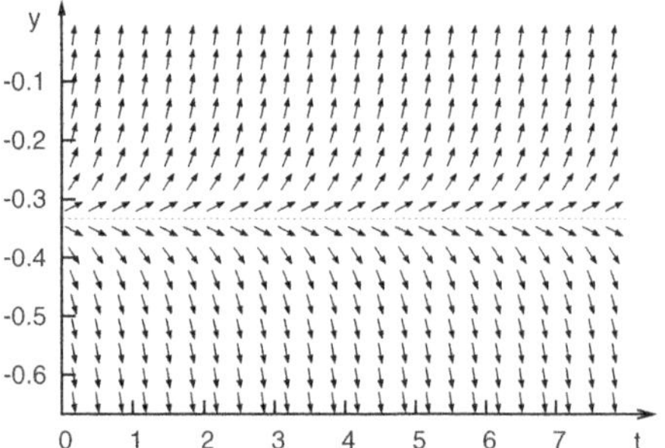

Figure 1.18: Direction field for Problem 41.

34. Show that the initial value problem $y' = 4x\sqrt{y}$, $y(0) = 0$ has infinitely many solutions.

The following problems require utilization of a computer package.

35. For the following two initial value problems

    (a) $y' = y(1-y)$, $y(0) = \frac{1}{2}$; **(b)** $y' = y(1-y)$, $y(0) = 2$;

    show that $y_a(x) = \left(1+e^{-x}\right)^{-1}$ and $y_b = 2\left(2-e^{-x}\right)^{-1}$ are their solutions, respectively. What are their long-term behaviors?

Consider the following list of differential equations, some of which produced the direction fields shown in Figures 1.13 through 1.18. In each of Problems 36 through 41, identify the differential equation that corresponds to the given direction field.

(a)	$\dot{y} = y^{2.5}$;	**(b)**	$\dot{y} = 3y - 1$;	**(c)**	$\dot{y} = y(y+3)^2$;	**(d)**	$\dot{y} = y^5(y+1)$;
(e)	$\dot{y} = y(y+3)^2$;	**(f)**	$\dot{y} = y^2(y+3)$;	**(g)**	$\dot{y} = (y+3)^2$;	**(h)**	$\dot{y} = y+3$;
(i)	$\dot{y} = y-3$;	**(j)**	$\dot{y} = y-3$;	**(k)**	$\dot{y} = -y+3$;	**(l)**	$\dot{y} = 3y+1$.

36. The direction field of Figure 1.13.
37. The direction field of Figure 1.14.
38. The direction field of Figure 1.15.
39. The direction field of Figure 1.16.
40. The direction field of Figure 1.17.
41. The direction field of Figure 1.18.

1.6 Existence and Uniqueness

An arbitrary differential equation of the first order $y' = f(x, y)$ does not necessarily have a solution that satisfies it. Therefore, the existence of a solution is an important problem for both the theory of differential equations and their applications.

If some phenomenon is modeled by a differential equation, then the equation should have a solution. If it does not, then presumably there is something wrong with the mathematical modeling and the simulation needs improvement. So, an engineer or a scientist would like to know whether a differential equation has a solution before investing time, effort, and computer applications in a vain attempt to solve it. An application of a software package may fail to provide a solution to a given differential equation, but this doesn't mean that the differential equation doesn't have a solution.

Whenever an initial value problem has been formulated, there are three questions that could be asked before finding a solution:

1. Does a solution of the differential equation satisfying the given conditions exist?
2. If one solution satisfying the given conditions exists, can there be a different solution that also satisfies the conditions?
3. What is the reason to determine whether an initial value problem has a unique solution if we won't be able to explicitly determine it?

A positive answer for the first question is our hunting license to go looking for a solution. In practice, one wishes to find the solution of a differential equation satisfying the given conditions to less than a finite number of decimal places. For example, if we want to draw the solution, our eyes cannot distinguish two functions which have values that differ by less than 1%. Therefore, for printing applications, the knowledge of three significant figures in the solution is admissible accuracy. This may be done, for instance, with the aid of available software packages.

In general, existence or uniqueness of an initial value problem cannot be guaranteed. For example, the initial value problem $y' = y^2$, $x < 1$, $y(1) = -1$ has a solution $y(x) = -x^{-1}$, which does not exist for $x = 0$. On the other hand, Example 1.4.2 on page 9 shows that the initial value problem may have two (or more) solutions.

For most of the differential equations in this book, there are unique solutions that satisfy certain prescribed conditions. However, let us consider the differential equation

$$xy' - 5y = 0,$$

which arose in a certain problem. Suppose a scientist has drawn an experimental curve as shown on the left side of Fig.1.19 (page 22).

The general solution of the given differential equation is $y = C\,x^5$ with an arbitrary constant C. From the initial condition $y(1) = 2$, it follows that $C = 2$ and $y = 2\,x^5$. Thus, the theoretical and experimental graphs agreed for $x > 0$, but disagree for $x < 0$.

If the scientist had erroneously assumed that a unique solution exists, s/he may decide that the mathematics was wrong. However, since the differential equation has a singular point $x = 0$, its general solution contains two arbitrary constants, A and B, one for domain $x > 0$ and another one for $x < 0$. So

$$y(x) = \begin{cases} A\,x^5, & x \geqslant 0, \\ B\,x^5, & x \leqslant 0. \end{cases}$$

Therefore, the experimental graph corresponds to the case $A = 2$ and $B = 0$.

Now suppose that for the same differential equation $xy' = 5y$ we have the initial condition at the origin: $y(0) = 0$. Then any function $y = Cx^5$ satisfies it for arbitrary C and we have infinitely many solutions to the given initial value problem (IVP). On the other hand, if we want to solve the given equation with the initial condition $y(0) = 1$, we are out of luck. There is no solution to this initial value problem!

In this section, we discuss two fundamental theorems for first order ordinary differential equations subject to initial conditions that prove the existence and the uniqueness of their solutions. These theorems provide *sufficient conditions* for the existence and uniqueness of a solution; that is, if the conditions hold, then uniqueness and/or existence are guaranteed. However, the conditions are not *necessary conditions* at all; there may still be a unique solution if these conditions are not met. The following theorem guarantees the uniqueness and existence for linear differential equations.

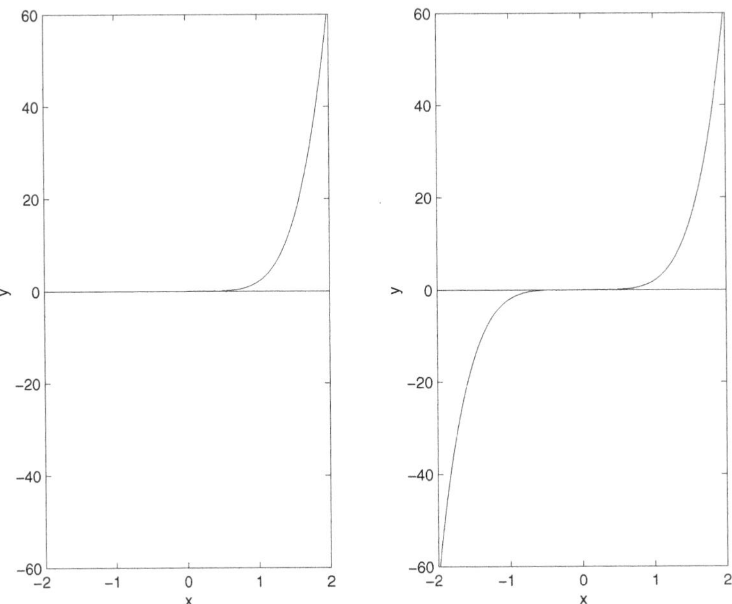

Figure 1.19: Experimental curve at the left and modeled solution at the right.

Theorem 1.1: Let us consider the initial value problem for the linear differential equation

$$y' + a(x)y = f(x), \tag{1.6.1}$$

$$y(x_0) = y_0, \tag{1.6.2}$$

where $a(x)$ and $f(x)$ are known functions and y_0 is an arbitrary prescribed initial value. Assume that the functions $a(x)$ and $f(x)$ are continuous on an open interval $\alpha < x < \beta$ containing the point x_0. Then the initial value problem (1.6.1), (1.6.2) has a unique solution $y = \phi(x)$ on the same interval (α, β).

PROOF: In §2.5 we show that if Eq. (1.6.1) has a solution, then it must be given by the following formula:

$$y(x) = \mu^{-1}(x)\left[\int \mu(x)f(x)\,\mathrm{d}x + C\right], \quad \mu(x) = \exp\left\{\int a(x)\,\mathrm{d}x\right\}. \tag{1.6.3}$$

When $\mu(x)$ is a nonzero differentiable function on the interval (α, β), we have from Eq. (2.5.2), page 86, that

$$\frac{\mathrm{d}}{\mathrm{d}x}\left[\mu(x)y(x)\right] = \mu(x)f(x).$$

Since both $\mu(x)$ and $f(x)$ are continuous functions, its product $\mu(x)f(x)$ is integrable, and formula (1.6.3) follows from the latter. Hence, from Eq. (1.6.3), the function $y(x)$ exists and is differentiable over the interval (α, β). By substituting the expression for $y(x)$ into Eq. (1.6.1), one can verify that this expression is a solution of Eq. (1.6.1). Finally, the initial condition (1.6.2) determines the constant C uniquely.

If we choose the lower limit to be x_0 in all integrals in the expression (1.6.3), then

$$y(x) = \frac{1}{\mu(x)}\left[\int_{x_0}^{x} \mu(s)f(s)\,\mathrm{d}s + y_0\right], \qquad \mu(x) = \exp\left\{\int_{x_0}^{x} a(s)\,\mathrm{d}s\right\}$$

is the solution of the initial value problem (1.6.1), (1.6.2). ■

In 1886, Giuseppe Peano[9] gave sufficient conditions that only guarantee the existence of a solution for initial value problems (IVPs).

[9]Giuseppe Peano (1858–1932) was a famous Italian mathematician who worked at the University of Turin. The existence theorem was published in his article [36].

The Peano existence theorem can be viewed as a generalization of the fundamental theorem of calculus, which makes the same assertion for the first order equation $y' = f(x)$. Geometrical intuition suggests that a solution curve, if any, of the equation $y' = f(x,y)$ can be obtained by threading the segments of the direction field. We may also imagine that a solution is a trajectory or path of a particle moving under the influence of the force field. Physical intuition asserts the existence of such trajectories when that field is continuous.

Theorem 1.2: [Peano] Suppose that the function $f(x,y)$ is continuous in some rectangle:

$$\Omega = \{(x,y) : x_0 - a \leqslant x \leqslant x_0 + a, \quad y_0 - b \leqslant y \leqslant y_0 + b\}, \tag{1.6.4}$$

Let

$$M = \max_{(x,y)\in\Omega} |f(x,y)|, \quad h = \min\left\{a, \frac{b}{M}\right\}. \tag{1.6.5}$$

Then the initial value problem

$$y' = f(x,y), \qquad y(x_0) = y_0, \tag{1.6.6}$$

has a solution in the interval $[x_0 - h, x_0 + h]$.

Corollary 1.1: If the continuous function $f(x,y)$ in the domain $\Omega = \{(x,y) : \alpha < x < \beta, \ -\infty < y < \infty\}$ satisfies the inequality $|f(x,y)| \leqslant a(x)|y| + b(x)$, where $a(x)$ and $b(x)$ are positive continuous functions, then the solution to the initial value problem (1.6.1), (1.6.2) exists in the interval $\alpha < x < \beta$. ■

In most of today's presentations, Peano's theorem is proved with the help of either the Arzela–Ascoli compactness principle for function sequences or Banach's fix-point theorem, which are both beyond the scope of this book.

In 1890, Peano showed that the solution of the nonlinear differential equation $y' = 3y^{2/3}$ subject to the initial condition $y(0) = 0$ is not unique. He discovered, and published, a method for solving linear differential equations using successive approximations. However[10], Emile Picard[11] had independently rediscovered this method and applied it to show the existence and uniqueness of solutions to the initial value problems for ordinary differential equations. His result, known as Picard's theorem, imposes a stronger condition[12] on $f(x,y)$ to prevent the equation $y' = f(x,y)$ from having singular solutions.

Theorem 1.3: [Picard] Let $f(x,y)$ be a continuous function in a rectangular domain Ω containing the point (x_0, y_0). If $f(x,y)$ satisfies the Lipschitz condition

$$|f(x,y_1) - f(x,y_2)| \leqslant L|y_1 - y_2|$$

for some positive constant L (called the Lipschitz constant) and any x, y_1, and y_2 from Ω, then the initial value problem (1.6.6) has a unique solution in some interval $x_0 - h \leqslant x \leqslant x_0 + h$, where h is defined in Eq. (1.6.5).

PROOF: We cannot guarantee that the solution $y = \phi(x)$ of the initial value problem (1.6.6) exists in the interval $(x_0 - a, x_0 + a)$ because the integral curve $y = \phi(x)$ can exist outside of the rectangle Ω. For example, if there exists x_1 such that $x_0 - a < x_1 < x_0 + a$ and $y_0 + b = \phi(x_1)$, then for $x > x_1$ (if $x_1 > x_0$) the solution $\phi(x)$ cannot be defined.

We definitely know that the solution $y = \phi(x)$ is in the range $y_0 - b \leqslant \phi(x) \leqslant y_0 + b$ when $x_0 - h \leqslant x \leqslant x_0 + h$ with $h = \min\{a, \ b/M\}$ since the slope of the graph of the solution $y = \phi(x)$ is at least $-M$ and at most M. If the graph of the solution $y = \phi(x)$ crosses the lines $y = y_0 \pm b$, then the points of intersection with the abscissa are $x_0 \pm b/M$. Therefore, the abscissa at the point where the integral curve goes out of the rectangle Ω is less than or equal to $x_0 + b/M$ and is greater than or equal to $x_0 - b/M$.

[10] In 1838, Joseph Liouville first used the method of successive approximations in a special case.

[11] Charles Emile Picard (1856–1941) was one of the greatest French mathematicians of the nineteenth century. In 1899, Picard lectured at Clark University in Worcester, Massachusetts. Picard and his wife had three children, a daughter and two sons, who were all killed in World War I.

[12] It is called the Lipschitz condition in honor of the German mathematician Rudolf Lipschitz (1832–1903), who introduced it in 1876 when working out existence proofs for ordinary differential equations.

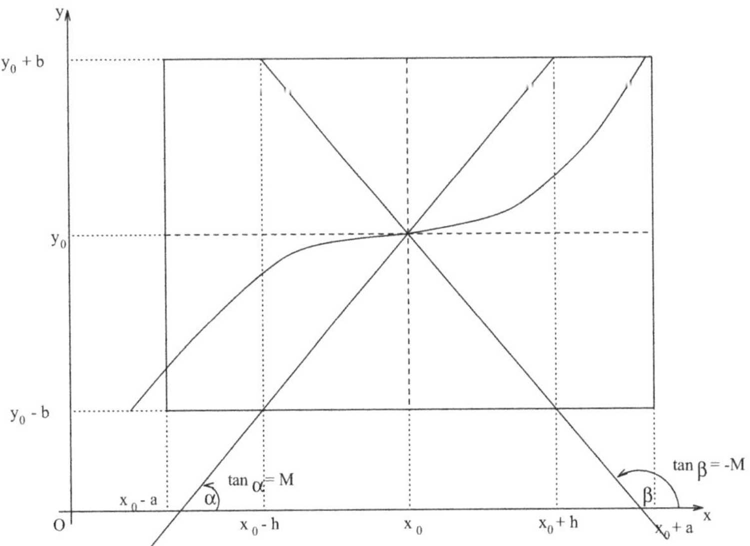

Figure 1.20: The domain of existence.

To prove the theorem, we transform the initial value problem (1.6.6) into an integral equation. After integrating both sides of Eq. (1.6.6) from the initial point x_0 to an arbitrary value of x, we obtain

$$y(x) = y_0 + \int_{x_0}^{x} f(s, y(s))\, \mathrm{d}s. \tag{1.6.7}$$

Since the last equation contains an integral of the unknown function $y(x)$ it is called an **integral equation**. More precisely, the equation of the form (1.6.7) is called a *Volterra integral equation of the second kind*. This integral equation is equivalent to the initial value problem (1.6.6) in the sense that any solution of one is also a solution of the other.

We prove the existence and uniqueness of Eq. (1.6.7) using **Picard's iteration method** or the **method of successive approximations**. We start by choosing an initial function ϕ_0, either arbitrarily or to approximate solution of Eq. (1.6.6) in some way. The simplest choice is

$$\phi_0(x) = y_0.$$

Then this (constant) function satisfies the initial condition $y(x_0) = y_0$. The next approximation ϕ_1 is obtained by substituting ϕ_0 for $y(s)$ in the right side of Eq. (1.6.7), namely,

$$\phi_1(x) = y_0 + \int_{x_0}^{x} f(s, \phi_0)\, \mathrm{d}s = y_0 + \int_{x_0}^{x} f(s, y_0)\, \mathrm{d}s.$$

Let us again substitute the first order approximation in the right-hand side of Eq. (1.6.7) to obtain

$$\phi_2(x) = y_0 + \int_{x_0}^{x} f(s, \phi_1(s))\, \mathrm{d}s.$$

Each successive substitution into Eq. (1.6.7) results in a sequence of functions. In general, if the n-th approximation $\phi_n(s)$ has been obtained in this way, then the $(n+1)$-th approximation is taken to be the result of substituting ϕ_n in the right-hand side of Eq. (1.6.7). Therefore,

$$\phi_{n+1}(x) = y_0 + \int_{x_0}^{x} f(s, \phi_n(s))\, \mathrm{d}s. \tag{1.6.8}$$

All terms of the sequence $\{\phi_n(x)\}$ exist because

$$\left| \int_{x_0}^{x} f(s, \phi(s))\, \mathrm{d}s \right| \leqslant \max |f(x,y)| \left| \int_{x_0}^{x} \mathrm{d}s \right| = M|x - x_0| \leqslant b$$

for $|x - x_0| \leqslant b/M$, where $M = \max_{(x,y)\in\Omega} |f(x,y)|$.

The method of successive approximations gives a solution of Eq. (1.6.7) if and only if the successive approximations ϕ_{n+1} from Eq. (1.6.8) approach uniformly to a certain limit as $n \to \infty$. Then the sequence $\{\phi_n(x)\}$ converges to a true solution $y = \phi(x)$ as $n \to \infty$, which in fact is the unique solution of Eq. (1.6.7):

$$y = \phi(x) = \lim_{n\to\infty} \phi_n(x). \tag{1.6.9}$$

We can identify each element $\phi_n(x)$ on the right-hand side of Eq. (1.6.9),

$$\phi_n(x) = \phi_0 + [\phi_1(x) - \phi_0] + [\phi_2(x) - \phi_1(x)] + \cdots + [\phi_n(x) - \phi_{n-1}(x)],$$

as the n-th partial sum of the telescopic series

$$\phi(x) = \phi_0 + \sum_{n=1}^{\infty} [\phi_n(x) - \phi_{n-1}(x)]. \tag{1.6.10}$$

The convergence of the sequence $\{\phi_n(x)\}$ is established by showing that the series (1.6.10) converges. To do this, we estimate the magnitude of the general term $|\phi_n(x) - \phi_{n-1}(x)|$.

We start with the first iteration:

$$|\phi_1(x) - \phi_0| = \left| \int_{x_0}^{x} f(s, y_0)\, \mathrm{d}s \right| \leqslant M \int_{x_0}^{x} \mathrm{d}s = M|x - x_0|.$$

For the second term we have

$$|\phi_2(x) - \phi_1(x)| \leqslant \int_{x_0}^{x} |f(s,\phi_1(s)) - f(s,\phi_0(s))|\, \mathrm{d}s \leqslant L \int_{x_0}^{x} M(s - x_0)\, \mathrm{d}s = \frac{ML(x - x_0)^2}{2},$$

where L is the Lipschitz constant for the function $f(x,y)$, that is, $|f(x,y_1) - f(x,y_2)| \leqslant L|y_1 - y_2|$. For the n-th term, we have

$$\begin{aligned}
|\phi_n(x) - \phi_{n-1}(x)| &\leqslant \int_{x_0}^{x} |f(s,\phi_{n-1}(s)) - f(s,\phi_{n-2}(s))|\, \mathrm{d}s \\
&\leqslant L \int_{x_0}^{x} |\phi_{n-1}(s) - \phi_{n-2}(s)|\, \mathrm{d}s \leqslant L \int_{x_0}^{x} \frac{ML^{n-2}(s - x_0)^{n-1}}{(n-1)!}\, \mathrm{d}s \\
&= \frac{ML^{n-1}|x - x_0|^n}{n!} \leqslant \frac{ML^{n-1}h^n}{n!}, \qquad h = \max|x - x_0|.
\end{aligned}$$

Substituting these results into the finite sum

$$\phi_n(x) = \phi_0 + \sum_{k=1}^{n} [\phi_k(x) - \phi_{k-1}(x)],$$

we obtain

$$\begin{aligned}
|\phi_n - \phi_0| &\leqslant M|x - x_0| + \frac{ML|x - x_0|^2}{2} + \cdots + \frac{ML^{n-1}|x - x_0|^n}{n!} \\
&= \frac{M}{L} \left[L|x - x_0| + \frac{L^2|x - x_0|^2}{2} + \cdots + \frac{L^n|x - x_0|^n}{n!} \right].
\end{aligned}$$

When n approaches infinity, the sum

$$L|x - x_0| + \frac{L^2|x - x_0|^2}{2} + \cdots + \frac{L^n|x - x_0|^n}{n!}$$

approaches $e^{L|x-x_0|} - 1$. Thus, we have

$$|\phi_n(x) - y_0| \leqslant \frac{M}{L} \left[e^{L|x-x_0|} - 1 \right]$$

for any n. Therefore, by the Weierstrass M-test, the series (1.6.10) converges absolutely and uniformly on the interval $|x - x_0| \leqslant h$. It follows that the limit function (1.6.9) of the sequence (1.6.8) is a continuous function on the interval $|x - x_0| \leqslant h$. Sometimes a sequence of continuous functions converges to a limit function that is discontinuous, as Exercise 26 on page 35 shows. This may happen only when the sequence of functions converges pointwise, but not uniformly.

Next we will prove that $\phi(x)$ is a solution of the initial value problem (1.6.6). First of all, $\phi(x)$ satisfies the initial condition. In fact, from (1.6.8),

$$\phi_n(x_0) = y_0, \quad n = 0, 1, 2, \ldots$$

and taking limits of both sides as $n \to \infty$, we find $\phi(x_0) = y_0$. Since $\phi(x)$ is represented by a uniformly convergent series (1.6.10), it is a continuous function on the interval $x_0 - h \leqslant x \leqslant x_0 + h$.

Allowing n to approach ∞ on both sides of Eq. (1.6.8), we get

$$\phi(x) = y_0 + \lim_{n\to\infty} \int_{x_0}^{x} f(s, \phi_n(s))\,\mathrm{d}s. \tag{1.6.11}$$

Recall that the function $f(x, y)$ satisfies the Lipschitz condition for $|s - x_0| \leqslant h$:

$$|f(s, \phi_n(s)) - f(s, \phi(s))| \leqslant L|\phi_n(s) - \phi(s)|.$$

Since the sequence $\phi_n(s)$ converges uniformly to $\phi(s)$ on the interval $|s - x_0| \leqslant h$, it follows that the sequence $f(s, \phi_n(s))$ also converges uniformly to $f(s, \phi(s))$ on this interval. Therefore, we can interchange integration with the limiting operation on the right-hand side of Eq. (1.6.11) to obtain

$$\phi(x) = y_0 + \int_{x_0}^{x} \lim_{n\to\infty} f(s, \phi_n(s))\,\mathrm{d}s = y_0 + \int_{x_0}^{x} f(s, \phi(s))\,\mathrm{d}s.$$

Thus, the limit function $\phi(x)$ is a solution of the integral equation (1.6.7) and, consequently, a solution of the initial value problem (1.6.6). In general, taking the limit under the sign of integration is not permissible, as Exercise 27 (page 35) shows. However, it is true for a uniform convergent sequence. Differentiating both sides of the last equality with respect to x and noting that the right-hand side is a differentiable function of the upper limit, we find

$$\phi'(x) = f(x, \phi(x)).$$

This completes the proof that the limit function $\phi(x)$ is a solution of the initial value problem (1.6.6). ■

Uniqueness. Finally we will prove that $\phi(x)$ is the only solution of the initial value problem (1.6.6). To start, we assume the existence of another solution $y = \psi(x)$. Then

$$\phi(x) - \psi(x) = \int_{x_0}^{x} [\,f(s, \phi(s)) - f(s, \psi(s))\,]\,\mathrm{d}s$$

for $|x - x_0| \leqslant h$. Setting $U(x) = |\phi(x) - \psi(x)|$, we have

$$U(x) \leqslant \int_{x_0}^{x} |f(s, \phi(s)) - f(s, \psi(s))|\,\mathrm{d}s \leqslant L \int_{x_0}^{x} U(s)\,\mathrm{d}s.$$

By differentiating both sides with respect to x, we obtain $U'(x) - LU(x) \leqslant 0$. Multiplying by the integrating factor e^{-Lx} reduces the latter inequality to the following one:

$$\left[e^{-Lx}\, U(x)\right]' \leqslant 0.$$

The function $e^{-Lx}\, U(x)$ has a nonpositive derivative and therefore does not increase with x. After integrating from x_0 to x $(> x_0)$, we obtain

$$e^{-Lx}\, U(x) \leqslant 0$$

since $U(0) = 0$. The absolute value of any number is positive; hence, $U(x) \geqslant 0$. Thus, $U(x) \equiv 0$ for $x > x_0$. The case $x < x_0$ can be treated in a similar way. Therefore $\phi(x) \equiv \psi(x)$.

Corollary 1.2: If the functions $f(x, y)$ and $\partial f/\partial y$ are continuous in a rectangle (1.6.4), then the initial value problem $y' = f(x, y)$, $y(x_0) = y_0$ has a unique solution in the interval $|x - x_0| \leqslant h$, where h is defined in Eq. (1.6.5) and the Lipschitz constant is $L = \max|\partial f(x, y)/\partial y|$.

Corollary 1.3: If the functions $f(x, y)$ and $\partial f/\partial x$ are continuous in a neighborhood of the point (x_0, y_0) and $f(x_0, y_0) \neq 0$, then the initial value problem $y' = f(x, y)$, $y(x_0) = y_0$ has a unique solution.

The above statements follow from the mean value relation

$$f(x, y_1) - f(x, y_2) = f_y(x, \xi)(y_1 - y_2),$$

where $\xi \in [y_1, y_2]$ and $f_y = \partial f/\partial y$. The proof of Corollary 1.3 is based on conversion of the original problem to its reciprocal counterpart: $\partial x/\partial y = 1/f(x, y)$. ∎

The proof of Picard's theorem is an example of a constructive proof that includes an iterative procedure and an error estimate. When iteration stops at some step, it gives an approximation to the actual solution. With the availability of a computer algebra system, such an approximation can be exactly found for many analytically defined slope functions. The disadvantage of Picard's method is that it only provides a solution locally—in a small neighborhood of the initial point.

Usually, the solution to the IVP exists in a wider region (see, for instance, Example 1.6.3) than Picard's theorem guarantees. Once the solution $y = \phi(x)$ of the given initial value problem is obtained, we can consider another initial condition at the point $x = x_0 + \Delta x$ and set $y_0 = \phi(x_0 + \Delta x)$. Application of Picard's theorem to this IVP may allow us to extend the solution to a larger domain. By continuing in such a way, we could extend the solution of the original problem to a bigger domain until we reach the boundary (this domain could be unbounded; in this case we define the function in the interval $x_0 - h \leqslant x < \infty$). Similarly, we can extend the solution to the left end of the initial interval $x_0 - h \leqslant x \leqslant x_0 + h$. Therefore, we may obtain some open interval $p < x < q$ (which could be unbounded) on which the given IVP has a unique solution. Such an approach is hard to call constrictive.

The ideal existence theorem would assure the existence of a solution in a longest possible interval, called the **validity interval**. It turns out that another method is known (see, for example, [24]) to extend the existence theorem that furnishes a solution in a validity interval. It was invented in 1768 by Leonhard Euler (see §3.2). However, the systematic method was developed by the famous French mathematician Augustin-Louis Cauchy (1789–1857) between the years 1820 and 1830. Later, in 1876, Rudolf Lipschitz substantially improved it. The Cauchy–Lipschitz method is based on the following fundamental inequality. This inequality can be used not only to prove the existence theorem by linear approximations, but also to find estimates produced by numerical methods that are discussed in Chapter 3.

Theorem 1.4: [Fundamental Inequality] Let $f(x, y)$ be a continuous function in the rectangle $[a, b] \times [c, d]$ and satisfying the Lipschitz condition

$$|f(x, y_1) - f(x, y_2)| \leqslant L|y_1 - y_2|$$

for some positive constant L and all pairs y_1, y_2 uniformly in x. Let $y_1(x)$ and $y_2(x)$ be two continuous piecewise differentiable functions satisfying the inequalities

$$|y_1'(x) - f(x, y_1(x))| \leqslant \epsilon_1, \qquad |y_2'(x) - f(x, y_2(x))| \leqslant \epsilon_2$$

with some positive constants $\epsilon_{1,2}$. If, in addition, these functions differ by a small amount $\delta > 0$ at some point:

$$|y_1(x_0) - y_2(x_0)| \leqslant \delta,$$

then

$$|y_1(x) - y_2(x)| \leqslant \delta\, e^{L|x-x_0|} + \frac{\epsilon_1 + \epsilon_2}{L}\left(e^{L|x-x_0|} - 1\right). \tag{1.6.12}$$

The Picard theorem can be extended for non-rectangular domains, as the following theorem[13] states.

[13] In honor of Sergey Mikhailovich Lozinskii/Lozinsky (1914–1985), a famous Russian mathematician who made an important contribution to the error estimation methods for various types of approximate solutions of ordinary differential equations.

Theorem 1.5: [**Lozinsky**] Let $f(x,y)$ be a continuous function in some domain Ω and $M(x)$ be a nonnegative continuous function on some finite interval I ($x_0 \leqslant x \leqslant x_1$) inside Ω. Let $|f(x,y)| \leqslant M(x)$ for $x \in I$ and $(x,y) \in \Omega$. Suppose that the closed finite domain Q, defined by inequalities

$$x_0 \leqslant x \leqslant x_1, \qquad |y - y_0| \leqslant \int_{x_0}^{x} M(u)\,\mathrm{d}u,$$

is a subset of Ω and there exists a nonnegative integrable function $k(x)$, $x \in I$, such that

$$|f(x,y_2) - f(x,y_1)| \leqslant k(x)\,|y_2 - y_1|, \qquad x_0 \leqslant x \leqslant x_1, \quad (x,y_2),(x,y_1) \in Q.$$

Then formula (1.6.8) on page 24 defines the sequence of functions $\{\phi_n(x)\}$ that converges to a unique solution of the given IVP (1.6.1), (1.6.2) provided that all points $(x,\phi_n(x))$ are included in Q when $x_0 \leqslant x \leqslant x_1$. Moreover,

$$|y(x) - \phi_n(x)| \leqslant \frac{1}{n!}\int_{x_0}^{x} M(u)\,\mathrm{d}u \left[\int_{x_0}^{x} k(u)\,\mathrm{d}u\right]^n. \tag{1.6.13}$$

Actually, the Picard theorem allows us to determine the accuracy of the n-th approximation:

$$\begin{aligned}
|\phi(x) - \phi_n(x)| &\leqslant \left|\int_{x_0}^{x} [f(x,\phi(x)) - f(x,\phi_{n-1}(x))]\,\mathrm{d}x\right| \\
&\leqslant \int_{x_0}^{x} |f(x,\phi(x)) - f(x,\phi_{n-1}(x))\,\mathrm{d}x| \leqslant \int_{x_0}^{x} L|\phi(x) - \phi_{n-1}(x)|\,\mathrm{d}x \\
&\leqslant \int_{x_0}^{x} L\,\mathrm{d}x \int_{x_0}^{x} L|\phi(x) - \phi_{n-2}(x)|\,\mathrm{d}x \leqslant \ldots .
\end{aligned}$$

Therefore,

$$|\phi(x) - \phi_n(x)| \leqslant \frac{M}{L}\,\frac{(L|x-x_0|)^{n+1}}{n!} \leqslant \frac{ML^n}{n!}\,h^{n+1} = \frac{M}{L}\,\frac{(Lh)^{n+1}}{n!}, \tag{1.6.14}$$

which is in agreement with the inequality (1.6.13).

Example 1.6.1: Find the global maximum of the continuous function $f(x,y) = (x+2)^2 + (y-1)^2 + 1$ in the domain $\Omega = \{(x,y):\ |x+2| \leqslant a,\ |y-1| \leqslant b\}$.

Solution. The function $f(x,y)$ has the global minimum at $x = -2$ and $y = 1$. This function reaches its maximum values in the domain Ω at such points that are situated furthest from the critical point $(-2,1)$. These points are vertices of the rectangle Ω, namely, $(-2 \pm a, 1 \pm b)$. Since the values of the function $f(x,y)$ at these points coincide, we have

$$\max_{(x,y)\in\Omega} f(x,y) = f(-2+a, 1+b) = a^2 + b^2 + 1.$$

Example 1.6.2: Show that the function

$$f(y) = \begin{cases} y\ln|y|, & \text{if } y \neq 0, \\ 0, & \text{if } y = 0 \end{cases}$$

is not a Lipschitz function on the interval $[-b,b]$, where b is a positive real number.

Solution. We prove by contradiction, assuming that $f(y)$ is a Lipschitz function. Let y_1 and y_2 be arbitrary points from this interval. Then $|f(y_1) - f(y_2)| \leqslant L|y_1 - y_2|$ for some constant L. We set $y_2 = 0$ ($\lim_{y\to 0} y\ln|y| = 0$), making $|f(y_1)| = |y_1 \ln|y_1|| \leqslant L|y_1|$ or $|\ln|y_1|| \leqslant L$ for all $y_1 \in [0,b]$, which is impossible. Indeed, for small values of an argument y, the function $\ln y$ is unbounded. Thus, the function $f(y)$ does not satisfy the Lipschitz condition.

Example 1.6.3: Let us consider the initial value problem for the Riccati equation

$$y' = x^2 + y^2, \quad y(0) = 0$$

in the rectangle $\{(x,y) : |x| \leqslant a, \quad |y| \leqslant b\}$. The maximum value of h is $\max\min\left\{a,\ \dfrac{b}{a^2+b^2}\right\} = 2^{-1/2}$, which does not exceed 1. According to Picard's theorem, the solution of the given IVP exists (and could be found by successive approximations) within the interval $|x| < h < 1$, but cannot be extended on whole line. The best we can get from the theorem is the existence of the solution on the interval $|x| \leqslant \frac{1}{\sqrt{2}} \approx .7071067810$.

This result can be improved with the aid of the Lozinsky theorem (on page 28). We consider the domain Ω defined by inequalities (containing positive numbers x_1 and A to be determined)

$$0 \leqslant x \leqslant x_1, \qquad 0 \leqslant |y| \leqslant A\,x^3 .$$

Then $M(x) = \max_{(x,y)\in\Omega}\left(x^2+y^2\right) = x^2 + A^2x^6$, $0 \leqslant x \leqslant x_1$, and the set Q is defined by inequalities

$$0 \leqslant x \leqslant x_1, \qquad 0 \leqslant |y| \leqslant \frac{x^3}{3} + A^2\,\frac{x^7}{7} .$$

In order to have $Q \subset \Omega$, the following inequality must hold:

$$\frac{1}{3} + A^2\,\frac{x_1^4}{7} \leqslant A .$$

This is equivalent to the quadratic equation $A^2\,\frac{x_1^4}{7} - A + \frac{1}{3} = 0$, which has two roots: $A = \dfrac{1 \pm \sqrt{1 - \frac{4\,x_1^4}{21}}}{2\,x_1^4/7}$. Hence, the quadratic equation has two real roots when x_1 satisfies enequality $x_1 \leqslant \sqrt[4]{\frac{21}{4}} \approx 1.513700052$. Therefore, Lozinsky's theorem gives a larger interval of existence (more than double) than Picard's theorem.

As shown in Example 2.6.10, page 103, this Riccati equation $y' = x^2 + y^2$ has the general solution of the ratio $-u'/(2u)$, where u is a linear combination of Bessel functions (see §4.9):

$$y(x) = \frac{1}{2x} - x\,\frac{J_{1/4}'(x^2/2) + C\,Y_{1/4}'(x^2/2)}{J_{1/4}(x^2/2) + C\,Y_{1/4}(x^2/2)} .$$

A constant C should be chosen to satisfy the initial condition $y(0) = 0$ (see Eq. (2.7.1) on page 114 for the explicit expression). As seen from this formula, the denominator has zeroes, the smallest of them is approximately 2.003, so the solution has the asymptote $x = h$, where $h < 2.003$.

The situation changes when we consider another initial value problem (IVP):

$$y' = x^2 - y^2, \qquad y(0) = 0.$$

The slope function $x^2 - y^2$ attains its maximum in a domain containing lines $y = \pm x$:

$$\max|x^2 - y^2| = \begin{cases} x^2, & \text{if } x^2 \geqslant y^2, \\ y^2, & \text{if } y^2 \geqslant x^2. \end{cases}$$

For the function $x^2 - y^2$ in the rectangle $|x| \leqslant a$, $|y| \leqslant b$, the Picard theorem guarantees the existence of a unique solution within the interval $|x| \leqslant h$, where $h = \min\left\{a,\ \dfrac{b}{\max\{a^2, b^2\}}\right\} \leqslant 1$. To extend the interval of existence, we apply the Lozinsky theorem.

First, we consider the function $x^2 - y^2$ in the domain Ω bounded by inequalities $0 \leqslant x \leqslant x_p^*$ and $|y| \leqslant Ax^p$, where A and p are some positive constants, and x_p^* will be determined shortly. Then

$$|x^2 - y^2| \leqslant M(x) \equiv \max_{(x,y)\in\Omega} |x^2 - y^2| = \begin{cases} x^2, & \text{if } x^2 \geqslant (Ax^p)^2, \\ (Ax^p)^2 = A^2x^{2p}, & \text{if } (Ax^p)^2 \geqslant x^2. \end{cases}$$

Now we define the domain Q by inequalities $0 \leqslant x \leqslant x_p^*$, $|y| \leqslant \int_0^x M(u)\,\mathrm{d}u$, where

$$\begin{aligned} \int_0^x M(u)\,\mathrm{d}u &= \int_0^{A^{1/(p-1)}} u^2\,\mathrm{d}u + \int_{A^{1/(p-1)}}^x A^2u^{2p}\,\mathrm{d}u \\ &= \frac{1}{3}\,A^{3/(p-1)} - \frac{1}{2p+1}\,A^{4+3/(p-1)} + \frac{A^2}{2p+1}\,x^{2p+1}. \end{aligned}$$

In order to guarantee inclusion $Q \subset \Omega$, the following inequality should hold: $\int_0^x M(u)\,\mathrm{d}u \leqslant Ax^p$. It is valid in the interval $\epsilon < x < x_p^*$, where x_p^* is the root of the equation $\int_0^x M(u)\,\mathrm{d}u = Ax^p$ and ϵ is a small number. When $A \to +0$ and $p \to 1+0$, the root, x_p^*, could be made arbitrarily large. For instance, when $A = 0.001$ and $p = 1.001$, the root is $x_p^* \approx 54.69$. Therefore, the given IVP has a solution on the whole line $-\infty < x < \infty$.

Example 1.6.4: (Example 1.4.2 revisited) Let us reconsider the initial value problem (page 9):

$$\frac{dy}{dx} = 2\,y^{1/2}, \quad y(0) = 0.$$

Peano's theorem (page 23) guarantees the existence of the initial value problem since the slope function $f(x, y) = 2\,y^{1/2}$ is a continuous function. The critical point $y \equiv 0$ is obviously a solution of the initial value problem. We show that $f(x, y) = 2\,y^{1/2}$ is not a Lipschitz function by assuming the opposite. That is to say suppose there exists a positive constant L such that

$$|y_1^{1/2} - y_2^{1/2}| \leqslant L|y_1 - y_2|.$$

Setting $y_2 = 0$, we have

$$|y_1^{1/2}| \leqslant L|y_1| \quad \text{or} \quad 1 \leqslant L\,|y_1^{1/2}|.$$

The last inequality does not hold for small y_1; therefore, $f(y) = 2\,y^{1/2}$ is not a Lipschitz function. In this case, we cannot apply Picard's theorem (page 23), and the given initial value problem may have multiple solutions. According to Theorem 2.2 (page 48), since the integral

$$\int_0^y \frac{dy}{2\sqrt{y}}$$

converges, the given initial value problem doesn't have a unique solution. Indeed, for arbitrary $x_0 > 0$, the function

$$y = \varphi(x) \stackrel{\text{def}}{=} \begin{cases} 0, & \text{for } -\infty < x \leqslant x_0, \\ (x - x_0)^2, & \text{for } x > x_0, \end{cases}$$

is a singular solution of the given initial value problem. Note that $y \equiv 0$ is the envelope of the one-parameter family of curves, $y = (x - C)^2$, $x \geqslant C$, corresponding to the general solution.

Example 1.6.5: Consider the autonomous equation

$$y' = |y|.$$

The slope function $f(x, y) = |y|$ is not differentiable at $x = 0$, but it is a Lipschitz function, with $L = 1$. According to Picard's theorem, the initial value problem with the initial condition $y(0) = y_0$ has a unique solution:

$$y(x) = \begin{cases} y_0\,e^{x}, & \text{for } y_0 > 0, \\ 0, & \text{for } y_0 = 0, \\ y_0\,e^{-x}, & \text{for } y_0 < 0. \end{cases}$$

Since an exponential function is always positive, the integral curves never meet or cross the equilibrium solution $y \equiv 0$.

Example 1.6.6: Does the initial value problem

$$\frac{dy}{dx} = 1 + \frac{3}{2}\,(y - x)^{1/3}, \qquad y(0) = 0,$$

have a singular solution? Find all solutions of this differential equation.

Solution. Changing the dependent variable to $y - x = u$, we find that the differential equation with respect to u is

$$u' = 3\,u^{1/3}/2.$$

The derivative of its slope function $f(x, u) = f(u) = \frac{3}{2}\,u^{1/3}$ is $f'(u) = 1/2\,u^{-2/3}$, which is unbounded at $u = 0$. In this case, Picard's theorem (page 23) is not valid and the differential equation $u' = \frac{3}{2}\,u^{1/3}$ may have a singular solution. Since the equation for u is autonomous, we apply Theorem 2.2 (page 48). The integral

$$\int_0^u \frac{du}{f(u)} = \frac{3}{2}\int_0^u u^{-1/3}\,du = u^{2/3}\big|_{u=0}^{u} = u^{2/3}$$

converges. Hence, there exists another solution besides the general one, $y = \eta(x + C)^{3/2}$, where $\eta = \pm 1$ and C is an arbitrary constant. Notice that the general solution of this nonlinear differential equation depends on more than the single constant of integration; it also depends on the discrete parameter η. Returning to the variable y, we get the singular solution $y = x$ (which corresponds to $u \equiv 0$) and the general solution $y = x + \eta(x - C)^{3/2}$, $x \geqslant C$.

Example 1.6.7: Let us find the solution of the problem

$$y' = y^2, \quad y(0) = 1,$$

by the method of successive approximations. We choose the first approximation $\phi_0 = 1$ according to the initial condition $y(0) = 1$. Then, from formula (1.6.8), we find

$$\phi_{n+1}(x) = 1 + \int_0^x \phi_n^2(s)\,\mathrm{d}s \qquad (n = 0, 1, 2, \ldots).$$

Hence, we have

$$\begin{aligned}
\phi_1(x) &= 1 + x; \\
\phi_2(x) &= 1 + \int_0^x (1+s)^2\,\mathrm{d}s = 1 + x + x^2 + \frac{x^3}{3}; \\
\phi_3(x) &= 1 + \int_0^x \left(1 + s + s^2 + \frac{s^3}{3}\right)^2 \mathrm{d}s \\
&= 1 + x + x^2 + x^3 + \frac{2}{3}x^4 + \frac{1}{3}x^5 + \frac{1}{63}x^7,
\end{aligned}$$

and so on. The limit function is

$$\phi(x) = 1 + x + x^2 + \cdots + x^n + \cdots = \frac{1}{1-x}.$$

To check our calculations, we can use either *Mathematica:*

```
Clear[phi]
T[phi_] := Function[x, 1 + Integrate[phi[t]^2, {t, 0, x}]];
f[x_] = 1;          (* specify the initial function *)
Nest[T, f, 5][x]    (* Find the result of 5th iterations *)
```

or *Maple:*

```
y0:=1:  T(phi,x):=(phi,x)->y0+eval(int(phi(t)^2,t=0..x)):
y:=array(0..n):   Y:=array(0..n):
y[0]:=x->y0:
for i from 1 to n do
    y[i]:=unapply(T(y[i-1],x),x):
    Y[i]:=plot(y[i](x),x=0..1):
od:
display([seq(Y[i],i=1..n)]);
seq(eval(y[i]),i=1..n);
```

Example 1.6.8: Using the Picard method, find the solution of the initial value problem

$$y' = x + 2y, \quad y(0) = 1.$$

Solution. From the recursion relation

$$\phi_{n+1}(x) = 1 + \int_0^x (t + 2\,\phi_n(t))\,\mathrm{d}t = 1 + \frac{x^2}{2} + 2\int_0^x \phi_n(t)\,\mathrm{d}t, \quad n = 0, 1, 2, \ldots,$$

where $\phi_0(x) \equiv 1$ is the initial function, we obtain

$$\begin{aligned}
\phi_1(x) &= 1 + \int_0^x (t+2)\,\mathrm{d}t = 1 + 2x + \frac{x^2}{2}, \\
\phi_2(x) &= 1 + \int_0^x (t + 2\,\phi_1(t))\,\mathrm{d}t = 1 + 2x + \frac{5}{2}x^2 + \frac{x^3}{3}, \\
\phi_3(x) &= 1 + \int_0^x (t + 2\,\phi_2(t))\,\mathrm{d}t = 1 + 2x + \frac{5}{2}x^2 + \frac{5}{3}x^3 + \frac{x^4}{6},
\end{aligned}$$

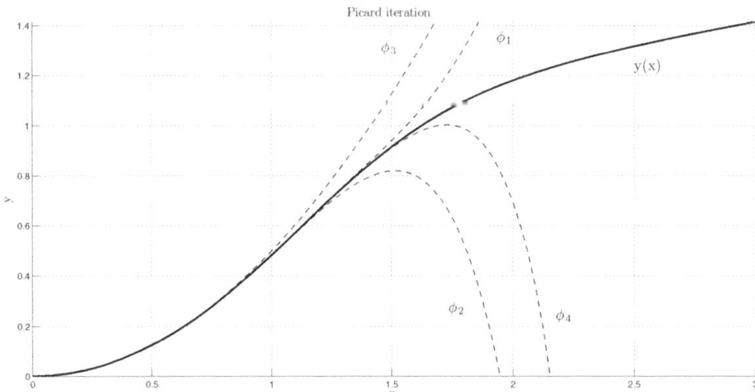

Figure 1.21: First four Picard approximations, plotted with MATLAB.

and so on.

To check calculations, we compare these approximations with the exact solution

$$\begin{aligned}
y(x) &= -\frac{1}{4} - \frac{x}{2} + \frac{5}{4}\,e^{2x} \\
&= -\frac{1}{4} - \frac{x}{2} + \frac{5}{4}\left[1 + 2x + \frac{4x^2}{2} + \frac{2^3\,x^3}{3!} + \frac{2^4\,x^4}{4!} + \frac{2^5\,x^5}{5!} + \cdots\right] \\
&= -\frac{1}{4} - \frac{x}{2} + \frac{5}{4} + \frac{5}{4}\,2x + \frac{5}{4}\,4x^2 + \frac{5}{4}\,\frac{8}{6}\,x^3 + \frac{5}{4}\,\frac{16}{24}\,x^4 + \frac{5}{4}\,\frac{32}{24\cdot 5}\,x^5 + \cdots \\
&= 1 + 2x + \frac{5}{2}\,x^2 + \frac{5}{3}\,x^3 + \frac{5}{6}\,x^4 + \frac{1}{3}\,x^5 + \cdots .
\end{aligned}$$

Maple can also be used:

```
restart:  picard:=proc(f,x0,y0,n)  # n is the number of iterations
local s,y;      y:=y0;             # y0 is the initial value at x=x0
for s from 1 to n do
y:=y0+int(f(a,subs(x=a,y)),a=x0..x);
y:=sort(y,x,ascending);
od;           return(y);     end
```

Example 1.6.9: Find successive approximations to the initial value problem

$$y' = x - y^3, \quad y(0) = 0$$

in the square $|x| \leqslant 1$, $|y| \leqslant 1$. On what interval does Picard's theorem guarantee the convergence of successive approximations? Determine the error in the third approximation.

Solution. The slope function $f(x, y) = x - y^3$ has continuous partial derivatives; therefore, this function satisfies the conditions of Theorem 1.2 with the Lipschitz constant

$$L = \max_{\substack{|x|\leqslant 1\\ |y|\leqslant 1}} \left|\frac{\partial f}{\partial y}\right| = \max_{|y|\leqslant 1} |-3y^2| = 3.$$

Calculations show that

$$M = \max_{\substack{|x|\leqslant 1\\ |y|\leqslant 1}} |f(x,y)| = \max_{\substack{|x|\leqslant 1\\ |y|\leqslant 1}} |x - y^3| = 2, \quad h = \min\left\{a, \frac{b}{M}\right\} = \min\left\{1; \frac{1}{2}\right\} = \frac{1}{2}.$$

The successive approximations converge at least on the interval $[-1/2, 1/2]$. From Eq. (1.6.8), we get

$$\phi_{n+1} = \int_0^x (t - \phi_n^3(t))\,\mathrm{d}t = \frac{x^2}{2} - \int_0^x \phi_n^3(t)\,\mathrm{d}t, \quad n = 0, 1, 2, \ldots, \qquad \phi_0(x) \equiv 0.$$

For $n = 0, 1, 2$, we have

$$\begin{aligned}
\phi_1(x) &= \int_0^x (t-0)\,\mathrm{d}t = \frac{x^2}{2},\\
\phi_2(x) &= \int_0^x \left(t - \frac{t^6}{8}\right)\mathrm{d}t = \frac{x^2}{2} - \frac{x^7}{56},\\
\phi_3(x) &= \int_0^x \left[t - \left(\frac{t^2}{2} - \frac{t^7}{56}\right)^3\right]\mathrm{d}t = \frac{x^2}{2} - \frac{x^7}{56} + \frac{x^{12}}{896} - \frac{3x^{17}}{106,624} + \frac{x^{22}}{3,963,552}.
\end{aligned}$$

Their graphs along with the exact solution are presented in Fig. 1.21. The absolute value of the error of the third approximation can be estimated as follows:

$$|y(x) - \phi_3(x)| \leqslant \frac{ML^3}{3!}\,h^4 = \frac{2\cdot 3^3}{6}\cdot\frac{1}{2^4} = \frac{9}{16} = 0.5625.$$

Of course, the estimate could be improved if the exact solution were known. We plot the four Picard approximations along with the solution using the following MATLAB commands (which call for **quadl**, adaptive Gauss–Lobatto rules for numerical integral evaluation):

```
t = linspace(time(1),time(2),npts); % Create a discrete time grid
y = feval(@init,t,y0);              % Initialize y = y0
window = [time,space];
for n = 1:N                         % Perform N Picard iterations
    [t,y] = picard(@f,t,y,npts);    % invoke picard.m
    plot(t,y,'b--','LineWidth',1);  % Plot the nth iterant
    axis(window);  drawnow;  hold on;
end
[t,y] = ode45(@f,time,y0);          % Solve numerically the ODE
plot(t,y,'k','LineWidth',2);        % Plot the numerical solution
hold off;
axis([min(t) max(t) min(y) max(y)])
function [t,y] = picard(f,t,y,n)    % picard.m
tol = 1e-6;                         % Set tolerance
phi = y(1)*ones(size(t));           % Initialize
for j=2:n
    phi(j) = phi(j-1)+quadl(@fint,t(j-1),t(j),tol,[],f,t,y);
end         y = phi;
```

Example 1.6.10: Let

$$f(x,y) = \begin{cases} 0, & \text{if } x \leqslant 0,\ -\infty < y < \infty;\\ x, & \text{if } 0 < x \leqslant 1,\ -\infty < y < 0;\\ x - \frac{2y}{x}, & \text{if } 0 < x \leqslant 1,\ 0 \leqslant y \leqslant x^2;\\ -x, & \text{if } 0 < x \leqslant 1,\ x^2 < y < \infty. \end{cases}$$

Prove that the Picard iterations, ϕ_0, ϕ_1, ..., for the solution of the initial value problem $y' = f(x,y)$, $y(0) = 0$, do not converge on $[0,\varepsilon]$ for any $0 < \varepsilon \leqslant 1$.

Solution. It is not hard to verify that the function $f(x,y)$ is continuous and bounded in the domain $\Omega = \{\,0 \leqslant x \leqslant 1,\ -\infty < y < \infty\,\}$. Moreover, $|f(x,y)| \leqslant 1$. Hence, the conditions of Theorem 1.2 are satisfied and the initial value problem $y' = f(x,y)$, $y(0) = 0$ has a solution. Let us find Picard's approximations for $0 \leqslant x \leqslant 1$, starting with $\phi_0 \equiv 0$:

$$\begin{aligned}
\phi_1 &= \int_0^x f(t,0)\,\mathrm{d}t = \frac{x^2}{2};\\
\phi_2 &= \int_0^x f(t,\phi_1(t))\,\mathrm{d}t = \int_0^x f\left(t, \frac{t^2}{2}\right)\mathrm{d}t \equiv 0;\\
\phi_3 &= \int_0^x f(t,\phi_2(t))\,\mathrm{d}t = \int_0^x f(t,0)\,\mathrm{d}t = \frac{x^2}{2};
\end{aligned}$$

and so on. Hence, $\phi_{2m}(x) \equiv 0$ and $\phi_{2m+1}(x) = \frac{x^2}{2}$, $m = 0, 1, \ldots$. Thus, the sequence $\phi_n(x)$ has two accumulation points (0 and $x^2/2$) for every $x \neq 0$. Therefore, Picard's approximation does not converge. This example shows that continuity of the function $f(x, y)$ is not enough to guarantee the convergence of Picard's approximations. □

There are times when the slope function $f(x, y)$ in Eq. (1.6.6), page 23, is piecewise continuous. We will see such functions in Example 2.5.4 and Problems 5 and 13 (§2.5). When abrupt changes occur in mechanical and electrical applications, the corresponding mathematical models lead to differential equations with intermittent slope functions. So, solutions to differential equations with discontinuous forcing functions may exist; however, the conditions of Peano's theorem are not valid for them. Such examples serve as reminders that the existence theorem 1.2 provides only sufficient conditions needed to guarantee a solution to the first order differential equation.

Computer-drawn pictures can sometimes make uniqueness misleading. Human eyes cannot distinguish drawings that differ within 1% to 5%. For example, solutions of the equation $x' = x^2 - t^2$ in Fig. 1.11(b) on page 16 and Fig. 2.32(b) on page 111 appear to merge; however, they are only getting very close.

Problems

1. Show that $|y(x)| = \begin{cases} C_1 x^2, & \text{for } x < 0, \\ C_2 x^2, & \text{for } x > 0. \end{cases}$ is the general solution of the differential equation $x\,y' - 2y = 0$.
2. Prove that no solution of $x^3\,y' - 2x^2\,y = 4$ can be continuous at $x = 0$.
3. Show that the functions $y \equiv 0$ and $y = x^4/16$ both satisfy the differential equation $y' = x\,y^{1/2}$ and the initial conditions $y(0) = 0$. Do the conditions of Theorem 1.2 (page 23) hold?
4. Show that the hypotheses of Theorem 1.3 (page 23) do not hold in a neighborhood of the line $y = 1$ for the differential equation $y' = |y - 1|$. Nevertheless, the initial value problem $y' = |y - 1|$, $y(1) = 1$ has a unique solution; find it.
5. Does the initial value problem $y' = \sqrt{|y|}$, $x > 0$, $y(0) = 0$ have a unique solution? Does $f(y) = \sqrt{|y|}$ satisfy the Lipschitz condition?
6. Show that the function $f(x, y) = y^2 \cos x + e^{2x}$ is a Lipschitz function in the domain $\Omega = \{\,(x, y) : |y| < b\,\}$ and find the least Lipschitz constant.
7. Show that the function $f(x, y) = (2 + \sin x) \cdot y^{2/3} - \cos x$ is not a Lipschitz function in the domain $\Omega = \{\,(x, y) : |y| < b\,\}$. For what values of α and β is this function a Lipschitz function in the domain $\Omega_1 = \{\,(x, y) : 0 < \alpha \leqslant y \leqslant \beta\,\}$?
8. Could the Riccati equation $y' = a(x)\,y^2 + b(x)\,y + c(x)$, where $a(x)$, $b(x)$, and $c(x)$ are continuous functions for $x \in (-\infty, \infty)$, have a singular solution?
9. Find the global maximum of the continuous function $f(x, y) = x^2 + y^2 + 2(2 - x - y)$ in the domain $\Omega = \{\,(x, y) : |x - 1| \leqslant a,\ |y - 1| \leqslant b\,\}$.
10. Consider the initial value problem $y' = -2x + 2(y + x^2)^{1/2}$, $\quad y(-1) = -1$.
    (a) Find the general solution of the given differential equation. *Hint:* use the substitution $y(x) = -x^2 + u(x)$.
    (b) Derive a particular solution from the general solution that satisfies the initial value.
    (c) Show that $y = 2Cx + C^2$, where C is an arbitrary positive constant, satisfies the differential equation $y' + 2x = 2\sqrt{x^2 + y}$ in the domain $x > -C/2$.
    (d) Verify that both $y_1(x) = 2x + 1$ and $y_2(x) = -x^2$ are solutions of the given initial value problem.
    (e) Show that the function $y_2(x) = -x^2$ is a singular solution.
    (f) Explain why the existence of three solutions to the given initial value problem does not contradict the uniqueness part of Theorem 1.3 (page 23).
11. Determine a region of the xy-plane for which the given differential equation would have a unique solution whose graph passes through a given point in the region.
    (a) $(y - y^2)\,y' = x$, $y(1) = 2$; **(b)** $y' = y^{2/3}$, $y(0) = 1$;
    (c) $(y - x)\,y' = y + x^2$, $y(1) = 2$; **(d)** $y' = \sqrt{x}\,y$, $y(0) = 1$;
    (e) $(x^2 + y^2)\,y' = \sin y$, $y(\pi/2) = 0$; **(f)** $x\,y' = y$, $y(2) = 3$;
    (g) $(1 + y^3)\,y' = x$, $y(1) = 0$; **(h)** $y' - y = x^2, y(2) = 1$.
12. Solve each of the following initial value problems and determine the domain of the solution.
    (a) $(1 + \sin x)\,y' + \cot(x)\,y = \cos x$, $y(\pi/2) = 1$. (b) $(1 - x^3)\,y' - 3x^2\,y = 4x^3$, $y(0) = 1$.
    (c) $(\sin x)\,y' + \cos x\,y = \cot x$, $y(3\pi/4) = 2$. (d) $x^2\,y' + xy/\sqrt{1 - x^2} = 0$, $y(\sqrt{3}/2) = 1$.
    (e) $(x \sin x)\,y' + (\sin x + x \cos x)\,y = e^x$, $y(-0.5) = 0$. (f) $y' = 2xy^2$, $\quad y(0) = 1/k^2$.

13. Determine (without solving the problem) an interval in which the solution of the given initial value problem is certain to exist.
    (a) $(\ln t)\,y' + ty = \ln^2 t, \quad y(2) = 1;$ **(b)** $x(x-2)\,y' + (x-1)y + x^3 y = 0, \ x(1) = 2;$
    (c) $y' + (\cot x)y = x, \quad y(\pi/2) = 9;$ **(d)** $(t^2-9)\,y' + ty = t^4, \quad y(-1) = 2;$
    (e) $(t^2-9)\,y' + ty = t^4, \quad y(4) = 2;$ **(f)** $(x-1)\,y' + (\sin x)y = x^3, \quad x(2) = 1.$

14. Prove that two distinct solutions of a first order linear differential equation cannot intersect.

15. For each of the following differential equations, find a singular solution and the general solution.
    (a) $y' = 2x + 3(y-x^2)^{2/3};$ **(b)** $y' = \frac{1}{2y} + \frac{6x^2}{y}\,(y^2-x)^{3/4};$ **(c)** $y' = x^2 - \sqrt{x^3-3y};$
    (d) $y' = \frac{\sqrt{y-1}}{x};$ **(e)** $y' = \sqrt{(y-1)(x+2)};$ **(f)** $y' = x\sqrt{2x+y} - 2;$
    (g) $y' = \sqrt{y^2-1};$ **(h)** $y' = \frac{\sqrt{y^3-x^2+2x}}{3y^2};$ **(i)** $y' = \sqrt{\frac{y-1}{y(1+x)}};$
    (j) $y' = (y^2-1)\sqrt{y};$ **(k)** $(y')^2 + xy' = y;$ **(l)** $y\,y' = \left(1-y^2\right)^{1/3}.$

16. Compute the first two Picard iterations for the following initial-value problems.
    (a) $y' = 1 - (1+x)y + y^2, \quad y(0) = 1;$ **(b)** $y' = x - y^2, \quad y(0) = 1;$
    (c) $y' = 1 + x\,\sin y, \quad y(\pi) = 2\pi;$ **(d)** $y' = 1 + x - y^2, \quad y(0) = 0.$

17. Compute the first three Picard iterations for the following initial value problems. On what interval does Picard's theorem guarantee the convergence of successive approximations? Determine the error of the third approximation.
    (a) $y' = xy, \quad y(1) = 1;$ **(b)** $y' = x - y^2, \quad y(0) = 0;$
    (c) $x\,y' = y - y^2, \quad y(1) = 1;$ **(d)** $y' = 3y^2 + 4x^2, \quad y(0) = 0;$
    (e) $x\,y' = y^2 - 2y, \quad y(1) = 2;$ **(f)** $y' = \sin x - y, \quad y(0) = 0.$

18. Compute the first four Picard iterators for the differential equation $y' = x^2 + y^2$ subject to the initial condition $y(0) = 0$ and then another initial condition $y(1) = 2$. Estimate the error of the fourth approximation for each.

19. Find the general formula for n-th Picard's approximation, $\phi_n(x)$, for the given differential equation subject to the specified initial condition.
    (a) $y' = 3\,e^{-2x} + y, \quad y(1) = 0;$ **(b)** $y' = e^{2x} - y, \quad y(0) = 1;$
    (c) $y' = x + y, \quad y(0) = 1;$ **(d)** $y' = -y^2, \quad y(0) = 1.$

20. Let $f(x,y)$ be a continuous function in the domain $\Omega = \{\,(x,y)\,:\, x_0 \leqslant x \leqslant x_0 + \varepsilon, \ \ |y - y_0| \leqslant b\,\}$. Prove the uniqueness for the initial value problem $y' = f(x,y)$, $x_0 \leqslant x \leqslant x_0 + \varepsilon, \ \ 0 < \varepsilon \leqslant a, \ \ y(x_0) = y_0$ if $f(x,y)$ does not increase in y for each fixed x.

21. For which nonnegative values of α does the uniqueness theorem for the differential equation $y' = |y|^{\alpha}$ fail?

22. For the following IVPs, show that there is no solution satisfying the given initial condition. Explain why this lack of solution does not contradict Peano's theorem.
    (a) $xy' - y = x^2, \ y(0) = 1;$ **(b)** $xy' = 2y - x^3, \ y(0) = 1.$

23. Convert the given initial value problem into an equivalent integral equation and find the solution by Picard iteration:

$$y' = 6x - 2xy, \qquad y(0) = 1.$$

24. Prove that an initial value problem for the differential equation $y' = 2\sqrt{|y|}$ has infinitely many solutions.

25. Does the equation $y' = y^{3/4}$ have a unique solution through every initial point (x_0, y_0)? Can solution curves ever intersect for this differential equation?

26. Show that the sequence of continuous functions $y_n(x) = x^n$ for $0 \leqslant x \leqslant 1$ converges as $x \to 0$ pointwise to a discontinuous function.

27. Find a sequence of functions for which $\lim_{n\to\infty} \int_0^1 f_n(x)\,\mathrm{d}x \ne \int_0^1 \lim_{n\to\infty} f_n(x)\,\mathrm{d}x$.

28. Find successive approximations $\phi_0(x)$, $\phi_1(x)$, and $\phi_2(x)$ of the initial value problem for the Riccati equation $y' = 1 - (1+x)y + y^2, \quad y(0) = 1$. Estimate the difference between $\phi_2(x)$ and the exact solution $y(x)$ on the interval $[-0.25, 0.25]$.

29. Does the initial value problem $y' = x^{-1/2}$, $y(0) = 1$ have a unique solution?

30. Consider the initial value problem $y' = \sqrt{x} + \sqrt{y}$, $y(0) = 0$. Using substitution $y = t^6 v^2$, $x = t^4$, derive the differential equation and the initial conditions for the function $v(t)$. Does this problem have a unique solution?

Review Questions for Chapter 1

1. Eliminate the arbitrary constant (denoted by C) to obtain a differential equation.
   (a) $y^2 = \frac{1}{2\sin x + C\cos x}$; **(b)** $\sqrt{y - x} - \ln Cx = 0$; **(c)** $x^2 + Cy^2 = 1$;
   (d) $y + 1 = C\tan x$; **(e)** $C\,e^x + e^y = 1$; **(f)** $C\,e^x = \frac{1}{\sin x} + \frac{1}{\sin y}$;
   (g) $y\,e^{Cx} = 1$; **(h)** $C\,e^x = \arcsin x + \arcsin y$; **(i)** $C\,y - \ln x = 0$;
   (j) $\sqrt{1+x} = 1 + C\sqrt{1+y}$; **(k)** $y = C(x - C)^2$; **(l)** $y(x+1) + Ce^{-x} + e^x = 0$.
2. In each of the following problems, verify whether or not the given function is a solution of the given differential equation and specify the interval or intervals in which it is a solution; C always denotes a constant.
   (a) $(y')^2 - xy + y = 0, \quad 3\sqrt{y} = (x-1)^{3/2}$; **(b)** $y\,y' = x, \quad y^2 - x^2 = C$;
   (c) $y' = 2x\,\cos(x^2)\,y, \quad y = C\,e^{\sin x^2}$; **(d)** $y' + y^2 = 0, \quad y(x) = 1/x$;
   (e) $y' + 2xy = 0, \quad y(x) = C\,e^{-x^2}$; **(f)** $xy' = 2y, \quad y = C\,x^2$;
   (g) $y' + x/y = 0, \quad y(x) = \sqrt{C^2 - x^2}$; **(h)** $y' = \cos^2 y, \quad \tan y = x + C$.
3. The curves of the one-parameter family $x^3 + y^3 = 3Cxy$, where C is a constant, are called **folia of Descartes**. By eliminating C, show that this family of graphs is an implicit solution to
   $$\frac{dy}{dx} = \frac{y(y^3 - 2x^3)}{x(2y^3 - x^3)}.$$
4. Show that the functions in parametric form satisfy the given differential equation.
   (a) $x\,y' = 2y, \quad x = t^2,\ y = t^4$; **(b)** $a^2 y\,y' = b^2 x, \quad x = a\sinh t,\ y = b\cosh t$.
5. A particle moves along the abscissa in such a way that its instantaneous velocity is given as a function of time t by $v(t) = 4 - 3t^2$. At time $t = 0$, it is located at $x = 2$. Set up an initial value problem describing the motion of the particle and determine its position at any time $t > 0$.
6. Determine the values of λ for which the given differential equation has a solution of the form $y = e^{\lambda t}$.
   (a) $y'' - y' - 6y = 0$; **(b)** $y''' + 3y'' + 3y' + y = 0$.
7. Determine the values of λ for which the given differential equation has a solution of the form $y = x^{\lambda}$.
   (a) $2x^2y'' - 3xy' + 3y = 0$; **(b)** $2x^2y'' + 5xy' - 2y = 0$.
8. Find a singular solution for the given differential equation
   (a) $y' = 3x\sqrt{1 - y^2}$; **(b)** $y' = \frac{1}{x}\sqrt{2x + y} - 2$;
   (c) $(x^2 + 1)\,y' = \sqrt{4y - 1}$; **(d)** $y' = -\sqrt{y^2 - 4}$.
9. Verify that the function $y = \sqrt{(x^2 - 2x + 3)/(x^2 + x - 2)}$ is a solution of the differential equation $2yy' = (9 - 3x^2)/(x^2 + x - 2)^2$ on some interval. Give the largest intervals of definition of this solution.
10. The sum of the x- and y-intercepts of the tangent line to a curve in the xy-plane is a constant regardless of the point of tangency. Find a differential equation for the curve.
11. Derive a differential equation of the family of circles having a tangent $y = 0$.
12. Derive a differential equation of the family of unit circles having centers on the line $y = 2x$.
13. Show that x^3 and $(x^{3/2} + 5)^2$ are solutions of the nonlinear differential equation $(y')^2 - 9xy = 0$ on $(0, \infty)$. Is the sum of these functions a solution?
14. In each of the following problems, verify whether or not the given function is a solution of the given differential equation and specify the interval or intervals in which it is a solution; C always denotes a constant.
    (a) $x\,y' = y + x, \quad y(x) = x\,\ln x + Cx$; **(b)** $y'\,\cos x + y\,\sin x = 1, \quad y = C\,\cos x + \sin x$;
    (c) $y' + y\,e^x = 0, \quad C = e^x + \ln y$; **(d)** $y' + 2xy = e^{-x^2}, \quad y = (x + C)\,e^{-x^2}$;
    (e) $xy' + y = y\,\ln|xy|, \quad \ln|xy| = Cx$; **(h)** $y' = \sec(y/x) + y/x, \quad y = x\arcsin(\ln x)$;
    (f) $\dfrac{y'}{\sqrt{1-y^2}} + \dfrac{1}{\sqrt{1-x^2}} = \arcsin x + \arcsin y, \quad C\,e^x = \arcsin x + \arcsin y$;
    (g) $t^2\,y' + 2ty - t + 1 = 0, \quad y = \frac{1}{2} - 1/t + C/t^2$.
15. Show that $y' = y^a$, where a is a constant such that $0 < a < 1$ has the singular solution $y = 0$ and the general solution is $y = [(1-a)(x + C)]^{1/(1-a)}$. What is the limit of the general solution as $a \to 1$?
16. Which straight lines through the origin are solutions of the following differential equations?
    (a) $y' = \dfrac{x^2 - y^2}{3xy}$; **(b)** $y' = xy$; **(c)** $y' = \dfrac{5x + 4y}{4x + 5y}$; **(d)** $y' = \dfrac{x + 3y}{3x + y}$.

17. Show that $y = Cx - C^2$, where C is an arbitrary constant, is an equation for the family of tangent lines for the parabola $y = x^2/4$.

18. Show that in general it is not possible to write every solution of $y' = f(x)$ in the form $y(x) = \int_a^x f(t)\,dt$ and compare this result with the fundamental theorem of calculus.

19. Show that the differential equation $y^2\,(y')^2 + y^2 = 1$ has the general solution family $(x+C)^2 + y^2 = 1$ and also singular solutions $y = \pm 1$.

20. The charcoal from a tree killed in the volcanic eruption that formed Crater Lake in Oregon contained 44.5% of carbon-14 found in living matter. The half-life of C^{14} is 5730 ± 40 years. About how old is Crater Lake?

21. Derive a differential equation for uranium with half-life 4.5 billion years.

22. With time measured in years, the value of λ in Eq. (1.1.1) for cobalt-60 is about 0.13. Estimate the half-life of cobalt-60.

23. The general solution of $1 - y^2 = (y\,y')^2$ is $(x-c)^2 + y^2 = 1$, where c is an arbitrary constant. Does there exist a singular solution?

24. Use a computer graphic routine to display the direction field for each of the following differential equations. Based on the slope field, determine the behavior of the solution as $x \to +\infty$. If this behavior depends on the initial value at $x = 0$, describe the dependency.
    (a) $y' = x^3 + y^3, \quad -2 \leqslant x \leqslant 2$; **(b)** $y' = \cos(x+y), \quad -4 \leqslant x \leqslant 4$;
    (c) $y' = x^2 + y^2, \quad -3 \leqslant x \leqslant 3$; **(d)** $y' = x^2 y^2, \quad -2 \leqslant x \leqslant 2$.

25. Using a software solver, estimate the validity interval of each of the following initial value problems.
    (a) $y' = y^3 - xy + 1,\ y(0) = 1$; **(b)** $y' = y^2 + xy + 1,\ y(0) = 1$;
    (c) $y' = x^2 y^2 - x^3,\ y(0) = 1$; **(d)** $y' = 1/(x^2 + y^2),\ y(0) = 1$.

26. Using a software solver, draw a direction field for each of the given differential equations of the first order $y' = f(x,y)$. On the same graph, plot the graph of the curve defined by $f(x,y) = 0$; this curve is called the nullcline.
    (a) $y' = y^3 - x$; **(b)** $y' = y^2 + xy + x^2$; **(c)** $y' = y - x^3$;
    (d) $y' = xy/(x^2 + y^2)$; **(e)** $y' = xy^{1/3}$; **(f)** $y' = \sqrt{|xy|}$.

27. A body of constant mass m is projected away from the earth in a direction perpendicular to the earth's surface. Let the positive x-axis point away from the center of the earth along the line of motion with $x = 0$ lying on the earth's surface. Suppose that there is no air resistance, but only the gravitational force acting on the body given by $w(x) = -k(x+R)^{-2}$, where k is a constant and R is the radius of the earth. Derive a differential equation for modeling body's motion.

28. Find the maximum interval for which Picard's theorem guarantees the existence and uniqueness of the initial value problem $y' = (x^2 + y^2 + 1)\,e^{1-x^2-y^2}, \quad y(0) = 0$.

29. Find all singular solutions to the differential equation $y' = y^{2/3}\,(y^2 - 1)$.

30. Under what condition on C does the solution $y(x) = y(x,C)$ of the initial value problem $y' = ky(1+y^2)$, $y(0) = C$ exist on the whole interval $[0,1]$?

31. Determine whether Theorem 1.3 (page 23) implies that the given initial value problem has a unique solution.
    (a) $y' + yt = \sin^2 t,\ y(\pi) = 1$; **(b)** $y' = x^3 + y^3,\ y(0) = 1$;
    (c) $y' = x/y,\ y(2) = 0$; **(d)** $y' = y/x,\ y(2) = 0$;
    (e) $y' = x - \sqrt[3]{y-1},\ y(5) = 1$; **(f)** $y' = \sin y + \cos y,\ y(\pi) = 0$;
    (g) $y' = x\,\sqrt{y},\ y(0) = 0$. **(h)** $y' = x\,\ln|y|,\ y(1) = 0$.

32. Convert the given IVP, $y' = x(y-1)$, $y(0) = 1$, into an equivalent integral equation and determine the first four Picard iterations.

33. Consider the initial value problem $y' = y^2 + 4\sin^2 x$, $y(0) = 0$. According to Picard's theorem, this IVP has a unique solution in any rectangle $D = \{(x,y) : |x| < a,\ |y| < b/M\}$, where M is the maximum value of $|f(x,y)|$. Show that the unique solution exists at least on the interval $[-h,h]$, $h = \min\left\{a, \frac{b}{b^2+4}\right\}$.

34. Use the existence and uniqueness theorem to prove that $y = 2$ is the only solution to the IVP $y' = \frac{4x}{x^2+9}\,(y^2 - 4)$, $\quad y(0) = 2$.

35. Prove that the differential equation $y' = x - 1/y$ has a unique solution on $(0,\infty)$.

36. Determine (without solving the problem) an interval in which the solution of the given initial value problem is certain to exist.
    (a) $y' + \frac{x}{1-x^2}\,y = x,\ y(0) = 0$; **(b)** $(1+x^3)\,y' + xy = x^2,\ y(0) = 1$;
    (c) $y' + (\tan t)y = t^2,\ y(\pi/2) = 9$; **(d)** $(x^2+1)\,y' + xy = x^2,\ y(0) = 0$;
    (e) $t\,y' + t^2\,y = t^4,\ y(\pi) = 1$; **(f)** $(\sin^2 t)\,y' + y = \cos t,\ y(1) = 1$.

37. Compute the first two Picard iterations for the following initial-value problems.
(a) $y' = x^2 + y^2, \quad y(0) = 0$; (b) $y' = (x^2 + y^2)^2, \quad y(0) = 1$;
(c) $y' = \sin(xy), \quad y(\pi) = 2$; (d) $y' = e^{xy}, \quad y(0) = 1$.

38. Compute the first three Picard iterations for the following initial value problems. On what interval does Picard's theorem guarantee the convergence of successive approximations? Determine the error of the third approximation when the given function is defined in the rectangle $|x - x_0| < a, \ |y - y_0| < b$.
(a) $y' = x^2 + xy^2, \ y(0) = 1$; (b) $x\,y' = y^2 - 1, \ y(1) = 2$;
(c) $y' = xy^2, \ y(1) = 1$; (d) $y' = x^2 - y, \ y(0) = 0$;
(e) $xy' = y^2, \ y(1) = 1$; (f) $y' = 2y^2 + 3x^2, \ y(0) = 0$;
(g) $xy' = 2y^2 - 3x^2, \ y(0) = 0$; (h) $y' = y + \cos x, \ y(\pi/2) = 0$.

39. Compute the first four Picard iterations for the given initial value problems. Estimate the error of the fourth approximation when the given function is defined in the rectangle $|x - x_0| < a, \ |y - y_0| < b$.
(a) $y' = 3x^2 + xy^2 + y, \ y(0) = 0$; (b) $y' = y - e^x + x, \ y(0) = 0$;
(c) $y' + y = 2\sin x, \ y(0) = 1$; (d) $y' = -y - 2t, \ y(0) = 1$.

40. Find the general formula for n-th Picard's approximation, $\phi_n(x)$, for the given differential equation subject to the specified initial condition.
(a) $y' = y - e^{2x}, \ y(0) = 1$; (b) $y' = 2y + x, \ y(1) = 0$;
(c) $y' = y - x^2, \ y(0) = 1$; (d) $y' = y^2, \ y(1) = 1$.

41. Find the formula for the eighth Picard's approximation, $\phi_8(x)$, for the given differential equation subject to the specified initial condition. Also find the integral of the given initial value problem and compare its Taylor's series with $\phi_8(x)$.
(a) $y' = 2y + \cos x, \ y(0) = 0$; (b) $y' = \sin x - y, \ y(0) = 1$;
(c) $y' = x^2 y + 1, \ y(0) = 0$; (d) $y' + 2y = 3x^2, \ y(1) = 1$.

42. Sometimes the quadratures that are required to carry further the process of successive approximations are difficult or impossible. Nevertheless, even the first few approximations are often quite good. For each of the following initial value problems, find the second Picard's approximation and compare it with the exact solution at points $x_k = k/4$, $k = -1, 0, 1, 2, 3, 4, 5, 6, 7, 8$.
(a) $y' = 2\sqrt{y}, \quad y(0) = 1$; (b) $y' = \frac{2 - e^{-y}}{1 + 2x}, \quad y(0) = 0$.

43. The accuracy of Picard's approximations depends on the choice of the initial approximation, $\phi_0(x)$. For the following problems, calculate the second Picard's approximation for two given initial approximations, $\phi_0(x) = 1$ and $y_0(x) = x$, and compare it with the exact solution at points $x_k = k/4$, $k = -1, 0, 1, 3, 4, 5, 6, 7, 8$.
(a) $y' = y^2, \ y(1) = 1$; (b) $y' = y^{-2}, \ y(1) = 1$.

44. **The Grönwall[14] inequality:** Let x, g, and h be real-valued continuous functions on a real t-interval I: $a \leqslant t \leqslant b$. Let $h(t) \geqslant 0$ on I, $g(t)$ be differentiable, and suppose that for $t \in I$,

$$x(t) \leqslant g(t) + \int_a^t h(\tau) x(\tau)\, \mathrm{d}\tau.$$

Prove that on I

$$x(t) \leqslant g(t) + \int_a^t h(\tau) g(\tau) \exp\left\{ \int_\tau^t h(s)\, \mathrm{d}s \right\} \mathrm{d}\tau.$$

Hint: Differentiate the given inequality and use an integrating factor $\mu(t) = \exp\left\{ -\int_a^t h(\tau)\, \mathrm{d}\tau \right\}$.

45. For a positive constant $k > 0$, find the general solution of the differential equation $\dot{y} = \sqrt{|y|} + k$. Show that while the slope function $\sqrt{|y|} + k$ does not satisfy a Lipschitz condition in any region containing $y = 0$, the initial value problem for this equation has a unique solution.

[14] In honor of the Swedish mathematician Thomas Hakon Grönwall (1877–1932), who proved this inequality in 1919.

Chapter 2

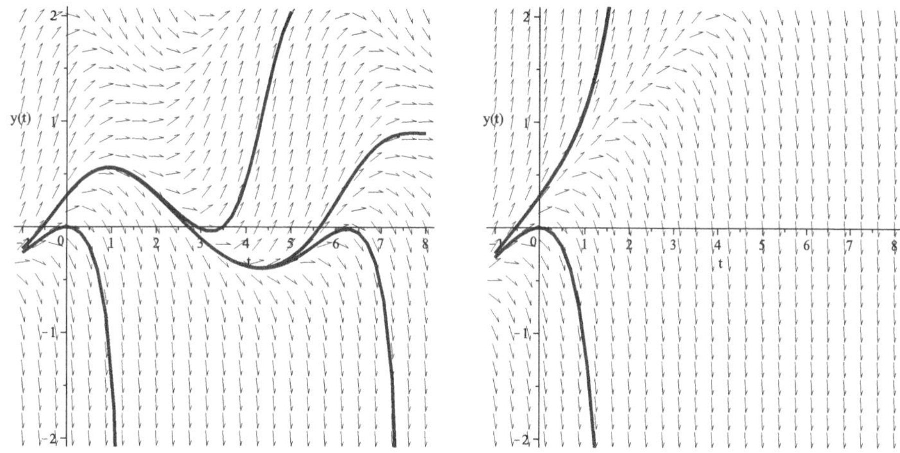

Linearization of the equation $y' = 2y - y^2 - \sin(t)$ **at the right**

First Order Equations

This chapter presents some wide classes of first order differential equations that can be solved at least implicitly. The last section is devoted to the introduction of qualitative analysis. All pictures are generated with one of the following software packages: MATLAB®, *Mathematica*®, *Maple*™, or *Maxima*.

2.1 Separable Equations

In this section, we consider a class of differential equations for which solutions can be determined by integration (called quadrature). However, before doing this, we introduce the definition that is applicable for arbitrary differential equations in normal form $y' = f(x, y)$.

Definition 2.1: A constant solution $y \equiv y^*$ of the differential equation $y' = f(x, y)$ is called the **equilibrium** or **stationary solution**, if $f(x, y^*) \equiv 0$ for all x.

In calculus, points where the derivative vanishes are called the critical points. Therefore, sometimes equilibrium solutions are called the **critical points** of the slope function.

Definition 2.2: We say that the differential equation

$$y' = f(x, y) = p(x)\, q(y) \tag{2.1.1}$$

is **separable** if $f(x, y)$ is a product of a function of x times a function of y: $f(x, y) = p(x)\, q(y)$, where $p(x)$ and $q(y)$ are continuous functions on some intervals I_x and I_y, respectively.

Let us consider the separable equation $y' = p(x)q(y)$. For values of y from interval I_y (which is abbreviated as $y \in I_y$) where $q(y)$ is not zero, we can divide both sides by $q(y)$ and rewrite the equation as

$$\frac{y'}{q(y)} = p(x), \quad q(y) \neq 0 \quad \text{for } y \in I_y.$$

When we integrate both sides, the corresponding integrals will be the same up to an additive constant (denoted by C):

$$\int \frac{1}{q(y(x))}\left(\frac{\mathrm{d}y}{\mathrm{d}x}\right)\mathrm{d}x = \int p(x)\,\mathrm{d}x + C.$$

If we substitute in the left-hand side integral $u = y(x)$, then $\mathrm{d}u = \frac{\mathrm{d}y}{\mathrm{d}x}\,\mathrm{d}x$ and

$$\int \frac{1}{q(u)}\,\mathrm{d}u = \int p(x)\,\mathrm{d}x + C.$$

Replacing u again by y, we obtain the general solution in implicit form:

$$\int \frac{\mathrm{d}y}{q(y)} = \int p(x)\,\mathrm{d}x + C, \quad q(y) \neq 0, \tag{2.1.2}$$

where C is a constant of integration. We separate variables such that the multiplier of $\mathrm{d}x$ is a function of only one variable x, and the multiplier of $\mathrm{d}y$ is a function of another variable y. That is why we call Eq. (2.1.1) a **separable equation**. Each member of the equation $q^{-1}(y)\,\mathrm{d}y = p(x)\,\mathrm{d}x$ is now a differential, namely, the differential of its own[15] quadrature. Generally speaking, formula (2.1.2) only gives an implicit general solution of Eq. (2.1.1). An explicit solution is obtained by solving Eq. (2.1.2) for y as a function of x. Note that the equilibrium solutions (if any) of the equation $y' = p(x)q(y)$ which satisfy $q(y) = 0$ do not appear in Eq. (2.1.2).

Definition 2.3: When the differential equation is given in an irreducible differential form

$$M(x,y)\,\mathrm{d}x + N(x,y)\,\mathrm{d}y = 0, \tag{2.1.3}$$

we say that it is separable if each function $M(x,y)$ and $N(x,y)$ is a product of a function of x by a function of y: $M(x,y) = p_1(x)\,q_1(y)$ and $N(x,y) = p_2(x)\,q_2(y)$. Hence, a separable differential equation is an equation of the form

$$p_1(x)q_1(y)\,\mathrm{d}x + p_2(x)q_2(y)\,\mathrm{d}y = 0. \tag{2.1.4}$$

A differential equation may have either horizontal or vertical constant solutions. Their determination becomes easier when a differential equation is written in the differential form[16] $M\,\mathrm{d}x + N\,\mathrm{d}y = 0$, with some continuous functions $M(x,y)$ and $N(x,y)$. It is assumed that functions $p_1(x)$, $p_2(x)$, and $q_1(y)$, $q_2(y)$ are defined and continuous on some intervals I_x and I_y, respectively, where neither p_1, p_2 nor q_1, q_2 are zeroes.

The equation $p_1(x)q_1(y)\,\mathrm{d}x + p_2(x)q_2(y)\,\mathrm{d}y = 0$ can be solved in a similar way:

$$\frac{p_1(x)}{p_2(x)}\,\mathrm{d}x + \frac{q_2(y)}{q_1(y)}\,\mathrm{d}y = 0 \quad \Longrightarrow \quad \int \frac{q_2(y)}{q_1(y)}\,\mathrm{d}y = -\int \frac{p_1(x)}{p_2(x)}\,\mathrm{d}x + C. \tag{2.1.5}$$

It is assumed that there exist intervals I_x and I_y such that the functions $p_2(x)$ and $q_1(y)$ are not zero for all values of $x \in I_x$ and $y \in I_y$, respectively. Therefore, solutions to a separable differential equation (2.1.4) reside inside some rectangular domain bounded left-to-right by consecutive vertical lines where $p_2(x) = 0$ (or by $\pm\infty$) and top-to-bottom by consecutive horizontal lines where $q_1(y) = 0$ (or by $\pm\infty$).

Definition 2.4: A point (x_0, y_0) on the (x,y)-plane is called a **singular point** of the differential equation (2.1.3), written in an irreducible differential form, if $M(x_0,y_0) = 0$ and $N(x_0,y_0) = 0$.

---

[15]The word quadrature means an expression via an indefinite integral.

[16]Modeling physical problems, mechanical ones in particular, usually leads to differential equations in the form (2.1.3) rather than (2.1.1). Because of that, these equations are historically called "differential equations" rather than "derivative equations."

If $M(x, y^*) = 0$ but $N(x, y^*) \neq 0$, then the given differential equation has the horizontal equilibrium solution $y = y^*$. On the other hand, if there exists a point $x = x^*$ such that the function $M(x^*, y) \neq 0$ but $N(x^*, y) = 0$, then the solution to the differential equation $M(x, y)\,\mathrm{d}x + N(x, y)\,\mathrm{d}y = 0$ (or $y' = -M(x, y)/N(x, y)$) has an infinite slope at that point. Since numerical determination of integral curves in the neighborhood of a point with large (infinite) slope becomes problematic, it is convenient to switch to a reciprocal form and consider the equation $\mathrm{d}x/\mathrm{d}y = -N(x, y)/M(x, y)$ for which $x = x^*$ is an equilibrium solution. Then y is assumed to be an independent variable. A special case when both functions, $M(x, y)$ and $N(x, y)$, vanish at the origin is considered in §2.2.2.

From a geometrical point of view, there is no difference between horizontal lines ($y = y^*$) and vertical lines ($x = x^*$), so these two kinds of points should share the same name. However, the vertical line is not a function as it is defined in mathematics. Another way to get around this problem is to introduce an auxiliary variable, say t, and rewrite the given differential equation (2.1.3) in an equivalent form as the system of equations

$$\mathrm{d}x/\mathrm{d}t = N(x, y), \qquad \mathrm{d}y/\mathrm{d}t = -M(x, y).$$

When solving a differential equation, try to find all constant solutions first because they may be lost while reducing it to the differential form (2.1.3) and following integration. For instance, dividing both sides of the equation $y' = p(x)q(y)$ by $q(y)$ assumes that $q(y) \neq 0$. If there is a value y^* at which $q(y^*) = 0$, the equation $y' = p(x)q(y)$ has a constant solution $y = y^*$ since $y' = 0$ or $\mathrm{d}y = 0$. Sometimes, such constant solutions (horizontal $y = y^*$ or vertical $x = x^*$), if they exist, cannot be obtained from the general solution (2.1.5), but sometimes they can. These constant solutions may or may not be singular. It depends on whether an integral curve from the general solution touches the singular solution or not.

We would like to classify equilibrium solutions further, but we will use only descriptive definitions without heavy mathematical tools. When every solution that starts "near" a critical point moves away from it, we call this constant solution an **unstable equilibrium solution** or unstable critical point. We refer to it as a *repeller* or *source*. If every solution that starts "near" a critical point moves toward the equilibrium solution, we call it an **asymptotically stable equilibrium solution**. We refer to it as an *attractor* or *sink*. When solutions on one side of the equilibrium solution move toward it and on the other side of the constant solution move away from it, we call the equilibrium solution semi-stable.

Example 2.1.1: (Population growth) The growth of the U.S. population (excluding immigration) during the 20 years since 1980 may be described, to some extent, by the differential equation

$$\frac{\mathrm{d}P}{\mathrm{d}t} = \frac{t}{t^2 + 1}\, P(t)\,,$$

where $P(t)$ is the population at year t beginning in 1980. To solve the given equation, we separate variables to obtain

$$\frac{\mathrm{d}P}{P} = \frac{t\,\mathrm{d}t}{t^2 + 1}.$$

Integration yields

$$\ln P = \frac{1}{2}\ln(t^2 + 1) + \ln C = \ln\sqrt{t^2 + 1} + \ln C = \ln C\sqrt{t^2 + 1},$$

where C is an arbitrary (positive) constant. Raising to an exponent, we get $P = C\,(t^2 + 1)^{1/2}$. The given differential equation has the unstable equilibrium solution $P \equiv 0$, which corresponds to $C = 0$. To check our answer, we ask *Mathematica* for help:

```
DSolve[{y'[t] == t*y[t]/(t*t + 1)}, y[t], t]
Plot[{Sqrt[1 + t*t], 4*Sqrt[1+t*t], 0.4*Sqrt[1+t*t], 8*Sqrt[1+t*t]},
{t, 0, 5},  PlotLabel -> "Solutions to Example 1.2.1"]
```

Example 2.1.2: Find the general solution of the equation

$$2x(y - 1)\,\mathrm{d}x + (x^2 - 1)\,\mathrm{d}y = 0. \tag{2.1.6}$$

Solution. The variables in Eq. (2.1.6) can be separated:

$$\frac{2x\,\mathrm{d}x}{x^2 - 1} = -\frac{\mathrm{d}y}{y - 1} \quad \text{for } x \neq \pm 1,\ y \neq 1.$$

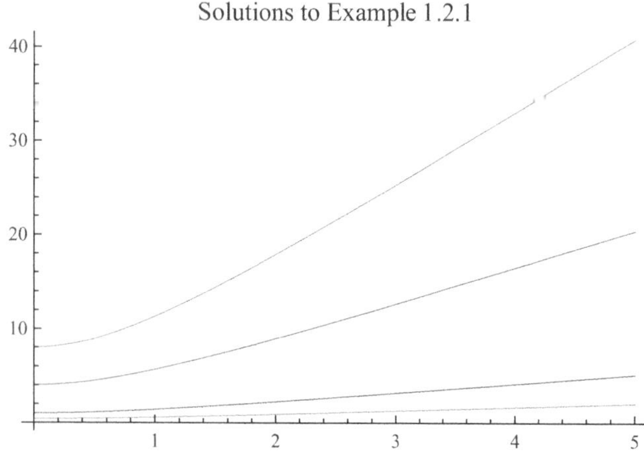

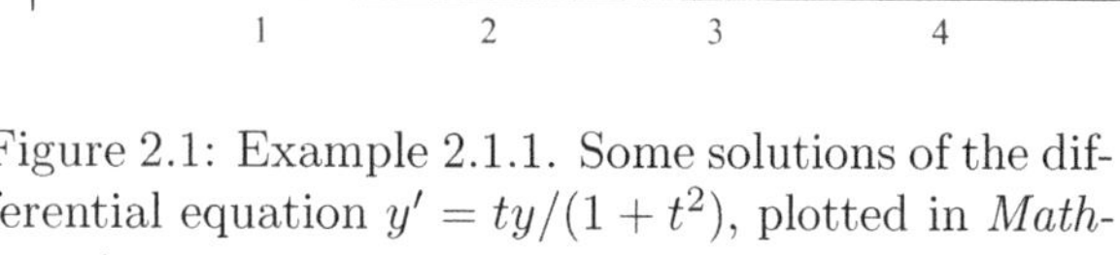

Figure 2.1: Example 2.1.1. Some solutions of the differential equation $y' = ty/(1+t^2)$, plotted in *Mathematica.*

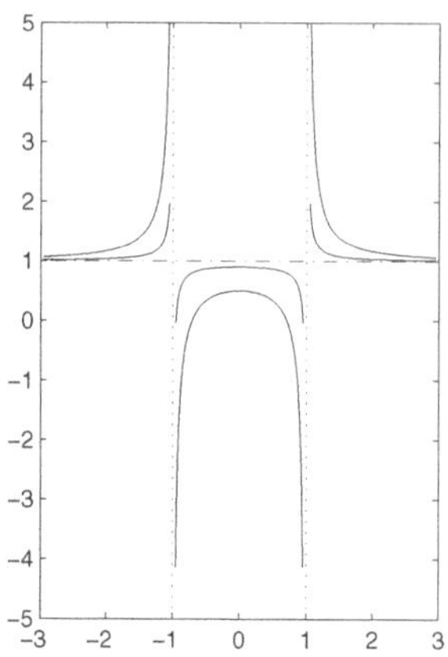

Figure 2.2: Example 2.1.2. Solutions $(x^2-1)(y-1) = C$ of Eq. (2.1.6), plotted with MATLAB.

The expression in the left-hand side is the ratio where the numerator is the derivative of the denominator. Recall the formula from calculus:

$$\int \frac{v'(x)}{v(x)}\,\mathrm{d}x = \ln|v(x)| + C = \ln Cv(x), \qquad v(x) \neq 0. \tag{2.1.7}$$

Integration of both sides yields $\ln|x^2-1| = -\ln|y-1| + C$, where C is an arbitrary constant. Therefore,

$$\ln|x^2-1| + \ln|y-1| = C \qquad \text{or} \qquad \ln|(x^2-1)(y-1)| = C.$$

Taking an exponential of both sides, we obtain

$$|(x^2-1)(y-1)| = e^C \qquad \text{or} \qquad (x^2-1)(y-1) = \pm e^C,$$

where sign plus is chosen when the product $(x^2-1)(y-1)$ is positive, otherwise it is negative. Since C is an arbitrary constant, clearly $\pm e^C$ is also an arbitrary nonzero constant. For economy of notation, we can still denote it by C, where C is any constant, including infinity and zero. Thus, the general integral of Eq. (2.1.6) in implicit form is $(x^2-1)(y-1) = C$. The general solution in explicit form becomes $y = 1 + C\left(x^2-1\right)^{-1}$, $x \neq \pm 1$.

Finally, we must examine what happens when $x = \pm 1$ and when $y = 1$. Going back to the original equation (2.1.6), we see that the vertical lines $x^* = \pm 1$ are solutions of the given differential equation. The horizontal line $y = 1$ is an equilibrium solution of Eq. (2.1.6), which is obtained from the general solution by setting $C = 0$. Therefore, the critical point $y^* = 1$ is a nonsingular stationary solution. MATLAB codes to plot solutions are as follows.

```
C1 = 1; C2 = 10; C3 = -2;
x0 = 1.1; x = x0:.01:x0 + 2.2;
x2d = 1./(x.^2 - 1);
y1=1+C1.*x2d; y2=1+C2.*x2d; y3=1+C3.*x2d;
plot(x,y1,'r',x,y2,'k-.',x,y3,'b--','linewidth',2)
legend('C = 1','C = 10','C = -2');
```

Example 2.1.3: Solve the initial value problem first analytically and then using an available software package:

$$(1+e^y)\,\mathrm{d}x - e^{2y}\sin^3 x\,\mathrm{d}y = 0, \quad y(\pi/2) = 0.$$

Solution. The given differential equation has infinitely many constant solutions $x = x^* = n\pi$ $(n = 0, \pm 1, \pm 2, \ldots)$, each of them is a vertical asymptote for other solutions. Outside these points, division by $(1+e^y)\sin^3 x$ reduces this differential equation to a separable one, namely,

$$\frac{\mathrm{d}x}{\sin^3 x} = \frac{e^{2y}}{1+e^y}\,\mathrm{d}y \qquad \text{or} \qquad y' = \frac{e^{-2y}+e^{-y}}{\sin^3 x} \quad (x \neq n\pi),$$

where n is an integer. Integrating both sides, we obtain the general solution in implicit form,

$$\int \frac{dx}{\sin^3 x} = \int \frac{e^{2y}}{1+e^y}\, dy + C,$$

where C is an arbitrary constant. To bypass this integration problem, we use the available software packages:
Maple: `int((sin(x))^(-3),x);` *Mathematica:* `Integrate[(Sin[x])^(-3),x]`
Maxima: `integrate((sin(x))^(-3),x);`
and *Sage:* `from sage.symbolic.integration.integral import indefinite_integral`
`indefinite_integral(1/(sin(x))^3, x)`

It turns out that all computer algebra systems give different forms of the antiderivative; however, all corresponding expressions are equivalent. In *Maple*, we define the function of two variables:

```
restart; with(plots): with(DEtools):
f:=(x,y) -> (1+exp(y) )/(exp(2*y)*sin(x)*sin(x)*sin(x));
```

After that we solve and plot the solution in the following steps:

```
dsolve( diff(y(x),x) = f(x,y(x)), y(x));  # find the general solution
dsolve( {diff(y(x),x) = f(x,y(x)), y(Pi/2) =0}, y(x));    # or
soln := dsolve( {diff(y(x),x) = f(x,y(x)), y(Pi/2) =0}, y(x), numeric);
odeplot(soln, [x,y(x)], 0..2);              # plot solution
fieldplot([x,soln], x=0..2, soln=-2..2);  # direction field
```

Example 2.1.4: Draw the integral (solution) curves of the differential equation

$$\frac{dy}{dx} = \frac{\sin y}{\sin x}.$$

Solution. The function on the right-hand side is unbounded at the points $x_n^* = n\pi$, $n = 0, \pm 1, \pm 2, \ldots$. The constant functions $y = k\pi$, $k = 0, \pm 1, \pm 2, \ldots$, are equilibrium solutions (either stable or unstable depending on the parity of k) of the given differential equation except points x_n^* $(n = 0, \pm 1, \pm 2, \ldots)$. At the singular points $(n\pi, k\pi)$ where $\sin y = 0$ and $\sin x = 0$ simultaneously, the slope function $\sin y / \sin x$ is undefined. For other points, we have

$$\frac{dy}{\sin y} = \frac{dx}{\sin x} \qquad \text{or} \qquad \int \frac{dy}{\sin y} = \int \frac{dx}{\sin x}.$$

Integration leads to

$$\ln\left|\tan\frac{y}{2}\right| = \ln\left|\tan\frac{x}{2}\right| + \ln C = \ln\left|C\tan\frac{x}{2}\right|.$$

Hence, the general solution is $\tan\frac{y}{2} = C\tan\frac{x}{2}$ or $y = 2\arctan\left(C\tan\frac{x}{2}\right)$ if $\left|C\tan\frac{x}{2}\right| < \frac{\pi}{2}$. Using the following MATLAB commands, we plot a few solutions.

```
C = [1,.02,.2,2,8,22]; C = [C,-C];
epsilon = .001; x0 = pi - epsilon;
x = -x0:0.01:x0;
for i = 1:numel(C)
y(:,i) = 2*atan(C(i)*tan(0.5*x));
end
plot(x,y,'k','linewidth',2);
```

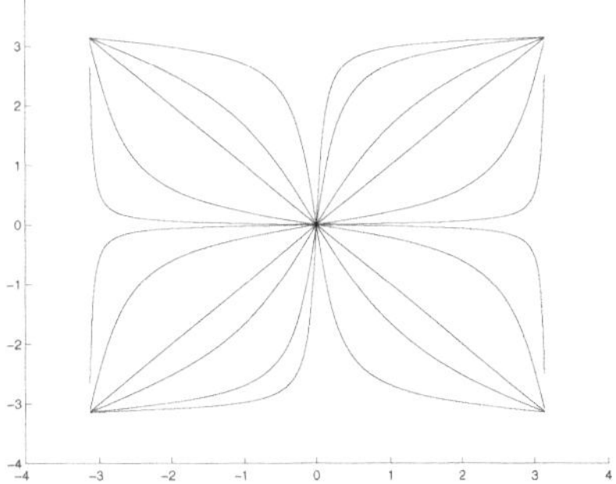

Figure 2.3: Example 2.1.4. The solutions of the initial value problem, plotted with MATLAB.

Several integral curves are shown in Fig. 2.3. If we set $C = 0$ in the general solution, then the equation $\tan\frac{y}{2} = 0$ has infinitely many constant solutions $y = 2k\pi$ $(k = 0, \pm 1, \pm 2 \ldots)$. The other constant solutions, $y = (2k+1)\pi$, are obtained by setting $C = \infty$ in the general formula. Therefore, all equilibrium solutions are not singular because they can be obtained from the general solution. On the other hand, the vertical lines $x = n\pi$ $(n = 0, \pm 1, \pm 2, \ldots)$ that we excluded as singular points for the given differential equation $dy/dx = \sin y / \sin x$ become equilibrium solutions for the reciprocal equation $dx/dy = \sin x / \sin y$.

Example 2.1.5: (Young leaf growth) A young leaf of the vicuna plant has approximately a circular shape. Its leaf area has a growth rate that is proportional to the leaf radius and the intensity of the light beam hitting the leaves. This intensity is proportional to the leaf area and the cosine of the angle between the direction of a beam of light and a line perpendicular to the ground. Express the leaf area as a function of time if its area was $0.16\,\mathtt{cm}^2$ at 6 a.m. and $0.25\,\mathtt{cm}^2$ at 6 p.m. Assume for simplicity that sunrise was at 6 a.m.[17] and sunset was at 6 p.m. that day.

Solution. Let $S = S(t)$ be the area of the leaf at instant t. We set $t = 0$ for 6 a.m.; hence, $S(0) = 0.16\,\mathtt{cm}^2$. It is given that $\mathrm{d}S/\mathrm{d}t = \alpha r I$, where α is a coefficient, $r = r(t)$ is the radius of the leaf at time t, and I is the intensity of light. On the other hand, $I(t) = \beta S(t)\cos\theta$, where β is a constant, and θ is the angle between the light beam and the vertical line (see Fig. 2.4(a)). Assuming that at midday the sun is in its highest position, and the angle θ is approximately a linear function with respect to time t, we have

$$\theta(t) = at + b, \quad \theta(0) = -\frac{\pi}{2}, \quad \theta(6) = 0, \quad \theta(12) = \frac{\pi}{2}.$$

Therefore, $a = \pi/12$, $b = -\pi/2$, and $\theta(t) = \pi\,(t-6)/12$. Then we find the intensity:

$$I(t) = \beta S(t)\,\cos\frac{\pi}{12}(t-6).$$

Since $S(t) = \pi r^2(t)$ and $r = \sqrt{S/\pi}$, we obtain the differential equation for $S(t)$:

$$\frac{\mathrm{d}S}{\mathrm{d}t} = k\,S\sqrt{S}\,\cos\frac{\pi}{12}(t-6),$$

where $k = \alpha\beta/\sqrt{\pi}$. Separating variables $\dfrac{\mathrm{d}S}{S\sqrt{S}} = k\,\cos\dfrac{\pi}{12}(t-6)\,\mathrm{d}t$ which, when integrated, yields

$$-\frac{2}{\sqrt{S}} = k\,\frac{12}{\pi}\,\sin\frac{\pi}{12}(t-6) + C.$$

We use the conditions $S(0) = 0.16$ and $S(12) = 0.25$ to determine the values of the unknown coefficients k and C to be $k = \pi/24$ and $C = -9/2$. Hence

$$S(t) = \frac{16}{\left[\sin\frac{\pi}{12}(t-6) - 9\right]^2}.$$

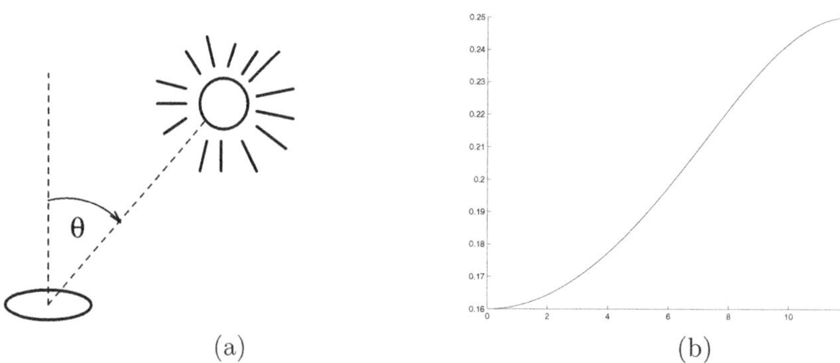

Figure 2.4: Example 2.1.5; (a) The growth rate of a vicuna leaf, and (b) integral curve $S(t)$.

Example 2.1.6: (Curve on the plane) Let us consider a curve in the positive half plane $y \geqslant 0$ that passes through the origin. Let $M(x, y)$ be a point on the curve and $\mathtt{OBMA}$ be the rectangle (see Fig. 2.5, page 45) with the vertices at M and at the origin O. The curve divides this rectangle into two parts; the area under the curve is half the area above or 1/3 the total area. Find the equation of the curve.

[17]It is a custom to use a.m. for abbreviation of the Latin expression *ante meridiem*, which means *before noon*; similarly, p.m. corresponds to the Latin expression *post meridiem* (after noon).

Solution. Suppose $y = y(x)$ is the equation of the curve to be determined. The area of the rectangle OBMA is $A_{\text{OBMA}} = xy$ since the curve passes through the point $M(x, y)$. The area under the curve is $Q = \int_0^x y(t)\,\mathrm{d}t$. We know that $A_{\text{OBMA}} - Q = 2Q$ or $A_{\text{OBMA}} = 3Q$. Thus we have

$$xy = 3\int_0^x y(t)\,\mathrm{d}t.$$

Differentiation yields

$$\frac{\mathrm{d}}{\mathrm{d}x}(xy) = y + x\,y' = 3\frac{\mathrm{d}}{\mathrm{d}x}\int_0^x y(t)\,\mathrm{d}t = 3y(x).$$

Hence,

$$y + x\,y' = 3y \qquad \text{or} \qquad x\,y' = 2y.$$

This is a separable differential equation. Since the curve is situated in the positive half plane $y \geqslant 0$, we obtain the required solution to be $y(x) = C\,x^2$. □

Next, we present several applications of separable differential equations in physics and engineering. We refer the reader to [14], which contains many supplementary examples.

Example 2.1.7: (Temperature of an iron ball) Let us consider a hollow iron ball in steady state heat; that is, the temperature is independent of time, but it may be different from point to point. The inner radius of the sphere is 6 cm and the outer radius is 10 cm. The temperature on the inner surface is 200°C and the temperature on the outer surface is 100°C. Find the temperature at the distance 8 cm from the center of the ball.

Solution. Many experiments have shown that the quantity of heat Q passing through a surface of the area A is proportional to A and to the temperature gradient (see [14]), namely,

$$Q = -k\,A\,\nabla T = -k\,A\,\frac{\mathrm{d}T}{\mathrm{d}r},$$

where T is the temperature, ∇ is the gradient operator, k is the thermal conductivity coefficient ($k = 50\,\mathrm{W/(m{\cdot}K)}$ for iron), and r is the distance to the center. For a sphere of radius r, the surface area is $A = 4\pi r^2$. In steady heat state, Q is a constant; hence,

$$-4\pi k\,r^2\,\frac{\mathrm{d}T}{\mathrm{d}r} = Q.$$

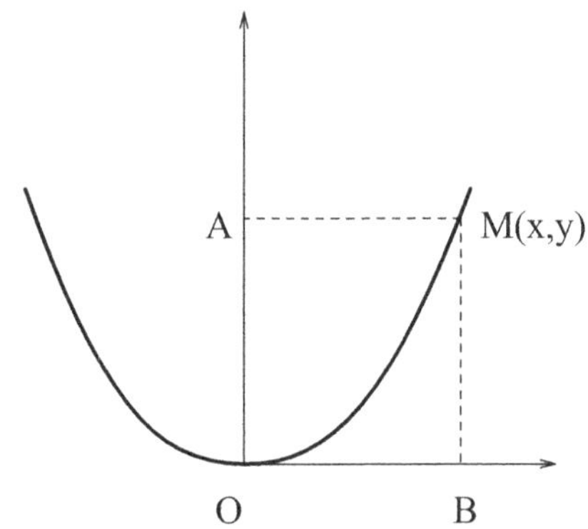

Figure 2.5: Example 2.1.6. The curve to be determined.

Separation of variables and integration yields $-4\pi k\int \mathrm{d}T = Q\int \frac{\mathrm{d}r}{r^2}$ or $4\pi k\,K - 4\pi k\,T = -\frac{Q}{r}$ for some constant K. Thus, $T = \frac{Q}{4\pi k\,r} + K$. From the boundary conditions $T(6) = 200$ and $T(10) = 100$, we obtain the system of algebraic equations

$$200 = \frac{Q}{4\pi k \cdot 6} + K, \quad 100 = \frac{Q}{4\pi k \cdot 10} + K.$$

Solving for Q and K, we get

$$Q = 6000\,\pi\,k \quad \text{and} \quad K = -50.$$

Therefore, $T(r) = 1500/r - 50$ and, in particular, $T(8) = 137.5$°C.

Example 2.1.8: (Light passing through water) A light beam is partially absorbed by water as it passes through. Its intensity decreases at a rate proportional to its intensity and the depth. At 1 m, the intensity of a light beam directed downward into water is three quarters of its surface intensity. What part of a light beam will reach the depth of h meters?

Solution. Let us denote the intensity of the beam at the depth h by $Q(h)$. Passing through a water layer of depth dh, the rate of absorption dQ equals $-k\,Q\,\mathrm{d}h$ with some coefficient k. Hence, $Q(h) = C\,e^{-kh}$.

Suppose that at the surface $Q(0) = Q_0$. Then $Q(h) = Q_0\,e^{-kh}$. We know that at the depth $h = 1$, $Q(1) = \frac{3}{4}\,Q_0$. Thus,

$$\frac{3}{4}\,Q_0 = Q_0\,e^{-k} \qquad \Longrightarrow \qquad e^{-k} = \frac{3}{4}.$$

Therefore, $Q(h) = Q_0\,\left(\frac{3}{4}\right)^h$.

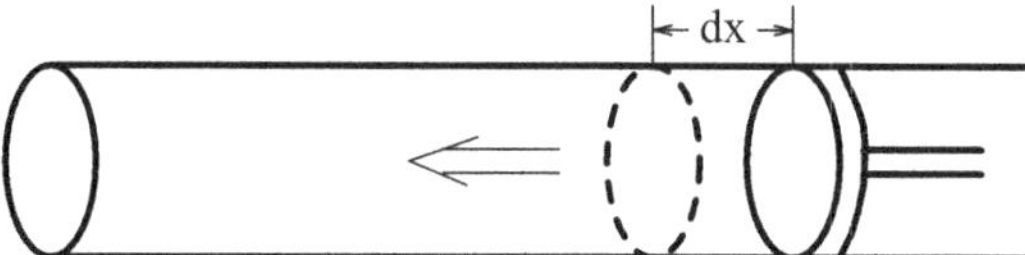

Figure 2.6: The infinitesimal work done by the system during a small compression dx.

Example 2.1.9: (Work done by compression) In a cylinder of volume V_0, air is adiabatically (no heat transfer) compressed to the volume V. Find the work done by this process.

Solution. Suppose A is the cross-sectional area of the cylinder. We assume that the pressure exerted by the movable piston face is p. When the piston moves an infinitesimal distance dx, the work dW done by this force is $\mathrm{d}W = -pA\,\mathrm{d}x$.

On the other hand, $A\,\mathrm{d}x = \mathrm{d}V$, where d$V$ is the infinitesimal change of volume of the system. Thus, we can express the work done by the movable pistol as $\mathrm{d}W = -p\,\mathrm{d}V$. For an adiabatic compression of an ideal gas,

$$p\,V^{\gamma} = \text{constant}, \quad \gamma = \frac{R}{C_V} + 1,$$

where C_V is the molar heat capacity at constant volume and R is the ideal[18] gas constant (≈ 8.3145 `J/mol·K`).

From the initial condition, it follows that

$$p\,V^{\gamma} = p_0\,V_0^{\gamma} \qquad \Longrightarrow \qquad p = p_0\,V_0^{\gamma}\,V^{-\gamma}.$$

Substituting this expression into the equation $\mathrm{d}W = -p\,\mathrm{d}V$, we obtain $\mathrm{d}W = -p_0\,V_0^{\gamma}\,\frac{\mathrm{d}V}{V^{\gamma}}$. Since $W(V_0) = 0$, integration yields

$$W(V) = \frac{p_0\,V_0^{\gamma}}{(\gamma-1)\,V^{\gamma-1}} - \frac{p_0 V_0}{\gamma-1}.$$

Example 2.1.10: (Fluid motion) Let us consider the motion of a fluid in a straight pipeline of radius R. The velocity v of the flow pattern increases as it approaches the center (axis) of the pipeline. Find the fluid velocity $v = v(r)$ as a function of the distance r from the axis of a cylinder.

Solution. From a course in fluid mechanics, it is known that the fluid velocity v and its distance from the axis of a cylinder are governed by the separable differential equation:

$$\mathrm{d}v = -\frac{\rho\Delta}{2\mu}\,r\,\mathrm{d}r,$$

where ρ is the density, μ is the viscosity, and Δ is the pressure difference. The minus sign shows that the fluid velocity decreases as the flow pattern approaches the boundary of a cylinder. Integration yields

$$v(t) = -\frac{\rho\Delta}{4\mu}\,r^2 + C$$

with an arbitrary constant C. We know that on the boundary $r = R$ the velocity is zero, so

$$v(R) = -\frac{\rho\Delta}{4\mu}\,R^2 + C = 0 \qquad \Longrightarrow \qquad v(r) = \frac{\rho\Delta}{4\mu}\,\left(R^2 - r^2\right).$$

[18]$\gamma = 5/3$ for every gas whose molecules can be considered as points.

2.1.1 Autonomous Equations

Definition 2.5: A differential equation in which the independent variable does not appear explicitly, i.e., an equation of the form

$$\dot{y} \stackrel{\text{def}}{=} \mathrm{d}y/\mathrm{d}t = f(y) \tag{2.1.8}$$

is called **autonomous** (or **self-regulating**); here $y = y(t)$ is a function of variable t.

For example, a simple autonomous equation $y' = y$ is a topic in calculus; it leads to the definition of an exponential function with base e. Autonomous equations arise naturally in the study of conservative mechanical systems. For instance, the equations used in celestial mechanics are autonomous. The book [14] contains numerous examples of autonomous equations.

As we discussed previously (page 41), the constant, or **equilibrium** solution, is particularly important to autonomous equations. The constant solution is called an *asymptotically stable equilibrium* if all sufficiently small disturbances away from it die out. Conversely, an *unstable equilibrium* is a constant solution for which small perturbations grow with t rather than die out. By definition, stability is a local characteristic of a critical point because it is based on small disturbances, but certain large perturbations may fail to decay. Sometimes we observe **global stability** when all trajectories approach a critical point.

To determine stability of a critical point y^*, it is convenient to apply the *Stability Test:* If $f'(y^*) > 0$, then the equilibrium solution is unstable. However, if $f'(y^*) < 0$, then the equilibrium solution is stable.

The implicit solution of the Eq. (2.1.8) can be determined by separation of variables:

$$\frac{\mathrm{d}y}{f(y)} = \mathrm{d}t \qquad \Longrightarrow \qquad \int_{y_0}^{y} \frac{\mathrm{d}y}{f(y)} = t - t_0,$$

if the function $f(y) \neq 0$ on some interval. Otherwise there exists a point y^* such that $f(y^*) = 0$, and the horizontal line $y(t) \equiv y^*$ is an equilibrium solution of Eq. (2.1.8).

Example 2.1.11: (Example 1.6.7 revisited) Solve the initial value problem

$$\dot{y} = y^2, \quad y(t_0) = y_0.$$

Solution. The given differential equation has one equilibrium solution $y \equiv 0$ for which the stability test is inconclusive. Let us show that the solution is

$$y(t) = \frac{y_0}{1 - y_0(t - t_0)}. \tag{2.1.9}$$

It is easy to see that after separation of variables we have

$$\frac{dy}{y^2} = \mathrm{d}t, \qquad \text{or} \qquad -y^{-1} = t + C,$$

which is a solution in implicit form. Some algebra work is required to obtain an explicit solution:

$$y(t) = \frac{-1}{t + C}. \tag{2.1.10}$$

Substituting the initial point $t = t_0$, $y = y_0$ into Eq. (2.1.10), we find the value of an arbitrary constant C to be $-y_0^{-1} = t_0 + C$ or

$$C = -t_0 - y_0^{-1} = -\frac{y_0 t_0 + 1}{y_0}.$$

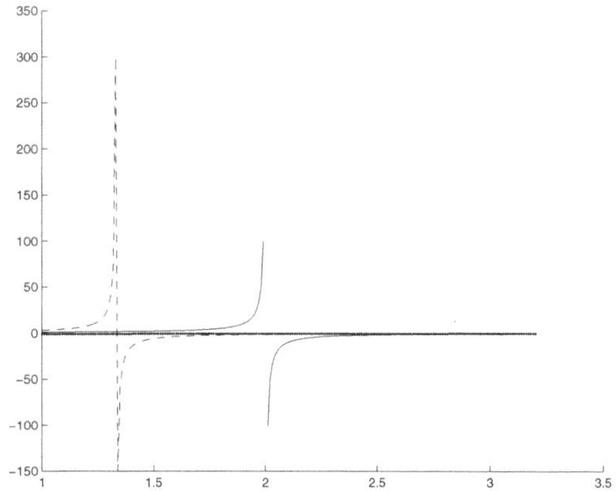

Figure 2.7: Example 2.1.11. Solutions (2.1.9) of the autonomous differential equation $\dot{y} = y^2$, plotted with MATLAB.

If $y_0 \neq 0$, then the integral curve (2.1.10) is a hyperbola. If $y_0 = 0$, then we get the integral curve to be nonsingular equilibrium solution $y = 0$, which is the abscissa (horizontal axis). Indeed, assuming the opposite is also true, we should have $y(t_1) = y_1 \neq 0$ at some point t_1 and also $y(t_0) = 0$. According to formula (2.1.9), the initial value problem $\dot{y} = y^2$, $y(t_1) = y_1$ has the solution $y(t) = \frac{y_1}{1-y_1(t-t_1)}$ that should go through the point $y(t_0) = 0$. Since $y(t) \neq 0$ for any t, it is impossible to satisfy the condition $y(t_0) = 0$ and we get a contradiction. Therefore such a point (t_1, y_1), $y_1 \neq 0$, does not exist and the integral curve should be $y(t) \equiv 0$. □

Example 2.1.11 shows that the solution (2.1.9) may reach infinity in finite time. In fact, if $y_0 > 0$, then the solution (2.1.9) reaches infinity at $t = t_0 + 1/y_0$.

The next theorem shows that solutions of an autonomous differential equation may reach infinity at the endpoints of an interval if the slope function in Eq. (2.1.8) is positive or negative on this interval and vanishes at the endpoints.

Theorem 2.1: Let $f(y)$ be a continuously differentiable function on the closed interval $[a, b]$, where $f(y)$ is positive except for the only two endpoints, $f(a) = f(b) = 0$. If the initial value, y_0, is chosen within the open interval (a, b), the initial value problem

$$\dot{y} = f(y), \quad y(0) = y_0 \in (a, b)$$

has a solution $y(t)$ on $(-\infty, \infty)$ such that $\lim_{t\to-\infty} y(t) = a$ and $\lim_{t\to+\infty} y(t) = b$.

PROOF: The constant functions $y(t) \equiv a$ and $y(t) \equiv b$ are solutions of the autonomous equation (2.1.8) since $f(a) = f(b) = 0$. Let $y = \varphi(t)$ be a solution of the given initial value problem. This function monotonically increases if $f(y) > 0$ and it monotonically decreases if $f(y) < 0$ since $\dot{\varphi} = f(\varphi)$. The function $\varphi(t)$ cannot pass through b (or a, respectively). We assume the opposite: that there exists $t = t^*$ such that $\varphi(t^*) = b$ for some $t = t^*$. This leads to a contradiction of the uniqueness of the solution (see Theorem 1.3 on page 23) because we have two solutions $y \equiv b$ and $y = \varphi(t)$ that pass through the point $t = t^*$. Therefore, the bounded solution $y = \varphi(t)$ monotonically increases; hence, there exists the limit $\lim_{t\to\infty} \varphi(t) = C$.

We have to prove that $C = b$ ($C = a$, respectively). Suppose the opposite: that $C \neq b$ ($C \neq a$, respectively). Then $C < b$ ($C > a$) and $f(C) > 0$ ($f(C) < 0$). Since $\varphi(t) \to C$ as $t \to \infty$, we have $\varphi'(t) \to 0$ as $t \to \infty$. Thus

$$0 = \lim_{t\to\infty} \frac{d\varphi(t)}{dt} = \lim_{t\to\infty} f(\varphi(t)) = f(C).$$

This contradiction proves that $C = b$.

Theorem 2.2: Let $f(y)$ be a continuous function on the closed interval $[a, b]$ that has only one null $y^* \in (a, b)$, namely, $f(y^*) = 0$ and $f(y) \neq 0$ for all other points $y \in (a, b)$. If the integral

$$\int_y^{y^*} \frac{dy}{f(y)} \tag{2.1.11}$$

diverges, then the initial value problem for the autonomous differential equation

$$\dot{y} \stackrel{\text{def}}{=} dy/dt = f(y), \quad y(t_0) = y^* \tag{2.1.12}$$

has the unique constant solution $y(t) \equiv y^*$. If the integral (2.1.11) converges, then the initial value problem (2.1.12) has multiple solutions.

PROOF: Let $y = \varphi(t)$ be a solution of the differential equation $\dot{y} = f(y)$ such that $\varphi(t_0) = y_0 < y^*$. Suppose $f(y) > 0$ for $y \in [y_0, y^*)$. Hence, separation of variables and integration yield that the function $y = \varphi(t)$ satisfies the equation

$$t - t_0 = \int_{y_0}^{\varphi(t)} \frac{dy}{f(y)}.$$

The integral

$$\int_{y_0}^{\varphi(t)} \frac{dy}{f(y)}$$

diverges as $\varphi(t)$ approaches y^*. Therefore, the variable t grows with no limit. This means that the integral curve of the solution $y = \varphi(t)$ asymptotically approaches the line $y = y^*$ as t increases infinitely, but does not cross the line. Thus, the initial value problem (2.1.12) has a unique solution.

Let us consider the case when the integral

$$\int_{y_0}^{y^*} \frac{\mathrm{d}y}{f(y)} = A < \infty$$

converges. Then the curve $y = \varphi(t)$ goes through the point $(t_0 + A, y^*)$ on the line $y = y^*$. Hence, at this point, the initial value problem (2.1.12) has no unique solution since the function $y = \varphi(t - t_0)$ is the solution of $\dot{y} = f(y)$ for an arbitrary constant $t_0 \in \mathbb{R}$. Thus, $y = y^*$ is a singular solution.

Note: As can be seen from the proof, Theorem 2.2 can be extended for separable equations $\dot{y} = f(y)g(t)$ if $g(t)$ is a monotonically increasing continuous function without bounds. ■

The following examples present an array of physical models and mathematical equations where autonomous differential equations occur.

Example 2.1.12: (Cell dynamics) The bone marrow of a human constantly produces red blood cells (erythrocytes) because they have a finite life span (approximately 120 days). Let us denote the number of erythrocytes in the blood at time t by $u(t)$. This number may be modeled by the solution of the autonomous differential equation (see [1])

$$\frac{\mathrm{d}u}{\mathrm{d}t} = -a\,u(t) + \frac{b}{u(t)},$$

with positive constants a and b. The coefficient a determines the rate of natural loss of erythrocytes. The oxygen level in kidney tissue determines the production of erythropoietin (Epo), and the level of Epo is proportional to the rate of red blood cell production. This means that the oxygen level is inversely proportional to the rate of erythrocyte production.

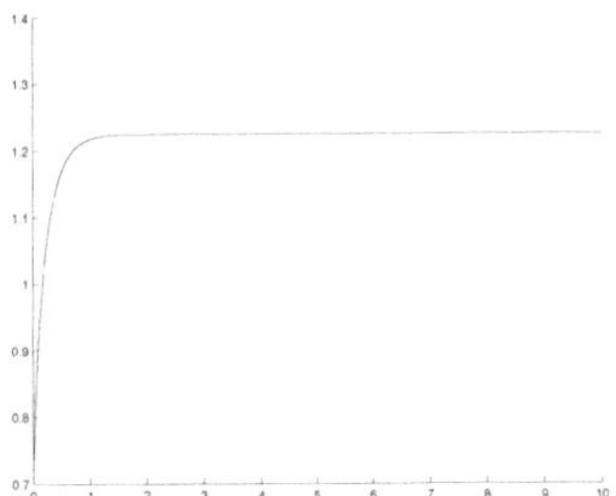

Figure 2.8: Example 2.1.12. Erythrocyte dynamics, plotted with MATLAB.

To find the general solution of the equation, we separate variables to obtain

$$\frac{\mathrm{d}u}{-au + b/u} = \mathrm{d}t.$$

Integration yields

$$t + C_1 = \int \frac{u\,\mathrm{d}u}{b - a\,u^2} = -\frac{1}{2a}\int \frac{\mathrm{d}(b - au^2)}{b - au^2} = -\frac{1}{2a}\ln|b - au^2|,$$

with an arbitrary constant C_1. Raising both sides of the equation $-2a(t + C_1) = \ln|b - au^2|$ to an exponent, we get

$$b - a\,u^2 = aC\,e^{-2at},$$

with another arbitrary constant $C = e^{-2aC_1}/a$. Hence,

$$u(t) = \left[\frac{b}{a} - C\,e^{-2at}\right]^{1/2}$$

is the general solution of the differential equation that serves as a mathematical model of the red blood cell system (see its graph in Figure 2.8).

Example 2.1.13: (Investment) Suppose you learned that the U.S. government[19] borrowed from one of your ancestors \$24 two hundred years ago at an interest rate of 8% compounded continuously. How much should you get for this investment? Find the amount after 300 years.

[19]This example is based on a story about George Washington (1732–1799), who borrowed \$24 on behalf of the U.S. government.

Solution. Let $s(t)$ be the amount of money (capital plus interest) at time t measured in years. Initially, $s(0) = \$24$. The balance $s(t)$ grows by the accumulated interest, which is proportional to the interest rate of 8%, or 0.08 in absolute numbers. Thus,

$$\frac{ds(t)}{dt} = 0.08\, s(t).$$

This differential equation is separable and its solution is

$$s(t) = C\, e^{0.08t}, \quad \text{where} \quad C = s(0) = \$24.$$

At $t = 200$, we have $s(200) = 24\, e^{16} \approx \$2.1326665 \times 10^8$, which is more than 200 million dollars.

The amount after 300 years will be $s(300) = 24\, e^{24} \approx \$6.357398 \times 10^{11}$ (in the short scale, one billion is 10^9), which is about half the property value of New York City.

Example 2.1.14: (Population growth) The simplest mathematical model of population growth is obtained by assuming that the rate of population increase at any time is proportional to the size of the population at that time:

$$\dot{P} = \mathrm{d}P/\mathrm{d}t = rP, \tag{2.1.13}$$

where $P(t)$ is the population at an instant t. For other population models, see [14].

If r is a positive constant, then Eq. (2.1.13) is called the *Malthusian*[20] *growth model*, which gives a reasonably accurate description of the population of certain algae, bacteria, and cell cultures. This differential equation is easily solved by simple integration. Collecting like variables in Eq. (2.1.13) gives

$$\frac{\mathrm{d}P}{P} = r\,\mathrm{d}t \qquad \Longrightarrow \qquad P(t) = P(0)\, e^{rt}.$$

The time taken for a population to double in size is called the **doubling time**. For the Malthusian model, this time, t_d, is determined from the algebraic equation $P(t_d) = 2P(0)$ which leads to

$$P(0)\, e^{rt_d} = 2P(0) \quad \text{and hence} \quad e^{rt_d} = 2.$$

Thus, the doubling time is $t_d = (\ln 2)/r$.

Problems

1. Determine which of the following differential equations are separable.
   (a) $(4 + y^2)\, x\, \mathrm{d}x + y\, \mathrm{d}y = 0$; **(b)** $x^2\, y' = 3 - y^2$; **(c)** $(x^2 + y^2)\, \mathrm{d}x - 2xy\, \mathrm{d}y = 0$;
   (d) $y' = \cos y$; **(e)** $y' = \cos(xy)$; **(f)** $y' = x - 2xy - y + 4$;
   (g) $x\, y' - y = xy^2$; **(h)** $y' = \cos(x + y)$; **(i)** $y\, y' = x^2 + y^2$;
   (j) $y\, y' = e^{x+y}$; **(k)** $x\, y' - y = x^3$; **(l)** $(1 + y^2)\, \cos x\, \mathrm{d}x = (1 - \sin^2 x)\, y\, \mathrm{d}y$.
2. Find the general solution of the following differential equations. Check your answer by substitution.
   (a) $2y' + (1 + 4x)\, y^3 = 0$; **(b)** $y\, y' = \sin^2 \omega x$; **(c)** $x^3 y^2\, y' = e^y$;
   (d) $e^{y^2}\, \mathrm{d}x + 2x^2 y\, \mathrm{d}y = 0$; **(e)** $4\, y' = y^3\, \sin x$; **(f)** $y' = y^2\, \cos x$;
   (g) $\cos x\, \cos y\, y' = \sin x\, \sin y$; **(h)** $y' = y\, \tanh x$; **(i)** $y' = \cos^2 x \cos y$;
   (j) $y' = (1 + 2x)(1 + 4y^2)$; **(k)** $y' = \sqrt{1 - y^2}$; **(l)** $y' = 4 - y^2$;
   (m) $\mathrm{d}r = r \cos\theta\, \mathrm{d}\theta - \sin\theta\, \mathrm{d}r$; **(n)** $(x^2 + 4)\, y' = 2xy$; **(o)** $y' = x^2 y$;
   (p) $y' = xy^2 - 3y^2 + x - 3$; **(q)** $\sqrt{1 - x^2}\, y' + x = 0$; **(r)** $y' + xy = x/y$;
   (s) $y' = (\tan y)/(x + 1)$; **(t)** $y' = (4 + x)y^3$.
3. Solve the following separable equations
   (a) $e^{-x} \cos y\, \mathrm{d}x + \sec y\, \mathrm{d}y = 0$; **(b)** $(y + 1)^2\, \mathrm{d}x = x^2\, \mathrm{d}y$; **(c)** $y' = (x - 1)(y + 1)^{2/3}$;
   (d) $(1 + x^2)\, \mathrm{d}y - e^{-y}\, \mathrm{d}x = 0$; **(e)** $(x + 1)\, y' = xy$; **(f)** $(3y^2 + 1)\, y' = \cos x$;
   (g) $\mathrm{d}y = \cos^2 x \cos^2(2y)\, \mathrm{d}x$; **(h)** $x\, \mathrm{d}x + ye^{-x}\, \mathrm{d}y = 0$; **(i)** $y^2 y' = x/\sqrt{1 + x^2}$.
4. Solve each of the following equations
   (a) $\frac{\mathrm{d}y}{\mathrm{d}x} = \frac{(y+1)(x-1)(y-2)}{(x+1)(y-1)(x-2)}$; **(b)** $\frac{\mathrm{d}y}{\mathrm{d}x} = \frac{y+1}{\sqrt{x} + \sqrt{xy}}$; **(c)** $y' = 2^{x+y}$;
   (d) $x\, \frac{\mathrm{d}y}{\mathrm{d}x} - y = x^2 - y$; **(e)** $y' = \frac{4x - e^{-x}}{2y + e^y}$; **(f)** $x\, y' = 1 + y^2$;
   (g) $y' = \sqrt{y^2 + 1}/(x^2 + 1)$; **(h)** $y' = \frac{x^2}{y^2(1+x^3)}$; **(i)** $y' = 4x^3\, (1 + y^2)^{3/2}$.

[20] The British economist Thomas Malthus (1766–1834) discovered this law at the end of the 18th century.

5. Find all solutions (including singular) of the following differential equations.
   (a) $x^2y^2y' + 1 = y$; **(b)** $xyy' = (x^2+1)(1-y^2)$.
6. Obtain the particular solution satisfying the initial condition.
   (a) $y\,y' = 1,\quad y(1) = 1$; **(b)** $y' = x^2\,e^{-y},\quad y(1) = 0$;
   (c) $6xy\,y' = 1 + y^2,\quad y(4) = 1$; **(d)** $y' = y^2 + y - 6,\quad y(1) = -1/2$;
   (e) $y' = y(y+2),\quad y(2) = 4$; **(f)** $y' = 1 + y^2,\quad y(0) = 0$;
   (g) $4x\,dx + 9y\,dy = 0,\quad y(3) = 0$; **(h)** $y' = 2y/x,\quad y(1) = -2$;
   (i) $y' = (1-x)/(1+y),\quad y(1) = 0$; **(j)** $3x^{-1}y^2\,dx - 2y\,dy = 0,\quad y(4) = 8$;
   (k) $y' + y\sin 2x = 0,\quad y(0) = 1$; **(l)** $y' = y\ln y,\quad y(0) = e$;
   (m) $y' = (y+1)^2,\quad y(0) = 0$; **(n)** $y' = (2y+1)^2,\quad y(0) = -1$;
   (o) $y' = e^{y-x^2+1},\quad y(0) = 0$; **(p)** $y' = y\sec x,\quad y(0) = 1$;
   (q) $2xyy' = 1 + y^2,\quad y(1) = 2$; **(r)** $xy^2\,dx + e^x\,dy = 0,\ y \to 1/2$ when $x \to \infty$.
7. Solve the given initial value problems and determine the interval of definition for the explicit solution. *Hint:* To find the validity interval (= the longest interval where the solution is defined), look for points where the integral curve has vertical tangent.
   (a) $\frac{dy}{dx} = \frac{2x-x^3}{1+y^3}$, $\quad y(1) = 1$; **(b)** $\frac{dy}{dx} = \frac{4x-x^5}{3+y^5}$, $\quad y(1) = 1$.
8. Find the most general function that has the property that its square plus the square of its derivative equals 1.
9. The initial population in a village is 1,000. After five years, it reaches 1,848. Using the logistic population model $P' = 0.00025(k-P)P$, determine the population after 15 years.
10. The initial population in a town is 10,000. After five years, it reaches 13,564, and after 10 years it equals 14,929. Is there a solution to the logistic equation $P' = r(k-P)P$ that fits this data?
11. Six of the 1,000 passengers, crew, and staff that board a cruise ship have the flu. After one day of sailing, the number of infected people rises to 12. Assuming that the rate at which the flu virus spreads is proportional to the product of the number of infected individuals times the number of people not yet infected, determine how many people will have the flu at the end of the 7-day cruise.
12. The population of insects in a certain area increases at a rate proportional to the current population and, in the absence of other factors, the doubling time is one month. There are 500 insects in the area initially, and predators eat 100 insects/month. Determine the population of insects in the area at any time.
13. Suppose that a certain population has the growth rate $(1+\cos(0.5t))\,P(t)$, where $P(t)$ is the population at time t. Determine the population at any time.
14. A population of birds increases at a rate proportional to the number present. If the number of birds doubles in eight years, how long will it take to triple the initial number?
15. The half-life of radioactive white lead is 22 years. Find the general formula modeling its radioactive decay.
16. In winter, the thickness of the ice on a lake increases at a rate proportional to the square root of the time t. Write a differential equation for the thickness at any time.
17. Observations show that the rate of change of atmospheric pressure with altitude is proportional to the pressure. At 6 `km` (about 18,000 `ft`), the pressure is half of its value at sea level. Find the formula for the pressure at any height.
18. Solve the differential equation (that is, find its general solution and equilibrium solutions)

$$\frac{dx}{dt} = k\,(a-x)(b-x), \quad k > 0,\ a > 0,\ b > 0,$$

    which occurs in chemistry (see [14]). What value does x approach as $t \to \infty$?
19. For the given differential equation $2yy' = (1+y^2)\cos t$,
    (a) find its general solution;
    (b) determine a particular solution subject to $y(\pi/2) = -1$ and identify its interval of existence.
20. A person invested some amount of money at the rate 4.8% per year compounded continuously. After how many years would he/she double this sum?
21. What constant interest rate compounded annually is required if an initial deposit invested into an account is to double its value in seven years? Answer the same question if an account accrues interest compounded continuously.
22. In a car race, driver A had been leading archrival B for a while by a steady 5 `km` (so that they both had the same speed). Only 1 `km` from finish, driver A ran out of gas and decelerated thereafter at a rate proportional to the square of his remaining speed. Half a kilometer later, driver A's speed was exactly halved. If driver B's speed remained constant, who won the race?

23. For a positive number $a > 0$, find the solution to the initial value problem $y' = xy^3$, $y(0) = a$, and then determine the domain of its solution. Show that as a approaches zero the domain approaches the entire real line $(-\infty, \infty)$ and as a approaches $+\infty$ the domain shrinks to a single point $x = 0$.

24. For each of the following autonomous differential equations, find all equilibrium solutions and determine their stability or instability.
(a) $\dot{x} = x^2 - 4x + 3$; (b) $\dot{x} = x^2 - 6x + 5$; (c) $\dot{x} = 4x^2 - 1$;
(d) $\dot{x} = 2x^2 - x - 3$; (e) $\dot{x} = (x-2)^2$; (f) $\dot{x} = (1+x)(x-2)^2$.

25. For each of the following initial value problems, determine (at least approximately) the interval in which the solution is valid and positive, identifying all singular points. *Hint:* Solve the equation and plot its solution.
(a) $y' = (2+3x^2)/(5y^4 - 4y)$, $y(0) = 1$; (b) $y' = (1-y^2)/(x^2+1)$, $y(0) = 0$;
(c) $y' = (2x-3)/(y^2-1)$, $y(1) = 0$; (d) $y' = \sinh x/(1+2y)$, $y(0) = 1$;
(e) $y' = 9\cos(3t)/(2+6y)$, $y(0) = 1$; (f) $y' = (3-e^x)/(2+3y)$, $y(0) = 0$.

26. Solve the equation $y' = y^p/x^q$, where p and q are positive integers.

27. In each of the following equations, determine how the long-term behavior of the solution depends on the initial value $y(0) = a$.
(a) $\dot{y} = ty(3-y)/(2+t)$; (b) $\dot{y} = (1+y)/(y^2+3)$; (c) $\dot{y} = (1+t^2)/(4y^2-9)$;
(d) $\dot{y} = y^2 - 4t^2$; (e) $\dot{y} = (1+t)(4+y^2)$; (f) $\dot{y} = \sin t/(3+4y)$.

28. Suppose that a particle moves along a straight line with velocity that is inversely proportional to the distance covered by the particle. At time $t = 0$ the particle was 10 m from the origin and its velocity was $v_0 = 20$ m/sec. Determine the distance and the velocity of the particle after 12 seconds.

29. Show that the initial value problems, $\dot{y} = \cos t$, $y(0) = 0$ and $\dot{y} = \sqrt{1-y^2}$, $y(0) = 0$, have solutions that coincide for $0 \leqslant t \leqslant \frac{\pi}{2}$.

30. In positron emission tomography, a patient is given a radioactive isotope, usually technetium-99m, that emits positrons as it decays. When a positron meets an electron, both are annihilated and two gamma rays are given off. These are detected and a tomographic image (of a tumor or cancer) is created.
Technetium-99m decays in accordance with the equation $dy/dt = -ky$, with $k = 0.1155$/hour. The short half-life of technetium-99m has the advantage that its radioactivity does not endanger the patient. A drawback is that the isotope must be manufactured in a cyclotron, which is usually located far away from the hospital. It therefore needs to be ordered in advance from medical suppliers.
Suppose a dosage of 7 millicuries (mCi) of technetium-99m is to be administrated to a patient. Estimate the amount of the radionuclide that the manufacturer should produce 24 hours before its use in the hospital treatment room.

31. Torricelli's equation is an equation created by Evangelista Torricelli (1608–1647) to find the final velocity of an object moving with a constant acceleration without having a known time interval. It is based on expressing the velocity $v = gt$ of a falling liquid drop in terms of distance: $v = \sqrt{2gh}$, where g is acceleration due to gravity, and h is the distance.
Suppose a tank has an outlet hole near the bottom. Let $h = h(t)$ be the height of a fluid above the outlet at time t, and let a be the area of the outlet hole. For a tank with cross-sectional area $A(h)$ at height h, the fluid volume V up to height h is given by an integral $V(h) = \int_0^h A(y)\,dy$. This volume $V(t)$, remaining after time t, satisfies *Torricelli's equation:*
$$\frac{dV}{dt} = -ka\sqrt{2gh} \qquad \Longrightarrow \qquad A(h)\,\frac{dh}{dt} = -ka\sqrt{2gh},$$
where the flow constant k $(0 < k < 1)$ depends on properties of the fluid such as viscosity and on the shape of the hole.
A tank in the shape of a circular cone has radius $r_0 = 0.3$ meters and vertical height $h_0 = 1.5$ meters. Hence the radius r at height h satisfies $r = h/5$. A circular outlet has area $a = \pi/100$. Find the time to empty the tank if $k \approx 0.7$.

32. An upright hemispherical bowl of radius 1 meter (m) has an outlet hole near the bottom of radius $r = 0.05$ m. Assuming flow constant $k \approx 0.5$, about how long would it take for the full bowl to empty under the influence of gravity?

33. Stefan's Law of radiation states that the radiation energy of a body is proportional to the fourth power of the absolute temperature T (in the Kelvin scale) of a body. The rate of change of this energy in a surrounding medium of absolute temperature M is thus
$$\frac{dT}{dt} = \sigma\left(M^4 - T^4\right),$$
where σ is a positive constant when $T > M$. Find the general solution of Stefan's equation assuming M to be a constant.

34. The antioxidant activity (of crude hsian-tsao leaf gum extracted by sodium bicarbonate solutions and precipitated by 70% ethanol) $A(c)$ depending on its concentration c can be modeled by the following initial value problem[21] that you aresed to solve:
$$\frac{dA}{dc} = k\left[A^* - A(c)\right] \quad k > 0, \qquad A(0) = 0.$$

[21] Lih-Shiun Lai et al., *Journal of Agricultural and Food Chemistry*, **49**, 963–968, 2001.

2.2 Equations Reducible to Separable Equations

One of the best techniques used to solve ordinary differential equations is to change one or both of the variables. Many differential equations can be reduced to separable equations by substitution. We start with a nonlinear equation of the form

$$y' = F(ax + by + c), \qquad b \neq 0, \tag{2.2.1}$$

where $F(v)$ is a given continuous function of a variable v, and a, b, c are some constants. Equation (2.2.1) can be transformed into a separable differential equation by means of the substitution $v = ax + by + c$, $b \neq 0$. Then $v' = a + by'$, and we obtain from $y' = F(v)$ that

$$v' = a + by' = a + b\,F(v) \qquad \text{or} \qquad \frac{\mathrm{d}v}{a + bF(v)} = \mathrm{d}x,$$

which is a separable differential equation.

For example, the differential equation $y' = (x + y + 1)^2$ is not a separable one. By setting $v = x + y + 1$, we make it separable: $v' = 1 + y' + 0 = 1 + v^2$. Separation of variables yields

$$\frac{\mathrm{d}v}{1 + v^2} = \mathrm{d}x \qquad \Longrightarrow \qquad \arctan(v) = x + C \qquad \text{or} \qquad v = \tan(x + C).$$

Therefore, the general solution of the given equation is $\arctan(x + y + 1) = x + C$, where C is an arbitrary constant. Equation (2.2.1) is a particular case of more general equations discussed in §2.2.3. ■

There is another class of nonlinear equations that can be reduced to separable equations. Let us consider the differential equation

$$x\,y' = y\,F(xy), \tag{2.2.2}$$

where $F(v)$ is a given function. By setting $v = xy$, we reduce Eq. (2.2.2) to the following one:

$$v' = y + y\,F(v) = \frac{v}{x}\,(1 + F(v))$$

since $v' = y + xy'$ and $y = v/x$. Separation of variables and integration yields the general solution in implicit form

$$\int \frac{\mathrm{d}v}{v(1 + F(v))} = x + C, \qquad v = xy.$$

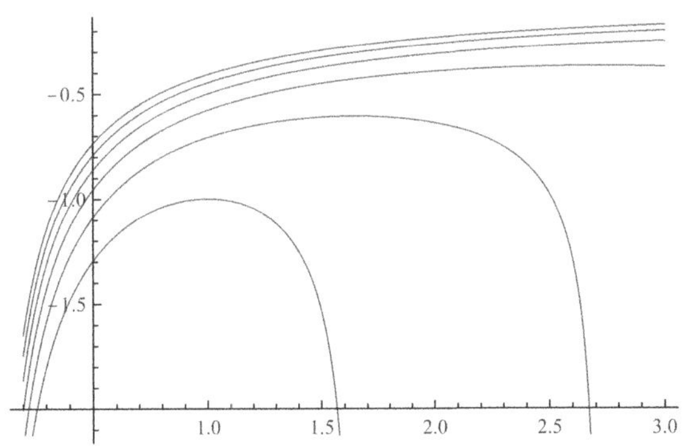

Figure 2.9: Example 2.2.1. Solutions for different values of C, plotted with *Mathematica*.

Example 2.2.1: Solve the differential equation

$$x\,y' = x^2y^3 - y.$$

Solution. Since the equation is of the form (2.2.2), with $F(v) = v^2 - 1$, we use substitution $v = xy$, which yields

$$v' = y + xy' = y + x^2y^3 - y = x^2y^3 = v^3/x.$$

We separate variables v and x to obtain

$$\frac{\mathrm{d}v}{v^3} = \frac{\mathrm{d}x}{x} \qquad \text{or} \qquad -\frac{1}{2v^2} = \ln|x| + C.$$

Since

$$v^2 = \frac{1}{-2(\ln|x| + C)} = \frac{1}{C_1 - \ln x^2},$$

where $C_1 = -2C$ is an arbitrary constant (we denote it again by C), we get the general solution

$$y = \frac{1}{x\sqrt{C - \ln x^2}}, \qquad \ln x^2 < C.$$

To find the general solution in *Mathematica*, use the command `DSolve`. However, there are two options in its application:

```
DSolve[y'[x] == x*(y[x])^3-y[x]/x,y[x],x]      or
DSolve[y'[x] == x*(y[x])^3-y[x]/x,y,x]
```

The former does not give values for `y'[x]` or even `y[1]`. The latter will return `y` as a pure function, so you can evaluate it at any point. Say, to find the value at $x = 1$, you need to execute the command `y[1]/.%`
To plot solutions in *Mathematica* (see Fig. 2.9 on page 53), we type:

```
solution = DSolve[y'[x] == x*(y[x])^3 - y[x]/x, y[x], x]
g[x_] = y[x] /. solution[[1]]
t[x_] = Table[g[x] /. C[1] -> j, {j, 1, 6}]
Plot[t[x], {x, 0.2, 3}]
```

However, since the general solution is known explicitly, we can plot the family of streamlines with one *Mathematica* command:

```
Plot[1/(x Sqrt[#1 - 2 Log[x]] &) /@ {1, 2, 3, 4}, {x, .3, 10} ]
```

2.2.1 Equations with Homogeneous Coefficients

Definition 2.6: Let r be a real number. A function $g(x, y)$ is called **homogeneous** (*with accent on syllable "ge"*) of degree r if $g(\lambda x, \lambda y) = \lambda^r g(x, y)$ for any nonzero constant λ.

For example, the functions x/y, $x - y$, $x^2 + \sqrt{x^4 + y^4}$, $x^3 + y^3 + xy^2$ are homogeneous of degree zero, one, two, and three, respectively. Usually homogeneous functions of zero degree are referred to as homogeneous. Obviously, if f is a function of the ratio y/x, i.e.,

$$f(x, y) = F\left(\frac{y}{x}\right), \qquad x \neq 0,$$

then f is a homogeneous function[22] (of degree 0). We avoid the trivial case when $f(x, y) = y/x$. Let us consider the differential equation with homogeneous right-hand side function:

$$\frac{dy}{dx} = F\left(\frac{y}{x}\right), \qquad x \neq 0. \tag{2.2.3}$$

To solve this equation, we set the new dependent variable $v(x) = y(x)/x$ or $y = vx$. Then Eq. (2.2.3) becomes

$$\frac{dy}{dx} = F(v).$$

Applying the product rule, we find the derivative dy/dx to be

$$\frac{dy}{dx} = \frac{d}{dx}(vx) = x\frac{dv}{dx} + v = xv' + v.$$

Substituting this expression into the equation $y' = F(v)$, we get

$$x\frac{dv}{dx} + v = F(v) \qquad \text{or} \qquad x\frac{dv}{dx} = F(v) - v.$$

We now can separate the variables v and x, finding

$$\frac{dv}{F(v) - v} = \frac{dx}{x}, \quad F(v) \neq v.$$

Integration produces a logarithm of x on the right-hand side:

$$\int \frac{dv}{F(v) - v} = \ln C + \ln|x| = \ln C|x| = \ln Cx \quad (v = y/x),$$

[22]Because the ratio y/x is the tangent of the argument in the polar system of coordinates, the function $f(y/x)$ is sometimes called "homogeneous-polar."

where C is an arbitrary constant so that the product Cx is positive. Thus, the last expression defines the general solution of Eq. (2.2.3) in implicit form.

Sometimes it is more convenient to consider x as a new dependent variable. Since $\frac{dy}{dx} = 1/(dx/dy)$, we can rewrite Eq. (2.2.3) as

$$\frac{dx}{dy} = \frac{1}{F(y/x)}.$$

Using substitution $x = uy$ and separation of variables, we find

$$\frac{dx}{dy} = y\,\frac{du}{dy} + u = \frac{1}{F(u^{-1})} \qquad \Longrightarrow \qquad \ln Cy = \int \frac{du}{1/F(u^{-1}) - u}\bigg|_{u=x/y}.$$

Example 2.2.2: Solve the equation

$$\frac{dy}{dx} = \frac{x^2+y^2}{xy}, \qquad x \ne 0,\ y \ne 0.$$

Solution. Ruling out the singular lines $x = 0$ and $y = 0$, we make substitution $y = vx$. This leads to a new equation in v:

$$x\,\frac{dv}{dx} + v = \frac{v^2x^2 + x^2}{xvx} = \frac{v^2+1}{v} = v + \frac{1}{v} \qquad (x \ne 0, \quad v \ne 0).$$

Canceling v from both sides, we get $\frac{dv}{dx} = \frac{1}{xv}$. The equation can be solved by separation of variables:

$$v\,dv = \frac{dx}{x} \qquad \Longrightarrow \qquad \frac{1}{2}\,v^2 = \ln|x| + C,$$

where C is a constant of integration. If we denote $C = \ln C_1$, then the right-hand side can be written as $\ln|x| + C = \ln|x| + \ln C_1 = \ln C_1|x|$. We can drop the index of C_1 and write down this expression as $\ln C|x| = \ln Cx$, where arbitrary constant C takes care of the product Cx to guarantee its positiveness. That is, $C > 0$ when x is positive and $C < 0$ when x is negative. The last step is to restore the original variables by reversing the substitution $v = y/x$:

$$\frac{y^2}{2x^2} = \ln C|x| = \ln Cx \qquad \text{or} \qquad \frac{y^2}{x^2} = 2\ln C|x| = \ln C^2x^2,$$

because C^2 is again an arbitrary positive constant and we can denote it by the same letter. The general solution can be written explicitly as $y = \pm x\sqrt{\ln Cx^2}$ with a positive constant C.

Now we rewrite the given equation in a reciprocal form:

$$\frac{dx}{dy} = \frac{xy}{x^2+y^2}, \qquad x \ne 0, \quad y \ne 0.$$

Setting $x = uy$ and separating variables, we obtain

$$\left(\frac{1}{u} + \frac{1}{u^3}\right) du = -\frac{dy}{y} \qquad (u \ne 0,\ y \ne 0).$$

Integration yields $u^2 = \frac{x^2}{y^2} = \frac{1}{\ln Cx^2}$, so $y = \pm x\sqrt{\ln Cx^2}$. To check our solution, we use the following *Maple* commands:

```
dsolve(D(y)(x)=(x*x+y(x)*y(x))/x*y(x),y(x))
```

To plot a particular solution we write

```
p:=dsolve({D(y)(x)=(x*x+y(x)*y(x))/x*y(x),y(1)=3},numeric)
odeplot(p,0.4..5)
```

We can reduce the given equation to a separable one by typing in *Mathematica:*

```
ode[x_, y_] = (x^2 + y^2)*dx == x*y*dy
ode[x, v*x] /. {dy -> x*dv + v*dx}
Map[Cancel, Map[Function[q, q/x^2], %]]
Map[Function[u, Collect[u, {dx, dv}]], %]
```

□

Now suppose that $M(x,y)$ and $N(x,y)$ are both homogeneous functions of the same degree α, and consider the differential equation written in the differential form:

$$M(x,y)\,\mathrm{d}x + N(x,y)\,\mathrm{d}y = 0. \tag{2.2.4}$$

Equation (2.2.4) can be converted into a separable equation by substitution $y = vx$. This leads to

$$M(x,vx)\,\mathrm{d}x + N(x,vx)\,(v\mathrm{d}x + x\mathrm{d}v) = 0,$$

which, because of the homogeneity of M and N, can be rewritten as

$$x^{\alpha}\,M(1,v)\,\mathrm{d}x + x^{\alpha}\,N(1,v)\,(v\mathrm{d}x + x\mathrm{d}v) = 0$$

or

$$M_1(v)\,\mathrm{d}x + N_1(v)\,(v\mathrm{d}x + x\mathrm{d}v) = 0,$$

where $M_1(v) = M(1,v)$ and $N_1(v) = N(1,v)$. We separate variables to get

$$[M_1(v) + v\,N_1(v)]\,\mathrm{d}x + N_1(v)x\,\mathrm{d}v = 0 \qquad \text{or} \qquad \frac{\mathrm{d}x}{x} + \frac{N_1(v)}{M_1(v) + vN_1(v)}\,\mathrm{d}v = 0.$$

Since the latter is already separated, we integrate both sides and obtain a solution of the given differential equation in implicit form.

Example 2.2.3: Solve the differential equation

$$\frac{\mathrm{d}y}{\mathrm{d}x} = \frac{y + \sqrt{x^2 - y^2}}{x}, \qquad x \ne 0 \quad \text{and} \quad x^2 \geqslant y^2.$$

Solution. When an equation contains a square root, it is assumed that the value of the root is always positive (positive branch of the root). We consider first the case when $x > 0$. Let us substitute $y = vx$ into the given equation; then it becomes

$$x\,v' + v = v + \sqrt{1 - v^2} \qquad \text{or} \qquad x\,v' = \sqrt{1 - v^2}.$$

Separation of variables and integration yield $\arcsin v = \ln|x| + C_1$, or

$$v(x) = \sin\left(\ln|x| + C_1\right) = \sin\left(\ln Cx\right), \qquad |\ln Cx| \leqslant \frac{\pi}{2},$$

where $C_1 = \ln C$ is a constant of integration. The constant functions, $v = 1$ and $v = -1$, are equilibrium solutions of this differential equation that cannot be obtained from the general one for any choice of C. These two critical points are singular solutions because they are touched by integral curves corresponding to the general solution ($\arcsin(\pm 1) = \pm\frac{\pi}{2}$). Moreover, the constants $v = \pm 1$ are envelopes of the one-parameter family of integral curves corresponding to the general solution. Since $v = y/x$, we obtain the one-parameter family of solutions

$$y = x\,\sin\left(\ln Cx\right), \qquad |\ln Cx| \leqslant \frac{\pi}{2}.$$

Now let us consider the case where $x < 0$. Setting $x = -t$, $t > 0$, we get

$$\frac{\mathrm{d}y}{\mathrm{d}x} = \frac{\mathrm{d}y}{\mathrm{d}t}\cdot\frac{\mathrm{d}t}{\mathrm{d}x} = -\frac{\mathrm{d}y}{\mathrm{d}t} \qquad \text{and} \qquad \frac{y + \sqrt{x^2 - y^2}}{x} = \frac{y + \sqrt{t^2 - y^2}}{-t}.$$

Equating these two expressions, we obtain the same differential equation. Therefore, the solutions are symmetrical with respect to the vertical y-axis, and the function $y = y(t) = y(-x) = t\,\sin\left(\ln Ct\right)$ is its general solution. The given differential equation has two equilibrium solutions that correspond to $v = \pm 1$. One of them, $y = x$, is stable, and the other one, $y = -x$, is unstable (see Fig. 2.11 on page 57).

The given differential equation can be rewritten in the following equivalent form:

$$xy' - y = \sqrt{x^2 - y^2}.$$

By squaring both sides, we get the nonlinear equation without the radical:

$$x^2(y')^2 - 2xyy' + 2y^2 - x^2 = 0. \tag{2.2.5}$$

As usual, raising both sides of an equation to the second power may lead to an equation with an additional solution. In our case, Eq. (2.2.5) combines two equations: $xy' = y \pm \sqrt{x^2 - y^2}$. This additional solution,

$$y = -x \sin(\ln Cx),$$

is the general solution to the equation $xy' = y - \sqrt{x^2 - y^2}$. To plot the direction field, we ask *Maple:*

```
DEplot(x*diff(y(x),x)=y(x)-sqrt(x*x-y(x)*y(x)),y(x),
x=-10..10,y=-10..10,[y(4)=1,y(12)=1,y(-4)=1,y(-12)=1],
scaling=constrained, dirgrid=[30,30], color=blue, linecolor=black,
title="x*dydx=y-sqrt(x*x-y*y)");
```

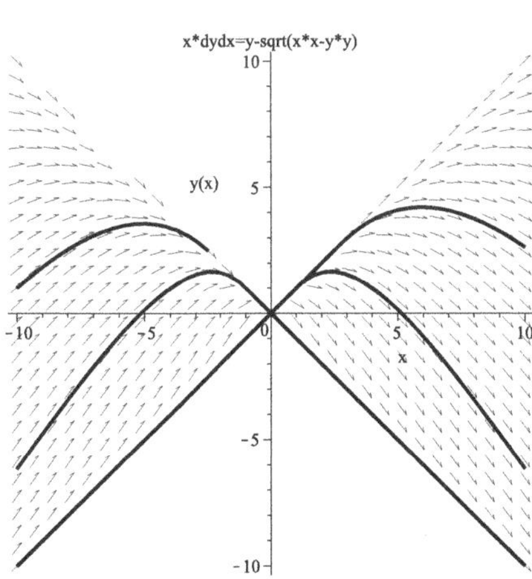

Figure 2.10: Example 2.2.3: direction field and solutions of the equation $xy' = y - \sqrt{x^2 - y^2}$, plotted with *Maple*.

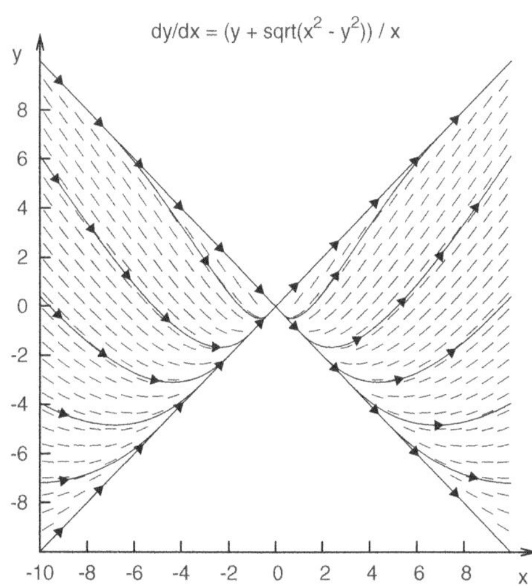

Figure 2.11: Example 2.2.3: direction field and solutions of the equation $xy' = y + \sqrt{x^2 - y^2}$, plotted with *Maxima*.

2.2.2 Equations with Homogeneous Fractions

Let us consider the behavior of the solutions of the differential equation

$$M(x, y)\,\mathrm{d}x + N(x, y)\,\mathrm{d}y = 0$$

near a singular point (x^*, y^*), where $M(x^*, y^*) = N(x^*, y^*) = 0$. To make our exposition as simple as possible, we assume that the singular point is $(0, 0)$, and consider the direction field near the origin (locally). When $M(x, y)$ is a smooth function in the neighborhood of $(0, 0)$ such that $M(0, 0) = 0$, it has a Maclaurin series expansion of the form

$$M(x, y) = ax + by + cx^2 + dxy + ey^2 + \cdots,$$

and $N(x, y)$ has a similar expansion. If we neglect terms of the second and higher degrees, which are small near the origin, we are led to an equation of the form

$$\frac{\mathrm{d}y}{\mathrm{d}x} = \frac{ax + by}{Ax + By} \qquad \text{or} \qquad (ax + by)\,\mathrm{d}x = (Ax + By)\,\mathrm{d}y. \tag{2.2.6}$$

Such an operation, replacing the given differential equation with a simpler one (2.2.6), is called *linearization* (see §2.5). In what follows, it is assumed that $aB \neq Ab$; otherwise, the numerator $ax + by$ will be proportional to the denominator: $Ax + By = \frac{B}{b}\,(ax+by)$. Setting $y = xv$ $(x \neq 0)$, where $v = v(x)$ is the function to be determined, we get

$$y' = v + xv' = \frac{a+bv}{A+Bv} \qquad \Longrightarrow \qquad xv' = \frac{a+bv-Av-Bv^2}{A+Bv}. \tag{2.2.7}$$

The obtained differential equation for v may have critical points that are solutions of the quadratic equation

$$Bv^2 + (A-b)v - a = 0. \tag{2.2.8}$$

Since Eq. (2.2.7) is a separable one, we can integrate it to obtain

$$\int \frac{A+Bv}{Bv^2+(A-b)v-a}\,\mathrm{d}v = -\int \frac{\mathrm{d}x}{x} = -\ln Cx = \ln Cx^{-1} \qquad (x \neq 0).$$

The explicit expression of the integral on the left-hand side depends on the roots of the quadratic equation (2.2.8). If $b = -A$, then the numerator is the derivative of the denominator (up to a factor 1/2), and we can apply the formula from calculus:

$$\int \frac{f'(v)}{f(v)}\,\mathrm{d}v = \ln|f(v)| + C_1 = \ln Cf(v), \tag{2.2.9}$$

where $C_1 = \ln C$ is a constant of integration. This yields

$$\int \frac{A+Bv}{Bv^2+2Av-a}\,\mathrm{d}v = \frac{1}{2}\int \frac{\mathrm{d}(Bv^2+2Av-a)}{Bv^2+2Av-a} = \frac{1}{2}\ln(Bv^2+2Av-a) = \ln Cx^{-1}.$$

By exponentiating, we have $Bv^2 + 2Av - a = Cx^{-2}$. Substituting $v = y/x$ leads to the following general solution:

$$By^2 + 2Axy - ax^2 = C,$$

which is a family of similar quadratic curves centered at the origin and having common axes. They are ellipses (or circles) if $Ab - aB > 0$, and hyperbolas together with their common asymptotes if $Ab - aB < 0$. The case $b + A = 0$ is also treated in §2.3.

In general, the family of integral curves of Eq. (2.2.6) appears unchanged when the figure is magnified or shrunk. The behavior of the integral curves for Eq. (2.2.6) in the neighborhood of the origin is one of the following three types according to

Case I:	$(A-b)^2 + 4aB > 0$	**(a)**	$aB - Ab < 0$,	semi-stable,
		(b)	$aB - Ab > 0$,	hyperbola.
Case II:	$(A-b)^2 + 4aB = 0.$			
Case III:	$(A-b)^2 + 4aB < 0.$			

When $(A-b)^2 + 4aB > 0$, the roots of the equation $Bv^2 + (A-b)v - a = 0$ are real and distinct, that is, $Bv^2 + (A-b)v - a = B(v-v_1)(v-v_2)$. Partial fraction decomposition then yields

$$\int \frac{A+Bv}{Bv^2+2Av-a}\,\mathrm{d}v = \int \left[\frac{D}{v-v_1} + \frac{E}{v-v_2}\right]\mathrm{d}v = D\,\ln|v-v_1| + E\,\ln|v-v_2|,$$

with some constants D and E (of different signs if $aB < Ab$). When $(A-b)^2 + 4aB = 0$, the equation $Bv^2 + (A-b)v - a = 0$ has one double root $v_1 = v_2 = -\frac{A-b}{2B}$. Then

$$\int \frac{A+Bv}{Bv^2+2Av-a}\,\mathrm{d}v = \int \frac{A+Bv}{B(v-v_1)^2}\,\mathrm{d}v = -\frac{A+Bv_1}{B(v-v_1)} + \ln|v-v_1|.$$

When $(A-b)^2 + 4aB < 0$, the roots of the equation $Bv^2 + (A-b)v - a = 0$ are complex conjugates, and the integral $\int \frac{A+Bv}{Bv^2+2Av-a}\,\mathrm{d}v$ can be expressed through the arctan function, as seen in the next example.

Example 2.2.4: (Case III) Solve the equation $5x\,\mathrm{d}x + 2\,(x+y)\,\mathrm{d}y = 0$, $x + y \neq 0$ and $x \neq 0$.

Solution. Excluding the line $x + y = 0$ where the slope function is unbounded and setting $y = vx$, we calculate the differential of y to be $\mathrm{d}y = x\,\mathrm{d}v + v\,\mathrm{d}x$. Therefore, the given differential equation becomes:

$$5x\,\mathrm{d}x + 2(vx+x)(x\,\mathrm{d}v + v\,\mathrm{d}x) = 0 \qquad (v \neq -1 \text{ and } x \neq 0),$$

from which a factor $x \neq 0$ can be removed at once. That done, we have to solve

$$5\,\mathrm{d}x + 2(v+1)(x\,\mathrm{d}v + v\,\mathrm{d}x) = 0 \qquad \text{or} \qquad [5 + 2v(v+1)]\,\mathrm{d}x + 2(v+1)x\,\mathrm{d}v = 0.$$

Since $2v(v+1)+5 > 0$, the differential equation in v has no equilibrium solution. Separation of variables yields

$$\int \frac{2(v+1)\,\mathrm{d}v}{2v^2+2v+5} = -\int \frac{\mathrm{d}x}{x} = -\ln|x| + C/2,$$

where C is an arbitrary constant. Since $\mathrm{d}(2v^2+2v+5) = 4v+2$, we have

$$\begin{aligned}
\int \frac{2(v+1)\,\mathrm{d}v}{2v^2+2v+5} &= \frac{1}{2}\int \frac{(4v+4)\,\mathrm{d}v}{2v^2+2v+5} = \frac{1}{2}\int \frac{(4v+2)\,\mathrm{d}v}{2v^2+2v+5} + \frac{1}{2}\int \frac{2\,\mathrm{d}v}{2v^2+2v+5} \\
&= \frac{1}{2}\int \frac{\mathrm{d}(2v^2+2v+5)}{2v^2+2v+5} + \int \frac{\mathrm{d}v}{2v^2+2v+5} \\
&= \frac{1}{2}\ln(2v^2+2v+5) + \int \frac{\mathrm{d}v}{2v^2+2v+5}.
\end{aligned}$$

To find the antiderivative of the latter integral, we "complete the squares" in the denominator as follows:

$$\begin{aligned}
2v^2+2v+5 &= 2\left(v^2+v+\frac{5}{2}\right) = 2\left[v^2 + 2\cdot\frac{1}{2}\cdot v + \left(\frac{1}{2}\right)^2 - \left(\frac{1}{2}\right)^2 + \frac{5}{2}\right] \\
&= 2\left[\left(v+\frac{1}{2}\right)^2 - \frac{1}{4} + \frac{5}{2}\right] = 2\left[\left(v+\frac{1}{2}\right)^2 + \frac{9}{4}\right] = 2\left[\left(v+\frac{1}{2}\right)^2 + \left(\frac{3}{2}\right)^2\right].
\end{aligned}$$

Using the identity

$$\int \frac{\mathrm{d}u}{u^2+a^2} = \frac{1}{a}\arctan\frac{u}{a} + C,$$

we obtain

$$\int \frac{\mathrm{d}v}{2v^2+2v+5} = \frac{1}{2}\int \frac{\mathrm{d}v}{\left(v+\frac{1}{2}\right)^2 + \left(\frac{3}{2}\right)^2} = \frac{1}{3}\arctan\left(\frac{2v+1}{3}\right).$$

Substitution yields the general solution

$$\frac{1}{2}\ln(2v^2+2v+5) + \frac{1}{3}\arctan\left(\frac{2v+1}{3}\right) = -\ln|x| + C/2$$

that is defined in two domains $v > -1$ and $v < -1$. Using properties of logarithms, we get

$$\ln\left[x^2(2v^2+2v+5)\right] + \frac{2}{3}\arctan\left(\frac{2v+1}{3}\right) = C \qquad (v \neq -1 \ \text{ and } \ x \neq 0).$$

To check the integration, we use *Maxima:*

```
integrate((2*(v+1))/(2*v^2+2*v+5), v);
```

Since $v = y/x$, we obtain the integral of the given differential equation to be

$$C = \ln[2y^2 + 2xy + 5x^2] + \frac{2}{3}\arctan\frac{2y+x}{3x} \qquad (y \neq -x \ \text{ and } \ x \neq 0).$$

To draw the direction field and some solutions, we type in *Maxima:*

```
drawdf(-(5*x)/(2*(x+y)),
makelist( implicit(C=log(2*y^2+2*x*y+5*x^2)+2/3*atan((2*y+x)/(3*x)),
     x,-10,10, y,-10,10), C, [1, 3, 5, 10]))$
```

Example 2.2.5: (Case II) Consider the differential equation

$$\frac{\mathrm{d}y}{\mathrm{d}x} = \frac{-3x+8y}{2x+3y}, \quad 3y \neq -2x.$$

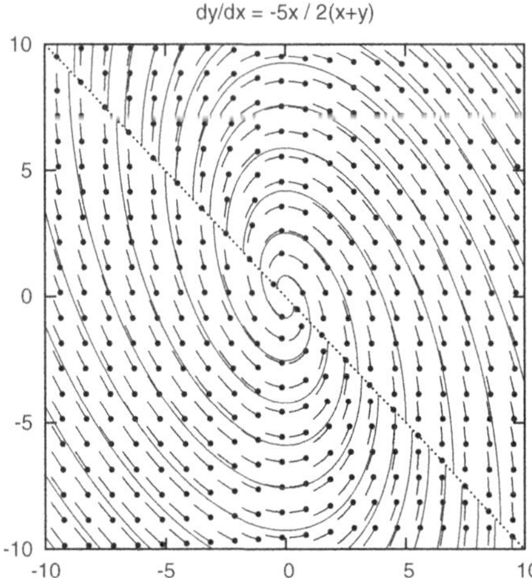

Figure 2.12: Example 2.2.4 (Case III): direction field and solutions to $5x\,\mathrm{d}x+2(x+y)\,\mathrm{d}y=0$, plotted with *Maxima.*

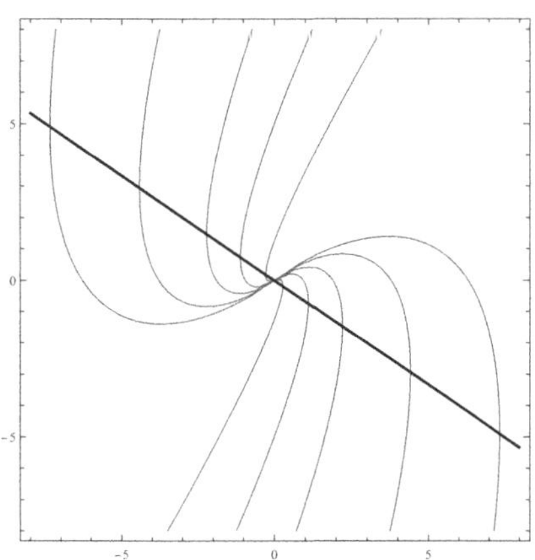

Figure 2.13: Example 2.2.5 (Case II): solutions to $y'=(8y-3x)/(2x+3y)$, plotted with *Mathematica.*

Excluding the line $3y=-2x$ where the slope function is unbounded, we make the substitution $y=xv$ that results in

$$\frac{2+3v}{(v-1)^2}\,\mathrm{d}v=-3\,\frac{\mathrm{d}x}{x}\qquad\Longrightarrow\qquad\left(\frac{5}{v-1}+3\,\ln|v-1|+3\,\ln Cx\right)_{v=y/x}=0.$$

Intersections of the integral curves from the general solution with the line $2x+3y=0$ give the points where tangent lines are vertical (see Fig. 2.13). The given differential equation has the singular solution $y=x$ (not a member of the general solution). To plot the direction field and solutions for the equation, we use the following *Maple* commands:

```
ode:=diff(y(x),x)=(a*x+b*y(x))/(A*x+B*y(x));                inc:=y(0)=1;
Y:=unapply(rhs(dsolve({eval(ode,{a=-3,b=8,A=2,B=3}), ics}, y(x))), x);
plot(Y, 0 .. 3*Pi, numpoints = 1000, scaling=constrained);
dfieldplot(eval(ode,{a=-3,b=8,A=2,B=3}), ics}, y(x), x=-9..9,y=-8..8)
```

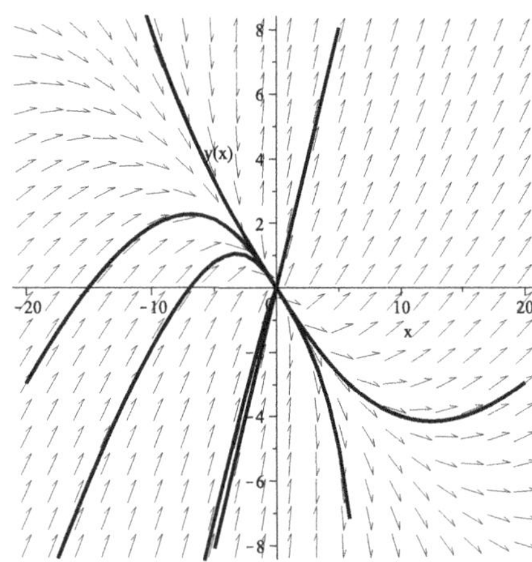

Figure 2.14: Case Ia: direction field and solutions to $(x+3y)\,\mathrm{d}x=(2x+y)\,\mathrm{d}y$, plotted with *Maple.*

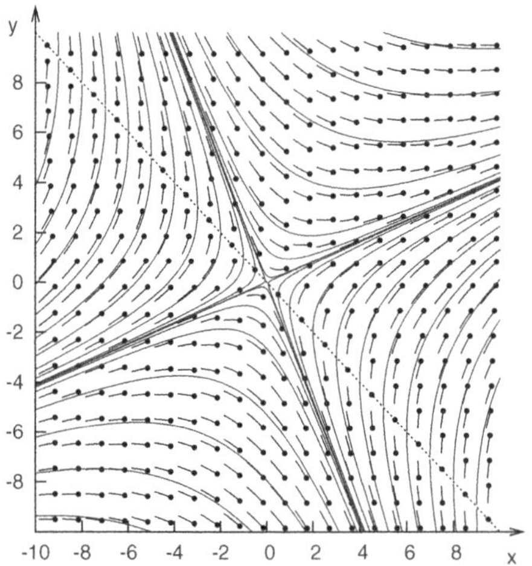

Figure 2.15: Example 2.2.6 (Case Ib): direction field and solutions to $y'=(x-y)/(x+y)$, plotted with *Maxima.*

Example 2.2.6: (Case I) Solve the equation $\dfrac{\mathrm{d}y}{\mathrm{d}x}=\dfrac{x-y}{x+y},\qquad y\neq -x\ \text{ and }\ x\neq 0.$

Solution. The slope function $(x-y)/(x+y)$ is a homogeneous function because

$$\frac{x-y}{x+y} = \frac{1-y/x}{1+y/x} \qquad (x \neq 0).$$

Thus, we set $y = vx$ $(x \neq 0)$ and get

$$\frac{\mathrm{d}y}{\mathrm{d}x} = x\frac{\mathrm{d}v}{\mathrm{d}x} + v = \frac{1-v}{1+v} \qquad \text{or} \qquad x\frac{\mathrm{d}v}{\mathrm{d}x} = \frac{1-v}{1+v} - v \qquad (v \neq -1).$$

We simplify this expression algebraically and separate variables to obtain

$$x\frac{\mathrm{d}v}{\mathrm{d}x} = -\frac{v^2+2v-1}{1+v} \qquad \Longrightarrow \qquad \frac{1+v}{v^2+2v-1}\,\mathrm{d}v = -\frac{\mathrm{d}x}{x},$$

or, after integration,

$$\int \frac{1+v}{v^2+2v-1}\,\mathrm{d}v = -\int \frac{\mathrm{d}x}{x} = -\ln Cx \quad (v \neq -1 \pm \sqrt{2},\ x \neq 0).$$

The derivative of the denominator $\frac{\mathrm{d}}{\mathrm{d}v}(v^2+2v-1) = 2v+2 = 2(v+1)$ equals twice the numerator. Thus,

$$\int \frac{1+v}{v^2+2v-1}\,\mathrm{d}v = \frac{1}{2}\int \frac{\mathrm{d}(v^2+2v-1)}{v^2+2v-1} = \frac{1}{2}\ln\left|v^2+2v-1\right|.$$

From these relations, we get the general solution

$$\frac{1}{2}\ln\left|v^2+2v-1\right| = -\ln C - \ln|x| \qquad \Longrightarrow \qquad \ln\left|v^2+2v-1\right| + \ln x^2 = \ln C.$$

Using the property of the logarithm function, $\ln(ab) = \ln a + \ln b$, we rewrite the general solution in a simpler form: $\ln x^2|v^2+2v-1| = \ln C$. After substituting $v = y/x$ $(x \neq 0)$ into the equation, we obtain

$$\left|y^2+2yx-x^2\right| = C \geqslant 0 \qquad \text{or} \qquad y^2+2yx-x^2 = c,$$

where c is an arbitrary constant (not necessarily positive). This is the general solution (which is a family of hyperbolas) in an implicit form. We solve the equation with respect to y to obtain the general solution in the explicit form:

$$y = -x \pm \sqrt{2x^2+c}\,,$$

which is valid for all real x where $2x^2+c \geqslant 0$. The given differential equation, $y' = (x-y)/(x+y)$, also has two equilibrium solutions $y = (-1 \pm \sqrt{2})x$ that are the asymptotes of the solutions (hyperbolas). In fact, integration yields

$$\int \frac{1+v}{v^2+2v-1}\,\mathrm{d}v = \int \frac{u\,\mathrm{d}u}{u^2-2}, \qquad \text{where } v = \frac{y}{x} \quad \text{and} \quad u = v+1 = \frac{x+y}{x},$$

subject to $v^2+2v-1 \neq 0$ (or in original variables, $y^2+2xy-x^2 \neq 0$). Upon equating the quadratic function to 0, we obtain the critical points $v = -1 \pm \sqrt{2}$, which correspond to $y = (-1 \pm \sqrt{2})x$. The line where the slope function is unbounded, $y = -x$, separates the direction field and all solutions into two parts. As Fig. 2.15 shows, one part of the equilibrium solutions is stable, and the other part is unstable. □

The following equation:

$$\frac{\mathrm{d}y}{\mathrm{d}x} = F\left(\frac{ax+by}{Ax+By}\right) \tag{2.2.10}$$

is an example of a differential equation with a homogeneous forcing function, where a, b, A, and B are some constants. This means that the function $F(\omega)$ on the right side of Eq. (2.2.10) remains unchanged if x is replaced by αx, and y by αy, with any constant $\alpha \neq 0$ because

$$F\left(\frac{a\alpha x+b\alpha y}{A\alpha x+B\alpha y}\right) = F\left(\frac{ax+by}{Ax+By}\right).$$

In what follows, we present an example for $F(\omega) = \sqrt{\omega}$.

Example 2.2.7: Solve the differential equation $y' = \sqrt{\dfrac{x+y}{2x}}$, $x > 0$, under two sets of initial conditions: $y(1) = 1$ and $y(1) = 3$.

Solution. Setting $y = vx$ in the given equation, we transform it to a separable equation $v'\,x + v = \sqrt{(1+v)/2}$. For positive root branch, the slope function has the critical point: $v = 1$, which corresponds to the equilibrium solution: $y = x$. Upon separation of variables, we obtain

$$\frac{dv}{\sqrt{(1+v)/2} - v} = \frac{dx}{x} \qquad \Longrightarrow \qquad \int \frac{dv}{v - \sqrt{(1+v)/2}} = -\int \frac{dx}{x} = -\ln C|x|,$$

where absolute value sign in the right-hand side can be dropped because we assume that $x > 0$. In the left-hand side integral, we change the variable by setting $v = 2u^2 - 1$. Then

$$u^2 = \frac{1+v}{2} \quad \text{and} \quad v - \sqrt{(1+v)/2} = 2u^2 - u - 1, \quad dv = 4u\,du.$$

Hence,

$$\int \frac{dv}{v - \sqrt{(1+v)/2}} = \int \frac{4u\,du}{2u^2 - u - 1}.$$

Partial fraction decomposition yields

$$\frac{4u}{2u^2 - u - 1} = \frac{4/3}{u-1} + \frac{2/3}{u+1/2}, \qquad u \ne 1, \quad u \ne -1/2.$$

Note that u cannot be negative because $u = \sqrt{(x+y)/2x} \geqslant 0$. Integrating, we have

$$\frac{4}{3}\ln|u-1| + \frac{2}{3}\ln\left|u + \frac{1}{2}\right| = -\ln C\,x \qquad \text{or} \qquad 2\ln|u-1| + \ln\left|u + \frac{1}{2}\right| = -\frac{3}{2}\ln C\,x\,.$$

Thus,

$$\ln(u-1)^2|u+1/2| = \ln C|x|^{-3/2}.$$

Exponentiation yields the general solution

$$(u-1)^2|u+1/2| = C|x|^{-3/2}.$$

Substituting back $u = \sqrt{(1+v)/2} = \sqrt{(x+y)/2x}$, we find the general solution of the given differential equation to be

$$\left[\sqrt{\frac{y+x}{2x}} - 1\right]^2 \left|\sqrt{\frac{y+x}{2x}} + \frac{1}{2}\right| = C|x|^{-3/2}. \tag{2.2.11}$$

As usual, we assume that the square root is nonnegative. Setting $x = 1$ and $y = 1$ in Eq. (2.2.11), we get $C = 0$ from the initial condition and

$$\left[\sqrt{\frac{y+x}{2x}} - 1\right]^2 \left|\sqrt{\frac{y+x}{2x}} + \frac{1}{2}\right| = 0.$$

It is known that the product of two terms is zero if and only if one of the multipliers is zero. Hence,

$$\sqrt{\frac{y+x}{2x}} = 1 \qquad \text{or} \qquad \sqrt{\frac{y+x}{2x}} = -\frac{1}{2}.$$

Since the square root cannot be equal to a negative number $(-1/2)$, we disregard the latter equation and get the required solution $y + x = 2x$ or $y = x$. To check our answer, we type in *Maple:*

```
dsolve({diff(y(x),x)=sqrt((x+y)/(2*x)),y(1)=1},y(x))}
```

We may use the general solution to determine an arbitrary constant C in order to satisfy the initial condition, $y(1) = 3$. Namely, we set $x = 1$ and $y = 3$ in Eq. (2.2.11) to obtain $C = 2\sqrt{2} - 1/2$. Thus, the solution (in the implicit form) of the given initial value problem is

$$\left[\sqrt{\frac{y+x}{2}} - \sqrt{x}\right]^2 \left|\sqrt{\frac{y+x}{2}} + \frac{1}{2}\sqrt{x}\right| = \left(2\sqrt{2} - \frac{1}{2}\right).$$

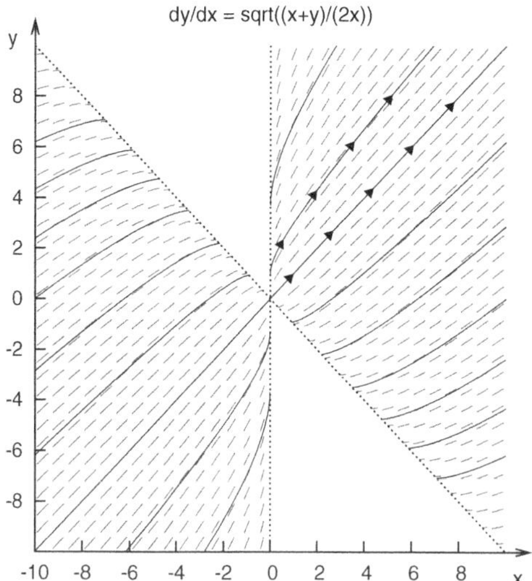

Figure 2.16: Direction field and solutions in Example 2.2.7, plotted with *Maxima*.

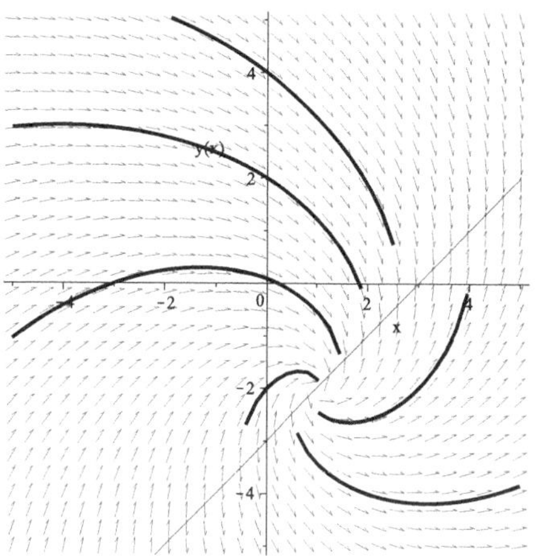

Figure 2.17: Example 2.2.8: direction field and solutions to $y' = (x + y + 1)/(x - y - 3)$, plotted with *Maple*.

2.2.3 Equations with Linear Coefficients

We are now able to handle differential equations of the form

$$\frac{dy}{dx} = F\left(\frac{ax + by + c}{Ax + By + C}\right), \tag{2.2.12}$$

where a, b, c, A, B, and C are constants. When $c = C = 0$, the right-hand side is a homogeneous function. Equation (2.2.12) coincides with Eq. (2.2.1) if $A = B = 0$ and $C = 1$.

Each of the equations $ax + by + c = 0$ and $Ax + By + C = 0$ defines a straight line in the xy-plane. They may be either parallel or intersecting. If we wish to find the point of intersection, we have to solve simultaneously the system of algebraic equations

$$\begin{cases} ax + by + c = 0, \\ Ax + By + C = 0. \end{cases} \tag{2.2.13}$$

This system has a unique solution if and only if the determinant of the corresponding matrix,

$$\begin{bmatrix} a & b \\ A & B \end{bmatrix},$$

does not vanish (consult §7.2.1); namely, $aB - Ab \neq 0$. Otherwise ($aB = Ab$), these two lines are parallel. We consider below these two cases.

Case 1. Suppose that $aB - bA \neq 0$. Then two lines $ax + by + c = 0$ and $Ax + By + C = 0$ have a unique intersection. In this case, the constants c and C can be eliminated from the coefficients by changing the variables. The right-hand side of Eq. (2.2.12) can be made into a homogeneous function by shifting variables:

$$x = X + \alpha \quad \text{and} \quad y = Y + \beta,$$

with constants α and β to be chosen appropriately. Since a derivative is the slope of a tangent line, its value is not changed by shifting the system of coordinates. Then

$$\frac{dy}{dx} = \frac{dY}{dX},$$

and

$$ax + by + c = aX + bY + (a\alpha + b\beta + c), \quad Ax + By + C = AX + BY + (A\alpha + B\beta + C).$$

Now choose α and β so that

$$a\alpha + b\beta + c = 0, \quad A\alpha + B\beta + C = 0. \tag{2.2.14}$$

Since the determinant of this system of algebraic equations is equal to $aB - bA \neq 0$, there exist an α and a β that satisfy these equations. With these choices, we get the equation

$$\frac{\mathrm{d}Y}{\mathrm{d}X} = F\left(\frac{aX + bY}{AX + BY}\right)$$

with a homogeneous slope function in X and Y. The resulting equation can be solved by the method previously discussed in §2.2.2 (using substitution $Y = vX$). Its solution in terms of X and Y gives a solution of the original equation (2.2.12) upon resetting $X = x - \alpha$ and $Y = y - \beta$.

Case 2. Suppose that $aB - bA = 0$. The two lines $ax + by + c = 0$ and $Ax + By + C = 0$ in Eq. (2.2.13) are parallel. In this case we can transfer the differential equation (2.2.12) into a separable one by changing the dependent variable y with a new one that is proportional to the linear combination $ax + by$. That is, the numerator $ax + by$ is proportional to the denominator $Ax + By$, and the fraction $(ax + by)/(Ax + By)$ will depend only on this linear combination. Since $b = aB/A$, we have

$$ax + by = ax + \frac{aB}{A}y = \frac{a}{A}(Ax + By).$$

Therefore, we set

$$av = ax + by \quad \text{(or we can let } Av = Ax + By\text{)}.$$

Then

$$a\frac{\mathrm{d}v}{\mathrm{d}x} = a + b\frac{\mathrm{d}y}{\mathrm{d}x} = a + bF\left(\frac{ax + by + c}{Ax + By + C}\right)$$

and, after dividing both sides by a, we obtain

$$\frac{\mathrm{d}v}{\mathrm{d}x} = 1 + \frac{b}{a}F\left(\frac{av + c}{Av + C}\right).$$

This is a separable equation that can be solved according to the method described in §2.1.

Example 2.2.8: (Case 1) Solve the differential equation (see its direction field in Fig. 2.17 on page 63)

$$(x - y - 3)\,\mathrm{d}y = (x + y + 1)\,\mathrm{d}x, \qquad x - y - 3 \neq 0.$$

Solution. We rewrite the slope function in the form

$$\frac{x + y + 1}{x - y - 3} = \frac{(x - 1) + (y + 2)}{(x - 1) - (y + 2)},$$

which suggests the substitution $X = x - 1$ and $Y = y + 2$. This leads to the following differential equation:

$$\frac{\mathrm{d}Y}{\mathrm{d}X} = \frac{X + Y}{X - Y}, \qquad X \neq Y,$$

with the homogeneous right-hand side function $(X + Y)/(X - Y)$. To solve the equation, we make substitution $x - 1 = X = r\cos\theta$, $y + 2 = Y = r\sin\theta$, with r, θ being new variables. Then

$$\mathrm{d}X = \cos\theta\,\mathrm{d}r - r\sin\theta\,\mathrm{d}\theta, \qquad \mathrm{d}Y = \sin\theta\,\mathrm{d}r + r\cos\theta\,\mathrm{d}\theta.$$

Substituting these expressions into the given differential equation, we get, after some cancellations, that

$$\mathrm{d}r = r\,\mathrm{d}\theta.$$

Then its general solution is $\ln r = \theta + C$, where C is a constant of integration. So, in terms of $X = x - 1$, $Y = y + 2$, we have

$$\ln(X^2 + Y^2) - 2\arctan(Y/X) = C \quad \text{or} \quad \ln\left[(x - 1)^2 + (y + 2)^2\right] - 2\arctan\frac{y + 2}{x - 1} = C.$$

The given differential equation has no equilibrium solution.

Example 2.2.9: (Case 2) Solve the equation

$$\frac{dy}{dx} = \frac{x+2y-5}{2x+4y+7}, \qquad 2x+4y+7 \neq 0. \tag{2.2.15}$$

Solution. This differential equation is of the form (2.2.12), where $a=1$, $b=2$, $c=-5$, $A=2$, $B=4$, $C=7$, and $F(z)=z$. In this case, $aB-bA=0$, and we set $v=x+2y$. Then the slope function becomes

$$\frac{x+2y-5}{2x+4y+7} = \frac{v-5}{2v+7}, \qquad v \neq -7/2 = -3.5.$$

This leads to

$$\frac{dv}{dx} = 1+2\frac{dy}{dx} = 1+2\frac{v-5}{2v+7} = \frac{4v-3}{2v+7} \qquad \text{or} \qquad (2v+7)\,dv = (4v-3)\,dx.$$

Synthetic division yields the separable equation

$$\left(\frac{2v+7}{4v-3}\right) dv \equiv \left(\frac{1}{2}+\frac{1}{2}\times\frac{17}{4v-3}\right) dv = dx.$$

After integrating, we find the general solution

$$\frac{v}{2}+\frac{17}{8}\ln|4v-3| = x+C, \qquad \text{where } v=x+2y,$$

with the potential function

$$C = 8y+3x+17\,\ln|4x+8y-3|.$$

The differential equation $(2v+7)\,dv=(4v-3)\,dx$ has the critical point $v^*=3/4$ and the singular point $v=-7/2$ to which correspond lines $y=(3-4x)/8$ and $y=-(7+2x)/4$, respectively. The unstable equilibrium solution $y=(3-4x)/8$ is nonsingular.

If we try to find the solution subject to the initial condition $y(0)=0$ using a software package, we will obtain the formula expressed via a special function. For instance, MATLAB code

```
solution=dsolve('(2*x+4*y+7)*Dy=x+2*y-5')
```

gives the output expressed through the omega function:

```
exp((2*C3)/3 + (2*t)/3 + (2*x)/3 - 2/3)/(2*exp(wrightOmega((2*C3)/3
+ (2*t)/3 + (2*x)/3 + log(2/3) - 2/3))) - x/2 + 1/2
subs(solution, 'C3', 0);
```

Similar output gives *Maple:*

```
deq1:=diff(y(x),x)=(x+2*y(x)-5)/(2*x+4*y(x)+7):
Y:=rhs(dsolve({deq1,y(0)=0},y(x)));
```

which presents the solution via the Lambert function (another name for the omega function):

$$-\frac{1}{2}x+\frac{3}{8}+\frac{17}{8}\,\text{LambertW}\left(\frac{3}{17}\,e^{\frac{1}{17}\,I(-81x+3I+17\pi)}\right)$$

Similarly, *Mathematica*'s output for

```
DSolve[{y'[x] == (x + 2*y[x] - 5)/(2*x + 4*y[x] + 7), y[0] == 0},
 y[x], x]
```

is $\quad y[x] \to \frac{1}{8}\left(3-4x+17\,\text{ProductLog}\left[-\frac{3}{17}\,e^{-\frac{3}{17}+\frac{8x}{17}}\right]\right)$

On the other hand, we can plot the solution of the initial value problem subject to $y(0)=0$ along with the direction field using *Maple*'s commands

```
implicitplot(17*log(3)=8*y-4*x+17*log(abs(4*x+8*y-3)), x=0..2,y=-6.4..0);
DEplot(deq1,y(x),x=0..1,y=-1..0,dirgrid=[16,16],color=blue)
```

□

An equation of the form

$$\frac{dy}{dx} = \frac{y}{x} + f\left(\frac{y}{x}\right) g(x), \qquad x \neq 0, \tag{2.2.16}$$

can also be reduced to a separable one. If we set $y=vx$, then

$$x\frac{dv}{dx}+v = v+f(v)\,g(x), \quad \text{or} \quad x\frac{dv}{dx} = f(v)g(x).$$

The latter is a separable equation, and we obtain its solution in implicit form by integration

$$\int \frac{dv}{f(v)} = \int g(x)\,\frac{dx}{x}.$$

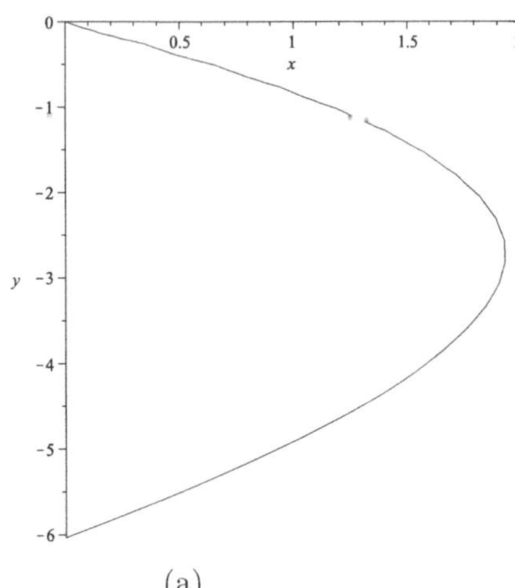

(a)

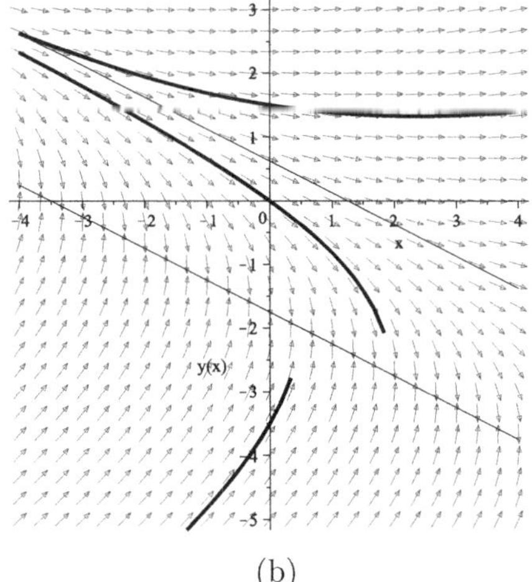

(b)

Figure 2.18: Example 2.2.9: (a) the implicit solution of Eq. (2.2.15) subject to $y(0) = 0$; (b) the direction field along with three solutions, the equilibrium solution $8y = 5 - 4x$, and the singular line $2x + 4y + 7 = 0$, plotted with *Maple*. The nullcline $x + 2y = 5$ is not shown.

Example 2.2.10: Solve the differential equation

$$\frac{\mathrm{d}y}{\mathrm{d}x} = 2\left(\frac{y+1}{x+y-2}\right)^2.$$

Solution. This is not an equation with a homogeneous function, but it is of the form under consideration (2.2.12), with $a = 0,\ b = 1,\ c = 1,\ A = 1,\ B = 1,\ C = -2$, and $F(\omega) = \omega^2$. Since $aB \neq bA$, we set $x = X + \alpha$ and $y = Y + \beta$ to get

$$\frac{\mathrm{d}y}{\mathrm{d}x} = 2\left(\frac{Y+\beta+1}{X+\alpha+Y+\beta-2}\right)^2.$$

We choose α and β so that

$$\beta + 1 = 0 \quad \text{and} \quad \alpha + \beta - 2 = 0.$$

Then $\beta = -1$ and $\alpha = 3$. With this in hand, we reduce the given differential equation to

$$\frac{\mathrm{d}Y}{\mathrm{d}X} = 2\left(\frac{Y}{X+Y}\right)^2 = 2\left(\frac{Y/X}{1+Y/X}\right)^2.$$

Setting $Y = y + 1 = vX = v(x-3)$ gives us

$$X\,\frac{\mathrm{d}v}{\mathrm{d}X} + v = 2\left(\frac{v}{1+v}\right)^2 \qquad \text{or} \qquad X\,\frac{\mathrm{d}v}{\mathrm{d}X} = \frac{2\,v^2 - v(1+v)^2}{(1+v)^2}.$$

Separation of variables yields

$$\frac{(1+v)^2\,\mathrm{d}v}{v+v^3} + \frac{\mathrm{d}X}{X} = 0 \qquad (X = x - 3 \neq 0,\ v \neq 0).$$

Using partial fraction decomposition,

$$\frac{(1+v)^2}{v+v^3} = \frac{1}{v} + \frac{2}{1+v^2} \qquad (v \neq 0),$$

we integrate to obtain $\ln|v| + 2\arctan v + \ln|X| = \ln C$, or $C = vX\,e^{2\arctan v}$. Substitution of $X = x - 3$, $Y = y + 1$, and $v = (y+1)/(x-3)$ yields the general solution in an implicit form:

$$C = (y+1)\,\exp\left\{2\,\arctan\frac{y+1}{x-3}\right\}.$$

The given differential equation has the (nonsingular) equilibrium solution, $y \equiv -1$. □

More advanced material is presented next and may be omitted on the first reading.

Example 2.2.11: Solve the initial value problem for $x > 0$:

$$\frac{dy}{dx} = \frac{y}{x} + \frac{4x^3 \cos(x^2)}{y}, \qquad y(\sqrt{\pi}) = 1.$$

Solution. We set $y = vx$. Then

$$x\frac{dv}{dx} + v = v + \frac{4x^2 \cos(x^2)}{v}.$$

This equation can be simplified by multiplying both sides by v/x, yielding

$$vv' = 4x \cos(x^2).$$

After integration, we obtain its solution in implicit form to be $v^2/2 = 2\sin(x^2) + C$. Since $v = y/x$, this gives

$$y = vx = \pm 2x \sqrt{\sin(x^2) + C/2}.$$

From the initial condition, $y(\sqrt{\pi}) = 1$, it follows that we should choose sign "+," which results in $C/2 = 1/\pi$. Hence, the answer is

$$y = x\sqrt{4 \sin(x^2) + 1/\pi}.$$

□

Definition 2.7: A function $g(x, y)$ is said to be **quasi-homogeneous** (or **isobaric**) with weights α and β if

$$g(\lambda^\alpha x, \lambda^\beta y) = \lambda^\gamma g(x, y) \tag{2.2.17}$$

for some real numbers α, β, and γ, and every nonzero constant λ.

A differential equation with a quasi-homogeneous slope function can be reduced to a separable one by setting $y = z^{\beta/\alpha}$ or $y = u\, x^{\beta/\alpha}$. We demonstrate two examples of differential equations with quasi-homogeneous right-hand side functions.

Example 2.2.12: Find the general solution of the differential equation

$$\frac{dy}{dx} = f(x, y), \quad \text{where} \quad f(x, y) = \frac{6x^6 - 3y^4}{2x^4 y}, \qquad x \neq 0, \quad y \neq 0.$$

Solution. Calculations show that

$$f(\lambda^\alpha x, \lambda^\beta y) = \frac{6\lambda^{6\alpha} x^6 - 3\lambda^{4\beta} y^4}{2\lambda^{4\alpha+\beta} x^4 y} = 3\,\lambda^{6\alpha-4\alpha-\beta}\,\frac{x^2}{y} - \lambda^{4\beta-4\alpha-\beta}\,\frac{3}{2}\,\frac{y^3}{x^4}.$$

The function $f(x, y)$ is a quasi-homogeneous one if and only if the numbers

$$6\alpha - 4\alpha - \beta = 2\alpha - \beta \quad \text{and} \quad 4\beta - 4\alpha - \beta = -4\alpha + 3\beta$$

are equal. This is valid only when $3\alpha = 2\beta$.

Therefore, for these values of α and β, the given differential equation can be reduced to a separable one by setting $y = v\,x^{3/2}$. From the product rule, it follows that $y' = v'\,x^{3/2} + v\,\frac{3}{2}\,x^{1/2}$. Substitution $y = v\,x^{3/2}$ reduces the given differential equation into a separable one:

$$y' = x^{3/2}\,v' + \frac{3}{2}\,x^{1/2}\,v = \frac{3x^2}{x^{3/2}\,v} - \frac{3v^3 x^{9/2}}{2x^4}$$

or

$$x\,v' = \frac{3}{v} - \frac{3(v + v^3)}{2} = \frac{3}{2v}\left[2 - v^2(1 + v^2)\right].$$

Integration gives

$$\int \frac{2v\,dv}{2 - v^2(1 + v^2)} = 3 \int \frac{dx}{x}.$$

Since

$$\frac{2v}{2 - v^2(1 + v^2)} = \frac{2v}{3(v^2 + 2)} - \frac{2v}{3(v^2 - 1)}, \qquad v \neq \pm 1,$$

we have

$$\ln \frac{v^2+2}{v^2-1} = 9\,\ln|x| + \ln C = \ln C\,x^9 \qquad \text{or}$$

$\frac{v^2+2}{v^2-1} = C\,x^9$. Back substitution $v = y\,x^{-3/2}$ leads to the general solution

$$\frac{y^2+x^3}{y^2-x^3} = C\,x^9.$$

The given differential equation has two equilibrium solutions $y = \pm x^{3/2}$ (see Figure at right).

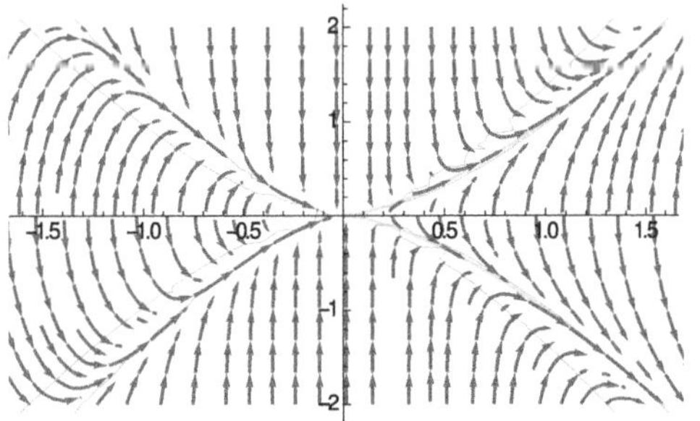

Example 2.2.13: Under which values of p and q is the slope function of the differential equation

$$y' = \mathrm{d}y/\mathrm{d}x = a\,x^p + b\,y^q$$

quasi-homogeneous? For appropriate values of p, solve the equation $y' = -6\,x^p + y^2$.

Solution. The given equation is quasi-homogeneous if

$$a\,\lambda^{\alpha p}\,x^p + b\lambda^{\beta q}\,y^q = \lambda^{\beta-\alpha}\,(a\,x^p + b\,y^q)$$

for some constants α and β. This is true if and only if

$$p\alpha = q\beta = \beta - \alpha, \qquad \text{that is,} \qquad p = q(p+1) \qquad \text{or} \qquad \frac{1}{q} - \frac{1}{p} = 1.$$

Hence, the function $-6\,x^p + y^2$ is quasi-homogeneous if and only if $p = -2$. For this value of p, we substitute $y = u \cdot x^{p/q} = u \cdot x^{-1}$ into the given differential equation to obtain

$$x^{-1}\,u' - u\,x^{-2} = -6\,x^{-2} + x^{-2}\,u^2 \qquad \text{or} \qquad x\,u' = u^2 + u - 6.$$

This is a separable differential equation. The roots $u = 2$ and $u = -3$ of the quadratic equation $u^2 + u - 6 = 0$ are critical points of this differential equation, and we must preclude $u = 2$ ($y = 2x^{-1}$) and $u = -3$ ($y = -3x^{-1}$) in the following steps. Other solutions can be found by separation of variables. Integration yields

$$\int \frac{\mathrm{d}u}{u^2+u-6} = \int \frac{\mathrm{d}x}{x} = \ln Cx, \qquad u \ne -3, \quad u \ne 2.$$

Since $u^2 + u - 6 = (u+3)(u-2)$, we have

$$\begin{aligned}\int \frac{\mathrm{d}u}{u^2+u-6} &= \int \frac{\mathrm{d}u}{(u+3)(u-2)} = \frac{1}{5}\int \left[\frac{1}{u-2} - \frac{1}{u+3}\right] \mathrm{d}u \\ &= \frac{1}{5}\,[\ln|u-2| - \ln|u+3|] = \frac{1}{5}\,\ln\frac{|u-2|}{|u+3|}.\end{aligned}$$

Substituting $u = xy$, we get the general solution

$$\frac{1}{5}\,\ln\left|\frac{xy-2}{xy+3}\right| = \ln Cx \qquad \text{or} \qquad \ln\left|\frac{xy-2}{xy+3}\right| = \ln Cx^5.$$

Raising to the exponent, we obtain all solutions of the given differential equation:

$$\text{general:} \quad \frac{xy-2}{xy+3} = C\,x^5, \quad \text{and equilibrium:} \quad y = \frac{2}{x}, \; y = -\frac{3}{x}.$$

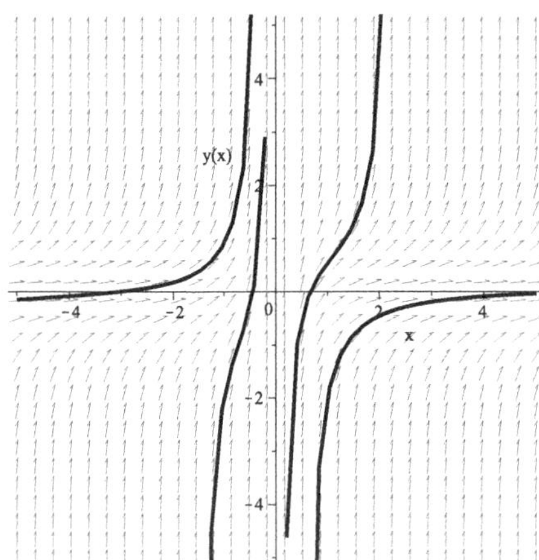

Figure 2.19: Direction field and solutions for Example 2.2.13, plotted with *Maple*.

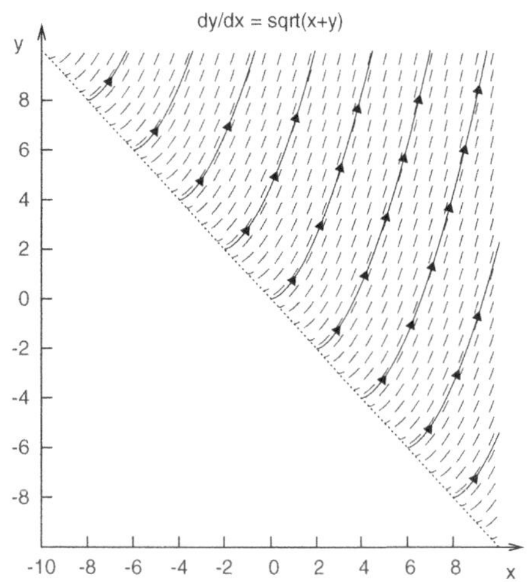

Figure 2.20: Direction field and solutions in Example 2.2.14, plotted with *Maxima*.

Example 2.2.14: Solve the differential equation $y' = \sqrt{x+y}$.

Solution. The given equation is of the form (2.2.12), page 63, with $a = 1$, $b = 1$, $c = 0$, $A = 0$, $B = 0$, $C = 1$, and $F(v) = \sqrt{v}$. Since $aB = bA = 0$, we might try a change of variable y to v given by $v = x + y$. Then $v' = 1 + y' = 1 + \sqrt{v}$, which can be solved by separation of variables. We rewrite the differential equation as

$$\frac{\mathrm{d}v}{1+\sqrt{v}} = \mathrm{d}x.$$

Integration leads to

$$\int \frac{\mathrm{d}v}{1+\sqrt{v}} = x + C.$$

In the left-hand side integral, we change the variable of integration by setting $p^2 = v$; therefore,

$$\begin{aligned}\int \frac{\mathrm{d}v}{1+\sqrt{v}} &= \int \frac{2p\,\mathrm{d}p}{1+p} = 2\int \frac{(p+1-1)\,\mathrm{d}p}{1+p} \qquad (\mathrm{d}v = 2p\,\mathrm{d}p)\\ &= 2\int \mathrm{d}p - 2\int \frac{\mathrm{d}p}{1+p} = 2p - 2\,\ln(p+1)\\ &= 2\sqrt{v} - 2\,\ln(1+\sqrt{v}) = 2\,\sqrt{x+y} - 2\,\ln(1+\sqrt{x+y}).\end{aligned}$$

Thus,

$$2\,\sqrt{x+y} - 2\,\ln(1+\sqrt{x+y}) = x + C$$

is the general solution in implicit form, which contains an arbitrary constant C.

Example 2.2.15: (Swan–Solow economic model) It may be said that the neoclassical approach has played a central role in modern economic growth theory. Most of these models are extensions and generalizations of the pioneering works published in 1956 by T. W. Swan and R. Solow. In this growth[23] model, the production is described by some smooth function

$$Y = F(K, L),$$

where Y is the output flow attainable with a given amount of capital, K, and labor, L. In the neoclassical approach, a production function F is considered to be a homogeneous function of degree one: $F(rK, rL) = rF(K, L)$, for all nonnegative r. In order to have a growing economy, it is assumed that a constant fraction s of the total output flow is saved and set aside to be added to the capital stock. Neglecting depreciation of the capital, we have

$$\dot{K} = sY = s\,F(K, L), \qquad K(0) > 0.$$

[23]The model was developed independently by an American economist Robert Merton Solow (born in 1924) and an Australian economist Trevor Winchester Swan (1918–1989) in 1956.

Introducing the ratio $k = K/L$, the above equation can be reduced to the differential equation

$$\frac{\mathrm{d}k}{\mathrm{d}t} = s f(k) - nk, \tag{2.2.18}$$

where $f(k) = F(K, L)/L = F(k, 1)$ and n is the coefficient of the labor growth. Recall that for the Malthusian model, n is a constant; however, for another model it may be a function (like the logistic equation). In the Solow model (2.2.18), once the capital per labor function is determined, all other variables, such as consumption, savings, wages, K, and L can be calculated accordingly. For example, the wage rate w and the profit rate r are given, respectively, by $w = f(k) - kf'(k)$ and $r = f'(k)$.

Example 2.2.16: Consider the differential equation

$$y' - \sqrt{1 + y/x} - 1 = 0.$$

We demonstrate *Sage* code that could be used to find the general solution:

```
y = function('y')(x)
de = diff(y,x) -sqrt(1+y/x)-1
h = desolve(de, y); h
c*x == e^(-2/3*log(sqrt((x + y(x))/x) + 1) - 4/3*log(sqrt((x + y(x))/x) - 2))
```

Problems

1. In each exercise, transform the given differential equation to a separable one and solve it.
   (a) $y' = e^{x+y-2} - 1$; (b) $y' = (2x + y - 1)^2 - 1$; (c) $y' = \sin(x + y)$;
   (d) $y' = e^{x+y}/(x + y) - 1$; (e) $yy' + xy^2 = x$; (f) $y' = \sqrt{4x + y}$.
   Hint: $\int \frac{dv}{1+\sin v} = \tan\left(\frac{v}{2} - \frac{\pi}{4}\right) = -\frac{2}{1+\tan(v/2)}$, $\int \frac{du}{4+\sqrt{u}} = 2\sqrt{u} - 8\ln(4 + \sqrt{u})$.
2. In each exercise, solve the given differential equation of the form $x\,y' = y\,F(xy)$ by using transformation $v = xy$.
   (a) $y' = y^2$; (b) $x\,y' = e^{xy} - y$; (c) $x\,y' = y/(xy + 1)$;
   (d) $x^2\,y' = \cos^2(xy) - xy$; (e) $xy' = y\,e^{xy}$; (f) $xy' = x^2y^3 + 2xy^2$.
3. In each exercise, determine whether or not the function is homogeneous. If it is homogeneous, state the degree of the function.

   (a) $5x^2 - 2xy + 3y^2$;
   (b) $y + \sqrt{x^2 + 2y^2}$;
   (c) $\sqrt{x^2 + xy - y^2}$;
   (d) $x\cos(y/x) - y\sin(x/y)$;
   (e) $(x^2 + y^2)\exp\left(\frac{2x}{y}\right)$;
   (f) $\sqrt{x - y}$;
   (g) $x^3 - xy + y^3$;
   (h) $\ln|x| - \ln|y|$;
   (i) $\dfrac{x^2 + 2xy}{x + y}$;
   (j) $\tan\dfrac{x}{2y}$;
   (k) e^x;
   (l) $(x^2 + y^2)^{3/2}$;
   (m) $\sqrt{x^3 + y^3}$.

4. In each exercise, determine whether or not the equation has a homogeneous right-hand side function. If it does, find the general solution of the equation.
   (a) $y' = \frac{y}{x} + 2\sqrt{\frac{x}{y}}$; (b) $(y^3 - yx^2)\,dx = 2xy^2\,dy$;
   (c) $y' = \frac{2x+y}{x}$; (d) $(x^2 + y^2)\,dx + xy\,dy = 0$;
   (e) $2xy\,y' = x^2 + y^2$; (f) $(y + \sqrt{x^2 + y^2})\,dx - x\,dy = 0$;
   (g) $x^2y' = 2y^2 + xy$; (h) $\left(\frac{y}{x} + 1\right)dx + \left(\frac{x}{y} + 1\right)dy = 0$;
   (i) $x^2y' = y^2 + 2xy - 6x^2$; (j) $(x^4 + y^4)\,dx - 2xy^3\,dy = 0$;
   (k) $y' = \frac{\sqrt{x+y}+\sqrt{x-y}}{\sqrt{x+y}-\sqrt{x-y}}$; (l) $y(y/x - 1)\,dx + x(x/y + 1)\,dy = 0$;
   (m) $x^2y' = y^2 + 3xy + x^2$; (n) $(2y^2 + xy)\,dx + (2xy + y^2)\,dy = 0$;
   (o) $x^3\,y' + 3xy^2 + 2y^3 = 0$; (p) $(2x\sin(y/x) - y\cos(y/x))\,dx + x\cos(y/x)\,dy = 0$.
5. Solve the given differential equation with a homogeneous right-hand side function. Then determine an arbitrary constant that satisfies the auxiliary condition.

   (a) $xy\,dx + 2(x^2 + 2y^2)\,dy = 0$, $y(1) = 1$;
   (b) $(y + \sqrt{x^2 + y^2})\,dx - 2x\,dy = 0$, $y(1) = 0$;
   (c) $(x - y)\,dx + (3x + y)\,dy = 0$, $y(3) = -2$;

(d) $(1 + 2e^{x/y})\,dx + 2e^{x/y}(1 - x/y)\,dy = 0, \quad y((1) = 1;$

(e) $(y^2 + xy)\,dx - 3x^2\,dy = 0, \quad y(1) = 1;$

(f) $(y^2 + 7xy + 16x^2)\,dx + x^2\,dy = 0, \quad y(1) = 1;$

(g) $(3x^2 - 2y^2)\,y' = 2xy, \quad y(1) = 1;$

(h) $y^2\,dx + (x^2 + 3xy + 4y^2)\,dy = 0, \quad y(2) = 1;$

(i) $y(3x + 2y)\,dx - x^2\,dy = 0, \quad y(1) = 1;$

(j) $x(y + x)^2\,dy = (y^3 + 2xy^2 + x^2y + x^3)\,dx, \quad y(1) = 2;$

(k) $(2x - 3y)\,dy = (4y - x)\,dx, \quad y(-1) = 2;$ **(l)** $(x^3 + y^3)\,dx - 2x^2y\,dy = 0, \quad y(1) = 1/2.$

6. Solve the following equations with linear coefficients.

(a) $(3x + y - 1)\,dx = (y - x - 1)\,dy;$ **(b)** $(x + 2y + 2)\,dx + (2x + 3y + 2)\,dy = 0;$
(c) $(4x + 2y - 8)\,dx + (2x - y)\,dy = 0;$ **(d)** $(x - 4y - 9)\,dx + (4x + y - 2)\,dy = 0;$
(e) $(5y - 10)\,dx + (2x + 4)\,dy = 0;$ **(f)** $(2x + y - 8)\,dx = (-2x + 9y - 12)\,dy;$
(g) $(x - 1)\,dx - (3x - 2y - 5)\,dy = 0;$ **(h)** $(2x - y - 5)\,dx = (x - 2y - 1)\,dy;$
(i) $(x + y - 2)\,dx + (x + 1)\,dy = 0;$ **(j)** $(2x + y - 2)\,dx - (x - y + 4)\,dy = 0;$
(k) $(3x + 2y - 5)\,dx = (2x - 3y + 1)\,dy;$ **(l)** $(2x - y - 3)\,dx + (3x + y + 3)\,dy = 0;$
(m) $(4y - 5x - 1)\,dx = (2x + y + 3)\,dy;$ **(n)** $(3x + y + 1)\,dx + (x + 3y + 11)\,dy = 0.$

7. Find the indicated particular solution for each of the following equations.

(a) $(x + 5y + 5)\,dx = (7y - x + 7)\,dy, \quad y(1) = 1;$

(b) $(x - y - 3)\,dx + (3x + y - 1)\,dy = 0, \quad y(2) = 2;$

(c) $(y - x + 1)\,dx = (x + y + 3)\,dy, \quad y(0) = \sqrt{3} - 2;$

(d) $(y + 2)\,dx - (x + y + 2)\,dy = 0, \quad y(1) = -1;$

(e) $(4x + 2y - 2)\,dx = (2x - y - 3)\,dy, \quad y(3/2) = -1;$

(f) $y\,e^{x/y}\,dx + (y - x\,e^{x/y})\,dy = 0, \quad y(1) = 1;$

(g) $(4x + 3y + 2)\,dx + (2x + y)\,dy = 0, \quad y(0) = 0;$

(h) $(x + y\,e^{y/x} + y\,e^{-y/x})\,dx = x(e^{-y/x} + e^{y/x})\,dy, \quad y(1) = 0;$

(i) $(4x + y)^2\,dx = xy\,dy, \quad y(1) = -1;$ **(j)** $(x + 1)\,dx = (x + 2y + 3)\,dy, \quad y(0) = 1.$

8. In order to get some practice in using these techniques when the variables are designated as something other than x and y, solve the following equations.

(a) $\dfrac{dp}{dq} = \dfrac{q + 4p + 5}{3q + 2p + 5};$ **(b)** $\dfrac{dr}{d\theta} = \dfrac{2r - \theta + 2}{r + \theta + 7};$ **(c)** $\dfrac{dw}{dt} = \dfrac{t - w + 5}{t - w + 4};$

(d) $\dfrac{dw}{dv} = \dfrac{w + v - 4}{3v + 3w - 8};$ **(e)** $\dfrac{dk}{ds} = \dfrac{3k + 2s - 2}{k + 4s + 1};$ **(f)** $\dfrac{dz}{du} = \dfrac{5u - 3z + 8}{z + u};$

(g) $\dfrac{dt}{dx} = \dfrac{3x + t + 4}{5t - x + 4};$ **(h)** $\dfrac{dx}{dt} = \dfrac{x + 3t + 3}{t - 2};$ **(i)** $\dfrac{dz}{dt} = \dfrac{4t + 11z - 42}{9z - 11t + 37}.$

9. Solve the following equations.

(a) $\dfrac{dy}{dx} = \left(\dfrac{x - y + 1}{2x - 2y}\right)^2;$ **(b)** $\dfrac{dy}{dx} = \left(\dfrac{x - y + 2}{x + 1}\right)^2;$ **(c)** $y' = \dfrac{1}{3x + y};$

(d) $y' = (4x + y)^2;$ **(e)** $y' = \dfrac{6x + 3y - 5}{2x + y};$ **(f)** $y' = \dfrac{2x - 4y + 1}{x - 2y + 1}.$

10. For any smooth functions $f(z)$ and $g(z)$, show that $yf(xy) + xg(xy)\,y'$ is isobaric, i.e., by setting λx, $\lambda^{\alpha} y$, and $\lambda^{\alpha-1} y'$ instead of x, y, and y', respectively, the equation remains the same up to the multiple of λ^{-1}.

11. Solve the following equations with quasi-homogeneous right-hand side functions.

(a) $y' = 2y/x + x^3/y;$ **(b)** $y' = y^2/x^3;$ **(c)** $y' = 3y/x^2 + x^4/y;$ **(d)** $y' = y^2/x^4.$

12. Find the general solution of the differential equation $(x - y)\,y' = x + y$. Use a software package to draw the solution curves.

13. Use the indicated change of variables to solve the following problems.

(a) $x^2y\,dx = (x^2y - 2y^5)\,dy; \quad x = t\sqrt{u}, \quad y = \sqrt{u}.$

(b) $(1 + y^2)\,dx + [(1 + y^2)(e^y + x^2\,e^{-y}) - x]\,dy = 0; \quad x = t\,e^y.$

(c) $(2yx^2 + 2x^4y^2 - e^{x^2y})\,\mathrm{d}x + x^3(1 + x^2y)\,\mathrm{d}y = 0; \qquad u = x^2y.$

(d) $(1 + 2x + 2y)\,\mathrm{d}x + (2x + 2y + 1 + 2\,e^{2y})\,\mathrm{d}y = 0; \qquad u = x + y.$

(e) $\left(2y + x^2y^2 + \frac{2x^2y}{1+x^2}\right)\mathrm{d}x + (2x + x^2y)\,\mathrm{d}y = 0; \qquad u = xy.$

(f) $(y - 2xy\ln x)\,\mathrm{d}x + x\ln x\,\mathrm{d}y = 0; \qquad u = y\ln x.$

(g) $\left(2y - \sqrt{1 - x^2y^2}\right)\mathrm{d}x + 2x\,\mathrm{d}y = 0; \qquad y = u/x.$

(h) $\left(3y + 3x^3y^2 - \frac{1}{x^3}\,e^{-x^3y}\right)\mathrm{d}x + (x + x^4y)\,\mathrm{d}y = 0; \qquad u = yx^3.$

(i) $(2x + y)\,y' = 6x + 3y + 5; \qquad u = 2x + y.$

(j) $(x - y^2)\,\mathrm{d}x + y(1 - x)\,\mathrm{d}y = 0; \qquad u = y^2.$

14. In each exercise, find a transformation of variables that reduces the given differential equation to a separable one. *Hint:* The equation $y' = f(x, y)$ with an isobaric function $f(x, y)$ can be reduced to a separable equation by changing variables $u = y/x^m$, where m is determined upon substitution $f(\lambda x, \lambda^m y) = \lambda^r f(x, y)$.

(a) $2xyy' = y^2 + \sqrt{y^4 - 9x^2}$;
(b) $3x^2y' = x^2y^2 - 3xy - 4$;
(c) $(x + x^2y)y' + 2y + 3xy^2 = 0$;
(d) $x^3y' + x^2y + 3 = 0$;
(e) $3x^3y' - 2x^2y + y^2 = 0$;
(f) $x^3\,y' = 2 + \sqrt{4 + 6x^2y}$.

15. Find the general solution by making an appropriate substitution.

(a) $y' = 4y^2 + 4(x + 3)\,y + (x + 3)^2$;
(b) $y' = \frac{x^2+y^2}{2y}$;
(c) $3xy^2\,y' + y^3 + x^3 = 0$;
(d) $y' = \sec y + x\tan y$;
(e) $(x^2 + y^2 + 3)\,y' = 2x(2y - x^2/y)$;
(f) $y' = \frac{2x^3+3xy^2-7x}{3x^2y+2y^3-8y}$.

16. In each of the following initial value problems, find the specific solution and show that the given function $y_s(x)$ is a singular solution.

(a) $y' = (3x + y - 5)^{1/4} - 6x, \qquad y(1) = 2; \qquad y_s(x) = 5 - 3x^2.$

(b) $y' = \sqrt{\frac{y+2}{x-3}}, \qquad y(4) = 2; \qquad y_s(x) = 2.$

(c) $3y' = 1 + \sqrt{1 - 3y/x}, \qquad y(3) = 1; \qquad y_s(x) = \frac{x}{3}.$

(d) $y' = \sqrt{x(x^3 + 2y)} - \frac{3}{2}\,x^2, \qquad y(1) = -\frac{1}{2}; \qquad y_s(x) = -\frac{x^3}{2}.$

17. Solve the initial value problems.

(a) $(2x + y)\,y' = y, \quad y(1) = 2$;
(b) $x\,y' = x + 4y, \quad y(1) = 1$;
(c) $(x + 2y)\,y' = x + y, \quad y(1) = 0$;
(d) $(5xy^2 - 3x^3)\,y' = x^2y + y^3, \quad y(2) = 1$;
(e) $x^2\,y' = xy + y^2, \quad y(1) = -1$;
(f) $xy^2\,y' = x^3 + y^3, \quad y(1) = 0$;
(g) $x = e^{y/x}\,(xy' - y), \quad y(1) = 0$;
(h) $y' = 2xy/(x^2 - y^2), \quad y(1) = 1$;
(i) $y^2 = x(y - x)y', \quad y(1) = e$;
(j) $y' = (2xy + y^2)/(3x^2), \quad y(1) = 1/2$;
(k) $xy\,y' = x^2 + y^2, \quad y(e) = e\sqrt{2}$;
(l) $(y'\,x - y)\cos\frac{y}{x} = -x\,\sin\frac{y}{x}, \quad y(1) = \frac{\pi}{2}$.

18. Use the indicated change of variables to solve the following equations.

(a) $y' = 2y/x + x^3/y, \quad y = ux^2$.
(b) $\left(3 + 9x + 3y + \frac{1/2}{1+x}\right)\mathrm{d}x + (1 + 3x + y)\,\mathrm{d}y = 0, \quad u = 3x + y$.
(c) $y' = y + 7x^3\sqrt{y}, \quad y = u^2x$.
(d) $(2y + 1 - x^2y^2)\,\mathrm{d}x + 2x\,\mathrm{d}y = 0, \quad u = xy$.
(e) $2xyy' = y^2 + \sqrt{y^4 - 4x^2}, \quad y = ux^{1/2}$.
(f) $2x^3y' = 1 + \sqrt{1 + 4x^2y}, \quad u = x^2y$.

19. Let $P(x) = ax^2 + bx + c$ be a polynomial of the second degree. The separable differential equation

$$\frac{\mathrm{d}x}{\sqrt{P(x)}} + \frac{\mathrm{d}y}{\sqrt{P(y)}} = 0 \qquad (P(x) = ax^2 + bx + c, \quad P(y) = ay^2 + by + c)$$

was first considered by L. Euler for a polynomial $P(\cdot)$ of the fourth degree. Find the general solution of the above differential equation using substitution $x = k\,u + h$, for some constants k and h, assuming that $b^2 - 4ac > 0$.

20. Suppose that a small (tennis) ball of mass m is thrown straight up from the ground into earth's atmosphere with the initial velocity v_0. It is known from elementary physics that the presence of air influences the ball's motion: it experiences now two forces acting on it—the force of gravity mg and the air resistance force, which can be assumed to be proportional to the velocity, $-k|v|$. Here v denotes the ball's velocity and g the acceleration due to gravity. Make a numerical experiment by choosing some numerical values for m, v_0, and $k < 1$ to show that the time for traveling up is not equal to the time traveling down.

2.3 Exact Differential Equations

In this section, it will be convenient to use the language of differential forms. A **differential form** with two variables x and y is an expression of the type

$$M(x,y)\,\mathrm{d}x + N(x,y)\,\mathrm{d}y,$$

where M and N are functions of x and y. The simple forms $\mathrm{d}x$ and $\mathrm{d}y$ are **differentials** and their ratio $\mathrm{d}y/\mathrm{d}x$ is a derivative. A differential equation $y' = f(x,y)$ can be written as $\mathrm{d}y = f(x,y)\,\mathrm{d}x$. This is a particular case of a general equation

$$M(x,y)\,\mathrm{d}x + N(x,y)\,\mathrm{d}y = 0 \tag{2.3.1}$$

written in a differential form, which suppresses the distinction between independent and dependent variables. Note that the differential equation (2.3.1) can have a constant solution $y = y^*$ (called the critical point or equilibrium solution) if $M(x, y^*) \equiv 0$.

Let $\psi(x,y)$ be a continuous function of two independent variables, with continuous first partial derivatives in a simply connected region Ω, whose boundary is a closed curve with no self-intersections; this means that the region Ω has no hole in its interior. The level curves $\psi(x,y) = C$ of the surface $z = \psi(x,y)$ can be considered as particular solutions to some differential equation. This point of view is very important when solutions to the differential equations are composed of two or more functions (see Fig. 2.18 on page 66). Taking the differential from both sides of the equation $\psi(x,y) = C$, we obtain

$$\mathrm{d}\psi \equiv \psi_x\,\mathrm{d}x + \psi_y\,\mathrm{d}y = 0 \qquad \text{or} \qquad \frac{\mathrm{d}y}{\mathrm{d}x} = -\frac{\psi_x(x,y)}{\psi_y(x,y)} \equiv -\frac{\partial\psi/\partial x}{\partial\psi/\partial y}. \tag{2.3.2}$$

Thus, the equation $\psi(x,y) = C$ defines the general solution of Eq. (2.3.2) in implicit form. Furthermore, the function $\psi(x,y)$ is a constant along any solution $y = \phi(x)$ of the differential equation (2.3.2), which means that $\psi(x,\phi(x)) \equiv C$ for all $x \in \Omega$. For example, let us consider a family of parabolas $\psi(x,y) = C$ for $\psi(x,y) = y - x^2$. Then

$$\mathrm{d}\psi = \mathrm{d}y - 2x\,\mathrm{d}x.$$

Equating $\mathrm{d}\psi$ to zero, we have

$$\mathrm{d}y - 2x\,\mathrm{d}x = 0 \qquad \text{or} \qquad \frac{\mathrm{d}y}{\mathrm{d}x} \stackrel{\text{def}}{=} y' = 2x.$$

The latter equation has the general solution $y - x^2 = C$ for some constant C.

Definition 2.8: A differential equation $M(x,y)\,\mathrm{d}x + N(x,y)\,\mathrm{d}y = 0$ is called **exact** if there is a function $\psi(x,y)$ whose total differential is equal to $M\,\mathrm{d}x + N\,\mathrm{d}y$, namely,

$$\mathrm{d}\psi(x,y) = M(x,y)\,\mathrm{d}x + N(x,y)\,\mathrm{d}y,$$

or

$$\frac{\partial\psi}{\partial x} = M(x,y), \quad \frac{\partial\psi}{\partial y} = N(x,y). \tag{2.3.3}$$

The function $\psi(x,y)$ is called a **potential function** of the differential equation (2.3.1).

The exact differential equation (2.3.1) can be written as $\mathrm{d}\psi(x,y) = 0$. By integrating, we immediately obtain the general solution of (2.3.1) in the implicit form

$$\psi(x,y) = \text{constant}. \tag{2.3.4}$$

An exact differential equation is equivalent to reconstructing a potential function $\psi(x,y)$ from its gradient $\nabla\psi = \langle \psi_x, \psi_y \rangle$. The potential function, $\psi(x,y)$, of the differential equation (2.3.1) is not unique because an arbitrary constant may be added to it. Some practical applications of exact equations are given in [14].

Example 2.3.1: The differential equation

$$\left(xy + \frac{y^2}{2}\right)\mathrm{d}x + \left(\frac{x^2}{2} + xy\right)\mathrm{d}y = 0$$

is exact because $\psi(x,y) = xy(x+y)/2$ is its potential function having partial derivatives

$$\psi_x \stackrel{\text{def}}{=} \frac{\partial\psi}{\partial x} = xy + \frac{y^2}{2}, \qquad \psi_y \stackrel{\text{def}}{=} \frac{\partial\psi}{\partial y} = \frac{x^2}{2} + xy.$$

Surely, for any constant c, $\psi(x,y)+c$ is also a potential function for the given differential equation. Therefore, this equation has the general solution in an implicit form: $xy(x+y) = C$, where C is an arbitrary constant. To check our answer, we ask *Mathematica:*

```
psi := x*y*(x + y)/2
factored = Factor[Dt[psi]] /. {Dt[x] -> dx, Dt[y] -> dy}
Collect[factored[[2]], {dx, dy}]
```

□

The beauty of exact equations is immediate—they are solvable by elementary integration methods. To see how we can solve exact equations, we need to continue our theoretical development.

Theorem 2.3: Suppose that continuous functions $M(x,y)$ and $N(x,y)$ are defined and have continuous first partial derivatives in a rectangle $R = [a,b] \times [c,d]$ or any domain bounded by a closed curve without self-intersections. Then Eq. (2.3.1) is an exact differential equation in R if and only if

$$\frac{\partial M(x,y)}{\partial y} = \frac{\partial N(x,y)}{\partial x} \tag{2.3.5}$$

at each point of R. That is, there exists a function $\psi(x,y)$ such that $\psi(x,y)$ is a constant along solutions to Eq. (2.3.1) if and only if $M(x,y)$ and $N(x,y)$ satisfy the relation (2.3.5).

PROOF: Computing partial derivatives $M_y = \partial M/\partial y$ and $N_x = \partial N/\partial x$ from Eqs. (2.3.3), we obtain

$$\frac{\partial M(x,y)}{\partial y} = \frac{\partial\psi(x,y)}{\partial x\partial y}, \qquad \frac{\partial N(x,y)}{\partial x} = \frac{\partial\psi(x,y)}{\partial y\partial x}.$$

Since M_y and N_x are continuous, it follows that second derivatives of ψ are also continuous in the domain R. This guarantees[24] the equality $\psi_{xy} = \psi_{yx}$, and Eq. (2.3.5) follows.

To prove that condition (2.3.5) satisfies the requirements for exactness of Eq. (2.3.3), we should show that it implies existence of a potential function. Choose any point (x_0, y_0) in the domain R where the functions M, N, and their partial derivatives $M_y = \partial M/\partial y$, $N_x = \partial N/\partial x$ are continuous. We try to restore $\psi(x,y)$ from its partial derivative $\psi_x = M(x,y)$:

$$\psi(x,y) = \int_{x_0}^{x} M(\xi,y)\,\mathrm{d}\xi + h(y),$$

where $h(y)$ is an arbitrary function of y, and ξ is a dummy variable of integration. We see that knowing one partial derivative is not enough to determine $\psi(x,y)$, and we have to use another partial derivative. Differentiating $\psi(x,y)$ with respect to y, we obtain

$$\psi_y(x,y) = \frac{\partial}{\partial y}\int_{x_0}^{x} M(x,y)\,dx + h'(y) = \int_{x_0}^{x} \frac{\partial M(\xi,y)}{\partial y}\,\mathrm{d}\xi + h'(y).$$

Setting $\psi_y = N(x,y)$ and solving for $h'(y)$ gives

$$h'(y) = N(x,y) - \int_{x_0}^{x} \frac{\partial M(x,y)}{\partial y}\,\mathrm{d}x. \tag{2.3.6}$$

The right-hand side of this equation does not depend on x because its derivative with respect to x is identically zero:

$$N_x(x,y) - M_y(x,y) \equiv 0 \qquad \text{for all } (x,y) \text{ in } R.$$

Now we can find $h(y)$ from Eq. (2.3.6) by a single integration with respect to y, treating x as a constant. This yields

$$h(y) = \int_{y_0}^{y} \left[N(x,y) - \int_{x_0}^{x} \frac{\partial M(x,y)}{\partial y}\,\mathrm{d}x \right] \mathrm{d}y.$$

[24] This condition is necessary for multi-connected domains, whereas for simple connected domains it is also sufficient.

Note that an arbitrary point (x_0, y_0) serves as a constant of integration. Substitution for $h(y)$ gives the potential function

$$\psi(x,y) = \int_{x_0}^{x} M(x,y)\,\mathrm{d}x + \int_{y_0}^{y} \left[N(x,y) - \int_{x_0}^{x} \frac{\partial M(x,y)}{\partial y}\,\mathrm{d}x \right] \mathrm{d}y. \tag{2.3.7}$$

To verify that the function $\psi(x,y)$ is the integral of Eq. (2.3.1), it suffices to prove that

$$M_y = N_x \quad \text{implies that} \quad \psi_x = M, \quad \psi_y = N.$$

In fact, from Eq. (2.3.7), we have

$$\begin{aligned}\psi_x(x,y) &= M(x,y) + \frac{\partial}{\partial x}\int_{y_0}^{y} N(x,y)\,\mathrm{d}y - \int_{y_0}^{y} \mathrm{d}y\, \frac{\partial}{\partial x}\int_{x_0}^{x} \frac{\partial M(x,y)}{\partial y}\,\mathrm{d}x \\ &= M(x,y) + \int_{y_0}^{y} M_y(x,y)\,\mathrm{d}y - \int_{y_0}^{y} M_y(x,y)\,\mathrm{d}y = M(x,y).\end{aligned}$$

Similarly, we can show that $\psi_y = N(x,y)$.

Example 2.3.2: Given $\psi(x,y) = \arctan x + \ln(1+y^2)$, find the exact differential equation $\mathrm{d}\psi(x,y) = 0$.

Solution. The derivative of $\psi(x,y)$ is (note that y is a function of x)

$$\frac{\mathrm{d}}{\mathrm{d}x}\psi(x,y) = \frac{\partial \psi}{\partial x} + \frac{\partial \psi}{\partial y}\,y'(x) = \frac{1}{1+x^2} + \frac{2yy'}{1+y^2} = 0.$$

Thus the function $\psi(x,y)$ is the potential for the exact differential equation

$$(1+y^2)\,\mathrm{d}x + 2y(1+x^2)\,\mathrm{d}y = 0 \qquad \text{or} \qquad y' = -\frac{1+y^2}{2y(1+x^2)}\,. \qquad \square$$

The simplest way to define the function $\psi(x,y)$ is to take a line integral of $M(x,y)\,\mathrm{d}x + N(x,y)\,\mathrm{d}y$ between some fixed point (x_0, y_0) and an arbitrary point (x,y) along any path:

$$\psi(x,y) = \int_{(x_0,y_0)}^{(x,y)} M(x,y)\,\mathrm{d}x + N(x,y)\,\mathrm{d}y. \tag{2.3.8}$$

The value of this line integral does not depend on the path of integration, but only on the initial point (x_0, y_0) and terminal point (x,y). Therefore, we can choose a curve of integration as we wish. It is convenient to integrate along some piecewise linear curve, as indicated in Fig. 2.21. Integration along the horizontal line first and then along the vertical line yields

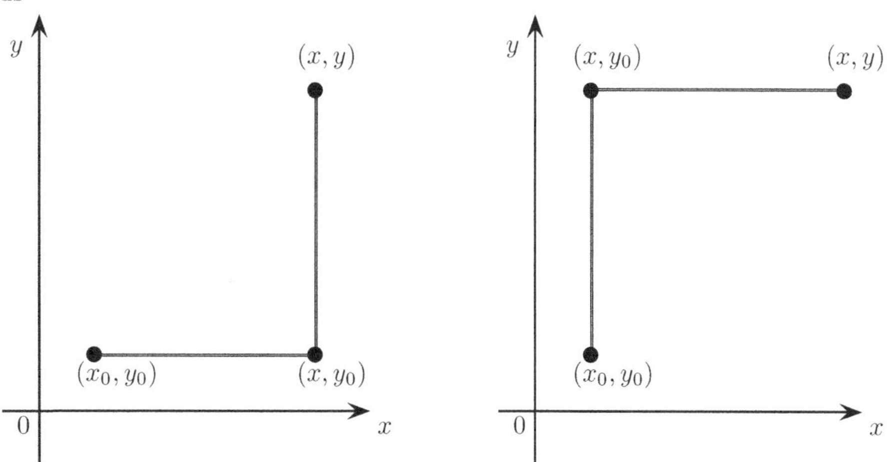

Figure 2.21: Lines of integration, plotted with `pstricks`.

$$\psi = \int_{(x_0,y_0)}^{(x,y)} M(x,y)\,\mathrm{d}x + N(x,y)\,\mathrm{d}y = \int_{(x_0,y_0)}^{(x,y_0)} M(x,y_0)\,\mathrm{d}x + \int_{(x,y_0)}^{(x,y)} N(x,y)\,\mathrm{d}y. \tag{2.3.9}$$

Similarly, if we integrate along the vertical line from (x_0, y_0) to (x_0, y) and then along the horizontal line from (x_0, y) to (x, y), we obtain

$$\psi = \int_{(x_0,y_0)}^{(x,y)} M(x,y)\,\mathrm{d}x + N(x,y)\,\mathrm{d}y = \int_{(x_0,y_0)}^{(x_0,y)} N(x_0,y)\,\mathrm{d}y + \int_{(x_0,y)}^{(x,y)} M(x,y)\,\mathrm{d}x. \tag{2.3.10}$$

If the straight line connecting the initial point (x_0, y_0) with an arbitrary point (x, y) belongs to the domain where functions $M(x, y)$ and $N(x, y)$ are defined and integrable, the potential function can be obtained by integration along the line (for simplicity, we set $x_0 = y_0 = 0$):

$$\psi(x,y) = \int_0^1 [x\,M(xt,yt) + y\,N(xt,yt)]\,\mathrm{d}t. \tag{2.3.11}$$

We may further simplify the integration in Eq. (2.3.8) by choosing the point (x_0, y_0) judiciously. The line integral representation (2.3.8) for $\psi(x, y)$ is especially convenient for dealing with an initial value problem when the initial condition $y(x_0) = y_0$ is specified. Namely, to define the solution explicitly, the equation $\psi(x, y) = 0$ has to be solved for $y = y(x)$ as a function of x, which is the solution of Eq. (2.3.1) satisfying the initial condition $y(x_0) = y_0$. If it is impossible to resolve the equation $\psi(x, y) = 0$ with respect to y, the equation $\psi(x, y) = 0$ defines the solution implicitly.

Note that the point (x_0, y_0) in the line integral has to be chosen in agreement with the initial condition $y(x_0) = y_0$. Then the solution of the initial value problem

$$M(x,y)\,\mathrm{d}x + N(x,y)\,\mathrm{d}y = 0, \qquad y(x_0) = y_0$$

becomes $\psi(x, y) = 0$, where the potential function $\psi(x, y)$ is defined by the line integral (2.3.8).

While line integration is useful for solving initial value problems, its well-known variation, the method of "integration by parts," is beneficial for a large class of exact differential equations. For example, let us take the exact equation

$$(3x^2y - 2y^2 - 1)\,\mathrm{d}x + (x^3 - 4xy + 2y)\,\mathrm{d}y = 0.$$

Its integration yields

$$\int (3x^2y - 2y^2 - 1)\,\mathrm{d}x + \int (x^3 - 4xy + 2y)\,\mathrm{d}y = C,$$

where C is an arbitrary constant. We integrate the "parts" terms having more than one variable. Since $\int x^3\mathrm{d}y = x^3y - \int 3x^2y\,\mathrm{d}x$ and $-\int 2y^2\,\mathrm{d}x = -2xy^2 + \int 4xy\,\mathrm{d}x$, it follows

$$\int 3x^2y\,\mathrm{d}x - 2xy^2 + \int 4xy\,\mathrm{d}x - x + x^3y - \int 3x^2y\,\mathrm{d}x - \int 4xy\,\mathrm{d}y + y^2 = C.$$

The remaining integrals cancel, giving the solution

$$x^3y - 2xy^2 - x + y^2 = C.$$

For visualization, we use *Mathematica*'s command `ContourPlot` (see Fig. 2.22):

```
ContourPlot[x*x*x*y - 2*x*y*y - x + y*y, {x, -3, Pi}, {y, -3, Pi}]
```

Solution. This is an exact equation because for $M(x, y) = 2x + y + 1$ and $N(x, y) = x - 3y + 4$ we have

$$\frac{\partial(2x+y+1)}{\partial y} = \frac{\partial(x-3y^2+4)}{\partial x} = 1.$$

Hence there exists a potential function $\psi(x, y)$ such that $\psi_x = M(x, y) = 2x+y+1$ and $\psi_y = N(x, y) = x - 3y^2 + 4$. We integrate the former with respect to x, treating y as a constant to obtain

$$\psi(x,y) = x^2 + xy + x + h(y),$$

where $h(y)$ is an unknown function to be determined. Differentiation of $\psi(x, y)$ with respect to y yields

$$\frac{\partial\psi}{\partial y} = N(x,y) = x + h'(y) = x - 3y^2 + 4,$$

Example 2.3.3: Consider a separable equation

$$y' = p(x)q(y).$$

We rewrite this equation in the form $M(x)\,dx + N(y)\,dy = 0$, where $M(x) = p(x)$ and $N(y) = -q^{-1}(y)$. Since $M_y = 0$ and $N_x = 0$ for any functions $M(x)$ and $N(y)$, the relation (2.3.5) is valid and this equation is an exact differential equation. Therefore, any separable differential equation is an exact equation.

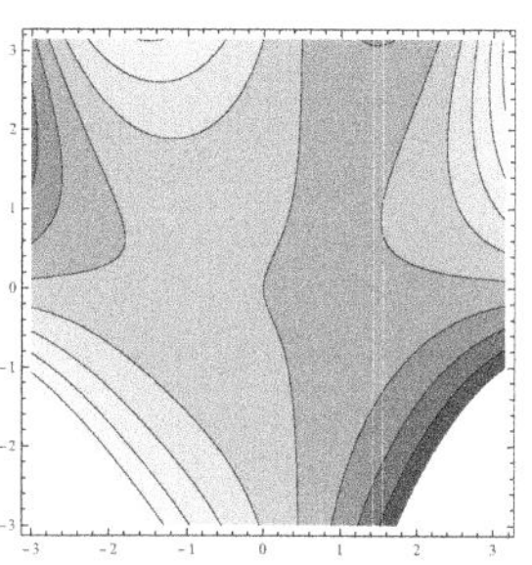

Figure 2.22: Contour curves $x^3y - 2xy^2 - x + y^2 = C$.

Example 2.3.4: Solve the equation

$$(2x + y + 1)\mathrm{d}x + (x - 3y^2 + 4)\mathrm{d}y = 0.$$

so we get

$$h'(y) = -3y^2 + 4 \qquad \Longrightarrow \qquad h(y) = -y^3 + 4y + c,$$

where c is a constant of integration. Substituting $h(y)$ back into $\psi(x, y)$ gives the exact expression for the potential function

$$\psi(x, y) = x^2 + xy + x - y^3 + 4y + c.$$

Of course, we can drop the constant c in the above equation because in the general solution $C = \psi(x, y)$ the constant C can absorb c. Instead, we apply the formula (2.3.11) and choose the initial point of integration as $(0, 0)$. Then Eq. (2.3.11) defines the potential function ψ:

$$\psi(x, y) = \int_0^1 \left[x(2xt + yt + 1) + y(xt - 3y^2t^2 + 4)\right] \mathrm{d}t = x^2 + xy + x - y^3 + 4y.$$

If we choose another point, say $(1, 2)$, we can apply, for instance, the line integration along the horizontal line $y = 2$ first and then along the vertical line:

$$\int_{(1,2)}^{(x,2)} (2x + 2 + 1)\,\mathrm{d}x + \int_{(x,2)}^{(x,y)} (x - 3y^2 + 4)\,\mathrm{d}y = x^2 - 1 + 3x - 3 + xy - 2x - y^3 + 3 \cdot 2^3 + 4y - 8,$$

which differs from the previously obtained potential function $\psi(x, y) = x^2 + xy + x - y^3 + 4y$ by the constant 12. In MATLAB, we plot the direction field along with the contour plot:

```
[x,y] = meshgrid(-2:.1:2,-1:.1:1);
z = x.^2 + x.*y + x - y.^3 + 4*y;
[DX,DY] = gradient(z,.2,.2);
quiver(x,y,DX,DY), hold on
title('Direction field and level curves of z=x^2+xy+x-y^3+4y')
[c,h] = contour(x,y,z,10,'linewidth',2);
h.LevelList = round(h.LevelList,1); clabel(c,h), hold off
```

$$\psi(x, y) = \int_{(0,1)}^{(x,y)} (1 + y^2 + xy^2)\,\mathrm{d}x + (x^2y + y + 2xy)\,\mathrm{d}y.$$

We choose the path of integration along the coordinate axes; first we integrate along the vertical straight line from $(0, 1)$ to $(0, y)$ (along this line $\mathrm{d}x = 0$ and $x = 0$) and then we integrate along the horizontal straight line from $(0, y)$ to (x, y) ($\mathrm{d}y = 0$ along this line); this gives us the function

$$\psi(x, y) = \int_1^y y\,\mathrm{d}y + \int_0^x (1 + y^2 + xy^2)\,\mathrm{d}x = \frac{1}{2}\left(y^2 - 1\right) + \left(2x + 2xy^2 + x^2y^2\right).$$

The potential function $\psi(x, y)$ can be determined by choosing another path of integration, but the result will be the same because the line integral does not depend on the curve of integration. For example, we may integrate along

Example 2.3.5: Let us consider the initial value problem

$$(1+y^2+xy^2)\,\mathrm{d}x + (x^2y+y+2xy)\,\mathrm{d}y = 0, \quad y(0)=1.$$

First, we can check the exactness with *Mathematica:*

```
MM[x_, y_] = 1 + y^2 + xy^2
NN[x_, y_] = yx^2 + y + 2 xy
Simplify[D[MM[x, y], y] == D[NN[x, y], x]]
```

The line integral (2.3.8) defines the potential function for this differential equation:

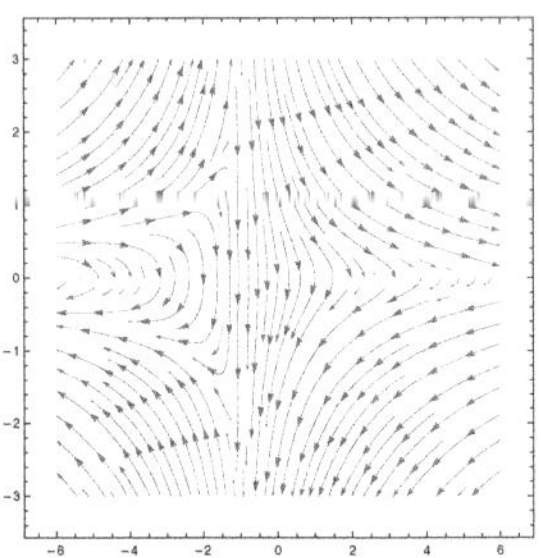

Figure 2.23: Direction field for Example 2.3.5, plotted with *Mathematica.*

the horizontal line first and then along the vertical line or use the straight line formula (2.3.11):

$$\begin{aligned}\psi(x,y) &= \int_0^x \left(1+1^2+x\cdot 1^2\right)\mathrm{d}x + \int_1^y \left(x^2y+y+2xy\right)\mathrm{d}y \\ &= \int_0^1 \left[x\left(1+(x+1)(1+t(y-1))^2\right)+(y-1)(1+t(y-1))^2\left(x^2t^2+1+2xt\right)\right]\mathrm{d}t.\end{aligned}$$

To check our answer, we type in *Mathematica:*

```
psi[X_, Y_] = Integrate[MM[x, y0], {x, x0, X}] + Integrate[NN[X, y], {y, y0, Y}]
psi[X, Y] == 0
```

From the equation $\psi(x,y)=0$, we find the solution of the given initial value problem to be $y=\phi(x)=\sqrt{(1-2x)/(1+x)^2}$. □

Equation (2.3.1) is a particular case of the more general equation (functions M, N, and f are given)

$$M(x,y)\,\mathrm{d}x + N(x,y)\,\mathrm{d}y = f(x)\,\mathrm{d}x, \tag{2.3.12}$$

called a nonhomogeneous equation. If the left-hand side is the differential of a potential function $\psi(x,y)$, we get the **exact** equation $\mathrm{d}\psi(x,y)=f(x)\mathrm{d}x$. After integrating, we obtain the general solution of Eq. (2.3.12) in implicit form:

$$\psi(x,y) = \int f(x)\,\mathrm{d}x + C.$$

Remark 1. We can obtain the potential function for Eq. (2.3.1) from the second equation (2.3.3), that is, $\psi_y(x,y)=N(x,y)$. Then integration with respect to y yields

$$\psi(x,y) = \int_{y_0}^y N(x,y)\,\mathrm{d}y + g(x),$$

where g is an arbitrary function of x. Differentiating with respect to x and setting $\psi_x = M(x,y)$, we obtain the ordinary differential equation for $g(x)$:

$$g'(x) = M(x,y) - \int_{y_0}^y N(x,y)\,\mathrm{d}y.$$

Remark 2. Most exact differential equations cannot be solved explicitly for y as a function of x. While this may appear to be very disappointing, with the aid of a computer it is very simple to compute $y(x)$ from the equation $\psi(x,y)=C$ up to any desired accuracy. ■

If Eq. (2.3.1) is not exact, then we cannot restore the function $h(y)$ from Eq. (2.3.6) as the following example shows.

Example 2.3.6: The differential equation

$$(1+y^2)\,\mathrm{d}x + (x^2y+y+2xy)\,\mathrm{d}y = 0$$

is not exact because $M_y = 2y \ne 2xy + 2y = N_x$. If we try the described procedure, then

$$\psi_x = 1 + y^2 \qquad \text{and} \qquad \psi(x,y) = x + xy^2 + h(y),$$

where $h(y)$ is an arbitrary function of y only. Substitution $\psi(x,y)$ into the equation $\psi_y = N(x,y) \equiv x^2y + y + 2xy$ yields

$$2xy + h'(y) = x^2y + y + 2xy \qquad \text{or} \qquad h'(y) = x^2y + y.$$

It is impossible to solve this equation for $h(y)$ because the right-hand side function depends on x as well as on y.

Problems

1. Given a potential function $\psi(x,y)$, find the exact differential equation $\mathrm{d}\psi(x,y) = 0$.
   (a) $\psi(x,y) = x^2 + y^2$; **(b)** $\psi(x,y) = \exp(xy^2)$;
   (c) $\psi(x,y) = \ln(x^2y^2)$; **(d)** $\psi(x,y) = (x + y - 2)^2$;
   (e) $\psi(x,y) = \tan(9x^2 + y^2)$; **(f)** $\psi(x,y) = \sinh(x^3 - y^3)$;
   (g) $\psi(x,y) = x/y^2$; **(h)** $\psi(x,y) = \sin(x^2y)$;
   (i) $\psi(x,y) = x^2y + 2y^2$; **(j)** $\psi(x,y) = x + 3xy^2 + x^2y$.
2. Show that the following differential equations are exact and solve them
   (a) $x^2\,y' + 2yx = 0$; **(b)** $y\,(e^{xy} + y)\,\mathrm{d}x + x\,(e^{xy} + 2y)\,\mathrm{d}y = 0$;
   (c) $(2x + e^y)\,\mathrm{d}x + xe^y\,\mathrm{d}y = 0$; **(d)** $(2xy - 3x^2)\,\mathrm{d}x + (x^2 + y)\,\mathrm{d}y = 0$;
   (e) $(y - 1)\,\mathrm{d}x + (x - 2)\,\mathrm{d}y = 0$; **(f)** $(6xy + 2y^2 - 4)\,\mathrm{d}x = (1 - 3x^2 - 4xy)\,\mathrm{d}y$;
   (g) $(\cot y + x^3)\,\mathrm{d}x = x\,\csc^2 y\,\mathrm{d}y$; **(h)** $\frac{\mathrm{d}x}{(1-xy)^2} + [y^2 + x^2(1 - xy)^{-2}]\,\mathrm{d}y = 0$;
   (i) $2xy\,\mathrm{d}x + (x^2 + 4y)\,\mathrm{d}y = 0$; **(j)** $(\cos xy - \sin xy)\,(y\,\mathrm{d}x + x\,\mathrm{d}y) = 0$;
   (k) $y^3\,\mathrm{d}x + 3xy^2\,\mathrm{d}y = 0$; **(l)** $(y/x + x)\,\mathrm{d}x + (\ln x - 1)\,\mathrm{d}y = 0$;
   (m) $e^{-\theta}\,\mathrm{d}r - r\,e^{-\theta}\,\mathrm{d}\theta = 0$; **(n)** $2x(1 + \sqrt{x^2 - y})\,\mathrm{d}x = \sqrt{x^2 - y}\,\mathrm{d}y$;
   (o) $4x^{-1}\,\mathrm{d}x + \mathrm{d}y = 0$; **(p)** $(\frac{y}{x^2} - \frac{1}{y^2})\,\mathrm{d}x + (\frac{2x}{y^3} - \frac{1}{x})\,\mathrm{d}y = 0$.
3. Solve the following exact equations.
   (a) $(2y + 3x^2)\,\mathrm{d}x + 2(x - y)\,\mathrm{d}y = 0$; **(b)** $\frac{y}{x}\,\mathrm{d}x + (y^2 + \ln x)\,\mathrm{d}y = 0$;
   (c) $\left(\frac{1}{y} + \frac{y}{x^2}\right)\mathrm{d}x - \left(\frac{1}{x} + \frac{x}{y^2}\right)\mathrm{d}y = 0$; **(d)** $\dfrac{3x^2 + y^2}{y^2}\,\mathrm{d}x - \dfrac{2x^3 + 4y}{y^3}\,\mathrm{d}y = 0$;
   (e) $\ln|y|\,\sinh x\,\mathrm{d}x + \dfrac{\cosh x}{y}\,\mathrm{d}y = 0$; **(f)** $\left(\frac{y}{x} + y + \frac{1}{y}\right)\mathrm{d}x + \left(x + 2y + \ln x - \frac{x}{y^2}\right)\mathrm{d}y = 0$;
   (g) $\dfrac{y\,\mathrm{d}x + x\,\mathrm{d}y}{1 + x^2y^2} = 0$; **(h)** $\left(\dfrac{y}{(x-y)^2} - \dfrac{1}{2\sqrt{1 - x^2}}\right)\mathrm{d}x - \dfrac{x}{(y-x)^2}\,\mathrm{d}y = 0$;
   (i) $\left(\frac{3}{x} - y\right)\mathrm{d}x = \left(x - \frac{3}{y} + 1\right)\mathrm{d}y$; **(j)** $\left[y^2 - \frac{1}{x^2y}\right]\mathrm{d}x + \left[3y^2 - \frac{1}{xy^2} + 2xy\right]\mathrm{d}y = 0$.
4. Are the following equations exact? Solve the initial value problems.
   (a) $\cos \pi x\, \cos 2\pi y\,\mathrm{d}x = 2\sin \pi x\, \sin 2\pi y\,\mathrm{d}y$, $y(3/2) = 1/2$;
   (b) $2xy\,\mathrm{d}y + (x^2 + y^2)\,\mathrm{d}x = 0$, $y(3) = 4$;
   (c) $(2xy - 4)\,\mathrm{d}x + (x^2 + 4y - 1)\,\mathrm{d}y = 0$, $y(1) = 2$;
   (d) $\sin(\omega y)\,\mathrm{d}x + \omega x \cos(\omega y)\,\mathrm{d}y = 0$, $y(1) = \pi/2\omega$;
   (e) $(\cos\theta - 2r\cos^2\theta)\,\mathrm{d}r + r\sin\theta(2r\cos\theta - 1)\,\mathrm{d}\theta = 0$, $r(\pi/4) = 1$;
   (f) $(2r + \sin\theta - \cos\theta)\,\mathrm{d}r + r(\sin\theta + \cos\theta)\,\mathrm{d}\theta = 0$, $r(\pi/4) = 1$;
   (g) $x^{-1}e^{-y/x}\,\mathrm{d}y - x^{-2}ye^{-y/x}\,\mathrm{d}x = 0$, $y(-2) = -2$;
   (h) $(2x - 3y)\,\mathrm{d}x + (2y - 3x)\,\mathrm{d}y = 0$, $y(0) = 1$;
   (i) $3y(x^2 - 1)\,\mathrm{d}x + (x^3 + 8y - 3x)\,\mathrm{d}y = 0$, $y(0) = 1$;
   (j) $(xy^2 + x - 2y + 5)\,\mathrm{d}x + (x^2y + y^2 - 2x)\,\mathrm{d}y = 0$, $y(1) = 3$;
   (k) $2y^2\sin^2 x\,\mathrm{d}x - y\sin 2x\,\mathrm{d}y = 0$, $y(\pi/2) = 1$;
   (l) $(3x^2\sin 2y - 2xy)\,\mathrm{d}x + (2x^3\cos 2y - x^2)\,\mathrm{d}y = 0$, $y(1) = 0$;
   (m) $(xy^2 + y)\,\mathrm{d}x + (x^2y + x - 3y^2)\,\mathrm{d}y = 0$, $y(0) = -1$;
   (n) $(\sin y + y\cos x)\,\mathrm{d}x + (\sin x + x\cos y)\,\mathrm{d}y = 0$, $y(\pi) = 2$;
   (o) $(\sin xy + xy\cos xy)\,\mathrm{d}x + x^2\cos xy\,\mathrm{d}y = 0$, $y(1) = 1$;
5. Solve an exact equation $2x\left[3x^2 + y - ye^{-x^2}\right]\mathrm{d}x + \left[x^2 + 3y^2 + e^{-x^2}\right]\mathrm{d}y = 0$.

2.4 Simple Integrating Factors

The idea of the integrating factor method is quite simple. Suppose that we are given a first order differential equation of the form

$$M(x,y)\,\mathrm{d}x + N(x,y)\,\mathrm{d}y = 0 \tag{2.4.1}$$

that is not exact, namely, $M_y \neq N_x$. (It is customary to use a subscript for the partial derivative, for example, $M_y = \partial M/\partial y$.) If Eq. (2.4.1) is not exact, we may be able to make the equation exact by multiplying it by a suitable function $\mu(x,y)$:

$$\mu(x,y)M(x,y)\mathrm{d}x + \mu(x,y)N(x,y)\mathrm{d}y = 0. \tag{2.4.2}$$

Such a function (other than zero) is called an **integrating factor** of Eq. (2.4.1) if Eq. (2.4.2) is exact. The integrating factor method was introduced by the French mathematician Alexis Clairaut (1713–1765). If an integrating factor exists, then it is not unique and there exist infinitely many functions that might be used for this purpose. Indeed, let $\mu(x,y)$ be an integrating factor for the first order differential equation (2.4.1) and $\psi(x,y)$ be the corresponding potential function, that is, $\mathrm{d}\psi = \mu M\,\mathrm{d}x + \mu N\,\mathrm{d}y$. Then any integrating factor for Eq. (2.4.1) is expressed by $m(x,y) = \mu(x,y)\phi(\psi(x,y))$ for an appropriate function $\phi(z)$ having continuous derivatives.

Using this claim, we can sometimes construct an integrating factor in the following way. Suppose that the differential equation (2.4.1) can be broken into two parts:

$$M_1(x,y)\,\mathrm{d}x + N_1(x,y)\,\mathrm{d}y + M_2(x,y)\,\mathrm{d}x + N_2(x,y)\,\mathrm{d}y = 0 \tag{2.4.3}$$

and suppose that $\mu_1(x,y)$, $\psi_1(x,y)$ and $\mu_2(x,y)$, $\psi_2(x,y)$ are integrating factors and corresponding potential functions to $M_1(x,y)\,\mathrm{d}x + N_1(x,y)\,\mathrm{d}y = 0$ and $M_2(x,y)\,\mathrm{d}x + N_2(x,y)\,\mathrm{d}y = 0$, respectively. Then all integrating factors for equations $M_k\,\mathrm{d}x + N_k\,\mathrm{d}y = 0$ $(k = 1,2)$ are expressed via the formula $\mu(x,y) = \mu_k(x,y)\phi_k(\psi_k(x,y))$ $(k = 1,2)$, where ϕ_1 and ϕ_2 are arbitrary smooth functions. If we can choose them so that $\mu_1(x,y)\phi_1(\psi_1(x,y)) = \mu_2(x,y) \times \phi_2(\psi_2(x,y))$, then $\mu(x,y) = \mu_1(x,y)\phi_1(\psi_1(x,y))$ is an integrating factor.

It should be noted that the differential equation is changed when multiplied by a function. Therefore, multiplication of Eq. (2.4.1) by an integrating factor may introduce new discontinuities in the coefficients or may also introduce extraneous solutions (curves along which the integrating factor is zero). It may result in either the loss of one or more solutions of Eq. (2.4.1), or the gain of one or more solutions, or both of these phenomena. Generally speaking, careful analysis is needed to ensure that no solutions are lost or gained in the process.

Example 2.4.1: The equation

$$2\,\mathrm{d}x + \left(x^3 y + \frac{x}{y}\right)\mathrm{d}y = 0$$

is not exact because

$$0 = \frac{\partial M}{\partial y} \neq \frac{\partial N}{\partial x} = 3x^2 y + y^{-1}$$

for $M = 2$ and $N = x^3 y + x\,y^{-1}$. If we multiply the given equation by $\mu(x,y) = x^{-3}y^{-1}$ $(x \neq 0,\ y \neq 0)$, we obtain

$$2\,x^{-3}y^{-1}\,\mathrm{d}x + \left(1 + x^{-2}y^{-2}\right)\mathrm{d}y = 0,$$

which is now exact since

$$\frac{\partial M}{\partial y} = \frac{\partial}{\partial y}\left(2\,x^{-3}y^{-1}\right) = -2\,x^{-3}y^{-2} = \frac{\partial N}{\partial x} = \frac{\partial}{\partial x}\left(1 + x^{-2}y^{-2}\right).$$

Therefore, the potential function is $\psi(x,y) = y - x^{-2}y^{-1}$ and the general solution becomes $\psi(x,y) = C$, for some constant C. □

If an integrating factor exists, then it should satisfy the partial first order differential equation

$$(\mu M)_y = (\mu N)_x \qquad \text{or} \qquad M\mu_y - N\mu_x + (M_y - N_x)\mu = 0 \tag{2.4.4}$$

by Theorem 2.3, page 74. In general, we don't know how to solve the partial differential equation (2.4.4) other than transfer it to the ordinary differential equation (2.4.1). In principle, the integrating factor method is a powerful tool for solving differential equations, but in practice, it may be very difficult, perhaps impossible, to find an integrating factor. Usually, an integrating factor can be found only under several restrictions imposed on M and N. In this section, we will consider two simple classes when μ is a function of one variable only.

Case 1. Assume that an integrating factor is a function of x alone, $\mu = \mu(x)$. Then $\mu_y = 0$ and, instead of Eq. (2.4.4), we have

$$(\mu N)_x = \mu M_y, \qquad \text{or} \qquad \frac{\mathrm{d}\mu}{\mathrm{d}x} = \frac{M_y - N_x}{N}\,\mu. \tag{2.4.5}$$

If $(M_y - N_x)/N$ is a function of x only, we can can solve Eq. (2.4.5) because it is a separable differential equation:

$$\frac{\mathrm{d}\mu}{\mu} = \frac{M_y - N_x}{N}\,\mathrm{d}x.$$

Therefore,

$$\mu(x) = \exp\left\{\int \frac{M_y - N_x}{N}\,\mathrm{d}x\right\}. \tag{2.4.6}$$

Case 2. A similar procedure can be used to determine an integrating factor when μ is a function of y only. Then $\mu_x = 0$ and we have

$$\frac{\mathrm{d}\mu}{\mathrm{d}y} = -\frac{M_y - N_x}{M}.$$

If $(M_y - N_x)/M$ is a function of y alone, we can find the integrating factor explicitly:

$$\mu(y) = \exp\left\{-\int \frac{M_y - N_x}{M}\,\mathrm{d}y\right\}. \tag{2.4.7}$$

Example 2.4.2: Consider the equation $\sinh x\,\mathrm{d}x + \frac{\cosh x}{y}\,\mathrm{d}y = 0$, where

$$\sinh x = \frac{1}{2}\,e^{x} - \frac{1}{2}\,e^{-x}, \qquad \cosh x = \frac{1}{2}\,e^{x} + \frac{1}{2}\,e^{-x} \geqslant 1.$$

This is an equation of the form (2.4.1) with $M(x, y) = \sinh x$ and $N(x, y) = \frac{\cosh x}{y}$. Since

$$M_y = 0 \quad \text{and} \quad N_x = \frac{\sinh x}{y},$$

this equation is not exact. The ratio

$$\frac{M_y - N_x}{N} = -\frac{(\sinh x)\,y}{y\,\cosh x} = -\frac{\sinh x}{\cosh x}$$

is a function on x only. Therefore, we can find an integrating factor as a function of x:

$$\mu(x) = \exp\left\{\int \frac{M_y - N_x}{N}\,\mathrm{d}x\right\} = \exp\left\{-\int \frac{\sinh x}{\cosh x}\,\mathrm{d}x\right\}.$$

We don't need to find all possible antiderivatives but only one; thus, we can drop the arbitrary constant to obtain

$$\int \frac{\sinh x}{\cosh x}\,\mathrm{d}x = \int \frac{\mathrm{d}(\cosh x)}{\cosh x} = \ln(\cosh x).$$

Substitution yields

$$\mu(x) = \exp\{-\ln\cosh x\} = \exp\left\{\ln(\cosh x)^{-1}\right\} = \frac{1}{\cosh x}.$$

Now we multiply the given equation by $\mu(x)$ to obtain the exact equation

$$\frac{\sinh x}{\cosh x}\,\mathrm{d}x + \frac{1}{y}\,\mathrm{d}y = 0.$$

This equation is exact because

$$\frac{\partial}{\partial y}\left(\frac{\sinh x}{\cosh x}\right) = 0 \quad \text{and} \quad \frac{\partial}{\partial x}\left(\frac{1}{y}\right) = 0.$$

Since this differential equation is exact, there exists a function $\psi(x, y)$ such that

$$\frac{\partial\psi}{\partial x} = \psi_x = \frac{\sinh x}{\cosh x} \quad \text{and} \quad \frac{\partial\psi}{\partial y} \equiv \psi_y = \frac{1}{y}.$$

We are free to integrate either equation, depending on which we deem easier. For example, integrating the latter leads to

$$\psi(x, y) = \ln|y| + k(x),$$

where $k(x)$ is an unknown function of x only and it should be determined later. We differentiate the equation $\psi(x, y) = \ln|y| + k(x)$ with respect to x and get

$$\psi_x = k'(x) = \frac{\sinh x}{\cosh x}.$$

Integration leads to $k(x) = \ln\cosh x + C_1$ with some constant C_1 and the potential function becomes

$$\psi(x, y) = \ln|y| + \ln\cosh x = \ln(|y| \cosh x).$$

We drop the constant C_1 because a potential function is not unique and it is defined up to an arbitrary constant. Thus, the given differential equation has the general solution in implicit form: $\ln(|y| \cosh x) = C$ for some constant C.

Another way to approach this problem is to look at the ratio

$$\frac{M_y - N_x}{M} = \frac{\frac{\partial}{\partial y}(\sinh x) - \frac{\partial}{\partial x}\left(\frac{\cosh x}{y}\right)}{\sinh x} = -\frac{\sinh x}{y \sinh x} = -\frac{1}{y},$$

which is a function of y only. Therefore, there exists another integrating factor,

$$\mu(y) = \exp\left\{-\int \frac{M_y - N_x}{M}\, \mathrm{d}y\right\} = \exp\left\{\int \frac{\mathrm{d}y}{y}\right\} = \exp\{\ln|y|\} = y,$$

as a function of y. Multiplication by $\mu(y) = y$ yields the exact equation $y \sinh x\, \mathrm{d}x + \cosh x\, \mathrm{d}y = 0$ for which a potential function $\psi(x, y)$ exists. Since

$$\mathrm{d}\psi = y \sinh x\, \mathrm{d}x + \cosh x\, \mathrm{d}y \quad \Longrightarrow \quad \frac{\partial\psi}{\partial x} = y \sinh x \quad \text{and} \quad \frac{\partial\psi}{\partial y} = \cosh x.$$

Integration of $\psi_x = y \sinh x$ yields

$$\psi(x, y) = y \cosh x + h(y),$$

where $h(y)$ is a function to be determined. We differentiate $\psi(x, y)$ with respect to y and equate the result to $\cosh x$. This leads to $h'(y) = 0$ and $h(y)$ is a constant, which can be chosen as zero. So the potential function is $\psi(x, y) = y \cosh x$, and we obtain the general solution in implicit form:

$$y \cosh x = C,$$

where C is an arbitrary constant. This equation can be solved with respect to y and we get the explicit solution

$$y(x) = \frac{C}{\cosh x} = C \operatorname{csch} x.$$

Example 2.4.3: Find the general solution to the following differential equation

$$(y + 2x)\, \mathrm{d}x = (x + 1)\, \mathrm{d}y.$$

Solution. We begin by looking for an integrating factor as a function of only one variable. The following equations

$$\frac{\partial}{\partial y} M(x, y) \stackrel{\text{def}}{=} M_y = \frac{\partial(y + 2x)}{\partial y} = 1 \quad \text{and} \quad \frac{\partial}{\partial x} N(x, y) \stackrel{\text{def}}{=} N_x = \frac{\partial(-x - 1)}{\partial x} = -1,$$

show that the given differential equation is not exact. However, the ratio

$$\frac{M_y - N_x}{N} = \frac{2}{-(x+1)}$$

is a function of x alone. Therefore, there exists an integrating factor:

$$\mu(x) = \exp\left\{-\int \frac{2}{x+1}\,\mathrm{d}x\right\} = \exp\left\{-2\ln|x+1|\right\} = \exp\left\{\ln(x+1)^{-2}\right\} = \frac{1}{(x+1)^2}.$$

Clearly, multiplication by $\mu(x)$ makes the given differential equation exact:

$$\frac{y+2x}{(x+1)^2}\,\mathrm{d}x - \frac{1}{x+1}\,\mathrm{d}y = 0, \qquad x \neq -1.$$

Hence, there exists a potential function $\psi(x,y)$ such that

$$\mathrm{d}\psi(x,y) = \psi_x\,\mathrm{d}x + \psi_y\,\mathrm{d}y, \quad \text{with} \quad \psi_x = \frac{y+2x}{(x+1)^2}, \quad \psi_y = -\frac{1}{x+1}.$$

Integrating the latter equation with respect to y, we obtain

$$\psi(x,y) = \frac{-y}{x+1} + k(x)$$

for some function $k(x)$. By taking the partial derivative with respect to x of both sides, we get

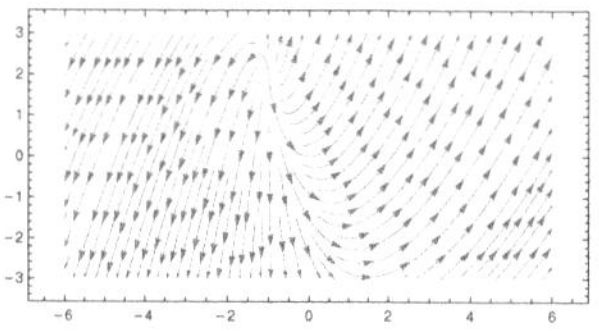

$$\psi_x = \frac{y}{(x+1)^2} + k'(x) = \frac{y+2x}{(x+1)^2}.$$

Figure 2.24: Direction field for Example 2.4.3, plotted with *Mathematica*.

Therefore, $k'(x) = 2x(x+1)^{-2}$. Its integration yields the general solution

$$C = \frac{2-y}{x+1} + \ln(x+1)^2.$$

With the integrating factor $\mu(x) = (x+1)^{-2}$, the vertical line solution $x = -1$ has been lost.

Example 2.4.4: We consider the equation $(x^2y - y^3 - y)\,\mathrm{d}x + (x^3 - xy^2 + x)\,\mathrm{d}y = 0$, which we break into the form (2.4.3), where

$$M_1(x,y) = y(x^2 - y^2), \quad N_1(x,y) = x(x^2-y^2), \quad M_2(x,y) = -y, \quad N_2(x,y) = x.$$

First, we exclude from our consideration the equilibrium solutions $x = 0$ and $y = 0$. Using an integrating factor $\mu_1 = (x^2-y^2)^{-1}$, we reduce the equation $M_1(x,y)\,\mathrm{d}x + N_1(x,y)\,\mathrm{d}y = 0$ to an exact equation with the potential function $\psi_1(x,y) = xy$. Therefore, all its integrating factors are of the form $\mu_1(x,y) = (x^2-y^2)^{-1}\phi_1(xy)$, where ϕ_1 is an arbitrary smooth function. The second equation, $M_2\,\mathrm{d}x + N_2\,\mathrm{d}y = 0$, has an integrating factor $\mu_2(x,y) = (xy)^{-1}$; hence, its potential function is $\psi_2(x,y) = y/x$. All integrating factors for the equation $x\,\mathrm{d}y - y\,\mathrm{d}x = 0$ are of the form $\mu_2(x,y) = (xy)^{-1}\phi_2(y/x)$, where $\phi_2(\cdot)$ is an arbitrary function.

Now we choose these functions, ϕ_1 and ϕ_2, so that the following equation holds:

$$\frac{1}{x^2-y^2}\,\phi_1(xy) = \frac{1}{xy}\,\phi_2\left(\frac{y}{x}\right).$$

If we take $\phi_1(z) = 1$ and $\phi_2(z) = z^{-1}(1-z^2)^{-1}$, the above equation is valid and the required integrating factor for the given differential equation becomes $\mu(x,y) = \mu_1(x,y) = (x^2-y^2)^{-1}$. After multiplication by $\mu(x,y)$, we get the exact equation with the potential function

$$\psi(x,y) = xy - \frac{1}{2}\ln\left|\frac{x-y}{x+y}\right| \quad (y \neq \pm x).$$

□

The next examples demonstrate the integrating factor method for solving initial value problems.

Example 2.4.5: Solve the initial value problem

$$\mathrm{d}x + (x + y + 1)\,\mathrm{d}y = 0, \quad y(0) = 1.$$

Solution. This equation is not exact because $M_y \neq N_x$, where $M(x, y) = 1$ and $N(x, y) = x + y + 1$. Although

$$\frac{\partial M}{\partial y} - \frac{\partial N}{\partial x} = -1 \qquad \Longrightarrow \qquad \frac{1}{M}\left[\frac{\partial M}{\partial y} - \frac{\partial N}{\partial x}\right] = -1$$

is independent of x. Therefore, from Eq. (2.4.7), we have $\mu(y) = e^y$. Multiplying both sides of the given differential equation by the integrating factor $\mu(y) = e^y$, we obtain the exact equation:

$$e^y\,\mathrm{d}x + e^y\,(x + y + 1)\,\mathrm{d}y = 0.$$

To find the potential function $\psi(x, y)$, we use the equations $\psi_x = e^y$ and $\psi_y = e^y\,(x + y + 1)$. Hence, integrating the former yields $\psi(x, y) = xe^y + h(y)$. The partial derivative of ψ with respect to y is

$$\psi_y = xe^y + h'(y) = (x + y + 1)e^y.$$

From this equation, we get

$$h'(y) = (y + 1)e^y \qquad \Longrightarrow \qquad h(y) = ye^y + C_1,$$

where C_1 is a constant. Substituting back, we have $\psi(x, y) = (x + y)\,e^y + C_1$. We may drop C_1 because the potential function is defined up to an arbitrary constant and therefore is not unique. Recall that $\psi(x, y) = C$ for some constant C defines the general solution in implicit form. Hence, adding a constant C_1 does not change the general form of a solution. This allows us to find the potential function explicitly: $\psi(x, y) = (x + y)e^y$ and the general solution becomes

$$(x + y)e^y = C.$$

From the initial condition $y(0) = 1$, we get the value of the constant C, namely, $C = e$. So $(x + y)\,e^y = e$ is the solution of the given differential equation in implicit form.

We can use line integral method instead since we know the differential of the potential function

$$\mathrm{d}\psi(x, y) = e^y\,\mathrm{d}x + e^y\,(x + y + 1)\,\mathrm{d}y.$$

We choose the path of integration from the initial point $(0, 1)$ to an arbitrary point (x, y) on the plane along the coordinate axis. So

$$\psi(x, y) = \int_0^x e^1\,\mathrm{d}x + \int_1^y e^y\,(x + y + 1)\,\mathrm{d}y = (x + y)e^y - e.$$

To find the solution of the given initial value problem, we just equate $\psi(x, y)$ to zero to obtain $e = (x + y)e^y$.

There is another approach to solving the given initial value problem using substitution (see §2.3). To do this, we rewrite the differential equation as $y' = -(x + y + 1)^{-1}$. We change the dependent variable by setting $v = x + y + 1$. Then $v' = 1 + y'$. Since $y' = -v^{-1}$, we have

$$v' = (x + y + 1)' = 1 - \frac{1}{v} = \frac{v - 1}{v}.$$

Separation of variables and integration yields

$$\int \frac{v}{v - 1}\,\mathrm{d}v = \int \left(1 + \frac{1}{v - 1}\right)\mathrm{d}v = \int \mathrm{d}x = x + C$$

or

$$v + \ln|v - 1| = x + C.$$

We substitute $v = x + y + 1$ back into the above equation to obtain

$$x + y + 1 + \ln|x + y + 1 - 1| = x + C \qquad \text{or} \qquad y + 1 + \ln|x + y| = C.$$

Now we can determine the value of an arbitrary constant C from the initial condition $y(0) = 1$. We set $x = 0$ and $y = 1$ in the equation $y + 1 + \ln|x + y| = C$ to obtain $C = 2$, and, hence,

$$y + \ln|x + y| = 1$$

is the solution of the given initial value problem in implicit form.

Example 2.4.6: Solve the initial value problem

$$3y\,\mathrm{d}x + 2x\,\mathrm{d}y = 0, \quad y(-1) = 2.$$

Solution. For $M = 3y$ and $N = 2x$, we find out that $M_y \neq N_x$ and the equation is not exact. Since $M_y - N_x = 1$, the expression $(M_y - N_x)/M = 1/(3y)$ is a function of y alone and the ratio $(M_y - N_x)/N = 1/(2x)$ is a function of x only. Thus, the given differential equation can be reduced to an exact equation by multiplying it by a suitable integrating factor. We can choose the integrating factor either as a function of x or as a function of y alone. Let us consider the case when the integrating factor is $\mu = \mu(y)$.

Integrating $(M_y - N_x)/M$ with respect to y, we obtain

$$\mu(y) = \exp\left\{-\int \frac{1}{3y}\,\mathrm{d}y\right\} = \exp\left\{\ln|y|^{-1/3}\right\} = |y|^{-1/3}.$$

Multiplication by $\mu(y)$ yields

$$3y|y|^{-1/3}\,\mathrm{d}x + 2x\,|y|^{-1/3}\,\mathrm{d}y = 0.$$

This is an exact equation with the potential function

$$\psi(x,y) = 3xy|y|^{-1/3} = 3xy^{2/3}$$

since the variable y is positive in the neighborhood of 2 (see the initial condition $y(-1) = 2$). Therefore, the general solution of the differential equation is

$$3xy^{2/3} = C \qquad \text{or} \qquad y = C\,|x|^{-3/2}.$$

From the initial condition, we find the value of the constant $C = 2$. Hence, the solution of the initial value problem is $y = 2(-x)^{-3/2}$ for $x < 0$ since

$$|x| = \begin{cases} x, & \text{for positive } x, \\ -x, & \text{for negative } x. \end{cases}$$

On the other hand, the integrating factor as a function of x is

$$\mu(x) = \exp\left\{\int \frac{\mathrm{d}x}{2x}\right\} = \exp\left\{\ln|x|^{1/2}\right\} = |x|^{1/2}.$$

With this in hand, we get the exact equation

$$3y|x|^{1/2}\,\mathrm{d}x + 2x|x|^{1/2}\,\mathrm{d}y = 0.$$

For negative x, we have

$$3y(-x)^{1/2}\,\mathrm{d}x + 2x(-x)^{1/2}\,\mathrm{d}y = 0.$$

The potential function is $\psi(x,y) = 2xy(-x)^{1/2}$ and the general solution (in implicit form) becomes $2xy(-x)^{1/2} = c$. The constant c is determined by the initial condition to be $c = -4$.

Problems

1. Show that the given equations are not exact, but become exact when multiplied by the corresponding integrating factor. Find an integrating factor as a function of x only and determine a potential function for the given differential equations (a and b are constants).

(a)	$y' + y(1+x) = 0$;	**(b)**	$x^3\,y' = xy + x^2$;
(c)	$ay\,\mathrm{d}x + bx\,\mathrm{d}y = 0$;	**(d)**	$5\,\mathrm{d}x - e^{y-x}\,\mathrm{d}y = 0$;
(e)	$(yx^3e^{xy} - 2y^3)\,\mathrm{d}x + (x^4e^{xy} + 3xy^2)\,\mathrm{d}y = 0$;	**(f)**	$\frac{1}{2}y^2\,\mathrm{d}x + (e^x - y)\,\mathrm{d}y = 0$;
(g)	$(x^2y^2 - y)\,\mathrm{d}x + (2x^3y + x)\,\mathrm{d}y = 0$;	**(h)**	$xy^2\,\mathrm{d}y - (x^2 + y^3)\,\mathrm{d}x = 0$;
(i)	$(x^2 - y^2 + y)\,\mathrm{d}x + x(2y-1)\,\mathrm{d}y = 0$;	**(j)**	$(5 - 6y + 2^{-2x})\,\mathrm{d}x = \mathrm{d}y$;
(k)	$(x^2 - 3y)\,\mathrm{d}x + x\,\mathrm{d}y = 0$;	**(l)**	$(e^{2x} + 3y - 5)\,\mathrm{d}x = \mathrm{d}y$.

2. Find an integrating factor as a function of y only and determine the general solution for the given differential equations (a and b are constants).

(a)	$(y+1)\,\mathrm{d}x - (x-y)\,\mathrm{d}y = 0$;	**(b)**	$\left(\frac{y}{x} - 1\right)\mathrm{d}x + \left(2y^2 + 1 + \frac{x}{y}\right)\mathrm{d}y = 0$;
(c)	$ay\,\mathrm{d}x + bx\,\mathrm{d}y = 0$;	**(d)**	$y(x+y+1)\,\mathrm{d}x + x(x+3y+2)\,\mathrm{d}y = 0$;
(e)	$(2xy^2 + y)\,\mathrm{d}x - x\,\mathrm{d}y = 0$;	**(f)**	$2(x+y)\,\mathrm{d}x - x(x+2y-2)\,\mathrm{d}y = 0$;
(g)	$xy^2\,\mathrm{d}x - (x^2 + y^2)\,\mathrm{d}y = 0$;	**(h)**	$(x + xy)\,\mathrm{d}x + (x^2 + y^2 - 3)\,\mathrm{d}y = 0$;
(i)	$(y+1)\,\mathrm{d}x = (2x+5)\,\mathrm{d}y$;	**(j)**	$4y(x^3 - y^5)\,\mathrm{d}x = (3x^4 + 8xy^5)\,\mathrm{d}y$.

2.5 First-Order Linear Differential Equations

Differential equations are among the main tools that scientists use to develop mathematical models of real world phenomena. Most models involve nonlinear differential equations, which are often difficult to solve. It is unreasonable to expect a breakthrough in solving and analyzing nonlinear problems. The transition of a nonlinear problem to a linear problem is called **linearization**. To be more specific, the differential equation $y' = f(x, y)$ is linearized by replacing the slope function $f(x, y)$ with its first-order approximation

$$y' = f(x, y_0) + f_y(x, y_0)\,(y - y_0)\,, \tag{2.5.1}$$

where y is restricted to some small interval containing a value y_0 of particular interest. In practice, y_0 is usually a critical point of the slope function where $f(x, y_0) = 0$. Before computers and corresponding software were readily available (circa 1960), engineers spent much of their time linearizing their problems so they could be solved analytically by hand.

For instance, consider a nonlinear equation $y' = 2y - y^2 - \sin(t)$. Its linear version in a proximity of the origin is obtained by dropping the square y^2 and substituting t instead of $\sin t$ when t is small. We plot in *Maple* the corresponding direction fields and some solutions (see the figures on page 39):

```
ode6:=diff(y(t),t)=2*y(t)-y(t)*y(t)-sin(t);
inc:=[y(0)=0, y(0)=0.3, y(0)=0.298177, y(0)=0.298174];
DEplot(ode6,y(t),inc,t=-1..8,y=-2..2, color=black, linecolor=blue, dirgrid=[25,25]);
```

A **first-order linear differential equation** is an equation of the form

$$a_1(x)\,y'(x) + a_0(x)y(x) = g(x), \tag{2.5.2}$$

where $a_0(x)$, $a_1(x)$, and $g(x)$ are given continuous functions of x on some interval. Values of x for which $a_1(x) = 0$ are called **singular points** of the differential equation. Equation (2.5.2) is said to be **homogeneous** (with accent on "ge") if $g(x) \equiv 0$; otherwise it is **nonhomogeneous** (also called inhomogeneous or driven).

A homogeneous (undriven) linear differential equation $a_1(x)y'(x) + a_0(x)y(x) = 0$ is separable, so it can be solved implicitly. In an interval where $a_1(x) \neq 0$, we divide both sides of Eq. (2.5.2) by the leading coefficient $a_1(x)$ to reduce it to a more useful form, usually called the **standard form** of a linear equation:

$$y' + a(x)y = f(x), \tag{2.5.3}$$

where $a(x) = a_0(x)/a_1(x)$ and $f(x) = g(x)/a_1(x)$ are given continuous functions of x in some interval. As we saw in §1.6 (Theorem 1.1, page 22), the initial value problem for Eq. (2.5.3) always has a unique solution in the interval where both functions $a(x)$ and $f(x)$ are continuous. Moreover, a linear differential equation has no singular solution.

The solutions to Eq. (2.5.3) have the property that they can be written as the sum of two functions: $y = y_h + y_p$, where y_h is the general solution of the associated homogeneous equation $y' + a(x)y = 0$ and y_p is a particular solution of the nonhomogeneous equation $y' + a(x)y = f(x)$. The function y_h is usually referred to as the **complementary function** (which contains an arbitrary constant) to Eq. (2.5.3).

In engineering applications, the independent variable often represents time and conventionally is denoted by t. Then Eq. (2.5.2) can also be rewritten in the form with isolated y:

$$p(t)\,\frac{\mathrm{d}y}{\mathrm{d}t} + y = F(t) \qquad \text{or} \qquad p(t)\dot{y} + y = F(t), \tag{2.5.4}$$

where the term $F(t)$ is called input, and the dependent variable $y(t)$ corresponds to a measure of physical quantity; in particular applications, y may represent a measure of temperature, current, charge, velocity, displacement, or mass. In such circumstances, a particular solution of Eq. (2.5.4) is referred to as output or response.

We shall generally assume that $p(t) \neq 0$, so that the equations (2.5.3) and (2.5.4) are equivalent. If $p(t)$ were 0, then y would be exactly $F(t)$. Therefore, the term $p(t)\dot{y}$ describes an obstacle that prevents y from equating $F(t)$. For instance, consider an experiment of reading a thermometer when it is brought outside from a heated house. The output will not agree with the input (outside temperature) until some transient period has elapsed.

Example 2.5.1: The differential equation

$$y' + y = 5$$

is a particular example of a linear equation (2.5.4) or (2.5.3) in which $a(x) = 1$ and $f(x) = 5$. Of course, this is a separable differential equation, so

$$\frac{\mathrm{d}y}{y-5} = -\mathrm{d}x \qquad \Longrightarrow \qquad \int \frac{\mathrm{d}y}{y-5} = -\int \mathrm{d}x.$$

Then after integration, we obtain

$$\ln|y-5| = -x + \ln C,$$

where C is a positive arbitrary constant. The general solution of the given equation becomes

$$|y-5| = Ce^{-x} \quad \text{or} \quad y - 5 = \pm Ce^{-x}.$$

We can denote the constant $\pm C$ again as C and get the general solution to be $y = 5 + Ce^{-x}$.

There is another way to determine the general solution. Rewriting the given equation in the form $M(y)\,\mathrm{d}x + N\,\mathrm{d}y = 0$ with $M(y) = y - 5$ and $N(x) = 1$, we can find an integrating factor as a function of x:

$$\mu(x) = \exp\left\{\int \frac{M_y - N_x}{N}\,\mathrm{d}x\right\} = \exp\left\{\int \mathrm{d}x\right\} = e^x.$$

Hence, after multiplication by $\mu(x) = e^x$, we get the exact equation

$$e^x(y-5)\,\mathrm{d}x + e^x\,\mathrm{d}y = 0$$

with the potential function $\psi(x,y) = e^x(y-5)$. Therefore, the general solution is $e^x(y-5) = C$ (in implicit form).
□

Generally speaking, Eq. (2.5.3) or Eq. (2.5.4) is neither separable nor exact, but can always be reduced to an exact equation with an integrating factor $\mu = \mu(x)$. There are two methods to solve the linear differential equation: the Bernoulli method and the integrating factor method. We start with the latter.

Integrating Factor Method. In order to reduce the given linear differential equation (2.5.3) to an exact one, we need to find an integrating factor. If in some domain an integrating factor is not zero, the reduced equation will be equivalent to the original one—no solution is lost and no solution is added. Multiplying both sides of Eq. (2.5.3) by a nonzero function $\mu(x)$, we get

$$\mu(x)y'(x) + \mu(x)a(x)y(x) = \mu(x)f(x).$$

Adding and then subtracting the same value $\mu'(x)y(x)$, we obtain

$$\mu(x)y'(x) + \mu'(x)y(x) - \mu'(x)y(x) + \mu(x)a(x)y(x) = \mu(x)f(x).$$

By regrouping terms, we can rewrite the equation in the equivalent form

$$\frac{\mathrm{d}(\mu y)}{\mathrm{d}x} = \left[\mu'(x) - a(x)\mu(x)\right]y(x) + \mu(x)f(x)$$

since $(\mu y)' = \mu y' + \mu' y$ according to the product rule. If we can find a function $\mu(x)$ such that the first term in the right-hand side is equal to zero, that is,

$$\mu'(x) - a(x)\mu(x) = 0, \tag{2.5.5}$$

then we will reduce Eq. (2.5.3) to an exact equation:

$$\frac{\mathrm{d}}{\mathrm{d}x}\left[\mu(x)y(x)\right] = \mu(x)f(x).$$

Excluding singular points where $\mu(x) = 0$, we obtain after integration

$$\mu(x)y(x) = \int \mu(x)f(x)\,\mathrm{d}x + C,$$

where C is a constant of integration and the function $\mu(x)$ is the solution of Eq. (2.5.5): $\mu(x) = \exp\left\{\int a(x)\,\mathrm{d}x\right\}$. It is obvious that $\mu(x)$ is positive for all values of x where $a(x)$ is continuous, so $\mu(x) \neq 0$. Hence, the general solution of the nonhomogeneous linear differential equation (2.5.3) is

$$y(x) = \frac{C}{\mu(x)} + \frac{1}{\mu(x)} \int \mu(x) f(x)\,\mathrm{d}x, \quad \mu(x) = \exp\left\{\int a(x)\,\mathrm{d}x\right\}. \tag{2.5.6}$$

Once we know an integrating factor, we can solve Eq. (2.5.3) by changing the dependent variable

$$w(x) = y(x)\mu(x) = y(x) \exp\left\{\int a(x)\,\mathrm{d}x\right\}.$$

This transforms Eq. (2.5.3) into a separable differential equation

$$w'(x) = y'(x)\mu(x) + a(x)y(x)\mu(x) \quad \text{or} \quad w' = [y' + ay]\mu$$

because

$$\frac{\mathrm{d}}{\mathrm{d}x} \exp\left\{\int a(x)\,\mathrm{d}x\right\} = a(x) \exp\left\{\int a(x)\,\mathrm{d}x\right\}.$$

From Eq. (2.5.3), it follows that $y' + ay = f$; therefore, $w'(x) = f(x)\mu(x)$. This is a separable differential equation having the general solution

$$w(x) = C + \int f(x)\mu(x)\,\mathrm{d}x,$$

which leads to Eq. (2.5.6) for $y(x)$. ■

Bernoulli's[25] Method. We are looking for a solution of Eq. (2.5.3) in the form of the product of two functions: $y(x) = u(x)v(x)$. Substitution into Eq. (2.5.3) yields

$$\frac{\mathrm{d}u}{\mathrm{d}x}\, v(x) + u(x)\, \frac{\mathrm{d}v}{\mathrm{d}x} + a(x)\, u(x)v(x) = f(x).$$

If we choose $u(x)$ so that it is a solution of the homogeneous (also separable) equation $u' + au = 0$, then the first and third terms on the left-hand side drop out leaving an equation easily seen to be separable with respect to v: $u(x)\, v' = f(x)$. After division by u, it can be integrated to give $v(x) = \int f(x)\, u^{-1}(x)\,\mathrm{d}x$. All you have to do is to multiply $u(x)$ and $v(x)$ together to get the solution.

Let us perform all the steps in detail. First, we need to find a solution of the homogeneous linear differential equation (which is actually a separable equation)

$$u'(x) + a(x)\, u(x) = 0. \tag{2.5.7}$$

Since we need just one solution, we may pick $u(x)$ as

$$u(x) = e^{-\int a(x)\,\mathrm{d}x}. \tag{2.5.8}$$

Then $v(x)$ is a solution of the differential equation

$$u(x)\, \frac{\mathrm{d}v}{\mathrm{d}x} = f(x) \qquad \text{or} \qquad \frac{\mathrm{d}v}{\mathrm{d}x} = \frac{f(x)}{u(x)} = f(x)\, e^{\int a(x)\,\mathrm{d}x}.$$

Integrating, we obtain

$$v(x) = C + \int f(x)\, e^{\int a(x)\,\mathrm{d}x}\,\mathrm{d}x$$

with an arbitrary constant C. Multiplication of u and v yields the general solution (2.5.6). ■

If coefficients in Eq. (2.5.2) are constants, then the equation $a_1 y' + a_0 y = g(x)$ has an explicit solution

$$y(x) = \frac{1}{a_1}\, e^{-a_0 x/a_1} \int g(x)\, e^{a_0 x/a_1}\,\mathrm{d}x + C\, e^{-a_0 x/a_1}. \tag{2.5.9}$$

[25] It was first formalized by Johannes/Johann/John Bernoulli (1667–1748) in 1697.

Example 2.5.2: Find the general solution to

$$y' + \frac{y}{x} = x^2, \quad 0 < x < \infty.$$

Solution. **Integrating Factor Method**. An appropriate integrating factor $\mu(x)$ for this differential equation is a solution of $x\,\mu' = \mu$. Separation of variables yields

$$\frac{\mathrm{d}\mu}{\mu} = \frac{\mathrm{d}x}{x}$$

and after integration we get $\mu(x) = x$. Multiplying by $\mu(x)$ both sides of the given differential equation, we obtain

$$x\,y' + y = x^3 \qquad \text{or} \qquad \frac{\mathrm{d}}{\mathrm{d}x}\,(xy) = x^3.$$

Integration yields (with a constant C)

$$xy = \frac{x^4}{4} + C \qquad \text{or} \qquad y = \frac{x^3}{4} + \frac{C}{x}.$$

Bernoulli's Method. Let $u(x) = 1/x$ be a solution of the homogeneous differential equation $u' + u/x = 0$. Setting $y(x) = u(x)v(x) = v(x)/x$, we obtain

$$y' = v'(x)/x - v(x)/x^2 + v(x)/x^2 = x^2 \qquad \text{or} \qquad v'(x) = x^3.$$

Simple integration gives us

$$v(x) = \frac{x^4}{4} + C,$$

where C is a genuine constant. Thus, from the relation $y(x) = v(x)/x$, we get the general solution

$$y(x) = \frac{C}{x} + \frac{x^3}{4}.$$

Example 2.5.3: Consider the initial value problem for a linear differential equation with a discontinuous coefficient

$$y' + p(x)\,y = 0, \qquad y(0) = y_0,$$

where

$$p(x) = \begin{cases} 2, & \text{if } 0 \leqslant x \leqslant 1, \\ 0, & \text{if } x > 1. \end{cases}$$

Solution. The problem can be solved in two steps by combining together two solutions on the intervals $[0, 1]$ and $[1, \infty)$. First, we solve the equation $y' + 2y = 0$ subject to the initial condition $y(0) = y_0$ in the interval $[0, 1]$. Its solution is $y(x) = y_0\,e^{-2x}$. Then we find the general solution of the equation $y' = 0$ for $x \geqslant 1$: $y = C$.

Now we glue these two solutions together in such a way that the resulting function becomes continuous. We choose the constant C from the condition $y_0\,e^{-2} = C$. Hence, the continuous solution of the given problem is

$$y(x) = \begin{cases} y_0\,e^{-2x}, & \text{if } 0 \leqslant x \leqslant 1; \\ y_0\,e^{-2}, & \text{if } 1 \leqslant x < \infty. \end{cases}$$

Example 2.5.4: Let us consider the initial value problem with a piecewise continuous forcing function:

$$y' - 2y = f(x) \equiv \begin{cases} 4x, & x \geqslant 0, \\ 0, & x < 0, \end{cases} \qquad y(0) = \lim_{x \to +0} y(x) = 0.$$

Solution. The homogeneous equation $y' - 2y = 0$ has the general solution

$$y(x) = Ce^{2x} \qquad (-\infty < x < \infty),$$

where C is an arbitrary constant. Now we consider the given equation for positive values of the independent variable x:

$$\phi' - 2\phi = 4x, \qquad x \geqslant 0.$$

Solving the differential equation for an integrating factor, $\mu' + 2\mu = 0$, we get $\mu(x) = e^{-2x}$. Multiplication by $\mu(x)$ reduces the given equation to an exact equation:

$$\mu\phi' - 2\phi\mu = 4x\mu \qquad \text{or} \qquad (\mu\phi)' = 4x\,e^{-2x}.$$

Integration yields

$$\mu(x)\phi(x) = 4\int x\,e^{-2x}\,\mathrm{d}x + K = 4\int_{x_0}^{x} x\,e^{-2x}\,\mathrm{d}x + K,$$

where K is a constant of integration. Here x_0 is a fixed positive number, which can be chosen to be zero without any loss of generality. Indeed,

$$\int_{x_0}^{x} x\,e^{-2x}\,\mathrm{d}x = \int_{x_0}^{0} x\,e^{-2x}\,\mathrm{d}x + \int_{0}^{x} x\,e^{-2x}\,\mathrm{d}x = \int_{0}^{x} x\,e^{-2x}\,\mathrm{d}x + C_1,$$

where $C_1 = \int_{x_0}^{0} x\,e^{-2x}\,\mathrm{d}x$ can be added to an arbitrary constant K:

$$\mu(x)\phi(x) = e^{-2x}\,\phi(x) = 4\int_{0}^{x} x\,e^{-2x}\,\mathrm{d}x + K.$$

To perform integration in the right-hand side (but avoid integration by parts), we consider the auxiliary integral

$$F(k) = \int_0^x e^{kx}\,\mathrm{d}x = \frac{1}{k}\,e^{kx} - \frac{1}{k}$$

for an arbitrary parameter k. Differentiation with respect to k leads to the relation:

$$F'(k) = \frac{\mathrm{d}}{\mathrm{d}k}\left(\frac{1}{k}\,e^{kx} - \frac{1}{k}\right) = \frac{1}{k^2} - \frac{1}{k^2}\,e^{kx} + \frac{x}{k}\,e^{kx} = \frac{\mathrm{d}}{\mathrm{d}k}\int_0^x e^{kx}\,\mathrm{d}x.$$

Changing the order of integration and differentiation, we get

$$\frac{1}{k^2} - \frac{1}{k^2}\,e^{kx} + \frac{x}{k}\,e^{kx} = \int_0^x \frac{\mathrm{d}}{\mathrm{d}k}\,e^{kx}\,\mathrm{d}x = \int_0^x x\,e^{kx}\,\mathrm{d}x.$$

Setting in the latter equation $k = -2$ to fit our problem, we obtain

$$1 - e^{-2x} - 2x\,e^{-2x} = 4\int_0^x x\,e^{-2x}\,\mathrm{d}x.$$

Using this value of the definite integral, we get the solution

$$e^{-2x}\,\phi(x) = 1 - e^{-2x} - 2x\,e^{-2x} + K = (K+1) - e^{-2x} - 2x\,e^{-2x}.$$

Hence, the general solution of the given differential equation becomes

$$y(x) = \begin{cases} (K+1)\,e^{2x} - 1 - 2x, & x \geqslant 0, \\ C\,e^{-2x}, & x < 0. \end{cases}$$

This function is continuous when the limit from the right, $\lim_{x\to+0} y(x) = K + 1 - 1 = K$, is equal to the limit from the left, $\lim_{x\to-0} y(x) = C$. Therefore $C = K = 0$ in order to satisfy the initial condition $y(0) = 0$. This yields the solution

$$y(x) = \begin{cases} e^{2x} - 1 - 2x, & x \geqslant 0, \\ 0, & x < 0. \end{cases}$$

To check our calculation, we use *Maple:*

```
f:=x->piecewise(x>=0,4*x);
dsolve({diff(y(x),x)-2*x=f(x),y(0)=0},y(x))
```

Example 2.5.5: Find a solution of the differential equation

$$(\sin x)\,\frac{dy}{dx} - y\,\cos x = -\frac{2\sin^2 x}{x^3} \qquad (x \neq k\pi, \quad k = 0, \pm 1, \pm 2, \ldots)$$

such that $y(x) \to 0$ as $x \to \infty$.

Solution. Using Bernoulli's method, we seek a solution as the product of two functions: $y(x) = u(x)\,v(x)$. Substitution into the given equation yields

$$\frac{du}{dx}\,v(x)\sin x + \frac{dv}{dx}\,u(x)\,\sin x - u(x)v(x)\,\cos x = -\frac{2\sin^2 x}{x^3}.$$

If $u(x)$ is a solution of homogeneous equation $u'\,\sin x - u\,\cos x = 0$, say $u(x) = \sin x$, then $v(x)$ must satisfy the following equation:

$$\frac{dv}{dx}\,\sin^2 x = -\frac{2\sin^2 x}{x^3} \qquad \text{or} \qquad \frac{dv}{dx} = -\frac{2}{x^3}.$$

The function $v(x)$ is not hard to determine by simple integration. Thus, $v(x) = x^{-2} + C$ and

$$y(x) = u(x)\,v(x) = C\,\sin x + \frac{\sin x}{x^2}.$$

The first term on the right-hand side does not approach zero as $x \to \infty$; therefore, we have to set $C = 0$ to obtain $y(x) = \sin x / x^2$ (see Fig. 2.25).

Example 2.5.6: Let the function $f(x)$ approach zero as $x \to \infty$ in the linear differential equation $y' + a(x)\,y(x) = f(x)$ with $a(x) \geqslant c > 0$ for some positive constant c. Show that every solution of this equation goes to zero as $x \to \infty$.

Solution. Multiplying both sides of this equation by an integrating factor yields

$$\frac{d}{dx}\,[\mu(x)y(x)] = \mu(x)f(x),$$

where $\ln\mu(x) = a(x)$. Hence,

$$y(x)\,\exp\left\{\int_{x_0}^{x} a(t)\,dt\right\} - y_0 = \int_{x_0}^{x} \exp\left\{\int_{x_0}^{\tau} a(t)\,dt\right\} f(\tau)\,d\tau,$$

and

$$y(x) = \exp\left\{-\int_{x_0}^{x} a(t)\,dt\right\} \left[y_0 + \int_{x_0}^{x} \exp\left\{\int_{x_0}^{\tau} a(t)\,dt\right\} f(\tau)\,d\tau\right].$$

The first term vanishes as $x \to \infty$ because

$$\exp\left\{-\int_{x_0}^{x} a(t)\,dt\right\} \leqslant \exp\left\{-\int_{x_0}^{x} c\,dx\right\} = \exp\{-c(x - x_0)\} \quad \text{for all } x.$$

We can estimate the second term as follows.

$$\begin{aligned}
I &\stackrel{\text{def}}{=} \left|\exp\left\{-\int_{x_0}^{x} a(t)\,dt\right\} \int_{x_0}^{x} \exp\left\{\int_{x_0}^{\tau} a(t)\,dt\right\} f(\tau)\,d\tau\right| \\
&= \left|\int_{x_0}^{x} \exp\left\{-\int_{x_0}^{x} a(t)\,dt + \int_{x_0}^{\tau} a(t)\,dt\right\} f(\tau)\,d\tau\right| \\
&= \left|\int_{x_0}^{x} \exp\left\{-\int_{\tau}^{x} a(t)\,dt\right\} f(\tau)\,d\tau\right| \leqslant \int_{x_0}^{x} \exp\{-c(x-\tau)\}\,|f(\tau)|\,d\tau \\
&= \int_{x_0}^{x/2} \exp\{-c(x-\tau)\}\,|f(\tau)|\,d\tau + \int_{x/2}^{x} \exp\{-c(x-\tau)\}\,|f(\tau)|\,d\tau \\
&\leqslant \max_{x_0 \leqslant t \leqslant x/2} |f(\tau)|\,\frac{1}{c}\,\exp\{-c(x-\tau)\}\Big|_{\tau=x_0}^{\tau=x/2} + \max_{x/2 \leqslant t \leqslant x} |f(\tau)|\frac{1}{c}\,\exp\{-c(x-\tau)\}\Big|_{\tau=x/2}^{\tau=x} \\
&= \max_{x_0 \leqslant t \leqslant x/2} |f(\tau)|\,\frac{1}{c}\,e^{-x/2}\left[1 - e^{-c(x/2 - x_0)}\right] + \frac{1}{c}\max_{x/2 \leqslant t \leqslant x} |f(\tau)|\,\left[1 - e^{-cx/2}\right] \to 0
\end{aligned}$$

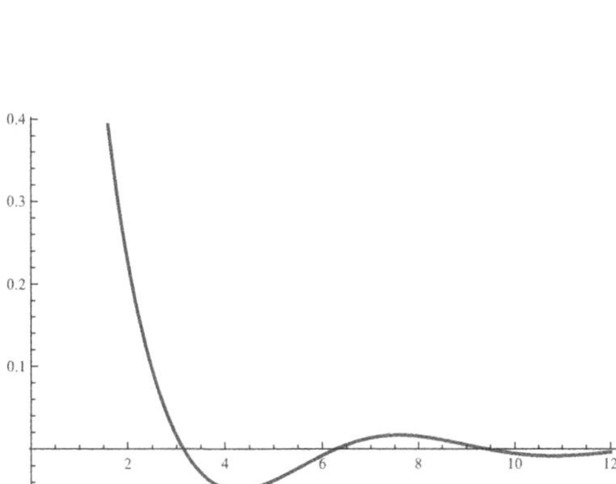

Figure 2.25: Example 2.5.5. A solution plotted with *Mathematica*.

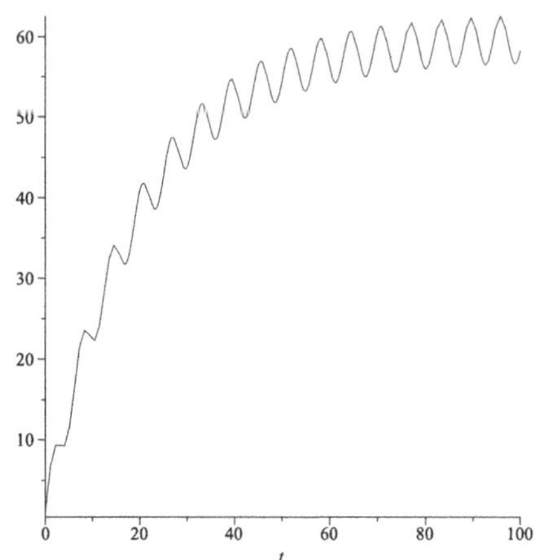

Figure 2.26: Example 2.5.7. The solution of Eq. (2.5.10) when $a = 0.05$, $A = 3$, and $p(0) = 1$.

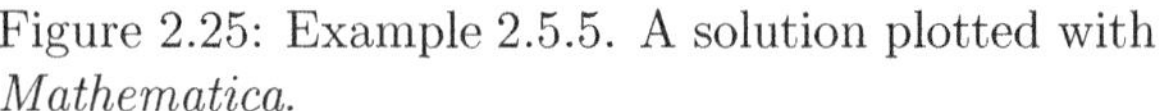

as $x \to \infty$ since $\left[1 - e^{-c(x/2 - x_0)}\right]$ and $\max_{x_0 \leqslant t \leqslant x/2} |f(\tau)|$ are bounded and $\displaystyle\max_{x/2 \leqslant t \leqslant x} |f(\tau)| \to 0$. Thus, both terms approach zero as $x \to \infty$ and therefore $y(x)$ also approaches zero.

Example 2.5.7: (Pollution model) Let us consider a large lake formed by damming a river that initially holds 200 million (2×10^8) liters of water. Because a nearby chemical plant uses the lake's water to clean its reservoirs, 1,000 liters of brine, each containing $100(1 + \cos t)$ kilograms of dissolved pollution, run into the lake every hour. Let's make the simplifying assumption that the mixture, kept uniform by stirring, runs out at the rate of 1,000 liters per hour and no additional spraying causes the lake to become even more contaminated. Find the amount of pollution $p(t)$ in the lake at any time t.

Solution. The time rate of pollution change $\dot{p}(t)$ equals the inflow rate minus the outflow rate. The inflow of pollution is given, which is $100(1 + \cos t)$. We determine the outflow, which is equal to the product of the concentration and the output rate of mixture. The lake always contains $200 \times 10^6 = 2 \times 10^8$ liters of brine. Hence, $p(t)/(2 \times 10^8)$ is the pollution content per liter (concentration), and $1000\, p(t)/(2 \times 10^8)$ is the pollution content in the outflow per hour. The time rate of change $\dot{p}(t)$ is the balance:

$$\dot{p}(t) = 100(1 + \cos t) - 1000\, p(t)/(2 \times 10^8)$$

or

$$\dot{p}(t) + a\, p(t) = A\,(1 + \cos t), \tag{2.5.10}$$

where $a = 5 \times 10^{-6}$ and $A = 100$. This is a nonhomogeneous linear differential equation. With the aid of the integrating factor $\mu(t) = e^{a\,t}$, we find the general solution of Eq. (2.5.10) to be

$$p(t) = p(0)\, e^{-at} + A\, \frac{a\, \cos t + \sin t}{a^2 + 1} + A\, \frac{1 + a^2}{a(a^2 + 1)} - A\, \frac{2a^2 + 1}{a(a^2 + 1)}\, e^{-at}.$$

Example 2.5.8: (Iodine radiation) Measurements showed the ambient radiation level of radioactive iodine in the Chernobyl area (in Russia) after the nuclear disaster to be five times the maximum acceptable limit. These radionuclides tend to decompose into atoms of a more stable substance at a rate proportional to the amount of radioactive iodine present. The proportionality coefficient, called the **decay constant**, for radioactive iodine is about 0.004. How long will it take for the site to reach an acceptable level of radiation?

Solution. Let $Q(t)$ be the amount of radioactive iodine present at any time t and Q_0 be the initial amount after the nuclear disaster. Thus, Q satisfies the initial value problem

$$\frac{\mathrm{d}Q}{\mathrm{d}t} = -0.004\, Q(t), \quad Q(0) = Q_0.$$

Since this equation is a linear homogeneous differential equation and also separable, its general solution is $Q(t) = C\,e^{-0.004t}$, where C is an arbitrary constant. The initial condition requires that $C = Q_0$.

We are looking for the period of time t when $Q(t) = Q_0/5$. Hence, we obtain the equation

$$Q_0\,e^{-0.004t} = \frac{Q_0}{5} \qquad \text{or} \qquad e^{-0.004t} = \frac{1}{5}.$$

Its solution is $t = 250\ \ln 5 \approx 402$ years.

Problems

1. Find the general solution of the following linear differential equations with constant coefficients. (Recall that $\sinh x = 0.5\,e^{x} - 0.5\,e^{-x}$ and $\cosh x = 0.5\,e^{x} + 0.5\,e^{-x}$).
   (a) $y' + 4y = 17\ \sin x$; (b) $y' + 4y = 2\,e^{-2x}$; (c) $y' + 4y = e^{-4x}$;
   (d) $y' - 2y = 4$; (e) $y' - 2y = 2 + 4x$; (f) $y' - 2y = 3\,e^{-x}$;
   (g) $y' - 2y = e^{2x}$; (h) $y' - 2y = 5\ \sin x$; (i) $y' + 2y = 4$;
   (j) $y' + 2y = 4\,e^{2x}$; (k) $y' + 2y = e^{-2x}$; (l) $y' + 2y = 3\ \cosh x$;
   (m) $y' + 2y = 3\ \sinh x$; (n) $y' - y = 4\ \sinh x$; (o) $y' - y = 4\ \cosh x$;
   (p) $y' = 2y + x^2 + 3$.
2. Find the general solution of the following linear differential equations with variable coefficients.
   (a) $y' + xy = x$; (b) $x\,y' + (3x+1)\,y = e^{-3x}$; (c) $x^2y' + xy = 1$;
   (d) $x\,y' + (2x+1)\,y = 4x$; (e) $(1-x^2)\,y' - xy + x(1-x^2) = 0$; (f) $y' + (\cot x)\,y = x$;
   (g) $x \ln x\,y' + y = 9x^3\ \ln x$; (h) $(2 - y\ \sin x)\,dx = \cos x\,dy$; (i) $(\cos x)^2\,y' + y = \tan x$.
3. Use the Bernoulli method to solve the given differential equations.
   (a) $x\,y' = y + x^2\,e^x$; (b) $y' + 2x\,y = 4x$;
   (c) $y' = (y-1)\ \tan x$; (d) $(1+x)\,y' = xy + x^2$;
   (e) $x\ \ln x\,y' + y = x$; (f) $(1+x^2)\,y' + (1-x^2)\,y - 2x\,e^{-x} = 0$;
   (g) $y' = y/x + 4x^2\ \ln x$; (h) $y' + 2xy/(1-x^2) + 4x = 0\ (|x| < 1)$.
4. Solve the given initial value problems.
   (a) $y' + 2y = 10,\quad y(0) = 8$; (b) $x^2y' + 2xy - x + 1 = 0,\quad y(1) = 0$;
   (c) $y' = y + 6x^2,\quad y(0) = -2$; (d) $x^2\,y' + 4xy = 2\,e^{-x},\quad y(1) = 1$;
   (e) $x\,y' = y + 2x^2,\quad y(5) = 5$; (f) $x\,y' + (x+2)\,y = 2\sin x,\quad y(\pi) = 1$;
   (g) $xy' + 2y = x,\quad y(3) = 1$; (h) $xy' + 2y = 2\sin x,\quad y(\pi) = 1$;
   (i) $y' + y = e^x,\quad y(0) = 1$; (j) $x^2\,y' - 4xy = 4x^3,\quad y(1) = 1$;
   (k) $y' + y = e^{-x},\quad y(0) = 1$; (l) $y' + y/(x^2-1) = 1 + x,\quad y(0) = 1$.
5. Find a continuous solution of the following initial value problems.
   (a) $y' + 2y = f_k(x),\quad y(0) = 0$; (b) $y' + y = f_k(x),\quad y(0) = 0$;
   (c) $y' - y = f_k(x),\quad y(0) = 1$; (d) $y' - 2y = f_k(x),\quad y(0) = 1$;

   for each of the following functions ($k = 1, 2$):

$$f_1(x) = \begin{cases} 1, & x \leqslant 3, \\ 0, & x > 3; \end{cases} \qquad f_2(x) = \begin{cases} 1, & x \leqslant 1, \\ 2-x, & x > 1. \end{cases}$$

6. Show that the integrating factor $\mu(x) = K\,e^{\int a(x)\,dx}$, where $K \neq 0$ is an arbitrary constant, produces the same general solution.
7. Find a general solution of a linear differential equation of the first-order when two solutions of this equation are given.
8. Which nonhomogeneous linear ordinary differential equations of first-order are separable?
9. One of the main contaminants of the nuclear accident at Chernobyl is strontium-90, which decays at a constant rate of approximately 2.47% per year. What percent of the original strontium-90 will still remain after 100 years?
10. In a certain RL-series circuit, $L = 1 + t^2$ henries, $R = 1 - t$ ohms, $V(t) = t$ volts, and $I(0) = 1$ ampere. Compute the value of the current at any time.
11. In a certain RL-series circuit, $L = t$ henries, $R = 2t + 1$ ohms, $V(t) = 4t$ volts, and $I(1) = 2$ amperes. Compute the value of the current at any time.
12. Find the charge, $q(t)$, $t \geqslant 1$, in a simple RC-series circuit with electromotive force $E(t) = t$ volts. It is given that $R = t$ ohms, $C = (1+t)^{-1}$ farads, and $q(1) = 1$ coulomb.

13. An electric circuit, consisting of a capacitor, resistor, and an electromotive force (see [14] for details) can be modeled by the differential equation $R\,\dot{q} + \frac{1}{C}\,q = E(t)$, where R (resistance) and C (capacitance) are constants, and $q(t)$ is the amount of charge on the capacitor at time t. If we introduce new variables $q = CQ$ and $t = RC\tau$, then the differential equation for $Q(\tau)$ becomes $Q' + Q = E(\tau)$. Assuming that the initial charge on the capacitor is zero, solve the initial value problems for $Q(\tau)$ with the following piecewise electromotive forces.

   (a) $E(\tau) = \begin{cases} 1, & 0 < \tau < 2, \\ 0, & 2 < \tau < \infty; \end{cases}$ (b) $E(\tau) = \begin{cases} \tau, & 0 < \tau < 1, \\ 1/\tau, & 1 < \tau < \infty; \end{cases}$

   (c) $E(\tau) = \begin{cases} 0, & 0 < \tau < 2, \\ 2, & 2 < \tau < \infty; \end{cases}$ (d) $E(\tau) = \begin{cases} \tau, & 0 < \tau < 1, \\ 1, & 1 < \tau < \infty. \end{cases}$

14. Solve the initial value problems and estimate the initial value a for which the solution transitions from one type of behavior to another (such a value of a is called the **critical value**).

   (a) $y' - y = 10\,\sin(3x)$, $y(0) = a$; (b) $y' - 3y = 5\,e^{2x}$, $y(0) = a$.

15. Solve the initial value problem and determine the critical value of a for which the solution behaves differently as $t \to 0$.

   (a) $t^2\,\dot{y} + t(t-2)\,y + 3\,e^{-t}$, $y(1) = a$; (b) $(\sin t)\,\dot{y} + (\cos t)y = e^{2t/\pi}$, $y(\frac{\pi}{2}) = a$;

   (c) $t\dot{y} + y = t\cos t$, $y(\pi) = a$; (d) $\dot{y} + y\,\tan t = \cos t/(t+1)$, $y(0) = a$.

16. Find the value of a for which the solution of the initial value problem $y' - 2y = 10\,\sin(4x)$, $y(0) = a$ remains finite as $t \to \infty$.

17. Solve the following problems by considering x as a dependent variable.

   (a) $\dfrac{dy}{dx} = \dfrac{y}{x + y^2}$; (b) $\dfrac{dy}{dx} = \dfrac{y - y^2}{x}$; (c) $\dfrac{dy}{dx} = \dfrac{y^2}{2x+1}$.

18. Solve the initial value problem: $\dot{y} = ay + \int_0^b y(t)\,dt$, $y(0) = c$, where a, b, and c are constants. *Hint:* Set $k = \int_0^b y(t)\,dt$, solve for y in terms of k, and then determine k.

19. *CAS Problem.* Consider the initial value problem

$$x\,y' + 2y = \sin x, \qquad y(\pi/2) = 1.$$

   (a) Find its solution with a computer algebra system.

   (b) Graph the solution on the intervals $0 < x \leqslant 2$, $1 \leqslant x \leqslant 10$, and $10 \leqslant x \leqslant 100$. Describe the behavior of the solution near $x = 0$ and for large values of x.

20. A tank contains 100 liters of pure water. Brine with 30 grams of salt per liter flows in at the rate of 2 liters per minute. The thoroughly stirred mixture then flows out at the rate of 3 liters per minute.

   (a) Find the amount of salt in the tank when the brine in it has been reduced to 50 liters.

   (b) When is the amount of salt in the tank greatest?

21. Consider the ordinary differential equation $y' = e^{2y}/2(y - x\,e^{2x})$. None of the methods developed in the preceding discussion can be used for solving this equation. However, by considering x as a function of y, it can be rewritten in the form $dx/dy = -2x + 2y\,e^{-2y}$, which is a linear equation. Find its general solution.

22. Suppose that the moisture loss of a wet sponge in open air is directly proportional to the moisture content; that is, the rate of change of the moisture M, with respect to the time t, is proportional to M. Write an ODE modeling this situation. Solve for $M(t)$ as a function of t given that $M = 1,000$ for $t = 0$ and that $M = 100$ when $t = 10$ hours.

23. According to an *absorption law*[26] derived by Lambert, the rate of change of the amount of light in a thin transparent layer with respect to the thickness h of the layer is proportional to the amount of light on the layer. Formulate and solve a differential equation describing this law.

In each of Problems 14 through 19, determine the long-term behavior as $t \mapsto +\infty$ of solutions to the given linear differential equation.

14. $\dot{y} + 2y = e^{-2t}\,\sin t$;
15. $\dot{y} + 2y = e^{2t}\,\cos t$;
16. $\dot{y} + 2y = e^{2t}\,\sin t$;
17. $t\,\dot{y} + 2y = 4t$;
18. $\dot{y} + 2ty = 2t$;
19. $\dot{y} + 2ty = t + t^2$.

[26] Johann Heinrich Lambert (1728–1777) was a famous French/Prussian scientist who published this law in 1760, although Pierre Bouguer discovered it earlier. By the way, Lambert also made the first systematic development of hyperbolic functions. A few years earlier they had been studied by the Italian mathematician and physicist Vincenzo Riccati (1707–1775).

2.6 Special Classes of Equations

This section contains more advanced material, which discusses the issue of reducing given differential equations to simpler ones. We present several cases when such simplification is possible. Other methods are given in §4.1.

When the solution of a differential equation is expressed by a formula involving one or more integrals, it is said that the equation is solvable[27] by *quadrature*, and the corresponding formula is called a *closed-form solution*. So far, we have discussed methods that allow us to find (explicitly or implicitly) solutions of differential equations in a closed form. Only a few types of differential equations can be solved by quadrature.

2.6.1 The Bernoulli Equation

Certain nonlinear differential equations can be reduced to linear or separable forms. The most famous of these is the **Bernoulli[28] equation:**

$$y' + p(x)y = g(x)y^{\alpha}, \quad \alpha \text{ is a real number.} \tag{2.6.1}$$

If $\alpha = 0$ or $\alpha = 1$, the equation is linear. Otherwise, it is nonlinear and it always has the equilibrium solution $y = 0$. We present two methods to solve Eq. (2.6.1): substitution (discovered by Leibniz in 1695) and the Bernoulli method. We start with the former.

1. (Leibniz substitution) The Bernoulli equation (2.6.1) can be reduced to a linear one by setting

$$u(x) = [y(x)]^{1-\alpha}. \tag{2.6.2}$$

We differentiate both sides and substitute $y' = gy^{\alpha} - py$ from Eq. (2.6.1) to obtain

$$\begin{aligned} u' &= (1-\alpha)y^{-\alpha}y' = (1-\alpha)y^{-\alpha}(g(x)y^{\alpha} - p(x)y) \\ &= (1-\alpha)(g(x) - p(x)y^{1-\alpha}). \end{aligned}$$

Since $y^{1-\alpha} = u$, we get the linear differential equation for u:

$$u' + (1-\alpha)p(x)u = (1-\alpha)g(x).$$

2. (Bernoulli method) This equation (2.6.1) can also be solved using the Johann's Bernoulli method (1697). Suppose there exists a solution of the form

$$y(x) = u(x)\,v(x), \tag{2.6.3}$$

where $u(x)$ is a solution of the linear homogeneous equation (which is also separable)

$$u'(x) + p(x)\,u(x) = 0 \qquad \Longrightarrow \qquad u(x) = \exp\left\{-\int p(x)\,dx\right\}.$$

Substituting $y = uv$ leads to the separable equation with respect to $v(x)$:

$$u(x)\,v'(x) = g(x)\,u^{\alpha}(x)\,v^{\alpha}(x) \qquad \text{or} \qquad \frac{dv}{v^{\alpha}} = g(x)\,u^{\alpha-1}(x)\,dx.$$

Integration yields

$$\int \frac{dv}{v^{\alpha}} = \frac{1}{1-\alpha}\left(v^{1-\alpha} - C\right) = \int g(x)\,u^{\alpha-1}(x)\,dx.$$

Thus,

$$y(x) = \exp\left\{-\int p(x)\,dx\right\}\left[C + (1-\alpha)\int g(x)\,u^{\alpha-1}(x)\,dx\right]^{1/(1-\alpha)}.$$

■

Example 2.6.1: (Logistic equation) A particular Bernoulli equation of the form

$$y' - Ay = -By^2 \qquad \text{or} \qquad y' = y\,(A - By),$$

[27] The term *quadrature* has its origin in plane geometry.

[28] This equation was proposed for solution in 1695 by Jakob (=Jacobi = Jacques = James) Bernoulli (1654–1705), a Swiss mathematician from Basel, where he was a professor of mathematics until his death.

where A and B are positive constants, is called the **Verhulst equation** or the **logistic equation**. This equation has many applications in population models, which are discussed in detail in [14]. This is a Bernoulli equation with $\alpha = 2$; hence, we reduce the logistic equation to the linear equation by setting $u = y^{-1}$. Then we obtain

$$u' + Au = B.$$

Its general solution is $u = C_1\, e^{-Ax} + B/A$, where C_1 is an arbitrary constant, so

$$y(x) = \frac{1}{(B/A) + C_1\, e^{-Ax}} = \frac{A}{B + C\, e^{-Ax}},$$

where $C = AC_1$ is an arbitrary constant. On the other hand, using the Bernoulli method, we seek a solution of the logistic equation as the product $y = uv$, where u is a solution of the linear equation $u' = Au$ (so $u = e^{Ax}$) and v is the solution of the separable equation $v' = -Buv^2$.

A similar equation $\dot{y} + a(t)\, y = b(t)\, y^2$ occurs in various problems in solid mechanics, where it is found to describe a propagation of acceleration waves in nonlinear elastic materials [7].

We can plot some selected solutions with *Mathematica*.

```
curves = Flatten[Table[{A/(B + C*Exp[-A*x])}, {C, 1/4, 5/2, 1/2}]]
Plot[Evaluate[curves], {x, 0, 4}, PlotRange -> All]
```

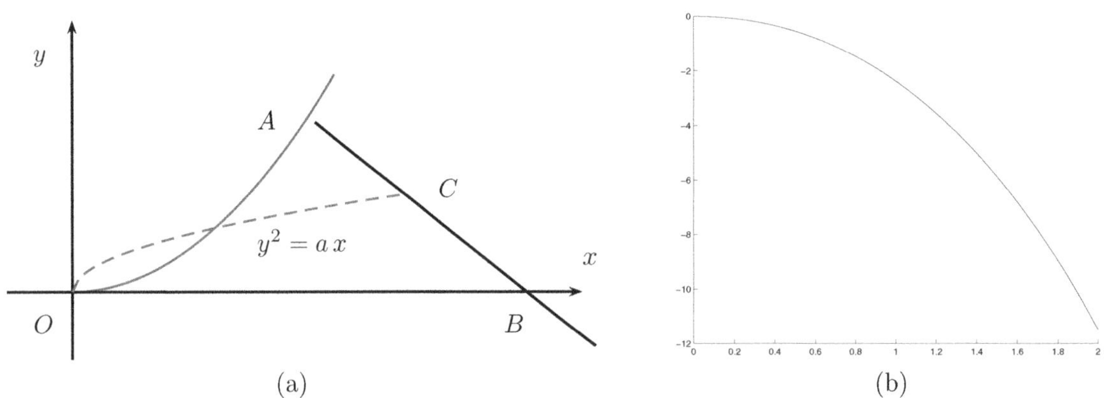

Figure 2.27: (a) Example 2.6.2 and (b) its solution, plotted with MATLAB.

Example 2.6.2: Let us consider the curve $y = y(x)$ that goes through the origin (see Fig. 2.27). Let the middle point of the normal to an arbitrary point on the curve and horizontal line (abscissa) belong to parabola $y^2 = ax$. Find the equation of the curve.

Solution. The slope of the tangent line to the curve is $y'(x)$ at any point x. The slope of the normal line is $k = -1/y'(x)$. With this in hand, we find the equation of the normal $y = kx + b$ that goes through an arbitrary point $A(x, y)$ on the curve and crosses the abscissa at $B(X, 0)$. Since X belongs to the normal, we obtain

$$X = -\frac{b}{k} \quad \text{from the equation} \quad 0 = kX + b.$$

Substituting $y - kx$ instead of b yields

$$X = -\frac{y - kx}{k} = x - \frac{y}{k} = x + y \cdot y'$$

because $k = -1/y'$. The middle point C of the segment AB has coordinates (X_1, Y_1) that are mean values of the corresponding coordinates $A(x, y)$ and $B(x + y\,y', 0)$, that is,

$$X_1 = \frac{x + x + y\,y'}{2} = x + \frac{y}{2}\,y' \quad \text{and} \quad Y_1 = \frac{y}{2}.$$

The coordinates of the point C satisfy the equation of the parabola $y^2 = ax$; thus,

$$Y_1^2 = \left(\frac{y}{2}\right)^2 = aX_1 \qquad \text{or} \qquad \frac{\mathrm{d}y}{\mathrm{d}x} - \frac{y}{2a} = -\frac{2x}{y}.$$

This is a Bernoulli equation (2.6.1) with

$$p = -\frac{1}{2a}, \quad g(x) = -2x, \quad \alpha = -1.$$

Setting $u(x) = y^{1-\alpha} = y^2$, we obtain the linear differential equation with constant coefficients

$$u'(x) - \frac{1}{a}u = -4x.$$

From Eq. (2.5.9), page 88, it follows that

$$y^2(x) = u(x) = 4ax + 4a^2 + C\,e^{x/2}.$$

Since the curve goes through the origin, we have $0 = 4a^2 + C$. Therefore, the required equation of the curve becomes

$$y^2(x) = 4ax + 4a^2\left(1 - e^{x/2}\right).$$

To check our answer, we ask *Mathematica:*

```
BerEq[x_, y_] := D[y[x], x] == -2*x*y[x]^(-1)  + y[x]/(2*a)
alpha = -1
ypRule = Flatten[Solve[u'[x] == D[y[x]^(1 - alpha), x], y'[x]]]
yRule = {y[x] -> u[x]^(1/(1 - alpha))}
BerEq[x, y] /. ypRule /. yRule // Simplify
term = PowerExpand[%]
NewLHS = Simplify[term[[1]]/u[x]^(alpha/(1 - alpha))]
uRule = DSolve[NewLHS == 0, u[x], x]
yRule /.uRule // Flatten
```

Example 2.6.3: Solve $xy' + y = y^2 \qquad (x \ne 0)$.

Solution. This equation is not linear because it contains y^2. It is, however, a Bernoulli differential equation of the form in Eq. (2.6.1), with $p(x) = g(x) = 1/x$ and $\alpha = 2$. Instead of making the Leibniz substitution suggested by (2.6.2), we apply the Bernoulli method by setting $y(x) = u(x)v(x)$, where $u(x)$ is a solution of the linear differential equation $xu' + u = 0$. This is a separable (and exact) equation $\mathrm{d}u/u = -\mathrm{d}x/x$; hence, we may take $u(x) = 1/x$. Therefore,

$$y(x) = \frac{v(x)}{x} \qquad \text{and} \qquad y'(x) = \frac{v'}{x} - \frac{v(x)}{x^2}, \qquad x \ne 0.$$

Substituting these formulas into the given differential equation, we get $v' = x^{-2}v^2$. This is again a separable equation with the general solution

$$v(x) = \frac{x}{1 + Cx} \qquad \text{and} \qquad y(x) = \frac{1}{1 + Cx},$$

with a constant of integration C. We check our answer using *Mathematica:*

```
Urule = Flatten[DSolve[x u'[x] + u[x] == 0, u[x], x]] /. C[1] -> 1
Vrule = Flatten[DSolve[x u[x] v'[x] == u[x]^2 v[x]^2 /. Urule, v[x], x]]
y[x_] = u[x] v[x] /. Urule /. Vrule
```

Example 2.6.4: Solve $xy' + y = y^2 \ln x$.

Solution. This is a Bernoulli equation (2.6.1) with $\alpha = 2$. By making the substitution (2.6.2) $u(x) = [y(x)]^{-1}$, we get

$$y(x) = \frac{1}{u(x)} \quad \text{and} \quad y'(x) = -\frac{u'}{u^2}.$$

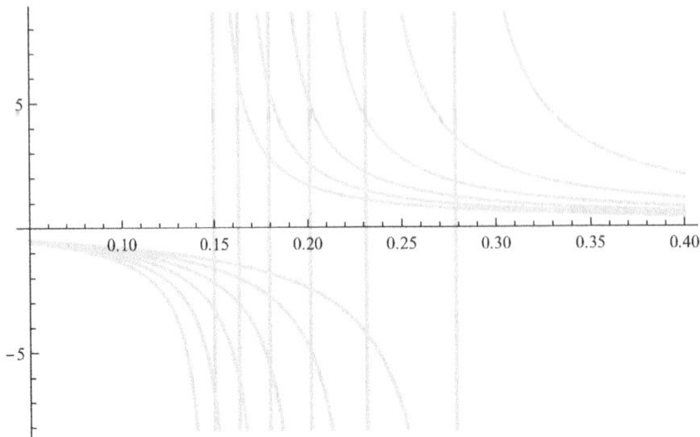

Figure 2.28: Family of solutions from Example 2.6.4, plotted with *Mathematica*.

A direct substitution of $y = u^{-1}$ and $y' = -u^{-2}u'$ into the given differential equation yields

$$-x\,\frac{u'}{u^2(x)} + \frac{1}{u(x)} = \frac{\ln x}{u^2(x)} \qquad \text{or} \qquad -\,x\,u' + u = \ln x.$$

This is a linear differential equation; thus,

$$-x^2\frac{\mathrm{d}}{\mathrm{d}x}\left(\frac{u}{x}\right) = \ln x \qquad \text{or} \qquad \frac{\mathrm{d}}{\mathrm{d}x}\left(\frac{u}{x}\right) = -\frac{\ln x}{x^2}.$$

Integrating both sides by parts, we obtain

$$\frac{u}{x} = -\int \frac{\ln x}{x^2}\,\mathrm{d}x = \int \ln x\,\mathrm{d}(x^{-1}) = \frac{\ln x}{x} - \int \frac{\mathrm{d}x}{x^2} = \frac{\ln x}{x} + \frac{1}{x} + C.$$

Therefore, $u(x) = \ln x + 1 + Cx$, and the solution of the given differential equation becomes

$$y(x) = \frac{1}{\ln x + 1 + Cx}.$$

Example 2.6.5: Solve the Bernoulli equation

$$y' + ay = y^2\,e^{bx}.$$

Solution. We use the Bernoulli substitution (2.6.3) by setting $y(x) = u(x)v(x)$, where $u(x) = e^{-ax}$ is a solution of the homogeneous linear differential equation $u' + au = 0$. Substitution of the derivative $y'(x) = (uv)' = -a\,e^{-ax}\,v(x) + e^{-ax}\,v'(x)$ into the equation leads to

$$e^{-ax}\,v'(x) = e^{bx}\,e^{-2ax}\,v^2(x) \qquad \text{or} \qquad v' = v^2\,e^{(b-a)x}.$$

This is a separable differential equation, and we have

$$\frac{\mathrm{d}v}{v^2} = e^{(b-a)x}\,\mathrm{d}x.$$

Integrating, we obtain

$$\frac{1}{v} = \begin{cases} \frac{C+e^{(b-a)x}}{a-b} & \text{if } a \ne b, \\ -x + C & \text{if } a = b. \end{cases}$$

Therefore,

$$y(x) = \begin{cases} \frac{(a-b)\,e^{-ax}}{C+e^{(b-a)x}} = \frac{a-b}{C\,e^{ax}+e^{bx}} & \text{if } a \ne b, \\ \frac{e^{-ax}}{C-x} & \text{if } a = b. \end{cases}$$

2.6.2 The Riccati Equation

Let us consider a first-order differential equation in the normal form: $y' = f(x,y)$. If we approximate $f(x,y)$, while x is kept constant, we get

$$f(x,y) = P(x) + yQ(x) + y^2R(x) + \cdots .$$

If we stop at y^2 and drop higher powers of y, we will have

$$y' + p(x)y = g(x)y^2 + h(x). \tag{2.6.4}$$

This equation is called the **Riccati**[29] **equation**. It occurs frequently in dynamic programming, continuous control processes, the Kolmogorov–Wiener theory of linear prediction, and many physical applications. The reciprocal of any solution of Eq. (2.6.4), $v = 1/y$, is a solution of the associated Riccati equation

$$v' - p(x)v = -g(x) - h(x)v^2. \tag{2.6.5}$$

When $h(x) \equiv 0$, we have a Bernoulli equation. The Riccati equation has much in common with linear differential equations; for example, it has no singular solution. Except special cases that we will discuss later, the Riccati equation cannot be solved analytically using elementary functions or quadratures, and the most common way to obtain its solution is to represent it in series. Moreover, the Riccati equation can be reduced to the second order linear differential equation by substitution:

$$u(x) = \exp\left\{-\int g(x)y(x)\,\mathrm{d}x\right\} \quad \text{or} \quad y(x) = -\frac{u'(x)}{g(x)u(x)}.$$

Indeed, the derivatives of $u(x)$ are

$$u'(x) = -g(x)y(x)\,\exp\left\{-\int g(x)y(x)\,\mathrm{d}x\right\} = -g(x)y(x)u(x),$$
$$u''(x) = -g'(x)\,y(x)u(x) - g(x)\,y'(x)\,u(x) - g(x)y(x)\,u'(x).$$

Since

$$u' = -gyu \quad \text{and} \quad y' = -py + gy^2 + h,$$

we have

$$u''(x) = -g'(x)\,y(x)u(x) - g(x)\left[-p(x)y(x) + g(x)y^2(x) + h(x)\right]u(x) + g^2(x)y^2(x)u(x)$$

and

$$u''(x) = -g'(x)\,y(x)u(x) + g(x)y(x)p(x)u(x) - g(x)h(x)u(x).$$

Eliminating the product $yu = -u'/g$ from the above equation, we obtain

$$u'' + a(x)\,u'(x) + b(x)\,u(x) = 0, \tag{2.6.6}$$

where

$$a(x) = p(x) - \frac{g'(x)}{g(x)}, \quad b(x) = g(x)h(x). \tag{2.6.7}$$

This linear second order homogeneous differential equation (2.6.6) may sometimes be easier to solve than the original one (2.6.4).

Conversely, every linear homogeneous differential equation (2.6.6) with variable coefficients can be reduced to the Riccati equation

$$y' + y^2 + a(x)y + b(x) = 0$$

by the substitution

$$u(x) = \exp\left\{\int y(x)\,\mathrm{d}x\right\}.$$

It is sometimes possible to find a solution of a Riccati equation by guessing. One may try such functions as $y = a\,x^b$ or $y = a\,e^{bx}$, with undetermined constants a and b, or some other functions that the differential equation suggests.

[29] In honor of the Italian mathematician Jacopo Francesco (Count) Riccati (1676–1754), who studied equations of the form $y' + by^2 = cx^m$.

Without knowing a solution of the nonlinear differential equation (2.6.4), there is absolutely no chance of finding its general solution explicitly. Later we will consider some special cases when the Riccati equations are solvable in quadratures. Indeed, if one knows a solution of a Riccati equation $\varphi(x)$, it could be reduced to a Bernoulli equation by setting $w = y - \varphi$. Then for $w(x)$ we have the equation

$$w' + p(x)\,w = g(x)\,w^2 + 2g(x)\varphi(x)\,w.$$

In practice, one may also try the substitution $y = \varphi + 1/v$ (or $v = (y-\varphi)^{-1}$), which leads to the linear differential equation

$$v' + [2\varphi(x)\,g(x) - p(x)]v = -g(x).$$

Example 2.6.6: Solve the second order linear differential equation with variable coefficients:

$$x\,u'' + (x - x^2 - 1)\,u' - x^2\,u = 0, \qquad x > 0.$$

Solution. To reduce the given differential equation to a first-order one, we make the substitution $y = u'/u$ and divide by x to obtain

$$y' + y^2 + \left(1 - x - \frac{1}{x}\right)\,y - x = 0.$$

This Riccati equation has a solution $y = x$ (which can be determined by trial). By setting $y = x + 1/v$, we get the linear differential equation for v:

$$xv' + (1 - x - x^2)v - x = 0.$$

Multiplying by an integrating factor $\mu(x) = e^{-x-x^2/2}$, we reduce the equation to the exact equation:

$$\frac{\mathrm{d}}{\mathrm{d}x}\left[v(x)\,x\,e^{-x-x^2/2}\right] = x\,e^{-x-x^2/2}.$$

Integration yields

$$v(x)\,x\,e^{-x-x^2/2} = \int x\,e^{-x-x^2/2}\,\mathrm{d}x + C \quad\Longrightarrow\quad v(x) = \frac{1}{x}\,e^{x+x^2/2}\left[\int x\,e^{-x-x^2/2}\,\mathrm{d}x + C\right]$$

and

$$y(x) = x + \frac{1}{v(x)} = x + x\,e^{-x-x^2/2}\left[\int x\,e^{-x-x^2/2}\,\mathrm{d}x + C\right]^{-1}.$$

To check our answer, we type in *Mathematica:*

```
Ricc[x_, y_] = y'[x] + y[x]^2 + (1 - x - 1/x)*y[x] - x
Expand[v[x]^2 Ricc[x, Function[t, t + 1/v[t]]]]
DSolve[% == 0, v[x], x]
```

Example 2.6.7: Let us consider the differential equation

$$y' = y^2 + x^2.$$

The substitution $u'(x) = \exp[-\int y(x)\,\mathrm{d}x]$ reduces this equation to $u'' + x^2 u = 0$. By setting $y = -u'/u$, we get

$$y' = \frac{-u''\,u + (u')^2}{u^2}.$$

Therefore, from the equation $y' = y^2 + x^2$, it follows that

$$\frac{-u''\,u + (u')^2}{u^2} = \left(-\frac{u'}{u}\right)^2 + x^2.$$

Multiplication by u^2 yields

$$-u''\,u + (u')^2 = (u')^2 + x^2\,u^2 \qquad \text{or} \qquad -u'' = x^2\,u.$$

□

The main property of a Riccati equation is that it has no singular solution as it follows from Picard's Theorem 1.3 (see §1.6 for details). Therefore, the initial value problem for Eq. (2.6.4) has a unique solution. By a linear transformation, the Riccati equation can be reduced to the canonical form

$$y' = \pm y^2 + R(x). \tag{2.6.8}$$

Theorem 2.4: Let $R(x)$ be a periodic function with period T, $R(x+T) = R(x)$. If the Riccati equation $y' = y^2 + R(x)$ has two different periodic solutions with period T, say $y_1(x)$ and $y_2(x)$, then

$$\int_0^T [y_1(x) + y_2(x)]\, \mathrm{d}x = 0.$$

PROOF: For two solutions $y_1(x)$ and $y_2(x)$, we have

$$y_1' = y_1^2 + R(x) \quad \text{and} \quad y_2' = y_2^2 + R(x).$$

Subtracting one equation from another, we have

$$\frac{\mathrm{d}}{\mathrm{d}x}(y_1 - y_2) = y_1^2 - y_2^2 = (y_1 + y_2)(y_1 - y_2).$$

Let $u = y_1 - y_2$; then the above equation can be rewritten as $u' = (y_1 + y_2)u$ or $u^{-1}\mathrm{d}u = (y_1 + y_2)\,\mathrm{d}x$. Its integration leads to

$$y_1(x) - y_2(x) = \exp\left\{\int_0^x [y_1(t) + y_2(t)]\,\mathrm{d}t\right\} [y_1(0) - y_2(0)].$$

Setting $x = T$, we obtain

$$y_1(T) - y_2(T) = \exp\left\{\int_0^T [y_1(t) + y_2(t)]\,\mathrm{d}t\right\} [y_1(0) - y_2(0)].$$

The functions $y_1(x)$ and $y_2(x)$ are periodic with period T; therefore,

$$\exp\left\{\int_0^T [y_1(t) + y_2(t)]\,\mathrm{d}t\right\} = 1, \quad \text{which leads to} \quad \int_0^T [y_1(t) + y_2(t)]\,\mathrm{d}t = 0.$$

Example 2.6.8: Integrate the Riccati equation

$$y' + y(y - 2x) = 2.$$

Solution. It is easy to verify that $y(x) = 2x$ is a solution of this differential equation. We set $y(x) = 2x + u(x)$; then $y' = 2 + u'(x)$ and we obtain

$$u'(x) + (2x + u)\, u(x) = 0 \qquad \text{or} \qquad u' + 2xu = -u^2.$$

This is a Bernoulli equation, so we seek a solution of the above equation in the form $u = v\,w$, where w is a solution of the homogeneous equation $w' + 2xw = 0$. Separating variables yields

$$\frac{\mathrm{d}w}{w} = -2x\,\mathrm{d}x$$

and its solution becomes $\ln w = -x^2 + \ln C$. Hence, $w = C\,e^{-x^2}$. We set $C = 1$ and get $u(x) = v(x)\,e^{-x^2}$. Then the function $v(x)$ satisfies the separable equation

$$v'(x)\,e^{-x^2} = -v^2(x)\,e^{-2x^2} \qquad \text{or} \qquad \frac{v'}{v^2} = -e^{-x^2}.$$

Integration yields $-\frac{1}{v(x)} = -\int e^{-x^2}\,\mathrm{d}x + C$; hence, $v(x) = \frac{1}{\int e^{-x^2}\,\mathrm{d}x - C}$. Thus,

$$u(x) = \frac{e^{-x^2}}{\int e^{-x^2}\,\mathrm{d}x - C} \qquad \text{and} \qquad y(x) = 2x + u(x) = 2x + \frac{e^{-x^2}}{\int e^{-x^2}\,\mathrm{d}x - C}.$$

Example 2.6.9: Solve the Riccati equation

$$y' = e^{2x} + (1 - 2\,e^{x})\,y + y^2.$$

Solution. If we try $y = a\,e^{bx}$, we find that $\varphi(x) = e^x$ is a solution. Setting $y = \varphi(x) + 1/v$, we obtain the linear differential equation for v to be $v' + v + 1 = 0$. Its general solution becomes

$$v(x) = C\,e^{-x} - 1. \qquad \text{Hence,} \qquad y(x) = e^x + \frac{1}{C\,e^{-x} - 1}$$

is the general solution of the given Riccati equation. □

Generally speaking, it is impossible to find solutions of the Riccati equation using elementary functions, though there are many cases when this can happen. We mention several[30] of them here. Except for these and a few other rare elementary cases, the Riccati equation is not integrable by any systematic method other than that of power series.

1. The Riccati equation
$$y' = a\,\frac{y^2}{x} + \frac{y}{2x} + c, \quad a^2 + c^2 \neq 0,$$
is reduced to a separable equation by setting $y = v\sqrt{x}$.

2. The Riccati equation
$$y' = ay^2 + \frac{b}{x}\,y + \frac{c}{x^2}$$
can also be simplified by substituting $y = u/x$.

3. The equation
$$y' = y^2 - a^2x^2 + 3a$$
has a particular solution $y_a(x) = ax - 1/x$.

4. The Riccati equation
$$y' = ay^2 + bx^{-2}$$
has the general solution of the form
$$y(x) = \frac{\lambda}{x} - x^{2a\lambda}\left(\frac{ax}{2a\lambda + 1}\,x^{2a\lambda} + C\right)^{-1},$$
where C is an arbitrary constant and λ is a root of the quadratic equation $a\lambda^2 + \lambda + b = 0$.

5. The **Special Riccati Equation** is the equation of the form
$$y' = ay^2 + bx^{\alpha},$$
and can be integrated in closed form[31] if and only if[32]
$$\frac{\alpha}{2\alpha + 4} \quad \text{is an integer.}$$
The special Riccati equation can be reduced to a linear differential equation $u'' + abx^{\alpha}\,u = 0$ by substituting $y(x) = -u'/(au)$, which in turn has a solution expressed through Bessel functions J_ν and Y_ν (see §4.9):
$$u(x) = \sqrt{x}\begin{cases} C_1\,J_{1/(2q)}\left(\frac{\sqrt{ab}}{q}\,x^q\right) + C_2\,Y_{1/(2q)}\left(\frac{\sqrt{ab}}{q}\,x^q\right), & \text{if } ab > 0, \\ C_1\,I_{1/(2q)}\left(\frac{\sqrt{-ab}}{q}\,x^q\right) + C_2\,K_{1/(2q)}\left(\frac{\sqrt{-ab}}{q}\,x^q\right), & \text{if } ab < 0, \end{cases} \tag{2.6.9}$$
where $q = 1 + \alpha/2 = (2 + \alpha)/2$.

[30] The reader could find many solutions of differential equations and Riccati equations in [37].

[31] Here *closed form* means that all integrations can be explicitly performed in terms of elementary functions.

[32] This result was proved by the French mathematician Joseph Liouville (1809–1882) in 1841.

Remark. The function $u(x)$ contains two arbitrary constants since it is the general solution of a second order linear differential equation. The function $y(x) = -u'/(au)$, however, depends only on one arbitrary constant $C = C_1/C_2$ ($C_2 \neq 0$), the ratio of arbitrary constants C_1 and C_2.

Example 2.6.10: (Example 2.6.7 revisited) For the equation

$$y' = x^2 + y^2,$$

it is impossible to find an exact solution expressed with elementary functions since $\alpha = 2$ and the fraction $\alpha/(2\alpha+4) = 1/4$ is not an integer. This equation has the general solution $y(x) = -u'/u$, where

$$u(x) = \sqrt{x}\left[C_1 J_{1/4}(x^2/2) + C_2 Y_{1/4}(x^2/2)\right].$$

The function $u(x)$ is expressed via Bessel functions, $J_{1/4}(z)$ and $Y_{1/4}(z)$, which are presented in §4.9. Similarly, the Riccati equation

$$y' = x^2 - y^2$$

has the general solution $y = u'/u$, which is expressed through modified Bessel functions: $u(x) = \sqrt{x}\left[C_1 I_{1/4}(x^2/2) + C_2 K_{1/4}(x^2/2)\right]$.

Example 2.6.11: Solve the Riccati equation (see its direction field in Fig. 2.29)

$$y' = xy^2 + x^2 y - 2x^3 + 1.$$

Solution. By inspection, the function $y_1(x) = x$ is a solution of this equation. After the substitution $y = x + 1/u$, we have $u' + 3x^2 u = -x$, which leads to the general solution

$$u(x) = e^{-x^3}\left(C - \int x e^{x^3}\,\mathrm{d}x\right).$$

Example 2.6.12: Solve the differential equation $y' + y^2 = 6\,x^{-2}$ (see its direction field in Fig. 2.30).

Solution. This Riccati equation has a particular solution of the form $y(x) = k/x$. Substituting this expression into the differential equation, we obtain

$$-\frac{k}{x^2} + \frac{k^2}{x^2} = \frac{6}{x^2} \qquad \text{or} \qquad k^2 - k = 6.$$

The quadratic equation has two solutions, $k_1 = 3$ and $k_2 = -2$, and we can pick either of them. For example, choosing the latter one, we set $y(x) = u(x) - 2/x$. Substitution leads to

$$u'(x) + \frac{2}{x^2} + u^2(x) + \frac{4}{x^2} - \frac{4}{x}\,u(x) = \frac{6}{x^2} \qquad \text{or} \qquad u'(x) - \frac{4}{x}\,u(x) + u^2(x) = 0.$$

This is a Bernoulli differential equation of the form of (2.6.1), page 95, with $p(x) = -4/x$, $g(x) = -1$, and $\alpha = 2$. According to the Bernoulli method, we make the substitution $u(x) = v(x)\,\varphi(x)$, where $\varphi(x)$ is a solution of its linear part: $\varphi'(x) - 4\varphi(x)/x = 0$, so $\varphi(x) = x^4$. Thus, $u(x) = x^4\,v(x)$ and we obtain the separable differential equation for $v(x)$:

$$v'(x) + x^4\,v^2(x) = 0 \qquad \text{or} \qquad -\frac{\mathrm{d}v}{v^2} = x^4\,\mathrm{d}x.$$

Integration yields

$$\frac{1}{v} = \frac{x^5}{5} + \frac{C}{5} \qquad \text{or} \qquad v(x) = \frac{5}{x^5 + C}.$$

Thus,

$$y(x) = \frac{5\,x^4}{x^5 + C} - \frac{2}{x} = \frac{3x^5 - 2C}{x(x^5 + C)}.$$

Similarly, we can choose $k_1 = 3$ and set $y(x) = u(x) + 3/x$. Then we obtain the solution $y(x) = \frac{3Kx^5+2}{x(Kx^5-1)}$, which is equivalent to the previously obtained one because $K = -1/C$ is an arbitrary constant.

Example 2.6.13: The equation

$$y' = y^2 + x^{-4}$$

can be integrated in elementary functions since $\alpha = -4$ and $\alpha/(2\alpha+4) = 1$. We make the substitution $y = u/x^2$ to obtain

$$\frac{u'}{x^2} - 2\frac{u}{x^3} = \frac{u^2}{x^4} + \frac{1}{x^4} \qquad \text{or} \qquad u' - \frac{2}{x}\,u = \frac{u^2}{x^2} + \frac{1}{x^2}.$$

Setting $u = v - x$ in this equation, we get $v' = (v^2+1)/x^2$. Separation of variables yields

$$\frac{\mathrm{d}v}{v^2+1} = \frac{\mathrm{d}x}{x^2} \qquad \text{or} \qquad \int \frac{\mathrm{d}v}{v^2+1} = \int \frac{\mathrm{d}x}{x^2} = -\frac{1}{x} + C.$$

Thus, the general solution becomes

$$\arctan v = \frac{Cx-1}{x} \qquad \text{and} \qquad x^2\,y(x) = \tan\frac{Cx-1}{x} - x.$$

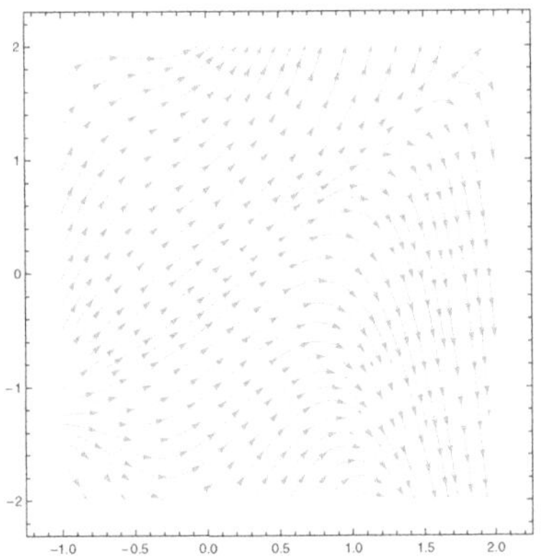

Figure 2.29: Example 2.6.11 (*Mathematica*).

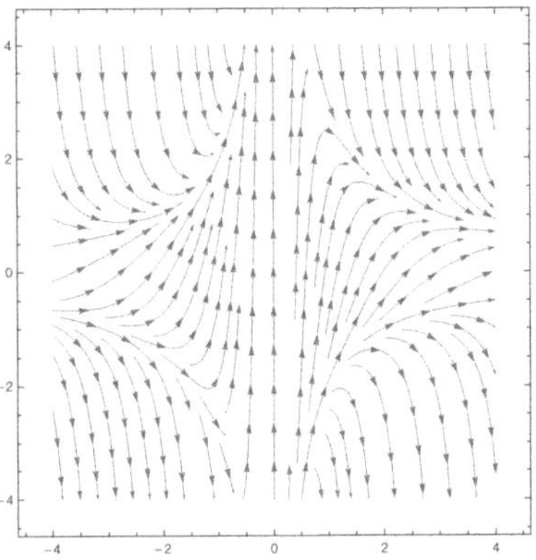

Figure 2.30: Example 2.6.12 (*Mathematica*).

Example 2.6.14: (A rod bombarded by neutrons) Suppose that a thin rod is bombarded from its right end by a beam of neutrons, which can be assumed without any loss of generality to be of unit intensity. Considering only a simple version of the problem, we assume that the particles only move in one direction and that the rod is also in one dimension. On average, in each unit length along the rod a certain fraction p of the neutrons in the beam will interact with the atoms of the rod. As a result of such an interaction, this neutron is replaced by two neutrons, one moving to the right and one moving to the left.

The neutrons moving to the right constitute a reflection beam, whose intensity we denote by $y(x)$, where x is the rod length. For small h (which is an abbreviation of common notation Δx), consider the portion of the rod between $x-h$ and x. Since every neutron that interacts with the rod in this interval is replaced by a neutron moving to the left, the right-hand end of the portion of the rod from 0 to $x-h$ is struck by a beam of intensity one. This produces a reflected beam of intensity $y(x-h)$. Since the neutrons interact within the interval $[x, x-h]$, it causes an additional reflected beam of intensity ph, proportional to its size. In addition, because of interaction that occurs between the neutrons in either of these reflected beams, we have a contribution to $y(x)$ of magnitude $ph + y(x-h)$.

Moreover, since some of the neutrons in each of the reflected beams just described will interact with the rod, each beam will give rise to a secondary beam moving to the left, which in turn will produce additional contributions to the reflected beam when it strikes the right-hand end of the portion of the rod between 0 and $x-h$. The intensities of these secondary beams are, respectively, $ph\,y(x-h)$ and $(ph)^2$, and they will contribute $ph\,y^2(x-h)$ and $(ph)^2\,y(x-h)$ to the reflected beam. Repetitions of this argument yield a third and ternary contributions, and so forth.

To derive the equation, we neglect terms containing h^2 because of its small size, which results in the equation:

$$y(x) = ph + y(x-h) + ph\,y^2(x-h).$$

Assuming that the function $y(x)$ is differentiable, we approximate further $y(x-h)$ along the tangent line: $y(x-h) = y(x) - h\,y'(x)$. This substitution produces the equation

$$y(x) = ph + y(x) - hy'(x) + ph(y - hy')^2,$$

which again is simplified to the Riccati equation

$$y'(x) = p + py^2(x),$$

because of neglecting terms with h^2. Solving the equation using the initial condition $y(0) = 0$ yields $y(x) = \tan(px)$.

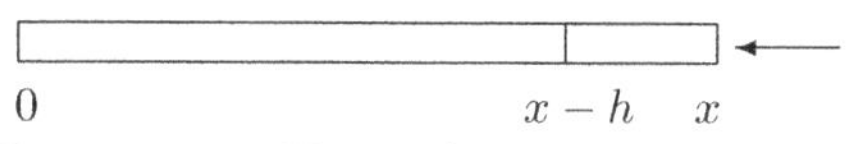

Figure 2.31: Example 2.6.14.

2.6.3 Equations with the Dependent or Independent Variable Missing

If the second order differential equation is of the form $y'' = f(x, y')$, then substituting $p = y'$, $p' = y''$ leads to the first-order equation of the form $p' = f(x, p)$. If this first-order equation can be solved for p, then y can be obtained by integrating $y' = p$. In general, equations of the form

$$F(x, y^{(k)}, y^{(k+1)}, \ldots, y^{(n)}) = 0$$

can be reduced to the equation of the $(n-k)$-th order using the substitution $y^{(k)} = p$. In particular, the equation

$$y^{(n)} = F(x, y^{(n-1)}),$$

containing only two consecutive derivatives, can be reduced to a first-order equation by means of the transformation $p = y^{(n-1)}$.

Example 2.6.15: Compute the general solution of the differential equation

$$y''' - \frac{1}{x}\,y'' = 0.$$

Solution. Setting $p = y''$, the equation becomes

$$p' - \frac{1}{x}\,p = 0,$$

which is separable. The general solution is seen to be $p(x) = Cx$ (C is an arbitrary constant). Thus, $y'' = Cx$. Integrating with respect to x, we obtain $y' = Cx^2/2 + C_2$. Another integration yields the general solution $y = C_1x^3 + C_2x + C_3$ with arbitrary constants $C_1 = C/6$, C_2, and C_3. □

Definition 2.9: A differential equation in which the independent variable is not explicitly present is called **autonomous**.

We consider a particular case of the second order autonomous equation

$$\ddot{y} = f(y, \dot{y}),$$

where dot stands for derivative with respect to t. A solution to such an equation is still a function of the independent variable (which we denote by t), but the equation itself does not contain t. Therefore, the rate function $f(y, \dot{y})$ is independent of t, and shifting a solution graph right or left along the abscissa will always produce another solution graph. Identifying the "velocity" $\dot{y}$ with the letter v, we get the first-order equation for it: $\ddot{y} = \dot{v} = f(y, v)$. Since this equation contains three variables (t, y, and v), we use the chain rule:

$$\dot{v} = \frac{\mathrm{d}v}{\mathrm{d}t} = \frac{\mathrm{d}v}{\mathrm{d}y}\,\frac{\mathrm{d}y}{\mathrm{d}t} = v\,\frac{\mathrm{d}v}{\mathrm{d}y}.$$

Replacing $\dot{v}$ by $v\,(\mathrm{d}v/\mathrm{d}y)$ enables us to eliminate explicit use of t in the velocity equation $\dot{v} = f(y, v)$ to obtain

$$v\,\frac{\mathrm{d}v}{\mathrm{d}y} = f(y, v).$$

If we are able to solve this equation implicitly, $\Phi(y, v, C_1) = 0$, with some constant C_1, then we can use the definition of v as $v = \mathrm{d}y/\mathrm{d}t$, and try to solve the first-order equation $\Phi(y, \dot{y}, C_1) = 0$, which is called a **first integral** of the given equation $\ddot{y} = f(y, \dot{y})$. In some lucky cases, the first integral can be integrated further to produce a solution of the given equation. The following examples clarify this procedure.

Example 2.6.16: Solve the equation

$$y\,\ddot{y} - (\dot{y})^2 = 0.$$

Solution. Setting $\dot{y} = v$ and $\ddot{y} = v(\mathrm{d}v/\mathrm{d}y)$, we get the separable differential equation

$$yv\,\frac{\mathrm{d}v}{\mathrm{d}y} - v^2 = 0.$$

The general solution of the latter equation is $v = C_1 y$ or $\mathrm{d}y/\mathrm{d}t = C_1 y$. This is also a separable equation and we obtain the general solution by integration: $\ln|y| = C_1 t + \ln C_2$ or $y = C_2 \exp(C_1 t)$.

Example 2.6.17: Solve the initial value problem

$$\ddot{y} = 2y\,\dot{y}, \quad y(0) = 1, \quad \dot{y}(0) = 2.$$

Solution. Setting $v = \dot{y}$, we get the first-order equation

$$v\,\frac{\mathrm{d}v}{\mathrm{d}y} = 2yv \qquad \Longrightarrow \qquad \mathrm{d}v = 2y\,\mathrm{d}y.$$

Integration yields the first integral:

$$\dot{y} = v = y^2 + C_1 \qquad \text{or after separation of variables} \qquad \frac{\mathrm{d}y}{y^2+1} = \mathrm{d}t$$

because $v(0) = y(0)^2 + C_1 = 1 + C_1 = 2$. Integration gives

$$\arctan y = t + C_2 \qquad \Longrightarrow \qquad y(t) = \tan(t + C_2).$$

The value of the constant C_2 is determined from the first initial condition $y(0) = \tan(C_2) = 1$, so $C_2 = \pi/4$, and we get the solution

$$y(t) = \tan\left(t + \frac{\pi}{4}\right).$$

□

Another type of autonomous equations constitute second order equations where both the independent variable and velocity are missing:

$$\ddot{y} = f(y). \tag{2.6.10}$$

If antiderivative $F(y)$ of the function $f(y)$ is known (see §9.4), so $\mathrm{d}F/\mathrm{d}y = f(y)$, then the above equation can be reduced to the first-order equation by multiplying both sides by $\dot{y}$:

$$\ddot{y}\cdot\dot{y} = f(y)\cdot\dot{y} \qquad \Longleftrightarrow \qquad \frac{1}{2}\,\frac{\mathrm{d}}{\mathrm{d}t}\,(\dot{y})^2 = \frac{\mathrm{d}}{\mathrm{d}t}\,F(y). \tag{2.6.11}$$

Then the first integral becomes

$$(\dot{y})^2 = 2\,F(y) + C_1 \qquad \Longleftrightarrow \qquad \dot{y} = \pm\sqrt{2\,F(y) + C_1}.$$

Since the latter is an autonomous first-order equation, we separate variables and integrate:

$$\frac{\mathrm{d}y}{\sqrt{2\,F(y) + C_1}} = \pm\mathrm{d}t \qquad \Longrightarrow \qquad \int \frac{\mathrm{d}y}{\sqrt{2\,F(y) + C_1}} = \pm t + C_2.$$

It should be noted that while the autonomous equation (2.6.10) may have a unique solution, Eq. (2.6.11) may not. For instance, the initial value problem

$$\ddot{y} + y = 0, \qquad y(0) = 0, \quad \dot{y}(0) = 1$$

can be reduced to $(\dot{y})^2 + y^2 = 1$, $y(0) = 0$. However, the latter has at least three distinct solutions:

$$y_1 = \sin t; \qquad y_2 = \begin{cases} \sin t, & \text{for } 0 \leqslant t \leqslant \frac{\pi}{2}, \\ 1, & \text{for } \frac{\pi}{2} \leqslant t < \infty; \end{cases} \qquad y_3 = \begin{cases} \sin t, & \text{for } 0 \leqslant t \leqslant \frac{\pi}{2}, \\ 1, & \text{for } \frac{\pi}{2} \leqslant t < T, \\ \cos(t-T), & \text{for } T \leqslant t < \infty. \end{cases}$$

Example 2.6.18: (Pendulum) Consider the pendulum equation (see also Example 9.4.1 on page 515)

$$\ddot{\theta} + \omega^2 \sin\theta = 0,$$

where θ is the angle of deviation from the downward vertical position, $\ddot{\theta} = \mathrm{d}^2\theta/\mathrm{d}t^2$, $\omega^2 = g/\ell$, ℓ is the length of the pendulum, and g is acceleration due to gravity. Multiplying both sides of the pendulum equation by $\mathrm{d}\theta/\mathrm{d}t$, we get

$$\frac{\mathrm{d}^2\theta}{\mathrm{d}t^2} \cdot \frac{\mathrm{d}\theta}{\mathrm{d}t} + \omega^2 \sin\theta\, \frac{\mathrm{d}\theta}{\mathrm{d}t} = 0 \qquad \Longleftrightarrow \qquad \frac{1}{2}\frac{\mathrm{d}}{\mathrm{d}t}\left(\frac{\mathrm{d}\theta}{\mathrm{d}t}\right)^2 = \omega^2 \frac{\mathrm{d}}{\mathrm{d}t} \cos\theta.$$

This leads to the first integral:

$$\left(\frac{\mathrm{d}\theta}{\mathrm{d}t}\right)^2 = 2\omega^2 \cos\theta + C_1 \qquad \Longleftrightarrow \qquad \frac{\mathrm{d}\theta}{\mathrm{d}t} = \pm\sqrt{2\omega^2 \cos\theta + C_1},$$

where the value of constant C_1 is determined by the initial conditions $\left.\frac{\mathrm{d}\theta}{\mathrm{d}t}\right|_{t=0} = \left.\dot{\theta}\right|_{t=0} = \pm \left.\sqrt{2\omega^2\cos\theta + C_1}\right|_{t=0}$, and the sign $\pm$ depends to what side (left or right) the bob of the pendulum was moved initially. Next integration yields

$$\frac{\mathrm{d}\theta}{\sqrt{2\omega^2\cos\theta + C_1}} = \pm \mathrm{d}t \qquad \Longrightarrow \qquad t = \pm\int_{\theta_0}^{\theta} \frac{\mathrm{d}\tau}{\sqrt{2\omega^2\cos\tau + C_1}}$$

because initially at $t = 0$ the pendulum was in position θ_0. Unfortunately, the integral in the right-hand side cannot be expressed through elementary functions, but can be reduced to the incomplete elliptic integral of the first kind:

$$F(\varphi, k) = \int_0^{\varphi} \frac{\mathrm{d}t}{\sqrt{1 - k^2\sin^2 t}}.$$

2.6.4 Equations Homogeneous with Respect to Their Dependent Variable

Suppose that in the differential equation

$$F(x, y, y', y'', \ldots, y^{(n)}) = 0 \tag{2.6.12}$$

the function $F(x, y, p, \ldots, q)$ is homogeneous with respect to all variables except x, namely, $F(x, ky, kp, \ldots, kq) = k^n F(x, y, p, \ldots, q)$. The order of Eq. (2.6.12) can then be reduced by the substitution

$$y = \exp\left\{\int v(x)\,\mathrm{d}x\right\} \qquad \text{or} \qquad v = \frac{y'}{y}, \tag{2.6.13}$$

where v is an unknown dependent variable.

The same substitution reduces the order of the differential equation of the form

$$y'' = f(x, y, y')$$

with homogeneous function $f(x, y, y')$ of degree one in y and y', that is, $f(x, ky, ky') = kf(x, y, y')$. If the function $F(x, y, y', y'', \ldots, y^{(n)})$ in Eq. (2.6.12) is homogeneous in the extended sense, namely,

$$F(kx, k^m y, k^{m-1}y', k^{m-2}y'', \ldots, k^{m-n}y^{(n)}) = k^{\alpha}\, F(x, y, y', y'', \ldots, y^{(n)}),$$

then we make the substitution $x = e^t$, $y = z\,e^{mt}$, where $z = z(t)$ and t are new dependent and independent variables, respectively. Such a substitution leads to a differential equation for z that does not contain an independent variable. Calculations show that

$$y' = \frac{\mathrm{d}y}{\mathrm{d}x} = \frac{\mathrm{d}y}{\mathrm{d}t}\,\frac{\mathrm{d}t}{\mathrm{d}x} = \frac{\mathrm{d}y}{\mathrm{d}t}\,\frac{1}{\frac{\mathrm{d}x}{\mathrm{d}t}} = \frac{\mathrm{d}y}{\mathrm{d}t}\,\frac{1}{e^t} = \frac{\mathrm{d}y}{\mathrm{d}t}\,e^{-t}.$$

With this in hand, we differentiate $y = z\,e^{mt}$ with respect to t to obtain

$$\frac{\mathrm{d}y}{\mathrm{d}t} = \left(\frac{\mathrm{d}z}{\mathrm{d}t} + mz\right) e^{mt} \quad \text{and} \quad y' = \frac{\mathrm{d}y}{\mathrm{d}x} = \left(\frac{\mathrm{d}z}{\mathrm{d}t} + mz\right) e^{(m-1)t}.$$

The second differentiation yields

$$\begin{aligned} y'' &= \frac{\mathrm{d}^2y}{\mathrm{d}x^2} = e^{-t}\,\frac{\mathrm{d}}{\mathrm{d}t}\left(\frac{\mathrm{d}z}{\mathrm{d}t} + mz\right) e^{(m-1)t} \\ &= \left(\frac{\mathrm{d}^2z}{\mathrm{d}t^2} + 2m\,\frac{\mathrm{d}z}{\mathrm{d}t} + m^2 z\right) e^{(m-2)t}. \end{aligned}$$

Substituting the first and second derivatives into the second order differential equation $F(x, y, y', y'') = 0$, we obtain

$$F\left(e^t, z\,e^{mt}, (z' + mz)\,e^{(m-1)t}, (z'' + 2mz' + m^2)\,e^{(m-2)t}\right) = 0,$$

or

$$e^{mt}\,F\left(1, z, (z' + mz), (z'' + 2mz' + m^2)\right) = 0.$$

Example 2.6.19: Reduce the following nonlinear differential equation of the second order to an equation of the first order:

$$x^2\,y'' = (y')^2 + x^2.$$

Solution. To find the degree of homogeneity of the function $F(x, y', y'') = x^2y'' - (y')^2 - x^2$, we equate it to $F(kx, k^{m-1}y', k^{m-2}y'')$, which leads to the algebraic equation $m = 2m - 2 = 2$. Therefore, $m = 2$ and we make the substitution $y = z\,e^{2t}$ that leads to the equation

$$z'' + 4z' + 4z = (z' + 2z)^2 + 1.$$

This is an equation without an independent variable, so we set $p = z'$ and $z'' = p\,p'$. Thus, we obtain the first-order differential equation with respect to p:

$$p\,p' + 4p + 4z = (p + 2z)^2 + 1.$$

Example 2.6.20: Find the general solution of the differential equation

$$yy'' - (y')^2 = 6xy^2.$$

Solution. Computing the derivatives of the function (2.6.13), we have

$$y' = v\exp\left\{\int v(x)\,\mathrm{d}x\right\}, \quad y'' = v'\exp\left\{\int v(x)\,\mathrm{d}x\right\} + v^2\exp\left\{\int v(x)\,\mathrm{d}x\right\}.$$

Substituting into the given equation, we obtain $v' = 6x$. Hence $v(x) = 3x^2 + C_1$ and the general solution becomes

$$y(x) = C_2\exp\left\{\int [3x^2 + C_1]\,\mathrm{d}x\right\} = C_2\exp\left\{x^3 + C_1x\right\}.$$

2.6.5 Equations Solvable for a Variable

We present another method of attacking first-order equations when it is possible to solve for a variable. Suppose that the equation $F(x, y, y') = 0$ is solved for y, it is then in the form $y = \Phi(x, y')$. Setting $p = y'$ and differentiating the equation $y = \Phi(x, p)$ with respect to x, we obtain

$$y' = p = \frac{\partial \Phi}{\partial p}\frac{\mathrm{d}p}{\mathrm{d}x} + \frac{\partial \Phi}{\partial x},$$

which contains only $y' = p$, x, and $\mathrm{d}p/\mathrm{d}x$, and is a first-order differential equation in $p = y'$. If it is possible to solve this equation for p, then we can integrate it further to obtain a solution. We demonstrate this approach in the following example.

Example 2.6.21: Consider the equation $y' = (x + y)^2$, which, when solved for y, gives

$$y = \sqrt{y'} - x = \sqrt{p} - x \quad \Longrightarrow \quad y' = p = \frac{1}{2}\,p^{-1/2}\,\frac{\mathrm{d}p}{\mathrm{d}x} - 1.$$

Separating variables yields

$$\frac{\mathrm{d}p}{2\sqrt{p}(p+1)} = \mathrm{d}x \quad \Longrightarrow \quad y' = p = \tanh^2(C + x).$$

Since $y' = p = (x + y)^2$, we get the general solution to be $(x + y)^2 = \tanh^2(C + x)$. □

A similar method is possible by reducing the equation $F(x, y, y') = 0$ to the form $x = \Phi(x, y')$. Differentiating, we obtain the equation

$$1 = \frac{\partial \Phi}{\partial y}\,y' + \frac{\partial \Phi}{\partial y'}\,\frac{\mathrm{d}y'}{\mathrm{d}x},$$

which contains only the variables y and $p = y'$, and the derivative $\frac{\mathrm{d}y'}{\mathrm{d}x}$ and which may be written as $\frac{\mathrm{d}y'}{\mathrm{d}x} = \frac{\mathrm{d}p}{\mathrm{d}x} = p\,\frac{\mathrm{d}p}{\mathrm{d}y}$. That is, the process above has replaced x with a new variable, $p = y'$. If it's possible to solve it for p and then integrate, we may get the general solution.

Example 2.6.22: Reconsider the equation $y' = (x + y)^2$, which, when solved for x, gives

$$x = \sqrt{p} - y, \quad \text{or after differentiation,} \quad 1 = \frac{1}{2}\,p^{-1/2}\,\frac{\mathrm{d}p}{\mathrm{d}y} - p.$$

Since the integration by separation of variables is almost the same as in Example 2.6.21, we omit the details.

Problems

1. Solve the following Bernoulli equations.
   (a) $y' + 2y = 4y^2$; **(b)** $(1 + x^2)\,y' - 2xy = 2\sqrt{y(1 + x^2)}$; **(c)** $y' - 2y = y^{-1}$;
   (d) $4x^3\,y' = y(y^2 + 5x^2)$; **(e)** $y' + y = e^x\,y^3$; **(f)** $2xy' + y = y^2(x^5 + x^3)$;
   (g) $y' - 4y = 2y^2$; **(h)** $2x\,y' - y + y^2\,\sin x = 0$; **(i)** $x^2\,y' + axy^2 = 0$;
   (j) $x^3\,y' = xy + y^2$; **(k)** $y' + y = e^{2x}\,y^3$; **(l)** $y' + y/2 = (1 - 2x)\,y^4/2$;
   (m) $y\,y' = x - xy^2$; **(n)** $(x - 1)y' - 2y = \sqrt{y(x^2 - 1)}$; **(o)** $x^2y' - xy = y^3$.
2. Find the particular solution required.
   (a) $y' = \frac{x+1}{2x}\,y - 2y^3, \quad y(1) = 1$; **(b)** $xy' + y^2 = 2y; \quad y(1) = 4$;
   (c) $y' + 2y/x = x^2y^2, \quad y(1) = 1$; **(d)** $x\,y' + y = x^4y^3, \quad y(1) = -1$;
   (e) $y' + \frac{y}{2}\,\cot x = \frac{y^3}{\sin x}, \quad y(\pi/2) = 1$; **(f)** $y' - \frac{2}{3x}\,y = \frac{1 - 2\ln x}{3xy}, \quad y(1) = 1$.
3. Using the Riccati substitution $u' = yu$, reduce each second order linear differential equation to a Bernoulli equation.
   (a) $u'' + 9u = 0$; **(b)** $u'' + 2u' + u = 0$; **(c)** $u'' - 2u' + 5u = 0$;
   (d) $u'' - 4u' + 4u = 0$; **(e)** $u'' - 5u' + 6u = 0$; **(f)** $u'' - 2u' + u = 0$.
4. Reduce the given Riccati equations to linear differential equations of the second order.
   (a) $y' - \frac{2}{x+1}\,y = (x + 1)^2y^2 + \frac{4}{(x+1)^2}$; **(b)** $y' = y\,\cot x + y^2\,\sin x + (\sin x)^{-1}$;
   (c) $y' = 2\,\frac{1-x}{x}\,y + x^2y^2 + x^{-2}$; **(d)** $y' + \left(2 + \frac{1}{x}\right)\,y = y^2/x + 5x$;
   (e) $y' + 2y = y^2\,e^{2x} + 4e^{-2x}$; **(f)** $y' + (2 - 3x)y = x^3\,y^2 + x^{-3}$.
5. By the use of a Riccati equation, find the general solution of the following differential equations with variable coefficients of the second order.

(a) $u'' - 2x^2 u' - 4xu = 0$. Hint: $y' + y^2 - 2x^2 y - 4x = 0$, try $y = 2x^2$.

(b) $u'' - \left(2 + \frac{1}{x}\right) u' - x^2 e^{4x} u = 0$. Hint: $y' + y^2 - \left(2 + \frac{1}{x}\right) y - x^2 e^{4x} = 0$, try $y = x\, e^{2x}$.

(c) $u'' + (\tan x - 2x \cos x)\, u' + (x^2 \cos^2 x - \cos x) u = 0$. Hint: $y' + y^2 + (\tan x - 2x \cos x)\, y + (x^2 \cos^2 x - \cos x) = 0$, try $y = x \cos x$.

(d) $u'' - 3u' - e^{6x} u = 0$. Hint: $y' + y^2 - 3y - e^{6x} = 0$, try $y = e^{3x}$.

(e) $u'' - \frac{1}{x} u' - (1 + x^2 \ln^2 x) u = 0$. Hint: $y' + y^2 - \frac{1}{x} y - (1 + x^2 \ln^2 x) = 0$, try $y = x \ln x$.

(f) $x \ln x\, u'' - u' - x \ln^3 x\, u = 0$. Hint: $y' + y^2 - \frac{1}{x \ln x} - \ln^2 x = 0$, try $y = \ln x$.

(g) $x\, u'' + (1 - \sin x)\, u' - \cos x\, u = 0$. Hint: $y' + y^2 + \frac{1 - \sin x}{x}\, y - \frac{\cos x}{x} = 0$, try $y = \frac{\sin x}{x}$.

(h) $u'' - \frac{2}{x^2} u' + \frac{4}{x^3} u = 0$. Hint: $y' + y^2 - \frac{2}{x^2} y + \frac{4}{x^3} = 0$, try $y = 2\, x^{-2}$.

6. Given a particular solution, solve each of the following Riccati equations.
   (a) $y' + xy = y^2 + 2(1 - x^2)$, $y_1(x) = 2x$; (b) $xy' - 3y + y^2 = 4x^2 + 4x$, $y_1(x) = -2x$;
   (c) $y' = y^2 + 2x - x^4$, $y_1(x) = x^2$; (d) $y' = 0.5\, y^2 + 0.5 x^{-2}$, $y_1(x) = -1/x$;
   (e) $y' + y^2 + \frac{y}{x} - \frac{4}{x^2} = 0$, $y_1(x) = 2/x$; (f) $y' + 2y = y^2 + 2 - x^2$, $y_1(x) = (x + 1)$.

7. Find a particular solution in the form $y = kx$ of the following Riccati equation and then determine the general solution.
   (a) $y' = 1 + 3\frac{y}{x} + \frac{y^2}{3x^2}$; (b) $y' + 2y(y - x) = 1$; (c) $y' = 4x^3 + y/x - x\, y^2$.

8. Find a particular solution of each Riccati equation satisfying the given condition.
   (a) $y' + y^2 = 4$, $y(0) = 6$. (b) $y' = \frac{y^2}{x^2} - 2\frac{y}{x} + 2$, $y(1) = 1$. Try $y = 2x$.
   (c) $y' + y^2 = 1$, $y(0) = -1$. (d) $y' = \frac{y^2}{x^2} - y/x^2 - 2/x^2$, $y(3) = 7/2$. Try $y = 2$.
   (e) $y' + y^2 + y - 2 = 0$, $y(0) = 4$. (f) $x\, y' = x\, y^2 + (1 - 2x)\, y + x - 1$, $y(1) = 3$. Try $y = 1 - 2/x$.
   (g) $xy\, y' - y^2 + x^2 = 0$, $y(1) = 2$. (h) $y' = e^x y^2 + y - 3e^{-x}$, $y(0) = 0$. Try $y = e^{-x}$.
   (i) $y' = \frac{y^2}{2x} + \frac{y}{2x} - \frac{1}{x}$, $y(1) = 0$. (j) $y' = y^2 - \frac{2}{x^2} y - \frac{2}{x^3} + \frac{1}{x^4}$, $y(2) = 1$.

9. Let us consider the initial value problem $y' + y = y^2 + g(t)$, $y(0) = a$. Prove that $|y(t)|$ is bounded for $t > 0$ if $|a|$ and $\max_{t>0} |g(t)|$ are sufficiently small.

10. Prove that a linear fractional function in an arbitrary constant C,
$$y(x) = \frac{C\, \varphi_1(x) + \varphi_2(x)}{C\, \psi_1(x) + \psi_2(x)}$$
is the general solution of the Riccati equation, provided that $\varphi_1(x)$, $\varphi_2(x)$, $\psi_1(x)$, and $\psi_2(x)$ are some smooth functions.

11. Let y_1, y_2, and y_3 be three particular solutions of the Riccati equation (2.6.4). Prove that the general solution of Eq. (2.6.4) is
$$C = \frac{y - y_2}{y - y_1} \cdot \frac{y_3 - y_1}{y_3 - y_2}.$$

12. Solve the equations with the dependent variable missing.
   (a) $x\, y'' + y' = x$; (b) $x\, y'' + y' = f(x)$; (c) $x^2\, y'' + 2y' = 4x$;
   (d) $y'' + y' = 4 \sinh x$; (e) $x\, y'' = (1 + x^2)(y' - 1)$; (f) $y'' + x(y')^3 = 0$;
   (g) $x\, y'' + x(y')^2 + y' = 0$; (h) $y'' + (2y')^3 = 2y'$; (i) $4y'' - (y')^2 + 4 = 0$;
   (j) $y''' - y'' = 1$; (k) $yx''' - 2y'' = 0$; (l) $2x^2\, y'' + (y')^3 - 2xy' = 0$.

13. Solve the equations with the independent variable missing.
   (a) $2y'' + 3y^2 = 0$; (b) $y'' + \omega^2 y = 0$; (c) $y'' - \omega^2 y = 0$;
   (d) $y^2\, y'' + y' + 2y(y')^2 = 0$; (e) $y'' - 2(y')^2 = 0$; (f) $(1 + y^2) yy'' = (3y^2 + 1)(y')^2$;
   (g) $y\, y'' + 2(y')^2 = 0$; (h) $y'' + 2y(y')^3 = 0$; (i) $4y'' + y = 0$;
   (j) $2yy'' = y^2 + (y')^2$; (k) $y^2 y'' = (y')^3$; (l) $y\, y'' + (y')^2 = 2y^3\, (y')^3$.

14. Reduce the order of the differential equations with homogeneous functions and solve them.
   (a) $y\, y'' = (y')^2 + 28 y^2 \sqrt[3]{x}$; (b) $y\, y'' - (y')^2 = (x^2 + 1)\, yy'$;
   (c) $x(y\, y'' - (y')^2) + y\, y' = y^2$; (d) $y'' = y\, x^{-8/3}$;
   (e) $y\, y'' = 2(y')^2 + y^2\, x^{-4}$; (f) $y\, y'' = (y')^2 + 28 y^2\, x^{1/3}$;
   (g) $x^2 y\, y'' - x^2 (y')^2 = yy'$; (h) $y\, y'' = x(y')^2$;
   (i) $xy\, y'' - x(y')^2 = (x^2 + 1) yy'$; (j) $y\, y'' = (y')^2$;
   (k) $x^2 y\, y'' = (y - xy')^2$; (l) $xyy'' = yy' + (y')^2 x/2$;
   (m) $y(x\, y'' + y') = (y')^2 (x - 1)$; (n) $yy'' = (n + 1) x^n (y')^2$.

2.7 Qualitative Analysis

In the previous sections, we concentrated our attention on developing tools that allow us to solve certain types of differential equations (§§2.1–2.6) and discussed existence and uniqueness theorems (§1.6). However, most nonlinear differential equations cannot be explicitly integrated; instead, they define functions as solutions to these equations. Even when solutions to differential equations are found (usually in implicit form), it is not easy to extract valuable information about their behavior.

This conclusion pushes us in another direction: the development of qualitative methods that provide a general idea of how solutions to given equations behave, how solutions depend on initial conditions, whether there are curves that attract or repel solutions, whether there are vertical, horizontal, or oblique asymptotes, determination of periodic solutions, and any bounds. Realizing that qualitative methods fall into a special course on differential equations, we nevertheless motivate the reader to look deeper into this subject by exploring some tools and examples. More information can be found in [2, 24]. We start with a useful definition.

Definition 2.10: The **nullclines** of the differential equation $y' = f(x, y)$ are curves of zero inclination that are defined as solutions of the equation $f(x, y) = 0$.

Nullclines are usually not solutions of the differential equation $y' = f(x, y)$ unless they are constants (see Definition 2.1 on page 39). The curves of zero inclination divide the xy-plane into regions where solutions either rise or fall. The solution curves may cross nullclines only when their slope is zero. This means that critical points of solution curves are points of their intersection with nullclines. We clarify the properties of nullclines in the following four examples.

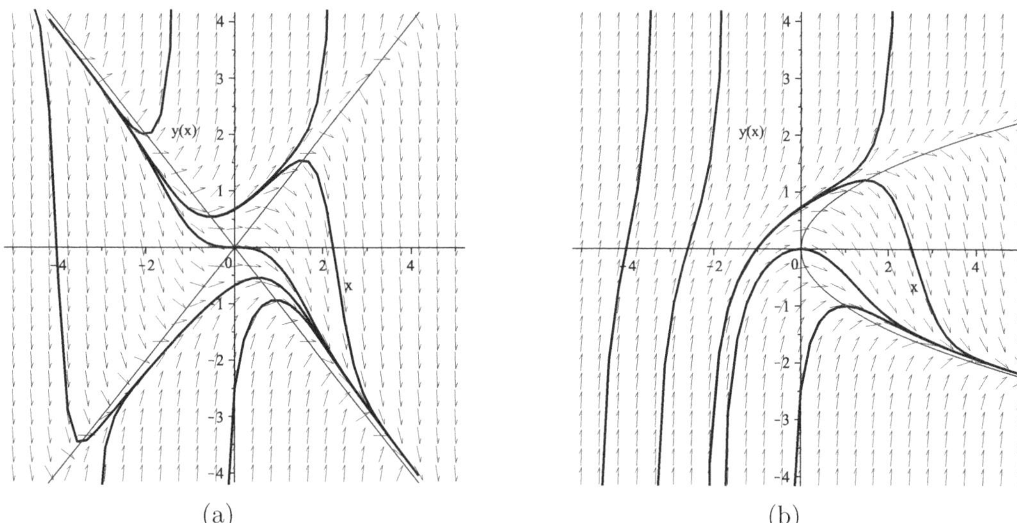

Figure 2.32: Example 2.7.2: Direction fields, nullclines, and streamlines for (a) $y' = y^2 - x^2$ and (b) $y' = y^2 - x$, plotted with *Maple*. Points where solution curves intersect with nullclines are their critical points (local maxima or minima).

Example 2.7.1: Consider the linear differential equation $a\dot{y} + ty = b$ with some nonzero constants a and b. Using the integrating factor method, we find its general solution (see Eq. (2.5.9) on page 88) to be

$$y = C\, e^{-t^2/(2a)} + \frac{b}{a}\, e^{-t^2/(2a)} \int_0^t e^{s^2/(2a)}\, \mathrm{d}s,$$

where C is an arbitrary constant. The first term, $C\, e^{-t^2/(2a)}$, decays exponentially for any value of $C \neq 0$ if $a > 0$. The second term can be viewed as the ratio of two functions:

$$y_p \stackrel{\text{def}}{=} \frac{b}{a}\, e^{-t^2/(2a)} \int_0^t e^{s^2/(2a)}\, \mathrm{d}s = \frac{f(t)}{g(t)},$$

where $f(t) = \frac{b}{a}\int_0^t e^{s^2/(2a)}\,ds$ and $g(t) = e^{t^2/(2a)}$. To find the limit of $y_p(t)$ when $t \mapsto +\infty$, we apply l'Hôpital's rule:

$$\lim_{t\to+\infty} y_p(t) = \lim_{t\to+\infty}\frac{f(t)}{g(t)} = \lim_{t\to+\infty}\frac{f'(t)}{g'(t)} = \lim_{t\to+\infty}\frac{b}{a}\,e^{t^2/(2a)}\Big/\frac{2t}{2a}\,e^{t^2/(2a)} = \frac{b}{t} \mapsto 0.$$

Therefore, every solution of the given linear differential equation approaches the nullcline $y = b/t$ as $t \to +\infty$.

Example 2.7.2: Let us consider two Riccati equations that, as we know from §2.6.2, have no singular solutions. The equation $y' = y^2 - x^2$ has two nullclines $y = \pm x$ that are not solutions of the given equation. When $|y| < |x|$, the slope is negative and the solutions decline in this region. In contrast, the solutions increase in the region $|y| > |x|$. As can be seen from Figure 2.32(a), both nullclines $y = \pm x$ are semistable because half of them attract solutions whereas the other half repel solutions.

Consider another differential equation $y' = y^2 - x$, which also cannot be integrated using elementary functions (neither explicitly nor implicitly). It has two nullclines $y = \sqrt{x}$ and $y = -\sqrt{x}$ that are not solutions of the given differential equation. Slope fields and some plotted streamlines presented in Figure 2.32(b) strongly indicate that there are two types of solutions: those that approach ∞ for large x, and those that follow the nullcline $y = -\sqrt{x}$ and ultimately go to $-\infty$. Moreover, these two types of solutions are separated by an exceptional solution, which can be estimated from the graphs. The same observation is valid for the equation $y' = y^2 - x^2$.

Example 2.7.3: Consider a Bernoulli equation $t^2\dot{y} = y(y^2 - 2t)$. Its general solution is $y = \phi(t) = \sqrt{t}/\sqrt{2 + Ct^5}$. The equation has the equilibrium solution $y = 0$ and two nullclines $y = \pm\sqrt{2t}$. If $C \neq 0$, solutions approach the critical point $y = 0$ as $t \to +\infty$. If $C = 0$, then the solution $\phi(t) = \sqrt{t/2}$ follows the nullcline $y = \sqrt{2t}$ as $t \to +\infty$. The following MATLAB codes allow us to plot the solutions.

```
s=dsolve('t^2*Dy = y^3- 2*t*y')    /* invoke Maple's solver */
for cval=0:5             hold on    /* assign 6 integer values to cval */
ezplot(subs(s(2), 'C11', cval));    /* C11 is a constant from dsolve */
hold off                    end
axis([0 5,0 2 ]);
title('solution to t^2*Dy = y^3- 2*t*y');
xlabel('t'); ylabel('y');
```
□

While Examples 2.7.1 and 2.7.2 may give an impression that solutions either approach or repel the nullclines, generally speaking, they do not. There could be many differential equations with distinct solutions that share the same nullclines. As an example, we recommend plotting direction fields with some solutions for the differential equations $y' = (y^2 - x^2)/(x^4 + 1)$ or $y' = (y^2 - x)/(x^2 + 1)$, which have the same nullclines as the differential equations in Example 2.7.2. Nevertheless, nullclines provide useful information about the behavior of solutions: they identify domains where solutions go down or up, or where solutions attain their local maximum or minimum. For instance, solutions for the differential equation $y' = y^2 - x$ decrease inside the parabola $y^2 - x = 0$ because their slopes are negative. Outside this parabola all solutions increase. Similar observations can be made for the equation $y' = y^2 - x^2$: solutions decrease inside the region $|y| < |x|$.

The first partial derivative $f_y = \partial f/\partial y$ of the rate function $f(x, y)$ gives more information about disintegration of solutions. Solutions to $y' = f(x, y)$ pull apart when f_y is large and positive; on the other hand, solutions tend to stay together when f_y is close to zero. As an example, consider the equation $y' = f(x, y)$ with the slope function $f = y^2 - x^2$. Its partial derivative $f_y = 2y$ is positive when $y > 0$, so solutions fly apart in the upper half plane, and the nullclines $y = \pm x$ cannot be stable in this region.

Example 2.7.4: Let us consider the differential equation $y' = f(x, y)$, where $f(x, y) = (y - 1)(y - \sin x)$. The horizontal line $y = 1$ in the xy-plane is its equilibrium solution, but the nullcline $y = \sin x$ is not a solution. The former is unstable for $y > 1$; however, when $y < 1$, it is neither attractive nor repulsive. Since $f(x, y)$ contains the periodic function $\sin x$, the slope field is also periodic in x.

None of the solutions can cross the equilibrium point $y = 1$ because the given equation is a Riccati equation that has no singular solution. The solutions do not approach the nullcline $y = \sin x$, but "follow" it in the sense that they have the same period. □

Drawing a slope field for a particular differential equation may provide some additional information about its solutions. Sometimes certain types of symmetry are evident in the direction field. If the function $f(x, y)$ is odd with respect to the dependent variable y, $f(x, -y) = -f(x, y)$, then solutions of the differential equation $y' = f(x, y)$

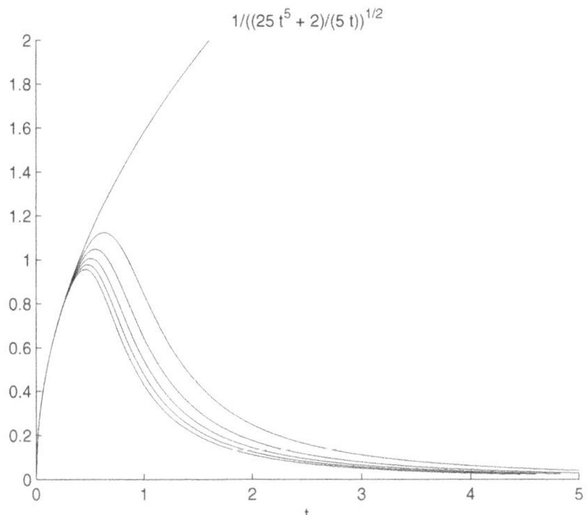

Figure 2.33: Example 2.7.3: solutions to the equation $t^2 y' = y^3 - 2ty$, plotted in MATLAB.

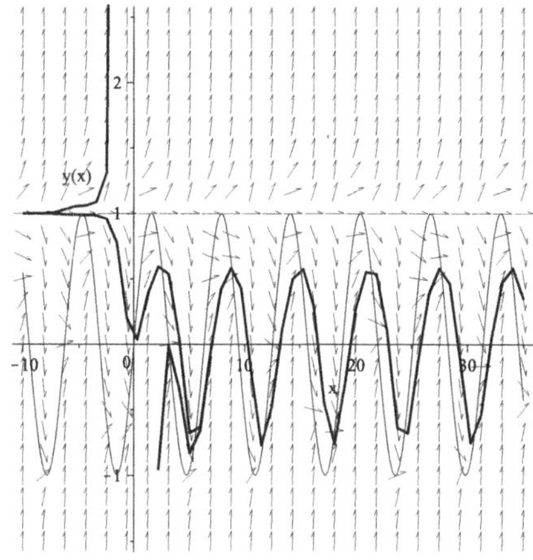

Figure 2.34: Example 2.7.4. Black lines are solutions, and blue ones are nullclines (plotted with *Maple*).

are symmetric with respect to abscissa $y = 0$. So if $y(x)$ is a solution, then $z(x) = -y(x)$ is also a solution to the equation $y' = f(x, y)$. Similarly, if $f(x, y) = -f(-x, y)$, then solutions are symmetric with respect to the vertical y-axis (see Fig. 2.3, page 43).

Also, we have already discovered that solutions of autonomous differential equations are symmetric under translations in the x direction. In general, a differential equation may possess other symmetries. Unfortunately, we cannot afford this topic for presentation, but we refer the reader to [3].

Many solutions to differential equations have *inflection points*, where the second derivative is zero: $y'' = 0$. To find y'', you need to differentiate $f(x, y)$ with respect to x and substitute $f(x, y)$ for y'. This expression for y'' gives information on the concavity of solutions. For example, differentiating both sides of the equation $y' = y^2 - 1$ with respect to x, we obtain $y'' = 2yy' = 2y(y^2 - 1)$. Therefore, solutions to the given differential equation have three inflection points (which are actually horizontal lines): $y = 0$ and $y = \pm 1$.

As we will see in Chapter 3, numerical methods usually provide useful information about solutions to the differential equations in the "short" range, not too far from the starting point. This leads us to the next question: "how do solutions behave in the long range (as $x \to +\infty$)?" More precisely, "what are the asymptotes (if any) of the solutions?" Sometimes it is not always the right question because a solution may fly up in finite time, that is, has a vertical asymptote. We can answer these questions in a positive way at least for solvable differential equations (see §2.1 – 2.6). For instance, the following statement assures the existence of vertical asymptotes for a certain class of autonomous equations.

Corollary 2.1: Nonzero solutions to the autonomous differential equation $y' = A\, y^a$ (A is a constant) have vertical asymptotes for $a > 1$.

However, what are we supposed to do when our differential equation is not autonomous, or if it is autonomous, but is not of the form $y' = A\, y^a$? The natural answer is to use estimates, which are based on the following **Chaplygin** [33] **inequality**.

Theorem 2.5: [Chaplygin] Suppose the functions $u(x)$ and $v(x)$ satisfy the differential inequalities

$$u'(x) - f(x, u) > 0, \qquad v'(x) - f(x, v) < 0$$

on some closed interval $[a, b]$. If $u(x_0) = v(x_0) = y_0$, where $x_0 \in [a, b]$, then the solution $y(x)$ of the differential equation $y' = f(x, y)$ that passes through the point (x_0, y_0) is bounded by u and v; that is, $v(x) < y(x) < u(x)$ on the interval $[a, b]$. The functions u and v are called the **upper fence** and the **lower fence**, respectively, for the solution $y(x)$.

[33] The Russian mathematician Sergey Alexeyevich Chaplygin (1869–1942) proved this inequality in 1919.

Let $y = \phi(x)$ be a solution for the differential equation $y' = f(x,y)$ and $v(x)$ be its lower fence on the interval $|a,b|$; then the lower fence pushes solutions up. Similarly, the upper fence pushes solutions down.

Example 2.7.5: Consider the Riccati equation $y' = y^2 + x^2$. On the semi-open interval $[0,1)$, the slope function $f(x,y) = x^2 + y^2$ can be estimated as

$$x^2 < x^2 + y^2 < y^2 + 1 \qquad (0 \leqslant x < 1)$$

because $y \equiv 0$ cannot be a solution of the given equation (check it). If $\alpha(x)$ is a solution to the equation $\alpha'(x) = x^2$, so $\alpha(x) = x^3/3 + C$, then $\alpha(x)$ is a lower fence for the given differential equation on the interval $(0,1)$. Hence, every solution of $y' = x^2 + y^2$ grows faster than $x^3/3 + C$ on $[0,1)$. To be more specific, consider a particular solution of the above equation that goes through the origin, $y(0) = 0$. To find its formula, we use *Maple:*

```
deq1:=dsolve({diff(y(x),x)=x*x+y(x)*y(x),y(0)=0},y(x));
```

which gives the solution

$$\phi(x) = x\,\frac{Y_{-3/4}\left(\frac{x^2}{2}\right) - J_{-3/4}\left(\frac{x^2}{2}\right)}{J_{1/4}\left(\frac{x^2}{2}\right) - Y_{1/4}\left(\frac{x^2}{2}\right)}, \tag{2.7.1}$$

where $J_\nu(z)$ and $Y_\nu(z)$ are Bessel functions (see §4.9). Its lower fence is, for instance, the function $\alpha(x) = x^3/3 - 0.1$, which is the solution to the initial value problem $\alpha' = x^2$, $\alpha(0) = -0.1$. To find an upper fence on the interval $(0,1)$, we solve the initial value problem $\beta' = \beta^2 + 1$, $\beta(0) = 0.1$, which gives $\beta(x) = \tan(x + c)$, where $c = \tan(0.1) \approx 0.1003346721$. Since at the origin we have estimates $\alpha(0) = -0.1 < 0 = \phi(0) < 0.1 = \beta(0)$, we expect the inequalities $\alpha(x) < \phi(x) < \beta(x)$ to hold for all $x \in [0,1]$. Therefore, the solution $y = \phi(x)$ to the initial value problem ($y' = y^2 + x^2$, $y(0) = 0$) is a continuous function on the interval $[0,1]$.

To find a lower estimate for $x > 1$, we observe that $y^2 + 1 < y^2 + x^2$ $(1 < x < \infty)$. Since we need the value of the solution $\phi(x)$ at $x = 1$, we use *Maple* again:

```
phi:=dsolve({diff(y(x),x)=x*x+y(x)*y(x),y(0)=0},y(x),numeric);  phi(1)
```

This provides the required information: $\phi(1) \approx 0.3502319798054$. Actually, this value was obtained by solving the IVP numerically—we do not need to know the exact formula (2.7.1). To find the lower fence in the interval $x > 1$, we solve the initial value problem: $\alpha' = \alpha^2 + 1$, $\alpha(1) = 0.35$. This gives us $\alpha(x) = \tan(x + C)$, where $C = \arctan(0.35) - 1 \approx -0.6633$. At the point $x = 1$, we have the inequality $0.35 = \alpha(1) < \phi(1) \approx 0.35023$. Since the tangent function has vertical asymptotes, the lower fence $\alpha(x) = \tan(x + C)$ blows up at $x + C = \frac{\pi}{2}$, when $x = \frac{\pi}{2} - C \approx 2.2341508$. Therefore, we expect that the solution $\phi(x)$ has a vertical asymptote somewhere before $x < 2.234$. Indeed, the solution $y = \phi(x)$ has a vertical asymptote around $x \approx 2.003147359$—the first positive root of the transcendent equation $J_{1/4}\left(\frac{x^2}{2}\right) = Y_{1/4}\left(\frac{x^2}{2}\right)$. To plot fences for the given problem (see Fig. 2.35), we type in *Maple*

```
cm:=arctan(0.35)-1.0;
ym:=plot(tan(x+cm),x=1..2.1,color=blue);
dd:=DEplot(deq1,y(x),x=0..2,y=0..2,[y(0)=0],dirgrid=[16,16],
   color=blue,linecolor=black);
display(ym,dd)
```

□

Let us consider the differential equation $y' = f(x,y)$ under two different initial conditions $y(x_0) = y_{01}$ and $y(x_0) = y_{02}$. If y_{01} is close to y_{02}, do the corresponding solutions stay close to each other? This question is of practical importance because, in mathematical modeling, the initial condition is usually determined experimentally and is subject to experimental error. In other words, we are interested in how close the solution to the initial value problem with incorrect initial data will be to the exact solution. A similar problem is important in numerical calculations and is discussed in Example 3.4.2, page 171. Actually, the fundamental inequality (1.6.12) provides us with useful information.

Corollary 2.2: Suppose the function $f(x,y)$ satisfies the conditions of Picard Theorem 1.3 and that $y_1(x)$ and $y_2(x)$ are solutions to the equation $y' = f(x,y)$ subject to the initial conditions $y(x_0) = y_{01}$ and $y(x_0) = y_{02}$, respectively. Then

$$|y_1(x) - y_2(x)| \leqslant |y_{01} - y_{02}|\, e^{L|x-x_0|}, \tag{2.7.2}$$

where L is the Lipschitz constant for the slope function. ◁

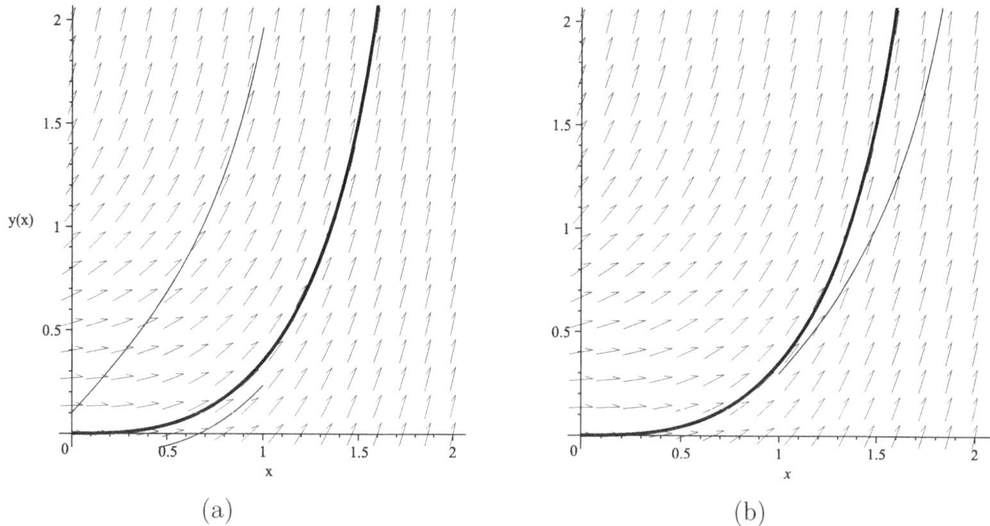

Figure 2.35: Example 2.7.5: direction field along with (a) lower and upper fences for the solution on the interval $(0, 1)$ and (b) lower fence on the interval $(1, 2)$, plotted with *Maple*.

A similar result is valid when the initial values x_0 and y_0 in the initial condition $y(x_0) = y_0$ are continuous functions of a parameter.

As follows from the inequality (2.7.2), the Lipschitz constant or $f_y = \partial f/\partial y$ evaluated at a particular point (if it exists) measures the stability of solutions. If f_y is small, then error in initial conditions is less critical; however, if $|f_y|$ is large, then solutions fly apart exponentially. However, inequality (2.7.2) only provides an upper bound to the difference between solutions. This worst case scenario may not always occur.

As we learned from the numerous examples, many solutions to the given differential equations belong to different classes that behave in similar ways. Some of them may tend to infinity, some asymptotically approach a certain curve, and some may approach a particular value. Classes of solutions which behave similarly are often separated by solutions with exceptional behavior. For instance, the differential equation $y' = y(y^2 - 9)$ considered in Example 2.7.15, page 120, has two unstable critical points: $y = \pm 3$. Such equilibrium solutions are exceptional solutions from which all solutions tend to go away. If the initial conditions are close to these two unstable critical points, the perturbed solutions may fly away in different directions exponentially, as inequality (2.7.2) predicts.

Example 2.7.6: Consider the initial value problem for the following linear differential equation

$$y' = y + 3 - t, \qquad y(0) = a.$$

The general solution to the above equation is known to be (see §2.5)

$$y(t) = C\,e^t + t - 2,$$

where C is an arbitrary constant. This is an exponentially growing function approaching $+\infty$ if $C > 0$ and $-\infty$ if $C < 0$. However, when $C = 0$, the solution grows only linearly as $t \to \infty$. Therefore, $\phi(t) = t - 2$ is an exceptional solution that separates solutions that satisfy the initial condition $y(0) = a > -2$ from solutions that satisfy the initial condition $y(0) = a < -2$. Therefore, this exceptional solution $\phi(t)$ is called a separatrix.

Example 2.7.7: Consider the differential equation

$$y' = \sin(\pi x y).$$

Since the slope function is an odd function, it is sufficient to consider solutions in the positive quadrant $(x, y \geqslant 0)$. The nullclines $xy = k$ $(k = 1, 2, \ldots)$, shown in Fig. 2.36, divide the first quadrant into regions where solutions to the equation $y' = \sin(\pi x y)$ are eventually either increasing or decreasing once they reach the domain $y < x/\pi$ (marked with a dashed line). Moreover, a simple calculus argument shows that, once a solution enters a decreasing region, it cannot get out provided the solution enters to the right of the line $y = x/\pi$. As seen from Fig. 2.37, solutions can be very sensitive to the initial conditions. The given differential equation has a stable equilibrium solution $y = 0$.

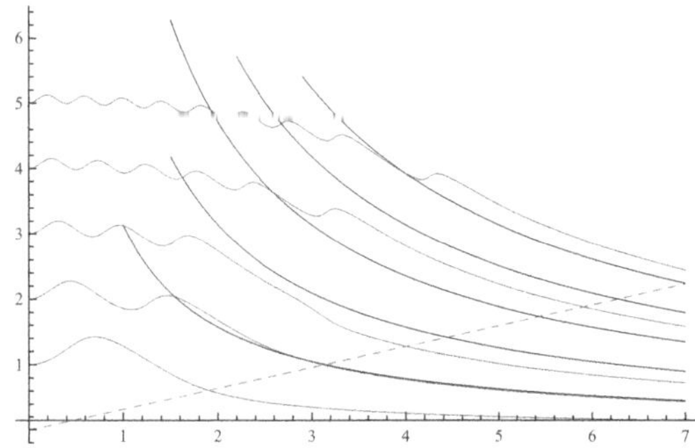

Figure 2.36: Example 2.7.7. Nullclines (in black) and solutions, plotted with *Mathematica*.

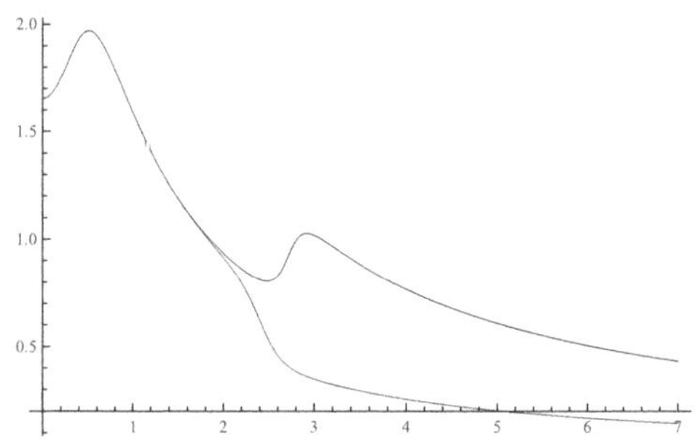

Figure 2.37: Two solutions to the equation $y' = \sin(\pi xy)$ subject to the initial conditions $y(0) = 1.652$ and $y(0) = 1.653$.

2.7.1 Bifurcation Points

A differential equation used to model real-world problems usually contains a parameter, a numerical value that affects the behavior of the solution. As a rule, the corresponding solutions change smoothly with small changes of the parameter. However, there may exist some exceptional values of the parameter when the solution completely changes its behavior.

Example 2.7.8: Consider a linear differential equation

$$y' = a - 2ay + e^{-ax},$$

where a is a number (parameter). The equation has one nullcline: $y = \frac{1}{2a}\left(a + e^{-ax}\right)$. All solutions approach this nullcline when the parameter a is positive, and repel when it is negative. From Fig. 2.38, you may get the impression that solutions merge to $y = 0.5 = \lim \frac{1}{2a}\left(a + e^{-ax}\right)$ when $x \to +\infty$ if $a > 0$ and when $x \to -\infty$ if $a < 0$; however, we know from §2.5 that a linear differential equation has no singular solutions. When $a = 0$, the slope becomes 1, and the equation $y' = 1$ has a totally different general solution $y = x + C$.

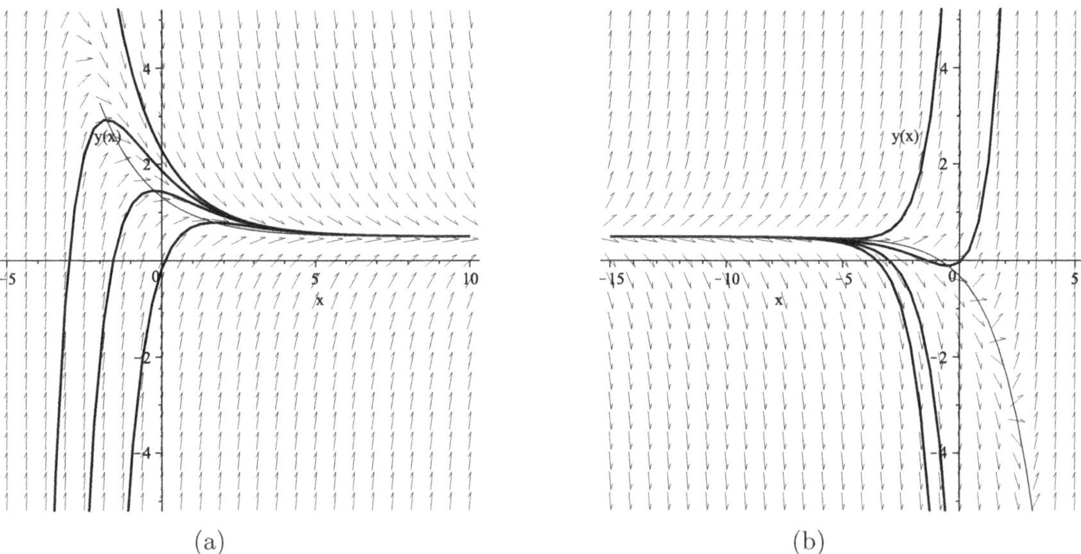

Figure 2.38: Example 2.7.8: direction fields, nullclines, and streamlines for (a) $a > 0$ and (b) $a < 0$, plotted with *Maple*.

Definition 2.11: An autonomous differential equation $y' = f(y, \alpha)$, depending on a parameter α, would have a solution that also depends on the parameter. In particular, if such an equation has a critical point y^*, then it also depends on the parameter $y^* = y^*(\alpha)$. The value of the parameter $\alpha = \alpha^*$ is called the **bifurcation point** of the differential equation if in its neighborhood the solutions exhibit a sudden qualitative change in their behavior.

Bifurcation points are of great interest in many applications because the nature of solutions is drastically altered around them. After a bifurcation occurs, critical points may either come together or separate, causing equilibrium solutions to be lost or gained. For example, when an aircraft exceeds the speed of sound, it undergoes a different behavior: supersonic motion is modeled by a hyperbolic equation, while subsonic motion requires an elliptic equation. More examples can be found in [28].

For visualization, it is common to draw a bifurcation diagram that shows the possible equilibria as a function of a bifurcation parameter. For an autonomous differential equation $y' = f(y, \alpha)$ with one parameter α, there are four common ways in which bifurcation occurs.

- A **saddle-node bifurcation** is a collision and disappearance of two equilibria. This happens if the following conditions hold:
$$\left.\frac{\partial^2 f(y,\alpha)}{\partial y^2}\right|_{y=y^*,\alpha=\alpha^*} \ne 0, \qquad \left.\frac{\partial f(y,\alpha)}{\partial \alpha}\right|_{y=y^*,\alpha=\alpha^*} \ne 0, \tag{2.7.3}$$
where y^* is the equilibrium solution and α^* is the bifurcation value. A typical equation is $y' = \alpha - y^2$.

- A **transcritical bifurcation** occurs when two equilibria change stability. Analytically, this means that
$$\left.\frac{\partial^2 f(y,\alpha)}{\partial y^2}\right|_{y=y^*,\alpha=\alpha^*} \ne 0, \qquad \left.\frac{\partial^2 f(y,\alpha)}{\partial y \partial \alpha}\right|_{y=y^*,\alpha=\alpha^*} \ne 0. \tag{2.7.4}$$
A typical equation is $y' = \alpha y - y^2$.

- **Pitchfork bifurcation: supercritical.** A pitchfork bifurcation is observed when critical points tend to appear and disappear in symmetrical pairs. In the αy-plane, it looks like a pitchfork, with three tines and a single handle. A supercritical bifurcation occurs when two additional equilibria are stable. A typical equation is $y' = \alpha y - y^3$.

- **Pitchfork bifurcation: subcritical** occurs when two additional equilibria are unstable. A typical equation is $y' = \alpha y + y^3$. In the case of pitchfork bifurcation (either supercritical or subcritical), the slope function $f(y, \alpha)$ satisfies the following conditions at the critical point $y = y^*$, $\alpha = \alpha^*$:
$$\left.\frac{\partial^3 f(y,\alpha)}{\partial y^3}\right|_{y=y^*,\alpha=\alpha^*} \ne 0, \qquad \left.\frac{\partial^2 f(y,\alpha)}{\partial y \partial \alpha}\right|_{y=y^*,\alpha=\alpha^*} \ne 0. \tag{2.7.5}$$

Example 2.7.9: (Saddle-node bifurcation) Consider the differential equation
$$y' = \alpha - y^2,$$
containing a parameter α. Equating $\alpha - y^2$ to zero, we find the equilibrium solutions to be $y = \pm\sqrt{\alpha}$. When $\alpha < 0$, there is no equilibrium solution. When $\alpha = 0$, the only equilibrium solution is $y \equiv 0$. If $\alpha > 0$, we have two equilibrium solutions: $y = \pm\sqrt{\alpha}$. One of them, $y = \sqrt{\alpha}$, is stable, while another, $y = -\sqrt{\alpha}$, is unstable. Hence, $\alpha = 0$ is a bifurcation point, classified as a saddle-node because of a sudden appearance of stable and unstable equilibrium points.

The function $f(y, \alpha) = \alpha - y^2$ satisfies the conditions (2.7.3) at the point $y = 0$, $\alpha = 0$: $f_{yy}(0, 0) = -2$, $f_\alpha(0, 0) = 1$.

Example 2.7.10: (Transcritical bifurcation) The logistic equation
$$y' = ry - y^2 = y(r - y),$$
has two equilibrium solutions: $y = 0$ and $y = r$. When $r = 0$, these two critical points merge; hence, we claim that $r = 0$ is a bifurcation point. The function $f(y, r) = ry - y^2$ satisfies the conditions (2.7.4) at the point $y = 0$, $r = 0$: $f_{yy}(0, 0) = -2$, $f_{yr}(0, 0) = 1$. The equilibrium solution $y = 0$ is stable when $r < 0$, but becomes unstable when the parameter $r > 0$.

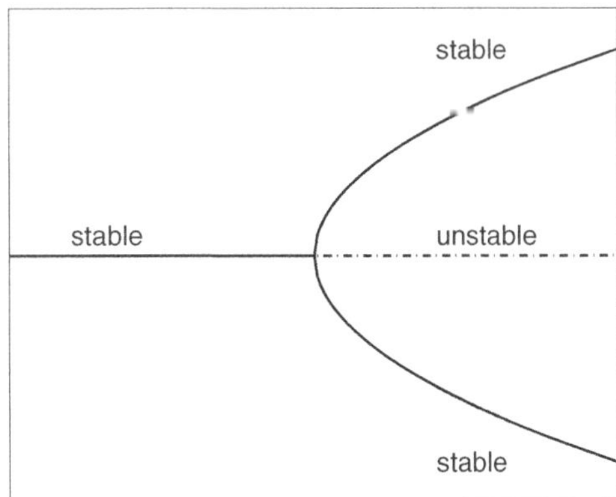

Figure 2.39: Supercritical pitchfork bifurcation diagram, plotted with MATLAB.

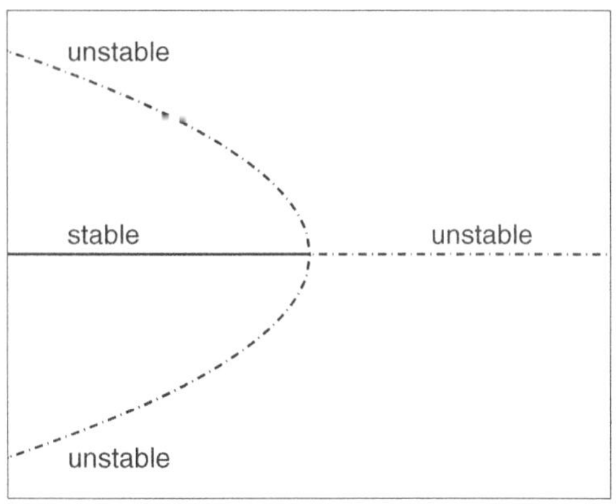

Figure 2.40: Subcritical pitchfork bifurcation diagram, plotted with MATLAB.

Example 2.7.11: (Pitchfork bifurcation, supercritical) Consider the one-parameter first order differential equation

$$y' = \alpha y - y^3.$$

Obviously, $\alpha = 0$ is a bifurcation point because the function $f(y) = \alpha y - y^3$ can be factored: $f(y) = y(\alpha - y^2)$. When $\alpha \leqslant 0$, there is only one critical point $y = 0$. However, when $\alpha > 0$, there are two additional equilibrium solutions $y = \pm\sqrt{\alpha}$. The slope function $f(y) = \alpha y - y^3$ satisfies the conditions (2.7.5) at the point $y = 0$, $\alpha = 0$: $\left.\frac{\partial^3 f(y,\alpha)}{\partial y^3}\right|_{y=0,\alpha=0} = -6$, $\left.\frac{\partial^2 f(y,\alpha)}{\partial y \partial \alpha}\right|_{y=0,\alpha=0} = 1$.
The equilibrium $y = 0$ is stable for every value of the parameter $\alpha < 0$. However, this critical point becomes unstable for $\alpha > 0$. In addition, two branches of unstable equilibria ($y = \pm\sqrt{\alpha}$) emerge.

Example 2.7.12: (Neuron activity) Let $y(t)$ denote the level of activity of a single neuron (nerve cell) at time t, normalized to be between 0 (lowest activity) and 1 (highest activity). Since a neuron receives external input from surrounding cells in the brain and feedback from its own output, we model its activity with a differential equation of the form

$$\dot{y} = S(y + E - \theta) - y \qquad (\dot{y} = \mathrm{d}y/\mathrm{d}t),$$

where S is a response function, E is the level of input activity from surrounding cells, and θ is the cell's threshold. The response function $S(z)$ is a monotonically increasing function from 0 to 1 as $z \to +\infty$. By choosing it as $(1 + e^{-az})^{-1}$ with an appropriate positive constant a, we obtain a simple model for neutron activity to be

$$\frac{\mathrm{d}y}{\mathrm{d}t} = \frac{1}{1 + e^{-a(y+E-\theta)}} - y(t). \tag{2.7.6}$$

The equilibrium solutions of the differential equation are determined by solving the transcendent equation

$$y = \frac{1}{1 + e^{-a(y+E-\theta)}},$$

with respect to y; however, its solution cannot be expressed through elementary functions. Because the response function is always between 0 and 1, the critical points must be within the interval $(0, 1)$ as well.

Assuming that the level of activity is a positive constant, we draw graphs $y = x$ and $y = \left(1 + e^{-a(x+E-\theta)}\right)^{-1}$ to determine their points of intersection. As seen from Fig. 2.41, for small threshold values there will be one equilibrium solution near $x = 1$, and for large θ there will be one critical point near $x = 0$. However, in a middle range there could be three equilibrium solutions.

Using the numerical values $E = 0.1$, $a = 8$, and $\theta = 0.55$, we find three critical points: $y_1^* \approx 0.03484$, $y_2^* \approx 0.4$, and $y_3^* \approx 0.9865$. (Note that these equilibrium solutions will be different for other numerical values of parameters E, a, and θ.) The slope function $f(y) = \left(1 + e^{-a(x+E-\theta)}\right)^{-1} - y$ is positive if $y < y_1^*$, negative if $y_1^* < y < y_2^*$, positive when y is between y_2^* and y_3^*, and negative if $y > y_3^*$. Therefore, two critical points y_1^* and y_3^* are stable, while y_2^* is the unstable equilibrium solution.

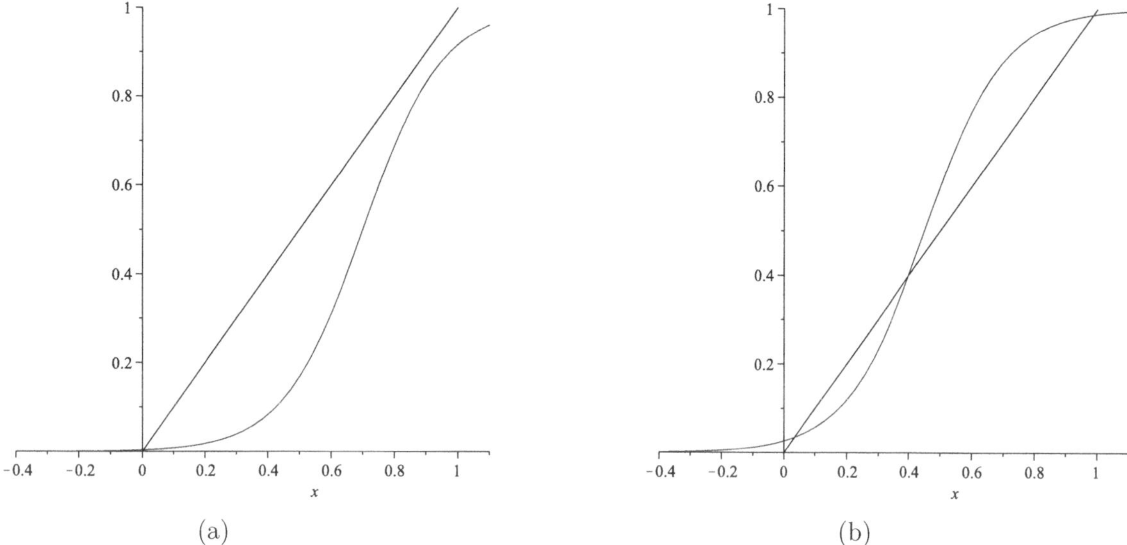

Figure 2.41: Example 2.7.12: (a) one bifurcation point when $\theta = 0.8$ and (b) three bifurcation points when $\theta = 0.55$, plotted with *Maple*.

To find the bifurcation points, we must determine the values of θ for which the graph of $y = \left(1 + e^{-a(x+E-\theta)}\right)^{-1}$ is tangent to the line $y = x$. This point of tangency is a solution of the following system of equations:

$$\begin{cases} 1 + e^{-a(x+E-\theta)} = \frac{1}{x}, \\ \frac{\mathrm{d}}{\mathrm{d}x}\left(\frac{1}{1+e^{-a(x+E-\theta)}}\right) = a\,\frac{e^{-a(x+E-\theta)}}{\left(1+e^{-a(x+E-\theta)}\right)^2} = 1. \end{cases} \tag{2.7.7}$$

From the first equation, we express the exponential function as $e^{-a(x+E-\theta)} = \frac{1}{x} - 1 = \frac{1-x}{x}$. Hence, the second equation is reduced to the quadratic equation with respect to x:

$$ax^2\,\frac{1-x}{x} = 1 \qquad \Longrightarrow \qquad x_{1,2} = \frac{1}{2} \pm \frac{1}{2}\sqrt{1 - \frac{4}{a}} \quad (a > 4).$$

Now the bifurcation points are found from the transcendental equation $e^{-a(x_{1,2}+E-\theta)} = \dfrac{1 - x_{1,2}}{x_{1,2}} = \dfrac{1}{ax_{1,2}^2}$, which leads to

$$\theta = x_{1,2} + E + \frac{1}{a}\ln\frac{1 - x_{1,2}}{x_{1,2}} \approx E + \begin{cases} 0.523 & \text{for } x_1 = \frac{2+\sqrt{2}}{4} \text{ and } a = \frac{16}{3}, \\ 0.477 & \text{for } x_2 = \frac{2-\sqrt{2}}{4} \text{ and } a = \frac{16}{3}. \end{cases}$$

2.7.2 Validity Intervals of Autonomous Equations

The existence and uniqueness theorems discussed in §1.6 do not determine the largest interval, called the **validity interval**, on which a solution to the given initial value problem can be defined. They also do not describe the behavior of solutions at the endpoints of the maximum interval of existence. The largest intervals of existence for the linear equation $y' + a(x)y = f(x)$ are usually bounded by points (if any) where the functions $a(x)$ and $f(x)$ are undefined or unbounded. Such points, called singular points or singularities, do not depend on the initial conditions. In contrast, solutions to nonlinear differential equations, in addition to having fixed singular points, may also develop movable singularities whose locations depend on the initial conditions.

In this subsection, we address these issues to autonomous (for simplicity) equations. Actually, Theorems 2.1 and 2.2 on page 48 describe the behavior of solutions at critical points of the differential equation. Before extending these theorems for the general case, we present some examples.

Example 2.7.13: Using the well-known trigonometric identity $(\sin x)' = \cos x = \sqrt{1 - \sin^2 x}$, we see that the sine function should be a solution of the differential equation

$$y' = \sqrt{1 - y^2}.$$

After separation of variables and integration, we get its general solution to be $\arcsin(y) = x + C$, where C is a constant of integration (see Fig. 2.42). To have an explicit form, we apply the sine function to both sides:

$$y(x) = \sin(x + C), \quad \text{if} \quad -\frac{\pi}{2} - C \leqslant x \leqslant \frac{\pi}{2} - C.$$

Every member of the general solution is defined only on the finite interval $\left[-\frac{\pi}{2} - C, \frac{\pi}{2} - C\right]$, depending on C. Hence, these functions do not intersect. The envelope of singular solutions consists of two equilibrium solutions $y = \pm 1$. Since for $f(y) = \sqrt{1 - y^2}$, the integral (2.1.11) converges, so we expect the uniqueness of the initial value problem to fail. Indeed, every member of the one-parameter family of the general solution touches the equilibrium singular solutions $y = \pm 1$. Moreover, we can construct infinitely many singular solutions by extending the general solution:

$$\varphi(x) = \begin{cases} 1, & \text{if } x \geqslant \frac{\pi}{2} - C, \\ \sin(x + C), & \text{if } -\frac{\pi}{2} - C < x < \frac{\pi}{2} - C, \\ -1, & \text{if } x \leqslant -\frac{\pi}{2} - C. \end{cases}$$

Example 2.7.14: Consider the initial value problem

$$y' = y^2 - 9, \qquad y(0) = 1.$$

The differential equation has two equilibrium solutions: $y = \pm 3$. One of them, $y = 3$, is unstable, while another one, $y = -3$, is stable. Using *Mathematica*

```
DSolve[{y'[x] == y[x]*y[x] - 9, y[0] == 1}, y[x], x]
```

we find its solution to be

$$y(x) = \frac{3(e^{6x} - 2)}{e^{6x} + 2}.$$

The maximum interval of existence is $(-\infty, \infty)$; $y(x)$ approaches 3 as $x \to +\infty$, and $y(x)$ approaches -3 as x goes to $-\infty$. If we change the initial condition to be $y(0) = 5$, the solution becomes

$$y(x) = \frac{3(4 + e^{6x})}{4 - e^{6x}}.$$

The maximum interval is $(-\infty, \frac{1}{3}\ln 2)$; $y(x)$ blows up at $\frac{1}{3}\ln 2$, and $y(x)$ approaches -3 as x goes to $-\infty$. We ask *Maple* to find the solution subject to another initial condition, $y(0) = -5$:

```
dsolve({diff(y(x),x)= y(x)*y(x) - 9, y(0)=-5}, y(x));
```

$$y(x) = \frac{3(e^{-6x} + 4)}{e^{-6x} - 4}.$$

The maximum interval of existence becomes $(-\frac{1}{3}\ln 2, \infty)$; $y(x)$ approaches $-\infty$ as $x \to -\frac{1}{3}\ln 2 \approx -0.2310490602$ (from the left), and $y \mapsto -3$ as $x \to +\infty$.

The initial value problem for the given equation with the initial condition $y(0) = 3$ has the unique constant solution $y = 3$ in accordance with Theorem 2.2, page 48.

Example 2.7.15: Consider the initial value problem

$$y' = y(y^2 - 9), \qquad y(0) = 1.$$

This equation has three equilibrium solutions, one of which, $y = 0$, is stable, while the two others, $y = \pm 3$, are unstable. The solution $y(x) = \dfrac{3}{\sqrt{1 + 8\,e^{18x}}}$ exists for all $x \in \mathbb{R}$; $y \mapsto 0$ as $x \to +\infty$ and $y \mapsto 3$ as $x \to -\infty$. To plot streamlines (see Fig. 2.43), we use *Mathematica:*

```
StreamPlot[{1, y (y^2 - 9)}, {x, -2, 2}, {y, -5, 5}, StreamStyle -> Blue]
```

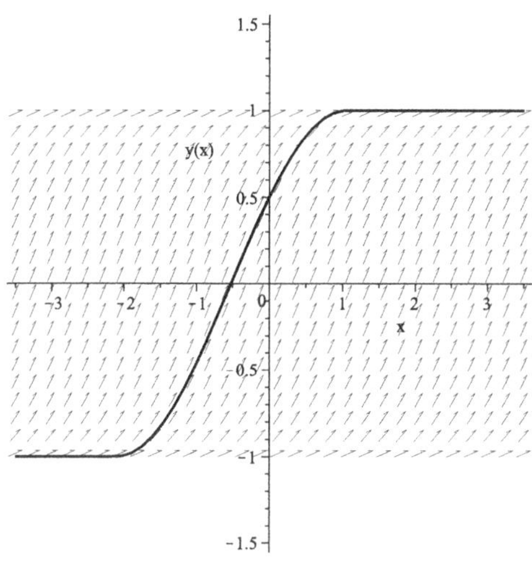

Figure 2.42: Example 2.7.13: direction field, plotted with *Maple*.

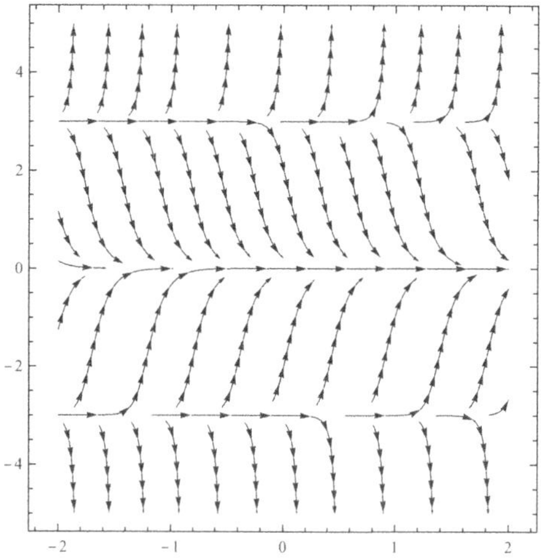

Figure 2.43: Streamlines of the differential equation $y' = y(y^2 - 9)$, plotted with *Mathematica*.

For another initial condition, $y(0) = 5$, we get the solution, $y(x) = \dfrac{15}{\sqrt{25 - 16\, e^{18x}}}$, which has the maximum interval of existence $\left(-\infty, \frac{1}{6}\ln\frac{5}{4}\right)$; the function $y(x)$ approaches $+\infty$ as $x \to \frac{1}{18}\ln\frac{25}{16} = \frac{1}{9}\ln\frac{5}{4} \approx 0.02479372792$, while $y(x) \mapsto 3$ as $x \to -\infty$.

Again, changing the initial condition to be $y(0) = -5$, we get $y(x) = -\dfrac{15}{\sqrt{25 - 16\, e^{18x}}}$, with the maximum interval of existence $\left(-\infty, \frac{1}{6}\ln\frac{5}{4}\right)$; the function $y(x)$ approaches $-\infty$ as $x \to \frac{1}{6}\ln\frac{5}{4}$, while $y(x) \mapsto -3$ as $x \to -\infty$.

The slope function $f(y) = y(y^2 - 9)$ has three zeroes, $y = 0$, $y = \pm 3$, that are critical points of the given differential equation. Since $1/f(y)$ is not integrable in the neighborhood of any of these points, the equilibrium solutions $y^* = 0$, 3, or -3 are unique solutions of the given equation subject to the corresponding initial conditions. Therefore, these critical points are not singular solutions.

Example 2.7.16: Solve the initial value problem

$$\frac{\mathrm{d}y}{\mathrm{d}x} = \frac{6x^2 + 4x + 1}{2y - 1}, \qquad y(0) = 2,$$

and determine the longest interval (the validity interval) in which the solution exists.

Solution. Excluding the horizontal line $2y = 1$, where the slope is unbounded, we rewrite the given differential equation in the differential form as

$$(6x^2 + 4x + 1)\,\mathrm{d}x = (2y - 1)\,\mathrm{d}y.$$

The line $2y = 1$ separates the range of solutions into two parts. Since the initial condition $y(0) = 2$ belongs to the half plane $y > 0.5$, we integrate the right-hand side with respect to y in this range. Next, integrating the left-hand side with respect to x, we get $y^2 - y = 2x^3 + 2x^2 + x + C$, where C is an arbitrary constant. To determine the required solution that satisfies the prescribed initial condition, we substitute $x = 0$ and $y = 2$ into the above equation to obtain $C = 2$. Hence the solution of the given initial value problem is given implicitly by $y^2 - y = 2x^3 + 2x^2 + x + 2$.

Since this equation is quadratic in y, we obtain the explicit solution:

$$y = \frac{1}{2} \pm \frac{1}{2}\sqrt{9 + 4x + 8x^2 + 8x^3}.$$

This formula contains two solutions of the given differential equation, only one of which, however, belongs to the range $y > 0.5$ and satisfies the initial condition:

$$y = \phi(x) = \frac{1}{2} + \frac{1}{2}\sqrt{9 + 4x + 8x^2 + 8x^3}.$$

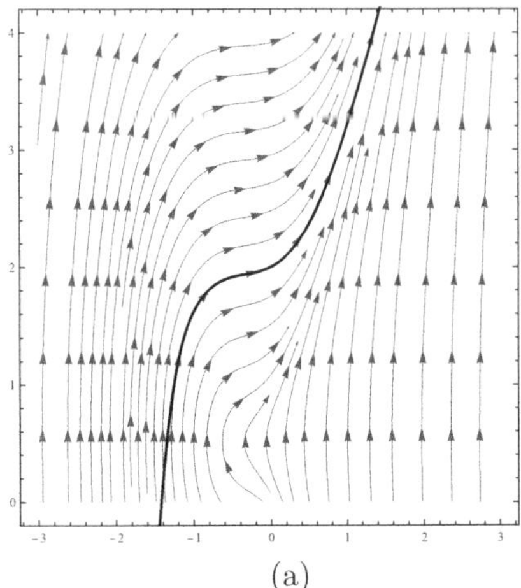

(a)

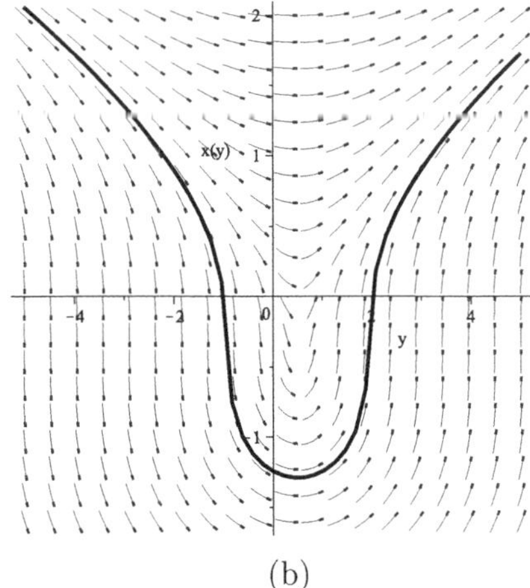

(b)

Figure 2.44: Example 2.7.16. (a) The streamlines for the equation $\frac{\mathrm{d}y}{\mathrm{d}x} = \frac{6x^2+4x+1}{2y-1}$ and the solution $y = \phi(x)$, plotted with *Mathematica;* (b) the direction field and solution for the reciprocal equation $\frac{\mathrm{d}x}{\mathrm{d}y} = \frac{2y-1}{6x^2+4x+1}$, plotted with *Maple.*

The form of solution indicates that it consists of two branches. Therefore, it is more natural to consider the reciprocal equation

$$\frac{\mathrm{d}x}{\mathrm{d}y} = \frac{2y-1}{6x^2+4x+1},$$

considering y as the independent variable. As can be seen from Fig. 2.44, the solution $y = \phi(x)$ is only part of the solution $x = x(y)$. Finally, to determine the interval in which the solution $y = \frac{1}{2} + \frac{1}{2}\sqrt{9+4x+8x^2+8x^3}$ is valid, we must find the interval in which the quantity under the radical is positive. Solving numerically the equation $9+4x+8x^2+8x^3 = 0$, we see that it has only one real root, $x_0 \approx -1.28911$, where the slope is infinite. The desired interval is $x > x_0$. On the other hand, the reciprocal equation has the solution $x(y)$, which exists for any y. □

Is it possible to determine the maximum interval of existence of the solution to $y' = f(y)$, and the behavior of the solution near the ends of that interval without actually solving the problem? To answer this question, we observe that the zeroes of the slope function f, known as the critical points of the differential equation, divide the real line into subintervals so that f has a constant sign on each interval. This allows us to formulate the following results.

Theorem 2.6: Let $f(y)$ in the autonomous equation $y' = f(y)$ be a continuous function on some interval (a, b), $a < b$. Suppose that $f(y)$ has a finite number of simple zeroes $y_1, y_2, \ldots, y_n$ on this interval, so that $f(y_k) = 0$ and $f'(y_k) \neq 0$, $k = 1, 2, \ldots, n$. These points, called equilibrium solutions, divide the interval into subintervals where f has a constant sign.

If on subinterval (y_k, y_{k+1}) the function $f(y)$ is positive, then any solution of the equation $y' = f(y)$ subject to the initial condition $y(x_0) = y_0 \in (y_k, y_{k+1})$ must remain in (y_k, y_{k+1}) and be an increasing function. This makes the equilibrium point $y = y_{k+1}$ stable, and the critical point $y = y_k$ unstable (if the function $f(y)$ is negative outside this subinterval).

If on subinterval (y_k, y_{k+1}) the function $f(y)$ is negative, then any solution of the equation $y' = f(y)$ with an initial value $y(x_0) = y_0$ within this subinterval is also decreasing. The point $y = y_{k+1}$ is an unstable critical point, while $y = y_k$ is stable (if the function $f(y)$ is positive outside this subinterval).

Theorem 2.6 can be extended for multiple nulls of odd degrees; it allows us to consider an autonomous equation $y' = f(y)$ on the interval bounded by critical points.

Theorem 2.7: Suppose that a continuously differentiable function $f(y)$ has no zeroes on an open interval (a, b). Then the maximum interval on which the solution to the initial value problem

$$y' = f(y), \qquad y(x_0) = y_0 \in (a, b),$$

can be defined is either

$$\left(x_0 + \int_{y_0}^{a} \frac{du}{f(u)},\ x_0 + \int_{y_0}^{b} \frac{du}{f(u)}\right) \qquad \text{if } f(y) \text{ is positive,} \tag{2.7.8}$$

or

$$\left(x_0 + \int_{y_0}^{b} \frac{du}{f(u)},\ x_0 + \int_{y_0}^{a} \frac{du}{f(u)}\right) \qquad \text{if } f(y) \text{ is negative.} \tag{2.7.9}$$

PROOF: It is similar to the proof of Theorem 2.1, page 48. See details in [10]. ∎

At the end of this section, we reconsider previous examples along with two new ones.

Example 2.7.17: (Previous examples revisited) In Example 2.7.13, the domain of the function $f(y) = \sqrt{1-y^2}$ is the closed interval $[-1, 1]$. The corresponding initial value problem has two equilibrium singular solutions $y = \pm 1$. To find the largest interval on which a solution to the initial value problem

$$y' = \sqrt{1-y^2}, \qquad y(0) = 0,$$

can be defined, we apply Theorem 2.7. Since

$$\int_0^{-1} \frac{dy}{\sqrt{1-y^2}} = -\frac{\pi}{2} \qquad \text{and} \qquad \int_0^{1} \frac{dy}{\sqrt{1-y^2}} = \frac{\pi}{2},$$

we conclude that the given initial value problem has the unique solution, $y = \sin x$, on the interval $\left(-\frac{\pi}{2}, \frac{\pi}{2}\right)$.

In Example 2.7.14, the function $f(y) = y^2 - 9 = (y-3)(y+3)$ has two zeroes: $y = \pm 3$. These equilibrium solutions are not singular because solutions approach/leave them, but do not touch them. In the interval $(-3, 3)$, the function $f(y)$ is positive. According to Theorem 2.1, page 48, any initial value problem with the initial condition within this interval has the validity interval $(-\infty, \infty)$. The initial condition $y(0) = 5 \in (3, \infty)$ is outside the interval $(-3, 3)$ where it is negative. Since

$$\int_5^{3} \frac{dy}{y^2-9} = -\infty \qquad \text{and} \qquad \int_5^{\infty} \frac{dy}{y^2-9} = \frac{1}{3}\ln 2 \approx 0.2310490602,$$

the maximum interval of existence for such a problem is $\left(-\infty, \frac{1}{3}\ln 2\right)$. Similarly, we can determine the validity interval for the initial value problem with the initial condition $y(0) = -5$ to be $\left(-\frac{1}{3}\ln 2, \infty\right)$. In general, the validity interval for the initial value problem

$$y' = y^2 - 9, \qquad y(0) = y_0,$$

depends on the value of y_0:

$$\begin{array}{ll} \left(-\frac{1}{6}\ln\left|\frac{y_0-3}{y_0+3}\right|, \infty\right), & \text{if } y_0 < 3; \\ (-\infty, \infty), & \text{if } -3 < y_0 < 3; \\ \left(-\infty, \frac{1}{6}\ln\left|\frac{y_0+3}{y_0-3}\right|\right), & \text{if } y_0 > 3. \end{array}$$

In Example 2.7.15, the function $f(y) = y(y^2-9) = y(y-3)(y+3)$ has three equilibrium (nonsingular) solutions $y = \pm 3$ and $y = 0$. The initial condition $y(0) = 1$ belongs to the interval $(0, 3)$ where $f(y)$ is negative. Using Theorem 2.7, we conclude that the maximum interval of existence is $(-\infty, \infty)$ because

$$\int_1^{3} \frac{dy}{y(y^2-9)} = -\infty \qquad \text{and} \qquad \int_1^{0} \frac{dy}{y(y^2-9)} = \infty.$$

If the initial condition $y(0) = y_0 \in (3, \infty)$, $f(y_0) > 0$, the maximum interval of existence is determined by Eq. (2.7.8):

$$\left(\int_{y_0}^{3} \frac{dy}{y(y^2-9)}, \int_{y_0}^{\infty} \frac{dy}{y(y^2-9)} = \frac{1}{18}\ln\left|1 - \frac{9}{y^2}\right|_{y=y_0}^{y=\infty}\right) = \left(-\infty, -\frac{1}{18}\ln\left|1 - \frac{9}{y_0^2}\right|\right).$$

Example 2.7.18: Find the maximum interval of existence of the solution to the initial value problem

$$y' = y^3 + 3y^2 + 4y + 2, \qquad y(0) = 2.$$

Since $y^3 + 3y^2 + 4y + 2 = (y+1)(y^2 + 2y + 2)$, the equation has only the one equilibrium solution $y = -1$. The function $f(y) = y^3 + 3y^2 + 4y + 2$ is positive for $y > -1$; therefore, the maximum interval of existence will be

$$\left(\int_2^{-1} \frac{du}{f(u)}, \int_2^{\infty} \frac{du}{f(u)} \right) = \left(-\infty, \frac{1}{2} \ln \frac{10}{9} \right) \approx (-\infty, 0.05268).$$

For the initial condition $y(0) = -2$, the validity interval will be

$$\left(\int_{-2}^{-1} \frac{du}{f(u)}, \int_{-2}^{-\infty} \frac{du}{f(u)} \right) = \left(-\infty, \frac{1}{2} \ln 2 \right) \approx (-\infty, 0.34657).$$

Example 2.7.19: Consider the initial value problem

$$y' = y^3 - 2y^2 - 11y + 12, \qquad y(-1) = y_0.$$

The slope function $f(y) = y^3 - 2y^2 - 11y + 12 = (y+3)(y-1)(y-4)$ has three zeroes: $y = -3$, $y = 1$, and $y = 4$. These are equilibrium solutions, two of which ($y = -3$ and $y = 4$) are unstable and one ($y = 1$) is stable. If the initial value $y_0 < -3$, the largest interval on which the solution exists is

$$\left(-1 + \int_{y_0}^{-3} \frac{du}{f(u)}, -1 + \int_{y_0}^{-\infty} \frac{du}{f(u)} \right) = \left(-\infty, -1 - \ln \frac{|y_0 - 4|^{1/21} \, |y_0 + 3|^{1/28}}{|y_0 - 1|^{1/12}} \right).$$

If the initial value y_0 is within the interval $(-3, 1)$, the validity interval becomes

$$\left(-1 + \int_{y_0}^{-3} \frac{du}{f(u)}, -1 + \int_{y_0}^{1} \frac{du}{f(u)} \right) = (-\infty, \infty).$$

If the initial value $y_0 \in (1, 4)$, then the maximum interval of existence is

$$\left(-1 + \int_{y_0}^{4} \frac{du}{f(u)}, -1 + \int_{y_0}^{1} \frac{du}{f(u)} \right) = (-\infty, \infty).$$

Finally, when the initial value exceeds 4, the validity interval becomes

$$\left(-1 + \int_{y_0}^{4} \frac{du}{f(u)}, -1 + \int_{y_0}^{\infty} \frac{du}{f(u)} \right) = \left(-\infty, -1 - \ln \frac{|y_0 - 4|^{1/21} \, |y_0 + 3|^{1/28}}{|y_0 - 1|^{1/12}} \right).$$

Problems

1. Consider the equation $y' = 4\sqrt{1 - y^2}$.
   (a) Find its general solution.
   (b) Show that the constant functions $y(t) \equiv 1$ and $y(t) \equiv -1$ are solutions for all t.
   (c) Show that the following function is a singular solution:
   $$\varphi(t) = \begin{cases} 1, & \text{for } t \geqslant \frac{\pi}{8}, \\ \sin(4t), & \text{for } -\frac{\pi}{8} \leqslant t \leqslant \frac{\pi}{8}, \\ -1, & \text{for } t \leqslant -\frac{\pi}{8}. \end{cases}$$
   (d) What is the maximum interval of existence for the IVP $y' = 4\sqrt{1 - y^2}$, $y(0) = 0$?

2. For each of the following autonomous differential equations, draw the slope field and label each equilibrium solution as a sink, source, or node.

   (a) $y' = y(1-y)(y-3)$; **(b)** $y' = y^2(1-y)(y-3)$;
   (c) $y' = y(1-y)^2(y-3)^2$; **(d)** $y' = y^2(1-y)^2(y-3)^2$;
   (e) $y' = y(1-y)^3(y-3)$; **(f)** $y' = y^3(1-y)^2(y-3)^3$.

3. Find the general solution $y(t)$ to the autonomous equation $y' = A\,y^a$ $(a > 1)$ and determine where it blows up.
4. For each of the following differential equations, find its nullclines. By plotting their solutions, determine which nullclines the solutions follow.
   (a) $t^2\dot{y} = y^3 - 4ty$; **(b)** $\dot{y} = y - t^2$; **(c)** $\dot{y} = \sin(y)\sin(t)$;
   (d) $\dot{y} = y^2 - \frac{t^2}{t^2+1}$; **(e)** $\dot{y} = \sin(y - t^2)$; **(f)** $\dot{y} = y^2 - t^3$.
5. Consider the initial value problem $\dot{y} - t\,y = 1$, $y(0) = y_0$. Using any available solver, find approximately (up to 5 significant places) the value of y_0 that separates solutions that grow positively as $t \to +\infty$ from those that grow negatively. How does the solution that corresponds to the exceptional value of y_0 behave as $t \to +\infty$?
6. Consider the differential equations:

$$\text{(a) } \dot{y} = y^2; \qquad \text{(b) } \dot{y} = y^2 + 1; \qquad \text{(c) } \dot{y} = y^2 - 1.$$

   (a) Using any available solver, draw a direction field and some solutions for each equation.
   (b) Which solutions have vertical/horizontal asymptotes?

In each of Problems 7 through 12,

(a) Draw a direction field for the given linear differential equation. How do solutions appear to behave as x becomes large? Does the behavior depend on the choice of the initial value a? Let a_0 be the value of a for which there exists an exceptional solution that separates one type of behavior from another one. Estimate the value of a_0.
(b) Solve the initial value problem and find the critical value a_0 exactly.
(c) Describe the behavior of the solution corresponding to the initial value a.

7. $y' - y = 2\sin x$, $y(0) = a$;
8. $y' = 2y + t\,e^{-2t}$, $y(0) = a$;
9. $y' = x + 3\,y$, $y(0) = a$;
10. $y' = 3\,y + t^2$, $y(0) = a$;
11. $y' = x^2 + x\,y$, $y(0) = a$;
12. $y' = 3t^2\,y + t^2$, $y(0) = a$.

In each of Problems 13 through 18, determine whether solutions to the given differential equation exhibit any kind of symmetry. Is it symmetrical with respect to the vertical and/or horizontal axis?

13. $y' = 2y$;
14. $y' = y^3 + x$;
15. $y' = 2y + 1$;
16. $y' + 2y = e^{-t}\,y^3$;
17. $y' = \sin(xy)$;
18. $y' = \cos(xy)$.

In each of Problems 19 through 24, prove that a vertical asymptote exists for the given differential equation.

19. $y' = y^4 - x^2$;
20. $y' = y^3 - yx^2$;
21. $y' = e^{-y} + y^3$;
22. $y' = y^3 + 2y^2 - y - 2$;
23. $y' = y^2 + \sin(x)$;
24. $y' = y - 4y^2$.

In each of Problems 25 through 30, find the maximum interval of existence of the solution to the given initial value problem.

25. $\dot{y} = (y-2)(y^2+1)(y^2+2y+2) \equiv y^5 - y^3 - 4y^2 - 2y - 4$, $y(0) = 0$;
26. $\dot{y} = (y^2+2y+3)(y^2-1) \equiv y^4 + 2y^3 + 2y^2 - 2y - 3$, $y(0) = 0$;
27. $\dot{y} = (y^3-1)(y^4+y^2+1) \equiv y^7 + y^5 + y^3 - y^4 - y^2 - 1$, $y(0) = 0$;
28. $\dot{y} = (y^3-1)(y^2+2y+4) \equiv y^5 + 2y^4 + 4y^3 - y^2 - 2y - 4$, $y(0) = 2$;
29. $\dot{y} = (y^4-1)(y^2+y+1) \equiv y^6 + y^5 + y^4 - y^2 - y - 1$, $y(0) = 0$;
30. $\dot{y} = (y^3+y+2)(y^3+y+10) \equiv y^6 + 2y^4 + 12y^3 + y^2 + 12y + 20$, $y(0) = 0$.
31. Consider the differential equation $y' = -x/y$.
    (a) Use Picard's theorem to show that there is a unique solution going through the point $y(1) = -0.5$.
    (b) Show that for $|x| < 1$, $y(x) = -\sqrt{1-x^2}$ is a solution of $y' = -x/y$.
    (c) Using the previous result, find a lower fence of the solution in part (a).
    (d) Solve the initial value problem in part (a) analytically. Find the exact x-interval on which the solution is defined.
32. Suppose that $y = y^*$ is a critical point of an autonomous differential equation $\dot{y} = f(y)$, where $f(y)$ is a smooth function. Show that the constant equilibrium solution $y = y^*$ is asymptotically stable if $f(y) < 0$ (> 0) and unstable if $f(y) > 0$ (< 0) for $y > y^*$ $(y < y^*)$.

In each of Problems 33 through 36, find the solution to the given initial value problem. Then plot the solution defined implicitly and determine or estimate the validity interval.

33. $y' = 3x^2/(3y^2 - 3.4), \quad y(1) = 0;$
34. $y' = (y^2 - 2y)/(2 + 6x^2), \quad y(0) = 1;$
35. $y' = (2 + 6x^2)/(y^2 - 2y), \quad y(0) = 1;$
36. $y' = (y^2 + 2y)/(x^2 - 4x), \quad y(1) = 1.$

Summary for Chapter 2

1. The differential equation $\mathrm{d}y/\mathrm{d}x = f(x, y)$ is a **separable** equation if the slope function $f(x, y)$ is a product of two functions of dependent and independent variables only, that is, $f(x, y) = p(x)\,q(y)$. The integral

$$C = \int p(x)\,\mathrm{d}x - \int \frac{\mathrm{d}y}{q(y)}$$

defines the general solution in implicit form.

2. A separable differential equation $y' = f(y)$ is called **autonomous**.
3. A nonlinear differential equation $y' = F(ax + by + c)$ can be transformed to a separable one by the substitution $v = ax + by + c$, $b \neq 0$.
4. A nonlinear differential equation $x\,y' = y\,F(xy)$ can be transformed to a separable one by the substitution $v = xy$.
5. The differential equation $y' = F\left(\frac{y}{x}\right)$, $x \neq 0$, can be reduced to a separable differential equation for a new dependent variable $v = v(x)$ by setting $y = vx$. This equation may have a singular solution of the form $y = kx$, where k is (if any) a root of the equation $k = F(k)$.
6. A function $g(x, y)$ is called a *quasi-homogeneous* function with weights α and β if

$$g(\lambda^\alpha x, \lambda^\beta x) = \lambda^{\beta-\alpha}\, g(x, y),$$

for some real numbers α and β. If $\alpha = \beta$, the function is called homogeneous. By setting $y = z^{\beta/\alpha}$ or $y = u\,x^{\beta/\alpha}$, the differential equation $y' = g(x, y)$ with a quasi-homogeneous (or isobaric) equation $g(x, y)$ is reduced to a separable one.

7. The differential equation

$$\frac{\mathrm{d}y}{\mathrm{d}x} = F\left(\frac{ax + by}{Ax + By}\right)$$

is an example of a differential equation with a homogeneous slope function.

8. The differential equation

$$\frac{\mathrm{d}y}{\mathrm{d}x} = F\left(\frac{ax + by + c}{Ax + By + C}\right)$$

has a slope function which is not a homogeneous function, if either C or c is not zero. The solution of this differential equation is case dependent.

(a) **Case 1**. If $aB \neq bA$, then this equation is reduced to a differential equation with a homogeneous slope function by shifting variables

$$x = X + \alpha \quad \text{and} \quad y = Y + \beta.$$

To obtain this, one should determine the constants α and β that satisfy the following system of algebraic equations:

$$a\alpha + b\beta + c = 0, \qquad A\alpha + B\beta + C = 0.$$

With this in hand, we get the differential equation with a homogeneous rate function

$$\frac{\mathrm{d}Y}{\mathrm{d}X} = F\left(\frac{aX + bY}{AX + BY}\right)$$

in terms of the new variables X and Y.

(b) **Case 2**. If $aB = bA$, then we set

$$v = k\,(ax + by)$$

for an arbitrary nonzero constant k. Substituting a new variable v into Eq. (2.2.12), we obtain an autonomous differential equation

$$v' = ka + kb\,F\left(\frac{av + kac}{Av + kaC}\right).$$

9. A differential equation

$$M(x, y)\,\mathrm{d}x + N(x, y)\,\mathrm{d}y = 0 \tag{2.3.1}$$

is called **exact** if and only if $\frac{\partial M}{\partial y} = \frac{\partial N}{\partial x}$. In this case, there exists a **potential function** $\psi(x, y)$ such that

$$\mathrm{d}\psi(x, y) = M(x, y)\,\mathrm{d}x + N(x, y)\,\mathrm{d}y, \quad \text{or} \quad \frac{\partial \psi}{\partial x} = M(x, y), \quad \frac{\partial \psi}{\partial y} = N(x, y).$$

10. An exact differential equation (2.3.1) has the general solution in implicit form $\psi(x,y) = C$, where ψ is a potential function and C is an arbitrary constant.

11. A nonhomogeneous equation
$$M(x,y)\,\mathrm{d}x + N(x,y)\,\mathrm{d}y = f(x)\,\mathrm{d}x \tag{2.3.12}$$
is also called **exact** if there exists a function $\psi(x,y)$ such that $\mathrm{d}\psi(x,y) = M(x,y)\,\mathrm{d}x + N(x,y)\,\mathrm{d}y$. This equation (2.3.12) has the general solution in implicit form
$$\psi(x,y) = \int f(x)\,\mathrm{d}x + C.$$

12. A differential equation of the first order $M(x,y)\,\mathrm{d}x + N(x,y)\,\mathrm{d}y = 0$ can sometimes be reduced to an exact equation. If after multiplication by a function $\mu(x,y)$ the resulting equation
$$\mu(x,y)M(x,y)\,\mathrm{d}x + \mu(x,y)N(x,y)\,\mathrm{d}y = 0$$
is an exact one, we call $\mu(x,y)$ the **integrating factor**.

13. In general it is almost impossible to find such a function $\mu(x,y)$. Therefore, we consider two of the easiest cases when an integrating factor is a function of only one of the variables, either x or y. Thus,
$$\mu(x) = \exp\left\{\int \frac{M_y - N_x}{N}\,\mathrm{d}x\right\}, \quad \mu(y) = \exp\left\{-\int \frac{M_y - N_x}{M}\,\mathrm{d}y\right\}.$$
The integrating factors in these forms exist if the fractions $(M_y - N_x)/N$ and $(M_y - N_x)/M$ are functions of x and y alone, respectively.

14. If the coefficients M and N of Eq. (2.3.1) satisfy the relation $M_y(x,y) - N_x(x,y) = p(x)N(x,y) - q(y)M(x,y)$, an integrating factor has the form $\mu(x,y) = \exp\left\{\int p(x)\,\mathrm{d}x + \int q(y)\,\mathrm{d}y\right\}$.

15. There are two known methods to solve linear differential equations (2.5.3) of the first order in **standard form**,
$$y'(x) + a(x)\,y(x) = f(x), \tag{2.5.3}$$
namely, the **Bernoulli method** and the **integrating factor method**.

    (a) **Bernoulli's method**. First, find a function $u(x)$ which is a solution of the homogeneous equation
    $$u'(x) + a(x)\,u(x) = 0.$$
    Since it is a separable one, its solution is
    $$u(x) = e^{-\int a(x)\,\mathrm{d}x}.$$
    Then, we seek the solution of the given nonhomogeneous equation in the form $y(x) = u(x)v(x)$, where $v(x)$ is a solution of another separable equation $u(x)v'(x) = f(x)$, which has the general solution
    $$v(x) = \int \frac{f(x)}{u(x)}\,\mathrm{d}x + C.$$
    Multiplying $u(x)$ and $v(x)$, we obtain the general solution of Eq. (2.5.3):
    $$y(x) = e^{-\int a(x)\,\mathrm{d}x}\left[C + \int f(x)\,e^{\int a(x)\,\mathrm{d}x}\,\mathrm{d}x\right].$$

    (b) **Integrating factor method**. Let $\mu(x)$ be a solution of the homogeneous differential equation
    $$\mu'(x) - a(x)\,\mu(x) = 0 \qquad \Longrightarrow \qquad \mu(x) = e^{\int a(x)\,\mathrm{d}x}.$$
    Upon multiplication of both sides of Eq. (2.5.3) by μ, the given differential equation is reduced to the exact equation
    $$\frac{\mathrm{d}}{\mathrm{d}x}[\mu(x)\,y(x)] = \mu(x)\,f(x),$$
    and, after integrating, we obtain
    $$\mu(x)\,y(x) = \int \mu(x)\,f(x)\,\mathrm{d}x \qquad \text{or} \qquad y(x) = \frac{1}{\mu(x)}\int \mu(x)\,f(x)\,\mathrm{d}x.$$

Review Questions for Chapter 2

Section 2.1 of Chapter 2 (Review)

1. Which of the following equations are separable?
   (a) $y' = y\,e^{x+y}\,(x^2+1)$; (b) $x^2y' = 1+y^2$; (c) $y' = \sin(xy)$; (d) $x(e^y+4)\,dx = e^{x+y}\,dy$;
   (e) $y' = \cos(x+y)$; (f) $xy' + y = xy^2$; (g) $y' = t\,\ln(y^{2t}) + t^2$; (h) $y' = x\,e^{y^2-x}$;
   (i) $y' = \ln|xy|$; (j) $y' = \tan y$.
2. Solve the following separable equations.
   (a) $x(y+1)^2\,dx = (x^2+1)ye^y\,dy$; (b) $(y-1)\,dx - (x+1)\,dy = 0$; (c) $xy\,y' = (y+1)(2-x)$;
   (d) $\tan x\,dx + \cot y\,dy = 0$; (e) $(x^2-1)\,dy + (y+1)\,dx = 0$; (f) $x\,dy + \sqrt{1+y^2}\,dx = 0$;
   (g) $(x+1)\sin y\,y' + 2x\cos y = 0$; (h) $\sqrt{1+y^2}\,dx + xy\,dy = 0$; (i) $xy\ln y\,dx = \sec x\,dy$;
   (j) $(4x^2+1)y' = y^2$; (k) $(1-x)\,y' = 3y^2 - 2y$; (l) $y\,y' = x$.
3. Find the general solution of the following differential equations. Check your answer by substitution.
   (a) $(y+1)\,dx - (x+2)\,dy = 0$; (b) $y\,dx - x(y-1)\,dy = 0$;
   (c) $\cos y\,dx - x\,\sin y\,dy = 0$; (d) $x\,dx + e^{x-y}\,dy = 0$;
   (e) $(y-1)(1-x)\,dx + xy\,dy = 0$; (f) $xy\,dx - (x^2+1)\,dy = 0$;
   (g) $(y+1)\,dx + (x^2-4)\,dy = 0$; (h) $\sin 2x\,dx + \cos 3y\,dy = 0$.
4. Obtain the particular solution satisfying the initial condition.
   (a) $2xyy' = 1+y$, $y(1) = 1$; (b) $\tan y\,y' + x\cos^2 y = 0$, $y(0) = 0$;
   (c) $(1-y^2)y' = \ln|x|$, $y(1) = 0$; (d) $\cos y\,dx + x\sin y\,dy = 0$, $y(1) = 0$;
   (e) $xy' = \sqrt{1-y^2}$, $y(1/2) = 0$; (f) $xy\ln y\,dx = \sec x\,dy$, $y(0) = e$;
   (g) $y' = xy$, $y(0) = 2$; (h) $\sqrt{1+y^2}\,dx + xy\,dy = 0$, $y(1) = 0$;
   (i) $y' = -y^2\sin x$, $y(0) = 1$; (j) $x\,dx + ye^{-x}\,dy = 0$, $y(0) = 1$.
5. At time $t = 0$, a new version of WebCT software is introduced to n faculty members of a college. Determine a differential equation governing the number of people $P(t)$ who have adapted the software for education purposes at time t if it is assumed that the rate at which the invention spreads through the faculty members is jointly proportional to the number of instructors who have adopted it and the number of faculty members who have not adopted it.
6. Separate the equation $dy/dx = (4x-3)(y-2)^{2/3}$ to derive the general solution. Show that $y = 2$ is also a solution of the given equation, but it cannot be found from the general solution. Thus, the singular solution $y = 2$ is lost upon division by $(y-2)^{2/3}$.
7. Solve the equation $\dfrac{dy}{dx} = \dfrac{1}{x} - \dfrac{1}{y^2+2} - \dfrac{1}{x(y^2+2)} + 1$.
8. Replace $y = \sin u$ in the equation $\dfrac{dy}{dx} = \dfrac{1}{x}\sqrt{1-y^2}\arcsin y$ and solve it.
9. A gram of radium takes 10 years to diminish to 0.997 gm. What is its half-life?
10. For a positive number $a > 0$, find the solution to the initial value problem $y' = x^2y^4$, $y(0) = a$, and then determine the domain of its solution. Show that as a approaches zero the domain approaches the entire real line $(-\infty, \infty)$, and as a approaches $+\infty$ the domain contains all points except $x = 0$.
11. Bacteria in a colony are born and die at rates proportional to the number present, so that its growth obeys Eq. (1.1.1) with $\lambda = k_1 - k_2$, where k_1 corresponds to the birth rate and k_2 to the death rate. Find k_1 and k_2 if it is known that the colony doubles in size every 48 hours and that its size would be halved in 12 hours if there were no births.
12. Are the following equations separable? If so, solve them.
    (a) $\frac{dr}{d\phi} = \frac{2\cos\phi + e^r\cos\phi}{e^{3r} + e^{3r}\sin\phi/2}$; (b) $4x^2\,e^{3x^3-y^2}\,dx = 3y^5\,e^{y^2-x^3}\,dy$.
13. For each of the following autonomous differential equations, find all equilibrium solutions and determine their stability or instability.
    (a) $\dot{x} = 3x^2 + 8x - 3$; (b) $\dot{x} = x^2 - 8x + 7$; (c) $\dot{x} = 2x^2 - 9x + 4$; (d) $\dot{x} = 3x^2 - 5x + 2$.
14. Sketch a direction field of the differential equation $\dot{y} = y(1-2t)$ and determine the maximum value of a particular solution subject to the initial condition $y(0) = 1$.
15. Suppose that a hole is available through the earth (it has an equatorial diameter of 12,756 km, and a polar diameter of 12713.6 km). If an object were dropped down a hole through the center of the earth, it would be attracted toward the center with the force directly proportional to the distance from the center. Newton's universal law of attraction gives the differential equation for the velocity v of the ball as $\dfrac{dv}{dt} = -\dfrac{g\,r}{R}$, where $R \approx 6{,}371$ km is the radius of the earth, $g \approx 9.8$ m/sec is acceleration due to gravity, and r is the distance of the ball from the center of the earth. Find the time to reach the other end of the hole.

16. Consider a conical reservoir 20 meters deep with an open top that has radius 60 meters. Initially, the reservoir is empty, but water is added at a rate of $r\,\text{m}^3/\text{hr}$. Water evaporates from the tank at a rate proportional to the surface area, the constant of proportionality being $1/(9\pi)$. Convert the differential equation that describes the volume V of water having height h
$$\frac{\mathrm{d}V}{\mathrm{d}t} = r - \frac{h^2}{9\pi}$$
into the differential equation, containing h and solve it.

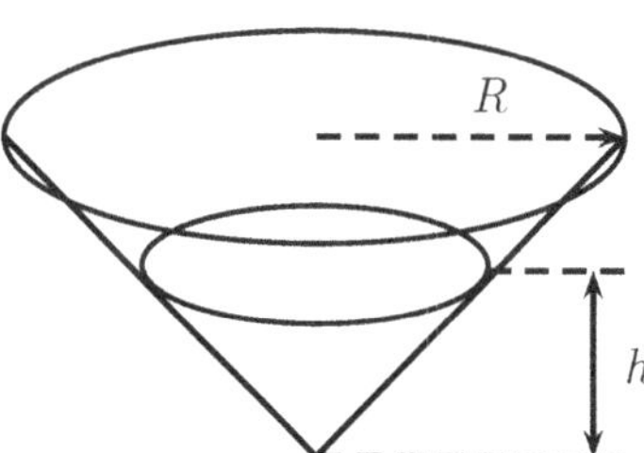

Section 2.2 of Chapter 2 (Review)

1. Use appropriate substitution to solve the given equations.

(a) $y' = \sqrt{4x+2y-1}$; (b) $y' = -\frac{x+2y-5}{x+2y}$; (c) $y' = (2x+3y+1)^2 - 2$;
(d) $y' = \frac{6x+3y-10}{2x+y}$; (e) $y' = \sqrt{2x+3y}$; (f) $y' = (x+y+2)^2 - (x+y-2)^2$;
(g) $y' = (2x+3y)^{-1}$; (h) $y' = \sqrt{2x+y} - 2$; (i) $y' = y - x + 1 + (x-y+2)^{-1}$.

2. Determine whether or not the function is homogeneous. If it is homogeneous, state the degree of the function.

(a) $x^2 + 4xy + 2\,y^2$; (b) $\ln|y| - \ln|x|$; (c) $\frac{x}{y-1}$; (d) $\frac{\sqrt{x^2+4y^2}}{3x+y}$;
(e) $\frac{x+2}{y} + \frac{5y-4}{2y}$; (f) $\sin(y+x)$; (g) $\sin(y/x)$; (h) $\frac{\sqrt{x+y}}{x-y}$.

3. Solve the given differential equations with homogeneous right-hand side functions.

(a) $yy' = x + y$; (b) $(x^2y + xy^2 - y^3)\,\mathrm{d}x + (xy^2 - x^3)\,\mathrm{d}y = 0$;
(c) $(x+y)\,y' = y$; (d) $y\sqrt{x^2+y^2}\,\mathrm{d}x = (x\sqrt{x^2+y^2} + y^2)\,\mathrm{d}y$;
(e) $y' = e^{y/x} + y/x$; (f) $y' = (4xy^2 - 2x^3)/(4y^3 - x^2y)$;
(g) $y' = (y^3 + x^3)/xy^2$; (h) $x^2y' = 4x^2 + 7xy + 2y^2$;
(i) $y' = 2x/(x+y)$; (j) $(6x^2 - 5xy - 2y^2)\,\mathrm{d}x = (8xy - y^2 - 6x^2)\,\mathrm{d}y$;
(k) $x^3y' = y^3 + 2x^2y$; (l) $\left(\frac{1}{r^2} - \frac{3\theta^2}{r^4}\right)\frac{\mathrm{d}r}{\mathrm{d}\theta} + \frac{2\theta}{r^3} = 0$.

4. Solve the given differential equation with a homogeneous right-hand side function. Then determine an arbitrary constant that satisfies the auxiliary condition.

(a) $y' = \frac{x^2y - 4y^3}{2x^3 - 4xy^2}$, $y(1) = 1$; (b) $x^2y' = y^2 + 2xy$, $y(1) = 1$;
(c) $y' = \frac{y^3 + 3x^3}{xy^2}$, $y(1) = \ln 2$; (d) $y' = \frac{y}{x}\left(\ln\frac{y}{x} + 1\right)$, $y(1) = e$.

5. Solve the following equations with linear coefficients.

(a) $(x-y)\,y' = x + y + 2$; (b) $(4x+5y-5)\,y' = 4 - 5x - 4y$;
(c) $y' = x - y + 3$; (d) $(x+2y)\,\mathrm{d}x + (2x+3y+1)\,\mathrm{d}y = 0$;
(e) $(x+2y)\,y' = 1$; (f) $(x+4y-3)\,\mathrm{d}x - (2x+6y-2)\,\mathrm{d}y = 0$;
(g) $(2x+y)\,y' = 6x + 3y - 1$; (h) $(3x+y-1)\,\mathrm{d}x - (6x+2y-3)\,\mathrm{d}y = 0$;
(i) $(x+3)\,y' = x + y + 4$; (j) $(x+y+1)\,\mathrm{d}x + (3y-3x+9)\,\mathrm{d}y = 0$;
(k) $(x-y)\,y' = x + y - 2$; (l) $(2x+y-4)\,\mathrm{d}x + (x-y+1)\,\mathrm{d}y = 0$;
(m) $(x-y)\,\mathrm{d}x + (3x+y)\,\mathrm{d}y = 0$; (n) $(x-3y+8)\,\mathrm{d}x - (x+y)\,\mathrm{d}y = 0$;
(o) $(x+y)y' = 2 - x - y$; (p) $(2x+y+1)y' = 4x + 2y + 6$.

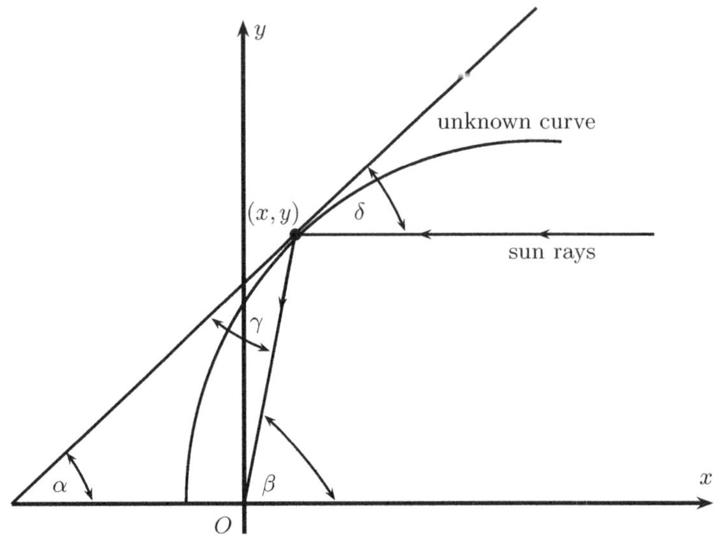

6. Solve the following equations subject to the given initial conditions.

(a) $y' = \dfrac{x-2y+5}{y-2x-4}$, $y(1)=1$; (b) $y' = \dfrac{2-2x-3y}{1-2x-3y}$, $y(1)=-2/3$;

(c) $y' = \dfrac{x+3y-5}{x-y-1}$, $y(2)=2$; (d) $y' = \dfrac{4y-3x+1}{2x-y+1}$, $y(0)=1$;

(e) $y' = \dfrac{y-4x+6}{x-y-3}$, $y(2)=-3$; (f) $y' = \dfrac{4x+11y-15}{11x-4y-7}$, $y(2)=1$;

(g) $y' = \dfrac{4x+3y+2}{3x+y-1}$, $y(1)=-1$; (h) $y' = \dfrac{3y-5x-1}{y-3x+1}$, $y(1)=1$;

(i) $y' = \dfrac{7x+3y+1}{y-3x+5}$, $y(1)=1$; (j) $y' = \dfrac{2x+y-8}{-2x+9y-12}$, $y(3)=7/3$;

(k) $y' = \dfrac{3x-2y-5}{2x+7y+5}$, $y(1)=1$; (l) $y' = -\dfrac{4x+y+7}{3x+y+5}$, $y(-1)=6$.

7. Consider a solar collector that redirects the sun rays into a fixed point. Its shape is a surface of revolution obtained by revolving a curve $y=\phi(x)$ about the $x-$axis. Without loss of generality, you can assume that the rays are parallel to the $x-$axis and that they are focused at the origin (see above picture). Show that the curve $y=\phi(x)$ satisfies the differential equation

$$\frac{dy}{dx} = \frac{\sqrt{x^2+y^2}-x}{y}.$$

Then find its solution. *Hint:* The law of reflection says: $\angle\alpha = \angle\gamma = \angle\delta$ and $\angle\beta = 2\angle\alpha$.

8. Show that the nonseparable differential equation $xy' = -y + F(x)/G(xy)$ becomes separable on changing the dependent variable $v = xy$. Using this result, solve the equation $(4x^3 + y\cos(xy))\,dx + x\cos(xy)\,dy = 0$.

9. Solve the differential equation $x(y^2+y)\,dx = (x^2-y^3)\,dy$ by changing to polar coordinates $x = r\cos\theta$, $y = r\sin\theta$.

10. An airplane flies horizontally with constant airspeed v km per hour, starting at distance a east of a fixed beacon and always pointing directly at the beacon. There is a crosswind from the south with constant speed $w < v$.

Locate the positive x axis along the east-west line through the beacon and with the origin at the beacon. Show that the graph $y = y(x)$ of the airplane's path satisfies

$$\frac{dy}{dx} = \frac{y}{x} - \frac{w}{v}\sqrt{1+\left(\frac{y}{x}\right)^2}.$$

Using substitution $y = x\,u$, solve the airplane equation.

11. By making the substitution $y = v\,x^n$ and choosing a convenient value of n, show that the following differential equations can be transformed into equations with separable variables, and then solve them.

(a) $y' = \dfrac{1-x\,y^2}{10\,x^2 y}$; (b) $y' = \dfrac{1-xy^3}{6x^2y^2}$; (c) $y' = \dfrac{y-xy^2}{x+3x^2y}$.

12. Find a particular solution for the initial value problem $y' = y^2/x$, $\quad y(1) = 1$ with the isobaric slope function (see definition on page 67).

13. In each of the following initial value problems, find the specific solution and show that the given function $y_s(x)$ is a singular solution.

    (a) $y' = (x^2 + 3x - y)^{2/3} + 2x + 3, \qquad y(1) = 4; \qquad y_s(x) = x^2 + 3x.$

    (b) $y' = \sqrt{\frac{y-1}{x+3}}, \qquad y(1) = 1; \qquad y_s(x) = 1.$

    (c) $9y' = 4\left[x - \sqrt{x^2 - 9y}\right], \qquad y(3) = 1; \qquad y_s(x) = \frac{x^2}{9}.$

    (d) $y' = \sqrt{2 + y/x} - 2, \quad y(1) = -2; \qquad y_s(x) = -2x.$

Section 2.3 of Chapter 2 (Review)

1. Write the given differential equation in the form $M(x, y)\,dx + N(x, y)\,dy = 0$ and test for exactness. Solve the equation if it is exact.

   (a) $\dfrac{dy}{dx} = \dfrac{y}{4y^3 - x}$; **(b)** $\dfrac{dy}{dx} = -\dfrac{3y + 3x^2y^2}{3x + 2x^3y}$; **(c)** $y' + \dfrac{\frac{5}{2}x^{3/2} + 14y^3}{\frac{3}{2}y^{1/2} + 42xy^2} = 0$;

   (d) $y' = \dfrac{6xy - 2/x^2y}{-3x^2 + 2/xy^2}$; **(e)** $y' = \dfrac{\cos x\,\cos y + 2x}{\sin x\,\sin y - 2y}$; **(f)** $y' = \dfrac{4x^3 - e^x\,\sin y}{e^x\,\cos y + y^{-1/3}}$;

   (g) $y' = \dfrac{x - 3}{y^2}$; **(h)** $y' = -\dfrac{3x^2 + 2xy + y^2}{x^2 + 2xy + 3y^2}$; **(i)** $y' = \dfrac{e^x + 4x^3 + \cos y}{5 + x\,\sin y}$.

2. Solve the exact differential equations by finding a potential function.

   (a) $[y\sinh(xy) + 1]\,dx + [2 + x\sinh(xy)]\,dy = 0$; **(b)** $2r\theta^2\,dr + [2r^2\theta + \cos\theta]\,d\theta = 0$;

   (c) $[2r\cos(r\theta) - r^2\theta\sin(r\theta)]\,dr - r^3\sin(r\theta)\,d\theta = 0$; **(d)** $(y + e^{2x})\,dx + (x + y^2)\,dy = 0$;

   (e) $(4xy + x^2)\,dx + (2x^2 + y)\,dy = 0$; **(f)** $2xye^{x^2}\,dx + e^{x^2}\,dy = 0$;

   (g) $\left(\ln x + \frac{1}{x}\right)\,dx + \left(\frac{x}{y} + 2y\right)\,dy = 0$; **(h)** $y^2\sin x\,dx = (2y\cos x + 1)\,dy$;

   (i) $(2x^2 + 3y)\,dx + (4y^2 + 3x)\,dy = 0$; **(j)** $(x^2/2 + \ln y)\,y' + xy + \ln x = 0$;

   (k) $\left(\frac{x}{y} + \ln x\right)\,dy + \left(\frac{y}{x} + \ln y\right)\,dx = 0$; **(l)** $(x^{-2} + y)\,dx + (y^2 + x)\,dy = 0$;

   (m) $\left(y^3 - \frac{1}{x^2y}\right)\,dx = \left(\frac{1}{xy^2} - 3xy^2 - 1\right)\,dy$; **(n)** $(2xy + 1)\,dx + x^2\,dy = 0$.

3. Solve the exact differential equation and then find a constant to satisfy the initial condition.

   (a) $\dfrac{2x - 1}{t}\,dx + \dfrac{x - x^2}{t^2}\,dt = 0, \quad x(1) = 2$; **(b)** $\left(\dfrac{x^2}{y} - \dfrac{y^3}{3x^2}\right)dx = \left(\dfrac{x^3}{3y^2} - \dfrac{y^2}{x}\right)dy,\ y(1) = 1$;

   (c) $ye^{xy}\,dx + (xe^{xy} + 4y)\,dy = 0,\ y(0) = 2$; **(d)** $2xy\,dx + (x^2 - y^2)\,dy = 0,\ y(0) = 3$;

   (e) $\left(y + \dfrac{1}{y}\right)dx = \left(\dfrac{x}{y^2} - 1 - x\right)dy,\ y(0) = 1$; **(f)** $(x^2 + y)\,dx + (x + e^y)\,dy = 0,\ y(0) = 0$;

   (g) $dx + 2e^y\,dy = 0, \quad y(0) = 1$; **(h)** $(2xy + 6x)\,dx + (x^2 + 2)\,dy = 0,\ y(0) = 1$;

   (i) $\frac{x}{y^2}\,dx - \frac{x^2}{y^3}\,dy = 0, \quad y(1) = 1$; **(j)** $\left(\frac{y}{x} + x^2\right)dx + (\ln|x| - 2y)dy = 0,\ y(1) = 1$;

   (k) $2x^2\,dx - 4y^3\,dy = 0, \quad y(1) = 0$; **(l)** $e^y\,dx - (2x + y)\,dy = 0, \quad y(1) = 2$.

4. Using the line integral method, solve the initial value problems.

   (a) $\left(6xy + \frac{1}{\sqrt{y^4 - x^2}}\right)dx = \left(\frac{2x}{y\sqrt{y^4 - x^2}} - 3x^2\right)dy, \quad y(1) = 1.$

   (b) $[e^x\sin(y^2) + xe^x\sin(y^2)]\,dx + [2xye^x\cos(y^2) + y]\,dy = 0, \quad y(0) = 1.$

   (c) $(2 + \ln(xy))\,dx + \left(1 + \frac{x}{y}\right)\,dy = 0, \quad y(1) = 1.$

   (d) $\left(\frac{1}{y} - xy^3\right)dx = \left(\frac{3x^2y^2}{2} + \frac{x}{y^2} + y\right)dy, \quad y(1) = 1.$

   (e) $y(y^3 - 4x^3)\,dx + x(4y^3 - x^3)\,dy = 0, \quad y(1) = 1.$

   (f) $(2x\sin y + 3x^2y)\,dx + (x^2\,\cos y + x^3)\,dy = 0, \quad y(1) = \pi/6.$

   (g) $x(2 + x)y' + 2(1 + x)y = 1 + 3x^2, \quad y(1) = 1.$

   (h) $\frac{3x^2}{2\sqrt{x^3 + y^2}}\,dx + \left(\frac{2}{\sqrt{x^3 + y^2}} + 6y^2\right)dy = 0, \quad y(0) = 1.$

   (i) $\left(x + \frac{1}{y}\right)dx = \left(\frac{x}{y^2} - 2y\right)dy, \quad y(0) = 1.$

   (j) $(e^x\sin y + e^{-y})\,dx = (x\,e^{-y} = e^x\cos y)\,dy, \quad y(0) = \pi/2.$

5. For each of the following equations, find the most general function $N(x,y)$ so that the equation is exact.

(a)	$(\cos(x+y)+y^2)\,dx+N(x,y)\,dy=0;$	(b)	$(y\sin(xy)+e^{2x})\,dx+N(x,y)\,dy;$
(c)	$(e^x\sin y+x^2)\,dx+N(x,y)\,dy=0;$	(d)	$\left(2x+\frac{y}{1+x^2y^2}\right)dx+N(x,y)\,dy=0,$
(e)	$(2xy+1)\,dx+N(x,y)\,dy=0;$	(f)	$(3x+2y)\,dx+N(x,y)\,dy=0;$
(g)	$(e^y+x)\,dx+N(x,y)\,dy=0;$	(h)	$(2y^{3/2}+1)\,x^{-1/2}\,dx+N(x,y)\,dy=0;$
(i)	$(2+y^2\cos 2x)\,dx+N(x,y)\,dy=0;$	(j)	$(3x^2+2y)\,dx+N(x,y)\,dy=0;$
(k)	$x\,e^{-y}\,dx+N(x,y)\,dy=0;$	(l)	$2xy^{-3}\,dx+N(x,y)\,dy=0.$

6. Solve the equation $xf(x^2+y^2)\,dx+yf(x^2+y^2)\,dy=0$, where f is a differentiable function in some domain.

7. Prove that if the differential equation $M\,dx+N\,dy=0$ is exact, and if N (or, equivalently, M) is expressible as a sum $\sum_{i=1}^{n}a_i(x)b_i(y)$ of separated arbitrary (differentiable and integrable) functions $a_i(x)$ and $b_i(y)$, then this equation may be integrated by parts.

8. Determine p so that the equation $\dfrac{x^2+2xy+3y^2}{(x^2+y^2)^p}\,(y\,dx-x\,dy)$ is exact.

9. For an exact equation $M\,dx+N\,dy=0$, show that the potential function can be obtained by integration

$$\psi=\int_0^1[(x-x_0)M(\xi,\eta)+(y-y_0)N(\xi,\eta)]\,dt,\quad \xi=x_0+t(x-x_0),\quad \eta=y_0+t(y-y_0),$$

provided that the functions $M(x,y)$ and $N(x,y)$ are integrable along the straight line connecting (x_0,y_0) and (x,y).

10. Solve the nonhomogeneous exact equations.

(a)	$[2xy^2+y\,e^y]\,dx+[x(1+y)\,e^y+2yx^2]\,dy=2x\,dx;$	(b)	$y^2\,dx+2xy\,dy=\cos x\,dx;$
(c)	$\cos(xy)\,[y\,dx+x\,dy]=2x\,e^{x^2}\,dx;$	(d)	$\dfrac{x\,dy-y\,dx}{xy}=\dfrac{x\,dy+y\,dx}{\sqrt{1+x^2y^2}}.$

11. Consider the differential equation $(y+e^y)\,dy=\left(x-e^{-x}\right)dx$.

(a) Solve it using a computer algebra system. Observe that the solution is given implicitly in the form $\psi(x,y)=C$.

(b) Use the `contourplot` command from *Maple* or the `ContourPlot` command from *Mathematica* to see what the solution curves look like. For your x and y ranges you might use $-1\leqslant x\leqslant 3$ and $-2\leqslant y\leqslant 2$.

(c) Use `implicitplot` command (consult Example 1.3.2, page 6) to plot the solution satisfying the initial condition $y(1.5)=0.5$. Your plot should show two curves. Indicate which one corresponds to the solution.

12. Find the potential function $\psi(x,y)$ to the exact equation $x\,dx+y^{-2}\,(y\,dx-x\,dy)=0$, and then plot some streamlines $\psi(x,y)=C$ for positive and negative values of the parameter C.

13. Given a differential equation $y'=p(x,y)/q(x,y)$ in the rectangle $R=\{(x,y)\,:\,a<x<b,\ c<y<d\}$, suppose that $\psi(x,y)$ is its potential function. Show that each statement is true.

(a) For any positive integer n, $\psi^n(x,y)$ is also a potential function.

(b) If $\psi(x,y)\neq 0$ in R and if n is any positive integer, then $\psi^{-n}(x,y)$ is a potential function.

(c) If $F(t)$ and $F'(t)$ are continuous functions on the entire line $-\infty<t<\infty$ and do not vanish for every t except $t=0$, then $F(\psi(x,y))$ is a potential function whenever $\psi(x,y)$ is a potential function for a differential equation.

14. If the differential equation $M(x,y)\,dx+N(x,y)\,dy=0$ is exact, prove that

$$[M(x,y)+f(x)]\,dx+[N(x,y)+g(y)]\,dy=0$$

is also exact for any differentiable functions $f(x)$ and $g(y)$.

15. Show that the linear fractional equation

$$\frac{dy}{dx}=\frac{ax+by+c}{Ax+By+C}$$

is exact if and only if $A+b=0$.

16. In each of the following equations, determine the constant λ such that the equation is exact, and solve the resulting exact equation:

(a)	$(x^2+3xy)\,dx+(\lambda x^2+4y-1)\,dy=0;$	(b)	$(2x+\lambda y^2)\,dx-2xy\,dy=0;$
(c)	$\left(\dfrac{1}{x^2}+\dfrac{1}{y^2}\right)dx+\dfrac{\lambda x+1}{y^3}\,dy=0;$	(d)	$\left(\dfrac{\lambda y}{x^3}+\dfrac{y}{x^2}\right)dx=\left(\dfrac{1}{x}-\dfrac{1}{x^2}\right)dy;$
(e)	$(1-\lambda x^2y-2y^2)\,dx=(x^3+4xy+4)\,dy;$	(f)	$(\lambda xy+y^2)\,dx+(x^2+2xy)\,dy=0;$
(g)	$(e^x\sin y+3y)\,dx+(\lambda e^x\cos y+3x)\,dy=0;$	(h)	$\left(\frac{y}{x}+6x\right)dx+(\ln\lvert x\rvert+\lambda)\,dy=0.$

17. In each of the following equations, determine the most general function $M(x, y)$ such that the equation is exact:
   (a) $M(x,y)\,dx + (2xy - 4)\,dy = 0$; (b) $M(x,y)\,dx + (2x^2 - y^2)\,dy = 0$;
   (c) $M(x,y)\,dx + (2x + y)\,dy = 0$; (d) $M(x,y)\,dx + (4y^3 - 2x^3 y)\,dy = 0$;
   (e) $M(x,y)\,dx + (3x^2 + 4xy - 6)\,dy = 0$; (f) $M(x,y)\,dx + (y^3 + \ln|x|)\,dy = 0$;
   (g) $M(x,y)\,dx + \sqrt{3x^2 - y}\,dy = 0$; (h) $M(x,y)\,dx + 2y\,\sin x\,dy = 0$;
   (i) $M(x,y)\,dx + (\sec^2 y + x/y)\,dy = 0$; (j) $M(x,y)\,dx = (e^{-y} - \sin x\cos y - 2xy)\,dy$;
   (k) $M(x,y)\,dx + (xe^{xy} + x/y^2)\,dy = 0$; (l) $M(x,y)\,dx + \left(\frac{1}{x} - \frac{x}{y}\right)dy = 0$.

Section 2.4 of Chapter 2 (Review)

1. Find an integrating factor as a function of x only and determine the general solution for the given differential equations.
   (a) $y(1+x)\,dx + x^2\,dy = 0$; (b) $xy^2\,dy - (x^3 + y^3)\,dx = 0$;
   (c) $(x^2 - 3y)\,dx + x\,dy = 0$; (d) $(y - 2x)\,dx - x\,dy = 0$;
   (e) $(3x^2 + y + 3x^3 y)\,dx + x\,dy = 0$; (f) $(2x^2 + 2xy^2 + 1)y\,dx + (3y^2 + x)\,dx = 0$;
   (g) $\sin y\,dx + \cos y\,dy = 0$; (h) $(3y^2\cot x + \frac{1}{2}\sin 2x)\,dx = 2y\,dy$.
2. Find an integrating factor as a function of y only and determine the general solution for the given differential equations.
   (a) $y(y^2+1)\,dx + x(y^2 - 1)\,dy = 0$; (b) $(y^2 - 3x^2 + xy + 8y + 2x - 5)\,dy + (y - 6x + 1)\,dx = 0$;
   (c) $y(y^3+2)\,dx + x(y^3 - 4)\,dy = 0$; (d) $(2x + 2xy^2)\,dx + (x^2 y + 1 + 2y^2)\,dy = 0$;
   (e) $e^{-y}\,dx + (x/y)\,e^{-y}\,dy = 0$; (f) $(3x^2 y - x^{-2}y^5)\,dx = (x^3 - 3x^{-1}y^4)\,dy$;
   (g) $x\,dx = (x^2 y + y^3)\,dy$; (h) $(\frac{y}{x}\sec y - \tan y)\,dx = (x - \sec y\,\ln x)\,dy$.
3. Use an integrating factor to find the general solution of the given differential equations.
   (a) $(y^3 + y)\,dx + (xy^2 - x)\,dy = 0$; (b) $y\cos x\,dx + y^2\,dy = 0$;
   (c) $\left(e^{2x} + y - 1\right)\,dx - dy = 0$; (d) $dx + (x/y + \cos y)\,dy = 0$;
   (e) $(4x^3 y^{-2} + 3y^{-1})\,dx + (3xy^{-2} + 4y)\,dy = 0$; (f) $y\,dx + (2xy - e^{-y})\,dy = 0$;
   (g) $(1 + y^2)\,dx + y(x + y^2 + 1)\,dy = 0$; (h) $(1 + xy)\,dx + x^2\,dy = 0$;
   (i) $(1 + 2x^2 + 4xy)\,dx + 2\,dy = 0$; (j) $y\,dx + x^2\,dy = 0$;
   (k) $y(1 + y^3)\,dx + x(y^3 - 2)\,dy = 0$; (l) $(x^4 + y^4)\,dx + 2xy^3\,dy = 0$.
4. Using an integrating factor method, solve the given initial value problems.
   (a) $4y\,dx + 5x\,dy = 0,\ y(1) = 1$. (b) $(2x\,e^x - y^2)\,dx + 2y\,dy = 0,\ y(0) = \sqrt{2}$.
   (c) $3x^{-1}y^2\,dx - 2y\,dy = 0,\ y(4) = 8$. (d) $dx + (x + 2y + 1)\,dy = 0,\ y(0) = 1$.
   (e) $(y + 1)\,dx + x^2\,dy = 0,\ y(1) = 2$. (f) $xy^3\,dx + (x^2y^2 - 2)\,dy = 0,\quad y(1) = 1$.
   (g) $y' = e^x + 2y,\quad y(0) = -1$. (h) $2\cos y\,dx = \tan 2x\,\sin y\,dy,\ y(\pi/4) = 0$.
   (i) $\left(x^3 - y^3\right)dx = xy^2 dy,\quad y(1) = 1$. (j) $\left(2y - 9x^2\right)dx + 6x\left(1 - x^2y^{-2}\right)dy = 0\ y(1) = 1$.
5. Show that if the ratio $(M_y - N_x)\,/\,(y\,N - x\,M)$ is a function $g(z)$ of the product $z = xy$, then $\mu(xy) = \exp\left\{\int g(z)\,dz\right\}$ is an integrating factor for the differential equation (2.4.1).

Section 2.5 of Chapter 2 (Review)

1. Solve the given linear differential equation with variable coefficients.
   (a) $t\,y' + (t+1)\,y = 1$; (b) $(1 + x^2)y' + xy = 2x(1 + x^2)^{1/2}$;
   (c) $xy' = y + 4x^3\ln x$; (d) $y' + \frac{2xy}{1-x^2} = 2x \quad (|x| < 1)$;
   (e) $2y' + xy = 2x$; (f) $x(1 - x^2)y' = y + x^3 - 2yx^2$.
2. Solve the given initial value problems.
   (a) $(x^2 + 1)\,y' + 2xy = x,\quad y(2) = 1$; (b) $y' + 2y = x^2,\quad y(0) = 1$;
   (c) $y' + y = x\,\cos x,\quad y(0) = 1$; (d) $y' - y = x\,\sin x,\quad y(0) = 1$;
   (e) $x^3 y' + 4x^2 y + 1 = 0,\quad y(1) = 1$; (f) $x^2 y' + xy = 2,\quad y(1) = 0$.
3. Solve the initial value problems and estimate the initial value a for which the solution transits from one type of behavior to another.
   (a) $y' + 5y = 3\,e^{-2x},\quad y(0) = a$; (b) $y' + 5y = 29\cos 2x,\quad y(0) = a$.
4. Consider the differential equation $y' = ay + b\,e^{-x}$, subject to the initial condition $y(0) = c$, where a, b, and c are some constants. Describe the asymptotic behavior as $x \to \infty$ for different values of these constants.
5. Suppose that a radioactive substance decomposes into another substance which then decomposes into a third substance. If the instantaneous amounts of the substances at time t are given by $x(t)$, $y(t)$, and $z(t)$, respectively, find $y(t)$ as follows.
   The rate of change of the first substance is proportional to the amount present so that $\frac{dx}{dt} = \dot{x} = -\alpha x$. The rate of increase of the second substance is equal to the rate at which it is formed less the rate at which it decomposes. Thus, $\dot{y} = \alpha x(t) - \beta y(t)$. Eliminate $x(t)$ from this equation and solve the resulting equation for $y(t)$.

6. Let $a(x)$, $f_1(x)$, and $f_2(x)$ be continuous for all $x \geqslant 0$, and let y_1 and y_2 be solutions of the following IVPs.

$$y' + a(x)y = f_1(x), \quad y(0) = c_1 \qquad \text{and} \qquad y' + a(x)y = f_2(x), \quad y(0) = c_2,$$

respectively.

(a) Show that if $f_1(x) \equiv f_2(x)$ and $c_1 > c_2$, then $y_1(x) > y_2(x)$ for all $x \geqslant 0$.

(b) Show that if $f_1(x) > f_2(x)$ for all $x > 0$, $f_1(0) = f_2(0)$, and $c_1 = c_2$, then $y_1(x) > y_2(x)$.

7. For what values of constants a, $b \neq 0$, and c, a solution to the initial value problem $y' = ax + by + c$, $y(0) = 1$ is bounded.

8. Formula (2.5.6) shows that the general solution of a first order linear differential equation (2.5.3) is a family of curves of the form

$$y(x) = C \cdot k(x) + h(x).$$

Show, conversely, that the differential equation of any such family is a first order linear differential equation.

9. Show that the differential equation $y' + p(x)\,y = q(x)\,y \ln y$ can be reduced to a linear equation by substitution $v = \ln y$. Apply this method to solve the equation

$$x\,y' = x^2 y + xy \ln y.$$

10. Show that **Newton's law of cooling**, $\dot{T} = k(M - T)$, can be regarded as the linearization of Stefan's law, $\dot{T} = -\sigma\left(T^4 - M^4\right)$, near the equilibrium solution $T(t) = M$. Here the constant of proportionality k is positive if $T > M$.

Section 2.6 of Chapter 2 (Review)

1. Solve the following Bernoulli equations.

(a)	$6x^2\, y' = y(x + 4y^3)$;	**(b)**	$y' + y\sqrt{x} = \frac{2}{3}\sqrt{\frac{x}{y}}$;
(c)	$y' + 2y = e^{-x}\, y^2$;	**(d)**	$(1 + x^2)y' - 2xy = 4\sqrt{y(1 + x^2)}\,\arctan x$;
(e)	$y' + 2y = e^{x}\, y^2$;	**(f)**	$\dot{y} = (a\,\cos t + b)y - y^3$;
(g)	$(t^2 + 1)\dot{y} = ty + y^3$;	**(h)**	$y' = y/(x + 1) + (x + 1)^{1/2} y^{-1/2}$;
(i)	$\dot{y} + 2ty + ty^4 = 0$;	**(j)**	$xy' + y = xy^3$;
(k)	$t^2\dot{y} = 2ty - y^2$;	**(l)**	$2\dot{y} - y/t + y^3\,\cos t = 0$;
(m)	$y' + \frac{y}{x} + y^{-3}\,\sin x = 0$;	**(n)**	$y - y'\,\cos x = y^2\,\cos x\,(1 - \sin x)$.

2. Find the particular solution required.

(a)	$y' + y + xy^3 = 0$, $y(0) = 1$;	**(b)**	$y' + 4xy = xy^3$, $y(0) = 1$;
(c)	$xy' + t = 2xy^{1/2}$, $y(1) = 1$;	**(d)**	$xy' - 2y = 4xy^{3/2}$, $y(1) = 1$;
(e)	$y' - y = 3\,e^{2x}$, $y(0) = 1/2$;	**(f)**	$xy' = 2y - 5x^3 y^2$, $y(1) = 1/2$.

3. Solve the equations with the independent variable missing.

(a)	$y\,y'' = 4(y')^2 - y'$;	**(b)**	$y\,y'' + 3\,(y')^3 = 0$;	**(c)**	$y\,y'' - 2(y')^3 = 0$;
(d)	$(1 + y^2)yy'' = (3y^2 - 1)(y')^2$;	**(e)**	$y'' + (y')^2 = 2e^{-2y}$;	**(f)**	$y'' - 2(y')^2 = 0$;
(g)	$y'' + y' - 2y^3 = 0$;	**(h)**	$y'' = y - (y')^2$;	**(i)**	$y^2 y'' + y(y')^2 = 1/2$.

4. Find validity interval for the solution of the initial value problem:

$$y'\sqrt{1 + x^2} = x\,y^3, \qquad y(0) = 1.$$

5. Solve the equations with the dependent variable missing.

(a) $y'' = 1 + (y')^2$; **(b)** $x^2\,y'' = 2xy' - (y')^2$; **(c)** $y'' - x = 2y'$; **(d)** $y'' + y' = e^{-2t}$.

6. Find the general solution of the given equations by solving for one of the variables x or y.

(a) $(y')^2 + 2y' - x = 0$; **(b)** $x(y')^2 + 2y' = 0$; **(c)** $2(y')^3 + (y')^2 - y = 0$; **(d)** $(x^2 - 4)y'^2 = 1$.

7. Consider a population $P(t)$ of a species whose dynamics are described by the logistic equation with a constant harvesting rate:

$$\dot{P} = r\,(1 - P)\,P - k,$$

where r and k are positive constants. By solving a quadratic equation, determine the equilibrium population(s) of this differential equation. Then find the bifurcation points that relay the values of parameters r and k.

8. Assume that $y = x + \sqrt{9 - x^2}$ is an explicit solution of the following initial value problem

$$(y + ax)\,y' + ay + bx = 0, \qquad y(0) = y_0.$$

Determine values for the constants a, b, and y_0.

9. A block of mass m is pulled over a frictionless smooth surface by a string having a constant tension F (it has units of force, e.g. newtons). The block starts from rest at a horizontal distance d from the base of the pulley. Using Newton's law of motion, derive the (horizontal) velocity $v(x)$ of the block as a function of position x.

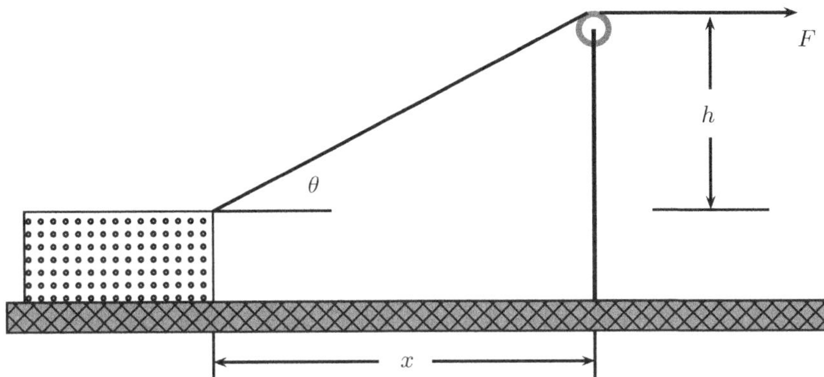

10. Reduce the order of the differential equations with homogeneous functions (some of them are homogeneous in the extended sense) and solve them.

(a)	$2x^2y'' = x^2 + 8y;$	**(b)**	$3\frac{y^2}{x^2} + 2(y')^2 = y\,y'' + 3\frac{yy'}{x^2};$
(c)	$(y')^2 - 2yy'' = x^2;$	**(d)**	$y\,y'' + (y')^2 = 2yy';$
(e)	$y\,y'' = (y')^2 + y\sqrt{x}/2;$	**(f)**	$y'' + 3\frac{y'}{x} = \frac{y}{x^2} + 2\,(y')^2;$
(g)	$yy'' = (y')^2 + x^2;$	**(h)**	$xy\,y'' + y^2/x = 2yy' + x/2;$

11. Solve the equation $(x^3 \sin y - x)y' + 2y = 0$ by reducing it to the Bernoulli equation. *Hint:* Consider x as a dependent variable.

12. A model for the variation of a finite amount of stock $y(t)$ in a warehouse as a function of the time t caused by the supply of fresh stock and its renewal by demand is

$$y' = a\,y(t)\,\cos\omega t - b\,y^2(t),$$

with the positive constants a, b, and ω, where $y(0) = y_0$. Solve for $y(t)$ and plot its solution using one of the available software packages.

13. Stefan–Boltzmann law for the Kelvin-scale temperature $T(t)$ of a body in an environment with ambient temperature M states that $\dot{T} = \sigma\left(M^4 - T^4\right)$ for some positive constant σ (with units `K`$^{-3}$`/sec`). Using second order Taylor's series approximation for the slope function $f(T) = \sigma\left(M^4 - T^4\right)$ around the equilibrium temperature $T = M$, derive a Bernoulli equation for $T(t)$ and then solve it.

14. In Example 2.6.14, show that when second order effects are taken into consideration the intensity of the reflected beam must satisfy the differential equation $h\,y''/2 - (1 + 2phy)y' + p + py^2 + p^2h(y + y^3) = 0$.

Section 2.7 of Chapter 2 (Review)

1. Show that the solution to the initial value problem $y' = x^2 + y^2$, $y(0) = 1$ blows up at some point within the interval $[0, 1]$.

In each of Problems 2 through 7, find the maximum interval of existence of the solution to the given initial value problem.

2. $\dot{y} = (y^3 + 1)(y^4 + y^2 + 2) \equiv y^7 + y^5 + y^4 + y^2 + 2y^3 + 2, \quad y(0) = 0;$
3. $\dot{y} = (y^4 + y^2 + 1)(y - 1)^2 \equiv y^6 - 2y^5 + 2y^4 - 2y^3 + 2y^2 - 2y + 1, \quad y(0) = 0;$
4. $\dot{y} = ((y^4 + y^2 + 1)(y - 1)^3 \equiv y^7 - 3y^6 + 4y^5 - 4y^4 + 4y^3 - 4y^2 + 3y - 1, \quad y(0) = 0;$
5. $\dot{y} = (y^3 + 1)(y^2 + 2y + 3) \equiv y^5 + 2y^4 + 3y^3 + y^2 + 2y + 3, \quad y(0) = 2;$
6. $\dot{y} = (y^4 - 1)(y^2 + 2y + 2) \equiv y^6 + 2y^5 + 2y^4 - y^2 - 2y - 2, \quad y(0) = 0;$
7. $\dot{y} = (y^3 + 2y + 12)(y^3 + 2y + 3) \equiv y^6 + 4y^4 + 15y^3 + 4y^2 + 30y + 36, \quad y(0) = 0.$

In each of Problems 8 through 19, find the bifurcation value of the parameter α and determine its type (saddle node, transcritical, or pitchfork).

8. $\dot{y} = 1 + \alpha y + y^2$;
9. $\dot{y} = y - \alpha y/(1 + y^2)$;
10. $\dot{y} = \alpha - y - e^{-y}$;
11. $\dot{y} = \alpha y - y(1 - y)$;
12. $\dot{y} = \alpha y - \sin(y)$;
13. $\dot{y} = \alpha \tanh y - y$.
14. $\dot{y} = \alpha y + y/4 - y/(1 + y)$;
15. $\dot{y} = \alpha - y - e^{-y}$;
16. $\dot{y} = y(1 - y^2) - \alpha(1 - e^{-y})$;
17. $\dot{y} = \alpha \ln(1 + y) + y$;
18. $\dot{y} = \alpha y + y^3 - y^5$;
19. $\dot{y} = \sin y(\cos y - \alpha)$.

20. Consider the differential equation

$$\frac{dy}{dx} = f(x, y) = \begin{cases} \frac{4x^3 y}{x^4 + y^2}, & (x, y) \neq (0, 0), \\ 0, & (x, y) = (0, 0), \end{cases}$$

subject to the initial condition $y(0) = 0$. Verify that the slope function $f(x, y)$ is continuous at the origin but does not satisfy the Lipschitz condition. Show that the initial value problem has infinitely many solutions

$$y(x) = c^2 - \sqrt{x^4 + c^4} \qquad \text{for a real constant } c.$$

Chapter 3

The graph on the right-hand side is obtained by plotting the solution to the initial value problem

$$\frac{\mathrm{d}y}{\mathrm{d}x} = \frac{x + 2y - 1}{2x + 4y + 1}, \qquad y(0) = 0,$$

using the standard Runge–Kutta algorithm implemented in MATLAB®. As seen from the graph, the numerical procedure became unstable where the slope is close to the vertical direction.

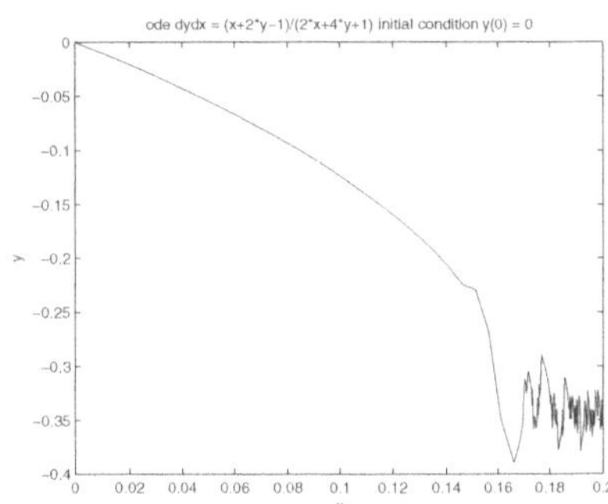

Numerical Methods

As shown in the previous chapters, there are some classes of ordinary differential equations that can be solved implicitly or, in exceptional cases, explicitly. Even when a formula for the solution is obtained, a numerical solution may be preferable, especially if the formula is very complicated. In practical work, it is an enormous and surprising bonus if an equation yields a solution by the methods of calculus. Generally speaking, the differential equation itself can be considered as an expression describing its solutions, and there may be no other way of defining them. Therefore, we need computational tools to obtain quantitative information about solutions that otherwise defy analytical treatment.

A complete analysis of a differential equation is almost impossible without exploiting the capabilities of computers that involve numerical methods. Because of the features of numerical algorithms, their implementations approximate (with some accuracy) only a single solution of the differential equation and usually in a "short" run from the starting point. In this chapter, we concentrate our attention on presenting some simple discrete numerical algorithms for solutions of the first order differential equation in the normal form subject to the initial condition

$$y' = f(x, y), \qquad y(x_0) = y_0,$$

assuming that the given initial value problem has a unique solution in the interval of interest. Discrete numerical methods are the procedures that can be used to calculate a table of approximate values of $y(x)$ at certain discrete points called grid, nodal, net, or mesh points. In this case, a numerical solution is not a continuous function at all, but rather an array of discrete pairs, i.e., points. When these points are connected, we get a polygonal curve consisting of line segments that approximates the actual solution. Moreover, a numerical solution is a discrete approximation to the actual solution, and therefore is not 100% accurate.

In two previous chapters, we demonstrated how available computer packages can be successfully used to solve and plot solutions to various differential equations. They may give an impression that finding or plotting these solutions is no more challenging than finding a root or logarithm. Actually, any numerical solver can fail with an appropriate initial value problem. The objective of this chapter is three-fold:

- to present the main ideas used in numerical approximations to solutions of first order differential equations;
- to advise the reader on programming an algorithm;
- to demonstrate the most powerful technique in applied mathematics—iteration.

An in-depth treatment of numerical analysis requires careful attention to error bounds and estimates, the stability and convergence of methods, and machine error introduced during computation. We shall not attempt to treat all these topics in one short chapter. Instead, we have selected a few numerical methods that are robust and for which algorithms for numerical approximations can be presented without a great deal of background. Furthermore, these techniques serve as general approaches that form a foundation for further theoretical understanding. To clarify the methods, numerical algorithms are accompanied with scripts written in popular computer algebra systems and MATLAB.

Since both differentiation and integration are infinite processes involving limits that cannot be carried out on computers, they must be discretized instead. This means that the original problem is replaced by some finite system of equations that can be solved by algebraic methods. Most of them include a sequence of relatively simple steps related to each other—called recurrence. Numerical algorithms define a sequence of discrete approximate values to the actual solution recursively or iteratively. Therefore, the opening section is devoted to the introduction of recurrences related to numerical solutions of the initial value problems.

3.1 Difference Equations

Many mathematical models that attempt to interpret physical phenomena often can be formulated in terms of the rate of change of one or more variables and as such naturally lead to differential equations. Such models are usually referred to as continuous models since they involve continuous functions. Numerical treatment of problems for differential equations requires discretization of continuous functions. So a numerical algorithm is dealing with their values at a discrete number of grid points, although there are some cases where a discrete model may be more natural.

Equations for sequences of values arise frequently in many applications. For example, let us consider the **amortization loan** problem for which the loan amount plus interest must be paid in a number of equal monthly installments. Suppose each installment is applied to the accrued interest on the debt and partly retires the principal. We introduce the following notations. Let

A	be the amount of the principal borrowed,
m	be the amount of each monthly payment,
r	be the monthly interest rate, and
p_n	be the outstanding balance at the end of month n.

Since interest in the amount rp_n is due at the end of the n-th month, the difference $m - rp_n$ is applied to reduce the principal, and we have

$$p_{n+1} = p_n - (m - rp_n) \quad \text{or} \quad p_{n+1} - (1+r)p_n = -m \quad (n = 1, 2, \ldots).$$

This equation and the initial condition $p_1 = A$ constitute the initial value problem for the so-called first order constant coefficient *linear difference equation* or *recurrence*.

So our object of interest is a sequence (either finite or infinite) of numbers rather than a function of a continuous variable. Traditionally, a sequence is denoted by $\mathbf{a} = \{a_n\}_{n\geqslant 0} = \{a_0, a_1, a_2, \ldots\}$ or simply $\{a_n\}$. An element a_n of a sequence can be defined by giving an explicit formula, such as $a_n = n!$, $n = 0, 1, 2, \ldots$. More often than not, the elements of a sequence are defined implicitly via some equation.

Definition 3.1: A **recurrence** is an equation that relates different members of a sequence of numbers $\mathbf{y} = \{y_0, y_1, y_2, \ldots\}$, where y_n $(n = 0, 1, 2, \ldots)$ are the values to be determined.

A solution of a recurrence is a sequence that satisfies the recurrence throughout its range.

Definition 3.2: The **order** of a recurrence relation is the difference between the largest and smallest subscripts of the members of the sequence that appear in the equation. The general form of a recurrence relation (in normal form) of order p is $y_n = f(n, y_{n-1}, y_{n-2}, \ldots, y_{n-p})$ for some function f. A recurrence of a finite order is usually called a **difference equation**.

For example, the recurrence $p_{n+1} - (1+r)p_n = -m$ from the amortization loan problem is a difference equation of the first order. In general, a first order difference equation is of the form $\Phi(y_n, y_{n-1}) = 0$, $n = 1, 2, \ldots$. In what follows, we consider only first order difference equations in normal form, namely, $y_{n+1} = f(n, y_n)$. When the function f does not depend on n, such difference equations are referred to as autonomous: $y_n = f(y_{n-1})$.

Definition 3.3: If in the difference equation $y_n = f(y_{n-1}, y_{n-2}, \ldots, y_{n-p})$ of order p, the function f is linear in all its arguments, then the equation is called **linear**. The first and second order linear difference equations have the following form:

$$a_n y_{n+1} + b_n y_n = f_n \qquad \text{first order linear equation,}$$
$$a_n y_{n+1} + b_n y_n + c_n y_{n-1} = f_n \qquad \text{second order linear equation,}$$

where $\{f_n\}$, $\{a_n\}$, $\{b_n\}$, and $\{c_n\}$ are given (known) sequences of coefficients. When all members of these sequences do not depend on n, the equation is said to have **constant coefficients**; otherwise, these difference equations (recurrences) are said to have **variable coefficients**.

The sequence $\{f_n\}$ is called a **nonhomogeneous sequence**, or **forcing sequence** of the difference equation. If all members of $\{f_n\}$ are zero, then the linear difference equation is called a **homogeneous equation**; otherwise, we call it nonhomogeneous (or inhomogeneous).

A difference equation usually has infinitely many solutions. In order to pin down a solution (which is a sequence of numbers), we have to know one or more of its elements. If we have a difference equation of order p, we need to specify p sequential values of the sequence, called the *initial conditions*. So, for first order recurrences we have to specify only one element of the sequence, say the first one $y_0 = a$; for the second order difference equations we have to know two elements: $y_0 = a$ and $y_1 = b$; and so forth. It should be noted that the unique solution of a recurrence may be specified by imposing restrictions other than initial conditions.

There are some examples of difference equations:

$y_{n+1} - y_n^2 = 0$	(first order, nonlinear, homogeneous, autonomous);
$y_{n+1} + y_n - n\,y_{n-1} = 0$	(linear second order, variable coefficients, homogeneous);
$y_{n+1} - y_{n-1} = n^2$	(linear second order, constant coefficients, nonhomogeneous);
$F_{n+1} = F_n + F_{n-1}$	(Fibonacci recurrence is a linear second order, constant coefficients, homogeneous, autonomous difference equation).

Much literature has been produced over the last two centuries on the properties of difference equations due to their prominence in many areas of applied mathematics, numerical analysis, and computer science.

Let us consider a couple of examples of difference equations. For instance, the members of the sequence of factorials, $\{0!, 1!, 2!, \ldots\}$ can be related via either a simple first order variable coefficients homogeneous equation $y_{n+1} = (n+1)y_n$ $(n = 0, 1, 2\ldots)$, $y_0 = 0! = 1$, or a second order variable coefficients homogeneous equation $y_{n+1} = n[y_n + y_{n-1}]$, subject to the initial conditions $y_0 = 1$, $y_1 = 1$.

Example 3.1.1: Consider now a problem of evaluating the integrals

$$I_n = \int_0^1 x^n\, e^x\, \mathrm{d}x \qquad \text{and} \qquad S_n = \int_0^\pi x^n \sin x\, \mathrm{d}x, \qquad n \geqslant 0.$$

Assuming that a computer algebra system is not available, solving such a problem becomes very tedious. Integrating I_{n+1} and S_{n+2} by parts, we get

$$\begin{aligned} I_{n+1} &= \left[x^{n+1}\, e^x\right]\Big|_{x=0}^{x=1} - \int_0^1 (n+1)x^n\, e^x\, \mathrm{d}x, \\ S_{n+2} &= \left[-x^{n+2}\cos x + (n+2)x^{n+1}\sin x\right]\Big|_{x=0}^{x=\pi} - (n+2)(n+1)\int_0^1 x^n \sin x\, \mathrm{d}x, \end{aligned}$$

or

$$I_{n+1} = e - (n+1)I_n, \qquad I_0 = \int_0^1 e^x\, \mathrm{d}x = e - 1; \tag{3.1.1}$$

$$S_{n+2} = \pi^{n+2} - (n+2)(n+1)S_n, \qquad S_0 = 2, \quad S_1 = \pi. \tag{3.1.2}$$

Equation (3.1.1) is a first order variable coefficient inhomogeneous difference equation. It can be used to generate the sequence of integrals I_n, $n = 1, 2, \ldots$. For instance,

$$I_1 = e - I_0 = 1, \quad I_2 = e - 2I_1 = e - 2, \quad I_3 = e - 3I_2 = 6 - 2e, \quad I_4 = 9e - 24, \ldots.$$

Equation (3.1.2) is a second order variable coefficient inhomogeneous difference equation. However, it is actually a first order difference equation for even and odd indices because it relays entries with indices n and $n+2$. Similarly, we can generate its elements using the given recurrence:

$$S_2 = \pi^2 - 4, \quad S_3 = \pi^3 - 6\pi, \quad S_4 = \pi^4 - 12\pi^2 + 48, \ldots. \qquad \square$$

The problems studied in the theory of recurrences not only concern their existence and analytical representation of some or all solutions, but also the behavior of these solutions, especially when n tends to infinity, and computational stability.

In this section, we consider only first order recurrences. We start with a constant coefficient case:

$$y_{n+1} = f(y_n), \qquad n = 0, 1, 2, \ldots, \tag{3.1.3}$$

where the given function f does not depend on n. If the initial value y_0 is given, then successive terms in the sequence $\{y_0, y_1, y_2, \ldots\}$ can be found from the recurrence (3.1.3):

$$y_1 = f(y_0), \quad y_2 = f(y_1) = f(f(y_0)), \quad y_3 = f(y_2) = fff(y_0), \ldots.$$

The quantity $f(f(y_0))$ or simply $ff(y_0)$ is called the second iterate of the difference equation and is denoted by $f^2(y_0)$. Similarly, the n-th iterate y_n is $y_n = f^n(y_0)$. So such an iterative procedure allows us to determine all values of the sequence $\{y_n\}$.

It may happen that the recurrence (3.1.3) has a constant solution (which does not depend on n). In this case this constant solution satisfies the equation $y_n = f(y_n)$ and we call it the **equilibrium solution** because $y_{n+1} = f(y_n) = y_n$.

The general linear recurrence relation of the first order can be written in the form

$$y_{n+1} = p_n y_n + q_n, \qquad n = 0, 1, 2, \ldots, \tag{3.1.4}$$

where p_n and q_n are given sequences of numbers. If $q_n = 0$ for all n, the recurrence is said to be homogeneous and its solution can easily be found by iteration:

$$y_1 = p_0 y_0, \quad y_2 = p_1 y_1 = p_1 p_2\, y_0, \quad y_3 = p_2 p_1 p_2\, y_0, \quad \ldots, \quad y_n = p_{n-1} \cdots p_1 p_2\, y_0.$$

As no initial condition is specified, we may select y_0 as we wish, so y_0 can be considered as an arbitrary constant.

If the coefficient p_n in Eq. (3.1.4) does not depend on n, the constant coefficient homogeneous difference equation

$$y_{n+1} = p\, y_n, \qquad n = 0, 1, 2, \ldots,$$

has the general solution $y_n = p^n\, y_0$. Then the limiting behavior of y_n is easy to determine

$$\lim_{n\to\infty} y_n = \lim_{n\to\infty} p^n\, y_0 = \begin{cases} 0, & \text{if } |p| < 1; \\ y_0, & \text{if } p = 1; \\ \text{does not exist}, & \text{otherwise}. \end{cases}$$

Now we consider a constant coefficient first order linear nonhomogeneous difference equation

$$y_{n+1} = p y_n + q_n, \qquad n = 0, 1, 2, \ldots. \tag{3.1.5}$$

Iterating in the same manner as before, we get

$$\begin{aligned} y_1 &= p y_0 + q_0, \\ y_2 &= p y_1 + q_1 = p(p y_0 + q_0) + q_1 = p^2\, y_0 + p q_0 + q_1, \\ y_3 &= p y_2 + q_2 = p^3\, y_0 + p^2 q_0 + p q_1 + q_2, \end{aligned}$$

and so on. In general, we have

$$y_n = p^n\, y_0 + \sum_{k=0}^{n-1} p^{n-1-k} q_k. \tag{3.1.6}$$

In the special case where $q_n = q \neq 0$ for all n, the difference equation (3.1.5) becomes

$$y_{n+1} = p y_n + q, \qquad n = 0, 1, 2, \ldots,$$

and from Eq. (3.1.6), we find its solution to be

$$y_n = p^n\, y_0 + (1 + p + p^2 + \cdots + p^{n-1}) q. \tag{3.1.7}$$

The geometric polynomial $1 + p + p^2 + \cdots + p^{n-1}$ can be written in the more compact form:

$$1 + p + p^2 + \cdots + p^{n-1} = \begin{cases} \frac{1-p^n}{1-p}, & \text{if } p \neq 1; \\ n, & \text{if } p = 1. \end{cases} \tag{3.1.8}$$

The limiting behavior of y_n follows from Eq. (3.1.8). If $|p| < 1$ then $y_n \mapsto q/(1-p)$ since $p^n \to 0$ as $n \to \infty$. If $p = 1$ then y_n has no limit as $n \to \infty$ since $y_n = y_0 + nq \to \infty$. If $|p| > 1$ or $p = -1$ then y_n has no limit unless the right-hand side in Eq. (3.1.7) approaches a constant, that is,

$$p^n\, y_0 + (1 + p + p^2 + \cdots + p^{n-1}) q \to \text{ constant } \quad (p \neq 1) \quad \text{ as } n \to \infty.$$

Since

$$p^n\, y_0 + (1 + p + p^2 + \cdots + p^{n-1}) q = p^n\, y_0 + \frac{1-p^n}{1-p}\, q = \frac{q}{1-p} + p^n \left(y_0 - \frac{q}{1-p} \right),$$

we conclude that $y_n = q/(1-p)$ is the equilibrium solution when $y_0 = q/(1-p)$. ■

Solution (3.1.6) can be simplified in some cases. For example, let $q_n = \alpha q^n$ for a constant α and $q \neq p$. We seek a particular solution of Eq. (3.1.5) in the form $y_n = A\, q^n$, where A is some constant to be determined. Substitution into the equation $y_{n+1} = p\, y_n + \alpha q^n$ yields

$$A q^{n+1} = p A q^n + \alpha q^n \quad \text{or} \quad A q = p A + \alpha,$$

because $y_{n+1} = A q^{n+1}$. Solving for A, we get

$$A = \frac{\alpha}{q-p} \qquad (q \neq p)$$

and the general solution becomes

$$y_n = p^n\, C + \frac{\alpha}{q-p}\, q^n \qquad (q \neq p), \qquad n = 0, 1, 2, \ldots,$$

for some constant C. If y_0 is given, then

$$y_n = p^n\, y_0 + \frac{\alpha}{q-p}\, (q^n - p^n) \qquad (q \neq p), \qquad n = 0, 1, 2, \ldots. \tag{3.1.9}$$

Now we consider the difference equation

$$y_{n+1} = p y_n + \alpha p^n, \qquad n = 0, 1, 2, \ldots.$$

We are looking for its particular solution in the form $y_n = A n\, p^n$ with unknown constant A. Substitution into the given recurrence yields

$$A(n+1) p^{n+1} = p n A p^n + \alpha p^n \quad \text{or} \quad A p = \alpha.$$

So the general solution becomes

$$y_n = p^n\, y_0 + \alpha n\, p^{n-1}, \qquad n = 0, 1, 2, \ldots. \tag{3.1.10}$$

If q_n is a polynomial in n, that is, $q_n = \alpha_m n^m + \alpha_{m-1} n^{m-1} + \cdots + \alpha_0$, then the solution of Eq. (3.1.6) has the form

$$y_n = p^n\, y_0 + \beta_m n^m + \beta_{m-1} n^{m-1} + \cdots + \beta_0 \qquad (p \neq 1), \tag{3.1.11}$$

where coefficients β_i are determined by substituting (3.1.11) into the given difference equation.

The general linear difference equation of the first order (3.1.4) can be solved explicitly by an iterative procedure:

$$y_n = \pi_n \left(y_0 + \frac{q_0}{\pi_1} + \frac{q_2}{\pi_2} + \cdots + \frac{q_{n-1}}{\pi_n} \right), \qquad n = 0, 1, 2, \ldots, \tag{3.1.12}$$

where $\pi_0 = 1$, $\pi_1 = p_0$, $\pi_2 = p_0 p_1$, $\ldots$, $\pi_n = p_0 p_1 \cdots p_{n-1}$.

Example 3.1.2: Suppose we are given a sequence of numbers $\{p_0,\ p_1,\ \ldots,\ p_n, \ldots\}$. For any first n elements of the sequence, we define the polynomial

$$P_n(x) = p_0 x^n + p_1 x^{n-1} + \ldots + p_n.$$

Now suppose we are given a task to evaluate the polynomial $P_n(x)$ and some of its derivatives at a given point $x = t$.

Let $y_n \stackrel{\text{def}}{=} y_n^{(0)} = P_n(t)$. We calculate the numbers y_n recursively by the relations

$$y_n = t y_{n-1} + p_n, \quad n = 1, 2, \ldots, \qquad y_0 = p_0. \tag{3.1.13}$$

The given recurrence has the solution (3.1.12), where $\pi_n = t^n$; so

$$y_n \stackrel{\text{def}}{=} y_n^{(0)} = p_0 t^n = p_1 t^{n-1} + \cdots + p_n.$$

Let $y_n^{(k)} = P_n^{(k)}(t)$ be the value of k-th derivative of the polynomial $P_n(x)$ at a given point $x = t$. For $k = 1, 2, \ldots$, we generate the sequence $\{y_n^{(k)}\}$ recursively using the difference equation

$$y_0^{(k)} = y_0^{(k-1)}, \qquad y_n^{(k)} = t y_{n-1}^{(k)} + y_n^{(k-1)}, \qquad n = 0, 1, 2, \ldots. \tag{3.1.14}$$

Note that the sequence $y_n^{(k)}$ terminates when k exceeds n.

As an example, we consider the polynomial $P_5(x) = 8x^5 - 6x^4 + 7x^2 + 3x - 5$, and suppose that we need to evaluate $P_5(x)$ and its derivative $P_5'(x)$ at $x = 0.5$. Using algorithm (3.1.13) with $t = 0.5$, we obtain

$$\begin{aligned}
y_0 &= 8; \\
y_1 &= t y_0 + p_1 = 0.5\, y_0 - 6 = 4 - 6 = -2; \\
y_2 &= t y_1 + p_2 = 0.5\, y_1 + 0 = -1; \\
y_3 &= t y_2 + p_3 = 0.5\, y_2 + 7 = 6.5; \\
y_4 &= t y_3 + p_4 = 0.5\, y_3 + 3 = 3.25 + 3 = 6.25; \\
y_5 &= t y_4 + p_5 = 0.5\, y_4 - 5 = -1.875.
\end{aligned}$$

So $P_5(0.5) = -1.875$. To find $P_5'(x)$, we use the algorithm (3.1.14):

$$\begin{aligned}
y_0^{(1)} &= y_0 = 8; \\
y_1^{(1)} &= t y_0^{(1)} + y_1 = 8t + y_1 = 4 - 2 = 2; \\
y_2^{(1)} &= t y_1^{(1)} + y_2 = 1 - 1 = 0; \\
y_3^{(1)} &= t y_2^{(1)} + y_3 = y_3 = 6.5; \\
y_4^{(1)} &= t y_3^{(1)} + y_4 = 3.25 + 6.25 = 9.5.
\end{aligned}$$

Therefore, $P_5'(0.5) = 9.5$.

Example 3.1.3: Compute values for $I_{12} = \int_0^1 x^{12}\, e^x\, dx$ using recurrence (3.1.1) with different "approximations" for the initial value $I_0 = e - 1$. Using the *Maple*™ command

```
rsolve( {a(n+1) = e-(n+1) * a(n), a(0) =e-1 }, {a});
```

we fill out Table 143, where

$$e_n(x) \stackrel{\text{def}}{=} 1 + x + \frac{x^2}{2!} + \frac{x^3}{3!} + \cdots + \frac{x^n}{n!} \tag{3.1.15}$$

is the *incomplete exponential function*. This example demonstrates the danger in applying difference equations without sufficient prior analysis. □

As we can see in Example 3.1.3, a small change in the initial value for I_0 produces a large change in the solution. Such problems are said to be **ill-conditioned**. This ill-conditioning can be inherent in the problem itself or induced by the numerical method of solution.

Consider the general first order difference equation (3.1.4) subject to the initial condition $y_0 = a$. Let z_n be a solution of the same recurrence relation (3.1.4) with the given initial value $z_0 = y_0 + \varepsilon$. The iterative procedure yields

$$\begin{aligned} z_1 &= p_0 z_0 + q_0 = p_0(y_0 + \varepsilon) + q_0 = y_1 + p_0\varepsilon, \\ z_2 &= p_1 z_1 + q_1 = p_1(y_1 + p_0\varepsilon) + q_1 = p_1 y_1 + q_1 + p_1 p_0 \varepsilon = y_2 + p_1 p_0 \varepsilon, \end{aligned}$$

and in general

$$z_n = y_n + p_{n-1}p_{n-2}\cdots p_0\varepsilon.$$

Clearly, after n applications of the recurrence, the original error ε will be amplified by a factor $p_{n-1}p_{n-2}\cdots p_0$. Hence, if $|p_k| \leqslant 1$ for all k, the difference $|y_n - z_n|$ remains small when ε is small and the difference equation is said to be **absolutely stable**; otherwise we call it **unstable**. If the *relative error* $|y_n - z_n|/|y_n|$ remains bounded, then the difference equation is said to be **relatively stable**.

Table 143: Solution of Eq. (3.1.1) for initial conditions rounded to various decimal places. (a) Exact value expressed with 7 decimal places; (b) 6 places; (c) 5 places, and (d) 3 places.

	Exact	(a)	(b)	(c)	(d)
I_0	$e-1$	1.7182818	1.718282	1.7183	1.718
I_1	1	1.0	1.000000828	1.000081828	1.008281828
I_2	$e-2$	.718281828	.718280172	.718118172	.701718172
I_3	$6-2e$	.563436344	.563441312	.563927312	.613127312
⋮	⋮	⋮	⋮	⋮	⋮
I_6	$265\,e - 720$	.344684420	.344088260	.285768260	-5.618231740
⋮	⋮	⋮	⋮	⋮	⋮
I_{12}	$12!\,[I_0 - e + e\cdot e_{12}(-1)]$	.114209348	-396.4991155	-39195.62872	$-.3967\times 10^7$

Example 3.1.4: We reconsider the recurrence (3.1.1), page 139, and let $\{z_n\}$ be its solution subject to the initial condition $z_0 = I_0 + \varepsilon$, where ε is a small number (perturbation). Then $z_n = I_n + (-1)^n\, n!\,\varepsilon$ and the difference $|z_n - I_n| = n!\,\varepsilon$ is unbounded. Therefore, the difference equation (3.1.1) is unstable.

The general solution of the recurrence relation (3.1.1) is

$$I_n = (-1)^n\, n!\,[I_0 + e(e_n(-1) - 1)] = (-1)^n\, n!\left[I_0 - e\left(1 - \frac{1}{2!} + \frac{1}{3!} - \cdots \pm \frac{1}{n!}\right)\right],$$

which can be verified by substitution. Here $e_n(x)$ is the incomplete exponential function (3.1.15). Hence, the relative error

$$E = \frac{Z_n - I_n}{I_n} = \frac{(-1)^n\, n!\,\varepsilon}{I_n} = \frac{\varepsilon}{I_0 + e(e_n(-1) - 1)} \quad \to \quad \frac{\varepsilon}{I_0 + 1 - e}$$

because $e_n(x) \to e^x$ as $n \to \infty$. Since $I_0 = e - 1$, we conclude that the recurrence relation (3.1.1) is absolutely unstable.

Example 3.1.5: Consider the following first order nonlinear difference equation

$$u_{k+1} - u_k = (b - a u_k) u_k, \qquad k = 0, 1, 2, \ldots, \qquad u_0 \text{ is given.} \tag{3.1.16}$$

This is a discrete logistic equation that describes the population of species u_k at the k-th year when a, b are positive numbers. For instance, the logistic model that fits the population growth in the U.S. for about hundred years until 1900 is as follows:

$$u_{k+1} = 1.351\, u_k - 1.232 \times 10^{-9} u_k^2.$$

However, this model cannot be used after 1930, which indicates that human population dynamics is more complicated. Actually Eq. (3.1.16) is the discrete Euler approximation of the logistic differential equation (see §2.6, page 96) when the derivative with respect to time is replaced by the finite difference $\dfrac{du}{dt} \sim \dfrac{u(t_{k+1}) - u(t_k)}{t_{k+1} - t_k}$ at point $t = t_k$. It is more convenient to rewrite the logistic equation (3.1.16) in the following form:

$$y_{k+1} = r y_k (1 - y_k), \qquad k = 0, 1, 2, \ldots, \tag{3.1.17}$$

where $r = b + 1$ and $a u_k = (b+1) y_k$. Equation (3.1.17) turns out to be a mathematical equation with extraordinary complex and interesting properties. Its equilibrium solutions can be found from the equation $r(1-y)y = y$, so there are two such solutions: $y = 0$ and $y = 1 - 1/r$.

Because of its nonlinearity, the discrete logistic equation is impossible to solve explicitly in general. Two particular cases are known when its solution can be obtained: for $r = 2$ and $r = 4$. The behavior of the solution to Eq. (3.1.17) depends on the initial condition, y_0, that together with the value of the parameter r eventually determines the trend of the population. Recurrence (3.1.16) manifests many trademark features in nonlinear dynamics. There are known three different types of possible behavior of solutions to the discrete logistic equation:

Fixed: The population approaches a stable value either from one side or asymptotically from both sides.

Periodic: The population alternates between two or more fixed values.

Chaotic: The population will eventually visit every neighborhood in a subinterval of $(0, 1)$.

The equilibrium solution $y_k = 0$ is stable for $|r| < 1$, and another constant solution $y_k = 1 - 1/r$ is stable for $r \in (1, 3)$. A stable 2-cycle begins at $r = 3$ followed by a stable 4-cycle at $r = 1 + \sqrt{6} \approx 3.449489743$. The period continues doubling over even shorter intervals until around $r = 3.5699457\ldots$, where the chaotic behavior takes over. Within the chaotic regime, there are interspersed various windows with periods other than powers of 2, most notably a large 3-cycle window beginning at $r = 1 + \sqrt{8} \approx 3.828427125$. When the growth rate exceeds 4, all solutions zoom to infinity, which makes this model unrealistic.

Problems

1. For the recurrence relation (3.1.2),
   (a) find exact expressions for $S_5, S_6, \ldots, S_{12}$;
   (b) express the sequence $\{z_n\}$ via $\{S_n\}$ where z_n satisfies (3.1.2) subject to the initial conditions $z_0 = 2 + \varepsilon_0$, $z_1 = \pi + \varepsilon_1$;
   (c) based on part (b), determine the stability of the recurrence (3.1.2). Is it ill-conditioned or not?

2. Calculate $P(x)$, $P'(x)$, and $P''(x)$ for $x = 1.5$, where
   (a) $P(x) = 4x^5 + 5x^4 - 6x^3 - 7x^2 + 8x + 9$; **(b)** $P(x) = x^5 - 2x^4 + 3x^3 - 4x^2 + 5x - 6$;
   (c) $P(x) = 6x^5 - 4x^4 + 8x^3 - 3x^2 + 6x + 3$; **(d)** $P(x) = x^5 + 6x^4 - 3x^3 - 8x^2 + 3x - 1$.

3. Solve the given first order difference equations in terms of the initial value a_0. Describe the behavior of the solution as $n \to \infty$.
   (a) $y_{n+1} = 0.5\, y_n$; **(b)** $y_{n+1} = \frac{n+4}{n+2}\, y_n$; **(c)** $y_{n+1} = \sqrt{\frac{n+3}{n+4}}\, y_n$;
   (d) $y_{n+1} = (-2)^{n+2}\, y_n$; **(e)** $y_{n+1} = 0.8\, y_n + 20$; **(f)** $y_{n+1} = -1.5\, y_n - 1$.

4. Using a computer solver, show that the first order recurrence $x_{n+1} = 1 + x_n - x_n^2/4$ subject to the initial condition $x_0 = 7/4$ converges to 2. Then show that a similar recurrence $x_{n+1} = 1 + x_n - x_n^2/4 + e^{x_n - 2}/2$ under the same initial condition diverges.

5. Using a computer solver, show that the first order recurrence $x_{n+1} = x_n/2 + 2/x_n$ subject to the initial condition $x_0 = 7/4$ converges to 2. Then show that a similar recurrence $x_{n+1} = x_n/2 + 2/x_n - 4\left(1 + 10^{12}(x_n - 2)^2\right)^{-1}$ under the same initial condition converges to -2.

6. An investor deposits \$250,000 in an account paying interest at a rate of 6% compounded quarterly. She also makes additional deposits of \$4,000 every quarter. Find the account balance in 5 years.
7. A man takes a second mortgage of \$200,000 for 30-years period. What monthly payment is required if the interest rate is 4%?
8. Find the effective annual percentage yield of a bank account that pays an interest rate of 2% compounded weekly.
9. A college student borrows \$40,000 for a flashy car. The lender charges an annual interest rate of 5.5%. What monthly payment is required to payoff the loan in 10 years? What is the total amount paid during the term of loan?
10. If the interest rate on 15-years mortgage is fixed at 3.375% and \$500 is the maximum monthly payment the borrower can afford, what is the maximum mortgage loan possible?
11. A man wants to purchase a yacht for \$200,000, so he wishes to borrow that amount at the interest rate 6.275% for 10 years. What would be his monthly payment?
12. A home-buyer wishes to finance a mortgage of \$250,000 with a 15-year term. What is the maximum interest rate the buyer can afford if the monthly payment is not to exceed \$2,000?
13. Due to natural evaporation, the amount of water in a bowl or aquarium decreases with time. This leads to an increase of sodium concentration (as ordinary table salt, $NaCl$) that is present in almost every water supply. The amount of salt is unchanged, and eventually its concentration will exceed the survivable level for the fish living in the fishbowl. To avoid an increase of sodium concentration, a certain amount of water is periodically removed and replaced with more fresh water to compensate for its evaporation and deliberate removal. Let x_n be the amount of salt present in the aquarium at the moment of the n-th removal of water. Then x_n satisfies the recurrence $x_{n+1} = x_n + r - \delta x_n$. Solve this first order difference equation assuming that parameters r and δ are constants, and the initial amount of salt is known.
14. Let the elements of a sequence $\{w_n\}_{n\geqslant 0}$ satisfy the inequalities
$$w_{n+1} \leqslant (1+a)w_n + B, \qquad n = 0, 1, 2, \ldots,$$
where a and B are certain positive constants, and let $w_0 = 0$. Prove (by induction) that
$$w_n \leqslant \frac{B}{a}\left(e^{na} - 1\right), \qquad n = 0, 1, 2, \ldots.$$
15. Consider the discrete logistic equation (3.1.16). For what positive value of u_k is $u_{k+1} = 0$? What positive value of u_k provides the maximum value of u_{k+1}? What is this maximum value?
16. Let $g(y) = ry(1-y)$ be the right-hand side of the logistic equation (3.1.17).
    (a) Show that $g(g(y)) = r^2 y(1-y)(1-ry+y^2)$.
    (b) By solving the equation $y = g(g(y))$, show that for any $r > 3$, there are exactly two initial conditions $y_0 = (1 + r \pm \sqrt{r^2 - 2r - 3})/(2r)$ within the unit interval $(0,1)$ for which the discrete logistic recurrence (3.1.17) has nontrivial 2-cycles.
17. Determine the first order recurrence and the initial condition for which the given sequence of numbers is its solution.
$$\text{(a)}\ a_n = \frac{1}{4^n\, n!}; \qquad \text{(b)}\ a_n = 1 + \frac{1}{2} + \cdots + \frac{1}{n}; \qquad \text{(c)}\ a_n = \frac{n+1}{n}.$$
18. Starting at $x_0 = 0.5$, how many iterations are needed to find the root of the quadratic equation $x^2 + x - 2 = 0$ with 7 exact decimal places using each of the following recurrences.
    (a) $x_{n+1} = 2 - x_n^2$;
    (b) $x_{n+1} = \dfrac{2}{x_n} - 1$;
    (c) $x_{n+1} = \dfrac{2 + x_n^2}{1 + 2x_n}$;
    (d) $x_{n+1} = x_n - \dfrac{x_n^2 + x_n - 2}{x_n^4 + 4x_n^3 + 8x_n^2 + 9x_n + 2}$;
    (e) $x_{n+1} = x_n - \dfrac{x_n^4 + 4x_n^3 - 6x_n + 1}{(2x_n + 1)^2}$.
19. Consider Tower of Hanoi recurrence
$$h_n = 2\,h_{n-1} + 1 \quad (n = 1, 2, \ldots), \qquad h_0 = 0.$$
For this sequence of (integer) numbers $\{h_n\}_{n\geqslant 0}$ we can assign an infinite series $H(z) = \sum_{n\geqslant 0} h_n z^n$, called the generating function. Find the generating function for the sequence of the Tower of Hanoi. *Note:* the Tower of Hanoi puzzle was introduced by the French mathematician Edouard Lucas (1842–1891) in 1889.

3.2 Euler's Methods

In this section, we will discuss the numerical algorithms to approximate the solution of the initial value problem

$$\frac{dy}{dx} = f(x, y), \qquad y(x_0) = y_0 \tag{3.2.1}$$

assuming that the problem has a unique solution, $y = \phi(x)$ (see §1.6 for details), on some interval $|a, b|$ including x_0, where usually the left endpoint a coincides with x_0.

A numerical method frequently begins by imposing a partition of the form $a = x_0 < x_1 < x_2 < \cdots < x_{N-1} < x_N = b$ of the x-interval $[a, b]$. For simplicity, these points, called the **mesh points**, are assumed to be uniformly distributed:

$$x_1 = x_0 + h,\ x_2 = x_0 + 2h, \ldots, x_n = x_0 + nh = x_{n-1} + h, \quad n = 0, 1, \ldots, N, \tag{3.2.2}$$

where $h = \frac{b-a}{N}$ is the **step size**. Note that in practice, the uniform grid is used very rerely. The number N of subintervals is related to the step size by the identity $b - a = Nh$. Therefore, the uniform partition of the interval $[a, b]$ is uniquely identified by specifying either the step size h or the number of mesh points N. The value h is called the *discretization parameter.* At each mesh point x_n, the numerical algorithm generates an approximation y_n to the actual solution $y = \phi(x)$ of the initial value problem (3.2.1) at that point, so we expect $y_n \approx \phi(x_n)$ $(n = 1, 2, \ldots, N)$. Note that the initial condition provides us an exact starting point (x_0, y_0).

A preeminent mathematician named Leonhard Euler was the first to numerically solve an initial value problem. Leonhard Euler was born on April 15, 1707 in Basel, Switzerland and died September 18, 1783 in St. Petersburg, Russia. He left Basel when he was 20 years old and never returned. Leonhard Euler was one of the greatest mathematicians of all time. After his death, the St. Petersburg Academy of Science (Russia) continued to publish Euler's unpublished work for nearly 50 more years and has yet to publish all his works. His name is pronounced "oiler," not "youler."

In 1768, he published (St. Petersburg) what is now called the **tangent line method**, or more often, the **Euler method**. This is a variant of a **one-step** method (or single-step method) that computes the solution on a step-by-step basis iteratively. That is why a one-step method is usually called a memory-free method: it performs computation of the solution's next value based only on the previous step and does not retain the information in future approximations. In the one-step method, we start from the given $y_0 = y(x_0)$ and advance the solution from x_0 to x_1 using y_0 as the initial value. Since the true value of the solution at the point $x = x_1$ is unknown, we approximate it by y_1 according to a special rule. Next, to advance from x_1 to x_2, we discard y_0 and employ y_1 as the new initial value. This allows us to find y_2, approximate value of the solution at $x = x_2$, using only information at the previous point $x = x_1$. And so on, we proceed stepwise, computing approximate values $\{y_n\}$ of the solution $y = \phi(x)$ at the mesh points $\{x_n\}_{n \geqslant 0}$.

In the initial value problem (3.2.1), the slope of the solution is known at every point, but the values of the solution are not. Any one-step method is based on a specific rule or algorithm that approximates the solution at the right end of a mesh interval using slope values from the interval. From the geometric point of view, it defines the slope of advance from (x_n, y_n) to (x_{n+1}, y_{n+1}) over the mesh interval $[x_n, x_{n+1}]$. Its derivation becomes more clear when we replace the given initial value problem by its integral counterpart. If we integrate both sides of Eq. (3.2.1) with respect to x, we reduce the initial value problem to the integral equation

$$y(x) = y_0 + \int_{x_0}^{x} f(s, y(s))\,ds. \tag{3.2.3}$$

After splitting the integral by the mesh points x_n, $n = 1, 2, \ldots$, we obtain

$$\begin{aligned}
y(x_1) &= y_0 + \int_{x_0}^{x_1} f(s, y(s))\,ds, \\
y(x_2) &= y_0 + \int_{x_0}^{x_2} f(s, y(s))\,ds \\
&= y_0 + \int_{x_0}^{x_1} f(s, y(s))\,ds + \int_{x_1}^{x_2} f(s, y(s))\,ds = y_1 + \int_{x_1}^{x_2} f(s, y(s))\,ds, \\
y(x_3) &= y_0 + \int_{x_0}^{x_3} f(s, y(s))\,ds \\
&= y_0 + \int_{x_0}^{x_2} f(s, y(s))\,ds + \int_{x_2}^{x_3} f(s, y(s))\,ds = y_2 + \int_{x_2}^{x_3} f(s, y(s))\,ds,
\end{aligned}$$

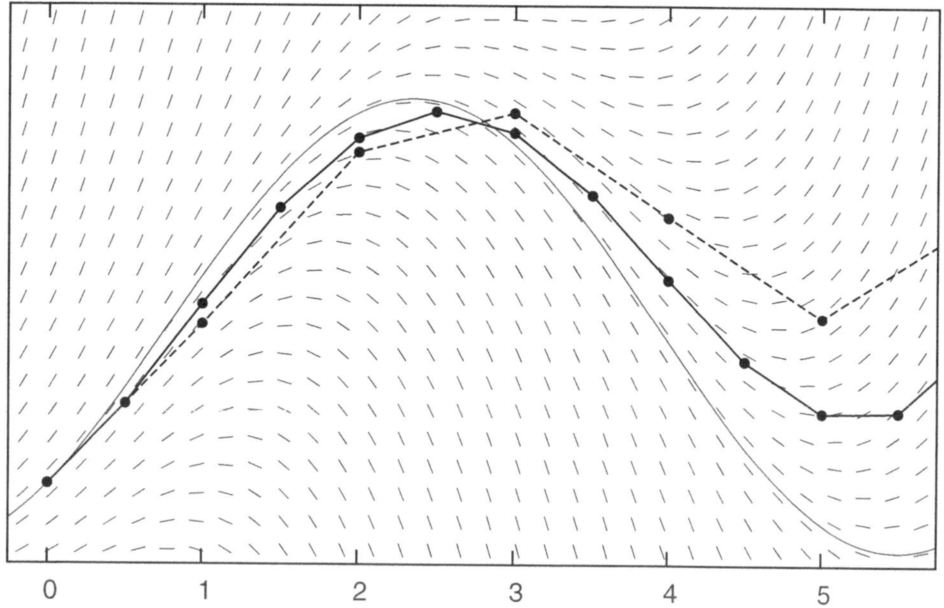

Figure 3.1: Two Euler semi-linear approximations (in black) calculated for two step sizes $h = 1$ and $h = 1/2$ along with the exact solution (in blue) to the linear differential equation $\dot{y} = 2\,\cos(t) + y(t) - 1$ subject to $y(0) = 0$, plotted with *Maxima*.

and so on. In general, we have

$$y(x_{n+1}) = y(x_n) + \int_{x_n}^{x_{n+1}} f(s, y(s))\,\mathrm{d}s. \tag{3.2.4}$$

This equation cannot be used computationally because the actual function $y(s)$ is unknown on the partition interval $[x_n, x_{n+1}]$ and therefore the slope function $f(s, y(s))$ is also unknown. Suppose, however, that the step length h is small enough so that the slope function is nearly constant over the interval $x_n \leqslant s \leqslant x_{n+1}$. In this case, the crude approximation (rectangular rule) of the integral in the right-hand side of Eq. (3.2.4) gives us $\int_{x_n}^{x_{n+1}} f(s, y(s))\,\mathrm{d}s \approx h\,f(x_n, y(x_n))$. Using this approximation, we obtain the **Euler rule:**

$$y_{n+1} = y_n + (x_{n+1} - x_n) f(x_n, y_n) \qquad \text{or} \qquad y_{n+1} = y_n + h f_n, \tag{3.2.5}$$

where the following notations are used: $h = x_{n+1} - x_n$, $f_n = f(x_n, y_n)$, and y_n denotes the approximate value of the actual solution $y = \phi(x)$ at the point x_n $(n = 1, 2, \ldots)$. A new value of y at $x = x_{n+1}$ is predicted using the slope at the left end x_n to extrapolate linearly over a mesh interval of size h. This means that the Euler formula is asymmetrical because it uses derivative information, $y' = f(x_n, y_n)$, only at the beginning of the interval $[x_n, x_{n+1}]$. Note that the step size h can be either positive or negative, giving approximations to the right of the initial value or to the left, respectively.

Equation (3.2.5) is the **difference equation** or **recurrence** of the first order associated with the Euler method. Solving recurrences numerically may lead to instability and further serious analysis is usually required [13]. For some simple slope functions, the recurrence (3.2.5) can be solved explicitly. Sometimes it is beneficial to transfer the difference equation to another form that is more computationally friendly. To compute numerically the integral on the right-hand side of Eq. (3.2.3), we apply the left-point Riemann sum approximation to obtain the so-called *quadrature form* of the Euler algorithm:

$$y_{n+1} = y_0 + h \sum_{j=0}^{n} f(x_j, y_j), \qquad n = 0, 1, \ldots. \tag{3.2.6}$$

Let us estimate the computational cost of Euler's algorithms (3.2.5) and (3.2.6). The dominant contribution of these techniques is n_f, the number of arithmetic operations required for evaluating $f(x, y)$, which depends on the complexity of the slope function. At each step, the difference equation (3.2.5) uses 1 addition (A), 1 multiplication (M) by h, and n_f operations to evaluate the slope. Therefore, to calculate y_n, we need $n(\mathrm{A} + \mathrm{M} + n_f)$ arithmetic operations, where A stands for addition and M for multiplication. On the other hand, the full-history recurrence (3.2.6) requires much more effort: $\mathrm{M} + \frac{1}{2}\,n(n-1)\mathrm{A} + \frac{1}{2}\,n(n-1)\,n_f$ arithmetic operations. However, with clever coding the quadratic term $n(n-1)/2$ can be reduced to the linear one:

$$s_{k+1} = s_k + f(x_{k+1}, y_0 + hs_k), \qquad s_0 = f(x_0, y_0) \qquad (k = 0, 1, \ldots),$$

where $s_k = \sum_{i=0}^{k} f(x_i, y_i)$ is the partial sum in the quadrature form (3.2.6). Evaluation of s_{k+1} requires 2 additions, 1 multiplication by h, and 1 value of the slope function, so the total cost to evaluate the partial sum for y_n is $(n-1)(2\mathrm{A}+1\mathrm{M}+n_f)+n_f$. Using this approach, the full-history recurrence (3.2.6) can be solved using $(2n-1)\mathrm{A} + n(\mathrm{M}+n_f)$ arithmetic operations, which requires only $(n-1)$ more additions compared to the Euler rule (3.2.5).

The Euler algorithm, either (3.2.5) or (3.2.6), generates a sequence of points (x_0, y_0), (x_1, y_1), ... on the plane that, when connected, produces a polygonal curve consisting of line segments. When the step size is small, the naked eye cannot distinguish the individual line segments constituting the polygonal curve. Then the resulting polygonal curve looks like a smooth graph representing the solution of the differential equation. Indeed, this is how solution curves are plotted by computers.

Example 3.2.1: Let us start with the linear differential equation

$$\dot{y} = 2\,\cos(t) + y(t) - 1,$$

having the oscillating slope function. First, we find its general solution by typing in *Maple*
`dsolve(D(y)(t) = 2*cos(t)+y(t)-1,y(t))`
This yields $y = \phi(t) = 1 + \sin t - \cos t + C\,e^t$, where $C = y(0)$ is an arbitrary constant. We consider three initial conditions: $y(0) = 0.01$, $y(0) = 0$, and $y(0) = -0.01$, and let $y = \phi_1(t)$, $y = \phi_0(t)$, and $y = \phi_{-1}(t)$ be their actual solutions, respectively. The function $\phi_0(t) = 1 + \sin t - \cos t$ separates solutions (with $y(0) > 0$) that grow unboundedly from those (with $y(0) < 0$) that decrease unboundedly. *Maple* can help to find and plot the solutions:

```
dsolve({D(y)(t) = 2*cos(t)-y(t)+1,y(0)=0.01},y(t))
u := unapply(rhs(%), t); plot(u(t), t = 0 .. 1)
```

Now we can compare true values with approximate values obtained with the full-history recurrence (3.2.6):

$$y_{n+1} = 0.01 - h(n+1) + 2h\sum_{j=0}^{n}\cos(jh) + h\sum_{j=0}^{n} y_j, \qquad y_0 = 0.01,$$

where y_n is an approximation of the actual solution $y = \phi(t)$ at the mesh point $t = t_n = nh$ with fixed step size h. For instance, the exact solution has the value $\phi(1) = 1.328351497$ at the point $t = 1$, whereas the quadrature formula approximates this value as $y_{10} \approx 1.2868652685$ when $h = 0.1$ and $y_{100} \approx 1.3240289197$ when $h = 0.01$. The Euler algorithm (3.2.5) provides almost the same answer, which differs from the quadrature output in the ninth decimal place (due to round-off error): $y_{n+1} = y_n + h(2\cos t_n + y_n - 1)$.

The following MATLAB code helps to find the quadrature approximations:

```
y0=0.01;   tN=2*pi;  h=.01;     % final point is tN
N=round(tN/h);     % rounded number of steps
y(N)=0;            % memory allocation
n=0;        s1=cos(n);  % sum cos(jh) for n=0
s2=y0;             % sum of y(j) for j=0
y(1)=y0-h*(n+1)+2*h*s1+h*s2;
for n=1:N-1
  s1=s1+cos(n*h); % sum of cos(jh), j=0,1,...,n
  s2=s2+y(n);     % sum of y(j), j=0,1,...,n
  y(n+1)=y0-h*(n+1)+2*h*s1+h*s2;
end
plot((0:1:N)*h,[y0 y]);
```

Example 3.2.2: Let us apply the Euler method to solve the initial value problem

$$2y' + y = 3x, \qquad y(0) = 0.$$

The given differential equation is linear, so it has the unique solution $y = \phi(x) = 3x - 6 + 6e^{-x/2}$. The integral equation (3.2.4) for $f(x, y) = (3x - y)/2$ becomes

$$y(x_{n+1}) = y(x_n) + \frac{1}{2}\int_{x_n}^{x_{n+1}} (3s - y(s))\,\mathrm{d}s,$$

and the Euler algorithm can be written as

$$y_{n+1} = \frac{3}{2}\,hx_n + \left(1 - \frac{h}{2}\right) y_n, \qquad y_0 = 0, \quad x_n = nh, \qquad n = 0, 1, 2, \ldots. \tag{3.2.7}$$

According to Eq. (3.1.6), this linear first order constant coefficient difference equation has the unique solution

$$y_n = 6\left(1 - \frac{h}{2}\right)^n - 6 + 3hn, \qquad n = 0, 1, 2, \ldots. \tag{3.2.8}$$

The natural question to address is how good is its approximation, namely, how close is it to the exact solution $\phi(x)$? A couple of the first Euler approximations are not hard to obtain:

$$y_1 = 0, \qquad y_2 = \frac{3}{2}\,h^2, \qquad y_3 = \frac{9}{2}\,h^2 - \frac{3}{4}\,h^3, \qquad y_4 = 9h^2 - 3h^3 + \frac{3}{8}\,h^4.$$

In general,

$$y_n = \frac{3}{4}\,n(n-1)\,h^2 - \frac{1}{8}\,n(n-1)(n-2)\,h^3 + \cdots.$$

Using the Maclaurin series for the exponential function, $e^{-t} = 1 - t + \frac{t^2}{2!} - \frac{t^3}{3!} + \cdots$, we get

$$\phi(x) = 3x - 6 + 6e^{-x/2} = 6\sum_{k=2}^{\infty} \frac{1}{k!}\left(-\frac{x}{2}\right)^k = \frac{3}{4}\,x^2 - \frac{1}{8}\,x^3 + \frac{1}{64}\,x^4 - \cdots. \tag{3.2.9}$$

So, for $x = x_n = nh$ $(n = 2, 3, 4, \ldots)$, we have

$$\phi(nh) = \frac{3}{4}\,n^2h^2 - \frac{n^3}{8}\,h^3 + \frac{n^4}{64}\,h^4 - \cdots.$$

Therefore, we see that the error of such an approximation, $\phi(x_n) - y_n$, is an alternating series in h starting with h^2. So we expect this error to be small when h is small enough.

Our next question is: can we choose h arbitrarily? Unfortunately, the answer is no. Let us look at the general solution (3.2.8) of the Euler algorithm that contains the power $(1 - h/2)^n$. If $1 - h/2 < -1$, that is $4 < h$, the power function x^n for $x = 1 - h/2 < -1$ oscillates and the approximate solution diverges. □

Because Euler's method is so simple, it is possible to apply it "by hand." However, it is better to use a software package and transfer this job to a computer. Comparisons of the exact solution with its approximations obtained by the Euler method are given in Figure 3.2, plotted using the following *Mathematica*® script:

```
f[x_, y_] = (3*x - y)/2
phi[x_] := 3*x - 6 + 6*Exp[-x/2];
x[0] = 0; y[0] = 0; h = 0.1; n = 10;
Do[x[k + 1] = x[k] + h; y[k + 1] = y[k] + f[x[k], y[k]]*h; , {k, 0, n}]
data = Table[{x[k], y[k]}, {k, 0, n}]
Show[Plot[phi[x], {x, 0, 1}], ListPlot[data, PlotMarkers -> Automatic]]
```

Similar output can be obtained with the following *Maple* commands:

```
h:=0.1:  n:=10: xx:=0.0:  yy:=0.0: # initiation
points:=array(0..n) points[0]:=[xx,yy]
for i from 1 to n do
  yy:=evalf(xx+h*f(xx,yy))  xx:=xx+h
  points[i]:=[xx,yy]
od
plotpoints := [seq(points[i], i = 0 .. n)]
plot1 := plot(plotpoints, style = point, symbol = circle)
plot2 := plot(3*x-6+6*exp(-.5*x), x = 0 .. 1)
plots[display](plot1, plot2, title = 'Your Name')
```

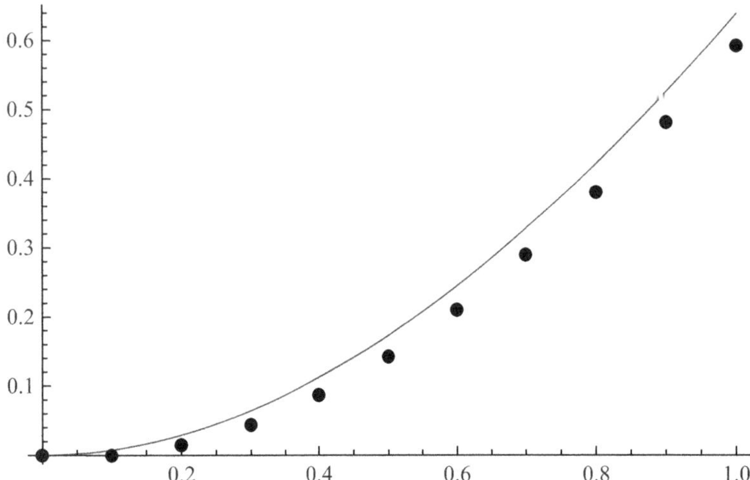

Figure 3.2: Example 3.2.2, a comparison of the exact solution $y = 3x - 6 + 6e^{-x/2}$ with the results of numerical approximations using the Euler method with step size $h = 0.1$.

Sometimes the Euler formulas can be improved by partial integration over those terms that do not contain an unknown solution on the right-hand side of Eq. (3.2.3). For instance, in our example, the term $3s/2$ can be integrated, leading to the integral equation

$$y(x_{n+1}) = y(x_n) + \frac{3}{4}\left(x_{n+1}^2 - x_n^2\right) - \frac{1}{2}\int_{x_n}^{x_{n+1}} y(s)\,\mathrm{d}s,$$

Application of the right rectangular rule to the integral yields

$$y_{n+1} = \frac{3}{4}h^2(2n+1) + \left(1 - \frac{h}{2}\right)y_n, \tag{3.2.10}$$

which has the unique solution

$$y_n = 3\left(2 - \frac{h}{2}\right)\left(1 - \frac{h}{2}\right)^n + 3hn - 3\left(2 - \frac{h}{2}\right), \qquad n = 0, 1, 2, \ldots. \tag{3.2.11}$$

Comparing the value $y_{10} = 0.59242$ obtained by the algorithm (3.2.7) and the value $y_{10} = 0.652611$ obtained by the algorithm (3.2.10) for $h = 0.1$ and $n = 10$, we see that the latter is closer to the true value $\phi(1.0) = 0.63918$. □

The previous example shows that there exists a critical step size beyond which numerical instabilities become manifest. Such partial stability is typical for forward Euler technique (3.2.5). To overcome stringent conditions on the step size, we present a simple implicit **backward Euler formula**:

$$y_{n+1} = y_n + (x_{n+1} - x_n)f(x_{n+1}, y_{n+1}) = y_n + hf_{n+1}. \tag{3.2.12}$$

Since the quantity $f_{n+1} = f(x_{n+1}, y_{n+1})$ contains the unknown value of y_{n+1}, one should solve the (usually nonlinear with respect to y_{n+1}) algebraic equation (3.2.12). If $f(x, y)$ is a linear or quadratic function with respect to y, Eq. (3.2.12) can be easily solved explicitly. (Cubic and fourth order algebraic equations also can be solved explicitly but at the expense of complicated formulas.) Otherwise, one must apply a numerical method to find its solution, making this method more computationally demanding. Another drawback of the backward Euler algorithm consists in the nonuniqueness of y_{n+1} in the equation $y_{n+1} = y_n + h\,f(x_{n+1}, y_{n+1})$: it may have multiple solutions and the numerical solver should be advised which root it should choose.

Example 3.2.3: Consider the initial value problem for the nonlinear equation

$$y' = \frac{1}{x - y + 2} = (x - y + 2)^{-1}, \qquad y(0) = 1. \tag{3.2.13}$$

Application of algorithm (3.2.12) yields

$$y_{n+1} = y_n + \frac{h}{x_{n+1} - y_{n+1} + 2} \qquad \Longleftrightarrow \qquad y_{n+1}^2 - y_{n+1}\,(2 + x_{n+1} + y_n) + y_n(2 + x_n) + h = 0.$$

This is a quadratic equation in y_{n+1}, and we instruct *Maple* to choose one root:

```
f:=(x,y)-> (x-y+2)^(-1); x[0]:= 0: y[0]:= 1: h:= 0.1:
 for k from 0 to 9 do
   x[k+1]:= x[k]+h:  bb := solve(b=y[k]+h*f(x[k+1],b),b);
   y[k+1]:= evalf(min(bb));
 end do;
a := [seq(x[i], i = 0 .. 10)]; b := [seq(y[i], i = 0 .. 10)];
pair:= (x, y) -> [x, y]; dplot := zip(pair, a, b);
plot(dplot, color = blue)
```

In the code above, the command `solve` is applied because the quadratic equation in `y[k+1]` has an explicit solution. In general, the `fsolve` command should be invoked instead to find the root numerically. Similarly, *Maxima* can be used for calculations:

```
load(newton1)$
f(x,y) := 1/(x-y+2); x[0]:0$ y[0]:1$ h:0.1$
  for k:0 thru 9 do (
    x[k+1]: x[k]+h,
    y[k+1]: newton(-b+y[k]+h*f(x[k+1],b), b, y[k], 1e-6) );
```

Fortunately, the given IVP has the unique solution (consult §1.6) $y = x + 1$. However, when the initial condition is different from 1, say $y(0) = 0$, the corresponding actual solution can be expressed only in implicit form: $y = \ln|x - y + 1|$. □

Table 151: A comparison of the results for the numerical solution of $y' = (x - y + 2)^{-1}$, $y(0) = 0$, using the backward Euler rule (3.2.12) for different step sizes h.

x	Exact	$h = 0.1$	$h = 0.05$	$h = 0.025$	$h = 0.01$
1.0	0.442854401	0.436832756	0.439848517	0.441352740	0.442254051
2.0	0.792059969	0.781058081	0.786559812	0.789310164	0.790960127
3.0	1.073728938	1.058874274	1.066296069	1.070011211	1.072241553

There is too much inertia in Euler's method: it keeps the same slope over the whole interval of length h. A better formula can be obtained if we use a more accurate approximation of the integral in the right-hand side of Eq. (3.2.4) than the rectangular rule as in Eq. (3.2.5). If we replace the integrand by the average of its values at the two end points, we come to the **trapezoid rule:**

$$y_{n+1} = y_n + \frac{h}{2}[f(x_n, y_n) + f(x_{n+1}, y_{n+1})]. \tag{3.2.14}$$

This equation defines implicitly y_{n+1} and it must be solved to determine the value of y_{n+1}. The difficulty of this task depends entirely on the nature of the slope function $f(x, y)$.

Example 3.2.4: (Example 3.2.2 revisited) When the differential equation is linear, the backward Euler algorithm and the trapezoid rule can be successfully implemented. For instance, application of the backward Euler method to the initial value problem from Example 3.2.2 yields

$$y_{n+1} = y_n + \frac{h}{2}(3x_{n+1} - y_{n+1}) \quad \Longrightarrow \quad y_{n+1} = \frac{y_n}{1 + h/2} + \frac{3hx_{n+1}/2}{1 + h/2} = \frac{2y_n}{2 + h} + \frac{3h^2(n+1)}{2 + h}$$

since $x_n = nh$ $(n = 0, 1, 2, \ldots)$. At each step, the backward algorithm (3.2.12) requires 10 arithmetic operations, whereas the forward formula (3.2.7) needs only 6 arithmetic operations. Solving the recurrence, we obtain

$$y_n = 6\left(\frac{2}{2+h}\right)^n + 3hn - 6 = \frac{3}{4}n(n-1)h^2 - \frac{nh^3}{8}(n^2 + 3n + 2) + \cdots$$

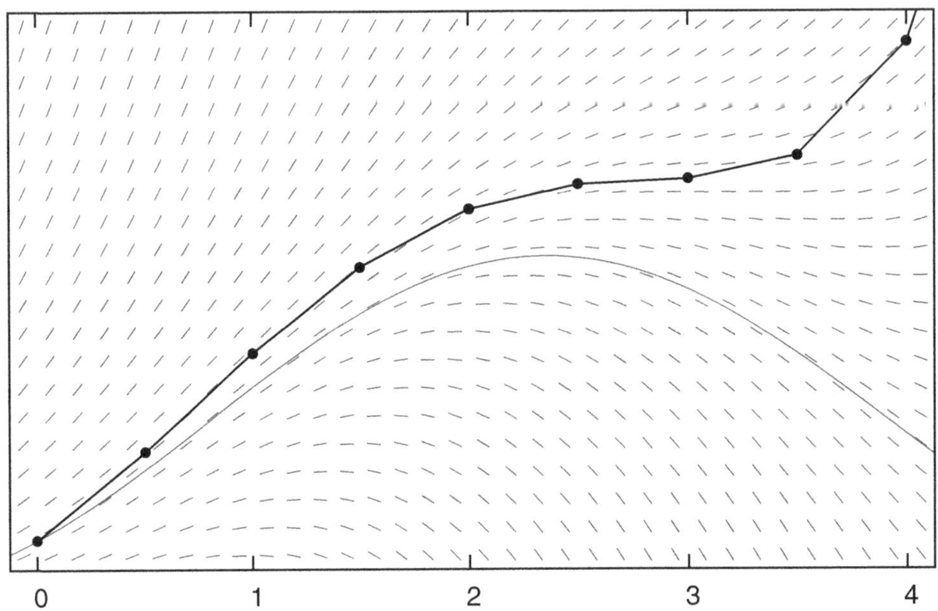

Figure 3.3: The backward Euler approximations (in black) along with the exact solution (in blue) to the linear differential equation $\dot{y} = 2\,\cos(t) + y(t) - 1$, plotted with *Maxima*.

Comparison with the exact solution (3.2.9) shows that the error using backward Euler formula of this approximation, $\phi(x_n) - y_n$, is again proportional to h^2.

Now we consider the trapezoid rule, and implement the corresponding *Mathematica* script:

```
x = 0; y = 0; X = 1; h = 0.1; n = (X - x)/h;
For[i = 1, i < n + 1, i++, x = x + h;
 y = Y /. FindRoot[Y == y + (h/4)*(3*(2*x - h) + y + Y), {Y, y}]]
Print[y]  (* to see outcome of computations *)
```

Table 152: A comparison of the results for the numerical solution of $2y' + y = 3x$, $y(0) = 0$ using the implicit trapezoid procedure (3.2.14) for different step sizes h.

x	Exact	$h = 0.1$	$h = 0.05$	$h = 0.025$	$h = 0.01$
0.1	0.00737654	0.0073170	0.07361682	0.00737283	0.00737595
0.5	0.17280469	0.1725612	0.17274384	0.17278948	0.17280226
1.0	0.63918395	0.6388047	0.63908918	0.63916026	0.63918016

The algorithm (3.2.14) leads to the following (implicit) recurrence:

$$y_{n+1} = y_n + \frac{h}{4}(3x_n - y_n) + \frac{h}{4}(3x_{n+1} - y_{n+1}), \qquad n = 0, 1, 2, \ldots.$$

Solving for y_{n+1}, we get

$$y_{n+1} = \frac{(1 - h/4)y_n + 3h(x_n + x_{n+1})/4}{1 + h/4} = \frac{4-h}{4+h}\,y_n + \frac{3h^2(2n+1)}{4+h},$$

which requires 12 arithmetic operations. If we calculate y_n for the first few values of n,

$$y_1 = \frac{3h^2}{4+h}, \quad y_2 = \frac{(4-h)3h^2}{(4+h)^2} + \frac{9h^2}{4+h}, \quad y_3 = \frac{(4-h)^2 3h^2}{(4+h)^3} + \frac{(4-h)9h^2}{(4+h)^2} + \frac{15h^2}{4+h},$$

then an unmistakable pattern emerges:

$$y_n = 6\left(\frac{4-h}{4+h}\right)^n + 3hn - 6 = \frac{3}{4}(nh)^2 - \frac{nh^3}{16}(2n^2+3n-1) + \cdots .$$

Hence, the difference $\phi(nh) - y_n$ between the true value and its approximation is proportional to h^3 because $\phi(nh) = \frac{3}{4}(nh)^2 - \frac{n^3}{8}h^3 + \cdots$. □

There are several reasons that Euler's approximation is not recommended for practical applications, among them, (i) the method is not very accurate when compared to others, and (ii) it may be highly erratic (see Example 3.2.2 on page 148) when a step size is not small enough.

Since, in practice, the Euler rule requires a very small step to achieve sufficiently accurate results, we present another explicit method that allows us to find an approximate solution with better accuracy. It is one of a class of numerical techniques known as **predictor-corrector** methods. First, we approximate y_{n+1} using Euler's rule (3.2.5): $y^*_{n+1} = y_n + h\,f(x_n, y_n)$. It is our intermediate prediction, which we distinguish with a superscript $*$. Then we correct it by substituting y^*_{n+1} instead of the unknown value of y_{n+1} into the trapezoid formula (3.2.14) to obtain

$$y_{n+1} = y_n + \frac{h}{2}[f(x_n, y_n) + f(x_{n+1}, y_n + hf(x_n, y_n))], \quad n = 0, 1, 2, \ldots . \tag{3.2.15}$$

This formula is known as the **improved Euler formula** or the average slope method. Equation (3.2.15), which is commonly referred to as the Heun[34] formula, gives an explicit formula for computing y_{n+1} in terms of the data at x_n and x_{n+1}. Figure 3.4 gives a geometrical explanation of how the algorithm (3.2.15) works on each mesh interval $[x_n, x_{n+1}]$. First, it finds the Euler point (x_{n+1}, y^*_{n+1}) as the intersection of the vertical line $x = x_{n+1}$ and the tangent line with slope $k_1 = f(x_n, y_n)$ to the solution at the point $x = x_n$. Next, the algorithm evaluates the slope $k_2 = f\left(x_{n+1}, y^*_{n+1}\right)$ at the Euler point. At the end, the improved Euler formula finds the average of two slopes $k = \frac{1}{2}(k_1 + k_2)$ and draws the straight line with the slope k to the intersection with the vertical line $x = x_{n+1}$. This is the Heun point that the algorithm chooses as y_{n+1}, the approximation of the solution at $x = x_{n+1}$ (see Fig. 3.4).

Example 3.2.5: Let us consider the initial value problem (3.2.1) with the linear slope function $f(x, y) = \frac{1}{2}(3x + y)$ and the initial condition $y(0) = 0$. Its analytic solution is

$$y = \phi(x) = 6\,e^{x/2} - 3x - 6 = 6\sum_{k=2}^{\infty}\frac{1}{k!}\left(\frac{x}{2}\right) = \frac{3}{4}x^2 + \frac{1}{8}x^3 + \frac{1}{64}x^4 + \cdots .$$

In our case, the improved Euler formula (3.2.15) becomes

$$y_{n+1} = y_n\left(1 + \frac{h}{2} + \frac{h^2}{8}\right) + \frac{3h}{4}(2x_n + h) + \frac{3h^2}{8}x_n = y_n\left(1 + \frac{h}{2} + \frac{h^2}{8}\right) + \frac{3h^2}{8}(4+h)n + \frac{3h^2}{4}$$

because $x_n = nh$, $x_{n+1} = (n+1)h$, $y_n + hf_n = y_n + h(\frac{3}{2}x_n + \frac{1}{2}y_n) = y_n(1 + \frac{h}{2}) + \frac{3h}{2}x_n$, $f(x_{n+1}, y_n + hf_n) = \frac{3}{2}x_{n+1} + \frac{1}{2}y_n(1 + \frac{h}{2}) + \frac{3h}{4}x_n$. To advance on each mesh interval, the average slope method requires 11 arithmetic operations while the Euler rule (3.2.7) uses about half as many arithmetic operations—just 6.

To facilitate understanding, let us write the two first steps of the improved Euler algorithm explicitly. Using a uniform partition with a step size h, we start with $x_0 = 0$, $y_0 = 0$. Then

$$x_1 = h, \quad y_1 = y_0 + \tfrac{h}{2}f(x_0, y_0) + \tfrac{h}{2}f(x_1, y_0 + hf(x_0, y_0)) = \tfrac{3}{4}h^2;$$

$$x_2 = 2h \quad y_2 = y_1 + \tfrac{3}{4}h(x_1 + x_2) + \tfrac{h}{2}y_1 + \tfrac{h^2}{8}(3x_1 + y_1)$$

$$= y_1 + \tfrac{3}{4}h(3h) + \tfrac{h}{2}y_1 + \tfrac{h^2}{8}3h + \tfrac{h^2}{8}y_1 = 3h^2 + \tfrac{3}{4}h^3 + \tfrac{3}{32}h^4.$$

Since the slope function is linear, the corresponding Heun recurrence can be solved explicitly:

$$y_n = 6\left(1 + \frac{h}{2} + \frac{h^2}{8}\right)^n - 3hn - 6, \qquad n = 0, 1, 2, \ldots .$$

[34]Karl Heun (1859–1929) was a German mathematician best known for the Heun differential equation that generalizes the hypergeometric differential equation. Approximation (3.2.15) was derived by C. Runge.

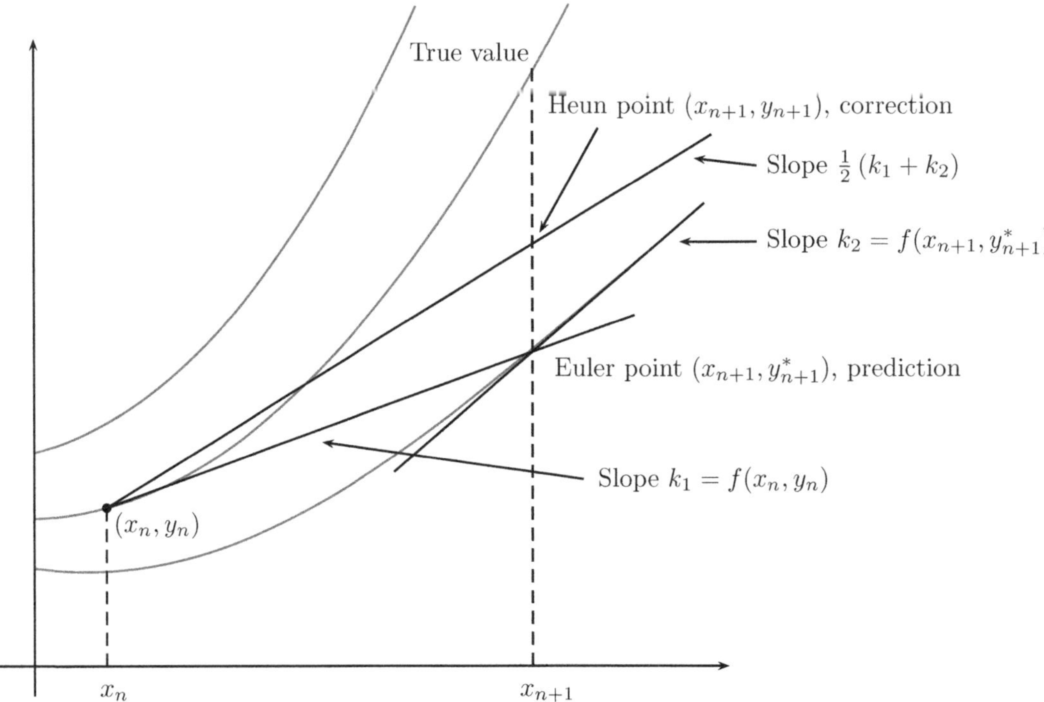

Figure 3.4: The improved Euler approximation for the one step to the solution of the linear differential equation $\dot{y} = 2\cos(t) + y(t) - 1$ subject to the initial condition $y(0) = y_0 > 0$.

Using the `series` command from *Maple*, we get y_n as a function of h:

$$y_n = \frac{3}{4}\,n^2h^2 + \frac{n(n^2-1)}{8}\,h^3 + \frac{n(n-1)(n^2+n-3)}{64}\,h^4 + \cdots, \qquad n = 0, 1, 2, \ldots.$$

Therefore, the difference $\phi(nh) - y_n$ is the series in h starting with h^3. □

When a slope function is nonlinear in y, it is generally impossible to find an exact formula for y_n. In this case one can use a software solver. With *Mathematica*, we have

```
For[i = 1, i < n + 1, i = i + 1; k1 = F[t,y];
k2 = F[t+h,y+h*k1]; y = y+h(k1+k2)/2; t = t+h];
Print[y]
```

MATLAB does a similar job:

```
function y=heun(h,t0,y0,T),t=t0; y=y0; n=round(T-t0)/h;
for j=1:n
k1 = f(t,y); k2 = f(t+h,y+h*k1);
y = y+h*(k1+k2)/2; t =t0+j*h; disp([t,y])
end
```

Similarly with *Maple:*

```
impeu:=proc(f,a,b,A,N)
local n, xn, yn, h, k1, k2, w;
h:=evalf((b-a)/N); xn:=evalf(a); yn:=evalf(A); w[0]:=[xn,yn];
for n from 1 to N do
k1:=evalf(f(xn,yn)); k2:=evalf(f(xn+h,yn+h*k1));
yn:=evalf(yn+h*(k1+k2)/2); xn:=evalf(xn+h); w[n]:=[xn,yn]; end do;
w;
vv:=impeu(f,0,1,0,N);
for n from 0 to N do
print(x||n=vv[n][1],ye||n=vv[n][2]); od;
```

Table 154: A comparison of the results for the numerical solution of the initial value problem $\dot{y} = 2\pi\, y\, \cos(2\pi t)$, $y(0) = 1$, using the improved Euler method (Heun formula) and the forward Euler rule for different step sizes h. The actual solution is $y = \phi(t) = e^{\sin(2\pi t)}$.

t	Exact	$h = 0.1$	$h = 0.05$	$h = 0.025$	$h = 0.01$
	Heun:				
0.5	1.0	0.9803425392	0.9982313528	0.9998019589	0.9999877448
1.0	1.0	0.9610714964	0.9964658340	0.9996039591	0.9999754899
1.5	1.0	0.9421792703	0.9947034358	0.9994059972	0.9999632284
3.0	1.0	0.8877017706	0.9894349280	0.9988123511	0.9999264662
	Euler:				
0.5	1.0	1.162054041	1.071163692	1.034443454	1.013584941
1.0	1.0	0.3082365926	0.5988071290	0.7795371389	0.9058853939
1.5	1.0	0.3581875773	0.6414204548	0.8063870900	0.9181917922
3.0	1.0	0.02928549567	0.2147142589	0.4737076882	0.7433952325

Example 3.2.6: The actual solution of the separable (linear) differential equation $\dot{y} = 2\pi\, y\, \cos(2\pi t)$ subject to the initial condition $y(0) = 1$ is the periodic function $y = \phi(t) = e^{\sin(2\pi t)}$. Table 154 shows that the improved Euler method (3.2.15) gives good approximations even for relatively large step sizes. On the other hand, the forward Euler method (3.2.5) has considerable difficulty keeping up with the actual solution, and requires a very small step size to give a reasonable approximation. □

Another modification of Euler's method is known. Using the midpoint rule for the evaluation of the integral on the right-hand side of Eq. (3.2.4), we obtain the so-called **modified Euler formula**, or *explicit midpoint rule*, or midpoint Euler algorithm:

$$y_{n+1} = y_n + hf\left(x_n + \frac{h}{2}\,,\ y_n + \frac{h}{2}\, f(x_n, y_n)\right), \qquad n = 0, 1, 2, \ldots. \tag{3.2.16}$$

This formula reevaluates the slope halfway through the line segment by taking two trial slopes, $k_1(x, y) = f(x, y)$ and $k_2(x, y; h) = f\left(x + h/2, y + hk_1/2\right)$, and then using the latter as the final slope. Therefore, the midpoint rule is another example of the predictor-corrector method: it extrapolates the value of y at the midpoint: $y_{n+1/2} = y_n + h\, f(x_n, y_n)/2$. Then this predicted value is used to calculate a slope at the midpoint.

Example 3.2.7: (Example 3.2.5 revisited) The modified Euler algorithm for $f(x, y) = \frac{1}{2}\,(3x + y)$ becomes

$$\begin{aligned} y_{n+1} &= y_n + \frac{h}{2}\left[3\left(x_n + \frac{h}{2}\right) + y_n + \frac{h}{2}\frac{1}{2}\,(3x_n + y_n)\right] \\ &= y_n\left(1 + \frac{h}{2} + \frac{h^2}{8}\right) + \frac{3h}{2}\, x_n + \frac{3h^2}{8}\, x_n + \frac{3h^2}{4}. \end{aligned}$$

Table 155: A comparison of the results for the numerical solution of the IVP $y' = 2\,\cos(t) + y(t) - 1$, $y(0) = 0$ using the Euler method, the trapezoid method, Heun's formula, and midpoint Euler algorithm for step size $h = 0.05$.

t	Exact	Euler	Trapezoid	Heun	Midpoint
1.0	1.301168679	1.280605284	1.3007147307	1.300202716	1.301122644
2.0	2.325444264	2.303749072	2.3238172807	2.323267203	2.325921001
2.5	2.399615760	2.386084742	2.3969091047	2.396552146	2.400687151
3.0	2.131112505	2.129626833	2.1267185085	2.126646709	2.133104265

Comparison with the previous example reveals a remarkable result: it is the same recurrence as with the improved Euler formula. Indeed, these two formulas (3.2.15) and (3.2.16) coincide for a linear slope function $f(x, y)$. However, if the function $f(x, y)$ is not linear, then these formulas lead to different results. For example, let $f(x, y) = 4x^2 + 2y$.

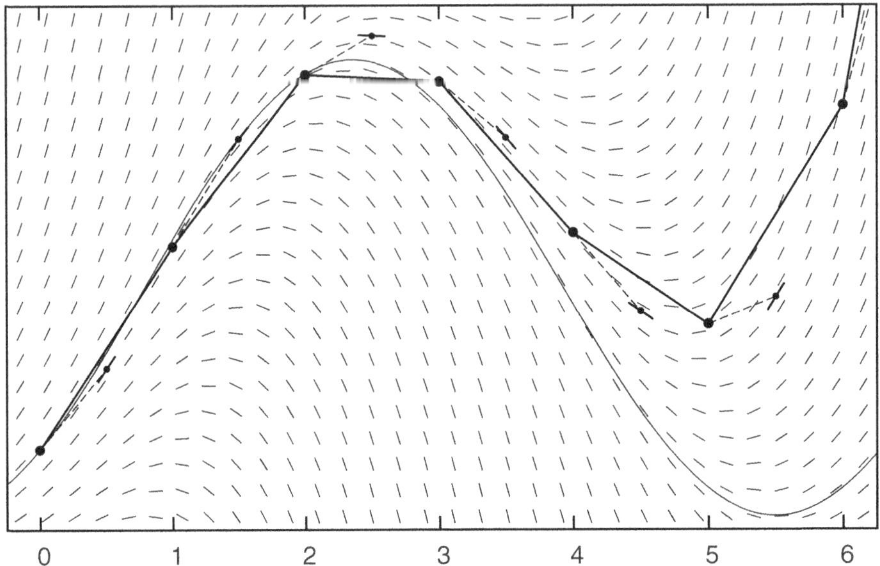

Figure 3.5: Modified Euler approximations (in black) along with the exact solution (in blue) to the linear differential equation $\dot{y} = 2\cos(t) + y(t) - 1$, plotted with *Maxima*.

Then we have the following approximations (with 11 and 13 arithmetic operations at each step, respectively):

$$y_{n+1} = y_n(1 + 2h + 2h^2) + 2h(x_n^2 + x_{n+1}^2) + 4h^2 x_n^2, \qquad \text{improved Euler,}$$

$$y_{n+1} = y_n(1 + 2h + 2h^2) + 4h\left(x_n + \frac{h}{2}\right)^2 + 4h^2 x_n^2, \qquad \text{modified Euler.}$$

Table 155 gives comparisons of numerical calculations based on the methods discussed in this section.

Problems

In all problems, use a uniform grid of points: $x_k = x_0 + kh$, $k = 0, 1, 2, \ldots, n$, where $h = (x_n - x_0)/n$ is a fixed step size. A numerical method works with a discrete set of mesh points $\{x_k\}$ and the sequence of values $y_0, y_1, \ldots, y_n$, such that each y_k is approximately equal to the true value of the actual solution $\phi(x)$ at that point x_k, $k = 0, 1, \ldots, n$. Throughout, primes denote derivatives with respect to x, and a dot stands for the derivative with respect to t.

1. A projectile of mass $m = 0.2\,\texttt{kg}$ shot vertically upward with the initial velocity $v(0) = 8\,\texttt{m/sec}$ (see [14] for details) is slowed due to the gravity force $F_g = mg$ ($g = 9.81\,\texttt{m/sec}^2$) and due to the air resistance force $F_r = -kv|v|$, where $k = 0.002\,\texttt{kg/m}$. Use Euler's methods (3.2.5), (3.2.15), and (3.2.16) with $h = 0.1$ to approximate the velocity after 2.0 seconds.
2. Given $y' = y$, $y(0) = 1$, find $y(1)$ numerically using all five numerical methods presented in the section by taking $h = 0.1$ and compare the results with its exact value. How many correct decimal places did you get? How can you use your results to compute e?
3. Solve the initial value problem $(1 + x^2)\,y' = 1$, $y(0) = 0$, using the Euler method (3.2.5) to find $y(1)$ with step sizes $h = 0.1,\ 0.05,\ 0.01$. How can you use your results to compute π?
4. As we have learned from §2.1.1, solutions to autonomous differential equations may blow up at a point. Consider the initial value problem $\dot{y} = y^2$, $y(0) = 1$, and pretend that you don't know its solution, $y = \phi(t) = (1 - t)^{-1}$. Use Euler's numerical method to attempt estimating the value of the solution at $t = 1$. How are you going to verify that the actual solution does not exist on the closed interval $[0, 1]$?
5. Consider the initial value problem $3xy' + y = 0$, $y(-1) = 1$. Show that it has the actual solution $y = \phi(x) = -x^{-1/3}$, which exists for $x < 0$, having discontinuity at $x = 0$. Apply Euler's algorithm (3.2.5) with step size $h = 0.15$ to approximate this solution on the interval $[-1, 0.5]$. Show that the numerical method "jumps across discontinuity" to another solution for $x > 0$. Repeat calculations for $h = 0.03$, but keeping results at the previous mesh points. Would you now suspect a discontinuity at $x = 0$?
6. Solve the following differential equation that describes the amount $x(t)$ of potassium hydroxide (KOH) after time t:
$$\frac{\mathrm{d}x}{\mathrm{d}t} = k\left(n_1 - \frac{x}{2}\right)^2 \left(n_2 - \frac{x}{2}\right)^2 \left(n_3 - \frac{3x}{4}\right)^3,$$

where k is the velocity constant of the reaction. If $k = 6.22 \times 10^{-19}$, $n_1 = n_2 = 2 \times 10^3$, and $n_3 = 3 \times 10^3$, how many units of KOH will have been formed after 0.3 sec? Use the modified Euler method (3.2.16) with uniform step size $h = 0.01$ to calculate the approximate value, and then compare it with the true value.

7. Consider the following autonomous differential equations subject to the initial condition $y(0) = 1$ at the specified point $x = 0$.

   (a) $\dot{y} = e^{-y}$; **(d)** $\dot{y} = \frac{y}{8}(1 - y/4)$; **(g)** $\dot{y} = 1/(2y + 3y^2)$;
   (b) $\dot{y} = \frac{1}{6}(1 + y^2)$; **(e)** $\dot{y} = (y^2 + 1)/y$; **(h)** $\dot{y} = (y - 5)^2$;
   (c) $\dot{y} = 7y(1 - y/100)$; **(f)** $\dot{y} = y^2 - y/10$; **(i)** $\dot{y} = y^{3/2}$.

   Use the Euler rule (3.2.5), improved Euler formula (3.2.15), and modified Euler method (3.2.16) to obtain an approximation at $x = 1$. First use $h = 0.1$ and then use $h = 0.05$. Then compare numerical values with the true value.

8. Answer the following questions for each linear differential equation **(a)** – **(f)** subject to the initial condition $y(0) = 1$ at the specified point $x = 0$.

   (a) $y' = y - 2$; **(b)** $y' = 2 + x - y$; **(c)** $y' = 3 - 9x + 3y$;
   (d) $y' + y = 2x$; **(e)** $(x^2 + 1)y' + 2xy = 6x$; **(f)** $(x + 1)y' + y = x^2$.

   (i) Find the actual solution $y = \phi(x)$.

   (ii) Use Euler's method $y_{k+1} = y_k + hf_k$ to obtain the approximation y_k at the mesh point x_k.

   (iii) Use the backward Euler method (3.2.12) to obtain the approximation y_k at the mesh point x_k.

   (iv) Use the trapezoid rule (3.2.14) to obtain the approximation y_k at the mesh point x_k.

   (v) Use the improved (or modified) Euler formula (3.2.15) to obtain the approximation y_k at the mesh point x_k.

   (vi) For $h = 0.1$, compare numerical values y_4 found in the previous parts with the true value $\phi(0.4)$.

9. Consider the following separable differential equations subject to the initial condition $y(0) = 1$ at the specified point $x_0 = 0$.

   (a) $y' = xy^2$; **(b)** $y' = (2x + 1)(y^2 + 1)$; **(c)** $y^3 y' = \cos x$;
   (d) $y' = 2xy + x/y$; **(e)** $(1 + x)\, y' = y$; **(f)** $y' = xy^2 - 4x$;
   (g) $y' = 2x \cos^2 y$; **(h)** $y' = (2x/y)^2$; **(i)** $y' = 3x^2 + y$.

   Use the Euler method (3.2.5), improved Euler formula (3.2.15), and modified Euler algorithm (3.2.16) on the uniform mesh grid to obtain an approximation at the point $x = 1$. First use $h = 0.1$, and then use $h = 0.05$. Compare numerical values at $x = 1$ with the true value.

10. Consider the following initial value problems for the Bernoulli equations.

    (a) $y' = y(2x + y)$, $y(0) = 1/5$. **(b)** $xy' + y = x^3 y^{-2}$, $y(1) = 1$.
    (c) $y' = xy^2 + y/(2x)$, $y(1) = 1$. **(d)** $\dot{y} + y = y^2 e^t$, $y(1) = 1$.
    (e) $\dot{y} = y + ty^{1/3}$, $y(0) = 1$. **(f)** $y' = 2xy(1 - y)$, $y(0) = 2$.

    First find the actual solutions and then calculate approximate solutions at the point $x = 1.5$ using the Euler method (3.2.5), improved Euler formula (3.2.15), and modified Euler method (3.2.16) on the uniform grid. First use $h = 0.1$ and then $h = 0.05$.

11. Consider the following initial value problems for which analytic solutions are not available.

    (a) $y' = x + \sqrt{y}$, $y(0) = 1$. **(b)** $y' = x^2 - y^3$, $y(0) = 1$.
    (c) $y' = x + y^{1/3}$, $y(0) = 1$. **(d)** $y' = e^x + x/y$, $y(0) = 1$.
    (e) $y' = \sin(x) - y^2$, $y(0) = 1$. **(f)** $y' = x^2 - x^3 \cos(y)$, $y(0) = 1$.

    Find approximate solutions at the point $x = 1.4$ for each problem using the Euler method (3.2.5), improved Euler formula (3.2.15), and modified Euler method (3.2.16) on the uniform grid. First use $h = 0.1$ and then $h = 0.05$.

12. Apply the backward Euler algorithm to solve the following differential equations involving cubic and fourth powers. In particular, find the approximate value of the solution at the point $x = 0.5$ with the uniform step size $h = 0.1$ and $h = 0.05$.

    (a) $y' = y^3 - 5x$, $y(0) = 1$; **(b)** $\dot{y} = y^4 - 40\,t^2$, $y(0) = 1$.

13. Make a computational experiment: in the improved Euler algorithm (3.2.15), use the corrector value as the next prediction, and only after the second iteration use the trapezoid rule. Derive an appropriate formula and apply it to solve the following differential equations subject to the initial condition $y(0) = 1$. Compare your answers with the true values at $x = 2$ and the approximate values obtained with the average slope algorithm (3.2.15) with $h = 0.1$.

    (a) $y' = (1 + x)\sqrt{y}$; **(b)** $y' = (2x + y + 1)^{-1}$; **(c)** $y' = (y - 4x)^2$.

14. Consider the **generalized logistic equation**

$$\frac{\mathrm{d}P(t)}{\mathrm{d}t} = k\,P^{\alpha}\left(1 - \frac{P^{\beta}}{K}\right).$$

Find numerical approximations to the solution in the range $0 \leqslant t \leqslant 10$ for the parameter pairs $(\alpha, \beta) = (0.5, 1)$, $(0.5, 2)$, $(1.5, 1)$, $(1.5, 2)$, $(2, 2)$ and the values of parameters $k = 1$, $K = 5$, and the initial condition $P(0) = 1$ using

(a) Euler's rule (3.2.5) with $h = 0.1$;
(b) Euler's rule (3.2.5) with $h = 0.01$;
(c) improved Euler's method (3.2.15) with $h = 0.1$;
(d) improved Euler's method (3.2.15) with $h = 0.01$;
(e) modified Euler's method (3.2.16) with $h = 0.1$;
(f) modified Euler's method (3.2.16) with $h = 0.01$;

15. Repeat the previous exercise with another initial condition $P(0) = 6$.

16. Consider the initial value problem for the generalized logistic equation

$$\frac{\mathrm{d}P(t)}{\mathrm{d}t} = P(t)\left(1 - \frac{P(t)}{K(t)}\right), \qquad P(0) = \frac{1}{2},$$

where $K(t)$ depends on time variable t. Find numerical approximations to the solution in the range $0 \leqslant t \leqslant 10$ for the following functions

(A) $K(t) = 3 + \sin t$; **(B)** $K(t) = \ln(2 + t)$; **(C)** $K(t) = t + 1$;

using

(a) Euler's rule (3.2.5) with $h = 0.1$;
(b) Euler's rule (3.2.5) with $h = 0.01$;
(c) improved Euler's method (3.2.15) with $h = 0.1$;
(d) improved Euler's method (3.2.15) with $h = 0.01$;
(e) modified Euler's method (3.2.16) with $h = 0.1$;
(f) modified Euler's method (3.2.16) with $h = 0.01$;

17. Consider the initial value problem $y' = x(1 + y^2)$, $y(0) = 1$. Use Euler's rule (3.2.5) and the Heun method (3.2.15) to find the approximate value of the exact solution $\phi(x) = \tan\left(\frac{x^2}{2} + \frac{\pi}{4}\right)$ at the point $x = 1.5$. Use several step decrements, starting with $h = 0.1$, and then $h = 0.01$, $h = 0.001$. Do you observe any problem in achieving this task?

18. Consider the initial value problem $y' = (x + y - 3/2)^2$, $y(0) = 2$. Use the improved Euler method (3.2.15) with $h = 0.1$ and $h = 0.05$ to obtain approximate values of the solution at $x = 0.5$. At each step compare the approximate value with the true value of the actual solution.

19. Which of two numerical methods, average slope algorithm (3.2.15) or midpoint rule (3.2.16), requires fewer arithmetic operations?

20. Let a and b be some real numbers. Apply Heun's formula and the midpoint Euler method with a fixed positive step size h to the initial value problem $y' = ax + by$, $y(x_0) = y_0$. Show that in either case these methods lead to the same difference equation, which has the solution

$$y_n = \left(1 + bh + \frac{b^2h^2}{2}\right)^n \left[y_0 + \frac{ax_0}{b} + \frac{a}{b^2}\right] - \frac{anh}{b} - \frac{ax_0}{b} - \frac{a}{b^2}.$$

In each problem, the given iteration is the result of applying Euler's method, the trapezoid rule, improved Euler formula, or the midpoint Euler rule to an initial value problem $y' = f(x, y)$, $y(x_0) = y_0$ on the interval $a = x_0 \leqslant x \leqslant b$. Identify the numerical method and determine $a = x_0$, b, and $f(x, y)$.

14. $y_{n+1} = y_n + hx_n^2 + hy_n^2$, $y_0 = 1$, $x_n = 1 + nh$, $h = 0.01$, $n = 0, 1, \ldots, 100$.

15. $y_{n+1} = y_n + hx_n + hx_{n+1} + hy_{n+1}$, $y_0 = 1$, $x_n = 1 + nh$, $h = 0.02$, $n = 0, 1, \ldots, 100$.

16. $y_{n+1} = y_n + 3h + hx_n y_n^2 + h^2\left(x_n^2 y_n^3 + 3x_n y_n + \frac{1}{2}\,y_n^2\right) + h^3\left(3x_n^2 y_n^2 + \frac{1}{2}\,x_n^3 y_n^4 + \frac{9}{2}\,x_n + x_n y_n^3 + 3y_n\right) + h^4\left(3x_n y_n^2 + \frac{9}{2}\right)$, $y_0 = 2$, $x_n = 1 + nh$, $h = 0.05$, $n = 0, 1, \ldots, 40$.

17. $y_{n+1} = y_n + hx_n + \frac{h^2}{2} + h\left(y_n + \frac{h}{2}\left(x_n + y_n^2\right)^2\right)$, $y_0 = 1$, $x_n = nh$, $h = 0.01$, $n = 0, 1, \ldots, 100$.

18. $y_{n+1} = y_n + \frac{h}{2}\left(y_n^2 + y_{n+1}^2\right)$, $y_0 = 0$, $x_n = 1 + nh$, $h = 0.05$, $n = 0, 1, \ldots, 100$.

3.3 The Polynomial Approximation

One of the alternative approaches to the discrete numerical methods discussed in §3.2 is the Taylor series method. If the slope function $f(x, y)$ in the initial value problem

$$y' = f(x, y), \qquad y(x_0) = y_0, \tag{3.3.1}$$

is sufficiently differentiable or is itself given by a power series, then the numerical integration by Taylor's expansion is possible. In what follows, it is assumed that $f(x, y)$ possesses continuous derivatives with respect to both x and y of all orders required to justify the analytical operations to be performed.

There are two ways in which a Taylor series can be used to construct an approximation to the solution of the initial value problem. One can either utilize the differential equation to generate Taylor polynomials that approximate the solution, or one can use a Taylor polynomial as a part of a numeric integration scheme similar to Euler's methods. We are going to illustrate both techniques.

It is known from calculus that a smooth function $g(x)$ can be approximated by its Taylor series polynomial of the order n:

$$p_n(x) = g(x_0) + g'(x_0)(x - x_0) + \cdots + \frac{g^{(n)}(x_0)}{n!}(x - x_0)^n, \tag{3.3.2}$$

which is valid in some neighborhood of the point $x = x_0$. How good this approximation is depends on the existence of the derivatives of the function $g(x)$ and their values at $x = x_0$, as the following statement shows (consult an advanced calculus course).

Theorem 3.1: [**Lagrange**] Let $g(x) - p_n(x)$ measure the accuracy of the polynomial approximation (3.3.2) of the function $g(x)$ that possesses $(n + 1)$ continuous derivatives on an interval containing x_0 and x, then

$$g(x) - p_n(x) = \frac{g^{(n+1)}(\xi)}{(n+1)!}(x - x_0)^{n+1},$$

where ξ, although unknown, is guaranteed to be between x_0 and x.

Therefore, one might be tempted to find the solution to the initial value problem (3.3.1) as an infinite Taylor series:

$$\begin{aligned} y(x) &= \sum_{n \geqslant 0} a_n (x - x_0)^n \qquad\qquad (3.3.3)\\ &= y(x_0) + y'(x_0)(x - x_0) + \frac{y''(x_0)}{2!}(x - x_0)^2 + \frac{y'''(x_0)}{3!}(x - x_0)^3 + \cdots . \end{aligned}$$

This series provides an explicit formula for the solution, which can be computed to any desired accuracy. If a numerical method based on a power series expansion is to yield efficient, accurate approximate solutions to problems in ordinary differential equations, it must incorporate accurate determinations of the radii of convergence of the series. However, the series representation (3.3.3) may not exist, and when it does exist, the series may not converge in the interval of interest.

Usually solutions of nonlinear differential equations contain many singularities that hinder the computation of numerical solutions. The determination of singular points or estimation of the radius of convergence requires quite a bit of effort. Without knowing a radius of convergence (which is the distance to the nearest singularity) and the rate of decrease of coefficient magnitudes, power series solutions are rather useless because we don't know how many terms in the truncated series must be kept to achieve an accurate approximation. In general, the Taylor series approximation of a smooth function is often not so accurate over an interval of interest, but may provide a reliable power series solution when some information about its convergence is known, for example, when its coefficients are determined, or satisfy a recurrence, or some pattern is evident.

Since the coefficients $a_n = y^{(n)}(x_0)/n!$ in series (3.3.3) are expressed through derivatives of the (unknown) function $y(x)$, they could be determined by differentiation of both sides of the equation $y' = f(x, y)$ as follows:

$$\begin{aligned} y' &= f(x, y), \\ y'' &= f' = f_x + f_y y' = f_x + f\, f_y = \left(\frac{\partial}{\partial x} + f\,\frac{\partial}{\partial y}\right) f(x, y), \\ y''' &= f'' = f_{xx} + 2 f_{xy} f + f_{yy} f^2 + f_y f_x + f_y^2 f = \left(\frac{\partial}{\partial x} + f\,\frac{\partial}{\partial y}\right)^2 f(x, y). \end{aligned}$$

Continuing in this manner, one can express any derivative of an unknown function y in terms of $f(x,y)$ and its partial derivatives. However, we are most likely forced to truncate the infinite series after very few terms because the exact expressions for derivatives of f rapidly increase in complexity unless f is a very simple function or some of its derivatives become identically zero (for example, if $f(x,y) = (2y/x) - 1$, then $f'' \equiv 0$).

Since formulas for successive derivatives can be calculated by repeatedly differentiating the previous derivative and then replacing the derivative y' by $f(x, y(x))$, we delegate this job to a computer algebra system. For instance, the following *Maple* code allows us to find derivatives of the function $y(x)$:

```
derivs:=proc(n)
option remember;
if n=0 then y(x)
elif n=1 then f(x,y(x))
else simplify(subs(diff(y(x),x)=f(x,y(x)),diff(derivs(n-1),x)))
end if; end;
```

Similarly, we can find derivatives using *Mathematica:*

```
f[x_,y_] = x^2 + y^2;        (* function f is chosen for illustration *)
y1[x_,y_] = D[f[x,y[x]],x]/.{y'[x]->f[x,y],y[x]->y};
y2[x_,y_] = D[y1[x,y[x]],x]/.{y'[x]->f[x,y],y[x]->y};          (* or *)
f2[x_,y_]:= D[f[x, y], x] + D[f[x, y], y] f[x, y]
f3[x_,y_]:= D[f2[x, y], x] + D[f2[x, y], y] f[x, y]
```

Example 3.3.1: Consider the initial value problem

$$y' = x - y + 2, \qquad y(0) = 1.$$

Since the given differential equation is linear, the problem has a unique solution on the real line (see Theorem 1.1 on page 22). Therefore, we try to find its solution as the power series

$$y(x) = 1 + \sum_{n\geqslant 1} c_n\, x^n,$$

where $n!\, c_n = y^{(n)}(0)$, $n = 1, 2, \ldots$. This series obviously satisfies the initial condition $y(0) = 1$ provided that it converges in a neighborhood of the origin. Differentiating both sides of the equation $y' = x - y + 2$ and setting $x = 0$, we obtain

$$\begin{aligned} c_1 &= y'(0) = (x - y + 2)_{x=0} = -y(0) + 2 = 1, \\ 2\, c_2 &= y''(0) = (x - y + 2)'_{x=0} = (1 - y')_{x=0} = 0, \\ 3!\, c_3 &= y'''(0) = (x - y + 2)''_{x=0} = \frac{\mathrm{d}}{\mathrm{d}x}(1 - y')\bigg|_{x=0} = -y''(0) = 0. \end{aligned}$$

All coefficients c_n in the series representation $y(x) = 1 + x + \sum_{n\geqslant 2} c_n\, x^n$ vanish and we get the solution $y = 1 + x$.

Now we consider the same equation subject to a different initial condition $y(1) = 3$. We seek its solution in the form

$$y(x) = 3 + \sum_{n\geqslant 1} c_n\, (x-1)^n, \qquad n!\, c_n = y^{(n)}(1).$$

Using the same approach, we obtain

$$\begin{aligned} c_1 &= y'(1) = (x - y + 2)_{x=1} = -y(1) + 3 = 0, \\ 2\, c_2 &= y''(1) = (x - y + 2)'_{x=1} = (1 - y')_{x=1} = 1, \\ 3!\, c_3 &= y'''(1) = (x - y + 2)''_{x=1} = \frac{\mathrm{d}}{\mathrm{d}x}(1 - y')\bigg|_{x=1} = -y''(1) = -1, \\ 4!\, c_4 &= y^{(4)}(1) = -y'''(1) = 1, \end{aligned}$$

and so forth. Hence, $c_n = \frac{(-1)^n}{n!}$, $n = 2, 3, \ldots$, and the solution becomes

$$y(x) = 3 + \sum_{n\geqslant 2} \frac{(-1)^n}{n!}\,(x-1)^n = 3 + \sum_{n\geqslant 0} \frac{(-1)^n}{n!}\,(x-1)^n - 1 + (x-1) = 1 + x + e^{1-x}.$$

Note that the series for $y(x)$ converges everywhere.

Example 3.3.2: (Example 3.2.3 revisited) Consider the initial value problem for the following nonlinear equation:

$$y' = \frac{1}{x-y+2} = (x-y+2)^{-1}, \qquad y(0) = 1.$$

Let us try the series method: $y(x) = 1 + x + \sum_{n\geqslant 2} a_n x^n$, where $n!\,a_n = y^{(n)}(0)$. Here we have used our knowledge that $y(0) = 1$ and $y'(0) = \left.\frac{1}{x-y+2}\right|_{\substack{x=0\\y=1}} = 1$. Now we need to find further derivatives:

$$\begin{aligned} y'' &= \frac{\mathrm{d}}{\mathrm{d}x}(x-y+2)^{-1} = (-1)(x-y+2)^{-2}(1-y') \quad\Longrightarrow\quad y''(0) = 0, \\ y'' &= 2(x-y+2)^{-3}(1-y')^2 - (x-y+2)^{-2}(-y'') \quad\Longrightarrow\quad y'''(0) = 0, \end{aligned}$$

and so on. Hence, all derivatives of the function y greater than 2 vanish and the required solution becomes $y = x+1$.

If we would like to solve the same differential equation under another initial condition $y(x_0) = y_0$, we cannot choose the initial point (x_0, y_0) arbitrarily; it should satisfy $x_0 - y_0 + 2 \neq 0$, otherwise the slope function $f(x,y) = (x-y+2)^{-1}$ is undefined. For instance, we cannot pick $y(1) = 3$ simply because this point $(1,3)$ is outside of the domain of the slope function. So we consider the same equation subject to the initial condition $y(1) = 1$ and seek its solution in the form: $y(x) = 1 + \sum_{n\geqslant 1} c_n (x-1)^n$. Then the coefficients are evaluated as follows:

$$\begin{aligned} c_1 &= y'(1) = (x-y+2)^{-1}_{x=1} = 1/2, \\ 2\,c_2 &= y''(1) = \left.\frac{\mathrm{d}}{\mathrm{d}x}(x-y+2)^{-1}\right|_{x=1} = -(x-y+2)^{-2}(1-y')_{x=1} = -\frac{1}{2^3}, \\ 3!\,c_3 &= y'''(1) = -\left.\frac{\mathrm{d}}{\mathrm{d}x}(x-y+2)^{-2}(1-y')\right|_{x=1} \\ &= 2\,(x-y+2)^{-3}(1-y')^2_{x=1} + (x-y+2)^{-2}y''|_{x=1} = \frac{1}{2^5}, \\ 4!\,c_4 &= -\left.\frac{6(1-y')^3}{(x-y+2)^4}\right|_{x=1} - \left.\frac{6(1-y')y''}{(x-y+2)^3}\right|_{x=1} + \left.\frac{y'''}{(x-y+2)^2}\right|_{x=1} = \frac{1}{2^7}, \end{aligned}$$

and so on. Substituting back, we get

$$y = 1 + \frac{x-1}{2} - \left(\frac{x-1}{4}\right)^2 + \frac{1}{3}\left(\frac{x-1}{4}\right)^3 + \frac{1}{12}\left(\frac{x-1}{4}\right)^4 - \frac{13}{60}\left(\frac{x-1}{4}\right)^5 + \cdots.$$

To verify our calculations, we ask *Maple* for help.

```
ode:=D(y)(x)=1/(x-y(x)+2);
dsolve({ode, y(1) = 1}, y(x), 'series', x = 1)
```

The series solution seems to converge for small $|x-1|$, but we don't know its radius of convergence because we have no clue how the coefficients decrease with n in general. It is our hope that the truncated series formula gives a good approximation to the "exact" solution $y = \ln|x-y+2| + 1 - \ln 2$, but hopes cannot be used in calculations. We are lucky that the solution can be found in an implicit form or even as an inverse function: $x = y - 1 + e^{y-1}$. Nevertheless, the truncated series approximation is close to the actual solution in some neighborhood of the point $x = 1$, as seen from Fig. 3.6. Using the following *Maple* commands, we can evaluate the solution at any point, say at $x = 2$, to get $y(2) \approx 1.442854401$.

```
soln := dsolve({ode, y(1) = 1}) assign(soln); y(x);
evalf(subs(x = 2, y(x)))
```

Now we plot the actual solution using *Mathematica:*

```
ans1 = DSolve[{y'[x] == 1/(x - y[x] + 2), y[0] == 1}, y[x], x]
Plot[y[x] /. ans1, {x, -1, 1}, PlotRange -> {0, 2}, AspectRatio ->1]
ans2 = DSolve[{x'[y] == x[y] - y + 2, x[1] == 0}, x[y], y] //Simplify
ParametricPlot[Evaluate[{x[y], y} /. ans2], {y, 0, 2},
 PlotRange -> {{-1, 1}, {0, 2}}, AspectRatio -> 1]
```

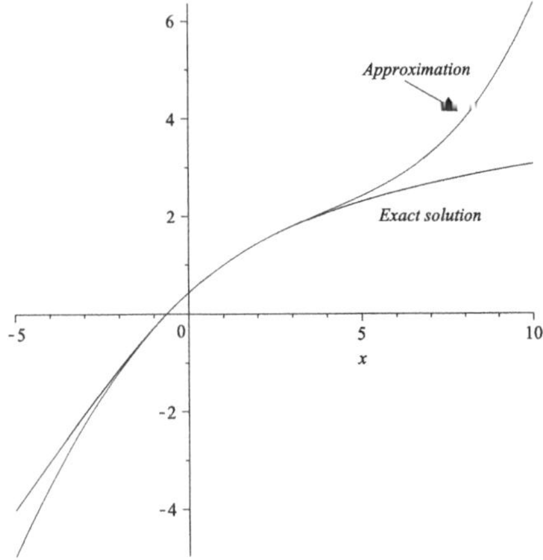

Figure 3.6: Example 3.3.2: fourth Taylor approximation to the exact solution, plotted with *Maple*.

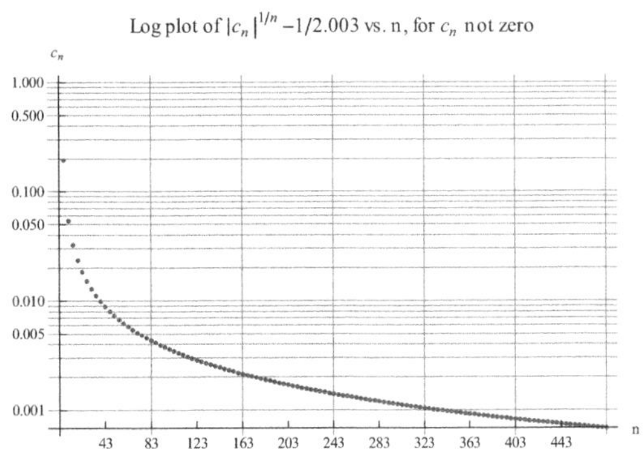

Figure 3.7: The first 100 roots $\sqrt[4n+3]{|c_{4n+3}|}$ of coefficients of power series (3.3.4), plotted with *Mathematica* using a logarithm scale.

Example 3.3.3: Let $f(x,y) = x^2 + y^2$ and consider the equation $y' = f(x,y)$ subject to the initial condition $y(0) = 0$. The amount of work to find sequential derivatives of $y^{(n)}$ accumulates like a rolling snowball with n; however, first two derivatives of f are easy to obtain

$$\begin{aligned} y'' &= f' = f_x + f_y f = 2(x + x^2 y + y^3), \\ y''' &= f'' = 2(1 + 2xy + x^4 + 4x^2y^2 + 3y^4), \end{aligned}$$

which allow us to find the first few coefficients in its power series representation

$$y(x) = \sum_{n\geqslant 0} c_n x^n = \frac{x^3}{3} + \frac{x^7}{63} + \frac{2\,x^{11}}{2079} + \frac{13\,x^{15}}{218295} + \frac{46\,x^{19}}{12442815} + \cdots . \tag{3.3.4}$$

Since no simple pattern seems to be emerging for the coefficients, it is hard to say what the radius of convergence is for this series. We know from Example 2.7.5, page 114, that this radius should be around 2 because the solution (3.3.4) blows up around $x \approx 2.003147$. Actually, computer algebra systems are so powerful that we may try to estimate the radius R of convergence using the root test $R^{-1} = \limsup_{n\to\infty} \sqrt[n]{|c_n|} = \lim_{n\to\infty} \sqrt[4n+3]{|c_{4n+3}|}$, where lim sup is the upper limit. Indeed, Fig. 3.7, plotted with *Mathematica*, unambiguously indicates that the reciprocal of the radius of convergence is around 0.5.

```
DSolve[{y'[x] == y[x]^2 + x^2, y[0] == 0}, y[x], x]
mylist = CoefficientList[y'[x] - ( y[x]^2 + x^2) /.
    {y[x] -> Sum[c[n] x^n, {n, 0, m}],
     y'[x] -> D[Sum[c[n] x^n, {n, 0, m}], x]}, x];
c[0] = 0.;       (* m is the number of terms in truncated polynomial *)
Sum[c[n] x^n, {n, 0, m}] /.
 Solve[Take[mylist, {1, m}] == Table[0., {n, 1, m}],
   Table[c[n], {n, 1, m}]][[1]]
CoefficientList[%, x]
Table[%[[i]]^(1/i), {i, 1, Length[%]}]       (* find i-th root of c_i *)
ListPlot[%, Frame -> True, GridLines -> {None, {0.5}}]
```
□

As seen from the previous examples, practical applications of the power series method are quite limited. Nevertheless, polynomial approximations can be used locally in a step-by-step procedure. For example, if the series (3.3.3) is terminated after the first two terms and $\phi(x_n)$ and $\phi(x_{n+1})$ are replaced by their approximate values y_n and y_{n+1}, respectively, then we obtain the Euler method. Viewed in this light, we might retain more terms in the Taylor series to obtain a better approximation. Considering a uniform grid (3.2.2), page 146, it is common to calculate approximation y_{n+1} of the exact solution at the next step point x_{n+1} using either the **Taylor Series Method of Order 2**

$$y_{n+1} = y_n + hf_n + \frac{h^2}{2}\left[f_x + f_y\, f\right]_{\substack{x=x_n \\ y=y_n}} = y_n + h\Phi_2(x_n, y_n; h), \tag{3.3.5}$$

where

$$\Phi_2(x,y;h) = f(x,y) + \frac{h}{2}\left(f_x + f_y f\right),$$

or the **Taylor Series Method of Order 3** (or a three-term polynomial approximation)

$$y_{n+1} = y_n + h\Phi_2(x_n,y_n;h) + \frac{h^3}{3!}\left[f_{xx} + 2f_{xy}\,f + f_{yy}f^2 + f_x f_y + f_y^2 f\right]\Big|_{\substack{x=x_n\\ y=y_n}}. \tag{3.3.6}$$

In order to formalize the procedure, we first introduce, for any positive integer p, the increment operator acting on a function f:

$$\Phi_p f \stackrel{\text{def}}{=} \Phi_p(x,y;h) = f(x,y) + \frac{h}{2!}\,f'(x,y) + \cdots + \frac{h^{p-1}}{p!}\,f^{(p-1)}(x,y), \tag{3.3.7}$$

where the derivatives $f^{(k)}$ can be calculated by means of the recurrence

$$f^{(k+1)} = \left(\frac{\partial}{\partial x} + f\,\frac{\partial}{\partial y}\right)^k f(x,y(x)), \qquad f^{(0)} = f, \qquad k = 0,1,2,\ldots. \tag{3.3.8}$$

As always, the approximate value of the actual solution at $x = x_n$ is denoted by y_n, where $x_n = x_0 + nh$, $n = 0,1,2,\ldots$. If it is desired to have a more accurate method, then it should be clear that more computational work will be required. In general, we have **Taylor's algorithm of order** p to find an approximate solution of the differential equation $y' = f(x,y)$ subject $y(a) = y_0$ over an interval $[a,b]$:

1. For simplicity, choose a uniform partition with $N+1$ points $x_k = a + kh$, $k = 0,1,\ldots,N$, and set the step size $h = (b-a)/N$.

2. Generate approximations y_k at mesh points x_k to the actual solution of the initial value problem $y' = f(x,y)$, $y(x_0) = y_0$ from the recurrence

$$y_{k+1} = y_k + h\Phi_p(x_k,y_k;h), \qquad k = 0,1,2,\ldots,N-1. \tag{3.3.9}$$

When x_n and $y(x_n)$ are known exactly, then Eq. (3.3.9) could be used to compute y_{n+1} with an error

$$\frac{h^{p+1}}{(p+1)!}\,y^{(p+1)}(\xi_n) = \frac{h^{p+1}}{(p+1)!}\,f^{(p)}(\xi_n, y(\xi_n)), \qquad x_n < \xi_n < x_{n+1},$$

where ξ_n is an unknown parameter within the interval (x_n, x_{n+1}). If the number of terms to be included in (3.3.7) is fixed by the permissible error ε and the series is truncated by p terms, then

$$\frac{h^{p+1}}{(p+1)!}\left|f^{(p)}(\xi_n, y(\xi_n))\right| < \varepsilon. \tag{3.3.10}$$

For given h, the inequality (3.3.10) will determine p; otherwise, if p is specified, it will give an upper bound on h. Actually, *Maple* has a special subroutine based on the Taylor approximation:

```
Digits := 10; dsolve(ode, numeric, method = taylorseries, output =
Array([0, .2, .4, .6, .8, 1]), abserr = 1.*10^(-10))
```

Here the `Array` option allows us to obtain the output at specified points. There is another alternative approach of the polynomial approximation of great practical importance. The *Mathematica* function `NDSolve` approximates the solution by cubic splines, a sequence of cubic polynomials that are defined over mesh intervals $[x_n, x_{n+1}]$ ($n = 0,1,\ldots$), and adjacent polynomials agree at the end mesh points.

Example 3.3.4: (Example 3.3.3 revisited) For the slope function $f(x,y) = x^2 + y^2$, the Taylor series methods of order 2 and 3 lead to the following formulas:

$$y_{n+1} = y_n + h(x_n^2 + y_n^2) + h^2(x_n + x_n^2 y_n + y_n^3),$$
$$y_{n+1} = y_n + h(x_n^2 + y_n^2) + h^2(x_n + x_n^2 y_n + y_n^3) + \frac{h^3}{3}\left(1 + 2x_n y_n + x_n^4 + 4x_n^2 y_n^2 + 3y_n^4\right),$$

respectively. The operators $\Phi_2(x, y; h)$ and $\Phi_3(x, y; h)$ become

$$\Phi_2(x, y; h) = x^2 + y^2 + h(x + x^2 y + y^3),$$
$$\Phi_3(x, y; h) = x^2 + y^2 + h(x + x^2 y + y^3) + \frac{h^2}{3}(1 + 2xy + x^4 + 4x^2y^2 + 3y^4).$$

Assuming that x and y have been evaluated previously, the operator $\Phi_2(x, y; h)$ requires 9 arithmetic operations, whereas $\Phi_3(x, y; h)$ needs 23 arithmetic operations at each step.

Example 3.3.5: We consider the initial value problem

$$y' = x^2 + y, \qquad y(0) = 1.$$

Since the equation is linear, the given problem has the explicit solution $y = 3\,e^x - 2 - 2x - x^2$. The higher order derivatives of $y(x)$ can be calculated by successively differentiating the equation $y' = f(x, y)$, where $f(x, y) = x^2 + y$:

$$y' = f(x, y) = x^2 + y,$$
$$y'' = f' = f_x + f_y f = 2x + x^2 + y,$$
$$y''' = 2 + 2x + x^2 + y,$$

and for the derivative of arbitrary order,

$$y^{(r)} = 2 + 2x + x^2 + y, \qquad r = 3, 4, \ldots.$$

Hence, we obtain the p-th order approximation:

$$y_p(x) = 1 + x + \frac{x^2}{2!} + 3\left(\frac{x^3}{3!} + \frac{x^4}{4!} + \cdots + \frac{x^p}{p!}\right).$$

To get results accurate up to $\varepsilon = 10^{-7}$ on every step of size h, we have from Eq. (3.3.10)

$$\frac{h^{p+1}}{(p+1)!}\left|f^{(p)}(\xi_n, y(\xi_n))\right| = \frac{h^{p+1}}{(p+1)!}\left|2 + 2\xi_n + \xi_n^2 + y(\xi_n)\right| < \varepsilon.$$

Since $|y(\xi_n)| < 3e$ on the interval $[0, 1]$, the error of approximation should not exceed

$$\frac{h^{p+1}}{(p+1)!}(5 + 3e) \leqslant \varepsilon \quad \text{or after rounding} \quad \frac{h^{p+1}}{(p+1)!}(5 + 3e) \leqslant 5 \times 10^{-8}.$$

For $h = 0.1$ this inequality gives $p \approx 7$. So about 7 terms are required to achieve the accuracy ε in the range $0 \leqslant x \leqslant 1$. Indeed, $|y_7(x) - y(x)| = 3\sum_{k\geqslant 8}\frac{x^k}{k!} \leqslant 0.000083579$ for $0 \leqslant x \leqslant 1$.

Example 3.3.6: (Example 3.3.2 revisited) We consider the initial value problem

$$y' = (x - y + 2)^{-1}, \qquad y(1) = 1.$$

The Euler approximation $y_{n+1} = y_n + h(x_n - y_n + 2)^{-1}$ requires 5 arithmetic operations at each step. The second order Taylor series approximation (3.3.5) gives

$$y_{n+1} = y_n + \frac{h}{x_n - y_n + 2} + \frac{h^2}{2}\,\frac{y_n' - 1}{(x_n - y_n + 2)^2} = y_n + h\,y_n' + \frac{h^2\,(y_n')^2}{2}\,(y_n' - 1),$$

where $y_n' = (x_n - y_n + 2)^{-1}$. Such a second order approximation requires 11 arithmetic operations. The third order approximation (3.3.6) yields

$$y_{n+1} = y_n + h\,y_n' + \frac{h^2\,(y_n')^2}{2}\,(y_n' - 1) + \frac{h^3}{6}\left[y_n''(y_n')^2 + 2(y_n' - 1)^2(y_n')^3\right],$$

which requires 22 arithmetic operations.

Problems

1. Determine the first three nonzero terms in the Taylor polynomial approximations for the given initial value problems.
   (a) $y' = 2 + 3y^2$, $y(0) = 1$. **(b)** $y' = 2x^2 + 3y^2$, $y(0) = 1$.
   (c) $y' = 2x^2 + 3y$, $y(0) = 1$. **(d)** $y' = y^2 + \sin x^2$, $y(0) = 1$.
   (e) $y' = 1/(x - y)$, $y(0) = 1$. **(f)** $y' = y - y^3/3$, $y(0) = 1$.
   (g) $y' = (y - 5)^2$, $y(0) = 1$. **(h)** $y' = x - y^2$, $y(0) = 1$.
2. Find four-term Taylor approximations for the given differential equations subject to the initial condition $y(0) = 1$.
   (a) $y' = 1 + y^2$; **(b)** $y' = 3x + 2y$; **(c)** $y' = 2x^2 + 2y^2$; **(d)** $y' = 4 - y^2$;
   (e) $y' = 1/(x + y)$; **(f)** $y' = 3x^2 - 2y$; **(g)** $y' = 2 + x + y$; **(h)** $y' = 1 + xy$;
   (i) $y' = (2y - y^2)/(x - 3)$; **(j)** $y' = y - 3\,e^x$; **(k)** $y' = (1 - 3x + y)^{-1}$; **(l)** $y' = 2xy$.
3. Suppose that the operator $T_p(x, y; h)$ defined by (3.3.7) satisfies the Lipschitz condition: $|T_p(x, y; h) - T_p(x, z; h)| \leqslant L|y - z|$ for any y, z and all h and x (x, $x + h \in [a, b]$), and the $(p + 1)$-st derivative of the exact solution of the IVP $y' = f(x, y)$, $y(a) = y_0$ is continuous on the closed interval $[a, b]$. Show that the local truncation error $z_n = y_n - u(x_n)$ between y_n, the approximate value by the Taylor algorithm, and $u(x_n)$, the reference solution, is as follows:
$$|z_n| \leqslant h^p \,\frac{M}{L\,(p+1)!}\,\left[e^{L(x_n - a)} - 1\right], \qquad M = \max_{x\in[a,b]} |u^{(p+1)}(x)|.$$
4. Obtain the Taylor series solution of the initial value problem $y' = x + 2xy$, $y(0) = 0$ and determine:
   (a) x when the error in $y(x)$ obtained from four terms only is to be less than 10^{-7} after rounding.
   (b) The smallest number of terms in the series needed to find results correct to 10^{-7} for $0 \leqslant x \leqslant 1$.
5. Obtain the Taylor series solution of the initial value problem
$$y' + 2xy = 2x, \qquad y(0) = 0$$
   and determine:
   (a) x when the error in $y(x)$ obtained from four terms only is to be less than 10^{-7} after rounding.
   (b) The number of terms in the series to find results correct to 10^{-7} for $0 \leqslant x \leqslant 1$.
6. Solve the IVP $y' = x - y^2$, $y(0) = 1/2$ by the Taylor algorithm of order $p = 3$, using the steps $h = 0.5$ and $h = 0.1$, and compare the values of the numerical solution at $x = 1$ with the true value of the actual solution.
7. Repeat the previous exercise for the IVP: $y' = x - y^2$, $y(0) = 1$.

In Exercises 8 through 14

(a) Obtain an actual solution and, from it, an accurate true value at the indicated terminal point.

(b) At the initial point, expand the actual solution in a Taylor polynomial of degree $p = 5$.

(c) Build a Taylor series numerical method of degree p and use it to calculate the value at the terminal point, using a stepsize of $h = 0.1$. Determine the error at the terminal point.

(d) Using $h = 0.05$, recalculate the value at the terminal point and again determine the error.

8. $y' = \dfrac{(36x^3 + 10x)\,y}{4 - 5x^2 - 9x^4}$, $y(-2) = 1$; terminal point at $x = -1$, and degree $p = 3$.
9. $y' = \dfrac{4\,x^3}{14\,y^3 + 3y}$, $y(0) = 1$; terminal point at $x = 1$, and degree $p = 2$.
10. $y' = \dfrac{8 + 18\,x^2}{27\,y^2}$, $y(2) = 2$; terminal point at $x = 3$, and degree $p = 2$.
11. $y' = \dfrac{3x}{6\,y + 7x}$, $y(1) = 1$; terminal point at $x = 2$, and degree $p = 2$.
12. $y' = \dfrac{xy}{y^2 - x^2}$, $y(2) = 3$; terminal point at $x = 3$, and degree $p = 3$.
13. $y' = \dfrac{7\,x^2 - 5xy}{y^2 + x^2}$, $y(1) = 2$; terminal point at $x = 2$, and degree $p = 2$.
14. $y' = \dfrac{3\,xy - 10x^2}{7\,y^2 + 6\,x^2}$, $y(1) = 1$; terminal point at $x = 2$, and degree $p = 2$.

3.4 Error Estimates

Mathematical models are essential tools in our understanding of the real world; however, they are usually not accurate because it is impossible to take into account everything that affects the phenomenon under consideration. Also, almost every measurement is not accurate and is subject to human perception. The errors associated with both calculations and measurements can be characterized with regard to their accuracy and precision. *Accuracy* means how closely a computed or measured value agrees with the true value. *Precision* refers to how closely individual computed or measured values agree with each other.

Once models are established, they are used for quantitative and qualitative analysis. Available computational tools allow many people, some of them with rather limited background in numerical analysis, to solve initial value problems numerically. Since digital computers cannot represent some quantities exactly, this section addresses concerns about the reliability of numbers that computers produce.

Definition 3.4: The initial value problem

$$y' = f(x,y), \quad (a < x < b) \quad y(x_0) = y_0 \quad (a \leqslant x_0 < b) \tag{3.4.1}$$

is said to be a **well-posed problem** if a unique solution to the given problem exists and for any positive ε there exists a positive constant k such that whenever $|\varepsilon_0| < \varepsilon$ and $|\delta(x)| < \varepsilon$, a unique solution, $z(x)$, to the problem

$$z' = f(x,z) + \delta(x), \qquad z(x_0) = y_0 + \varepsilon_0, \tag{3.4.2}$$

exists with $\quad |z(x) - y(x)| < k\varepsilon \qquad$ for all x within the interval $[a,b]$.

The problem specified by Eq. (3.4.2) is often called a **perturbed problem** associated with the given problem (3.4.1). Usually, we deal with a perturbed problem due to inaccurate input values. Therefore, numerical methods are almost always applied to perturbed problems. If the slope function in Eq. (3.4.1) satisfies the conditions of Theorem 1.3 (page 23), the initial value problem (3.4.1) is well-posed.

At the very least, we insist on $f(x,y)$ obeying the Lipschitz condition

$$|f(x,y_1) - f(x,y_2)| \leqslant L|y_1 - y_2| \quad \text{for all } y_1, y_2 \in [\alpha,\beta], \ a \leqslant x \leqslant b.$$

Here $L > 0$, called a Lipschitz constant, is independent of the choice of y_1 and y_2. This condition guarantees that the initial value problem (3.4.1) possesses a unique solution (Theorem 1.3 on page 23). The Lipschitz condition is not an easy one to understand, but there is a relatively simple criterion that is sufficient to ensure a function satisfies the Lipschitz condition within a given domain $\Omega = [a,b] \times [\alpha,\beta]$ containing (x_0, y_0). If the function $f(x,y)$ has a continuous partial derivative with respect to y throughout $[\alpha,\beta]$, then it obeys the Lipschitz condition within Ω. If $f_y = \partial f/\partial y$ exists and is bounded in the domain Ω, then for some θ $(y_1 \leqslant \theta \leqslant y_2)$ $f(x,y_1) - f(x,y_2) = (y_1 - y_2) f_y(x,\theta)$.

Taking a stronger requirement, we may stipulate that the rate function f in the problem (3.4.1) has as many derivatives as needed by Taylor expansions. By choosing a constant step size h for simplicity, we generate a uniform set of mesh points: $x_n = x_0 + nh$, $n = 0, 1, 2, \ldots$. The quantity h, called the *discretization parameter*, measures the degree to which the discrete algorithm represents the original problem. This means that as h decreases, the approximate solution should come closer and closer to the actual solution. In what follows, we denote the actual solution of the initial value problem (3.4.1) by $\phi(x)$ and its approximation at the mesh point $x = x_n$ by y_n.

The proof of the convergence $y_n \mapsto \phi(x_n)$ as $h \to 0$ is the ultimate goal of the numerical analysis. However, to establish such convergence and estimate its speed, we need to know a good deal about the exact solution (including its derivatives). In most practical applications, this is nearly impossible to achieve. When we leave the realm of textbook problems, the requirements of most convergence analysis are too restrictive to be applicable. In practice, numerical experiments are usually conducted for several values of a discretization parameter until the numerical solution has "settled down" in the first few decimal places. Unfortunately, it does not always work; for instance, this approach cannot detect a multiplication error.

Recall that a "numerical solution" to the initial value problem (3.4.1) is a finite set of ordered pairs of rational numbers, (x_n, y_n), $n = 0, 1, 2, \ldots, N$, whose first coordinates, x_n, are distinct mesh points and whose second coordinates, y_n, are approximations to the actual solution $y = \phi(x)$ at these grid points. Numerical algorithms that we discussed so far generate a sequence of approximations $\{y_n\}_{n\geqslant 0}$ by operating within one grid interval $[x_n, x_{n+1}]$ of size $h = x_{n+1} - x_n$ and use the values of the slope function on this interval along with only one approximate value y_n at the left end x_n. They define the approximation y_{next} to the solution at the right end x_{n+1} by the formula

$$y_{next} = y + h\Phi(x, y, y_{next}; h) \qquad \text{or} \qquad y_{n+1} = y_n + h\Phi(x_n, y_n, y_{n+1}; h), \tag{3.4.3}$$

where the function $\Phi(x, y, y_{next}; h)$ may be thought of as the approximate increment per unit step h and it is defined by the applied method. For example, $\Phi(x, y; h) = f(x, y)$ in the Euler method, but in the trapezoid method (3.2.14), Φ depends on y_{next}, that is, $\Phi = \frac{1}{2} f(x, y) + \frac{1}{2} f(x + h, y_{next})$.

Definition 3.5: The big Oh notation, $O(h^n)$, also known as the Landau symbol, means a set of functions that are bounded when divided by h^n, that is, $|h^{-n} f(h)| \leqslant K$ for every function f from $O(h^n)$ and some constant $K = K(f)$ independent of h.

It is customary to denote the big Oh relation as either $f = O(h^n)$ or $f \in O(h^n)$ when $h^{-n} f(h)$ is bounded. The Landau symbol is used to estimate the **truncation errors** or **discretization errors**, which are the errors introduced by algorithms in replacing an infinite or infinitesimal quantity by something finite.

Definition 3.6: The one-step difference method (3.4.3) is said to have the **truncation error** given by

$$T_n(h) = \frac{u(x_n + h) - u(x_n)}{h} - \Phi(x_n, y_n; h), \qquad n = 0, 1, 2, \ldots, \tag{3.4.4}$$

where $u(x)$ is the true solution of the local initial value problem

$$\frac{\mathrm{d}u}{\mathrm{d}x} = f(x, u), \qquad x_n \leqslant x \leqslant x_n + h, \quad u(x_n) = y_n.$$

This function $u(x)$ on the mesh interval $[x_n, x_n + h]$ is called the *reference solution*. The difference $u(x_n + h) - y_{n+1} = h\,T_n(h)$ is called the **local truncation error**. We say that the truncation error is of order $p > 0$ if $T_n(h) = O(h^p)$, that is, $|T_n(h)| \leqslant K\,h^p$ for some positive constant K (not depending on h but on the slope function).

A local truncation error (from the Latin *truncare*, meaning to *cut off*) measures the accuracy of the method at a specific step, assuming that the method was exact at the previous step:

$$hT_n(h) = (u(x_{n+1}) - u(x_n)) - (y_{n+1} - y_n),$$

where $u(x)$ is the reference solution on the interval $[x_n, x_{n+1}]$. A numerical method is of order p if at each mesh interval of length h its local truncation error is $O(h^p)$. The truncation error can be defined as

$$T_n(h) = \frac{1}{h}\,[u(x_n + h) - y_{next}] \tag{3.4.5}$$

because of the assumption that $u(x_n) = y_n$. This shows that the local truncation error $hT_n(h)$ is the difference between the exact and the approximate increment per unit step because the reference solution is an exact solution within every step starting at (x_n, y_n). Truncation errors would vanish if computers were infinitely fast and had infinitely large memories such that there was no need to settle for finite approximations.

As a rule, for a given amount of computational effort, a method with a higher order of convergence tends to give better results than one with lower order. However, using a high-order method does not *always* mean that you get greater accuracy. This is balanced by the fact that higher order methods are usually more restrictive in their applicability and are more difficult to implement. When a higher order method is carried out on a computer, the increased accuracy repays the extra effort for more work to be performed at each step.

Let $\phi(x)$ be the actual solution to the initial value problem (3.4.1). The difference

$$E_n = \phi(x_n) - y_n, \qquad n = 0, 1, 2, \ldots, \tag{3.4.6}$$

is known as the **global truncation error** (also called the accumulative error), which measures how far away the approximation y_n is from the true value $\phi(x_n)$. To find this approximation, y_n, we need to make $n - 1$ iterations

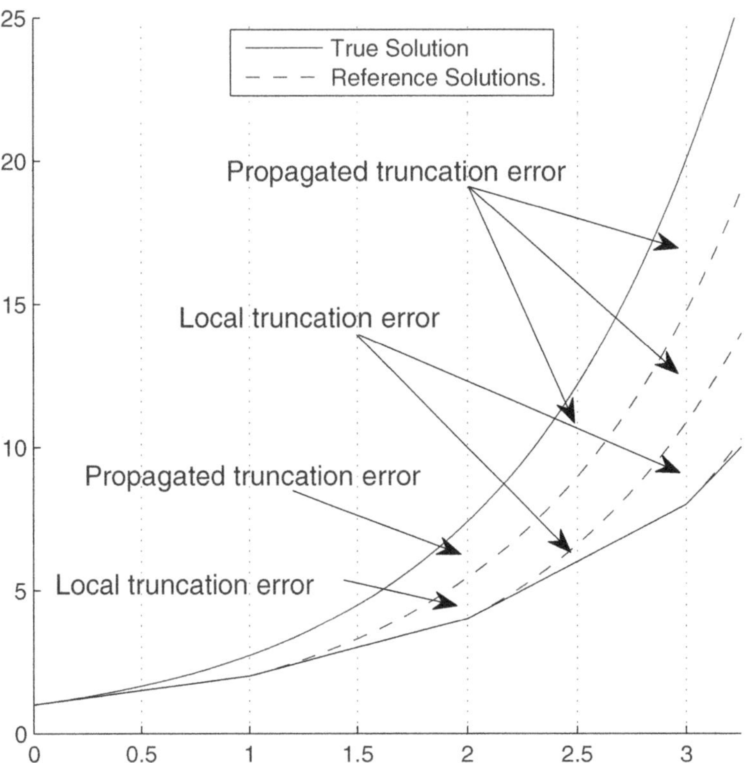

Figure 3.8: Local and propagated errors in Euler's approximations along with the exact solution (solid) and reference solutions (dashed), plotted with MATLAB.

according to the one-step algorithm (3.4.3), and every step is based on the solution from the previous step. The input data at each step are only approximately correct since we don't know the true value $\phi(x_k)$, but only its approximation y_k $(k = 0, \ldots, n-1)$. Therefore, at each step, we numerically solve a perturbed problem by introducing a local truncation error that arises from the use of an approximate formula. The essential part of the global error (not included in the local truncation error), accumulated with iteration, is called the **propagated truncation error** (see Fig. 3.8).

Definition 3.7: The one-step method (3.4.3) is considered consistent if its truncation error

$$T_n(h) \to 0 \quad \text{as} \quad h \to 0$$

uniformly for (x, y). A method is said to be convergent if

$$\lim_{h \to +0} \max_n |y_n - \phi(x_n)| = 0$$

for the actual solution $y = \phi(x)$ of the initial value problem (3.4.1) with a Lipschitz slope function $f(x, y)$.

Hence, convergence means that the numerical solution tends to the values of the actual solution at mesh points as the grid becomes increasingly fine. From Eq. (3.4.3), we have consistency if and only if $\Phi(x, y; 0) = f(x, y)$. Every one-step method of order $p > 0$ is consistent.

■

To derive the local truncation error for the Euler method, we use Taylor's theorem:

$$\phi(x_{n+1}) = \phi(x_n) + (x_{n+1} - x_n)\phi'(x_n) + \frac{(x_{n+1} - x_n)^2}{2}\,\phi''(\xi_n)$$

for some number $\xi_n \in (x_n, x_{n+1})$. Since $x_{n+1} - x_n = h$, $\phi'(x_n) = f(x_n, \phi(x_n))$, and $\phi(x_n)$ is assumed to be equal to

y_n, we obtain the local truncation error $h\,T_n(h) = \phi(x_n) - y_n$ to be

$$h\,T_n(h) = \frac{h^2}{2}\,\phi''(\xi_n) = \frac{h^2}{2}\,[f_x(\xi_n,\phi_n) + f_y(\xi_n,\phi_n)\,f(\xi_n,\phi_n)] = O(h^2),$$

where $\phi_n = \phi(\xi_n)$ is the value of the actual solution $\phi(x)$ at the point $x = \xi_n$. So the local truncation error for the Euler algorithm (either forward or backward) is of order $p = 1$, and we would expect the error to be small for values of h that are small enough. If the initial value $y_0 = y(a)$ is given and we seek the solution value at the point b, then there are $N = (b-a)/h$ steps involved in covering the entire interval $[a,b]$. The cumulative error after n steps will be proportional to h, as the following calculations show.

$$\begin{aligned}
E_{n+1} &= \phi(x_{n+1}) - y_{n+1} = \phi(x_n) + h\phi'(x_n) + \frac{h^2}{2}\,\phi''(\xi_n) - y_{n+1}\\
&= \phi(x_n) + h\phi'(x_n) + \frac{h^2}{2}\,\phi''(\xi_n) - y_n - hf(x_n,y_n)\\
&= \phi(x_n) - y_n + h\,[\phi'(x_n) - f(x_n,y_n)] + \frac{h^2}{2}\,\phi''(\xi_n)\\
&= E_n + h[f(x_n,\phi(x_n)) - f(x_n,y_n)] + \frac{h^2}{2}\,\phi''(\xi_n)\\
&\leqslant |E_n| + hL|\phi_n - y_n| + Mh^2 = |E_n|\,(1+hL) + Mh^2,
\end{aligned}$$

where $M = \max_x |\phi''(x)|/2$ and L is the Lipschitz (positive) constant for the function f, that is, $|f(x,y_1) - f(x,y_2)| \leqslant L|y_1 - y_2|$. If the slope function is differentiable in a rectangular domain containing the solution curve, then constants L and M can be estimated as $L = \max_{(x,y)\in R}|f_y(x,y)|$ and $2M = \max_{(x,y)\in R}|f_x(x,y) + f(x,y)\,f_y(x,y)|$. Now we claim that from the inequality

$$|E_{n+1}| \leqslant |E_n|(1+hL) + Mh^2, \tag{3.4.7}$$

it follows that

$$|E_n| \leqslant \frac{M}{L}\,h\,[(1+hL)^n - 1], \qquad n = 0,1,2,\ldots. \tag{3.4.8}$$

We prove this statement by induction on n. When $n = 0$, the error is zero since at $x_0 = a$ the numerical solution matches the initial condition.

For arbitrary $n \geqslant 0$, we assume that inequality (3.4.8) is true up to n; using (3.4.7), we have

$$\begin{aligned}
|E_{n+1}| &\leqslant |E_n|(1+hL) + Mh^2 \leqslant \frac{M}{L}\,h\,[(1+hL)^n - 1]\,(1+hL) + Mh^2\\
&= \frac{M}{L}\,h\,(1+hL)^{n+1} - \frac{M}{L}\,h\,(1+hL) + Mh^2\\
&= \frac{M}{L}\,h\,(1+hL)^{n+1} - \frac{M}{L}\,h = \frac{M}{L}\,h\,\left[(1+hL)^{n+1} - 1\right].
\end{aligned}$$

This proves that the inequality (3.4.8) is valid for all n. The constant hL is positive, hence $1 + hL < e^{hL}$, and we deduce that $(1+hL)^n \leqslant e^{nhL}$. The index n is allowed to range through $\left\{0,1,\ldots,\frac{b-a}{h}\right\}$; therefore, $(1+hL)^n \leqslant e^{L(b-a)}$. Substituting into the inequality (3.4.8), we obtain the estimate

$$|\phi(x_{n+1}) - y_{n+1}| \leqslant \frac{M}{L}\,h\,\left(e^{L(b-a)} - 1\right), \qquad n = 0,1,\ldots,\frac{b-a}{h}. \tag{3.4.9}$$

Since $M\left(e^{L(b-a)} - 1\right)/L$ is independent of h, the global truncation error of the Euler rule (3.2.5) is proportional to h and tends to zero as $h \to 0$. ■

By scanning the rows in Table 151, we observe that for each mesh point x_n, the accumulative error $E_n = \phi(x_n) - y_n$ decreases when the step size is reduced. But by scanning the columns in this table, we see that the error increases as x_n gets further from the starting point. This reflects the general observation: the smaller the step size, the less the truncation error and the approximate solution is closer to the actual solution. On the other hand, the exponential term in the inequality (3.4.9) can be quite large for large intervals, which means that we should not trust numerical approximations over long intervals.

The inequality (3.4.9) should not be used to build bridges or airplanes, but our goal is to demonstrate the general form of the procedure. The bound (3.4.9) is too crude to be used in practical estimations of numerical errors because it is based only on two constants—the Lipschitz constant L (which is actually an estimation of f_y) and the upper bound M of the second derivative of the actual solution (which is usually unknown). In the majority of applications, the cumulative error rarely reaches the upper bound (3.4.9), but sometimes it does.

Example 3.4.1: In the initial value problem

$$y' = -10\,y, \qquad y(0) = 1,$$

for the linear equation, we have the slope function $f(x,y) = -10y$, with the Lipschitz constant $L = 10$. For the second derivative, $y''(x) = 100\,e^{-10x}$ of the exact solution $y(x) = e^{-10x}$ on the interval $[0,1]$, we have the estimate $M = \frac{1}{2}\max_{x\in[0,1]}|y''(x)| = 50$. Thus, we find the upper bound in the inequality (3.4.9) to be

$$|E_{n+1}| = |\phi(x_{n+1}) - y_{n+1}| \leqslant 5h\,(e^{10} - 1) \approx 110127.329\,h.$$

On the other hand, the Euler algorithm for the given initial value problem produces $y_{n+1} = y_n - 10hy_n = y_0\,(1 - 10h)^{n+1}$, which leads to the estimate at the point $x = 1$:

$$\begin{aligned} E_{n+1} &= e^{-10} - (1-10h)^{1/h} = \left(50h - \frac{2750}{3}\,h^2 + \frac{20000}{3}\,h^3 - \frac{38750}{9}\,h^4 + \cdots\right) e^{-10} \\ &\leqslant 50\,e^{-10}\,h \approx 0.00227\,h. \end{aligned}$$

If h were chosen so that $1 - 10h < -1$ ($h > 0.5$), the corresponding difference solution y_n would bear no resemblance to the solution of the given differential equation. This phenomenon is called **partial instability**. □

The truncation error for the improved Euler method is

$$T_n(h) = \frac{u_{n+1} - u_n}{h} - \frac{1}{2}\,f_n - \frac{1}{2}\,f(x_{n+1}, y_n + hf_n), \quad f_n = f(x_n, y_n), \quad u_n = u(x_n) = y_n.$$

The function $f(x_n + h, y_n + hf_n)$ can be estimated based on the Taylor theorem:

$$f(x_n + h, y_n + hf_n) = f_n + hf_x(x_n, y_n) + hf_n\,f_y(x_n, y_n) + E_n,$$

where the error E_n at each step is proportional to h^2. It is very difficult to predict whether such an approximation is an underestimate or otherwise. Hence,

$$T_n(h) = \frac{u_{n+1} - u_n}{h} - f_n - \frac{h}{2}\,[f_x(x_n, y_n) + f_n\,f_y(x_n, y_n)] + O(h^2).$$

On the other hand,

$$\begin{aligned} u_{n+1} &= u_n + hu'(x_n) + \frac{h^2}{2}\,u''(x_n) + \frac{h^3}{3!}\,u'''(\xi_n) \\ &= u_n + hf(x_n, u_n) + \frac{h^2}{2}\,u''(x_n) + O(h^3). \end{aligned}$$

Therefore,

$$\frac{u_{n+1} - u_n}{h} = f(x_n, u_n) + \frac{h}{2}\,u''(x_n) + O(h^2), \qquad u_n = y_n.$$

Substituting this representation into the above expression for $T_n(h)$, we obtain

$$T_n(h) = \frac{h}{2}\,u''(x_n) - \frac{h}{2}\,[f_x(x_n, y_n) + f_n\,f_y(x_n, y_n)] + O(h^2) = O(h^2)$$

since

$$u''(x_n) = \left.\frac{\mathrm{d}f(x,u)}{\mathrm{d}x}\right|_{x=x_n} = f_x(x_n, y_n) + f_n\,f_y(x_n, y_n) + O(h^2).$$

Thus, the truncation error for the improved Euler method is of the second order. Similarly, it can be shown (see Problems 3 and 4) that both the trapezoid and modified Euler methods are of the second order. ■

One of the popular approaches to maintain the prescribed error tolerance per unit step is based on interval halving or doubling. The idea consists of computing the approximate value y_{n+1} at the grid point x_{n+1} by using the current step $h = x_{n+1} - x_n$ and then recomputing its value using two steps of length $h/2$. One expects to reduce the error by approximately $(1/2)^p$, where p is the order of the truncation error. If the results of such calculations differ within the tolerance, the step size h is kept unchanged; otherwise, we halve it and repeat the procedure. The same approach is used to double the step size.

Example 3.4.2: Consider the differential equation $\dot{y} = y^2 \sin(y(t)) + t$ subject to three initial conditions: $y(0) = -0.8075$, $y(0) = -0.8076$, and $y(0) = -0.8077$. By plotting their solutions with *Maple*,

```
y3 := diff(y(t), t) = y(t)^2*sin(y(t))+t
DEplot(y3, y(t), t = 0 .. 6, [[y(0) = -.8075], [y(0) = -.8076],
[y(0) = -.8077]],arrows=medium,y= -3 .. 2,color = black,linecolor = blue)
```

we see (Fig. 3.9 on page 171) that solutions are very sensitive to the initial conditions. This is usually the case when there is an exceptional solution that separates solutions with different qualitative behavior. If the initial condition identifies the exceptional solution, any numerical method will eventually flunk (see Exercise 14 on page 184).

Euler's accumulative error $E_n(h) = \phi(x_n) - y_n$ is $O(h)$, that is, $|E_n(h)| \leqslant C\,h$, where C is a constant that depends on the behavior of the slope function within the mesh interval, but independent of step size h. To perform actual calculations, we consider the equation $\dot{y} = y^2 \sin(y(t)) + t$ subject to the initial condition: $y(0) = -0.8075$. By calculating the ratios E_n/h for different grid points, we see that they remain approximately unchanged for different step sizes:

h	Approx. at $t = 1$	$\lvert E(h)\rvert/h$	Approx. at $t = 2$	$\lvert E(h)\rvert/h$
2^{-5}	−1.2078679834365	0.30429340837767	−1.67796064616645	5.67692722580626
2^{-6}	−1.2033363156891	0.31856008092308	−1.59929791928236	6.31943993103098
2^{-7}	−1.2009071155484	0.32618254383425	−1.55268398278824	6.67229599081492
2^{-8}	−1.1996483656937	0.33012512485624	−1.52733905657636	6.85629087138864
2^{-9}	−1.1990075067446	0.33213046777075	−1.51413104023359	6.95007737527897

where $\phi(1) = -1.19835888124895$, $\phi(2) = -1.50055667035633$ are true values at $t = 1$ and $t = 2$, respectively (calculated with the tolerance 2^{-14}). This table tells us that the ratio $E(h)/E(h/2)$ of two errors evaluated for the same mesh point will be close to 2. Hence, the error goes down by approximately one-half when h is halved. If we repeat similar calculations using the Heun method (of order 2), we will get the table of values:

h	Approx. at $t = 1$	$\lvert E(h)\rvert/h^2$	Approx. at $t = 2$	$\lvert E(h)\rvert/h^2$
2^{-5}	−1.1987431613123	0.3935712129344	−1.50867192833994	8.31002417145396
2^{-6}	−1.1984558487773	0.3974527082992	−1.50261140901733	8.41620954044174
2^{-7}	−1.1983831848709	0.3992853900927	−1.50107317886035	8.46247526969455
2^{-8}	−1.1983649205765	0.4001727624709	−1.50068612220823	8.48375632541138
2^{-9}	−1.1983603426266	0.4006089575705	−1.50058907206927	8.49391367618227

In the case of a second order numerical method (Heun or midpoint), the ratio of cumulative errors $E(h)/E(h/2)$ will be close to $4 = 2^2$, which is clearly seen from Fig. 3.9(b).

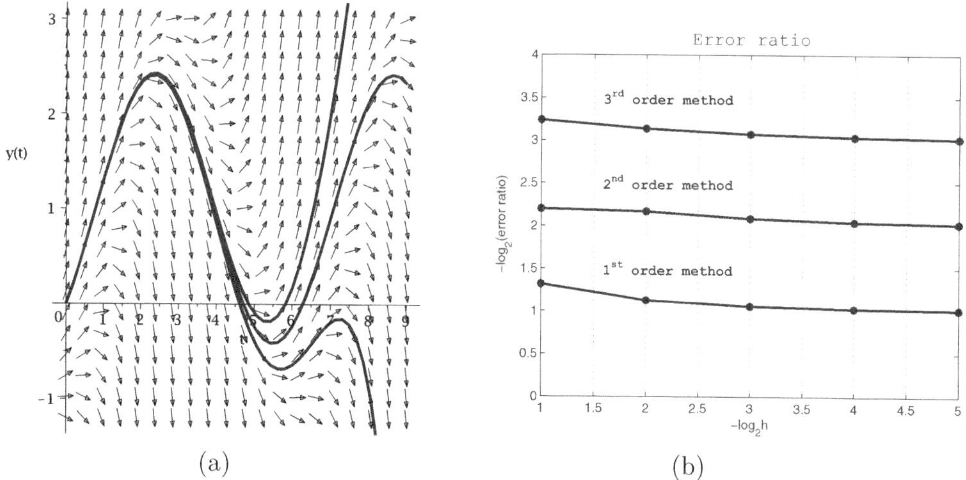

Figure 3.9: Example 3.4.2. (a) Solutions are sensitive to the initial conditions, plotted in *Maple*; and (b) error ratios for three different numerical methods, plotted with MATLAB.

The previous observations show that the smaller the step size, the better the approximation we obtain. But what should we do when we cannot use a small step size? For instance, when the slope function is given at particular mesh points, and it is unknown at other points. One of the approaches to improve approximations is to use iteration, as the following example shows.

Example 3.4.3: Consider the following "improvement" of the Heun method. For the initial value problem $y' = f(x, y)$, $y(x_0) = y_0$, to find the approximate value y_{n+1} of the actual solution at the right endpoint $x = x_{n+1}$ of the mesh interval $[x_n, x_{n+1}]$ of size $h = x_{n+1} - x_n$, the iterative algorithm can be expressed concisely as

$$\begin{array}{ll}\textbf{predictor}: & y^0_{k+1} = y_k + h\, f(x_k, y_k), \\ \textbf{corrector}: & y^j_{k+1} = y_k + \frac{h}{2}\, f(x_k, y_k) + \frac{h}{2}\, f(x_{k+1}, y^{j-1}_{k+1}), \quad j = 1, 2, \ldots m.\end{array}$$

The value m defines the number of iterations, which may depend on the mesh interval. Note that $m = 1$ corresponds to the improved Euler method. This value can be either fixed through the computations or variable, with a termination criterion

$$\left| \frac{y^j_{k+1} - y^{j-1}_{k+1}}{y^j_{k+1}} \right| \times 100\% < \varepsilon,$$

where ε is a tolerance, and y^{j-1}_{k+1} and y^j_{k+1} are results from the prior and present iteration. It should be understood that the iterative process does not necessarily converge to the true answer. To see that a priori a fixed number of iterations is not the best option, we reconsider the initial value problem from Example 3.2.6 on page 155: $\dot{y} = 2\pi\, y\, \cos(2\pi t)$, $y(0) = 1$. Numerical experimentation reveals the following results summarized in Table 172. The data show that the next iteration may not give a better approximation; therefore, it is preferable to use a variable termination criterion rather than a fixed number of iterations.

□

Table 172: A comparison of the results for the numerical approximations of the solution to the initial value problem $\dot{y} = 2\pi\, y\, \cos(2\pi t)$, $y(0) = 1$ at the point $t = 2$, using the iterative Heun formula and the improved Euler method (3.2.15) for different step sizes h.

		Iterative		**Heun**		
Exact	m	Approximation	m	Approximation	m	Approximation
1.00	1	$y_{20} \approx 0.92366$	2	$y_{40} \approx 0.88179$	3	$y_{60} \approx 0.99523$
1.00	4	$y_{80} \approx 0.99011$	5	$y_{100} \approx 0.99961$	6	$y_{200} \approx 0.99916$

		Heun				
Exact	h	Approximation	h	Approximation	h	Approximation
1.00	0.1	$y_{20} \approx 0.92366$	0.05	$y_{40} \approx 0.99294$	0.025	$y_{80} \approx 0.99921$

So far we have always used a uniform step size in deriving approximate formulas. The main reason for such an approach is to simplify the derivations and make them less cumbersome. However, in practical applications, the step size is usually adjusted to calculations because there is no reason to keep it fixed. Of course, we cannot choose a step size too large because the numerical approximation may be inconsistent with an error tolerance, which is usually prescribed in advance. Therefore, the numerical method chosen for computations imposes some restrictions on the step size to maintain its stability and accuracy.

The major disadvantage of halving and doubling intervals consists in the substantial computational effort required. Contemporary general-purpose initial value problem solvers provide an automatic step size control to produce the computed accuracy within the given tolerance. This approach gives a user confidence that the problem has been solved in a meaningful way. On the other hand, the step size should not be taken too small because the finer the set of mesh points, the more steps needed to cover a fixed interval.

When arithmetic operations are performed by computers, every input number is transformed into its floating-point form and represented in its memory using the binary system. For instance, the number π is represented in floating-point form as 0.31415926×10^1, where its fractional part is called the *mantissa*, and the exponential part is called the *characteristic*. Due to the important property of computers to carry out only a finite number of digits or characters, the difference between a number stored in a computer memory, Y_n, and its exact (correct) value, y_n, is called the **round-off error** (also called the **rounding error**):

$$R_n = y_n - Y_n, \qquad n = 0, 1, 2, \ldots .$$

The rounding errors would disappear if computers had infinite capacity to operate with numbers precisely. Round-off errors can affect the final computed result either in accumulating error during a sequence of millions of operations or in catastrophic cancellation. In general, the subject of round-off propagation is poorly understood and its effect on computations depends on many factors. If too many steps are required in the calculation, then eventually round-off error is likely to accumulate to the point that it deteriorates the accuracy of the numerical procedure. Therefore, the round-off error is proportional to the number of computations performed and it is inversely proportional to some power of the step size.

One-step methods do not usually exhibit any numerical instability for sufficiently small h. However, one-step methods may suffer from round-off error accumulated during computation because higher order methods involve considerably more computations per step.

Let us go to an example that explains round-off errors. Suppose you make calculations keeping 4 decimal places in the process. Say you need to add two fractions: $\frac{2}{3} + \frac{2}{3} = \frac{4}{3}$. First, you compute 2/3 to get 0.6667. Next you add these rounded numbers to obtain $0.6667 + 0.6667 = 1.3334$, while the answer should be equal to $\frac{4}{3} = 1.3333$. The difference $0.001 = 1.3334 - 1.3333$ is an example of round-off error. A similar problem arises when two close numbers are subtracted, causing cancellation of decimal digits. Moreover, the floating-point arithmetic operations do not commute.

Of course, a computer uses more decimal places and performs computing with much greater accuracy, but the principle is the same. When a lot of computations are being made, the round-off error could accumulate. This is well observed when a reduction in step size is made in order to decrease discretization error. This necessarily increases the number of steps and so introduces additional rounding error.

Another problem usually occurs when a numerical method has multiple scales. For instance, the Euler rule $y_{n+1} = y_n + h\, f(x_n, y_n)$ operates on two scales: there is y_n and another term with a multiple of h. If the product $h\, f_n$ is small compared to y_n, then the Euler method loses sensitivity, and the quadrature rule (3.2.6), page 147, should be used instead.

For instance, consider an initial value problem

$$y' = \cos x, \qquad y(0) = 10^7,$$

on the interval $[0, 1]$. Its exact solution is trivial: $y = 10^7 + \sin(x)$. However, the standard Euler algorithm computes $y_0 + h\, \cos(0) + h\, \cos(x_1) + \cdots$, whereas the quadrature rule provides $y_0 + h(y(0) + y(x_1) + y(x_2) + \cdots)$, which is not computationally equivalent to the former. Indeed, if h is small enough, the product $h\, \cos(x)$ may be within the machine epsilon[35] when added to $y_0 = 10^7$; then it will be dropped, causing the numerical solution to be unchanged from the initial condition.

Problems

1. Show that the cumulative error $E_n = \phi(x_n) - y_n$, where ϕ is an exact solution and y_n is its approximation at $x = x_n$, in the trapezoid rule (3.2.14) satisfies the inequality

$$|E_{n+1}| \leqslant \left(\frac{1 + hL/2}{1 - hL/2}\right) |E_n| + Mh^3 (1 - Lh)^{-1}$$

for some positive constant M, where L is the Lipschitz constant of the slope function f.

2. Suppose that $f(x, y)$ satisfies the Lipschitz condition $|f(x, y_1) - f(x, y_2)| \leqslant L|y_1 - y_2|$ and consider the Euler θ-method for approximating the solution of the initial value problem (3.2.1):

$$y_{n+1} = y_n + h[(1-\theta) f_n + \theta f_{n+1}], \qquad n = 0, 1, 2, \ldots,$$

where $f_n = f(x_n, y_n)$ and $f_{n+1} = f(x_{n+1}, y_{n+1})$. For $\theta = 0$ this is the standard Euler method; for $\theta = 1/2$ this is the trapezoid method, and for $\theta = 1$ this is the backward Euler's method. By choosing different values of θ $(0 < \theta < 1)$, numerically verify that the θ-method is of order 2 for $f(x, y) = 2\,(3x - y + 2)^{-1}$ and $x_0 = 0$, $y_0 = 1$.

3. Show that the trapezoid method (3.2.14) is of order 2.
4. Show that the midpoint Euler method (3.2.16) is of order 2.
5. Consider the initial value problem $y' = x + y + 1$, $y(0) = 1$, which has the analytic solution $y = \phi(x) = 3\,e^x - x - 2$.
   (a) Approximate $y(0.1)$ using Euler's method with $h = 0.1$.
   (b) Find a bound for the local truncation error in part (a).
   (c) Compare the actual error in y_1, the first approximation, with your error bound.
   (d) Approximate $y(0.1)$ using Euler's method with $h = 0.05$.
   (e) Verify that the global truncation error for Euler's method is $O(h)$ by comparing the errors in parts (a) and (d).

[35] For the IEEE 64-bit floating-point format, the machine epsilon is about 2.2×10^{-16}.

3.5 The Runge–Kutta Methods

In §3.2 we introduced the Euler algorithms for numerical approximations of solutions to the initial value problem $y' = f(x, y)$, $y(x_0) = y_0$ on a uniform mesh $x_{n+1} = x_n + h$, $n = 0, 1, \dots, N-1$. The main advantage of Euler's methods is that they all use the values of the slope function in at most two points at each step. However, the simplicity of these methods is downgraded by the relatively high local truncated errors, which leads to a requirement of small step size h in actual calculations, and as consequence, to longer computations where round-off errors may affect the accuracy.

In this section, we discuss one of the most popular numerical procedures used in obtaining approximate solutions to initial value problems—the **Runge–Kutta method** (RK for short). It is an example of so-called self-starting (or one-step) numerical methods that advance from one point x_n ($n = 0, 1, \dots, N$) to the next mesh point $x_{n+1} = x_n + h$, and use only one approximate value y_n of the solution at x_n and some values of the slope function from the interval $[x_n, x_{n+1}]$. Recall that slopes are known at any point of the plane where $f(x, y)$ is defined, but the values of the actual solution $y = \phi(x)$ are unknown. The accuracy of a self-starting method depends on how close the next value y_{n+1} will be to the true value $\phi(x_{n+1})$ of the solution $y = \phi(x)$ at the point $x = x_{n+1}$. This can be viewed as choosing the slope of the line connecting the starting point $S(x_n, y_n)$ and the next point $P(x_{n+1}, y_{n+1})$ as close as possible to the slope of the line connecting the starting point S with the true point $A\,(x_{n+1}, \phi(x_{n+1}))$.

The main idea of the Runge–Kutta method (which is actually a family of many algorithms including Euler's methods) was proposed in 1895 by Carle Runge[36] who intended to extend Simpson's quadrature formula to ODEs. Later Martin Kutta[37], Karl Heun, then John Butcher [6], and others generalized these approximations and gave a theoretical background. These methods effectively approximate the true slope of SA on each step interval $[x_n, x_{n+1}]$ by the weighted average of slopes evaluated at some sampling points within this interval. Mathematically this means that the truncation of the Taylor series expansion

$$y_{n+1} = y_n + h y'_n + \frac{h^2}{2}\, y''_n + \frac{h^3}{6}\, y'''_n + \cdots \tag{3.5.1}$$

is in agreement with an approximation of y_{n+1} calculated from a formula of the type

$$\begin{aligned} y_{n+1} = y_n &+ h[\alpha_0 f(x_n, y_n) + \alpha_1 f(x_n + \mu_1 h, y_n + b_1 h) \\ &+ \alpha_2 f(x_n + \mu_2 h, y_n + b_2 h) + \cdots + \alpha_p f(x_n + \mu_p h, y_n + b_p h)]. \end{aligned} \tag{3.5.2}$$

Here the α's, μ's, and b's are so determined that, if the right-hand expression of Eq. (3.5.2) were expanded in powers of the spacing h, the coefficients of a certain number of the leading terms would agree with the corresponding coefficients in Eq. (3.5.1). The number p is called the order of the method. An explicit Runge–Kutta method (3.5.2) can be viewed as a "staged" sampling process. That is, for each i ($i = 1, 2, \dots, p$), the value μ_i is chosen to determine the x-coordinate of the i-th sampling point. Then the y-coordinate of the i-th sampling point is determined using prior stages.

The backbone of the Runge–Kutta formulas is based on the Taylor series in two variables:

$$f(x + r, y + s) = \sum_{k \geqslant 0} \frac{1}{k!} \left(r\,\frac{\partial}{\partial x} + s\,\frac{\partial}{\partial y} \right)^k f(x, y). \tag{3.5.3}$$

This series is analogous to the Taylor series in one variable. The mysterious-looking terms in Eq. (3.5.3) are understood as follows:

$$\begin{aligned} \left(r\,\frac{\partial}{\partial x} + s\,\frac{\partial}{\partial y} \right)^0 f(x, y) &= f(x, y), \\ \left(r\,\frac{\partial}{\partial x} + s\,\frac{\partial}{\partial y} \right)^1 f(x, y) &= r\,\frac{\partial f}{\partial x} + s\,\frac{\partial f}{\partial y}, \\ \left(r\,\frac{\partial}{\partial x} + s\,\frac{\partial}{\partial y} \right)^2 f(x, y) &= r^2\,\frac{\partial^2 f}{\partial x^2} + 2rs\,\frac{\partial^2 f}{\partial xy} + s^2\,\frac{\partial^2 f}{\partial y^2}, \end{aligned}$$

[36] Carle David Tolmé Runge (1856–1927) was a famous applied mathematician from Germany. Runge was always a fit and active man and on his 70th birthday he entertained his grandchildren by doing handstands.

[37] Martin Wilhelm Kutta (1867–1944) is best known for the Runge–Kutta method (1901) for solving ordinary differential equations numerically and for the Zhukovsky–Kutta airfoil. (The letter u in both names, Runge and Kutta, is pronounced as u in the word "rule.")

and so on. To simplify notations, we use subscripts to denote partial derivatives. For example, $f_{xx} = \partial^2 f/\partial x^2$. Since the actual derivation of the formulas (3.5.2) involves substantial algebraic manipulations, we consider in detail only the very simple case $p = 1$, which may serve to illustrate the procedure in the more general case.

Thus, we proceed to determine α_0, α_1, μ, and b such that

$$y_{n+1} = y_n + h[\alpha_0 f_n + \alpha_1 f(x_n + \mu h, y_n + bh)]. \tag{3.5.4}$$

First, we expand $f(x_n + \mu h, y_n + bh)$ using Taylor's series (3.5.3) for a function of two variables and drop all terms in which the exponent of h is greater than two:

$$f(x_n + \mu h, y_n + bh) = f_n + f_x \mu h + f_y bh + \frac{h^2}{2}\left(\mu^2 f_{xx} + 2\mu b f_{xy} + b^2 f_{yy}\right) + O(h^3),$$

where $f_n \equiv f(x_n, y_n)$, $f_x \equiv f_x(x_n, y_n)$, and so forth. The nomenclature $O(h^3)$ means an expression that is proportional to the step size h raised to the third power. Here we used the first few terms of the two-variable Taylor series:

$$\begin{aligned} f(x + r, y + s) &= f(x, y) + r f_x(x, y) + s f_y(x, y) + r^2 f_{xx}(x, y)/2 \\ &\quad + rs f_{xy}(x, y) + s^2 f_{yy}(x, y)/2 + O\left[(|r| + |s|)^3\right]. \end{aligned} \tag{3.5.5}$$

Hence, Eq. (3.5.4) can be reduced to

$$y_{n+1} = y_n + h(\alpha_0 + \alpha_1) f_n + h^2 \alpha_1 [\mu f_x + b f_y] + \frac{h^3}{2}\alpha_1[\mu^2 f_{xx} + 2\mu b f_{xy} + b^2 f_{yy}] + O(h^4). \tag{3.5.6}$$

By chain rule differentiation,

$$\frac{\mathrm{d}f(x, y)}{\mathrm{d}x} = \frac{\partial f}{\partial x} + \frac{\partial f}{\partial y}\frac{\mathrm{d}y}{\mathrm{d}x}, \qquad \frac{\mathrm{d}y}{\mathrm{d}x} = f(x, y);$$

with the same abbreviated notation, we obtain

$$y' = f_n, \quad y'' = (f_x + f f_y)_n, \quad y''' = \left[f_{xx} + 2f f_{xy} + f^2 f_{yy} + f_y(f_x + f f_y)\right]_n,$$

where all functions are evaluated at $x = x_n$. Therefore, Eq. (3.5.1) becomes

$$\begin{aligned} y_{n+1} &= y_n + h f_n + \frac{h^2}{2}(f_x + f f_y)_n \\ &\quad + \frac{h^3}{6}\left[f_{xx} + 2f f_{xy} + f^2 f_{yy} + f_y(f_x + f f_y)\right]_n + O(h^4). \end{aligned} \tag{3.5.7}$$

Finally, we equate terms in like powers of h and h^2 in (3.5.6) and (3.5.7) to obtain the three conditions

$$\alpha_0 + \alpha_1 = 1, \quad \mu\alpha_1 = \frac{1}{2}, \quad b\alpha_1 = \frac{f_n}{2}.$$

From the preceding equations, we find the parameters

$$\alpha_0 = 1 - C, \quad \mu = \frac{1}{2C}, \quad b = \frac{f_n}{2C},$$

where $C = \alpha_1$ is an arbitrary nonzero constant, which clearly cannot be determined. Substituting these formulas into Eq. (3.5.4), we obtain a one-parameter family of numerical methods (one-parameter Runge–Kutta algorithms of the second order):

$$y_{n+1} = y_n + (1 - C)h f(x_n, y_n) + Ch f\left(x_n + \frac{h}{2C}, y_n + \frac{h}{2C} f(x_n, y_n)\right) + hT_n,$$

where hT_n is the local truncation error that contains the truncation error term

$$T_n = \left(\frac{h^2}{6} - \frac{h^2}{8C}\right)\left(\frac{\partial}{\partial x} + f\frac{\partial}{\partial y}\right)^2 f + \frac{h^2}{6} f_y \left(\frac{\partial}{\partial x} + f\frac{\partial}{\partial y}\right) f + O(h^3), \tag{3.5.8}$$

or, after some simplification,

$$T_n = \frac{h^2}{24C}\left[(4C-3)y_n''' + 3f_y(x_n, y_n)y_n''\right] + O(h^3).$$

There are two common choices: $C = 1/2$ and $C = 1$. For $C = 1/2$, we have

$$y_{n+1} = y_n + \frac{h}{2}f(x_n, y_n) + \frac{h}{2}f(x_n + h, y_n + hf(x_n, y_n)) + hT_n,$$

where $T_n = \frac{h^2}{12}[3f_y(x_n, y_n)y_n'' - y_n'''] + O(h^3)$. This one-step algorithm is recognized as the *improved Euler method* (3.2.15), page 153. For $C = 1$, we have

$$y_{n+1} = y_n + hf(x_n + h/2, y_n + hf(x_n, y_n)/2) + hT_n,$$

where $T_n = \frac{h^2}{24}[y_n''' + 3f_y(x_n, y_n)y_n''] + O(h^3)$. This one-step algorithm is the *modified Euler's method* (3.2.16). Note that other choices of C are possible, which provide different versions of the second order methods. For instance, when $C = 3/4$, the truncation error term (3.5.8) has the simplest form. However, none of the second order Runge–Kutta algorithms is widely used in actual computations because of their local truncation error being of order $O(h^3)$. ■

The higher order Runge–Kutta methods are developed in a similar way. For example, the increment function for the third order method is

$$y_{n+1} = y_n + h(ak_1 + bk_2 + ck_3),$$

where the coefficients k_1, k_2, k_3 are sample slopes, usually called *stages*, that are determined sequentially:

$$\begin{aligned} k_1 &= f(x_n, y_n), \\ k_2 &= f(x_n + ph, y_n + phk_1), \\ k_3 &= f(x_n + rh, y_n + shk_1 + (r-s)hk_2). \end{aligned}$$

To obtain the values of the constants $a, b, c, p, r,$ and s, we first expand k_2 and k_3 about mesh point (x_n, y_n) in a Taylor series (the analog of Eq. (3.5.5)):

$$\begin{aligned} k_2 &= f_n + phf_x + phk_1f_y + (ph)^2\left[\frac{1}{2}f_{xx} + k_1f_{xy} + \frac{k_1^2}{2}f_{yy}\right] + O(h^3), \\ k_3 &= f_nrh[f_x + k_1f_y] + h^2p(r-s)(f_x + k_1f_y)f_y + (rh)^2\left[\frac{1}{2}f_{xx} + k_1f_{xy} + \frac{k_1^2}{2}f_{yy}\right]. \end{aligned}$$

The function $y(x+h)$ is expanded in a Taylor series as before in Eq. (3.5.6). Coefficients of like powers of h through h^3 terms are equated to produce a formula with a local truncation error of order h^4. Again, we obtain four equations with six unknowns:

$$a + b + c = 1,\ bp + cr = 1/2,\ bp^2 + cr^2 = 1/3,\ cp(r-s) = 1/6.$$

Two of the constants $a, b, c, p, r,$ and s are arbitrary. For one set of constants, selected by Kutta, the **third-order method** (or three-stage algorithm) becomes

$$y_{n+1} = y_n + \frac{h}{6}(k_1 + 4k_2 + k_3), \tag{3.5.9}$$

where

$$\begin{aligned} k_1 &= f(x_n, y_n), \\ k_2 &= f(x_n + h/2, y_n + k_1h/2), \\ k_3 &= f(x_n + h, y_n + 2hk_2 - hk_1). \end{aligned}$$

Note that if the slope function in Eq. (3.2.1) does not depend of y, then the formula (3.5.9) becomes the Simpson approximation of the integral in Eq. (3.2.3), page 146.

Table 176: Example 3.5.1: a comparison of the results for the numerical solution of the initial value problem $\dot{y} = y - 1 + \cos t$, $y(0) = 0$, using the third-order Kutta method (3.5.9) for different step sizes h. The true values at the points $t = 1$ and $t = 2$ are $\phi(1) = 1.3011686789397567$ and $\phi(2) = 2.325444263372824$, respectively.

h	Approx. at $t = 1$	$\lvert E\rvert/h^3$	Approx. at $t = 2$	$\lvert E\rvert/h^3$
$1/2^5$	1.3011675995209735	0.0353704	2.3254337871513067	0.343285
$1/2^6$	1.3011685445838836	0.0352206	2.3254430282232250	0.323787
$1/2^7$	1.3011686621826817	0.0351421	2.3254441138396460	0.313594
$1/2^8$	1.3011686768475141	0.0351020	2.3254442449916843	0.308384
$1/2^9$	1.3011686786783776	0.0350817	2.3254442610948014	0.305751

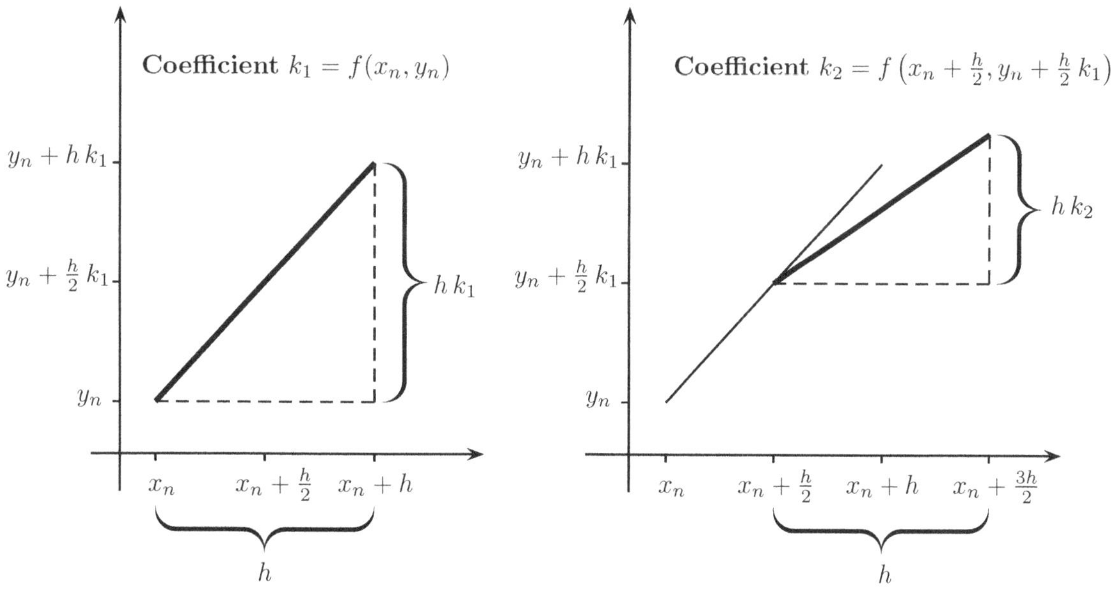

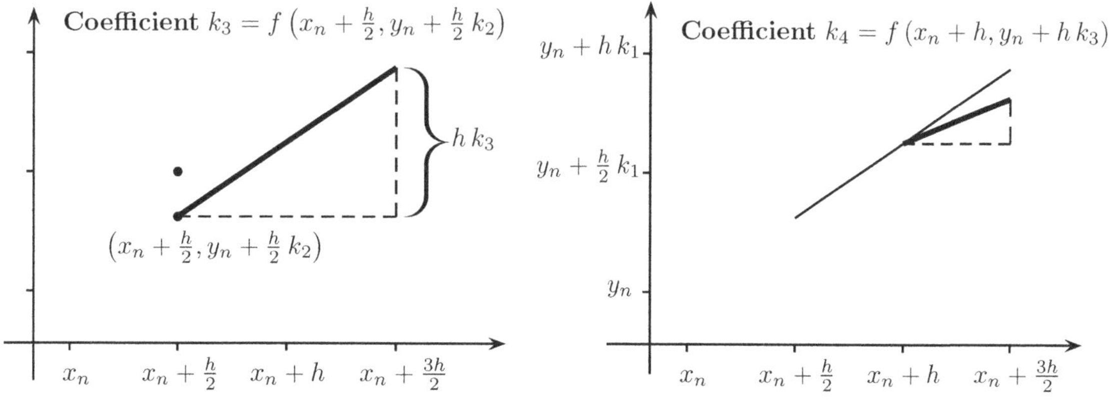

Figure 3.10: Classical Runge–Kutta algorithm.

Example 3.5.1: (Example 3.2.1 revisited) Let us start with the initial value problem for a linear differential equation where all calculations become transparent. For the slope function $f(t, y) = y - 1 + 2\cos t$ from Example 3.2.1, page 148, the stages on each mesh interval $[t_n, t_{n+1}]$ of length $h = t_{n+1} - t_n$ in the Kutta algorithm (3.5.9) have the following expressions:

$$k_1 = y_n - 1 + 2\cos(t_n),$$

$$k_2 = y_n\left(1+\frac{h}{2}\right) - \left(1+\frac{h}{2}\right) + h\,\cos t_n + 2\cos\left(t_n+\frac{h}{2}\right),$$

$$k_3 = (1+h+h^2)\,(y_n-1) + 2(h^2-h)\,\cos(t_n) + 4h\,\cos\left(t_n+\frac{h}{2}\right) + 2\cos(t_{n+1}).$$

For this slope function, the Kutta algorithm (3.5.9) requires 24 arithmetic operations and 3 evaluations of the cosine function at each step. On the other hand, the improved Euler method (3.2.15), which is of the second order, needs 13 arithmetic operations and 3 evaluations of the cosine function. The Taylor series approximation of the third order,

$$y_{n+1} = y_n(1+h) + h\,(2\cos t_n - 1) + \frac{h^2}{2}\,(y_n - 1 + 2\cos t_n - 2\sin t_n) + \frac{h^3}{6}\,(y_n - 1 - 2\sin t_n),$$

requires 13 arithmetic operations and 2 evaluations of the trigonometric functions on each step. The following table confirms that the Kutta method (3.5.9) is of order 3.

All the **fourth-order** algorithms are of the form

$$y_{n+1} = y_n + h(a\,k_1 + b\,k_2 + c\,k_3 + d\,k_4),$$

where k_1, k_2, k_3, and k_4 are derivative values (slopes) computed at some sample points within the mesh interval $[x_n, x_{n+1}]$, and a, b, c, and d are some coefficients. The **classical fourth order Runge–Kutta method** (see Fig. 3.10) is

$$y_{n+1} = y_n + \frac{h}{6}\,(k_1 + 2k_2 + 2k_3 + k_4), \tag{3.5.10}$$

where

$$\begin{aligned} k_1 &= f_n = f(x_n, y_n), \\ k_2 &= f\left(x_n + \tfrac{h}{2}, y_n + \tfrac{h}{2}k_1\right), \\ k_3 &= f\left(x_n + \tfrac{h}{2}, y_n + \tfrac{h}{2}k_2\right), \\ k_4 &= f(x_n + h, y_n + hk_3). \end{aligned}$$

Example 3.5.2: Consider the nonlinear differential equation $y' = (x+y-4)^2$ subject to $y(0) = 4$. This initial value problem has the exact solution $y = 4 + \tan(x) - x$. Using the third-order Kutta method (3.5.9), we obtain the following values of coefficients:

$$k_1 = (x_n + y_n - 4)^2, \quad k_2 = \left(x_n + \frac{h}{2} + y_n + \frac{h}{2}(x_n + y_n - 4)^2 - 4\right)^2,$$

$$k_3 = (x_n + h + y_n + 2hk_2 - hk_1 - 4)^2.$$

In particular, for $n = 0$ (first step), we have $y_1 = 4 + \frac{h}{6}\left[h^2 + \left(h + \frac{h^3}{2}\right)^2\right]$ since $k_1 = 0$, $4k_2 = h^2$, $k_3 = \left(h + \frac{h^3}{2}\right)^2$.

The algorithm (3.5.9) requires 22 arithmetic operations at each step. The Taylor series approximation of the third order,

$$y_{n+1} = y_n + h\,(x_n + y_n - 4)^2 + h^2\,(x_n + y_n - 4)\,[1 + (x_n + y_n - 4)^2],$$

requires 11 arithmetic operations. On the other hand, the classical fourth order Runge–Kutta method (3.5.10) asks to calculate four slopes:

$$k_1 = (x_n + y_n - 4)^2, \qquad k_2 = \left(x_n + y_n - 4 + \frac{h}{2} + \frac{h}{2}\,k_1\right)^2,$$

$$k_3 = \left(x_n + y_n - 4 + \frac{h}{2} + \frac{h}{2}\,k_2\right)^2, \qquad k_4 = (x_n + y_n - 4 + h + h\,k_3)^2.$$

Therefore, the algorithm (3.5.10) requires 25 arithmetic operations at each step. □

Although the Runge–Kutta methods look a little bit cumbersome, they are well suited for programming needs, as the following codes show. For example, in *Maple* the classical fourth order formula (3.5.10) can be written as

```
x[0]:=0.0:  y[0]:=0: h:=0.15:  # initial condition
for i from 0 to 9 do
    x[i+1] := x[i]+h:
    k1:=f(x[i] , y[i]);
    k2:=f(x[i]+h/2 , y[i]+h*k1/2);
    k3:=f(x[i]+h/2 , y[i]+h*k2/2);
    k4:=f(x[i+1] , y[i]+h*k3);
    y[i+1]:=y[i]+h*(k1+2*k2+2*k3+k4)/6:
  od:
ptsapp1 := seq([x[i],y[i]],i=1..10):
ptssol1 := seq([x[i],fsol(x[i])],k=1..10):
p1app:=pointplot({ptsapp1}):
display(psol,p1app);
errorplot := pointplot( {seq([x[i],fsol(x[i])-y[i]],i=1..10)}):
display(errorplot);
```

Maxima and MATLAB have a special solver based on scheme (3.5.10), called `rk` (load `dynamics`) and `ode45` (actually, it is a more sophisticated routine), respectively. Then the script to solve the differential equation $y' = f(x, y)$ (the slope function $f(x, y)$ is labeled as `FunctionName`) subject to the initial condition $y($x0$)=$a0 on the interval [x0,xf] and to plot the solution is

```
[x,y] = ode45(@FunctionName, [x0,xf], a0); plot (x,y)
```

For example, if the slope function is piecewise smooth

$$f(x, y) = \begin{cases} y\,\sin(\pi x/3), & \text{if } 0 \leqslant x \leqslant 3, \\ 0, & \text{otherwise,} \end{cases}$$

we define it with the following MATLAB script:

```
function f=FunctionName(x,y)
    f=y*sin(pi*x/3).*(x<=3);
```

We present two other popular fourth order Runge–Kutta type algorithms. The first one, ascribed to Kutta, is

$$y_{n+1} = y_n + \frac{h}{8}\,(k_1 + 3k_2 + 3k_3 + k_4), \tag{3.5.11}$$

where

$$\begin{aligned} k_1 &= f_n = f(x_n, y_n), \\ k_2 &= f\left(x_n + \tfrac{h}{3}, y_n + \tfrac{h}{3}\,k_1\right), \\ k_3 &= f\left(x_n + \tfrac{2h}{3}, y_n - \tfrac{h}{3}\,k_1 + hk_2\right), \\ k_4 &= f(x_n + h, y_n + hk_1 - hk_2 + hk_3). \end{aligned}$$

Another widely used fourth-order method is one credited to S. Gill [17],

$$y_{n+1} = y_n + \frac{h}{6}\left[k_1 + 2\left(1 - \frac{1}{\sqrt{2}}\right)k_2 + 2\left(1 + \frac{1}{\sqrt{2}}\right)k_3 + k_4\right], \tag{3.5.12}$$

where

$$\begin{aligned} k_1 &= f(x_n, y_n), \\ k_2 &= f\left(x_n + \tfrac{h}{2}, y_n + \tfrac{h}{2}\,k_1\right), \\ k_3 &= f\left(x_n + \tfrac{h}{2}, y_n - \left(\tfrac{1}{2} - \tfrac{1}{\sqrt{2}}\right)hk_1 + \left(1 - \tfrac{1}{\sqrt{2}}\right)hk_2\right), \\ k_4 &= f\left(x_n + h, y_n - \tfrac{1}{\sqrt{2}}hk_2 + \left(1 + \tfrac{1}{\sqrt{2}}\right)hk_3\right). \end{aligned}$$

Example 3.5.3: (Example 3.2.6 revisited) We compare the results of numerical calculations for different Runge–Kutta methods: the third-order (3.5.9), the classical fourth order Runge–Kutta method (3.5.10), the Kutta formula (3.5.11), and Gill's method (3.5.12). For this purpose we consider the linear differential equation with oscillating slope function $\dot{y} = 2\pi\, y\,\cos(2\pi t)$ subject to the initial condition $y(0) = 1$. The actual solution of this initial value problem is $y = \phi(t) = e^{\sin(2\pi\, t)}$.

Gill's algorithm for this initial value problem requires evaluation of the slope function at four sample points.

$$\begin{aligned} k_1 &= f_n = f(x_n, y_n) = 2\pi y_n\,\cos(2\pi t_n), \\ k_2 &= 2\pi y_n\left[1 + \pi h\cos(2\pi t_n)\right]\cos(2\pi t_n + \pi h), \\ k_3 &= 2\pi\left[y_n - \tfrac{h}{2}\,k_1 + hk_2 + \tfrac{h}{\sqrt{2}}\,(k_1 - k_2)\right]\cos(2\pi t_n + \pi h), \\ k_4 &= 2\pi\left[y_n - \tfrac{h}{2}\,k_2 + \left(1 + \tfrac{1}{\sqrt{2}}\right)hk_3\right]\cos(2\pi t_n + 2\pi h). \end{aligned}$$

The approximate value y_{n+1} of the solution at the next step $x_{n+1} = x_n + h$ is obtained according to formula (3.5.12), which requires 45 arithmetic operations and 3 evaluations of cosine function. The classical fourth order Runge–Kutta method (3.5.10) is less costly—it requires only 30 arithmetic operations and the same number of cosine evaluations.

At the end, we present results of numerical calculations with constant step size $h = 0.1$ for the IVP $y' = (2/x)\,(y - \ln x)^{1/2} + 1/x$, $y(1) = 0$, where the actual solution $y = \phi(x) = (\ln x)^2 + \ln x$ is a monotonically increasing function, using the third-order Kutta algorithm (3.5.9), the Kutta formula (3.5.11), the classical fourth order Runge–Kutta method (3.5.10), and Gill's method (3.5.12).

Table 180: A comparison of the results for the numerical solutions of the initial value problem $\dot{y} = 2\pi\, y\, \cos(2\pi t)$, $y(0) = 1$ using the third-order and fourth order Kutta algorithms (3.5.9) and (3.5.11), the classical Runge–Kutta method (3.5.10), and Gill's rule (3.5.12) with step size $h = 0.1$.

x	Exact	Kutta-3	Kutta-4	Classical	Gill
1.0	1.000	1.0143670292	0.9930807904	0.9992873552	0.9992873552
2.0	1.000	1.0289404699	0.9862094562	0.9985752184	0.9985752184
3.0	1.000	1.0437232876	0.9793856663	0.9978635890	0.9978635890
4.0	1.000	1.0587184905	0.9726090915	0.9971524667	0.9971524667

Problems

In all problems, use a uniform mesh obtained by a partition with grid points $x_n = x_0 + nh$, $n = 0, 1, \ldots$. As usual, denote by y_n the approximate value of the actual solution at $x = x_n$. Throughout, primes denote derivatives with respect to x, while dots denote derivatives with respect to t.

1. For the given initial value problems, use the classical Runge–Kutta method (3.5.10) with $h = 0.1$ to obtain an approximation to $y(2.5)$, the value of the unknown function at $x = 2.5$. Compare your result with the true value obtained from the actual solution.

   (a) $y' = (y + x - 4)^2$, $y(0) = 2$. **(b)** $y' = y^2/(x+1)$, $y(0) = 1/2$.
   (c) $y' = (x + y - 3)/(x - y + 1)$, $y(2) = 2$. **(d)** $y' = 2y + 4x^2$, $y(0) = 0$.
   (e) $\dot{y} = y\cos t + 2\cos^3 t$, $y(0) = 1$. **(f)** $\dot{y} = y + 2\cos t$, $y(0) = 0$.

2. Redo Problem 1 using the optimal second order Runge–Kutta scheme:
$$y_{n+1} = y_n + \frac{h}{4}\, f(x_n, y_n) + \frac{3h}{4}\, f\left(x_n + \frac{2h}{3}, y_n + \frac{2h}{3}\, f(x_n, y_n)\right), \qquad n = 0, 1, 2, \ldots. \tag{3.5.13}$$

3. Redo Problem 1 using Ralson's method (which corresponds $C = 2/3$ in Eq. (3.5.8)):
$$y_{n+1} = y_n + \frac{h}{3}\, f(x_n, y_n) + \frac{2h}{3}\, f\left(x_n + \frac{3h}{4}, y_n + \frac{3h}{4}\, f(x_n, y_n)\right), \qquad n = 0, 1, 2, \ldots. \tag{3.5.14}$$

4. Redo Problem 1 using each of the following third order approximations:

   (a) the third order Kutta method (3.5.9);

   (b) the Nystrom method:
$$y_{n+1} = y_n + \frac{h}{8}\,(2k_1 + 3k_2 + 3k_3), \qquad n = 0, 1, 2, \ldots, \tag{3.5.15}$$
   where
$$k_1 = f(x_n, y_n),\ k_2 = f\left(x_n + \frac{2h}{3}, y_n + \frac{2h}{3}\, k_1\right),\ k_3 = f\left(x_n + \frac{2h}{3}, y_n + \frac{2h}{3}\, k_2\right);$$

   (c) the nearly optimal method (which is used in MATLAB function `ode23`):
$$y_{n+1} = y_n + \frac{h}{9}\,(2k_1 + 3k_2 + 4k_3), \qquad n = 0, 1, 2, \ldots, \tag{3.5.16}$$
   in which the stages are computed according to
$$k_1 = f(x_n, y_n), \quad k_2 = f\left(x_n + \frac{h}{2}, y_n + \frac{h}{2}\, k_1\right), \quad k_3 = f\left(x_n + \frac{3h}{4}, y_n + \frac{3h}{4}\, k_2\right);$$

   (d) Heun's algorithm:
$$y_{n+1} = y_n + \frac{h}{4}\,(k_1 + 3k_3), \qquad n = 0, 1, 2, \ldots, \tag{3.5.17}$$
   where
$$k_1 = f(x_n, y_n),\ k_2 = f\left(x_n + \frac{h}{3}, y_n + \frac{h}{3}\, k_1\right),\ k_3 = f\left(x_n + \frac{2h}{3}, y_n + \frac{2h}{3}\, k_2\right).$$

5. Solve numerically the following initial value problems using the classical Runge–Kutta method with $h = 0.1$ and $h = 0.01$. Determine the numerical value of the solution 2 units from the initial position at $x = 0$, and then compare with the true value.
   (a) $y' = y^2 + 4, \quad y(0) = 0.$ **(b)** $y' = 3x^2(y - y^2), \quad y(0) = 1/2.$
   (c) $y' = y + 4x - 1, \quad y(0) = 0.$ **(d)** $y' = x - y, \quad y(0) = 1/2.$
   (e) $y' = \sqrt{4x + y - 1}, \quad y(0) = 1.$ **(f)** $y' = (2xy + y^2)/(1 + x^2), \quad y(0) = 1.$
   (g) $y' = 3y - 2x + 5, \quad y(0) = 1.$ **(h)** $y' = \cos(y - x + \pi/2), \quad y(0) = 0.$
6. Consider the following initial value problems for which closed form solution formulas are not available. Apply the classical Runge–Kutta method (3.5.10) to these problems with $h = 0.1$, $h = 0.05$, and $h = 0.025$ to see whether the numerical solutions settle down. Pick the interval of interest to be 2 units from the initial position at $x = 0$ and find approximations at $x = 2$.
   (a) $y' = x^2 - x\cos(y), \quad y(0) = 1.$ **(b)** $y' = \sin y + \sin x, \quad y(0) = 0.$
   (c) $y' = x^2 + y^{-2}, \quad y(0) = 1.$ **(d)** $y' = y\sqrt{x^2 + y^2 + 3}, \quad y(0) = 1.$
   (e) $y' = 2x\,e^{-y} + \sin(xy), \quad y(0) = 0.$ **(f)** $y' = \cos\left(x^2 + y^2\right), \quad y(0) = 1.$
   (g) $y' = y^{-3} - 2x^2, \quad y(0) = 1.$ **(h)** $y' = e^{-\sqrt{x^2+y^2}}, \quad y(0) = 0.$
7. Redo the previous problem using the Kutta method (3.5.11).
8. Redo Problem 5 using the Gill method (3.5.12).
9. Using a second order implicit method

$$y_{n+1} = y_n + \frac{h}{2}\left(k_1 + k_2\right), \tag{3.5.18}$$

where

$$k_1 = f\left(x_n + \frac{\sqrt{3}-1}{2\sqrt{3}}\,h,\ y_n + \frac{h}{4}\,k_1 + \frac{\sqrt{3}-2}{4\sqrt{3}}\,hk_2\right),$$
$$k_2 = f\left(x_n + \frac{\sqrt{3}+1}{2\sqrt{3}}\,h,\ y_n + \frac{\sqrt{3}+2}{4\sqrt{3}}\,hk_1 + \frac{h}{4}\,k_2\right),$$

find a four-decimal-place approximation to $y(1.0)$ with step size $h = 0.1$ of the following IVPs.
   (a) $y' = y + x^2, \quad y(0) = -1.$ **(b)** $\dot{y} = -4t^3 y, \quad y(0) = 2.$
   (c) $y' = 2xy - \sqrt{y}, \quad y(0) = 1.$ **(d)** $\dot{y} = 2ty + 1, \quad y(0) = 2.$
   Compare the numerical results with the exact solution. Express the exact solution to part (d) in terms of the error function, $\text{erf}(t) = \frac{2}{\sqrt{\pi}}\int_0^t e^{-x^2}\,dx.$
10. Redo Problem 5 using Butcher's (1964) fifth order RK method

$$y_{k+1} = y_k + \frac{h}{90}\left(7k_1 + 32k_3 + 12k_4 + 32k_5 + 7k_6\right), \qquad k = 0, 1, \ldots, \tag{3.5.19}$$

where $k_1 = f(x_k, y_k)$, $k_2 = f\left(x_k + h/4, y_k + hk_1/4\right)$, $k_3 = f\left(x_k + h/4, y_k + k_1 h/8 + k_2 h/8\right)$, $k_4 = f\left(x_k + h/2, y_k - k_2 h/2 + k_3 h\right)$, $k_5 = f\left(x_k + 3h/4, y_k + 3k_1 h/16 + 9k_4 h/16\right)$, $k_6 = f\left(x_k + h, y_k - 3k_1 h/7 + 2k_2 h/7 + 12k_3 h/7 - 12k_4 h/7 + 8k_5 h/7\right)$.
11. Select one of the following initial value problems and compare the numerical solutions obtained with the classical fourth-order RK formula (3.5.10) and the fourth-order Taylor series. Use different values of step size $h = 2^{-n}$, for $n = 2, 3, \ldots, 7$, to compare with the true value at the point $x = 2$.
   (a) $y' = 1 - x/(4y), \ y(1) = 1.$ **(b)** $y' = 5xy/(x^2 - y^2), \ y(1) = 2.$
   (c) $xy' = y + \sqrt{x^2 + y^2}, \ y(1) = 1.$ **(d)** $y' = 2\left(\frac{y}{x+y}\right)^2, \ y(1) = 1.$
   (e) $xy' = y + 10\sqrt{x^2 - y^2}, \ y(1) = 1.$ **(f)** $y' = y - \sqrt{y}, \ y(1) = 1/4.$
12. Stiffness is a subtle, difficult, and important concept in the numerical solution of ordinary differential equations. It depends on the differential equation, the initial conditions, and the numerical method. A problem is stiff if the solution being sought varies slowly, but there are nearby solutions that vary rapidly. The following example, credited to Larry Shampine, shows that a practical model of flame propagation is stiff.

    If you light a match, the ball of flame grows rapidly until it reaches a critical size. Then it remains at that size because the amount of oxygen being consumed by the combustion in the interior of the ball balances the amount available through the surface. Let $y(t)$ represent the radius of the ball at time t, and let $\dot{y}$ be its derivative with respect to t. The simple model is $\dot{y} = y^2 - y^3$, $y(0) = \delta$, where $0 \leqslant t \leqslant 2/\delta$, and terms y^2 and y^3 come from the surface area and the volume. The critical parameter is the initial radius δ.

    First, solve the given initial value problem analytically since the slope function is autonomous. Second, use a computer solver based on the Runge–Kutta method to solve this problem numerically for two different values, $\delta = 0.01$ and $\delta = 0.00001$, to observe the stiffness. For instance, MATLAB has a special subroutine `ode23s` to solve some stiff problems:

```
delta = 0.01; F = @(t,y) y^2 - y^3;
opt=odeset('stats','on');figure;title({'ode45, \delta = ',delta});hold on
disp('**Statistics for ode45**');ode45(F,[0 2/delta],delta,opt);
figure;title({'ode23s, \delta = ',delta});hold on;
disp('**Statistics for ode23s**');ode23s(F,[0 2/delta],delta,opt);
```

13. Solve the initial value problem $y' = y\sqrt{y^2-4}$ with $y(0) = \sqrt{5}$ by the Runge–Kutta method (3.5.10) on the interval $[0, 0.55]$, and account for any difficulties. Then, using negative h, solve the same differential equation on the interval with the initial value $y(0.55) = 280$.
14. Consider the IVP: $y' = x^2 + y^2$, $y(0) = 0$. To solve the problem numerically, apply the Runge–Kutta methods of order 4: (3.5.10), (3.5.11), and (3.5.12). Calculate the number of arithmetic operations required per step.

Summary for Chapter 3

1. A recurrence is an equation that relates different members of a sequence of numbers $\{y_0, y_1, y_2, \ldots\}$, where the values of y_n of the sequence are unknown quantities that we wish to determine.
2. The order of a recurrence relation is the difference between the largest and smallest subscripts of the members of the sequence that appear in the equation. The general form of a recurrence relation (in normal form) of order p is $y_n = f(y_{n-1}, y_{n-2}, \ldots, y_{n-p})$ for some function f. A recurrence of a finite order is usually called the difference equation.
3. A constant coefficient recurrence relation of the first order $y_n = f(y_{n-1})$ has an equilibrium solution $y = y_n$ (which does not depend on n) if $y = f(y)$.
4. A constant coefficient linear nonhomogeneous difference equation $y_{n+1} = py_n + q_n$ has the general solution (3.1.6). If $q_n = \alpha q^n$, its exact solution is given by formulas (3.1.9), (3.1.10).
5. The first order variable coefficient difference equation $y_{n+1} = p_n y_n + q_n$ has a solution (3.1.12).
6. If a small change in the initial value has produced a large change in the solution, we call such a problem (recurrence) ill-conditioned.
7. The Euler method, or the tangent line method, is the simplest numerical method for solving the initial value problem $y' = f(x, y)$, $y(x_0) = y_0$. For computing approximate values y_n of the solution $\phi(x_n)$ at the mesh points $x_n = x_0 + nh$, $(n = 0, 1, 2, \ldots)$, the Euler approximation is defined by Eq. (3.2.5) (recall that $f_n = f(x_n, y_n)$).
8. The backward Euler method is defined in Eq. (3.2.12).
9. The trapezoid rule (3.2.14) is another example of an implicit one-step method.
10. The improved Euler formula, or the Heun formula, or the average slope method, is defined by Eq. (3.2.15).
11. The modified Euler formula is given in Eq. (3.2.16).
12. The **round-off error** is the difference between an approximation, y_n, and its computed numerical value using the floating-point presentation, Y_n.
13. The **global truncation error** (also called accumulative error) at the n-th step is the difference $E_n = \phi(x_n) - y_n$ between the true value $\phi(x_n)$ and its approximation.
14. The **local truncation error** for the difference one-step method $y_{n+1} = y_n + h\,\Phi(x_n, y_n, y_{n+1}; h)$, $(n = 0, 1, 2, \ldots)$ is the error that the increment function, $h\,\Phi$, causes during a single iteration, assuming perfect knowledge of the true solution at the previous iteration (called the reference solution).
15. If in the IVP $y' = f(x, y)$, $y(a) = y_a$, the function $f(x, y)$ is sufficiently differentiable, then calculate its derivatives by means of the recurrence
$$f^{(0)} = f, \qquad f^{(k+1)} = f_x^{(k)} + f_y^{(k)}\, f, \qquad k = 0, 1, 2, \ldots.$$
We define the operator $T_p(x, y; h)$ by Eq. (3.3.7) to advance from y_n to $y_{n+1} = y_n + h\,T_p(x, y; h)$.
16. **Taylor's algorithm of order** p to find an approximate solution of the differential equation $y' = f(x, y)$ subject to $y(a) = y_a$ over an interval $[a, b]$ is outlined on page 163.
17. The methods associated with the names of Runge, Kutta, and others are based on the following approximation of the initial value problem $y' = f(x, y)$, $y(x_0) = y_0$:
$$y_{n+1} = y_n + h[\alpha_0 f(x_n, y_n) + \alpha_1 f(x_n + \mu_1 h, y_n + b_1 h) + \alpha_2 f(x_n + \mu_2 h, y_n + b_2 h) + \cdots + \alpha_p f(x_n + \mu_p h, y_n + b_p h)].$$
18. The third order Kutta method is given in Eq. (3.5.9).
19. There are several fourth order Runge–Kutta methods. We present three of the most popular algorithms: the classical Runge–Kutta method (3.5.10), the Kutta scheme (3.5.11), and Gill's algorithm (3.5.12).

Review Questions for Chapter 3

Section 3.1

1. Solve the given first order difference equations in terms of the initial value a_0. Describe the behavior of the solution as $n \to \infty$.

   (a) $y_{n+1} = -1.2\, y_n$; **(b)** $y_{n+1} = \frac{n+5}{n+1}\, y_n$; **(c)** $y_{n+1} = \sqrt{\frac{n+2}{n+1}}\, y_n$;
   (d) $y_{n+1} = (0.5)^{n-1}\, y_n$; **(e)** $y_{n+1} = 1.8\, y_n - 8$; **(f)** $y_{n+1} = -0.4\, y_n + 7$.

2. An investor deposits \$2,500 in an account paying interest at a rate of 4% compounded quarterly. She also makes additional deposits of \$50 every quarter. Find the account balance in 5 years.
3. A man takes a second mortgage of \$250,000 for a 30-year period. What monthly payment is required if the interest rate is 5.75%?
4. Find the effective annual yield of a bank account that pays interest at a rate of 4% compounded weekly.
5. A college student borrows \$40,000 for a flashy car. The lender charges annual interest at a rate of 6.0%. What monthly payment is required to pay off the loan in 10 years? What is the total amount paid during the term of the loan?
6. If the interest rate on a 15-year mortgage is fixed at 7.0% and \$700 is the maximum monthly payment the borrower can afford, what is the maximum mortgage loan possible?
7. A man wants to purchase a yacht for \$300,000, so he wishes to borrow that amount at the interest rate 6.0% for 10 years. What would be his monthly payment?
8. A home-buyer wishes to finance a mortgage of \$300,000 with a 15-year term. What is the maximum interest rate the buyer can afford if the monthly payment is not to exceed \$2,500?

Section 3.2 of Chapter 3 (Review)

In all problems, use a uniform grid of points $x_k = x_0 + kh$, $k = 0, 1, 2, \ldots, n$, where $h = (x_n - x_0)/n$ is a fixed step size. A numerical method works with a discrete set of mesh points $\{x_k\}$ and the sequence of values $y_0, y_1, \ldots, y_n$, such that each y_k is approximately equal to the true value of the actual solution $\phi(x)$ at that point x_k, $k = 0, 1, \ldots, n$. Throughout, primes denote derivatives with respect to x, and a dot stands for the derivative with respect to t.

1. Consider the following autonomous differential equations subject to the initial condition $y(0) = 1$ at the specified point $t = 0$.

   (a) $\dot{y} = e^{y/2}$; **(d)** $\dot{y} = y\,(1 - y/5)$; **(g)** $\dot{y} = 1/(4y^3 + 3y^2)$;
   (b) $\dot{y} = 4 + y^4$; **(e)** $\dot{y} = (y^3 + 1)/y^2$; **(h)** $\dot{y} = (y/2 - 1)^3$;
   (c) $\dot{y} = y^{3/4}$; **(f)** $\dot{y} = 3y^2 - y/2$; **(i)** $\dot{y} = 5y\,(1 - y/3)$.

   Use the Euler rule (3.2.5), Heun's formula (3.2.15), and modified Euler method (3.2.16) to obtain an approximation at $t = 1.2$. First use $h = 0.1$ and then use $h = 0.05$. Then compare numerical values with the true value.
2. In each problem **(a)** through **(i)**, use an appropriate method to find the actual solution $y = \phi(x)$ to a given differential equation subject to the initial condition $y(0) = 1$. Calculate the true value $\phi(1.5)$ of the actual solution at the point $x = 1.5$. Then apply the average slope algorithm (3.2.15) and the midpoint rule (3.2.16) to determine its approximate value using a uniform grid with a constant step size h. Obtain at least a four-decimal place approximation to the indicated value. First use $h = 0.1$ and then use $h = 0.05$. Which of these two numerical methods gives the better approximation?

   (a) $y' = (x - 4y + 5)^2$; **(d)** $y' = 2x\left(y - x^2 - 0.5\right)^2$; **(g)** $y' = \sqrt{x + 2y}$;
   (b) $y' = 4x + y$; **(e)** $y' = (y - x)/(x + 1)$; **(h)** $y\,y' = 6x + 6 + y$;
   (c) $y' = 5 + (y - x)^2$; **(f)** $y' = (4x - y)^2$; **(i)** $y' = \cos(y + x)$.

3. The initial value problem $\dot{y} = \sqrt{y}$, $y(0) = 0$ has two distinct solutions for $x \geqslant 0$: $y_1(x) \equiv 0$ and $y_2(x) = x^2/4$. Which of the five numerical methods (Euler, backward Euler, trapezoid, Heun, and midpoint) may produce these two solutions?
4. Answer the following questions for each linear differential equation **(a)** – **(f)** subject to the initial condition $y(0) = 1$ at $x = 0$.

   (a) $y' = 3 - y$; **(c)** $y' - 5\,y + 8x = 10$; **(e)** $(4 + x^2)\,y' = 4xy - 8x$;
   (b) $y' = 2\,y + 8x + 4$; **(d)** $y' + 2y + 8x = 0$; **(f)** $y' = x(x^2 + y)$.

   (i) Find the actual solution $y = \phi(x)$.

(ii) Use Euler's method $y_{k+1} = y_k + hf_k$ to obtain the approximation y_k at the mesh point x_k.

(iii) Use the backward Euler method to obtain the approximation y_k at the mesh point x_k.

(iv) Use the trapezoid rule to obtain the approximation y_k at the mesh point x_k.

(v) Use the improved and modified Euler formula to obtain the approximation y_k at the mesh point x_k.

(vi) For $h = 0.1$, compare numerical values y_6 found in the previous parts with the true value $\phi(0.6)$.

5. Consider the following separable differential equations subject to the initial condition $y(x_0) = y_0$ at the specified point x_0.

   (a) $e^x\, y' + \cos^2 y = 0, \quad y(0) = 1.$ **(b)** $y' = e^y\,(\cos x - 2x)\,, \quad y(0) = e.$
   (c) $x^2(2 - y)y' = (x - 2)y, \quad y(1) = 1.$ **(d)** $y' = 3x^2 y - y/x, \quad y(1) = 1.$
   (e) $y' = e^{2x-y}, \quad y(0) = 1.$ **(f)** $y' + 3x^2(y + 1/y) = 0, \quad y(0) = 4.$
   (g) $y' = y\,\sin x, \quad y(0) = 1.$ **(h)** $t^2\dot{y} + \sec y = 0, \quad y(1) = 0.$

   Use the Euler method (3.2.5), Heun formula (3.2.15), and modified Euler method (3.2.16) to obtain a four-decimal approximation at $x_0 + 1.2$. First use $h = 0.1$ and then use $h = 0.05$. Then compare numerical values with the true value.

6. Consider the following initial value problems for the Bernoulli equations.

   (a) $y' = y(8x - y), \quad y(0) = 1.$ **(b)** $y' = 4x^3 y - x^2 y^2, \quad y(0) = 1.$
   (c) $y' + 3y = xy^{-2}, \quad y(1) = 1.$ **(d)** $y' + 6xy = y^{-2}, \quad y(0) = 1.$
   (e) $2(x - y^4)y' = y, \quad y(0) = 4.$ **(f)** $(y - x)\,y' = y, \quad y(0) = 1.$

   First find the actual solutions and then calculate approximate solutions for $y(1.5)$ at the point $x = 1.5$ using the Euler method (3.2.5), trapezoid method (3.2.14), Heun formula (3.2.15), and modified Euler method (3.2.16). First use $h = 0.1$ and then $h = 0.05$.

7. Consider the following initial value problems for the Riccati equation.

   (a) $y' = 2 - 2y/x + y^2/x^2, \quad y(1) = 7/3.$ **(b)** $y' = 3x^3 + y/x + 3xy^2, \quad y(1) = 1.$

   Use the Euler rule (3.2.5), trapezoid method (3.2.14), Heun formula (3.2.15), and midpoint rule (3.2.16) to obtain a four-decimal approximation at $x = 2.0$. First use $h = 0.1$ and then use $h = 0.05$.

8. Consider the following initial value problems for which analytic solutions are not available.

   (a) $y' = \sin(y^2 - x), \quad y(0) = 1.$ **(b)** $y' = x^{1/3} + y^{1/3}, \quad y(0) = 1.$
   (c) $y' = (y - x)x^2 y^2, \quad y(0) = 1.$ **(d)** $y' = y^2/10 - x^3, \quad y(0) = 1.$
   (e) $y' = y^3/10 + \sin(x), \quad y(0) = 1.$ **(f)** $y' = x^{1/3} - \cos(y), \quad y(0) = 1.$

   Find approximate solutions for $y(1.5)$ at the point $x = 1.5$ for each problem using the Euler rule (3.2.5), trapezoid method (3.2.14), Heun formula (3.2.15), and midpoint rule (3.2.16). First use $h = 0.1$ and then $h = 0.05$.

9. Consider the initial value problem $y' = (4 + x^2)^{-1} - 2y^2$, $y(0) = 0$, which has the actual solution $y = \phi(x) = x/(4 + x^2)$.

   (a) Approximate $\phi(0.1)$ using Euler method (3.2.5) with one step.

   (b) Find the error of this approximation.

   (c) Approximate $\phi(0.1)$ using Euler method with two steps.

   (d) Find the error of this approximation.

10. Repeat the previous problem using the improved Euler's method (3.2.15).

11. Repeat Problem 9 using the modified Euler's method (3.2.16).

12. Which of two second order numerical methods, improved Euler (3.2.15) or modified Euler (3.2.16), requires fewer arithmetic operations performed at each step when applied to the IVP $y' = x^2 + y^2$, $y(0) = 0$?

13. Show that Euler's method can be run forward or backward, according to whether the step size is positive or negative.

14. Apply the Heun method with iteration (see Example 3.4.3, page 172) to integrate $y' = 5\,e^{0.5\,x} - y$ from $x = 0$ to $x = 1$ with a step size of $h = 0.1$. The initial condition at $x = 0$ is $y = 2$. Employ a stopping criterion of 0.000001% to terminate the correct iteration.

15. Can Euler's method be used to solve the initial value problem

$$\dot{y} = 4 + y^2, \qquad y(0) = 0$$

over interval $[0, 1]$?

16. In psychology, the Weber–Fechner law was first published in 1860 by a German philosopher, physicist, and experimental psychologist Gustav Theodor Fechner (1801–1887), a student of Ernst Heinrich Weber (1795–1878) from Leipzig University, one of the founders of experimental psychology. This law states that the rate of change $\mathrm{d}R/\mathrm{d}S$ of the reaction R is inversely proportional to the stimulus S. Use Heun's method with step size $h = 0.1$ to solve the initial value problem
$$\frac{\mathrm{d}R}{\mathrm{d}S} = \frac{1}{S}, \qquad R(0.1) = 0$$
over interval $[0.1, 5, 2]$.

Section 3.3 of Chapter 3 (Review)

1. Solve the IVP $y' = 1 - y^2$, $y(0) = 0$ by the Taylor algorithm of order $p = 3$, using the steps $h = 0.5$ and $h = 0.1$, and compare the values of the numerical solution at $x_n = 1$ with the true value of the exact solution.
2. This exercise addresses the problem of numerical approximations of the following irrational numbers $e \approx 2.718281828$, $\ln 2 \approx 0.6931471806$, $\pi \approx 3.141592654$, and the golden ratio $(1 + \sqrt{5})/2 \approx 1.618033988$. In each case, apply the second and the third order polynomial approximation with the step size $h = 0.01$ to determine how many true decimal places you can obtain with such approximations.
   - **(a)** The number $e = y(1)$, where $y(x)$ is the solution of the IVP $y' = y$, $y(0) = 1$.
   - **(b)** The number $\ln 2 = y(2)$, where $y(x)$ is the solution of the IVP $y' = x^{-1}$, $y(1) = 0$.
   - **(c)** The number $\pi = y(1)$, where $y(x)$ is the solution of the IVP $y' = 4/(1 + x^2)$, $y(0) = 0$.
   - **(d)** The number $(1 + \sqrt{5})/2 = y(2)$, where $y(t) = \frac{\sqrt{3+t}\left(5-19\sqrt{5}+20\sqrt{3+t}\right)}{10\sqrt{3-t}}$ is the solution of the IVP $y' = \frac{3y}{9-t^2} + \frac{1}{\sqrt{3-t}}$, $y(0) = \frac{1}{2} + 2\sqrt{3} - \frac{19}{2\sqrt{5}}$.
3. Consider Taylor series approximation of order $n = 4$:
$$y_{n+1} = y_n + h\,\Phi_4(x_n, y_n; h) = y_n + h \sum_{k=0}^{3} \frac{h^k}{(k+1)!} \left(\frac{\partial}{\partial x} + f\,\frac{\partial}{\partial y} \right)^k f(t, y) \Big|_{\substack{x=x_n \\ y=y_n}},$$
where h is step length and $x_{n+1} = x_n + h$ $(n = 0, 1, \ldots)$ are the mesh points.
The Richardson improvement method can be used in conjunction with Taylor's method. Let the interval of interest be $[0, b]$, where $b = 1$ is the right end. If Taylor's method of order $n = 4$ is used with step size h, then $y(b) \approx y_h + C\,h^4$ for some constant C. If Taylor's method of order $n = 4$ is used with step size $2h$, then $y(b) \approx y_{2h} + 16C\,h^4$. The terms involving $C\,h^4$ can be eliminated to obtain an improved approximation for $y(b)$:
$$y(b) \approx \frac{16\,y_h - y_{2h}}{15}.$$
Use this improvement scheme to obtain better approximation to $y(1)$ with $h = 1/2$ and $h = 1/4$, where $y(x)$ is the solution of the initial value problem
$$y' = 9x + 3y \qquad \text{with} \qquad y(0) = 1.$$

Section 3.4 of Chapter 3 (Review)

1. For the initial value problem $y' = 2x - y$, $y(0) = 0$,
   - (a) find the actual solution $y = \phi(x)$;
   - (b) using the Maclaurin expansion of the exponential function $e^z = 1 + z + \frac{z^2}{2!} + \frac{z^3}{3!} + \frac{z^4}{4!} + \cdots$, expand $\phi(x)$ in a power series with respect to x;
   - (c) for a uniform grid with step size h, verify that the Euler method and the backward Euler method have the local truncation error $O(h^2)$, and the Heun formula, the modified Euler method, and the trapezoid rule all have the local truncation error $O(h^3)$.
2. Numerically estimate the step size that is needed for the Euler method (3.2.5) to make the local truncation error less than 0.0025 at the first step.
   (a) $\dot{y} = 2t^2 - 3y^2$, $y(0) = 1$; **(b)** $\dot{y} = t - 2\sqrt{y}$, $y(0) = 1$;
   (c) $\dot{y} = \sqrt{t + y^2}$, $y(0) = 1$; **(d)** $\dot{y} = t - e^{-ty}$, $y(0) = 1$.
3. For each of the following problems, estimate the local truncation error, $T_n(h)$, for the Euler method in terms of the reference solution $y = \phi(t)$. Obtain a bound for $T_n(h)$ in terms of t and $\phi(t)$ that is valid on the closed interval $0 \leqslant t \leqslant 1$. Find the actual solution and obtain a more accurate error bound for $T_n(h)$. Compute a bound for the error $T_4(h)$ in the fourth step for $h = 0.1$ and compare it with the true error.
   (a) $\dot{y} = 1 - 2y$, $y(0) = 2$; **(b)** $2\dot{y} = t + y$, $y(0) = 1$; **(c)** $\dot{y} = t^2 - y$, $y(0) = 0$.

4. Redo Exercise 3 using the Heun formula (3.2.15) instead of the Euler method (3.2.5).
5. Estimate the largest stepsize for which Euler's rule (3.2.5) remains stable on the interval $[0, 1]$ for the initial value problem: $y' + 3x^2 y = x^2$, $y(0) = 2$.
6. Some efficient algorithms for integration of initial value problems may produce instability. When such unstable schemes are employed, we may observe numerical solutions that are qualitatively different from the true solutions. Such solutions are called "ghost solutions."

   Consider the initial value problem for the logistic equation

$$\dot{y} = y\,(1 - y)\,, \qquad y(0) = y_0,$$

   where $y_0 = 0.5$ is given. In order to integrate the logistic equation, employ the central difference scheme

$$\frac{y_{n+1} - y_{n-1}}{2h} = y_n\,(1 - y_n)$$

   with initial conditions y_0 (given exactly to be 0.5) and $y_1 = y_0 + hy_0\,(1 - y_0)$, computed by Euler's algorithm.

   Compute a numerical solution by the central difference scheme using fixed time-mesh length $h = 0.1$ and make at least 500 iterations. Then plotting the results, observe a ghost solution.

Section 3.5 of Chapter 3 (Review)

1. For the given initial value problems, use the classical Runge–Kutta method (3.5.10) with $h = 0.1$ to obtain a four-decimal-place approximation to $y(0.5)$, the value of the unknown function at $x = 0.5$. Compare your result with the actual solution.

   (a) $y' = (y + 4x - 1)^2$, $y(0) = 1$. **(b)** $y' = 3y + 27x^2$, $y(0) = 0$.
   (c) $y' = y^{2/3}$, $y(0) = 1$. **(d)** $y' = x/y$, $y(0) = 1$.
2. Consider the following initial value problems.

   (a) $y' = y + 2x^2$, $y(1) = 2$. **(b)** $y' = (x^2 + y^2)/(x - y)$, $y(3) = 1$.
   (c) $y' = 3y - x + 5$, $y(1) = 2$. **(d)** $y' = 4xy/(x^2 - y^2)$, $y(1) = 3$.
   (e) $y' = x^2 + \sqrt{y}$, $y(0) = 4$. **(f)** $y' = (xy + 4y^2)/(y^2 + 4)$, $y(0) = 1$.
   (g) $y' = 2y + x^2$, $y(0) = 0$. **(h)** $y' = 2y - 5x + 3$, $y(3) = 4$.

   Using the classical Runge–Kutta method (3.5.10) with step size $h = 0.1$, compute approximations at 2 units from the initial point.
3. Find the actual solutions to the following Bernoulli equations subject to specified initial conditions.

   (a) $2y' = y - x/y$, $y(1) = 2$. **(b)** $y' = 3y - 3x\,y^3$, $y(0) = 1$.
   (c) $y' + 2xy + 2xy^4 = 0$, $y(0) = 1$. **(d)** $xy' + y = x^2y^2$, $y(1) = 1/3$.
   (e) $y' + y\tan x = y^3/2$, $y(0) = 1$. **(f)** $xy' - y = x^2y^2$, $y(1) = 1/9$.
   (g) $y' + 2xy = 2xy^{-3}$, $y(0) = 2$. **(h)** $xy' + y = x^2y^2\ln x$, $y(1) = 1/2$.

   Compare the true values evaluated at two units away from the initial position with approximations found with the aid of the following Runge–Kutta methods with fixed step size $h = 0.1$.

   (a) The fifth order Butcher's scheme (3.5.19).
   (b) The second order implicit method (3.5.18).
   (c) The fourth order classical method (3.5.10).
   (d) The fourth order Gill's method (3.5.12).
   (e) The fourth order Kutta method (3.5.11).
4. Consider the first-order integro-differential equation subject to the initial condition:

$$y'(x) = 2.3\,y - 0.01\,y^2 - 0.1\,y\int_0^x y(x)\,\mathrm{d}x, \qquad y(0) = 50.$$

   Use the Heun method with $h = 0.001$ over the interval $[0, 2.0]$, and the trapezoidal rule to find an approximate solution to the problem.

Chapter 4

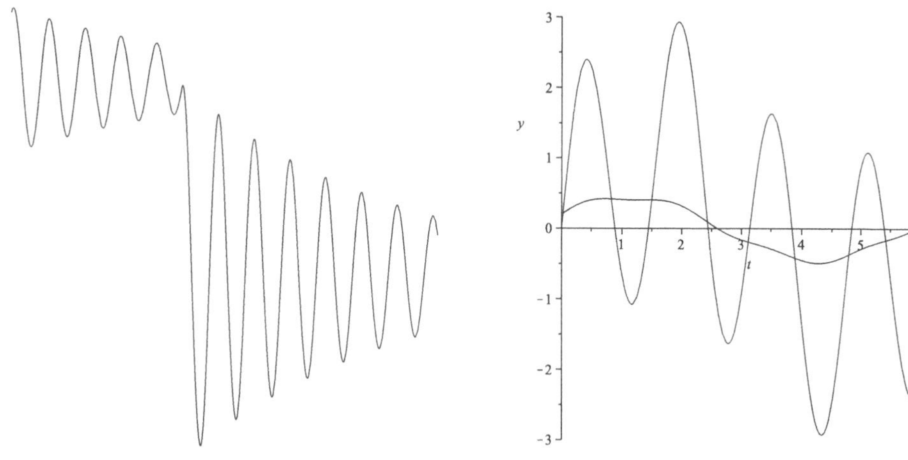

Response for square pulse **Filter output for periodic input**

Second and Higher Order Linear Differential Equations

Ordinary differential equations (ODEs) may be divided into two classes: linear equations and nonlinear equations. The latter have a richer mathematical structure than linear equations and are generally more difficult to solve in closed form. As we saw in Chapters 1 and 2, first order differential equations have an unusual combination of features: they are important in applications and some of them can be solved implicitly or even explicitly. Unfortunately, the techniques applicable for solving second order (and higher order) *nonlinear* ODEs are not available at an undergraduate level. Therefore, we consider only *linear* equations.

This chapter is an introduction to the elegant theory of second and higher order linear differential equations. There are two main reasons for concentrating on equations of the second order. First, they have important applications in mechanics, electric circuit theory, and physics. First and second derivatives have well-understood interpretations: velocity for order one and acceleration for order two. On the other hand, derivatives of order more than two are usually not easy to interpret conceptually. Second, their theory resembles that of linear differential equations of any order, using a constant interplay of ideas from calculus and analysis. Therefore, we often formulate the definitions and theorems only for second order differential equations, leaving generalizations for higher order differential equations to the reader. Furthermore, a substantial part of the theory of higher order linear differential equations is understandable at a fairly elementary mathematical level.

The text is very selective in presenting general statements about linear n-th order differential equations because it has a different goal. The theorems are usually proved only for second order equations. For complete proofs and deeper knowledge about this subject, the curious reader is advised to consult more advanced books (see, for instance, [9]).

4.1 Second and Higher Order Differential Equations

Recall from calculus that derivatives of functions $u(x)$ and $y(t)$ are denoted as $u'(x)$ or $\mathrm{d}u/\mathrm{d}x$ and $y'(t) = \mathrm{d}y/\mathrm{d}t$ or $\dot{y}$, respectively. Newton's dot notation ($\dot{y}$) is usually used to represent the derivative with respect to time. The notation x or t stands for the independent variable and will be widely used in this chapter. Higher order derivatives have similar notation; for example, f'' or $\mathrm{d}^2 f/\mathrm{d}x^2$ denotes the second derivatives.

A second order differential equation in the normal form is as follows:

$$\frac{\mathrm{d}^2 y}{\mathrm{d}x^2} = F\left(x, y, \frac{\mathrm{d}y}{\mathrm{d}x}\right) \quad \text{or} \quad y'' = F\left(x, y, y'\right), \tag{4.1.1}$$

where $F(x, y, p)$ is some given function of three variables. If the function $F(x, y, p)$ is linear in variables y and p (that is, $F(x, ay_1 + by_2, p) = a\,F(x, y_1, p) + b\,F(x, y_2, p)$ for any constants a, b, and similar for variable p), then Eq. (4.1.1) is called linear. For example, the equation $y'' = \cos x + 5y^2 + 2\,(y')^2$ is a second order nonlinear differential equation, while the equation $y'' = (\cos x)\,y$ is a linear one.

A function $y = \phi(x)$ is a solution of (4.1.1) in some interval $a < x < b$ (perhaps infinite), having derivatives up to the second order throughout the interval, if $\phi(x)$ satisfies the differential equation identically in the interval (a, b), that is,

$$\frac{\mathrm{d}^2\phi}{\mathrm{d}x^2} = F\left(x, \phi(x), \frac{\mathrm{d}\phi}{\mathrm{d}x}\right) \qquad \text{for all } x \in (a, b).$$

For many of the differential equations to be considered, it will be found that solutions of Eq. (4.1.1) can be included in one formula, either explicit

$$y = \phi(x, C_1, C_2)$$

or implicit

$$\Phi(x, y, C_1, C_2) = 0,$$

where C_1 and C_2 are arbitrary constants. Such a solution is referred to as the **general solution** of the differential equation of the second order in either explicit or implicit form. Choosing specific values of the constants C_1 and C_2, we obtain a particular solution of Eq. (4.1.1). All solutions can be so found except, possibly, singular and/or equilibrium solutions.

Second order differential equations are widely used in science and engineering to model real world problems. The most famous second order differential equation is Newton's second law of motion, $m\,\ddot{y} = F\,(t, y, \dot{y})$, which describes a one-dimensional motion of a particle of mass m moving under the influence of a force F. In this equation, $y = y(t)$ is the position of a particle at a time t, $\dot{y} = \mathrm{d}y/\mathrm{d}t$ is its velocity, $\ddot{y} = \mathrm{d}^2 y/\mathrm{d}t^2$ is its acceleration, and F is the total force acting on the particle.

For given two numbers y_0 and y_1, we impose the **initial conditions** on $y(x)$ in the form

$$y(x_0) = y_0, \quad y'(x_0) = y_1. \tag{4.1.2}$$

The differential equation (4.1.1) with the initial conditions (4.1.2) is called the **initial value problem** (IVP for short) or the Cauchy problem. Now we can state the following theorem, which is a direct extension of Picard's Theorem 1.3, page 23, for first order equations.

Theorem 4.1: Suppose that F, $\partial F/\partial y$, and $\partial F/\partial y'$ are continuous in a closed 3-dimensional domain Ω in xyy'-space, and the point (x_0, y_0, y_0') belongs to Ω. Then the initial value problem (4.1.1), (4.1.2) has a unique solution $y = \phi(x)$ on an x-interval in Ω containing x_0.

The general **linear differential equation of the second order** is an equation that can be written as

$$a_2(x)\frac{\mathrm{d}^2 y}{\mathrm{d}x^2} + a_1(x)\frac{\mathrm{d}y}{\mathrm{d}x} + a_0(x)y(x) = g(x). \tag{4.1.3}$$

The given functions $a_0(x), a_1(x), a_2(x)$, and $g(x)$ are independent of variable y. When Eq. (4.1.1) cannot be written in the form (4.1.3), then it is called **nonlinear**. If the function $a_2(x)$ in Eq. (4.1.3) has no zeroes on some interval, then we can divide both sides of the equation by $a_2(x)$ to obtain its normalized form:

$$y''(x) + p(x)\,y'(x) + q(x)\,y(x) = f(x). \tag{4.1.4}$$

The points where the coefficients of Eq. (4.1.4) are discontinuous or undefined are called the **singular** points of the equation. These points are usually not used in the initial conditions except some cases (see Exercise 10). For example, the equation $(x^2 - 1)\, y'' + y = 1$ has two singular points $x = 1$ and $x = -1$ that must be excluded from consideration. If in opposite, the initial condition $y(1) = y_0$ is imposed, then the differential equation dictates that $y_0 = 1$; otherwise, it has no solution.

Theorem 4.2: Let $p(x)$, $q(x)$, and $f(x)$ be continuous functions on an open interval $a < x < b$. Then, for each $x_0 \in (a, b)$, the initial value problem

$$y'' + p(x)y' + q(x)y = f(x), \qquad y(x_0) = y_0, \quad y'(x_0) = y_1$$

has a unique solution for arbitrary specified real numbers y_0, y_1.

PROOF: Since every linear second order differential equation can be reduced to the first order (nonlinear) Riccati equation (see Problem 17), Theorem 4.2 follows from Picard's theorem (page 23). ■

Equation (4.1.3) is a particular case of the general linear differential equation of the n-th order

$$a_n(x)\frac{\mathrm{d}^n y}{\mathrm{d}x^n} + a_{n-1}(x)\frac{\mathrm{d}^{n-1} y}{\mathrm{d}x^{n-1}} + \cdots + a_0(x)y(x) = g(x). \tag{4.1.5}$$

If $g(x)$ is identically zero, equations (4.1.3) or (4.1.5) are said to be **homogeneous** (the accent is on the syllable "ge"); if $g(x)$ is not identically zero, equations (4.1.3) and (4.1.5) are called **nonhomogeneous** (or inhomogeneous or driven) and the function $g(x)$ is referred to as the *nonhomogeneous term*, which is variously also called the *input function* or *forcing function*. The functions $a_0(x), a_1(x), \ldots, a_n(x)$ are called the *coefficients* of differential equation (4.1.5). If all the coefficients of Eq. (4.1.5) are constants, then we speak of this equation as a linear, n-th order differential equation with *constant coefficients*.

The *existence and uniqueness theorem* for the initial value problem for linear differential equations of the form (4.1.5) is valid. This theorem, which we state below, guarantees the existence of only one solution for the initial value problem. Also, Eq. (4.1.5) does not have a singular solution (i.e., a solution not obtained from the general solution). Therefore, the initial value problem for a linear equation always has a unique solution.

Theorem 4.3: Let functions $a_0(x)$, $a_1(x)$, ..., $a_n(x)$ and $f(x)$ be defined and continuous on the closed interval $a \leqslant x \leqslant b$ with $a_n(x) \ne 0$ for $x \in [a, b]$. Let x_0 be such that $a \leqslant x_0 \leqslant b$ and let $y_0, y_0', \ldots, y_0^{(n-1)}$ be any constant. Then in the closed interval $[a, b]$, there exists a unique solution $y(x)$ satisfying the initial value problem:

$$a_n(x)y^{(n)} + a_{n-1}(x)y^{(n-1)} + \cdots + a_1(x)y' + a_0(x)y = g(x),$$

$$y(x_0) = y_0, \; y'(x_0) = y_0', \; \ldots, \; y^{(n-1)}(x_0) = y_0^{(n-1)}.$$

Is a similar theorem valid for boundary value problems? No, as shown in §10.1, where a boundary value problem can have many, one, or no solutions.

According to Theorem 4.2, any initial value problem for the second order differential equation (4.1.4) has a unique solution in some interval $|a, b|$ (where $a < b$). In particular, if a solution and its derivative vanish at some point in the interval $|a, b|$, then such a solution is identically zero. Therefore, if a nontrivial (not identically zero) solution $y(x)$ is zero at some point $x_0 \in |a, b|$, then $y'(x_0) \ne 0$, and hence the solution changes its sign at x_0 when it goes through it (which means that the graph of $y(x)$ crosses the abscissa at x_0).

Example 4.1.1: Let us consider the initial value problem

$$x(x^2 - 4)\, y'' + (x + 2)\, y' + x^2\, y = \sin x, \qquad y(x_0) = y_0, \quad y'(x_0) = y_0'.$$

To determine the interval of existence, we divide both sides of the differential equation by $x(x^2 - 4) = x(x-2)(x+2)$ to obtain $y'' + p(x)y' + q(x)y = f(x)$ or

$$y'' + \frac{1}{x(x-2)}\, y' + \frac{x}{(x-2)(x+2)}\, y = \frac{\sin x}{x} \cdot \frac{1}{(x-2)(x+2)}.$$

The coefficient $p(x) = 1/x(x-2)$ is not defined at two singular points $x = 0$ and $x = 2$. Similarly, the functions $q(x) = x/(x^2-4)$ and $f(x) = \frac{\sin x}{x} \cdot \frac{1}{(x-2)(x+2)}$ fail to be continuous at singular points $x = \pm 2$. So we don't want to choose the initial point x_0 as 0 or ± 2. For example, if $x_0 = 1$, the given initial value problem has a unique solution in the open interval $(0, 2)$. If $x_0 = 3$, the given initial value problem has a unique solution in the interval $2 < x < \infty$. The behavior of a solution at these singularities requires additional analysis.

4.1.1 Linear Operators

With a function $y = y(x)$ having two derivatives, we associate another function, which we denote $(Ly)(x)$ (or $L[y]$ or simply Ly), by the relation

$$(Ly)(x) = a_2(x)\, y''(x) + a_1(x) y'(x) + a_0(x) y(x), \tag{4.1.6}$$

where $a_2(x)$, $a_1(x)$, and $a_0(x)$ are given functions, and $a_2(x) \neq 0$. In mathematical terminology, L is an **operator**[38] that operates on functions; that is, there is a prescribed recipe for associating with each function $y(x)$ a new function $(Ly)(x)$. An operator L assigns to each function $y(x)$ having two derivatives a new function called $(Ly)(x)$. Therefore, the concept of an operator coincides with the concept of a "function of a function."

> **Definition 4.1:** By an **operator** we mean a transformation that maps a function into another function. A linear operator L is an operator such that $L[af + bg] = aLf + bLg$ for any functions f, g and any constants a, b.

In other words, a linear operator is an operator that satisfies the following two properties:
Property 1: $L[cy] = cL[y]$, for any constant c.
Property 2: $L[y_1 + y_2] = L[y_1] + L[y_2]$.
All other operators are nonlinear. another function. Analysis of operators, their properties, and corresponding techniques are called **operator methods**.

Example 4.1.2: The operator $L[y] = y'' + x^2 y$ is a linear operator. If $y(x) = \sin x$, then

$$(Ly)(x) = (\sin x)'' + x^2 \sin x = (x^2 - 1)\sin x,$$

and if $y(x) = x^4$, then

$$(Ly)(x) = (x^4)'' + x^2 x^4 = 12x^2 + x^6.$$

Example 4.1.3: An operator

$$Ly = y^2 + y''$$

is a nonlinear differential operator because for any constant c we have

$$L[cy] = (cy)^2 + cy'' = c^2 y^2 + cy'' \neq cLy = c(y^2 + y''). \qquad \square$$

Differentiation gives us an example of a linear operator. Let $\texttt{D}$ denote differentiation with respect to the independent variable (x in our case), that is,

$$\texttt{D}y = y' = \frac{\mathrm{d}y}{\mathrm{d}x} \quad \left(\text{so } \texttt{D} = \frac{\mathrm{d}}{\mathrm{d}x} \right).$$

Sometimes we may use a subscript $(\texttt{D}_x)$ to emphasize the differentiation with respect to x. Then $\texttt{D}$ is a linear operator transforming a function $y(x)$ (assumed differentiable) into its derivative $y'(x)$. For example,

$$\texttt{D}(x^3) = 3x^2, \quad \texttt{D}e^{2x} = 2e^{2x}, \quad \texttt{D}(\cos x) = -\sin x.$$

Applying $\texttt{D}$ twice, we obtain the second derivative $\texttt{D}(\texttt{D}y) = \texttt{D}y' = y''$. We simply write $\texttt{D}(\texttt{D}y) = \texttt{D}^2 y$, so that

$$\texttt{D}y = y', \quad \texttt{D}^2 y = y'', \quad \texttt{D}^3 y = y''', \cdots \quad \left(\texttt{D}^2 = \frac{\mathrm{d}^2}{\mathrm{d}x^2}, \ \texttt{D}^3 = \frac{\mathrm{d}^3}{\mathrm{d}x^3}, \ldots \right).$$

Note that $\texttt{D}^0$ is the identity operator (which we omit). With this in mind, the following definition is natural.

[38]The word "operator" was introduced in mathematics by the famous Polish mathematician Stefan Banach (1892–1945) who published in 1932 the first textbook *Théorie des opérations linéaires* (Theory of Linear Operations) on operator theory.

Definition 4.2: The expression

$$L[x,\mathtt{D}] = a_n(x)\mathtt{D}^n + a_{n-1}(x)\mathtt{D}^{n-1} + \cdots + a_1(x)\mathtt{D} + a_0(x) \tag{4.1.7}$$

is called an n-th order linear differential operator, where $a_0(x)$, $a_1(x)$, ..., $a_n(x)$ are real-valued functions.

Then homogeneous and nonhomogeneous equations can be rewritten as $L[x,\mathtt{D}]y = 0$ and $L[x,\mathtt{D}]y = f$, respectively. Note that we write the operator in the form where the coefficient $a_k(x)$ precedes the derivative operator. For instance, $x^2\mathtt{D}$ operates on a function f as $x^2\mathtt{D}f = x^2 f'(x)$, while $\mathtt{D}x^2$ produces a different output: $\mathtt{D}x^2 f(x) = 2xf(x) + x^2 f'(x)$. If all coefficients in Eq. (4.1.7) are constants, we drop x and denote such an operator as $L[\mathtt{D}]$. It is a custom, which we follow throughout the book, to drop the identity operator $\mathtt{D}^0$ (that is, $\mathtt{D}^0 y(x) = y(x)$ for any function $y(x)$) in expressions containing the derivative operator $\mathtt{D}$.

Theorem 4.4: [Principle of Superposition for homogeneous equations] If each of the functions y_1, y_2, ..., y_m is a solution to the same linear homogeneous differential equation,

$$a_n(x)y^{(n)} + a_{n-1}(x)y^{(n-1)} + \cdots + a_1(x)y' + a_0(x)y = 0, \tag{4.1.8}$$

then for every choice of the constants c_1, c_2, ..., c_m the linear combination

$$y = c_1 y_1 + c_2 y_2 + \cdots + c_m y_m$$

is also a solution of Eq. (4.1.8).

PROOF: Since the left-hand side of this equation is a linear operator, we have

$$Ly = L[c_1y_1 + c_2y_2 + \cdots + c_my_m] = c_1Ly_1 + c_2Ly_2 + \cdots + c_mLy_m = 0$$

because $Ly_j = 0,\ j = 1, 2, \ldots, m$.

4.1.2 Exact Equations and Integrating Factors

We define a certain class of differential equations that can be reduced to lower order equations by simple integration.

Definition 4.3: An **exact** differential equation is a derivative of an equation of lower order. If this equation is written in the operator form $\mathtt{A}\,y = f$, then it is exact if and only if there exists a differential operator $\mathtt{B}$ such that $\mathtt{A} = \mathtt{D}\,\mathtt{B}$, where $\mathtt{D}$ stands for the derivative operator.

In particular, a linear exact second order differential equation $\frac{\mathrm{d}}{\mathrm{d}x}[P(x)y'] + \frac{\mathrm{d}}{\mathrm{d}x}[Q(x)y] = f(x)$, where $P(x)$ and $Q(x)$ are given continuously differentiable functions, can be integrated immediately:

$$P(x)\,y' + Q(x)y = \int f(x)\,\mathrm{d}x + C.$$

Since this equation is a linear first order differential equation, the function $y(x)$ can be determined in an explicit form (see §2.5).

Theorem 4.5: A linear differential equation $a_2(x)\,y'' + a_1(x)\,y' + a_0 y = f(x)$ is exact if and only if its coefficients satisfy the equation

$$a_1'(x) = a_2''(x) + a_0(x). \tag{4.1.9}$$

PROOF: For the equation $a_2\,y'' + a_1\,y' + a_0\,y = f(x)$ to be exact, there should exist functions $P(x)$ and $Q(x)$ such that

$$a_2(x) = P', \quad a_1 = P' + Q, \quad a_0 = Q'(x).$$

By eliminating Q, we obtain the necessary and sufficient condition (4.1.9). ■

We may try to reduce the differential equation $L[x, \mathtt{D}]y = f(x)$, where $L[x, \mathtt{D}] = a_2(x)\,\mathtt{D}^2 + a_1(x)\,\mathtt{D} + a_0$, to an exact one by multiplying it by an appropriate integrating factor $\mu(x)$:

$$\mu a_2 y'' + \mu a_1 y' + \mu a_0 y = \mu f.$$

Using the condition (4.1.9), we conclude that μ should satisfy the following equation

$$(\mu a_1)' = (\mu a_2)'' + \mu a_0.$$

Reducing parentheses, we see that μ is a solution of the so-called **adjoint equation**

$$L'\mu = 0, \qquad \text{where} \qquad L' = \mathtt{D}^2 a_2 - \mathtt{D} a_1 + a_0.$$

Example 4.1.4: Find the general solution of the equation

$$xy'' + (2 + x^2)y' + 3xy = 0.$$

Solution. An integrating factor $\mu = \mu(x)$ is a solution of the adjoint equation $x\mu'' - x^2\mu' + x\mu = 0$, which has a solution $\mu = x$. So, after multiplication by μ, we obtain an exact equation $\frac{\mathrm{d}}{\mathrm{d}x}\left[x^2 y' + x^3 y\right] = 0$, or after integration, we get $x^2 y' + x^3 y = C_1$. Division by x^2 yields $y' + xy = C_1/x^2$. Upon multiplication by an integrating factor, we get $\frac{\mathrm{d}}{\mathrm{d}x}\left[y\, e^{x^2/2}\right] = C_1 x^{-2}\, e^{x^2/2}$. Next integration gives us the general solution:

$$y\, e^{x^2/2} = C_1 \int x^{-2}\, e^{x^2/2}\, \mathrm{d}x + C_2\,. \qquad \square$$

The definition of exactness can be extended for nonlinear equations in a more general form:

$$\left[\sum_{i=0}^{n} a_i(x,y)(y')^i\right] y'' + \sum_{j=0}^{m} b_j(x,y)(y')^j = f(x). \tag{4.1.10}$$

For the first set of terms, we use the relation $uv' = (uv)' - vu'$. Letting $u = a_i$ and $v' = (y')^i y'' = \left(\frac{1}{i+1}\,(y')^{i+1}\right)'$, we reduce the above equation to the following:

$$\left(\sum_{i=0}^{n} \frac{a_i(x,y)(y')^{i+1}}{i+1}\right)' - \sum_{i=0}^{n} \frac{(y')^{i+1}}{i+1}\left[\frac{\partial a_i}{\partial x} + \frac{\partial a_i}{\partial y}\, y'\right] + \sum_{j=0}^{m} b_j(x,y)(y')^j = f(x).$$

Since the first term is exact, the remainder must also be exact; but the first order exact equation is necessarily linear in y'. Hence, the terms of higher degree in y' must vanish. Collecting terms of order 0 and 1 and equating them to zero, we get $b_0 + b_1 y' - \frac{\partial a_0}{\partial x}\, y' = 0$. Using the exactness criterion (2.3.5), page 74, we obtain

$$\frac{\partial}{\partial x}\left(b_1 - \frac{\partial a_0}{\partial x}\right) = \frac{\partial b_0}{\partial y}. \tag{4.1.11}$$

The necessary and sufficient conditions of exactness for Eq. (4.1.10) consist of the exactness condition (4.1.11) and the conditions for the vanishing of higher powers:

$$b_k = \frac{1}{k}\frac{\partial a_{k-1}}{\partial x} + \frac{1}{k-1}\frac{\partial a_{k-2}}{\partial y}, \qquad 2 \leqslant k \leqslant \max(m, n+2), \tag{4.1.12}$$

where we let $a_i = 0$ and $b_j = 0$ for $i > n$ and $j > m$, respectively.

Example 4.1.5: Let us consider the equation

$$(x^2 y + 2x^3 y')y'' + y^2 + 4xyy' + 4x^2(y')^2 = 2x,$$

which is of the form (4.1.10). Here $a_0 = x^2 y$, $a_1 = 2x^3$, $n = 1$ and $b_0 = y^2$, $b_1 = 4xy$, $b_2 = 4x^2$, $m = 2$. It is easily checked that conditions (4.1.11), (4.1.12) are satisfied. We have $\int (x^2 y)\, y''\, \mathrm{d}x = x^2 yy' - \int (x^2 y' + 2xy) y'\, \mathrm{d}x$, $\int (x^3)(2y'y''\, \mathrm{d}x) = x^3 (y')^2 - 3\int x^2 (y')^2\, \mathrm{d}x$, and $\int y^2\, \mathrm{d}x = y^2 x - \int 2yy'x\, \mathrm{d}x$ so that, upon multiplying by $\mathrm{d}x$ and integrating, the given equation is reduced to

$$x^2 yy' + x^3 (y')^2 + xy^2 = \int 2x\, \mathrm{d}x = x^2 + C. \qquad \square$$

Let us consider the general second order differential equation in normal form:

$$R(x,y,y')\,y'' + S(x,y,y') = 0.$$

Introducing a new variable $p = y' = \mathrm{d}y/\mathrm{d}x$, we rewrite the given equation in differential form:

$$R(x,y,p)\,\mathrm{d}p + S(x,y,p)\,\mathrm{d}x = 0. \tag{4.1.13}$$

Treating x, y as constants, we define the quadrature

$$\int R(x,y,p)\,\mathrm{d}p = \varphi(x,y,p).$$

Then, using the equation $\mathrm{d}\varphi(x,y,p) = R(x,y,p)\,\mathrm{d}p + \varphi_x(x,y,p)\,\mathrm{d}x + \varphi_y(x,y,p)\,\mathrm{d}y$, we replace $R(x,y,p)\,\mathrm{d}p = \mathrm{d}\varphi - \varphi_x\,\mathrm{d}x - \varphi_y\,\mathrm{d}y$ in Eq. (4.1.13) to obtain

$$\mathrm{d}\varphi(x,y,p) + [S(x,y,p) - \varphi_x]\,\mathrm{d}x - \varphi_y\,\mathrm{d}y = 0. \tag{4.1.14}$$

This equation is exact if and only if the first order differential equation

$$[S(x,y,p) - \varphi_x]\,\mathrm{d}x - \varphi_y\,\mathrm{d}y = 0 \tag{4.1.15}$$

is exact. If this is the case, there exists a potential function $\psi(x,y)$ such that $\mathrm{d}\psi = [S(x,y,p)\ -\varphi_x]\,\mathrm{d}x - \varphi_y\,\mathrm{d}y$. Therefore, the first integral of Eq. (4.1.14) has the form

$$\varphi(x,y,p) + \psi(x,y) = C_1 \qquad \text{with} \quad p = y', \tag{4.1.16}$$

which is a first order differential equation. So we reduced the second order nonlinear differential equation (4.1.13) to a first order equation, which may be solved using one of the known methods (see Chapter 2).

Example 4.1.6: Integrate the nonlinear differential equation

$$(3y^2 - 1)\,y'' + 6y(y')^2 + y' - 3y^2y' + 1 - x^2 = 0.$$

Solution. In Eq. (4.1.13), we have $R(x,y,p) = 3y^2 - 1$ and $\varphi(x,y,p) = (3y^2-1)p$, and hence Eq. (4.1.15) becomes

$$\left[6yp^2 + p - 3y^2p + 1 - 3x^2\right]\mathrm{d}x - 6yp\,\mathrm{d}y = 0.$$

Replacing p by $\mathrm{d}y/\mathrm{d}x$, we obtain the equation

$$(1 - 3y^2)\,\mathrm{d}y + (1 - 3x^2)\,\mathrm{d}x = 0,$$

which is exact and has the potential function $\psi(x,y) = y - y^3 + x - x^3$. From Eq. (4.1.16), we get the first order equation

$$(3y^2 - 1)\,y' + y - y^3 + x - x^3 = C_1\,.$$

The change of variable $u = y^3 - y$ transforms the left-hand side into $u' - u + x - x^3$, which is linear. The general integral of the given equation is found to be

$$y^3 - y = C_2\,e^{-x} - C_1 - 5 - 5x - 3x^2 - x^3.$$

4.1.3 Change of Variables

We introduce some transformations that reduce the homogeneous second order equation

$$y'' + p(x)y' + q(x)y = 0 \tag{4.1.17}$$

to other canonical forms. Let us start with the Bernoulli substitution $y = \varphi(x)v(x)$, where $\varphi(x)$ is some known nonvanishing function. The transformed equation for v is then

$$\varphi(x)v'' + (2\varphi'(x) + p\varphi)v' + (\varphi'' + p\varphi' + q\varphi)v = 0.$$

This new equation may be integrated for some choice of $\varphi(x)$. However, there is no rule for such a fortunate choice. The substitution

$$y = v(x)\, \exp\left\{-\frac{1}{2}\int_{x_0}^{x} p(t)\,\mathrm{d}t\right\} \tag{4.1.18}$$

reduces Eq. (4.1.17) to the form $v'' + \left(q - p^2/4 - p'/2\right) v = 0$, which does not contain the first derivative of an unknown function. If $q - p^2/4 - p'/2$ is equal to a constant divided by x^2, the equation for v can be integrated explicitly. If this expression is a constant, then it will be shown how to integrate it in §4.5. For a second order constant coefficient differential equation $ay'' + by' + cy = 0$, substitution

$$y = e^{-bx/(2a)}\, v(x) \tag{4.1.19}$$

reduces the given equation to the following one:

$$v'' + \frac{1}{a}\left(c - \frac{b^2}{4a}\right) v = 0. \tag{4.1.20}$$

Theorem 4.6: The second order homogeneous linear differential equation

$$y'' - q(x)\, y = 0, \tag{4.1.21}$$

where $q(x)$ is a continuous function for $x \geqslant 0$ and $q(x) > q_0 > 0$, has a solution that is unbounded and another one that is bounded.

PROOF: The given equation (4.1.21) with positive initial conditions $y(0) = y_0 > 0$ and $y'(0) = y_1 > 0$ has the unbounded solution $y(x)$ that also has an unbounded derivative because its graph is concave up. Indeed, integration of Eq. (4.1.21) yields

$$y'(x) - y'(0) = \int_0^x q(s)\, y(s)\,\mathrm{d}s > q_0 \int_0^x y(s)\,\mathrm{d}s \qquad \Longrightarrow \qquad y'(x) > y_1 + q_0 \int_0^x y(s)\,\mathrm{d}s.$$

Therefore, $y'(x)$ is positive and the solution is growing. Taking the lower bound, we get

$$y'(x) > y_1 + x q_0 y_0 > y_1 \qquad \Longrightarrow \qquad y(x) > y_1 x + y_0.$$

Using this unbounded solution, we construct a new bounded solution

$$u(x) \stackrel{\text{def}}{=} Cy(x) - y(x)\int_0^x y(s)^{-2}\,\mathrm{d}s \qquad \left(C = \int_0^\infty \frac{\mathrm{d}t}{y^2(t)} \leqslant \int_0^\infty \frac{\mathrm{d}t}{(y_0 + t y_1)^2} = \frac{1}{y_0 y_1}\right).$$

Since C is finite, we have $u(0) = Cy(0) = Cy_0 > 0$ and $u(x)$ is always positive by our construction. At infinity, $u(x)$ tends to zero as the following limit shows, evaluated with the aid of l'Hôpital's rule,

$$\lim_{x\to\infty} u(x) = \lim_{x\to\infty} \left(\int_x^\infty \frac{\mathrm{d}s}{y^2(s)} \Big/ \frac{1}{y(x)}\right) = \lim_{x\to\infty} \frac{-y^{-2}(x)}{\left(-y'(x) y^{-2}(x)\right)} = \lim_{x\to\infty} \frac{1}{y'(x)} = 0.$$

With a little pencil pushing, it can be checked that $u(x)$ is also a solution of Eq. (4.1.21). ■

The substitution $y' = yw$ reduces the equation $y'' + Q(x)y = 0$ to the Riccati equation

$$w' + w^2 + Q(x) = 0$$

since

$$w' = \frac{\mathrm{d}}{\mathrm{d}x}\left(\frac{y'}{y}\right) = \frac{y''y - y'y'}{y^2} = \frac{-Q(x)y^2 - (yw)^2}{y^2} = -w^2 - Q(x).$$

In §2.6.2 we discussed the reduction of the Riccati equation to a second order linear differential equation. Now we consider the inverse procedure. Namely, the second order linear differential equation

$$a_2(x)y'' + a_1(x)y' + a_0(x)y = 0$$

has a solution

$$y(x) = \exp\left\{\int \sigma(x)v(x)\,\mathrm{d}x\right\}, \tag{4.1.22}$$

where $v(x)$ is any solution of the Riccati equation

$$v' + \frac{a_0(x)}{\sigma(x)a_2(x)} + \left[\frac{a_1(x)}{a_2(x)} + \frac{\sigma'(x)}{\sigma(x)}\right] v + \sigma(x)v^2 = 0 \tag{4.1.23}$$

on an interval upon which $a_2(x) \neq 0$ and $\sigma(x) \neq 0$.

Example 4.1.7: Solve the initial value problem:

$$y'' + 2y'/x + y = 0, \qquad y(1) = 1, \quad y'(1) = 0.$$

Solution. In this case, we have $p(x) = 2/x$, $q(x) = 1$, and $x_0 = 1$. Hence, substitution (4.1.18) becomes

$$y(x) = v(x)\,\exp\left\{-\int_1^x \frac{\mathrm{d}t}{t}\right\} = v(x)\,\exp\{-\ln x\} = \frac{v(x)}{x}.$$

Then its derivatives are

$$y' = \frac{v'}{x} - \frac{v}{x^2}, \quad y'' = \frac{v''}{x} - \frac{2v'}{x^2} + \frac{2v}{x^3}.$$

This yields the equation for v: $v'' + v = 0$. It is not hard to find its solution (consult §4.5): $v(x) = C_1\,\cos x + C_2\,\sin x$, where C_1 and C_2 are arbitrary constants. Thus, the general solution of the given equation becomes

$$y(x) = \frac{C_1\,\cos x + C_2\,\sin x}{x}.$$

After substituting this expression into the initial conditions, we get

$$y(x) = \frac{\cos x\,(\cos 1 - \sin 1) + \sin x\,(\cos 1 + \sin 1)}{x}.$$

Problems In all problems, $\mathtt{D}$ denotes the derivative operator, that is, $\mathtt{D}y = y'$; its powers are defined recursively: $\mathtt{D}^n = \mathtt{D}(\mathtt{D}^{n-1})$, $n = 1, 2, \ldots$, and $\mathtt{D}^0$ is the identical operator, which we usually drop.

1. Classify the differential equation as being either linear or nonlinear. Furthermore, classify the linear ones as being homogeneous or nonhomogeneous, with constant coefficients or with variable coefficients, and state the order.

   (a) $y'' + x^2 y = 0$; **(b)** $y''' + xy = \sin x$; **(c)** $y'' + yy' = 1$;
   (d) $y^{(5)} - y^{(4)} + y' = 2x^2 + 3$; **(e)** $y'' + yy^{(4)} = 1$; **(f)** $y''' + xy = \cosh x$;
   (g) $(\cos x)y' + y\,e^{x^2} = \sinh x$; **(h)** $y''' + xy = \cosh x$; **(i)** $y \cdot y' = 1$;
   (j) $(\sinh x)(y')^2 + 3y = 0$; **(k)** $5y' - xy = 0$; **(l)** $y'^2\,y^{1/2} = \sin x$;
   (m) $2y'' + 3y' + 4x^2 y = 1$; **(n)** $y''' - 1 = 0$; **(o)** $x^2\,y'' - y = \sin^2 x$.

2. For each of the following differential equations, state the order of the equation.

   (a) $y'' = x^2 + y$; **(b)** $y''' + xy'' - y^2 = \sin x$; **(c)** $(y')^2 + yy'xy' = \ln x$;
   (d) $\sin(y'') + yy^{(4)} = 1$; **(e)** $(\sinh x)(y')^2 + y'' = xy$; **(f)** $y \cdot y'' = 1$;
   (g) $y^{(5)} - xy = 0$; **(h)** $(y''')^2 + y^{1/2} = \sin x$; **(i)** $y'' + y^2 = x^2$.

3. Using the symbol $\mathtt{D} = \mathrm{d}/\mathrm{d}t$, rewrite the given differential equation in the operator form.

   (a) $y'' + 4y' + y = 0$; **(b)** $y''' - 5y'' + y' - y = 0$; **(c)** $2y'' - 3y' - 2y = 0$;
   (d) $3y^{(4)} - 2y'' + y' = 0$; **(e)** $y''' - (\sin x)\,y'' + y = x$; **(f)** $7y^{(4)} + 8y''' - 9y'' = 0$.

4. Write the differential equation corresponding to the given operator.

   (a) $\mathtt{D}^2 - 2\,\mathtt{D} + \pi^2 + 1$; **(b)** $(\mathtt{D} + 1)^2$; **(c)** $\mathtt{D}^2 + 3\,\mathtt{D} - 10$;
   (d) $4\,\mathtt{D}^4 - 8\,\mathtt{D}^3 - 7\,\mathtt{D} + 6$; **(e)** $3\,\mathtt{D}^2 + 2\,\mathtt{D} + 1$; **(f)** $(\mathtt{D} + 1)^3$.

5. In each of the following initial value problems, determine, without solving the problem, the longest interval in which the solution is certain to exist.

   (a) $(x - 3)y'' + (\ln x)y = x^2$, $y(1) = 1$, $y'(1) = 2$;

   (b) $y'' + (\tan x)y' + (\cot x)y = 0$, $y(\pi/4) = 1$, $y'(\pi/4) = 0$;

   (c) $(x^2 + 1)y'' + (x - 1)y' + y = 0$, $y(0) = 0$, $y'(0) = 1$;

(d) $xy'' + 2x^2y' + y\sin x = \sinh x, \quad y(0) = 1, \ y'(0) = 1;$

(e) $\sin x\, y'' + xy' + 7y = 1, \quad y(1) = 1, \ y'(1) = 0;$

(f) $y'' - (x-1)y' + x^2\, y = \tan x, \quad y(0) = 0, \ y'(0) = 1;$

(g) $(x^2 - 4)\, y'' + x\, y' + (x+2)\, y = 0, \quad y(0) = 1, \ y'(0) = -1.$

6. Evaluate (the symbol D stands for the derivative $\mathrm{d}/\mathrm{d}x$) the following expressions.
   (a) $(\mathtt{D}-2)(x^3+2x)$; (b) $(\mathtt{D}-1)(\mathtt{D}+2)(\sin 2x)$; (c) $(\mathtt{D}^3 - 5\,\mathtt{D}^2 + 11\,\mathtt{D} - 1)(x^2-2)$;
   (d) $(\mathtt{D}-x)(x^2-2x)$; (e) $(\mathtt{D}-2)(\mathtt{D}^2+4)\cos 2x$; (f) $(\mathtt{D}^3 - \mathtt{D}^2 + 2\,\mathtt{D})\sin 3x$.

7. A particle moves along the abscissa so that its instantaneous acceleration is given as a function of time t by $2 - 3t^2$. At times $t = 0$ and $t = 2$, the particle is located at $x = 0$ and $x = -10$, respectively. Set up a differential equation and associated conditions describing the motion. Is the problem an initial or boundary value problem? Determine the position and velocity of the particle at $t = 5$.

8. Let $y = Y_1(x)$ and $y = Y_2(x)$ be two solutions of the differential equation $y'' = x^2 + y^2$. Is $y = Y_1(x) + Y_2(x)$ also a solution?

9. (a) Show that $\dfrac{\mathrm{d}^2y}{\mathrm{d}x^2} = -\dfrac{\mathrm{d}^2x}{\mathrm{d}y^2}\left[\left(\dfrac{\mathrm{d}x}{\mathrm{d}y}\right)^3\right]^{-1}$. *Hint:* Differentiate both sides of $\mathrm{d}y/\mathrm{d}x = 1/(\mathrm{d}x/\mathrm{d}y)$ with respect to x.

   (b) Use the result in (a) to solve the differential equation $\frac{\mathrm{d}^2x}{\mathrm{d}y^2} + x\left(\frac{\mathrm{d}x}{\mathrm{d}y}\right)^3 = 0$.

10. Show that $y_1 = x$ and $y_2 = e^x$ are solutions of the linear differential equation $(x-1)y'' - xy' + y = 0$.

11. Find the general solution
    (a) $\dfrac{\mathrm{d}}{\mathrm{d}x}\left(x^3\dfrac{\mathrm{d}y}{\mathrm{d}x}\right) = 0$; (b) $\dfrac{1}{r}\dfrac{\mathrm{d}}{\mathrm{d}r}\left(r\dfrac{\mathrm{d}u}{\mathrm{d}r}\right) = 0$.

12. Replace r in the Bessel equation $\frac{\mathrm{d}^2u}{\mathrm{d}r^2} + \frac{1}{r}\frac{\mathrm{d}u}{\mathrm{d}r} + \nu^2\, u = 0$ by $r = e^t$.

13. Show that $x^3/9$ and $(x^{3/2}+1)^2/9$ are solutions of the nonlinear differential equation $(y')^2 - xy = 0$ on $(0,\infty)$. Is the sum of these functions a solution?

14. Show that a nontrivial (not identically zero) solution of the second order linear differential equation $y''+p(x)y'+q(x)y = 0$, $x \in |a,b|$, with continuous coefficients $p(x)$ and $q(x)$, cannot vanish at an infinite number of points on any subinterval $[\alpha,\beta] \subset |a,b|$.

15. Given two differential equations

$$(P(x)u')' + Q(x)u = f(x) \qquad \text{and} \qquad (P(x)v')' + Q(x)v = g(x).$$

Prove the Green formula

$$\int_{x_0}^{x} [g(\tau)u(\tau) - f(\tau)v(\tau)]\,\mathrm{d}\tau = [P(t)(u'(t)v'(t) - u'(t)v(t)]\Big|_{t=x_0}^{t=x}.$$

Hint: Derive the Lagrange identity: $[P(x)(uv' - u'v)]' = Q(x)u(x) - f(x)v(x)$.

16. Using the substitution $y = \varphi(x)u(x)$, transfer the differential equation $x^2\, y'' - 4x^2\, y' + (x^2+1)\, y = 0$ to the equation without u'.

17. Show that the change of the dependent variable $u = P(x)y'/y$ transfers the self-adjoint differential equation $(P(x)y')' + Q(x)y = 0$ into the Riccati equation $u' + \dfrac{u^2}{P(x)} + Q(x) = 0$.

18. Show that for any constant C, the function $y(x) = x\left(C + \dfrac{1}{x}\sqrt{\dfrac{25k}{6}}\right)^{2/5}$ is a solution to the nonlinear differential equation $y'' + k\,x\,y^{-4} = 0 \quad (k > 0)$.

19. Using the substitution $y' = uy$, show that any linear constant coefficient differential equation $ay'' + by' + cy = 0$ has an explicit solution.

20. For a given second order differential operator $L = a_1\mathtt{D}^2 + a_1\mathtt{D} + a_0$, the operator $L' = \mathtt{D}^2 a_1 - a_1\mathtt{D} + a_0$ is called the adjoint operator to L. Find an adjoint operator to the given one.

    (a) $x\mathtt{D}^2 + x^2\mathtt{D} + 1$; (b) $x^2\mathtt{D}^2 + x\mathtt{D} - 1$; (c) $\mathtt{D}^2 + 2x\mathtt{D}$; (d) $(1-x^2)\mathtt{D}^2 - 2x\mathtt{D} + 3$.

21. Find the adjoint of the given differential equation (where α is a real number and n is a positive integer).

(a) $y'' - y' - y = 0$, Fibonacci equation; (b) $\left(1 - x^2\right) y'' - 2x\, y' + n(n+1)\, y = 0$, Legendre's equation;
(c) $y'' - x\, y = 0$, Airy's equation; (d) $\left(x^2 - 4\alpha\right) y'' + x\, y' - n^2 y = 0$, Dickson's equation.

22. Solve the following linear exact equations.
(a) $y'' + 2xy' + 2y = 0$; (b) $xy'' + (\sin x)y' + (\cos x)y = 0$;
(c) $y'' + 2x^2y' + 4xy = 2x$; (d) $(1 - x^2)y'' + (1 - x)y' + y = 1 - 2x$;
(e) $y'' + 4xy' + (2 + 4x^2)y = 0$; (f) $x^2y'' + x^2y' + 2(1 - x)y = 0$;
(g) $y'' + x^2y' + 2xy = 2x$; (h) $y'' \ln(x^2 + 1) + \frac{4x}{1+x^2}\, y' + y\, \frac{1(1-x^2)}{(1+x^2)^2} = 0$;
(i) $xy'' + x^2y' + 2xy = 0$; (j) $y'' + (\sin x)y' + (\cos x)y = \cos x$;
(k) $y'' + y' \cot x - y \csc^2 x = \cos x$; (l) $y''x \ln x + 2y' - y/x = 1$.

23. Solve the following nonlinear exact equations.
(a) $xy'' + (6xy^2 + 1)y' + 2y^3 + 1 = 0$; (b) $(1 + y)x\, y'' + yy' - x(y')^2 + y' = (1 + y)^2 x \sin x$;
(c) $yy'' \sin x + [y' \sin x + y \cos x]y' = \cos x$; (d) $(x \cos y + \sin x)y'' - x(y')^2 \sin y + 2(\cos y + \cos x)y' = y \sin x$;
(e) $(1 - y)y'' - (y')^2 = 0$; (f) $(\cos y - y \sin y)y'' - (y')^2(2 \sin y + y \cos y) = \sin x$.

24. Find an integrating factor that reduces each given differential equation to an exact equation and use it to determine the equation's general solution.
(a) $y'' + \frac{2x}{2x-1}\, y' - \frac{4x}{(2x-1)^2}\, y = 0$; (b) $(2x + x^2)y'' + (10 + x + x^2)y' = (25 - 6x)y$;
(c) $y'' + \frac{y'}{1+x} - \frac{2+x}{x^2(1+x)}\, y = 0$; (d) $(x^2 - x)y'' + (2x^2 + 4x - 3)y' + 8xy = 1$;
(e) $\frac{x^2-1}{x}\, y'' + \frac{3x+1}{x}\, y' + \frac{y}{x} = 3x$; (f) $(2 \sin x - \cos x)y'' + (7 \sin x + 4 \cos x)y' + 10y \cos x = 0$;
(g) $y'' + \frac{x-1}{x}\, y' + \frac{y}{x^3} = \frac{1}{x^3}\, e^{-1/x}$; (h) $y'' + (2x + 5)y' + (4x + 8)y = e^{-2x}$.

25. Use the integrating factor method to reduce the second order (pendulum) equation $\ell\ddot{\varphi} = g \sin \varphi$ to the first order equation. *Note:* the reduced first order equation cannot be solved in elementary functions.

26. By substitution, reduce the coefficient of y in $y'' - x^n\, y = 0$ to negative unity.

27. Differential equations may have stationary solutions, also called **equilibrium solutions**. Find all equilibrium solutions of the given differential equations.

(a) $\ddot{y} + y = y^3$; (b) $\ddot{y} + 4y = 8$; (c) $\ddot{y} + 4\dot{y} = 0$; (d) $\ddot{y} + y^2 = 1$.

28. Formulate a differential equation governing the motion of an object with mass $m = 1$ kg that stretches a vertical spring 6 cm when attached and experiences a resistive forth whose magnitude is one-sixteenth of the object's speed.

29. Let u be a function of the variables x and y. Show that the operator A defined by

$$\mathrm{A}u = \int_0^1 \mathrm{d}\xi \int_0^1 \mathrm{d}\eta\, u(\xi, \eta)\, \sqrt{(x - \xi)^2 + (y - \eta)^2}$$

is a linear operator.

30. Derive a differential equation whose solution is a family of circles $(x - a)^2 + (y - b)^2 = 1$.

31. What differential equation does the general solution $(ax + b)y = C_1 + C_2\, x\, e^{ax/b}$ satisfy?

32. Show that the change of the dependent variable $y' = uy$ transfers the differential equation $y'' + p(x)\, y' + q(x)\, y = 0$ into the Riccati equation $u' + u^2 + p(x)u + q(x) = 0$.

33. Find a necessary condition for a differential operator $L = a_2\texttt{D}^2 + a_1\texttt{D} + a_0$ to be self-adjoint.

34. By changing the independent variable $x = x(t)$, rewrite the differential equation $y'' + p(x)y' + g(x)y = f(x)$ in new variables, y and t.

35. In the differential equation $y'' + p(x)y' + g(x)y = f(x)$, make a change of independent variable, $x = x(t)$, so that the equation in the new variable does not contain the first derivative.

36. Prove that the second order variable coefficient differential equation $y'' + p(x)\, y' + q(x)\, y = 0$ sometimes could be reduced to a constant coefficient equation by setting $t = \psi(x)$ if $\psi(x) = \int \sqrt{kq(x)}\, \mathrm{d}x$, where $k \neq 0$ is a constant. *Note* that this is only a necessary condition. Derive the sufficient condition on the coefficients $p(x)$ and $q(x)$ under which the transformation $t = \psi(x)$ reduces the given equation to a constant coefficient differential equation.

4.2 Linear Independence and Wronskians

For a finite set of functions $f_1(x)$, $f_2(x)$, ..., $f_m(x)$, its **linear combination** is defined as

$$\alpha_1 f_1(x) + \alpha_2 f_2(x) + \ldots + \alpha_m f_m(x),$$

where α_1, α_2, ..., α_m are some constants. We begin this section with a very important definition:

Definition 4.4: A set of m functions f_1, f_2, ..., f_m, each defined and continuous on the interval $|a, b|$, is said to be **linearly dependent** on $|a, b|$ if there exist constants $\alpha_1, \alpha_2, \ldots, \alpha_m$, not all of them zero, such that

$$\alpha_1 f_1(x) + \alpha_2 f_2(x) + \cdots + \alpha_m f_m(x) = 0, \qquad x \in |a, b|,$$

for every x in the interval $|a, b|$. Otherwise, the functions $f_1, f_2, \ldots, f_m$ are said to be **linearly independent** on this interval.

A set of functions f_1, f_2, ..., f_m is linearly dependent on an interval if at least one of these functions can be expressed as a linear combination of the remaining functions. Two functions $f_1(x)$ and $f_2(x)$ are linearly dependent if and only if there exists a constant k such that $f_1(x) = k\, f_2(x)$.

The interval on which functions are defined plays a crucial role in this definition. The set of functions can be linearly independent on some interval, but can become dependent on another one (see Example 4.2.3).

Example 4.2.1: The n $(n \geqslant 2)$ functions $f_1(x) = 1 = x^0$, $f_2(x) = x, \ldots, f_n(x) = x^{n-1}$ are linearly independent on the interval $(-\infty, \infty)$ (and on any interval). Indeed, the relation

$$\alpha_1 \cdot 1 + \alpha_2 x + \cdots + \alpha_n x^{n-1} \equiv 0$$

fails to be valid because a polynomial cannot be identically zero.

Example 4.2.2: Any two of the four functions

$$f_1(x) = e^x, \;\; f_2(x) = e^{-x}, \;\; f_3(x) = \sinh x, \;\; f_4(x) = \cosh x$$

are linearly independent, but any three of them are linearly dependent. The last statement follows from the formulas

$$\sinh x = \frac{1}{2}\left(e^x - e^{-x}\right), \quad \cosh x = \frac{1}{2}\left(e^x + e^{-x}\right).$$

The equation

$$\alpha_1 e^x + \alpha_2 e^{-x} = 0$$

with nonzero constants α_1 and α_2 cannot be true for all x because, after multiplying both sides by e^x, we obtain

$$\alpha_1 e^{2x} + \alpha_2 = 0.$$

The last equation is valid only for $x = \frac{1}{2}\ln\left(-\frac{\alpha_2}{\alpha_1}\right)$, but not for all $x \in (-\infty, \infty)$.

Example 4.2.3: Consider two functions $f_1(x) = x$ and $f_2(x) = |x|$. They are linearly independent on any interval containing zero, but they are linearly dependent on any interval $|a, b|$ when either $0 < a < b$ or $a < b < 0$.

Example 4.2.4: The functions $f_1(x) = \sin^2 x$, $f_2(x) = \cos^2 x$, and $f_3(x) = 1$ are linearly dependent on any finite interval. This follows from the identity

$$\sin^2 x + \cos^2 x - 1 \equiv 0. \qquad \square$$

Recall that a **matrix** is a rectangular array of objects or entries, written in rows and columns. The properties of square matrices and their determinants are given in §6.2 and §7.2.

Definition 4.5: Let $f_1, f_2, \ldots, f_m$ be m functions that together with their first $m-1$ derivatives are continuous on an interval $|a,b|$ $(a < b)$. The **Wronskian** or the **Wronskian determinant** of $f_1, f_2, \ldots, f_m$, evaluated at $x \in |a,b|$ is denoted by $W[f_1, f_2, \ldots, f_m](x)$ or $W(f_1, f_2, \ldots, f_m; x)$ or simply by $W(x)$ and is defined to be the determinant

$$W[f_1, f_2, \ldots, f_m](x) = \det \begin{bmatrix} f_1 & f_2 & \cdots & f_m \\ f_1' & f_2' & \cdots & f_m' \\ f_1'' & f_2'' & \cdots & f_m'' \\ \vdots & \vdots & \ddots & \vdots \\ f_1^{(m-1)} & f_2^{(m-1)} & \cdots & f_m^{(m-1)} \end{bmatrix}. \tag{4.2.1}$$

Each of the functions appearing in this determinant is to be evaluated at $x \in |a,b|$.

For the special case $m = 2$, the Wronskian[39] takes the form

$$W[f_1, f_2](x) = \begin{vmatrix} f_1(x) & f_2(x) \\ f_1'(x) & f_2'(x) \end{vmatrix} = f_1(x)\, f_2'(x) - f_1'(x)\, f_2(x).$$

In the practical evaluation of a Wronskian, the following lemma may be very helpful. We leave its proof for the reader (see Exercise 15).

Lemma 4.1: *For $g_0 = 1$ and arbitrary functions f, g_1, $\ldots$, g_{n-1}, the Wronskian determinants satisfy the equation*

$$\det \left[\frac{\mathrm{d}^j (f g_k)}{\mathrm{d}x^j} \right]_{j,k=0,1,\ldots,n-1} = f^n \det \left[\frac{\mathrm{d}^j g_k}{\mathrm{d}x^j} \right]_{j,k=0,1,\ldots,n-1}.$$

In particular,

$$W[f, fg] = \det \begin{bmatrix} f & fg \\ f' & f'g + fg' \end{bmatrix} = f^2 \begin{vmatrix} 1 & g \\ 0 & g' \end{vmatrix} = f^2\, g' = f^2\, W[1, g],$$

$$W[f, fg_1, fg_2] = \begin{vmatrix} f & fg_1 & fg_2 \\ f' & (fg_1)' & (fg_2)' \\ f'' & (fg_1)'' & (fg_2)'' \end{vmatrix} = f^3\, W[1, g_1, g_2].$$

Example 4.2.5: Let us find the Wronskian for the given functions: $f_1(x) = x$, $f_2(x) = x \cos x$, $f_3(x) = x \sin x$. From the definition, we have

$$W[f_1, f_2, f_3](x) = \det \begin{bmatrix} x & x\cos(x) & x\sin(x) \\ 1 & \cos(x) - x\sin(x) & \sin(x) + x\cos(x) \\ 0 & -2\sin(x) - x\cos(x) & 2\cos(x) - x\sin(x) \end{bmatrix} = x^3,$$

which could be verified after tedious calculations without a computer algebra system. On the other hand, using Lemma 4.1, we get

$$W[f_1, f_2, f_3](x) = x^3 \det \begin{bmatrix} 1 & \cos(x) & \sin(x) \\ 0 & -\sin(x) & \cos(x) \\ 0 & -\cos(x) & -\sin(x) \end{bmatrix} = x^3 \det \begin{bmatrix} -\sin(x) & \cos(x) \\ -\cos(x) & -\sin(x) \end{bmatrix} = x^3.$$

All computer algebra systems (*Maple*, *Mathematica*, *Maxima*, *Sage*, SymPy, and MuPad from MATLAB) have a dedicated command to calculate a Wronskian:

```
with(VectorCalculus): Wronskian([x,x*sin(x),x*cos(x)],x)  # Maple
Wronskian[{x,x Sin[x], x Cos[x]},x]               (* Mathematica *)
load(functs)$   wronskian([f1(x), f2(x) , f3(x)],x); /* Maxima, Sage, MuPad, and SymPy */
```

[39]Wronskian determinants are named after the Polish philosopher Jósef Maria Höené-Wronski (1776–1853). He was born Höené to a Czech emigrant, but in 1792 Jósef ran away from home and changed his name; he served in the Russian army (1795–1797) and later worked mostly in France. The term "Wronskian" was coined by the Scottish mathematician Thomas Muir (1844–1934) in 1882.

Theorem 4.7: Let $f_1, f_2, \ldots, f_m$ be m functions that together with their first $m-1$ derivatives are continuous on an open interval $a < x < b$. If their Wronskian $W[f_1, f_2, \ldots, f_m](x_0)$ is not equal to zero at some point $x_0 \in (a, b)$, then these functions $f_1, f_2, \ldots, f_m$ are linearly independent on (a, b). Alternatively, if $f_1, f_2, \ldots, f_m$ are linearly dependent and they have $m-1$ first derivatives on the open interval (a, b), then their Wronskian $W[f_1, f_2, \ldots, f_m](x) \equiv 0$ for every x in (a, b).

PROOF: We prove this theorem only for the case $m = 2$ by contradiction. Let the Wronskian of two functions f_1 and f_2 be not zero, and suppose the contrary is true, namely, functions $f_1(x)$ and $f_2(x)$ are linearly dependent. Then there exist two constants α_1 and α_2, at least one not equal to zero, such that

$$\alpha_1 f_1(x) + \alpha_2 f_2(x) = 0$$

for any $x \in (a, b)$. Evaluating the derivative at x_0, we obtain

$$\alpha_1 f_1'(x_0) + \alpha_2 f_2'(x_0) = 0.$$

Thus, we have obtained a linear system of algebraic equations with respect to α_1, α_2. The right-hand side of this system is zero. It is known that a homogeneous system of algebraic equations has nontrivial (that is, not identically zero) solutions if and only if the determinant of the corresponding matrix is zero. The determinant of coefficients of the last two equations with respect to α_1, α_2 is precisely $W[f_1, f_2](x_0)$, which is not zero by the hypothesis. Therefore, $\alpha_1 = \alpha_2 = 0$ and this contradiction proves that functions $f_1(x)$ and $f_2(x)$ are linearly independent on (a, b).

The second part of this theorem follows immediately from the first one. Thus, if functions $f_1(x)$ and $f_2(x)$ are linearly dependent on (a, b), then $f_2(x) = \beta f_1(x)$ for some constant β. In this case, the Wronskian becomes

$$\begin{aligned} W[f_1, f_2](x) &= \begin{vmatrix} f_1(x) & f_2(x) \\ f_1'(x) & f_2'(x) \end{vmatrix} = \begin{vmatrix} f_1(x) & \beta f_1(x) \\ f_1'(x) & \beta f_1'(x) \end{vmatrix} \\ &= \beta \begin{vmatrix} f_1(x) & f_1(x) \\ f_1'(x) & f_1'(x) \end{vmatrix} \equiv 0. \end{aligned}$$

Example 4.2.6: Show that the functions $f(x) = x^m$ and $g(x) = x^n$ $(n \neq m)$ are linearly independent on any interval (a, b) that does not contain zero.

Solution. The Wronskian $W(x)$ of these two functions,

$$W(x) = \begin{vmatrix} x^m & x^n \\ m x^{m-1} & n x^{n-1} \end{vmatrix} = (n-m) x^{m+n-1}, \quad n \neq m,$$

is not equal zero at any point $x \neq 0$. So, from Theorem 4.7, it follows that these functions are linearly independent. It is also clear that there does not exist a constant k such that $x^n = k\, x^m$. □

Theorem 4.7 gives us only the necessary condition of linear dependency, but not the sufficient one, as the following example shows: linearly independent functions may have an identically zero Wronskian!

Example 4.2.7: Consider two functions

$$f_1(x) = \begin{cases} x^2, & \text{for } x \geqslant 0, \\ 0, & \text{for } x \leqslant 0, \end{cases} \qquad f_2(x) = \begin{cases} 0, & \text{for } x \geqslant 0, \\ x^2, & \text{for } x \leqslant 0. \end{cases}$$

Then $W(x) \equiv 0$ for all x, but the functions are linearly independent. If this were not true we would have two nonzero constants α_1 and α_2 such that

$$\alpha_1 f_1(x) + \alpha_2 f_2(x) = \begin{cases} \alpha_1 x^2, & \text{for } x \geqslant 0, \\ \alpha_2 x^2, & \text{for } x \leqslant 0, \end{cases} \quad = \quad x^2 \begin{cases} \alpha_1, & \text{for } x \geqslant 0, \\ \alpha_2, & \text{for } x \leqslant 0, \end{cases} = 0.$$

Since this relation holds only when constants α_1 and α_2 are both zero, we get a contradiction and the claim about linear dependency of functions f_1 and f_2 is void.

Theorem 4.8: A finite set of linearly independent holomorphic (represented by convergent power series) functions has a nonzero Wronskian.

Theorem 4.9: Let $y_1, y_2, \ldots, y_n$ be the solutions of the n-th order differential homogeneous equation

$$y^{(n)} + a_{n-1}(x)y^{(n-1)} + \cdots + a_1(x)y' + a_0(x)y = 0$$

with continuous coefficients $a_0(x)$, $a_1(x)$, ..., $a_{n-1}(x)$, defined on an open interval $a < x < b$. Then functions $y_1, y_2, \ldots, y_n$ are linearly independent on this interval (a, b) if and only if their Wronskian never vanishes on (a, b). Alternatively, these functions $y_1, y_2, \ldots, y_n$ are linearly dependent on this interval (a, b) if and only if $W[y_1, y_2, \ldots, y_n](x)$ is zero for all $x \in (a, b)$.

PROOF: We prove the theorem only for the case $n = 2$. We wish to determine the sufficient conditions in order for the solutions y_1, y_2 to be linearly dependent. The necessary conditions were found in Theorem 4.7. These functions will be linearly dependent if we can find constants α_1 and α_2 to be not all zero, such that

$$\alpha_1 y_1(x) + \alpha_2 y_2(x) = 0.$$

Suppose the Wronskian $W[y_1, y_2](x)$ is zero at some point $x = x_0$. Then the system of algebraic equations

$$C_1 y_1(x_0) + C_2 y_2(x_0) = 0, \quad C_1 y_1'(x_0) + C_2 y_2'(x_0) = 0$$

has a nontrivial solution $C_1 = \alpha_1$ and $C_1 = \alpha_2$. In fact, the corresponding matrix

$$\begin{bmatrix} y_1(x_0) & y_2(x_0) \\ y_1'(x_0) & y_2'(x_0) \end{bmatrix}$$

is singular, that is, its determinant is zero. Only in this case the corresponding system of algebraic equations has a nontrivial (not identically zero) solution (see §7.2.1). Then the function

$$\tilde{y} = \alpha_1 y_1(x) + \alpha_2 y_2(x)$$

is a solution of the differential equation $y'' + a_1 y' + a_0 y = 0$. Moreover, $\tilde{y}$ also satisfies the initial conditions

$$\tilde{y}(x_0) = 0, \quad \left.\frac{\mathrm{d}\tilde{y}}{\mathrm{d}x}\right|_{x=x_0} = 0.$$

However, only the trivial function $\tilde{y}(x) \equiv 0$ satisfies the homogeneous linear differential equation and the homogeneous initial conditions (Theorem 4.2 on page 189).

Theorem 4.10: (Abel) If $y_1(x)$ and $y_2(x)$ are solutions of the differential equation

$$y'' + p(x)y' + q(x)y = 0, \tag{4.2.2}$$

where $p(x)$ and $q(x)$ are continuous on an open interval (a, b), then the Wronskian $W(x) = W[y_1, y_2](x)$ is given by

$$W[y_1, y_2](x) = W[y_1, y_2](x_0) \exp\left\{-\int_{x_0}^{x} p(t)\,\mathrm{d}t\right\}. \tag{4.2.3}$$

Here x_0 is any point from the interval (a, b).

Note: The formula (4.2.3) was derived by the greatest Norwegian mathematician Niels Henrik Abel (1802–1829) in 1827 for the second order differential equation. In the general case, namely, for the equation $y^{(n)} + p_{n-1}(x)y^{(n-1)} + \cdots + p_1(x)y' + p_0(x)y = 0$, Joseph Liouville (1809–1882) and Michel Ostrogradski (1801–1861) independently showed in 1838 that

$$W(x) = W(x_0) \exp\left\{-\int_{x_0}^{x} p_{n-1}(t)\,\mathrm{d}t\right\} \quad \Longleftrightarrow \quad W'(x) + p_{n-1}(x)W(x) = 0. \tag{4.2.4}$$

PROOF: Each of the functions $y_1(x)$ and $y_2(x)$ satisfies Eq. (4.2.2). Therefore,

$$y_1'' + p(x)y_1' + q(x)y_1 = 0, \quad y_2'' + p(x)y_2' + q(x)y_2 = 0.$$

Solving this system of algebraic equations with respect to p and q, we obtain

$$p(x) = -\frac{y_1y_2'' - y_1''y_2}{y_1y_2' - y_1'y_2} = -\frac{W'(x)}{W(x)}, \qquad q(x) = \frac{y_1'y_2'' - y_1''y_2'}{y_1y_2' - y_1'y_2} = \frac{1}{W(x)}W[y_1', y_2'], \tag{4.2.5}$$

where $W(x) = y_1y_2' - y_1'y_2$ is the Wronskian of the functions y_1, y_2. The first equation can be rewritten in the form

$$W'(x) + p(x)W(x) = 0. \tag{4.2.6}$$

This is a separable differential equation of the first order and its solution is given by the formula (4.2.3). Abel's formula shows that the Wronskian for any set of solutions is determined up to a multiplicative constant by the differential equation itself.

Corollary 4.1: If the Wronskian $W(x)$ of solutions of Eq. (4.2.2) is zero at one point $x = x_0$ of an interval (a, b) where coefficients are continuous, then $W(x) \equiv 0$ for all $x \in (a, b)$.

Corollary 4.2: If the Wronskian $W(x)$ of solutions of Eq. (4.2.2) at one point $x = x_0$ is not zero, then it is not zero at every point of an interval, where coefficients $p(x)$ and $q(x)$ are continuous.

Example 4.2.8: Given the differential equation

$$y^{(4)} - y = 0.$$

Find the Wronskian of solutions

$$y_1(x) = e^x, \quad y_2(x) = e^{-x}, \quad y_3(x) = \sinh x, \quad y_4(x) = \cosh x.$$

Solution. We calculate derivatives of these functions to obtain

$$\begin{array}{llll} y_1'(x) = y_1(x), & y_2'(x) = -y_2(x), & y_3'(x) = \cosh x, & y_4'(x) = \sinh x, \\ y_1''(x) = e^x, & y_2''(x) = e^{-x}, & y_3''(x) = \sinh x, & y_4''(x) = \cosh x, \\ y_1'''(x) = y_1(x), & y_2'''(x) = -y_2(x), & y_3'''(x) = \cosh x, & y_4'''(x) = \sinh x. \end{array}$$

To evaluate the Wronskian $W(x)$, we use the Abel formula (4.2.4). Its value at $x = 0$ is

$$W(0) = \det \begin{bmatrix} 1 & 1 & 0 & 1 \\ 1 & -1 & 1 & 0 \\ 1 & 1 & 0 & 1 \\ 1 & -1 & 1 & 0 \end{bmatrix} \equiv \begin{vmatrix} 1 & 1 & 0 & 1 \\ 1 & -1 & 1 & 0 \\ 1 & 1 & 0 & 1 \\ 1 & -1 & 1 & 0 \end{vmatrix}.$$

We add the second column to the first one to obtain

$$W(0) = \det \begin{bmatrix} 2 & 1 & 0 & 1 \\ 0 & -1 & 1 & 0 \\ 2 & 1 & 0 & 1 \\ 0 & -1 & 1 & 0 \end{bmatrix} = 2 \det \begin{bmatrix} 1 & 1 & 0 & 1 \\ 0 & -1 & 1 & 0 \\ 1 & 1 & 0 & 1 \\ 0 & -1 & 1 & 0 \end{bmatrix}.$$

Since the latter matrix has two equal columns (the first and the fourth), its determinant is zero. Hence, at any point, the Wronskian $W(x) = W(0) \equiv 0$ and the given functions are linearly dependent. To check our answer, we type in *Mathematica:*

```
y = {Exp[x], Exp[-x], Sinh[x], Cosh[x]}
w = {y, D[y, x], D[y, {x,2}], D[y, {x, 3}]}     Det[w]
```

Since the output is zero, the functions are linearly dependent. □

The Wronskian unambiguously determines whether functions are linearly dependent or independent if and only if these functions are solutions of the same differential equation. But what should we do if we don't know whether the functions are solutions of the same differential equation? If these functions are represented as convergent power series, then Theorem 4.8 assures us that evaluating the Wronskian is sufficient. If they are not, there is another advanced tool to determine whether or not a set of functions is linearly independent or dependent, called the Gramm determinant, but it is out of the scope of this book.

Problems

1. Show that the system of functions $\{f_1, f_2, \ldots, f_n\}$ is linearly dependent if one of the functions is identically zero.
2. Which polynomials can be expressed as linear combinations of the set of functions $\{(x-2),\ (x-2)^2,\ (x-2)^3\}$?
3. Determine whether the given pair of functions is linearly independent or linearly dependent.
   (a) $f_1(x) = x^2,\ f_2(x) = (x+1)^2$; **(b)** $f_1(x) = x^2,\ f_2(x) = x^2+1$;
   (c) $f_1(x) = 2x^2+3x,\ f_2(x) = 2x^2-3x$; **(d)** $f_1(x) = x|x|,\ f_2(x) = x^2$;
   (e) $f_1(x) = 1+\cos x,\quad f_2(x) = \cos^2 x$; **(f)** $f_1(x) = \cos x,\quad f_2(x) = \sin\left(x - \frac{\pi}{2}\right)$;
   (g) $f_1(x) = \sin 2x,\quad f_2(x) = \cos x \sin x$; **(h)** $f_1(x) = \tan x,\ f_2(x) = \cot x$;
   (i) $f_1(x) = e^x,\ f_2(x) = e^{2x}$; **(j)** $f_1(x) = e^x,\ f_2(x) = e^{x+1}$;
   (k) $f_1(x) = e^x,\ f_2(x) = x\,e^x$; **(l)** $f_1(x) = e^x,\ f_2(x) = 1+e^x$.
4. Obtain the Wronskian of the following three functions.
   (a) $f_1(x) = e^x,\quad f_2(x) = x\,e^x,\quad$ and $\quad f_3(x) = (2x-1)\,e^x$.
   (b) $f_1(x) = e^x,\quad f_2(x) = e^x \sin x,\quad$ and $\quad f_3(x) = e^x \cos x$.
   (c) $f_1(x) = x,\quad f_2(x) = x\,e^x,\quad$ and $\quad f_3(x) = 2x\,e^{-x}$.
   (d) $f_1(x) = e^x \cos x,\quad f_2(x) = e^{2x} \cos x,\quad$ and $\quad f_3(x) = e^{3x} \cos x$.
5. Suppose the Wronskian $W[f,g]$ of two functions f and g is known; find the Wronskian $W[u,v]$ of $u = af + bg$ and $v = \alpha f + \beta g$, where a, b, α, and β are some constants.
6. Are the following functions linearly independent or dependent on $[1,\infty)$?
   (a) x^2 and $x^2 \ln x$; **(b)** $x^2 \ln x$ and $x^2 \ln 2x$;
   (c) $\ln x$ and $\ln(x^4)$; **(d)** $\ln x$ and $(\ln x)^2$.
7. Are the functions $f_1(x) = x|x|$ and $f_2(x) = x^2$ linearly independent or dependent on the following intervals?
   (a) $0 \leqslant x \leqslant 1$; **(b)** $-1 \leqslant x \leqslant 0$; **(c)** $-1 \leqslant x \leqslant 1$.
8. Determine whether the following three functions are linearly dependent or independent.
   (a) $f_1(x) = \sqrt{x}+2,\quad f_2(x) = \sqrt{x}+2x,\quad$ and $\quad f_3(x) = x-1$;
   (b) $1,\quad \sin x,\quad$ and $\quad \cos x$; **(c)** $1,\quad \cos^3 x,\quad$ and $\quad \cos 3x + 3\cos x$.
9. Let α, β, and γ be distinct real numbers. Show that x^α, x^β, and x^γ are linearly independent on any subinterval of the positive x-axis.
10. The Wronskian of two functions is $W(x) = x^2 - 1$. Are the functions linearly independent or linearly dependent? Why?
11. Find the Wronskian of the two solutions of the given differential equation without solving the equation.
    (a) $(\sin x)y'' + (\cos x)y' + \tan x y = 0$; **(b)** $x\,y'' - (2x+1)\,y' + \sin 2x\, y = 0$;
    (c) $x\,y'' + y' + x\,y = 0$; **(d)** $(1-x^2)\,y'' - 2x\,y' + 2y = 0$;
    (e) $x^2\,y'' - x(x+1)\,y' + (x+1)^2\,y = 0$; **(f)** $\sin(2x)\,y'' - 2y' + xy = 0$.
12. Derive the Abel formula for the self-adjoint operator $\mathtt{D}P(x)\mathtt{D} + Q(x)$, where $\mathtt{D} = \mathrm{d}/\mathrm{d}x$.
13. Prove that for any real number α and any positive integer n, the functions $e^{\alpha x}$, $x\,e^{\alpha x}$, $\ldots$, $x^{n-1}\,e^{\alpha x}$ are linearly independent on any interval.
14. For what coefficients of the second order differential equation $a_2(x)y'' + a_1(x)y' + a_0(x)y = 0$ is the Wronskian a constant?
15. Prove Lemma 4.1 on page 199.
16. Prove the following claim:
    If two solutions of the differential equation $y'' + p(x)y' + q(x)y = 0$ on some interval $|a,b|$ vanish at a point from $|a,b|$, then these solutions are linearly dependent.
    The same claim is valid when their derivatives are zero at a point from $|a,b|$.
17. Let f and g be two continuously differentiable functions on some interval $|a,b|$, $a < b$, and suppose that g never vanishes in it. Prove that if their Wronskian is identically zero on $|a,b|$, then f and g are linearly dependent. *Hint:* Differentiate $f(x)/g(x)$.

4.3 The Fundamental Set of Solutions

Let us consider a homogeneous linear differential equation of the second order

$$y'' + p(x)y' + q(x)y = 0 \tag{4.3.1}$$

in some interval $x \in |a, b|$. Any two solutions of Eq. (4.3.1) are said to form a **fundamental set of solutions** of this equation if they are linearly independent on the interval $|a, b|$. This definition can be extended to the n-th order differential equation ($\texttt{D} = \mathrm{d}/\mathrm{d}x$)

$$L[x, \texttt{D}]y = 0, \quad \text{where } Ly = y^{(n)} + a_{n-1}(x)y^{(n-1)} + \cdots + a_1(x)y' + a_0(x)y. \tag{4.3.2}$$

Namely, any set of solutions $y_1, y_2, \ldots, y_n$ is a fundamental set of solutions of Eq. (4.3.2) if they are linearly independent on some interval. Note that the number of functions in the fundamental set of solutions coincides with the order (the highest derivative in the equation) of the equation.

According to Theorem 4.9, the functions y_1 and y_2 form a fundamental set of solutions of Eq. (4.3.1) if and only if their Wronskian $W[y_1, y_2](x)$ is not zero.

The family $y = \varphi(x, C_1, C_2)$ of solutions of Eq. (4.3.1), which depends on two arbitrary constants C_1 and C_2, is called the **general solution** of Eq. (4.3.1) because it encompasses every solution. If we know $\{y_1, y_2\}$, a fundamental set of solutions of Eq. (4.3.1), then we can construct the general solution of the homogeneous equation (4.3.1) as

$$y(x) = C_1 y_1(x) + C_2 y_2(x), \tag{4.3.3}$$

where C_1 and C_2 are arbitrary constants. The function (4.3.3) is also a solution of Eq. (4.3.1) as it follows from the Principle of Superposition (Theorem 4.4, page 191).

Theorem 4.11: Let $y_1, y_2, \ldots, y_n$ be n linearly independent solutions (a fundamental set of solutions) of the linear differential equation of the n-th order (4.3.2). Let y be any other solution. Then there exist constants $C_1, C_2, \ldots, C_n$ such that

$$y(x) = C_1 y_1(x) + C_2 y_2(x) + \cdots + C_n y_n(x), \tag{4.3.4}$$

PROOF: As usual, we prove the theorem only for $n = 2$. Let $y(x)$ be any solution of Eq. (4.3.1). We must show that the linear combination $C_1 y_1(x) + C_2 y_2(x)$ is equal to $y(x)$ for some choice of coefficients C_1 and C_2. Let x_0 be a point in the interval (a, b). Since the solutions y_1 and y_2 are linearly independent, their Wronskian $W(y_1, y_2; x) \neq 0$ for any $x \in (a, b)$. Consequently $W(x_0) \neq 0$. We know that the solution of Eq. (4.3.2) is uniquely defined by its initial conditions at any point of interval (a, b), so the values $y_0 = y(x_0)$ and $y_0' = y'(x_0)$ uniquely define the function $y(x)$. After differentiation, we obtain the system of algebraic equations

$$\left.\begin{array}{rcl} y_0 & = & C_1\, y_1(x_0) \;+\; C_2\, y_2(x_0), \\ y_0' & = & C_1\, y_1'(x_0) \;+\; C_2\, y_2'(x_0). \end{array}\right\}$$

This system has a unique solution because the determinant of the corresponding system of algebraic equations is the Wronskian $W(x_0) \neq 0$. Thus, the function y is the general solution of Eq. (4.3.2). Note that Theorem 4.11 can be proved by mathematical induction (Exercise 9) for any $n \geqslant 2$.

Corollary 4.3: Let $y_1, y_2, \ldots, y_n$ be solutions to the linear differential equation (4.3.2). Then the family of solutions (4.3.4) with arbitrary coefficients $C_1, C_2, \ldots, C_n$ includes every solution if and only if their Wronskian does not vanish.

Theorem 4.12: Consider Eq. (4.3.1) whose coefficients $p(x)$ and $q(x)$ are continuous on some interval (a, b). Choose some point x_0 from this interval. Let y_1 be the solution of Eq. (4.3.1) that also satisfies the initial conditions

$$y(x_0) = 1, \quad y'(x_0) = 0,$$

and let y_2 be another solution of Eq. (4.3.1) that satisfies the initial conditions

$$y(x_0) = 0, \quad y'(x_0) = 1.$$

Then y_1 and y_2 form a fundamental set of solutions of Eq. (4.3.1).

PROOF: To show that functions y_1 and y_2 are linearly independent, we should calculate their Wronskian at one point

$$W(y_1, y_2; x_0) = \begin{vmatrix} y_1(x_0) & y_2(x_0) \\ y_1'(x_0) & y_2'(x_0) \end{vmatrix} = \begin{vmatrix} 1 & 0 \\ 0 & 1 \end{vmatrix} = 1.$$

Since their Wronskian is not zero at the point x_0, it is not zero anywhere on the interval (a, b) (see Corollary 4.2 on page 202). Therefore, these functions are linearly independent (by Theorem 4.9, page 201). ■

This theorem has a natural generalization, which we formulate below.

Theorem 4.13: There exists a fundamental set of solutions for the homogeneous linear n-th order differential equation (4.3.2) on an interval where coefficients are all continuous.

Theorem 4.14: Let $y_1, y_2, \ldots, y_n$ be n functions that together with their first $n-1$ derivatives are continuous on an open interval (a, b). Suppose their Wronskian $W(y_1, y_2, \ldots, y_n; x) \neq 0$ for $a < x < b$. Then there exists the unique linear homogeneous differential equation of the n-th order for which the collection of functions $y_1, y_2, \ldots, y_n$ forms a fundamental set of solutions.

PROOF: We prove the statement only for a second order differential equation of the form (4.3.1). Let y_1 and y_2 be linearly independent functions on interval (a, b) with $W(y_1, y_2; x) \stackrel{\text{def}}{=} W(x) \neq 0,\ x \in (a, b)$. Then we have

$$y_1'' + p(x)y_1' + q(x)y_1 = 0, \quad y_2'' + p(x)y_2' + q(x)y_2 = 0.$$

Solving this system of algebraic equations for $p(x)$ and $q(x)$, we arrive at the equations (4.2.5), which determine the functions $p(x)$ and $q(x)$ uniquely. The formula for the general case is given in #12 of Summary, page 255. ■

To solve an initial value problem for the linear n-th order differential equation (4.3.2), one must choose the constants C_j $(j = 1, 2, \ldots, n)$ in the general solution (4.3.4) so that the initial conditions $y(x_0) = y_0$, $y'(x_1) = y_1$, $\ldots$, $y^{(n-1)}(x_0) = y_{n-1}$ are satisfied. These constants are determined upon substitution of the solution (4.3.4) into the initial conditions, which leads to n simultaneous algebraic equations

$$\sum_{j=1}^{n} C_j\, y_j^{(k)}(x_0) = y_k, \qquad k = 0, 1, \ldots, n-1.$$

Since the determinant of the above system is the Wronskian evaluated at the point $x = x_0$, the constants are determined uniquely. Does it mean that a nonvanishing Wronskian at a point implies a unique solution of the corresponding initial value problem? Unfortunately no because it is a necessary but not a sufficient condition. (For instance, the Wronskian of $y_1 = x^{-1}$ and $y_2 = x^2$, the fundamental set of solutions for the equation $x^2y'' - 2y = 0$, is constant everywhere, including the singular point $x = 0$.) On the other hand, if the Wronskian is zero or infinite at the point where the initial conditions are specified, the initial value problem may or may not have a solution or, if the solution exists, it may not be unique.

Example 4.3.1: Let us consider the differential equation

$$y'' + y = 0.$$

The function $y_1(x) = \cos x$ is a solution of this equation and it satisfies the initial conditions $y(0) = 1,\ y'(0) = 0$. Another linearly independent solution of this equation is $y_2(x) = \sin x$, which satisfies $y(0) = 0,\ y'(0) = 1$. Therefore, the couple $\{y_1, y_2\}$ is the fundamental set of solutions of the equation, and the general solution becomes

$$y(x) = C_1\, y_1(x) + C_2\, y_2(x) = C_1 \cos x + C_2 \sin x.$$

To find the values of constants C_1, C_2 so that the function satisfies the initial conditions $y(\pi/4) = 3\sqrt{2}$ and $y'(\pi/4) = -2\sqrt{2}$, we ask *Mathematica* to solve the system of algebraic equations

$$C_1 \cos(\pi/4) + C_2 \sin(\pi/4) = 3, \qquad -C_1 \sin(\pi/4) + C_2 \cos(\pi/4) = -2.$$

```
y[x_] = c1 Cos[x] + c2 Sin[x];  y''[x] + y[x] == 0 (* to check solution *)
eqns = {y[Pi/4] == 3*Sqrt[2], y'[Pi/4] == -2*Sqrt[2]}
c1c2 = Solve[eqns, {c1, c2}]       soln[x_] = y[x] /. c1c2[[1]]
```

This yields the solution $y = 5\cos x + \sin x$.

Example 4.3.2: Given two functions $y_1(x) = x^3$ and $y_2(x) = x^2$. These functions are linearly independent on the interval $(-\infty, \infty)$. Their Wronskian $W(x) = -x^4$ vanishes at $x = 0$ because x^3 and x^2 are tangent there. Using formulas (4.2.5), page 202, we find

$$p(x) = -\frac{4}{x}, \qquad q(x) = \frac{6}{x^2}, \quad x \neq 0.$$

Therefore, the corresponding differential equation is

$$y'' - \frac{4}{x}\,y' + \frac{6}{x^2}\,y = 0 \qquad \text{or} \qquad x^2y'' - 4xy' + 6y = 0.$$

The coefficients of this equation are discontinuous at $x = 0$. Hence, functions $y_1(x)$ and $y_2(x)$ constitute a fundamental set of solutions for this equation on each of the intervals $(-\infty, 0)$ and $(0, \infty)$.

An initial value problem has a unique solution when the initial conditions are specified at any point other than $x = 0$. When these initial conditions are replaced with $y(0) = 2$, $y'(0) = 3$, the problem has no solution. However, with the initial conditions $y(0) = 0$, $y'(0) = 0$, the problem has infinitely many solutions $y = C_1x^3 + C_2x^2$, with arbitrary constants C_1 and C_2.

Example 4.3.3: For a positive integer n and a real number a, let us consider n functions $y_1 = e^{ax}$, $y_2 = x\,e^{ax}$, ..., $y_n = x^{n-1}\,e^{ax}$. Using Lemma 4.1 on page 199, the Wronskian of these functions becomes

$$W(y_1, y_2, \ldots, y_n; x) = e^{anx}\,W(1, x, \ldots, x^{n-1}; x) = e^{anx}\prod_{k=0}^{n-1}(k!) \neq 0, \tag{4.3.5}$$

where $\prod_{k=0}^{n-1} a_k$ denotes the product $a_0a_1a_2\cdots a_{n-1}$, similar to $\sum$ for summation. Each function $y_{k+1} = x^k\,e^{ax}$, $k = 0, 1, \ldots, n-1$, is annihilated by the operator $(\texttt{D} - a)^{k+1}$ and, therefore, by any operator $(\texttt{D} - a)^m$ for $m > k$, where $\texttt{D} = \text{d}/\text{d}x$. This claim can be shown by induction and it is based on the formula $(\texttt{D} - a)x^k\,e^{ax} = kx^{k-1}\,e^{ax}$. So every application of the operator $\texttt{D} - a$ to $x^k\,e^{ax}$ reduces the power x^k by 1. This observation allows us to conclude that the given functions y_1, y_2, ..., y_n form the fundamental set of solutions to the differential equation $(\texttt{D} - a)^n y = 0$.

Example 4.3.4: Find the value of the Wronskian, $W[y_1, y_2](2)$, at the point $x = 2$ given that its value at $x = 1$ is $W[y_1, y_2](1) = 4$, where y_1 and y_2 are linearly independent solutions of $xy'' + 2y' + xy\sin x = 0$.

Solution. First, we reduce the equation to the normal form (4.2.2):

$$y'' + \frac{2}{x}\,y' + (\sin x)\,y = 0,$$

with $p(x) = 2/x$ and $q(x) = \sin x$. From Eq. (4.2.3), we find

$$\begin{aligned} W(2) &= W(1)\exp\left\{-\int_1^2 p(t)\,\text{d}t\right\} = W(1)\exp\{-2(\ln 2 - \ln 1)\} \\ &= W(1)\exp\{\ln 2^{-2}\} = W(1)\cdot 2^{-2} = 4\cdot 2^{-2} = 1. \end{aligned}$$

Problems

1. Verify that $y_1(x) = \cos 3x\,e^{2x}$ and $y_2(x) = \sin 3x\,e^{2x}$ both satisfy $y'' - 4y' + 13y = 0$ and compute the Wronskian of the functions y_1 and y_2.
2. Verify that the functions $y_1(x) = x^2$ and $y_2(x) = x^{-1}$ form a fundamental set of solutions to some differential equation on an interval that does not contain $x = 0$.
3. For the given pairs of functions, construct the linear differential equation for which they form a fundamental set of solutions.

(a)	$y_1 = x, \quad y_2 = x^3$;	**(b)**	$y_1 = \sin x, \quad y_2 = \cos x$;	**(c)**	$y_1 = \sin x, \quad y_2 = \tan x$;
(d)	$y_1 = x^{-1}, \; y_2 = x^3$;	**(e)**	$y_1 = x^{1/2}, \quad y_2 = x^{-1/2}$;	**(f)**	$y_1 = \sin 2x, \; y_2 = x\,y_1$;
(g)	$y_1 = x^2, \quad y_2 = x^5$;	**(h)**	$y_1 = x^{1/4}, \quad y_2 = x^{3/4}$;	**(i)**	$y_1 = \tan x, \quad y_2 = \cot x$;
(j)	$y_1 = x, \quad y_2 = 1/x$;	**(k)**	$y_1 = x + 1, \quad y_2 = x - 1$;	**(l)**	$y_1 = x + 1, \quad y_2 = e^x$.

4. Prove that if two solutions $y_1(x)$ and $y_2(x)$ of the differential equation $y''+p(x)y'+q(x)y=0$ are zero at the same point in an interval (a,b), where the coefficients $p(x)$ and $q(x)$ are continuous functions, then they cannot form a fundamental set of solutions.

5. Answer `true` or `false` for each of the following statements.

   (a) If $y_1(x)$ and $y_2(x)$ are linearly independent solutions to the second order linear differential equation (4.3.1) on interval $|a,b|$, then they are linearly independent on a smaller interval $|\alpha,\beta| \subset |a,b|$.

   (b) If $y_1(x)$ and $y_2(x)$ are linearly dependent solutions to the second order linear differential equation (4.3.1) on interval $|a,b|$, then they are linearly dependent on the larger interval $|\alpha,\beta| \supset |a,b|$.

6. Let $\phi_1(x)$ and $\phi_2(x)$ be nontrivial solutions to the linear equation (4.3.1), where the functions $p(x)$ and $q(x)$ are continuous at x_0. Show that there exists a constant K such that $\phi_1(x)=K\phi_2(x)$ if $\phi_1'(x_0)=\phi_2'(x_0)=0$.

7. Prove the **Sturm Theorem**:
   If t_1 and t_2 ($t_1<t_2$) are zeroes of a solution $x(t)$ to the equation $\ddot{x}(t)+Q(t)x(t)=0$ and
$$Q(t)<Q_1(t) \qquad \text{for all } t\in[t_1,t_2],$$
   then any solution $y(t)$ to the equation $\ddot{y}(t)+Q_1(t)y(t)=0$ has at least one zero in the closed interval $[t_1,t_2]$.

8. Find the longest interval in which the initial value problem is certain to have a unique (twice differentiable) solution.

(a)	$t(t-5)\ddot{y}+3t\,\dot{y}=3,\quad y(3)=\dot{y}(3)=1;$	**(b)**	$t(t-5)\ddot{y}+3t\,\dot{y}=3,\quad y(6)=\dot{y}(6)=1;$
(c)	$t(t-5)\ddot{y}+4y=3,\quad y(-1)=y'(-1)=1;$	**(d)**	$(t^2-1)\ddot{y}+\dot{y}+y=3,\quad y(0)=\dot{y}(0)=1;$
(e)	$(t^2-1)\ddot{y}+\dot{y}+y=3,\quad y(2)=\dot{y}(2)=1;$	**(f)**	$(t^2+1)\ddot{y}+\dot{y}+y=3,\quad y(0)=\dot{y}(0)=1;$
(g)	$(t^2+4)\ddot{y}+4y=4,\quad y(-2)=\dot{y}(-2)=1;$	**(h)**	$\sin t\,\ddot{y}+t\,y=t^2,\quad y(1)=\dot{y}(1)=2;$
(i)	$(t^2+t-6)\ddot{y}+\dot{y}=t,\quad y(0)=\dot{y}(0)=1;$	**(j)**	$(t^2+2t-3)\ddot{y}+y=2,\quad y(0)=\dot{y}(0)=1.$

9. Prove Theorem 4.11 (page 204) by mathematical induction.

10. Show that the functions x^α, x^β, and x^γ form a fundamental set of solutions to the differential equation
$$x^3y'''+(3-\alpha-\beta-\gamma)y''+(1-\alpha-\beta-\gamma+\alpha\beta+\alpha\gamma+\beta\gamma)x\,y'-\alpha\beta\gamma\,y=0$$
    if α, β, and γ are distinct real numbers.

11. If y_1 and y_2 are linearly independent solutions of $xy''+y'+xy\cos x=0$, and if $W[y_1,y_2](1)=2$, find the value of $W[y_1,y_2](2)$.

12. Does the initial value problem $x^2y''-2y=0$, $y(0)=y'(0)=1$ have a solution?

13. Verify that each of the sets, $\{x^{-2},x^{-3}\}$ and $\{2x^{-2}-5x^{-3},5x^{-2}+2x^{-3}\}$, is a fundamental set of solutions to the differential equation $x^2y''+6x\,y'+6\,y=0$.

14. Show that the functions $u(x)=x^{-1/2}$ and $v(x)=x^2$ are solutions of the nonlinear differential equation $x^2y\,y''+(xy'-y)^2=3\,y^2$, but no linear combination of them is a solution, although $w(x)=\sqrt{c_1x^{-1}+c_2x^4}$ is a solution for arbitrary constants c_1, c_2.

15. Verify that $\sin^3 x$ and $\sin x-\frac{1}{3}\sin 3x$ are solutions of $y''+(\tan x-2\cot x)y'=0$ on any interval where $\tan x$ and $\cot x$ are both defined. Are these solutions linearly independent?

16. From the given functions, choose a fundamental set of solutions to the differential equation.

    (a) $2\,e^{-x}$, $\cosh x$, $-\sinh x$ on $(-\infty,\infty)$ for the equation $y''-y=0$.

    (b) x^3, $3x^3\ \ln x$, $x^3\,(\ln x-5)$ on $(0,\infty)$ for the equation $x^2\,y''-5xy'+9y=0$.

    (c) $2\,\sin 2x$, $-\cos 2x$, $\cos(2x+7)$ on $(-\infty,\infty)$ for the equation $y''+4y=0$.

    (d) $4x$, $\frac{x}{2}\ln\frac{x+1}{1-x}-2$, $\frac{x}{2}$ on $(-1,1)$ for the Legendre equation $(1-x^2)y''-2xy'+2y=0$.

    (e) e^x, $e^x\ln x$, $x\,e^x$ on $(0,\infty)$ for the equation $x\,y''+(1-x)\,y'+x\,y=0$.

    (f) x, $x\,e^x$, $x\,\ln x$ on $(0,\infty)$ for the equation $x^2y''-x(2+x)\,y'+(x+2)\,y=0$.

    (g) x, $\frac{1}{2}\left(x^2-2\right)\sqrt{x^2+1}+\frac{3x}{2}\operatorname{arcsinh}(x)$, $x\,\ln x$ on $(0,\infty)$ for the equation $\left(x^2+1\right)y''-3x\,y'+3\,y=0$.

    (h) x^3-3x, $\frac{(x^2-1)\sqrt{4-x^2}}{x(x^2-3)}\left(x^3-3x\right)$, $x^3\sqrt{4-x^2}$ on $|x|<2$ for the Dickson equation $\left(x^2-4\right)y''+x\,y'-9\,y=0$.

17. For the given pairs of functions, construct the linear differential equation for which they form the fundamental set of solutions.

(a)	$y_1=x,\quad y_2=x^4;$	**(b)**	$y_1=\sin 2x,\ y_2=\cos 2x;$	**(c)**	$y_1=x^{-1},\ y_2=x^{-3};$
(d)	$y_1=x^{1/2},\ y_2=x^{1/3};$	**(e)**	$y_1=\sin^3 x,\ y_2=\sin^{-2} x;$	**(f)**	$y_1=x,\quad y_2=\ln\lvert x\rvert;$
(g)	$y_1=x^3,\quad y_2=x^4;$	**(h)**	$y_1=x^{1/4},\quad y_2=x^{-1/4};$	**(i)**	$y_1=x,\quad y_2=\sin x;$
(j)	$y_1=x^2,\quad y_2=x^{-2};$	**(k)**	$y_1=x^2+1,\ y_2=x^2-1;$	**(l)**	$y_1=\ln\lvert x\rvert,\ y_2=x\,y_1;$
(m)	$y_1=x,\ y_2=e^{-x};$	**(n)**	$y_1=1,\ y_2=1/x;$	**(o)**	$y_1=e^{x^2},\ y_2=x\,y_1.$

4.4 Homogeneous Equations with Constant Coefficients

Consider a linear differential equation $L[\mathtt{D}]y = 0$, where $L[\mathtt{D}]$ is the linear differential constant coefficient operator

$$L[\mathtt{D}] = a_n \mathtt{D}^n + a_{n-1}\mathtt{D}^{n-1} + \cdots + a_1 \mathtt{D} + a_0, \qquad a_n \neq 0, \tag{4.4.1}$$

containing the derivative operator $\mathtt{D} = \mathrm{d}/\mathrm{d}x$, and some given constants a_k, $k = 0, 1, \ldots, n$. As usual, we drop $\mathtt{D}^0$, the identical operator. This differential equation $L[\mathtt{D}]y = 0$ can be rewritten explicitly as

$$a_n \frac{\mathrm{d}^n y}{\mathrm{d}x^n} + a_{n-1}\frac{\mathrm{d}^{n-1} y}{\mathrm{d}x^{n-1}} + \cdots + a_1 \frac{\mathrm{d}y}{\mathrm{d}x} + a_0\, y(x) = 0. \tag{4.4.2}$$

In particular, when $n = 2$, we have the equation

$$ay'' + by' + cy = 0, \tag{4.4.3}$$

whose coefficients a, b, and c are constants. No doubt, such constant coefficient equations are the simplest of all differential equations. They can be handled entirely within the context of linear algebra, and form a substantial class of equations of order greater than one which can be explicitly solved. Surprisingly, such equations arise in a wide variety of physical problems, including mechanical and electrical applications. A remarkable result about linear constant coefficient equations follows from Theorem 4.3 on page 189, which we formulate below.

Theorem 4.15: For any real numbers a_k $(k = 0, 1, 2, \ldots, n)$, $a_n \neq 0$, and y_i $(i = 0, 1, \ldots, n-1)$, there exists a unique solution to the initial value problem

$$a_n\, y^{(n)} + a_{n-1}\, y^{(n-1)} + \cdots + a_0 y(x) = 0, \quad y(x_0) = y_0,\ y'(x_0) = y_1,\ \ldots,\ y^{(n-1)}(x_0) = y_{n-1}.$$

The solution is valid for all real numbers $-\infty < x < \infty$. ■

The main idea (proposed by L. Euler) of how to find a solution of Eq. (4.4.3) or Eq. (4.4.2) is based on a property of the exponential function. Namely, since for a constant λ and a positive integer k,

$$\mathtt{D}^k e^{\lambda x} = \frac{\mathrm{d}^k}{\mathrm{d}x^k}\, e^{\lambda x} = \lambda^k e^{\lambda x},$$

it is easy to find the effect an operator $L[\mathtt{D}]$ has on $e^{\lambda x}$. That is,

$$\begin{aligned} L[\mathtt{D}]e^{\lambda x} &= a_n \mathtt{D}^n e^{\lambda x} + a_{n-1}\mathtt{D}^{n-1} e^{\lambda x} + \cdots + a_1 \mathtt{D} e^{\lambda x} + a_0 e^{\lambda x} \\ &= a_n \lambda^n\, e^{\lambda x} + a_{n-1}\lambda^{n-1}\, e^{\lambda x} + \cdots + a_1 \lambda\, e^{\lambda x} + a_0 e^{\lambda x} = L(\lambda) e^{\lambda x}. \end{aligned}$$

In other words, the differential operator $L[\mathtt{D}]$ acts as an operator of multiplication by a polynomial on functions of the form $e^{\lambda x}$. Therefore, the substitution of an exponential instead of $y(x)$ into Eq. (4.4.3) or Eq. (4.4.2) reduces these differential equations into algebraic ones, which are more simple to solve.

For the second order differential equation (4.4.3), we have

$$a\,\frac{\mathrm{d}^2}{\mathrm{d}x^2} e^{\lambda x} + b\,\frac{\mathrm{d}}{\mathrm{d}x} e^{\lambda x} + c\, e^{\lambda x} = \left(a\mathtt{D}^2 + b\mathtt{D} + c\right) e^{\lambda x} = \left(a\lambda^2 + b\lambda + c\right) e^{\lambda x}.$$

Hence, the exponential function $e^{\lambda x}$ is a solution of Eq. (4.4.3) if λ is a solution of the quadratic equation

$$a\lambda^2 + b\lambda + c = 0 \qquad \Longrightarrow \qquad \lambda_{1,2} = \frac{-b \pm \sqrt{b^2 - 4ac}}{2a}. \tag{4.4.4}$$

The quadratic equation (4.4.4) is called the **characteristic equation** for the differential equation of the second order (4.4.3). This definition can be extended for the general linear differential equation $L[\mathtt{D}]y = 0$ (where $L[\mathtt{D}]$ is defined in Eq. (4.4.1)) using the relationship

$$e^{-\lambda x}\, L[\mathtt{D}]e^{\lambda x} = a_n \lambda^n + a_{n-1}\lambda^{n-1} + \cdots + a_0.$$

Definition 4.6: Let $L[\mathtt{D}] = a_n \mathtt{D}^n + a_{n-1}\mathtt{D}^{n-1} + \cdots + a_0 \mathtt{D}^0$ be the differential operator of order n with constant coefficients. The polynomial $L(\lambda) = a_n \lambda^n + a_{n-1}\lambda^{n-1} + \cdots + a_0$ is called the **characteristic polynomial**, and the corresponding algebraic equation

$$a_n \lambda^n + a_{n-1}\lambda^{n-1} + \cdots + a_0 = 0 \tag{4.4.5}$$

is called the characteristic equation.

Assuming for simplicity that the roots of the characteristic equation (4.4.5) are distinct, let them be denoted by $\lambda_1, \lambda_2, \ldots, \lambda_n$. Then the n solutions

$$y_1(x) = e^{\lambda_1 x}, \quad y_2(x) = e^{\lambda_2 x}, \quad \ldots \quad , \quad y_n(x) = e^{\lambda_n x} \tag{4.4.6}$$

are linearly independent because their Wronskian is not zero. In fact, after pulling out all exponentials from the columns, the Wronskian becomes

$$W(y_1, y_2, \ldots, y_n; x) \stackrel{\text{def}}{=} W(x) = e^{(\lambda_1+\lambda_2+\cdots+\lambda_n)x}\, V(\lambda_1, \lambda_2, \ldots, \lambda_n),$$

where

$$V(\lambda_1, \lambda_2, \ldots, \lambda_n) = \begin{vmatrix} 1 & 1 & \ldots & 1 \\ \lambda_1 & \lambda_2 & \ldots & \lambda_n \\ \lambda_1^2 & \lambda_2^2 & \ldots & \lambda_n^2 \\ \vdots & \vdots & \ddots & \vdots \\ \lambda_1^{n-1} & \lambda_2^{n-1} & \ldots & \lambda_n^{n-1} \end{vmatrix}$$

is the so-called **Vandermonde**[40] **determinant**. It can be shown that V equals

$$V(\lambda_1, \lambda_2, \ldots, \lambda_n) = \prod_{1\leqslant i<j\leqslant n} (\lambda_j - \lambda_i) = (-1)^{n(n-1)/2} \prod_{1\leqslant i<j\leqslant n} (\lambda_i - \lambda_j),$$

which is the product of all differences $\lambda_j - \lambda_i$ with $j > i$. For instance, when $n = 3$ we obtain $V = (\lambda_2 - \lambda_1)(\lambda_3 - \lambda_2)(\lambda_3 - \lambda_1) = -(\lambda_1 - \lambda_2)(\lambda_1 - \lambda_3)(\lambda_2 - \lambda_3)$. This shows that the Wronskian of exponential functions is not zero if and only if all roots of the characteristic equation are different.

The general solution of Eq. (4.4.2) is a linear combination of the functions from the set (4.4.6), that is,

$$y = C_1 e^{\lambda_1 x} + C_2 e^{\lambda_2 x} + \cdots + C_n e^{\lambda_n x},$$

in which $C_1, C_2, \ldots, C_n$ are arbitrary constants. In particular, the general solution for Eq. (4.4.3) is

$$y = C_1 e^{\lambda_1 x} + C_2 e^{\lambda_2 x}, \tag{4.4.7}$$

where λ_1 and λ_2 are distinct roots of the characteristic equation $a\lambda^2 + b\lambda + c = 0$. Actually, this result follows from the statement below.

Theorem 4.16: If $m_1, m_2, \ldots, m_n$ are all distinct numbers, and $P_1, P_2, \ldots, P_n$ are polynomials, then

$$P_1(x)e^{m_1 x} + P_2(x)e^{m_2 x} + \cdots + P_n(x)e^{m_1 x} \tag{4.4.8}$$

cannot vanish identically unless all the $P_i(x)$ $(i = 1, 2, \ldots, n)$ vanish identically.

PROOF: We use mathematical induction to prove this claim. The basis case $n = 1$ is obviously true. Assuming that this is true for a given n; we prove it is true for $n + 1$. Let $m_1, m_2, \ldots, m_n, m_{n+1}$ be all distinct numbers, and $P_1, P_2, \ldots, P_n, P_{n+1}$ be polynomials such that

$$P_1(x)e^{m_1 x} + P_2(x)e^{m_2 x} + \cdots + P_n(x)e^{m_n x} + P_{n+1}(x)e^{m_{n+1} x} = 0.$$

If $P_{n+1}(x) \equiv 0$, then all the other polynomials are zero by the inductive hypothesis. If not, we can divide the equation by $P_{n+1}(x)e^{m_{n+1}x}$, obtaining

$$R_1(x)e^{r_1 x} + R_2(x)e^{r_2 x} + \cdots + R_n(x)e^{r_n x} + 1 = 0,$$

[40] Named after the French musician, mathematician, and chemist Alexandre-Théophile Vandermonde (1735–1796).

where the R_i are rational, $r_i = m_i - m_{n+1}$. Differentiating, we get

$$(R_1' + r_1 R_1)e^{r_1 x} + \cdots + (R_n' + r_n R_n)e^{r_n x} = 0.$$

If we clear fractions, the left-hand side becomes an expression of form (4.4.8), which vanishes identically. Since the common denominator does not vanish identically, the hypothesis of induction shows that for all i

$$R_i' + r_i R_i = 0, \qquad R_i = C_i\, e^{-r_i x}, \quad i = 1, 2, \ldots, n.$$

This contradicts the fact that all $R_i(x)$ are rational. Hence, all $P_i(x)$ are identically 0.

Example 4.4.1: Solve the initial value problem $y'' - 4y = 0, \quad y(0) = 0, \ y'(0) = 3$.

Solution. The characteristic equation is $\lambda^2 - 4 = 0$, with roots $\lambda_1 = 2$ and $\lambda_2 = -2$. Hence, the general solution of the differential equation is

$$y = C_1 e^{2x} + C_2 e^{-2x}.$$

It remains to enforce the conditions at $x = 0$. Now we have

$$0 = C_1 + C_2, \quad 3 = 2C_1 - 2C_2.$$

Simultaneously solving these equations for C_1 and C_2, we conclude that $C_1 = 3/4$ and $C_2 = -3/4$. Thus,

$$y = \frac{3}{4}\,e^{2x} - \frac{3}{4}\,e^{-2x} \qquad \text{or} \qquad y = \frac{3}{2}\,\sinh(2x)$$

since $\sinh\alpha = \frac{1}{2}\left(e^{\alpha} - e^{-\alpha}\right)$.

Example 4.4.2: Solve the equation $\dfrac{\mathrm{d}^3y}{\mathrm{d}x^3} - 4\dfrac{\mathrm{d}^2y}{\mathrm{d}x^2} + \dfrac{\mathrm{d}y}{\mathrm{d}x} + 6y = 0$.

Solution. First we write the characteristic equation

$$\lambda^3 - 4\lambda^2 + \lambda + 6 = 0.$$

Its roots $\lambda = -1, 2, 3$ may be obtained by synthetic division if one of the roots is known (or guessed). Suppose we find one root $\lambda = -1$. Then the characteristic polynomial can be factored as $\lambda^3 - 4\lambda^2 + \lambda + 6 = (\lambda+1)(\lambda^2 - 5\lambda + 6) = (\lambda+1)(\lambda-2)(\lambda-3)$. The general solution is seen to be $y = C_1 e^{-x} + C_2 e^{2x} + C_3 e^{3x}$. We check our answer with *Mathematica:*

```
L[x_, y_] = y'''[x] - 4 y''[x] + y'[x] + 6 y[x]
CharPoly[lambda_] = Coefficient[L[x, Exp[lambda #] &], Exp[lambda x]]
roots = lambda /. Solve[CharPoly[lambda] == 0, lambda]
Solns = Map[Exp[# x] &, roots]
AllSolns[x_] = Solns/.{c1, c2, c3}
Simplify[L[x, AllSolns] == 0]
```

□

Now suppose that we want to find a particular member of the set of solutions (4.4.7) that satisfies the initial conditions $y(x_0) = y_0, \ y'(x_0) = y_0'$. To answer this question, we substitute the general solution $y = C_1\, e^{\lambda_1 x} + C_2\, e^{\lambda_2 x}$ into the given initial conditions. This will lead to the following algebraic system of equations with respect to C_1, C_2:

$$C_1 e^{\lambda_1 x_0} + C_2 e^{\lambda_2 x_0} = y_0, \qquad C_1 \lambda_1\, e^{\lambda_1 x_0} + C_2 \lambda_2\, e^{\lambda_2 x_0} = y_0'.$$

The determinant of these algebraic equations for C_1 and C_2 is the Wronskian $W\left[e^{\lambda_1 x}, e^{\lambda_2 x}; x\right] = (\lambda_2 - \lambda_1)e^{(\lambda_1+\lambda_2)x} \neq 0$. Therefore, we can find C_1 and C_2 to be

$$C_1 = \frac{y_0' - y_0\lambda_2}{\lambda_1 - \lambda_2}\, e^{-\lambda_1 x_0}, \qquad C_2 = \frac{y_0\lambda_1 - y_0'}{\lambda_1 - \lambda_2}\, e^{-\lambda_2 x_0}. \tag{4.4.9}$$

Problems In all problems, D stands for the derivative, while $\mathtt{D}^0$, the identity operator, is omitted. The fourth derivative of the function y is denoted by $y^{(4)}$, whereas all previous derivatives are denoted by primes; so y''' is the third derivative of y. The derivatives with respect to t are denoted by dots.

1. Would the function $y(x) = e^{\lambda x}$ be a solution of the differential equation $x^2\, y'' + 2x\, y' + y = 0$ for some value λ?

2. Write out the characteristic equation for the given differential equation.

(a)	$y'' - y' + 2y = 0;$	(b)	$2y''' + y' - 4y = 0;$	(c)	$y^{(4)} - y = 0;$
(d)	$2y'' + y' - 5y = 0;$	(e)	$y''' - y = 0;$	(f)	$y^{(4)} + 2y'' + y = 0;$
(g)	$3y'' + 4y' + y = 0;$	(h)	$3y''' - y'' + y = 0;$	(i)	$y^{(4)} - 2y''' + y'' - 2y' = 0.$

3. The characteristic equation for a certain differential equation is given. State the order of the differential equation and give the form of the general solution.

(a)	$\lambda^2 + \lambda - 6 = 0;$	(b)	$\lambda^2 - 3\lambda + 2 = 0;$	(c)	$\lambda^2 + 2\lambda - 3 = 0;$
(d)	$\lambda^2 - 1 = 0;$	(e)	$\lambda^3 - 3\lambda^2 - 2 = 0;$	(f)	$3\lambda^2 + 2\lambda^2 - 1 = 0;$
(g)	$\lambda^3 - 2\lambda^2 - \lambda + 2 = 0;$	(h)	$2\lambda^2 - \lambda^2 - 1 = 0;$	(i)	$4\lambda^2 - 1 = 0.$

4. Write the general solution to the given differential equations (D stands for the derivative).

(a)	$8y'' - 6y' + y = 0;$	(b)	$y''' + y'' - 2y' = 0;$	(c)	$4y''' - 21y' - 10y = 0;$
(d)	$y''' + 30y = 19y';$	(e)	$y''' - 14y' + 8y = 0;$	(f)	$(\mathtt{D}^3 + 3\,\mathtt{D}^2 - 4\,\mathtt{D})y = 0;$
(g)	$(\mathtt{D}^2 + 2\,\mathtt{D})y = 0;$	(h)	$(\mathtt{D}^2 + \mathtt{D} - 6)y = 0;$	(i)	$(\mathtt{D}^3 - 3\,\mathtt{D}^2 - 10\,\mathtt{D})y = 0;$
(j)	$(\mathtt{D}^2 + 2\,\mathtt{D})y = 3y;$	(k)	$(\mathtt{D}^2 - 5\,\mathtt{D} + 6)y = 0;$	(l)	$y''' + 6y'' + 11y' + 6y = 0;$
(m)	$4y'' - 4y' - 3y = 0;$	(n)	$4y'' - 4y' - 5y = 0;$	(o)	$y''' + 3y'' - 4y' - 12y = 0.$

5. Solve the initial value problems.

(a) $y'' - 2y' - 3y = 0, \quad y(0) = 0,\ y'(0) = -4.$

(b) $y'' - y' - 6y = 0, \quad y(0) = 0,\ y'(0) = 5.$

(c) $4y'' - 4y' - 3y = 0, \quad y(0) = 0,\ y'(0) = 2.$

(d) $y'' + 3y' - 10y = 0, \quad y(0) = 2,\ y'(0) = -3.$

(e) $y'' - 6y' + 8y = 0, \quad y(0) = 2,\ y'(0) = 6.$

(f) $2y'' - 5y' + 2y = 0, \quad y(0) = 3,\ y'(0) = 3.$

6. Find the solutions to the given initial value problems for equations of degree greater than 2.

(a) $y''' - 4y' = 0, \quad y(0) = 0,\ y'(0) = 1,\ y''(0) = 8.$

(b) $y''' - 2y'' - 5y' + 6y = 0, \quad y(1) = 0,\ y'(1) = 1,\ y''(1) = -1.$

(c) $4y''' - 4y'' - 23y' + 30y = 0, \quad y(0) = 0,\ y'(0) = -4,\ y''(0) = 4.$

(d) $y''' - 2y'' - y' + y = 0, \quad y(0) = 0,\ y'(0) = 5,\ y''(0) = 3.$

(e) $y''' - 3y'' - 25y' - 21y = 0, \quad y(0) = 5,\ y'(0) = -2,\ y''(0) = 8.$

7. Construct the general form of the solution that is bounded as $x \to \infty$.

(a)	$y'' - y = 0;$	(b)	$y'' = 0;$	(c)	$y'' + y' - 6y = 0;$	(d)	$2y'' + 3y' - 2y = 0;$
(e)	$y'' + y = 0;$	(f)	$y'' = 9;$	(g)	$y''' + 108y = 9y';$	(h)	$3y'' + 17y' - 6y = 0.$

8. Consider a constant coefficient differential equation $y'' + a_1\,y' + a_0\,y = 0$, whose characteristic equation has two distinct real roots λ_1 and λ_2. What conditions on the coefficients a_1 and a_0 guarantee that every solution to the given ODE satisfies $\lim_{x\to\infty} y(x) = 0$?

9. Find the fundamental set of solutions for the following equations specified by Theorem 4.12, page 204.

(a) $\ddot{y} = y;$ (b) $\ddot{y} + \dot{y} - 6y = 0;$ (c) $\ddot{y} - 3\dot{y} + 2y = 0;$ (d) $2\ddot{y} + \dot{y} - y = 0.$

10. Let D denote the derivative operator. Find the fundamental set of solutions of the following third order differential equations by extending Theorem 4.12.

(a) $(3\,\mathtt{D}^3 + 5\,\mathtt{D}^2 - 2\,\mathtt{D})y = 0;$ (b) $(\mathtt{D}^3 - 3\,\mathtt{D}^2 + 4)y = 0;$ (c) $(\mathtt{D}+1)(\mathtt{D}+2)\mathtt{D}y = 0.$

11. Find the solution of the initial value problem

$$y'' - 4y = 0, \qquad y(0) = 3, \quad y'(0) = 2.$$

Plot the solution for $0 \leqslant t \leqslant 1$ and determine its minimum value.

12. Find the solution of the initial value problem

$$2y'' - 5y' + 2y = 0, \qquad y(0) = 1, \quad y'(0) = -1.$$

Then determine the point where the solution is zero.

13. Solve the initial value problem $2y'' + 5y' - 3y = 0,\ y(0) = 1,\ y'(0) = \beta$. Then find the smallest positive value of β for which the solution has no minimum point.

4.5 Complex Roots

Consider an equation $L[\mathtt{D}]y = 0$ for the linear constant coefficient differential operator (4.4.1), page 208. Suppose that the corresponding characteristic equation has complex roots. To construct the fundamental set of solutions, we need to lay down a definition of $\exp(z)$ for the complex values of z. First, we recall some basic properties of complex[41] numbers.

Definition 4.7: A *complex number* is an ordered pair of two real numbers, often written as $z = (x, y)$. The first coordinate (abscissa) x is called the *real part* of the complex number z. Historically, the real number (ordinate) y is called the *imaginary part* of the complex number z. For these components, the following notations are usually used: $\Re z = \operatorname{Re} z = x$ and $\Im z = \operatorname{Im} z = y$.

Two complex numbers $z_1 = (x_1, y_1)$ and $z_2 = (x_2, y_2)$ are equal if and only if their real and imaginary components are separately equal, that is, $x_1 = x_2$ and $y_1 = y_2$. The set of all ordered pairs can be visualized as the set of points on the plane $\mathbb{R}^2$ or the set of vectors on the plane starting at the origin. However, we distinguish the set complex numbers from two former sets by introducing the following arithmetic operations:

1° **Addition:** $z_1 + z_2 = (x_1 + x_2, y_1 + y_2)$.

2° **Subtraction:** $z_1 - z_2 = (x_1 - x_2, y_1 - y_2)$.

3° **Multiplication:** $z_1 z_2 = (x_1 x_2 - y_1 y_2, x_1 y_2 + x_2 y_1)$.

4° **Division:** $\dfrac{z_1}{z_2} = \left(\dfrac{x_1 x_2 + y_1 y_2}{x_2^2 + y_2^2}, \dfrac{x_2 y_1 - x_1 y_2}{x_2^2 + y_2^2} \right)$.

Definition 4.8: The set of all ordered pairs with such operations of addition, subtraction, multiplication, and division is called the field of complex numbers and it is denoted as $\mathbb{C}$.

Unfortunately, there is no unique notation for geometric representation of complex numbers—actually, there are two of them. Euler suggested to denote by $\mathbf{1} = (1, 0)$ and $\mathbf{i} = (0, 1)$ the unit vectors of abscissa (x-axis) and ordinate (y-axis), respectively. Mathematicians still follow this genius. On the other hand, in engineering and computer science, it is a custom to use another notation, $\mathbf{j} = (0, 1)$, for the unit vector in the positive vertical direction, while keeping $\mathbf{i} = (1, 0)$ for the unit vector in the horizontal direction. Historically, the x-axis is called the **axis of reals** and the y-axis is called the **axis of imaginaries**. Such representation of an arbitrary complex number z in terms of its projections $x = \Re z$ and $y = \Im z$,

$$z = x\,\mathbf{1} + y\,\mathbf{i} \qquad \text{or} \qquad z = x\,\mathbf{i} + y\,\mathbf{j}, \tag{4.5.1}$$

is called the **Cartesian** (or rectangular) form of a complex number. It is a custom to drop the unit vector in the horizontal direction and use only one vector along the ordinate, either $\mathbf{i}$ in mathematics or $\mathbf{j}$ in engineering. Henceforth, we shall denote the complex number $(a, 0)$ simply as a. Since addition and multiplication are commutative, we have $x + y\mathbf{j} = y\mathbf{j} + x$ and $y\mathbf{j} = \mathbf{j}y$; for brevity, we write x instead of $x + \mathbf{j}0$ and $\mathbf{j}y$ instead of $0 + \mathbf{j}y$. The former complex number is said to be *purely real* and the latter is said to be *purely imaginary*.

The representation in **polar** coordinates proves a useful alternative to the Cartesian form (4.5.1). We take the pole at the origin of the Cartesian axis and the polar axis along the positive axis of reals. Let r, θ denote the polar coordinates of the point $z = (x, y)$ on the complex plane. Then

$$x = r\cos\theta, \quad y = r\sin\theta$$

and (4.5.1) becomes

$$z = x + \mathbf{j}y = r(\cos\theta + \mathbf{j}\sin\theta).$$

The last relation gives the **trigonometric form** of the complex number z. The quantities r and θ are called the **modulus** (absolute value) and **argument** (or amplitude) of z, respectively. The modulus $r = |z| = \sqrt{x^2 + y^2}$ is the

[41]Complex numbers were named by Carl Friedrich Gauss (1777–1855), but they were used many years previously, starting from the work on cubic equations by the Italian mathematician Gerolamo Cardano (1501–1576). The word "complex" means a union of two (real) numbers.

distance from the origin to the point $z = x + \mathbf{j}y$. In polar coordinates, infinitely many values of θ represent the same point on the plane, which depend on the number of whole revolutions. Therefore, the argument is not a function because it has an infinite number of values differing from one another by arbitrary multiples of 2π. To single out one of them, we restrict the range of θ to obtain a function $\theta = \arg z$, called the **branch of the argument** of z, if the range of $\arg z$ coincides with an interval of the length 2π. If $-\pi < \theta \leqslant \pi$, then such a value of θ is called the **chief amplitude** of z or the **principal branch** of the argument of z.

Definition 4.9: The complex number $x - \mathbf{j}y$ is called the **conjugate** of the complex number $z = x + \mathbf{j}y$ and is denoted by $\overline{z} = x - \mathbf{j}y$. Their product $z \cdot \overline{z} = r^2$ is modulus squared.

Definition 4.10: The operation of raising the number $e \approx 2.71828\ldots$ to the power $z = x + \mathbf{j}y$ is defined by the equation

$$e^{x+\mathbf{j}y} = e^x \, e^{\mathbf{j}y} = e^x \,(\cos y + \mathbf{j} \sin y).$$

Putting $x = 0$ and $y = \theta$, we obtain **Euler's formula**:

$$e^{\mathbf{j}\theta} = \cos\theta + \mathbf{j}\sin\theta \qquad (\mathbf{j}^2 = -1). \tag{4.5.2}$$

From elementary algebra, it is known that complex roots of a polynomial with real coefficients always come in conjugate pairs. Therefore, if $\lambda = \alpha + \mathbf{j}\beta$ is a root of the characteristic equation $L(\lambda) = 0$, then $\overline{\lambda} = \alpha - \mathbf{j}\beta$ is also a root of $L(\lambda) = 0$. It must be kept in mind that this result is a consequence of the real coefficients in the equation $L(\lambda) = 0$. Complex roots do not necessarily appear in pairs in an algebraic equation whose coefficients involve imaginary numbers.

Example 4.5.1: Consider a differential equation $y'' + \pi^2 y = 0$, for which the characteristic equation becomes $\lambda^2 + \pi^2 = 0$. Obviously, it has two complex conjugate roots: $\lambda = \pm\mathbf{j}\pi$, to which correspond two linearly independent solutions $y_{1,2} = e^{\pm\mathbf{j}\pi x} = \cos(\pi x) \pm \mathbf{j}\sin(\pi x)$. Some values of these functions are

$$e^{\mathbf{j}\pi/2} = \cos\frac{\pi}{2} + \mathbf{j}\sin\frac{\pi}{2} = \mathbf{j}. \qquad e^{\mathbf{j}\pi} = \cos\pi + \mathbf{j}\sin\pi = -1.$$

$$e^{\mathbf{j}3\pi/2} = \cos\frac{3\pi}{2} + \mathbf{j}\sin\frac{3\pi}{2} = -\mathbf{j}. \qquad e^{\mathbf{j}2\pi} = \cos 2\pi + \mathbf{j}\sin 2\pi = 1.$$

$$e^{\mathbf{j}\pi/4} = \cos\frac{\pi}{4} + \mathbf{j}\sin\frac{\pi}{4} = \frac{1}{\sqrt{2}} + \mathbf{j}\frac{1}{\sqrt{2}}.$$

□

Now we construct, in a usable form, solutions of the homogeneous equation $L[\texttt{D}]y = 0$, where $L[\texttt{D}]$ is a constant coefficient differential operator (4.4.1), page 208, whose characteristic equation has a complex root. Let us for simplicity consider the second order differential equation

$$ay'' + by' + cy = 0 \tag{4.5.3}$$

with real constant coefficients. Suppose now that the discriminant $b^2 - 4ac$ is negative. Then, the roots of the characteristic equation, $a\lambda^2 + b\lambda + c = 0$, are complex conjugate numbers; we denote them by

$$\lambda_1 = \alpha + \mathbf{j}\beta, \quad \lambda_2 = \alpha - \mathbf{j}\beta \quad \left(\alpha = -\frac{b}{2a}, \; \beta = \frac{\sqrt{4ac - b^2}}{2a}, \quad 4ac > b^2\right).$$

The corresponding complex conjugate functions

$$y_1(x) = e^{(\alpha+\mathbf{j}\beta)x}, \qquad y_2(x) = e^{(\alpha-\mathbf{j}\beta)x}$$

form the fundamental set of solutions for Eq. (4.5.3) because their Wronskian $W[y_1, y_2](x) = -2\mathbf{j}\beta\, e^{2\alpha x} \neq 0$ for any x. According to Theorem 4.11 on page 204, the linear combination

$$y(x) = C_1 y_1(x) + C_2 y_2(x) = C_1\, e^{\alpha x + \mathbf{j}\beta x} + C_2\, e^{\alpha x - \mathbf{j}\beta x} \tag{4.5.4}$$

is also a solution of Eq. (4.5.3). The constants C_1 and C_2 in the above linear combination cannot be chosen arbitrary—neither real nor complex. Indeed, if C_1 C_2 are arbitrary real numbers, then the solution (4.5.4) is a complex-valued function. However, we are after a real-valued solution. Since y_1 and y_2 are complex conjugate functions, the coefficients C_1 and C_2 must also be complex conjugate numbers $C_1 = \overline{C}_2$. Only in this case their linear combination will define a real-valued function containing two real arbitrary constants. Using Euler's formula (4.5.2), we obtain

$$y(x) = C_1\, e^{\alpha x}\,(\cos \beta x + \mathbf{j} \sin \beta x) + C_2\, e^{\alpha x}\,(\cos \beta x - \mathbf{j} \sin \beta x). \tag{4.5.5}$$

We can rewrite the last expression as

$$y(x) = (C_1 + C_2)\, e^{\alpha x}\, \cos \beta x + \mathbf{j}(C_1 - C_2)\, e^{\alpha x}\, \sin \beta x.$$

Finally, let $C_1 + C_2 = C_1 + \overline{C}_1 = C_3$ and $\mathbf{j}(C_1 - C_2) = \mathbf{j}(C_1 - \overline{C}_1) = C_4$, where C_3 and C_4 are new real arbitrary constants. Then the linear combination (4.5.5) is seen to provide the real-valued general solution

$$y(x) = C_3\, e^{\alpha x}\, \cos \beta x + C_4 e^{\alpha x}\, \sin \beta x = e^{\alpha x}\,(C_3\, \cos \beta x + C_4\, \sin \beta x)\,, \tag{4.5.6}$$

corresponding to the two complex conjugate roots $\lambda_1 = \alpha + \mathbf{j}\beta$, and $\lambda_2 = \alpha - \mathbf{j}\beta$, $(\beta \neq 0)$ of the characteristic equation. We would prefer to have a real-valued solution (4.5.6) instead of the complex form (4.5.5). In order to guarantee a real value of the expression (4.5.5), we must choose arbitrary constants C_1 and C_2 to be conjugate numbers. It is more convenient to use real numbers for C_3 and C_4 to represent a general solution as a real-valued function. Whenever a pair of simple conjugate complex roots of the characteristic equation appears, we write down at once the general solution corresponding to those two roots in the form given on the right-hand side of Eq. (4.5.6).

We can rewrite Eq. (4.5.3) in the operator form

$$a\left[(\texttt{D} - \alpha)^2 + \beta^2\right] y = 0, \tag{4.5.7}$$

where $\alpha = \frac{b}{2a}$, $\beta = \frac{4ac-b^2}{4a^2}$, and $\texttt{D}$ stands for the derivative operator. This is obtained by completing the squares:

$$\begin{aligned} a\texttt{D}^2 + b\texttt{D} + c &= a\left[\texttt{D}^2 + 2 \cdot \texttt{D} \cdot \frac{b}{2a} + \left(\frac{b}{2a}\right)^2 - \left(\frac{b}{2a}\right)^2 + \frac{c}{a}\right] \\ &= a\left[\left(\texttt{D} + \frac{b}{2a}\right)^2 + \frac{c}{a} - \frac{b^2}{4a^2}\right] = a\left[\left(\texttt{D} + \frac{b}{2a}\right)^2 + \frac{4ac - b^2}{4a^2}\right]. \end{aligned}$$

The general solution of Eq. (4.5.7) has the form (4.5.6). If in Eq. (4.5.7) we make the substitution: $y = e^{\alpha t}\, v$, then v will satisfy the canonical equation

$$v'' + \beta^2 v = 0 \qquad \Longrightarrow \qquad v(x) = C_1\, \cos \beta x + C_2\, \sin \beta x. \tag{4.5.8}$$

Example 4.5.2: Find the general solution of

$$y'' + 4y = 0.$$

Solution. The characteristic equation is $\lambda^2 + 4 = 0$, with the roots $\lambda = \pm 2\mathbf{j}$. Thus, the real part of λ is $\Re\lambda = 0$ and the imaginary part of λ is $\Im\lambda = \pm 2$. The general solution becomes

$$y(x) = C_1 \cos 2x + C_2 \sin 2x.$$

Note that if the real part of the roots is zero, as in this example, then there is no exponential term in the solution.

Example 4.5.3: Solve the equation

$$y'' - 2y' + 5y = 0.$$

Solution. We can rewrite this equation in operator form as

$$\left[\texttt{D}^2 - 2\,\texttt{D} + 5\right] y = 0 \qquad \text{or} \qquad \left[(\texttt{D} - 1)^2 + 4\right] y = 0,$$

where $\texttt{D} = \mathrm{d}/\mathrm{d}x$. Then, the general solution is obtained according to the formula (4.5.6):

$$y(x) = C_1\, e^x\, \cos 2x + C_2\, e^x\, \sin 2x = e^x\,(C_1\, \cos 2x + C_2\, \sin 2x)\,.$$

Note that the same answer can be obtained with the aid of substitution (4.1.19), page 194, $y = e^x v(x)$, where v satisfies the two-term (canonical) equation $v'' + 4v = 0$.

If the initial position is specified, $y(0) = 2$, we get $C_1 = 2$, so the solution becomes

$$y(x) = 2\, e^x\, \cos 2x + C_2\, e^x\, \sin 2x.$$

This one-parameter family of curves is plotted with the following *Mathematica* script:

```
DSolve[{y''[x] - 2 y'[x] + 5 y[x] == 0, y[0] == 2}, y[x], x]
soln[x_] = Expand[y[x]] /. %[[1]] /. C[1] -> c1
curves = Table[soln[x], {c1, -2, 2}]
Plot[Evaluate[curves], {x, -1, 2.5}, PlotRange -> {-16, 15},
 PlotStyle -> {{Thick, Blue}, {Thick, Black}, {Thick, Blue}, {Thick, Black}}]
```

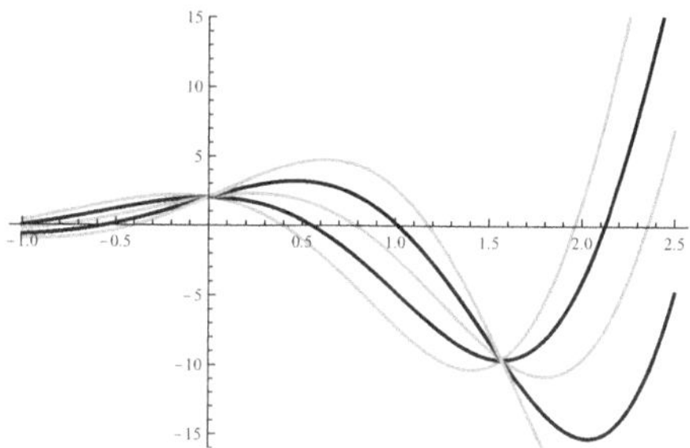

Figure 4.1: Example 4.5.3. The family of solutions subject to the initial condition $y(0) = 2$, plotted with *Mathematica*.

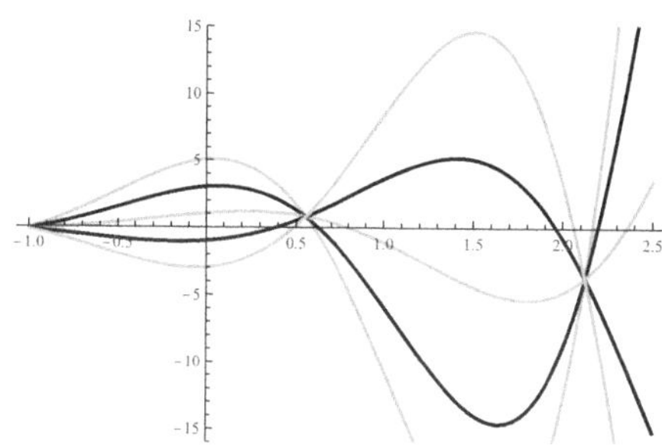

Figure 4.2: Example 4.5.3. The family of solutions subject to the initial condition $y'(0) = 1$, plotted with *Mathematica*.

Note that all solutions meet at the discrete number of points $x = \frac{\pi}{2} + n\pi$, $n = 0, \pm1, \pm2, \ldots$. If the initial velocity is specified, $y'(0) = 1$, we obtain another one-parameter family of solutions (see Fig. 4.2)

$$y(x) = (1 - C_2)\, e^x \, \cos 2x + C_2\, e^x \, \sin 2x.$$

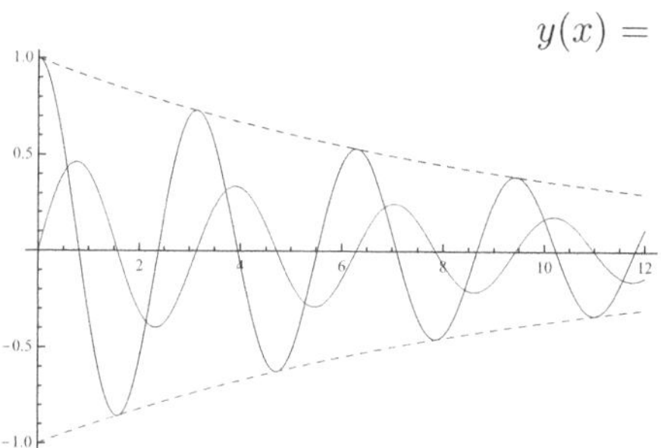

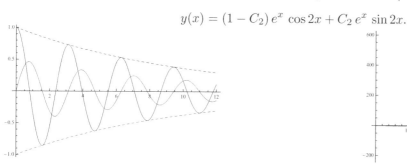

Figure 4.3: Two solutions of the equation $y'' + 0.2y' + 4.01\, y = 0$, plotted with *Mathematica*.

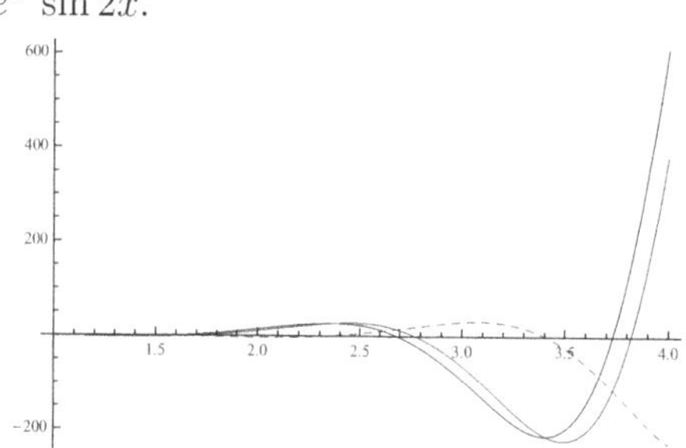

Figure 4.4: Example 4.5.4. Three solutions, plotted with *Mathematica*.

Example 4.5.4: Solve the equation

$$y''' - 3y'' + 9y' + 13y = 0.$$

Solution. The corresponding characteristic equation

$$\lambda^3 - 3\lambda^2 + 9\lambda + 13 = 0 \qquad \text{or} \qquad (\lambda + 1)\left((\lambda - 2)^2 + 9\right) = 0$$

has one real root, $\lambda_1 = -1$, and two complex conjugate roots $\lambda_{2,3} = 2 \pm 3\mathbf{j}$. They can be found with the aid of the *Mathematica* or *Maple*, as well as MuPad, *Maxima*, *Sage*, SymPy commands `Solve` or `solve`, respectively. Hence, the general solution of the differential equation is

$$y(x) = C_1\, e^{-x} + C_2\, e^{2x} \cos 3x + C_3\, e^{2x} \sin 3x.$$

Problems In all problems, `D` stands for the derivative operator, while $\mathtt{D}^0$, the identity operator, is omitted. The derivatives with respect to t are denoted by dots.

1. The characteristic equation for a certain homogeneous differential equation is given. Give the form of the general solution.

(a)	$\lambda^2 + 1 = 0;$	**(b)**	$\lambda^2 - 2\lambda + 2 = 0;$	**(c)**	$\lambda^2 + 2\lambda + 5 = 0;$
(d)	$\lambda^2 - 4\lambda + 5 = 0;$	**(e)**	$\lambda^2 - 2\lambda + 9 = 0;$	**(f)**	$9\lambda^2 - 6\lambda + 37 = 0;$
(g)	$4\lambda^2 + 4\lambda + 65 = 0;$	**(h)**	$9\lambda^2 + 6\lambda + 82 = 0;$	**(i)**	$\lambda^2 - 2\lambda + 26 = 0;$
(j)	$4\lambda^2 + 16\lambda + 17 = 0;$	**(k)**	$9\lambda^2 + 54\lambda + 82 = 0;$	**(l)**	$4\lambda^2 - 48\lambda + 145 = 0.$

2. Write the general solution of the following differential equations.

(a)	$y'' + 9y = 0;$	(b)	$y'' + 6y' + 13y = 0;$	(c)	$y''' + 4y'' + 5y' = 0;$
(d)	$y'' - 6y' + 25y = 0;$	(e)	$y'' + 10y' + 29y = 0;$	(f)	$4y'' - 4y' + 5y = 0;$
(g)	$2y'' + 10y' + 25y = 0;$	(h)	$y'' + 2y' + 10y = 0;$	(i)	$4y'' - 4y' + 17y = 0.$

3. Find the general solution of the differential equation of order 3, where $\texttt{D} = \mathrm{d}/\mathrm{d}x$.

$$\text{(a) } (\texttt{D}^3 + 8\,\texttt{D}^2 + 25\,\texttt{D})y = 0; \quad \text{(b) } 16y''' - 16y'' + 5y' = 0; \quad \text{(c) } (\texttt{D}^3 - 3\,\texttt{D} - 2)y = 0.$$

4. Find the solution of the initial value problems.

   (a) $y'' + 2y' + 2y = 0, \quad y(0) = 1,\ y'(0) = 1.$

   (b) $y'' + 4y' + 5y = 0, \quad y(0) = 0,\ y'(0) = 9.$

   (c) $y'' - 2y' + 2y = 0, \quad y(0) = -3,\ y'(0) = 0.$

   (d) $4y'' + 8y' + 5y = 0,\ y(0) = 1,\ y'(0) = 0.$

   (e) $\ddot{x} + 2\sqrt{k^2 - b^2}\,\dot{x} + k^2 x = 0,\ k > b > 0; \quad x = 0$ and $\dot{x} = \mathrm{d}x/\mathrm{d}t = v_0$ when $t = 0$.

   (f) $y'' + 2\alpha y' + (\alpha^2 + 1)y = 0, \quad y(0) = 1,\ y'(0) = 0.$ **(g)** $y'' + 49\,y = 0, \quad y(0) = 2,\ y'(0) = 1.$

5. Find the solution of the given IVPs for equations of degree greater than 2.

   (a) $y''' - 2y'' - 5y' + 6y = 0, \quad y(1) = 5,\ y'(1) = 0,\ y''(1) = 0.$

   (b) $y''' + y' = 0, \quad y(1) = 1,\ y'(1) = 1,\ y''(1) = 0.$

   (c) $y''' - 3y'' + 9y' + 13y = 0, \quad y(0) = 1,\ y'(0) = 0,\ y''(0) = 5.$

   (d) $y''' + 5y'' + 17y' + 13y = 0, \quad y(0) = 2,\ y'(0) = 0,\ y''(0) = -15.$

   (e) $y''' - y'' + y' - y = 0, \quad y(0) = 0,\ y'(0) = 1,\ y''(0) = 2.$

   (f) $4y''' + 28y'' + 61y' + 37y = 0, \quad y(0) = 1,\ y'(0) = 0,\ y''(0) = -5.$

   (g) $2y^{(4)} + 11y''' - 4y'' - 69y' + 34y = 0, \quad y(0) = 1,\ y'(0) = 3,\ y''(0) = -4,\ y'''(0) = 55.$

   (h) $y^{(4)} - 2y'' + 16y' - 15y = 0, \quad y(0) = 0,\ y'(0) = 0,\ y''(0) = -3,\ y'''(0) = -11.$

6. Replace $y = u/x$ in the equation $\dfrac{\mathrm{d}^2y}{\mathrm{d}x^2} + \dfrac{2}{x}\dfrac{\mathrm{d}y}{\mathrm{d}x} + y = 0$ and solve it.

7. Consider a constant coefficient differential equation $y'' + a_1 y' + a_0 y = 0$; its characteristic equation has two complex conjugate roots λ_1 and λ_2. What conditions on the coefficients, a_1 and a_0, guarantee that every solution to the given ODE satisfies $\lim_{x\to\infty} y(x) = 0$?

8. Find the solution to the initial value problem

$$4y'' - 4y' + 5y = 0, \qquad y(0) = 0, \quad y'(0) = 2,$$

   and determine the first time at which $|y(t)| = 3.4$.

9. Using the Taylor series $e^x = 1 + x + \frac{x^2}{2!} + \frac{x^3}{3!} + \cdots$, $\cos x = 1 - \frac{x^2}{2!} + \frac{x^4}{4!} - \cdots$, and $\sin x = x - \frac{x^3}{3!} + \frac{x^5}{5!} - \cdots$, prove the Euler formula (4.5.2).

10. As another way of obtaining Euler's formula (4.5.2), consider two functions $y_1(\theta) = e^{\mathbf{j}\theta}$ and $y_2(\theta) = \cos\theta + \mathbf{j}\sin\theta$. Show that these two functions are both solutions of the complex-valued initial value problem

$$y' = \mathbf{j}y, \qquad y(0) = 1.$$

11. Using a computer solver, plot on a common set of axes over the interval $-5 \leqslant x \leqslant 5$ some solutions to the differential equation subject to one initial condition $y(0) = 2$. At what point do all solutions meet?

    (a) $y'' - 2y' + 10y = 0;$ **(b)** $y'' - 4y' + 13y = 0.$

12. Solve the initial value problems.

(a) $4\,y'' - 12\,y' + 13y = 0,$	$y(0) = 0, \quad y'(0) = 1.$
(b) $4\,y'' - 20\,y' + 41y = 0,$	$y(0) = 2, \quad y'(0) = 5.$
(c) $y'' - 10\,y' + 29y = 0,$	$y(0) = 0, \quad y'(0) = 2.$
(d) $9\,y'' - 12\,y' + 40y = 0,$	$y(0) = 3, \quad y'(0) = 2.$

4.6 Repeated Roots. Reduction of Order

Suppose that in a constant coefficient linear differential equation

$$L[\mathtt{D}]y = 0, \tag{4.6.1}$$

the operator $L[\mathtt{D}] = a_n\mathtt{D}^n + a_{n-1}\mathtt{D}^{n-1} + \cdots + a_1\mathtt{D} + a_0$, with $\mathtt{D} = \mathrm{d}/\mathrm{d}x$, has repeated factors. That is, the characteristic equation $L(\lambda) = 0$ has repeated roots. For example, the second order constant coefficient linear operator $L[\mathtt{D}] = a\,\mathtt{D}^2 + b\,\mathtt{D} + c$ has a repeated factor if and only if the corresponding characteristic equation $a\lambda^2 + b\lambda + c = 0$ has a double root

$$\lambda_1 = \lambda_2 = -b/(2a)\,.$$

In other words, the quadratic polynomial can be factored $\lambda^2 + b\lambda + c = a(\lambda - \lambda_1)^2$ if and only if its discriminant $b^2 - 4ac$ is zero. In this case we have only one solution of exponential form:

$$y_1(x) = e^{-bx/(2a)}$$

for the differential equation

$$a\,y'' + b\,y' + c\,y = 0, \qquad b^2 = 4ac. \tag{4.6.2}$$

To find another linearly independent solution of Eq. (4.6.2), we use the method of reduction of order (see §2.6), credited to Jacob Bernoulli. Setting

$$y = v(x)\,y_1(x) = v(x)\,e^{-bx/(2a)}, \tag{4.6.3}$$

we have

$$\begin{aligned} y' &= v'(x)\,y_1(x) + v(x)\,y_1'(x) = v'(x)\,y_1(x) - \frac{b}{2a}v(x)\,y_1(x), \\ y'' &= v''(x)\,y_1(x) - \frac{b}{a}v'(x)\,y_1(x) + \left(\frac{b}{2a}\right)^2 v(x)\,y_1(x). \end{aligned}$$

By substituting these expressions into Eq. (4.6.2), we obtain

$$a\left[v''(x)\,y_1(x) - \frac{b}{a}v'(x)\,y_1(x) + \left(\frac{b}{2a}\right)^2 v(x)\,y_1(x)\right] + b\left[v'(x)\,y_1(x) - \frac{b}{2a}v(x)\,y_1(x)\right] + cv(x)\,y_1(x) = 0.$$

After collecting terms, the above expression is simplified to

$$av''(x)\,y_1(x) = 0.$$

The latter can be divided by the nonzero term $ay_1(x) = a\,e^{-bx/(2a)}$. This yields $v''(x) = 0$. After integrating, we obtain $v'(x) = C_1$ and $v(x) = C_1x + C_2$, where C_1 and C_2 are arbitrary constants. Finally, substituting for $v(x)$ into Eq. (4.6.3), we obtain the general solution of Eq. (4.6.2) as

$$y(x) = (C_1x + C_2)e^{-bx/(2a)}.$$

Functions $e^{\gamma x}$ and $x\,e^{\gamma x}$ are linearly independent since their Wronskian

$$W(x) = \begin{vmatrix} e^{\gamma x} & x\,e^{\gamma x} \\ \gamma\,e^{\gamma x} & (x\gamma + 1)\,e^{\gamma x} \end{vmatrix} = e^{2\gamma x} \neq 0.$$

Therefore, these functions form a fundamental set of solutions for Eq. (4.6.2), whatever the constant γ is.

Theorem 4.17: Let a_0, a_1, ..., a_n be n real (or complex) numbers with $a_n \neq 0$, and $y(x)$ be a n times continuously differentiable function on some interval $|a, b|$. Then $y(x)$ is a solution of the n-th order linear differential equation with constant coefficients

$$L[\mathtt{D}]y \stackrel{\text{def}}{=} \sum_{k=0}^{n} a_k\mathtt{D}^k y = a_n y^{(n)} + a_{n-1}y^{(n-1)} + \cdots + a_0 y(x) = 0, \quad \mathtt{D} = \mathrm{d}/\mathrm{d}x,$$

if and only if

$$y(x) = \sum_{j=1}^{m} e^{\lambda_j x} P_j(x), \tag{4.6.4}$$

where $\lambda_1, \lambda_2, \ldots, \lambda_m$ are distinct roots of the characteristic polynomial $\sum_{k=0}^{n} a_k \lambda^k = 0$ with multiplicities $m_1, m_2, \ldots, m_m$, respectively, and $P_k(x)$ is a polynomial of degree $m_k - 1$.

PROOF: We prove this statement by induction. For $n = 1$, the characteristic polynomial $L(\lambda) = a_1\lambda + a_0$ has the null $\lambda_1 = -a_0/a_1$. So the general solution is $y = C\, e^{\lambda_1 x}$, where $P_1(x) = C$ is a polynomial of degree $0 = m_1 - 1$.

Now, let $n \geqslant 1$ and assume that the assertion of the theorem is valid for the differential equation of order $k < n$. If λ_1 is a root of $L(\lambda) = 0$ with multiplicity m_1, then $L(\lambda) = P(\lambda)(\lambda - \lambda_1)$, where $P(\lambda)$ is a polynomial of degree $n - 1$ whose nulls are $\lambda_1, \lambda_2, \ldots, \lambda_m$ with multiplicities $m_1 - 1, m_2, \ldots, m_m$, respectively. Then the equation $P[\mathtt{D}]y = 0$ has a solution of the form (4.6.4) according to the induction hypothesis. Therefore, the equation $L[\mathtt{D}]y = P(\mathtt{D})(\mathtt{D} - \lambda_1)y = (\mathtt{D} - \lambda_1)P(\mathtt{D})y = 0$ is equivalent to $(\mathtt{D} - \lambda_1)y = \sum_k e^{\lambda_k x} Q_k(x)$ by the induction hypothesis, where $Q_k(x)$ are polynomials in x with degree $m_k - 1$ if $k > 1$ and degree $m_1 - 2$ if $k = 1$. Since $(\mathtt{D} - \lambda_1)y = \mathtt{D}\left(e^{-\lambda_1 x} y\right)$, we get the required result.

Example 4.6.1: Let us consider the differential equation

$$y'' + 4y' + 4y = 0.$$

The characteristic equation $\lambda^2 + 4\lambda + 4 = 0$ has a repeated root $\lambda = -2$. Hence, we get one exponential solution $y_1(x) = e^{-2x}$. Starting with $y(x) = v(x) y_1(x) = v(x)\, e^{-2x}$, we have

$$\begin{aligned} y'(x) &= v'(x)e^{-2x} - 2v(x)e^{-2x} \\ y''(x) &= v''(x)e^{-2x} - 4v'(x)e^{-2x} + 4v(x)e^{-2x}. \end{aligned}$$

Then,

$$\begin{aligned} y'' + 4y' + 4y &= v''(x)e^{-2x} - 4v'(x)e^{-2x} + 4v(x)e^{-2x} + 4v'(x)e^{-2x} \\ &\quad - 8v(x)e^{-2x} + 4v(x)e^{-2x} = v''(x)e^{-2x} = 0. \end{aligned}$$

Therefore, $v'' = 0$ and its solution is $v(x) = C_1 x + C_2$, Thus, the general solution of the equation becomes

$$y(x) = (C_1 x + C_2)e^{-2x}.$$ □

In general, suppose that the linear differential operator $L[\mathtt{D}]$ in Eq. (4.6.1) has repeated factors; that is, it can be written as

$$L[\mathtt{D}] = l[\mathtt{D}]\,(\mathtt{D} + \gamma)^m,$$

where $l[\mathtt{D}]$ is a linear operator of lower order, and m is a positive integer. This would lead us to the conclusion that any solution of the equation

$$(\mathtt{D} + \gamma)^m y = 0 \tag{4.6.5}$$

is also a solution of Eq. (4.6.1). The corresponding characteristic equation $L(\lambda) = 0$ would have a repeated factor $(\lambda + \gamma)^m$. Equation (4.6.5) would have an exponential solution

$$y(x) = e^{-\gamma x}.$$

Calculations show that

$$(\mathtt{D} + \gamma)[x^k\, e^{-\gamma x}] = kx^{k-1}\, e^{-\gamma x} - \gamma x^k\, e^{-\gamma x} + \gamma x^k\, e^{-\gamma x} = kx^{k-1}\, e^{-\gamma x}.$$

Then,

$$(\mathtt{D} + \gamma)^2[x^k\, e^{-\gamma x}] = x(\mathtt{D} + \gamma)[x^{k-1}\, e^{-\gamma x}] = k(k-1)x^{k-2}\, e^{-\gamma x}.$$

Repeating the operation, we are led to the formula

$$(\mathtt{D}+\gamma)^n[x^k\,e^{-\gamma x}] = \begin{cases} k(k-1)\cdots(k-n+1)\,e^{-\gamma x} = k^{\underline{n}}\,e^{-\gamma x}, & \text{if } n \leqslant k,\\ 0, & \text{if } k < n. \end{cases} \tag{4.6.6}$$

Here, $k^{\underline{n}} = k(k-1)\cdots(k-n+1)$ abbreviates the n-th falling factorial. We know that the function $e^{-\gamma x}$ is a solution of the equation $(\mathtt{D}+\gamma)y = 0$; therefore, for all $n > k$,

$$(\mathtt{D}+\gamma)^n[x^k\,e^{-\gamma x}] = 0, \qquad k = 0, 1, 2, \ldots, (n-1). \tag{4.6.7}$$

From Eq. (4.6.7), we find the general solution for Eq. (4.6.5)

$$y(x) = (C_1 + C_2 x + \ldots + C_n x^{n-1})\,e^{-\gamma x}$$

because functions $e^{-\gamma x}$, $x\,e^{-\gamma x}$, $\ldots$, $x^{k-1}\,e^{-\gamma x}$ are linearly independent (Example 4.3.3, page 206).

Example 4.6.2: Let us consider the differential equation

$$(\mathtt{D}^4 - 7\,\mathtt{D}^3 + 18\,\mathtt{D}^2 - 20\,\mathtt{D} + 8)y = 0, \quad \mathtt{D} = \mathrm{d}/\mathrm{d}x.$$

Since the corresponding characteristic polynomial is $L(\lambda) = (\lambda-1)(\lambda-2)^3$, the operator can be factored as

$$L[\mathtt{D}] \stackrel{\text{def}}{=} \mathtt{D}^4 - 7\,\mathtt{D}^3 + 18\,\mathtt{D}^2 - 20\,\mathtt{D} + 8 = (\mathtt{D}-1)(\mathtt{D}-2)^3 = (\mathtt{D}-2)^3(\mathtt{D}-1)$$

because constant coefficient linear differential operators commute. Then the general solution of the equation is a sum of two general solutions corresponding to $(\mathtt{D}-1)y = 0$ and $(\mathtt{D}-2)^3 y = 0$, respectively. Therefore,

$$y(x) = (C_1 + C_2 x + C_3 x^2)e^{-2x} + C_4 e^{-x}.$$

4.6.1 Reduction of Order

So far we have discussed constant coefficient linear differential equations. In this subsection, we show that the Bernoulli method and Abel's theorem are applicable to reduce the order of equations with variable coefficients.

Let us consider (for simplicity) a second order linear differential equation

$$y'' + p(x)\,y' + q(x)\,y = 0, \tag{4.6.8}$$

where $p(x)$ and $q(x)$ are continuous functions on the x-interval of interest. Suppose that one solution $y_1(x)$ of Eq. (4.6.8) is obtained by inspection (which is just a dodge to hide the fact that the process was one of trial and error). According to Bernoulli, we seek the second unknown member of the fundamental set of solutions, $y_2(x)$, in the product form

$$y_2(x) = v(x)\,y_1(x).$$

In order y_1 and y_2 to be linearly independent on some interval, the ratio y_2/y_1 must be nonconstant. Differentiating $y_2 = vy_1$ twice with respect to x yields

$$y_2' = v'y_1 + vy_1' \qquad \text{and} \qquad y_2'' = v''y_1 + 2v'y_1' + vy_1''.$$

Substituting into Eq. (4.6.8) gives $v''y_1 + 2v'y_1' + vy_1'' + p(x)(v'y_1 + vy_1') + qvy_1 = 0$ or

$$v''y_1 + v'(2y_1' + py_1) + v(y_1'' + py_1' + qy_1) = 0.$$

The coefficient of v in this expression vanishes because $y_1(x)$ is a solution of Eq. (4.6.8). Therefore, the product $y_2 = vy_1$ solves Eq. (4.6.8) provided $v(x)$ satisfies

$$v''y_1 + v'(2y_1' + py_1) = 0, \tag{4.6.9}$$

which is a first order separable (and linear) differential equation for $u = v'$. Separating the variables in Eq. (4.6.9), we get

$$\frac{u'}{u} = -\frac{2y_1' + py_1}{y_1} = -2\,\frac{y_1'}{y_1} - p = -2\,(\ln y_1)' - p.$$

Formal integration yields

$$\ln u = -2\ln y_1 - \int_{x_0}^{x} p(t)\,\mathrm{d}t + \ln C_1,$$

which, upon exponentiation, gives

$$u = v' = \frac{C_1}{y_1^2}\exp\left\{-\int_{x_0}^{x} p(t)\,\mathrm{d}t\right\},$$

where C_1 is a constant of integration provided that $y_1(x) \ne 0$ on some interval $|a,b|$ $(a < b)$. One more integration results in

$$v = C_1\int y_1^{-2}\exp\left\{-\int_{x_0}^{x} p(t)\,\mathrm{d}t\right\}\mathrm{d}x + C_2, \qquad x \in |a,b|.$$

Since $y_2 = vy_1$, another linearly independent solution becomes

$$y_2(x) = y_1(x)\int y_1^{-2}(x)\exp\left\{-\int_{x_0}^{x} p(t)\,\mathrm{d}t\right\}\mathrm{d}x, \quad y_1(x) \ne 0 \text{ on } |a,b|. \tag{4.6.10}$$

Actually, another linearly independent solution y_2 can be found from Abel's formula (4.2.3), page 201, and Exercise 12 asks you to generalize it for n-th order equations. Let $y_1(x)$ be a known solution of Eq. (4.6.5), then for another solution we have Abel's relation (4.2.3)

$$y_1y_2' - y_1'y_2 = W_0\exp\left\{-\int_{x_0}^{x} p(t)\,\mathrm{d}t\right\},$$

where W_0 is a constant. If y_2 is linearly independent from y_1, then this constant W_0 is not zero. The above differential equation can be solved with respect to y_2 using the integrating factor $\mu(x) = y_1^{-2}(x)$ to obtain the exact equation

$$\frac{\mathrm{d}}{\mathrm{d}x}\left[\frac{y_2(x)}{y_1(x)}\right] = \frac{W_0}{y_1^2(x)}\exp\left\{-\int_{x_0}^{x} p(t)\,\mathrm{d}t\right\}.$$

Since we are looking for just one linearly independent solution, we can set W_0 to be any number we want. For instance, we can choose $W_0 = 1$, and the next integration leads us to the formula (4.6.10). The following result generalizes the reduction of order for n-th order equations.

Theorem 4.18: If $y_1(x)$ is a known solution of the linear differential equation

$$a_n(x)y^{(n)} + a_{n-1}(x)y^{(n-1)} + \cdots + a_1(x)y' + a_0(x)y = 0,$$

which has the property that it never vanishes in the interval of definition of the differential equation, then the change of dependent variable $y = v\,y_1$ produces a linear differential equation of order $n-1$ for v.

Example 4.6.3: For a positive integer n, let us consider the differential equation

$$x\,y'' - (x+n)\,y' + ny = 0. \tag{4.6.11}$$

By inspection, we find an exponential solution, $y_1 = e^x$. To determine another solution, we use the Bernoulli substitution $y = v(x)\,e^x$. Substituting this function into Eq. (4.6.11) and solving the separable equation (4.6.9) for $u = v'$, we obtain $u = v' = x^n\,e^{-x}$ and the fundamental set of solutions for Eq. (4.6.11) consists of

$$y_1 = e^x \qquad \text{and} \qquad y_2 = e^x\int x^n e^{-x}\,\mathrm{d}x.$$ □

Let $w = v''/v' = u'/u$; we rewrite Eq. (4.6.9) in the following form:

$$y_1' + \frac{1}{2}\left(p(x) + \frac{v''}{v'}\right)y_1 = 0.$$

Assuming $w = v''/v'$ is known, we obtain the fundamental set of solutions for Eq. (4.6.8) given by

$$y_1 = \exp\left\{-\frac{1}{2}\int (p(x)+w)\,\mathrm{d}x\right\} \quad \text{and} \quad y_2 = v\exp\left\{-\frac{1}{2}\int (p(x)+w)\,\mathrm{d}x\right\}. \tag{4.6.12}$$

We substitute y_1 and its derivatives

$$y_1' = -\frac{1}{2}(p+w)\exp\left\{-\frac{1}{2}\int(p(x)+w)\,\mathrm{d}x\right\} = -\frac{1}{2}(p+w)\,y_1,$$
$$y_1'' = -\frac{1}{2}(p'+w')\exp\left\{-\frac{1}{2}\int(p(x)+w)\,\mathrm{d}x\right\} + \frac{1}{4}(p+w)^2\exp\left\{-\frac{1}{2}\int(p(x)+w)\,\mathrm{d}x\right\}$$

into Eq. (4.6.8) to obtain the Riccati equation (see §2.6.2 for details) $w^2 - 2w' = 2p' + p^2 - 4q$. If we know its solution, the fundamental set of solutions of Eq. (4.6.8) is given by the formula (4.6.12). In what follows, the function $D(x) = 2p'(x) + p^2(x) - 4q(x)$ is called the **discriminant** of Eq. (4.6.8). From §2.6.2, we know that the Riccati equation

$$w^2 - 2w' = D(x) \tag{4.6.13}$$

has a solution expressed in quadratures when $D(x)$ is a function of special form. For example, when $D(x) = k$ or $D(x) = kx^{-2}$, where k is a constant, the Riccati equation (4.6.13) is explicitly integrable. It is also known (see §2.6.2) that if $D(x) = kx^n$ when $n = (-4m)/(2m \pm 1)$ for some positive integer m, then the Riccati equation can be solved in terms of standard functions.

We consider first the case when $D(x)$ is a constant k. Then Eq. (4.6.13) becomes separable

$$\int \frac{2}{w^2 - k}\,\mathrm{d}w = \int \mathrm{d}x$$

and its integration depends on the sign of the coefficient k. Since $v''/v' = w$, the next integration yields

$$v(x) = \begin{cases} \dfrac{c_1 + c_2\tan(\alpha x)}{c_3 + c_4\tan(\alpha x)}, & \text{if } k = -4\alpha^2 < 0, \\[2ex] \dfrac{c_1 + c_2 x}{c_3 + c_4 x}, & \text{if } k = 0, \\[2ex] \dfrac{c_1 + c_2\,e^{x\sqrt{k}}}{c_3 + c_4\,e^{x\sqrt{k}}}, & \text{if } k > 0, \end{cases}$$

where c_1, c_2, c_3, and c_4 are real constants with $c_1c_4 \ne c_2c_3$.

Now we consider the case when $D(x) = kx^{-2}$ for some constant $k \ne 0$. The Riccati equation $w^2 - 2w' = kx^{-2}$ has a solution $w = \dfrac{\gamma}{x} + \dfrac{2(\gamma+1)x^\gamma}{C_1 - x^{\gamma+1}}$, where γ is a root of the quadratic equation $\gamma^2 + 2\gamma = k$ and C_1 is a constant of integration. Solving $v''/v' = w$, we obtain

$$v(x) = \int e^{\int w\,\mathrm{d}x}\,\mathrm{d}x = C_2\ln(C_1 - x^{\gamma+1}).$$

Example 4.6.4: Consider the linear equation

$$y'' + 2y' + \left(1 - \frac{3}{4x^2}\right)y = 0.$$

Since the discriminant is $D(x) = 3x^{-2}$, the equation for w becomes $w^2 - 2w' = 3x^{-2}$. It has a solution $w = 1/x$, so, according to Eq. (4.6.12), one of the solutions is

$$y_1 = \exp\left\{-\frac{1}{2}\int\left(2 + \frac{1}{x}\right)\mathrm{d}x\right\} = x^{-1/2}\,e^{-x}.$$

We find another linearly independent solution y_2 as a product $y_2 = v(x)y_1(x)$, where v is a solution of Eq. (4.6.9), i.e., $v''/v' = x^{-1}$. Integration gives $v = x^2$ and the second linearly independent solution becomes

$$y_2 = x^{3/2}\,e^{-x}.$$

4.6.2 Euler's Equations

An equation with variable coefficients

$$a\,x^2y'' + b\,x\,y' + c\,y = 0, \qquad x > 0, \tag{4.6.14}$$

where a, b, and c are real numbers, is called an **Euler's equation** (it is also known as the equidimensional equation). It is a particular case of more general equation

$$x^n a_n y^{(n)} + x^{n-1}a_{n-1}y^{(n-1)} + \cdots + a_1 x\,y' + a_0 y = 0 \tag{4.6.15}$$

of n-th order that is considered in §8.1.1. We present two approaches to deriving the general solution of the Euler equation: one is just guessing a solution in the form $y = x^r$, and another one is based on the change of independent variable $t = \ln x$. These two approaches are closely related.

The Euler equation can be reduced to an algebraic problem if we seek its solution in the form $y = x^r$, where r is a parameter to be determined. Upon differentiation, $y' = r\,x^{r-1}$ and $y'' = r(r-1)\,x^{r-2}$, we get from Eq. (4.6.14) that

$$a\,x^2\,r(r-1)\,x^{r-2} + bx\,r\,x^{r-1} + c\,x^r = 0,$$

or collecting similar terms

$$a\,r(r-1)\,x^r + br\,x^r + c\,x^r = 0, \qquad x > 0.$$

Factoring out x^r, we get the algebraic equation

$$a\,r^2 + (b-a)\,r + c = 0. \tag{4.6.16}$$

If this quadratic equation has two real distinct roots r_1 and r_2, then we have two linearly independent functions

$$y_1 = x^{r_1} \qquad \text{and} \qquad y_2(x) = x^{r_2}$$

that form the fundamental set of solutions to Eq. (4.6.14).

When equation (4.6.16) has either repeated roots or complex roots, it is more convenient to make substitution

$$t = \ln x \qquad \Longrightarrow \qquad \frac{\mathrm{d}}{\mathrm{d}x} = \frac{1}{x}\frac{\mathrm{d}}{\mathrm{d}t}, \qquad \frac{\mathrm{d}^2}{\mathrm{d}x^2} = -\frac{1}{x^2}\frac{\mathrm{d}}{\mathrm{d}t} + \frac{1}{x^2}\frac{\mathrm{d}^2}{\mathrm{d}t^2}.$$

Then the Euler equation (4.6.14) is converted into constant coefficient differential equation with respect to variable t:

$$a\,\ddot{y} + (b-a)\,\dot{y} + c\,y = 0. \tag{4.6.17}$$

Example 4.6.5: Solve the Euler equation

$$x^2y'' - x\,y' + 5\,y = 0, \qquad x > 0.$$

Solution. In the given differential equation, we make substitution $t = \ln x$, which yields the constant coefficient equation:

$$\ddot{y} - 2\,\dot{y} + 5\,y = 0.$$

The corresponding characteristic equation $\lambda^2 - 2\,\lambda + 5 = 0$ has two complex conjugate roots $\lambda = 1 \pm 2\mathbf{j}$. Therefore, this differential equation has the general solution

$$y(t) = e^t\left[C_1\cos 2t + C_2\sin 2t\right].$$

In original variables, we have

$$y(x) = x\left[C_1\cos\left(2\,\ln x\right) + C_2\sin\left(2\,\ln x\right)\right].$$

Problems In all problems, $\mathtt{D}$ stands for the derivative operator, while $\mathtt{D}^0$, the identity operator, is omitted. The derivatives with respect to t are denoted by dots.

1. The factored form of the characteristic equation for certain homogeneous differential equations is given. State the order of the differential equation and write down the form of its general solution.

(a)	$(\lambda-2)^2$;	**(b)**	$(\lambda-1)(\lambda+2)^2$;	**(c)**	$(\lambda^2-2\lambda+2)^2$;
(d)	$(\lambda+4)^3$;	**(e)**	$(\lambda-1)^2(\lambda+3)^3$;	**(f)**	$(\lambda-3)(\lambda^2+1)^2$;
(g)	$(\lambda^2+1)^2$;	**(h)**	$(\lambda^2+9)(\lambda^2-2\lambda+2)^2$;	**(i)**	$(\lambda+1)(\lambda^2+4)^3$;
(j)	$(\lambda^2+16)^3$;	**(k)**	$(\lambda^2+1)^3(\lambda^2-4\lambda+5)$;	**(l)**	$(\lambda^2+6\lambda+13)^2$.

2. Write the general solution of the second order differential equation

(a) $y'' - 6y' + 9y = 0$; (b) $y'' - 8y' + 16y = 0$; (c) $4y'' - 4y' + y = 0$;
(d) $y'' - 2y' + y = 0$; (e) $25y'' - 10y' + y = 0$; (f) $16y'' + 8y' + y = 0$;
(g) $(2\texttt{D} + 3)^2 y = 0$; (h) $(\texttt{D} + 1)^2 y = 0$; (i) $(3\texttt{D} - 2)^2 y = 0$.

3. Write the general solution of the differential equation of order larger than 2.

(a) $y^{(4)} - 6y''' + 9y'' = 0$; (b) $y''' - 8y'' + 16y' = 0$; (c) $4y''' + 4y'' + y' = 0$;
(d) $y''' - 2y'' + y' = 0$; (e) $(\texttt{D}^4 + 18\texttt{D}^2 + 81)y = 0$; (f) $9y''' + 6y'' + y' = 0$;
(g) $y^{(4)} + 6y''' + 9y'' = 0$; (h) $(\texttt{D}^2 + 1)^2 y = 0$; (i) $(\texttt{D}^2 + \texttt{D} - 6)^2\, y = 0$;
(j) $9y^{(4)} + 6y''' + y'' = 0$; (k) $4y''' + 12y'' + 9y' = 0$; (l) $y^{(4)} - 4y'' = 0$.

4. Find the solution of the initial value problem for the second order equation.

(a) $y'' + 4y' + 4y = 0$, $y(0) = 1$, $y'(0) = 1$. (b) $4y'' - 20y' + 25y = 0$, $y(0) = 1$, $y'(0) = 1$.
(c) $4y'' - 4y' + y = 0$, $y(1) = 0$, $y'(1) = 1$. (d) $4y'' + 28y' + 49y = 0$, $y(0) = 1$, $y'(0) = 1$.

5. Find the solution of the initial value problem for the third order equation.

(a) $y''' + 3y'' + 3y' + y = 0$, $y(0) = y'(0) = 0$, $y''(0) = 2$.

(b) $y''' + y'' - 5y' + 3y = 0$, $y(0) = 2$, $y'(0) = 5$, $y''(0) = 8$.

(c) $y''' + 5y'' + 7y' + 3y = 0$, $y(0) = 2$, $y'(0) = 0$, $y''(0) = \frac{3}{2}$.

(d) $y''' + 2y'' = 0$, $y(0) = y''(0) = 0$, $y'(0) = 4$.

(e) $y''' - 3y' - 2y = 0$, $y(0) = 1$, $y'(0) = 0$, $y''(0) = 8$.

(f) $8y''' - 4y'' - 2y' + y = 0$, $y(0) = 4$, $y'(0) = 0$, $y''(0) = 3$.

6. Consider the initial value problems.

(a) $25y'' - 20y' + 4y = 0$, $y(0) = 5$, $y'(0) = b$. (b) $4y'' - 12y' + 9y = 0$, $y(0) = 2$, $y'(0) = b$.

Find the solution as a function of b and then determine the critical value of b that separates negative solutions from those that are always positive.

7. Find the fundamental set of solutions for the following differential equations.

(a) $y'' + 3y' + \frac{1}{4}(9 + 1/x^2)y = 0$; (b) $y'' - 4y' \tan(2x) - 5y = 0$;
(c) $y'' + \frac{2}{1+x}\, y' + 9y = 0$; (d) $y'' + \left(1 + \frac{1}{x}\right) y' + \left(\frac{1}{4} + \frac{1}{2x}\right) y = 0$;
(e) $y'' + 2xy' + x^2 y = 0$; (f) $xy'' + (2 - 3x)y' + (2x - 4)y = 0$.

8. Determine the values of the constants a_0, a_1, and a_2 such that $y(x) = a_0 + a_1 x + a_2 x^2$ is a solution to the given differential equations

(a) $(1 + 9x^2)y'' = 18y$; (b) $(4 + x^2)\, y'' = 2y$; (c) $(x + 3)y'' + (2x - 5)y' - 4y = 0$.

Use the reduction of order technique to find a second linearly independent solution.

9. One solution y_1 of the differential equation is given, find another linearly independent solution.

(a) $(1 - x^2)y'' - 2xy' + 2y = 0$, $y_1 = x$. (b) $y'' - 2xy' + 2y = 0$, $y_1 = x$.
(c) $y'' + \tan x\, y' = 6\cot^2 x\, y$, $y_1 = \sin^3 x$. (d) $\frac{\text{d}}{\text{d}x}\left(x\frac{\text{d}y}{\text{d}x}\right) + \frac{4x^2-1}{4x}\, y = 0$, $y_1 = \frac{\cos x}{\sqrt{x}}$.
(e) $(1 + x^2)y'' = xy' + 2y/x^2$, $y_1 = x^2$. (f) $xy'' + y = (x - 1)y'$, $y_1 = 1 - x$.
(g) $(x + 2)y'' = (3x + 2)y' + 12y$, $y_1 = e^{3x}$. (h) $y'' - \frac{1}{x}\, y' + 4x^2\, y = 0$, $y_1 = \sin(x^2)$.

10. Find the discriminant of the given differential equation and solve it.

(a) $xy'' + 2y' + y/(4x) = 0$; (b) $x^2 y'' + x(2 - x)y = 0$.

11. The Legendre polynomial $P_0(x) = 1$ is clearly a solution of the differential equation $\left(1 - x^2\right) y'' - 2x\, y' = 0$. Find another linearly independent solution of this equation.

12. The Chebyshev polynomial $T_0(x) = 1$ of the first kind is clearly a solution of the differential equation $\left(1 - x^2\right) y'' - x\, y' = 0$. Find another linearly independent solution of this equation.

13. The Fibonacci polynomial $F_1(x) = 1$ is clearly a solution of the differential equation $\left(4 + x^2\right) y'' + 3x\, y' = 0$. Find another linearly independent solution of this equation.

14. The Lucas polynomial $L_1(x) = 1$ is clearly a solution of the differential equation $\left(4 + x^2\right) y'' + x\, y' = 0$. Find another linearly independent solution of this equation.

4.7 Nonhomogeneous Equations

In this section, we turn our attention from homogeneous equations to nonhomogeneous linear differential equations:

$$a_n(x)\frac{\mathrm{d}^n y}{\mathrm{d}x^n} + a_{n-1}(x)\frac{\mathrm{d}^{n-1} y}{\mathrm{d}x^{n-1}} + \cdots + a_0(x)y(x) = f(x), \tag{4.7.1}$$

where $f(x)$ and coefficients $a_n(x)$, ..., $a_0(x)$ are given real-valued functions of x in some interval. The associated (auxiliary) homogeneous equation is

$$a_n(x)\, y^{(n)} + a_{n-1}(x)\, y^{(n-1)} + \cdots + a_0(x)y(x) = 0, \tag{4.7.2}$$

where $y^{(n)}$ stands for the n-th derivative of $y(x)$. It is convenient to use operator notation for the left-hand side expression in equations (4.7.1) and (4.7.2): $L[x, \texttt{D}]$ or simply L, where

$$L[x, \texttt{D}] = a_n(x)\texttt{D}^n + a_{n-1}(x)\texttt{D}^{n-1} + \cdots + a_0(x), \tag{4.7.3}$$

and $\texttt{D}$ stands for the derivative operator $\texttt{D} = \mathrm{d}/\mathrm{d}x$. To find an integral of the equation $L[x, \texttt{D}]y = f$, we must determine a function of x such that, when $L[x, \texttt{D}]$ operates on it, the result is $f(x)$. The following theorem relates a nonhomogeneous differential equation to a homogeneous one and gives us a plan for solving Eq. (4.7.1).

Theorem 4.19: The difference between two solutions of the nonhomogeneous equation (4.7.1) on some interval $|a, b|$ is a solution of the homogeneous equation (4.7.2).

The sum of a particular solution of the driven equation (4.7.1) on an interval $|a, b|$ and a solution of the homogeneous equation (4.7.2) on $|a, b|$ is a solution of the nonhomogeneous equation (4.7.1) on the same interval.

PROOF: When the differential operator $L[x, \texttt{D}]$ acts on a function y, we write it as $L[x, \texttt{D}]y$ or $L[y]$ or simply Ly. Let y_1 and y_2 be two solutions of the nonhomogeneous equation (4.7.1):

$$Ly_1 = f(x), \qquad Ly_2 = f(x).$$

Since L is a linear operator (see Properties 1 and 2 on page 190), we obtain for their difference

$$L[y_1 - y_2] = Ly_1 - Ly_2 = f(x) - f(x) \equiv 0.$$

Similarly, for any particular solution $Y(x)$ of Eq. (4.7.1) and for a solution $y(x)$ of the homogeneous equation (4.7.2), we have

$$L[Y + y] = L[Y] + L[y] = f(x) + 0 = f(x).$$

Theorem 4.20: A **general solution** of the nonhomogeneous equation (4.7.1) on some open interval (a, b) can be written in the form

$$y(x) = y_h(x) + y_p(x), \tag{4.7.4}$$

where

$$y_h(x) = C_1 y_1(x) + C_2 y_2(x) + \cdots + C_n y_n(x)$$

is the general solution of the associated homogeneous equation (4.7.2) on (a, b), which is frequently referred to as a **complementary function**, and $y_p(x)$ is a particular solution of Eq. (4.7.1) on (a, b). Here C_j represents an arbitrary constant for $j = 1, 2, \ldots, n$, and $\{y_1(x), y_2(x), \ldots, y_n(x)\}$ is the fundamental set of solutions of Eq. (4.7.2).

PROOF: This theorem is a simple corollary of the preceding theorem.

Theorem 4.21: [**Superposition Principle for nonhomogeneous equations**] The general solution of the differential equation

$$L[x, \texttt{D}]y = f_1(x) + f_2(x) + \cdots + f_m(x), \quad \texttt{D} = \mathrm{d}/\mathrm{d}x,$$

on an interval $|a, b|$ is

$$y(x) = y_h(x) + y_{p1}(x) + y_{p2}(x) + \cdots + y_{pm}(x),$$

where $L[x, \texttt{D}]$ is a linear differential operator (4.7.3), $y_{pj}(x)$, $j = 1, 2, \ldots, m$, are particular solutions of $L[x, \texttt{D}]y = f_j(x)$ on $|a, b|$, and $y_h(x)$ is a general solution of the homogeneous equation $L[x, \texttt{D}]y_h = 0$.

PROOF: This follows from the fact that L is a linear differential operator. ■

Now we illustrate one of the remarkable properties of linear differential equations that, in some cases, it is possible to find the general solution of a nonhomogeneous equation $Ly = f$ without knowing all the solutions of the associated homogeneous equation $Ly = 0$. Consider a nonhomogeneous second order linear differential equation

$$y'' + p(x)y' + q(x)y = f(x) \tag{4.7.5}$$

on some interval where the coefficients $p(x)$, $q(x)$ and the forcing function $f(x)$ are continuous. Suppose that we know one solution $y_1(x)$ of the associated homogeneous equation $y'' + p(x)y' + q(x)y = 0$. Following Bernoulli, we seek a solution of Eq. (4.7.5) as a product $y = u(x)\, y_1(x)$, with some yet unknown function $u(x)$. Substituting $y = u\, y_1$ into Eq. (4.7.5) yields

$$y_1\, u'' + (2y_1' + py_1)\, u' = f,$$

which is a first order linear equation in u'. Therefore, it can be solved explicitly. Once u' has been obtained, $u(x)$ is determined by integration.

Example 4.7.1: Find the general solution of $x^2y'' + 2xy' - 2y = 4x^2$ on the interval $(0, \infty)$.

Solution. The associated homogeneous equation $x^2y'' + 2xy' - 2y = 0$ clearly has $y_1 = x$ as a solution. To find the general solution of the given nonhomogeneous equation, we substitute $y = u(x)\, x$, with a function $u(x)$ to be determined, into the given nonhomogeneous equation. Using the product rule of differentiation, we obtain

$$x^2\,(xu'' + 2u') + 2x\,(xu' + u) - 2xu = 4x^2,$$

which reduces to $x^3u'' + 4x^2u' = 4x^2$. Since the latter becomes exact upon multiplication by x, we get $\frac{\mathrm{d}}{\mathrm{d}x}\left(x^4\, u'\right) = 4x^3$. The next integration of $x^4\, u' = x^4 + C_1$ yields the general solution $u = x + C_1\, x^{-3} + C_2$; multiplying it by $y_1 = x$, we obtain: $y = x^2 + C_1\, x^{-2} + C_2\, x$.

4.7.1 The Annihilator

The relation between homogeneous (4.7.2) and nonhomogeneous differential equation (4.7.1) is better understood when the language of differential operators is involved. Upon introducing the derivative operator $\mathtt{D} = \mathrm{d}/\mathrm{d}x$, a differential equation (4.7.1) can be written in compact form $L[x, \mathtt{D}]\, y(x) = f(x)$. Correspondingly, the associated homogeneous equation (4.7.2) has the form $L[x, \mathtt{D}]\, y(x) = 0$. It is natural to introduce the following definition.

Definition 4.11: The linear differential operator $L[x, \mathtt{D}]$ is said to annihilate a function $y(x)$ if

$$L[x, \mathtt{D}]\, y(x) \equiv 0$$

for all x where the function $y(x)$ is defined. In this case, $L[x, \mathtt{D}]$ is called an **annihilator** of $y(x)$.

Note that the annihilator is not unique and can be multiplied (from left) by any linear differential operator. For instance, if L is an annihilator for $y(x)$, then $LL = L^2$ is also an annihilator of the function $y(x)$.

Now we turn our attention to constant coefficient linear differential operators, written as $L[\mathtt{D}] = a_n\mathtt{D}^n + a_{n-1}\mathtt{D}^{n-1} + \cdots + a_1\mathtt{D} + a_0$. It is a custom to drop the identity operator $\mathtt{D}^0$ and write the coefficient a_0 alone. From the equation $\mathtt{D}\, e^{\lambda x} = \lambda e^{\lambda x}$, it follows that $\mathtt{D}^n e^{\lambda x} = \lambda^n\, e^{\lambda x}$ and

$$L[\mathtt{D}]e^{\lambda x} = L(\lambda)e^{\lambda x}. \tag{4.7.6}$$

Eq. (4.7.6) shows that $\mathtt{D} - \gamma$ annihilates $e^{\gamma x}$. Therefore, if the operator $L[\mathtt{D}]$ annihilates the exponential function $e^{\gamma x}$, then $L[\mathtt{D}]$ has a factor $(\mathtt{D} - \gamma)$. This means that its characteristic polynomial $L(\lambda)$ is divisible by $\lambda - \gamma$.

A solution of the constant coefficient equation $L[\mathtt{D}]\, y = 0$ contains a term like x^n, with $n = 0, 1, 2, \ldots$, only when the operator $L[\mathtt{D}]$ has a factor $\mathtt{D}^{n+1}$. Similarly, solutions of homogeneous equations with constant coefficients comprise terms such as $e^{\alpha x}\cos\beta x$ or $e^{\alpha x}\sin\beta x$ only when the operator $L[\mathtt{D}]$ has a factor $((\mathtt{D} - \alpha)^2 + \beta^2)$. Say for the function $e^{\alpha x}\sin\beta x$, we have

$$[(\mathtt{D} - \alpha)^2 + \beta^2]\, e^{\alpha x}\sin\beta x = (\mathtt{D}^2 - 2\alpha\mathtt{D} + \alpha^2 + \beta^2]\, e^{\alpha x}\sin\beta x.$$

Substitution yields

$$\begin{aligned}[(\mathtt{D}-\alpha)^2+\beta^2]\,e^{\alpha x}\,\sin\beta x-\mathtt{D}^2 e^{\alpha x}\,\sin\beta x-2\alpha\mathtt{D}e^{\alpha x}\,\sin\beta x+(\alpha^2+\beta^2)\,e^{\alpha x}\,\sin\beta x\\ =(\alpha^2-\beta^2)\,e^{\alpha x}\,\sin\beta x+2\alpha\beta e^{\alpha x}\,\cos\beta x\\ -2\alpha^2 e^{\alpha x}\,\sin\beta x-2\alpha\beta e^{\alpha x}\,\cos\beta x+(\alpha^2+\beta^2)\,e^{\alpha x}\,\sin\beta x\equiv 0.\end{aligned}$$

Let $P_k(x)=p_k x^k+p_{k-1}x^{k-1}+\cdots+p_0$ and $Q_k(x)=q_k x^k+q_{k-1}x^{k-1}+\cdots+q_0$ be polynomials in x of order k $(p_k^2+q_k^2>0)$, then the function

$$P_k(x)\,e^{\alpha x}\,\cos\beta x+Q_k(x)\,e^{\alpha x}\,\sin\beta x,\quad p_k^2+q_k^2>0, \tag{4.7.7}$$

is a solution of the constant-coefficient differential equation

$$L[\mathtt{D}]\,f(x)=0,\qquad \mathtt{D}=\mathrm{d}/\mathrm{d}x,$$

if and only if its characteristic polynomial $L(\lambda)$ has a multiple $[(\lambda-\alpha)^2+\beta^2]^{k+1}$, that is,

$$L(\lambda)=[(\lambda-\alpha)^2+\beta^2]^{k+1}\,L_1(\lambda),$$

where $L_1(\lambda)$ is a polynomial of order $n-2k-2\geqslant 0$. Now we consider a nonhomogeneous differential equation with constant coefficients

$$L\,[\mathtt{D}]\,y(x)\stackrel{\text{def}}{=}\left[a_n\mathtt{D}^n+a_{n-1}\mathtt{D}^{n-1}+\cdots+a_1\mathtt{D}+a_0\right]y(x)=f(x), \tag{4.7.8}$$

assuming that we know an annihilator $\psi[\mathtt{D}]$ of the driving term $f(x)$, that is, $\psi\,[\mathtt{D}]\,f(x)\equiv 0$ for all x. Then the driven equation (4.7.8) can be reduced to a homogeneous equation, as the following statements assure us.

Theorem 4.22: Let p, q, and L be polynomials such that $L(\lambda)=p(\lambda)q(\lambda)$ and q is relatively prime to ψ, the annihilator of the forcing function $f(x)$ in the right-hand side of Eq. (4.7.8). Then there exists a solution of the differential equation $L[\mathtt{D}]y=f$, which is also a solution of

$$\psi[\mathtt{D}]\,L\,[\mathtt{D}]\,y(x)=\psi\,[\mathtt{D}]\,f(x)=0. \tag{4.7.9}$$

Such a solution can be obtained by applying a polynomial differential operator to any solution u of the equation $p[\mathtt{D}]u=f$.

PROOF: Since q and ψ are relatively prime, there exist polynomials h and g such that

$$gq+h\psi=1. \tag{4.7.10}$$

Now, let u be any solution of the equation $p[\mathtt{D}]u=f$ and set $y=g[\mathtt{D}]u$. To complete the proof, we need only to show that y is a solution of both (4.7.9) and (4.7.8). First, we multiply (4.7.10) by p and apply the corresponding differential operators to the function u. This gives

$$\begin{aligned} g[\mathtt{D}]\,q[\mathtt{D}]\,p[\mathtt{D}]u+h[\mathtt{D}]\,\psi[\mathtt{D}]\,p[\mathtt{D}]u &= p[\mathtt{D}]u,\\ L[\mathtt{D}]\,g[\mathtt{D}]u+h[\mathtt{D}]\,\psi[\mathtt{D}]\,p[\mathtt{D}]u &= p[D]u,\qquad \text{since } g[\mathtt{D}]u=y,\\ L[\mathtt{D}]y &= f,\qquad \text{by } \psi[\mathtt{D}]f=0\quad\text{and}\quad p[\mathtt{D}]u=f.\end{aligned}$$

Thus, y is a solution of Eq. (4.7.8) and we have

$$\psi[\mathtt{D}]\,p[\mathtt{D}]y=\psi[\mathtt{D}]\,p[\mathtt{D}]\,g[\mathtt{D}]u=g[\mathtt{D}]\,\psi[\mathtt{D}]f=0$$

since $p[\mathtt{D}]u=f$ and $\psi[\mathtt{D}]f=0$. So the equation (4.7.9) is also fulfilled.

Corollary 4.4: If polynomials L and ψ are relatively prime and $\psi(\mathtt{D})$ is an annihilator of f, then a solution of the equation $L[\mathtt{D}]\,y=f$ can be obtained by applying a polynomial differential operator to f.

Now we look into an opposite direction: given a set of linearly independent functions $\{y_1, y_2, \dots, y_n\}$, find the linear differential operator of minimum order that annihilates these functions. We are not going to treat this problem in general (for two functions y_1, y_2, the answer was given in §4.2); instead, we consider only constant coefficient differential equations:

$$L[\mathtt{D}]y = f \quad \text{or} \quad a_n y^{(n)} + a_{n-1}\, y^{(n-1)} + \cdots + a_1\, y' + a_0\, y = f. \tag{4.7.11}$$

As shown in §§4.4–4.6, a solution of a linear homogeneous differential equation $L[\mathtt{D}]y = 0$, for some linear constant coefficient differential operator L, is one of the following functions: exponential, sine, cosine, polynomial, or sums or products of such functions. That is, a solution can be a finite sum of the functions of the form (4.7.7).

We assign to a function of the form (4.7.7) a complex number $\sigma = \alpha + \mathbf{j}\beta$, called the **control number**. Its complex conjugate $\sigma = \alpha - \mathbf{j}\beta$ is also a control number. Actually, the control number of the function (4.7.7) is a root of the characteristic equation $\psi(\lambda) = 0$ associated with the annihilator $\psi[\mathtt{D}]$ of $f(x)$. Since we consider only differential equations with constant real coefficients, the complex roots (if any) of the corresponding characteristic equation may appear only in pairs with their complex conjugates.

The control number guides us about the form of a solution for an equation (4.7.11). Here $f(x)$ has a dual role. It is not just the forcing function of our nonhomogeneous differential equation $L[\mathtt{D}]\, y = f$, but it is also a solution to some unspecified, *homogeneous* differential equation with constant coefficients $\psi[\mathtt{D}]\, f = 0$. We use $f(x)$ to deduce the roots (control numbers) of the characteristic equation $\psi(\sigma) = 0$ corresponding to our unspecified, homogeneous differential equation $\psi[\mathtt{D}]\, f = 0$. This gives us the form of a particular solution for the nonhomogeneous differential equation whose solution is what we are really after. Such an approach is realized in §4.7.2.

Before presenting examples of control numbers, we recall the definition of the multiplicity. Let $L(\lambda) = a_n\lambda^n + a_{n-1}\lambda^{n-1} + \cdots + a_0$ be a polynomial or entire function and λ_0 is its null, that is, a root of the equation $L(\lambda) = 0$. The **multiplicity** of this null (or the root of the equation $L(\lambda) = 0$) is the number of its first derivative at which it is not equal to zero; that is, λ_0 is a null of the multiplicity m if and only if

$$L(\lambda_0) = 0,\ L'(\lambda_0) = 0, \dots, L^{(m-1)}(\lambda_0) = 0, \quad \text{but} \quad L^{(m)}(\lambda_0) \ne 0.$$

From the fundamental theorem of algebra, it follows that a polynomial of degree n has n (generally speaking, complex) nulls, $\lambda_1, \lambda_2, \cdots, \lambda_n$, some of which may be equal. If all coefficients of a polynomial are real numbers, the complex nulls (if any) appear in pairs with their complex conjugate numbers. The characteristic polynomial can be written in the product form

$$L(\lambda) = a_n(\lambda - \lambda_1)^{m_1}(\lambda - \lambda_2)^{m_2} \dots (\lambda - \lambda_k)^{m_k},$$

where $\lambda_1, \lambda_2, \dots, \lambda_k$ are distinct roots of the equation $L(\lambda) = 0$. The power m_j in this representation is called the **multiplicity** of zero (or null) λ_j $(j = 1, 2, \dots, k)$.

Example 4.7.2: Find the control number of the given functions.
(a) $f(x) = 2\mathbf{j} + 1$; **(b)** $f(x) = x^2 + 2x + \mathbf{j}$; **(c)** $f(x) = 2\mathbf{j}\, e^{-x}$;
(d) $f(x) = 2\, e^{2\mathbf{j}x}$; **(e)** $f(x) = e^{-x}\cos 2x$; **(f)** $f(x) = (x+2)e^{x}\sin 2x$.

Solution. In case **(a)**, the function $f(x)$ is a constant; therefore the control number is zero.
(b) The function $f(x)$ is a polynomial of the second order; therefore, the control number $\sigma = 0$. **(c)** $\sigma = -1$.
(d) The control number for the function

$$e^{-2\mathbf{j}x} = \cos 2x - \mathbf{j}\sin 2x \quad \text{(Euler's formula)}$$

is $\sigma = -2\mathbf{j}$ (or $2\mathbf{j}$, it does not matter because both are complex conjugate control numbers).
(e) The control number for the function $e^{-x}\cos 2x$ is $\sigma = -1 + 2\mathbf{j}$ or $\sigma = -1 - 2\mathbf{j}$ (it does not matter which sign is chosen). **(f)** The control number is $\sigma = 1 + 2\mathbf{j}$ (or $1 - 2\mathbf{j}$).

Example 4.7.3: Find a linear differential operator that annihilates the function $x^2\, e^{3x}\sin x$.

Solution. Since the control number of the given function is $3 + \mathbf{j}$, the operator that annihilates $e^{3x}\sin x$ is

$$(\mathtt{D} - 3)^2 + 1 = \mathtt{D}^2 - 6\,\mathtt{D} + 10\,\mathtt{D}^0 = \mathtt{D}^2 - 6\,\mathtt{D} + 10.$$

The multiple x^2 indicates that we need to apply the above operator three times, and the required differential operator is

$$\psi[\mathtt{D}] = [(\mathtt{D} - 3)^2 + 1]^3.$$

To check the answer, we ask *Maple* for help. First, we verify that $[(\mathtt{D} - 3)^2 + 1]^2 = \mathtt{D}^4 - 12\,\mathtt{D}^3 + 56\,\mathtt{D}^2 - 120\,\mathtt{D} + 100$ does not annihilate the given function:

```
f:=x->x*x*exp(3*x)*sin(x);
diff(f(x),x$4)-12*diff(f(x),x$3)+56*diff(f(x),x$2)-120*diff(f(x),x)+100*f(x)
```

Since the answer is $-8\,e^{3x}\,\sin x$, we conclude that $[(\mathtt{D}-3)^2+1]^3$ annihilates the given function.

Example 4.7.4: Find the linear homogeneous differential equation of minimum order having the solution $y = C_1\,\cos x + C_2\,\sin x + C_3 x$, where C_1, C_2, C_3 are arbitrary constants.

Solution. To answer the question, we ask *Mathematica* for help.

```
expr = (y[x] == C1 Cos[x]+C2 Sin[x]+C3 x +C4)
equations=Table[D[expr,{x,k}],{k,0,4}] (* evaluate 4 derivatives *)
the4deriv=Simplify[Solve[equations, y'''[x], {C1,C2,C3,C4}]]
de[x_, y_] = (y''''[x] == (y''''[x] /. the4deriv[[1]]))
```

This gives the required equation $y^{(4)} + y'' = 0$ of fourth order.

4.7.2 The Method of Undetermined Coefficients

We know from previous sections 4.4–4.6 how to find the general solution of a homogeneous equation with constant coefficients. The task of this subsection is to discuss methods for finding a particular solution of a nonhomogeneous equation. Various methods for obtaining a particular solution of Eq. (4.7.1) are known; some are more complicated than others. A simple technique of particular practical interest is called the **method of undetermined coefficients**. In this subsection, we present a detailed algorithm of the method and various examples of its application.

The method of undetermined coefficients can only be applied to linear differential equations with real constant coefficients:

$$a_n y^{(n)} + a_{n-1} y^{(n-1)} + \cdots + a_1\, y' + a_0 y = f(x), \tag{4.7.11}$$

when the nonhomogeneous term $f(x)$ is of special form. It is assumed that $f(x)$ is, in turn, a solution of some homogeneous differential equation with constant coefficients $\psi[\mathtt{D}]f = 0$ for some linear differential operator $\psi[\mathtt{D}]$. Namely, the function $f(x)$ should be of the special form (4.7.7), page 226, for the method of undetermined coefficients to be applicable.

The idea of the method of undetermined coefficients becomes crystal clear when the language of differential operators is used. Denoting by $L[\mathtt{D}]y$ the left-hand side expression in Eq. (4.7.11), we rewrite it as $L[\mathtt{D}]y = f$. If f is a solution of another linear constant coefficient homogeneous differential equation $\psi[\mathtt{D}]f = 0$, then applying ψ to both sides of Eq. (4.7.11), we obtain

$$\psi[\mathtt{D}]\,L[\mathtt{D}]y = \psi[\mathtt{D}]f = 0. \tag{4.7.12}$$

Since $\psi[\mathtt{D}]\,L[\mathtt{D}]y = 0$ is a homogeneous linear differential equation with constant coefficients, its general solution has a standard form (Theorem 4.17, page 217)

$$y(x) = \sum_j e^{\lambda_j x}\,P_j(x), \tag{4.7.13}$$

where $P_j(x)$ are polynomials in x with arbitrary coefficients and λ_j are roots of the algebraic equation $\psi(\lambda)L(\lambda) = 0$. Now we extract from the expression (4.7.13) those terms that are included in the general solution of the homogeneous solution $L[\mathtt{D}]\,y = 0$. The rest contains only terms corresponding to the roots of the equation $\psi(\sigma) = 0$:

$$y_p = \sum_j e^{\sigma_j x}\,Q_j(x),$$

which is your particular solution. To ensure that y_p is a required solution, we determine the coefficients of $Q_j(x)$ from the equation

$$L[\mathtt{D}]\sum_j e^{\sigma_j x}\,Q_j(x) = f(x).$$

Before presenting specific examples, let us review the rules for the method of undetermined coefficients.

Superposition Rule. If the driving term $f(x)$ in Eq. (4.7.11) can be broken down into the sum $f(x) = f_1(x) + f_2(x) + \cdots + f_m(x)$, where each term $f_j(x)$, $j = 1, 2, \ldots, m$, is a function of the form (4.7.7), apply the rules described below to obtain a particular solution $y_{pj}(x)$ of the nonhomogeneous equation with each forcing term $f_j(x)$. Then the sum of these particular solutions $y_p(x) = y_{p1}(x) + \cdots + y_{pm}(x)$ is a particular solution of Eq. (4.7.11) for $f(x) = f_1(x) + f_2(x) + \cdots + f_m(x)$.

Basic Rule: *When the control number is not a root of the characteristic equation for the associated homogeneous equation.* If $f(x)$ in Eq. (4.7.11) is the function of the form (4.7.7), page 226, and its control number is **not** a root of the characteristic equation $L(\lambda) = 0$, choose the corresponding particular solution in the **same form** as $f(x)$. In particular, if $f(x)$ is one of the functions in the first column in Table 229 times a constant, and it is not a solution of the homogeneous equation $L[\mathtt{D}]y = 0$, choose the corresponding function y_p in the third column as a particular solution of Eq. (4.7.11) and determine its coefficients (which are denoted by Cs and Ks) by substituting y_p and its derivatives into Eq. (4.7.11).

Modification Rule: *When the control number of the right-hand side function in Eq. (4.7.11)* **is** *a root of multiplicity r of the characteristic equation*, choose the corresponding particular solution in the same form multiplied by x^r. In other words, if a term in your choice for y_p happens to be a solution of the corresponding homogeneous equation $L[\mathtt{D}]y = 0$, then multiply your choice of y_p by x to the power that indicates the multiplicity of the corresponding root of the characteristic equation.

Table 229: Method of undetermined coefficients. σ is the control number of the function in the first column. All Cs and Ks in the last column are constants.

Term in $f(x)$	σ	Choice for $y_p(x)$
$e^{\gamma x}$	γ	$Ce^{\gamma x}$
x^n $(n = 0, 1, \ldots)$	0	$C_n x^n + C_{n-1}x^{n-1} + \cdots + C_1 x + C_0$
$x^n\, e^{\gamma x}$	γ	$(C_n x^n + C_{n-1}x^{n-1} + \cdots + C_1 x + C_0)e^{\gamma x}$
$\cos ax$	$\mathbf{j}a$	$C_1 \cos ax + K_2 \sin ax$
$\sin ax$	$\mathbf{j}a$	$C_1 \cos ax + K_2 \sin ax$
$x^n\ \cos ax$	$\mathbf{j}a$	$(C_n x^n + C_{n-1}x^{n-1} + \cdots + C_1 x + C_0) \cos ax +$ $+(K_n x^n + K_{n-1}x^{n-1} + \cdots + K_1 x + K_0) \sin ax$
$x^n\ \sin ax$	$\mathbf{j}a$	$(C_n x^n + C_{n-1}x^{n-1} + \cdots + C_1 x + C_0) \cos ax +$ $+(K_n x^n + K_{n-1}x^{n-1} + \cdots + K_1 x + K_0) \sin ax$
$e^{\alpha x} \cos \beta x$	$\alpha + \mathbf{j}\beta$	$e^{\alpha x}(C_1 \cos \beta x + K_2 \sin \beta x)$
$e^{\alpha x} \sin \beta x$	$\alpha + \mathbf{j}\beta$	$e^{\alpha x}(C_1 \cos \beta x + K_2 \sin \beta x)$
$x^n\, e^{\alpha x} \cos \beta x$	$\alpha + \mathbf{j}\beta$	$e^{\alpha x}\ \cos \beta x(C_n x^n + C_{n-1}x^{n-1} + \cdots + C_1 x + C_0) +$ $+e^{\alpha x}\ \sin \beta x(K_n x^n + K_{n-1}x^{n-1} + \cdots + K_1 x + K_0)$
$x^n\, e^{\alpha x} \sin \beta x$	$\alpha + \mathbf{j}\beta$	$e^{\alpha x}\ \cos \beta x(C_n x^n + C_{n-1}x^{n-1} + \cdots + C_1 x + C_0) +$ $+e^{\alpha x}\ \sin \beta x(K_n x^n + K_{n-1}x^{n-1} + \cdots + K_1 x + K_0)$

The method of undetermined coefficients allows us to guess a particular solution of a specific form with the coefficients left unspecified. If we cannot determine the coefficients, this means that there is no solution of the form that we assumed. In this case, we may either modify the initial assumption or choose another method.

Remark. The method of undetermined coefficients works well for driving terms of the form

$$f(x) = P_k(x)\, e^{\alpha x}\ \cos(\beta x + \varphi) + Q_k(x)\, e^{\alpha x}\ \sin(\beta x + \varphi).$$

Such functions often occur in electrical networks. ■

Let us consider in detail the differential equation

$$a_n y^{(n)} + a_{n-1} y^{(n-1)} + \cdots + a_0 y = h\, e^{\sigma x}, \tag{4.7.14}$$

where h and σ are constants. If σ is not a root of the characteristic equation $L(\lambda) \stackrel{\text{def}}{=} a_n \lambda^n + a_{n-1}\lambda^{n-1} + \cdots + a_0 = 0$, then we are looking for a particular solution of Eq. (4.7.14) in the form

$$y(x) = A\, e^{\sigma x}$$

with a constant A to be determined later. Since

$$\frac{\mathrm{d}}{\mathrm{d}x}\, e^{\sigma x} = \sigma e^{\sigma x}, \quad \frac{\mathrm{d}^2}{\mathrm{d}x^2}\, e^{\sigma x} = \sigma^2 e^{\sigma x}, \quad \ldots, \quad \frac{\mathrm{d}^n}{\mathrm{d}x^n}\, e^{\sigma x} = \sigma^n e^{\sigma x},$$

we have

$$L[\mathtt{D}]\, e^{\sigma x} = L(\sigma)\, e^{\sigma x} \quad \text{with} \quad \mathtt{D} = \mathrm{d}/\mathrm{d}x.$$

Hence, $A = h/L(\sigma)$ and a particular solution of Eq. (4.7.14) is

$$y_p(x) = \frac{h}{L(\sigma)}\, e^{\sigma x}.$$

If σ is a simple root of the characteristic equation $L(\lambda) = 0$, then Table 229 suggests the choice for a particular solution as $y_p(x) = Ax\, e^{\sigma x}$. Its derivatives are

$$\begin{aligned}
y_p'(x) &= A\, e^{\sigma x} + \sigma\, y_p(x), \\
y_p''(x) &= A\sigma\, e^{\sigma x} + \sigma\, y_p'(x) = 2A\,\sigma\, e^{\sigma x} + \sigma^2\, y_p(x), \\
y_p'''(x) &= 2A\,\sigma^2\, e^{\sigma x} + \sigma^2\, y_p'(x) = 3A\,\sigma^2\, e^{\sigma x} + \sigma^3\, y_p(x), \\
& \quad \text{and so on;} \\
y_p^{(n)}(x) &= nA\sigma^{n-1}\, e^{\sigma x} + \sigma^n\, y_p(x).
\end{aligned}$$

Substitution of $y_p(x)$ and its derivatives into Eq. (4.7.14) yields

$$\begin{aligned}
& A\left[a_n\, nA\sigma^{n-1}\, e^{\sigma x} + a_n\sigma^n\, y_p + a_{n-1}(n-1)\sigma^{n-2}\, e^{\sigma x} + a_{n-1}\sigma^{n-1}\, y_p + \cdots + a_0 y_p\right] \\
&= AL(\sigma)y_p(x) + A\left[a_n\, nA\sigma^{n-1} + a_{n-1}(n-1)\sigma^{n-2} + \cdots + a_1\right]\, e^{\sigma x} \\
&= A\, L(\sigma)y_p(x) + A\left(\frac{\mathrm{d}}{\mathrm{d}\sigma}\, L(\sigma)\right)\, e^{\sigma x} = AL'(\sigma)\, e^{\sigma x} = h\, e^{\sigma x}
\end{aligned}$$

since $L(\sigma) = 0$ and $L'(\sigma) \ne 0$. From the latter relation, we determine A to be $h/L'[\sigma]$, and a particular solution becomes

$$y_p(x) = \frac{hx}{L'(\sigma)}\, e^{\sigma x}.$$

In general, if σ is a root of the characteristic equation of multiplicity r, then a particular solution of Eq. (4.7.14) is

$$y_p(x) = \frac{h\, x^r}{L^{(r)}(\sigma)}\, e^{\sigma x}. \tag{4.7.15}$$

This follows from the product rule

$$\frac{\mathrm{d}^k}{\mathrm{d}x^k}\,(u(x)v(x)) = \sum_{s=0}^{k} \binom{k}{s} u^{(k-s)}(x)v^{(s)}(x), \qquad \binom{k}{s} = \frac{k!}{(k-s)!s!}$$

and the equation

$$L[\mathtt{D}]x^r\, e^{\sigma x} = L^{(r)}(\sigma)\, e^{\sigma x}.$$

■

Now we present various examples to clarify the application of this method.

Example 4.7.5: (Basic rule) Find a particular solution of the nonhomogeneous differential equation

$$L[\mathtt{D}]\, y \stackrel{\text{def}}{=} \left(\mathtt{D}^2 - 2\,\mathtt{D} - 3\right) y = 3\, e^{2x} \qquad \text{or} \qquad y'' - 2y' - 3y = 3\, e^{2x}.$$

Solution. The characteristic equation corresponding to the homogeneous equation $L[\mathtt{D}]\, y = 0$,

$$\lambda^2 - 2\lambda - 3 = 0,$$

has two real roots $\lambda_1 = -1$ and $\lambda_2 = 3$. The associated homogeneous equation has the general solution $y_h = C_1\, e^{-x} + C_2\, e^{3x}$ with some arbitrary constants C_1 and C_2. Thus, the function $f(x) = 3\, e^{2x}$ is not a solution of the homogeneous differential equation—its control number, $\sigma = 2$, does not match the roots of the characteristic equation. Therefore, we are looking for a particular solution in the form $y(x) = A\, e^{2x}$, with unknown coefficient A. We find two derivatives of $y_p(x)$ that are used in the equation so that $y'(x) = 2Ae^{2x} = 2y_p$, $y''(x) = 4Ae^{2x} = 4y_p$. Now we substitute the functions into the original equation to obtain

$$A\,(4 - 2\cdot 2 - 3)\, e^{2x} = 3\, e^{2x}.$$

Canceling out the exponent, we get $-3A = 3$, hence $A = -1$. Our general solution is, therefore,

$$y(x) = -e^{2x} + C_1\, e^{-x} + C_2\, e^{3x}.$$

We can find a particular solution of the given differential equation using Eq. (4.7.15). In our case, $L[\mathtt{D}] = \mathtt{D}^2 - 2\,\mathtt{D} - 3$, $r = 0$ (since $\lambda = 2$ is not a root of the characteristic equation) and $L(2) = -3$. Therefore

$$y_p(x) = \frac{3}{L(2)}\, e^{2x} = -\frac{3}{3}\, e^{2x} = -e^{2x}.$$

Example 4.7.6: (Basic rule) Find a particular solution of the differential equation

$$y'' - 2y' - 3y = -3x^2 + 2/3.$$

Solution. The polynomial on the right-hand side is not a solution of the homogeneous equation $y'' - 2y' - 3y = 0$ because its control number $\sigma = 0$ is not a root of the characteristic equation $\lambda^2 - 2\lambda - 3 = 0$. Since the right-hand side function $-3x^2 + 2/3$ in the given differential equation is a polynomial of the second degree, we guess a particular solution in the same form, namely, as a quadratic function

$$y_p(x) = Ax^2 + Bx + C.$$

Substituting the function $y_p(x)$ and its derivatives into the differential equation, we obtain

$$2A - 2(2Ax + B) - 3(Ax^2 + Bx + C) = -3x^2 + 2/3.$$

Equating the coefficients of the like powers of x, we have

$$-3A = -3, \quad -4A - 3B = 0, \quad 2A - 2B - 3C = 2/3.$$

The solution of the last system of algebraic equations is $A = 1$, $B = -C = -4/3$. Thus, a particular solution of the nonhomogeneous differential equation has the form

$$y_p(x) = x^2 - \frac{4}{3}(x - 1).$$

Example 4.7.7: (Basic rule) Find a particular solution of the differential equation

$$y'' - 2y' - 3y = -65\,\cos 2x$$

and determine the coefficients in this particular solution.

Solution. The characteristic equation $\lambda^2 - 2\lambda - 3 = 0$ corresponding to the homogeneous differential equation $y'' - 2y' - 3y = 0$ has two real roots $\lambda_1 = -1$ and $\lambda_2 = 3$. The control number of the function $f(x) = -65\,\cos 2x$ is $\sigma = \pm 2\mathbf{j}$. Therefore, the right-hand side is not a solution of the homogeneous differential equation, so we choose the function

$$y_p(x) = A\cos 2x + B\sin 2x$$

as a particular solution with undetermined coefficients A and B. Its derivatives are

$$y_p'(x) = -2A\,\sin 2x + 2B\,\cos 2x, \quad y_p''(x) = -4A\,\cos 2x - 4B\,\sin 2x = -4y_p.$$

Substituting the function y_p into the given differential equation $L[\mathtt{D}]\,y = -65\,\cos 2x$, we obtain

$$4A\sin 2x - 4B\cos 2x - 7A\cos 2x - 7B\sin 2x = -65\cos 2x.$$

Collecting similar terms, we get

$$(4A - 7B)\sin 2x - (4B + 7A)\cos 2x = -65\cos 2x.$$

Functions $\sin 2x$ and $\cos 2x$ are linearly independent because they are solutions of the same differential equation $y'' + 4y = 0$ and their Wronskian equals 1, see Theorem 4.9 on page 201. Recall that two functions $f(x)$ and $g(x)$ are linearly dependent if and only if $f(x)$ is a constant multiple of the other function $g(x)$. Therefore, we must equate the corresponding coefficients to zero. This leads us to

$$4A - 7B = 0, \quad 4B + 7A = 65.$$

Solving for A and B, we obtain the particular solution to be

$$y_p(x) = 7\cos 2x + 4\sin 2x.$$

We can derive the same result directly from Eq. (4.7.15). Since $\cos 2x$ is the real part of $e^{2\mathbf{j}x}$, namely, $\cos 2x = \Re e^{2\mathbf{j}x}$, a particular solution of the given differential equation is the real part of its complex solution:

$$y_p(x) = \Re\frac{-65}{L(2\mathbf{j})}\, e^{2\mathbf{j}x} = \Re\frac{-65}{(2\mathbf{j})^2 - 2(2\mathbf{j}) - 3}\, e^{2\mathbf{j}x}.$$

Since

$$\frac{-65}{(2\mathbf{j})^2 - 2(2\mathbf{j}) - 3} = \frac{-65}{-4 - 4\mathbf{j} - 3} = \frac{65}{7 + 4\mathbf{j}} = \frac{65(7 - 4\mathbf{j})}{7^2 + 4^2} = 7 - 4\mathbf{j},$$

we have

$$y_p(x) = \Re\,(7 - 4\mathbf{j})\, e^{2\mathbf{j}x} = \Re(7 - 4\mathbf{j})\,(\cos 2x + \mathbf{j}\sin 2x) = 7\cos 2x + 4\sin 2x.$$

Example 4.7.8: (Basic rule) Find a particular solution of the differential equation

$$y'' - 2y' - 3y = 65x\sin 2x - 7\,\cos 2x.$$

Solution. Table 229 suggests the choice for a particular solution

$$y_p(x) = (Ax + B)\sin 2x + (Cx + D)\cos 2x$$

with coefficients A, B, C, and D to be determined because the control number of the right-hand side function $\sigma = \pm 2\mathbf{j}$ does not match the roots of the characteristic equation. The first two derivatives of the function y_p are

$$\begin{aligned} y_p'(x) &= A\sin 2x + 2(Ax + B)\cos 2x + C\cos 2x - 2(Cx + D)\sin 2x \\ &= (A - 2Cx - 2D)\sin 2x + (2Ax + 2B + C)\cos 2x, \\ y_p''(x) &= -2C\,\sin 2x + 2(A - 2Cx - 2D)\,\cos 2x + 2A\,\cos 2x - 2(2Ax + 2B + C)\,\sin 2x \\ &= -4(Ax + B + C)\,\sin 2x + 4(A - Cx - D)\,\cos 2x. \end{aligned}$$

Substituting into the nonhomogeneous equation $L[\mathtt{D}]\,y = 65x\sin 2x - 7\,\cos 2x$ yields

$$\begin{aligned} L[\mathtt{D}]\,y = &-4(Ax + B + C)\,\sin 2x + 4(A - Cx - D)\,\cos 2x - 2(A - 2Cx - 2D)\,\sin 2x \\ &- 2(2Ax + 2B + C)\,\cos 2x - 3(Ax + B)\sin 2x - 3(Cx + D)\cos 2x \\ = &\ 65x\sin 2x - 7\,\cos 2x \end{aligned}$$

or, after collecting similar terms,

$$\begin{aligned} L[\mathtt{D}]\,y = &-[(2A + 7B + 4C - 4D) + x(7A - 4C)]\sin 2x \\ &- [(-4A + 4B + 2C + 7D) + x(7C + 4A)]\cos 2x = 65x\sin 2x - 7\,\cos 2x. \end{aligned}$$

Equating the coefficients of $\sin 2x$ and $\cos 2x$, we obtain

$$\begin{aligned} -(2A + 7B + 4C - 4D) - x(7A - 4C) &= 65x, \\ (-4A + 4B + 2C + 7D) + x(7C + 4A) &= 7. \end{aligned}$$

Two polynomials are equal if and only if their coefficients of like powers of x coincide, so

$$4C - 7A = 65, \quad 2A + 7B + 4C - 4D = 0, \quad 4A + 7C = 0, \quad -4A + 4B + 2C + 7D = 7.$$

Solving for A, B, C, and D, we determine a particular solution

$$y_p(x) = -(2 + 7x)\,\sin 2x + (4x - 3)\,\cos 2x.$$

The tedious problem of coefficient determination can be solved using *Mathematica:*

```
eq = y''[x] - 2 y'[x] - 3 y[x] == 65*x*Sin[2 x] - 7*Cos[2 x]
assume[x_] = (a*x + b)*Sin[2 x] + (c*x + d)*Cos[2 x]
seq = eq /.{y-> assume}
system = Thread[((Coefficient[#1, {Cos[2 x], x*Cos[2 x], Sin[ 2 x], x*Sin[2 x]}]) &) /@ seq]
cffs = Solve[system, {a, b, c, d}]
assume[x] /. cffs
```

Example 4.7.9: (Modification rule) Compute a particular solution of the differential equation

$$y'' - 2y' - 3y = 4\,e^{-x}.$$

Solution. As we know from Example 4.7.5, the right-hand side term $4\,e^{-x}$ is a solution to the corresponding homogeneous equation. Therefore, we try the function $y_p(x) = Ax\,e^{-x}$ as a solution because $\lambda = -1$ is a simple root of the characteristic equation $\lambda^2 - 2\lambda - 3 = 0$. Its derivatives are

$$y_p' = A\,e^{-x} - Ax\,e^{-x}, \quad y_p'' = -2A\,e^{-x} + Ax\,e^{-x}.$$

We substitute the function $y_p(x) = Ax\,e^{-x}$ and its derivatives into the differential equation to obtain

$$A\left[-2 + x - 2 + 2x - 3x\right]\,e^{-x} = 4\,e^{-x}.$$

The xe^{-x}-terms cancel each other out, and $-4A\,e^{-x} = 4\,e^{-x}$ remains. Hence, $A = -1$ and a particular solution becomes $y_p(x) = -x\,e^{-x}$.

If we want to obtain a particular solution from Eq. (4.7.15), we set $L(\lambda) = \lambda^2 - 2\lambda - 3$. Since $\lambda = -1$ is a simple root (of multiplicity 1) of $L(\lambda) = 0$, we set $r = 1$ in formula (4.7.15) and, therefore,

$$L^{(r)}(\lambda) = L'(\lambda) = 2\lambda - 2 = 2(\lambda - 1).$$

From Eq. (4.7.15), it follows that

$$y_p(x) = \frac{4x}{L'(-1)}\,e^{-x} = -x\,e^{-x}.$$

Example 4.7.10: (Modification rule) Find a particular solution of the differential equation

$$y'' - 2y' - 3y = 4\,x\,e^{-x}.$$

Solution. It is easy to check that $\sigma = -1$, the control number of the forcing function $4x\,e^{-x}$, is the root of the characteristic equation $\lambda^2 - 2\lambda - 3 = (\lambda - 3)(\lambda + 1) = 0$. Table 229 suggests choosing a particular solution as

$$y_p(x) = x(A + Bx)\,e^{-x},$$

where A and B are undetermined coefficients. The derivatives are

$$y_p' = -(Ax + Bx^2)\,e^{-x} + (A + 2Bx)\,e^{-x},$$
$$y_p'' = (Ax + Bx^2)\,e^{-x} - 2(A + 2Bx)\,e^{-x} + 2B\,e^{-x}.$$

Substituting y_p and its derivatives into the equation, we obtain

$$(2B - 4A - 8Bx)\,e^{-x} = 4x\,e^{-x}.$$

Multiplying both sides by e^x, we get the algebraic equation

$$2B - 4A - 8Bx = 4x.$$

Equating like power terms, we obtain the system of algebraic equations

$$2B - 4A = 0, \quad -8Bx = 4x.$$

Its solution is $B = -1/2$ and $A = -1/4$. Hence,

$$y_p(x) = -\frac{x}{4}(1 + 2x)\, e^{-x}.$$

Example 4.7.11: (Modification rule) Find the general solution of the differential equation

$$y'' - 4y' + 4y = 2\, e^{2x}.$$

Solution. The corresponding characteristic equation $(\lambda - 2)^2 = 0$ has the double root $\lambda = 2$ that coincides with the control number $\sigma = 2$ of the forcing function $2\, e^{2x}$. Hence, the general solution of the corresponding homogeneous equation is $y_h(x) = (C_1 + C_2 x)\, e^{2x}$, with arbitrary constants C_1 and C_2. As a particular solution, we choose the function

$$y_p(x) = Ax^2\, e^{2x}.$$

The power 2 in the multiplier x^2 appears because of the repeated root $\lambda = 2$ of the characteristic equation. Substituting this into the nonhomogeneous equation, we obtain

$$y_p'' - 4y_p' + 4y_p = 2A\, e^{2x},$$

where

$$\begin{aligned} y_p'(x) &= 2Ax\, e^{2x} + 2Ax^2\, e^{2x} = 2Ax\, e^{2x} + y_p(x), \\ y_p''(x) &= 2A\, e^{2x} + 4Ax\, e^{2x} + 2y_p'(x) = 2A\, e^{2x} + 8Ax\, e^{2x} + 4y_p(x). \end{aligned}$$

The expression $y_p'' - 4y_p' + 4y_p$ should be equal to $2\, e^{2x}$. Therefore, $A = 1$ and the general solution is the sum

$$y(x) = y_p(x) + y_h(x) = x^2\, e^{2x} + (C_1 + C_2 x)\, e^{2x}.$$

We can obtain the same result from Eq. (4.7.15), page 230. In our case, the characteristic polynomial $L(\lambda) = \lambda^2 - 4\lambda + 4 = (\lambda - 1)^2$ has double root $\lambda = 2$. Therefore, $r = 2$ and $L^{(r)}(\lambda) = L''(\lambda) = 2$. From Eq. (4.7.15), it follows

$$y_p(x) = \frac{2x^2}{L''(2)}\, e^{2x} = x^2\, e^{2x}.$$

Example 4.7.12: (Modification rule and superposition rule) Find a particular solution of

$$L[\mathtt{D}]y \equiv y'' - 2y' + 10y = 9e^x + 26 \sin 2x + 6e^x \cos 3x.$$

Solution. The right-hand side function is the sum of three functions having the control numbers 1, $\pm 2\mathbf{j}$, and $1 \pm 3\mathbf{j}$, respectively. Thus, we split up the equation as follows:

$$y'' - 2y' + 10y = 9e^x, \quad \text{or} \quad L[\mathtt{D}]y = 9e^x, \tag{a}$$

$$y'' - 2y' + 10y = 26 \sin 2x, \quad \text{or} \quad L[\mathtt{D}]y = 26 \sin 2x, \tag{b}$$

$$y'' - 2y' + 10y = 6e^x \cos 3x, \quad \text{or} \quad L[\mathtt{D}]y = 6e^x \cos 3x. \tag{c}$$

The nonhomogeneous terms of the first two equations (a) and (b) are not solutions of the corresponding homogeneous equation $y'' - 2y' + 10y = 0$ because their control numbers $\sigma = 1$ and $\sigma = 2\mathbf{j}$ do not match the roots $\lambda_{1,2} = 1 \pm 3\mathbf{j}$ of the characteristic equation $\lambda^2 - 2\lambda + 10 = 0$, but the latter one does.

We seek a particular solution of the equation (a) in the form

$$y_1(x) = A\,e^x,$$

where the coefficient A is yet to be determined. To find A we calculate

$$y_1'(x) = A\,e^x = y_1(x), \quad y_1''(x) = y_1(x)$$

and substitute for y_1, y_1', and y_1'' into the equation (a). The result is

$$y_1 - 2y_1 + 10y_1 = 9e^x, \quad \text{or} \quad 9Ae^x = 9e^x.$$

Hence, $A = 1$ and $y_1(x) = e^x$ is a particular solution of the equation (a). The value of the constant A can also be obtained from the formula (4.7.15) on page 230.

Table 229 suggests the choice for a particular solution of the equation (b) to be

$$y_2(x) = A\sin 2x + B\cos 2x,$$

where the coefficients A and B are to be determined. Then

$$y_2'(x) = 2A\cos 2x - 2B\sin 2x, \quad y_2''(x) = -4A\sin 2x - 4B\cos 2x = -4y_2(x).$$

Upon substitution these expressions for y_2, y_2', and y_2'' into the equation (b) and collecting terms, we have

$$\begin{aligned} L[\mathtt{D}]y &= -2(2A\cos 2x - 2B\sin 2x) + 6(A\sin 2x + B\cos 2x) \\ &= (6B - 4A)\cos 2x + (6A + 4B)\sin 2x = 26\sin 2x. \end{aligned}$$

The last equation is satisfied if we match the coefficients of $\sin 2x$ and $\cos 2x$ on each side of the equation

$$6B - 4A = 0, \quad 6A + 4B = 26.$$

Hence, $B = 2$ and $A = 3$, so a particular solution of the equation (b) becomes

$$y_2(x) = 3\sin 2x + 2\cos 2x.$$

To find a particular solution y_3 of the equation (c) we assume that

$$y_3(x) = x(A\cos 3x + B\sin 3x)\,e^x,$$

where coefficients A and B are to be determined numerically. We calculate the derivatives y_3' and y_3'' to obtain

$$y_3'(x) = (A\cos 3x + B\sin 3x)\,e^x + 3x(-A\sin 3x + B\cos 3x)\,e^x + y_3;$$

$$\begin{aligned} y_3''(x) &= 6(-A\sin 3x + B\cos 3x)\,e^x + (A\cos 3x + B\sin 3x)\,e^x - \\ &\quad -9x(A\cos 3x + B\sin 3x)\,e^x + 3x(-A\sin 3x + B\cos 3x)\,e^x + y_3'(x) \\ &= 6(-A\sin 3x + B\cos 3x)\,e^x + 2(A\cos 3x + B\sin 3x)\,e^x \\ &\quad -9x(A\cos 3x + B\sin 3x)\,e^x + 6x(-A\sin 3x + B\cos 3x)\,e^x + y_3(x). \end{aligned}$$

Substituting these expressions into the equation (c) yields

$$\begin{aligned} L[\mathtt{D}]y_3 &= -8y_3 + 2(-3A\sin 3x + 3B\cos 3x)\,e^x + 2(A\cos 3x + B\sin 3x)\,e^x \\ &\quad +2x(-3A\sin 3x + 3B\cos 3x)\,e^x - 2(A\cos 3x + B\sin 3x)\,e^x \\ &\quad -2x(-3A\sin 3x + 3B\cos 3x)\,e^x - 2y_3 + 10y_3 = 6e^x\,\cos 3x. \end{aligned}$$

Collecting similar terms, we obtain

$$6(-3A\sin 3x + 3B\cos 3x)\,e^x = 6e^x\,\cos 3x.$$

Equating coefficients of $\sin 3x$ and $\cos 3x$ yields $A = 0$ and $B = 1$. Hence,

$$y_3(x) = x\, e^x \sin 3x.$$

Since all these calculations are tedious, we may want to use Eq. (4.7.15), page 230, in order to determine a particular solution. Since the right-hand side of Eq. (c) is the real part of $6\, e^{(1+3\mathbf{j})x}$, we obtain a particular solution to be the real part of

$$\frac{6x}{L'(1+3\mathbf{j})}\, e^{(1+3\mathbf{j})x},$$

where $L(\lambda) = \lambda^2 - 2\lambda + 1$ and $L'(\lambda) = 2(\lambda - 1)$. Therefore,

$$y_p = \Re \frac{6x}{2(1+3\mathbf{j}-1)}\, e^{(1+3\mathbf{j})x} = \Re \frac{6x}{6\mathbf{j}}\, e^x\, e^{3\mathbf{j}x} = xe^x \sin 3x.$$

With this in hand, we are in a position to write down a particular solution y_p of the original nonhomogeneous equation as the sum of particular solutions of auxiliary equations (a), (b), and (c):

$$y_p(x) = y_1(x) + y_2(x) + y_3(x) = e^x + 3\sin 2x + 2\cos 2x + x\, e^x \sin 3x.$$

Example 4.7.13: (Modification rule and superposition rule) Solve the initial value problem

$$L[\mathtt{D}]\, y \stackrel{\text{def}}{=} y'' - 2y' + y = 6x\, e^x + x^2, \quad y(0) = 0,\ y'(0) = 1.$$

Solution. The general solution y_h of the corresponding homogeneous equation $y'' - 2y' + y = 0$ (also called the complementary function) is of the form

$$y_h(x) = (C_1 + C_2 x)\, e^x$$

because the characteristic equation $\lambda^2 - 2\lambda + 1 = (\lambda - 1)^2 = 0$ has the double root $\lambda = 1$. To find a particular solution of the nonhomogeneous equation, we split this equation into two

$$y'' - 2y' + y = 6x\, e^x \qquad \text{or} \qquad L[\mathtt{D}]\, y = 6x\, e^x, \tag{d}$$

where the forcing term, $6x\, e^x$, has the control number $\sigma = 1$, and

$$y'' - 2y' + y = x^2 \qquad \text{or} \qquad L[\mathtt{D}]\, y = x^2, \tag{e}$$

where x^2 has the control number $\sigma = 0$. We determine a particular solution $y_1(x)$ of the equation (d) according to Table 229 as

$$y_1(x) = x^2 (Ax + B)\, e^x$$

since $\lambda = 1$ is the double root of the characteristic equation $(\lambda - 1)^2 = 0$. That is why we multiplied the expression $(Ax + B)\, e^x$ by x^2, where the power 2 of x indicates the multiplicity of the root $\lambda = 1$. We calculate its derivatives

$$\begin{aligned} y_1'(x) &= (3Ax^2 + 2Bx)\, e^x + (Ax^3 + Bx^2)\, e^x, \\ y_1''(x) &= (6Ax + 2B + Ax^3 + Bx^2 + 6Ax^2 + 4Bx)\, e^x. \end{aligned}$$

Substituting these expressions into the equation (d) yields

$$(6Ax + 2B)\, e^x = 6x\, e^x.$$

Hence, $A = 1, B = 0$, and a particular solution of the equation (d) becomes $y_1(x) = x^3\, e^x$.

Let us consider the equation (e). The right-hand side function in this equation is not a solution of the corresponding homogeneous equation. Therefore, Table 229 suggests a particular solution y_2 in the same form:

$$y_2(x) = Ax^2 + Bx + C.$$

Its derivatives are

$$y_2'(x) = 2Ax + B, \quad y_2''(x) = 2A.$$

Upon substitution of these expressions into the equation (e), we have

$$2A - 2(2Ax + B) + Ax^2 + Bx + C = x^2 \quad \text{or} \quad Ax^2 + (B - 4A)x + 2A - 2B + C = x^2.$$

Equating the power like terms of x, we get the system of algebraic equations for A, B, and C:

$$A = 1, \quad B - 4A = 0, \quad 2A - 2B + C = 0.$$

Thus, $A = 1$, $B = 4$, $C = 6$ and a particular solution of the equation (e) becomes $y_2(x) = x^2 + 4x + 6$. Consequently, the general solution of the driven equation is the sum of particular solutions y_1 and y_2 plus the general solution of homogeneous equation $y_h(x)$, that is,

$$y(x) = y_h + y_1 + y_2 = x^2 + 4x + 6 + x^3\, e^x + (C_1 + C_2 x)\, e^x.$$

Substituting this expression into the initial condition, we obtain the relations allowing us to determine the unknown constants C_1 and C_2:

$$y(0) = 6 + C_1 = 0, \quad y'(0) = 4 + C_2 + C_1 = 1.$$

Hence, $C_1 = -6$, $C_2 = 3$ and

$$y(x) = x^2 + 4x + 6 + (-6 + 3x + x^3)\, e^x$$

is the solution of the given initial value problem.

Problems In all problems, D stands for the derivative operator, while $\mathtt{D}^0$, the identity operator, is omitted. The derivatives with respect to t are denoted by dots.

1. Find the control number of the following functions.
   (a) x; **(b)** $x^5 + 2x$; **(c)** e^{-2x};
   (d) $x^2\, e^x$; **(e)** e^{-ix}; **(f)** $\cos 2x$;
   (g) $x^2 \sin 3x$; **(h)** $e^{-2x} \sin x$; **(i)** $(x+1)^2\, e^x \cos 3x$;
   (j) $x^2\, e^{2ix}$; **(k)** $\sin 2x + \cos 2x$; **(l)** $e^{-x} (\cos 3x - 2x \sin 3x)$.

2. Obtain a linear differential equation with real, constant coefficients in the factorial form $a_n(\mathtt{D} - \lambda_1)^{m_1}(\mathtt{D} - \lambda_2)^{m_2} \cdots$ for which the given function is its solution.
   (a) $y = 2\, e^x + 3\, e^{-2x}$; **(b)** $y = 5 + 4\, e^{-3x}$; **(c)** $y = 2x + e^{4x}$;
   (d) $y = x^2 + e^{-2x}$; **(e)** $y = x^2 - 1 + \cos 3x$; **(f)** $y = 2e^{-x} \sin 2x$;
   (g) $y = (x+1) \cos 3x$; **(h)** $y = xe^{2x} \cos 3x$; **(i)** $y = x^2\, e^{2x} + \cos x$;
   (j) $y = x^2 + e^{-2x} \sin 2x$; **(k)** $y = 3x\, e^{2x} \cos 5x$; **(l)** $y = x^2 + e^{-x} \cos 2x$.

3. State the roots and their multiplicities of the characteristic equation for a homogeneous linear differential equation with real, constant coefficients and having the given function as a particular solution.
   (a) $y = 2x\, e^x$; **(b)** $y = x^2\, e^{-2x} + 2x$; **(c)** $y = e^{-2x} \sin 3x$;
   (d) $y = e^{-x} \sin 2x$; **(e)** $y = 4x + x\, e^{3x}$; **(f)** $y = 2 + 4\cos 2x$;
   (g) $y = 2x^3 - e^{-3x}$; **(h)** $y = 1 + 2x^2 + e^{-2x} \sin x$; **(i)** $y = \cos 3x$;
   (j) $y = 2 \cos 3x - 3 \sin 3x$; **(k)** $y = x \cos 3x - 3 \sin 2x$; **(l)** $y = e^{-x} \cos 2x$;
   (m) $y = e^{-x} (x + \sin 2x) + 3x^2$; **(n)** $y = \cos^2 x$; **(o)** $y = x\, e^{2x}$;
   (p) $y = x^2\, e^{2x} + 3e^{2x}$; **(q)** $y = \sin^3 x$; **(r)** $y = x^2 + e^{3x}$.

4. Find a homogeneous linear equation with constant coefficients that has a given particular solution.
   (a) $y_p(x) = \cos 2x$; **(b)** $y_p(x) = e^x + 2\, e^{-x}$; **(c)** $y_p(x) = e^x \sin 2x$;
   (d) $y_p(x) = 3x^2$; **(e)** $y_p(x) = \sin x + e^x$; **(f)** $y_p(x) = x \cos x$;
   (g) $y_p(x) = x\, e^{2x}$; **(h)** $y_p(x) = \sinh x$; **(i)** $y_p(x) = x \cosh x$;
   (j) $y_p(x) = x\, e^x \sin 2x$; **(k)** $y_p(x) = x^2 e^{-2x}$; **(l)** $y_p(x) = (\sinh x)^2$.

5. Write out the assumed form of a particular solution, but do not carry out the calculations of the undetermined coefficients.
   (a) $y'' + 4y = x$; **(b)** $y'' + 4y = (x+1) \cos 2x$;
   (c) $y'' + 4y = \cos x$; **(d)** $y'' + 4y = (x+1) \sin 2x + \cos 2x$;
   (e) $y'' + 4y = x \sin x$; **(f)** $y'' + 2y' - 3y = \cos 2x + \sin x$;
   (g) $y'' + 4y = e^{2x}$; **(h)** $y'' + 2y' - 3y = 20\, e^x \cos 2x$;
   (i) $y'' + 4y = (x-1)^2$; **(j)** $y'' + 2y' - 3y = x \sin x + \cos 2x$;
   (k) $y'' + 2y' - 3y = (x+1)^3$; **(l)** $y'' + 2y' - 3y = e^{-3x} \sin x$;
   (m) $y'' + 2y' - 3y = e^x$; **(n)** $y'' + 2y' - 3y = e^x + e^{-3x} \cos x$;
   (o) $y'' + 2y' - 3y = x\, e^x$; **(p)** $y'' - 2y' + 5y = x\, e^x \sin 2x$;
   (q) $y'' + 2y' - 3y = x^2\, e^x$; **(r)** $y'' - 2y' + 5y = e^x + \cos 2x$;
   (s) $y'' - 5\, y' + 6y = x^5\, e^{2x}$; **(t)** $y'' - 2y' + 5y = x\, e^x \cos 2x + e^x \sin 2x$;
   (u) $y'' - 2y' + 5y = \cos 2x$; **(v)** $2y'' + 7y' - 4y = e^{-4x}$.

6. Find the general solution of the following equations when the control number of the right-hand side function does not match a root of the characteristic equation.

(a)	$y'' + y = x^3$;	**(b)**	$y'' + y' - 6y = x^2 - 2x$;
(c)	$y'' + 2y' - 8y = 6e^{2x}$;	**(d)**	$y'' - 6y' + 9y = 4e^{x}$;
(e)	$y'' - 4y = 4e^{2x}$;	**(f)**	$y'' + 4y' + 4y = 9\sinh x$;
(g)	$y'' + 5y' + 6y = 12e^{x} + 6x^2 + 10x$;	**(h)**	$y'' + 3y' + 2y = 8x^3$;
(i)	$y'' - 2y' + 2y = 2e^{x}\cos x$,	**(j)**	$4y'' - 4y' + y = 8x^2 e^{x/2}$;
(k)	$y'' - 4y' + 3y = 24x\,e^{-3x} - 2e^{x}$;	**(l)**	$(\texttt{D}^2 + \texttt{D} + 1)y = 3x^2$;
(m)	$(\texttt{D}^2 + 5\,\texttt{D})y = 17 + 10x$;	**(n)**	$(\texttt{D}^2 + \texttt{D} - 6)y = 50\sin x$;
(o)	$(\texttt{D}^2 - \texttt{D} - 2)y = 4x + 16\,e^{-2x}$;	**(p)**	$(\texttt{D}^2 - 3\,\texttt{D} - 4)y = 36x\,e^{x}$;
(q)	$(\texttt{D}^4 - 1)y = 4\sin x$;	**(r)**	$(\texttt{D}^2 - 4\,\texttt{D} + 4)y = e^{x}$;
(s)	$(\texttt{D}^3 + \texttt{D}^2 - 4\,\texttt{D} - 4)y = 12\,e^{x} - 4x$;	**(t)**	$(\texttt{D}^2 + \texttt{D} + 13/4)y = -7\sin\left(\sqrt{3}x\right)$.

7. Find the general solution of the following equations when the control number of the right-hand side function matches a root of the characteristic equation.

(a)	$(\texttt{D}^2 + \texttt{D} - 6)\,y = 5\,e^{2x}$,	**(b)**	$(\texttt{D}^2 - \texttt{D} - 2)\,y = 18\,e^{-x}$,	**(c)**	$(\texttt{D}^2 - 10\,\texttt{D} + 26)y = e^{5x}\sin x$,
(d)	$(\texttt{D}^2 - 3\,\texttt{D} - 4)y = 50x\,e^{4x}$,	**(e)**	$(\texttt{D}^2 - 4\,\texttt{D} + 4)y = 2e^{2x}$,	**(f)**	$(\texttt{D}^2 - 6\,\texttt{D} + 13)y = 2e^{3x}\cos 2x$.

8. Solve the following initial value problems.

(a) $y'' + 4y' + 5y = 8\sin x, \quad y(0) = 0,\ y'(0) = 1$;

(b) $y'' - y = 4x\,e^{x}, \quad y(0) = 8,\ y'(0) = 1$;

(c) $y'' + y = 2x\,e^{-x}, \quad y(0) = 0,\ y'(0) = 2$;

(d) $y'' - y' - 2y = (16x - 40)\,e^{-2x} + 2x + 1, \quad y(0) = -3,\ y'(0) = 20$;

(e) $y'' - 7y' - 8y = -(14x^2 - 18x + 2)\,e^{x}, \quad y(0) = 0,\ y'(0) = 9$;

(f) $y'' + 2y' + 5y = 4e^{-x}, \quad y(0) = 0,\ y'(0) = 2$;

(g) $y'' + 2y' - 3y = 25x\,e^{2x}, \quad y(0) = -6,\ y'(0) = -3$;

(h) $y'' + 6\,y' + 10y = 4xe^{-3x}\sin x, \quad y(0) = 0,\ y'(0) = 1$;

(i) $y'' + 8y' + 7y = 12\sinh x + 7, \quad y(0) = 3/8,\ y'(0) = 3/8$;

(j) $y'' - 3y' - 5y = 39\sin(2x) + x\,e^{4x}, \quad y(0) = -3,\ y'(0) = -27$;

(k) $y'' - 2y' + 10y = 6\,e^{x}\cos 3x, \quad y(0) = 1,\ y'(0) = 1$;

(l) $y'' + 2y' + y = 2\,e^{-x}, \quad y(0) = 0,\ y'(0) = 1$.

9. In the following exercises, solve the initial value problems where the characteristic equation is of degree 3 and higher. At least one of its roots is an integer and can be found by inspection.

(a) $y''' + y' = 2x + 2\cos x, \quad y(0) = 1,\ y'(0) = 0,\ y''(0) = 2$;

(b) $y''' + 3y'' - 4y = 18x\,e^{x}, \quad y(0) = 0,\ y'(0) = -10/3,\ y''(0) = -5/3$;

(c) $y''' - 3y'' + 3y' - y = x^2, \quad y(0) = -3,\ y'(0) = -3,\ y''(0) = 0$;

(d) $y''' + y'' - 2y = \sin x - 3\cos x, \quad y(0) = 3,\ y'(0) = 0,\ y''(0) = 0$;

(e) $y''' + y'' - y' - y = -4\,e^{-x}, \quad y(0) = 0,\ y'(0) = 3,\ y''(0) = 0$;

(f) $y''' - 2y'' - 4y' + 8y = 8\,e^{2x}, \quad y(0) = 0,\ y'(0) = 5,\ y''(0) = 6$;

(g) $y''' - y'' - 4y' + 4y = 3\,e^{x}, \quad y(0) = 1,\ y'(0) = -4,\ y''(0) = -1$;

(h) $y''' - 2y'' + y' = 2x, \quad y(0) = 0,\ y'(0) = 4,\ y''(0) = 3$;

(i) $y^{(4)} - y'' = -12x^2, \quad y(0) = 1,\ y'(0) = 1,\ y''(0) = 24,\ y'''(0) = 2$;

(j) $(\texttt{D}^3 - 3\,\texttt{D}^2 + 4)y = 16\cos 2x + 8\sin 2x, \quad y(0) = 1,\ y'(0) = -2,\ y''(0) = -3$;

(k) $(\texttt{D}^3 - 3\,\texttt{D} - 2)y = 100\sin 2x, \quad y(0) = 7,\ y'(0) = 2,\ y''(0) = -27$;

(l) $(\texttt{D}^3 + 4\,\texttt{D}^2 + 9\,\texttt{D} + 10)y = 24\,e^{x}, \quad y(0) = 4,\ y'(0) = -3,\ y''(0) = 39$;

(m) $(\texttt{D}^3 + \texttt{D} - 10)\,y = 13\,e^{2x}, \quad y(0) = 2,\ y'(0) = 0,\ y''(0) = 9$;

(n) $y''' - 2y'' - y' + 2y = e^{x}, \quad y(0) = 2,\ y'(0) = 4,\ y''(0) = 3$.

4.8 Variation of Parameters

The method of undetermined coefficients is simple and has important applications, but it is applicable only to constant coefficient equations with a special forcing function. In this section, we discuss the *general method* (which is actually a generalization of the Bernoulli method presented in §2.6.1) credited[42] to **Lagrange**, known as the **method of variation of parameters**. If a fundamental set of solutions for a corresponding homogeneous equation is known, this method can be applied to find a solution of a nonhomogeneous linear differential equation with variable coefficients of any order. Let us start, for simplicity, with the second order differential equation

$$y'' + p(x)y' + q(x)y = f(x) \tag{4.8.1}$$

with given continuous coefficients $p(x), q(x)$, and a piecewise continuous (integrable) function $f(x)$ on some open interval (a, b). The method is easily extended to equations of order higher than two, but no essentially new ideas appear.

In the language of linear operators, the problem of finding a solution of Eq. (4.8.1) is equivalent to finding a right inverse operator for $L = \texttt{D}^2 + p(x)\texttt{D} + q(x)$, where $\texttt{D} = \mathrm{d}/\mathrm{d}x$ is the operator of differentiation. In other words, $y(x) = L^{-1}[f](x)$ and the existence of the inverse operator L^{-1} is guaranteed by Theorem 4.3 on page 189. The only problem is how to go about finding such a solution.

The continuity of $p(x)$ and $q(x)$ on the interval (a, b) implies that the associated homogeneous equation, $y'' + py' + qy = 0$, has the general solution

$$y_h(x) = C_1 y_1(x) + C_2 y_2(x), \tag{4.8.2}$$

where C_1 and C_2 are arbitrary constants (or parameters) and $\{y_1(x), y_2(x)\}$ is, of course, a known fundamental set of solutions that was found by some method or other. We call $y_h(x)$, a general solution (4.8.2) of the associated homogeneous differential equation to Eq. (4.8.1), a **complementary function** of Eq. (4.8.1). The method of variation of parameters involves "varying" the parameters C_1 and C_2, replacing them by functions $A(x)$ and $B(x)$ to be determined so that the resulting function

$$y_p(x) = A(x)y_1(x) + B(x)y_2(x) \tag{4.8.3}$$

is a particular solution of Eq. (4.8.1) on the interval (a, b). By differentiating Eq. (4.8.3), we obtain

$$y_p'(x) = A'(x)y_1(x) + A(x)y_1'(x) + B'(x)y_2(x) + B(x)y_2'(x).$$

The expression (4.8.3) contains two unknown functions $A(x)$ and $B(x)$; however, the requirement that y_p satisfies the nonhomogeneous equation imposes only one condition on these functions. Therefore, we expect that there are many possible choices of A and B that will meet our needs. Hence, we are free to impose one more condition upon A and B. With this in mind, let us enforce the second condition by demanding that

$$A'(x)y_1(x) + B'(x)y_2(x) = 0. \tag{4.8.4}$$

Thus,

$$y_p'(x) = A(x)y_1'(x) + B(x)y_2'(x),$$

so the derivative, y_p', is obtained from Eq. (4.8.3) by differentiating only y_1 and y_2 but not $A(x)$ and $B(x)$. Next derivative is

$$y_p''(x) = A(x)y_1''(x) + A'(x)y_1'(x) + B(x)y_2''(x) + B'(x)y_2'(x).$$

Finally, we substitute these expressions for y_p, y_p', and y_p'' into Eq. (4.8.1). After rearranging the terms in the resulting equation, we find that

$$A(x)[y_1''(x) + p(x)y_1'(x) + q(x)y_1(x)] + B(x)[y_2''(x) + p(x)y_2'(x) + q(x)y_2(x)] + A'(x)y_1'(x) + B'(x)y_2'(x) = f(x).$$

Since y_1 and y_2 are solutions of the homogeneous equation, namely,

$$y_j''(x) + p(x)y_j'(x) + q(x)y_j(x) = 0, \; j = 1, 2,$$

[42] Joseph-Louis Lagrange (1736–1813), born in Turin as Giuseppe Lodovico Lagrangia, was a famous mathematician and astronomer, who lived for 21 years (1766–1787) in Berlin (Prussia) and then in France (1787–1813).

each of the expressions in brackets in the left-hand side is zero. Therefore, we have

$$A'(x)y_1'(x) + B'(x)y_2'(x) = f(x). \tag{4.8.5}$$

Equations (4.8.4) and (4.8.5) form a system of two linear algebraic equations for the unknown derivatives $A'(x)$ and $B'(x)$. The solution is obtained by Cramer's rule or Gaussian elimination. Thus, multiplying Eq. (4.8.4) by $y_1'(x)$ and Eq. (4.8.5) by $-y_2(x)$ and adding, we get

$$A'(x)(y_1y_2' - y_2y_1') = -y_2(x)f(x).$$

The expression in parenthesis in the left-hand side is exactly the Wronskian of a fundamental set of solutions, that is, $W(y_1, y_2; x) = y_1y_2' - y_2y_1'$. Therefore, $A'(x)W(x) = -y_2(x)f(x)$. We then multiply Eq. (4.8.4) by $-y_1'(x)$ and Eq. (4.8.5) by $y_1(x)$ and add to obtain $B'(x)W(x) = y_1(x)f(x)$. The functions $y_1(x)$ and $y_2(x)$ are linearly independent solutions of the homogeneous equation $y'' + py' + qy = 0$; therefore, its Wronskian is not zero on the interval (a, b), where functions $p(x)$ and $q(x)$ are continuous. Now, division by $W(x) \neq 0$ gives

$$A'(x) = -\frac{y_2(x)f(x)}{y_1y_2' - y_2y_1'}, \qquad B'(x) = \frac{y_1(x)f(x)}{y_1y_2' - y_2y_1'}.$$

By integration, we have

$$A(x) = -\int \frac{y_2(x)f(x)}{W(y_1, y_2; x)}\,\mathrm{d}x + C_1, \quad B(x) = \int \frac{y_1(x)f(x)}{W(y_1, y_2; x)}\,\mathrm{d}x + C_2, \tag{4.8.6}$$

where C_1, C_2 are constants of integration. Substituting these integrals into Eq. (4.8.3), we obtain the general solution of the given inhomogeneous equation:

$$y(x) = -y_1(x)\int \frac{y_2(x)f(x)}{W(x)}\,\mathrm{d}x + y_2(x)\int \frac{y_1(x)f(x)}{W(x)}\,\mathrm{d}x + C_1y_1(x) + C_2y_2(x).$$

Since the linear combination $y_h(x) = C_1y_1(x) + C_2y_2(x)$ is a solution of the corresponding homogeneous equation, we can disregard it because our goal is to find a particular solution. It follows from Theorem 4.19 (page 224) that the difference between a particular solution of the nonhomogeneous equation and a solution of the homogeneous equation is a solution of the driven equation. Thus, a particular solution of the nonhomogeneous equation has the form

$$y_p(x) = \int_{x_0}^{x} G(x, \xi)\, f(\xi)\,\mathrm{d}\xi, \tag{4.8.7}$$

where function $G(x, \xi)$, called the **Green function of the linear operator**[43] $L = \mathtt{D}^2 + p(x)\mathtt{D} + q(x)$, is

$$G(x, \xi) = \frac{y_1(\xi)y_2(x) - y_2(\xi)y_1(x)}{y_1(\xi)y_2'(\xi) - y_2(\xi)y_1'(\xi)}. \tag{4.8.8}$$

The Green function depends only on the solutions y_1 and y_2 of the corresponding homogeneous equation $L[y] = 0$ and is independent of the forcing term. Therefore, $G(x, \xi)$ is completely determined by the linear differential operator L (not necessarily with constant coefficients).

Example 4.8.1: We consider the equation

$$\left(\mathtt{D}^2 - 3\,\mathtt{D} + 2\right) y = e^{-x}, \quad \mathtt{D} = \mathrm{d}/\mathrm{d}x.$$

The complementary function (the general solution of the associated homogeneous equation) is $y_h(x) = C_1e^x + C_2e^{2x}$, so we put

$$y_p(x) = A(x)\,e^x + B(x)\,e^{2x},$$

where $A(x)$ and $B(x)$ are unknown functions of x. Since $y_p' = A(x)\,e^x + A'(x)\,e^x + 2B(x)\,e^{2x} + B'(x)\,e^{2x}$, we impose the condition

$$A'(x)\,e^x + B'(x)\,e^{2x} = 0.$$

[43]George Green (1793–1841) was a British mathematical physicist whom we remember for Green's theorem from calculus.

Substituting y_p into the differential equation, we obtain the second condition for derivatives $A'(x)$ and $B'(x)$:

$$A'(x)\, e^{x} + 2B'(x)\, e^{2x} = e^{-x}.$$

From this system of algebraic equations with respect to $A'(x)$ and $B'(x)$, we have

$$A'(x) = -e^{-2x}, \quad B'(x) = e^{-3x}.$$

Integration yields

$$A(x) = \frac{1}{2}\, e^{-2x} + C_1, \quad B(x) = -\frac{1}{3}\, e^{-3x} + C_2,$$

where C_1 and C_2 are arbitrary constants. Therefore, the general solution becomes

$$y(x) = \frac{1}{2}\, e^{-x} - \frac{1}{3}\, e^{-x} + C_1 e^{x} + C_2 e^{2x}, \qquad \text{or} \qquad y(x) = \frac{1}{6}\, e^{-x} + C_1 e^{x} + C_2 e^{2x}.$$

The Green function for the operator $\texttt{D}^2 - 3\,\texttt{D} + 2$ is $G(x, \xi) = e^{2(x-\xi)} - e^{x-\xi}$.

Example 4.8.2: Solve the equation

$$(\texttt{D}^2 + 1)y = \csc x = 1/\sin x, \qquad \texttt{D} = \mathrm{d}/\mathrm{d}x.$$

Solution. The general solution of the corresponding homogeneous equation is

$$y_h(x) = C_1 \sin x + C_2 \cos x,$$

that is, $y_h(x)$ is the complementary function for the operator $\texttt{D}^2 + 1$. Let

$$y_p(x) = A(x)\, \sin x + B(x)\, \cos x$$

be a solution of the given inhomogeneous equation. Then two algebraic equations (4.8.4) and (4.8.5) for derivatives A' and B' become:

$$A'(x)\, \sin x + B'(x)\, \cos x = 0, \qquad A'(x)\, \cos x - B'(x)\, \sin x = \csc x.$$

This system may be resolved for $A'(x)$ and $B'(x)$, yielding

$$A'(x) = \cos x \csc x = \cot x, \quad B'(x) = -\sin x \csc x = -1.$$

Integrating, we obtain

$$A(x) = \int \cot x \,\mathrm{d}x + C_1 = \ln|\sin x| + C_1, \quad B(x) = -x + C_2.$$

A particular solution of the nonhomogeneous equation becomes

$$y_p(x) = \sin x \cdot \ln|\sin x| - x \cos x.$$

□

If we want to find the solution of the initial value problem for the inhomogeneous equation (4.8.1), it is sometimes more convenient to split the problem into two problems. It is known from Theorem 4.20 on page 224 that the general solution of a linear inhomogeneous differential equation is a sum of a complementary function, y_h, and a particular solution of the nonhomogeneous equation y_p. Thus, a particular solution can be found by solving the initial value problem

$$y_p'' + p(x)y_p' + q(x)y_p = f(x), \quad y_p(x_0) = 0, \; y_p'(x_0) = 0, \tag{4.8.9}$$

and the complementary function is the solution of the initial value problem

$$y_h'' + p(x)y_h' + q(x)y_h = 0, \quad y_h(x_0) = y_0, \; y_h'(x_0) = y_0'. \tag{4.8.10}$$

Example 4.8.3: Solve the initial value problem

$$y'' - 2y' + y = e^{2x}/(e^x + 1)^2, \quad y(0) = 3, \; y'(0) = 1.$$

Solution. We split this initial value problem into the two IVPs similar to (4.8.9) and (4.8.10):

$$y_p'' - 2y_p' + y_p = e^{2x}/(e^x + 1)^2, \quad y_p(0) = 0, \; y_p'(0) = 0$$

and

$$y_h'' - 2y_h' + y_h = 0, \quad y_h(0) = 3, \; y_h'(0) = 1.$$

Then the solution to the given initial value problem is the sum: $y = y_p + y_h$. We start with the homogeneous equation $y'' - 2y' + y = 0$. The corresponding characteristic equation $\lambda^2 - 2\lambda + 1 = 0$ has a double root $\lambda = 1$. Therefore, the complementary function is

$$y_h(x) = (C_1 + C_2 x)e^x.$$

Substituting this solution into the initial conditions $y(0) = 3$ and $y'(0) = 1$, we get

$$C_1 = 3, \quad C_2 = -2.$$

The next step is to integrate the nonhomogeneous equation using the variation of parameters method. Let $y_1(x) = e^x$ and $y_2(x) = xe^x$ be linearly independent solutions, then from Eq. (4.8.8) we find the Green function:

$$G(x,\xi) = \frac{y_1(\xi)y_2(x) - y_2(\xi)y_1(x)}{y_1(\xi)y_2'(\xi) - y_2(\xi)y_1'(\xi)} = (x - \xi)\, e^{x-\xi}.$$

According to the formula (4.8.7), a particular solution of the nonhomogeneous equation with homogeneous initial conditions becomes

$$y_p(x) = \int_0^x G(x,\xi)\, \frac{e^{2\xi}}{(e^\xi + 1)^2}\, \mathrm{d}\xi = e^x \int_0^x \frac{(x - \xi)e^\xi}{(e^\xi + 1)^2}\, \mathrm{d}\xi.$$

To evaluate the integral, we integrate by parts ($-\int v\, \mathrm{d}u = -vu + \int u\, \mathrm{d}v$) using substitutions:

$$v = x - \xi, \quad \mathrm{d}v = -\mathrm{d}\xi, \quad u = \left(1 + e^\xi\right)^{-1}, \quad \mathrm{d}u = -\frac{e^\xi}{(e^\xi + 1)^2}\, \mathrm{d}\xi.$$

The result is

$$y_p = x\, e^x - e^x \int_0^x \frac{\mathrm{d}\xi}{1 + e^\xi}.$$

In the last integral, we change the independent variable by setting $t = e^{\xi/2}$. Then

$$\begin{aligned} \int_0^x \frac{\mathrm{d}\xi}{1 + e^\xi} &= 2\int_1^{e^{x/2}} \frac{\mathrm{d}t}{t(1 + t^2)} = 2\int_1^{e^{x/2}} \frac{\mathrm{d}t}{t} - 2\int_1^{e^{x/2}} \frac{t\,\mathrm{d}t}{1 + t^2} \\ &= \ln t^2 - \ln(1 + t^2)\Big|_{t=1}^{e^{x/2}} = \ln\left(\frac{e^x}{1 + e^x}\right) - \ln\left(\frac{1}{2}\right) = x - \ln(1 + e^x) + \ln 2. \end{aligned}$$

Hence,

$$y_p(x) = -e^x \left[\ln(1 + e^x) - \ln(2)\right].$$

□

Now we extend the method of variation of parameters to linear equations of arbitrary order. Let us consider an equation in normal form:

$$y^{(n)} + a_{n-1}(x)y^{(n-1)} + \cdots + a_0(x)y = f(x) \qquad \text{or} \qquad L[x,\mathtt{D}]y = f, \tag{4.8.11}$$

defined on some interval I and assume that the complementary function $y_h = C_1 y_1(x) + C_2 y_2(x) + \cdots + C_n y_n(x)$ is known, where C_k are arbitrary constants and y_k, $k = 1, 2, \ldots, n$, are linearly independent solutions of the homogeneous equation $L[x,\mathtt{D}]y = 0$. Following Lagrange, we seek a particular solution of Eq. (4.8.11) in the form

$$y_p(x) = A_1(x)y_1(x) + A_2(x)y_2(x) + \cdots + A_n(x)y_n(x)\,, \tag{4.8.12}$$

where we impose the following n conditions on the unknown functions $A_k(x)$, $k = 1, 2, \ldots n$:

$$\begin{aligned} A_1'(x)y_1 + A_2'(x)y_2 + \cdots A_n'(x)y_n &= 0, \\ A_1'(x)y_1' + A_2'(x)y_2' + \cdots A_n'(x)y_n' &= 0, \\ &\cdots \\ A_1'(x)y_1^{(n-2)} + A_2'(x)y_2^{(n-2)} + \cdots A_n'(x)y_n^{(n-2)} &= 0, \\ A_1'(x)y_1^{(n-1)} + A_2'(x)y_2^{(n-1)} + \cdots A_n'(x)y_n^{(n-1)} &= f(x) \end{aligned} \tag{4.8.13}$$

for all $x \in I$. This is a system of n algebraic linear equations in the unknown derivatives $A_1'(x)$, ..., $A_n'(x)$, whose determinant is the Wronskian $W[y_1(x), y_2(x), \ldots, y_n(x)]$. Therefore this system has a unique solution and after integration, we obtain the functions $A_k(x)$ that can be substituted into Eq. (4.8.12) to get a particular solution, which in turn can be written in the integral form as

$$y_p(x) = \int_{x_0}^{x} G(x, \xi)\, f(\xi)\, \mathrm{d}\xi\,. \tag{4.8.14}$$

The function $G(x, \xi)$ is called the **Green function to the operator** $L[x, \mathtt{D}] = \mathtt{D}^n + a_{n-1}(x)\mathtt{D}^{n-1} + \cdots + a_0(x)$ and it is uniquely defined by the ratio of two determinants:

$$G(x, \xi) = \begin{vmatrix} y_1(\xi) & y_2(\xi) & \cdots & y_n(\xi) \\ y_1'(\xi) & y_2'(\xi) & \cdots & y_n'(\xi) \\ \vdots & \vdots & \ddots & \vdots \\ y_1^{(n-2)}(\xi) & y_2^{(n-2)}(\xi) & \cdots & y_n^{(n-2)}(\xi) \\ y_1(x) & y_2(x) & \cdots & y_n(x) \end{vmatrix} \Bigg/ W[y_1(\xi), y_2(\xi), \cdots, y_n(\xi)]\,. \tag{4.8.15}$$

The integral expression in the right-hand side of Eq. (4.8.14) defines a right inverse for the operator L. Actually this inverse operator is uniquely defined by the following conditions, which we summarize in the following statement.

Theorem 4.23: Let $G(x, \xi)$ be defined throughout the rectangular region R of the $x\xi$-plane and suppose that $G(x, \xi)$ and its partial derivatives $\partial^k G(x, \xi)/\partial x^k$, $k = 0, 1, 2, \ldots, n$, are continuous everywhere in R. Then $G(x, \xi)$ is the Green function to the linear differential operator $L[x, \mathtt{D}] = a_n(x)\mathtt{D}^n + a_{n-1}(x)\mathtt{D}^{n-1} + \cdots + a_0(x)$, $\mathtt{D} = \mathrm{d}/\mathrm{d}x$, if and only if it is a solution of the differential equation $L[x, \mathtt{D}]\, G(x, \xi) = 0$ subject to the following conditions

$$G(x, x) = 0, \quad \left.\frac{\partial G(x, \xi)}{\partial x}\right|_{\xi = x} = 0, \ldots, \left.\frac{\partial^{n-2} G(x, \xi)}{\partial x^{n-2}}\right|_{\xi = x} = 0, \quad \left.\frac{\partial^{n-1} G(x, \xi)}{\partial x^{n-1}}\right|_{\xi = x} = \frac{1}{a_n(x)}\,.$$

PROOF: is left as Problem 10, page 245. ■

Our next example demonstrates the direct extension of the variation of parameters method for solving driven differential equations of the order greater than two. All required steps are presented in the Summary, page 255.

Example 4.8.4: Let us consider a constant coefficient linear nonhomogeneous equation of the fourth order

$$y^{(4)} + 2y'' + y = 2\csc t.$$

The characteristic equation $(\lambda^2 + 1)^2 = 0$ of the associated homogeneous equation has two double roots: $\lambda_1 = \mathbf{j}$ and $\lambda_2 = -\mathbf{j}$. So, the fundamental set of solutions consists of four functions

$$y_1 = \cos t, \quad y_2 = \sin t, \quad y_3 = t\cos t, \quad y_4 = t\sin t.$$

The main idea of the method of variation of parameters is based on the assumption that a particular solution is of the following form:

$$y = A_1(t)\, y_1 + A_2(t)\, y_2 + A_3(t)\, y_3 + A_4(t)\, y_4,$$

To determine four functions, $A_1(t)$, $A_2(t)$, $A_3(t)$, and $A_4(t)$, we impose four conditions:

$$\begin{aligned} A_1'(t)\,y_1 + A_2'(t)\,y_2 + A_3'(t)\,y_3 + A_4'(t)\,y_4 &= 0,\\ A_1'(t)\,y_1' + A_2'(t)\,y_2' + A_3'(t)\,y_3' + A_4'(t)\,y_4' &= 0,\\ A_1'(t)\,y_1'' + A_2'(t)\,y_2'' + A_3'(t)\,y_3'' + A_4'(t)\,y_4'' &= 0,\\ A_1'(t)\,y_1''' + A_2'(t)\,y_2''' + A_3'(t)\,y_3''' + A_4'(t)\,y_4''' &= 2\csc t. \end{aligned}$$

Since finding a solution of a fourth order algebraic system of equations is quite time consuming, we ask *Maple* to help us with the following commands:

```
eq1:= a1*cos(t) + a2*sin(t) +a3*t*cos(t) + a4*t*sin(t);
eq2:= a2*cos(t) - a1*sin(t) +a3*(cos(t)-t*sin(t)) + a4*(sin(t)+t*cos(t));
eq3:= a1*cos(t) + a2*sin(t) + a3*(2*sin(t)+t*cos(t)) +a4*(t*sin(t)-2*cos(t));
eq4:= a1*sin(t) -a2*cos(t) +a3*(t*sin(t)-3*cos(t)) - a4*(3*sin(t)+t*cos(t));
solve({eq1=0, eq2=0, eq3=0, eq4=f}, {a1,a2,a3,a4});
```

This results in

$$A_1'(t) = t\,\cot t - 1, \quad A_2'(t) = \cot t - t, \quad A_3'(t) = -\cot t, \quad A_4'(t) = -1.$$

Integration yields

$$A_1 = t\,\ln|\sin t| - \int \ln|\sin t|\,dt, \quad A_2 = \ln|\sin t| + t^2/2, \quad A_3 = -\ln|\sin t|, \quad A_4 = -t.$$

Substituting the values of functions $A_1(t)$, $A_2(t)$, $A_3(t)$, and $A_4(t)$, we obtain a particular solution to be

$$y = \sin t\,(\ln|\sin t| - t^2/2) - \cos t \int \ln|\sin t|\,dt.$$

The Green function for the given fourth order differential operator is

$$G(x-\xi) = \frac{1}{2}\,\sin(x-\xi) - \frac{1}{2}\,(x-\xi)\,\cos(x-\xi)$$

because the function $G(x)$ is the solution of the following IVP: $G^{(4)} + 2G'' + G = 0 \quad G(0) = G'(0) = G''(0) = 0$, $G'''(0) = 1$.

Problems In all problems, D stands for the derivative operator, while D^0, the identity operator, is omitted. The derivatives with respect to t are denoted by dots.

1. Use the variation of parameters method to find a particular solution of the given differential equations.
   - **(a)** $y'' + y = \cot x$;
   - **(b)** $y'' + y = 3\csc x$;
   - **(c)** $y'' + 4y = 4\sec^2 2x$;
   - **(d)** $9y'' + y = 3\csc(x/3)$;
   - **(e)** $y'' + 4y = 8\,\sin^2 x$;
   - **(f)** $y'' + 9y = 9\,\sec 3x$;
   - **(g)** $y'' + 4y' + 5y = e^{-2x}\,\tan x$;
   - **(h)** $y'' - y = 2/(e^x - 1)$;
   - **(i)** $y'' - 9y = \frac{18}{1+e^{3x}}$;
   - **(j)** $y'' - y = \frac{4x^2+1}{x\sqrt{x}}$.
2. Given a complementary function, y_h, find a particular solution using the Lagrange method.
   - **(a)** $x^2\,y'' + xy' - y = x^4$, $\quad y_h(x) = C_1\,x + C_2\,x^{-1}$;
   - **(b)** $x^2\,y'' - 4xy' + 6y = 2x^4$, $\quad y_h(x) = C_1\,x^2 + C_2\,x^3$;
   - **(c)** $x^2\,y'' - 3xy' + 4y = 4x^4$, $\quad y_h(x) = C_1\,x^2 + C_2\,x^2\,\ln x$;
   - **(d)** $x^2\,y'' + xy' + y = \ln x$, $\quad y_h(x) = C_1\,\cos(\ln x) + C_2\,\sin(\ln x)$;
   - **(e)** $x^2\,y'' + 3x\,y' - 3y = 2x$, $\quad y_h(x) = C_1\,x + C_2\,x^{-3}$;
   - **(f)** $x^2\,y'' + 3x\,y' + y = 9\,x^2$, $\quad y_h(x) = C_1\,x^{-1} + C_2\,x^{-1}\,\ln x$
3. Use the variation of parameters method to find a particular solution to the given inhomogeneous differential equations with constant coefficients. Note that the characteristic equation to the associated homogeneous equation has double roots.
   - **(a)** $y'' + 4y' + 4y = \frac{2\,e^{-2x}}{x^2+1}$;
   - **(b)** $y'' + 6y' + 9y = \frac{e^{-3x}}{1+x^2} + 27x^2 + 18x$;
   - **(c)** $y'' - 4y' + 4y = 2e^{2x}\,\ln x + 25\,\sin x$;
   - **(d)** $y'' - 2y' + y = 4\,e^x\,\ln x$;
   - **(e)** $y'' - 6y' + 9y = x^{-2}\,e^{3x}$;
   - **(f)** $y'' + 6y' + 9y = \frac{e^{-3x}}{\sqrt{4-x^2}}$;
   - **(g)** $y'' + 2y' + y = e^{-x}/(1+x^2)$;
   - **(h)** $(D-1)^2 y = e^x/(1-x)^2$.

4. Use the Lagrange method to find a particular solution to the given inhomogeneous differential equations with constant coefficients. Note that the characteristic equation to the associated homogeneous equation has complex conjugate roots.
 (a) $4y'' - 8y' + 5y = e^x \tan^2 \frac{x}{2}$; **(b)** $y'' + y = 2\sec^2 x$; **(c)** $y'' + y = 6\sec^4 x$;
 (d) $9y'' - 6y' + 10y = 13\, e^x$; **(e)** $y''' + y' = \csc x$; **(f)** $y''' + 4y' = 8\tan 2x$.

5. Use the method of variation of parameters to determine the general solution of the given differential equation of the order higher than two.
 (a) $y''' + 4y' = \tan 2t$; **(b)** $y''' - y'' + y' - y = \csc t$;
 (c) $y^{(4)} - y = \sec t$; **(d)** $y''' - 4y'' + y' + 6y = 12/(t\, e^t)$.

6. Find the solution to

$$y'' + y = H(t)H(\pi - t), \qquad \text{where } H(t) = \begin{cases} 1, & \text{if } t > 0, \\ 0, & \text{if } t < 0, \end{cases}$$

subject to the initial conditions $y(0) = y'(0) = 0$ and that y and y' are continuous at $t = \pi$.

7. Compute the Green function $G(x, \xi)$ for the given differential operator by using (4.8.15).
 (a) $\texttt{D}^2(\texttt{D} - 2)$; **(b)** $\texttt{D}(\texttt{D}^2 - 9)$; **(c)** $\texttt{D}^3 - 6\,\texttt{D}^2 + 11\,\texttt{D} - 6$;
 (d) $4\,\texttt{D}^3 - \texttt{D}^2 + 4\,\texttt{D} - 1$; **(e)** $\texttt{D}^2(\texttt{D}^2 + 1)$; **(f)** $\texttt{D}^4 - 1$.

8. Find a particular solution $y'' + 4\,y = 2\sin^2 x$ in two ways.
 (a) Use variation of parameters.
 (b) Use the identity $2\sin^2 x = 1 - \cos 2x$ and the method of undetermined coefficients.

9. Prove that the function (4.8.14) satisfies the following initial conditions $y_p(x_0) = y_p'(x_0) = \cdots = y_p^{(n-1)}(x_0) = 0$.

10. Prove Theorem 4.23 on page 243.

11. Using the variation of parameters method, solve problems 6 and 7 on page 238.

12. Solve the initial value problem

$$L[y] \stackrel{\text{def}}{=} x\,y'' - (2x+1)\,y' + (x+1)\,y = 2x^2 e^x \ln x \quad (x > 0), \quad y(1) = 2,\ y'(1) = 4,$$

given that $y_1(x) = e^x$ is a solution of the homogeneous equation $L[y] = 0$.

13. Solve the initial value problem

$$L[y] \stackrel{\text{def}}{=} x\left(1 + 3x^2\right) y'' + 2\,y' - 6x\,y = \left(1 + 3x^2\right)^2 \quad (x > 0), \quad y(1) = 2,\ y'(1) = 4,$$

given that $y_1(x) = x^{-1}$ is a solution of the homogeneous equation $L[y] = 0$.

14. Solve the initial value problem

$$L[y] \stackrel{\text{def}}{=} x\,y'' - 2\,(x+1)\,y' + (x+2)\,y = 3x^3 e^{2x} \ln x \quad (x > 0), \quad y(1) = 3\,e^2,\ y'(1) = 6\,e^2,$$

given that $y_1(x) = e^x$ is a solution of the homogeneous equation $L[y] = 0$.

15. Find the continuous solution to the following differential equation with the given piecewise continuous forcing function subject to the homogeneous initial conditions ($y(0) = y'(0) = 0$).

$$y'' + 4y = \begin{cases} 4, & \text{if } 0 < t < \frac{\pi}{2}, \\ 0, & \text{otherwise.} \end{cases}$$

16. Find the Green function for the given linear differential operators.
 (a) $\texttt{D}^2 + 4$; **(b)** $\texttt{D}^2 - \texttt{D} - 2$; **(c)** $\texttt{D}^2 + 4\,\texttt{D} + 4$;
 (d) $4\,\texttt{D}^2 - 8\,\texttt{D} + 5$; **(e)** $\texttt{D}^2 + 3\,\texttt{D} - 4$; **(f)** $x^2\texttt{D}^2 - 2x\,\texttt{D} + 2$;
 (g) $x\,\texttt{D}^2 - (1 + 2x^2)\,\texttt{D}$; **(h)** $(1 - x^2)\,\texttt{D}^2 - 2x\,\texttt{D}$; **(i)** $\texttt{D}^3 + \frac{3}{2}\,\texttt{D}^2 - \texttt{D} - \frac{3}{2}$.

17. Find the general solution of the given differential equations.
 (a) $\dfrac{1}{r}\dfrac{\mathrm{d}}{\mathrm{d}r}\left(r\,\dfrac{\mathrm{d}u}{\mathrm{d}r}\right) = -1$; **(b)** $\dfrac{1}{r^2}\dfrac{\mathrm{d}}{\mathrm{d}r}\left(r^2\,\dfrac{\mathrm{d}u}{\mathrm{d}r}\right) = -1$;
 (c) $y'' + 3y' + 2y = 12\cosh t$; **(d)** $t^2 y'' + ty' - y = 1$. *Hint:* $y_1 = t$.

18. Use the Green function to determine a particular solution to the given differential equations.
 (a) $y'' + 4y' + 4y = f(x)$; **(b)** $y'' - 4y' + 5y = f(x)$;
 (c) $y'' + \omega^2 y = f(x)$; **(d)** $y'' - k^2 y = f(x)$.

4.9 Bessel Equations

The differential equation

$$x^2\,y'' + x\,y + (x^2 - \nu^2)\,y = 0, \tag{4.9.1}$$

where ν is a real nonnegative constant, is called the **Bessel**[44] **equation** of order ν. It has singular points at $x = 0$ and $x = \infty$, but only the former point is regular. We seek its solution in the form of a generalized power series:

$$y(x) = x^\alpha \sum_{n=0}^{\infty} a_n\,x^n = \sum_{n=0}^{\infty} a_n\,x^{n+\alpha},$$

where α is a real number to be determined later. Without any loss of generality we assume that $a_0 \neq 0$. Substituting $y(x)$ into Eq. (4.9.1), we obtain

$$\sum_{n=0}^{\infty}(n+\alpha)(n+\alpha-1)a_n\,x^{n+\alpha} + \sum_{n=0}^{\infty}(n+\alpha)a_n\,x^{n+\alpha} + \sum_{n=0}^{\infty} a_n\,x^{n+\alpha+2} - \nu^2\sum_{n=0}^{\infty} a_n\,x^{n+\alpha} = 0.$$

Setting the coefficients of each power of x equal to zero, we get the recurrence equation for the coefficients:

$$\left[(n+\alpha)(n+\alpha-1) + (n+\alpha) - \nu^2\right] a_n + a_{n-2} = 0 \qquad (n \geqslant 2)$$

or

$$\left[(n+\alpha)^2 - \nu^2\right] a_n + a_{n-2} = 0 \qquad (n = 2, 3, 4, \ldots). \tag{4.9.2}$$

We equate the coefficients of x^α and $x^{\alpha+1}$ to zero, which leads to equations

$$(\alpha^2 - \nu^2)a_0 = 0 \quad \text{and} \quad \left[(n+1)^2 - \nu^2\right] a_1 = 0.$$

Since $a_0 \neq 0$, it follows that

$$\alpha^2 - \nu^2 = 0 \quad \text{or} \quad \alpha = \pm\nu \tag{4.9.3}$$

and $a_1 = 0$. The recurrence relation (4.9.2) forces all coefficients with odd indices to be zero, namely, $a_{2k+1} = 0, \quad k = 0, 1, 2, \ldots$.

We try $\alpha = +\nu$ as a solution of Eq. (4.9.3). Then the recurrence equation (4.9.2) gives

$$a_n = -\frac{a_{n-2}}{n(n+2\nu)}, \qquad n \geqslant 2$$

and, therefore,

$$\begin{aligned}
a_2 &= -\frac{a_0}{2(2+2\nu)} = -\frac{a_0}{2^2\,(1+\nu)},\\
a_4 &= -\frac{a_2}{4(4+2\nu)} = -\frac{a_2}{4\cdot 2(2+\nu)} = \frac{a_0}{2^{2\cdot 2}2!\,(1+\nu)(2+\nu)},\\
a_6 &= -\frac{a_4}{6(6+2\nu)} = -\frac{a_0}{2^{2\cdot 3}3!\,(1+\nu)(2+\nu)(3+\nu)},\\
\cdots & \quad \cdots \quad \cdots \quad \cdots \quad \cdots \quad \cdots\\
a_{2k} &= (-1)^k\,\frac{a_0}{2^{2k}\,k!\,(1+\nu)(2+\nu)\cdots(k+\nu)}, \qquad k = 0, 1, 2, \ldots.
\end{aligned}$$

It is customary to set the reciprocal to the arbitrary constant a_0 to be $1/a_0 = 2^\nu\,\Gamma(\nu)\nu$, where

$$\Gamma(\nu) = \int_0^\infty t^{\nu-1}\,e^{-t}\,dt, \qquad \Re\,\nu > 0, \tag{4.9.4}$$

is the gamma function of Euler. Since $\Gamma(\nu)$ possesses the convenient property $\Gamma(\nu+1) = \nu\Gamma(\nu)$, we can reduce the denominator in a_{2n} to one term. With such a choice of a_0, we have the solution, which is usually denoted by $J_\nu(x)$:

$$J_\nu(x) = \sum_{k=0}^{\infty} \frac{(-1)^k\,\left(\frac{x}{2}\right)^{2k+\nu}}{k!\,\Gamma(\nu+k+1)} \tag{4.9.5}$$

$$a_n = -\frac{a_{n-2}}{n(n-2\nu)}, \quad n \geqslant 2$$

and gives another solution of Eq. (4.9.1) to be

$$J_{-\nu}(x) = \sum_{k=0}^{\infty} \frac{(-1)^k \left(\frac{x}{2}\right)^{2k-\nu}}{k!\,\Gamma(-\nu+k+1)}.$$

This is the Bessel function of the first kind of order $-\nu$.

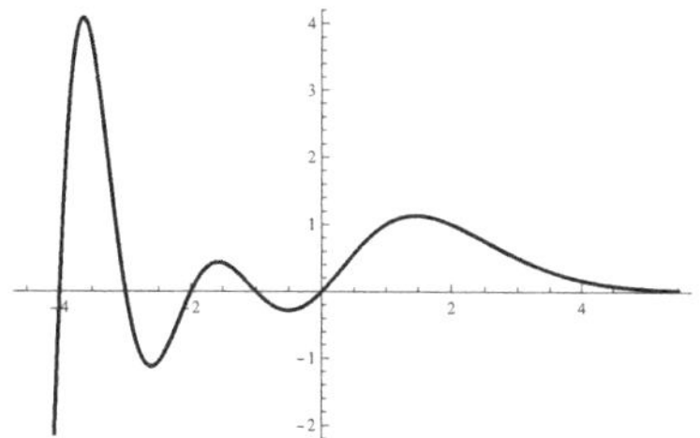

Figure 4.5: The reciprocal Γ-function.

called the **Bessel function of the first kind of the order** ν. The recurrence equation (4.9.2) with $\alpha = -\nu$ is reduced to

If ν is not an integer, the functions $J_\nu(x)$ and $J_{-\nu}(x)$ are linearly independent because they contain distinct powers of x. Thus, the general solution of the Bessel equation (4.9.1) is given by

$$y(x) = c_1\, J_\nu(x) + c_2\, J_{-\nu}(x), \tag{4.9.6}$$

with arbitrary constants c_1 and c_2.

When $\nu = n$, an integer, the denominators of those terms in the series for $J_{-n}(x)$ for which $-n+k+1 \leqslant 0$ contain the values of the gamma function at nonpositive integers. Since the reciprocal Γ-function is zero at these points (see Fig. 4.5) all terms of the series (4.9.6) for which $k \leqslant n-1$ vanish and

$$J_{-n}(x) = \sum_{k=n}^{\infty} \frac{(-1)^k \left(\frac{x}{2}\right)^{2k-n}}{k!\,\Gamma(-n+k+1)}.$$

We shift the dummy index k by setting $j = k - n$, then $k = j + n$ and we get

$$\begin{aligned} J_{-n}(x) &= \sum_{j=0}^{\infty} \frac{(-1)^{n+j} \left(\frac{x}{2}\right)^{2(j+n)-n}}{(j+n)!\,\Gamma(-n+j+n+1)} \\ &= (-1)^n \sum_{j=0}^{\infty} \frac{(-1)^j \left(\frac{x}{2}\right)^{2j+n}}{j!(j+n)!} = (-1)^n\, J_n(x) \end{aligned}$$

because $\Gamma(j+1) = j!$ for integers j. For example,

$$J_0(x) = \sum_{k=0}^{\infty} \frac{(-1)^k \left(\frac{x}{2}\right)^{2k}}{(k!)^2} = 1 - \frac{x^2}{2^2(1!)^2} + \frac{x^4}{2^2(2!)^2} - \frac{x^6}{2^6(3!)^2} + \cdots .$$

Since the Bessel functions $J_n(x)$ and $J_{-n}(x)$ are linearly dependent, we need to find another linearly independent integral of Bessel's equation to form its general solution. This independent solution can be obtained by using the Bernoulli change of variables (see §4.6.1) $y(x) = u(x)J_n(x)$. Substituting $y(x)$ and its derivatives into Eq. (4.9.1) with $\nu = n$ leads to

$$x^2(u''J_n + 2u'J_n' + J_n'') + x(u'J_n + uJ_n') + (x^2 - n^2)uJ_n = 0$$

or

$$x^2(u''J_n + 2u'J_n') + xu'J_n = 0$$

because $J_n(x)$ is an integral of Eq. (4.9.1). Letting $p = u'$, this equation may be rewritten as

$$x^2p'J_n + p(2x^2J_n' + xJ_n) = 0 \qquad \text{or} \qquad \frac{\mathrm{d}p}{\mathrm{d}x} = -p\left[2\,\frac{J_n'}{J_n} + \frac{1}{x}\right].$$

Separating the variables, we obtain

$$\frac{\mathrm{d}p}{p} = -\left[2\,\frac{J_n'(x)}{J_n(x)} + \frac{1}{x}\right]\mathrm{d}x.$$

[44]In honor of Friedrich Wilhelm Bessel (1784–1846), a German astronomer who in 1840 predicted the existence of a planet beyond Uranus. Equation (4.9.1) was first studied by Daniel Bernoulli in 1732.

After integration, we have

$$\ln p = -2 \ln J_n(x) - \ln x + \ln C_1 - \ln \frac{C_1}{xJ_n^2(x)} \qquad \text{or} \qquad p = \frac{du}{dx} = \frac{C_1}{x[J_n(x)]^2}$$

Next integration yields

$$u(x) = C_1 \int \frac{dx}{x[J_n(x)]^2} + C_2,$$

where C_1 and C_2 are arbitrary constants. Hence, the general solution of Eq. (4.9.1) becomes

$$y(x) = C_2\, J_n(x) + C_1\, J_n(x) \int \frac{dx}{x[J_n(x)]^2}.$$

The product

$$J_n(x) \int \frac{dx}{x[J_n(x)]^2}$$

defines another linearly independent solution. Upon multiplication by a suitable constant, this expression defines either $Y_n(x)$, the **Weber**[45] **function**, or $N_n(x)$, the **Neumann**[46] **function**. We call these two functions $N_n(x)$ and $Y_n(x) = \pi N_n(x)$ the **Bessel function of the second kind** of order n.

It is customary to define the Neumann function by means of a single expression

$$N_\nu(x) = \frac{\cos \nu\pi\, J_\nu(x) - J_{-\nu}(x)}{\sin \nu\pi}, \tag{4.9.7}$$

which is valid for all values of ν. If $\nu = n$, an integer, $\sin n\pi = 0$ and $\cos n\pi = (-1)^n$, then, by Eq. (4.9.7), $N_n(x)$ takes the indeterminate form 0/0. According to l'Hôpital's rule, we obtain

$$\begin{aligned} Y_n(x) &= \pi\, N_n(x) = 2J_n(x)\left(\gamma + \ln \frac{x}{2}\right) - \sum_{k=0}^{n-1} \frac{(n-k-1)!}{k!} \left(\frac{x}{2}\right)^{2k-n} \\ &- \sum_{k=1}^{\infty} \frac{(-1)^k \left(\frac{x}{2}\right)^{2k+n}}{k!\,(n+k)!} \left[2\left(1 + \frac{1}{2} + \frac{1}{3} + \cdots + \frac{1}{k}\right) + \frac{1}{k+1} + \frac{1}{k+2} + \cdots + \frac{1}{k+n}\right], \end{aligned}$$

where $\gamma = \Gamma'(1) = 0.5772156\ldots$ is the Euler[47] constant. For example,

$$Y_0(x) = \pi N_0(x) = 2J_0(x)\left(\ln \frac{x}{2} + \gamma\right) - 2\sum_{k=1}^{\infty} \frac{(-1)^k}{(k!)^2} \left(\frac{x}{2}\right)^{2k} \sum_{m=1}^{k} \frac{1}{m}.$$

Recall that the sum of reciprocals of the first n integers is called the n-th harmonic number, $H_n = \sum_{k=1}^{n} k^{-1}$. Obviously, the functions $Y_0(x)$ and $N_0(x)$ are unbounded when $x = 0$. The functions $Y_\nu(x)$ and $J_{-\nu}(x)$ approach infinity as $x \to 0$, whereas $J_\nu(x)$ remains finite as x approaches zero.

Hence, the general solution of a Bessel equation of any real order ν may be written

$$y(x) = C_1 J_\nu(x) + C_2 Y_\nu(x) \quad \text{or} \quad y(x) = C_1 J_\nu(x) + C_2 N_\nu(x)$$

with arbitrary constants C_1 and C_2.

---

[45] Heinrich Martin Weber (1842–1913) was a German mathematician whose main achievements were in algebra, number theory, and analysis. He introduced the function $Y_\nu(x)$ in 1873.

[46] Carl (also Karl) Gottfried Neumann (1832–1925), was a German mathematician who was born and studied in Königsberg (now Kaliningrad, Russia) University.

[47] Sometimes γ is also called the Euler–Mascheroni constant in honor of ordained Italian priest, poet, and teacher Lorenzo Mascheroni (1750–1800), who correctly calculated the first 19 decimal places of γ. It is not known yet whether the Euler constant is rational or irrational.

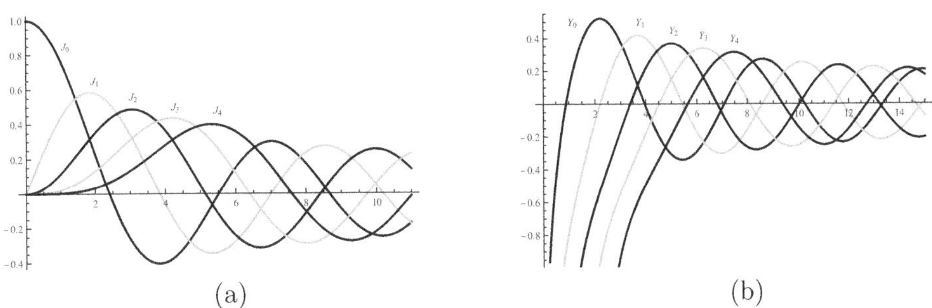

(a) (b)

Figure 4.6: (a) graphs of Bessel functions $J_n(x)$ for $n = 0, 1, 2, 3, 4$; and (b) graphs of Weber's functions $Y_n(x)$ for $n = 0, 1, 2, 3, 4$.

4.9.1 Parametric Bessel Equation

Another important differential equation is the **parametric Bessel equation**:

$$x^2 y'' + xy' + (\lambda^2 x^2 - \nu^2)y = 0. \tag{4.9.8}$$

This equation can be transformed into Eq. (4.9.1) by replacing the independent variable $t = \lambda x$. Then

$$\frac{\mathrm{d}y}{\mathrm{d}x} = \frac{\mathrm{d}y}{\mathrm{d}t} \cdot \frac{\mathrm{d}t}{\mathrm{d}x} = \lambda \cdot \frac{\mathrm{d}y}{\mathrm{d}t} = \lambda \dot{y}$$

and

$$\frac{\mathrm{d}^2 y}{\mathrm{d}x^2} = \frac{\mathrm{d}}{\mathrm{d}x}\left(\frac{\mathrm{d}y}{\mathrm{d}x}\right) = \frac{\mathrm{d}}{\mathrm{d}x}\left(\lambda\,\frac{\mathrm{d}y}{\mathrm{d}t}\right) = \lambda\,\frac{\mathrm{d}^2 y}{\mathrm{d}t^2} \cdot \frac{\mathrm{d}t}{\mathrm{d}x} = \lambda^2\,\frac{\mathrm{d}^2 y}{\mathrm{d}t^2} = \lambda^2\,\ddot{y}.$$

Therefore,

$$x\,y' = x\lambda\,\dot{y} = t\,\dot{y} = t\,\frac{\mathrm{d}y}{\mathrm{d}t}, \qquad x^2 y'' = x^2\lambda^2\,\ddot{y} = t^2\,\ddot{y} = t^2\,\frac{\mathrm{d}^2 y}{\mathrm{d}t^2}.$$

Substitution $t = \lambda x$ into Eq. (4.9.8) yields

$$t^2\,\ddot{y} + t\,\dot{y} + (t^2 - \nu^2)y = 0,$$

which is the regular Bessel equation. Thus, the general solution of Eq. (4.9.8) becomes

$$y(x) = C_1\,J_\nu(\lambda x) + C_2\,J_{-\nu}(\lambda x) \quad \text{for noninteger } \nu$$

or

$$y(x) = C_1\,J_\nu(\lambda x) + C_2\,Y_\nu(\lambda x) \quad \text{for arbitrary } \nu. \tag{4.9.9}$$

For $\lambda = \mathbf{j}$, namely, $\lambda^2 = -1$, we have

$$x^2\,y'' + x\,y' - (x^2 + \nu^2)\,y = 0. \tag{4.9.10}$$

This equation is called the **modified Bessel equation of order** ν. From (4.9.9), it follows that the **modified Bessel function of the first and second kinds**

$$I_\nu(x) = \mathbf{j}^{-\nu}\,J_\nu(\mathbf{j}x) = \sum_{k=0}^{\infty} \frac{\left(\frac{x}{2}\right)^{2k+\nu}}{k!\,\Gamma(k+\nu+1)}, \qquad K_\nu(x) = \frac{\pi}{2}\,\frac{I_{-\nu}(x) - I_\nu(x)}{\sin(\nu\pi)}$$

form a fundamental set of solution to Eq. (4.9.10). For example,

$$I_0(x) = 1 + \frac{x^2}{2^2} + \frac{x^4}{2^2\,4^2} + \frac{x^6}{2^2\,4^2\,6^2} + \cdots.$$

There are many amazing relations between Bessel functions of either the first or second kind. Some of them are given in the exercises.

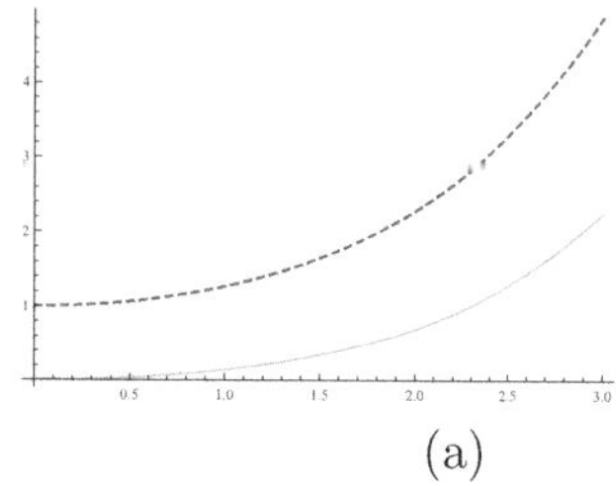
(a)

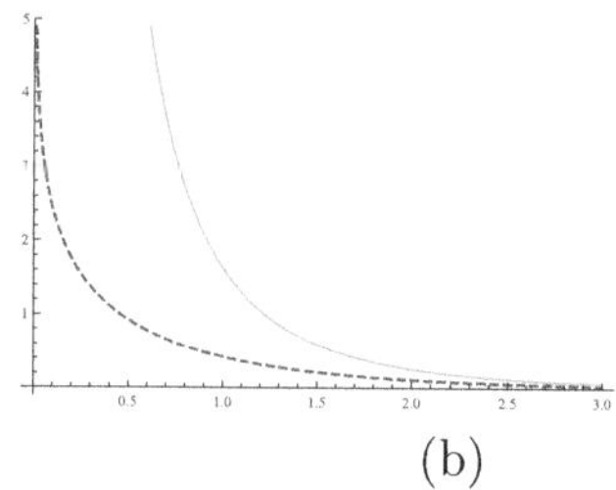
(b)

Figure 4.7: (a) Graphs of the modified Bessel functions of the first kind $I_0(x)$ (dashed line) and $I_2(x)$; (b) graphs of the modified Bessel functions of the second kind $K_0(x)$ (dashed line) and $K_2(x)$.

4.9.2 Bessel Functions of Half-Integer Order

The Bessel functions whose orders ν are odd multiples of $1/2$ are expressible in closed form via elementary functions. Let us consider the Bessel function of the first kind of order $1/2$:

$$J_{1/2}(x) = \sum_{k=0}^{\infty} \frac{(-1)^k}{k!\,\Gamma(k+1+1/2)} \left(\frac{x}{2}\right)^{2k+1/2}.$$

From the well-known relation

$$\Gamma(k+1+1/2) = \frac{(2k+1)!}{2^{2k+1}\,k!}\,\sqrt{\pi},$$

it follows that

$$\begin{aligned} J_{1/2}(x) &= \sum_{k=0}^{\infty} \frac{(-1)^k\, 2^{2k+1}\, k!}{k!\,(2k+1)!\,\sqrt{\pi}} \left(\frac{x^{2k+1/2}}{2^{2k+1/2}}\right) \\ &= \frac{2^{1/2}}{x^{1/2}\sqrt{\pi}} \sum_{k=0}^{\infty} \frac{(-1)^k\, x^{2k+1}}{(2k+1)!} = \sqrt{\frac{2}{\pi x}}\,\sin x. \end{aligned}$$

By a similar argument, it is possible to verify the following relations:

$$J_{-1/2}(x) = \sqrt{\frac{2}{\pi x}}\,\cos x, \quad I_{1/2}(x) = \sqrt{\frac{2}{\pi x}}\,\sinh x, \quad I_{-1/2}(x) = \sqrt{\frac{2}{\pi x}}\,\cosh x.$$

4.9.3 Related Differential Equations

Let $Z_\nu(x)$ denote either a solution of Bessel equation (4.9.1) or a solution of the modified Bessel equation (4.9.10). Let us introduce the function that depends on three parameters (p, q, and k)

$$u(x) = x^q Z_\nu\left(k\,x^p\right). \tag{4.9.11}$$

Then this function $u(x)$ is a solution of either the differential equation

$$x^2 u'' + x\,(1-2q)\,u' + \left(k^2 p^2 x^{2p} - p^2\nu^2 + q^2\right) u = 0 \tag{4.9.12}$$

or

$$x^2 u'' + x\,(1-2q)\,u' - \left(k^2 p^2 x^{2p} + p^2\nu^2 - q^2\right) u = 0 \tag{4.9.13}$$

depending on which Bessel equation the function $Z_\nu(t)$ satisfies.

Example 4.9.1: To find the general solution of the differential equation

$$y'' + a^2 x^2 y = 0,$$

we change variables $y = u\sqrt{x}$, $x = \sqrt{kt}$ (or $x^2 = kt$), where k will be determined shortly. Using the chain rule, we get

$$\begin{aligned} &\frac{\mathrm{d}t}{\mathrm{d}x} = \frac{2x}{k} \qquad \Longrightarrow \qquad \frac{\mathrm{d}}{\mathrm{d}x} = \frac{2x}{k}\,\frac{\mathrm{d}}{\mathrm{d}t}, \\ &\frac{\mathrm{d}y}{\mathrm{d}x} = \frac{\mathrm{d}}{\mathrm{d}x}\left(x^{1/2}\,u\right) = x^{1/2}\,\frac{\mathrm{d}u}{\mathrm{d}x} + \frac{1}{2}\,x^{-1/2}\,u = x^{1/2}\,\frac{\mathrm{d}u}{\mathrm{d}t}\left(\frac{2x}{k}\right) + \frac{1}{2}\,x^{-1/2}\,u = \frac{2}{k}\,x^{3/2}\dot{u} + \frac{1}{2}\,x^{-1/2}\,u, \end{aligned}$$

where dot stands for the derivative with respect to t. Now we find the second derivative:

$$\frac{\mathrm{d}^2y}{\mathrm{d}x^2} = \frac{\mathrm{d}}{\mathrm{d}x}\left(\frac{2}{k}\,x^{3/2}\,\dot{u} + \frac{1}{2}\,x^{-1/2}\,u\right) = \frac{2x}{k}\,\frac{\mathrm{d}}{\mathrm{d}t}\left(\frac{2}{k}\,x^{3/2}\,\dot{u} + \frac{1}{2}\,x^{-1/2}\,u\right)$$
$$= \frac{3}{k}\,x^{1/2}\,\dot{u} + \frac{4}{k^2}\,x^{5/2}\,\ddot{u} - \frac{1}{4}\,x^{-3/2}\,u + \frac{1}{k}\,x^{1/2}\,\dot{u}.$$

Substituting this expression for the second derivative into the given equation and multiplying by $x^{3/2}$, we obtain

$$\frac{4}{k^2}\,x^4\,\ddot{u} + \frac{4}{k}\,x^2\,\dot{u} - \frac{1}{4}\,u + a^2x^4\,u = 0 \quad\Longleftrightarrow\quad 4t^2\ddot{u} + 4\dot{u} + \left(a^2k^2t^2 - \frac{1}{4}\right)u = 0.$$

Now we set $k = 2/a$, and find that $u(t)$ is a solution of the Bessel equation

$$t^2\ddot{u} + \dot{u} + \left(t^2 - \frac{1}{16}\right)u = 0.$$

Therefore, the given differential equation has two linearly independent solutions

$$y_1(x) = x^{-1/2}J_{1/4}\left(\frac{ax^2}{2}\right) \quad \text{and} \quad y_2(x) = x^{-1/2}J_{-1/4}\left(\frac{ax^2}{2}\right).$$

Example 4.9.2: Let us consider the **Airy equation**[48]

$$y'' = xy \quad (-\infty < x < \infty). \tag{4.9.14}$$

Since this linear differential equation has no singular points, we seek its solution in the Maclaurin form

$$y(x) = a_0 + a_1x + a_2x^2 + \cdots = \sum_{n\geqslant 0} a_n\,x^n.$$

Substituting the series for $y(x)$ and for its second derivative $y''(x) = \sum_{n=0}^{\infty} a_{n+2}(n+2)(n+1)\,x^n$ into Airy's equation yields

$$\sum_{n=0}^{\infty} a_{n+2}(n+2)(n+1)\,x^n = \sum_{n=0}^{\infty} a_n\,x^{n+1} = \sum_{m=1}^{\infty} a_{m-1}\,x^m.$$

For two power series to be equal, it is necessary and sufficient that they have the same coefficients of all powers of x. This gives

$$a_2\,2\cdot 1 = 0, \qquad a_{n+2}(n+2)(n+1) = a_{n-1}, \quad n = 1, 2, 3, \ldots$$

or

$$a_{n+2} = \frac{a_{n-1}}{(n+2)(n+1)}, \qquad n = 1, 2, 3, \ldots; \qquad a_2 = 0. \tag{4.9.15}$$

Equation (4.9.15) is a recurrence relation of the third order. Since $a_2 = 0$, every third coefficient is zero, namely,

$$a_2 = 0,\ a_5 = 0,\ a_8 = 0, \ldots, a_{3k+2} = 0, \quad k = 0, 1, 2, \ldots.$$

Instead of solving the difference equation (4.9.15), we set $q = 1/2$, $p = 3/2$, $k = 2/3$, and $\nu = 1/3$ in Eq. (4.9.13) to obtain the modified Bessel equation that has two linearly independent solutions, known as **Airy functions**. They are denoted by $\mathrm{Ai}(x)$ and $\mathrm{Bi}(x)$, respectively. Airy functions commonly appear in physics, especially in optics, quantum mechanics, electromagnetics, and radioactive transfer. These functions are usually defined as

$$\mathrm{Ai}(x) = \frac{1}{\pi}\sqrt{\frac{x}{3}}\,K_{1/3}(z), \qquad z = \frac{2}{3}\,x^{3/2}, \tag{4.9.16}$$

$$\mathrm{Bi}(x) = \sqrt{\frac{x}{3}}\left[I_{1/3}(z) + I_{-1/3}(z)\right], \qquad z = \frac{2}{3}\,x^{3/2}, \tag{4.9.17}$$

[48]Sir George Biddell Airy (1801–1892) was an English astronomer and mathematician who was a director of the Greenwich Observatory from 1835 to 1881.

for positive x, and

$$\mathrm{Ai}(-x) = \frac{\sqrt{x}}{2}\left[J_{1/3}(z) - \frac{1}{\sqrt{3}}\,Y_{1/3}(z)\right] = \frac{\sqrt{x}}{3}\left[J_{1/3}(z) + J_{1/3}(z)\right], \tag{4.9.18}$$

$$\mathrm{Bi}(-x) = \frac{\sqrt{x}}{2}\left[\frac{1}{\sqrt{3}}\,J_{1/3}(z) + Y_{1/3}(z)\right] = \sqrt{\frac{x}{3}}\left[J_{-1/3}(z) - J_{1/3}(z)\right] \tag{4.9.19}$$

for the negative argument, where $z = \frac{2}{3}\,x^{3/2}$. Airy's functions are implemented in *Mathematica* as AiryAi[z] and AiryBi[z]. Their derivatives are denoted as AiryAiPrime[z] and AiryBiPrime[z]. In *Maple*, they have similar nomenclatures: AiryAi(x), AiryBi(x) and their derivatives AiryAi(1,x), AiryBi(1,x). *Maxima* and *Sage* share the notation `airy_ai` and `airy_bi`. MATLAB on prompt `airy(x)` or `airy(0,x)` returns the Airy function of the first kind Ai(x) and on prompt `airy(2,x)` returns the Airy function of the second kind, Bi(x). Its computer algebra package MuPad has special commands: `airyAi` and `airyBi`, respectively. SymPy uses `airyai` and `airybi` notations for these functions.

Figure 4.8: (a) Graph of Airy function Ai(x). (b) Graph of Airy function Bi(x).

Problems

1. Express $I_3(x)$ and $I_4(x)$ in terms of $I_0(x)$ and $I_1(x)$.
2. Show the following closed form formulas for Bessel functions.

   (a) $J_{3/2}(x) = \sqrt{\frac{2}{\pi x}}\left(\frac{\sin x}{x} - \cos x\right)$; **(b)** $J_{-3/2}(x) = -\sqrt{\frac{2}{\pi x}}\left(\frac{\cos x}{x} + \sin x\right)$;

   (c) $J_{5/2}(x) = \sqrt{\frac{2}{\pi x}}\left[\left(\frac{3}{x^2} - 1\right)\sin x - \frac{3}{x}\cos x\right]$; **(d)** $J_{-5/2}(x) = \sqrt{\frac{2}{\pi x}}\left[\left(\frac{3}{x^2} - 1\right)\cos x + \frac{3}{x}\sin x\right]$;

   (e) $J_{7/2}(x) = \sqrt{\frac{2}{\pi x}}\left[\left(\frac{30}{x^3} - \frac{11}{x}\right)\sin x - \left(\frac{30}{x^2} - 1\right)\cos x\right]$.
3. Find $J_1(0.1)$ to 5 significant figures.
4. Using generating functions
$$e^{\frac{x}{2}(z-z^{-1})} = \sum_{n=-\infty}^{\infty} J_n(x)\,z^n, \quad e^{\frac{x}{2}(z+z^{-1})} = \sum_{n=-\infty}^{\infty} I_n(x)\,z^n,$$
   prove the recurrences
$$\frac{2n}{x}\,J_n(x) = J_{n+1}(x) + J_{n-1}(x), \quad \frac{2n}{x}\,I_n(x) = I_{n-1}(x) - I_{n+1}(x), \quad n = 0, \pm 1, \ldots.$$
5. Prove the following identities.

   (a) $J_0''(x) = \frac{1}{2}\left[J_2(x) - J_0(x)\right]$; **(b)** $\frac{\mathrm{d}}{\mathrm{d}x}\left[x\,J_1(x)\right] = x\,J_0(x)$;

   (c) $\frac{J_2(x)}{J_1(x)} = \frac{1}{x} - \frac{J_0''(x)}{J_0'(x)}$; **(d)** $\frac{J_2(x)}{J_1(x)} = \frac{2}{x} - \frac{J_0(x)}{J_1(x)}$;

   (e) $\frac{\mathrm{d}}{\mathrm{d}x}\left[x^2\,J_1(x) + x\,J_0(x)\right] = x^2\,J_0(x) + J_0(x)$; **(f)** $J_1'(x) = J_0(x) - \frac{1}{x}\,J_1(x)$;

   (g) $\frac{\mathrm{d}}{\mathrm{d}x}\left[2x\,J_1(x) - x^2\,J_0(x)\right] = x^2\,J_1(x)$; **(h)** $J_0'(x) = -J_1(x)$.
6. Show that, if m is any positive integer,
$$\left(\frac{\mathrm{d}}{x\,\mathrm{d}x}\right)^m\left[x^{\nu}\,J_{\nu}(x)\right] = x^{\nu-m}\,J_{\nu-m}(x), \quad \left(\frac{\mathrm{d}}{x\,\mathrm{d}x}\right)^m\left[x^{-\nu}\,J_{\nu}(x)\right] = (-1)^m\,x^{-\nu-m}\,J_{\nu+m}(x).$$

7. Verify that the Bessel equation of the order one-half

$$x^2\,y'' + x\,y' + \left(x^2 - \frac{1}{4}\right) y = 0, \qquad x > 0$$

can be reduced to the linear equation with constant coefficients $u'' + u = 0$ by substitution $y = x^{-1/2}\,u(x)$.

8. Verify that when $x > 0$, the general solution of

$$x^2\,y'' + (1 - 2\alpha)\,x\,y' + \beta^2 x^2\,y = 0$$

is $y(x) = x^{\alpha}\,Z_{\alpha}(\beta x)$, where Z_{α} denotes the general solution of Bessel's equation of the order α and $\beta \neq 0$ is a positive constant.

9. Show that the equation $y'' + b^2 x^{a-2} y = 0$ has the general solution

$$y = \sqrt{x}\left(\frac{b}{a}\right)^{1/a}\left(c_1\Gamma\left(\frac{a+1}{a}\right) J_{1/a}\left(\frac{2bx^{1/a}}{a}\right) + c_2\Gamma\left(\frac{a-1}{a}\right) J_{-1/a}\left(\frac{2bx^{1/a}}{a}\right)\right),$$

where c_1, c_2 are arbitrary constants.

10. Express the general solution to the Riccati equation $y' + 4x^2 + y^2 = 0$ through Bessel functions. *Hint:* Make substitution $u' = uy$.

11. Show that a solution of the Bessel equation (4.9.1) can be represented in the form $J_{\nu}(x) = x^{-1/2}\,w(x)$, where $w(x)$ satisfies the differential equation

$$x^2\,w'' + \left(x^2 - \nu^2 + \frac{1}{4}\right) w = 0.$$

12. Determine the general solution of the following differential equations.

(a)	$xy'' + y' + xy = 0;$	**(b)**	$x^2 y'' + 2xy' + xy = 0;$
(c)	$xy'' + y' + 4xy = 0;$	**(d)**	$x^2 y'' + 3xy' + (1 + x)y = 0;$
(e)	$x^2 y'' + xy' + (2x^2 - 1)y = 0;$	**(f)**	$x^2 y'' + xy' + 2xy = 0;$
(g)	$x^2 y'' + xy' + (8x^2 - \frac{4}{9})y = 0;$	**(h)**	$x^2 y'' + 4xy' + (2 + x)y = 0;$
(i)	$xy'' + y' + \frac{1}{4}y = 0;$	**(j)**	$xy'' + 2y' + \frac{1}{4}y = 0;$
(k)	$xy'' + 2y' + \frac{1}{4}xy = 0;$	**(l)**	$x^2 y'' + xy' + (4x^4 - 16)y = 0.$

13. Find a particular solution to the initial value problems:

(a) $x^2\,y'' + xy' + (x^2 - 1)y = 0, \quad y(1) = 1.$

(b) $4x^2\,y'' + (x + 1)y = 0, \quad y(4) = 1.$

(c) $x\,y'' + 3y' + xy = 0, \quad y(0.5) = 1$; *Hint:* the function $x^{-1} J_1(x)$ satisfies this equation.

(d) $4x^2 y'' + 4xy' + (x - 1)y = 0, \quad y(4) = 1.$

(e) $x^4\,y'' + x^3\,y' + y = 0, \quad y(0.5) = 3$; *Hint:* use substitution $t = x^{-1}$.

(f) $x\,y'' - 3y' + xy = 0, \quad y(2) = 1.$

14. Show that $4J_{\nu}''(x) = J_{\nu-2}(x) - 2J_{\nu}(x) + J_{\nu+2}(x), \quad \nu \geqslant 2.$

15. Verify that the differential equation

$$x\,y'' + (1 - 2\nu)\,y' + xy = 0, \qquad x > 0$$

has a particular solution $y = x^{\nu}\,J_{\nu}(x)$.

16. Show that if $N_n(x) = \lim_{\nu\to n} N_{\nu}(x)$, $n = 0, 1, 2, \ldots$, where N_{ν} is defined by Eq. (4.9.7), then
$N_n(x) = \frac{1}{\pi}\left\{\frac{\partial}{\partial\nu}\left[J_{\nu}(x) - (-1)^n J_{-\nu}(x)\right]\right\}_{\nu=n} = \lim_{\nu\to n}\frac{1}{\pi}\frac{\partial}{\partial\nu}\left[J_{\nu}(x) - (-1)^n J_{-\nu}(x)\right].$

17. Bessel's equation of order 0 has a solution of the form $y_2(x) = J_0(x)\,\ln x + \sum_{k=1}^{\infty} a_k x^k$. Prove that the coefficients in the above series expansion are expressed as

$$y_2(x) = \sum_{k=1}^{\infty} \frac{(-1)^{k+1}}{(k!)^2}\left(1 + \frac{1}{2} + \cdots + \frac{1}{k}\right)\left(\frac{x}{2}\right)^{2k} + J_0(x)\,\ln x.$$

18. Find the general solution of $y'' + \dfrac{a+1}{x}\,y' + b^2 y = 0$, $x > 0$, where a and $b > 0$ are constants.

19. Show that if

$$\operatorname{ber} z = \sum_{n\geqslant 0} \frac{(-1)^n\, z^{4n}}{2^{4n}\,(2n!)^2}, \qquad \operatorname{bei} z = \sum_{n\geqslant 0} \frac{(-1)^n\, z^{4n+2}}{2^{4n+2}\,((2n+1)!)^2},$$

where $\mathbf{j}$ is the unit vector in the positive vertical direction, so $\mathbf{j}^2 = -1$. The functions $\operatorname{ber} z$ and $\operatorname{bei} z$ are named after William Thomson, 1st Baron Kelvin[49]; they are implemented in *Mathematica* and *Maple* as KelvinBer[ν, z] and KelvinBei[ν, z], respectively.

Show that the function $I_0(z\sqrt{\mathbf{j}})$ is a solution of the differential equation

$$\frac{\mathrm{d}^2 u}{\mathrm{d}z^2} + \frac{1}{z}\frac{\mathrm{d}u}{\mathrm{d}z} - \mathbf{j}\,u = 0.$$

20. Find the general solution of the differential equation

$$y'' + \left(1 - \frac{3}{4x^2}\right) y = 0$$

by making the change of variable $y = u\sqrt{x}$.

21. Find the general integral of the differential equation

$$x^2 y'' - xy' + \left(x^2 - \frac{7}{9}\right) y = 0$$

by making the change of variable $y = xu$.

22. Find the general integral of the differential equation

$$y'' + \left(e^x - \frac{9}{4}\right) y = 0$$

by making the change of independent variable $4\,e^x = t^2$.

23. Find the general solution of the differential equation

$$4y'' + 9x^4 y = 0.$$

24. Show that the Wronskian of Airy functions is $\mathrm{Ai}(x)\mathrm{Bi}'(x) - \mathrm{Ai}'(x)\mathrm{Bi}(x) = \dfrac{1}{\pi}$.

25. Show that the Airy functions satisfy the following initial conditions: $\mathrm{Ai}(0) = 3^{-2/3}/\Gamma\left(\frac{2}{3}\right)$, $\mathrm{Ai}'(0) = -3^{-1/3}/\Gamma\left(\frac{1}{3}\right)$ and $\mathrm{Bi}(0) = 3^{-1/6}/\Gamma\left(\frac{2}{3}\right)$, $\mathrm{Bi}'(0) = 3^{1/6}/\Gamma\left(\frac{1}{3}\right)$.

26. Use the transformations $u = x^a y$ and $t = b\,x^c$ to show that a solution of

$$x^2 u'' + (1-2a)\,x\,u' + (b^2 c^2 x^{2c} + a^2 - k^2 c^2)\,u = 0$$

is given by $u(x) = x^a J_k\,(b\,x^c)$.

27. By differentiating under the integral sign, show that the integral

$$y(x) \stackrel{\text{def}}{=} \int_0^{\pi} \cos\,(x\,\sin\theta)\,\mathrm{d}\theta$$

satisfies Bessel's equation of order zero.

28. Show that $y = \sqrt{x}\,J_2\left(4\,x^{1/4}\right)$ satisfies

$$x^{3/2} y'' + y = 0 \qquad \text{for} \quad x > 0.$$

[49]Lord Kelvin (1824–1907) was a Belfast-born British mathematical physicist and engineer.

Summary for Chapter 4

1. Since the general solution of a linear differential equation of n-th order contains n arbitrary constants, there are two common ways to specify a particular solution. If the unknown function and its derivatives are specified at a fixed point, we have the initial conditions. In contrast, the boundary conditions are imposed on the unknown function on at least two different points. The differential equation with the initial conditions is called the **initial value problem** or the Cauchy problem. The differential equation with the boundary conditions is called the **boundary value problem**.

2. The general linear differential equation of the n-th order ($\texttt{D} = \mathrm{d}/\mathrm{d}x$)

$$L[x, \texttt{D}]y \stackrel{\text{def}}{=} a_n(x)\, y^{(n)} + a_{n-1}(x)\, y^{(n-1)} + \cdots + a_0(x)\, y(x) = f(x). \tag{4.1.5}$$

is said to be **homogeneous** if $f(x)$ is identically zero. We call this equation **nonhomogeneous** (or inhomogeneous) if $f(x)$ is not identically zero.

3. Let functions $a_0(x)$, $a_1(x)$, ..., $a_n(x)$ and $f(x)$ be defined and continuous on the closed interval $a \leqslant x \leqslant b$ with $a_n(x) \ne 0$ for $x \in [a, b]$. Let x_0 be such that $a \leqslant x_0 \leqslant b$ and let y_0, y_1, ..., y_{n-1} be any constants. Then there exists a unique solution of Eq. (4.1.5) that satisfies the initial conditions $y^{(k)}(x_0) = y_k$ ($k = 0, 1, \ldots, n-1$).

4. A second order differential equation is called **exact** if it can be written in the form $\dfrac{\mathrm{d}}{\mathrm{d}x} F(x, y, y') = f(x)$. In particular, a linear exact equation is $\dfrac{\mathrm{d}}{\mathrm{d}x}\left[P(x)\, y' + Q(x)\, y\right] = f(x)$.

5. An **operator** transforms a function into another function. A linear operator L is an operator that possesses the following two properties:

$$L[y_1 + y_2] = L[y_1] + L[y_2], \quad L[cy] = cL[y], \quad \text{for any constant } c.$$

6. The adjoint to the linear differential operator $L[y](x) = a_2(x)\, y''(x) + a_1(x) y'(x) + a_0(x) y(x)$ is the operator $L' \equiv \texttt{D}^2 a_2(x) - \texttt{D} a_1(x) + a_0(x)$, where $\texttt{D}$ stands for the derivative. An operator is called self-adjoint (or Hermitian) if $L = L'$.

7. **Principle of Superposition:** If each of the functions $y_1, y_2, \ldots, y_n$ are solutions to the same linear homogeneous differential equation,

$$a_n(x) y^{(n)} + a_{n-1}(x) y^{(n-1)} + \cdots + a_1(x) y' + a_0(x) y = 0 \quad \text{or} \quad L[x, \texttt{D}]y = 0, \tag{4.1.8}$$

then for every choice of the constants $c_1, c_2, \ldots, c_n$ the linear combination $c_1 y_1 + c_2 y_2 + \cdots + c_n y_n$ is also a solution.

8. A set of m functions $f_1, f_2, \ldots, f_m$ each defined and continuous on the interval $|a, b|$ ($a < b$) is said to be **linearly dependent** on $|a, b|$ if there exist constants $\alpha_1, \alpha_2, \ldots, \alpha_m$, not all of which are zero, such that $\alpha_1 f_1(x) + \alpha_2 f_2(x) + \cdots + \alpha_m f_m(x) \equiv 0$ for every x in the interval $|a, b|$. Otherwise, the functions $f_1, f_2, \ldots, f_m$ are said to be **linearly independent** on this interval.

9. Let $f_1, f_2, \ldots, f_m$ be m functions that together with their first $m-1$ derivatives are continuous on the interval $|a, b|$. The **Wronskian** or the **Wronskian determinant** of $f_1, f_2, \ldots, f_m$, evaluated at $x \in |a, b|$, is denoted by $W[f_1, f_2, \ldots, f_m](x)$ or $W(f_1, f_2, \ldots, f_m; x)$ or simply by $W(x)$ and is defined to be the determinant (4.2.1), page 199.

10. Let $f_1, f_2, \ldots, f_m$ be m functions that together with their first $m-1$ derivatives are continuous on the interval $|a, b|$. If their Wronskian $W[f_1, f_2, \ldots, f_m](x_0)$ is not equal to zero at some point $x_0 \in |a, b|$, then these functions $f_1, f_2, \ldots, f_m$ are linearly independent on $|a, b|$. Alternatively, if $f_1, f_2, \ldots, f_m$ are linearly dependent and they have $m-1$ first derivatives on the interval $|a, b|$, then their Wronskian $W(f_1, f_2, \ldots, f_m; x) \equiv 0$ for every x in $|a, b|$.

11. Let $y_1, y_2, \ldots, y_n$ be the solutions of the n-th order differential homogeneous equation (4.1.8) with continuous coefficients $a_0(x)$, $a_1(x)$, ..., $a_n(x)$, defined on an interval $|a, b|$. Then functions $y_1, y_2, \ldots, y_n$ are linearly independent on this interval $|a, b|$ if and only if their Wronskian is never zero in $|a, b|$. Alternatively, these functions $y_1, y_2, \ldots, y_n$ are linearly dependent on this interval $|a, b|$ if and only if $W[y_1, y_2, \ldots, y_n](x)$ is zero for all $x \in |a, b|$.

12. For arbitrary set of linearly independent smooth functions $y_1, y_2, \ldots, y_n$, the linear differential operator of order n that annihilates these functions is

$$L[x, \texttt{D}]f = \frac{W[y_1, y_2, \ldots, y_n, f]}{W[y_1, y_2, \ldots, y_n]}.$$

13. The Wronskian of two solutions to the differential equation of the second order (4.2.2) satisfies the Abel identity (4.2.3), page 201.

14. Any set of solutions y_1, y_2, ..., y_n is called a fundamental set of solutions of (4.1.8) if they are linearly independent on some interval. If y_1, y_2, ..., y_n is a fundamental set of solutions of Eq. (4.3.2), then the formula

$$y(x) = C_1 y_1(x) + C_2 y_2(x) + \cdots + C_n y_n(x),$$

where $C_1, C_2, \ldots, C_n$ are arbitrary constants, gives the general solution of $L[x, \texttt{D}]y = 0$.

15. There exists a fundamental set of solutions for the homogeneous linear n-th order differential equation (4.3.2) on an interval where the coefficients are all continuous.

16. Let $y_1, y_2, \ldots, y_n$ be n functions that together with their first n derivatives are continuous on an open interval (a, b). Suppose their Wronskian $W(y_1, y_2, \ldots, y_n; x) \neq 0$ for $a < x < b$. Then there exists the unique linear homogeneous differential equation of the n-th order for which the collection of functions $y_1, y_2, \ldots, y_n$ is a fundamental set of solutions.

17. To any linear homogeneous differential equation with constant coefficients,

$$a_n\, y^{(n)} + a_{n-1}\, y^{(n-1)} + \cdots + a_0 y(x) = 0, \tag{4.4.2}$$

corresponds the algebraic equation $a_n \lambda^n + a_{n-1}\lambda^{n-1} + \cdots + a_0 = 0$, called the characteristic equation.

18. If all roots of the characteristic equation are different, the set of functions

$$y_1(x) = e^{\lambda_1 x}, \; y_2(x) = e^{\lambda_2 x}, \; \cdots, \; y_n(x) = e^{\lambda_n x}$$

constitutes the fundamental set of solutions.

19. If the characteristic polynomial $L(\lambda) = a_n \lambda^n + a_{n-1}\lambda^{n-1} + \cdots + a_1 \lambda + a_0$ of the differential operator $L[\texttt{D}] = a_n \texttt{D}^n + a_{n-1}\texttt{D}^{n-1} + \cdots + a_0$, $\texttt{D}$ stands for the derivative operator, has a complex null $\lambda = \alpha + \mathbf{j}\beta$, namely, $L(\alpha + \mathbf{j}\beta) = 0$, then

$$y_1 = e^{\alpha x} \cos \beta x \qquad \text{and} \qquad y_2 = e^{\alpha x} \cos \beta x$$

are two linearly independent solutions of the constant coefficient equation $L[\texttt{D}]y = 0$.

20. If the characteristic polynomial $L(\lambda) = a_n \lambda^n + a_{n-1}\lambda^{n-1} + \cdots + a_0 = a_n(\lambda - \lambda_1)^{m_1} \cdots (\lambda - \lambda_s)^{m_s}$ for constant coefficient differential operator $L[\texttt{D}] = a_n \texttt{D}^n + a_{n-1}\texttt{D}^{n-1} + \cdots + a_0$ has a null $\lambda = \lambda_k$ of multiplicity m_k, that is, $L(\lambda)$ contains a factor $(\lambda - \lambda_k)^{m_k}$, then the differential equation $L[\texttt{D}]y = 0$ has m_k linearly independent solutions:

$$e^{\lambda_k x}, \quad x\, e^{\lambda_k x}, \quad \ldots, \quad x^{m_k - 1}\, e^{\lambda_k x}.$$

21. If a solution $y_1(x)$ to the differential equation with variable coefficients $a_n(x)y^{(n)} + a_{n-1}y^{(n-1)} + \cdots + a_0 y = 0$ is known, then the Bernoulli substitution $y = v y_1$ reduces the given equation to an equation of order $n - 1$.

22. The expression $D(x) = 2p'(x) + p^2(x) - 4q(x)$ is called the **discriminant** of the differential equation $y'' + p(x)y' + q(x)y = 0$, which can be reduced to the Riccati equation $w^2 - 2w' = D(x)$, $w = y''/y'$.

23. Two linearly independent solutions, $y_1(x)$ and $y_2(x)$, to the second order differential equation $y'' + p(x)y' + q(x)y = 0$ can be expressed via w, the solution to the Riccati equation $w^2 - 2w' = 2p' + p^2 - 4q$:

$$y_1 = \exp\left\{ -\frac{1}{2} \int (p(x) + w)\, \mathrm{d}x \right\}, \quad y_2(x) = y_1(x) \int y_1^{-2}(x) \exp\left\{ - \int_{x_0}^{x} p(t)\, \mathrm{d}t \right\} \mathrm{d}x.$$

24. The linear differential equation

$$x^n a_n y^{(n)} + x^{n-1} a_{n-1} y^{(n-1)} + \cdots + a_1 x\, y' + a_0 y = 0, \tag{4.6.15}$$

called the Euler equation, can be reduced to a constant coefficient case either by substitution $x = \ln t$, or seeking a particular solution in the form $y = x^r$, for some power r.

25. The general solution of the nonhomogeneous linear differential equation (4.7.1) on some open interval (a, b) is the sum

$$y(x) = y_h(x) + y_p(x), \tag{4.7.4}$$

where $y_h(x)$ is the general solution of the associated homogeneous differential equation (4.7.2), and $y_p(x)$, which is a particular solution to a nonhomogeneous differential equation. The general solution $y_h(x)$ of the homogeneous equation (4.7.4) is frequently referred to as a complementary function.

26. The *method of undetermined coefficients* is a special method of practical interest that allows one to find a particular solution of a nonhomogeneous differential equation. It is based on guessing the form of the particular solution, but with coefficients left unspecified. This method can be applied if
    (a) Equation (4.7.11) is a linear differential equation with **constant coefficients**;
    (b) the nonhomogeneous term $f(x)$ can be broken down into the sum of either a *polynomial*, an *exponential*, a *sine* or *cosine*, or some product of these functions. In other words, $f(x)$ is a solution of a homogeneous differential equation with constant coefficients, $\psi[\texttt{D}]\, f = 0$.

27. There are several steps in the application of this method:

(a) For a given linear nonhomogeneous differential equation with constant coefficients

$$a_n y^{(n)}(x) + a_{n-1} y^{(n-1)}(x) + \cdots + a_0 y(x) = f(x),$$

find the roots of the characteristic equation $L(\lambda) = 0$, where

$$L(\lambda) = a_n \lambda^n + a_{n-1} \lambda^{n-1} + \cdots + a_0.$$

(b) Break down the nonhomogeneous term $f(x)$ into the sum $f(x) = f_1(x) + f_2(x) + \cdots + f_m(x)$ so that each term $f_j(x)$, $j = 1, 2, \ldots, m$, is one of the functions listed in the first column of Table 229. That is, each term $f_j(x)$ has the form

$$f_j(x) = P_k(x)\, e^{\alpha x} \cos\beta x \quad \text{or} \quad f_j(x) = P_k(x)\, e^{\alpha x} \sin\beta x,$$

where $P_k(x) = c_0 x^k + c_1 x^{k-1} + \cdots + c_k$ is a polynomial of the k-th order, and $\sigma = \alpha + \mathbf{j}\beta$ is a complex ($\beta \ne 0$) or real ($\beta = 0$) number, called the **control number**.

(c) Let the control number σ_j of the forcing term $f_j(x)$, $j = 1, 2, \ldots, m$, be the root of the characteristic equation $L(\lambda) = a_n \lambda^n + a_{n-1}\lambda^{n-1} + \cdots + a_0 = 0$ of the multiplicity $r > 0$. We set $r = 0$ if σ_j is not a root of the characteristic equation. It is said that there is a *resonance case* if $r > 0$, and a *nonresonance case* if $r = 0$, that is, the control number is not a null of the characteristic polynomial. Seek a particular solution $y_{pj}(x)$ in the form

$$y_{pj}(x) = x^r\, e^{\alpha x} \left[R_{1k}(x) \cos\beta x + R_{2k}(x) \sin\beta x\right],$$

where $R_{1k}(x)$ and $R_{2k}(x)$ are polynomials of the k-th order with coefficients to be determined.

(d) Substitute the assumed expression for $y_{pj}(x)$ into the left-hand side of the equation $L[\mathtt{D}]\, y_{pj} = f_j$ and equate coefficients of like terms, where $\mathtt{D} = d/dx$.

(e) Solve the resulting system of linear algebraic equations for the undetermined coefficients of polynomials $R_{1k}(x)$ and $R_{2k}(x)$ to determine $y_{pj}(x)$.

(f) Repeat steps (b) – (e) for each $j = 1, 2, \ldots, m$, and sum the functions $y_p(x) = y_{p1}(x) + y_{p2}(x) + \cdots + y_{pm}(x)$ to obtain a particular solution of the driven equation $L[\mathtt{D}]y = f$.

The method of variation of parameters (or Lagrange's method) for solving the inhomogeneous differential equation (4.1.5) requires knowledge of the fundamental set of solutions y_1, y_2, $\ldots$, y_n for the corresponding homogeneous equation (4.1.8). With this in hand, follow the following steps.

A. Form a particular solution as a linear combination of known linearly independent solutions

$$y_p(x) = A_1(x) y_1(x) + A_2(x) y_2(x) + \cdots + A_n(x) y_n(x) \tag{A}$$

with some yet unknown functions $A_1(x)$, $A_2(x)$, $\ldots$, $A_n(x)$.

B. Impose n auxiliary conditions on the derivatives of the unknown functions

$$\begin{aligned}
A_1' y_1(x) + A_2' y_2(x) + \cdots + A_n' y_n(x) &= 0, \\
A_1' y_1'(x) + A_2' y_2'(x) + \cdots + A_n' y_n'(x) &= 0, \\
\cdots &= 0 \\
A_1' y_1^{(n-2)}(x) + A_2' y_2^{(n-2)}(x) + \cdots + A_n' y_n^{(n-2)}(x) &= 0, \\
A_1' y_1^{(n-1)}(x) + A_2' y_2^{(n-1)}(x) + \cdots + A_n' y_n^{(n-1)}(x) &= f(x)/a_n(x).
\end{aligned}$$

C. Solve the obtained algebraic system of equations for unknown derivatives A_1', A_2', $\ldots$, A_n'.

D. Integrate the resulting expressions to obtain $A_1(x)$, $A_2(x)$, $\ldots$, $A_n(x)$ and substitute them into Eq. (A) to determine the general solution.

28. The **Green** function for the linear differential operator is uniquely defined either by the formula (4.8.15), page 243, or as the solution of the initial value problem formulated in Theorem 4.23 on page 243.

29. The differential equation

$$x^2\, y'' + x\, y + (x^2 - \nu^2)\, y = 0, \tag{4.9.1}$$

where ν is a real nonnegative constant, is called the **Bessel equation**.

30. The Bessel equation (4.9.1) has two linearly independent solutions

$$J_\nu(x) = \sum_{k=0}^{\infty} \frac{(-1)^k \left(\frac{x}{2}\right)^{2k+\nu}}{k!\, \Gamma(\nu + k + 1)}, \tag{4.9.5}$$

called the **Bessel function of the first kind of the order** ν, and

$$N_\nu(x) = \frac{\cos\nu\pi\, J_\nu(x) - J_{-\nu}(x)}{\sin\nu\pi}, \tag{4.9.7}$$

called the **Neumann function**. Its constant multiple $Y_\nu(x) = \pi\, N_\nu(x)$ is called the Weber function.

Review Questions for Chapter 4

In all problems, D is the derivative operator, that is, $\texttt{D}y = y'$; its powers are defined recursively: $\texttt{D}^n = \texttt{D}(\texttt{D}^{n-1})$, $n = 1, 2, \ldots$, with $\texttt{D}^0$ being the identical operator, which we drop.

Section 4.1

1. In each of the following initial value problems, determine, without solving the problem, the longest interval in which the solution is certain to exist.
   (a) $\cos x\, y'' + xy' + 7y = 1, \quad y(0) = 1,\ y'(0) = 0;$ **(b)** $(x-1)^2\, y'' + x\, y' - x^2\, y = \sin x, \quad y(2) = 1,\ y'(2) = 0;$
   (c) $(x^2 - x)\, y'' + xy = \sin x, \quad y(-1) = 0,\ y'(-1) = 1.$
   (d) $\sin x\, y'' + xy' + x^2\, y = x^3, \quad y(0.1) = 1,\ y'(0.1) = 2.$
   (e) $x\, y'' + x^2 y' + x^3\, y = \sin 2x, \quad y(1) = 1,\ y'(1) = 2.$ **(f)** $(x+1)(x-2)\, y'' + x\, y' + 5y = \cos x, \quad y(0) = 1,\ y'(0) = -1.$
   (g) $(x^2 - 4)\, y'' - x(x+2)\, y' + (x-2)\, y = 0, \quad y(0) = -1,\ y'(0) = 1.$
2. By changing the independent variable $t = \varphi(x)$, reduce the given differential equation to the equation without the first derivative.
   (a) $y'' - 2y' + e^{4x}\, y = 0;$ **(b)** $x^2 y'' + xy' + (x^2 - \nu^2) y = 0;$
   (c) $xy'' + 2y' + xy = 0;$ **(d)** $x^2 y'' - xy' + 9xy = 0.$
3. Changing the independent variable, reduce the given variable coefficient differential equation to a constant coefficient equation.
   (a) $y'' - y' + e^{2x} y = 0;$ **(b)** $(1 - x^2) y'' - xy' + 4y = 0;$
   (c) $(1 + x^2) y'' + xy' + 9y = 0;$ **(d)** $(x^2 - 1) y'' + xy - 16y = 0;$
   (e) $2xy'' + y' - 2y = 0;$ **(f)** $x^4 y'' + 2x^3 y' - 4y = 1/x;$
   (g) $xy'' - y' - 4x^3 y = 0;$ **(h)** $x^2 y'' + \left(x + x^3/(1 - x^2)\right) y' + 4(1 - x^2) y = 0;$
   (i) $(1 + x^2) y'' + xy' + 4y = 0;$ **(j)** $xy'' + (x^2 - 1) y' - 6x^3 y = 0.$
4. For what values of parameters p, q, and r does the differential equation $u'' + \frac{px}{ax+b}\, u' + \frac{qx+r}{(ax+b)^2}\, u = 0$ have an integrating factor $\mu(x) = (ax+b)^{1-pb/a^2}\, e^{px/a}$?
5. By the use of a Riccati equation (4.1.23), find an integral for each of the following differential equations.
   (a) $y'' - y'\left(2 + 4\, e^{2x}\right) - 5e^{4x}\, y = 0;$ **(b)** $xy'' \ln 2x - y' - 9xy(\ln 2x)^3 = 0;$
   (c) $x^3 y'' - xy' + 2y = 0;$ **(d)** $xy'' - 2y' - 4x^5 y = 0;$
   (e) $xy'' + (x^3 + x - 2) y' + x^3 y = 0;$ **(f)** $y'' - 2y' - 16e^{4x} y = 0.$
6. With the substitution $y = \varphi(x) v(x)$, transfer each of the following differential equations into an equation that does not contain the first derivative.
   (a) $(x^2 + 1) y'' + 4xy' + (4x^2 + 6) y = 0;$ **(b)** $x^2 y'' + (2x^2 - 2x) y' = (2x - 2 - 5x^2) y;$
   (c) $y'' - 6y' + (9 - 2/x) y = e^{3x};$ **(d)** $(x^2 + 9) y'' - 4xy' + 6y = 0;$
   (e) $y'' - 2y' \cot x + y(3 + 2\csc^2 x) = 4 \sin x;$ **(f)** $x^2 y'' - \frac{2x}{\ln x}\, y' = \left(2 - \frac{2}{\ln^2 x} - \frac{1}{\ln x}\right) y;$
   (g) $x^2 y'' - 4x^3 y' = (6 - 4x^4 + 2x^2) y;$ **(h)** $xy'' + 2y' + xy = 0.$
7. Solve the following linear exact equations.
   (a) $(x+1) y'' + 2y' + \sin x = 0;$ **(b)** $x(x+1) y'' + (4x + 2) y' + 2y = e^x;$
   (c) $x^3 y'' + 6x^2 y' + 6xy = 12x^2;$ **(d)** $\frac{y''}{1+x} - \frac{2y'}{(1+x)^2} + \frac{2y}{(1+x)^3} = 6x;$
   (e) $x^3 y'' + (3x^2 + x) y' + y = 0;$ **(f)** $y'' + \frac{2x}{1+x^2}\, y' + \frac{2(1-x^2)}{(1+x^2)^2}\, y = 4x;$
   (g) $y'' + \cot x\, y' - y \csc^2 x = 0;$ **(h)** $\sin x\, y'' + \sin x\, y = \cos x;$
   (i) $xy'' + \frac{1+x}{1+2x}\, y' - \frac{y}{(1+2x)^2} = 1;$ **(j)** $(1 + e^{2x})\, y''(2e^{2x} + x)\, y' + y = f(x);$
   (k) $x^4 y'' + (4x^3 + 2x) y' + 2y = 0;$ **(l)** $(\sin x) y'' + (\cos x + \sec x) y' + y \sec x \tan x = \cos x.$
8. Solve the following nonlinear exact equations.
   (a) $\frac{x^3 y''}{1 + x^2 y'} + \ln(1 + x^2 y') + \frac{2x^2}{1 + x^2 y'} = 0;$ **(b)** $x^2 yy'' + (xy')^2 + 4xyy' + y^2 = 0;$
   (c) $\frac{y''}{yy'} - \frac{y'}{y^2} \ln \frac{y'}{x} = \frac{1}{xy};$ **(d)** $xyy'' + x(y')^2 + yy' = 2x.$
9. Find an integrating factor that reduces the given differential equation to an exact equation and use it to determine its general solution.
   (a) $y'' - \frac{x+1}{x+2}\, y' = e^x.$ **(b)** $y'' + (1 + 1/x) y' + y/x = 3x.$
   (c) $y'' - (\sec x) y' = 1.$ **(d)** $x^4\, y'' + 2x^3\, y' = -k/x^4\ (k > 0).$
   (e) $(x^2 + x) y'' + (x^2 + 4x + 3) y' + 4xy = 5x^2 + 8x + 3.$
   (f) $(3x + x^2) y'' + (12 + 4x - x^2) y' = 4 + 4y + 5xy.$
   (g) $(x + \frac{1}{x}) y'' + 3y' + y/x = 1/\sqrt{1 + x^2}.$
   (h) $(y')^2 (1 + x^2 + y^2) + \frac{yy'}{x} (1 + y^2 - 3x^2) + yy''(1 + x^2 + y^2) = 2x^2.$

10. Reduce the coefficient of y in $y'' + xy = 0$ to unity.
11. In the answer to the preceding problem, reduce the coefficient of the first derivative to zero.
12. Reduce the coefficient of y in $y'' + x^n y = 0$ to unity.
13. Reduce the coefficient of the first derivative in the equation of the preceding problem to zero.
14. Suppose that $y_1(x)$ and $y_2(x)$ are solutions of an inhomogeneous equation $L[y] = f$, where L is a linear differential operator and $f(x)$ is not identically zero. Show that a linear combination $y = C_1 y_1(x) + C_2 y_2(x)$ of these solutions will also be a solution if and only if $C_1 + C_2 = 1$.
15. Solve the differential equations

(a) $\dfrac{\mathrm{d}}{\mathrm{d}r}\left(r\,\dfrac{\mathrm{d}u}{\mathrm{d}r}\right) + \dfrac{4a^2r^2 - 1}{4r}\,u = 0;$ (b) $\dfrac{\mathrm{d}}{\mathrm{d}r}\left(r^2\,\dfrac{\mathrm{d}u}{\mathrm{d}r}\right) + a^2r^2\,u = 0;$

using the change of variable $u = v(r)/\sqrt{r}$ and $u = v(r)/r$, respectively.
16. Consider the partial differential equation (called the Laplace equation): $\dfrac{\partial^2 u}{\partial x^2} + \dfrac{\partial^2 u}{\partial y^2} = 0$. Show that the substitution $u(x, y) = e^{x/\alpha} f(\alpha x - \beta y)$, where α and β are positive constants, reduces the Laplace equation to the ordinary differential equation $\dfrac{\mathrm{d}^2 f}{\mathrm{d}\xi^2} + 2p\,\dfrac{\mathrm{d}f}{\mathrm{d}\xi} + q\,f = 0$, with $p = 1/(\alpha^2 + \beta^2)$ and $q = p/\alpha^2$.
17. Formulate a differential equation governing the motion of an object with mass $m = 2$ kg that stretches a vertical spring 4 cm when attached and experiences a resistive force whose magnitude is three times the square of the speed.

Section 4.2 of Chapter 4 (Review)

1. Determine whether the given pair of functions is linearly independent or linearly dependent.
   (a) $2x^3$ & $-5x^3$; **(b)** e^{2x} & e^{-2x}; **(c)** t & t^{-1} $(t \neq 0)$;
   (d) $(x + 1)(x - 2)$ & $(2x - 1)(x + 2)$; **(e)** e^{3x} & $e^{3(x+3)}$; **(f)** t & $t \ln |t|$ $(t \neq 0)$.
2. Prove that any three polynomials of first degree must be linearly dependent.
3. Are the following functions linearly independent or dependent on the indicated interval?
   (a) $\arctan(x)$ and $\arctan(2x)$ on $(-\frac{\pi}{4}, \frac{\pi}{4})$. **(b)** x^2 and x^4 on $(0, \infty)$.
   (c) $\sqrt{1 - x^2}$ and x on $(-1, 1)$. **(d)** 1 and $\ln \frac{x-1}{x+1}$ on $(-\infty - 1)$.
   (e) $\ln x$ and $\log_{10} x$ on $[1, \infty)$. **(f)** $x + 1$ and $|x + 1|$ on $[-1, \infty)$.
4. Write Abel's identity (4.2.3) for the following equations.
   (a) $y'' - 4y' + 3y = 0;$ **(b)** $2y'' + 4y' - y = 0;$
   (c) $x^2 y'' - 2x\,y' + x^3 y = 0;$ **(d)** $(x^2 - 1)\,y'' + 4x\,y' + x^2 y = 0;$
   (e) $(2x^2 - 3x - 2)y'' + 5y' + xy = 0;$ **(f)** $(x^2 - 4)y'' + 2y' - 5y = 0.$
5. Find the Wronskian of the two solutions of the given differential equation without solving the equation.
   (a) $y'' + y'/x + y = 0;$ **(b)** $(x^2 + 1)\,y'' - x\,y' + (x + 2)\,y = 0.$
6. If y_1 and y_2 are two linearly independent solutions of the differential equation

$$a_2(x)\,y'' + a_1(x)\,y' + a_0(x)\,y = 0$$

with polynomial coefficients $a_0(x)$, $a_1(x)$, and $a_2(x)$ having no common factor other than a constant, prove that the Wronskian of these functions y_1 and y_2 is zero at only those points where $a_2(x) = 0$.
7. The Wronskian of two functions is $W(x) = x^2 + 2x + 1$. Are the functions linearly independent or linearly dependent? Why?
8. Suppose that the Wronskian $W[f, g]$ of two functions is known; find the Wronskian $W[u, v]$ of $u = af + bg$ and $v = \alpha f + \beta v$, where a, b, α, and β are some constants.
9. If the Wronskian of two functions f and g is $x \sin x$, find the Wronskian of the functions $u = 2f + g$ and $v = 2f - 3g$.
10. Prove that the functions e^{3x}, $x\,e^{3x}$, and $e^{3x}\cos x$ are linearly independent on any interval not containing $\frac{\pi}{2} + n\pi$, $n = 0, \pm 1, \pm 2, \ldots$.
11. Prove that if α and β are distinct real numbers, and n and m are positive integers, then the functions $e^{\alpha x}$, $x\,e^{\alpha x}$, ..., $x^{n-1}e^{\alpha x}$, and $e^{\beta x}$, $x\,e^{\beta x}$, ..., $x^{m-1}e^{\beta x}$ are linearly independent on any interval.
12. Are the functions $f_1(x) = |x|^3$ and $f_2(x) = x^3$ linearly independent or dependent on the following intervals?
   (a) $0 \leqslant x \leqslant 1$, (b) $-1 \leqslant x \leqslant 0$, (c) $-1 \leqslant x \leqslant 1$.

13. Prove the generalization of the previous exercise. Let f be an odd continuous differentiable function on an interval $|-a,a|$, $a>0$, (that is, $f(-x)=-f(x)$) and that $f(0)=f'(0)=0$. Show that the Wronskian $W[f,|f|](x)\equiv 0$ for $x\in|-a,a|$, but that f and $|f|$ are linearly independent on this interval.

14. What is the value of constant $W[y_1,y_2](x_0)$ in the formula (4.2.3) when two solutions y_1, y_2 satisfy the initial conditions

$$y_1(x_0)=a, \quad y_1'(x_0)=b, \quad y_2(x_0)=\alpha, \quad y_2'(x_0)=\beta?$$

15. Find the Wronskian of the solutions y_1, y_2 to the given differential equation that satisfy the given initial conditions.
    (a) $x^2y''+xy'+(1-x)y=0, \quad y_1(1)=0,\ y_1'(1)=1,\ y_2(1)=1,\ y_2'(1)=1.$
    (b) $(1-x^2)y''-2xy'+6y=0, \quad y_1(0)=3,\ y_1'(0)=3,\ y_2(0)=1,\ y_2'(0)=-1.$
    (c) $x^2y''-3xy'+(1+x)y=0, \quad y_1(1)=1,\ y_1'(-1)=1,\ y_2(-1)=0,\ y_2'(-1)=1.$
    (d) $y''-(\cos x)y'+4(\cot x)y=0, \quad y_1(0)=1,\ y_1'(0)=1,\ y_2(0)=0,\ y_2'(0)=1.$
    (e) $y''+2xy'+x^3y=0, \quad y_1(0)=1,\ y_1'(0)=1,\ y_2(0)=0,\ y_2'(0)=1.$
    (f) $\sqrt{1+x^3}\,y''-3x^2y'+y=0, \quad y_1(0)=1,\ y_1'(0)=1,\ y_2(0)=-1,\ y_2'(0)=1.$
    (g) $(1-x^2)\,y''-3x\,y'+8\,y=0, \quad y_1(0)=-1,\ y_1'(0)=0,\ y_2(0)=0,\ y_2'(0)=1.$

16. Let u_1, u_2, $\ldots$, u_n be linearly independent solutions to the n-th order differential equation $y^{(n)}+a_{n-1}(x)y^{(n-1)}+\cdots+a_0(x)y=0$. Show that the coefficient functions $a_k(x)$, $k=0,1,\ldots,n-1$, are uniquely determined by u_1, u_2, $\ldots$, u_n.

17. Let f be a continuous differentiable function on $|a,b|$, $a<b$, not equal to zero. Show that $f(x)$ and $xf(x)$ are linearly independent on $|a,b|$.

18. If the Wronskian $W[f,g]=t^3\,e^{2t}$ and if $f(t)=t^2$, find $g(t)$.

19. Verify that $y_1(x)=1$ and $y_2(x)=x^{1/3}$ are solutions to the differential equation $yy''+2(y')^2=0$. Then show that $C_1+C_2\,x^{1/3}$ is not, in general, a solution of this equation, where C_1 and C_2 are arbitrary constants. Does this result contradict Theorem 4.4 on page 191?

20. Find the Wronskian of the monomials $a_1x^{\lambda_1}$, $a_2x^{\lambda_2}$, $\ldots$, $a_nx^{\lambda_n}$.

21. Let $f_1(x)$, $f_2(x)$, $\ldots$, $f_n(x)$ be functions of x that at every point of the interval (a,b) have finite derivatives up to the order $n-1$. If their Wronskian $W[f_1,f_2,\ldots,f_n](x)$ vanishes identically but $W[f_1,f_2,\ldots,f_{n-1}](x)\neq 0$, prove that the set of functions $\{f_1,f_2,\ldots,f_n\}$ is linearly dependent.

22. Show that any two functions from the set $(\mathbf{j}^2=-1)$

$$f_1(x)=e^{\mathbf{j}x}, \quad f_2(x)=e^{-\mathbf{j}x}, \quad f_3(x)=\sin x, \quad f_4(x)=\cos x$$

are linearly independent, but any three of them are linearly dependent.

23. If $\{u(x),v(x)\}$ is known to be a fundamental set of solutions for a second order linear differential equation $y''+p(x)\,y'+q(x)\,y=0$, what conditions must a,b,c,a satisfy so that $\{a\,u+b\,v,c\,u+d\,v\}$ is also to be a fundamental set for the same equation?

24. Let $u(x)$ and $v(x)$ be differentiable on the interval $|a,b|$, and let $W[u,v]$ be their Wronskian. Show that $\frac{dW}{dx}=u\,v''-v\,u''$.

25. For given three linearly independent functions $y_1(x)$, $y_2(x)$, and $y_3(x)$, derive a linear differential equation

$$y'''+p(x)\,y''+q(x)\,y'+r(x)\,y=0$$

for which these three functions are its solutions. In other words, express three functions $p(x)$, $q(x)$, and $r(x)$ in terms of $\{y_1,y_2,y_3\}$ assuming that their Wronskian is not zero.

26. Let f and g be two continuously differentiable functions on some interval $|a,b|$, $a<b$, that have only finitely many zeroes in $|a,b|$ and no common zeroes. Prove that if $W[f,g](x)\equiv 0$ on $|a,b|$, then f and g are linearly dependent. *Hint:* Apply the result of the previous problem to the finite number of subintervals of $|a,b|$ on which f and g never vanish.

Section 4.3 of Chapter 4 (Review)

1. Prove that every solution to the differential equation $\ddot{x}(t)+Q(t)x(t)=0$ has at most one zero in the open interval (a,b) if $Q(t)\leqslant 0$ in $|a,b|$.

2. Prove that zeros of two linearly independent solutions to the equation $\ddot{x}(t)+Q(t)x(t)=0$ separate each other, that is, between two sequential zeros of one solution there is exactly one zero of another solution.

3. If y_1 and y_2 are linearly independent solutions of $(x^2+1)y''+xy'+y=0$, and if $W[y_1,y_2](1)=\sqrt{2}$, find the value of $W[y_1,y_2](2)$.

4. Suppose that two solutions, y_1 and y_2, of the homogeneous differential equation (4.3.1) have a common point of inflection; prove that they cannot form a fundamental set of solutions.

5. Let y_1 and y_2 be two linearly independent solutions of the second order differential equation (4.3.1). What conditions must be imposed on the constants a, b, c, and d for their linear combinations $y_3 = ay_1 + by_2$ and $y_4 = cy_1 + dy_2$ to constitute a fundamental set of solutions?

Section 4.4 of Chapter 4 (Review)

1. Write out the characteristic equation for the given differential equation.
   (a) $2y'' - y' + y = 0$; **(b)** $y''' + 2y' - 3y = 0$; **(c)** $4y^{(4)} + y = 0$;
   (d) $y'' + 5y' + y = 0$; **(e)** $3y''' + 7y = 0$; **(f)** $y^{(4)} - 3y'' + 5y = 0$;
   (g) $7y'' + 4y' - 2y = 0$; **(h)** $y''' - 3y'' + 4y = 0$; **(i)** $2y^{(4)} + y''' - 2y'' + y' = 0$.

2. The characteristic equation for a certain differential equation is given. State the order of the differential equation and give the form of the general solution.
   (a) $8\lambda^2 + 14\lambda - 15 = 0$; **(b)** $3\lambda^2 - 5\lambda + 2 = 0$; **(c)** $2\lambda^2 - \lambda - 3 = 0$;
   (d) $9\lambda^2 - 3\lambda - 2 = 0$; **(e)** $\lambda^2 - 4\lambda + 3 = 0$; **(f)** $\lambda^2 + 2\lambda - 8 = 0$;
   (g) $\lambda^3 + 3\lambda + 10 = 6\lambda^2$; **(h)** $2\lambda^3 - \lambda^2 - 5\lambda = 2$; **(i)** $3\lambda^3 + 11\lambda^2 + 5\lambda = 3$.

3. Write the general solution to the given differential equations.
   (a) $y''' + 3y'' - y' - 3y = 0$. **(b)** $4y''' - 13y' + 6y = 0$.
   (c) $2y''' + y'' - 8y' - 4y = 0$. **(d)** $y''' + 2y'' - 15y' = 0$.
   (e) $4y^{(4)} - 17y'' + 4y' = 0$. **(f)** $2y^{(4)} - 3y''' - 4y'' + 3y' + 2y = 0$.
   (g) $y''' + 5y'' - 8y' - 12y = 0$. **(h)** $y^{(4)} - 2y''' - 13y'' + 28y' - 24y = 0$.
   (i) $4y^{(4)} - 8y''' - y'' + 2y' = 0$. **(j)** $4y^{(4)} + 20y''' + 35y'' + 25y' + 6y = 0$.

4. Solve the initial value problems.

   (a) $\ddot{y} + 8\dot{y} - 9y = 0,\ y(0) = 0,\ \dot{y}(0) = 10$; **(b)** $3\ddot{y} - 8\dot{y} - 3y = 0,\ y(0) = 10,\ \dot{y}(0) = 0$;
   (c) $2\ddot{y} - 11\dot{y} - 6y = 0,\ y(0) = 0,\ \dot{y}(0) = 13$; **(d)** $3\ddot{y} - 17\dot{y} - 6y = 0,\ y(0) = 19,\ \dot{y}(0) = 0$;
   (e) $4\ddot{y} - 17\dot{y} + 4y = 0,\ y(0) = 1,\ \dot{y}(0) = 4$; **(f)** $5\ddot{y} - 24\dot{y} - 5y = 0,\ y(0) = 1,\ \dot{y}(0) = 5$.

5. Find the solutions to the given IVPs for equations of degree bigger than 2.

   (a) $y''' - 2y'' - 3y' = 0$, $y(0) = 0$, $y'(0) = 0$, $y''(0) = 12$.
   (b) $y''' + y'' - 16y' - 16y = 0$, $y(0) = 0$, $y'(0) = 8$, $y''(0) = 0$.
   (c) $y^{(4)} - 13y'' + 36y = 0$, $y(0) = 0$, $y'(0) = 0$, $y''(0) = -10$, $y'''(0) = 0$.
   (d) $y^{(4)} - 5y'' + 5y'' + 5y' = 6y$, $y(0) = 0$, $y'(0) = -1$, $y''(0) = -3$, $y'''(0) = -7$.
   (e) $10y''' + y'' - 7y' + 2y = 0$, $y(0) = 0$, $y'(0) = -6$, $y''(0) = 3$.
   (f) $4y^{(4)} - 15y'' + 5y' + 6y = 0$, $y(0) = 0$, $y'(0) = -16$, $y''(0) = -4$, $y'''(0) = -4$.
   (g) $3y''' - 7y'' - 7y' + 3y = 0$, $y(0) = 0$, $y'(0) = -12$, $y''(0) = -56$.

6. If $a > 0$, prove that all solutions to the equation $y'' + ay' = 0$ approach a constant value as $t \to +\infty$.

7. Find the fundamental set of solutions of the following equations specified by Theorem 4.12.
   (a) $\ddot{y} = 9y$; **(b)** $3\ddot{y} - 10\dot{y} + 3y = 0$; **(c)** $2\ddot{y} - 7\dot{y} - 4y = 0$; **(d)** $6\ddot{y} - \dot{y} - y = 0$.

8. Find the solution of the initial value problem

$$2y'' + 3y' - y = 0, \qquad y(0) = 3, \quad y'(0) = -1.$$

   Plot the solution for $0 \leqslant t \leqslant 1$ and determine its minimum value.

9. Find the solution of the initial value problem

$$3y'' - 10y' + 3y = 0, \qquad y(0) = 2, \quad y'(0) = -2.$$

   Then determine the point where the solution is zero.

10. Solve the initial value problem $3y'' + 17y' - 6y = 0$, $y(0) = 3$, $y'(0) = -\beta$. Then find the smallest positive value of β for which the solution has no minimum point.

11. Construct the general form of the solution that is bounded as $t \to \infty$.
    (a) $y'' - 4y = 0$; **(b)** $y''' = 0$; **(c)** $3y'' + 8y' - 3u = 0$; **(d)** $2y'' - 7y' - 4y = 0$.

12. Solve the initial value problem $2y'' - y' - 6y = 0$, $y(0) = \alpha$, $y'(0) = 2$. Then find the value of α so that the solution approaches zero as $t \to \infty$.
13. Solve the initial value problem $6y'' - y' - y = 0$, $y(0) = 3$, $y'(0) = \beta$. Then find the value of β so that the solution approaches zero as $t \to \infty$.
14. For the following differential equations, determine the values of γ, if any, for which all solutions tend to zero as $t \to \infty$. Also find the values of γ, if any, for which all (nonzero) solutions become unbounded as $t \to \infty$.
    (a) $y'' + 2(1-\gamma)y' - \gamma(2-\gamma)y = 0$; **(b)** $y'' - \gamma y' - (1+\gamma)y = 0$.
15. Solve the initial value problem $y'' - 4y' + 3y = 0$, $y(0) = 1$, $y'(0) = \beta$ and determine the coordinates x_m and y_m of the minimum point or the solution as a function of β.
16. Solve the initial value problem $y'' + y' - 2y = 0$, $y(0) = \alpha$, $y'(0) = 1$. Then find the value of α so that the solution approaches zero as $t \to \infty$.
17. Solve the initial value problem $3y'' + 4y' - 4y = 0$, $y(0) = 1$, $y'(0) = \beta$. Then find the value of β so that the solution approaches zero as $t \to \infty$.
18. For the following differential equations, determine the values of γ, if any, for which all solutions tend to zero as $t \to \infty$. Also find the values of γ, if any, for which all (nonzero) solutions become unbounded as $t \to \infty$.
    (a) $y'' - 2\gamma y' + (\gamma^2 - 1)y = 0$; **(b)** $y'' + \gamma(1+\gamma)y' - (\gamma+2)y = 0$.
19. Solve the initial value problem $y'' - 6y' + 8y = 0$, $y(0) = 1$, $y'(0) = \beta$ and determine the coordinates x_m and y_m of the minimum point or the solution as a function of β.
20. Consider the initial value problem

$$y'' + 5\,y' + 4\,y = 0, \qquad y(0) = 1, \quad y'(0) = \beta > 0.$$

   (a) Solve the initial value problem.

   (b) Determine the coordinates x_m and y_m of the maximum point of the solution as function of β.

   (c) Find the smallest value of β for which $y_m \geqslant 2$.

   (d) What is the behavior of x_m and y_m as $\beta \mapsto +\infty$?

Section 4.5 of Chapter 4 (Review)

1. The characteristic equation for a certain differential equation is given. Give the form of the general solution.

   (a) $\lambda^2 + 9 = 0$; **(b)** $\lambda^2 + 2\lambda + 2 = 0$; **(c)** $\lambda^2 + 4\lambda + 5 = 0$;
   (d) $4\lambda^2 - 4\lambda + 5 = 0$; **(e)** $16\lambda^2 + 8\lambda + 145 = 0$; **(f)** $9\lambda^2 - 6\lambda + 10 = 0$;
   (g) $\lambda^2 - 2\lambda + 5 = 0$; **(h)** $\lambda^2 + 4\lambda + 5 = 0$; **(i)** $4\lambda^2 - 4\lambda + 17 = 0$;
   (j) $9\lambda^2 + 6\lambda + 145 = 0$; **(k)** $\lambda^2 - 6\lambda + 13 = 0$; **(l)** $\lambda^2 - 6\lambda + 10 = 0$.

2. Write the general solution of the given differential equation (α is a parameter).

   (a) $4y'' - 4y' + 3y = 0$; **(b)** $64y'' - 48y' + 13y = 0$;
   (c) $y''' + 6y'' + 13y' = 0$; **(d)** $y''' + 3y'' + y' + 3y = 0$;
   (e) $y'' + 10y' + 26y = 0$; **(f)** $y'' - 2\alpha y' + (\alpha^2 + 9)y = 0$;
   (g) $4y'' - 24y' + 37y = 0$; **(h)** $9y'' - 18y' + 10y = 0$.

3. Solve the initial value problems (dot stands for the derivative with respect to t).

   (a) $9\ddot{y} - 24\dot{y} + 16y = 0$, $y(0) = 0$, $y'(0) = 1$; **(b)** $4\ddot{y} - 12\dot{y} + 9y = 0$, $y(0) = 2$, $y'(0) = 3$;
   (c) $\ddot{y} + 18\dot{y} + 81y = 0$, $y(0) = 0$, $y'(0) = 1$; **(d)** $9\ddot{y} - 42\dot{y} + 49y = 0$, $y(0) = 0$, $y'(0) = 1$.

4. Find the solutions to the given initial value problems for equations of the third order.

   (a) $y''' - 3y' - 2y = 0$, $\quad y(0) = 0$, $y'(0) = 0$, $y''(0) = 9$.
   (b) $y''' + y'' = 0$, $\quad y(0) = 0$, $y'(0) = 0$, $y''(0) = 1$.
   (c) $18y''' - 33y'' + 20y' - 4y = 0$, $\quad y(0) = 4$, $y'(0) = -1$, $y''(0) = -3$.
   (d) $4y''' + 8y'' - 11y' + 3y = 0$, $\quad y(0) = 1$, $y'(0) = 1$, $y''(0) = 13$.

5. Find the general solution of the given differential equations of order bigger than 2.

   (a) $y''' + y'' + 4y' + 4y = 0$. **(b)** $4y''' + 12y'' + 13y' + 10y = 0$.
   (c) $16y''' + 5y'' + 17y' + 13y = 0$. **(d)** $12y^{(4)} - 8y''' - 61y'' - 64y' - 15y = 0$.
   (e) $y''' - y'' + y' - y = 0$. **(f)** $y''' - 2y'' - 3y' + 10y = 0$.
   (g) $y''' - 2y'' + y' - 2y = 0$. **(h)** $8y^{(4)} + 44y''' + 50y'' + 3y' - 20y = 0$.

6. If a is a real positive number, prove that all solutions to the equation $y'' + a^2y = 0$ remain bounded as $x \to +\infty$.
7. Rewrite the function $y(t) = \sin t + \cos t$ in the form $y(t) = A\cos(\beta t - \delta)$.

The following problem requires access to a software package.

8. Find the solution to the initial value problem $4y'' - 12y' + 73y = 0$, $y(0) = 2$, $y'(0) = 3$, and determine the first time at which $|y(t)| = 1.0$.

9. Consider an electric circuit containing a resistor, an inductor, and a capacitor in series with a constant electromotive force (see figure at the right). Applying Kirchhoff's law (see §6.1.1), we obtain

$$L\frac{dI}{dt} + RI + \frac{q}{C} = E, \quad I = \frac{dq}{dt},$$

where $q(t)$ is electric charge, and $I(t)$ is the current in the circuit. Let $L = 2$ henries, $R = 248$ ohms, $C = 10^{-4}$ farads, and $E = 127$ volts. Determine the current $I(t)$ in the circuit assuming that no charge are present and no current is flowing at time $t = 0$ when E is applied.

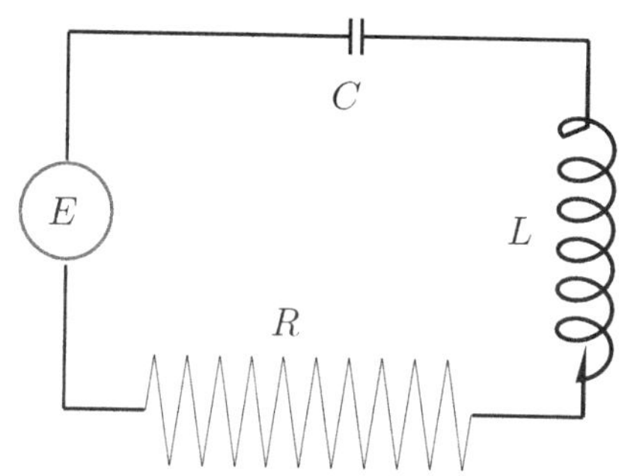

Section 4.6 of Chapter 4 (Review)

1. The factored form of the characteristic equation for certain differential equations is given. State the order of the differential equation and write down the form of its general solution.

(a)	$(\lambda - 3)^2$;	**(b)**	$(\lambda - 2)(\lambda + 1)^2$;	**(c)**	$(\lambda + 9)^3$;
(d)	$(\lambda^2 + 2\lambda + 2)^2$;	**(e)**	$(\lambda^2 + 4)^2$;	**(f)**	$(\lambda^2 - 2\lambda + 5)^2$;
(g)	$(\lambda^2 - 9)(\lambda^2 + 2\lambda + 5)^2$;	**(h)**	$(\lambda - 1)(\lambda^2 + 4)^2$;	**(i)**	$(\lambda - 4)(\lambda^2 + 4)^3$;
(j)	$(\lambda^2 + 1)^4(\lambda^2 - 4\lambda + 8)$;	**(k)**	$(2\lambda - 1)(4\lambda^2 + 1)^2$;	**(l)**	$(4\lambda^2 + 12\lambda + 10)^2$;
(m)	$(\lambda^2 - 1)(\lambda^2 + 4\lambda + 5)^2$;	**(n)**	$(\lambda + 1)^3(\lambda - 3)$;	**(o)**	$(\lambda^2 - 4)^2$.

2. Write the general solution of the given differential equation.

(a)	$9y'' - 12y' + 4y = 0$;	**(b)**	$2y'' - 2\sqrt{2}y' + y = 0$;	**(c)**	$9y'' + 6y' + y = 0$;
(d)	$4y'' - 12y' + 9y = 0$;	**(e)**	$y'' - 2\sqrt{5}y' + 5y = 0$;	**(f)**	$y''' + 3y'' - 4y = 0$;
(g)	$y^{(4)} - 8y'' + 16y = 0$;	**(h)**	$y^{(4)} + 18y'' + 81y = 0$;	**(i)**	$y^{(4)} + y''' + y'' = 0$;
(j)	$y^{(4)} - 4y''' + 6y'' = 4y' - y$;	**(k)**	$y^{(5)} + 2y''' + y' = 0$;	**(l)**	$y^{(4)} + 6y''' + 9y'' = 0$.

3. Find the solution of the initial value problem.

(a) $y'' - 6y' + 9y = 0, \quad y(0) = 0,\ y'(0) = 1.$

(b) $16y'' + 24y' + 9y = 0, \quad y(0) = 0,\ y'(0) = 4.$

(c) $4y'' + 12y' + 9y = 0, \quad y(0) = 2,\ y'(0) = 0.$

(d) $9y'' - 6y' + y = 0, \quad y(0) = 0,\ y'(0) = 3.$

4. Find the solutions to the given initial value problems for equations of the third order.

(a) $y''' + 6y'' + 12y' + 8y = 0, \quad y(0) = 0,\ y'(0) = 0,\ y''(0) = 2.$

(b) $y''' + y'' = 0, \quad y(0) = 0,\ y'(0) = 0,\ y''(0) = 1.$

(c) $18y''' - 33y'' + 20y' - 4y = 0, \quad y(0) = 4,\ y'(0) = -1,\ y''(0) = -3.$

(d) $4y''' + 8y'' - 11y' + 3y = 0, \quad y(0) = 1,\ y'(0) = 1,\ y''(0) = 13.$

(e) $4y''' - 27y'' + 27y = 0, \quad y(0) = -3,\ y'(0) = -7,\ y''(0) = 6.$

(f) $4y''' + 12y'' - 15y' + 4y = 0, \quad y(0) = 1,\ y'(0) = 0,\ y''(0) = 20.$

5. Find the solutions to the given initial value problems for equations of the fourth order.

(a) $y^{(4)} - 2y''' + 5y'' - 8y' + 4y = 0, \quad y(0) = 1,\ y'(0) = 1,\ y''(0) = 0,\ y'''(0) = 3.$

(b) $y^{(4)} + 3y''' - 4y' = 0, \quad y(0) = 0,\ y'(0) = 2,\ y''(0) = -3,\ y'''(0) = 13.$

6. Determine the values of the constants a_0, a_1, and a_2 such that $y(x) = a_0 + a_1x + a_2x^2$ is a solution to

$$(9 + 4x^2)y'' - 8y = 0$$

and use the reduction of order technique to find a second linearly independent solution.

7. Find the fundamental set of solutions for the following differential equations.
 (a) $y'' + y'/x - \frac{1}{4}(3 + 1/x^2)y = 0$; (b) $x^2y'' - 2y = 0$;
 (c) $xy'' - (x+2)y' + 2y = 0$; (d) $xy'' + (x+2)y' + y = 0$.
8. One solution y_1 of the differential equation is given; find another linearly independent solution.
 (a) $t^2y'' + (1-2a)ty' + a^2y = 0,\ y_1(t) = t^a$. (b) $y'' = (2\cot x - \tan x)\,y',\ y_1 = 1$.
 (c) $x^2y'' - 2xy' + (x^2+2)y = 0,\ y_1 = x\,\sin x$. (d) $y'' - e^x\,y' = 0,\quad y_1 = 1$.
 (e) $4x^2y'' + 4xy' + (4x^2-1)y = 0,\ y_1 = \frac{\sin x}{\sqrt{x}}$. (f) $xy'' - y' = 0,\quad y_1 = 1$.
 (g) $(t-1)^2\ddot{y} + 6y = 4(t-1)\dot{y},\ y_1 = (t-1)^2$. (h) $x\,y'' + y' = 0,\quad y_1 = \ln|x|$.
 (i) $x^2\,y'' - xy' + y = 0,\quad y_1 = x\,\ln|x|$. (j) $x^3y'' + 3xy' - 3y = 0,\ y_1 = x$.
9. Suppose that one solution, $y_1(x)$, to the differential equation $y'' + p(x)y' + q(x)y = 0$ is known. Then the inhomogeneous equation $y'' + p(x)y' + q(x)y = f(x)$ can be expressed in the following form: $[\texttt{D} + (p(x) + y_1'/y_1)][\texttt{D} - y_1'/y_1)]\,y = f(x)$. The required function $y(x)$ can be found by successive solutions of the equations

$$[\texttt{D} + (p(x) + y_1'/y_1)]\,v = f(x), \qquad [\texttt{D} - y_1'/y_1)]\,y = v(x).$$

Using the above approach, find the general integral of each of the following differential equations.
 (a) $xy'' + y' - \frac{4}{x}\,y = 4x + 12x^3;\ y_1 = x^2$. (b) $3xy'' - y' = x;\ y_1 = 1$.
 (c) $y'' - 4xy' + (4x^2-2)y = 2x;\ y_1 = e^{x^2}$. (d) $y'' + xy' = x;\ y_1 = 1$.
 (e) $xy'' - (2x+1)y' + 2y = 4x^2;\ y_1 = e^{2x}$. (f) $x^2y'' - 3xy' + 4y = 4x^4;\ y_1 = x^2$.
 (g) $xy'' + (1-2x)y' + (x-1)y = e^x;\ y_1 = e^x$. (h) $(1-x^2)y'' + 2xy' = 0;\ y_1 = 1$.
10. Use the factorization method to decompose the operator associated with the given differential equation as a product of first order operators and then find its general solution.
 (a) $x^4\,y'' - (2x+1)y = -4x^2\,e^{-1/x}$; (b) $4x^2\,y'' + y = 4x^{3/2}$.
11. Find the discriminant of the differential equation $(1-2x-x^2)y'' + 2(1+x)y' - 2y = 0$ and solve it.
12. For what values of parameters p, q, and r does the differential equation $u'' + \frac{px}{ax+b}\,u' + \frac{qx+r}{(ax+b)^2}\,u = 0$ have a solution $u = 1/(ax+b)$? Here a and b are two nonzero real constants.
13. Show that the solutions y_1 and y_2 defined by Eq. (4.6.10) are linearly independent.
14. By the use of Riccati's equation (4.1.23) (consult Problem 5 on page 258), find the general solution for each of the following differential equations.
 (a) $y'' - \frac{2}{x}\,y' - 16x^4y = 0$; (b) $y'' - \frac{1}{x^2}\,y' + \frac{2}{x^3}\,y = 0$;
 (c) $xy'' - (x^2e^x + 1)y' - x^2e^xy = 0$; (d) $y'' - y' - 9e^{2x}y = 0$;
 (e) $xy'' - (x^3 + x + 2)y' + x^3y = 0$; (f) $xy''\ln x - y' - xy\ln^3 x = 0$.
 (g) $xy'' + (1 - x\cos x)y' + (x\sin x - \cos x)y = 0$. (h) $y'' + (2\,e^x - 1)y' - 3\,e^{2x}\,y = 0$.
 (i) $y'' - (2x\sin x + \cot x)y' + (x^2\sin^2 x - \sin x)y = 0$.
15. Suppose that $y_1(x)$, $y_2(x)$, $\ldots$, $y_{n-1}(x)$ are linearly independent solutions of the n-th order variable coefficients differential equation $y^{(n)} + p_{n-1}(x)y^{(n-1)} + \cdots + p_1(x)y' + p_0(x)y = 0$. Find the fundamental set of solutions for the given equation.
16. The Dickson polynomial $E_3(x) = x^3 - 2x$ is a solution of the differential equation $\left(x^2-4\right)y'' + 3x\,y' - 15\,y = 0$. Find another linearly independent solution of this equation.
17. The Pell polynomial $P_2(x) = 2x$ is a solution of the differential equation $\left(x^2+1\right)y'' + 3x\,y' - 3\,y = 0$. Find another linearly independent solution of this equation.
18. The Hermite polynomial $H_2(x) = 4x^2 - 2$ is a solution of the differential equation $y'' - 2x\,y' + 4\,y = 0$. Find another linearly independent solution of this equation.
19. Show that if y is a solution of the linear equation

$$a(x)\,y'' + \left[a(x)b(x) - a'(x)\right]y' + a^2(x)\,c(x)\,y = 0,$$

then $u = y'/(a(x)\,y)$ is a solution of the Riccati equation $u' + a(x)\,u^2 + b(x)\,u + c(x) = 0$.
20. Consider the initial value problem

$$y'' + \gamma\,y' + y = 0, \qquad y(0) = 1, \quad y'(0) = 1,$$

where γ is a constant. Determine γ_{crit}, the damping constant value that makes the given differential equation critically damping (that is, the corresponding characteristic polynomial has a double root). Use a computational software to plot the solutions of the given IVP for $\gamma = 1, 2$, and 3 over a common time interval sufficiently large to display the main features of each solution.

21. Solve the Euler equations.
(a) $x^2y'' - x\,y' + 5y = 0$; (b) $x^2y'' - 2\,y' + 17y = 0$; (c) $x^2y'' - 2\,y = 0$;
(d) $x^2y'' - 2x\,y' + y = 0$; (e) $x^2y'' - 5x\,y' + 9y = 0$; (f) $x^2y'' + 11x\,y' + 25y = 0$;
(g) $x^2y'' - 3x\,y' + 29\,y = 0$; (h) $x^2y'' + 7x\,y' + 10\,y = 0$; (i) $x^2y'' + 4x\,y' + 2\,y = 0$.

22. Find a particular solution of the given initial value problem for the Euler equations.
(a) $x^2y'' - 2x\,y' + 2\,y = 0,\ y(1) = 1,\ y'(1) = -1$; (b) $4x^2y'' + 12x\,y' + 3\,y = 0,\ y(1) = 1,\ y'(1) = 5/2$;
(c) $x^2y'' + 3x\,y' - 3\,y = 0,\ y(2) = 8,\ y'(2) = 0$; (d) $x^2y'' - 9x\,y' + 25\,y = 0,\ y(1) = 1,\ y'(1) = 3$;
(e) $x^2y'' - 5x\,y' + 9\,y = 0,\ y(1) = 2,\ y'(1) = 5$; (f) $x^2y'' - 5x\,y' + 10\,y = 0,\ y(1) = 1,\ y'(1) = 0$;
(g) $x^2y'' - 3x\,y' + 13\,y = 0,\ y(1) = 2,\ y'(1) = 5$; (h) $x^2y'' - 12\,y = 0,\ y(1) = 1,\ y'(1) = 11$;
(i) $x^2y'' + x\,y' - 9\,y = 0,\ y(1) = 1,\ y'(1) = 9$; (j) $x^2y'' - 7x\,y' + 16\,y = 0,\ y(1) = 3,\ y'(1) = 4$.

23. Using a computer solver, plot solutions to the initial value problems in the interval $1 \leqslant x \leqslant 5$.
(a) $x^2y'' - 3x\,y' + 3\,y = 0,\ y(1) = 3,\ y'(1) = 1$; (b) $9x^2y'' + 3x\,y' - 8\,y = 0,\ y(1) = 3,\ y'(1) = -1/3$;
(c) $x^2y'' + x\,y' - 9\,y = 0,\ y(1) = 1,\ y'(1) = -1$; (d) $x^2y'' + x\,y' - y = 0,\ y(1) = 1,\ y'(1) = 2$;
(e) $x^2y'' + 7x\,y' + 8\,y = 0,\ y(1) = 2,\ y'(1) = -6$; (f) $x^2y'' - 3x\,y' + 5\,y = 0,\ y(1) = 1,\ y'(1) = 3$.

Section 4.7 of Chapter 4 (Review)

1. One solution y_1 of the associated homogeneous equation is given; find the general solution of the nonhomogeneous equation.
(a) $xy'' + (1-x)y' - y = e^{2x},\ y_1 = e^x$; (b) $(1-x^2)y'' - 2xy' + 2y = x,\ y_1 = x$.

2. Use the method of undetermined coefficients to find a particular solution of the following differential equations:
(a) $(\texttt{D}^2 - 4)y = 16x\,e^{2x}$; (b) $(\texttt{D}^2 - 1)y = 4x\,e^x$;
(c) $\texttt{D}(\texttt{D} + 1)y = 2\,\sin x$; (d) $(\texttt{D}^2 - 4\,\texttt{D} + 5)y = (x-1)^2$;
(e) $(\texttt{D}^2 + 4)y = \sin 3x$; (f) $(\texttt{D}^2 + 3\,\texttt{D})y = 28\,\cosh 4x$;
(g) $(\texttt{D}^2 - 1)y = 2\,e^x + 6\,e^{2x}$; (h) $(\texttt{D}^2 + 2\,\texttt{D} + 10)y = 25x^2 + 3$;
(i) $(\texttt{D}^2 - \texttt{D} - 2)y = 3\,e^{2x}$; (j) $(3\,\texttt{D}^2 + 10\,\texttt{D} + 3)y = 9x + 10\,\cos x$.

3. Use the method of undetermined coefficients to find the general solution of the following differential equations.
(a) $(\texttt{D}^2 + 2\,\texttt{D} + 2)y = 2x\,\cos x$; (b) $(\texttt{D} - 1)^3y = e^x$;
(c) $(\texttt{D} - 1)(\texttt{D}^2 + 4)y = e^x\,\sin 2x$; (d) $\texttt{D}(\texttt{D}^2 - 2\,\texttt{D} + 10)y = 6xe^x$;
(e) $(\texttt{D}^2 - 1)y = 2x\,\cos x$; (f) $(\texttt{D}^2 + 6\,\texttt{D} + 10)y = 2\cos x\,e^{-3x}$.

4. Using the method of undetermined coefficients find the general solution of the following equations.
(a) $y'' + 6y' + 9y = e^{-3x}$; (b) $y'' - y' - 2y = 1 + x + e^x$;
(c) $y'' + y' = x^2 + 2\,e^{-2x}$; (d) $y'' + y = 2\sin x + x$;
(e) $y''' - y'' = x + 2\sinh x$; (f) $y''' - 6y'' + 11y' - 6y = 2\,e^x$.

5. Solve the following initial value problems using the method of undetermined coefficients.

(a) $y'' - 9y = 1 + \cosh 3x, \quad y(0) = 1,\ y'(0) = 0.$

(b) $y'' + 4y = \sin 3x, \quad y(0) = 1,\ y'(0) = 0.$

(c) $6y'' - 5y'' + y = x^2, \quad y(0) = 0,\ y'(0) = -1.$

(d) $4y'' - 4y' + y = x^2\,e^{x/2}, \quad y(0) = 1,\ y'(0) = 0.$

(e) $4y'' - 12y' + 9y = x\,e^{3x/2}, \quad y(0) = 0,\ y'(0) = -1.$

(f) $y'' + \frac{3}{2}\,y' - y = 12x^2 + 6x^3 - x^4, \quad y(0) = 4,\ y'(0) = -8.$

(g) $y'' - 6y' + 13y = 4\,e^{3x}, \quad y(0) = 2,\ y'(0) = 4.$

(h) $y'' - 4y = e^{-2x} - 2x, \quad y(0) = 1,\ y'(0) = 0.$

(i) $y'' + 9y = 6\,\cos 3x, \quad y(0) = 1,\ y'(0) = 0.$

6. Write out the assumed form of a particular solution but do not carry out the calculations of the undetermined coefficients.
(a) $y^{(4)} + y^{(3)} = 12 - 6x^2\,e^{-x}$; (b) $y^{(4)} - y = e^x\,\cos x$;
(c) $y'' + 2y' + y = 3\,e^{-x}\,\sin x + 2x\,e^{-x}$; (d) $y^{(4)} - 5y''' + 6y'' = \cos 2x x^2\,e^{3x}$;
(e) $y^{(4)} - 4y''' + 5y'' = x^3\,\sin 2x + xe^x\,\cos 2x$; (f) $y''' + 6y'' + 10y' = \frac{2x\,\sin x}{e^{3x}} + x^2$;
(g) $y^{(5)} + y^{(4)} + 2y''' + 2y'' + y' + y = x^2$; (h) $y^{(4)} + 10y'' + 9y = \cos x$;
(i) $y''' - 6y'' + 11y' - 6y = 144x\,e^{-x}$; (j) $y''' - 7y'' + 16y' - 12y = 9\,e^{3x}$.

7. Use the form $y_p(t) = A\,\cos\omega t + B\,\sin\omega t$ to find a particular solution for the given differential equations.
(a) $\ddot{y} + 4y = 12\,\sin 2t$; (b) $\left(\texttt{D}^3 - 6\texttt{D}^2\right)y = 37\,\cos t - 12$;
(c) $\ddot{y} + 4y = \cos t$; (d) $\ddot{y} + 7\dot{y} + 6y = 100\,\cos 2t$;
(e) $\ddot{y} + 7\dot{y} + 10y = -43\,\sin 3t - 19\,\cos 3t$; (f) $\ddot{y} - 2\dot{y} - 15y = 377\,\cos 2t$.

8. The principle of decomposition ensures that the solution of the inhomogeneous equation $L[\mathtt{D}]y = f + g$, where $L[\mathtt{D}]$ is a linear operator (4.7.3), can be built up once the solutions of $L[\mathtt{D}]y = f$ and $L[\mathtt{D}]y = g$ are known. Can this process be inverted? Find conditions on the coefficients of a linear operator $L[\mathtt{D}]$ and functions f and g when the inverse statement of the decomposition principle is valid. That is, from the solution to $L[\mathtt{D}]y = f + g$ we can find solutions of equations $L[\mathtt{D}]y = f$ and $L[\mathtt{D}]y = g$.

9. In the following exercises solve the initial value problems where the characteristic equation is of degree 3 and higher. At least one of its roots is an integer and can be found by inspection.

   (a) $(\mathtt{D}^3 + \mathtt{D}^2 - 4\,\mathtt{D} - 4)y = 8x + 8 + 6\,e^{-x}, \quad y(0) = 2,\ y'(0) = -4,\ y''(0) = 12;$

   (b) $(\mathtt{D}^3 - \mathtt{D})y = 2x, \quad y(0) = 4,\ y'(0) = 1,\ y''(0) = 1;$

   (c) $(\mathtt{D}^3 - \mathtt{D}^2 + \mathtt{D} - 1)y = 4\sin x, \quad y(0) = 3,\ y'(0) = -2,\ y''(0) = -3;$

   (d) $(\mathtt{D}^3 + \mathtt{D}^2 - 4\,\mathtt{D} - 4)y = 3\,e^{-x} - 4x - 8, \quad y(0) = 1,\ y'(0) = 0,\ y''(0) = 2;$

   (e) $(\mathtt{D}^4 - 1)y = 7x^2, \quad y(0) = 2,\ y'(0) = -1,\ y''(0) = -14,\ y'''(0) = -1;$

   (f) $(\mathtt{D}^4 - 1)y = 4\,e^{-x}, \quad y(0) = 0,\ y'(0) = -1,\ y''(0) = 4,\ y'''(0) = -3;$

   (g) $y^{(4)} - 8y' = 16\,x\,e^{2x}, \quad y(0) = 1,\ y'(0) = \frac{4}{3},\ y''(0) = 2,\ y'''(0) = 4;$

   (h) $y''' + y = x^2 - x + 1, \quad y(0) = 2,\ y'(0) = -2,\ y''(0) = 3;$

   (i) $y''' - 7y'' + 6y = 6\,x^2, \quad y(0) = \frac{10}{3},\ y'(0) = 1,\ y''(0) = 3.$

10. Write the given differential equations in operator form $L[\mathtt{D}]y = f(x)$, where $L[\mathtt{D}]$ is a linear constant coefficient differential operator. Then factor $L[\mathtt{D}]$ into the product of first order differential operators and solve the equation by sequential integration.

(a)	$y'' - y' - 2y = 1/(e^x + 1);$	**(b)**	$y'' - y' - 2y = 13\sin 3x;$
(c)	$y'' - 4y' + 4y = 25\cos x + 2\,e^{2x};$	**(d)**	$y'' - y = \frac{2}{1+e^x};$
(e)	$y'' - 4y' + 5y = 25(x+1)^2;$	**(f)**	$y'' - y' + \frac{1}{4}\,y = x^2;$
(g)	$y'' - y' - 6y = 6x + 3\,e^{3x};$	**(h)**	$2y'' + 3y' - 2y = 4x;$
(i)	$y'' + y' - 2y = \frac{1}{x}\,e^{-2x};$	**(j)**	$y'' - \frac{3}{2}\,y' - y = 5\,e^{2x};$
(k)	$y'' - 4y' + 13y = 8e^{2x}\cos x;$	**(l)**	$y'' + y = 1 + 2x\,e^x;$
(m)	$y'' - y' = 2\sin x;$	**(n)**	$y''' - 3y'' + 3y' - y = 6\,e^x.$

11. Find a linear differential operator that annihilates the given functions.

(a)	$x^2 + \cos 2x;$	**(b)**	$e^x\sin 3x;$	**(c)**	$e^{5x}(1+2x);$	**(d)**	$x^2\,e^{2x};$
(e)	$\sin^2 x;$	**(f)**	$x\,e^{-x}\cos 2x;$	**(g)**	$x + \cos 3x;$	**(h)**	$x\cos 3x;$
(i)	$(x\,e^x)^3;$	**(j)**	$x^2 + \sin 2x;$	**(k)**	$(\cosh 3x)^2;$	**(l)**	$\sin x + \cos 2x;$
(m)	$x^2 + \cos^2 x;$	**(n)**	$x + e^x\cos 2x;$	**(o)**	$\sin x + e^x\cos x;$	**(p)**	$x + e^{-2x}\sin 3x;$
(q)	$x\,e^{-2x}\sin 3x;$	**(r)**	$\cos^3 2x;$	**(s)**	$\sin 2x + \cos 3x.$		

12. Find linearly independent functions that are annihilated by the given differential operator.

(a)	$(\mathtt{D}-1)^2 + 4;$	**(b)**	$\mathtt{D}^2 + 4;$	**(c)**	$[\mathtt{D}^2+4][(\mathtt{D}-1)^2+4];$
(d)	$[\mathtt{D}-5]^3;$	**(e)**	$\mathtt{D}^2[\mathtt{D}+2]^2;$	**(f)**	$[\mathtt{D}^2+1]^2;$
(g)	$[\mathtt{D}-1]^2[\mathtt{D}^2+1];$	**(h)**	$[\mathtt{D}^2-4];$	**(i)**	$[\mathtt{D}^2-1][\mathtt{D}^2+1];$
(j)	$\mathtt{D}^4 - 1;$	**(k)**	$\mathtt{D}^4 + 1;$	**(l)**	$2\,\mathtt{D}^2 + 3\,\mathtt{D} - 2;$
(m)	$\mathtt{D}(\mathtt{D}+4);$	**(n)**	$\mathtt{D}^2 + 4\,\mathtt{D} + 4;$	**(o)**	$(\mathtt{D}+4)^2;$
(p)	$(\mathtt{D}^2+4)^2;$	**(q)**	$2\,\mathtt{D}^2 - 3\,\mathtt{D} - 2;$	**(r)**	$3\,\mathtt{D}^2 + 2\,\mathtt{D} - 1.$

13. For the equation $y'' + 3y' = 9 + 27x$, find a particular solution that has at some point (to be determined) on the abscissa an inflection point with a horizontal tangent line.

14. Find a particular solution of the n-th order linear differential equation $a_n(x)y^{(n)} + a_{n-1}(x)y^{(n-1)} + \cdots + a_1(x)y' + a_0(x)y = f(x)$ when

   (a) $f(x) = mx + b$, $a_0 \ne 0$, a_1, m, and b are constants.

   (b) $f(x) = mx$, $a_0 = 0$, and a_1 is a nonzero constant.

15. For the equation $y''' + 2y'' = 48x$, find the solution whose graph has, at the origin, a point of inflection with the horizontal tangent line.

16. Consider an alternative method to determine a particular solution for the differential equation $(\mathtt{D} - \lambda_1)(\mathtt{D} - \lambda_2)y = f$, where $\mathtt{D}$ is the derivative operator and $\lambda_1 \ne \lambda_2$. Let y_1 and y_2 be solutions to the first-order differential equations

$$(\mathtt{D} - \lambda_1)\,y_1 = \frac{f}{\lambda_1 - \lambda_2}, \qquad (\mathtt{D} - \lambda_2)\,y_2 = \frac{f}{\lambda_2 - \lambda_1}.$$

Then $y = y_1 + y_2$ is a solution to the given equation $(\mathtt{D} - \lambda_1)(\mathtt{D} - \lambda_2)y = f$. Using this approach, find a particular solution to each of the following equations.

(a) $y'' + y' - 6y = 30\, e^x/(1 + e^x)$; (b) $y'' - 10y' + 21y = 12\, e^x$; (c) $y'' - 9y = 10 \cos x$;
(d) $y'' - 2y' + 5y = 125\, x^2$; (e) $y'' - 4y = 8/(1 + e^{2x})$; (f) $y'' - 2y' + 2y = 10 \cosh x$;
(g) $y'' - y' - 2y = 6\, e^{2x}/\,(e^x + 1)$; (h) $y'' - y = 2\,(1 + e^x)^{-1}$; (i) $2\,y'' + 7\,y' - 4\,y = e^x/\,(1 + e^x)$.

17. A uniform circular cylinder of height h, radius r, and mass m floats in a liquid of density ρ_0, with its axis being vertical. Suppose that cylinder is covered by the liquid to a height $h_0 < h$, which means that its density ρ is less than ρ_0. Archimedes' Principle tells us that the downward gravitational force of the cylinder must be counterbalanced by the buoyancy force of the liquid:

$$mg = \rho_0 \left(\pi r^2 h_0\right) g.$$

Solving for h_0 and using $m = \rho\left(\pi r^2 h\right)$ for the mass of the solid cylinder, we have $h_0 \rho_0 = h\,\rho$. If the cylinder is pushed downward x units with the periodic force, then the equation of motion becomes

$$m\,\ddot{x} = mg - \rho_0 \left(\pi r^2 (h_0 + x)\right) g + A\, \sin(\omega t),$$

where A is a constant magnitude. Find the general solution for two cases when the natural frequency $\omega_0 = \sqrt{\rho_0 g/(\rho h)}$ is different from ω and when it is equal to it.

Section 4.8 of Chapter 4 (Review)

1. Use the variation of parameters method to find a particular solution of the given differential equations.

(a) $y'' + 9y = \tan 3x$; (b) $y'' + 4y = 8 \sec^3 2x$;
(c) $y'' + 4y = 4\, \tan^2(2t)$; (d) $y'' + 9y = 9\, \tan^3(3t)$;
(e) $y'' + 4y = 4 \csc 2x + 4x(x - 1)$; (f) $y'' + 25y = 25 \sec(5x) + 130\, e^x$;
(g) $y'' - 2y' + y = \frac{2\, e^x}{4 + x^2}$; (h) $y'' - 6y' + 10y = e^{3x}\, \sec^2 x$;
(i) $y'' + 25y = \frac{100\sqrt{3}}{4 - \cos^2 5x}$; (j) $y'' - 4y' + 8y = \frac{4\sqrt{3}\, e^{2x}}{3 + \sin^2 2x}$.

2. Given a complementary function, y_h, find a particular solution using the Lagrange method.

(a) $x^3 y''' + x^2 y'' - 2xy' + 2y = 8\, x^3$; $y_h(x) = C_1\, x + C_2\, x^{-1} + C_3\, x^2$.
(b) $x^3 y''' - 3x^2 y'' + 6xy' - 6y = 6\, x^4$; $y_h(x) = C_1\, x + C_2\, x^2 + C_3\, x^3$.
(c) $(1 - x^2) y'' - 2xy' = 2x - 1$; $y_h = C_1 + C_2\, \ln \frac{1+x}{1-x}$, $-1 < x < 1$.
(d) $xy'' - (1 + 2x^2) y' = 8\, x^5\, e^{x^2}$; $y_h = C_1 + C_2\, e^{x^2}$.
(e) $(\sin 4x) y'' - 4(\cos^2 2x) y' = 16\, \tan x$; $y_h(x) = C_1 + C_2\, \cos 2x$.

3. Use the Lagrange method to find a particular solution to the given inhomogeneous differential equations with constant coefficients. Note that the characteristic equation to the associated homogeneous equation has a double root.

(a) $y'' - 2y' + y = \frac{x^2 + 2}{x^3}\, e^x$; (b) $y'' + 6y' + 9y = \frac{e^{-3x}}{1 + x^2} + 27x^2 + 18x$;
(c) $y'' - 6y' + 9y = 4x^{-3}\, e^{3x}\, \ln x$; (d) $y'' - 4y' + 4y = 2e^{2x}\, \ln x + 25\, \sin x$;
(e) $(\mathtt{D} + 2)^2 y = e^{-2x}/x$; (f) $(2\,\mathtt{D} + 1)^2 y = e^{-x/2}/x$;
(g) $(\mathtt{D} + 3)^2 y = e^{-3x}/x^2$; (h) $(3\,\mathtt{D} - 2)^2 y = e^{2x/3}/x^2$.

4. Use the method of variation of parameters to determine the general solution of the given differential equations of order higher than 2.

(a) $y''' + 4y' = 8\, \cot 2t$; (b) $y''' - 2y'' - y' + 2y = 6\, e^{-x}$;
(c) $y^{(4)} - 2y'' + y = 4\, \cos t$; (d) $y''' - y'' + y' - y = 5\, e^t\, \cos t$;
(e) $y''' + y' = \frac{\sin x}{\cos^2 x}$; (f) $y''' + y' = \tan x$.

Section 4.9 of Chapter 4 (Review)

1. Express the following functions as Taylor expansions.

(a) $J_0(\sqrt{x})$, (b) $\sqrt{x}\, J_1(\sqrt{x})$.

2. Show that the function $y(x) = \sqrt{x}\, J_0(\sqrt{x})$ satisfies the differential equation $4x^2\, y'' + (x + 1)\, y = 0$.

3. Using the substitution $t = \sqrt{x}$, convert the differential equation $4x^2\, y'' + 4x\, y' + (x - 1)y = 0$ into a Bessel equation.

4. Show that $8J_n'''(x) = J_{n-3}(x) - 3J_{n-1}(x) + 3J_{n+1}(x) - J_{n+3}(x)$.

5. Prove the recurrence relations

$$J_\nu'(x) = \frac{\nu}{x}\, J_\nu(x) - J_{\nu+1}(x), \qquad 2J_\nu'(x) = J_{\nu-1}(x) - J_{\nu+1}(x).$$

6. Establish the following identities.

   (a) $\displaystyle\int_0^x t\, J_0(t)\, \mathrm{d}t = x\, J_1(x);$ **(b)** $\displaystyle\int_0^x t^{-1}\, J_2(t)\, \mathrm{d}t = -x^{-1}\, J_1(x) + \frac{1}{2}.$

7. Verify that the function $y(x) = J_0\left(\frac{2}{\alpha}\sqrt{\frac{k}{m}}\, e^{-\alpha t/2}\right)$ is a particular solution to the differential equation $m\,\ddot{y}(t) + ke^{-\alpha t}\, y(t) = 0, \quad \alpha > 0.$

8. Using the change of variables $y = u\, x^{-1/2}$, transform the Bessel equation (4.9.1) into
$$u'' + \left[1 - \frac{1}{x^2}\left(\nu^2 - \frac{1}{4}\right)\right] u = 0.$$

9. Consider the modified Bessel equation $x^2 y'' + xy' - (x^2 + 1/4)y = 0$ for $x > 0$.
   (a) Define a new dependent variable $u(x)$ by substitution $y(x) = x^{-1/2} u(x)$. Show that $u(x)$ satisfies the differential equation $u'' - u = 0$.
   (b) Show that the given equation has a fundamental set of solutions $\frac{\cosh x}{\sqrt{x}}$ and $\frac{\sinh x}{\sqrt{x}}$, if $x > 0$.

10. Express the general solution to the Riccati equation $y' = 4x^2 - 9y^2$ through Bessel functions. *Hint:* Consult Example 2.6.10 on page 103.

11. By appropriate change of independent variable, find the general solution of
$$y'' + \left(\lambda^2 e^{2x} - \nu^2\right) y = 0.$$

12. Show that the Wronskian of $J_\nu(x)$ and $N_\nu(x)$ is $2/(\pi x)$.

13. (a) Prove that the Wronskian of J_ν and $J_{-\nu}$ satisfies the differential equation $\frac{\mathrm{d}}{\mathrm{d}x}\left[xW(J_\nu, J_{-\nu}\right] = 0$, and hence deduce that $W(J_\nu, J_{-\nu}) = C/x$, where C is a constant.
    (b) Use the series expansions for J_ν, $J_{-\nu}$, and their derivatives to conclude that whenever ν is not an integer $C = \dfrac{-2}{\Gamma(1-\nu)\Gamma(\nu)}$, where C is the constant in part (a).

14. Show that $\displaystyle\int x^2 J_0(x)\, \mathrm{d}x = x^2 J_1(x) + xJ_0(x) - \int J_0(x)\, \mathrm{d}x.$

15. By multiplying the power series for $e^{xt/2}$ and $e^{-x/(2t)}$, show that $e^{x(t-1/t)/2} = \displaystyle\sum_{n=-\infty}^{\infty} J_n(x)\, t^n$. The function $e^{x(t-1/t)/2}$ is known as the **generating function** for the $J_n(x)$.

Chapter 5

Oliver Heaviside (1850–1925)

Pierre-Simon Laplace (1749–1827)

Laplace Transforms

Between 1880 and 1887, Oliver Heaviside, a self-taught English electrical engineer, introduced his version of Laplace transforms to the world. This gave birth to the modern technique in mathematics and its applications, called the **operational method**. Although the famous French mathematician and astronomer Pierre Simon marquis de Laplace introduced the corresponding integral in 1782, the systematic use of this procedure in physics, engineering, and technical problems was stimulated by Heaviside's work. Unfortunately, Oliver's genius was acknowledged much later and he gained most of his recognition posthumously.

In mathematics, a *transform* is usually referred to as a procedure that changes one kind of operation into another. The purpose of such a change is that in the transformed state the object may be easier to work with, or we may get a problem where the solution is known. A familiar example is the logarithmic function, which allows us to replace multiplication by addition and exponentiation by multiplication. Another well-known example is the one-to-one correspondence between matrices and linear operators in a finite dimensional vector space.

Differential expressions and differential equations occur rather often in applications. Heaviside's ideas are based on transforming differential equations into algebraic equations. This allows us to define a function of the differential operator, $\texttt{D} = \mathrm{d}/\mathrm{d}t$, which acts on functions of a positive independent variable. That is, it transforms the operation of differentiation into the algebraic operation of multiplication. A transformation that assigns to an operation an algebraic operation of multiplication is called a **spectral representation** for the given operation. In other words, the Laplace transform is an example of a spectral representation for the differential operator $\texttt{D}$ acting in a space of smooth functions on a positive half-line.

Since the problems under consideration contain an independent variable that varies from zero to infinity, we denote it by t in order to emphasize the connection with time. Heaviside's idea was to eliminate the time-variable along with the corresponding differentiation from the problems to reduce their dimensions. It is no surprise that the operational method originated from electrical problems modeled by differential equations with discontinuous and/or periodic functions. For example, the impressed voltage on a circuit could be piecewise continuous and periodic. The Laplace transform is a powerful technique for solving initial value problems for constant coefficient ordinary and partial differential equations that have discontinuous forcing functions.

While the general solution gives us a set of all solutions to an ordinary differential equation, practical problems often require determination out of the infinity of solution curves only a specific solution that satisfies some auxiliary conditions like the initial conditions or boundary conditions. To solve the initial value problem, we have had to find the general solution first and then determine coefficients to fit the initial data. In contrast to techniques described in the previous chapter, the Laplace transform solves the initial value problems directly, without determining the general solution. Application of the Laplace transformation to constant coefficient linear differential equations includes the following steps:

1. Application of the Laplace transformation to the given problem.
2. Solving the transformed problem (algebraic equation).
3. Calculating the inverse Laplace transform to restore the solution.

5.1 The Laplace Transform

A pair of two integral equations

$$f(x) = \int_a^b K(x,y)\,\varphi(y)\,\mathrm{d}y, \qquad \varphi(y) = \int_c^d G(x,y)\,f(x)\,\mathrm{d}x$$

may be considered as an integral transformation and its inverse, respectively. In these equations, $K(x, y)$ and $G(x, y)$ are known functions, and we call them kernels of the integral transformation and its inverse, respectively. The Laplace transformation is a particular case of the general integral operator, which has proved its usefulness in numerous applications. The Laplace transform provides an example of a linear operator (see §4.1.1).

Definition 5.1: Let f be an arbitrary (complex-valued or real-valued) function, defined on the semi-infinite interval $[0, \infty)$; then the integral

$$f^L(\lambda) \equiv (\mathcal{L}f)(\lambda) = \int_0^\infty e^{-\lambda t} f(t)\,\mathrm{d}t \tag{5.1.1}$$

is said to be the *Laplace transform* of f if the integral (5.1.1) converges for some value $\lambda = \lambda_0$ of a parameter λ. Therefore, the Laplace transform of a function (if it exists) depends on a parameter λ, which could be either a real number or a complex number.

Saying that a function $f(t)$ has a Laplace transform $f^L(\lambda)$ means that the limit

$$f^L(\lambda) = \lim_{N\to\infty} \int_0^N f(t)\,e^{-\lambda_0 t}\,\mathrm{d}t$$

exists for some $\lambda = \lambda_0$. The integral on the right-hand side of Eq. (5.1.1) is an integral over an unbounded interval. Such integrals are called *improper integrals*, and they are defined as a limit of integrals over finite intervals. If such a limit does not exist, the improper integral is said to diverge.

From the definition of the integral, it follows that if the Laplace transform exists for a particular function, then it does not depend on the values of a function at a discrete number (finite or infinite) of points. Namely, we can change the values of a function at a finite number of points and its Laplace transform will still be the same.

The parameter λ in the definition of the Laplace transform is not necessarily a real number, and could be a complex number. Thus, $\lambda = \alpha + \mathbf{j}\beta$, where α is the real part of λ, denoted by $\alpha = \Re\lambda$, and β is an imaginary part of a complex number λ, $\beta = \Im\lambda$. The set of all complex numbers is denoted by $\mathbb{C}$, while the set of all real numbers is denoted by $\mathbb{R}$ (see §4.5).

Theorem 5.1: If a function f is absolutely integrable over any finite interval from $[0, \infty)$ and the integral (5.1.1) converges for some complex number $\lambda = \mu$, then it converges in the half-plane $\Re\lambda > \Re\mu$, i.e., in $\{\lambda \in \mathbb{C} : \Re\lambda > \Re\mu\}$.

PROOF: Since the integral (5.1.1) converges for $\lambda = \mu$, the integral

$$h(t, \mu) = \int_0^t f(\tau)\, e^{-\mu\tau}\, \mathrm{d}\tau$$

has a finite limit $h(\infty, \lambda)$ when $t \mapsto \infty$. Let λ be any complex number with $\Re\lambda > \Re\mu$. We have

$$\begin{aligned} f^L(\lambda) &= \int_0^\infty e^{-\lambda t}\, f(t)\, \mathrm{d}t \\ &= \int_0^\infty e^{-(\lambda-\mu)t}\, e^{-\mu t}\, f(t)\, \mathrm{d}t = \int_0^\infty e^{-(\lambda-\mu)t}\, \mathrm{d}h(t, \mu), \end{aligned}$$

where $\mathrm{d}h(t, \mu) = e^{-\mu t} f(t)\mathrm{d}t$. We apply the integration by parts formula

$$\int_a^b u(t)\, \mathrm{d}v(t) = u(b)\, v(b) - u(a)\, v(a) - \int_a^b v(t)\, \mathrm{d}u(t)$$

with

$$u(t) = e^{-(\lambda-\mu)t}, \ \mathrm{d}u(t) = -(\lambda-\mu)\, e^{-(\lambda-\mu)t}\mathrm{d}t, \quad \mathrm{d}v(t) = e^{-\mu t} f(t)\mathrm{d}t, \ v(t) = h(t, \mu)$$

to obtain

$$f^L(\lambda) = (\lambda - \mu) \int_0^\infty e^{(\lambda-\mu)t}\, h(t, \mu)\, \mathrm{d}t,$$

because $h(0, \mu) = 0$ and

$$\lim_{t \mapsto \infty} e^{-(\lambda-\mu)t}\, h(t, \mu) = h(\infty, \mu) \lim_{t \mapsto \infty} e^{-(\lambda-\mu)t} = 0.$$

The function $h(t, \mu)$ is bounded, meaning that there exists a positive number M such that $|h(t, \mu)| \leqslant M$ (so M is the maximum of the absolute values of the function $h(t, \mu)$). Therefore,

$$\begin{aligned} |f^L(\lambda)| &\leqslant |\lambda - \mu| \int_0^\infty e^{-t\Re(\lambda-\mu)}\, |h(t, \mu)|\, \mathrm{d}t \\ &\leqslant |\lambda - \mu| M \int_0^\infty e^{-t\Re(\lambda-\mu)}\, \mathrm{d}t = \frac{|\lambda - \mu|}{\Re(\lambda - \mu)} M. \end{aligned}$$

Consequently the integral (5.1.1) converges when $\Re\lambda > \Re\mu$. ∎

This theorem may be interpreted geometrically in the following manner. If the integral (5.1.1) converges for some complex number $\lambda = \mu$, then it converges everywhere in the region on the right-hand side of a straight line drawn through $\lambda = \mu$ and parallel to the imaginary axis (see Fig. 5.1). Thus, only the real part of λ is decisive for the convergence of the one-sided Laplace integral (5.1.1). The imaginary part of λ does not matter at all.

This implies that there exists some real value σ_c, called the **abscissa of convergence** of the function f, such that the integral (5.1.1) is convergent in the half-plane $\Re\lambda > \sigma_c$ and divergent in the half-plane $\Re\lambda < \sigma_c$. We cannot predict whether or not there are points of convergence on the line $\Re\lambda = \sigma_c$ itself.

We emphasize again that the imaginary part of the parameter λ in the definition of the Laplace transform does not affect the convergence of the integral (5.1.1) for real-valued functions. Therefore, for the question of existence of Laplace transforms, we may assume in the future that this parameter λ is a **real number** that is greater than σ_c, the abscissa of convergence (see Fig. 5.1 on page 272). Thus, for example, if we want to determine the convergence of the integral (5.1.1) for a particular function f, we may suppose that the parameter λ is a real number.

Theorem 5.2: The Laplace transform is a linear operator, that is,

$$(\mathcal{L}Cf)(\lambda) = C\,(\mathcal{L}f)(\lambda) \quad \text{and} \quad (\mathcal{L}(f+g))(\lambda) = (\mathcal{L}f)(\lambda) + (\mathcal{L}g)(\lambda),$$

where C is a constant and f and g are arbitrary functions for which their Laplace transforms exist.

PROOF: This result follows immediately from the definition of an improper integral. Thus, from the definition, we write

$$\begin{aligned}(\mathcal{L}Cf)(\lambda) &= \int_0^\infty e^{-\lambda t} Cf(t)\,\mathrm{d}t = \lim_{N\to\infty} \int_0^N Cf(t)\,e^{-\lambda t}\,\mathrm{d}t \\ &= C \lim_{N\to\infty} \int_0^N f(t)\,e^{-\lambda t}\,\mathrm{d}t = Cf^L(\lambda).\end{aligned}$$

Let us suppose that f and g are two functions whose Laplace transforms exist for $\lambda = \mu_f$ and $\lambda = \mu_g$, respectively. Let λ_0 be a complex number with its real part greater than the maximum of $\Re\mu_f$ and $\Re\mu_g$, that is, $\Re\lambda_0 > \Re\mu_f$ and $\Re\lambda_0 > \Re\mu_g$. Then for $\lambda = \lambda_0$ we have

$$\begin{aligned}(\mathcal{L}(f+g))(\lambda_0) &= \lim_{N\to\infty} \int_0^N (f+g)\,e^{-\lambda_0 t}\,\mathrm{d}t \\ &= \lim_{N\to\infty} \int_0^N f(t)\,e^{-\lambda_0 t}\,\mathrm{d}t + \lim_{N\to\infty} \int_0^N g(t)\,e^{-\lambda_0 t}\,\mathrm{d}t \\ &= \int_0^\infty f(t)\,e^{-\lambda_0 t}\,\mathrm{d}t + \int_0^\infty g(t)\,e^{-\lambda_0 t}\,\mathrm{d}t = f^L(\lambda_0) + g^L(\lambda_0).\end{aligned}$$

Hence, $(\mathcal{L}(f+g))(\lambda) = (\mathcal{L}f)(\lambda) + (\mathcal{L}g)(\lambda)$ for all λ such that $\Re\lambda \geqslant \Re\lambda_0$. ■

The next few examples demonstrate the application of the Laplace transformation to power functions and exponential functions.

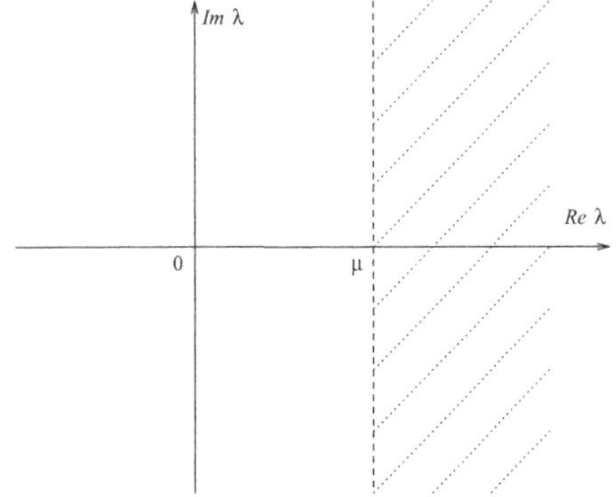

Figure 5.1: Region of convergence with the abscissa of convergence $\sigma_c = \mu$.

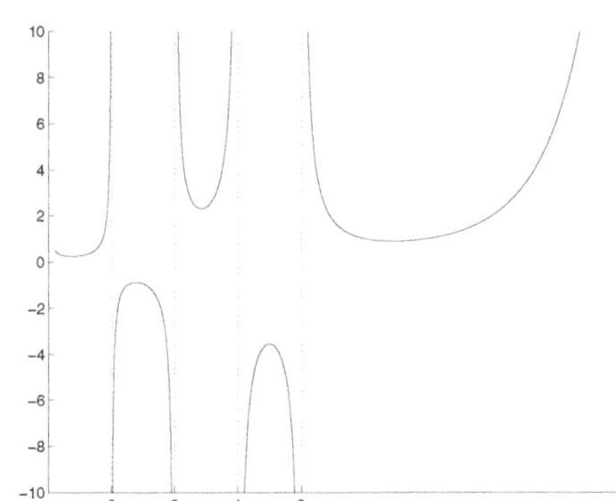

Figure 5.2: Example 5.1.4. The graph of the gamma function.

Example 5.1.1: Let $f(t) = e^{at}$, where a is a real constant. Find its Laplace transform.

Solution. We evaluate the integral over a semi-finite interval:

$$\begin{aligned}\int_0^N e^{at} e^{-\lambda t}\,\mathrm{d}t &= \int_0^N e^{-(\lambda-a)t}\,\mathrm{d}t = \left[-\frac{1}{\lambda-a}\,e^{-(\lambda-a)t}\right]_{t=0}^{t=N} \\ &= -\frac{1}{\lambda-a}\,e^{-(\lambda-a)N} + \frac{1}{\lambda-a}.\end{aligned}$$

We assume that $\lambda - a > 0$ for real λ and $\Re(\lambda - a) > 0$ for complex λ. Then the first term with the exponential multiple approaches zero as $N \to \infty$, leaving $1/(\lambda - a)$. Hence, we have

$$\mathcal{L}\left[e^{at}\right] = \int_0^\infty e^{at} e^{-\lambda t}\,\mathrm{d}t = \lim_{N\to\infty} \int_0^N e^{at} e^{-\lambda t}\,\mathrm{d}t = \frac{1}{\lambda-a}. \tag{5.1.2}$$

Example 5.1.2: Find the Laplace transform of the function $f(t) = t,\ t > 0$.

Solution. Since the Laplace transformation $(\mathcal{L}f)(\lambda) = \int_0^\infty e^{-\lambda t} t \, \mathrm{d}t$ is defined for $\Re\lambda > 0$, its abscissa of convergence is $\sigma_c = 0$. There are two options to determine the value of the integral. One of them calls for integration by parts, which yields

$$(\mathcal{L}f)(\lambda) = -\left.\frac{t}{\lambda} e^{-\lambda t}\right|_{t=0}^{\infty} + \frac{1}{\lambda}\int_0^\infty e^{-\lambda t}\,\mathrm{d}t = \frac{1}{\lambda^2}.$$

We remind that the expression

$$\left.\frac{t}{\lambda} e^{-\lambda t}\right|_{t=0}^{\infty}$$

means the difference of limits; that is,

$$\left.\frac{t}{\lambda} e^{-\lambda t}\right|_{t=0}^{\infty} = \lim_{t\to+\infty}\left(\frac{t}{\lambda} e^{-\lambda t}\right) - \lim_{t\to 0}\left(\frac{t}{\lambda} e^{-\lambda t}\right).$$

The upper limit in the right-hand side is zero because

$$\lim_{t\to+\infty} e^{-\lambda t} = \lim_{t\to+\infty} e^{-(\Re\lambda + \mathbf{j}\Im\lambda)t} = \lim_{t\to+\infty} e^{-(\Re\lambda)t} \lim_{t\to+\infty} e^{-\mathbf{j}(\Im\lambda)t}.$$

The first limit, $\lim_{t\to+\infty} e^{-(\Re\lambda)t}$, approaches zero since $\Re\lambda > 0$ and $t > 0$. The second multiplier, $\lim_{t\to+\infty} e^{-\mathbf{j}\Im\lambda t}$, has no limits, but is bounded because $|e^{-i\Im\lambda t}| = |\cos(\Im\lambda t) - \mathbf{j}\sin(\Im\lambda t)| = 1$.

There is another option to find the value of the integral. The integrand is equal to

$$e^{-\lambda t} t = -\frac{\mathrm{d}}{\mathrm{d}\lambda} e^{-\lambda t}.$$

Since λ and t are independent variables, we can interchange the order of integration and differentiation to obtain

$$\int_0^\infty e^{-\lambda t} t\,\mathrm{d}t = \int_0^\infty \left(-\frac{\mathrm{d}}{\mathrm{d}\lambda} e^{-\lambda t}\right)\mathrm{d}t = -\frac{\mathrm{d}}{\mathrm{d}\lambda}\int_0^\infty e^{-\lambda t}\,\mathrm{d}t.$$

It is not a problem to evaluate $\int_0^\infty e^{-\lambda t}\,\mathrm{d}t = \frac{1}{\lambda}$. So we have

$$\int_0^\infty e^{-\lambda t} t\,\mathrm{d}t = -\frac{\mathrm{d}}{\mathrm{d}\lambda}\left(\frac{1}{\lambda}\right) = \frac{1}{\lambda^2}.$$

Example 5.1.3: Using the result of Example 5.1.2, calculate the Laplace transform of $f(t) = t^n,\ t > 0$, where n is an integer.

Solution. The abscissa of convergence σ_c for this function is equal to zero. Therefore, according to Theorem 1.2, a parameter λ may be chosen as any positive number. To start, we consider $n = 2$. Again integrating by parts, we have

$$(\mathcal{L}f)(\lambda) = \int_0^\infty e^{-\lambda t} t^2\,\mathrm{d}t = -t^2 \left.\frac{1}{\lambda} e^{-\lambda t}\right|_{t=0}^{\infty} + \frac{1}{\lambda}\int_0^\infty e^{-\lambda t} 2t\,\mathrm{d}t$$

since

$$e^{-\lambda t}\,\mathrm{d}t = -\frac{1}{\lambda}\,\mathrm{d}\left(e^{-\lambda t}\right).$$

Therefore,

$$\left(\mathcal{L}t^2\right)(\lambda) = \frac{1}{\lambda}\int_0^\infty e^{-\lambda t} 2t\,\mathrm{d}t.$$

Using the result of Example 5.1.2, we obtain $\left(\mathcal{L}t^2\right)(\lambda) = \frac{2}{\lambda^3}$. In the general case, we have

$$\left(\mathcal{L}t^n\right)(\lambda) = \frac{n!}{\lambda^{n+1}}.$$

This formula can be obtained more easily by differentiation of the exponential function:

$$t^n e^{-\lambda t} = -\frac{\mathrm{d}}{\mathrm{d}\lambda}\left\{t^{n-1} e^{-\lambda t}\right\} = \left(-\frac{\mathrm{d}}{\mathrm{d}\lambda}\right)^n e^{-\lambda t}.$$

Thus, we have

$$\begin{aligned}(\mathcal{L}t^n)(\lambda) &= \int_0^\infty e^{-\lambda t} t^n \, \mathrm{d}t = \left(-\frac{\mathrm{d}}{\mathrm{d}\lambda}\right)^n \int_0^\infty e^{-\lambda t} \, \mathrm{d}t \\ &= \left(-\frac{\mathrm{d}}{\mathrm{d}\lambda}\right)^n \left(\frac{1}{\lambda}\right) = \frac{n!}{\lambda^{n+1}}.\end{aligned}$$

Example 5.1.4: Let p be any positive number (not necessarily an integer). Then the Laplace transform of the function $f(t) = t^p$, $t > 0$, is

$$\begin{aligned}(\mathcal{L}t^p)(\lambda) &= \int_0^\infty e^{-\lambda t} \, t^p \, \mathrm{d}t = \int_0^\infty e^{-\lambda t} \, (\lambda t)^p \lambda^{-p} \, \mathrm{d}t \\ &= \lambda^{-p-1} \int_0^\infty e^{-\tau} \, \tau^p \, \mathrm{d}\tau = \frac{\Gamma(p+1)}{\lambda^{p+1}},\end{aligned}$$

where

$$\Gamma(\nu) = \int_0^\infty e^{-\tau} \, \tau^{\nu-1} \, \mathrm{d}\tau \tag{5.1.3}$$

is Euler's **gamma function**. This improper integral converges for $\nu > 0$, and using integration by parts, we obtain

$$\Gamma(\nu+1) = \nu \, \Gamma(\nu). \tag{5.1.4}$$

Indeed, for $\nu > 0$, we have

$$\begin{aligned}\Gamma(\nu+1) &= \int_0^\infty e^{-\tau} \, \tau^\nu \, \mathrm{d}\tau = -\int_0^\infty \tau^\nu \, \mathrm{d}e^{-\tau} \\ &= -\tau^\nu \, e^{-\tau} \Big|_{\tau=0}^{\tau=\infty} + \nu \int_0^\infty e^{-\tau} \, \tau^{\nu-1} \, \mathrm{d}\tau = \nu \, \Gamma(\nu).\end{aligned}$$

The most remarkable property of the Γ-function is obtained when we set $\nu = n$, an integer. A comparison with the result of the previous example yields $\Gamma(n+1) = n!, \quad n = 0, 1, 2, \ldots$. The computation of $\mathcal{L}[t^{n/2}]$, where n is an integer, is based on the known special value $\Gamma\left(\frac{1}{2}\right) = \sqrt{\pi}$ and is left as Problem 7 on page 280.

Definition 5.2: A function f is said to be **piecewise continuous** (or **intermittent**) on a finite interval $[a, b]$ if the interval can be divided into finitely many subintervals so that $f(t)$ is continuous on each subinterval and approaches a finite limit at the end points of each subinterval from the interior. That is, there is a finite number of points $\{\alpha_j\}$, $j = 1, 2, \ldots, N$, where a function f has a *jump discontinuity* when both

$$\lim_{\substack{h \to 0 \\ h \geqslant 0}} f(\alpha_j + h) = f(\alpha_j + 0) \qquad \text{and} \qquad \lim_{\substack{h \to 0 \\ h \geqslant 0}} f(\alpha_j - h) = f(\alpha_j - 0)$$

exist but are different. A function is called piecewise continuous on an infinite interval if it is intermittent on every finite subinterval.

Recall that we have $f(t) = f(t+0) = f(t-0)$ for a continuous function f. If at some point $t = t_0$ this is not the case, then the function is discontinuous at $t = t_0$. In other words, the finite discontinuity occurs if the left-hand side and the right-hand side limits are finite and not equal.

Definition 5.3: The *Heaviside function* $H(t)$ is the unit step function, equal to zero for t negative and unity for t positive, with $H(0) = 1/2$, i.e.,

$$H(t) = \begin{cases} 1, & t > 0; \\ {}^1\!/_2, & t = 0; \\ 0, & t < 0. \end{cases} \tag{5.1.5}$$

Although *Mathematica*® has a built-in function `HeavisideTheta` (which is 1 for $t > 0$ and 0 for $t < 0$), it is convenient to define the Heaviside function directly:
`HVS[x_] := Piecewise[{{0, x < 0}, {1, x > 0}, {1/2, x==0}}]` or
`HVS[x_] := Piecewise[{{ 1, x > 0}, {1/2, x == 0}, {0, True}}]`
Maple™ uses the `Heaviside(t)` symbol to represent the Heaviside function. Both *Maple* and *Mathematica* leave the value at $t = 0$ undefined; however, *Mathematica* has a similar function `UnitStep` (which is 1 for $t \geqslant 0$ and 0 otherwise). It is also possible to modify the built-in symbol `UnitStep` in order to define it as 1/2 at $t = 0$:

```
Unprotect[UnitStep];     UnitStep[0] = 1/2;     Protect[UnitStep];
```

In *Maple*, we can either define the Heaviside function directly:

```
H:=x->piecewise(x<0,0,x>0,1,x=0,1/2)
```

or enforce the built-in function:
`Heaviside(0):=1/2` or type

```
NumericEventHandler(invalid_operation=‘Heaviside/EventHandler‘ (value_at_zero=0.5)):
```

Remark. Actually, the Laplace transformation is not sensitive to the values of the function at any finite number of points. Recall that a definite integral of a (positive) function is the area under its curve, which is further defined as the limit of small rectangles inscribed between the curve and the horizontal axis. Since the width of a point is zero, its product by the value of the function at that point is also zero, and one point cannot contribute to the area. Thus, a definite integral does not depend on the values of an integrated function (called the integrand) at a discrete number of points. So you can change the value of the function at any point to be any number, i.e., $\frac{1}{2}$ or 1,000,000 and its integral value will remain the same. □

Example 5.1.5: Since the Heaviside function is 1 for $t > 0$, its Laplace transform is

$$\int_0^\infty e^{-\lambda t}\,\mathrm{d}t = -\frac{1}{\lambda}\,e^{-\lambda t}\Big|_{t=0}^{\infty} = \frac{1}{\lambda}, \qquad \Re\lambda > 0.$$

The abscissa of convergence of this integral is equal to zero.

In the calculation, we never used the particular value of the Heaviside function at the point $t = 0$. Therefore, the Laplace transform of $H(t)$ does not depend on the value of the function at this point. □

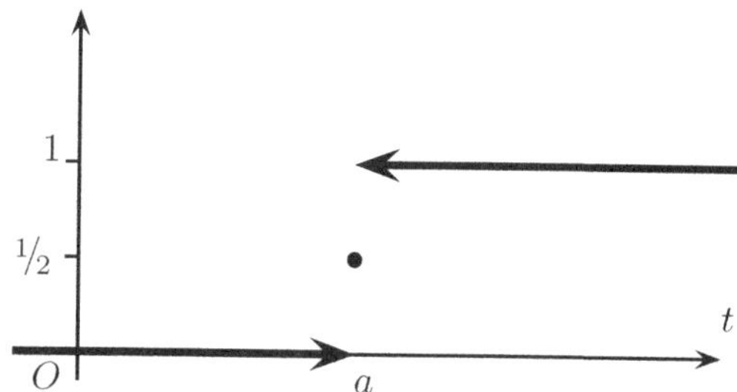

Figure 5.3: Shifted Heaviside function $H(t-a)$.

The Laplace transform of the shifted (sometimes referred to as retarded) Heaviside function $H(t-a)$ is

$$(\mathcal{L}H(t-a))\,(\lambda) = \int_0^\infty e^{-\lambda t}\,H(t-a)\,\mathrm{d}t = \int_a^\infty e^{-\lambda t}\,\mathrm{d}t = \frac{e^{-a\lambda}}{\lambda}.$$

In many cases, it is very difficult to find an exact expression for the Laplace transformation of a given function. However, if the Maclaurin series, $f(t) = \sum_{n\geqslant 0} f_n\,t^n$, for the function $f(t)$ is known, then using the formula $\mathcal{L}[t^n] = n!\,\lambda^{-n-1}$ found in Example 5.1.3, we sometimes are able to determine its Laplace transform as infinite series:

$$\mathcal{L}[f](\lambda) = \mathcal{L}\left[\sum_{n\geqslant 0} f_n\,t^n\right] = \sum_{n\geqslant 0} f_n\,\frac{n!}{\lambda^{n+1}} = \sum_{n\geqslant 0}\frac{f^{(n)}(0)}{\lambda^{n+1}} \tag{5.1.6}$$

provided the series converges.

Example 5.1.6: The function $\tan t$ is not piecewise continuous on any interval containing a point $t = \pi/2 + k\pi, k = 0, \pm 1, \pm 2, \ldots$, since it has infinite limits at these points. That is, for example, $\tan(\pi/2 - 0) = +\infty$ and $\tan(\pi/2 + 0) = -\infty$.

On the other hand, the function (see its graph in Fig. 5.4)

$$f(t) = \sum_{k=0}^{\infty} (-1)^k \, H(t - ka)$$

is piecewise continuous. This function is a bounded function; therefore, its Laplace transform exists. Thus,

$$f^L(\lambda) = \sum_{k=0}^{\infty} (-1)^k \, \mathcal{L}[H(t-ka)] = \frac{1}{\lambda} \sum_{k=0}^{\infty} (-1)^k \, e^{-ak\lambda} = \frac{1}{\lambda} \frac{1}{1+e^{-a\lambda}}$$

since we have $\dfrac{1}{1+z} = \sum_{k=0}^{\infty} (-1)^k \, z^k$ for $z = e^{-a\lambda}$.

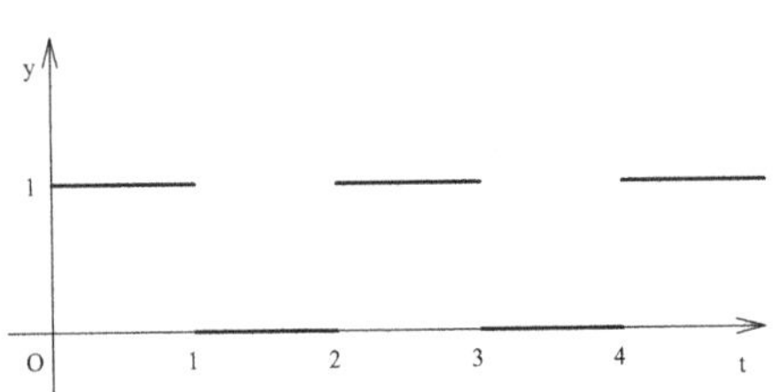

Figure 5.4: Example 5.1.6, the piecewise continuous function.

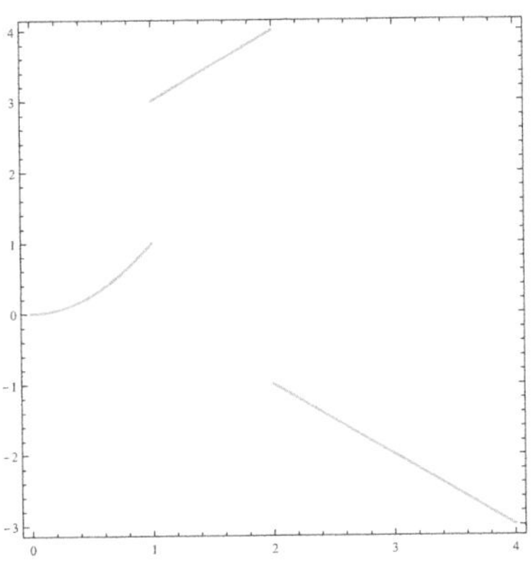

Figure 5.5: Example 5.1.7, the graph of the piecewise continuous function.

Example 5.1.7: Find the Laplace transform of the piecewise continuous function

$$f(t) = \begin{cases} t^2, & 0 \leqslant t \leqslant 1, \\ 2+t, & 1 < t \leqslant 2, \\ 1-t, & 2 < t < \infty. \end{cases}$$

Solution. This function is not continuous since it has two points of discontinuity, namely, $t = 1$ and $t = 2$. At these points we have $f(1-0) = 1$, $f(1+0) = 3$ and $f(2-0) = 4$, $f(2+0) = -1$. To plot the function, we use the following *Maple* commands:

```
f := piecewise(0<t < 1, t^2, 1<t < 2, 2+t, 2<t, 1-t);
plot(f, t = 0..4, discont=true);
```

To obtain the Laplace transform for a piecewise continuous function, just apply the integral transformation on intervals over which the function is continuous and sum the results. Therefore, the Laplace transform of the function $f(t)$ is

$$(\mathcal{L}f)(\lambda) = \int_0^1 t^2 \, e^{-\lambda t} \, \mathrm{d}t + \int_1^2 (2+t) \, e^{-\lambda t} \, \mathrm{d}t + \int_2^{\infty} (1-t) \, e^{-\lambda t} \, \mathrm{d}t.$$

Let us attack the integrals using integration by parts. Since the antiderivative of $e^{-\lambda t}$ is $-\lambda^{-1} \, e^{-\lambda t}$, we have

$$\begin{aligned} f^L(\lambda) &= \left[-\frac{t^2}{\lambda} \, e^{-\lambda t} - \frac{2t}{\lambda^2} \, e^{-\lambda t} - \frac{2}{\lambda^3} \, e^{-\lambda t} \right] \Bigg|_{t=0}^{t=1} \\ &\quad + \left[-\frac{2}{\lambda} \, e^{-\lambda t} - \frac{t}{\lambda} \, e^{-\lambda t} - \frac{1}{\lambda^2} \, e^{-\lambda t} \right] \Bigg|_{t=1}^{t=2} + \left[-\frac{1}{\lambda} + \frac{t}{\lambda} + \frac{1}{\lambda^2} \right] e^{-\lambda t} \Bigg|_{t=2}^{t=\infty} \\ &= \frac{2}{\lambda^3} + e^{-\lambda} \left[\frac{2}{\lambda} - \frac{1}{\lambda^2} - \frac{2}{\lambda^3} \right] - e^{-2\lambda} \left[\frac{5}{\lambda} + \frac{2}{\lambda^2} \right]. \end{aligned}$$

□

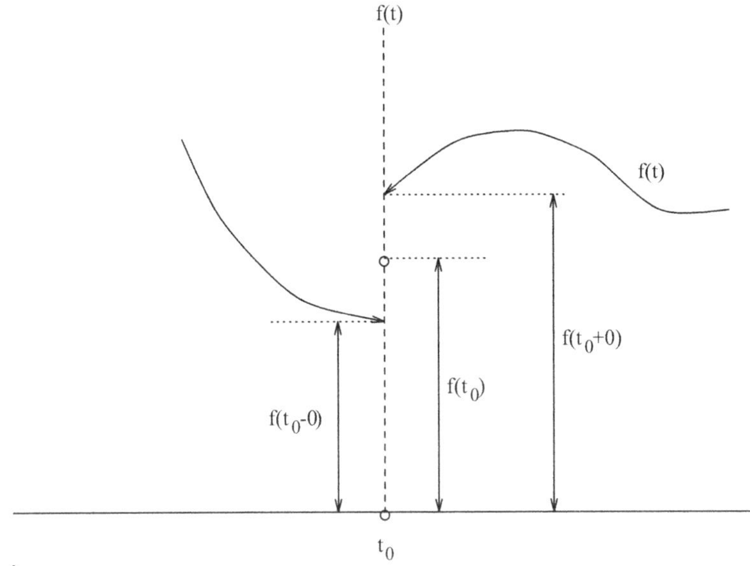

Figure 5.6: Mean value of a discontinuous function.

The last example shows that the Laplace transform of a discontinuous function is a holomorphic function (that possesses infinitely many derivatives) in half plane $\Re\lambda > \sigma_c$.

It soon becomes tiresome to test each function we encounter to determine whether the corresponding integral (5.1.1) exists in some range of values for λ. We therefore seek a fairly large class of functions for which the Laplace transform can easily be shown to exist. For this purpose, we introduce the following definition.

Definition 5.4: A function $f(t)$, $t \in [0, \infty)$ is said to be a **function-original** if it has a finite number of points of discontinuity (finite jumps) on every finite subinterval of $[0, \infty)$ and it grows not faster than an exponential, that is,

$$|f(t)| < M\, e^{ct} \quad (t > T) \tag{5.1.7}$$

for some constants c, M, and T, which may be very large. Moreover, we assume that at points of discontinuity the values of a function-original are equal to the corresponding *mean values*; thus,

$$f(t) = \frac{1}{2}\left[f(t+0) + f(t-0)\right] = \lim_{\substack{\varepsilon \to 0 \\ \varepsilon \geqslant 0}} \frac{f(t+\varepsilon) + f(t-\varepsilon)}{2}. \tag{5.1.8}$$

The Laplace transform of such a function is called the **image**.

A function-original can diverge to infinity as t tends to infinity. The restriction (5.1.7) tells us that a function-original can blow up at infinity as an exponent, but not any faster.

If a function is continuous at $t = t_0$, then $f(t_0) = f(t_0 + 0) = f(t_0 - 0)$, and the relation (5.1.8) is valid at that point. We impose condition (5.1.8) to guarantee the uniqueness of the inverse Laplace transform. In §5.4 we will see that the inverse Laplace transform uniquely restores a function from its image (Laplace transform) only if condition (5.1.8) holds at any point. So, there is a one-to-one correspondence between function-originals and their images.

Note that the Laplace transform may exist for a function that is not a function-original. A classic example is provided by the function $f(t) = t^{-1/2}$, which has an infinite jump at $t = 0$. Nevertheless, its Laplace transform exists and equals $\sqrt{\pi/\lambda}$. Therefore, function-originals constitute a subclass of functions for which the Laplace transformation exists.

Definition 5.5: We say that a function f is of *exponential order* if for some constants c, M, and T the inequality (5.1.7) holds. We abbreviate this as $f = O\left(e^{ct}\right)$ or $f \in O\left(e^{ct}\right)$. A function f is said to be of *exponential order* α, or $eo(\alpha)$ for abbreviation, if $f = O\left(e^{ct}\right)$ for any real number $c > \alpha$, but not when $c < \alpha$.

Any polynomial is of exponential order because the exponential function $e^{\varepsilon t}$ $(\varepsilon > 0)$ grows faster than any polynomial. If $f \in eo(\alpha)$, it may or may not be true that $f = O\left(e^{\alpha t}\right)$, as the following example shows.

Example 5.1.8: Both functions te^{2t} and $t^{-1}e^{2t}$ are of exponential order 2; the latter is $O\left(e^{2t}\right)$, but the former is not. According to the definition, a function $f(t)$ is of exponential order $f(t) = O(e^{ct})$ if and only if there exists a real number c such that the fraction $f(t)/e^{ct}$ is bounded for all $t > T$. Since the fraction $te^{2t}/e^{2t} = t$ is not bounded, the function te^{2t} does not belong to the class $O\left(e^{2t}\right)$. However, this function is of exponential order $2 + \varepsilon$ for any positive ε, namely, $te^{2t} = O(e^{(2+\varepsilon)t})$, because $te^{2t}/e^{(2+\varepsilon)t} = te^{-\varepsilon t}$ is a bounded function for large values of t, yet exponential functions decrease faster than any polynomial.

Example 5.1.9: The function $f(t) = e^{t^2}$, $t \in [0, \infty)$, is not a function-original because it grows faster than e^{ct} for any c; however, the function

$$f(t) = e^{\sqrt{t^4+1}-t^2}, \quad t \in [0, \infty)$$

does not grow faster than e^{ct} for any c. Therefore, it is a function-original. When t approaches infinity, the expression $\sqrt{t^4+1} - t^2$ approaches zero. This follows from the Taylor representation of the function

$$s(\varepsilon) \equiv \sqrt{1+\varepsilon} = s(0) + s'(0)\varepsilon + \frac{s''(0)}{2!}\varepsilon^2 + \cdots = 1 + \frac{1}{2}\varepsilon - \frac{1}{8}\varepsilon^2 + \cdots .$$

for small enough ε. If we let $\varepsilon = t^{-4}$, then we obtain

$$\sqrt{t^4+1} = t^2\sqrt{1+\frac{1}{t^4}} = t^2\sqrt{1+\varepsilon} = t^2\left[1 + \frac{1}{2t^4} + \cdots\right] = t^2 + \frac{1}{2t^2} + \cdots .$$

Hence,

$$\sqrt{t^4+1} - t^2 = \frac{1}{2t^2} - \frac{1}{8t^6} + \cdots$$

when t approaches infinity. The function e^{t^2} is not of exponential order since, for any constant λ,

$$\lim_{t\to\infty} e^{t^2} e^{-\lambda t} = \infty.$$

Definition 5.6: The integral (5.1.1) is said to be *absolutely convergent*, if the integral

$$\int_0^{\infty} e^{-(\Re\lambda)t}\,|f(t)|\,\mathrm{d}t \tag{5.1.9}$$

converges. The greatest lower bound σ_a of such numbers $\Re\lambda$ for which the integral (5.1.9) converges is called the *abscissa of absolute convergence*.

The following assessments follow from Definitions 5.1 and 5.5.

Theorem 5.3: If $|f(t)| \leqslant C$ for $t \geqslant T$, then the Laplace transform (5.1.1) converges absolutely for any λ with $\Re\lambda > 0$. In particular, the Laplace transform exists for any positive (real) λ.

Theorem 5.4: The integral (5.1.1) converges for any function-original. Moreover, if a function f is of exponential order α, then the integral (5.1.1) absolutely converges for $\Re\lambda > \alpha$. Furthermore, if f and g are piecewise continuous functions whose Laplace transforms exist and satisfy $(\mathcal{L}f) = (\mathcal{L}g)$, then $f = g$ at their points of continuity. Thus, if $F(\lambda)$ has a continuous inverse f, then f is unique.

Example 5.1.10: We have already shown in Example 5.1.1, page 272, that $\left(\mathcal{L}e^{\alpha t}\right)(\lambda) = (\lambda - \alpha)^{-1}$. With this in hand, we can find the Laplace transform of the trigonometric functions $\sin \alpha t$ and $\cos \alpha t$ using Euler's equations

$$\cos\theta = \Re e^{\mathrm{j}\theta} \qquad \text{and} \qquad \sin\theta = \Im e^{\mathrm{j}\theta}, \tag{5.1.10}$$

where $\Re z = \operatorname{Re} z = x$ denotes the real part of a complex number $z = x + \mathbf{j}y$ and $\Im z = \operatorname{Im} z = y$ is its imaginary part. Then assuming that λ is a real number, we find the Laplace transform of $\sin\alpha t$ as the imaginary part of the following integral:

$$(\mathcal{L}\sin\alpha t) = \int_0^\infty e^{-\lambda t}\,\Im e^{\mathbf{j}\alpha t}\,\mathrm{d}t = \Im\int_0^\infty e^{-\lambda t}\,e^{\mathbf{j}\alpha t}\,\mathrm{d}t = \Im\,\frac{1}{\lambda - \mathbf{j}\alpha} = \frac{\alpha}{\lambda^2+\alpha^2}.$$

Similarly, extracting the real part, we obtain

$$(\mathcal{L}\cos\alpha t) = \Re\int_0^\infty e^{-\lambda t}\,e^{i\alpha t}\,\mathrm{d}t = \Re\,\frac{1}{\lambda-\mathbf{j}\alpha} = \Re\,\frac{\lambda+\mathbf{j}\alpha}{(\lambda-\mathbf{j}\alpha)\,(\lambda+\mathbf{j}\alpha)} = \frac{\lambda}{\lambda^2+\alpha^2}.$$

Example 5.1.11: Since the hyperbolic functions are linear combinations of exponential functions,

$$\sinh\alpha t = \frac{1}{2}\left[e^{\alpha t} - e^{-\alpha t}\right], \quad \cosh\alpha t = \frac{1}{2}\left[e^{\alpha t} + e^{-\alpha t}\right],$$

their Laplace transformations follow from Eq. (5.1.2), page 272. For instance,

$$\begin{aligned}(\mathcal{L}\cosh\alpha t) &= \frac{1}{2}\int_0^\infty e^{-\lambda t}\,e^{\alpha t}\,\mathrm{d}t + \frac{1}{2}\int_0^\infty e^{-\lambda t}\,e^{-\alpha t}\,\mathrm{d}t \\ &= \frac{1}{2}\left[\frac{1}{\lambda-\alpha} + \frac{1}{\lambda+\alpha}\right] = \frac{\lambda}{\lambda^2-\alpha^2}.\end{aligned}$$

□

We summarize our results on the Laplace transforms as well as some relevant integrals in Table 280. In the previous examples, we found the Laplace transforms of some well known functions. It should be noted that all functions are considered only for positive values of the argument. For negative values of t, these functions are assumed to be identically zero; to ensure this property, we multiply all functions by the Heaviside step function, Eq. (5.1.5). For instance, in Example 5.1.3, the Laplace transform of the function $f(t) = t^2$ was found. However, we actually found the Laplace transform of the function $g(t) = H(t)t^2$, but not $f(t)$. Hence, we can consider the Laplace transform for $g(t)$ as the integral over the infinite interval, namely,

$$g^L(\lambda) = \int_{-\infty}^\infty g(t)\,e^{-\lambda t}\,\mathrm{d}t. \tag{5.1.11}$$

Such an integral, called a **two-sided** or **bilateral Laplace transform**, does not exist for the function $f(t) = t^2$ for any real value of λ.

Problems

1. Find the Laplace transform of the following functions.
   (a) $2-5t$; **(b)** $t\cos 2t$; **(c)** $e^{2t}\sin t$; **(d)** $e^{3t}\cosh 2t$;
   (e) $e^{2t}\sinh 4t$; **(f)** $e^t\cos 2t$; **(g)** t^4-2t; **(h)** $2t\sinh 2t$;
   (i) $\cos t+\sin 2t$; **(j)** $\sin t + e^{2t}$; **(k)** $t^2\cosh 2t$; **(l)** $t^2\sin t$;
   (m) $t\,e^{2t}$; **(n)** $t^2\,e^t$; **(o)** $t^2\sin 3t$; **(p)** $(t^2-1)\sin 2t$;
   (q) $t^2\cos 2t$; **(r)** $t^2\,e^{-3t}$.
2. Determine whether the given integral converges or diverges.
   (a) $\int_0^\infty (t^2+4)^{-1}\,\mathrm{d}t$; **(b)** $\int_0^\infty t^2\,e^{-2t}\,\mathrm{d}t$; **(c)** $\int_0^\infty t\cos 2t\,\mathrm{d}t$;
   (d) $\int_0^\infty \frac{\sin t}{2t}\,\mathrm{d}t$; **(e)** $\int_0^\infty \frac{1}{t+1}\,\mathrm{d}t$; **(f)** $\int_0^\infty \frac{\sqrt{t}}{t^2+1}\,\mathrm{d}t$.
3. Determine whether the given function is a function-original:
   (a) $\sin 2t$; **(b)** $e^t\cos 2t$; **(c)** $1/(1+t^2)$;
   (d) $\sin t^2$; **(e)** $t^{-1},\ t>0$; **(f)** $t^{-3},\ t>0$;
   (g) $\frac{\sin t}{t}$; **(h)** $\exp\left\{\sqrt{t^2+4}-t\right\}$; **(i)** $(1-\cos t)/t^2$;
   (j) $f(t) = \begin{cases} t, & 0\leqslant t\leqslant 1; \\ (1-t)^2+1, & 1<t<3; \\ 4, & 3\leqslant t; \end{cases}$ **(k)** $f(t) = \begin{cases} 1, & 0\leqslant t<1; \\ 2-t, & 1\leqslant t\leqslant 2; \\ t^2, & 2<t. \end{cases}$
4. Prove that if $f(t)$ is a function-original, then $\lim_{\lambda\to\infty} f^L(\lambda) = 0$.

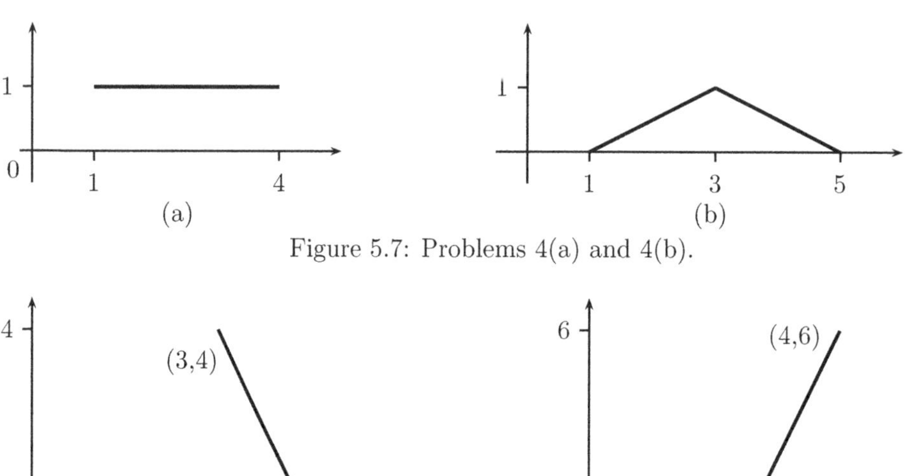

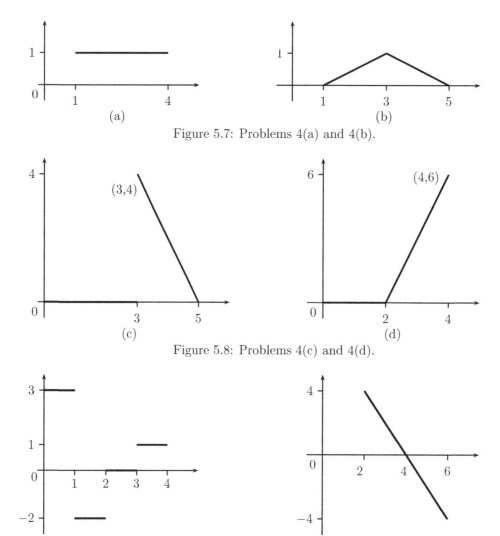

Figure 5.7: Problems 4(a) and 4(b).

4
(3,4)
0
3
5
(c)
6
(4,6)
0
2
4
(d)

Figure 5.8: Problems 4(c) and 4(d).

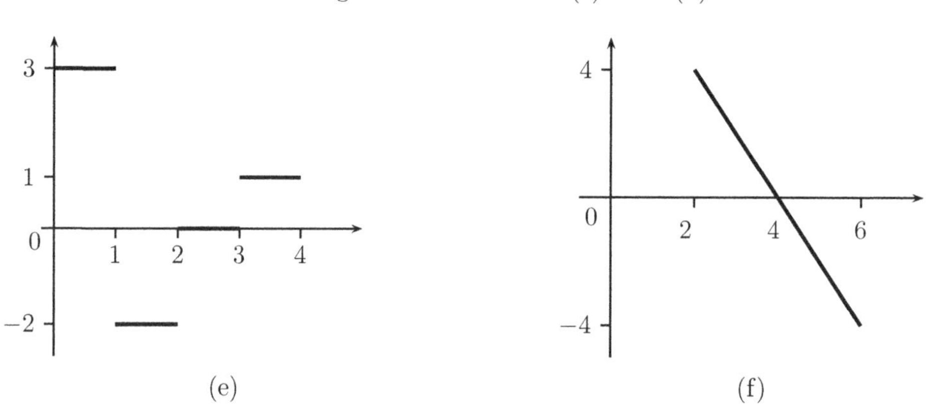

Figure 5.9: Problems 4(e) and 4(f).

5. The functions are depicted by the graphs in Fig. 5.7, 5.8, and 5.9, in which the rightmost segment extends to infinity. Construct an analytic formula for the function and obtain its Laplace transform.

6. Prove that if $f(t)$ and $g(t)$ are each of exponential order as $t \to \infty$, then $f(t) \cdot g(t)$ and $f(t) + g(t)$ are also of exponential order as $t \to \infty$.

7. Following Example 5.1.4 and using the known value $\Gamma(1/2) = \sqrt{\pi}$, find the Laplace transform of the following functions.

$$\text{(a)}\quad t^{1/2}; \qquad \text{(b)}\quad t^{-1/2}; \qquad \text{(c)}\quad t^{3/2}; \qquad \text{(d)}\quad t^{5/2}.$$

8. Which of the following functions are piecewise continuous on $[0, \infty)$ and which are not?

(a) $\ln(1 + t^2)$; **(b)** $\lfloor t \rfloor$, the greatest integer less than t;

(c) t^{-1}; **(d)** $f(t) = \begin{cases} 0, & t \text{ is an integer,} \\ 1, & \text{otherwise;} \end{cases}$

(e) $e^{1/t}$; **(f)** $f(t) = \begin{cases} 0, & t = 1/n, \quad n = 1, 2, \ldots, \\ 1, & \text{otherwise;} \end{cases}$

(g) $\sin \frac{1}{t}$; **(h)** $\dfrac{\sin t}{t}$.

Table 280: A table of elementary Laplace transforms. Note: each function in the left column is zero for negative t; that is, they must be multiplied by the Heaviside function $H(t)$. Here σ_c is the abscissa of convergence for the Laplace transform.

#	Function-original	Laplace transform	σ_c
1.	$H(t)$	$\dfrac{1}{\lambda}$	$\Re\lambda > 0$
2.	$H(t-a)$	$\dfrac{1}{\lambda}e^{-a\lambda}$	$\Re\lambda > 0$
3.	t	$\dfrac{1}{\lambda^2}$	$\Re\lambda > 0$
4.	$t^n, \ n = 1, 2, \ldots$	$\dfrac{n!}{\lambda^{n+1}}$	$\Re\lambda > 0$
5.	t^p	$\dfrac{\Gamma(p+1)}{\lambda^{p+1}}$	$\Re\lambda > 0$
6.	$e^{\alpha t}$	$\dfrac{1}{\lambda-\alpha}$	$\Re\lambda > \Re\alpha$
7.	$t^n\, e^{\alpha t}, \ n = 1, 2, \ldots$	$\dfrac{n!}{(\lambda-\alpha)^{n+1}}$	$\Re\lambda > \Re\alpha$
8.	$\sin\alpha t$	$\dfrac{\alpha}{\lambda^2+\alpha^2}$	$\Re\lambda > 0$
9.	$\cos\alpha t$	$\dfrac{\lambda}{\lambda^2+\alpha^2}$	$\Re\lambda > 0$
10.	$e^{\alpha t}\sin\beta t$	$\dfrac{\beta}{(\lambda-\alpha)^2+\beta^2}$	$\Re\lambda > \Re\alpha$
11.	$e^{\alpha t}\cos\beta t$	$\dfrac{\lambda-\alpha}{(\lambda-\alpha)^2+\beta^2}$	$\Re\lambda > \Re\alpha$
12.	$\sinh\beta t$	$\dfrac{\beta}{\lambda^2-\beta^2}$	$\Re\lambda > \Re\beta$
13.	$\cosh\beta t$	$\dfrac{\lambda}{\lambda^2-\beta^2}$	$\Re\lambda > \Re\beta$
14.	$t\sin\beta t$	$\dfrac{2\beta\lambda}{(\lambda^2+\beta^2)^2}$	$\Re\lambda > 0$
15.	$t\cos\beta t$	$\dfrac{\lambda^2-\beta^2}{(\lambda^2+\beta^2)^2}$	$\Re\lambda > 0$
16.	$e^{\alpha t}-e^{\beta t}$	$\dfrac{\alpha-\beta}{(\lambda-\alpha)(\lambda-\beta)}$	$\Re\lambda > \Re\alpha, \Re\beta$
17.	$e^{\alpha t}\left[\cos\beta t + \frac{\alpha}{\beta}\sin\beta t\right]$	$\dfrac{\lambda}{(\lambda-\alpha)^2+\beta^2}$	$\Re\lambda > \Re\alpha$
18.	$\frac{\sin\beta t}{2\beta^3} - \frac{t\cos\beta t}{2\beta^2}$	$\dfrac{1}{(\lambda^2+\beta^2)^2}$	$\Re\lambda > 0$
19.	$\frac{t\sin\beta t}{2\beta}$	$\dfrac{\lambda}{(\lambda^2+\beta^2)^2}$	$\Re\lambda > 0$
20.	$e^{\alpha t}\sinh\beta t$	$\dfrac{\beta}{(\lambda-\alpha)^2-\beta^2}$	$\Re\lambda > \Re(\alpha\pm\beta)$
21.	$e^{\alpha t}\cosh\beta t$	$\dfrac{\lambda-\alpha}{(\lambda-\alpha)^2-\beta^2}$	$\Re\lambda > \Re(\alpha\pm\beta)$

5.2 Properties of the Laplace Transform

The success of transformation techniques in solving initial value problems and other applications hinges on their *operational properties.* Rules that govern how operations in the time domain translate to operations in their transformation images are called *operational laws* or *rules.* In this section we present the 10 basic rules that are useful in applications of the Laplace transformation to differential equations. The justifications of these laws involve technical detail and require scrutiny that is beyond the scope of our text—we simply point to the books [8, 12, 47]. Let us start with the following definition.

Definition 5.7: The **convolution** $f * g$ of two functions f and g, defined on the positive half-line $[0, \infty)$, is the integral

$$(f * g)(t) = \int_0^t f(t-\tau) g(\tau)\,\mathrm{d}\tau = (g * f)(t).$$

Example 5.2.1: The convolution of two unit constants (which are actually two Heaviside functions) is

$$(H * H)(t) = \int_0^t H(t-\tau) H(\tau)\,\mathrm{d}\tau = \int_0^t \mathrm{d}\tau = t, \qquad t \geqslant 0. \qquad \square$$

Now we list the properties of the Laplace transform. All considered functions are assumed to be function-originals.

1° **The convolution rule**

The Laplace transform of the convolution of two functions is equal to the product of its images:

$$\mathcal{L}(f * g)(\lambda) = f^L(\lambda) g^L(\lambda). \tag{5.2.1}$$

PROOF: A short manipulation gives

$$\begin{aligned}\mathcal{L}(f * g)(\lambda) &= \int_0^\infty e^{-\lambda t}\,\mathrm{d}t \int_0^t f(t-\tau) g(\tau)\,\mathrm{d}\tau \\ &= \int_0^\infty g(\tau)\, e^{-\lambda\tau}\,\mathrm{d}\tau \int_\tau^\infty e^{\lambda\tau-\lambda t} f(t-\tau)\,\mathrm{d}(t-\tau) = g^L(\lambda) f^L(\lambda).\end{aligned}$$

2° **The derivative rule**

$$\mathcal{L}\left[f^{(n)}(t)\right](\lambda) = \lambda^n \mathcal{L}f(\lambda) - \sum_{k=1}^{n} \lambda^{n-k} f^{(k-1)}(+0). \tag{5.2.2}$$

Integration by parts gives us the equality (5.2.2). In particular,

$$\mathcal{L}\left[f'(t)\right](\lambda) = \lambda\, f^L(\lambda) - f(0); \tag{5.2.3}$$

$$\mathcal{L}\left[f''(t)\right](\lambda) = \lambda^2 f^L(\lambda) - \lambda f(0) - f'(0). \tag{5.2.4}$$

3° **The similarity rule**

$$\mathcal{L}[f(at)](\lambda) = \frac{1}{a} f^L\left(\frac{\lambda}{a}\right), \qquad \Re\lambda > a\sigma_c, \quad a > 0. \tag{5.2.5}$$

4° **The shift rule**

If we know $g^L(\lambda)$, the Laplace transform of $g(t)$, then the retarded function $f(t) = g(t-a)H(t-a)$ has the Laplace transform $g^L(\lambda)e^{-a\lambda}$, namely,

$$\mathcal{L}\left[H(t-a) g(t-a)\right](\lambda) = e^{-a\lambda} g^L(\lambda), \quad a > 0. \tag{5.2.6}$$

5° **The attenuation rule**

$$\mathcal{L}\left[e^{-at} f(t)\right](\lambda) = f^L(\lambda + a). \tag{5.2.7}$$

6° The integration rule

$$\mathcal{L}[t^{n-1} * f(t)](\lambda) = \mathcal{L} \int_0^t (t-\tau)^{n-1} f(\tau)\,\mathrm{d}\tau = \frac{(n-1)!}{\lambda^n} f^L(\lambda), \quad n = 1, 2, \ldots. \tag{5.2.8}$$

If $n = 1$, then

$$\frac{1}{\lambda} f^L(\lambda) = \mathcal{L} \int_0^t f(\tau)\,\mathrm{d}\tau. \tag{5.2.9}$$

PROOF: These equations (5.2.8) and (5.2.9) are consequences of the convolution rule and the equality

$$(\mathcal{L}t^p)(\lambda) = \int_0^\infty t^p\, e^{-\lambda t}\,\mathrm{d}t = \frac{\Gamma(p+1)}{\lambda^{p+1}}, \tag{5.2.10}$$

where $\Gamma(\nu)$ is Euler's gamma function (5.1.3), page 274. If ν is an integer (that is, $\nu = n$), then $\Gamma(n+1) = n!$ $n = 0, 1, 2, \ldots$. This relation has been proved with integration by parts in §5.1, page 274. Since the gamma function (see Fig. 5.2 on page 272) has the same value at $\nu = 1$ and at $\nu = 2$, that is, $\Gamma(1) = \Gamma(2) = 1$, we get $0! = 1! = 1$.

7° Rule for multiplicity by t^n or **the derivative of a Laplace transform**

$$\frac{\mathrm{d}^n}{\mathrm{d}\lambda^n} f^L(\lambda) = \mathcal{L}[(-1)^n t^n f(t)] \quad (n = 0, 1, \ldots). \tag{5.2.11}$$

8° Rule for division by t

$$\mathcal{L}\left[\frac{f(t)}{t}\right] = \int_\lambda^\infty f^L(\sigma)\,\mathrm{d}\sigma. \tag{5.2.12}$$

9° The Laplace transform of periodic functions

If $f(t) = f(t+\omega)$, then

$$f^L(\lambda) = \frac{1}{1 - e^{-\omega\lambda}} \int_0^\omega e^{-\lambda t} f(t)\,\mathrm{d}t. \tag{5.2.13}$$

10° The Laplace transform of anti-periodic functions

If $f(t) = -f(t+\omega)$, then

$$f^L(\lambda) = \frac{1}{1 + e^{-\omega\lambda}} \int_0^\omega e^{-\lambda t} f(t)\,\mathrm{d}t. \tag{5.2.14}$$

Theorem 5.5: If $f(t)$ is a continuous function such that $f'(t)$ is a function-original, then

$$f(0) = \lim_{\lambda \mapsto \infty} \lambda f^L(\lambda), \qquad f^L(\lambda) = \int_0^\infty f(t)\, e^{-\lambda t}\,\mathrm{d}t.$$

Remark 1. We can unite (5.2.5) and (5.2.7) into one formula:

$$\mathcal{L}\left[\frac{1}{a} e^{-bt/a} f\left(\frac{t}{a}\right)\right](\lambda) = f^L(a\lambda + b). \tag{5.2.15}$$

Remark 2. We do not formulate exact requirements on functions to guarantee the validity of each rule. A curious reader should consult [12, 47].

Example 5.2.2: (Convolution rule) The convolution of the function $f(t) = tH(t)$ with itself is

$$\begin{aligned} f * f = tH(t) * tH(t) &= \int_0^t \tau\,(t-\tau)\,\mathrm{d}\tau = t \int_0^t \tau\,\mathrm{d}\tau - \int_0^t \tau^2\,\mathrm{d}\tau \\ &= t\, \frac{\tau^2}{2}\bigg|_{\tau=0}^{\tau=t} - \frac{\tau^3}{3}\bigg|_{\tau=0}^{\tau=t} = \frac{t^3}{2} - \frac{t^3}{3} = \frac{t^3}{6} \qquad \text{for } t > 0. \end{aligned}$$

According to the convolution rule (5.2.1), we get its Laplace transform:

$$\mathcal{L}\,[t * t] = \mathcal{L}\left[\frac{t^3}{6}\right] = \mathcal{L}[t] \cdot \mathcal{L}[t] = \frac{1}{\lambda^2} \cdot \frac{1}{\lambda^2} = \frac{1}{\lambda^4}.$$

Example 5.2.3: (Convolution rule) The convolution product of the shifted Heaviside function $H(t-a)$ and t^2 is

$$H(t-a)*t^2 = \int_0^t H(t-\tau-a)\,\tau^2\,\mathrm{d}\tau = H(t-a)\int_0^{t-a}\tau^2\,\mathrm{d}\tau = H(t-a)\,\frac{(t-a)^3}{3}.$$

The same result is obtained with the convolution theorem:

$$\mathcal{L}\left[H(t-a)*t^2\right] = \mathcal{L}\left[H(t-a)\right]\cdot\mathcal{L}[t^2] = \frac{1}{\lambda}\,e^{-a\lambda}\cdot\frac{2}{\lambda^3} = \frac{2}{\lambda^4}\,e^{-a\lambda}.$$

The Laplace transformation is an excellent tool to determine convolution integrals. While we did not study the inverse Laplace transform, which is the topic of §5.4, we can find their inverses based on Table 280. Application of the shift rule (5.2.6) yields

$$\mathcal{L}^{-1}\left[\frac{2}{\lambda^4}\,e^{-a\lambda}\right] = H(t-a)\,\frac{2(t-a)^3}{3!} = H(t-a)\,\frac{(t-a)^3}{3}.$$

Example 5.2.4: (Derivative rule) Find the Laplace transform of $\cos at$.

Solution. We know that $-a^2\cos at = \mathrm{d}^2(\cos at)/\mathrm{d}t^2$. Let $f^L = \mathcal{L}\left[\cos at\right](\lambda)$ denote the required Laplace transform of $\cos at$. Using the initial values of sine and cosine functions, the derivative rule (5.2.2) yields

$$\mathcal{L}\left[-a^2\cos at\right](\lambda) = -a^2 f^L = \mathcal{L}\left[\frac{\mathrm{d}^2}{\mathrm{d}t^2}\,\cos at\right] = \lambda^2 f^L - \lambda.$$

Solving for f^L, we get $f^L = \mathcal{L}\left[\cos at\right](\lambda) = \lambda/(a^2+\lambda^2)$.

Example 5.2.5: (Similarity rule) Find the Laplace transform of $\cos\alpha t$.

Solution. We know that $\mathcal{L}[\cos t](\lambda) = \lambda(\lambda^2+1)^{-1}$. Then, using the similarity rule (5.2.5), we obtain

$$\mathcal{L}[\cos\alpha t](\lambda) = \frac{1}{\alpha}\,\frac{\dfrac{\lambda}{\alpha}}{1+\dfrac{\lambda^2}{\alpha^2}} = \frac{\lambda}{\lambda^2+\alpha^2}.$$

Now we find the same transform using the periodic property $\cos(t) = -\cos(t+\pi)$ for $0<t<\pi$. Indeed, Eq. (5.2.14) gives

$$\begin{aligned}\mathcal{L}[\cos t](\lambda) &= \frac{1}{1+e^{-\pi\lambda}}\int_0^{\pi} e^{-\lambda t}\,\cos(t)\,\mathrm{d}t \\ &= \frac{1}{1+e^{-\pi\lambda}}\,\frac{\lambda}{\lambda^2+1}\,\left(1+e^{-\lambda\pi}\right) = \frac{\lambda}{\lambda^2+1}.\end{aligned}$$

Example 5.2.6: Figure 5.10 shows the graphs of the functions $f_1(t) = H(t-a)\sin t$ and $f_2(t) = H(t-a)\sin(t-a)$.

These two functions f_1 and f_2 have different Laplace transforms. Indeed, using the trigonometric identity $\sin t = \sin(t-a+a) = \sin(t-a)\,\cos a + \cos(t-a)\,\sin a$, we obtain

$$f_1^L(\lambda) = \mathcal{L}\left\{H(t-a)\left[\sin(t-a)\,\cos a + \cos(t-a)\,\sin a\right]\right\} = \frac{e^{-\lambda a}}{\lambda^2+1}\,\left[\lambda\sin a + \cos a\right],$$

whereas $f_2^L(\lambda) = \frac{e^{-\lambda a}}{\lambda^2+1}$.

Example 5.2.7: (Shift rule) Evaluate $\mathcal{L}[H(t-a)](\lambda)$ (see graph of $H(t-a)$ on page 275). Then find the Laplace transform of the product $H(t-b)\cos\alpha(t-b)$.

Solution. The shift rule (5.2.6) gives

$$\mathcal{L}[H(t-a)](\lambda) = e^{-a\lambda}H^L(\lambda) = \frac{1}{\lambda}\,e^{-a\lambda}.$$

We again use the shift rule to obtain

$$\mathcal{L}[H(t-b)\cos\alpha(t-b)] = e^{-b\lambda}\,\frac{\lambda}{\lambda^2+\alpha^2}.$$

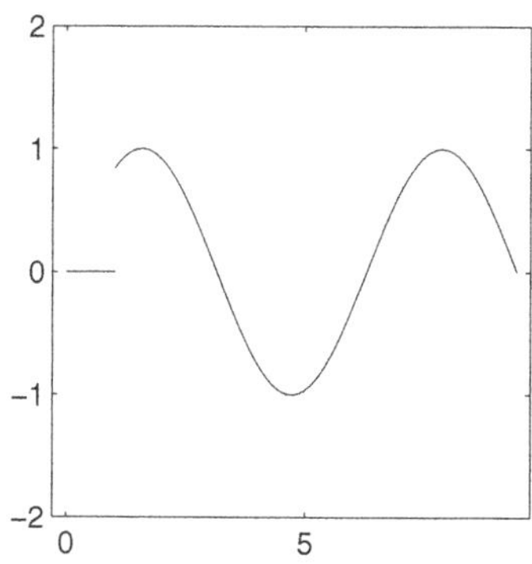

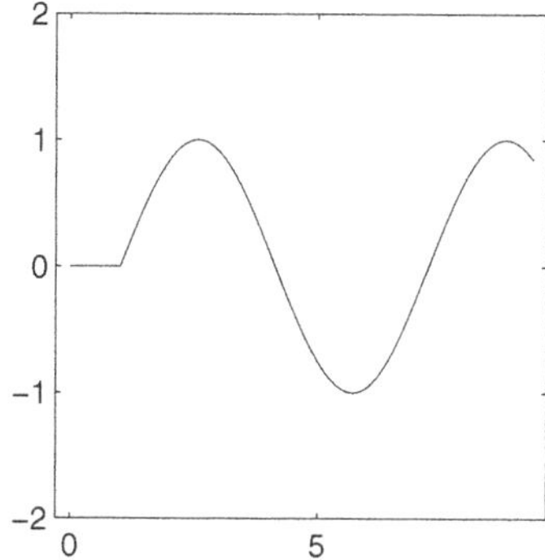

Figure 5.10: Example 5.2.6. The graph of the function $H(t-a)\sin t$ and $H(t-a)\sin(t-a)$, with $a=1$, plotted in MATLAB®.

Example 5.2.8: (Attenuation rule) Find the Laplace transform of $\cosh 2t$.

Solution. Using the attenuation rule Eq. (5.2.7), we have

$$\begin{aligned}\mathcal{L}\left[\cosh 2t\right] &= \mathcal{L}\left[\frac{1}{2}\,e^{2t}+\frac{1}{2}\,e^{-2t}\right]=\frac{1}{2}\,\mathcal{L}\left[e^{2t}\right]+\frac{1}{2}\,\mathcal{L}\left[e^{-2t}\right]\\ &= \frac{1}{2}\,\frac{1}{\lambda-2}+\frac{1}{2}\,\frac{1}{\lambda+2}=\frac{\lambda}{\lambda^2-4}.\end{aligned}$$

Example 5.2.9: (Attenuation rule) Find the Laplace transform of $t^3\sin 3t$.

Solution. We know that $\sin\theta=\Re\,e^{\mathbf{j}\theta}$ is the imaginary part of $e^{\mathbf{j}\theta}$, then the attenuation rule yields

$$\begin{aligned}\mathcal{L}\left[t^3\sin 3t\right] &= \mathcal{L}\left[t^3\,\Im e^{\mathbf{j}3t}\right]=\Im\mathcal{L}\left[t^3\,e^{\mathbf{j}3t}\right]\\ &= \Im\,\frac{3!}{(\lambda-3\mathbf{j})^4}=3!\,\Im\,\frac{(\lambda+3\mathbf{j})^4}{(\lambda-3\mathbf{j})^4(\lambda+3\mathbf{j})^4}=\frac{3!}{(\lambda^2+9)^4}\,\Im\,(\lambda+3\mathbf{j})^4\\ &= \frac{3!}{(\lambda^2+9)^4}\,\Im\,\left[\lambda^4+4\cdot\lambda^3(3\mathbf{j})+6\cdot\lambda^2\,(3\mathbf{j})^2+4\cdot\lambda(3\mathbf{j})^3+(3\mathbf{j})^4\right]\\ &= \frac{3!}{(\lambda^2+9)^4}\,\left[4\lambda^3\cdot 3-4\lambda\,3^3\right]=\frac{3!\,4\cdot 3}{(\lambda^2+9)^4}\,[\lambda^3-9\lambda].\end{aligned}$$

The same result can be obtained with the aid of the multiplication rule (5.2.11):

$$\mathcal{L}\left[t^3\,\sin 3t\right]=-\frac{\mathrm{d}^3}{\mathrm{d}\lambda^3}\mathcal{L}\left[\sin 3t\right]=-\frac{\mathrm{d}^3}{\mathrm{d}\lambda^3}\frac{3}{\lambda^2+9}=\frac{72\lambda(\lambda^2-9)}{(\lambda^2+9)^4}.$$

Example 5.2.10: (Integration rule) Find the Laplace transform of the integral

$$\int_0^t (t-\tau)\,\sin 2\tau\,\mathrm{d}\tau=\frac{t}{2}-\frac{1}{4}\,\sin 2t.$$

Solution. Using the integration rule (5.2.8), we obtain

$$\mathcal{L}\left[\int_0^t (t-\tau)\,\sin 2\tau\,\mathrm{d}\tau\right](\lambda)=\frac{\mathcal{L}[\sin 2t](\lambda)}{\lambda^2}=\frac{2}{\lambda^2+4}\cdot\frac{1}{\lambda^2}.$$

This integral is actually the convolution of two functions: $t*\sin 2t$. So, its Laplace transform is the product of Laplace transformations of the multipliers: $\frac{1}{\lambda^2}=\mathcal{L}[t]$ and $\frac{2}{\lambda^2+4}=\mathcal{L}[\sin 2t]$.

Example 5.2.11: (Multiplication by the t^n rule) Find the Laplace transform of $t\cos 2t$.

Solution. Using Eq. (5.2.11), we find its Laplace transform to be

$$\begin{aligned}
\mathcal{L}\left[t\,\cos 2t\right] &= \int_0^\infty t\,\cos 2t\, e^{-\lambda t}\,\mathrm{d}t = -\int_0^\infty \cos 2t\,\frac{\mathrm{d}}{\mathrm{d}\lambda}\left(e^{-\lambda t}\right)\,\mathrm{d}t \\
&= -\frac{\mathrm{d}}{\mathrm{d}\lambda}\int_0^\infty \cos 2t\, e^{-\lambda t}\,\mathrm{d}t = -\frac{\mathrm{d}}{\mathrm{d}\lambda}\,\mathcal{L}\left[\cos 2t\right] = -\frac{\mathrm{d}}{\mathrm{d}\lambda}\,\frac{\lambda}{\lambda^2+4} \\
&= -\left[\frac{\lambda^2+4-\lambda\cdot 2\lambda}{(\lambda^2+4)^2}\right] = \frac{\lambda^2-4}{(\lambda^2+4)^2}.
\end{aligned}$$

Alternatively, using Euler's formula (5.1.10), we have $\cos 2t = \Re\, e^{\mathbf{j}2t} = \mathrm{Re}\, e^{\mathbf{j}2t}$. Hence

$$\mathcal{L}\left[t\,\cos 2t\right] = \Re\,\mathcal{L}\left[t\,e^{\mathbf{j}2t}\right] \quad \text{and} \quad \mathcal{L}\left[t\right] = \frac{1}{\lambda^2}.$$

From the attenuation rule, it follows that

$$\begin{aligned}
\mathcal{L}\left[t\,\cos 2t\right] &= \Re\,\frac{1}{(\lambda-2\mathbf{j})^2} = \Re\,\frac{(\lambda+2\mathbf{j})^2}{(\lambda-2\mathbf{j})^2(\lambda+2\mathbf{j})^2} \\
&= \frac{1}{(\lambda^2+4)^2}\,\Re\,(\lambda+2\mathbf{j})^2 = \frac{1}{(\lambda^2+4)^2}\,\Re\,(\lambda^2+4\lambda\mathbf{j}-4) = \frac{\lambda^2-4}{(\lambda^2+4)^2}.
\end{aligned}$$

Example 5.2.12: (Periodic function) Find the Laplace transform of $\sin 2t$.

Solution. To find $\mathcal{L}[\sin 2t](\lambda)$, we apply the rule (5.2.13) because the sine is a periodic function with period 2π. The function $\sin 2t$ has the period π, that is, $\sin 2t = \sin 2(t+\pi)$. Hence,

$$\mathcal{L}[\sin 2t](\lambda) = \frac{1}{1-e^{-\pi\lambda}}\int_0^\pi e^{-\lambda t}\,\sin 2t\,dt.$$

The value of the integral can be determined using Euler's identity $\sin\theta = \Im e^{\mathbf{j}\theta}$, namely, $\sin\theta$ is the imaginary part of $e^{\mathbf{j}\theta}$:

$$\begin{aligned}
\int_0^\pi e^{-\lambda t}\,\sin 2t\,\mathrm{d}t &= \int_0^\pi e^{-\lambda t}\,\Im e^{2\mathbf{j}t}\,\mathrm{d}t = \Im\int_0^\pi e^{-\lambda t+2\mathbf{j}t}\,\mathrm{d}t \\
&= \Im\,\frac{e^{-\lambda t+2\mathbf{j}t}}{-\lambda+2\mathbf{j}}\bigg|_{t=0}^{t=\pi} = -\Im\left(\frac{e^{-\lambda\pi+2\mathbf{j}\pi}}{\lambda-2\mathbf{j}}+\frac{1}{\lambda-2\mathbf{j}}\right).
\end{aligned}$$

Multiplying the numerator and denominator by $\lambda+2\mathbf{j}$, the complex conjugate of $\lambda-2\mathbf{j}$, and using the identity $(\lambda+2\mathbf{j})(\lambda-2\mathbf{j}) = \lambda^2+4$, we obtain

$$\frac{1}{\lambda-2\mathbf{j}} - \frac{e^{-\lambda\pi+2\mathbf{j}\pi}}{\lambda-2\mathbf{j}} = \frac{\lambda+2\mathbf{j}}{\lambda^2+4}\left[1-e^{-\lambda\pi-2\mathbf{j}\pi}\right] = \frac{\lambda+2\mathbf{j}}{\lambda^2+4}\left[1-e^{-\lambda\pi}\right]$$

since $e^{2\pi\mathbf{j}} = e^{-2\pi\mathbf{j}} = 1$. Taking the imaginary part, we get

$$\mathcal{L}[\sin t](\lambda) = \frac{1}{1-e^{-\pi\lambda}}\,\frac{2}{\lambda^2+4}\,\left[1-e^{-\pi\lambda}\right] = \frac{2}{\lambda^2+4}.$$

Example 5.2.13: (Periodic function) Find the Laplace transforms of the periodic functions

$$f_a(t) = H(t) - H(t-a) + H(t-2a) - H(t-3a) + \cdots.$$

and

$$g(t) = \sin 2t f_{\pi/2}(t) = \sin 2t\sum_{k=0}^\infty (-1)^k\, H(t-\pi/2).$$

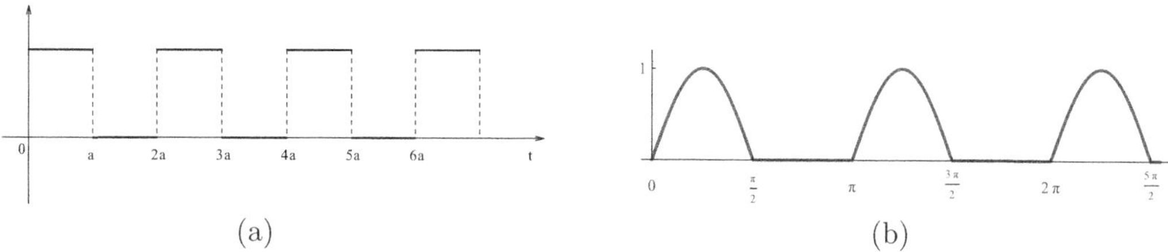

Figure 5.11: Example 5.2.13: (a) the piece-wise continuous function $f_a(t)$ (also called the square-wave function), and (b) the function $g(t) = \sin 2t f_{\pi/2}(t)$, plotted with *Mathematica.*

Solution. The function $f_a(t)$ has a period equal to $\omega = 2a$; therefore, we can apply Eq. (5.2.13) to find the Laplace transform of a periodic function. In fact, we have

$$\begin{aligned} f_a^L(\lambda) &= \frac{1}{\lambda} - \frac{1}{\lambda} e^{-a\lambda} + \frac{1}{\lambda} e^{-2a\lambda} - \frac{1}{\lambda} e^{-3a\lambda} + \cdots \\ &= \frac{1}{1-e^{-\omega\lambda}} \int_0^{\omega} e^{-\lambda t} f_a(t)\,\mathrm{d}t = \frac{1}{1-e^{-2a\lambda}} \int_0^a e^{-\lambda t}\,\mathrm{d}t \\ &= \frac{1-e^{-a\lambda}}{\lambda\,(1-e^{-2a\lambda})} = \frac{1-e^{-a\lambda}}{\lambda\,(1-e^{-a\lambda})\,(1+e^{-a\lambda})} = \frac{1}{\lambda\,(1+e^{-\lambda a})}. \end{aligned}$$

In turn, the function $g(t)$ has a period π and is zero for $t \in \left[\frac{\pi}{2}, \pi\right]$. Thus,

$$\mathcal{L}[g(t)](\lambda) = \frac{1}{1-e^{-\pi\lambda}} \int_0^{\pi/2} e^{-\lambda t}\,\sin(2t)\,\mathrm{d}t = \frac{2}{\lambda^2+4}\,\frac{1+e^{-\lambda\pi/2}}{1-e^{-\lambda\pi/2}}.$$

In the general case for the periodic function

$$f_\omega(t) = \begin{cases} \sin\omega t, & 0 \leqslant t \leqslant \pi/\omega, \\ 0, & \pi/\omega \leqslant t \leqslant 2\pi/\omega, \end{cases}$$

with the period $2\pi/\omega$, we can obtain its Laplace transform as follows:

$$\begin{aligned} f^L(\lambda) &= \frac{1}{1-e^{-2\pi\lambda/\omega}} \int_0^{\pi/\omega} \sin\omega t\, e^{-\lambda t}\,\mathrm{d}t \\ &= \frac{\omega}{\lambda^2+\omega^2}\,\frac{1}{1-e^{-2\pi\lambda/\omega}} \left[1+e^{-\pi\lambda/\omega}\right]. \end{aligned}$$

You can handle this function with a computer algebra system. For instance, in *Mathematica* it can be done as follows:

```
SawTooth[t_] := 2 t - 2 Floor[t] -1;
TriangularWave[t_] := Abs[2 SawTooth[(2 t -1)/4] -1;
SquareWave[t_] := Sign[ TriangularWave[t] ];
Plot[{ SawTooth[t], TriangularWave[t], SquareWave[t]}, {t, 0, 10}]
```

The function $g(t)$ can be plotted in *Mathematica* by executing the following code:

```
g[x_] := If[FractionalPart[x/Pi] < 0.5, Sin[2 x], 0]
Plot[g[t], {t, 0, 8}, PlotRange -> {0, 1.2}, AspectRatio -> 1/5,
 Ticks -> {Pi/2 Range[0, 8], {1}}, PlotStyle -> Thickness[0.006]]
```

Example 5.2.14: Figure 5.12(a) shows a pulsed periodic function $f(t)$ with a pulse repetition period $\tau = 4 + \pi$. Find its Laplace transform.

Solution. Thus, on the interval $[a, a+\tau]$, this function is defined as

$$f_a(t) = \begin{cases} \sin k(t-a), & a \leqslant t \leqslant a+\pi, \\ 0, & a+\pi \leqslant t \leqslant a+\tau. \end{cases}$$

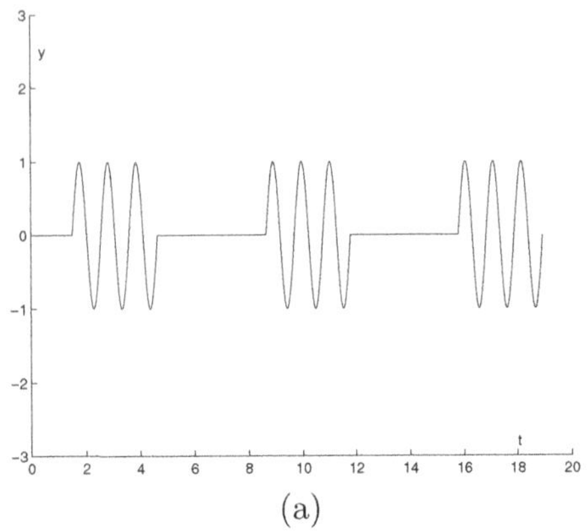

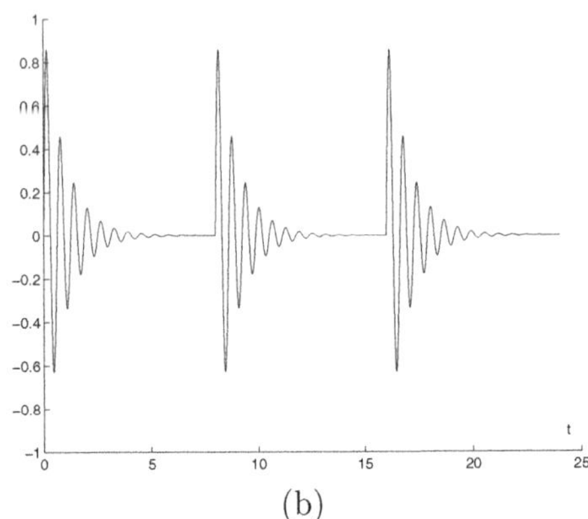

Figure 5.12: (a) Example 5.2.14: graph of the pulsed periodic function; (b) Example 5.2.15: graph of the damped oscillations $f(t) = e^{-\alpha t} \sin \omega t$ repeated every τ seconds; here $\alpha = 1.0$, $\omega = 10$, and $\tau = 8.0$, plotted with MATLAB.

To find the Laplace transform of this function, first we find the transformation of the function $f_0(t)$ when $a = 0$ and then we apply the shift rule (5.2.6). The function $f_0(t)$ is a periodic function with the period τ. Therefore, Eq. (5.2.13) yields

$$\begin{aligned} f_0^L(\lambda) &= \frac{1}{1-e^{-\tau\lambda}} \int_0^{\tau} f_0(t)\, e^{-\lambda t}\, dt = \frac{1}{1-e^{-\tau\lambda}} \int_0^{\pi} \sin kt\, e^{-\lambda t}\, dt \\ &= \frac{1}{1-e^{-\tau\lambda}} \frac{(-1)^{k+1}}{\lambda^2+k^2}\, e^{-\lambda\pi} \end{aligned}$$

if k is an integer. The shift rule gives us the general formula, that is,

$$f_a^L(\lambda) = \frac{(-1)^{k+1}}{\lambda^2+k^2} \frac{1}{1-e^{-\tau\lambda}}\, e^{-\lambda(\pi+a)}.$$

Example 5.2.15: Figure 5.12(b) shows a series of damped oscillations $f(t) = e^{-\alpha t} \sin \omega t$ repeated every τ sec. Assuming that α is large enough, show that its Laplace transform is

$$f^L(\lambda) = \left[\frac{\omega}{(\lambda+\alpha)^2+\omega^2}\right] \frac{1}{1-e^{-\tau\lambda}}.$$

Solution. According to Eq. (5.2.13) on page 283, we have

$$f^L(\lambda) = \frac{1}{1-e^{-\tau\lambda}} \int_0^{\tau} e^{-\alpha t} \sin \omega t\, e^{-\lambda t}\, dt = \left[\frac{\omega}{(\lambda+\alpha)^2+\omega^2}\right] \frac{1}{1-e^{-\tau\lambda}}$$

since the function $f(t)$ has the period τ.

Example 5.2.16: Figure 5.13 shows an anti-periodic function $f(t) = -f(t+a)$ defined as

$$f(t) = \begin{cases} \frac{E}{a}\, t, & \text{if } 0 < t < a; \\ -\frac{E}{a}\, t + E, & \text{if } a < t < 2a; \\ f(t) = f(t+2a), & \text{for all positive } t. \end{cases}$$

The Laplace transform of this function can be obtained according to Eq. (5.2.14) on page 283, that is,

$$\begin{aligned} f^L(\lambda) &= \frac{1}{1+e^{-a\lambda}} \int_0^{a} e^{-\lambda t} f(t)\, dt = \frac{1}{1+e^{-a\lambda}} \cdot \frac{E}{a} \int_0^{a} e^{-\lambda t}\, t\, dt \\ &= \frac{E}{a\lambda^2} \cdot \frac{1-e^{-a\lambda}}{1+e^{-a\lambda}} - \frac{E}{(1+e^{-a\lambda})\,\lambda}\, e^{-a\lambda}. \end{aligned}$$

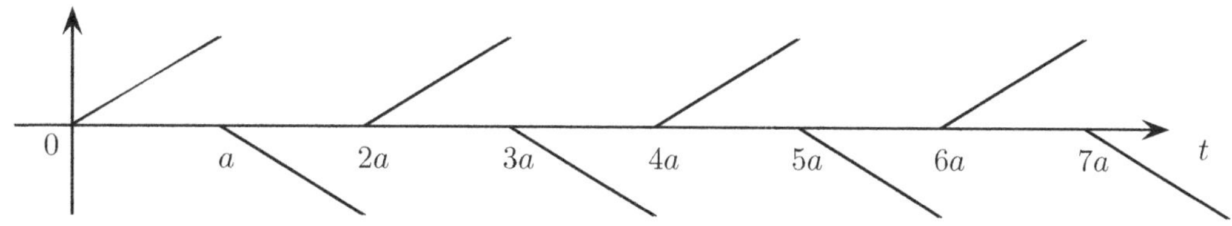

Figure 5.13: Example 5.2.16. Anti-periodic function.

Example 5.2.17: To find the Laplace transform of $\frac{\sin 3t}{t}$, we expand it into the Taylor series:

$$\frac{\sin 3t}{t} = \frac{1}{t}\sum_{n=0}^{\infty}(-1)^n \frac{(3t)^{2n+1}}{(2n+1)!} = \sum_{n=0}^{\infty}(-1)^n \frac{t^{2n}\,3^{2n+1}}{(2n+1)!}.$$

Interchanging the order of integration and summation, we obtain

$$\mathcal{L}\left[\frac{\sin 3t}{t}\right](\lambda) = \int_0^{\infty} \frac{\sin 3t}{t}\, e^{-\lambda t}\, \mathrm{d}t = \sum_{n=0}^{\infty}\int_0^{\infty} (-1)^n \frac{(3t)^{2n}}{(2n+1)!}\, e^{-\lambda t}\, \mathrm{d}t.$$

The fourth formula in Table 280 provides $\mathcal{L}\left[t^{2n}\right] = \frac{(2n)!}{\lambda^{2n+1}}$. Thus,

$$\mathcal{L}\left[\frac{\sin 3t}{t}\right](\lambda) = \sum_{n=0}^{\infty}(-1)^n \frac{(2n)!}{\lambda^{2n+1}}\,\frac{3^{2n+1}}{(2n+1)!} = \sum_{n=0}^{\infty}(-1)^n \frac{3^{2n+1}}{\lambda^{2n+1}(2n+1)}$$

since $(2n+1)! = (2n+1)\cdot(2n)!$.

Problems

1. Find the Laplace transform of the following functions.
   (a) $4t-2$; **(b)** t^6; **(c)** $\cos(2t-1)$;
   (d) $4t^2+\sin 2t$; **(e)** e^{2t+1}; **(f)** $\sin^2(2t-1)$;
   (g) $\cos^2(5t-1)$; **(h)** $(2t+3)^3$; **(i)** $2-4\,e^{4t}$;
   (j) $(2t-1)^3$; **(k)** $|\cos 2t|$; **(l)** $\sin 2t\,\cos 4t$.
2. Find the Laplace transform of the derivatives
   (a) $\frac{\mathrm{d}}{\mathrm{d}t}\left[t\,e^{2t}\right]$; **(b)** $\frac{\mathrm{d}}{\mathrm{d}t}\left[t^2\,e^{-t}\right]$; **(c)** $\frac{\mathrm{d}}{\mathrm{d}t}\left[\cos 2t - t\,e^{t}\right]$;
   (d) $\frac{\mathrm{d}}{\mathrm{d}t}\left[t^2\,e^{3t}\right]$; **(e)** $\frac{\mathrm{d}}{\mathrm{d}t}\left[t\,\sin 2t\right]$; **(f)** $\frac{\mathrm{d}}{\mathrm{d}t}\left[e^{-t}\,\cos 3t\right]$.
3. Use the attenuation rule and/or the similarity rule to find the Laplace transformations of the following functions.
   (a) $e^{2t}\,t^2$; **(b)** $e^{2t}\,t^3$; **(c)** $t\,e^{-4t}$;
   (d) $e^{t}\,\sinh t$; **(e)** $t(e^{-t}-e^{2t})^2$; **(f)** $e^{-10t}\,t$;
   (g) $e^{2t}\,\cosh t$; **(h)** $e^{t}\,\sinh t$; **(i)** $e^{-t}\,\sin 2t$;
   (j) $e^{-10t}\,t$; **(k)** $e^{2t}\,(5-2t+\cos t)$; **(l)** $e^{2t}\,(1+t)^2$;
   (m) $e^{-2t}\,\sin 4t$; **(n)** $(e^{-t}+3e^{2t})\,\sin t$; **(o)** $\cos(2t+1)$;
   (p) $\sin(3t-1)$; **(q)** $\sinh(3t+1)$; **(r)** $e^{2t}\,\cosh(3t+2)$.
4. Use the rules (5.2.2) and (5.2.11) to find the Laplace transform of the function $t\,\frac{\mathrm{d}}{\mathrm{d}t}\left(e^{t}\,\sin 2t\right)$.
5. Use the shift rule to determine the Laplace transform of the following functions.
   (a) $f(t-\pi)$ where $f(t) = \cos t\,H(t)$; (b) $f(t-1)$ where $f(t) = t\,H(t)$.
6. Use the rule (5.2.11) to find the Laplace transform of the following functions.
   (a) $t\,e^{2t}$; **(b)** $t^2\,\sin 2t$; **(c)** $t^2\,\cos(3t)$;
   (d) $t\,\sin(2t+7)$; **(e)** $t\,e^{2t-1}$; **(f)** $t^2\,\sin(2t-1)$;
   (g) $t^2\,e^{-2t}$; **(h)** $t^2\,\sin t$; **(i)** $t\,\cos(2t)$;
   (j) $t\,\sinh(2t)$; **(k)** $t^2\,\cosh(3t)$; **(l)** $t^3\,e^{-3t}$;
   (m) $-t\,e^{4t}\,\cosh(2t)$; **(n)** $t^3-t^4+t^6$; **(o)** $t\,e^{2t}\,\sinh(3t)$.

7. Find the Laplace transform of the following functions:

(a) $\frac{1}{t}\left(e^{2t}-1\right)$; (b) $\frac{1}{t^2}\left(\cos(2t)-1\right)$; (c) $\frac{1}{t}\left(2(1-\cos t)\right)$;

(d) $\frac{1}{t}\sin 2t$; (e) $\frac{1}{t}\left(\cosh 2t-1\right)$; (f) $\frac{1}{t}\left(\cos t-\cosh t\right)$.

8. Find the Laplace transform of the following periodic functions:

(a) $f(t)=\begin{cases}\sin t, & \text{if } 0<t<\pi,\\ -\sin t, & \text{if } \pi<t<2\pi;\end{cases}$ and $f(t)\equiv f(t+2\pi)$ for all positive t.

(b) $f(t)=\begin{cases}\cos t, & \text{if } 0<t<\pi,\\ 0, & \text{if } \pi<t<2\pi;\end{cases}$ and $f(t)\equiv f(t+2\pi)$ for all positive t.

(c) $f(t)=\begin{cases}1, & 0\leqslant t<1,\\ 0, & 1\leqslant t<2,\\ -1, & 2\leqslant t<3,\\ 0, & 3\leqslant t<4;\end{cases}$ and $f(t+4)\equiv f(t)$ for all $t\geqslant 0$.

(d) $f(t)=2t$ for $0\leqslant t<\frac{1}{2}$ and $f\left(t+\frac{1}{2}\right)\equiv f(t)$ for all t.

9. Find the Laplace transform of the periodic function: $f(t)=(-1)^{\lfloor at\rfloor}$, where a is a real positive number.

10. Use mathematical induction to justify $\mathcal{L}[t^n]=\frac{n}{\lambda}\,\mathcal{L}[t^{n-1}]$.

11. Find the Laplace transform of the triangular wave function $f(t)=\sum_{n\geqslant 0}[(t-2an)\,H(t-2an)+(2a+2an-t)\,H(2a+2an-t)]$.

12. Find the Laplace transform of the following integrals.

(a) $te^{-t}\int_0^t \frac{\mathrm{d}}{\mathrm{d}t}\left(e^{3t}\cos t\right)\,\mathrm{d}t$; (b) $\int_0^t te^{-2t}\cos t\,\mathrm{d}t$;

(c) $\frac{\mathrm{d}^2}{\mathrm{d}t^2}\int_0^t e^{2t}\cos t\,\mathrm{d}t$; (d) $\frac{\mathrm{d}}{\mathrm{d}t}\int_0^t e^{2t}\cos 3t\,\mathrm{d}t$.

13. Prove that

(a) $\mathcal{L}\left[t\,y'\right]=-y^L-\lambda\,\frac{\mathrm{d}y^L}{\mathrm{d}\lambda}$; (b) $\mathcal{L}\left[t\,y''\right]=-2\lambda y^L-\lambda^2\,\frac{\mathrm{d}y^L}{\mathrm{d}\lambda}$;

(c) $\mathcal{L}\left[t^2\,y''\right]=\lambda^2\,\frac{\mathrm{d}^2y^L}{\mathrm{d}\lambda^2}+4\lambda\,\frac{\mathrm{d}y^L}{\mathrm{d}\lambda}+2y^L(\lambda)$.

14. Show that $\mathcal{L}\left[H(t-a)g(t)\right](\lambda)=e^{-a\lambda}\,\mathcal{L}\left[g(t+a)\right](\lambda)$, $a>0$, and $\mathcal{L}\left[f(t+a)\right](\lambda)$ $=e^{a\lambda}\left\{f^L(\lambda)-\int_0^a e^{-\lambda t}\,f(t)\,\mathrm{d}t\right\}$, $a>0$.

15. Prove the identity: $\mathcal{L}[ty'']=y(+0)-2\lambda\,y^L-\lambda^2\,\frac{\mathrm{d}y^L}{\mathrm{d}\lambda}$.

16. Use any method to find the Laplace transform for each of the following functions.

(a) $\frac{2(1-\cosh t)}{t}$; (b) $\frac{1-e^{-2t}}{t}$; (c) $\frac{\sin kt}{\sqrt{t}}$; (d) $\frac{\sin kt}{t}$; (e) $\frac{1-\cos kt}{t}$.

17. Find the convolution of two functions $f(t)$ and $g(t)$, where

(a) $f(t)=\sin at$, $g(t)=\cosh at$; (b) $f(t)=\begin{cases}0, & 0\leqslant t\leqslant 1,\\ e^t, & 1<t\leqslant 2,\\ 0, & t\geqslant 2,\end{cases}$ $g(t)=t$;

(c) $t*\cosh t$; (d) $t*t\,e^{2t}$;
(e) $e^{at}*\sinh bt$; (f) $\cosh at*\cosh bt$;
(g) $\sin at*\sin at$; (h) $\cosh at*\cosh at$;
(i) $\cos at*\cos at$; (j) $\sinh at*\sinh at$;
(k) $t*t^2*t^3$; (l) $e^t*e^{2t}*e^{3t}$;
(m) $\sin at*\sin at*\sin at$; (n) $\sinh at*\sinh at*\sinh at$.

18. Use the rule (5.2.11) to find the Laplace transform of the following functions.

(a) $t\,e^{-3t}$; (b) $t\,e^{2t}\sin 3t$; (c) $t\,e^{2t}\cosh 3t$;
(d) $t^2\sinh(2t)$; (e) t^2-t^4; (f) $t^2\,e^{-2t}\sin(3t)$.

5.3 Discontinuous and Impulse Functions

In engineering applications, situations frequently occur in which there is an abrupt change in a system's behavior at specified values of time t. One common example is when a voltage is switched on or off in an electrical system. Many other physical phenomena can often be described by discontinuous functions. The value of $t = 0$ is usually taken as a convenient time to switch on or off the given electromotive force (*emf*, for short). The Heaviside function (5.1.5), page 274, or unit step function is an excellent tool to model the switching process mathematically. In many circuits, waveforms are applied at specified intervals. Such a function may be described using the shifted Heaviside function. A common situation in a circuit is for a voltage to be applied at a particular time (say at $t = a$) and removed later at $t = b$. Such a piecewise behavior is described by the rectangular window function:

$$W(t) = H(t-a) - H(t-b) = \begin{cases} 1, & \text{if } a < t < b, \\ 0, & \text{outside the interval } [a,b]. \end{cases}$$

This voltage across the terminals of the source of *emf* has strength 1 and duration $(b-a)$. For example, the LRC-circuit sketched in Fig. 6.1(a) on page 343 has one loop involving a resistance R, a capacitance C, an inductance L, and a time-dependent electromotive force $E(t)$. The charge $q(t)$ is modeled (see [14]) by the differential equation

$$L\,\frac{\mathrm{d}^2 q}{\mathrm{d}t^2} + R\,\frac{\mathrm{d}q}{\mathrm{d}t} + \frac{1}{C}\,q = E(t).$$

An initial charge q_0 and initial current I_0 are specified as the initial conditions: $q(0) = q_0$, $\dot{q}(0) = I_0$.

If the current is initially unforced but is plugged into an alternating electromotive force $E(t) = E_0 \sin\omega(t-T)$ at time T, then the piecewise continuous function $E(t)$ is defined by

$$E(t) = E_0 \sin\omega(t-T)\, H(t-T) = \begin{cases} 0, & t < T, \\ E_0 \sin\omega(t-T), & t > T. \end{cases}$$

Our primary tool to describe discontinuous functions is the Heaviside function (see Definition 5.3 on page 274). There are other known unit step functions that have different values at the point of discontinuity. For instance, a unit function can be defined as

$$H(t) = \frac{1}{2}\left(1 + \frac{t}{\sqrt{t^2}}\right), \qquad t \ne 0,$$

where the value at $t = 0$ is undefined, which means that it can be chosen arbitrarily. However, as will be clear later from §5.4, an application of the Laplace transformation calls for the Heaviside function. Many functions, including continuous and discontinuous, can be approximated as a sum of linear combinations of unit functions. For example, for small $\varepsilon > 0$, a continuous function $f(t)$ can be well approximated by a piecewise step-function $f_\varepsilon = f\left(\varepsilon \lfloor t/\varepsilon \rfloor\right)$, where $\lfloor A \rfloor$ is the floor of a real number A, namely, the largest integer that is less than or equal to A. It is not a surprise that the inverse statement holds: the Heaviside function is the limit of continuous functions containing a parameter when the parameter approaches a limit value. For example,

$$H(t) = \lim_{s\to\infty} \left\{ \frac{1}{2} + \frac{1}{\pi}\,\arctan(st) \right\}. \tag{5.3.1}$$

Related to the unit function is the function *signum* x, defined by

$$\operatorname{sign} x = \begin{cases} 1, & x > 0, \\ 0, & x = 0, \\ -1, & x < 0. \end{cases}$$

Maple and *Maxima* have the built-in symbol `signum`(x), while *Mathematica* uses the symbol `Sign[x]`, MATLAB, SymPy, and *R* utilize the nomenclature `sign(x)`, and *Sage* uses `sgn(x)`. The signum function can be expressed through the Heaviside function as $\operatorname{sign} t = 2H(t) - 1$, so

$$H(t) = \frac{1}{2}\left[\operatorname{sign} t + 1\right]. \tag{5.3.2}$$

The two representations, (5.3.1) and (5.3.2), of the Heaviside function are valid for all real values of the independent argument, including $t = 0$. Recall that $H(0) = \frac{1}{2}$. In the following, we will consider only such discontinuous functions

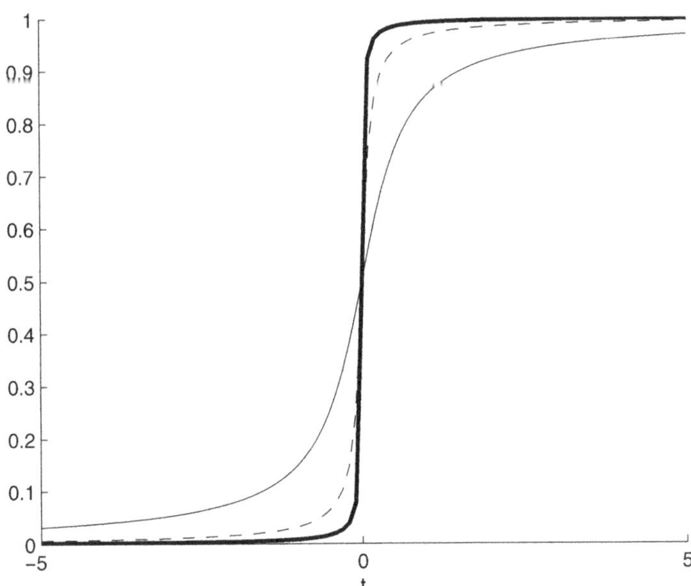

Figure 5.14: Approximations (5.3.1) to the Heaviside function for $s = 2$, $s = 10$ (dashed line), and $s = 40$ (solid line), plotted in MATLAB.

that satisfy the mean value property (5.1.8), page 277. That is, all functions under consideration have the value at a point of discontinuity to be equal to the average of limit values from the left and from the right.

As we saw in the first section, the Laplace transform of a piecewise continuous function is a smooth function (moreover, a holomorphic function in some half-plane). The Laplace transform is particularly beneficial for dealing with discontinuous functions.

Keep in mind that our ultimate goal is to solve differential equations with possible piecewise continuous forcing functions. Utilization of the Laplace transformation in differential equations involves two steps: the direct Laplace transform and the inverse Laplace transform. When applying the Laplace transform to an intermittent function, it does not matter what the values of the function at the points of discontinuity are—integration is not sensitive to the values of the function at a discrete number of points. On the other hand, the inverse Laplace transform defines the function that possesses a mean value property. For instance, the unit function

$$u(t) = 1 + \left\lfloor \frac{t}{t^2+1} \right\rfloor = \begin{cases} 1, & t \geqslant 0, \\ 0, & t < 0; \end{cases}$$

has the same Laplace transform as the Heaviside function: $\mathcal{L}[u] = \mathcal{L}[H] = 1/\lambda$. As we will see later in §5.4, the inverse Laplace transform of $1/\lambda$ is $\mathcal{L}^{-1}\left[\frac{1}{\lambda}\right] = H(t)$. Hence, $\mathcal{L}^{-1}\left[\mathcal{L}[u]\right] = \mathcal{L}^{-1}\mathcal{L}[H] = H(t)$, and the unit function cannot be restored from its Laplace transform.

If we need to cut out a piece of a function on the interval (a, b) and set it identically to zero outside this interval, we just multiply the function by the difference $H(t-a) - H(t-b)$, called the rectangular window. For example, the following discontinuous functions (see Fig. 5.15 on next page)

$$f(t) = \begin{cases} 1, & 0 < t < 2, \\ 0, & \text{elsewhere;} \end{cases} \qquad \text{and} \qquad g(t) = \begin{cases} 2, & 1 < t < 4, \\ 0, & \text{elsewhere;} \end{cases}$$

can be written as

$$f(t) = H(t) - H(t-2) \qquad \text{and} \qquad g(t) = 2H(t-1) - 2H(t-4),$$

respectively. According to our agreement, we do not pay attention to the values of the functions at the points of discontinuity and define piecewise functions only on open intervals. It is assumed that the functions under consideration satisfy the mean value property (5.1.8). In our example, we have $f(0) = f(2) = \frac{1}{2}$ and $g(1) = g(4) = 1$.

In general, if a function is defined on disjointed intervals via distinct formulas, we can represent this function as a sum of these expressions multiplied by the difference of the corresponding Heaviside functions. For instance,

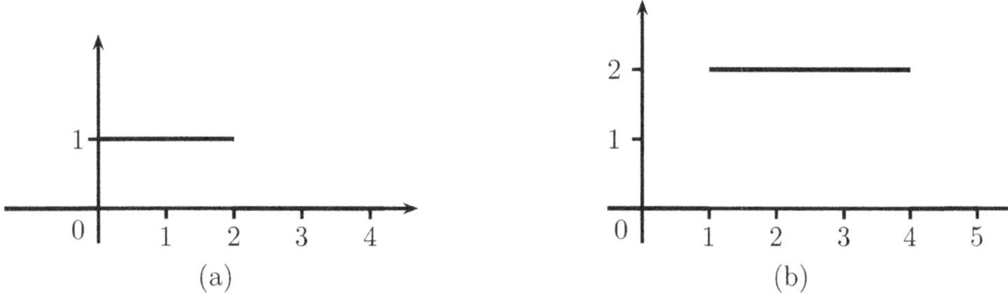

Figure 5.15: Graphs of the functions (a) $f(t) = H(t) - H(t-2)$, and (b) the function $g(t) = 2H(t-1) - 2H(t-4)$.

consider the function

$$f(t) = \begin{cases} t, & 0 < t < 1, \\ 1, & 1 < t < 3, \\ 4 - t, & 3 < t < 4, \\ 0, & t > 4. \end{cases}$$

This function is a combination of four functions defined on disjoint intervals. Therefore, we represent $f(t)$ as the sum:

$$\begin{aligned} f(t) &= t\,[H(t) - H(t-1)] + 1\,[H(t-1) - H(t-3)] + (4-t)\,[H(t-3) - H(t-4)] \\ &= t\,H(t) - (t-1)\,H(t-1) + (4-t-1)\,H(t-3) + (4-t)\,H(t-4) \\ &= t\,H(t) - (t-1)\,H(t-1) - (t-3)\,H(t-3) - (t-4)\,H(t-4). \end{aligned}$$

Using the shift rule, Eq. (5.2.6) on page 282, we obtain its Laplace transform

$$f^L = \frac{1}{\lambda^2} - \frac{1}{\lambda^2}\,e^{-\lambda} - \frac{1}{\lambda^2}\,e^{-3\lambda} - \frac{1}{\lambda^2}\,e^{-4\lambda}.$$

Example 5.3.1: (Example 5.1.7 revisited) We reconsider the discontinuous function from Example 5.1.7 on page 276. To find its Laplace transform, it is convenient to rewrite its formula using the Heaviside function:

$$f(t) = t^2[H(t) - H(t-1)] + (2+t)[H(t-1) - H(t-2)] + (1-t)[H(t-2).$$

In order to apply the shift rule, we need to do some extra work by adding and subtracting a number to the function that is multiplied by the shifted Heaviside function. Recall that the shift rule (5.2.6) requires the function to be shifted by the same value as the Heaviside function. For instance, instead of $t^2H(t-1)$ we use

$$t^2H(t-1) = (t-1+1)^2\,H(t-1) = \left[(t-1)^2 + 2(t-1) + 1\right]H(t-1).$$

This allows us to rewrite the given function as

$$f(t) = t^2H(t) + \left[2 - (t-1) - (t-1)^2\right]H(t-1) - [2(t-2) + 5]\,H(t-2).$$

Now the function is ready for application of the shift rule (5.2.6). Using formula (4) in Table 280 for the Laplace transform of the power function, we get

$$f^L = \frac{2}{\lambda^3} + \left[\frac{2}{\lambda} - \frac{1}{\lambda^2} - \frac{2}{\lambda^3}\right]e^{-\lambda} - \left[\frac{2}{\lambda^2} + \frac{5}{\lambda}\right]e^{-2\lambda}.$$

Example 5.3.2: Express the square wave function shown in Fig. 5.11 (Example 5.2.13 on page 286) and the so-called Meander function shown in Fig. 5.16 on page 294 in terms of the Heaviside function, and obtain their Laplace transforms.

Solution. It can be seen that the Meander function $f(t)$ is defined by the equation

$$f(t) = H(t) - 2H(t-a) + 2H(t-2a) - 2H(t-3a) + 2H(t-4a) + \cdots.$$

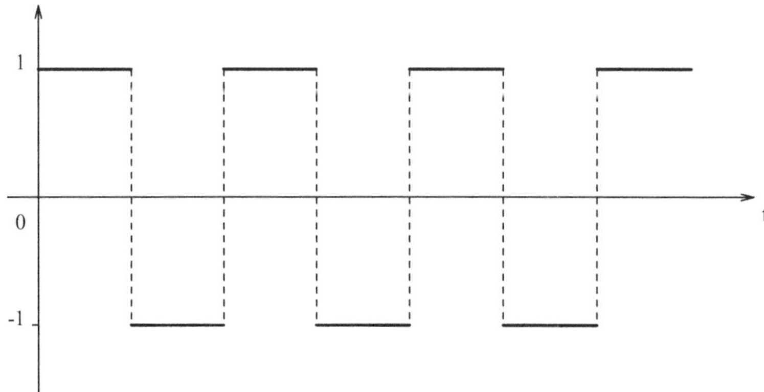

Figure 5.16: Example 5.3.2. The Meander function.

From formula 2 of Table 280, it follows that the Laplace transform of each term in the last representation of the Meander function is

$$\mathcal{L}[H(t-na)](\lambda) = \frac{1}{\lambda}\, e^{-an\lambda}, \quad n = 0, 1, 2, \ldots.$$

Therefore,

$$\begin{aligned} f^L(\lambda) &= \frac{1}{\lambda} - \frac{2}{\lambda}\, e^{-\lambda a} + \frac{2}{\lambda}\, e^{-2\lambda a} - \frac{2}{\lambda}\, e^{-3\lambda a} + \cdots \\ &= \frac{1}{\lambda}\left[1 - 2e^{-a\lambda}\left(1 - e^{-a\lambda} + e^{-2a\lambda} - e^{-3a\lambda} + \ldots\right)\right] \\ &= \frac{1}{\lambda}\left[1 - \frac{2e^{-a\lambda}}{1+e^{-a\lambda}}\right] = \frac{1}{\lambda}\left[\frac{1-e^{-a\lambda}}{1+e^{-a\lambda}}\right] = \frac{1}{\lambda}\tanh\left(\frac{a\lambda}{2}\right). \end{aligned}$$

Here, we used the geometric series

$$\frac{1}{1-z} = \sum_{k=0}^{\infty} z^k = 1 + z + z^2 + z^3 + \cdots \tag{5.3.3}$$

with $z = -e^{-\lambda a}$. On the other hand, the Laplace transform of the square wave function,

$$g(t) = H(t) - H(t-1) + H(t-3) - H(t-4) + \cdots,$$

is

$$\begin{aligned} g^L(\lambda) &= \frac{1}{\lambda} - \frac{1}{\lambda}\, e^{-\lambda} + \frac{1}{\lambda}\, e^{-2\lambda} - \frac{1}{\lambda}\, e^{-3\lambda} + \cdots \\ &= \frac{1}{\lambda}\left[1 - e^{-\lambda} + e^{-2\lambda} + e^{-3\lambda} + \cdots\right] \\ &= \frac{1}{\lambda}\sum_{n=0}^{\infty}\left(-e^{-\lambda}\right)^n = \frac{1}{\lambda}\,\frac{1}{1+e^{-\lambda}}. \end{aligned}$$

Definition 5.8: The **full-wave rectifier** of a function $f(t)$, $0 \leqslant t \leqslant T$, is a periodic function with period T that is equal to $f(t)$ on the interval $[0, T]$.

The **half-wave rectifier** of a function $f(t)$, $0 \leqslant t \leqslant T$, is a periodic function with period $2T$ that coincides with $f(t)$ on the interval $[0, T]$ and is identically zero on the interval $[T, 2T]$.

Example 5.3.3: Find the Laplace transform of the saw-tooth function

$$\begin{aligned} f(t) = {} & \frac{E}{a}\, t\, [H(t) - H(t-a)] + \frac{E}{a}\,(t-a)\,[H(t-a) - H(t-2a)] \\ & + \frac{E}{a}\,(t-2a)\,[H(t-2a) - H(t-3a)] + \frac{E}{a}\,(t-3a)\,[H(t-3a) - H(t-4a)] + \cdots, \end{aligned}$$

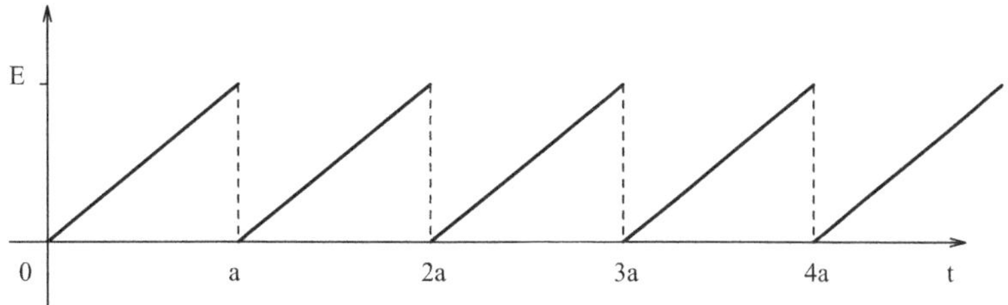

Figure 5.17: Example 5.3.3. The saw-tooth function.

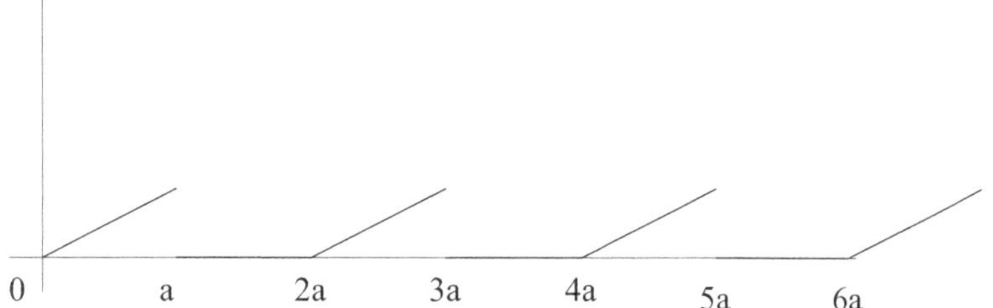

Figure 5.18: Example 5.3.4. The half-wave rectification of the function $f(t) = \frac{E}{a}\, t$, on the interval $[0, 2a]$.

which is a full-wave rectifier of the function $f(t) = \frac{E}{a}\, t$, on the interval $[0, a]$.

Solution. This is a periodic function with the period $\omega = a$. Applying Eq. (5.2.13), we obtain

$$\begin{aligned} f^L(\lambda) &= \frac{1}{1-e^{-a\lambda}}\,\frac{E}{a}\int_0^a t\,e^{-t\lambda}\,\mathrm{d}t \\ &= \frac{1}{1-e^{-a\lambda}}\,\frac{E}{a}\left[\frac{1}{\lambda^2} - \frac{1}{\lambda^2}\,e^{-a\lambda} - \frac{a}{\lambda}\,e^{-a\lambda}\right] = \frac{E}{a}\,\frac{1}{\lambda^2} - \frac{E}{\lambda}\,\frac{e^{-a\lambda}}{1-e^{-a\lambda}}. \end{aligned}$$

Example 5.3.4: Find the Laplace transform of the half-wave rectifier of the function $f(t) = \frac{E}{a}\, t$ on the interval $[0, 2a]$.

Solution. The half-wave rectification of the function $f(t)$ is the following function:

$$\begin{aligned} F(t) = {}& \frac{E}{a}\, t\, [H(t) - H(t-a)] + \frac{E}{a}\,(t-2a)\,[H(t-2a) - H(t-3a)] \\ & + \frac{E}{a}\,(t-4a)\,[H(t-4a) - H(t-5a)] + \cdots . \end{aligned}$$

From the shift rule (5.2.6), it follows that

$$\mathcal{L}\left[H(t-na)(t-na)\right] = e^{-na\lambda}\mathcal{L}[t] = e^{-na\lambda}\,\frac{1}{\lambda},$$

$$\begin{aligned} \mathcal{L}\left[H(t-na-a)(t-na)\right] &= \mathcal{L}\left[H(t-na-a)(t-na-a+a)\right] \\ &= \mathcal{L}\left[H(t-na-a)(t-na-a)\right] + a\,\mathcal{L}[H(t-na-a)] \\ &= e^{-(n+1)a\lambda}\,\frac{1}{\lambda^2} + \frac{a}{\lambda}\,e^{-(n+1)a\lambda}. \end{aligned}$$

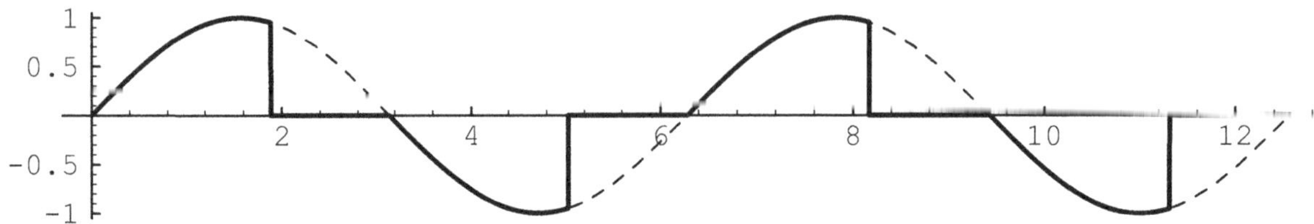

Figure 5.19: Example 5.3.5.

Hence, the Laplace transform of $F(t)$ is

$$\begin{aligned}
F^L(\lambda) &= \frac{E}{a\lambda^2}\left[1 - e^{-a\lambda} + e^{-2a\lambda} - e^{-3a\lambda} + e^{-4a\lambda} - e^{-5a\lambda} + \cdots\right] \\
&\quad - \frac{E}{\lambda}\left[e^{-a\lambda} + e^{-3a\lambda} + e^{-5a\lambda} + \cdots\right] \\
&= \frac{E}{a\lambda^2}\left[1 + e^{-2a\lambda} + e^{-4a\lambda} + e^{-6a\lambda} + \cdots\right] \\
&\quad - \frac{E}{a\lambda^2}\, e^{-a\lambda}\left[1 + e^{-2a\lambda} + e^{-4a\lambda} + e^{-6a\lambda} + \cdots\right] \\
&\quad - \frac{E}{\lambda}\, e^{-a\lambda}\left[1 + e^{-2a\lambda} + e^{-4a\lambda} + e^{-6a\lambda} + \cdots\right].
\end{aligned}$$

Setting $z = e^{-2a\lambda}$, we summarize the series using Eq. (5.3.3) to obtain

$$1 + e^{-2a\lambda} + e^{-4a\lambda} + \cdots = 1 + z + z^2 + \cdots = \frac{1}{1-z} = \frac{1}{1 - e^{-2a\lambda}}.$$

Thus,

$$F^L(\lambda) = \frac{E}{a\lambda^2}\,\frac{1 - e^{-a\lambda}}{1 - e^{-2a\lambda}} - \frac{E}{\lambda}\,\frac{e^{-a\lambda}}{1 - e^{-2a\lambda}}.$$

Example 5.3.5: Certain light dimmers produce the following type of function as an output voltage: a sine function that is cut off as shown in Fig. 5.19 by the solid line; the jumps are at a, $\pi + a$, $2\pi + a$, etc. Find the Laplace transform of this function.

Solution. The output is a periodic function $f(t)$ with the period $\omega = 2\pi$. According to Eq. (5.2.13) on page 283, its Laplace transform is

$$f^L(\lambda) = \frac{1}{1 - e^{-2\pi\lambda}} \int_0^{2\pi} e^{-\lambda t} f(t)\, dt,$$

where

$$f(t) = \begin{cases} \sin t, & 0 \leqslant t < a, \\ 0, & 0 < t \leqslant \pi, \\ \sin t, & \pi \leqslant t < \pi + a, \\ 0, & \pi + a < t \leqslant 2\pi. \end{cases}$$

Substituting this function under the integral sign yields

$$f^L(\lambda) = \frac{1}{1 - e^{-2\pi\lambda}} \left[\int_0^a \sin t\, e^{-\lambda t}\, dt + \int_\pi^{\pi + a} \sin t\, e^{-\lambda t}\, dt\right].$$

Using the relation $\sin t = \Im e^{\mathbf{j}t}$, the imaginary part of $e^{\mathbf{j}t}$, and multiplying both sides by $1 - e^{-2\pi\lambda}$, we obtain

$$\begin{aligned}
\left[1 - e^{-2\pi\lambda}\right] f^L(\lambda) &= \Im \int_0^a e^{\mathbf{j}t}\, e^{-\lambda t}\, \mathrm{d}t + \Im \int_\pi^{\pi+a} e^{\mathbf{j}t}\, e^{-\lambda t}\, \mathrm{d}t \\
&= \Im \left\{ - \frac{1}{\lambda - \mathbf{j}}\, e^{\mathbf{j}t - \lambda t} \Big|_{t=0}^{t=a} \right\} + \Im \left\{ - \frac{1}{\lambda - \mathbf{j}}\, e^{\mathbf{j}t - \lambda t} \Big|_{t=\pi}^{t=\pi+a} \right\} \\
&= \Im \left\{ - \frac{\lambda + \mathbf{j}}{\lambda^2 + 1}\, e^{\mathbf{j}a}\, e^{-\lambda a} + \frac{\lambda + \mathbf{j}}{\lambda^2 + 1} \right\} \\
&\quad + \Im \left\{ - \frac{\lambda + \mathbf{j}}{\lambda^2 + 1}\, e^{\mathbf{j}a + \mathbf{j}\pi}\, e^{-\lambda(\pi + a)} + \frac{\lambda + \mathbf{j}}{\lambda^2 + 1}\, e^{\mathbf{j}\pi}\, e^{-\lambda\pi} \right\} \\
&= \frac{1}{\lambda^2 + 1}\, \Im \left\{ -(\lambda + \mathbf{j})(\cos a + \mathbf{j} \sin a)\, e^{-\lambda a} + \lambda + \mathbf{j} \right\} \\
&\quad + \frac{e^{-\lambda\pi}}{\lambda^2 + 1}\, \Im \left\{ (\lambda + \mathbf{j})(\cos a + \mathbf{j} \sin a)\, e^{-\lambda a} - \lambda - \mathbf{j} \right\}
\end{aligned}$$

because $e^{\pi\mathbf{j}} = -1$ and $\frac{1}{\lambda - \mathbf{j}} = \frac{\lambda + \mathbf{j}}{(\lambda - i)(\lambda + \mathbf{j})} = \frac{\lambda + \mathbf{j}}{\lambda^2 + 1}$. Thus,

$$\begin{aligned}
\left[1 - e^{-2\pi\lambda}\right] f^L(\lambda) &= \frac{1}{\lambda^2 + 1} \left[1 - e^{-\lambda a} \left(\lambda \sin a + \cos a\right)\right] \\
&\quad + \frac{e^{-\lambda\pi}}{\lambda^2 + 1} \left[e^{-\lambda a} \left(\lambda \sin a + \cos a\right) - 1\right].
\end{aligned}$$

Therefore,

$$f^L(\lambda) = \frac{1 - e^{-\lambda\pi}}{\left(1 - e^{-2\lambda\pi}\right)\left(\lambda^2 + 1\right)} \left[\left(\lambda \sin a + \cos a\right) e^{-\lambda(a+\pi)} - e^{-a\lambda}\right].$$

Example 5.3.6: Find the Laplace transform of the ladder function

$$f(t) = \tau H(t) + \tau H(t - a) + \tau H(t - 2a) + \cdots .$$

Solution. We find the Laplace transform of the function by applying the shift rule: $\mathcal{L}H(t - a) = \lambda^{-1}\, e^{-\lambda a}$. Thus,

$$\begin{aligned}
f^L(\lambda) &= \frac{\tau}{\lambda} \left[1 + e^{-\lambda a} + e^{-2\lambda a} + e^{-3\lambda a} + \cdots\right] \\
&= \frac{\tau}{\lambda} \left[1 + e^{-\lambda a} + \left(e^{-\lambda a}\right)^2 + \left(e^{-\lambda a}\right)^3 + \left(e^{-\lambda a}\right)^4 + \cdots\right] \\
&= \frac{\tau}{\lambda} \frac{1}{1 - e^{-a\lambda}}. \qquad \square
\end{aligned}$$

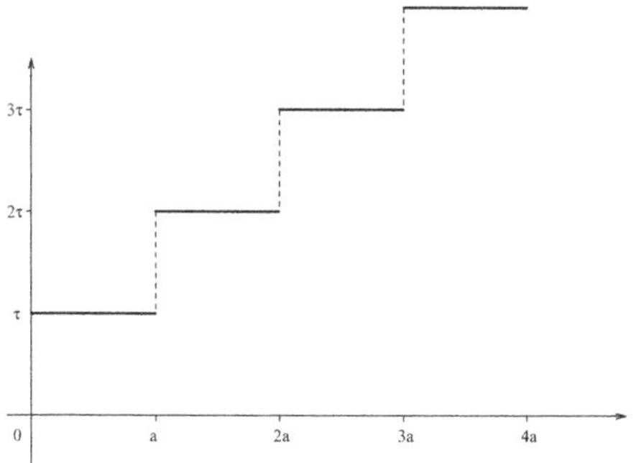

Figure 5.20: Example 5.3.6. The ladder function.

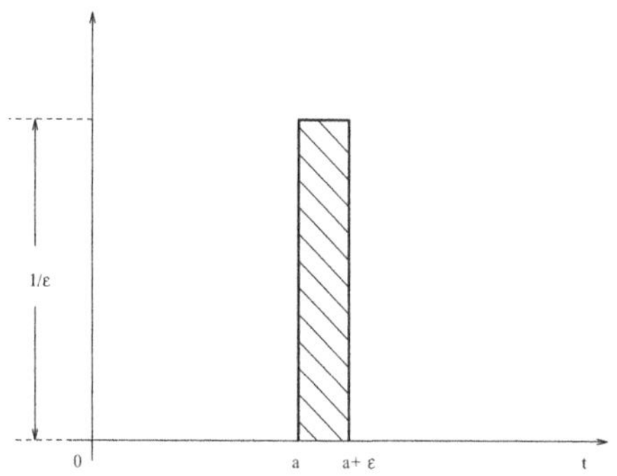

Figure 5.21: Approximation of the δ-function.

Mechanical systems are also often driven by an external force of large magnitude that acts for only very short periods of time. For example, the strike of a hammer exerts a relatively large force over a relatively short time.

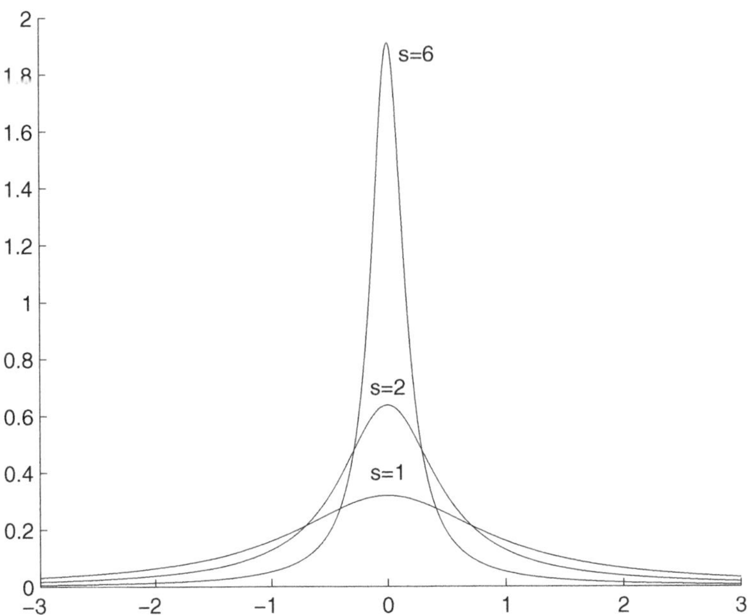

Figure 5.22: Approximations $\delta_2(t, s)$ to the δ-function for different values $s = 1$, $s = 3$, and $s = 6$, plotted with MATLAB.

Mathematical simulations of such processes involve differential equations with discontinuous or impulsive forcing functions.

Paul Dirac[50] introduced in 1926 his celebrated δ-function via the relation

$$u(x) = \int \delta(t - x)\, u(t)\, \mathrm{d}t, \tag{5.3.4}$$

where $\delta(x) = 0$ if $x \neq 0$. Such a "function" is zero everywhere except at the origin, where it becomes infinite in such a way as to ensure

$$\int_{-\infty}^{\infty} \delta(x)\, \mathrm{d}x = 1.$$

The Dirac delta-function $\delta(x)$ is not a genuine function in the ordinary sense, but it is a generalized function or distribution. Generalized functions were rigorously defined in 1936 by the Russian mathematician Sergei L'vovich Sobolev (1908–1989). Later in 1950 and 1951, the French mathematician Laurent Schwartz[51] published two volumes of "Theore des Distributions," in which he presented the theory of distributions.

Recall that Dirac introduced his δ-function in order to justify laws in quantum mechanics. We cannot see elementary particles like electrons, but we can observe the point where the electron strikes the screen. To describe this phenomenon mathematically, Dirac suggested using integration of two functions, one of which corresponds to a particle, and the other one, called the "probe" function, corresponds to the environment (as a screen). Hence, the δ-function operates on "probe" functions according to Eq. (5.3.4). The delta-function can be interpreted as the limit of a physical quantity that has a very large magnitude for a very short time, keeping their product finite (i.e., the strength of the pulse remains constant). For example,

$$\delta(t - a) = \lim_{\varepsilon \to 0} \delta_\varepsilon(t - a) \equiv \lim_{\varepsilon \to 0} \frac{1}{\varepsilon} \left[H(t - a) - H(t - a - \varepsilon) \right].$$

As $\varepsilon \mapsto 0$, the function $\delta_\varepsilon(t - a)$ approaches the unit impulse function or the Dirac delta-function. The right-hand side limit is the derivative of the Heaviside function $H(t - a)$ with respect to t, namely,

$$\delta(t - a) = \lim_{\varepsilon \to 0} \delta_\varepsilon(t - a) = H'(t - a), \tag{5.3.5}$$

[50]Paul Dirac (1902–1984), an English physicist, was awarded the Nobel Prize (jointly with Erwin Schrödinger) in 1933 for his work in quantum mechanics.

[51]Laurent Schwartz (1915–2002) received the most prestigious award in mathematics—The Fields Medal—for this work.

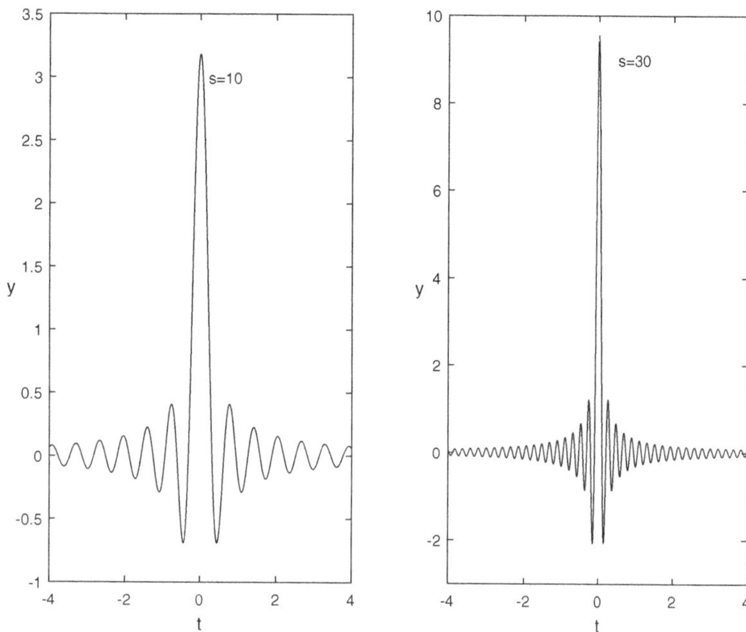

Figure 5.23: Approximations $\delta_4(t,s)$ to the δ-function for $s = 10$ and $s = 30$.

and hence

$$\int f(t)\, H'(t-a)\, \mathrm{d}t = -\int f'(t)\, H(t-a)\, \mathrm{d}t = \int f(t)\delta(t-a)\, \mathrm{d}t = f(a) \tag{5.3.6}$$

for any continuously differentiable function $f(t)$. Because of Eq. (5.3.4), the product of the delta-function $\delta(t-a)$ and a smooth function $g(t)$ has the same value as $g(a)\,\delta(t-a)$; that is,

$$g(t)\,\delta(t-a) = g(a)\,\delta(t-a). \tag{5.3.7}$$

It should be emphasized that the derivative of the Heaviside function is understood not in the ordinary sense, but as the derivative of a generalized function. With the exception of the point $t = 0$, the Heaviside function $H(t)$ permits differentiation anywhere, and its derivative vanishes in any region that does not contain the point $t = 0$. Although we cannot differentiate the Heaviside function because of its discontinuity at $t = 0$, we can approximate it by, for instance, formula (5.3.1) to obtain the derivative

$$\delta(t) = H'(t) = \lim_{s\to\infty} \frac{\mathrm{d}}{\mathrm{d}t}\left\{\frac{1}{2} + \frac{1}{\pi}\arctan(st)\right\} = \lim_{s\to\infty} \frac{s}{\pi(s^2t^2+1)}.$$

Therefore, the δ-function can be defined as the limit of another sequence of functions, namely,

$$\delta(t) = \lim_{s\to\infty} \delta_2(t,s), \quad \text{with} \quad \delta_2(t,s) = \frac{s}{\pi(s^2t^2+1)}.$$

There are many other known approximations of the delta-function, among which we mention two: $\delta_3(t,s) = \frac{s}{\sqrt{\pi}}\, e^{-s^2t^2}$ and $\delta_4(t,s) = \frac{\sin(st)}{t}$.

Using the shift rule (5.2.6) on page 282, we obtain

$$\delta(t-a) * f(t) = f(t-a). \tag{5.3.8}$$

The Laplace transform of the Dirac δ-function is

$$\mathcal{L}[\delta](\lambda) = \delta^L(\lambda) = \int_0^\infty \delta(t)\, e^{-\lambda t}\, \mathrm{d}t = e^{-\lambda 0} = 1. \tag{5.3.9}$$

The Dirac delta-function has numerous applications in theoretical physics, mathematics, and engineering problems. In electric circuits, the delta-function can serve as a model for voltage spikes. In a spike event, the electromotive function applied to a circuit increases by thousands of volts and then decreases to a normal level, all within nanoseconds (10^{-9} seconds). It would be nearly impossible to make measurements necessary to graph a spike. The best we can do is to approximate it by $m\delta(t-a)$, where $t-a$ is the spike time. The multiple m represents the magnitude of the impulse coursed by the spike.

The delta-function is also used in mechanical problems. For example, if we strike an object like a string with a hammer (this is commonly used in a piano) or beat a drum with a stick, a rather large force acts for a short interval of time. In [14], we discuss vibrations of a weighted spring system when the mass has been struck a sharp blow. This force can be approximated by the delta-function multiplied by some appropriate constant to make it equal to the total impulse of energy. Recall from mechanics that a force $\mathbf{F}$ acting on an object during a time interval $[0, t]$ is said to impact an **impulse**, which is defined by the integral $\int_0^t \mathbf{F}\,dt$. If $\mathbf{F}$ is a constant force, the impulse becomes the product $\mathbf{F}t$. When applied to mechanical systems, the impulse equals the change in momentum.

Remark. It should be pointed out that the approximations of the delta-function are defined not pointwise, but in the generalized sense: for any continuous and integrable function $f(t)$, the δ-function is defined as the limit

$$\int_{-\infty}^{\infty} \delta(t)\, f(t)\, dt = f(0) = \lim_{s\to\infty} \int_{-\infty}^{\infty} \delta(t,s)\, f(t)\, dt,$$

where $\delta(t,s)$ is one of the previous functions approximating the unit impulse function.

Problems

1. Rewrite each function in terms of the Heaviside function and find its Laplace transform.

   (a) $f(t) = \begin{cases} 1, & 0 \leqslant t < 2, \\ -1, & t \geqslant 2. \end{cases}$ **(b)** $f(t) = \begin{cases} 1, & 0 \leqslant t < 2, \\ 0, & t \geqslant 2. \end{cases}$

   (c) $f(t) = \begin{cases} t, & 0 \leqslant t < 3, \\ 0, & t \geqslant 3. \end{cases}$ **(d)** $f(t) = \begin{cases} 0, & 0 \leqslant t < 2, \\ t^2, & t \geqslant 2. \end{cases}$

   (e) $f(t) = \begin{cases} \sin t, & 0 \leqslant t < 2\pi, \\ 0, & t \geqslant 2\pi. \end{cases}$ **(f)** $f(t) = \begin{cases} 0, & 0 \leqslant t < 3\pi/4, \\ \sin(2t), & t \geqslant 3\pi/4. \end{cases}$

   (g) $f(t) = \begin{cases} 1, & 0 \leqslant t < 2, \\ 0, & 2 \leqslant t < 4, \\ -1, & 4 \leqslant t. \end{cases}$ **(h)** $f(t) = \begin{cases} 2, & 0 \leqslant t < 4, \\ t/2, & 4 \leqslant t < 6, \\ 3, & 6 \leqslant t. \end{cases}$

   (i) $f(t) = \begin{cases} \cos 3t, & 0 \leqslant t < \pi/2, \\ 0, & \pi/2 \leqslant t. \end{cases}$ **(j)** $f(t) = \begin{cases} \cos 3t, & 0 \leqslant t < \pi, \\ 0, & \pi \leqslant t. \end{cases}$

2. For $\varepsilon > 0$, find the Laplace transform of the piecewise-step function $f_\varepsilon(t) = f(\varepsilon \lfloor t/\varepsilon \rfloor)$, where $f(t) = t$.

3. Find the Laplace transform of the following functions.

   (a) $H(t) - H(t-1) + H(t-2) - H(t-3)$; **(b)** $\delta(t) - \delta(t-1)$;
   (c) $\sin t\,[H(t) - H(t-\pi)]$; **(d)** $e^t\, H(t-1)$.

4. Use the Heaviside function to redefine each of the following piecewise continuous functions. Then using the shift rule, find its Laplace transform.

   (a) $f(t) = \begin{cases} t, & 0 \leqslant t \leqslant 1, \\ 2-t, & 1 \leqslant t \leqslant 2, \\ 0, & 2 \leqslant t. \end{cases}$ **(b)** $f(t) = \begin{cases} 0, & 0 < t < 1, \\ t, & 1 < t < 4, \\ 1, & 4 < t. \end{cases}$

(c) $f(t) = \begin{cases} 0, & 0 \leqslant t \leqslant 1, \\ t-1, & 1 \leqslant t \leqslant 2, \\ 3-t, & 2 \leqslant t \leqslant 3, \\ 0, & 3 \leqslant t. \end{cases}$ (d) $f(t) = \begin{cases} t, & 0 \leqslant t \leqslant 1, \\ 1, & 1 \leqslant t \leqslant 2, \\ 3-t, & 2 \leqslant t \leqslant 3, \\ 0, & 3 \leqslant t. \end{cases}$

(e) $f(t) = \begin{cases} 1, & 0 < t < 2, \\ 2, & 2 < t < 3, \\ -2, & 3 < t < 4, \\ 3, & t > 4. \end{cases}$ (f) $f(t) = \begin{cases} 2, & 0 < t < 1, \\ t, & 1 < t < 2, \\ t-2, & 2 < t < 3, \\ 0, & t > 3. \end{cases}$

(g) $f(t) = \begin{cases} 1, & 0 < t < 2, \\ t, & 2 < t < 4, \\ 6-t, & 4 < t < 6, \\ t-8, & 6 < t < 8, \\ 0, & t > 8. \end{cases}$ (h) $f(t) = \begin{cases} 0, & 0 \leqslant t \leqslant 1, \\ t-1, & 1 \leqslant t \leqslant 2, \\ 1, & 2 \leqslant t \leqslant 3, \\ 4-t, & 3 \leqslant t \leqslant 4, \\ 0, & 4 \leqslant t. \end{cases}$

5. Find the Laplace transformation of the given periodic function.

(a) Full-wave rectifier of the function $f(t) = \sin t$ on the interval $[0, 2\pi]$.

(b) Full-wave rectifier of the function $f(t) = \sin 2t$ on the interval $[0, \pi]$.

(c) Full-wave rectifier of the function $f(t) = \cos t$ on the interval $[0, \pi/2]$.

(d) Full-wave rectifier of the function $f(t) = t^2$ on the interval $[0, 1]$.

(e) Half-wave rectifier of the function $f(t) = \sin t$ on the interval $[0, 2\pi]$.

(f) Half-wave rectifier of the function $f(t) = \sin 2t$ on the interval $[0, \pi]$.

(g) Half-wave rectifier of the function $f(t) = \cos t$ on the interval $[0, \pi/2]$.

(h) Half-wave rectifier of the function $f(t) = t^2$ on the interval $[0, 1]$.

(i) Periodic function $f(t) = f(t+2)$, where $f(t) = \begin{cases} t, & 0 \leqslant t \leqslant 1, \\ 2-t, & 1 \leqslant t \leqslant 2. \end{cases}$

(j) Periodic function $f(t) = f(t+3)$, where

$$f(t) = \begin{cases} t, & 0 \leqslant t \leqslant 1, \\ 1, & 1 \leqslant t \leqslant 2, \\ 3-t, & 2 \leqslant t \leqslant 3. \end{cases}$$

(k) Periodic function $f(t) = f(t+4)$, where

$$f(t) = \begin{cases} 1, & 0 \leqslant t \leqslant 1, \\ t^2, & 1 \leqslant t \leqslant 3, \\ 9, & 3 \leqslant t \leqslant 4. \end{cases}$$

(l) Periodic function $f(t) = f(t+5)$, where

$$f(t) = \begin{cases} 2-t, & 0 \leqslant t \leqslant 2, \\ 0, & 2 \leqslant t \leqslant 3, \\ (t-3)^2 - 2, & 3 \leqslant t \leqslant 5. \end{cases}$$

6. Find the Laplace transform of $\operatorname{sign}(\sin t)$.

7. Find the Laplace transforms of the following functions.

(a) $\delta(t-1) - \delta(t-3)$; (b) $(t-3)\delta(t-3)$; (c) $\sqrt{t^2+3t}\,\delta(t-1)$; (d) $\cos \pi t\,\delta(t-1)$.

8. Rewrite each function from Problem 5 in §5.1 (page 280) in terms of shifted Heaviside functions.

9. Show that the Laplace transform of the full-wave rectification of $\sin \omega t$ on the interval $[0, \pi/\omega]$ is $\frac{\omega}{\lambda^2+\omega^2} \coth \frac{\pi\lambda}{2\omega}$.

10. Find the Laplace transform of the given functions.

(a) $(t^2 - 25)H(t-5)$; (b) $(t-\pi)H(t-\pi) - (t-2\pi)H(t-2\pi)$;
(c) $e^{(t-3)}\,H(t-3)$; (d) $\cos(\pi t)[H(t) - H(t-2)]$;
(e) $(1+t)H(t-2)$; (f) $\sin(2\pi t)[H(t-1) - H(t-3/2)]$.

11. Can $\mathcal{L}^{-1}\left[\frac{\lambda}{\lambda+1}\right]$ be found by using the differential rule (5.2.2) $\mathcal{L}^{-1}\left[\frac{\lambda}{\lambda+1}\right] = \texttt{D}\mathcal{L}^{-1}\left[\frac{1}{\lambda+1}\right] = \texttt{D}\,e^{-t} = -e^{-t}$? Explain why or why not.

5.4 The Inverse Laplace Transform

We employ the symbol $\mathcal{L}^{-1}[F(\lambda)]$ to denote the inverse Laplace transformation of the function $F(\lambda)$, while $\mathcal{L}$ corresponds to the direct Laplace transform defined by Eq. (5.1.1), page 270. Thus, we have the Laplace pair

$$F(\lambda) = (\mathcal{L}f(t))(\lambda) = \int_0^\infty e^{-\lambda t} f(t)\,\mathrm{d}t, \qquad f(t) = \mathcal{L}^{-1}[F(\lambda)](t).$$

It has already been demonstrated that the Laplace transform $f^L(\lambda) = (\mathcal{L}f)(\lambda)$ of a given function $f(t)$ can be calculated either by direct integration or by using some straightforward rules. As we will see later, many practical problems, when solved using the Laplace transformation, provide us with $F(\lambda)$, the Laplace transform of some unknown function of interest. That is, we usually know $F(\lambda)$; however, we need to determine $f(t)$ such that $f(t) = \mathcal{L}^{-1}[F(\lambda)]$. The inverse Laplace transform was a challengeable problem, so it is not surprising that it took a while to discover exact formulas. In 1916, Thomas Bromwich[52] answered the question of how to find this function, $f(t)$, the inverse Laplace transform of a given function $F(\lambda)$. He expressed the inverse Laplace transform as the contour integral

$$\frac{1}{2}[f(t+0) + f(t-0)] = \frac{1}{2\pi\mathbf{j}} \int_{c-\mathbf{j}\infty}^{c+\mathbf{j}\infty} f^L(\lambda)\, e^{\lambda t}\,\mathrm{d}\lambda = \lim_{\omega\to\infty} \frac{1}{2\pi\mathbf{j}} \int_{c-\mathbf{j}\omega}^{c+\mathbf{j}\omega} f^L(\lambda)\, e^{\lambda t}\,\mathrm{d}\lambda, \tag{5.4.1}$$

where c is any number greater than the abscissa of convergence for $f^L(\lambda)$ and the integral takes its Cauchy principal[53] value. Note that the inverse formula (5.4.1) is valid only for $t \geqslant 0$, providing zero for negative t, that is, $\mathcal{L}^{-1}[F(\lambda)](t) = 0$ for $t < 0$.

Remark. From this formula (5.4.1), it follows that the inverse Laplace transform restores a function-original from its image in such a way that the value of a function-original at any point is equal to the mean of its right-hand side limit value and its left-hand side limit value. Algebra of arithmetic operations with function-originals requires reevaluations of the outcome function at the points of discontinuity. For example, this justifies the following relation for the Heaviside functions: $H^2(t) = H(t)$. ■

In this section, we will not use Eq. (5.4.1), as it is very complicated. Instead, we consider three practical methods to find the inverse Laplace transform: *Partial Fraction Decomposition*, the *Convolution Theorem*, and the *Residue Method*. The reader is free to use any of them or all of them. We will restrict ourselves to finding the inverse Laplace transform of rational functions or their products on exponentials; that is,

$$F(\lambda) = \frac{P(\lambda)}{Q(\lambda)} \quad \text{or} \quad F_\alpha(\lambda) = \frac{P(\lambda)}{Q(\lambda)}\, e^{-\alpha\lambda},$$

where $P(\lambda)$ and $Q(\lambda)$ are polynomials (or sometimes entire functions) without common factors. Such functions occur in most applications of the Laplace transform to differential equations with constant coefficients. In this section, it is always assumed that the degree of the denominator is larger than the degree of the numerator.

The case of the product of a rational function and an exponential can be easily reduced to the case without the exponential multiplier by the shift rule (5.2.6). In fact, suppose we know

$$f(t) = \mathcal{L}^{-1}[F(\lambda)](t),$$

the original of a rational function $F(\lambda) = P(\lambda)/Q(\lambda)$. Then according to the shift rule (5.2.6), page 282, we have

$$H(t-a)f(t-a) = \mathcal{L}^{-1}[F(\lambda)\, e^{-a\lambda}].$$

5.4.1 Partial Fraction Decomposition

The fraction of two polynomials in λ

$$F(\lambda) = \frac{P(\lambda)}{Q(\lambda)}$$

[52]Thomas John I'Anson Bromwich (1875–1929) was an English mathematician, and a Fellow of the Royal Society.

[53]We do not discuss the definitions of contour integration and Cauchy principal value because they require a solid knowledge of the theory of a complex variable. We refer the reader to other books, for example, [47].

can be expanded into elementary partial fractions, that is, P/Q can be represented as a linear combination of simple rational functions of the form $1/(\lambda-\alpha)$, $1/(\lambda-\alpha)^2$, and so forth. To do this, it is first necessary to find all nulls of the denominator $Q(\lambda)$ or, equivalently, to find all roots of the equation

$$Q(\lambda) = 0. \tag{5.4.2}$$

Then $Q(\lambda)$ can be factored as

$$Q(\lambda) = c_0(\lambda-\lambda_1)^{m_1}(\lambda-\lambda_2)^{m_2}\cdots(\lambda-\lambda_k)^{m_k},$$

where $\lambda_1, \lambda_2, \ldots, \lambda_k$ are the distinct roots of Eq. (5.4.2) and $m_1, m_2, \ldots, m_k$ are their respective **multiplicities**. A root of Eq. (5.4.2) is called **simple** if its multiplicity equals one. A root which appears twice is often called a **double root**. Recall that a polynomial of degree n has n roots, counting multiplicities, so $m_1+m_2+\cdots+m_k = n$. Thus, if Eq. (5.4.2) has a simple real root $\lambda=\lambda_0$, then the polynomial $Q(\lambda)$ has a factor $\lambda-\lambda_0$. In $F = P/Q$, this factor corresponds to the partial fraction decomposition of the form

$$\frac{A}{\lambda-\lambda_0},$$

where A is a constant yet to be found. The inverse Laplace transform of this fraction is (see Table 280, formula 6)

$$\mathcal{L}^{-1}\left[\frac{A}{\lambda-\lambda_0}\right] = A\,e^{\lambda_0 t}\,H(t),$$

where $H(t)$ is the Heaviside function (5.1.5), page 274.

The attenuation rule (5.2.7) gives us a clue about how to get rid of λ_0 in the denominator. Thus, formula 1 from Table 280 yields

$$\mathcal{L}^{-1}\left[\frac{A}{\lambda}\right] = AH(t) \qquad \text{(with a constant } A).$$

Therefore, $\mathcal{L}\left[e^{\lambda_0 t}\,H(t)\right] = (\lambda-\lambda_0)^{-1}$.

If a polynomial $Q(\lambda)$ has a repeated factor $(\lambda-\lambda_0)^m$, that is, if Eq. (5.4.2) has a root λ_0 with multiplicity m ($m \geqslant 1$), then the partial fraction decomposition of $F = P/Q$ contains a sum of m fractions

$$\frac{A_m}{(\lambda-\lambda_0)^m} + \frac{A_{m-1}}{(\lambda-\lambda_0)^{m-1}} + \cdots + \frac{A_1}{\lambda-\lambda_0}.$$

The inverse Laplace transform of each term is (see Table 280, formula 7)

$$\mathcal{L}^{-1}\left[\frac{A_m}{(\lambda-\lambda_0)^m}\right] = A_m\mathcal{L}^{-1}\left[\frac{1}{(\lambda-\lambda_0)^m}\right] = A_m\,\frac{t^{m-1}}{(m-1)!}\,e^{\lambda_0 t}\,H(t).$$

Suppose a polynomial $Q(\lambda)$ has an unrepeated complex factor $(\lambda-\lambda_0)(\lambda-\overline{\lambda_0})$, where $\lambda_0 = \alpha+\mathbf{j}\beta$, and $\overline{\lambda_0} = \alpha-\mathbf{j}\beta$ is the complex conjugate of λ_0. When the coefficients of $Q(\lambda)$ are real, complex roots occur in conjugate pairs. The pair of conjugate roots of Eq. (5.4.2) corresponding to this factor gives rise to the term

$$\frac{A\lambda+B}{(\lambda-\alpha)^2+\beta^2}$$

in the partial fraction decomposition since

$$(\lambda-\alpha-\mathbf{j}\beta)(\lambda-\alpha+\mathbf{j}\beta) = (\lambda-\alpha)^2+\beta^2.$$

The expansion of the fraction $F = P/Q$ can be rewritten as

$$\frac{A(\lambda-\alpha)+\alpha A+B}{(\lambda-\alpha)^2+\beta^2}.$$

From formulas 9 and 10, Table 280, and the shift rule (5.2.6), we obtain the inverse transform:

$$\mathcal{L}^{-1}\left[\frac{A\lambda+B}{(\lambda-\alpha)^2+\beta^2}\right] = e^{\alpha t}\left[A\cos\beta t + \frac{\alpha A+B}{\beta}\sin\beta t\right]H(t). \tag{5.4.3}$$

If the polynomial $Q(\lambda)$ has the repeated complex factor $[(\lambda - \lambda_0)(\lambda - \overline{\lambda}_0)]^2$, then the sum of the form

$$\frac{A\lambda + B}{[(\lambda-\alpha)^2+\beta^2]^2} + \frac{C\lambda + D}{(\lambda-\alpha)^2+\beta^2}$$

corresponds to this factor in the partial fraction decomposition of $F = P/Q$. The last fraction is in the form that appears in Eq. (5.4.3). To find the inverse Laplace transform of the first factor, we can use formulas 17 and 18 from Table 280 and the shift rule (5.2.6). This leads us to

$$\mathcal{L}^{-1}\left[\frac{A\lambda + B}{[(\lambda-\alpha)^2+\beta^2]^2}\right] = e^{\alpha t}\left[\frac{A}{2\beta}\,t\,\sin\beta t + \frac{\alpha A + B}{2\beta^2}\,(\sin\beta t - \beta t\cos\beta t)\right] H(t).$$

Example 5.4.1: To find the inverse Laplace transform of

$$\frac{1}{(\lambda-a)(\lambda-b)} \qquad (a \ne b),$$

where a and b are real constants, we expand the given fraction into elementary partial fractions:

$$\frac{1}{(\lambda-a)(\lambda-b)} = \frac{A}{(\lambda-a)} + \frac{B}{(\lambda-b)}.$$

The coefficients A and B are determined by multiplying each term by the lowest common denominator $(\lambda-a)(\lambda-b)$. This leads to the equation

$$A(\lambda-b) + B(\lambda-a) = 1,$$

from which we need to determine the unknown constants A and B. We equate the coefficients of like power terms of λ to obtain

$$A + B = 0, \qquad -(Ab + Ba) = 1.$$

Thus, $A = -B = 1/(a-b)$. Now we are in a position to find the inverse Laplace transform of this fraction:

$$\mathcal{L}^{-1}\left[\frac{1}{(\lambda-a)(\lambda-b)}\right] = \frac{1}{a-b}\,\mathcal{L}^{-1}\left[\frac{1}{\lambda-a} - \frac{1}{\lambda-b}\right] = \frac{1}{a-b}\,\left[e^{at} - e^{bt}\right] H(t),$$

where $H(t)$ is the Heaviside function (5.1.5). We multiply the last expression by $H(t)$ because the inverse Laplace transform vanishes for negative values of the argument t, but the functions e^{at} and e^{bt} are positive for all values of t. *Mathematica* confirms our calculations: `Apart[1/((lambda - a) (lambda - b))]`

Example 5.4.2: In a similar way, the fraction

$$\frac{1}{(\lambda^2+a^2)(\lambda^2+b^2)} \qquad (a \ne b)$$

can be expanded as

$$\frac{1}{(\lambda^2+a^2)(\lambda^2+b^2)} = \frac{1}{b^2-a^2}\left[\frac{1}{\lambda^2+a^2} - \frac{1}{\lambda^2+b^2}\right].$$

To check our answer, we type in *Maple:*

```
convert(1/((lambda^2 + a^2)*(lambda^2 + b^2)),parfrac,lambda)
```

Using formula 8 from Table 280, we obtain the inverse Laplace transform:

$$\begin{aligned}
\mathcal{L}^{-1}\left[\frac{1}{(\lambda^2+a^2)(\lambda^2+b^2)}\right] &= \frac{1}{b^2-a^2}\mathcal{L}^{-1}\left[\frac{1}{\lambda^2+a^2} - \frac{1}{\lambda^2+b^2}\right] \\
&= \frac{1}{b^2-a^2}\mathcal{L}^{-1}\left[\frac{1}{\lambda^2+a^2}\right] - \frac{1}{b^2-a^2}\mathcal{L}^{-1}\left[\frac{1}{\lambda^2+b^2}\right] \\
&= \frac{1}{b^2-a^2}\left[\frac{\sin at}{a} - \frac{\sin bt}{b}\right] H(t),
\end{aligned}$$

where $H(t)$ is the Heaviside function. The last expression is multiplied by $H(t)$ because the Laplace transform deals only with functions that are zero for negative values of an argument, but the functions $\sin at$ and $\sin bt$ are not zero for negative values of t.

Example 5.4.3: Find

$$\mathcal{L}^{-1}\left[\frac{\lambda+5}{\lambda^2+2\lambda+5}\right].$$

Solution. We first expand this ratio into partial fractions as follows:

$$\frac{\lambda+5}{\lambda^2+2\lambda+5}=\frac{\lambda+1+4}{(\lambda+1)^2+2^2}=\frac{\lambda+1}{(\lambda+1)^2+2^2}+2\frac{2}{(\lambda+1)^2+2^2}.$$

Since we know from Table 280 that $\mathcal{L}^{-1}\left[\frac{\lambda}{\lambda^2+2^2}\right]=\cos 2t$ and $\mathcal{L}^{-1}\left[\frac{2}{\lambda^2+2^2}\right]=\sin 2t$, the attenuation rule (5.2.7) yields

$$\mathcal{L}^{-1}\left[\frac{\lambda+5}{\lambda^2+2\lambda+5}\right]=\left[e^{-t}\cos 2t+2e^{-t}\sin 2t\right]H(t).$$

Example 5.4.4: Find the inverse Laplace transform of the function

$$F(\lambda)=\frac{4\lambda+5}{4\lambda^2+12\lambda+13}.$$

Solution. We rearrange the denominator to complete squares:

$$4\lambda^2+12\lambda+13=(2\lambda)^2+2\cdot 2\lambda\cdot 3+3^2-3^2+13=(2\lambda+3)^2+4.$$

Hence, we have

$$F(\lambda)=\frac{4\lambda+6-1}{(2\lambda+3)^2+4}=\frac{2(2\lambda+3)}{(2\lambda+3)^2+4}-\frac{1}{(2\lambda+3)^2+4}.$$

Setting $s=2\lambda+3$ gives

$$F(\lambda(s))=\frac{2s}{s^2+4}-\frac{1}{s^2+4}.$$

The inverse Laplace transform of the right-hand side is

$$2\cos 2t\,H(t)-\frac{1}{2}\sin 2t\,H(t).$$

Application of Eq. (5.2.15) yields

$$\begin{aligned}\mathcal{L}^{-1}[f(\lambda)](t) &= \frac{1}{2}e^{3t/2}\left[2\cos\left(2\frac{t}{2}\right)-\frac{1}{2}\sin\left(2\frac{t}{2}\right)\right]H\left(\frac{t}{2}\right)\\ &= e^{3t/2}\left[\cos t-\frac{1}{4}\sin t\right]H(t)\end{aligned}$$

because $H(t/2)=H(t)$.

Example 5.4.5: Find the inverse Laplace transfrom of the fraction

$$\frac{2\lambda^2+6\lambda+10}{(\lambda-1)(\lambda^2+4\lambda+13)}.$$

Solution. Since the denominator $(\lambda-1)(\lambda^2+4\lambda+13)=(\lambda-1)(\lambda+2)^2+3^2)$ has one real null $\lambda=1$ and two complex conjugate nulls $\lambda=-2\pm 3\mathbf{j}$, we expand the given function into partial fractions:

$$\frac{2\lambda^2+6\lambda+10}{(\lambda-1)(\lambda^2+4\lambda+13)}=\frac{1}{\lambda-1}+\frac{\lambda+2}{(\lambda+2)^2+3^2}+\frac{1}{(\lambda+2)^2+3^2}.$$

With the aid of Table 280 and the attenuation rule (5.2.7), it is easy to find the inverse Laplace transform:

$$\mathcal{L}^{-1}\left[\frac{2\lambda^2+6\lambda+10}{(\lambda-1)(\lambda^2+4\lambda+13)}\right]=e^{t}+e^{-2t}\cos 3t+\frac{1}{3}e^{-2t}\sin 3t,\qquad t>0.$$

As usual, we should multiply the right-hand side by the Heaviside function.

Example 5.4.6: Find the inverse Laplace transform of

$$F_2(\lambda) = \frac{\lambda+3}{\lambda^2(\lambda-1)}\,e^{-2\lambda}.$$

Solution. The attenuation rule (5.2.7) suggests that we first find the inverse Laplace transform of the fraction

$$F(\lambda) = \frac{\lambda+3}{\lambda^2(\lambda-1)}.$$

We expand $F(\lambda)$ into partial fractions:

$$\frac{\lambda+3}{\lambda^2(\lambda-1)} = \frac{A}{\lambda^2} + \frac{B}{\lambda} + \frac{C}{\lambda-1}.$$

To determine the unknown coefficients A, B, and C we combine the right-hand side into one fraction to obtain

$$\frac{\lambda+3}{\lambda^2(\lambda-1)} = \frac{A(\lambda-1)+B\lambda(\lambda-1)+C\lambda^2}{\lambda^2(\lambda-1)}.$$

Therefore, the numerator of the right-hand side must be equal to the numerator of the left-hand side. Equating the power like terms, we obtain

$$0 = C+B, \quad 1 = A-B, \quad 3 = -A.$$

Solving for A, B, and C in the system of algebraic equations, we get $A=-3$, $B=-4$, and $C=4$. To check our answer, we use the following *Maxima* command:

```
partfrac((lambda+3)/((lambda-1)*lambda^2)),lambda);
```

Thus, the inverse Laplace transform of F becomes

$$f(t) = \mathcal{L}^{-1}\left[\frac{\lambda+3}{\lambda^2(\lambda-1)}\right] = H(t)\left(4e^t - 3t - 4\right).$$

Now we are in a position to find the inverse Laplace transform of $F_2(\lambda)$. Using the shift rule (5.2.6), we have

$$\begin{aligned} f_2(t) &= \mathcal{L}^{-1}\left[F_2(\lambda)\right] = H(t-2)\left(4e^{t-2} - 3(t-2) - 4\right) \\ &= H(t-2)\left[4e^{t-2} - 3t + 2\right]. \end{aligned}$$

5.4.2 Convolution Theorem

Suppose that a function $F(\lambda)$ is represented as a product of two other functions: $F(\lambda) = F_1(\lambda)\cdot F_2(\lambda)$. Assume that we know the inverse Laplace transforms $f_1(t)$ and $f_2(t)$ of these functions $F_1(\lambda)$ and $F_2(\lambda)$, respectively. Then the inverse Laplace transform of the product $F_1(\lambda)\cdot F_2(\lambda)$ can be defined according to the convolution rule (5.2.1), page 282:

$$\mathcal{L}^{-1}\left[F_1(\lambda)\,F_2(\lambda)\right](t) = \mathcal{L}^{-1}\left[F_1(\lambda)\right] * \mathcal{L}^{-1}\left[F_2(\lambda)\right] = (f_1 * f_2)(t) = \int_0^t f_1(\tau) f_2(t-\tau)\,\mathrm{d}\tau.$$

It turns out that one can calculate the inverse of such a product in terms of the known inverses, with the convolution integral.

Example 5.4.7: Find the inverse Laplace transform of the function

$$F(\lambda) = \frac{1}{\lambda(\lambda-a)} = \frac{1}{\lambda}\cdot\frac{1}{\lambda-a}.$$

Solution. The function F is a product of two functions $F(\lambda) = F_1(\lambda)F_2(\lambda)$, where

$$F_1(\lambda) = \lambda^{-1}, \quad F_2(\lambda) = (\lambda-a)^{-1},$$

with known inverses (see formula 1 from Table 280 on page 281)

$$f_1(t) = \mathcal{L}^{-1}\left\{F_1(\lambda)\right\} = H(t), \quad f_2(t) = \mathcal{L}^{-1}\left\{F_2(\lambda)\right\} = H(t)\,e^{at},$$

where $H(t)$ is the Heaviside function (recall that $H(t) = 1$ for positive t, $H(t) = 0$ for negative t, and $H(0) = 1/2$). Then their convolution becomes

$$\begin{aligned} f(t) = (f_1 * f_2)(t) &= \int_0^t f_1(t-\tau) f_2(\tau)\,\mathrm{d}\tau = \int_0^t H(t-\tau)\,H(\tau)\,e^{a\tau}\,\mathrm{d}\tau \\ &= \int_0^t H(t-\tau)\,e^{a\tau}\,\mathrm{d}\tau = \int_0^t e^{a\tau}\,\mathrm{d}\tau = \frac{e^{at}-1}{a}\,H(t). \end{aligned}$$

As usual, we multiply the result by the Heaviside function.

Example 5.4.8: (Example 5.4.1 revisited) The function $F(\lambda) = (\lambda - a)^{-1}(\lambda - b)^{-1}$ is the product of two functions

$$F_1(\lambda) = \frac{1}{\lambda - a} \quad \text{and} \quad F_2(\lambda) = \frac{1}{\lambda - b}$$

with inverses known to be

$$f_1(t) = \mathcal{L}^{-1}\left[\frac{1}{\lambda - a}\right] = e^{at}\,H(t) \quad \text{and} \quad f_2(t) = \mathcal{L}^{-1}\left[\frac{1}{\lambda - b}\right] = e^{bt}\,H(t).$$

From Eq. (5.2.1) on page 282, it follows that

$$f(t) = \mathcal{L}^{-1}\left[\frac{1}{(\lambda - a)(\lambda - b)}\right] = (f_1 * f_2)(t).$$

Straightforward calculations show that

$$\begin{aligned} (f_1 * f_2)(t) &= \int_0^t f_1(t-\tau) f_2(\tau)\,\mathrm{d}\tau = \int_0^t e^{a(t-\tau)}\,e^{b\tau}\,\mathrm{d}\tau \qquad (t \geqslant 0) \\ &= e^{at} \int_0^t e^{b\tau - a\tau}\,\mathrm{d}\tau = \frac{1}{a-b}\left[e^{at} - e^{bt}\right] H(t). \end{aligned}$$

Example 5.4.9: Find

$$f(t) = \mathcal{L}^{-1}\left[\frac{1}{(\lambda^2 + a^2)(\lambda^2 + b^2)}\right].$$

Solution. The function $F(\lambda)$ is a product of two functions $F_1(\lambda) = (\lambda^2 + a^2)^{-1}$ and $F_2(\lambda) = (\lambda^2 + b^2)^{-1}$. Their inverse Laplace transforms are known from Table 280, namely,

$$f_1(t) = \mathcal{L}^{-1}\left[\frac{1}{\lambda^2 + a^2}\right] = \frac{1}{a}\sin at, \qquad f_2(t) = \mathcal{L}^{-1}\left[\frac{1}{\lambda^2 + b^2}\right] = \frac{1}{b}\sin bt, \quad t \geqslant 0.$$

Hence, their convolution gives us the required function

$$\begin{aligned} f(t) = (f_1 * f_2)(t) &= \frac{1}{ab}\int_0^t \sin a\tau\, \sin b(t-\tau)\,\mathrm{d}\tau \\ &= \frac{1}{2ab}\int_0^t \left[\cos(a\tau - bt + b\tau) - \cos(a\tau + bt - b\tau)\right]\mathrm{d}\tau \\ &= \frac{1}{2ab(a+b)}\int_0^t \mathrm{d}\,\sin(a\tau - bt + b\tau) - \frac{1}{2ab(a-b)}\int_0^t \mathrm{d}\,\sin(a\tau + bt - b\tau) \\ &= \frac{\sin(at - bt + bt) + \sin bt}{2ab(a+b)} - \frac{\sin(at + bt - bt) - \sin bt}{2ab(a-b)} \\ &= \frac{\sin at + \sin bt}{2ab(a+b)} - \frac{\sin at - \sin bt}{2ab(a-b)}, \qquad t \geqslant 0. \end{aligned}$$

Example 5.4.10: To find the inverse Laplace transform $\mathcal{L}^{-1}\left[\frac{1}{\lambda^2(\lambda^2+a^2)}\right]$, we use the convolution rule. Since

$$\mathcal{L}^{-1}\left[\lambda^{-2}\right] = t\,H(t), \qquad \mathcal{L}^{-1}\left[\frac{1}{\lambda^2+a^2}\right] = \frac{\sin at}{a}\,H(t),$$

we get

$$\begin{aligned}
f(t) = \mathcal{L}^{-1}\left[\frac{1}{\lambda^2(\lambda^2+a^2)}\right] &= t * \frac{\sin at}{a} = \frac{1}{a}\int_0^t (t-\tau)\,\sin a\tau\,\mathrm{d}\tau \\
&= \frac{t}{a}\int_0^t \sin a\tau\,\mathrm{d}\tau - \frac{1}{a}\int_0^t \tau\,\sin a\tau\,\mathrm{d}\tau \\
&= -\frac{t}{a}\,\frac{\cos a\tau}{a}\bigg|_{\tau=0}^{\tau=t} + \frac{1}{a^2}\,\tau\,\cos a\tau\bigg|_{\tau=0}^{\tau=t} - \frac{1}{a^2}\int_0^t \cos a\tau\,\mathrm{d}\tau \\
&= \frac{t}{a^2} - \frac{\sin a\tau}{a^3}\bigg|_{\tau=0}^{\tau=t} = \frac{t}{a^2} - \frac{\sin at}{a^3}, \qquad t \geqslant 0.
\end{aligned}$$

Example 5.4.11: Evaluate

$$f(t) = \mathcal{L}^{-1}\left[\frac{1}{(\lambda^2+a^2)^2}\right] = \frac{\sin at}{a} * \frac{\sin at}{a}.$$

According to the definition of convolution, we have

$$\begin{aligned}
f(t) &= \frac{1}{a^2}\int_0^t \sin a\tau\,\sin a(t-\tau)\,\mathrm{d}\tau \qquad (t \geqslant 0) \\
&= \frac{1}{2a^2}\int_0^t \cos(a\tau - at + a\tau)\,\mathrm{d}\tau - \frac{1}{2a^2}\int_0^t \cos(a\tau + at - a\tau)\,\mathrm{d}\tau \\
&= \frac{1}{4a^3}\,(\sin at + \sin at) - \frac{t\cos at}{2a^2} = \frac{1}{2a^3}\,(\sin at - at\cos at)\,H(t).
\end{aligned}$$

5.4.3 The Residue Method

Our presentation of the residue method does not have the generality that is given in more advanced books. Nevertheless, our exposition is adequate to treat the inverse Laplace transformations of the functions appearing in this course. The residue method can be tied to partial fraction decomposition [16], but our novel presentation of the material is independent.

Suppose a function $F(\lambda) = \dfrac{P(\lambda)}{Q(\lambda)}$ is a ratio of two irreducible polynomials (or entire functions). We denote by λ_j, $j = 1, 2, \ldots, N$, all nulls[54] of the denominator $Q(\lambda)$. Then the inverse Laplace transform of the function F can be found as

$$f(t) = \mathcal{L}^{-1}\left\{F(\lambda)\right\} = \sum_{j=1}^{N} \operatorname*{Res}_{\lambda=\lambda_j} F(\lambda)e^{\lambda t}, \tag{5.4.4}$$

where the sum ranges over all zeroes of the equation $Q(\lambda) = 0$, and residues of the function $F(\lambda)e^{\lambda t}$ at the point $\lambda = \lambda_j$ $(j = 1, 2, \ldots, N)$ are evaluated as follows.

If λ_j is a simple root of the equation $Q(\lambda) = 0$, then

$$\operatorname*{Res}_{\lambda_j} F(\lambda)e^{\lambda t} = \frac{P(\lambda_j)}{Q'(\lambda_j)}\,e^{\lambda_j t}. \tag{5.4.5}$$

If λ_j is a double root of $Q(\lambda) = 0$, then

$$\operatorname*{Res}_{\lambda_j} F(\lambda)e^{\lambda t} = \lim_{\lambda\mapsto\lambda_j} \frac{\mathrm{d}}{\mathrm{d}\lambda}\left\{(\lambda-\lambda_j)^2 F(\lambda)\,e^{\lambda t}\right\}. \tag{5.4.6}$$

[54] $N = \infty$ if $Q(\lambda)$ is an entire function.

In general, when λ_j is an n-fold root of $Q(\lambda) = 0$ (that is, it has multiplicity n), then

$$\operatorname*{Res}_{\lambda_j} F(\lambda)e^{\lambda t} = \lim_{\lambda \mapsto \lambda_j} \frac{1}{(n-1)!} \frac{\mathrm{d}^{n-1}}{\mathrm{d}\lambda^{n-1}} \left\{ (\lambda - \lambda_j)^n F(\lambda)\, e^{\lambda t} \right\}. \tag{5.4.7}$$

Note that if $\lambda_j = \alpha + \mathbf{j}\beta$ is a complex null of the denominator $Q(\lambda)$ (the root of the polynomial equation with real coefficients $Q(\lambda) = 0$), then $\overline{\lambda_j} = \alpha - \mathbf{j}\beta$ is also a null of $Q(\lambda)$. In this case, we don't need to calculate the residue (5.4.4) at $\alpha - \mathbf{j}\beta$ because it is known to be the complex conjugate of the residue at $\alpha + \mathbf{j}\beta$:

$$\operatorname*{Res}_{\alpha - \mathbf{j}\beta} \frac{P(\lambda)}{Q(\lambda)}\, e^{\lambda t} = \overline{\left(\operatorname*{Res}_{\alpha + \mathbf{j}\beta} \frac{P(\lambda)}{Q(\lambda)}\, e^{\lambda t} \right)}.$$

Therefore, if the denominator $Q(\lambda)$ has two complex conjugate roots $\lambda = \alpha \pm \mathbf{j}\beta$, then the sum of residues at these points is just double the value of the real part of one of them:

$$\operatorname*{Res}_{\alpha + \mathbf{j}\beta} \frac{P(\lambda)}{Q(\lambda)} + \operatorname*{Res}_{\alpha - \mathbf{j}\beta} \frac{P(\lambda)}{Q(\lambda)} = 2\, \Re \operatorname*{Res}_{\alpha + \mathbf{j}\beta} \frac{P(\lambda)}{Q(\lambda)}. \tag{5.4.8}$$

In the case of simple complex roots, we get

$$\operatorname*{Res}_{\alpha + \mathbf{j}\beta} F(\lambda)e^{\lambda t} + \operatorname*{Res}_{\alpha - \mathbf{j}\beta} F(\lambda)e^{\lambda t} = e^{\alpha t} \left\{ 2A \cos \beta t - 2B \sin \beta t \right\}, \tag{5.4.9}$$

where

$$A = \Re \operatorname*{Res}_{\alpha + \mathbf{j}\beta} F(\lambda) = \operatorname{Re} \frac{P(\alpha + \mathbf{j}\beta)}{Q'(\alpha + \mathbf{j}\beta)}, \qquad B = \Im \operatorname*{Res}_{\alpha + \mathbf{j}\beta} F(\lambda) = \operatorname{Im} \frac{P(\alpha + \mathbf{j}\beta)}{Q'(\alpha + \mathbf{j}\beta)}.$$

Example 5.4.12: We demonstrate the power of the residue method by revisiting Example 5.4.1. Consider the function

$$F(\lambda) = \frac{1}{(\lambda - a)(\lambda - b)}.$$

According to Eq. (5.4.4), we have

$$f(t) = \mathcal{L}^{-1} \left[\frac{1}{(\lambda - a)(\lambda - b)} \right] = \operatorname*{Res}_{\lambda = a} \frac{e^{\lambda t}}{(\lambda - a)(\lambda - b)} + \operatorname*{Res}_{\lambda = b} \frac{e^{\lambda t}}{(\lambda - a)(\lambda - b)}.$$

We denote $P(\lambda) = e^{\lambda t}/(\lambda - b)$ and $Q(\lambda) = \lambda - a$ when evaluating the residue at the point $\lambda = a$ because the function $P(\lambda)$ is well defined in a neighborhood of that point. Then from Eq. (5.4.4), it follows that

$$\operatorname*{Res}_{\lambda = a} \frac{e^{\lambda t}}{(\lambda - a)(\lambda - b)} = \frac{P(a)}{Q'(a)} = P(a) = \left. \frac{e^{\lambda t}}{\lambda - b} \right|_{\lambda = a} = \frac{1}{a - b}\, e^{at}$$

because $Q'(a) = 1$. Similarly we get

$$\operatorname*{Res}_{\lambda = b} \frac{e^{\lambda t}}{(\lambda - a)(\lambda - b)} = \left. \frac{e^{\lambda t}}{\lambda - a} \right|_{\lambda = b} = \frac{1}{b - a}\, e^{bt}.$$

Adding these expressions, we obtain

$$\mathcal{L}^{-1} \left[\frac{1}{(\lambda - a)(\lambda - b)} \right] = \frac{1}{a - b} \left[e^{at} - e^{bt} \right] H(t).$$

Note that we always have to multiply the inverse Laplace transformation by the Heaviside function to ensure that this function vanishes for negative t. However, sometimes we are too lazy to do this, so the reader is expected to finish our job.

Example 5.4.13: Consider the function

$$F(\lambda) = \frac{1}{(\lambda - a)^2(\lambda - b)}.$$

Using the residue method, we get its inverse Laplace transform

$$f(t) = \mathcal{L}^{-1}\left[\frac{1}{(\lambda-a)^2(\lambda-b)}\right] = \operatorname*{Res}_{\lambda=a} \frac{e^{\lambda t}}{(\lambda-a)^2(\lambda-b)} + \operatorname*{Res}_{\lambda=b} \frac{e^{\lambda t}}{(\lambda-a)^2(\lambda-b)}.$$

Since the $\lambda = a$ is a double root, we apply formula (5.4.6) and obtain

$$\operatorname*{Res}_{\lambda=a} \frac{e^{\lambda t}}{(\lambda-a)^2(\lambda-b)} = \frac{\mathrm{d}}{\mathrm{d}\lambda}\frac{e^{\lambda t}}{\lambda-b}\bigg|_{\lambda=a} = \frac{t\,e^{\lambda t}(\lambda-b) - e^{\lambda t}}{(\lambda-b)^2}\bigg|_{\lambda=a} = \frac{t}{a-b}\,e^{at} - \frac{1}{(a-b)^2}\,e^{at}.$$

The residue at $\lambda = b$ is evaluated according to Eq. (5.4.5):

$$\operatorname*{Res}_{\lambda=b} \frac{e^{\lambda t}}{(\lambda-a)^2(\lambda-b)} = \frac{e^{\lambda t}}{(\lambda-a)^2}\bigg|_{\lambda=b} = \frac{e^{bt}}{(b-a)^2}.$$

Therefore,

$$f(t) = \mathcal{L}^{-1}\left[\frac{1}{(\lambda-a)^2(\lambda-b)}\right] = \frac{t}{a-b}\,e^{at} - \frac{1}{(a-b)^2}\,e^{at} + \frac{e^{bt}}{(b-a)^2}, \qquad t > 0.$$

If one prefers to use the partial fraction decomposition, then we have to break the given function into the sum of simple terms:

$$\frac{1}{(\lambda-a)^2(\lambda-b)} = \frac{1}{a-b}\,\frac{1}{(\lambda-a)^2} - \frac{1}{(a-b)^2}\,\frac{1}{\lambda-a} + \frac{1}{a-b}\,\frac{1}{\lambda-b}.$$

Since the inverse Laplace transform of every term is known, we get the same result.

Application of the convolution rule yields:

$$\begin{aligned} f(t) &= \mathcal{L}^{-1}\left[\frac{1}{(\lambda-a)^2(\lambda-b)}\right] = \mathcal{L}^{-1}\left[\frac{1}{(\lambda-a)^2}\right] * \mathcal{L}^{-1}\left[\frac{1}{\lambda-b}\right] \\ &= t\,e^{at} * e^{bt} = \int_0^t \tau\,e^{a\tau}\,e^{b(t-\tau)}\,\mathrm{d}\tau = e^{bt}\int_0^t \tau\,e^{(a-b)\tau}\,\mathrm{d}\tau \\ &= e^{bt}\,\frac{1}{(a-b)^2}\left[(-1+(a-b)t)\,e^{(a-b)\tau}\right]\bigg|_{\tau=0}^{\tau=t} \\ &= e^{bt}\,\frac{1}{(a-b)^2}\left[(-1+(a-b)t)\,e^{(a-b)t}\right] + e^{bt}\,\frac{1}{(a-b)^2}. \end{aligned}$$

Example 5.4.14: (Example 5.4.10 revisited) Using the residue method, find the inverse Laplace transform

$$f(t) = \mathcal{L}^{-1}\left[\frac{1}{\lambda^2(\lambda^2+a^2)}\right].$$

Solution. The denominator $\lambda^2(\lambda^2+a^2)$ has one double null $\lambda = 0$ and two simple pure imaginary nulls $\lambda = \pm\mathbf{j}a$. The residue at the point $\lambda = 0$ can be found according to the formula (5.4.6):

$$\operatorname*{Res}_{0}\left\{\frac{e^{\lambda t}}{\lambda^2(\lambda^2+a^2)}\right\} = \lim_{\lambda\mapsto 0}\frac{\mathrm{d}}{\mathrm{d}\lambda}\left[\frac{e^{\lambda t}}{\lambda^2+a^2}\right] = \lim_{\lambda\mapsto 0}\frac{t\,e^{\lambda t}(\lambda^2+a^2) - 2\lambda\,e^{\lambda t}}{(\lambda^2+a^2)^2} = \frac{t}{a^2}.$$

To find the residues at the simple roots $\lambda = \pm\mathbf{j}a$, we set $P(\lambda) = \lambda^{-2}\,e^{\lambda t}$ and $Q(\lambda) = \lambda^2 + a^2$. Then with the benefit of the formula (5.4.5), we get

$$\operatorname*{Res}_{\pm\mathbf{j}a}\left\{\frac{e^{\lambda t}}{\lambda^2(\lambda^2+a^2)}\right\} = \lim_{\lambda\mapsto\pm\mathbf{j}a}\frac{e^{\lambda t}}{\lambda^2\cdot 2\lambda} = \frac{e^{\pm\mathbf{j}at}}{2(\pm\mathbf{j}a)^3} = -\frac{\cos at \pm \mathbf{j}\sin at}{\pm 2\mathbf{j}a^3}.$$

Summing all residues, we obtain the inverse Laplace transform:

$$f(t) = \frac{t}{a^2} - \frac{e^{\mathbf{j}at}}{2\mathbf{j}a^3} - \frac{e^{-\mathbf{j}at}}{-2\mathbf{j}a^3} = \frac{t}{a^2} - \frac{\sin at}{a^3}$$

because $\sin at = \frac{e^{\mathrm{j}at}}{2\mathrm{j}} - \frac{e^{-\mathrm{j}at}}{2\mathrm{j}}$.

Now we find $f(t)$ using partial fraction decomposition as follows:

$$\frac{1}{\lambda^2(\lambda^2+a^2)} = \frac{1}{a^2}\,\frac{1}{\lambda^2} - \frac{1}{a^2}\,\frac{1}{\lambda^2+a^2} = \frac{1}{a^2}\,\frac{1}{\lambda^2} - \frac{1}{a^3}\,\frac{a}{\lambda^2+a^2}.$$

The inverse Laplace transforms of each term are known from Example 5.1.2 (page 272) and Example 5.1.10 (page 278) to be

$$\mathcal{L}^{-1}\left[\frac{1}{\lambda^2(\lambda^2+a^2)}\right] = \frac{1}{a^2}\,\mathcal{L}^{-1}\left[\frac{1}{\lambda^2}\right] - \frac{1}{a^3}\,\mathcal{L}^{-1}\left[\frac{a}{\lambda^2+a^2}\right] = \frac{t}{a^2} - \frac{\sin at}{a^3}.$$

Example 5.4.15: Find the inverse Laplace transform

$$f(t) = \mathcal{L}^{-1}\left[\frac{\lambda^2+5\lambda-4}{\lambda^3+3\lambda^2+2\lambda}\right].$$

Solution. We start with the partial fraction decomposition method. The equation $\lambda^3+3\lambda^2+2\lambda = 0$ has three simple roots $\lambda = 0$, $\lambda = -1$, and $\lambda = -2$. Therefore,

$$\frac{\lambda^2+5\lambda-4}{\lambda^3+3\lambda^2+2\lambda} = \frac{A}{\lambda} + \frac{B}{\lambda+1} + \frac{C}{\lambda+2}$$

with three unknowns A, B, and C. Multiplying each term on the right-hand side of the last equation by the common denominator, $\lambda(\lambda+1)(\lambda+2)$, we get

$$A(\lambda+1)(\lambda+2) + B\lambda(\lambda+2) + C\lambda(\lambda+1) = \lambda^2+5\lambda-4.$$

We equate coefficients of like powers of λ to obtain $A = -2$, $B = 8$, and $C = -5$. Thus, the final decomposition has the form

$$\frac{\lambda^2+5\lambda-4}{\lambda^3+3\lambda^2+2\lambda} = -\frac{2}{\lambda} + \frac{8}{\lambda+1} - \frac{5}{\lambda+2}.$$

The inverse Laplace transform of each partial fraction can be found from Table 280. The result is

$$f(t) = \mathcal{L}^{-1}\left[-\frac{2}{\lambda} + \frac{8}{\lambda+1} - \frac{5}{\lambda+2}\right] = \left(-2 + 8\,e^{-t} - 5\,e^{-2t}\right)H(t).$$

Now we attack the inverse formula of the same fraction using the residue method. For simplicity, we denote the numerator, $\left(\lambda^2+5\lambda-4\right)e^{\lambda t}$, as $P(\lambda)$ and the denominator, $\lambda^3+3\lambda^2+2\lambda$, as $Q(\lambda)$. The derivative of the denominator is $Q'(\lambda) = 3\lambda^2+6\lambda+2$. Substituting into this formula $\lambda = 0, -1$, and -2 yields $Q'(0) = 2$, $Q'(-1) = -1$, and $Q'(-2) = 2$, respectively. The corresponding values of the numerator at these points are $P(0) = -4$, $P(-1) = -8\,e^{-t}$, and $P(-2) = -10\,e^{-2t}$. Since all roots of the equation $Q(\lambda) = 0$ are simple, Eq. (5.4.5) can be used to evaluate residues

$$f(t) = \frac{P(0)}{Q'(0)} + \frac{P(-1)}{Q'(-1)}\,e^{-t} + \frac{P(-2)}{Q'(-2)}\,e^{-2t} = \left(-2 + 8\,e^{-t} - 5\,e^{-2t}\right)H(t).$$

Example 5.4.16: Consider the function

$$F(\lambda) = \frac{e^{-c\lambda}}{(\lambda-a)(\lambda-b)}.$$

From Example 5.4.12, we know that

$$f(t) = \mathcal{L}^{-1}\left[\frac{1}{(\lambda-a)(\lambda-b)}\right] = \frac{1}{a-b}\left(e^{at} - e^{bt}\right)H(t).$$

Using the shift rule (5.2.6), page 282, we obtain

$$\mathcal{L}^{-1}\left[\frac{e^{-c\lambda}}{(\lambda-a)(\lambda-b)}\right] = f(t-c)H(t-c) = \frac{1}{a-b}\left[e^{a(t-c)} - e^{b(t-c)}\right]H(t-c).$$

On the other hand, if we apply the residue method, then formally we have

$$\begin{aligned}\mathcal{L}^{-1}\left[\frac{e^{-c\lambda}}{(\lambda-a)(\lambda-b)}\right] &= \operatorname*{Res}_{\lambda=a}\frac{e^{(t-c)\lambda}}{(\lambda-a)(\lambda-b)}+\operatorname*{Res}_{\lambda=b}\frac{e^{(t-c)\lambda}}{(\lambda-a)(\lambda-b)}\\ &= \frac{1}{a-b}\left[e^{a(t-c)}-e^{b(t-c)}\right];\end{aligned}$$

however, we obtain the correct answer only after multiplying this result by the shifted Heaviside function, $H(t-c)$. □

Finally, we present two examples to show that, in general, the considered methods (partial fraction decomposition, convolution, and the residue theorem) are not applicable directly; however, the inverse Laplace transform can be found with the aid of other techniques.

Example 5.4.17: Suppose we want to find the inverse Laplace transform $\mathcal{L}^{-1}\left[\ln\frac{\lambda+1}{\lambda+2}\right]$. No methods have been discussed so far for finding inverses of functions that involve logarithms. However, note that

$$\frac{\mathrm{d}}{\mathrm{d}\lambda}\ln\frac{\lambda+1}{\lambda+2}=\frac{\mathrm{d}}{\mathrm{d}\lambda}\left(\ln(\lambda+1)-\ln(\lambda+2)\right)=\frac{1}{\lambda+1}-\frac{1}{\lambda+2}$$

and that $\mathcal{L}^{-1}\left[\frac{1}{\lambda+1}-\frac{1}{\lambda+2}\right]=\left[e^{-t}-e^{-2t}\right]H(t)$. Then according to the multiplicity rule (5.2.11), page 283, we have

$$\mathcal{L}^{-1}\left[\ln\frac{\lambda+1}{\lambda+2}\right]=\frac{e^{-2t}-e^{-t}}{t}\,H(t)\,.$$

Example 5.4.18: In this example, we consider finding inverses of Laplace transforms of periodic functions. For instance, using the geometric series, we get

$$\begin{aligned}\mathcal{L}^{-1}\left[\frac{1-e^{-\lambda}}{\lambda(1+e^{-\lambda})}\right] &= \mathcal{L}^{-1}\left[\frac{1}{\lambda}\sum_{n=0}^{\infty}(-1)^n\,e^{-n\lambda}-\frac{1}{\lambda}\sum_{n=0}^{\infty}(-1)^n e^{-\lambda}e^{-n\lambda}\right]\\ &= \mathcal{L}^{-1}\left[\frac{1}{\lambda}\sum_{n=0}^{\infty}(-1)^n\,e^{-n\lambda}-\frac{1}{\lambda}\sum_{n=0}^{\infty}(-1)^n e^{-(n+1)\lambda}\right]\\ &= \mathcal{L}^{-1}\left[\frac{1}{\lambda}\sum_{n=0}^{\infty}(-1)^n\,e^{-n\lambda}-\frac{1}{\lambda}\sum_{k=1}^{\infty}(-1)^{k+1} e^{-k\lambda}\right]\\ &= \mathcal{L}^{-1}\left[\frac{1}{\lambda}+\frac{2}{\lambda}\sum_{n=1}^{\infty}(-1)^n\,e^{-n\lambda}\right]=H(t)+2\sum_{n=1}^{\infty}(-1)^n\,H(t-n).\end{aligned}$$

Thus, the inverse of the given function is the periodic function $f(t)=1$ for $0\leqslant t<1$ and $f(t)=-1$ for $1\leqslant t<2$, with the period 2: $f(t)=f(t+2)$.

Problems

1. Determine the inverse Laplace transforms by inspection.
   (a) $\frac{3}{\lambda+2}$; **(b)** $\frac{5-3\lambda}{\lambda^2-9}$; **(c)** $\frac{\lambda}{(\lambda-3)^2}$; **(d)** $\frac{4-8\lambda}{\lambda^2-4}$; **(e)** $\frac{5\lambda}{(\lambda+1)^2}$;
   (f) $\frac{2\lambda+6}{\lambda^2+4}$; **(g)** $\frac{2}{(\lambda-4)^3}$; **(h)** $\frac{2\lambda-3}{\lambda^2+9}$; **(i)** $\frac{\lambda+2}{\lambda^2+1}$; **(j)** $\frac{(\lambda+1)^3}{\lambda^4}$;
   (k) $\frac{18(\lambda+2)}{9\lambda^2-1}$; **(l)** $\frac{4\lambda}{4\lambda^2+1}$; **(m)** $\frac{(\lambda-2)^2}{\lambda^4}$; **(n)** $\frac{3}{9\lambda^2-1}$; **(o)** $\frac{9}{9\lambda^2+1}$.
2. Use partial fraction decomposition to determine the inverse Laplace transforms.
   (a) $\frac{8}{(\lambda+5)(\lambda-3)}$; **(b)** $\frac{1}{\lambda(\lambda^2+a^2)}$; **(c)** $\frac{20(\lambda^2+1)}{\lambda(\lambda+1)(\lambda-4)}$; **(d)** $\frac{9\lambda^2-36}{\lambda(\lambda^2-9)}$;
   (e) $\frac{5\lambda^2-\lambda+1}{\lambda(\lambda^2-1)}$; **(f)** $\frac{2\lambda+1}{(\lambda-7)(\lambda+3)}$; **(g)** $\frac{4\lambda}{(\lambda+2)(\lambda^2+4)}$; **(h)** $\frac{10\lambda}{(\lambda-1)(\lambda^2+9)}$;
   (i) $\frac{4\lambda-8}{\lambda(\lambda+1)(\lambda^2+1)}$; **(j)** $\frac{3\lambda-6}{\lambda^2(\lambda+1)}$; **(k)** $\frac{3\lambda-9}{\lambda(\lambda^2+9)}$; **(l)** $\frac{8}{\lambda^2-2\lambda-15}$;
   (m) $\frac{5}{\lambda^2+\lambda-6}$; **(n)** $\frac{2\lambda^2+2\lambda+4}{(\lambda-1)^2(\lambda+3)}$; **(o)** $\frac{4\lambda+1}{4\lambda^2+4\lambda+5}$; **(p)** $\frac{\lambda+2}{2\lambda^2+\lambda-15}$.
3. Use the attenuation rule (5.2.7) to find the inverse Laplace transforms.
   (a) $\dfrac{2\lambda+3}{\lambda^2-4\lambda+20}$. **(b)** $\dfrac{5}{\lambda^2+4\lambda+29}$. **(c)** $\dfrac{\lambda-1}{\lambda^2-2\lambda+5}$. **(d)** $\dfrac{4\lambda-2}{4\lambda^2-4\lambda+10}$.

4. Use the convolution rule to find the inverse Laplace transforms.

(a) $\frac{1}{\lambda^3}$; (b) $\frac{4}{\lambda^2(\lambda+2)}$; (c) $\frac{4}{(\lambda+3)(\lambda-1)}$; (d) $\frac{4}{\lambda(\lambda+2)^2}$;
(e) $\frac{4}{(\lambda-1)^2(\lambda+1)}$; (f) $\frac{54}{(\lambda-4)^2(\lambda+2)^2}$; (g) $\frac{13}{(\lambda+9)(\lambda-4)}$; (h) $\frac{24}{(\lambda-2)^5}$;
(i) $\frac{80\lambda}{(9\lambda^2+1)(\lambda^2+9)}$; (j) $\frac{82\lambda}{(9\lambda^2+1)(\lambda^2-9)}$; (k) $\frac{2\lambda}{(\lambda^2-1)^2}$; (l) $\frac{16}{\lambda(\lambda^2+4)^2}$;
(m) $\frac{24\lambda^2}{(\lambda-1)^5}$; (n) $\frac{63\lambda}{(\lambda^2+4)(16\lambda^2+1)}$; (o) $\frac{4(\lambda-1)}{\lambda^2(\lambda+2)}$; (p) $\frac{8\lambda+8}{(9\lambda^2+1)(\lambda^2-1)}$.

5. Use the residue method to determine the inverse Laplace transforms.

(a) $\frac{2\lambda+1}{(\lambda-1)(\lambda+2)}$; (b) $\frac{2-\lambda}{\lambda^2-4}$; (c) $\frac{\lambda^2+4}{\lambda(\lambda+1)(\lambda-4)}$;
(d) $\frac{2\lambda-8}{\lambda(\lambda^2+4)}$; (e) $\frac{2\lambda+3}{\lambda(\lambda+3)}$; (f) $\frac{13\lambda^2+2\lambda+126}{(\lambda-2)(\lambda^2+9)}$;
(g) $\frac{(\lambda+1)^2}{(\lambda-1)^3}$; (h) $\frac{(\lambda+1)^2+4}{(\lambda-2)(\lambda^2+9)}$; (i) $\frac{4\lambda+12}{(\lambda^2+6\lambda+13)^2}$;
(j) $\frac{4\lambda+16}{(\lambda^2+6\lambda+13)(\lambda-1)}$; (k) $\frac{4\lambda-16}{(\lambda^2+4)^2}$; (l) $\frac{8\lambda+16}{(\lambda^2-4)^2}$;
(m) $\frac{8\lambda-16}{(\lambda^2-4\lambda+13)(\lambda^2-4\lambda+5)}$; (n) $\frac{\lambda+2}{(\lambda^2-4)(\lambda^2-4\lambda+5)}$; (o) $\frac{\lambda+8}{\lambda^2+4\lambda+40}$.

6. Apply the residue method and then check your answer by using partial fraction decomposition to determine the inverse Laplace transforms.

(a) $\frac{2}{\lambda^2-6\lambda+10}$; (b) $\frac{\lambda-3}{\lambda^2-6\lambda+34}$; (c) $\frac{6\lambda+4}{\lambda^2+2\lambda+5}$;
(d) $\frac{4\lambda}{4\lambda^2-4\lambda+5}$; (e) $\frac{18\lambda+6}{(\lambda+1)^4}$; (f) $\frac{\lambda}{\lambda^2+8\lambda+16}$;
(g) $\frac{3\lambda+12}{(\lambda+2)(\lambda^2-1)}$; (h) $\frac{\lambda+1}{\lambda^2+2\lambda+10}$; (i) $\frac{6\lambda+4}{\lambda^2+2\lambda+5}$;
(j) $\frac{8\lambda}{4\lambda^2-4\lambda+5}$; (k) $\frac{16}{(\lambda^2-4\lambda+8)^2}$; (l) $\frac{39}{(\lambda^2-6\lambda+10)(\lambda^2+1)}$;
(m) $\frac{3(\lambda+1)}{(\lambda^2-4\lambda+13)(\lambda-2)}$; (n) $\frac{5\lambda-10}{(\lambda^2-4\lambda+13)(\lambda^2-4\lambda+8)}$; (o) $\frac{8(\lambda^3-33\lambda^2+90\lambda-11)}{(\lambda^2-6\lambda+13)(\lambda^2+4\lambda-12)}$.

7. Find the inverse Laplace transform.

(a) $\frac{20-(\lambda-2)(\lambda+3)}{(\lambda-1)(\lambda-2)(\lambda+3)}$; (b) $\frac{\lambda+9}{(\lambda-1)(\lambda^2+9)}$; (c) $\frac{(\lambda+2)(\lambda+1)-(\lambda-3)(\lambda-2)}{(\lambda-1)(\lambda+2)(\lambda-3)}$;
(d) $\frac{2\lambda^2+\lambda-28}{(\lambda+1)(\lambda-2)(\lambda+4)}$; (e) $\frac{12\lambda+2}{(\lambda^2+2\lambda+5)(\lambda-2)^2}$; (f) $\frac{17(\lambda+1)}{(\lambda^2-4\lambda+5)(\lambda+2)}$;
(g) $\frac{(\lambda+2)(\lambda+1)-3(\lambda-3)(\lambda-2)}{(\lambda-1)(\lambda+2)(\lambda-3)}$; (h) $\frac{5\lambda^2+5\lambda+10}{(\lambda-1)(\lambda-2)(\lambda+3)}$; (i) $\frac{5\lambda+29}{(\lambda^2+2\lambda+5)(\lambda-2)^2}$;
(j) $\frac{\lambda-1}{(\lambda^2-4\lambda+5)(\lambda-2)}$; (k) $\frac{\lambda+6}{(\lambda^2+4)(\lambda+2)}$; (l) $\frac{10\lambda}{(\lambda^2+9)(\lambda-1)}$;
(m) $\frac{2\lambda+8}{(\lambda^2-4\lambda+13)(\lambda-1)}$; (n) $\frac{7\lambda-11}{(\lambda^2-4\lambda+13)(\lambda+2)}$; (o) $\frac{6\lambda+6}{(4\lambda^2-4\lambda+10)(\lambda-2)}$.

8. Using each of the methods: partial fraction decomposition, convolution, and residue method, find the inverse Laplace transform of the function $1/(\lambda-a)(\lambda-b)(\lambda-c)$.

9. Find the inverse of each of the following Laplace transforms.

(a) $\ln(1+4/\lambda^2)$; (b) $\ln\frac{\lambda+a}{\lambda+b}$; (c) $\frac{2+e^{-\lambda}}{\lambda}$;
(d) $\frac{2\,e^{-3\lambda}}{2\lambda^2+1}$; (e) $\ln\frac{\lambda+3}{\lambda+2}$; (f) $\ln\frac{\lambda^2+1}{\lambda(\lambda+3)}$.

10. Using the shift rule (5.2.6), find the inverse Laplace transform of each of the following functions.

(a) $\frac{2\lambda^2-7\lambda+8}{\lambda(\lambda-2)^2}\,e^{-\lambda}$. (b) $\frac{2\lambda+1}{\lambda^2+\lambda-2}\,e^{-2\lambda}$. (c) $\frac{\lambda+3}{\lambda^2+6\lambda+10}\,e^{-\lambda}$.
(d) $\frac{2\lambda^3-4\lambda^2+10\lambda}{(\lambda-1)(\lambda^2-1)^2}\,e^{-2\lambda}$. (e) $\frac{2\lambda}{2\lambda^2+6\lambda+5}\,e^{-3\lambda}$. (f) $\frac{30}{\lambda(\lambda^2-2\lambda+10)}\,e^{-2\lambda}$.
(g) $\frac{4-4\,e^{-\pi\lambda}}{\lambda(\lambda^2+4)}$. (h) $\frac{\pi-\pi\,e^{-\lambda}}{\lambda^2+\pi^2}$. (i) $\frac{3+3\,e^{-\pi\lambda}}{\lambda^2+9}$.
(j) $\frac{9+9\,e^{-2\lambda}}{\lambda(\lambda^2+9)}$. (k) $\frac{3\lambda^2-2\lambda}{\lambda^4+5\lambda^2+4}\,e^{-3\lambda}$. (l) $\frac{2\lambda^3-\lambda^2}{(4\lambda^2-4\lambda+5)^2}\,e^{-\lambda}$.

11. Use the convolution formula to find the inverse Laplace transform of each of the following functions.

(a) $\frac{1}{\lambda^2+1}\,f^L$; (b) $\frac{2\,e^{-2\lambda}}{\lambda^3}\,f^L$; (c) $\frac{\lambda}{\lambda^2-4}\,f^L$; (d) $\frac{2}{(\lambda-1)^2+4}\,f^L$.

12. Show that for any integer $n \geqslant 1$ and any $a \neq 0$,

$$\mathcal{L}^{-1}\left[\frac{1}{(\lambda^2+a^2)^{n+1}}\right] = \frac{1}{2n}\int_0^t t\mathcal{L}^{-1}\left[\frac{1}{(\lambda^2+a^2)^n}\right]\,\mathrm{d}t.$$

13. Use the formula from the previous exercise to show that

$$\mathcal{L}^{-1}\left[\frac{1}{(\lambda^2+a^2)^{n+1}}\right] = \frac{1}{2^n an!}\underbrace{\int_0^t t\int_0^t t\cdots\int_0^t}_{n \text{ times}} t\sin at\,\mathrm{d}t\,\mathrm{d}t\cdots\mathrm{d}t.$$

5.5 Homogeneous Differential Equations

In Section 5.2, we established some properties about Laplace transforms. In particular, we got the differential rule, which we reformulate below as a theorem.

Theorem 5.6: Let a function-original $f(t)$ be defined on a positive half-line, have $n-1$ first continuous derivatives f', f'', ..., $f^{(n-1)}$, and a piecewise continuous derivative $f^{(n)}(t)$ on any finite interval $0 \leqslant t \leqslant t^* < \infty$. Suppose further that $f(t)$ and all its derivatives through $f^{(n)}(t)$ are of exponential order; that is, there exist constants T, s, M such that $|f(t)| \leqslant Me^{st}$, $|f'(t)| \leqslant Me^{st}$, ..., $|f^{(n)}(t)| \leqslant Me^{st}$ for $t > T$. Then the Laplace transform of $f^{(n)}(t)$ exists when $\Re\lambda > s$, and it has the following form:

$$\mathcal{L}\left[f^{(n)}(t)\right](\lambda) = \lambda^n \mathcal{L}[f](\lambda) - \sum_{k=1}^{n} \lambda^{n-k} f^{(k-1)}(+0). \tag{5.5.1}$$

Here, $f^{(k-1)}(+0)$ means the limit $f^{(k-1)}(+0) = \lim_{t\mapsto+0} f^{(k-1)}(t)$.

This is an essential feature of the Laplace transform when applied to differential equations. It transforms a derivative with respect to t into multiplication by a parameter λ (plus, possibly, some values of a function at the origin). Therefore, the Laplace transform is a very convenient tool to solve initial value problems for differential equations because it reduces the problem under consideration into an algebraic equation. Furthermore, initial conditions are automatically taken into account, and nonhomogeneous equations are handled in exactly the same way as homogeneous ones. When the initial values are not specified, we obtain the general solution. Also, this method has no restrictions on the order of a differential equation, and it can be successfully applied to higher order equations and to systems of differential equations (see §8.5).

We start with the initial value problem for a second order homogeneous linear differential equation with constant coefficients (a, b, and c)

$$ay''(t) + by'(t) + cy(t) = 0, \qquad y(0) = y_0, \quad y'(0) = y_1. \tag{5.5.2}$$

To apply the Laplace transform to the initial value problem (5.5.2), we multiply both sides of the differential equation by $e^{\lambda t}$ and then integrate the results with respect to t from zero to infinity to obtain the integral equation

$$a\int_0^\infty y''(t)\, e^{-\lambda t}\, \mathrm{d}t + b\int_0^\infty y'(t)\, e^{-\lambda t}\, \mathrm{d}t + c\int_0^\infty y(t)\, e^{-\lambda t}\, \mathrm{d}t = 0. \tag{5.5.3}$$

Integrating by parts (or using the differential rule (5.2.2)) yields

$$\begin{aligned}
\int_0^\infty y''(t)\, e^{-\lambda t}\, \mathrm{d}t &= \int_0^\infty e^{-\lambda t}\, \mathrm{d}\,(y') = y'\, e^{-\lambda t}\Big|_{t=0}^{\infty} + \lambda \int_0^\infty y'(t)\, e^{-\lambda t}\, \mathrm{d}t \\
&= -y'(0) + \lambda \int_0^\infty e^{-\lambda t}\, \mathrm{d}\,(y) \\
&= -y'(0) + \lambda\, y(t)\, e^{-\lambda t}\Big|_{t=0}^{\infty} + \lambda^2 \int_0^\infty y(t)\, e^{-\lambda t}\, \mathrm{d}t \\
&= -y'(0) - \lambda\, y(0) + \lambda^2 \int_0^\infty y(t)\, e^{-\lambda t}\, \mathrm{d}t = -y'(0) - \lambda\, y(0) + \lambda^2\, y^L(\lambda),
\end{aligned}$$

and analogously

$$\begin{aligned}
\int_0^\infty y'(t)\, e^{-\lambda t}\, \mathrm{d}t &= y\, e^{-\lambda t}\Big|_{t=0}^{\infty} + \lambda \int_0^\infty y(t)\, e^{-\lambda t}\, \mathrm{d}t \\
&= -y(0) + \lambda \int_0^\infty y(t)\, e^{-\lambda t}\, \mathrm{d}t = -y(0) + \lambda\, y^L(\lambda)
\end{aligned}$$

since the terms

$$y'(t)\, e^{-\lambda t} \quad \text{and} \quad y(t)\, e^{-\lambda t}$$

approach zero at infinity (when $t \to +\infty$); the general formula for integration by parts is

$$\int_a^b u(t)\,\mathrm{d}v(t) = v(b)\,u(b) - v(a)\,u(a) - \int_a^b u'(t)\,v(t)\,\mathrm{d}t.$$

From the integral equation (5.5.3), we get

$$a\left[\int_0^\infty y(t)\,e^{-\lambda t}\,\mathrm{d}t - \lambda y_0 - y_0'\right] + b\left[\int_0^\infty y(t)\,e^{-\lambda t}\,\mathrm{d}t - y_0\right] + c\int_0^\infty y(t)\,e^{-\lambda t}\,\mathrm{d}t = 0.$$

If we denote by

$$y^L(\lambda) = \int_0^\infty y(t)\,e^{-\lambda t}\,\mathrm{d}t$$

the Laplace transform of the unknown function, then the integral equation (5.5.3) can be written in compact form:

$$(a\lambda^2 + b\lambda + c)y^L = a(\lambda y_0 + y_0') + by_0. \tag{5.5.4}$$

Therefore, the Laplace transform reduces the initial value problem (5.5.2) to the algebraic (subsidiary) equation (5.5.4) for the Laplace transform of the unknown function, y^L. Solving Eq. (5.5.4) for y^L, we obtain

$$y^L(\lambda) = \frac{a(\lambda y_0 + y_0') + by_0}{a\lambda^2 + b\lambda + c}.$$

Notice the way in which the derivative rule (5.2.2), page 282, automatically incorporates the initial conditions into the Laplace transform of the solution to the given initial value problem. This is just one of many valuable features of the Laplace transform.

Thus, the Laplace transform y^L of the unknown function is represented as a fraction of two polynomials. The denominator of this fraction coincides with the characteristic polynomial of the homogeneous equation (5.5.2). The degree of the numerator is less than the degree of the denominator. Therefore, we can apply methods from the previous section to find the inverse Laplace transform of y^L.

To find the solution of the initial value problem, we apply the residue method to obtain the inverse of the function y^L. If both roots λ_1 and λ_2 of the characteristic equation $a\lambda^2 + b\lambda + c = 0$ are different (that is, the discriminant $b^2 - 4ac \ne 0$), then we apply the formula (5.4.5) to obtain

$$y(t) = \left[\frac{a(\lambda_1 y_0 + y_0') + by_0}{2a\lambda_1 + b}\,e^{\lambda_1 t} + \frac{a(\lambda_2 y_0 + y_0') + by_0}{2a\lambda_2 + b}\,e^{\lambda_2 t}\right] H(t). \tag{5.5.5}$$

If the characteristic equation has one double root $\lambda = \lambda_1$ (that is, $b^2 = 4ac$), then the characteristic polynomial will be $a\lambda^2 + b\lambda + c = a(\lambda - \lambda_1)^2$, where $\lambda_1 = -b/2a$. In this case, we apply the formula (5.4.6) and the solution becomes

$$\begin{aligned} y(t) &= \lim_{\lambda\mapsto\lambda_1} \frac{\mathrm{d}}{\mathrm{d}\lambda}\left\{(\lambda-\lambda_1)^2\,\frac{a(\lambda y_0 + y_0') + by_0}{a(\lambda-\lambda_1)^2}\,e^{\lambda t}\right\} = \frac{1}{a}\lim_{\lambda\mapsto\lambda_1}\frac{\mathrm{d}}{\mathrm{d}\lambda}\left\{[a(\lambda y_0 + y_0') + by_0]\,e^{\lambda t}\right\} \\ &= y_0\,e^{\lambda_1 t} + \frac{t}{a}[a(\lambda_1 y_0 + y_0') + by_0]\,e^{\lambda_1 t} = e^{-bt/2a}\left[y_0 + \frac{bt}{2a}\,y_0 + ty_0'\right] H(t). \end{aligned} \tag{5.5.6}$$

In formula (5.5.5), we use only our knowledge about the multiplicity of the roots of the characteristic equation, but do not acquire whether they are real or complex numbers. Hence, the formula (5.5.5) holds for distinct roots, which may be either real numbers or complex numbers.

Let us consider the initial value problem

$$L_n[\mathtt{D}]y(t) = 0, \qquad y(0) = y_0,\; y'(0) = y_0',\; \ldots,\; y^{n-1}(0) = y_0^{(n-1)}, \tag{5.5.7}$$

where $L_n[\mathtt{D}]$ is a linear constant coefficient differential operator of the n-th order, that is,

$$L_n[\mathtt{D}] = a_n\mathtt{D}^n + a_{n-1}\mathtt{D}^{n-1} + \cdots + a_0 = \sum_{k=0}^n a_k\,\mathtt{D}^k, \quad \mathtt{D} = \mathrm{d}/\mathrm{d}t,$$

and a_0, a_1, ..., a_n are some constants. To apply the Laplace transform, we multiply both sides of the differential equation $L_n[\texttt{D}]y(t) = 0$ by $e^{-\lambda t}$ and then integrate the result with respect to t from zero to infinity to obtain

$$a_n \int_0^\infty y^{(n)}(t)\, e^{-\lambda t}\, \mathrm{d}t + a_{n-1} \int_0^\infty y^{(n-1)}(t)\, e^{-\lambda t}\, \mathrm{d}t + \cdots + a_0 \int_0^\infty y(t)\, e^{-\lambda t}\, \mathrm{d}t = 0.$$

We integrate by parts all integrals that contain derivatives of the unknown function $y(t)$. This leads us to the algebraic (subsidiary) equation

$$[a_n\lambda^n + a_{n-1}\lambda^{n-1} + \cdots + a_1\lambda + a_0]y^L = P(\lambda),$$

where

$$y^L = \int_0^\infty y(t)\, e^{\lambda t}\, \mathrm{d}t, \qquad \text{and} \qquad P(\lambda) = \sum_{k=0}^{n} a_k \sum_{s=1}^{k} \lambda^{k-s}\, y^{(s-1)}(+0).$$

We can rewrite the equation in the following form:

$$L(\lambda)y^L = P(\lambda), \tag{5.5.8}$$

with the characteristic polynomial $L(\lambda) = a_n\lambda^n + a_{n-1}\lambda^{n-1} + \cdots + a_0$. Notice that Eq. (5.5.8) can also be obtained from the differential rule:

$$\begin{aligned} \mathcal{L}\left\{L[\texttt{D}]y\right\} &= \mathcal{L}\left[\sum_{k=0}^{n} a_k\, \texttt{D}^k y\right] = \sum_{k=0}^{n} a_k\, \mathcal{L}\left[\texttt{D}^k y\right] \\ &= \sum_{k=0}^{n} a_k \left[\lambda^k y^L(\lambda) - \sum_{s=1}^{k} \lambda^{k-s}\, y^{(s-1)}(+0)\right]. \end{aligned}$$

Equation (5.5.8) is a subsidiary algebraic equation, which is easy to solve:

$$y^L(\lambda) = \frac{P(\lambda)}{L(\lambda)}. \tag{5.5.9}$$

From this equation, we see that the Laplace transform y^L is represented as the fraction of two polynomials $P(\lambda)$ and $L(\lambda)$. Therefore, to find the solution of the initial value problem (5.5.7), we just need to apply the inverse Laplace transform to the right-hand side of Eq. (5.5.9). Various techniques to accomplish this goal are presented in the previous section.

Example 5.5.1: Solve the initial value problem

$$y'' + \omega^2 y = 0, \quad y(+0) = 1,\ y'(+0) = 2.$$

Solution. We transform the differential equation by means of the derivative rule (5.5.1), using the abbreviation y^L for the Laplace transform $y^L \stackrel{\text{def}}{=} \int_0^\infty y(t)\, e^{-\lambda t}\, \mathrm{d}t$ of the unknown solution $y(t)$. This gives

$$\lambda^2 y^L - \lambda y(+0) - y'(+0) + \omega^2 y^L = 0,$$

and, by collecting similar terms, $\left[\lambda^2 + \omega^2\right] y^L = \lambda + 2$. Division by $\lambda^2 + \omega^2$ yields the solution of the subsidiary equation:

$$y^L(\lambda) = \frac{\lambda + 2}{\lambda^2 + \omega^2}.$$

Thus, y^L is a fraction of two polynomials. To find its inverse Laplace transform, we apply the residue method. The equation $\lambda^2 + \omega^2 = 0$ has two simple roots $\lambda_1 = \mathbf{j}\omega$ and $\lambda_2 = -\mathbf{j}\omega$. Applying the formula (5.4.5) on page 308, we obtain the solution of the initial value problem

$$y(t) = \left.\frac{\lambda + 2}{2\lambda}\, e^{\lambda t}\right|_{\lambda = \mathbf{j}\omega} + \left.\frac{\lambda + 2}{2\lambda}\, e^{\lambda t}\right|_{\lambda = -\mathbf{j}\omega} = \left[\frac{1}{2} + \frac{1}{\mathbf{j}\,\omega}\right] e^{\mathbf{j}\omega t} + \left[\frac{1}{2} - \frac{1}{\mathbf{j}\,\omega}\right] e^{-\mathbf{j}\omega t}$$

since the derivative of $\lambda^2 + \omega^2$ is 2λ. Next step is to simplify the answer and show that it is a real-valued function. This can be done with the aid of the Euler formula (4.5.2), page 213; however, the fastest way is to apply Eq. (5.4.8), page 309, and extract the real part:

$$y(t) = 2\,\Re\left[\frac{1}{2} + \frac{1}{\mathbf{j}\omega}\right] e^{\mathbf{j}\omega t} = \Re\, e^{\mathbf{j}\omega t} + \Re\,\frac{2}{\mathbf{j}\omega}\, e^{\mathbf{j}\omega t} = \cos\omega t + \frac{2}{\omega}\,\sin\omega t, \qquad t > 0.$$

We can also find the solution using the partial fraction decomposition method. For this purpose, we rewrite the expression for y^L as

$$y^L(\lambda) = \frac{\lambda}{\lambda^2 + \omega^2} + \frac{2}{\omega}\,\frac{\omega}{\lambda^2 + \omega^2}.$$

Using formulas 7 and 8 from Table 280, we obtain the same expression.

Example 5.5.2: To solve the initial value problem

$$y'' + 5\,y' + 6y = 0, \quad y(0) = 1,\ y'(0) = 0,$$

by the Laplace transform technique, we multiply both sides of the given differential equation by $e^{\lambda t}$ and integrate with respect to t from zero to infinity. This yields

$$\left(\lambda^2 + 5\lambda + 6\right) y^L - \lambda - 5 = 0,$$

where y^L is the Laplace transform of an unknown solution $y(t)$. The solution of the subsidiary equation becomes

$$y^L(\lambda) = \frac{\lambda + 5}{\lambda^2 + 5\lambda + 6} = \frac{\lambda + 5}{(\lambda + 3)(\lambda + 2)}.$$

We expand the right-hand side into partial fractions

$$\frac{\lambda + 5}{(\lambda + 3)(\lambda + 2)} = \frac{A}{\lambda + 3} + \frac{B}{\lambda + 2}$$

with two unknowns A and B. Again, by combining the right-hand side into one fraction, we obtain the relation

$$\frac{\lambda + 5}{(\lambda + 3)(\lambda + 2)} = \frac{A(\lambda + 2) + B(\lambda + 3)}{(\lambda + 3)(\lambda + 2)}.$$

The denominators of these two fractions coincide; therefore, the numerators must be equal; that is,

$$\lambda + 5 = A(\lambda + 2) + B(\lambda + 3) = (A + B)\lambda + 2A + 3B.$$

In this equation, setting $\lambda = -2$ and then $\lambda = -3$, we get $A = -2$ and $B = 3$. Hence,

$$y^L(\lambda) = \frac{-2}{\lambda + 3} + \frac{3}{\lambda + 2}.$$

Applying the inverse Laplace transform, we obtain the solution of the initial value problem:

$$y(t) = 3H(t)\,e^{-2t} - 2H(t)\,e^{-3t}.$$

Following the previous example, the easiest way to find the solution is to apply the residue method. Since we know that the denominator has two simple roots $\lambda_1 = -2$ and $\lambda_2 = -3$, we immediately write down the explicit solution

$$y(t) = \frac{\lambda_1 + 5}{\lambda_1 + 3}\, e^{\lambda_1 t} + \frac{\lambda_2 + 5}{\lambda_2 + 2}\, e^{\lambda_2 t} = \left[3\,e^{-2t} - 2\,e^{-3t}\right] H(t).$$

We can also apply the convolution rule (5.2.1) if we rearrange y^L as follows:

$$y^L(\lambda) = \frac{\lambda + 5}{(\lambda + 3)(\lambda + 2)} = \frac{\lambda + 3 + 2}{(\lambda + 3)(\lambda + 2)} = \frac{1}{\lambda + 2} + \frac{2}{(\lambda + 3)(\lambda + 2)}.$$

Then using formula 6 from Table 280, we get

$$y(t) = \mathcal{L}^{-1}\left[\frac{1}{\lambda + 2}\right] + 2\,\mathcal{L}^{-1}\left[\frac{1}{\lambda + 3}\cdot\frac{1}{\lambda + 2}\right] = e^{-2t}H(t) + 2\,e^{-3t} * e^{-2t}\,H(t).$$

5.5.1 Equations with Variable Coefficients

The Laplace transformation can also be applied to differential equations with polynomial coefficients, as the following examples show.

Example 5.5.3: Let us start with the Euler differential equation

$$t^2 y'' + t\,y' - \frac{1}{4}\,y = 0.$$

Although we know from §4.6.2 that this equation has solutions expressed through power functions, it is instructive to apply the Laplace transform, which is based on the formulas from Problem 13 of §5.2, page 290. This yields the Euler equation in variable λ:

$$\lambda^2\,\frac{\mathrm{d}^2 y^L}{\mathrm{d}\lambda^2} + 3\lambda\,\frac{\mathrm{d}y^L}{\mathrm{d}\lambda} + \frac{3}{4}\,y^L = 0$$

because $\mathcal{L}[ty'] = -y^L + \lambda\mathcal{L}[ty]$ and $\mathcal{L}[t^2 y''] = 2y^L - 4\lambda\mathcal{L}[ty] + \lambda^2\mathcal{L}[t^2 y]$. We seek a solution in the form $y^L(\lambda) = \lambda^k$. Upon substitution into the differential equation, we obtain $k(k-1)+3k+\frac{3}{4} = 0$, which has two solutions $k_1 = -1/2$ and $k_2 = -3/2$. The obtained differential equation for y^L has two linearly independent solutions $y_1^L = \lambda^{-1/2}$ and $y_2^L = \lambda^{-3/2}$. Applying the inverse Laplace transform (see formula 5 in Table 280), we obtain two linearly independent solutions

$$y_1(t) = \mathcal{L}^{-1}\left[\frac{1}{\lambda^{1/2}}\right] = \frac{1}{\Gamma\left(\frac{1}{2}\right)}\,t^{-1/2}\,H(t), \quad y_2(t) = \mathcal{L}^{-1}\left[\frac{1}{\lambda^{3/2}}\right] = \frac{1}{\Gamma\left(\frac{3}{2}\right)}\,t^{1/2}\,H(t),$$

where $\Gamma\left(\frac{1}{2}\right) = \sqrt{\pi}$ and $\Gamma\left(\frac{3}{2}\right) = \sqrt{\pi}/2$.

Example 5.5.4: Solve the initial value problem for the Bessel equation of zero order:

$$t\,y'' + y' + t\,y = 0, \qquad y(0) = 1, \quad y'(0) = 0.$$

Solution. Using the derivative rule (5.2.11) on page 283, we find the Laplace transforms:

$$\begin{aligned}
\mathcal{L}\left[t\,y''\right] &= -\frac{\mathrm{d}}{\mathrm{d}\lambda}\,\mathcal{L}[y''] = -\frac{\mathrm{d}}{\mathrm{d}\lambda}\left[\lambda^2 y^L - \lambda\right] = -2\lambda\,y^L - \lambda^2\frac{\mathrm{d}y^L}{\mathrm{d}\lambda} + 1,\\
\mathcal{L}\left[y'\right] &= \lambda\,y^L - 1,\\
\mathcal{L}\left[t\,y\right] &= -\frac{\mathrm{d}}{\mathrm{d}\lambda}\,\mathcal{L}[y] = -\frac{\mathrm{d}}{\mathrm{d}\lambda}\,y^L.
\end{aligned}$$

This allows us to transfer the original initial value problem to the following separable equation for the Laplace transform y^L:

$$-2\lambda\,y^L - \lambda^2\frac{\mathrm{d}y^L}{\mathrm{d}\lambda} + 1 + \lambda\,y^L - 1 - \frac{\mathrm{d}}{\mathrm{d}\lambda}\,y^L = 0 \qquad \Longleftrightarrow \qquad (\lambda^2+1)\,\frac{\mathrm{d}y^L}{\mathrm{d}\lambda} = \lambda\,y^L.$$

Integration yields a solution $y^L = \left(\lambda^2+1\right)^{-1/2}$; its inverse Laplace transform is a special function, called the Bessel functions of zero order and denoted by $J_0(t)$. □

For a differential equation $a_2(t)\,y''(t) + a_1(t)\,y'(t) + a_0(t)\,y(t) = 0$ whose coefficients are linear in t, it is sometimes possible to find a solution in the form

$$y(t) = \int_{\lambda_1}^{\lambda_2} v(\lambda)\,e^{\lambda t}\,\mathrm{d}\lambda, \tag{5.5.10}$$

where λ_1 and λ_2 are constants and $v(\lambda)$ is a function to be determined. This form (5.5.10) resembles the inverse Laplace transform formula (5.4.1) on page 302.

Let the differential equation of interest be

$$(a_2 t + b_2)\,y'' + (a_1 t + b_1)\,y' + (a_0 t + b_0)\,y = 0\,. \tag{5.5.11}$$

After substituting (5.5.10) into (5.5.11), it becomes

$$t\int_{\lambda_1}^{\lambda_2} V(\lambda)\,e^{\lambda t}\,\mathrm{d}\lambda + \int_{\lambda_1}^{\lambda_2} R(\lambda)\,V(\lambda)\,e^{\lambda t}\,\mathrm{d}\lambda = 0, \tag{5.5.12}$$

where

$$V(\lambda) = (a_2\lambda^2 + a_1\lambda + a_0)\, v(\lambda) \tag{5.5.13}$$

and

$$R(\lambda) = \frac{b_2\lambda^2 + b_1\lambda + b_0}{a_2\lambda^2 + a_1\lambda + a_0}. \tag{5.5.14}$$

Since $t\, e^{\lambda t} = \frac{\mathrm{d}}{\mathrm{d}\lambda}\, e^{\lambda t}$, integration by parts changes (5.5.12) into

$$e^{\lambda t}\, V(\lambda)\Big|_{\lambda=\lambda_1}^{\lambda=\lambda_2} - \int_{\lambda_1}^{\lambda_2} e^{\lambda t}\, [V'(\lambda) - R(\lambda)V(\lambda)]\, \mathrm{d}\lambda = 0\,.$$

This equation is definitely fulfilled if $V'(\lambda) - R(\lambda)V(\lambda) = 0$ and if λ_1 and λ_2 are nulls of the function $V(\lambda)$. Separating the variables in the latter equation, we get

$$\frac{\mathrm{d}V}{V} = R(\lambda)\, \mathrm{d}\lambda \qquad \Longrightarrow \qquad \ln V(\lambda) = \int R(\lambda)\, \mathrm{d}\lambda\,. \tag{5.5.15}$$

Hence the function $v(\lambda)$ in Eq. (5.5.10) is expressed as

$$v(\lambda) = \frac{1}{a_2\lambda^2 + a_1\lambda + a_0}\, e^{\int R(\lambda)\, \mathrm{d}\lambda}\,.$$

Example 5.5.5: Let us solve the differential equation

$$t\, y'' - (2t - 3)\, y' - 4y = 0\,.$$

Solution. Formula (5.5.14) gives

$$R(\lambda) = \frac{3\lambda - 4}{\lambda^2 - 2\lambda} \qquad \Longrightarrow \qquad \int R(\lambda)\, \mathrm{d}\lambda = 2\ln\lambda + \ln(\lambda - 2) + \ln C = \ln C\lambda^2(\lambda - 2)\,,$$

where C is a constant of integration (which we can set equal to 1 because we seek just one of all possible solutions). Hence, by Eq. (5.5.15), $V(\lambda) = C\lambda^2(\lambda - 2)$ and from (5.5.13), we obtain

$$v(\lambda) = \frac{V(\lambda)}{\lambda^2 - 2\lambda} = \frac{\lambda^2(\lambda - 2)}{\lambda(\lambda - 2)} = \lambda\,.$$

Formula (5.5.10) gives the solution

$$y(t) = \int_0^2 \lambda\, e^{\lambda t}\, \mathrm{d}\lambda = \frac{1}{t^2} + e^{2t}\left(\frac{2}{t} - \frac{1}{t^2}\right).$$

Problems

1. Determine the initial value problem for which the following inverse Laplace transform is its solution.
   (a) $\frac{4\lambda - 1}{4\lambda^2 - 4\lambda + 1}$; **(b)** $\frac{9\lambda + 6}{9\lambda^2 + 6\lambda + 1}$; **(c)** $\frac{3\lambda + 5}{9\lambda^2 - 12\lambda + 4}$; **(d)** $\frac{4\lambda + 4}{4\lambda^2 + 12\lambda + 9}$.

Use the Laplace transform to solve the given initial value problems.

2. $y'' + 9y = 0, \quad y(0) = 2,\ y'(0) = 0.$
3. $4y'' - 4y' + 5y = 0, \quad y(0) = 2,\ y'(0) = 3.$
4. $y'' + 2y' + y = 0, \quad y(0) = -1,\ y'(0) = 2.$
5. $y'' - 4y' + 5y = 0, \quad y(0) = 0,\ y'(0) = 3.$
6. $y'' - y' - 6y = 0, \quad y(0) = 2,\ y'(0) = 1.$
7. $4y'' - 4y' + 37y = 0, \quad y(0) = 2,\ y'(0) = -3.$
8. $y'' + 3y' + 2y = 0, \quad y(0) = 2,\ y'(0) = 3.$
9. $y'' + 2y' + 5y = 0, \quad y(0) = 1,\ y'(0) = -1.$
10. $4y'' - 12y' + 13y = 0, \quad y(0) = 2,\ y'(0) = 3.$
11. $y'' + 4y' + 13y = 0, \quad y(0) = 1,\ y'(0) = -6.$

12. $y'' + 6y' + 9y = 0, \quad y(0) = 1,\ y'(0) = -3.$
13. $y^{(4)} + y = 0, \quad y(0) = y'''(0) = 0,\ y''(0) = y'(0) = 1/\sqrt{2}.$
14. $y'' - 2y' + 5y = 0, \quad y(0) = 0,\ y'(0) = -1.$
15. $y'' - 20y' + 51y = 0, \quad y(0) = 0,\ y'(0) = -14.$
16. $2y'' + 3y' + y = 0, \quad y(0) = 3,\ y'(0) = -1.$
17. $3y'' + 8y' - 3y = 0, \quad y(0) = 3,\ y'(0) = -4.$
18. $2y'' + 20y' + 51y = 0, \quad y(0) = 1,\ y'(0) = -5.$
19. $4y'' + 40y' + 101y = 0, \quad y(0) = 1,\ y'(0) = -5.$
20. $y'' + 6y' + 34y = 0, \quad y(0) = 3,\ y'(0) = 1.$
21. $y''' + 8y'' + 16y' = 0, \quad y(0) = y'(0) = 1,\ y''(0) = -8.$
22. $y''' + 6y'' + 13y' = 0, \quad y(0) = y'(0) = 1,\ y''(0) = -6.$
23. $y''' - 6y'' + 13y' = 0, \quad y(0) = y'(0) = 1,\ y''(0) = 6.$
24. $y''' + 4y'' + 29y' = 0, \quad y(0) = 1,\ y'(0) = 5,\ y''(0) = -20.$
25. $y''' + 6y'' + 25y' = 0, \quad y(0) = 1,\ y'(0) = 4,\ y''(0) = -24.$
26. $y''' - 6y'' + 10y' = 0, \quad y(0) = 1,\ y'(0) = 3,\ y''(0) = 8.$
27. $y^{(4)} + 13y'' + 36y = 0, \quad y(0) = 0, \quad y'(0) = -1, \quad y''(0) = 5, \quad y'''(0) = 19.$

The following second order differential equations with polynomial coefficients have at least one solution that is integrable in some neighborhood of $t = 0$. Apply the Laplace transform (for this purpose use rule (5.2.11) and Problem 13 in §5.2) to derive and solve the differential equation for $y^L = \mathcal{L}[y(t)]$, where $y(t)$ is a solution of the given differential equation. Then come back to the same equation and solve it using the formula (5.5.10).

28. $\ddot{y} - ty = 0;$
29. $t\ddot{y} + \dot{y} + ty = 0;$
30. $t\ddot{y} + (1-t)\dot{y} + 2y = 0;$
31. $\ddot{y} - 2t\dot{y} + 2ay = 0;$
32. $t\ddot{y} - (t-4)\dot{y} - 3y = 0;$
33. $t\ddot{y} + (1-t)\dot{y} - (2t+1)y = 0;$
34. $t\ddot{y} + \dot{y} + y = 0;$
35. $t\ddot{y} - (2t-1)\dot{y} - y = 0;$
36. $t\ddot{y} + 5\dot{y} - (t+3)y = 0;$
37. $t\ddot{y} + 3\dot{y} - ty = 0;$
38. $2t\ddot{y} + (2t+3)\dot{y} + y = 0;$
39. $t\ddot{y} - (4t-3)\dot{y} + (3t-7)y = 0.$

5.6 Nonhomogeneous Differential Equations

Let us examine a second order nonhomogeneous differential equation with constant coefficients:

$$a\,y''(t) + b\,y'(t) + cy(t) = f(t), \quad 0 < t < \infty, \tag{5.6.1}$$

subject to the initial conditions

$$y(0) = y_0, \quad y'(0) = y_0'. \tag{5.6.2}$$

Equation (5.6.1) together with the condition (5.6.2) is usually referred to as an initial value problem (IVP). We always assume that the unknown function $y(t)$ and the nonhomogeneous term $f(t)$ are function-originals, and, therefore, their Laplace transforms

$$y^L(\lambda) = \int_0^\infty y(t)\,e^{-\lambda t}\,\mathrm{d}t, \quad f^L(\lambda) = \int_0^\infty f(t)\,e^{-\lambda t}\,\mathrm{d}t \tag{5.6.3}$$

exist. From Theorem 4.20 (page 224), we know that the solution of the driven equation (5.6.1) is the sum of a particular solution of the nonhomogeneous equation and the general solution of the corresponding homogeneous equation. For simplicity, we split the problem under consideration into two problems, namely,

$$a\,y''(t) + b\,y'(t) + cy(t) = f(t), \quad y(0) = 0,\ y'(0) = 0 \tag{5.6.4}$$

and

$$a\,y''(t) + b\,y'(t) + cy(t) = 0, \quad y(0) = y_0,\ y'(0) = y_0'. \tag{5.6.5}$$

Then the solution $y(t)$ of the initial value problem (5.6.1) and (5.6.2) is the sum of $y_p(t)$, the solution of the problem (5.6.4), and $y_h(t)$, the solution of the problem (5.6.5) for the homogeneous equation, that is,

$$y(t) = y_p(t) + y_h(t). \tag{5.6.6}$$

First, we attack the problem (5.6.4) for the nonhomogeneous equation (5.6.1) because the IVP (5.6.5) was considered in the previous section. We know how we would have to apply the Laplace transform to find the solution of this initial value problem. We multiply both sides of Eq. (5.6.1) by $e^{-\lambda t}$ and integrate the result with respect to t from zero to infinity to obtain

$$\int_0^\infty \left[a\, y''(t) + b\, y'(t) + cy(t)\right] e^{-\lambda t}\, \mathrm{d}t = \int_0^\infty f(t)\, e^{-\lambda t}\, \mathrm{d}t.$$

Integrating by parts (or applying the derivative rule (5.2.2) on page 282) and using the notations (5.6.3), we get

$$a\lambda^2 y_p^L + b\lambda y_p^L + cy_p^L = f^L.$$

Solving the above algebraic equation for y_p^L, we obtain

$$y_p^L = f^L\, G^L(\lambda). \tag{5.6.7}$$

where

$$G^L(\lambda) = \frac{1}{a\lambda^2 + b\lambda + c} \tag{5.6.8}$$

is the reciprocal of the characteristic polynomial. Its inverse Laplace transform $G(t) = \mathcal{L}^{-1}\left[G^L(\lambda)\right]$ is called the **Green function** and it describes the impulse response of the system because $\mathcal{L}\left[\delta(t)\right] = 1$.

Note that the function $y_p^L(\lambda)$ in Eq. (5.6.7) is the product of two functions, $f^L(\lambda)$ and $G^L(\lambda)$. When the explicit formula for f^L is known, application of the inverse Laplace transform to $y_p^L = f^L\, G^L(\lambda)$ produces the solution $y_p(t)$ of the initial value problem (5.6.4). If an explicit formula for f^L is not available, we have no other option than to use the convolution rule:

$$y_p(t) = (f * G)(t) = \int_0^t f(\tau)\, G(t-\tau)\, \mathrm{d}\tau. \tag{5.6.9}$$

The Green function $G(t)$ can be defined directly without using the Laplace transformation. Recall that the Laplace transform of the δ-function is 1 since

$$\mathcal{L}[\delta](\lambda) = \int_0^\infty \delta(t)\, e^{-\lambda t}\, \mathrm{d}t = e^{-\lambda t}\Big|_{t=0} = e^{-\lambda 0} = 1.$$

Therefore, the Green function $G(t)$ for the given differential equation (5.6.1) is the solution of the following nonhomogeneous equation:

$$a\, G''(t) + b\, G'(t) + c\, G(t) = \delta(t).$$

Since the equation involves differentiation in the sense of distribution (recall that the δ-function is a distribution but not an ordinary function), it is more practical to define the Green function as the solution of the following initial value problem:

$$a\, G''(t) + b\, G'(t) + c\, G(t) = 0, \qquad G(0) = 0, \quad G'(0) = 1/a. \tag{5.6.10}$$

Therefore, the initial value problem (5.6.5) can be solved using techniques disclosed in §5.5. Application of the Laplace transform to the initial value problem (5.6.5) leads to the following algebraic equation for y_h^L, the Laplace transform of its solution:

$$a\lambda^2 y_h^L - a\lambda y_h(0) - ay_h'(0) + b\lambda y_h^L - by_h(0) + cy_h^L = 0.$$

Solving for y_h^L, we obtain

$$y_h^L(\lambda) = \frac{a\lambda y_0 + ay_0' + by_0}{a\lambda^2 + b\lambda + c}. \tag{5.6.11}$$

Any method presented in §5.4 can be used to determine the inverse Laplace transform of y_h^L, which gives the solution $y_h(t)$ of the problem (5.6.5). We demonstrate the Laplace transform technique in the following examples.

Example 5.6.1: Find the Green function for the given differential operator

$$4\mathtt{D}^4 + 12\mathtt{D}^3 + 5\mathtt{D}^2 - 16\mathtt{D} + 5, \qquad \mathtt{D} = \mathrm{d}/\mathrm{d}t.$$

Solution. The Green function for this operator is the solution of the following initial value problem:

$$4G^{(4)} + 12G''' + 5G'' - 16G' + 5G = 0, \quad G(0) = 0, \ G'(0) = 0, \ G''(0) = 0, \ G'''(0) = 1/4.$$

The application of the Laplace transform yields

$$\left(4\lambda^4 + 12\lambda^3 + 5\lambda^2 - 16\lambda + 5\right) G^L = 1.$$

Therefore, $G^L(\lambda)$, the Laplace transform of the Green function, is the reciprocal of the characteristic polynomial, that is,

$$G^L(\lambda) = \frac{1}{4\lambda^4 + 12\lambda^3 + 5\lambda^2 - 16\lambda + 5}.$$

The characteristic equation $Q(\lambda) \equiv 4\lambda^4 + 12\lambda^3 + 5\lambda^2 - 16\lambda + 5 = (2\lambda - 1)^2(\lambda^2 + 4\lambda + 5) = 0$ has one double root $\lambda_1 = 1/2$ and two complex conjugate roots $\lambda_{2,3} = -2 \pm \mathbf{j}$. Application of the residue method gives us

$$\begin{aligned} G(t) &= \left.\frac{\mathrm{d}}{\mathrm{d}\lambda}\frac{(\lambda - 1/2)^2\, e^{\lambda t}}{Q(\lambda)}\right|_{\lambda=\frac{1}{2}} + \sum_{k=2}^{3}\frac{e^{\lambda_k t}}{Q'(\lambda_k)} \\ &= \left.\frac{\mathrm{d}}{\mathrm{d}\lambda}\frac{e^{\lambda t}}{4(\lambda^2 + 4\lambda + 5)}\right|_{\lambda=\frac{1}{2}} + 2\,\Re\,\frac{e^{(-2\pm\mathbf{j})t}}{Q'(-2\pm\mathbf{j})}, \end{aligned}$$

where it does not matter which complex root λ_k $(k = 2, 3)$ will be chosen. Since

$$\frac{\mathrm{d}}{\mathrm{d}\lambda}\frac{e^{\lambda t}}{4(\lambda^2 + 4\lambda + 5)} = \frac{t\, e^{\lambda t}}{4(\lambda^2 + 4\lambda + 5)} - \frac{(2\lambda + 4)\, e^{\lambda t}}{4(\lambda^2 + 4\lambda + 5)^2}$$

and for $k = 2, 3$,

$$\begin{aligned} Q'(\lambda_k) &= (2\lambda_k - 1)^2\,(2\lambda_k + 4) = (2\lambda_k - 1)^2\, 2\,(\lambda_k + 2) = (-4 \pm 2\mathbf{j} - 1)^2\, 2\,(\pm\mathbf{j}) \\ &= (-5 \mp 2\mathbf{j})^2\, 2\,(\pm\mathbf{j}) = 2(25 - 4 \mp 20\mathbf{j})\,(\pm\mathbf{j}) = 2(20 \pm 21\mathbf{j}), \end{aligned}$$

$$\frac{1}{(2\lambda_k - 1)^2\,(2\lambda_k + 4)} = \frac{1}{2(20 \pm 21\mathbf{j})} = \frac{20 \mp 21\mathbf{j}}{2(21^2 + 20^2)} = \frac{10}{841} \mp \frac{21\mathbf{j}}{1682},$$

we have

$$G(t) = \left(\frac{t}{29} - \frac{20}{841}\right) e^{t/2}\, H(t) + \frac{1}{841}\, e^{-2t}\,(20\cos t + 21\sin t)\, H(t).$$

As usual, we have to multiply this result by the Heaviside function to guarantee vanishing of the Green function for negative t.

On the other hand, using partial fraction decomposition, we obtain

$$G^L(\lambda) = \frac{4}{29(2\lambda - 1)^2} - \frac{40}{841\,(2\lambda - 1)} + \frac{20\lambda + 61}{841\,(\lambda^2 + 4\lambda + 5)}.$$

Rewriting the last fraction as

$$\frac{20\lambda + 61}{\lambda^2 + 4\lambda + 5} = \frac{20(\lambda + 2) + 21}{(\lambda + 2)^2 + 1}$$

and using the attenuation rule (5.2.7), we get the same result produced by the residue method.

Example 5.6.2: Solve $y'' - y' - 6y = e^t$ subject to the given initial conditions $y(0) = 1,\ y'(0) = -1$.

Solution. The application of the Laplace transform gives us

$$\mathcal{L}[y''] - \mathcal{L}[y'] - 6\mathcal{L}[y] = \mathcal{L}[e^t].$$

We denote the Laplace transform $\int_0^\infty e^{-\lambda t}\, y(t)\,\mathrm{d}t$ of the unknown function $y(t)$ by $y^L = \mathcal{L}y$. Then from the differential rule (5.5.1), it follows that

$$\begin{aligned} \mathcal{L}[y'] &= \lambda y^L - y(0) = \lambda y^L - 1; \\ \mathcal{L}[y''] &= \lambda^2 y^L - \lambda y(0) - y'(0) = \lambda^2 y^L - \lambda + 1. \end{aligned}$$

The Laplace transform of the function $f(t) = e^t$ is $f^L(\lambda) = \mathcal{L}[e^t] = \frac{1}{\lambda - 1}$ (see formula 6 in Table 280). Therefore,

$$\lambda^2 y^L - \lambda + 1 - \lambda y^L + 1 - 6y^L = f^L \qquad \text{or} \qquad (\lambda^2 - \lambda - 6)y^L = f^L + \lambda - 2.$$

This algebraic equation has the solution $y^L = y_p^L + y_h^L$, where

$$y_p^L = \frac{1}{\lambda^2 - \lambda - 6}\cdot f^L \quad \text{and} \quad y_h^L = \frac{\lambda - 2}{\lambda^2 - \lambda - 6}.$$

Hence, we need to find the inverse Laplace transform of each of these functions, y_p^L and y_h^L, where y_p is the solution to the nonhomogeneous equation $y'' - y' - 6y = e^t$ subject to the homogeneous initial conditions $y(0) = y'(0) = 0$, and y_h is the solution of the homogeneous equation $y'' - y' - 6y = 0$ subject to the nonhomogeneous conditions $y(0) = 1$, $y'(0) = -1$. If we know the Green function $G(t) = \mathcal{L}^{-1}\left[\frac{1}{\lambda^2 - \lambda - 6}\right]$, we could find y_p as the convolution:

$$y_p(t) = G * f(t) = f * G(t) = \int_0^t G(t-\tau) f(\tau)\,\mathrm{d}\tau = \int_0^t G(\tau)\, e^{t-\tau}\,\mathrm{d}\tau.$$

On the other hand, the Green function can be obtained either with partial fraction decomposition:

$$G(t) = \mathcal{L}^{-1}\left[\frac{1}{\lambda^2 - \lambda - 6}\right] = \mathcal{L}^{-1}\left[\frac{1/5}{\lambda - 3} - \frac{1/5}{\lambda + 2}\right] = \frac{1}{5}\left[e^{3t} - e^{-2t}\right] H(t),$$

or with the convolution rule:

$$G(t) = \mathcal{L}^{-1}\left[\frac{1}{\lambda - 3} \cdot \frac{1}{\lambda + 2}\right] = e^{3t} * e^{-2t} = \left[\frac{1}{5}\, e^{3t} - \frac{1}{5}\, e^{-2t}\right] H(t),$$

or with the residue method:

$$\begin{aligned} G(t) &= \operatorname*{Res}_{\lambda=-2} \frac{e^{\lambda t}}{\lambda^2 - \lambda - 6} + \operatorname*{Res}_{\lambda=3} \frac{e^{\lambda t}}{\lambda^2 - \lambda - 6} \\ &= \left.\frac{e^{\lambda t}}{2\lambda - 1}\right|_{\lambda=-2} + \left.\frac{e^{\lambda t}}{2\lambda - 1}\right|_{\lambda=3} = \frac{1}{5}\left[e^{3t} - e^{-2t}\right] H(t). \end{aligned}$$

With this in hand, we define a particular solution of the nonhomogeneous equation as the convolution integral; that is,

$$y_p(t) = G * f = \frac{1}{5}\int_0^t \left[e^{3\tau} - e^{-2\tau}\right] e^{t-\tau}\,\mathrm{d}\tau = \left(\frac{1}{10}\, e^{3t} + \frac{1}{15}\, e^{-2t} - \frac{1}{6}\, e^{t}\right) H(t).$$

Note that the function $y_p(t)$ satisfies the homogeneous initial conditions $y_p(+0) = 0,\ y_p'(+0) = 0$.

The function $y_p(t)$ could also be found in a more straightforward manner since the Laplace transform of the forcing function is known:

$$y_p = \mathcal{L}^{-1}\left[\frac{1}{\lambda^2 - \lambda - 6} \cdot \frac{1}{\lambda - 1}\right] = \mathcal{L}^{-1}\left[\frac{1}{(\lambda - 3)(\lambda + 2)(\lambda - 1)}\right].$$

After application of the partial fraction decomposition method, we obtain

$$y_p = \mathcal{L}^{-1}\left[\frac{1}{15}\frac{1}{\lambda + 2} + \frac{1}{10}\frac{1}{\lambda - 3} - \frac{1}{6}\frac{1}{\lambda - 1}\right] = \frac{1}{15}\, e^{-2t} + \frac{1}{10}\, e^{3t} - \frac{1}{6}\, e^{t}, \quad t > 0.$$

The function y_h can be discovered using the residue method, rather painless, that

$$\begin{aligned} y_h(t) &= \mathcal{L}^{-1}\left[\frac{\lambda - 2}{\lambda^2 - \lambda - 6}\right] \\ &= \operatorname*{Res}_{\lambda=-2} \frac{(\lambda - 2)\, e^{\lambda t}}{\lambda^2 - \lambda - 6} + \operatorname*{Res}_{\lambda=3} \frac{(\lambda - 2)\, e^{\lambda t}}{\lambda^2 - \lambda - 6} \\ &= \left.\frac{(\lambda - 2)\, e^{\lambda t}}{2\lambda - 1}\right|_{\lambda=-2} + \left.\frac{(\lambda - 2)\, e^{\lambda t}}{2\lambda - 1}\right|_{\lambda=3} = \frac{4}{5}\, e^{-2t} + \frac{1}{5}\, e^{3t}, \quad t > 0. \end{aligned}$$

The function $y_h(t)$ satisfies the homogeneous equation $y'' - y' - 6y = 0$ and the given initial conditions $y(+0) = 1$, $y'(+0) = -1$. We check the answer with *Maple:*

```
sol := dsolve({ics, ode}, y(x), method = laplace) odetest(sol,[ode,ics])
```

5.6.1 Differential Equations with Intermittent Forcing Functions

In this subsection, we demonstrate the beauty and advantages of the Laplace transformation in solving nonhomogeneous differential equations with discontinuous forcing functions. This topic has many applications in engineering and numerical analysis when input functions are approximated by piecewise continuous or step-functions.

Example 5.6.3: We start with a simple problem for the first order differential equation

$$y' + ay = f(t), \qquad y(0) = 0,$$

where a is a real number, $f(t)$, called an input, is a given function; we will refer to the output as the solution of this problem. After applying the Laplace transformation, we get the algebraic equation for y^L, the Laplace transform of the unknown solution $y(t)$:

$$\lambda y^L + a y^L = f^L(\lambda) \qquad \Longrightarrow \qquad y^L = \frac{1}{\lambda + a}\, f^L(\lambda).$$

According to the convolution rule (5.2.1), the inverse Laplace transform yields the solution

$$y(t) = \mathcal{L}^{-1}\left[\frac{1}{\lambda + a}\, f^L(\lambda)\right] = \int_0^t e^{-a(t-\tau)}\, f(\tau)\,\mathrm{d}\tau. \tag{5.6.12}$$

Let us consider a periodic input modeled by function $f(t) = \sin t$. Then we get

$$y(t) = \mathcal{L}^{-1}\left[\frac{1}{\lambda + a}\,\frac{1}{\lambda^2 + 1}\right] = \frac{1}{a^2+1}\, e^{-at} + \frac{1}{a^2+1}\left(a\,\sin t - \cos t\right), \quad t \geqslant 0.$$

Now approximate $\sin t$ with a step-function so that the total energy (which is the integral) on each subinterval $[0, \pi]$ and $[\pi, 2\pi]$ remains the same:

$$f_\pi(t) = \frac{2}{\pi} \times \begin{cases} 1, & \text{if } 0 < t < \pi, \\ -1, & \text{if } \pi < t < 2\pi, \end{cases} \qquad f_\pi(t) = f_\pi(t + 2\pi),$$

Expanding $f_\pi(t)$ periodically with period 2π, we rewrite it using the Heaviside function: $f_\pi(t) = \frac{2}{\pi}\sum_{k\geqslant 0}(-1)^k \times [H(t - k\pi) - H(t - k\pi - \pi)]$. The Laplace transform of $f_\pi(t)$ is

$$\mathcal{L}[f_\pi] = \frac{2}{\pi\lambda}\sum_{k\geqslant 0}(-1)^k\left[e^{-\lambda k\pi} - e^{\lambda\pi(k+1)}\right] = \frac{2}{\pi\lambda}\,\frac{1 - e^{-\lambda\pi}}{1 + e^{-\lambda\pi}} = \frac{2}{\pi\lambda}\tanh\left(\frac{\pi\lambda}{2}\right).$$

From Eq. (5.6.12), we get the solution

$$y_\pi(t) = \mathcal{L}^{-1}\left[\frac{1}{\lambda + a}\, f_\pi^L\right] = \sum_{k\geqslant 0}(-1)^k\left[g(t - k\pi) - g(t - k\pi - \pi)\right],$$

where

$$g(t) = \frac{2}{a\pi}\left(1 - e^{-at}\right) H(t).$$

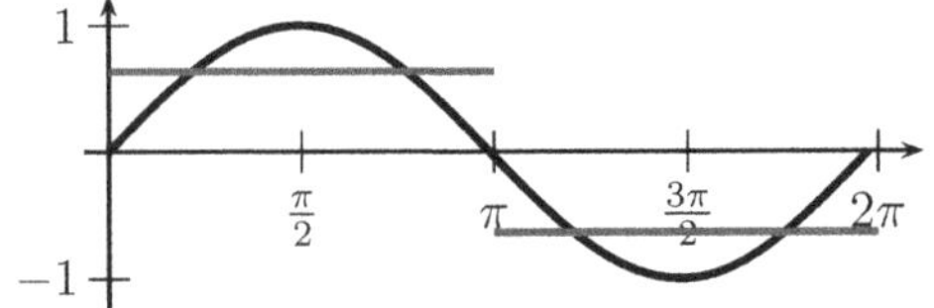

Figure 5.24: Example 5.6.3: approximation $f_\pi(t)$.

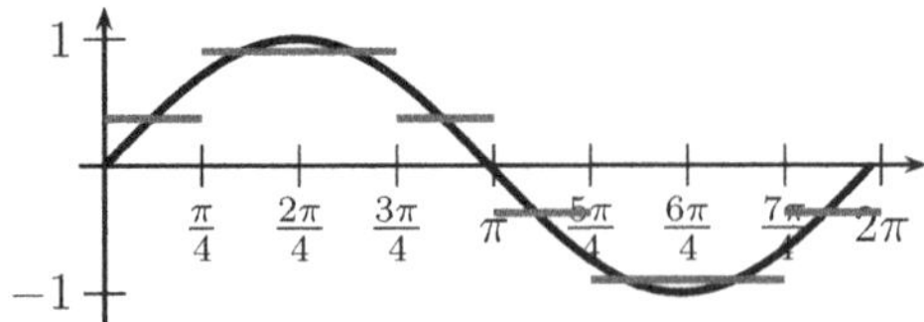

Figure 5.25: Example 5.6.3: approximation $f_{\pi/4}(t)$.

Similarly, we can truncate the interval $[0, 2\pi]$ into eight subintervals of length $\pi/4$ and approximate $\sin t$ on each interval keeping the integral unchanged:

$$f_{\pi/4}(t) = h_1 \left\{ 2\sum_{k\geqslant 1}(-1)^k H(t-k\pi) + \sum_{k\geqslant 0}(-1)^k \left[H\left(t - \frac{3\pi}{4} - k\pi\right) - H\left(t - \frac{\pi}{4} - k\pi\right)\right]\right\}$$
$$+ h_1 H(t) + h_2 \sum_{k\geqslant 0}(-1)^k \left[H\left(t - \frac{\pi}{4} - k\pi\right) - H\left(t - \frac{3\pi}{4} - k\pi\right)\right],$$

where $h_1 = \frac{4}{\pi}\int_0^{\pi/4} \sin t\,\mathrm{d}t = \frac{4-2\sqrt{2}}{\pi} \approx 0.37$ and $h_2 = \frac{2}{\pi}\int_{\pi/4}^{3\pi/2} \sin t\,\mathrm{d}t = \frac{2\sqrt{2}}{\pi} \approx 0.9$. As it is seen from Fig. 5.26(b), the corresponding solution $y_{\pi/4}(t)$ approximates the original solution (5.6.12) for $f(t) = \sin t$ better than $y_\pi(t)$.

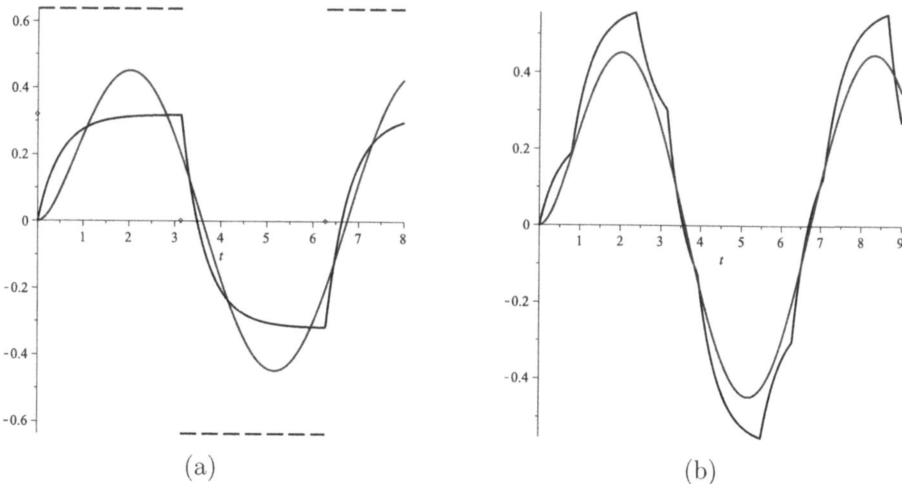

Figure 5.26: Example 5.6.3. (a) The output (black line) $y_\pi(t)$ to the input (dashed line) and (b) output $y_{\pi/4}(t)$, plotted with *Maple*.

Example 5.6.4: Use the Laplace transform to solve the initial value problem

$$y'' - y' - 6y = f(t) = \begin{cases} 50\sin 4t, & 0 \leqslant t \leqslant \pi, \\ 0, & \pi \leqslant t, \end{cases} \qquad y(0) = 1,\ y'(0) = -1.$$

Solution. Following the same procedure as in the previous example, we apply the Laplace transform to the given initial value problem to obtain

$$y^L = y_p^L + y_h^L,$$

where y^L is the Laplace transform of the unknown function $y(t)$ and

$$\mathcal{L}[y_p] \stackrel{\text{def}}{=} y_p^L = \frac{1}{\lambda^2 - \lambda - 6}\cdot f^L \quad \text{and} \quad \mathcal{L}[y_h] \stackrel{\text{def}}{=} y_h^L = \frac{\lambda - 2}{\lambda^2 - \lambda - 6}.$$

The function y_h was found in Example 5.6.2. We analyze here only the inverse of the function y_p^L. Since the Green function was defined in Example 5.6.2, we express y_p as the convolution integral

$$\begin{aligned} y_p(t) &= (G * f)(t) = \int_0^t G(t-\tau)\,f(\tau)\,\mathrm{d}\tau = \int_0^t G(\tau)\,f(t-\tau)\,\mathrm{d}\tau \\ &= \begin{cases} 10\int_0^t \left[e^{3\tau} - e^{-2\tau}\right]\sin 4(t-\tau)\,\mathrm{d}\tau, & \text{if } 0 \leqslant t \leqslant \pi, \\ 10\int_0^\pi \left[e^{3t-3\tau} - e^{-2t+2\tau}\right]\sin 4\tau\,\mathrm{d}\tau, & \text{if } \pi \leqslant t, \end{cases} \\ &= \begin{cases} \frac{8}{5}\,e^{3t} - \frac{11}{5}\sin 4t + \frac{2}{5}\cos 4t - 2\,e^{-2t}, & \text{if } 0 \leqslant t \leqslant \pi, \\ \frac{8}{5}\,e^{3t}\left[1 - e^{-3\pi}\right] + 2\,e^{2t}\left[e^{2\pi} - 1\right], & \text{if } \pi \leqslant t. \end{cases} \end{aligned}$$

However, we can find the solution y_p of the nonhomogeneous equation in another way. The Laplace transform of f is

$$f^L(\lambda) = \int_0^\infty e^{-\lambda t} f(t)\,\mathrm{d}t = 50 \int_0^\pi e^{-\lambda t} \sin 4t\,\mathrm{d}t = \frac{200}{\lambda^2+16}\left[1-e^{-\lambda\pi}\right].$$

Then

$$y_p^L = \frac{1}{\lambda^2-\lambda-6}\cdot f^L = \frac{1}{\lambda^2-\lambda-6}\cdot\frac{200}{\lambda^2+16}\left[1-e^{-\lambda\pi}\right] = x_0^L - x_\pi^L,$$

where

$$x_0^L = \frac{1}{\lambda^2-\lambda-6}\,\frac{200}{\lambda^2+16} = \frac{200}{(\lambda+2)(\lambda-3)(\lambda^2+16)}, \qquad x_\pi^L = x_0^L\,e^{-\lambda\pi}.$$

Since y_p^L is the difference of two similar expressions, one of which differs from the other only by the exponential multiplier, $e^{-\lambda\pi}$, it will be enough to find the inverse Laplace transform of one term, namely, $x_0(t) = \mathcal{L}^{-1}\left[x_0^L\right]$.

Of course we can apply the method of partial fraction decomposition, but the procedure to determine coefficients in the expansion

$$\frac{1}{\lambda^2-\lambda-6}\,\frac{200}{\lambda^2+16} = \frac{A}{\lambda-3} + \frac{B}{\lambda+2} + \frac{C\lambda+D}{\lambda^2+16}$$

becomes tedious (unless you use a computer algebra system).

The most straightforward way to determine the function $x_0(t)$ is to apply the residue method because the denominator (which we denote by $Q(t)$) has four simple different roots; that is,

$$Q(\lambda) = (\lambda^2-\lambda-6)(\lambda^2+16) = (\lambda-3)(\lambda+2)(\lambda+4\mathbf{j})(\lambda-4\mathbf{j}).$$

The derivative of $Q(\lambda)$ is

$$Q'(\lambda) = (\lambda+2)(\lambda^2+16) + (\lambda-3)(\lambda^2+16) + 2\lambda(\lambda-3)(\lambda+2).$$

Hence, the explicit expression for the function $x_0(t)$ becomes

$$\begin{aligned} x_0(t) &= \frac{200\,e^{3t}}{Q'(3)} + \frac{200\,e^{-2t}}{Q'(-2)} + \frac{200\,e^{4\mathbf{j}t}}{Q'(4\mathbf{j})} + \frac{200\,e^{-4\mathbf{j}t}}{Q'(-4\mathbf{j})} \\ &= \frac{8}{5}e^{3t} - 2\,e^{-2t} + \frac{25}{2(2-11\mathbf{j})}\,e^{4\mathbf{j}t} + \frac{25}{2(2+11\mathbf{j})}\,e^{-4\mathbf{j}t} \\ &= \left[\frac{8}{5}e^{3t} - 2\,e^{-2t} + \frac{2}{5}\cos 4t - \frac{11}{5}\sin 4t\right] H(t) \end{aligned}$$

because $\frac{1}{2-11\mathbf{j}} = \frac{2+11\mathbf{j}}{2^2+11^2} = \frac{1}{125}(2+11\mathbf{j})$. Note that instead of evaluating residues at two complex conjugate nulls $\lambda = \pm 4\mathbf{j}$, we can use the formula (5.4.8) on page 309. We must multiply the right-hand side by the Heaviside function $H(t)$ since $x_0(t)$ should be zero for $t<0$.

The inverse Laplace transform of the function $x_\pi^L = x_0^L\,e^{-\lambda\pi}$ can be found with the aid of the shift rule (5.2.6) on page 282; namely,

$$x_\pi(t) = \mathcal{L}^{-1}\left[x_0^L\,e^{-\lambda\pi}\right] = x_0(t-\pi)H(t-\pi).$$

Therefore, the solution $y_p(t)$ of the nonhomogeneous equation with the homogeneous initial conditions is

$$\begin{aligned} y_p(t) &= x_0(t) - x_\pi(t) = x_0(t)H(t) - x_0(t-\pi)H(t-\pi) \\ &= \left[\frac{8}{5}e^{3t} - 2\,e^{-2t}\right] H(t) - \left[\frac{8}{5}e^{3t-3\pi} - 2\,e^{-2t+2\pi}\right] H(t-\pi) \\ &\quad + \left[\frac{2}{5}\cos 4t - \frac{11}{5}\sin 4t\right]\left[H(t) - H(t-\pi)\right]. \end{aligned}$$

Example 5.6.5: Use the Laplace transform to solve the initial value problem

$$y'' + 4y = f(t) = H(t-\pi) - H(t-3\pi), \quad y(0)=0,\ y'(0)=0.$$

Solution. Since the input function $f(t)$ is the difference of two shifted Heaviside functions $f(t) = f_1(t) - f_2(t)$, with $f_1 = H(t-\pi)$ and $f_2 = H(t-3\pi)$, we apply the Laplace transform and obtain

$$\lambda^2 y^L + 4y^L = f^L = f_1^L - f_2^L,$$

where $y^L = \mathcal{L}[y]$, $f_1 = \mathcal{L}f_1$, $f_2 = \mathcal{L}f_2$, and

$$f^L(\lambda) = \int_0^\infty f(t)\,e^{-\lambda t}\,\mathrm{d}t = \int_\pi^\infty e^{-\lambda t}\,\mathrm{d}t - \int_{3\pi}^\infty e^{-\lambda t}\,\mathrm{d}t = \frac{1}{\lambda}\left[e^{-\lambda\pi} - e^{-3\lambda\pi}\right].$$

By straightforward algebraic operations we find the Laplace transform of the unknown solution to be

$$y^L(\lambda) = \frac{1}{\lambda^2+4}\cdot f^L = G^L\cdot f^L = \frac{1}{\lambda^2+4}\cdot\frac{1}{\lambda}\left[e^{-\lambda\pi} - e^{-3\lambda\pi}\right] = y_1^L - y_2^L,$$

where $y_1^L = \lambda^{-1}(\lambda^2+4)^{-1}e^{-\lambda\pi}$ and $y_2^L = \lambda^{-1}(\lambda^2+4)^{-1}e^{-3\lambda\pi}$. Let us introduce an auxiliary function $g(t)$, which we define through its Laplace transform:

$$g^L = \frac{1}{\lambda(\lambda^2+4)}.$$

Using the residue method, we find the explicit formula for $g(t)$:

$$\begin{aligned} g(t) &= \operatorname*{Res}_{\lambda=0}\frac{e^{\lambda t}}{\lambda(\lambda^2+4)} + 2\,\Re\operatorname*{Res}_{\lambda=2\mathbf{j}}\frac{e^{\lambda t}}{\lambda(\lambda^2+4)} \\ &= \frac{1}{4} + 2\,\Re\frac{e^{2\mathbf{j}t}}{(2\mathbf{j})(4\mathbf{j})} = \left(\frac{1}{4} - \frac{1}{4}\cos 2t\right)H(t). \end{aligned}$$

Since the Laplace transform of the unknown solution $y^L(\lambda)$ is expressed through $g^L(\lambda)$ as

$$y^L(\lambda) = g^L(\lambda)\,e^{-\lambda\pi} - g^L(\lambda)\,e^{-3\lambda\pi},$$

the above expression calls for application of the shift rule (5.2.6) on page 282. This yields

$$y(t) = g(t-\pi) - g(t-3\pi) = \frac{1}{4}(1-\cos 2t)[H(t-\pi) - H(t-3\pi)],$$

and $y_1(t) = g(t-\pi)$, $y_2(t) = g(t-3\pi)$. We check this answer by substituting $y(t)$ into the given equation and the initial conditions. First, we need to find its derivatives (using the product rule):

$$\begin{aligned} y' &= \frac{1}{4}(1-\cos 2t)'[H(t-\pi) - H(t-3\pi)] + \frac{1}{4}(1-\cos 2t)[H'(t-\pi) - H'(t-3\pi)] \\ &= \frac{1}{2}\sin 2t\,[H(t-\pi) - H(t-3\pi)] + \frac{1}{4}(1-\cos 2t)[\delta(t-\pi) - \delta(t-3\pi)] \\ &= \frac{1}{2}\sin 2t\,[H(t-\pi) - H(t-3\pi)] + \frac{1}{4}(1-\cos 2\pi)\delta(t-\pi) - \frac{1}{4}(1-\cos 6\pi)\delta(t-3\pi) \\ &= \frac{1}{2}\sin 2t\,[H(t-\pi) - H(t-3\pi)], \\ y'' &= \frac{1}{2}(\sin 2t)'\,[H(t-\pi) - H(t-3\pi)] + \frac{1}{2}\sin 2t\,[H'(t-\pi) - H'(t-3\pi)] \\ &= \cos 2t\,[H(t-\pi) - H(t-3\pi)] + \frac{1}{2}\sin 2t\,[\delta(t-\pi) - \delta(t-3\pi)] \\ &= \cos 2t\,[H(t-\pi) - H(t-3\pi)] \end{aligned}$$

because $H'(t-a) = \delta(t-a)$ and $f(t)\,\delta(t-a) = f(a)\delta(t-a)$ for any $a \geqslant 0$ and smooth function f (see Eqs. (5.3.5) and (5.3.7) on page 299). Then we substitute these derivatives into the given differential equation to verify that it is satisfied. Since the function $y(t) = y_1(t) - y_2(t)$ is identically zero for $t < \pi$, the initial conditions follow.

Example 5.6.6: Solve the initial value problem with intermittent forcing function

$$y'' - 5y' + 6y = f(t) = \begin{cases} 0, & 0 \leqslant t < 2, \\ t-2, & 2 \leqslant t < 3, \\ -(t-4), & 3 \leqslant t < 4, \\ 0, & 4 \leqslant t < \infty, \end{cases} \qquad y(0) = 0,\ y'(0) = 2.$$

Solution. We rewrite the function $f(t)$ via Heaviside functions:

$$\begin{aligned} f(t) &= (t-2)\left[H(t-2) - H(t-3)\right] - (t-4)\left[H(t-3) - H(t-4)\right] \\ &= (t-2)H(t-2) - 2(t-3)H(t-3) + (t-4)H(t-4). \end{aligned}$$

Using the shift rule (5.2.6), page 282, its Laplace transform becomes

$$f^L(\lambda) = \frac{1}{\lambda^2}\left[e^{-2\lambda} - 2\,e^{-3\lambda} + e^{-4\lambda}\right].$$

The application of the Laplace transform to the given initial value problem yields the algebraic equation $(\lambda^2 - 5\lambda + 6)y^L - 2 = f^L$ with respect to $y^L = \mathcal{L}y$. Hence, the Laplace transform of the unknown function $y(t)$ is

$$y^L(\lambda) = \frac{2}{\lambda^2 - 5\lambda + 6} + \frac{f^L}{\lambda^2 - 5\lambda + 6}.$$

Now we apply the inverse Laplace transform to obtain

$$y(t) = y_h(t) + y_p(t),$$

where

$$\begin{aligned} y_h(t) &= \mathcal{L}^{-1}\left[\frac{2}{\lambda^2 - 5\lambda + 6}\right] = \mathcal{L}^{-1}\left[\frac{2}{(\lambda-2)(\lambda-3)}\right] = 2\left(e^{3t} - e^{2t}\right)H(t), \\ y_p(t) &= \mathcal{L}^{-1}\left[\frac{1}{\lambda^2(\lambda-2)(\lambda-3)}\left(e^{-2\lambda} - 2\,e^{-3\lambda} + e^{-4\lambda}\right)\right]. \end{aligned}$$

According to the shift rule (5.2.6) on page 282, multiplication of the image by $e^{-a\lambda}$ causes shifting its function-original by a. Let $g(t)$ be the function defined by the inverse Laplace transform:

$$g(t) = \mathcal{L}^{-1}\left[\frac{1}{\lambda^2(\lambda-2)(\lambda-3)}\right].$$

Since the denominator, $\lambda^2(\lambda-2)(\lambda-3)$, has one double null $\lambda = 0$ and two simple nulls $\lambda = 2$ and $\lambda = 3$, we find its inverse Laplace transform using the residue method to be

$$g(t) = \operatorname*{Res}_0 \frac{e^{\lambda t}}{\lambda^2(\lambda-2)(\lambda-3)} + \operatorname*{Res}_2 \frac{e^{\lambda t}}{\lambda^2(\lambda-2)(\lambda-3)} + \operatorname*{Res}_3 \frac{e^{\lambda t}}{\lambda^2(\lambda-2)(\lambda-3)}.$$

From formula (5.4.6) on page 308, we get

$$\operatorname*{Res}_0 \frac{e^{\lambda t}}{\lambda^2(\lambda-2)(\lambda-3)} = \frac{\mathrm{d}}{\mathrm{d}\lambda}\left(\frac{e^{\lambda t}}{(\lambda-2)(\lambda-3)}\right)\bigg|_{\lambda=0} = \frac{t}{6} - \frac{(2\lambda-5)}{(\lambda^2-5\lambda+6)^2}\,e^{\lambda t}\bigg|_{\lambda=0} = \frac{t+5}{6}.$$

Residues at simple nulls are evaluated with the help of Eq. (5.4.5):

$$\begin{aligned} \operatorname*{Res}_2 \frac{e^{\lambda t}}{\lambda^2(\lambda-2)(\lambda-3)} &= \frac{e^{\lambda t}}{\lambda^2(\lambda-3)}\bigg|_{\lambda=2} = -\frac{1}{4}\,e^{2t}, \\ \operatorname*{Res}_3 \frac{e^{\lambda t}}{\lambda^2(\lambda-2)(\lambda-3)} &= \frac{e^{\lambda t}}{\lambda^2(\lambda-2)}\bigg|_{\lambda=3} = \frac{1}{9}\,e^{3t}, \end{aligned}$$

Combining all these formulas, we obtain $g(t) = \left[\frac{t+5}{6} - \frac{1}{4}\,e^{2t} + \frac{1}{9}\,e^{3t}\right]H(t)$. Therefore, the required solution becomes

$$y(t) = y_h(t) + g(t-2) - 2g(t-3) + g(t-4).$$

Example 5.6.7: A mass of 0.2 kg is attached to a spring with stiffness coefficient 10^4 newtons per meter. Suppose that at $t = 0$ the mass, which is motionless and in equilibrium position, is struck by a hammer of mass 0.25 kg, traveling at 40 m per sec. Find the displacement if the damping coefficient is approximately 0.02 kg/sec.

Solution. The equation of motion of this mechanical system is modeled by

$$my'' + cy' + ky = p\delta(t),$$

where $m = 0.2$ is the mass of the object, $c = 0.02$ is the damping coefficient due to air resistance, $k = 10^4$ is the string coefficient, and p is the momentum of the hammer to be determined. For more details, we recommend consulting [14].

The momentum of the hammer before the strike is $p_0 = mv = 0.25 \times 40 = 10$ kilogram meter per second. The law of conservation of momentum states that the total momentum is constant if the total external force acting on a system is zero. Hence $p_0 = p_w + p_h = m_w v_w + m v_h$, where p_w, v_w and p_h, v_h denote the momenta and velocities of the mass and hammer, respectively. The law of conservation of energy is also applicable: the total kinetic energy of the system after the hammer strikes is equal to the kinetic energy of the hammer before it struck. So $mv^2 = mv_h^2 + m_w v_w^2$, where v_h and v_w denote the velocity of the hammer and the mass after the strike. Excluding v_h and v_w from the above two equations, we obtain

$$p_w = \frac{2m_w}{m + m_w}\, p_0 \qquad \text{and} \qquad p_h = \frac{m - m_w}{m + m_w}\, p_0.$$

The momentum imparted to the mass by the collision is

$$p_w = \frac{2m_w}{m + m_w}\, p_0 = \frac{2 \times 0.2}{0.2 + 0.25}\, p_0 = \frac{0.4}{0.45} \times 10 = \frac{40}{4.5} = \frac{80}{9}.$$

Thus, the displacement y satisfies the IVP

$$0.2\, y'' + 0.01\, y' + 10^4\, y = \frac{80}{9}\, \delta(t), \qquad y(0) = y'(0) = 0.$$

Taking the Laplace transform, we get

$$y^L = \frac{400/9}{\lambda^2 + 0.1\lambda + 5 \times 10^4} = \frac{400/9}{(\lambda + 0.05)^2 + 5 \times 10^4 - 25 \times 10^{-4}} \approx \frac{0.2 \times 223.6}{(\lambda + 0.05)^2 + 223.6^2}.$$

It follows that the displacement of the mass is approximately $y(t) \approx 0.2\, e^{-0.05t} \sin(223.6t), \quad t > 0.$

Problems

In all problems, $\mathtt{D} = \mathrm{d}/\mathrm{d}t$ stands for the derivative operator, while $\mathtt{D}^0$, the identity operator, is omitted.

1. Obtain the Green function for the following differential operators.

(a)	$\mathtt{D}^2 + 2\mathtt{D} + 5$;	**(b)**	$\mathtt{D}^2 + 2\mathtt{D} + 1$;	**(c)**	$2\mathtt{D}^2 + \mathtt{D} - 1$;
(d)	$\mathtt{D}^2 + 6\mathtt{D} + 10$;	**(e)**	$\mathtt{D}^2 + 5\mathtt{D} + 6$;	**(f)**	$(\mathtt{D}^2 - 4\mathtt{D} + 20)^2$;
(g)	$\mathtt{D}^2 + 2\mathtt{D} + 17$;	**(h)**	$6\mathtt{D}^2 + 5\mathtt{D} + 1$;	**(i)**	$\mathtt{D}^4 + 1$;
(j)	$4\mathtt{D}^2 + 7\mathtt{D} + 3$;	**(k)**	$2\mathtt{D}^2 + 5\mathtt{D} - 3$;	**(l)**	$4\mathtt{D}^3 - \mathtt{D}$.

2. Solve the initial value problems using the Laplace transformation.

(a) $\ddot{y} + 2\dot{y} + 3y = 9t, \quad y(0) = 0, \; \dot{y}(0) = 1.$

(b) $4\ddot{y} + 16\dot{y} + 17y = 17t - 1, \quad y(0) = -1, \dot{y}(0) = 2.$

(c) $\ddot{y} + 5\dot{y} + 4y = 3\, e^{-t}, \quad y(0) = -\dot{y}(0) = -1.$

(d) $\ddot{y} - 4\dot{y} + 4y = t^2\, e^{2t}, \quad y(0) = 1, \quad \dot{y}(0) = 2.$

(e) $\ddot{y} + 9y = e^{-2t}, \quad y(0) = -\frac{2}{13}, \dot{y}(0) = \frac{1}{13}.$

(f) $2\ddot{y} - 3\dot{y} + 17y = 17t - 1, \quad y(0) = -1, \dot{y}(0) = 2.$

(g) $\ddot{y} + 2\dot{y} + y = e^{-t}, \quad y(0) = 1, \quad \dot{y}(0) = -1.$

(h) $\ddot{y} - 2\dot{y} + 5y = 2 + t, \quad y(0) = 4, \quad \dot{y}(0) = 1.$ **(i)** $2\dot{y} + y = e^{-t/2}, \quad y(0) = -1.$

(j) $\ddot{y} + 8\dot{y} + 20y = \sin 2t, \quad y(0) = 1, \dot{y}(0) = -4.$

(k) $4\ddot{y} - 4\dot{y} + y = t^2, \quad y(0) = -12, \dot{y}(0) = 7.$

(l) $2\ddot{y} + \dot{y} - y = 4\sin t, \quad y(0) = 0, \quad \dot{y}(0) = -4.$ **(m)** $\dot{y} - y = e^{2t}, \quad y(0) = 1.$

(n) $3\ddot{y} + 5\dot{y} - 2y = 7e^{-2t}, \quad y(0) = 3, \ \dot{y}(0) = 0.$

3. Use the Laplace transformation to solve the following initial value problems with intermittent forcing functions ($H(t)$ is the Heaviside function, page 274).

(a) $y' + y = H(t) - H(t-2), \quad y(0) = 1;$ **(b)** $y' - 2y = 4t[H(t) - H(t-2)], \quad y(0) = 1;$

(c) $y'' + 9y = 24 \sin t[H(t) + H(t-\pi)], \quad y(0) = 0, \ y'(0) = 0;$

(d) $y'' + 2y' + y = H(t) - H(t-1), \quad y(0) = 1, \ y'(0) = -1;$

(e) $y'' + 2y' + 2y = 5 \cos t[H(t) - H(t-\pi/4)], \quad y(0) = 1, \ y'(0) = -1;$

(f) $y'' + 5y' + 6y = 36t\,[H(t) - H(t-1)], \quad y(0) = -1, \ y'(0) = -2;$

(g) $y'' + 4y' + 13y = 39H(t) - 507(t-2)\,H(t-2), \quad y(0) = 3, \ y'(0) = 1;$

(h) $y'' + 4y = 3[H(t) - H(t-4)] + (2t-5)H(t-4), \quad y(0) = 3/4, \ y'(0) = 2;$

(i) $4y'' + 4y' + 5y = 25t[H(t) - H(t-\pi/2)], \quad y(0) = 2, \ y'(0) = 2;$

(j) $y'' + 4y' + 3y = H(t) - H(t-1) + H(t-2) - H(t-3), \quad y(0) = -2/3, \ y'(0) = 1;$

4. Use the Laplace transform to solve the initial value problems with piecewise continuous forcing functions.

(a) $y'' - 2y' = \begin{cases} 4, & 0 \leqslant t < 1, \\ 6, & t \geqslant 1, \end{cases} \qquad y(0) = -6, \quad y'(0) = 1.$

(b) $y'' - 3y' + 2y = \begin{cases} 0, & 0 \leqslant t < 1, \\ 1, & 1 \leqslant t < 2, \\ -1, & t \geqslant 2, \end{cases} \qquad y(0) = 3, \quad y'(0) = -1.$

(c) $y'' + 3y' + 2y = \begin{cases} 1, & 0 \leqslant t < 2, \\ -1, & t \geqslant 2, \end{cases} \qquad y(0) = 0, \quad y'(0) = 0.$

(d) $y'' + y = \begin{cases} t, & 0 \leqslant t < \pi, \\ -t, & t \geqslant \pi, \end{cases} \qquad y(0) = 0, \quad y'(0) = 0.$

(e) $y'' + 4y = \begin{cases} 8t, & 0 \leqslant t < \frac{\pi}{2}, \\ 8\pi, & t \geqslant \frac{\pi}{2}, \end{cases} \qquad y(0) = 0, \quad y'(0) = 0.$

5. Solve the IVPs involving the Dirac delta-function. The initial conditions are assumed to be homogeneous.

(a) $y'' + (2\pi)^2\, y = 3\delta(t - 1/3) - \delta(t-1).$ **(b)** $y'' + 2y' + 2y = 3\delta(t-1).$

(c) $y'' + 4y' + 29y = 5\delta(t-\pi) - 5\delta(t-2\pi).$ **(d)** $y'' + 3y' + 2y = 1 - \delta(t-1).$

(e) $4y'' + 4y' + y = e^{-t/2}\,\delta(t-1).$ **(f)** $y''' - 7y' + 6y = \delta(t-1).$

6. Consider an RC-circuit and associated equation $R\dot{q} + \frac{1}{C}\,q(t) = E(t)$, where $E(t)$ is the electromotive force function. Use the Laplace transform method to solve for the current $I(t) = \dot{q}$ and charge $q(t)$ in each of the following cases assuming homogeneous initial conditions.

(a) $R = 10$ ohms, $C = 0.01$ farads, $E(t) = H(t-1) - H(t-2)$.

(b) $R = 20$ ohms, $C = 0.001$ farads, $E(t) = (t-1)[H(t-1) - H(t-2)] + (3-t)[H(t-2) - H(t-3)]$.

(c) $R = 1$ ohm, $C = 0.01$ farads, $E(t) = t[H(t) - H(t-1)$.

(d) $R = 2$ ohms, $C = 0.001$ farads, $E(t) = (1 - e^{-t})[H(t) - H(t-2)]$.

7. Consider an LC-circuit and associated equation $L\,\dot{I} + \frac{1}{C}\,q(t) = E(t)$, where $E(t)$ is the electromotive force function. Assuming that $q(0) = 0$ and $I(0) = 0$, use Laplace transform methods to solve for $I(t) = \dot{q}$ and $q(t)$ in each of the following cases.

(a) $L = 1$ henry, $C = 1$ farad, $E(t) = H(t-1) - H(t-2)$.

(b) $L = 1$ henry, $C = 0.01$ farads, $E(t) = (t-1)[H(t-1) - H(t-2)] + (3-t)[H(t-2) - H(t-3)]$.

(c) $L = 2$ henries, $C = 0.05$ farads, $E(t) = t[H(t) - H(t-1)]$.

(d) $L = 2$ henries, $C = 0.01$ farads, $E(t) = (1 - e^{-t})[H(t) - H(t-\pi)]$.

8. Consider an RL-circuit and associated equation $L\,\dot{I} + R\,I = E(t)$. Assume that $I(0) = 0$, that L and R are constants, and that $E(t)$ is the periodic function whose description in one period is $E(t) = \begin{cases} 1, & 0 \leqslant t < 1, \\ -1, & 1 \leqslant t < 2. \end{cases}$ Solve for $I(t)$.

9. Suppose that a weight of constant mass m is suspended from a spring with spring constant k and that the air resistance is proportional to its speed. Then the differential equation of motion is $m\ddot{y} + c\dot{y} + ky = f(t)$, where $y(t)$ denotes the vertical displacement of the mass at time t and $f(t)$ is some force acting on the system. Assume that $c^2 = 4mk$ and that $y(0) = \dot{y}(0) = 0$. Find the response of the system under the following force: $f(t) = \begin{cases} \sin t, & 0 \leqslant t < \pi, \\ 0, & \pi \leqslant t. \end{cases}$

10. Suppose that in the preceding mass-spring problem $m = 1$, $c = 2$, $k = 10$; and $y(0) = 1$, $\dot{y}(0) = 0$; and $f(t)$ is a periodic function whose description in one period is $f(t) = \begin{cases} 10, & 0 \leqslant t < \pi, \\ -10, & \pi \leqslant t < 2\pi. \end{cases}$ Find $y(t)$.

11. A weight of 2 kilograms is suspended from a spring of stiffness 5×10^4 newtons per meter. Suppose that at $t = 2$ the mass, which is motionless and in its equilibrium position, is struck by a hammer of mass 0.25 kilogram, traveling at 20 meters per second. Find the displacement if the damping coefficient is approximately 0.02 `kg/sec`.

12. The values of mass m, spring constant k, dashpot resistance c, and external force $f(t)$ are given for a mass-spring-dashpot system:

    (a) $m = 1$, $k = 8$, $c = 4$, $f(t) = H(t) - H(t - \pi)$.

    (b) $m = 1$, $k = 4$, $c = 5$, $f(t) = H(t) - H(t - 2)$.

    (c) $m = 1$, $k = 8$, $c = 9$, $f(t) = 4 \sin t[H(t) - H(t - \pi)]$.

    (d) $m = 1$, $k = 4$, $c = 4$, $f(t) = t[H(t) - H(t - 2)]$.

    (e) $m = 1$, $k = 4$, $c = 5$, $f(t)$ is the triangular wave function with $a = 1$ (Problem 11, §5.2, page 290).

    Find the displacement which is the solution of the following initial value problem

$$m\ddot{x} + c\dot{x} + kx = f(t), \qquad x(0) = 0, \quad \dot{x}(0) = v_0.$$

13. Solve the initial value problems for differential equations of order higher than 2.

    (a) $y''' + y'' + 4y' + 4y = 8$, $y(0) = 4$, $y'(0) = -3$, $y''(0) = -3$.

    (b) $y''' - 2y'' - y' + 2y = 4t$, $y(0) = 2$, $y'(0) = -2$, $y''(0) = 4$.

    (c) $y''' - y'' + 4y' - 4y = 8\,e^{2t} - 5\,e^{t}$, $y(0) = 2$, $y'(0) = 0$, $y''(0) = 3$.

    (d) $y''' - 5y'' + y' - y = 2t - 10 - t^2$, $y(0) = 2$, $y'(0) = 0$, $y''(0) = 0$.

14. Use the Laplace transformation to solve the following initial value problems with intermittent forcing functions ($H(t)$ is the Heaviside function, page 274) for the differential equations of the order higher than 2.

    (a) $y^{(4)} - 5y'' + 4y = 12\,[H(t) - H(t - 1)]$, $y(0) = 0$, $y'(0) = 0$, $y''(0) = 0$, $y'''(0) = 0$;

    (b) $y^{(4)} - 16y = 32[H(t) - H(t - \pi)]$, $y(0) = 0$, $y'(0) = 0$, $y''(0) = 0$, $y'''(0) = 0$;

    (c) $(\mathtt{D} - 1)^3 y = 6\,e^t\,[H(t) - H(t - 1)]$, $y(0) = 0$, $y'(0) = 0$, $y''(0) = 0$;

    (d) $(\mathtt{D}^3 - \mathtt{D})y = H(t) - H(t - 2)$, $y(0) = 0$, $y'(0) = 0$, $y''(0) = 0$.

Summary for Chapter 5

1. The Laplace transform is applied to functions defined on the positive half-line. We always assume that such functions are zero for the negative values of its argument. In Laplace transform applications, we usually use well-known functions (such as $\sin t$ or t^3, for example) that originally are defined on the whole axis, but not only on the positive half-line. That is why we must multiply a function by the Heaviside function to guarantee that such a function is zero for negative values of the argument.

2. For a given function $f(t)$, $t \geqslant 0$, and some complex or real parameter λ , if the indefinite integral $\int_0^\infty f(t)\,e^{-\lambda t}\,dt$ exists, it is called the Laplace transform of the function $f(t)$ and denoted either by f^L or $\mathcal{L}[f]$. The imaginary part of the parameter λ does not affect the convergence of the Laplace integral. If the integral converges for some real value $\lambda = \mu$, then it converges for all real $\lambda > \mu$ as well as for complex numbers with a real part greater than μ.

3. There is a class of functions, called *function-originals*, for which the Laplace transformation exists. The Laplace transform establishes a one-to-one correspondence between a function-original and its *image*. Each function-original possesses the property (5.1.8), page 277, that is, at each point $t = t_0$, a function-original is equal to the mean of its left-hand side and right-hand side limit values.

4. A computer algebra system is useful for a quick calculation of Laplace transforms. For example, *Maple* can find the Laplace transform of the function $f(t) = e^{2t}\sin(3t)$ in the following steps:
```
with(inttrans): f:=exp(2*t)*sin(3*t);
F:=laplace(f,t,lambda); F:=simplify(expand(F));
```
Mathematica can do a similar job:
```
f = Exp[2*t]*Sin[3*t];
F = LaplaceTransform[f,t,lambda]
Simplify[Expand[ F ]]
```
When using *Maxima*, we type
```
laplace(exp(2*t)*sin(3*t),t,lambda)
```
Application of *Sage* yields
```
t,lambd,a = var('t,lambd,a')
f = exp(a*t)*sin(3*t)
f.laplace(t,lambd)
3/(a^2 - 2*a*lambd + lambd^2 + 9)
```

Below are the most important properties of the Laplace transformation:

1° **The convolution rule**, Eq. (5.2.1), page 282.

2° **The differential rule**, Eq. (5.2.2), page 282.

3° **The similarity rule**, Eq. (5.2.5), page 282.

4° **The shift rule** , Eq. (5.2.6), page 282.

5° **The attenuation rule**, Eq. (5.2.7), page 282.

6° **The integration rule**, Eq. (5.2.8), page 283.

7° **Rule for multiplicity by t^n**, Eq. (5.2.11), page 283.

8° **The Laplace transform of periodic functions:** If $f(t) = f(t+\omega)$, then

$$f^L(\lambda) = \frac{1}{1-e^{-\omega\lambda}} \int_0^{\omega} e^{-\lambda t} f(t)\,\mathrm{d}t. \tag{5.2.13}$$

5. The property

$$H(t) * f(t) = \int_0^t f(\tau)\,\mathrm{d}\tau$$

can be used to obtain an antiderivative of a given function $f(t)$, namely,

$$\int_0^t f(\tau)\,\mathrm{d}\tau = \mathcal{L}^{-1}\left[\frac{1}{\lambda} f^L(\lambda)\right].$$

6. The **full-wave rectifier** of a function $f(t)$, $0 \leqslant t \leqslant T$ is a periodic function with period T that coincides with $f(t)$ on the interval $[0, T]$.
The **half-wave rectifier** of a function $f(t)$, $0 \leqslant t \leqslant T$ is a periodic function with period $2T$ that coincides with $f(t)$ on the interval $[0, T]$ and is zero on the interval $[T, 2T]$.

7. The Dirac δ-function is a derivative of the Heaviside function, namely,

$$\lim_{\varepsilon\to 0} \frac{1}{\varepsilon} \left[H(t-a) - H(t-a-\varepsilon)\right],$$

where the limit is understood in generalized sense as $\int \delta(t-a)\, f(t)\,\mathrm{d}t = \lim_{\varepsilon\to 0} \frac{1}{\varepsilon} \int_a^{a+\varepsilon} f(t)\,\mathrm{d}t = f(a)$ for every smooth integrable function $f(t)$.

8. The Dirac delta-function has two remarkable properties: $\delta(t-a) * f(t) = f(t-a)$ and $\mathcal{L}[\delta](\lambda) = 1$.

We present three methods to determine the inverse Laplace transform, $\mathcal{L}^{-1}$, of rational functions $P(\lambda)/Q(\lambda)$, where $P(\lambda)$ and $Q(\lambda)$ are polynomials or entire functions.

A. The Partial Fraction Decomposition Method is based on the expansion of $P(\lambda)/Q(\lambda)$ into a linear combination of the simple rational functions

$$\frac{1}{\lambda-\alpha} \quad \text{or} \quad \frac{A\lambda + B}{(\lambda-\alpha)^2 + \beta^2},$$

or their powers. Table 280 suggests that

$$\mathcal{L}^{-1}\left[\frac{1}{\lambda-\alpha}\right] = e^{\alpha t}\,H(t),$$

$$\mathcal{L}^{-1}\left[\frac{A\lambda+B}{(\lambda-\alpha)^2+\beta^2}\right] = \left[A\,e^{\alpha t}\,\cos\beta t + \frac{A\alpha+\beta}{\beta}\,e^{\alpha t}\,\sin\beta t\right]\,H(t),$$

where $H(t)$ is the Heaviside (see Eq. (5.1.5) on page 274). The rule (5.2.11) of multiplicity by t^n may be used to find the inverse Laplace transform for powers of simple rational functions.

B. Convolution Theorem

$$\mathcal{L}^{-1}\left\{F_1(\lambda)\,F_2(\lambda)\right\} = (f_1 * f_2)(t),$$

where $F_1(\lambda) = \mathcal{L}[f_1] = f_1^L$ and $F_2(\lambda) = \mathcal{L}[f_2] = f_2^L$.

C. The Residue Method. For a function $F(\lambda) = P(\lambda)/Q(\lambda)$ which is a fraction of two entire functions

$$f(t) = \mathcal{L}^{-1}\left[F(\lambda)\right] = \sum_{j=1}^{N} \operatorname*{Res}_{\lambda_j} F(\lambda)e^{\lambda t},$$

where the sum covers all zeroes λ_j of the equation $Q(\lambda) = 0$. The residues $\mathrm{Res}_{\lambda_j}\, F(\lambda)e^{\lambda t}$ are evaluated as follows: If λ_j is a simple root of $Q(\lambda) = 0$, then

$$\operatorname*{Res}_{\lambda_j} F(\lambda)e^{\lambda t} = \frac{P(\lambda_j)}{Q'(\lambda_j)}\,e^{\lambda_j t}.$$

If λ_j is a double root of $Q(\lambda) = 0$, then

$$\operatorname*{Res}_{\lambda_j} F(\lambda)e^{\lambda t} = \lim_{\lambda\mapsto\lambda_j} \frac{\mathrm{d}}{\mathrm{d}\lambda}\left\{(\lambda-\lambda_j)^2 F(\lambda)\,e^{\lambda t}\right\}.$$

In general, when λ_j is an n-fold root of $Q(\lambda) = 0$, then

$$\operatorname*{Res}_{\lambda_j} F(\lambda)e^{\lambda t} = \lim_{\lambda\mapsto\lambda_j} \frac{1}{(n-1)!}\,\frac{\mathrm{d}^{n-1}}{\mathrm{d}\lambda^{n-1}}\left\{(\lambda-\lambda_j)^n F(\lambda)\,e^{\lambda t}\right\}.$$

9. To check your answer, you may want to use a computer algebra system:
Maple: `with(inttrans): invlaplace(function, lambda, t);` or
`with(MTM): ilaplace(function(lambda), lambda, t);`
Mathematica: `f = InverseLaplaceTransform[function,lambda,t] // Expand`
Maxima: `ilt(function,lambda,t);`
Sage: `f.inverse_laplace(lambd,t)`

Although the ideas of Chapter 4 could be used to solve the initial value problems, Laplace transform provides a convenient alternative solutions strategy. Applying the Laplace transform to the initial value problem for second order homogeneous linear differential equations with constant coefficients

$$a\,y''(t) + b\,y'(t) + c\,y(t) = 0, \qquad y(0) = y_0,\ y'(0) = y_1$$

includes the following steps:

1. Multiply both sides of the given differential equation by $e^{\lambda t}$ and then integrate the result with respect to t from zero to infinity. This yields the integral equation.
2. Introduce the shortcut notation for the Laplace transform of the unknown function, say y^L.
3. Integrate by parts the terms with derivatives y' and y'' or use the differential rule (5.2.2). This leads to the algebraic equation with respect to y^L:
$$(a\lambda^2 + b\lambda + c)y^L = a(\lambda y_0 + y_1) + by_0\,.$$
4. Solve this algebraic equation for y^L to obtain $y^L(\lambda) = \dfrac{a(\lambda y_0 + y_1) + by_0}{a\lambda^2 + b\lambda + c}$.
5. Apply the inverse Laplace transform to determine $y(t)$.

A similar procedure is applicable to initial value problems for n-th order differential equations.

The application of the Laplace transform to initial value problems for nonhomogeneous differential equations

$$ay''(t) + by'(t) + cy(t) = f(t),\ (t > t_0) \quad y(t_0) = y_0,\ y'(t_0) = y_0' \tag{$*$}$$

or, in the general case,

$$L[y] \stackrel{\text{def}}{=} a_n y^{(n)} + \cdots + a_0 y = f, \quad y(t_0) = y_0,\ \ldots,\ y^{(n-1)}(t_0) = y_0^{(n-1)}, \tag{$**$}$$

with linear constant coefficient differential operator L of the n-th order and given function f, consists of the following steps.

1. Set $t_0 = 0$ by making a shift and consider the initial value problem for this case.
2. Apply the Laplace transform using the differential rule (5.2.2); that is, substitute into Eq. (*) or Eq. (**) the algebraic expression
$$\mathcal{L}[y^{(k)}] = \lambda^k y^L - \lambda^{k-1} y_0 - \lambda^{k-2} y_0^{(1)} - \cdots - y_0^{(k-1)}.$$
for the k-th derivative of unknown function $y^{(k)}$. The result should look like this:
$$L(\lambda) y^L = f^L + \sum_{k=0}^{n} a_k \left[\lambda^{k-1} y_0 + \lambda^{k-2} y_0^{(1)} + \cdots - y_0^{(k-1)}\right].$$
3. Divide both sides of the last algebraic equation by the characteristic polynomial $L(\lambda) = a_n\lambda^n + \cdots + a_0$ to obtain
$$y^L = y_p^L + y_h^L,$$
where
$$y_p^L(\lambda) = G^L(\lambda) \cdot f^L \qquad \text{with} \qquad G^L(\lambda) = \frac{1}{L(\lambda)},$$
the Laplace transform of the Green function, and
$$y_h^L(\lambda) = \frac{1}{L(\lambda)} \left(\sum_{k=0}^{n} a_k \left[\lambda^{k-1} y_0 + \lambda^{k-2} y_0^{(1)} + \cdots - y_0^{(k-1)}\right]\right).$$
4. Determine $y_h(t)$, the solution of the homogeneous equation with the given initial conditions, as the inverse Laplace transform of $y_h^L(\lambda)$. For this purpose, use partial fraction decomposition or the residue method. For details, see §5.4.
5. To find $y_p(t)$, the solution of a nonhomogeneous equation with homogeneous initial conditions (at $t = 0$), there are two options:
   (a) Determine the explicit expression f^L of the Laplace transform of the function $f(t)$ by evaluating the integral
   $$f^L(\lambda) = \int_0^\infty f(t)\, e^{-\lambda t}\, \mathrm{d}t.$$
   For this purpose we can use any method that is most convenient, for instance, Table 280 can be utilized. With this in hand, find the function $y_p(t)$ as the inverse Laplace transform of the known function y_p^L (see §5.4).
   (b) If the explicit expression f^L is hard to determine, use the convolution rule. First, find the Green function $G(t) = \mathcal{L}^{-1}[G^L]$ (again use the methods of §5.4). Then apply the convolution rule to obtain
   $$y_p(t) = \int_0^t G(t-\tau)\, f(\tau)\, \mathrm{d}\tau = \int_0^t G(\tau)\, f(t-\tau)\, \mathrm{d}\tau.$$
6. Add the two solutions $y_p(t)$ and $y_h(t)$ to obtain the solution of the given initial value problem with initial conditions at $t = 0$.
7. Return to the original initial conditions at $t = t_0$ by using the shift rule (5.2.6). Namely, let $z(t)$ be the solution of the given initial value problem with the initial conditions at $t = 0$. Then the solution $y(t)$ for the same problem with the initial conditions at $t = t_0$ is $y(t) = H(t - t_0) z(t - t_0)$.

Review Questions for Chapter 5

Section 5.1.

1. Find the Laplace transform of the given functions.

   (a) $f(t) = \begin{cases} e^t, & t \leqslant 2, \\ e^2, & t \geqslant 2; \end{cases}$ (b) $f(t) = \begin{cases} t, & t \leqslant 2, \\ 2, & t \geqslant 2; \end{cases}$ (c) $f(t) = \begin{cases} 2t, & 0 \leqslant t \leqslant 2, \\ 4, & t \geqslant 2; \end{cases}$

   (d) $f(t) = \begin{cases} 1, & 0 \leqslant t \leqslant 1, \\ e^t, & 1 < t \leqslant 2, \\ 0, & t \geqslant 2; \end{cases}$ (e) $f(t) = \begin{cases} t^2, & 0 \leqslant t \leqslant 1, \\ 1, & 1 \leqslant t \geqslant 2, \\ (t-1)^2, & t \geqslant 2. \end{cases}$

2. Let a be a positive number. Find the Laplace transform of the given functions. A preliminary integration by parts may be necessary.

   (a) $e^{-t} \sinh at$; (b) $e^{at} \cosh t$; (c) $\sqrt{t} + t^2$; (d) $(t+1)^3$;
   (e) $\sin(2t)\, \cos(3t)$; (f) $\sinh^2 2t$; (g) $4t^{3/2} - 8t^{5/2}$; (h) $\cosh 3t \sinh 4t$;
   (i) $[\sin t + \cos t]^2$; (j) $\sin t\, \cosh(2t)$; (k) $(t+1)^2\, \cos 3t$; (l) $\sqrt{t}\, e^{at}$.

3. If $f(t)$ and $f'(t)$ are both Laplace transformable and $f(t)$ is continuous on $(0,\infty)$, then prove that $\lim\limits_{\lambda\to\infty} \lambda f^L(\lambda) = f(+0)$.
4. Find $\mathcal{L}\left[e^{-t}\sin t\right]$ by performing the integration $\int_0^\infty \left(e^{-t}\sin t\right) e^{-\lambda t}\,\mathrm{d}t$.
5. Which of the following functions have Laplace transforms?
   (a) $\frac{t}{t^2+1}$; **(b)** $\frac{1}{t+1}$; **(c)** $\lfloor t\rfloor !$; **(d)** $\frac{\sin(\pi t)}{t-\lfloor t\rfloor}$; **(e)** 3^t; **(f)** e^{-t^2}.
6. Which of the following functions are of exponential order on $[0,\infty)$?
   (a) $\ln(1+t^2)$; **(b)** $t^{1/2}$; **(c)** $t\ln t$; **(d)** $t^4\cos t$;
   (e) $e^{\sin t}$; **(f)** $\sin(t^2)$; **(g)** $\sinh(t^2)$; **(h)** $\cos\left(e^{t^2}\right)$.

Section 5.2 of Chapter 5 (Review)

1. Find the Laplace transform of the following functions.
   (a) $\sin^2 t$; **(b)** $\cos^2 t$; **(c)** $\sin(t+a)$; **(d)** $|\sin at|$; **(e)** $t^2\sin at$; **(f)** $t^2\cos at$.
2. Find the Laplace transform of the given periodic functions.
   (a) $f(t) = \begin{cases} 1, & t\leqslant 2,\\ 0, & 2\leqslant t\leqslant 4,\end{cases}$ $f(t+4)=f(t)$; **(b)** $f(t) = \begin{cases} t, & 0\leqslant t\leqslant 1,\\ 1, & 1\leqslant t\leqslant 2,\end{cases}$ $f(t+2)=f(t)$;
   (c) $f(t) = 1-t$, if $0\leqslant t\leqslant 2$, $f(t+2)=f(t)$.
   (d) $f(t) = \begin{cases} \sin(t/2), & 0\leqslant t< 2\pi,\\ 0, & 2\pi\leqslant t< 4\pi,\end{cases}$ and $f(t+4\pi)=f(t)$ for all $t\geqslant 0$.
3. Find the Laplace transform of the given functions.
   (a) $\frac{\mathrm{d}^2}{\mathrm{d}t^2}\sin 2t$; **(b)** $(t-3)\cos 2t$; **(c)** $t^2+3\cos 3t$;
   (d) $2t^2\cosh 3t$; **(e)** $\frac{\mathrm{d}}{\mathrm{d}t}(t^2-2t+5)\,e^{3t}$; **(f)** $[t+\cos t]^2$.
4. Use the attenuation rule and/or the similarity rule to find the Laplace transformations of the following functions.
   (a) $t\,e^{-2t}$; **(b)** $t^{3/2}\,e^t$; **(c)** $\sqrt{t}\,e^{2t}$;
   (d) $e^{\alpha t}\cos\beta t$; **(e)** $t^{200}\,e^{5t}$; **(f)** $e^{-2t}\,(t^2+3t-6)$.
5. Find the Laplace transform of the following integrals.
   (a) $\int_0^t t\cosh 2t\,\mathrm{d}t$; **(b)** $\int_0^t e^{2t}\sin 3t\,\mathrm{d}t$; **(c)** $\int_0^t te^{3t}\sin t\,\mathrm{d}t$;
   (d) $e^{-2t}\int_0^t t\sin 5t\,\mathrm{d}t$; **(e)** $t^2\int_1^t t\cos 2t\,\mathrm{d}t$; **(f)** $\frac{\mathrm{d}^2}{\mathrm{d}t^2}\int_0^t e^t\sin 2t\,\mathrm{d}t$;
   (g) $te^{-2t}\int_0^t \frac{\mathrm{d}}{\mathrm{d}t}\left(e^t\sin 3t\right)\mathrm{d}t$; **(h)** $\int_0^t t^2e^{-3t}\sin t\,\mathrm{d}t$; **(i)** $\frac{\mathrm{d}}{\mathrm{d}t}\int_0^t e^t\cosh 3t\,\mathrm{d}t$.
6. Compute the indicated convolution product (H is the Heaviside function)
   (i) by calculating the convolution integral;
   (ii) using the Laplace transform and the convolution rule (5.2.1).
   (a) $t * H(t-1)$; **(b)** $(t^2-1) * H(t-1)$;
   (c) $t^n * H(t-a)$; **(d)** $[H(t-2)-H(t-3)] * [H(t-1)-H(t-4)]$;
   (e) $H(t-1) * H(t-3)$; **(f)** $[H(t)-H(t-3)] * [H(t-1)-H(t-2)]$;
   (g) $t^2\sin t * H(t-a)$; **(h)** $2\sin t * \sin t[H(t)-H(t-\pi)]$.
7. Find the convolution of the following functions.
   (a) $H(t) * t$; **(b)** $t^2 * t$; **(c)** $t^2 * t^2$;
   (d) $t^2 * (t+1)^2$; **(e)** $t * e^{2t}$; **(f)** $t * \sinh bt$;
   (g) $e^{2t} * e^{2t}$; **(h)** $e^{at} * e^{bt}$, $a\neq b$; **(i)** $e^{at} * e^{at}$;
   (j) $\cos t * e^{at}$; **(k)** $\sin at * t$; **(l)** $\cos at * t$;
   (m) $\sin at * \cos at$; **(n)** $\sin at * \sin bt$; **(o)** $e^t * \cosh t$;
   (p) $e^{-t} * \cosh t$; **(q)** $t^n * t^m * t^k$; **(r)** $\sin at * \sin at * \cos at$.
8. Suppose that both $f(t)$ and $\int_a^t f(\tau)\,\mathrm{d}\tau$ are Laplace transformable, where a is any nonnegative constant. Show that

$$\mathcal{L}\left[\int_a^t f(\tau)\,\mathrm{d}\tau\right] = \frac{1}{\lambda}\,\mathcal{L}[f](\lambda) - \frac{1}{\lambda}\int_0^a f(\tau)e^{-\lambda\tau}\,\mathrm{d}\tau.$$

9. Let $f(t) = \lfloor t\rfloor$ be the floor function (the greatest integer that is less than or equal to t). Show that

$$\mathcal{L}\lfloor t\rfloor = \sum_{n\geqslant 0}\frac{n}{\lambda}\left(e^{-n\lambda} - e^{-(n+1)\lambda}\right).$$

10. Find the Laplace transform of the exponential function e^t by integrating term by term the Maclaurin series $e^t = \sum_{n\geqslant 0} \frac{t^n}{n!}$ and adding the results. *Hint:* The sum of the geometric series is $\sum_{n\geqslant 0} \lambda^n = \frac{1}{1-\lambda}$.

11. Find the Laplace transform of the function $\cos(2\sqrt{t})/\sqrt{\pi t}$ by expanding cosine into the Maclaurin series, $\cos u = \sum_{n\geqslant 0}(-1)^n \frac{u^{2n}}{(2n)!}$, and integrating term by term.

12. Let $x(t)$ be a piecewise continuous function. Show that the unweighted average of $x(t)$ over the interval $[t-a, t+a]$ can be written as
$$\frac{1}{2a}\int_{-a}^{a} x(t+\tau)\,\mathrm{d}\tau = \frac{1}{2a}\left[(H*x)(t+a) - (H*x)(t-a)\right].$$

13. Give a formula for the convolution of n Heaviside functions: $\underbrace{H * H * \ldots * H}_{n}$.

14. Evaluate each of the following convolutions using the Laplace transformation.
(a) $t * \cos at$; **(b)** $t * \sin at$; **(c)** $t^2 * \sinh 3t$; **(d)** $t^3 * \cosh 3t$.

Section 5.3 of Chapter 5 (Review)

1. Sketch the graph of the given function and determine whether it is continuous, piecewise continuous, or neither on the interval $0 \leqslant t \leqslant 4$.

(a) $f(t) = \begin{cases} 2t, & 0\leqslant t\leqslant 2,\\ 4-t, & 2<t\leqslant 3,\\ 3, & 3<t\leqslant 4.\end{cases}$ (b) $f(t) = \begin{cases} t^{-1}, & 0\leqslant t< 1,\\ (t-1)^{-1}, & 1<t\leqslant 3,\\ 1, & 3<t\leqslant 4.\end{cases}$

(c) $f(t) = \begin{cases} t^2, & 0\leqslant t< 2,\\ 4, & 2<t\leqslant 3,\\ 5-t, & 3<t\leqslant 4.\end{cases}$ (d) $f(t) = \begin{cases} t, & 0\leqslant t< 2,\\ 5-t, & 2<t\leqslant 3,\\ t-1, & 3<t\leqslant 4.\end{cases}$

2. Write each function in terms of the Heaviside function and find its Laplace transform.

(a) $f(t) = \begin{cases} 2, & 0\leqslant t<4,\\ -2, & t\geqslant 4.\end{cases}$ (b) $f(t) = \begin{cases} 0, & 0\leqslant t<3,\\ 1, & t\geqslant 3.\end{cases}$

(c) $f(t) = \begin{cases} t^2, & 0\leqslant t<1,\\ 0, & t\geqslant 1.\end{cases}$ (d) $f(t) = \begin{cases} 0, & 0\leqslant t<3,\\ t, & t\geqslant 3.\end{cases}$

(e) $f(t) = \begin{cases} \sin 2t, & 0\leqslant t<\pi,\\ 0, & t\geqslant \pi.\end{cases}$ (f) $f(t) = \begin{cases} 0, & 0\leqslant t<3\pi/4,\\ \cos(2t), & t\geqslant 3\pi/4.\end{cases}$

(g) $f(t) = \begin{cases} 2, & 0\leqslant t<1,\\ 0, & 1\leqslant t<2,\\ -2, & 2\leqslant t.\end{cases}$ (h) $f(t) = \begin{cases} 1, & 0\leqslant t<1,\\ t, & 1\leqslant t<2,\\ 2, & 2\leqslant t.\end{cases}$

3. Find the convolutions:

(a) $(e^t - 2) * \delta(t - \ln 3)$; (b) $\sin(\pi t) * \delta(t-3)$; (c) $H(t-1) * \delta(t-3)$.

4. Find expressions for the functions whose graphs are shown in Fig. 5.27, 5.28 in terms of shifted Heaviside functions, and calculate the Laplace transform of each.

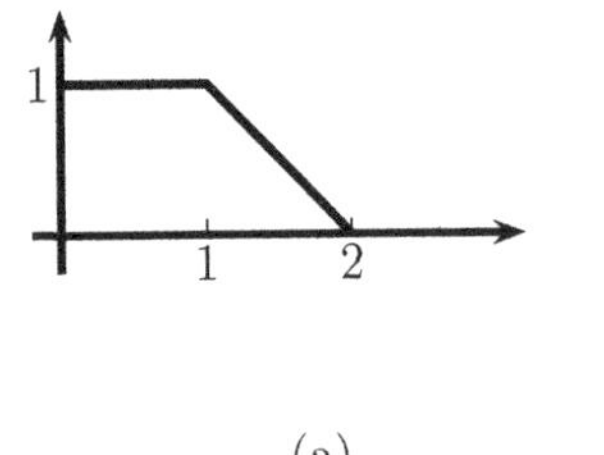

(a)

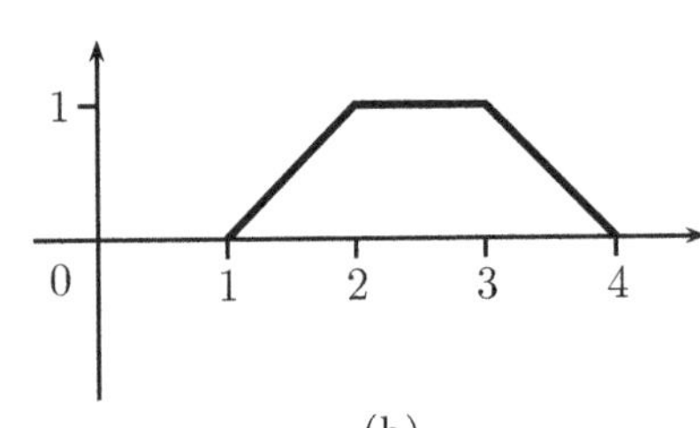

(b)

Figure 5.27: Problems 4(a) and 4(b).

Section 5.4 of Chapter 5 (Review)

1. Find the inverse Laplace transform of the given functions.

(a) $\frac{\lambda+4}{\lambda^2+4\lambda+8}$; (b) $\frac{\lambda+1}{\lambda^2-9}$; (c) $\frac{\lambda+1}{\lambda^2+9}$; (d) $\frac{65}{(\lambda^2+1)(\lambda^2+4\lambda+8)}$;

(e) $\frac{2}{2\lambda-3}$; (f) $\frac{2}{(\lambda-3)^3}$; (g) $\frac{\lambda+3}{\lambda^2+2\lambda+5}$; (h) $\frac{\lambda+2}{\lambda^3}$;

(i) $\frac{30\lambda}{3\lambda^2+8\lambda-3}$.

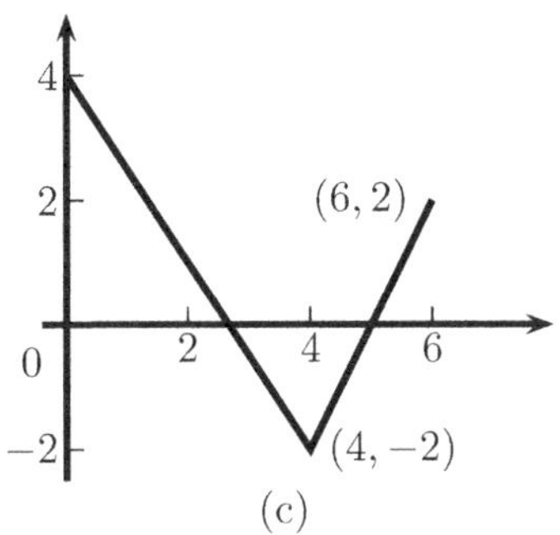

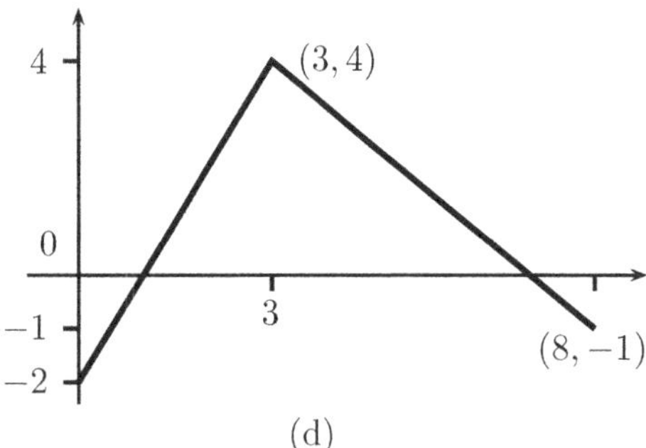

Figure 5.28: Problems 4(c) and 4(d).

2. Find the inverse of each of the following Laplace transforms.

   (a) $\ln \frac{\lambda^2+1}{\lambda(\lambda-1)}$; **(b)** $\ln \frac{(\lambda+a)^2}{(\lambda+b)^2}$; **(c)** $\ln \frac{\lambda-a}{\lambda+a}$; **(d)** $\arctan \frac{4}{\lambda+1}$; **(e)** $\ln \frac{\lambda^2+1}{(\lambda+3)(\lambda-2)}$; **(f)** $\ln\left(1+\frac{9}{\lambda+2}\right)$.

3. Use the convolution rule to find the inverse Laplace transforms.

   (a) $\frac{6}{\lambda^4}$; **(b)** $\frac{1}{\lambda^2(\lambda-1)}$; **(c)** $\frac{7}{(\lambda-2)(\lambda+5)}$;
   (d) $\frac{9}{\lambda(\lambda-3)^2}$; **(e)** $\frac{16}{(\lambda-2)^2(\lambda+2)}$; **(f)** $\frac{250}{(\lambda-9)^2(\lambda+1)^2}$;
   (g) $\frac{9}{(\lambda+4)(\lambda-5)}$; **(h)** $\frac{24}{(\lambda-3)^5}$; **(i)** $\frac{15\lambda}{(4\lambda^2+1)(\lambda^2+4)}$;
   (j) $\frac{17\lambda}{(4\lambda^2+1)(\lambda^2-4)}$; **(k)** $\frac{16\lambda}{(4\lambda^2-1)^2}$; **(l)** $\frac{2}{\lambda(\lambda^2+1)^2}$;
   (m) $\frac{6\lambda^2}{(\lambda^2+9)^2}$; **(n)** $\frac{8\lambda^3}{(\lambda^2+1)^3}$; **(o)** $\frac{82\lambda+246}{(\lambda^2-9)(9\lambda^2+1)}$.

4. Apply the residue method and then check your answer by using partial fraction decomposition to determine the inverse Laplace transforms.

   (a) $\frac{37\lambda-74}{(4\lambda^2+1)(\lambda^2-9)}$. **(b)** $\frac{8(\lambda+1)}{(\lambda^2-4\lambda+5)(\lambda^2-4\lambda+13)}$. **(c)** $\frac{5(\lambda+1)}{\lambda^2-6\lambda+34}$.
   (d) $\frac{28-2(\lambda-3)(\lambda-1)}{(\lambda+1)(\lambda-3)(\lambda+4)}$. **(e)** $\frac{3\lambda^2+4\lambda+9}{(\lambda^2+2\lambda+5)(\lambda+1)}$. **(f)** $\frac{3\lambda-7}{\lambda^2+8\lambda+13}$.
   (g) $\frac{3\lambda+1}{\lambda^2+4}$. **(h)** $\frac{3-(\lambda-1)(\lambda+1)}{(\lambda+4)(\lambda-1)(\lambda-2)}$. **(i)** $\frac{2\lambda+3}{(\lambda-1)^2+4}$.

5. Use the convolution formula to find the inverse Laplace transform of each of the following functions.

   (a) $\dfrac{\lambda}{\lambda^2+1}\, f^L$. **(b)** $\dfrac{2\,e^{-\pi\lambda}}{\lambda^2+4}\, f^L$. **(c)** $\dfrac{\lambda\,e^{-\lambda}}{(\lambda-2)^2}\, f^L$. **(d)** $\dfrac{2\,e^{-\lambda}}{\lambda^2+2\,\lambda+2}\, f^L$.

6. Show that for any integer $n \geqslant 1$ and any $a \neq 0$,

$$\mathcal{L}^{-1}\left[\frac{\lambda}{(\lambda^2+a^2)^{n+1}}\right] = \frac{t}{2n}\int_0^t \mathcal{L}^{-1}\left[\frac{\lambda}{(\lambda^2+a^2)^n}\right] dt.$$

7. Use the attenuation rule (5.2.7) to find the inverse Laplace transforms.

   (a) $\dfrac{3}{\lambda^2+4\lambda+13}$. **(b)** $\dfrac{6(\lambda+2)}{(\lambda^2+4\lambda+13)^2}$. **(c)** $\dfrac{\lambda}{\lambda^2+6\lambda+5}$.
   (d) $\dfrac{2}{4\lambda^2+12\lambda+10}$. **(e)** $\dfrac{8}{\lambda^2-10\lambda+29}$. **(f)** $\dfrac{9}{9\lambda^2+6\lambda+10}$.
   (g) $\dfrac{8}{4\lambda^2-4\lambda+17}$. **(h)** $\dfrac{3}{9\lambda^2-24\lambda+17}$. **(i)** $\dfrac{1}{\lambda^2-14\lambda+50}$.

8. Using the shift rule (5.2.6), find the inverse Laplace transform of each of the following functions.

   (a) $\frac{2\,e^{-2\lambda}}{\lambda^2+2\lambda+5}$. **(b)** $\frac{2\lambda\,e^{-2\lambda}}{\lambda^2+2\lambda+5}$. **(c)** $\frac{2\lambda^4+2\lambda^2+4}{\lambda^5+2\lambda^3}\,e^{-\lambda}$.
   (d) $\frac{4\lambda+2}{\lambda(\lambda^2+\lambda-2)}\,e^{-2\lambda}$. **(e)** $\frac{3}{\lambda^2-2\lambda+10}\,e^{-2\lambda}$. **(f)** $\left[\frac{1}{\lambda-1}+\frac{\lambda-1}{\lambda^2-2\lambda+5}\right]e^{-\lambda}$.
   (g) $\frac{2\,e^{-2\lambda}}{\lambda(\lambda+1)(\lambda+2)}$. **(h)** $\frac{\lambda}{\lambda^2+\pi^2}\,(1+e^{-5\lambda})$. **(i)** $\frac{\lambda}{\lambda^2+1}\,(1-e^{-2\lambda})$.
   (j) $\frac{\lambda}{\lambda^2+4}\,(e^{-\pi\lambda}-e^{-3\pi\lambda})$. **(k)** $\frac{\sqrt{3}\,e^{-2\lambda}}{3\lambda^2+1}$. **(l)** $\frac{2}{4\lambda^2+1}\,e^{-\pi\lambda}$.

9. Use the convolution theorem to find the inverse Laplace transform of the given functions.

   (a) $\dfrac{\lambda}{(\lambda^2+a^2)(\lambda^2+b^2)}$ $a \neq b$; **(b)** $\dfrac{\lambda}{(\lambda^2+a^2)^2}$.

Section 5.5 of Chapter 5 (Review)

1. Solve the initial value problems by the Laplace transform.

   (a) $\ddot{y}+4y=0$, $y(0)=2$, $\dot{y}(0)=2$. **(b)** $\ddot{y}+\dot{y}+y=0$, $y(0)=1$, $\dot{y}(0)=2$.

(c) $\ddot{y} + 4\dot{y} + 3y = 0, \quad y(0) = 3, \, \dot{y}(0) = 2.$

(d) $\ddot{y} + 4\dot{y} + 13y = 0, \quad y(0) = 1, \, \dot{y}(0) = -1.$

(e) $\ddot{y} + 3\dot{y} + 2y = 0, \quad y(0) = 1, \, \dot{y}(0) = 2.$

(f) $\ddot{y} + 6\dot{y} + 18y = 0, \quad y(0) = 2, \quad y'(0) = -3.$

(g) $\ddot{y} + 2\dot{y} + 10y = 0, \quad y(0) = 1, \, \dot{y}(0) = 2.$ (h) $\ddot{y} - 3\dot{y} + 2y = 0, \quad y(0) = 0, \, \dot{y}(0) = 1.$

(i) $\ddot{y} + 6\dot{y} + 8y = 0, \quad y(0) = 0, \, \dot{y}(0) = 2.$ (j) $2\ddot{y} - 11\dot{y} - 6y = 0, \quad y(0) = 3, \, \dot{y}(0) = 5.$

(k) $4\ddot{y} - 39\dot{y} - 10y = 0, \quad y(0) = 5, \, \dot{y}(0) = 9.$

(l) $2\ddot{y} - 27\dot{y} - 45y = 0, \quad y(0) = 3, \, \dot{y}(0) = 12.$

2. Let $\mathtt{D} = d/dt$. Solve the given differential equations of order greater than 2 by the Laplace transform.

(a) $\mathtt{D}^4 y - 5\,\mathtt{D}^2 y + 4y = 0, \quad y(0) = y'(0) = y''(0) = y'''(0) = 1.$

(b) $\mathtt{D}^3 y + 2\,\mathtt{D}^2 y - \mathtt{D}y - 2y = 0, \quad y(0) = y'(0) = y''(0) = 1.$

(c) $y^{(4)} + 8y'' + 16y = 0, \quad y(0) = y'(0) = y''(0) = 1, \quad y'''(0) = 16.$

(d) $y''' + y'' - 6y' = 0, \quad y(0) = y'(0) = 0, \quad y''(0) = 15.$

(e) $y''' - y'' + 9y' - 9y = 0, \quad y(0) = 1, \quad y'(0) = 0, \quad y''(0) = -9.$

(f) $y''' + y'' + 4y' + 4y = 0, \quad y(0) = 2, \quad y'(0) = 1, \quad y''(0) = -3.$

(g) $y''' - 2y'' - y' + 2y = 0, \quad y(0) = 3, \quad y'(0) = 2, \quad y''(0) = 6.$

3. The following second order differential equations with polynomial coefficients have at least one solution that is integrable in some neighborhood of $t = 0$. Apply the Laplace transform (for this purpose use rule (5.2.11) and Problem 13 in §5.2) to derive and solve the differential equation for $y^L = \mathcal{L}[y(t)]$, where $y(t)$ is a solution of the given differential equation. Then come back to the same equation and solve it using the formula (5.5.10).

(a) $t\ddot{y} + 2(t+1)\dot{y} + y = 0;$ (b) $t\ddot{y} + 3\dot{y} - 4ty = 0;$ (c) $(2t+1)\ddot{y} - t\dot{y} + 3y = 0;$
(d) $t\ddot{y} + (1+t)\dot{y} + 2y = 0;$ (e) $(t-1)\ddot{y} - \frac{5}{4}\,t\dot{y} = 0;$ (f) $(t-1)\ddot{y} - (t+1)y = 0.$

4. Determine the initial value problem for which the following inverse Laplace transform is the solution.

(a) $\frac{4\lambda}{4\lambda^2+4\lambda+1};$ (b) $\frac{5\lambda+3}{25\lambda^2-10\lambda+1};$ (c) $\frac{4\lambda+4}{16\lambda^2+8\lambda+1};$ (d) $\frac{\lambda+1}{4\lambda^2-12\lambda+9}.$

Section 5.6 of Chapter 5 (Review)

1. Let $\mathtt{D} = d/dt$ be the operator of differentiation. Obtain the Green function for the following differential operators.

(a) $\mathtt{D}^2 + 4\,\mathtt{D} + 13;$ (b) $\mathtt{D}^2 - 6\,\mathtt{D} + 25;$ (c) $\mathtt{D}^3 + 3\,\mathtt{D}^2 + \mathtt{D} - 5;$
(d) $\mathtt{D}^2 - 6\,\mathtt{D} + 18;$ (e) $\mathtt{D}^2 + \mathtt{D} - 12;$ (f) $(\mathtt{D}^2 - 2\,\mathtt{D} + 26)^2;$
(g) $\mathtt{D}^2 + 8\,\mathtt{D} + 17;$ (h) $6\,\mathtt{D}^2 + \mathtt{D} - 1;$ (i) $\mathtt{D}^4 + 16;$
(j) $4\,\mathtt{D}^2 - \mathtt{D} - 3;$ (k) $2\,\mathtt{D}^2 + \mathtt{D} - 3;$ (l) $9\,\mathtt{D}^3 - \mathtt{D}.$

2. Solve the initial value problems by Laplace transform.

(a) $\dot{y} + y = 2\,\sin t, \quad y(0) = 2.$ (b) $\ddot{y} + 4\dot{y} + 3y = 4\,e^{-t}, \quad y(0) = 0, \, \dot{y}(0) = 2.$

(c) $\dot{y} + 2y = 1, \quad y(0) = 3.$ (d) $\ddot{y} + 16y = 8\,\sin 4t, \quad y(0) = 1, \, \dot{y}(0) = -1.$

(e) $6\ddot{y} - \dot{y} - y = 3\,e^{2t}, \quad y(0) = 0, \, \dot{y}(0) = 0.$

(f) $\ddot{y} + 4\dot{y} + 13y = 120\sin t, \quad y(0) = 0, \, \dot{y}(0) = 3.$

(g) $\ddot{y} + y = 8\,\cos 3t, \quad y(0) = 0, \, \dot{y}(0) = -6.$

(h) $2\ddot{y} + 7\dot{y} + 5y = 27t\,e^{-t}, \quad y(0) = 0, \, \dot{y}(0) = 3.$

(i) $4\ddot{y} + y = t, \quad y(0) = 0, \, \dot{y}(0) = 2.$ (j) $\ddot{y} + 2\dot{y} + 5y = 5e^{-2t}, \quad y(0) = 0, \; \dot{y}(0) = 5.$

(k) $\ddot{y} + 5\dot{y} + 6y = 2e^{-t}, \quad y(0) = 0, \; \dot{y}(0) = 3.$

(l) $2\ddot{y} + \dot{y} = 68\,\sin 2t, \quad y(0) = -17, \, \dot{y}(0) = 0.$

(m) $4\ddot{y} - 4\dot{y} + 5y = 4\sin t, \quad y(0) = 0, \, \dot{y}(0) = 1.$

(n) $\ddot{y} + 2\dot{y} + y = 6\sin t - 4\cos t, \quad y(0) = -6, \, \dot{y}(0) = 2.$

3. Use the Laplace transformation to solve the following initial value problems with intermittent forcing functions ($H(t)$ is the Heaviside function, page 274).

(a) $y' + 2y = 4t[H(t) - H(t-2)], \quad y(0) = 1.$

(b) $4y'' - 12y' + 25y = H(t) - 2H(t-1), \quad y(0) = 1, \; y'(0) = 3.$

(c) $y'' + 25y = 13\sin t[H(t) + H(t - 5\pi)]$, $y(0) = 0$, $y'(0) = 0$.

(d) $y'' - 2y' + 2y = H(t) - H(t - \pi/2)$, $y(0) = 0$, $y'(0) = 1$.

(e) $y'' + 2y' + 10y = e^{-t}[H(t) - H(t - \pi/3)]$, $y(0) = 1$, $y'(0) = -2$.

(f) $y'' - 4y' + 53y = t[H(t - 1) - H(t - 2)]$, $y(0) = 1$, $y'(0) = 4$.

(g) $y'' + 4\pi y' + 4\pi^2 y = 4\pi^2[H(t) - H(t - 2)] + 2\pi^2(t - 2)H(t - 2)$, $y(0) = 1$, $y'(0) = -4\pi$.

(h) $y'' + 4y' + 3y = H(t - 2)$, $y(0) = 0$, $y'(0) = 1$.

(i) $y'' + 2y' + y = e^{2t}H(t - 1)$, $y(0) = 1$, $y'(0) = 1$.

(j) $y'' - y' - 6y = e^{t}H(t - 2)$, $y(0) = 0$, $y'(0) = 1$.

(k) $2y'' + 2y' + y = (\cos t)H(t - \pi/2)$, $y(0) = 0$, $y'(0) = 1$.

(l) $y'' + 2y' + \frac{5}{4}y = (t - 2)H(t - 2) - (t - 2 - k)H(t - 2 - k)$, $y(0) = 1$, $y'(0) = 0$.

4. Use the Laplace transformation to solve the given nonhomogeneous differential equations subject to the indicated initial conditions.

(a) $\ddot{y} + 4\dot{y} + 3y = f(t) \equiv \begin{cases} 6, & 0 \leqslant t \leqslant 1, \\ 0, & 1 < t \leqslant 2, \\ 6, & 2 \leqslant t, \end{cases}$ $y(0) = 1$, $y'(0) = -4$.

(b) $\ddot{y} + y = f(t) \equiv \begin{cases} 0, & 0 \leqslant t \leqslant 1, \\ 2, & 1 < t \leqslant 2, \\ 0, & t \geqslant 2, \end{cases}$ $y(0) = 1$, $y'(0) = 0$.

(c) $\ddot{y} - 4y = f(t) \equiv \begin{cases} 0, & 0 \leqslant t \leqslant 1, \\ t - 1, & 1 < t \leqslant 2, \\ 3 - t, & 2 \leqslant t < 3, \\ 0, & 3 \leqslant t, \end{cases}$ $y(0) = 0$, $y'(0) = 1$.

(d) $\ddot{y} - 5\dot{y} + 6y = f(t) \equiv \begin{cases} t, & 0 \leqslant t \leqslant 2, \\ 0, & 2 \leqslant t, \end{cases}$ $y(0) = 1$, $\dot{y}(0) = 0$.

(e) $3\ddot{y} - 5\dot{y} - 2y = f(t) \equiv \begin{cases} t^2, & 0 \leqslant t < 1, \\ e^{t-1}, & 1 < t < 2, \\ \cos(t - 2), & 2 < t, \end{cases}$ $y(0) = 1$, $\dot{y}(0) = 2$.

(f) $\ddot{y} - 4\dot{y} + 13y = f(t) \equiv \begin{cases} 0, & 0 \leqslant t < 1, \\ e^{t-1}, & 1 < t < 3, \\ t - 3, & 3 < t, \end{cases}$ $y(0) = 1$, $\dot{y}(0) = 2$.

(g) $\ddot{y} - 7\dot{y} + 6y = f(t) \equiv \begin{cases} 0, & 0 \leqslant t < \pi, \\ \sin 2t, & \pi < t < 2\pi, \\ t - 2\pi, & 2\pi < t, \end{cases}$ $y(0) = 1$, $\dot{y}(0) = 6$.

5. Solve the initial value problems for the differential equations of order higher than 2.

(a) $y''' + 6y'' + 12y' + 8y = 4913\cos\frac{t}{2}$, $y(0) = y'(0) = y''(0) = 0$.

(b) $y''' + 3y'' + 4y' + 2y = 10\sin t$, $y(0) = -3$, $y'(0) = -1$, $y''(0) = 3$.

(c) $y''' - 3y'' + 3y' - y = t^2 e^{2t}$, $y(0) = y'(0) = 0$, $y''(0) = 2$.

6. Solve the IVPs involving the Dirac delta-function. The initial conditions are assumed to be homogeneous.

(a) $y'' + (2\pi)^2 y = \delta(t - 1) - 3\delta(t - 1/2)$. **(b)** $y'' + 2y' + y = 3\delta(t - 2)$.

(c) $y''' + y'' + y' + y = \delta(t - 2\pi) - \delta(t - 4\pi)$. **(d)** $y^{(4)} + 4y' + 4y = \delta(t - \pi)$.

7. Use Laplace transforms to solve the IVP $\ddot{y} + 2\dot{y} + 5y = f(t)$, $y(0) = 0$, $\dot{y}(0) = 1$, where $f(t) \equiv \begin{cases} 10, & 0 \leqslant t < 1, \\ -10, & 1 \leqslant t < 2, \end{cases}$ and $f(t)$ is periodic on $[0, \infty)$ with period 2. *Hint:* $\mathcal{L}[f](\lambda) = 10[\tanh(\lambda/2)]/\lambda$.

8. Consider a simple electrical circuit with current satisfying $L\dot{I} + RI = \delta(t - 5)$, $I(0) = 0$. Use the Laplace transform method to find the response of the system.

9. Suppose that a weight of constant mass $m = 1$ is suspended from a spring with spring constant $k = 5$ and that the air resistance is proportional to the speed. Thus, the differential equation of motion is $\ddot{y} + 2\dot{y} + 5y = f(t)$, where $y(t)$ denotes the vertical displacement of the mass at time t and $f(t)$ is some periodic force acting on the system whose description in one period π is $f(t) = \sin t$, $0 \leqslant t < \pi$. Find $y(t)$ subject to the initial conditions $y(0) = y'(0) = 0$.

10. Consider a mass-spring-dashpot system that is modeled by the following differential equation:

$$y'' + 4y' + 20y = f(t) \equiv 20H(t) + 40 \sum_{n\geqslant 1} (-1)^n \, H(t - n\pi),$$

where $f(t)$ is an external force acting on the system and $y(t)$ is the displacement of the mass from the equilibrium position. Assuming that the system is initially at rest at equilibrium ($y(0) = y'(0) = 0$), find the position function $y(t)$.

11. Find the Laplace transform of $J_0(t)$ using the following steps.

(a) Apply the Laplace transform term-by-term to its Taylor expansion

$$\mathcal{L}\left[J_0(t)\right] = \mathcal{L}\left[\sum_{k=0}^{\infty} \frac{(-1)^k \left(\frac{1}{2}\right)^{2k} t^{2k}}{k!\, k!}\right] = \sum_{k\geqslant 0} \frac{(-1)^k \,(2k)!}{2^{2k}\, k!\, k!\, \lambda^{2k+1}}.$$

(b) Since $(2k)! = 2^k\, k!\,[1\cdot 3\cdot 5 \ldots (2k-1)] = 2^k\, k!\,(2k-1)!!$ conclude

$$\mathcal{L}\left[J_0(t)\right] = \frac{1}{\lambda}\left[1 + \sum_{k\geqslant 1} \frac{(-1)^k \,(2k-1)!!}{2^k\, k! \lambda^{2k}}\right] = \frac{1}{\lambda}\left(1 + \frac{1}{\lambda^2}\right)^{-1/2},$$

because

$$(1+z^2)^{-1/2} = \sum_{k\geqslant 0} \binom{-1/2}{k} z^{2k} = \sum_{k\geqslant 0} \frac{(-1)^k}{2^{2k}} \binom{2k}{k} z^{2k} = \sum_{k\geqslant 0} (-1)^k \frac{(2k-1)!!}{(2k)!!}\, z^{2k}.$$

(c) Hence

$$\mathcal{L}\left[J_0(t)\right] = \frac{1}{\sqrt{\lambda^2+1}}.$$

12. Use rule (5.2.12) for division by t to obtain the Laplace transform of the Bessel function $\mathcal{L}\left[J_0(t)\right] = \left(\lambda^2+1\right)^{-1/2}$.

Chapter 6

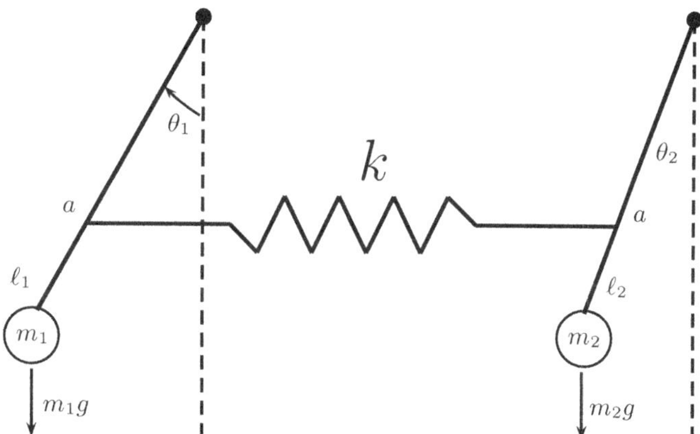

Two pendula connected with a spring

Introduction to Systems of ODEs

Many real-world problems can be modeled by differential equations containing more than one dependent variable to be considered. Mathematical and computer simulations of such problems can give rise to systems of ordinary differential equations. Another reason to study these systems is the fact that all higher order differential equations can be written as an equivalent system of first order equations. The inverse statement is not necessarily true. For this reason, most computer programs are written to approximate the solutions for a first order system of ordinary differential equations.

6.1 Some ODE Models

Differential equations might not be so important if their solutions never appeared in physical models. The electric circuit model and mechanical spring-mass model are two classical examples of modeling phenomena using a system of differential equations that almost every differential equation textbook presents. We cannot avoid a description of these models mostly because of tradition, their simplicity, and their importance in practice. Thus, we will start with these two models and then present other models of interest that also use systems of ordinary differential equations.

6.1.1 RLC-circuits

A typical RLC electric circuit consists of several loops or devices that can include resistors, inductors (coils), capacitors, and voltage sources (batteries). When charge, denoted by $q(t)$, flows through the circuit, it is said that there is a current—a flow of charge. The unit of electric charge is the coulomb[55] (abbreviated C). The conventional symbol for current[56] is I, which was used by the French scientist André-Marie Ampère (1775–1836), after whom the unit of electric current is named.

[55] Charles Augustin de Coulomb (1736–1806), French physicist and engineer who was best known for developing Coulomb's law that describes force interacting between static electrically charged particles.

[56] It originates from the French phrase *intensitè de courant* or in English *current intensity*.

The circuits we will consider involve these four elements (resistor, inductor, capacitor, and voltage source) and current, as flows through each of these elements causes a voltage drop. The equations needed to estimate the voltage drop across each circuit element are presented below. We are not going to discuss the conditions under which these equations are valid.

- In a **resistor**, the voltage drop is proportional to the current (Ohm's law, 1827):

$$\Delta V_{\text{res}} = R I_{\text{res}}.$$

The positive constant R is called the **resistance** and it is measured in ohms[57] (symbol: Ω) or milliohms (with scale factor 0.001).

- In a **capacitor**, the voltage drop is proportional to the charge difference between two plates:

$$\Delta V_{\text{cap}} = \frac{q_{\text{cap}}}{C}.$$

The positive constant C is the **capacitance** and is measured in farads[58] (F) or microfarads (symbolized by μF, which is equivalent to $0.000001 = 10^{-6}$ farads).

- In an **inductor** or **coil**, the voltage drop is proportional to the rate of change of the current:

$$\Delta V_{\text{ind}} = L\,\frac{\mathrm{d}I_{\text{ind}}}{\mathrm{d}t}.$$

The positive constant L is called the **inductance** of the coil and is measured in henries[59] (symbol H).

- A **voltage source** imposes an external voltage drop $V_{\text{ext}} = -V(t)$ and it is measured in volts[60] (V).

A current is a net charge flowing through the area per unit time; thus,

$$I = \texttt{D}\,q = \frac{\mathrm{d}q}{\mathrm{d}t} = \dot{q},$$

where the dot stands for the derivative with respect to t and $\texttt{D} = \mathrm{d}/\mathrm{d}t$. The differential equations that model voltage analysis of electric circuits are based on two fundamental laws[61] derived by Gustav R. Kirchhoff in 1845. **Kirchhoff's Current Law** states that the total current entering any point of a circuit equals the total current leaving it. **Kirchhoff's Voltage Law** states that the sum of the voltage drops around any loop in a circuit is zero.

Example 6.1.1: (A Simple LRC Circuit) In Fig. 6.1 on page 343, we sketch the simplest circuit involving a resistor, an inductor, a capacitor, and a voltage source in series. According to Kirchhoff's current law, the current is the same in each element, that is,

$$I_{\text{resistor}} = I_{\text{coil}} = I_{\text{capacitor}} = I_{\text{source}} = I.$$

Next we apply Kirchhoff's voltage law to obtain the following system of differential equations:

$$\begin{aligned} \frac{\mathrm{d}q}{\mathrm{d}t} &= I, \\ 0 &= V_{\text{resistor}} + V_{\text{coil}} + V_{\text{capacitor}} + V_{\text{source}} \\ &= RI + L\,\frac{\mathrm{d}I}{\mathrm{d}t} + \frac{q}{C} - V(t). \end{aligned}$$

57 George Simon Ohm (1789–1854) was a German physicist and mathematician who taught mathematics to Peter Dirichlet at the Jesuit Gymnasium in Cologne.

58 Michael Faraday (1791–1867) was an English scientist who contributed to the fields of electromagnetism and electrochemistry.

59 Joseph Henry (1797–1878) from the United States, who discovered electromagnetic induction independently of and at about the same time as Michael Faraday.

60 Named after Italian physicist Alessandro Volta (1745–1827), known for the invention of the battery in the 1800s.

61 German physicist Gustav R. Kirchhoff (1824–1887) who first demonstrated that current flows through a conductor at the speed of light. Both circuit rules can be directly derived through approximations from Maxwell's equations.

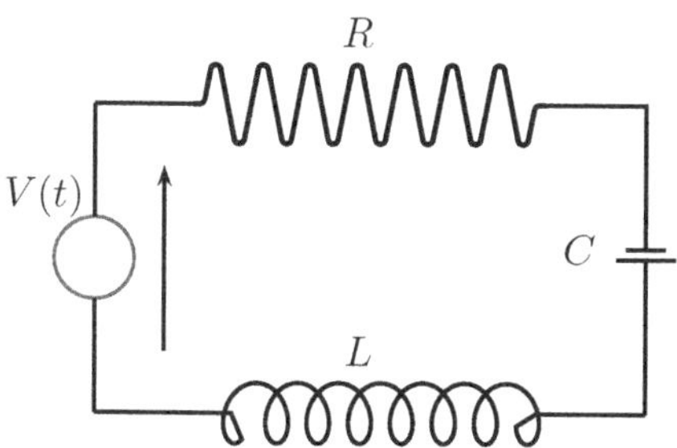

Figure 6.1: LRC-circuit.

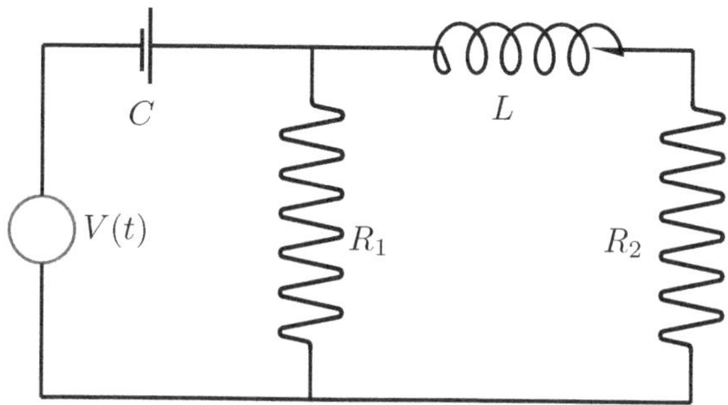

Figure 6.2: A two-loop circuit.

Example 6.1.2: (A Two-Loop Circuit) Figure 6.2 on page 343 shows a circuit that consists of two loops. We denote the current flowing through the left-hand loop by I_1 and the current in the right-hand loop by I_2 (both in the clockwise direction). The current through the resistor R_1 is $I_1 - I_2$. The voltage drops on each element are

$$\Delta V_{R_1} = R_1\,(I_1 - I_2), \quad \Delta V_{R_2} = R_2\,I_2$$

$$\Delta V_C = \frac{q_1}{C}, \quad \Delta V_L = L\,\frac{\mathrm{d}I_2}{\mathrm{d}t}.$$

The voltage analysis using Kirchhoff's law leads to the following equations:

$$\begin{aligned} V(t) &= \frac{q_1}{C} + R_1\,I_1 - R_1\,I_2 && \text{(left loop)},\\ 0 &= R_2\,I_2 + R_1\,I_2 + L\,\frac{\mathrm{d}I_2}{\mathrm{d}t} - R_1\,I_1 && \text{(right loop)},\\ \frac{\mathrm{d}q_1}{\mathrm{d}t} &= I_1. \end{aligned}$$

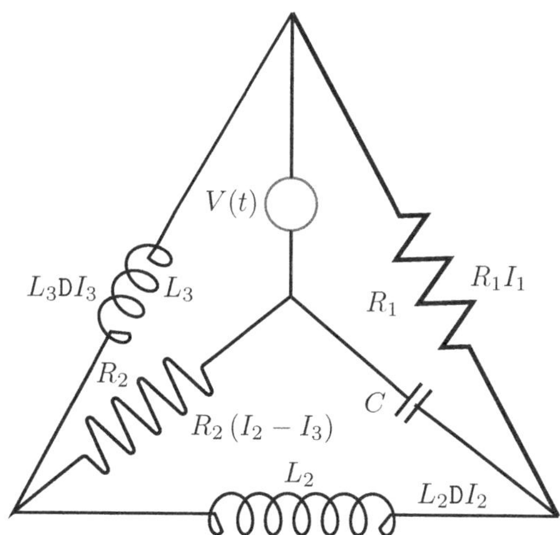

Figure 6.3: A three-loop circuit.

Example 6.1.3: (A Three-Loop Circuit) To handle the three-loop circuit sketched in Fig. 6.3, we apply Kirchhoff's laws to every loop:

$$\begin{aligned} -V(t) + R_1 I_1 + \frac{1}{C}\,q &= 0,\\ -\frac{1}{C}\,q + L_2\,(\mathtt{D}I_2) + R_2(I_2 - I_3) &= 0,\\ V(t) - R_2(I_2 - I_3) + L_3\,(\mathtt{D}I_3) &= 0, \end{aligned}$$

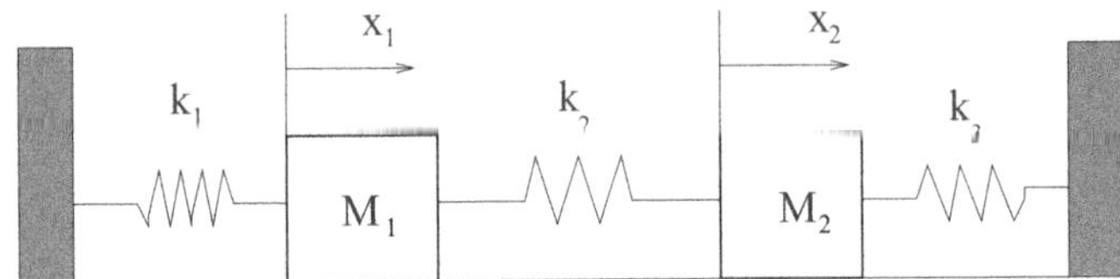

Figure 6.4: Spring-mass system.

where $\mathtt{D} = \mathrm{d}/\mathrm{d}t$ is the differential operator. From the first equation, it follows that

$$R_1 I_1 = V(t) - \frac{q}{C} \quad \Longrightarrow \quad I_1 = -\frac{q}{CR_1} + \frac{V(t)}{R_1}.$$

Hence,

$$\begin{aligned}
\mathtt{D}q \overset{\text{def}}{=} \dot{q} = \frac{\mathrm{d}q}{\mathrm{d}t} &= I_1 - I_2 = -\frac{q}{CR_1} + \frac{V(t)}{R_1} - I_2, \\
\mathtt{D}I_2 \overset{\text{def}}{=} \dot{I}_2 = \frac{\mathrm{d}I_2}{\mathrm{d}t} &= \frac{1}{CL_2}\, q - \frac{R_2}{L_2}\, I_2 + \frac{R_2}{L_2}\, I_3, \\
\mathtt{D}I_3 \overset{\text{def}}{=} \dot{I}_3 = \frac{\mathrm{d}I_3}{\mathrm{d}t} &= \frac{R_3}{L_3}\, I_2 - \frac{R_2}{L_3}\, I_3 - \frac{V(t)}{L_3}.
\end{aligned}$$

6.1.2 Spring-Mass Systems

When a system with masses connected by springs is in motion, the springs are subject to both elongation and compression. It is clear from experience that there is some force, called the **restoring force**, which tends to return the attached mass to its equilibrium position. By Hooke's law, the restoring force $\mathbf{F}$ has a direction opposite to the elongation (or compression) and is proportional to the distance of the mass from the equilibrium position. This can be expressed by $\mathbf{F} = -k\mathbf{x}$, where k is a constant of proportionality, called the **spring constant** or **force constant**, and $\mathbf{x}$ is the displacement. The former has units $\mathtt{kg/sec}^2$ in $\mathtt{SI}$ (abbreviation for the International System of Units).

In Figure 6.4, the two bodies of masses m_1 and m_2 are connected to three springs of negligible mass having spring constants k_1, k_2, and k_3, respectively. In turn, two of these three springs are attached to rigid supports. Let $x_1(t)$ and $x_2(t)$ denote the horizontal displacements of the bodies from their equilibrium positions. We assume that these two bodies move on a frictionless surface.

According to Newton's second law of motion, the time rate of change in the momentum[62] of a body is equal in magnitude and direction to the net force acting on the body and it has the same direction as the force.

Let us consider the left body of mass m_1. The elongation of the first spring is x_1, which therefore exerts a force of $-kx_1$ on mass m_1 by Hooke's law. The elongation of the second spring is $x_2 - x_1$ because it is subject to both an elongation and compression, so it exerts a force of $k_2(x_2 - x_1)$ on mass m_1. By Newton's second law, we have

$$m_1 \frac{\mathrm{d}v_1}{\mathrm{d}t} = -k_1\, x_1 + k_2\,(x_2 - x_1),$$

where v_1 is the velocity of the mass m_1.

Similarly, the net force exerted on the mass m_2 is due to the elongation of the second spring, namely, $x_2 - x_1$, and the compression of the third spring by x_2. Therefore, from Newton's second law, it follows that

$$m_2 \frac{\mathrm{d}v_2}{\mathrm{d}t} = -k_3\, x_2 - k_2\,(x_2 - x_1),$$

where v_2 is the velocity of the mass m_2.

[62]Momentum of a body of mass m is the product mv, where v is its velocity.

Since velocity is the derivative of displacement, we conclude that the motion of the coupled system is modeled by the following system of first order differential equations:

$$\begin{aligned} \frac{\mathrm{d}x_1}{\mathrm{d}t} &= v_1, \\ m_1 \frac{\mathrm{d}v_1}{\mathrm{d}t} &= -k_1\, x_1 + k_2\,(x_2 - x_1), \\ \frac{\mathrm{d}x_2}{\mathrm{d}t} &= v_2, \\ m_2 \frac{\mathrm{d}v_2}{\mathrm{d}t} &= -k_3\, x_2 - k_2\,(x_2 - x_1). \end{aligned} \tag{6.1.1}$$

6.1.3 The Euler–Lagrange Equation

Many interesting models originate from classical mechanical problems. The most general way to derive the corresponding systems of differential equations describing these models is the Euler–Lagrange equations. The method originated in works by Leonhard Euler in 1733; later in 1750, his student Joseph-Louis Lagrange substantially improved this method. There are two very important reasons for working with Lagrange's equations rather than Newton's. The first is that Lagrange's equations hold in any coordinate system, while Newton's is restricted to an inertial frame. The second is the ease with which we can deal with constraints in the Lagrangian system.

We demonstrate the Euler–Lagrange method in some elaborative examples from mechanics. We restrict ourselves to conservative systems. Let $\mathcal{L} = \mathrm{K} - \Pi$ be the Lagrange function (or Lagrangian) that is equal to the difference between the kinetic energy K and the potential energy Π.

For the spring-mass system from §6.1.2, the kinetic energy is given by

$$\mathrm{K} = \frac{1}{2}\, m_1 \dot{x}_1^2 + \frac{1}{2}\, m_2 \dot{x}_2^2,$$

where $\dot{x} = \mathrm{d}x/\mathrm{d}t$ is the derivative of x with respect to time. The potential energy is proportional to the square of the amount the spring is stretched or compressed, so

$$\Pi = \frac{1}{2}\, k_1 x_1^2 + \frac{1}{2}\, k_2 (x_1 - x_2)^2 + \frac{1}{2}\, k_3 x_2^2.$$

The Euler–Lagrange equations of the first kind for a system with two degrees of freedom are

$$\frac{\mathrm{d}}{\mathrm{d}t}\frac{\partial \mathcal{L}}{\partial \dot{x}_i} - \frac{\partial \mathcal{L}}{\partial x_i} = 0, \qquad i = 1, 2, \tag{6.1.2}$$

where x_1, x_2 are (generalized) coordinates. When the kinetic energy does not depend on displacements, and the potential energy does not depend on velocities, the Euler–Lagrange equations (6.1.2) become

$$\frac{\mathrm{d}}{\mathrm{d}t}\frac{\partial K}{\partial \dot{x}_i} + \frac{\partial \Pi}{\partial x_i} = 0, \qquad i = 1, 2. \tag{6.1.3}$$

Hence, $\dfrac{\partial \mathcal{L}}{\partial \dot{x}_i} = \dfrac{\partial \mathrm{K}}{\partial \dot{x}_i} = m_i\, \dot{x}_i$, $\quad \dfrac{\mathrm{d}}{\mathrm{d}t}\dfrac{\partial \mathcal{L}}{\partial \dot{x}_i} = m_i\, \ddot{x}_i$, $\quad i = 1, 2$; and we have $\dfrac{\partial \mathcal{L}}{\partial x_i} = -\dfrac{\partial \Pi}{\partial x_i}$. Since

$$\frac{\partial \Pi}{\partial x_1} = k_1\, x_1 + k_2\,(x_1 - x_2), \qquad \frac{\partial \Pi}{\partial x_2} = -k_2\,(x_1 - x_2) + k_3\, x_2,$$

the Euler–Lagrange equations are read as

$$m_1\, \ddot{x}_1 = -k_1\, x_1 - k_2\,(x_1 - x_2) \quad \text{and} \quad m_2\, \ddot{x}_2 = -k_3\, x_2 + k_2\,(x_1 - x_2),$$

which coincide with Newton's second law equations (6.1.1) found previously.

6.1.4 Pendulum

A bob of mass m is attached to one end of a rigid, but inextensible weightless rod (or shaft) of length ℓ. The other end of the shaft is attached to a support that allows it to rotate without friction. If the bob's oscillations take place within a plane, this system is called an **ideal pendulum**. The position of the bob is measured by the angle θ between the shaft and the downward vertical direction, with the counterclockwise direction taken as positive.

To analyze the plane motion of an oscillating pendulum, it is convenient to reformulate Newton's second law into its rotational equivalent. In science and engineering the usual unit of angle measurement is the radian, abbreviated by `rad`. It was abolished in 1995 and can be dropped—it is dimensionless in `SI` (from French: Système international d'unités).

Velocity is a measure of both the speed and direction that an object is traveling. When we consider only rotational motion, the direction of the object is known and its velocity is defined by its magnitude—speed, and its direction is determined by right hand rule. Therefore, we can drop vector notation and operate only with scalar quantities. The **angular velocity** (or instantaneous angular speed) of a body rotated about a fixed axis is the ratio of the angle traversed to the amount of time it takes to traverse that angle when the time tends to zero:

$$\omega = \lim_{\Delta t\to 0} \frac{\Delta\theta}{\Delta t} = \dot{\theta} = \frac{\mathrm{d}\theta}{\mathrm{d}t},$$

where θ is an angle in the cylindrical coordinate system and in which the axis of rotation is taken to be in the z direction. The unit of angular velocity is a radian per second (`rad/sec`, which is `1/sec`) and it is a measure of the angular displacement per unit time. Since the velocity vector is always tangent to the circular path, it is called the **tangential velocity**. Its magnitude $v = r\,\mathrm{d}\theta/\mathrm{d}t = r\omega$ is the linear velocity, where $r = |\mathbf{r}|$ and $\omega = |\boldsymbol{\omega}|$. In words, the tangential speed of a point on a rotating rigid object equals the product of the perpendicular distance of that point from the axis of rotation with the angular speed. Note that the tangential velocity and angular velocity only refer to its magnitude; no direction is involved. Although every point on the rigid body has the same angular speed, not every point has the same linear speed because r is not the same for all points on the object.

The **angular acceleration** is the rate of change of angular velocity with time:

$$\boldsymbol{\alpha} = \lim_{\Delta t\to 0} \frac{\Delta\omega}{\Delta t} = \frac{\mathrm{d}\,\omega}{\mathrm{d}t} = \frac{\mathrm{d}^2\theta}{\mathrm{d}t^2} = \ddot{\theta}.$$

It has units of radians per second squared (`rad/sec`^{2}), or just `sec`$^{-2}$. Note that $\boldsymbol{\alpha}$ is said to be positive when the rate of counterclockwise rotation is increasing or when the rate of clockwise rotation is decreasing. When rotating about a fixed axis, every particle on a rigid object rotates through the same angle and has the same angular speed and the same angular acceleration. That is, the quantities θ, ω, and α characterize the rotational motion of the entire rigid body.

Angular position (θ), angular speed (ω), and angular acceleration (α) are analogous to linear position (x), linear speed (v), and linear acceleration (a) but they differ dimensionally by a factor of unit length. The directions of angular velocity ($\boldsymbol{\omega}$) and angular acceleration ($\boldsymbol{\alpha}$) are along the axis of rotation. If an object rotates in the xy plane, the direction of $\boldsymbol{\omega}$ is out of the plane when the rotation is counterclockwise and into the plane when the rotation is clockwise.

A point rotating in a circular path undergoes a centripetal, or radial, acceleration $\mathbf{a}_r$ of magnitude v^2/r directed toward the center of rotation. Since $v = r\omega$ for a point on a rotating body, we can express the radial acceleration of that point as $a_r = r\omega^2$.

Suppose a body of mass m is constrained to move in a circle of radius r. Its position is naturally described by an angular displacement θ from some reference position. Such a body has **mass moment of inertia**[63]

$$J = m\,r^2 \quad [\mathtt{kg}\cdot\mathtt{m}^2].$$

It is a measure of an object's resistance to changes to its rotation rate—the angular counterpart to the mass. If a rotating rigid object consists of a collection of particles, each having mass m_i, then $J = \sum_i m_i r_i^2$. The rotational kinetic energy of a rigid body is

$$\mathrm{K} = \frac{1}{2}\sum_i m_i r_i^2 \omega^2 = \frac{1}{2}\,J\omega^2.$$

[63]We assume that the body has a point mass. The moment of inertia for a continuously distributed mass is the integral of density times the square of the radius.

Recall that the angular momentum of a body (particle) of mass m is

$$\mathbf{A} = \mathbf{r} \times m\mathbf{v} = J\boldsymbol{\omega},$$

where $\mathbf{v}$ is the velocity vector of the particle, $\mathbf{r}$ is the position vector of the particle relative to the origin, $m\mathbf{v}$ is the linear momentum of the particle, and $\times$ denotes the cross product. For a system of particles, the angular momentum is the vector sum of the individual angular momentums. The time derivative of the angular momentum is called torque:

$$\mathbf{M} = \frac{\mathrm{d}\mathbf{A}}{\mathrm{d}t} = \frac{\mathrm{d}\mathbf{r}}{\mathrm{d}t} \times m\mathbf{v} + \mathbf{r} \times m\,\frac{\mathrm{d}\mathbf{v}}{\mathrm{d}t} = \mathbf{r} \times m\,\frac{\mathrm{d}\mathbf{v}}{\mathrm{d}t}$$

because

$$\dot{\mathbf{r}} \times m\mathbf{v} = \mathbf{v} \times m\mathbf{v} \qquad \Longrightarrow \qquad |\mathbf{v} \times m\mathbf{v}| = m\,|v| \cdot |v| \sin 0^\circ = 0.$$

Torque or **moment** is a measure of how much a force acting on an object causes that object to rotate. Suppose that a force $\mathbf{F}$ acts on a wrench pivoted on the axis through the origin. Then torque is defined as

$$\mathbf{M} = \mathbf{r} \times \mathbf{F} \qquad \Longrightarrow \qquad M \stackrel{\text{def}}{=} |\mathbf{M}| = rF\sin\phi = F_{tan}\, r,$$

where F_{tan} is the component of the force tangent to the circle in the direction of an increasing θ in counterclockwise direction, r is the distance between the pivot point and the point of the application of $\mathbf{F}$, and ϕ is the angle between the line of action of the force $\mathbf{F}$ and the line of the radius r. The torque is defined only when a reference axis is specified.

Imagine pushing a door open. The force of your push ($\mathbf{F}$) causes the door to rotate about its hinges (the pivot point). How hard you need to push the door depends on the distance you are from the hinges (r). The closer you are to the hinges (i.e., the smaller r is), the harder it is to push.

Thus, Newton's second law when applied to rotational movement becomes

$$J\,\frac{\mathrm{d}^2\theta}{\mathrm{d}t^2} = M, \tag{6.1.4}$$

that is, the time rate of change of angular momentum about any point is equal to the resultant torque exerted by all external forces acting on the body about that point. The torque acting on the particle is proportional to its angular acceleration, and the proportionality constant is the moment of inertia: $M = J\alpha$. ■

For the pendulum, there are two forces (we neglect friction resistance force at the supported end of the shaft) acting on it: gravity, directed downward, and air resistance, directed opposite to the direction of motion. The latter force, denoted $F_{\text{resistance}}$, has only a tangential component. This damping force is assumed to be approximately proportional to the angular velocity, that is,

$$F_{\text{resistance}} = -\kappa\,\dot{\theta}$$

for some positive constant κ. We also assume here that θ and $\dot{\theta} = \mathrm{d}\theta/\mathrm{d}t$ are both positive when moved in counterclockwise direction. From this assumption, the moment of the damping force becomes

$$M_{\text{resistance}} = -\kappa\ell\,\dot{\theta}.$$

The tangential component of the gravitational force has magnitude $mg\sin\theta$ and acts in the clockwise direction. Therefore, its moment is

$$M_{\text{gravity}} = -\ell mg\,\sin\theta.$$

From Newton's second law (6.1.4), it follows that

$$m\ell^2\,\frac{\mathrm{d}^2\theta}{\mathrm{d}t^2} = -\kappa\ell\,\frac{\mathrm{d}\theta}{\mathrm{d}t} - \ell mg\,\sin\theta \qquad \text{or} \qquad \ddot{\theta} + \gamma\,\dot{\theta} + \omega^2\,\sin\theta = 0, \tag{6.1.5}$$

where

$$\gamma = \frac{\kappa}{m\ell} \qquad \text{and} \qquad \omega^2 = \frac{g}{\ell}.$$

If the damping force (due to air resistance and friction at the pivot) is small, it can be neglected; so we can set $\gamma = 0$ and the pendulum equation becomes

$$\ddot{\theta} + \omega^2\,\sin\theta = 0. \tag{6.1.6}$$

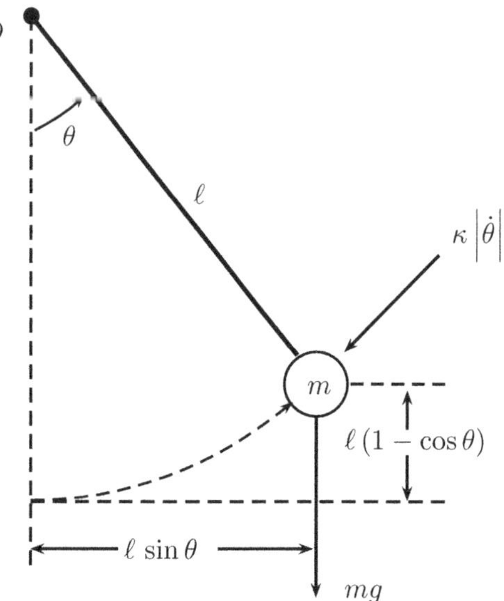

Figure 6.5: An oscillating pendulum.

Equation (6.1.6) has no solutions in terms of elementary functions. If θ remains small, say $|\theta| < 0.1$ radians, we may replace[64] $\sin\theta$ by θ in the pendulum equation and obtain the following linear differential equation:

$$\ddot{\theta} + \omega^2\,\theta = 0.$$

■

We can also derive the pendulum equation (6.1.5) using the Euler–Lagrange equation of motion (with n degrees of freedom)

$$\frac{\mathrm{d}}{\mathrm{d}t}\left(\frac{\partial \mathcal{L}}{\partial \dot{q}_k}\right) - \frac{\partial \mathcal{L}}{\partial q_k} = 0, \quad k = 1, 2, \ldots n, \tag{6.1.7}$$

where $\mathcal{L}$ is the Lagrange function, that is, $\mathcal{L} = \mathrm{K} - \Pi$, K is the kinetic energy, and Π is its potential energy. Here q_k represents the generalized coordinates of the system. For the pendulum $n = 1$ and $q_k = \theta$. Since the linear velocity of the bob is $\ell\dot{\theta}$, the kinetic energy is

$$\mathrm{K} = \frac{1}{2}\,m\left(\ell\,\dot{\theta}\right)^2.$$

For the potential energy expression, we assume that datum to be the lowest position of the bob, that is, when $\theta = 0$. When the pendulum is at some angle θ, the position of the mass above the datum is $\ell(1-\cos\theta)$. Hence the potential energy is

$$\Pi = mg\,\ell(1 - \cos\theta).$$

Therefore, the Lagrange function becomes

$$\mathcal{L} = \frac{1}{2}\,m\left(\ell\,\frac{\mathrm{d}\theta}{\mathrm{d}t}\right)^2 - mg\,\ell(1-\cos\theta) = \frac{1}{2}\,m\left(\ell\,\dot{\theta}\right)^2 - mg\,\ell(1-\cos\theta),$$

and we have

$$\begin{aligned} \frac{\partial \mathcal{L}}{\partial \dot{\theta}} &= m\ell^2\dot{\theta} \quad \text{and} \quad \frac{\mathrm{d}}{\mathrm{d}t}\left(\frac{\partial \mathcal{L}}{\partial \dot{\theta}}\right) = m\ell^2\ddot{\theta}, \\ \frac{\partial \mathcal{L}}{\partial \theta} &= -\frac{\partial \Pi}{\partial \theta} = -mg\ell\,\sin\theta, \end{aligned}$$

which leads to Eq. (6.1.5). □

---

[64] Since the Taylor series for $\sin\theta$ about $\theta = 0$ is the following sign alternating series

$$\sin\theta = \theta - \frac{\theta^3}{3!} + \frac{\theta^5}{5!} - \cdots,$$

the first term θ is a good approximation of the function $\sin\theta$ for small values of θ.

So far we have considered only an ideal pendulum that can oscillate within a plane. Now we relax this condition and consider a pendulum consisting of a compact mass m on the end of a light inextensible string of length ℓ. Suppose that the mass is free to move in any direction (as long as the string remains taut). Let the fixed end of the string be located at the origin of our coordinate system. We can define Cartesian coordinates, (x, y, z), such that the z-axis points vertically upward. We can also define spherical coordinates, (r, θ, ϕ), whose axis points along the $-z$-axis. The latter coordinates are the most convenient since r is constrained to always take the value $r = \ell$. However, the two angular coordinates, θ and ϕ, are free to vary independently. Hence, this is a two degrees of freedom system.

In spherical coordinates

$$x = \ell \sin\theta \cos\phi, \quad y = \ell \sin\theta \sin\phi, \quad z = -\ell \cos\theta,$$

the potential energy of the system becomes $\Pi = mg(\ell + z) = mg\ell\,(1 - \cos\theta)$. Since the velocity of the bob is

$$v^2 = \dot{x}^2 + \dot{y}^2 + \dot{z}^2 = \ell^2 \left(\dot{\theta}^2 + \sin^2\theta\dot{\phi}^2\right),$$

the Lagrangian of the system can be written as

$$\mathcal{L} = \mathrm{K} - \Pi = {}^1\!/_2\, m\ell^2 \left(\dot{\theta}^2 + \sin^2\theta\dot{\phi}^2\right) + mg\ell\,(\cos\theta - 1)\,.$$

The Euler–Lagrange equations give

$$\frac{\mathrm{d}}{\mathrm{d}t}\left(m\ell^2\dot{\theta}\right) - m\ell^2 \sin\theta\cos\theta\,\dot{\phi}^2 + mg\ell\,\sin\theta = 0,$$
$$\frac{\mathrm{d}}{\mathrm{d}t}\left(m\ell^2 \sin^2\theta\dot{\phi}\right) = 0.$$

Example 6.1.4: (Coupled pendula) Consider two pendula of masses m_1 and m_2 coupled by a Hookian spring with spring constant k (see figure on the front page 341). Suppose that the spring is attached to each rod at a distance a from their pivots, and that the pendula are far apart so that the spring can be assumed to be horizontal during their oscillations. Let θ_1/θ_2 and ℓ_1/ℓ_2 be the angle of inclination of the shaft with respect to the downward vertical line and the length of the shaft for each pendulum. Its kinetic energy is the sum of kinetic energies of two individual pendula:

$$\mathrm{K} = \frac{m_1}{2}\left(\ell_1\dot{\theta}_1\right)^2 + \frac{m_2}{2}\left(\ell_2\dot{\theta}_2\right)^2.$$

The potential energy is accumulated by the spring, which accounts for $\frac{k}{2}\,(a\,\sin\theta_1 - a\,\sin\theta_2)^2$, and by lifting both masses

$$\Pi = m_1 g\ell_1\,(1 - \cos\theta_1) + m_2 g\ell_2\,(1 - \cos\theta_2) + \frac{a^2 k}{2}\,(\sin\theta_1 - \sin\theta_2)^2\,.$$

Substituting these expressions into the Euler–Lagrange equations (6.1.3), we obtain the system of motion:

$$\begin{aligned} m_1\ell_1^2\ddot{\theta}_1 + m_1 g\ell_1 \sin\theta_1 + a^2 k\,(\sin\theta_1 - \sin\theta_2)\cos\theta_1 &= 0, \\ m_2\ell_2^2\ddot{\theta}_2 + m_2 g\ell_2 \sin\theta_2 - a^2 k\,(\sin\theta_1 - \sin\theta_2))\cos\theta_2 &= 0. \end{aligned} \tag{6.1.8}$$

6.1.5 Laminated Material

When a space shuttle enters the atmosphere to land on Earth, air friction causes its body to be heated to a high temperature. To prevent the shuttle from melting, its shell is covered with ceramic plates because there are no metals that can resist such a high temperature.

Let us consider the problem of describing the temperature in a laminated material. Suppose that the rate of heat flow between two objects in thermal contact is proportional to the difference in their temperatures and that the rate at which heat is stored in an object is proportional to the rate of change of its temperature. The net heat flow rate into an object must balance the heat storage rate. This balance for a three-layer material is described as

$$c_1\frac{\mathrm{d}T_1}{\mathrm{d}t} = p_{12}(T_2 - T_1) + p_{01}(T_0 - T_1),$$

$$c_2 \frac{\mathrm{d}T_2}{\mathrm{d}t} = p_{12}(T_1 - T_2) + p_{23}(T_3 - T_2),$$
$$c_3 \frac{\mathrm{d}T_3}{\mathrm{d}t} = p_{23}(T_2 - T_3) + p_{34}(T_4 - T_3),$$

where the c's and p's are constants of proportionality, and T_0 and T_4 are the temperatures at the two sides of the laminated material. Here T_k is the temperature in the k-th layer ($k = 1, 2, 3$).

6.1.6 Flow Problems

Examples of leaking tanks containing liquids constitute only a portion of flow problems; however, they model a whole host of familiar real-world phenomena. For example, "liquid" can also be heat flowing out of a cooling cup of coffee, a cold soda warming up to room temperature, an electric charge draining from a capacitor, or even a radioactive material leaking (decaying) to a stable material. The thermal counterparts of water/liquid volume, water height, tank base area, and drain area are heat, temperature, specific heat, and thermal conductivity. The corresponding electrical units are charge, voltage, capacitance, and the reciprocal of resistance.

We start with a simple example of two tanks coupled together. We assume that all tanks are of the same cylindrical shape and have a volume with base area 1; this allows us to track the tank's water volume using water height $y(t)$. If k denotes the cross-sectional area of its drain, we choose the units so that the water level y satisfies $y' = -ky$ when each tank is in isolation. To finish the physical setup, we put two tanks one above the other so that water drains from the top tank into the bottom one, and the pump returns to the top tank water that has gravity-drained from the bottom tank. This leads to the following system of differential equations:

$$\dot{y}_1 = -k_1 y_1 + k_2 y_2, \qquad \dot{y}_2 = k_1 y_1 - k_2 y_2. \tag{6.1.9}$$

By adding a third tank, the corresponding vector differential equation becomes

$$\begin{aligned} \dot{y}_1 &= -k_1 y_1 + 0 y_2 + k_3 y_3, \\ \dot{y}_2 &= k_1 y_1 - k_2 y_2 + 0 y_3, \\ \dot{y}_3 &= 0 y_1 + k_2 y_2 - k_3 y_3. \end{aligned} \tag{6.1.10}$$

A similar system of differential equations can be used to model a sequence of n connected tanks (n is a positive integer).

Example 6.1.5: (Cascading Tanks) Suppose there are three cascading tanks in sequence. Tank A contains 200 `liters` (`l`) of a salt solution with a concentration of 100 `grams` per `liter`. The next two tanks, B and C, contain pure water with volumes of 100 `liters` and 50 `liters`, respectively. At instant $t = 0$, fresh water is poured into tank A at a rate 10 `l/min`. The well-stirred mixture flows out of tank A at the same rate it flows into tank B. The well-stirred mixture is pumped from tank B into tank C at the same rate of 10 `l/min`. Tank C has a hole that allows brine to run out at a rate 10 `l/min`. Find the amount of salt in each tank at any time.

Solution. We denote the amount of salt in the containers A, B, and C by $Q^A(t)$, $Q^B(t)$, and $Q^C(t)$, respectively. Initially,

$$Q^A(0) = 200\,\texttt{liters} \times 0.1\,\texttt{kg/liter} = 20\,\texttt{kg}, \quad Q^B(0) = 0, \quad \text{and} \quad Q^C(0) = 0.$$

The `rate in` and `rate out` for each of the tanks are

$$R_{\text{in}}^A = 0 \frac{\texttt{kg}}{\texttt{min}}, \quad R_{\text{out}}^A = \left(\frac{Q^A(t)}{200} \frac{\texttt{kg}}{\texttt{liters}} \right) \times \left(10 \frac{\texttt{liters}}{\texttt{min}} \right) = \frac{Q^A(t)}{20} \frac{\texttt{kg}}{\texttt{min}},$$

$$R_{\text{in}}^B = R_{\text{out}}^A, \quad R_{\text{out}}^B = \left(\frac{Q^B(t)}{100} \frac{\texttt{kg}}{\texttt{liters}} \right) \times \left(10 \frac{\texttt{liters}}{\texttt{min}} \right) = \frac{Q^B(t)}{10} \frac{\texttt{kg}}{\texttt{min}},$$

$$R_{\text{in}}^C = R_{\text{out}}^B, \quad R_{\text{out}}^C = \left(\frac{Q^C(t)}{50} \frac{\texttt{kg}}{\texttt{liters}} \right) \times \left(10 \frac{\texttt{liters}}{\texttt{min}} \right) = \frac{Q^B(t)}{5} \frac{\texttt{kg}}{\texttt{min}}.$$

The balanced equation for each tank is

$$\frac{\mathrm{d}Q^A(t)}{\mathrm{d}t} = -\frac{Q^A(t)}{20} = -0.05\, Q^A(t), \tag{6.1.11}$$

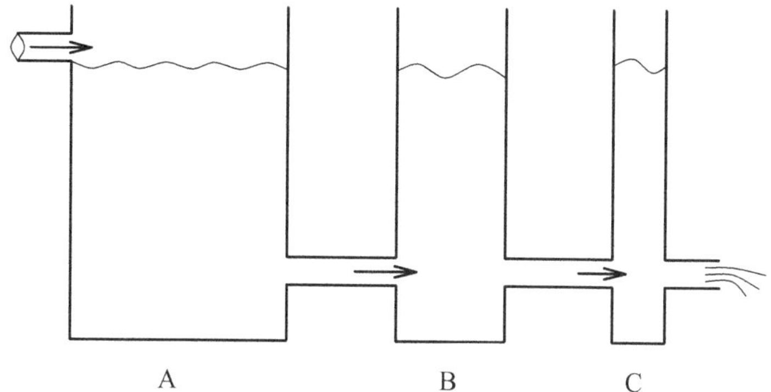

Figure 6.6: Cascading tanks.

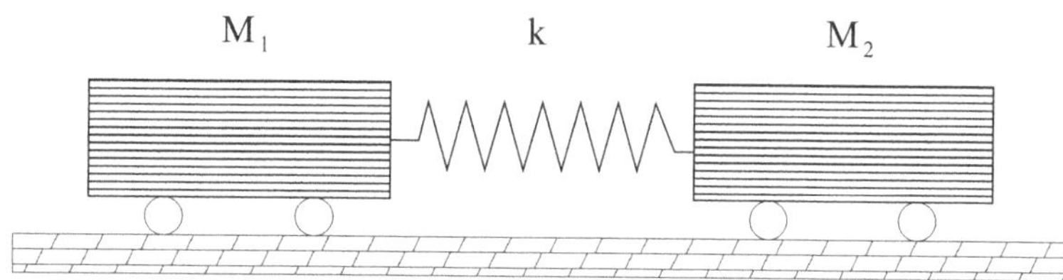

Figure 6.7: Problem 1.

$$\frac{\mathrm{d}Q^B(t)}{\mathrm{d}t} = 0.05\,Q^A(t) - 0.1\,Q^B(t), \tag{6.1.12}$$

$$\frac{\mathrm{d}Q^C(t)}{\mathrm{d}t} = 0.1\,Q^B(t) - 0.2\,Q^C(t), \tag{6.1.13}$$

Since Eq. (6.1.11) is a linear differential equation, the solution that satisfies the initial condition $Q^A(0) = 20$ is

$$Q^A(t) = 20\,e^{-0.05t}.$$

Equations (6.1.12) and (6.1.13) are nonhomogeneous linear differential equations. We recall from §2.5 that the initial value problem

$$\frac{\mathrm{d}Q}{\mathrm{d}t} + a\,Q(t) = f(t), \quad Q(0) = 0$$

has the explicit solution (see Eq. (2.5.9) on page 88)

$$Q(t) = \int_0^t f(\tau)\,e^{-a(t-\tau)}\,\mathrm{d}\tau.$$

With this in hand, the solutions of equations (6.1.12) and (6.1.13) that satisfy the homogeneous initial conditions are

$$Q^B(t) = 0.05\int_0^t \left[20\,e^{-0.05\tau}\right] e^{-0.1(t-\tau)}\,\mathrm{d}\tau = 20\left[e^{-0.05t} - e^{-0.1t}\right]$$

and

$$\begin{aligned} Q^C(t) &= 0.1\int_0^t 20\left[e^{-0.05\tau} - e^{-0.1\tau}\right] e^{-0.2(t-\tau)}\,\mathrm{d}\tau \\ &= \frac{40}{3}\left[e^{-0.05t} - e^{-0.2t}\right] - 20\left[e^{-0.1t} - e^{-0.2t}\right]. \end{aligned}$$

Problems

1. Two cars of masses m_1 and m_2 are connected by a spring of force constant k (Fig. 6.7). They are free to roll along the abscissa. Derive the equations of motion for each car.
2. Three cars of masses m_1, m_2, and m_3 are connected by two springs with spring constants k_1 and k_2, respectively (see Fig. 6.8). Set up the equations of motion for each car.

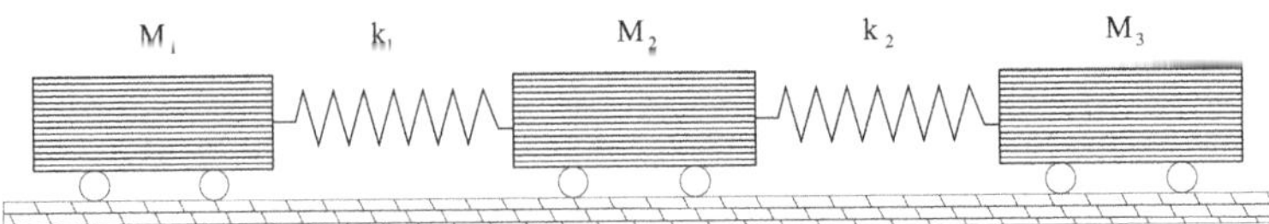

Figure 6.8: Problem 2.

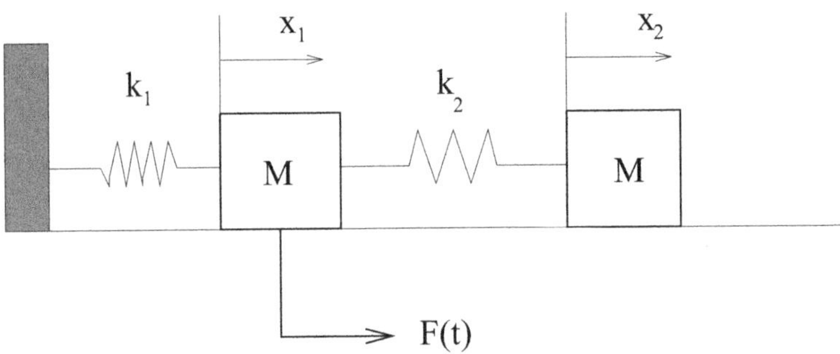

Figure 6.9: Problem 3, dynamic damper.

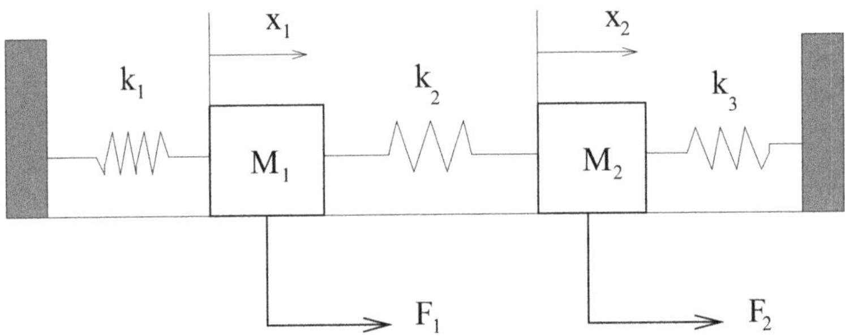

Figure 6.10: Problem 4.

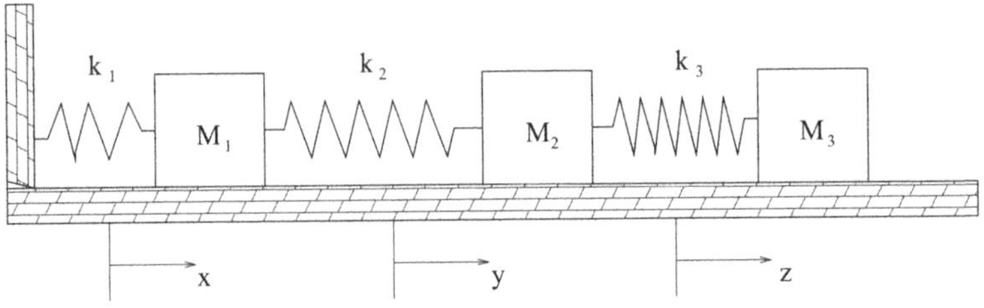

Figure 6.11: Problem 5.

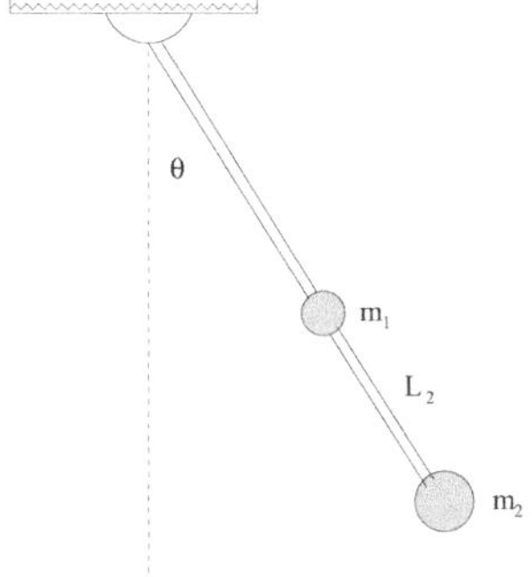

Figure 6.12: Problem 6.

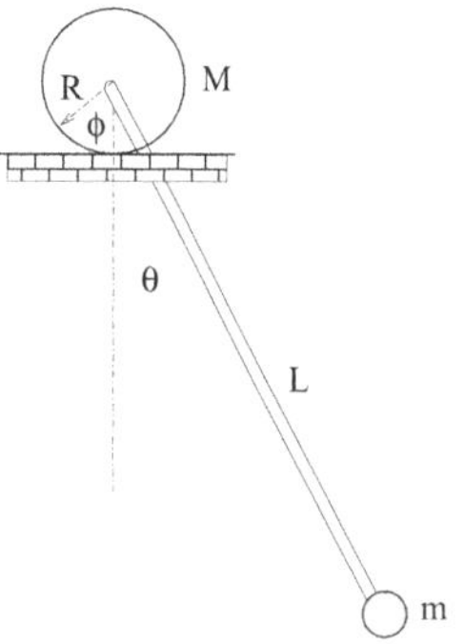

Figure 6.13: Problem 7.

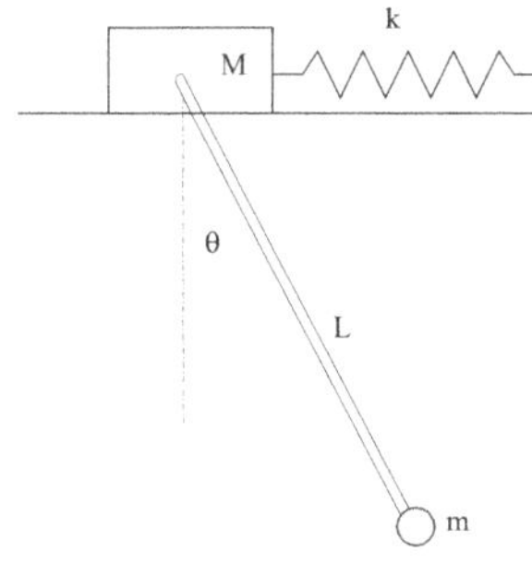

Figure 6.14: Problem 8.

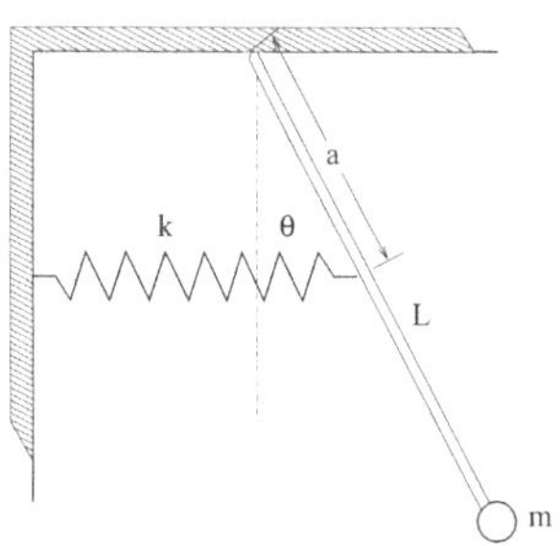

Figure 6.15: Problem 9.

3. Set up the system of differential equations that simulate a dynamic damper with $F(t) = \sin(\omega t + \alpha)$; see Fig. 6.9.
4. Derive the system of differential equations that simulate the mass-and-spring system shown in Fig. 6.10.
5. Consider the mechanical system of three springs with force constants k_1, k_2, and k_3 and three masses m_1, m_2, and m_3 (see Fig. 6.11). They are free to slide along the abscissa. The left spring is attached to the wall. Set up the equation of motion.
6. A pendulum consists of a rigid massless rod with two attached bobs of masses m_1 and m_2 (see Fig. 6.12). The distance between these bobs is L_2 and the bob of mass m_2 is attached to its end. The total length of the rod is L. Find the equation of motion when friction is neglected.
7. A pendulum consists of a bob of mass m attached to a pivot by a rigid and massless shaft of length L (see Fig. 6.13). The pivot, which is the axle of a homogeneous solid wheel of radius R and mass M, is free to roll horizontally. Set up the equation of motion for the system.
8. A pendulum consists of a bob of mass m attached to a pivot by a rigid and massless shaft of length L (see Fig. 6.14). The pivot is attached to a block of mass M that is free to slide along abscissa. The block is attached to a wall to the

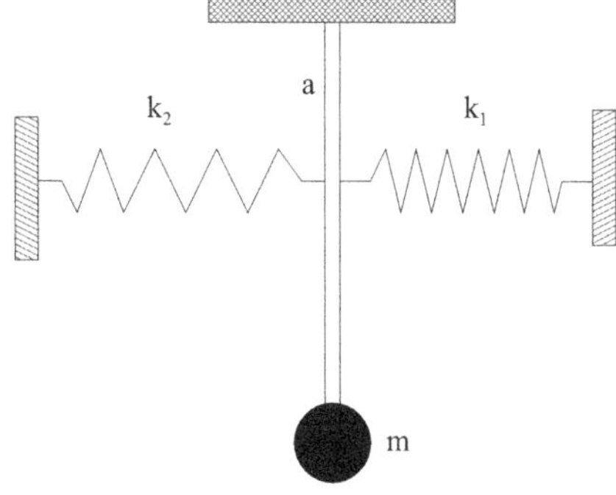

Figure 6.16: Problem 10.

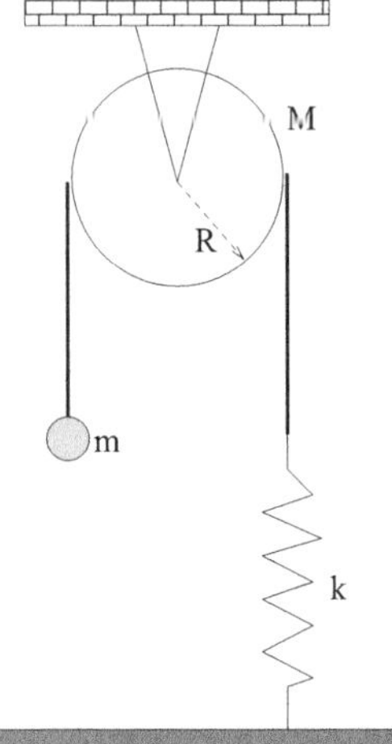

Figure 6.17: Problem 11.

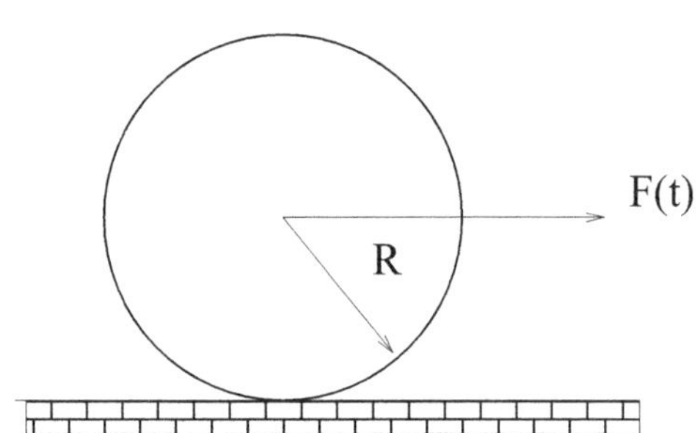

Figure 6.18: Problem 12.

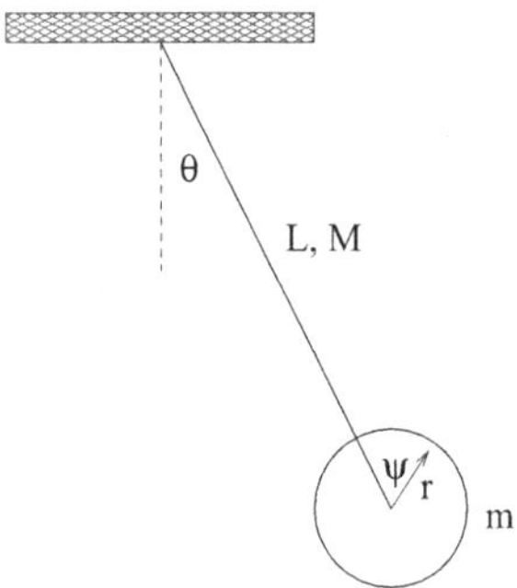

Figure 6.19: Problem 13.

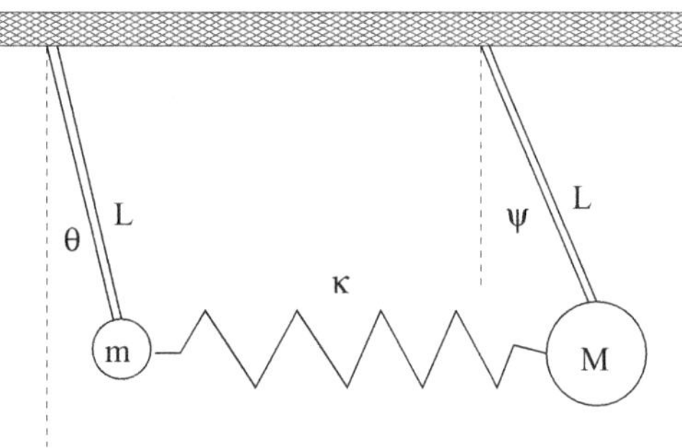

Figure 6.20: Problem 14.

left by a spring with a spring force constant k. Derive the equation of motion.

9. A pendulum consists of a bob of mass m attached to the lower end of a massless shaft of length L (see Fig. 6.15). If the spring has force constant k and is attached to the pendulum shaft at a distance $a < L$, find the equation of motion.

10. A pendulum consists of a rigid rod of length L, pivoted at its upper end and carrying a mass m at its lower end (see Fig. 6.16). Two springs with force constants k_1 and k_2, respectively, are attached to the rod at a distance a from the pivot and their other ends are rigidly attached to a wall. Find the equation of motion.

11. Find a system of equations for the mass m attached to a spring of force constant k via a pulley of radius R and mass M (see Fig. 6.17).

12. Derive an equation of motion for a uniformly thin disk of radius R that rolls without slipping on the abscissa under the action of a horizontal force F applied at its center (see Fig. 6.18). Let k be the coefficient of friction.

13. A uniform rod of length L and mass M is pivoted at one end and can swing in a vertical plane (see Fig. 6.19). A homogeneous disk of mass m and radius r is attached by a pivot at its center to the free end of the rod. Ignoring friction, set up the equation of motion for this system.

14. Two pendula of masses m and M, respectively, each of length L, are suspended from the same horizontal line (see Fig. 6.20). Their angular displacements are $\theta(t)$ and $\psi(t)$, respectively. The bobs are coupled by a spring of force constant k. Derive an equation of motion.

15. **(a)** Show that the functions

$$x(\theta) = 3\cos\theta + \cos 3\theta, \quad y(\theta) = 3\sin\theta - \sin 3\theta$$

satisfy the initial value problem

$$x'' - 2y' + 3x = 0, \quad y'' + 2x' + 3y = 0, \qquad x(0) = 4,\ y(0) = x'(0) = y'(0) = 0.$$

(b) Verify that these differential equations describe the trajectory $(x(t), y(t))$ of a particle moving in the plane along the *hypocycloid* traced by a point $P(x, y)$, fixed on the circumference of a circle of radius $r = 1$, and that rolls around

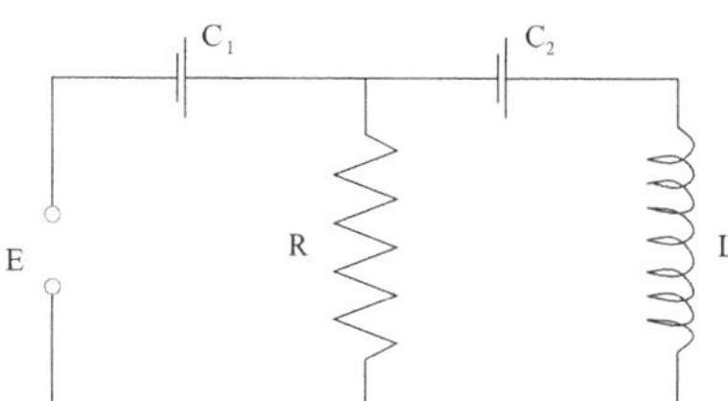

Figure 6.21: Problem 18.

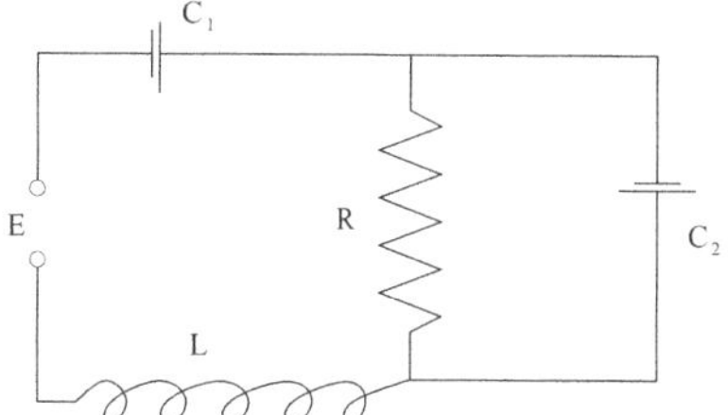

Figure 6.22: Problem 19.

inside a circle of radius $R = 4$. The parameter θ represents the angle measured in the counterclockwise direction from the abscissa to the line through the center of the small circle and the origin.

16. A ball of mass m and radius r rolls in a hemispherical bowl of radius R. Determine the Lagrangian equations of motion for the ball in spherical coordinates (θ and ϕ); denote by g the acceleration due to gravity. Find the period of oscillations. To avoid elliptic integrals, keep θ small and replace $\sin\theta$ by θ, and $\cos\theta$ by $1 - \theta^2/2$.
17. Solve the previous problem when the ball slides from the edge to the bottom. Find the time of descent.
18. An electrical circuit consists of two loops, as shown in Fig. 6.21. Set up the dynamic equations for the circuit.
19. An electrical circuit consists of two loops, as shown in Fig. 6.22. Set up the dynamic equations for the circuit.
20. Solve the problem from Example 6.1.5 when all three cascading tanks contain 200 liters of liquid.
21. A shaft carrying two disks is attached at each end to a wall. The distance between the disks and between each disk and the wall is ℓ. Each disk can turn about the shaft, but in so doing it exerts a torque on the shaft. The angular coordinates θ_1, θ_2 represent displacements from equilibrium, at which there is no torque. The total potential energy of the system is $\Pi = \dfrac{k}{2}\left[\theta_1^2 + (\theta_1 - \theta_2)^2 + \theta_2^2\right]$, where $k = C/\ell$ and C is the torsional stiffness of the shaft; the kinetic energy is K $= \frac{1}{2}\left[J_1\dot{\theta}_1^2 + J_2\dot{\theta}_2^2\right]$, where J_1 and J_2 are two moments of inertia. Set up the Lagrangian differential equations for the system.
22. The **Harrod–Domar model**[65] is used to explain the growth rate of developing countries' economies based on saving capital and using that capital as investment in productivity: $\dot{K} = s\,P(K, L)$, where K is capital, L is labor, $P(K, L)$ is output, which is assumed to be a homogeneous function $P(aK, aL) = a\,P(K, L)$, s is the fraction ($0 < s < 1$) of the output that is saved and the rest is consumed. Assuming that the labor force is growing according to the simple growth law $\dot{L} = rL$, derive a differential equation for the ratio $R(t) = K(t)/L(t)$.
23. Consider two interconnected tanks. The first one, which we call tank A, initially contains 50 `liters` of a solution with 50 `grams` of salt, and the second tank, which we call tank B, initially contains 100 `liters` of a solution with 100 `grams` of salt. Water containing 6 `gram/l` of salt flows into tank A at a rate of 10 `l/min`. The mixture flows from tank A to tank B at a rate of 6 `l/min`. Water containing 10 `gram/l` of salt also flows into tank B at a rate of 12 `l/min` from outside. The mixture drains from tank B at a rate of 7 `l/min`, of which some flows back into tank A at a rate of 3 `l/min`, while the remainder leaves the system.
    (a) Let $Q^A(t)$ and $Q^B(t)$, respectively, be the amount of salt in each tank at time t. Write down differential equations and initial conditions that model the flow process. Observe that the system of differential equations is nonhomogeneous.
    (b) Find the values of Q^A and Q^B for which the system is in equilibrium, that is, does not change with time. Let Q_E^A and Q_E^B be the equilibrium values. Can you predict which tank will approach its equilibrium state more rapidly?
    (c) Let $x_1(t) = Q^A(t) - Q_E^A$ and $x_2 = Q^B(t) - Q_E^B$. Determine an initial value problem for x_1 and x_2. Observe that the system of equations for x_1 and x_2 is homogeneous.
24. In 1963, American mathematician and meteorologist Edward Norton Lorenz (1917–2008) from Massachusetts Institute of Technology introduced a simplified mathematical model for atmospheric convection

$$\dot{x} = \sigma(y - x), \quad \dot{y} = \rho x - y - xz, \quad \dot{z} = xy - \beta z,$$

where σ, ρ, and β are constants. Use a computer solver to plot some solutions when $\sigma = 10$, $\rho = 28$, and $\beta = 8/3$.

[65]The model was developed independently by an English economist Roy F. Harrod (1900–1978) in 1939 and in 1946 by Evsey Domar, a Russian American economist Evsey David Domar/Domashevitsky (1914–1997), who immigrated to the US in 1936.

6.2 Matrices

Some quantities may be completely identified by a magnitude and a direction, as, for example, force, velocity, momentum, and acceleration. Such quantities are called **vectors**. This definition suggests the geometric representation of a vector as a directed line segment, or "arrow," where the length of the arrow is scaled according to the magnitude of the vector. Observe that the location of a vector in space is not specified; only its magnitude and direction are known. Fixing a Cartesian system of coordinates, we can move a vector so that it starts at the origin of a rectangular coordinate system. Then each vector can be characterized by a point corresponding to the end of the arrow. This endpoint is uniquely identified by its coordinates.

There are two ways to write coordinates of a vector: either as a column or as a row. Correspondingly, we get two kinds of vectors: column vectors, denoted by lower case letters in bold font (as $\mathbf{x}$), and row vectors, denoted by lower case letters with arrows above them (as $\vec{x}$). A transformation of a column vector into raw vector or vice versa is called **transposition**. In applications, it is a custom to identify a vector with its coordinates written in column form. For example, a vector $\mathbf{u}$ in fixed Cartesian 3-space can be written as a three-dimensional column vector:

$$\mathbf{u} = \begin{pmatrix} u_1 \\ u_2 \\ u_3 \end{pmatrix} \quad \text{or} \quad \mathbf{u} = \begin{bmatrix} u_1 \\ u_2 \\ u_3 \end{bmatrix} \quad \text{or} \quad \mathbf{u}^T = \langle u_1, u_2, u_3 \rangle,$$

where T denotes transposition (see Definition 6.3, page 358) and $\vec{u} = \langle u_1, u_2, u_3 \rangle$ is a row vector. In this text, we mostly use column vectors.

Definition 6.1: The **scalar** or **inner product** of two column vectors $\mathbf{x}^T = \langle x_1, x_2, \ldots, x_n \rangle$ and $\mathbf{y}^T = \langle y_1, y_2, \ldots, y_n \rangle$ of the same size n is a number (real or complex) denoted by $(\mathbf{x}, \mathbf{y})$ and is defined through the dot product:

$$(\mathbf{x}, \mathbf{y}) = \overline{\mathbf{x}} \cdot \mathbf{y}^T = \sum_{k=1}^{n} \overline{x}_k y_k,$$

where $\overline{\mathbf{x}}$ is a complex conjugate of the vector $\mathbf{x}$. The Euclidean **norm**, or **length**, or **magnitude** of a vector $\mathbf{x}$ is the positive number

$$\|\mathbf{x}\| = (\mathbf{x}, \mathbf{x})^{1/2} = \left(\overline{\mathbf{x}} \cdot \mathbf{x}^T\right)^{1/2} = \left[\sum_{k=1}^{n} \overline{x}_k x_k\right]^{1/2} = \left[\sum_{k=1}^{n} |x_k|^2\right]^{1/2}.$$

Thus, the Euclidean norm of a vector with real components is the square root of the sum of the squares of its components. If, for example, a vector $\mathbf{x}$ has complex components $\mathbf{x}^T = \langle x_1, x_2, \ldots, x_n \rangle$, $x_k = a_k + \mathbf{j} b_k$, $k = 1, 2, \ldots, n$, then its norm is

$$\|\mathbf{x}\| = \left[a_1^2 + b_1^2 + a_2^2 + b_2^2 + \cdots + a_n^2 + b_n^2\right]^{1/2}.$$

There are known other equivalent definitions of the norm; however, this one is the most useful in applications.

Example 6.2.1: In MATLAB®, we can define a vector of 20 random elements and find its Euclidean norm:

```
vect=rand(1,20);  norm(vect,2);
```

To find the sum of maximum entries (Manhattan norm), just type `norm(vect,1)`. *Maple*™ has a similar command `with(LinearAlgebra): VectorNorm( <2,-1>, 2);` which gives $\sqrt{2^2 + (-1)^2} = \sqrt{5}$. Similarly, the input `VectorNorm( <2,-1>, 1)` will give the Manhattan norm $3 = 2 + |-1|$ as output. *Mathematica*® also has a dedicated command: `Norm[vect]`. To operate with vectors in *Maxima*, one needs to load the package `load(vect)`. If your vector is $\langle -3, 2, 4 \rangle$ and you need to find its square norm, type in *Maxima*:

```
lsum(x^2 , x, [-3,2,4]);
```

Sage uses: `v.norm()` or `v.norm(2)` for the Euclidean norm of vector `v`.

Definition 6.2: Two vectors $\mathbf{x}$ and $\mathbf{y}$ are said to be **orthogonal** if their scalar product is zero:

$$(\mathbf{x}, \mathbf{y}) = \mathbf{0}.$$

A **matrix** is a rectangular array of objects or entries, written in rows and columns. The word originated in ancient Rome from Latin *matr-, mater* (womb or parent). The term "matrix" was first mentioned in the mathematical literature in a 1850 paper[66] by James Joseph Sylvester. In what follows, we use numbers (complex or real) or functions as entries, but generally speaking these objects can be anything you want. The rectangular array is usually enclosed either in square brackets

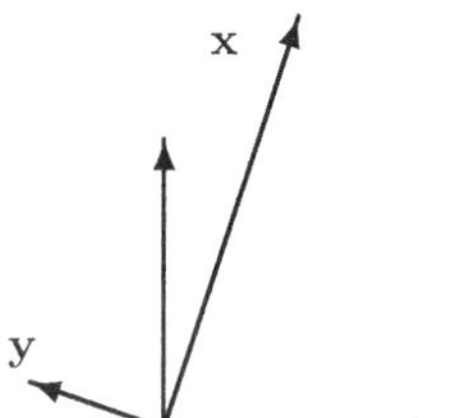

Figure 6.23: Orthogonal vectors

$$\mathbf{A} = [a_{ij}] = \begin{bmatrix} a_{11} & a_{12} & \cdots & a_{1n} \\ a_{21} & a_{22} & \cdots & a_{2n} \\ \vdots & \vdots & \ddots & \vdots \\ a_{m1} & a_{m2} & \cdots & a_{mn} \end{bmatrix}$$

or in parentheses

$$\mathbf{A} = (a_{ij}) = \begin{pmatrix} a_{11} & a_{12} & \cdots & a_{1n} \\ a_{21} & a_{22} & \cdots & a_{2n} \\ \vdots & \vdots & \ddots & \vdots \\ a_{m1} & a_{m2} & \cdots & a_{mn} \end{pmatrix},$$

and consists of m rows and n columns of mn objects (or entries) chosen from a given set. In this case we speak about an $m \times n$ matrix (pronounced "m by n"). In the symbol a_{ij}, which represents a typical entry (or element), the first subscript (i) denotes the row and the second subscript (j) denotes the column occupied by the entry.

The number of rows and the number of columns are called the **dimensions** of the matrix. When two matrices have the same dimensions, we say that they are of the same shape, of the same order, or of the same size. Unless otherwise indicated, matrices will be denoted by capital letters in a **bold font**.

Two matrices $\mathbf{A} = [a_{ij}]$ and $\mathbf{B} = [b_{ij}]$ are equal, written $\mathbf{A} = \mathbf{B}$, if and only if they have the same dimensions and their corresponding entries are equal. Thus, an equality between two $m \times n$ matrices $\mathbf{A}$ and $\mathbf{B}$ entails equalities between mn pairs of elements: $a_{11} = b_{11}$, $a_{12} = b_{12}$, and so on. Two matrices of the same size may be added or subtracted. We summarize the basic properties of arithmetic operations with matrices.

- **Equality:** Two matrices $\mathbf{A} = [a_{ij}]$ and $\mathbf{B} = [b_{ij}]$ are said to be equal if $a_{ij} = b_{ij}$ for all indices i, j.
- **Addition and Subtraction:** For matrices $\mathbf{A} = [a_{ij}]$ and $\mathbf{B} = [b_{ij}]$ of the same size, we have
$$\mathbf{A} \pm \mathbf{B} = [a_{ij} \pm b_{ij}].$$
- **Multiplication by a constant:** For any constant c, we have $c \cdot \mathbf{A} = [c \cdot a_{ij}]$.

Example 6.2.2: The equality between two 3×2 matrices

$$\begin{pmatrix} a_{11} & a_{12} \\ a_{21} & a_{22} \\ a_{31} & a_{32} \end{pmatrix} = \begin{pmatrix} 1 & 2 \\ 2 & 3 \\ 3 & 4 \end{pmatrix}$$

is equivalent to the six identities $a_{11} = 1$, $a_{12} = 2$, $a_{21} = 2$, $a_{22} = 3$, $a_{31} = 3$, $a_{32} = 4$.

Example 6.2.3: The sum and difference of two matrices

$$\mathbf{A} = \begin{bmatrix} 1 & 3 \\ 2 & 4 \\ 5 & 7 \end{bmatrix} \quad \text{and} \quad \mathbf{B} = \begin{bmatrix} -1 & 1 \\ 2 & -2 \\ 3 & 1 \end{bmatrix}$$

[66]Sylvester (1814–1897) was born James Joseph to a Jewish family in London, England; later he adopted the surname Sylvester. He invented a great number of other mathematical terms such as graph and discriminant.

are

$$\mathbf{A}+\mathbf{B}=\begin{bmatrix} 0 & 4 \\ 4 & 2 \\ 8 & 8 \end{bmatrix} \quad \text{and} \quad \mathbf{A}-\mathbf{B}=\begin{bmatrix} 2 & 2 \\ 0 & 6 \\ 2 & 6 \end{bmatrix} \qquad \square$$

The product of a scalar (number) α with an $m \times n$ matrix $\mathbf{A} = [a_{ij}]$ is an $m \times n$ matrix denoted by $\alpha\mathbf{A} = [\alpha a_{ij}]$. The scalar multiplies through each element of the matrix. Usually, we put the scalar multiplier on the left of the matrix, but $\mathbf{A}\alpha$ means the same as $\alpha\mathbf{A}$. Also, $(-1)\mathbf{A}$ is simply $-\mathbf{A}$ and is called the **negative** of $\mathbf{A}$. Similarly, $(-\alpha)\mathbf{A}$ is written $-\alpha\mathbf{A}$, and $\mathbf{A} + (-\mathbf{B})$ is written $\mathbf{A} - \mathbf{B}$.

> **Definition 6.3:** A matrix that is obtained from an $m \times n$ matrix $\mathbf{A} = [a_{ij}]$ by interchanging rows and columns is called the **transpose** of $\mathbf{A}$ and is usually denoted by $\mathbf{A}^T$ or $\mathbf{A}^t$ or even $\mathbf{A}'$. Thus $\mathbf{A}^T = [a_{ji}]$. A matrix is called **symmetric** if $\mathbf{A}^T = \mathbf{A}$; that is, $a_{ij} = a_{ji}$.

For any two matrices of the same size, $\mathbf{A}$ and $\mathbf{B}$, we have

$$\left(\mathbf{A}^T\right)^T = \mathbf{A}, \qquad (\mathbf{A}+\mathbf{B})^T = \mathbf{A}^T + \mathbf{B}^T.$$

If λ is a constant, then

$$(\lambda\mathbf{A})^T = \lambda\mathbf{A}^T.$$

Example 6.2.4: Let

$$\mathbf{A}=\begin{bmatrix} 1 & 3 \\ 2 & 4 \\ 5 & 7 \end{bmatrix} \quad \text{and} \quad \mathbf{B}=\begin{bmatrix} -1 & 1 \\ 2 & -2 \\ 3 & 1 \end{bmatrix}.$$

Then

$$\mathbf{A}^T=\begin{bmatrix} 1 & 2 & 5 \\ 3 & 4 & 7 \end{bmatrix} \quad \text{and} \quad \mathbf{B}^T=\begin{bmatrix} -1 & 2 & 3 \\ 1 & -2 & 1 \end{bmatrix}.$$

From Example 6.2.3, it follows that

$$(\mathbf{A}+\mathbf{B})^T=\begin{bmatrix} 0 & 4 \\ 4 & 2 \\ 8 & 8 \end{bmatrix}^T = \begin{bmatrix} 0 & 4 & 8 \\ 4 & 2 & 8 \end{bmatrix} = \mathbf{A}^T + \mathbf{B}^T.$$

> **Definition 6.4:** The complex conjugate of the matrix $\mathbf{A} = [a_{ij}]$, denoted by $\overline{\mathbf{A}}$, is the matrix obtained from $\mathbf{A}$ after replacing each element $a_{ij} = \alpha + \mathbf{j}\beta$ by its conjugate $\overline{a}_{ij} = \alpha - \mathbf{j}\beta$, where $\mathbf{j}^2 = -1$.

> **Definition 6.5:** The **adjoint** of the $m \times n$ matrix $\mathbf{A}$ is the transpose of its complex conjugate matrix and is denoted by $\mathbf{A}^*$ or $\mathbf{A}^H$, that is, $\mathbf{A}^* = \mathbf{A}^H = \overline{\mathbf{A}}^T$. A matrix is called **self-adjoint** or **Hermitian** if $\mathbf{A}^* = \mathbf{A}$; that is, $\overline{a}_{ij} = a_{ji}$.

Note that if $\mathbf{A}$ is a real-valued matrix, then its adjoint is just its transposition because the complex conjugate operation does not change real entries. If $\mathbf{A}$ and $\mathbf{B}$ are two $n \times n$ matrices and λ is a (complex) scalar, then the following properties hold:

- $\overline{(\overline{\mathbf{A}})} = \mathbf{A}, \quad (\mathbf{A}^T)^T = \mathbf{A}, \quad (\mathbf{A}^*)^* = \mathbf{A}.$
- $\overline{(\mathbf{A}+\mathbf{B})} = \overline{\mathbf{A}} + \overline{\mathbf{B}}, \quad (\mathbf{A}+\mathbf{B})^T = \mathbf{A}^T + \mathbf{B}^T, \quad (\mathbf{A}+\mathbf{B})^* = \mathbf{A}^* + \mathbf{B}^*.$
- $\overline{(\lambda\mathbf{A})} = \overline{\lambda}\,\overline{\mathbf{A}}, \quad (\lambda\mathbf{A})^T = \lambda\mathbf{A}^T, \quad (\lambda\mathbf{A})^* = \overline{\lambda}\,\mathbf{A}^*.$

Example 6.2.5: Consider the following two matrices with complex entries:

$$\mathbf{A}=\begin{bmatrix} 1+\mathbf{j} & \mathbf{j} \\ 2 & 3\mathbf{j} \end{bmatrix} \quad \text{and} \quad \mathbf{B}=\begin{bmatrix} 2 & \mathbf{j} \\ -\mathbf{j} & 3 \end{bmatrix}.$$

Their complex conjugate, transposed, and adjoint matrices can be written as follows:

$$\overline{\mathbf{A}} = \begin{bmatrix} 1-\mathrm{j} & -\mathrm{j} \\ 2 & -3\mathrm{j} \end{bmatrix}, \qquad \mathbf{A}^T = \begin{bmatrix} 1+\mathrm{j} & 2 \\ \mathrm{j} & 3\mathrm{j} \end{bmatrix}, \qquad \mathbf{A}^* = \begin{bmatrix} 1-\mathrm{j} & 2 \\ -\mathrm{j} & -3\mathrm{j} \end{bmatrix},$$

$$\overline{\mathbf{B}} = \begin{bmatrix} 2 & -\mathrm{j} \\ \mathrm{j} & 3 \end{bmatrix}, \qquad \mathbf{B}^T = \begin{bmatrix} 2 & -\mathrm{j} \\ \mathrm{j} & 3 \end{bmatrix}, \qquad \mathbf{B}^* = \begin{bmatrix} 2 & \mathrm{j} \\ -\mathrm{j} & 3 \end{bmatrix}.$$

Hence, the matrix **B** is self-adjoint, but the matrix **A** is not.

Definition 6.6: The square $n \times n$ matrix

$$\begin{bmatrix} 1 & 0 & 0 & \cdots & 0 \\ 0 & 1 & 0 & \cdots & 0 \\ \vdots & \vdots & \vdots & \ddots & \vdots \\ 0 & 0 & 0 & \cdots & 1 \end{bmatrix}$$

is denoted by the symbol **I** (or $\mathbf{I}_n$ when there is a need to emphasize the dimensions of the matrix) and is called the **identity matrix** (or unit matrix). The matrix with all entries being zero is denoted by **0** and is called the zero matrix.

The **product AB** of two matrices is defined whenever the number of columns of the first matrix **A** is the same as the number of rows of the second matrix **B**. If **A** is an $m \times n$ matrix and **B** is an $n \times r$ matrix, then the product $\mathbf{C} = \mathbf{AB}$ is an $m \times r$ matrix whose element c_{ij} in the i-th row and j-th column is defined as the inner or scalar product of the i-th row of **A** and j-th column of **B**. Namely,

$$c_{ij} = a_{i1}b_{1j} + a_{i2}b_{2j} + \cdots + a_{in}b_{nr} = \sum_{k=1}^{n} a_{ik}b_{kj}.$$

Every $n \times r$ matrix **B** can be represented in **column form** as

$$\mathbf{B} = [\mathbf{b}_1, \mathbf{b}_2, \ldots, \mathbf{b}_r]$$

with $\mathbf{b}_1$, ..., $\mathbf{b}_r$ being column vectors of size n: $\mathbf{b}_j = \langle b_{1j}, b_{2j}, \ldots, b_{nj}\rangle^T$. Similarly, the transpose $n \times m$ matrix $\mathbf{A}^T$ can also be rewritten in column form:

$$\mathbf{A}^T = [\mathbf{a}_1, \mathbf{a}_2, \ldots, \mathbf{a}_m]$$

with n-vectors $\mathbf{a}_k = \langle a_{k1}, a_{k2}, \ldots, a_{kn}\rangle^T$, $k = 1, 2, \ldots, m$. Then the product $\mathbf{A}\,\mathbf{B}$ of two matrices becomes

$$\mathbf{A}\,\mathbf{B} = [\,\mathbf{a}_k \cdot \mathbf{b}_j\,] \qquad (k = 1, 2, \ldots m,\ j = 1, 2, \ldots r).$$

The product of $m \times n$ matrix **A** and an n-column vector $\mathbf{x} = \langle x_1, x_2, \ldots, x_n\rangle^T$ (which is an $n \times 1$ matrix) can be written as

$$\mathbf{A}\,\mathbf{x} = x_1\mathbf{a}_1 + x_2\mathbf{a}_2 + \cdots + x_n\mathbf{a}_n,$$

where $\mathbf{A} = [\,\mathbf{a}_1, \mathbf{a}_2, \ldots, \mathbf{a}_n\,]$ is the matrix comprised of column vectors of size m.

It is easy to show that matrix multiplication satisfies the **associative law** $(\mathbf{AB})\mathbf{C} = \mathbf{A}(\mathbf{BC})$ and the **distributive law** $\mathbf{A}(\mathbf{B} + \mathbf{C}) = \mathbf{AB} + \mathbf{AC}$, but, generally speaking, it might not satisfy the **commutative law**, that is, it may happen that $\mathbf{AB} \ne \mathbf{BA}$.

Some properties of matrix multiplication differ from the corresponding properties of numbers, and we emphasize some of them below.

- The multiplication of matrices may not commute even for square matrices. Moreover, the product **AB** of two matrices **A** and **B** may exist but their inverse product **BA** may not. In general,

$$\mathbf{AB} \ne \mathbf{BA}.$$

- $\mathbf{AB} = \mathbf{0}$ does not generally imply $\mathbf{A} = \mathbf{0}$ or $\mathbf{B} = \mathbf{0}$ or $\mathbf{BA} = \mathbf{0}$.

- $\mathbf{AB} = \mathbf{AC}$ does not generally imply $\mathbf{B} = \mathbf{C}$.

The identity matrix $\mathbf{I}$ commutes with any other square matrix: $\mathbf{I} \cdot \mathbf{A} = \mathbf{A} \cdot \mathbf{I}$. However, there may exist a matrix $\mathbf{B} \neq \mathbf{I}$ such that $\mathbf{BA} = \mathbf{A}$ and $\mathbf{AB} = \mathbf{A}$ for a particular matrix $\mathbf{A}$.

Example 6.2.6: For the 2×2 matrices below, we have

$$\begin{bmatrix} 2 & 2 \\ 1 & 1 \end{bmatrix} \begin{bmatrix} 1 & -1 \\ -1 & 1 \end{bmatrix} = \begin{bmatrix} 0 & 0 \\ 0 & 0 \end{bmatrix}, \qquad \begin{bmatrix} 2 & 2 \\ 1 & 1 \end{bmatrix} \begin{bmatrix} 2 & 0 \\ 0 & 2 \end{bmatrix} = \begin{bmatrix} 4 & 4 \\ 2 & 2 \end{bmatrix} = 2 \begin{bmatrix} 2 & 2 \\ 1 & 1 \end{bmatrix},$$

and

$$\begin{bmatrix} 2 & 2 \\ 1 & 1 \end{bmatrix} \begin{bmatrix} 1 & 0 \\ 1 & 2 \end{bmatrix} = \begin{bmatrix} 4 & 4 \\ 2 & 2 \end{bmatrix} = 2 \begin{bmatrix} 2 & 2 \\ 1 & 1 \end{bmatrix}, \qquad \begin{bmatrix} 1 & 2 \\ 0 & 2 \end{bmatrix} \begin{bmatrix} 2 & 2 \\ 1 & 1 \end{bmatrix} = \begin{bmatrix} 4 & 4 \\ 2 & 2 \end{bmatrix} = 2 \begin{bmatrix} 2 & 2 \\ 1 & 1 \end{bmatrix}. \qquad \square$$

The following properties of matrix multiplication hold:

$$\overline{(\mathbf{AB})} = \overline{\mathbf{A}} \cdot \overline{\mathbf{B}}, \quad (\mathbf{AB})^T = \mathbf{B}^T \mathbf{A}^T, \quad (\mathbf{AB})^* = \mathbf{B}^* \mathbf{A}^*.$$

Definition 6.7: A square $n \times n$ matrix $\mathbf{A}$ is called **normal** if it commutes with its adjoint: $\mathbf{AA}^* = \mathbf{A}^*\mathbf{A}$.

Definition 6.8: The **trace** of an $n \times n$ matrix $\mathbf{A} = [\, a_{ij}\,]$, denoted by $\operatorname{tr}(\mathbf{A})$ or $\operatorname{tr}\mathbf{A}$, is the sum of its diagonal elements, that is, $\operatorname{tr}(\mathbf{A}) = a_{11} + a_{22} + \cdots + a_{nn}$.

Theorem 6.1: Whenever α is a number (complex or real) and $\mathbf{A}$ and $\mathbf{B}$ are square matrices of the same dimensions, the following identities hold:

- $\operatorname{tr}(\mathbf{A} + \mathbf{B}) = \operatorname{tr}(\mathbf{A}) + \operatorname{tr}(\mathbf{B})$;
- $\operatorname{tr}(\alpha\mathbf{A}) = \alpha \operatorname{tr}(\mathbf{A})$;
- $\operatorname{tr}(\mathbf{AB}) = \operatorname{tr}(\mathbf{BA})$.

Example 6.2.7: In *Maple*, you may define a matrix in a couple of different ways. However, before using matrix commands one should invoke one of the two linear algebra packages (or both): `LinearAlgebra` (recommended) or `linalg` (an old version, but still in use). A colon (:) at the end of each command is used when the user does not want to see the results of its execution on the screen. To minimize the possibility of mixing data from different problems, start every problem with
`restart:` `with(LinearAlgebra):` or/and `with(linalg):`

There are a few examples of entering the matrix with the same output:

```
with(LinearAlgebra):
A := Array(1..2, 1..3, [ [a,b,c],[d,e,f] ] );
```

or `A := Matrix( 2, 3, [ [a,b,c],[d,e,f] ] );` or `A := Matrix( 2, 3, [ a, b, c, d, e, f ] );`
which yields

$$\mathbf{A} := \begin{bmatrix} a & b & c \\ d & e & f \end{bmatrix}.$$

Note that the `linalg` package uses similar commands but they all start with lower case letters, for example, `matrix` instead of `Matrix`. The identity 3×3 matrix can be defined with the command:

```
M:=Matrix(3,3,shape=identity):
```

The entries of a matrix may be variables and not only numbers.

```
with(LinearAlgebra):
A := <<1,5,w>|<2,6,x>|<3,7,y>|<4,8,z>>;
```

or

```
linalg[matrix](2,3,[x,y,z,a,b,c]);
```

The `entermatrix` procedure prompts you for the values of matrix elements, one by one.

To find a transpose or Hermitian to the matrix **A**, type in

`transpose(A);`   `Transpose(A)` or `HermitianTranspose(A)`

The `evalm` function performs arithmetic on matrices: `evalm(matrix expression);` also, the dot as `A.B` or the command `Multiply(A,B)` is used within the `LinearAlgebra` package. When the package `linalg` is loaded, multiplication of two or more matrices can be obtained by executing the following command: `multiply(A,B)`. Recall that the matrix product `ABC` of three matrices may be entered within the `evalm` command as `A &* B &* C` or as `&*(A,B,C)`, the latter being more efficient. Automatic simplifications such as collecting constants and powers will be applied. Do not use the * to indicate purely matrix multiplication, as this will result in an error. The operands of `&*` must be matrices (or their names) with the exception of 0. Unevaluated matrix products are considered to be matrices. The operator `&*` has the same precedence as the * operator. So the `LinearAlgebra` package makes matrix operations more friendly, and its commands are similar to the ones of *Mathematica* (see Example 6.2.9).

In *Maple*, the trace of a matrix **A** can be found as `linalg[trace](A);` or just `trace(A);` if either of the packages `linalg` or `LinearAlgebra` was previously loaded. Actually, *Maple* has a dedicated command `MatrixFunction`, which is a part of the `LinearAlgebra` package:

```
with(LinearAlgebra):
A:=Matrix(2,2,[9,8,7,8]):
LinearAlgebra[MatrixFunction](A,cos(t*sqrt(lambda)),lambda)
MatrixFunction(A, exp(t*lambda), lambda)
MatrixExponential(A, t)
```

Example 6.2.8: In MATLAB, a matrix **A** can be entered as follows

$$\mathbf{A} = [1, 2, 3, 4; 3, 4, 5, 6; 5, 6, 7, 8] \quad \text{or} \quad \mathbf{A} = [1\ 2\ 3\ 4; 3\ 4\ 5\ 6; 5\ 6\ 7\ 8]$$

Then on the screen one can see

$$\mathbf{A} = \begin{bmatrix} 1 & 2 & 3 & 4 \\ 3 & 4 & 5 & 6 \\ 5 & 6 & 7 & 8 \end{bmatrix}.$$

If you type $A(:,1)$, then on the screen you will see the first column, namely, $\begin{matrix} 1 \\ 3 \\ 5 \end{matrix}$. The command $A(:,2)$ will invoke the second column, and so on. If you want to see the first row of the matrix **A**, type $A(1,:)$ to observe

1 2 3 4.

To display the diagonal of the matrix **A**, type `diag(A)` or `diag(A,0)` to observe [1 ; 4 ; 7]. The next diagonals can be retrieved by

`diag(A,1)` shows [2, 5, 8];   `diag(A,-1)` shows [3, 6].

To find the transpose matrix of matrix **A**, just type $A.'$ (with dot) at the MATLAB prompt. To get the adjoint matrix $\mathbf{A}^*$, you need to type A' (without dot); if the matrix **A** has only real entries, then A' gives you its transposed matrix.

Example 6.2.9: In *Mathematica*, a 2 × 3 matrix can be defined as follows:

```
{{1,2,3},{-2,0,3}}    (* or *)  {{1,2,3},{-2,0,3}} // MatrixForm
```

To multiply a matrix by a number, place the number in front of the matrix. A dot is used to define multiplication of matrices: (`A.B`). The operators `Transpose[A]` and `ConjugateTranspose[A]` are self-explanatory.

In *Mathematica*, the output of `IdentityMatrix[n]` gives the $n \times n$ identity matrix.

Example 6.2.10: In *Maxima* (or *Sage*), a matrix is defined as a collection of row vectors:

```
A: matrix([1,2],[-3,1]);
transpose(A);   /* to find a transpose matrix */
A.A;  /* or to square the matrix A */      A^^2;
mattrace(A);  /* trace of the matrix A */
```

6.3 Linear Systems of First Order ODEs

A system of differential equations is a set of equations involving more than one unknown function and their derivatives. The **order** of a system of differential equations is the highest derivative that occurs in the system. In this section, we consider only first order systems of differential equations. As we saw in §6.1, certain problems lead naturally to systems of nonlinear differential equations in **normal form**:

$$\begin{cases} dx_1/dt &= g_1(t, x_1, x_2, \ldots, x_n), \\ dx_2/dt &= g_2(t, x_1, x_2, \ldots, x_n), \\ \quad\vdots & \quad\vdots \\ dx_n/dt &= g_n(t, x_1, x_2, \ldots, x_n), \end{cases} \tag{6.3.1}$$

where $g_k(t, x_1, x_2, \ldots, x_n)$, $k = 1, 2, \ldots n$, is a given function of $n+1$ variables. Instead of dx/dt we will use either of shorter notations x' or the more customary notation $\dot{x}$ to denote a derivative of $x(t)$ with respect to variable t associated with time. Note that we consider only the case when the number of equations is equal to the number of unknown variables, which is called the **dimension** of the system. A system of dimension 2 is called a **planar** system. If the right-hand side functions g_1, g_2, …, g_n do not depend on t, then the corresponding system of equations is called **autonomous**.

When we say that we are looking for or have a solution of a system of first order linear differential equations, we mean a set of n continuously differentiable functions $x_1(t)$, $x_2(t)$, …, $x_n(t)$ that satisfy the system on some interval. In addition to the system, there may also be assigned initial conditions:

$$x_1(t_0) = x_{10}, \quad x_2(t_0) = x_{20}, \quad \ldots, \quad x_n(t_0) = x_{n0},$$

where t_0 is a specified value of t and x_{10}, x_{20}, …, x_{n0} are prescribed constants. The problem of finding a solution to a linear system of differential equations that satisfies the given initial conditions is called an **initial value problem** or a Cauchy problem.

When the functions $g_k(t, x_1, x_2, \ldots, x_n)$, $k = 1, 2, \ldots n$, in Eq. (6.3.1) are linear functions with respect to n dependent variables x_1, x_2, …, x_n, we obtain the general system of first order linear differential equations in normal form:

$$\begin{cases} \dot{x}_1(t) = p_{11}\, x_1(t) + p_{12}\, x_2(t) + \cdots + p_{1n}\, x_n(t) + f_1(t), \\ \dot{x}_2(t) = p_{21}\, x_1(t) + p_{22}\, x_2(t) + \cdots + p_{2n}\, x_n(t) + f_2(t), \\ \vdots \qquad \vdots \qquad\qquad \vdots \quad \vdots \qquad\qquad \vdots \quad \vdots \qquad \vdots \\ \dot{x}_n(t) = p_{n1}\, x_1(t) + p_{n2}\, x_2(t) + \cdots + p_{nn}\, x_n(t) + f_n(t). \end{cases} \tag{6.3.2}$$

In this system of equations (6.3.2), the n^2 **coefficients** $p_{11}(t)$, …, $p_{nn}(t)$ and the n functions $f_1(t)$, …, $f_n(t)$ are assumed to be known. If the coefficients p_{ij} are constants, then we have a constant coefficient system of differential equations. Otherwise, we have a linear system of differential equations with variable coefficients. The system is said to be **homogeneous** or **undriven** if $f_1(t) = f_2(t) = \ldots = f_n(t) \equiv 0$. If at least one of the components of the vector function $\mathbf{f}(t) = \langle f_1(t), \ldots, f_n(t)\rangle^T$ is not identically zero, the system is called **nonhomogeneous** or **driven** and the vector function $\mathbf{f}(t)$ is referred to as a **nonhomogeneous/driving term** or **forcing function** or **input**.

Example 6.3.1: The planar linear nonhomogeneous system of equations

$$\begin{cases} \dot{x}_1 = x_1 + 2\,x_2 + \sin(t), \\ \dot{x}_2 = 3\,x_1 + 4\,x_2 + t\,\cos(t) \end{cases}$$

is in normal form. However, the system

$$\begin{cases} \dot{x}_1 + 4\dot{x}_2 = x_1 + 2\,x_2 + \sin(t), \\ 2\dot{x}_1 - \dot{x}_2 = 2\,x_1 + 4\,x_2 + t\,\cos(t) \end{cases}$$

is not in normal form. The system

$$\dot{x}_1 = x_1 x_2, \qquad \dot{x}_2 = x_1 + x_2$$

is an example of a nonlinear system. □

We can rewrite the system (6.3.2) much more elegantly, in vector form. Let $\mathbf{x}(t)$ and $\mathbf{f}(t)$ be n-dimensional vectors, and $\mathbf{P}(t)$ denote the following square matrix:

$$\mathbf{x}(t) = \begin{bmatrix} x_1(t) \\ x_2(t) \\ \vdots \\ x_n(t) \end{bmatrix}, \quad \mathbf{P}(t) = \begin{bmatrix} p_{11}(t) & p_{12}(t) & \dots & p_{1n}(t) \\ p_{21}(t) & p_{22}(t) & \dots & p_{2n}(t) \\ \vdots & \vdots & \ddots & \vdots \\ p_{n1}(t) & p_{n2}(t) & \dots & p_{nn}(t) \end{bmatrix}, \quad \mathbf{f}(t) = \begin{bmatrix} f_1(t) \\ f_2(t) \\ \vdots \\ f_n(t) \end{bmatrix}.$$

Then the system of linear first order differential equations in normal form can be written as

$$\frac{\mathrm{d}}{\mathrm{d}t} \begin{bmatrix} x_1(t) \\ x_2(t) \\ \vdots \\ x_n(t) \end{bmatrix} = \begin{bmatrix} p_{11}(t) & p_{12}(t) & \dots & p_{1n}(t) \\ p_{21}(t) & p_{22}(t) & \dots & p_{2n}(t) \\ \vdots & \vdots & \ddots & \vdots \\ p_{n1}(t) & p_{n2}(t) & \dots & p_{nn}(t) \end{bmatrix} \begin{bmatrix} x_1(t) \\ x_2(t) \\ \vdots \\ x_n(t) \end{bmatrix} + \begin{bmatrix} f_1(t) \\ f_2(t) \\ \vdots \\ f_n(t) \end{bmatrix}$$

or simply

$$\dot{\mathbf{x}}(t) = \mathbf{P}(t)\mathbf{x}(t) + \mathbf{f}(t). \tag{6.3.3}$$

If the system is homogeneous, its vector form becomes

$$\dot{\mathbf{x}}(t) = \mathbf{P}(t)\mathbf{x}(t). \tag{6.3.4}$$

It is often a useful trade-off to replace a differential equation of order higher than one by a first order system at the expense of increasing the number of unknown functions. Such a transition to a system of differential equations in normal form (6.3.1) is important for numerical calculations as well as for geometrical interpretations. Any n-th order linear differential equation is equivalent to the system (6.3.3), meaning that their solutions are the same. If such an n-th order equation is given in normal form,

$$y^{(n)} = a_{n-1}\,y^{(n-1)} + a_{n-2}\,y^{(n-2)} + \cdots + a_1 y' + a_0\,y + g(t), \tag{6.3.5}$$

then introducing n variables regarding $y, y', \dots, y^{(n-1)}$ and renaming them

$$x_1(t) \stackrel{\text{def}}{=} y(t), \quad x_2(t) \stackrel{\text{def}}{=} y'(t), \quad x_2(t) \stackrel{\text{def}}{=} y''(t), \quad ,\dots, \quad x_n(t) \stackrel{\text{def}}{=} y^{(n-1)}(t),$$

we reduce Eq. (6.3.5) to the following vector form:

$$\frac{\mathrm{d}\mathbf{x}}{\mathrm{d}t} = \mathbf{P}\mathbf{x} + \mathbf{f}, \qquad \mathbf{P} = \begin{bmatrix} 0 & 1 & 0 & \dots & 0 \\ 0 & 0 & 1 & \dots & 0 \\ \vdots & \vdots & \vdots & \ddots & \vdots \\ a_0 & a_1 & a_2 & \dots & a_{n-1} \end{bmatrix}, \quad \mathbf{x} = \begin{bmatrix} x_1 \\ x_2 \\ \vdots \\ x_n \end{bmatrix}, \quad \mathbf{f} = \begin{bmatrix} 0 \\ 0 \\ \vdots \\ g(t) \end{bmatrix}. \tag{6.3.6}$$

Note that if the linear operator associated with the given n-th order equation, $L[\mathtt{D}] = \mathtt{D}^n - a_{n-1}\mathtt{D}^{n-1} - a_{n-2}\mathtt{D}^{n-2} - a_1\mathtt{D} - a_0$, where $\mathtt{D}$ is the derivative operator, has constant coefficients, then the determinant of the corresponding matrix $\lambda\mathbf{I} - \mathbf{P}$ is

$$\det(\lambda\mathbf{I} - \mathbf{P}) = \lambda^n - a_{n-1}\lambda^{n-1} - a_{n-2}\lambda^{n-2} - \cdots - a_1\lambda - a_0.$$

This means that the characteristic polynomial of the operator $L[\mathtt{D}]$ coincides with the characteristic polynomial of the constant matrix $\mathbf{P}$. (The definition of a determinant is given in §7.2, page 380.) The general case is considered in the following statement.

Theorem 6.2: Any ordinary differential equation or a system of ordinary differential equations in normal form can be expressed as a system of **first order** differential equations.

PROOF: Any n-th order differential equation, linear or nonlinear, can be written in the form

$$y^{(n)}(t) = F(t, y, y', \dots, y^{(n-1)}), \tag{6.3.7}$$

where derivatives are with respect to t. If we now rename the derivatives

$$u_1 = y, \quad u_2 = y' = u_1', \quad u_3 = y'' = u_2', \ldots, u_n = y^{(n-1)} = u^{(n-2)},$$

then Eq. (6.3.7) can be rewritten as

$$\frac{du_1}{dt} = u_2, \quad \frac{du_2}{dt} = u_3, \quad \ldots, \quad \frac{du_{n-1}}{dt} = u_n, \quad \frac{du_n}{dt} = F(t,\ u_1,\ u_2,\ \ldots,\ u_n),$$

which is a system of n first order differential equations. A similar proof is valid for any system of ordinary differential equations.

Example 6.3.2: Suppose that a mechanical system is modeled by the forced damped harmonic oscillator

$$\ddot{y}(t) - 2\beta\,\dot{y}(t) + \omega^2 y(t) = f(t),$$

where β, ω are positive constants and $f(t)$ is a given function. In order to convert this second order differential equation to a vector differential equation in standard form, we introduce a vector variable whose components are y and $\dot{y}$. In particular,

$$\mathbf{x}(t) \equiv \begin{bmatrix} x_1(t) \\ x_2(t) \end{bmatrix} = \begin{bmatrix} y(t) \\ \dot{y}(t) \end{bmatrix} \qquad \text{or} \qquad \mathbf{x}^T \equiv \langle x_1, x_2 \rangle^T = \langle y, \dot{y} \rangle^T.$$

Then its derivative is $\frac{d\mathbf{x}}{dt} = \begin{bmatrix} \dot{y} \\ \ddot{y} \end{bmatrix} = \begin{bmatrix} \dot{x}_1 \\ \dot{x}_2 \end{bmatrix}$. From the oscillator equation, we get $\ddot{y}$ to be $\ddot{y} = f(t) + 2\beta\,\dot{y} - \omega^2 y = f + 2\beta x_2 - \omega^2 x_1$. Now we can transform the given second order equation into standard normal form:

$$\begin{aligned} \frac{d\mathbf{x}}{dt} &= \frac{d}{dt}\begin{bmatrix} x_1 \\ x_2 \end{bmatrix} = \frac{d}{dt}\begin{bmatrix} y \\ \dot{y} \end{bmatrix} = \begin{bmatrix} \dot{y} \\ f + 2\beta\dot{y} - \omega^2 y \end{bmatrix} = \begin{bmatrix} x_2 \\ f + 2\beta x_2 - \omega^2 x_1 \end{bmatrix} \\ &= \begin{bmatrix} 0 & 1 \\ -\omega^2 & 2\beta \end{bmatrix} \mathbf{x}(t) + \begin{bmatrix} 0 \\ f(t) \end{bmatrix} = \begin{bmatrix} 0 & 1 \\ -\omega^2 & 2\beta \end{bmatrix} \begin{bmatrix} x_1 \\ x_2 \end{bmatrix} + \begin{bmatrix} 0 \\ f(t) \end{bmatrix}. \end{aligned}$$

Problems

1. Reduce the given differential equations to first order systems of equations. If the equation is linear, identify the matrix $\mathbf{P}(t)$ and the driving term $\mathbf{f}$ (if any) in Eq. (6.3.3) on page 363.
   (a) $t^2 y'' + 3t\,y' + y = t^7$; **(b)** $t^2 y'' - 6t\,y' + \sin 2t\,y = \ln t$;
   (c) $y'' + 3y' + y/t = t$; **(d)** $y'' + t\,y' - y\,\ln t = \cos 2t$;
   (e) $t^3 y'' - 2ty' + y = t^4$; **(f)** $y''' + 3t\,y'' + 3t^2\,y' + y = 7$;
   (g) $y''' = f(y) + g(t)$; **(h)** $t^3 y''' + 3t\,y' + y = \sin t$;
   (i) $t^2 y'' + 5t^{-1}\,y' - 2y\,\sin t = t$; **(j)** $y''(t) - t^2\,y'(t) + 2y(t) = \sin t$.

2. In each problem, reduce the given linear differential equation with constant coefficients to the vector form, $\dot{\mathbf{x}} = \mathbf{A}\mathbf{x} + \mathbf{f}$, and identify the matrix $\mathbf{A}$ and the forcing vector $\mathbf{f}$.
   (a) $y'' + 2y' + y = 1$; **(b)** $y'' - 2y' + 5y = e^t$;
   (c) $y'' - 3y' - 7y = 4$; **(d)** $y''' + 3y'' + 3y' + y = 5$;
   (e) $3y'' + 5y' - 2y = 3t^2$; **(f)** $y''' = 2y'' - 4y' + \sin t$.

3. Rewrite the system of differential equations in a matrix form $\dot{\mathbf{x}} = \mathbf{A}\mathbf{x}$ by identifying the vector $\mathbf{x}$ and the square matrix $\mathbf{A}$.

   (a) $\dot{x} = x - 2y, \quad \dot{y} = 3x - 4y$. **(b)** $\dot{x}_1 = \frac{5}{4}\,x_1 + \frac{3}{4}\,x_2, \quad \dot{x}_2 = \frac{1}{2}\,x_1 - \frac{3}{2}\,x_2$.

   (c) $x' - x + 2y = 0, \quad y' + y - x = 0$.

   (d) $x' + 5x - 2y = 0, \quad y' + 2x - y = 0$.

   (e) $x' - 3x + 2y = 0, \quad y' - x + 3y = 0$.

   (f) $x' + x - z = 0, \quad y' - y + x = 0, \quad z' + x + 2y - 3z = 0$.

   (g) $x' = -0.5\,x + 2\,y - 3\,z, \quad y' = y - 0.5\,z, \quad z' = -2\,x + z$.

4. Verify that the system
$$\dot{x} = -\alpha y - (1-\alpha)\sin t, \qquad \dot{y} = \alpha x - \alpha^2 t + (1-\alpha)\cos t$$
   has a solution of the form $x = \alpha t + \cos t + c\,\cos \alpha t$, $y = \sin t - 1 + c\,\sin \alpha t$ for some constant c.

5. Show that the system $\dot{x} = ty$, $\dot{y} = -tx$ has circular solutions of radius $r > 0$. *Hint:* Show that $x\dot{x} + y\dot{y} = 0$.

6.4 Reduction to a Single ODE

The Gauss elimination method for solving systems of algebraic equations can be adapted to systems of linear differential equations, not necessarily in normal form. In many cases, it is possible to eliminate all but one dependent variable in succession until there remains only a single differential equation containing only one dependent variable. When this single differential equation can be solved, other dependent variables can be found in turn, using the original system of equations. Such a procedure, called the **method of elimination**, provides an effective tool for solving systems of differential equations. The solution obtained may contain the sufficient number of constants of integration to identify it as the general solution. However, the eliminating procedure may not lead to an equivalent single differential equation, and some solutions could be missing. We shall illustrate the elimination method in examples, as it will enable us to anticipate the form of the solution.

Example 6.4.1: Suppose that there are two large interconnected tanks feeding each other; one of them we call tank A and the other one tank B. Suppose that initially tank A holds 60 liters of a brine solution, and tank B contains 40 liters of the same solution. Fresh water flows into tank A at a rate of 2 liters per minute (`l/min`), and fluid is drained out of tank B at the same rate. Also, 1 `l/min` of fluid are pumped from tank B to tank A, and 2 `l/min` from tank A to tank B. The liquids inside each tank are kept well stirred so that each mixture is homogeneous. If, initially, the brine solution in tank A contains x_0 `kg` of salt and that in tank B contains y_0 `kg` of salt, determine the amount of salt in each tank at time $t > 0$.

Solution. Note that the total volume of liquid remains constant at 100 liters because of the balance between inflow and outflow volume rates. However, the volume of liquid in tank B increases at a rate of 1 liter per minute, while the volume of liquid in tank A decreases at the same rate. Therefore, the volume of liquid in tank A at time t is $60 - t$, and the volume of liquid in tank B becomes $40 + t$. Let $x(t)$ denote the amount of salt in tank A at time t and $y(t)$ in tank B.

To formulate the equations for this system, we equate the rate of change of salt in each tank with the net rate at which salt is transferred to that tank. The salt concentration in tank A is $x(t)/(60 - t)$ `kg/l`, so the salt is carried out of tank A at a rate of $2x/(60 - t)$ `kg/min`. Similarly, the salt in tank B is transferred to tank A in the amount of $y/(40 + t)$ `kg/min`. The fresh water has no salt and its input just maintains the total volume. Since the difference between the input rate and output rate gives the net rate of change, we get the following system of equations:

$$\dot{x} \stackrel{\text{def}}{=} \texttt{D}x = \frac{\mathrm{d}x}{\mathrm{d}t} = \frac{y}{40+t} - \frac{2x}{60-t},$$
$$\dot{y} \stackrel{\text{def}}{=} \texttt{D}y = \frac{\mathrm{d}y}{\mathrm{d}t} = \frac{2x}{60-t} - \frac{3y}{40+t},$$

where D is the derivative operator. The system of linear differential equations obtained has variable coefficients. If the rate of exchange between the two tanks remains the same, say at 2 liters per minute, we get instead a constant coefficient system of differential equations:

$$\begin{aligned} \dot{x} &= \frac{2y}{40} - \frac{2x}{60} = \frac{y}{20} - \frac{x}{30}, \\ \dot{y} &= \frac{2x}{60} - \frac{4y}{40} = \frac{x}{30} - \frac{y}{10}, \end{aligned} \qquad \Longrightarrow \qquad \begin{aligned} \left(\texttt{D} + \frac{1}{10}\right)\left(\texttt{D} + \frac{1}{30}\right) x &= \frac{x}{600}, \\ \left(\texttt{D} + \frac{1}{10}\right)\left(\texttt{D} + \frac{1}{30}\right) y &= \frac{y}{600}. \end{aligned}$$

Example 6.4.2: Solve the following system of differential equations by reducing it to a single second order equation.

$$\begin{cases} x' = 3y, \\ y' = -x - 4y. \end{cases}$$

Solution. Differentiating the second equation yields

$$y'' = -x' - 4y' = -3y - 4y' \qquad \Longrightarrow \qquad y'' + 4y' + 3y = 0.$$

This homogeneous linear differential equation in y has the general solution $y = c_1\,e^{-t} + c_2\,e^{-3t}$. To find $x(t)$, we substitute $y(t)$ into $x' = 3y$ to obtain the equation $x' = 3c_1\,e^{-t} + 3c_2\,e^{-3t}$, which can be easily integrated: $x(t) = -3c_1\,e^{-t} - c_2\,e^{-3t} + c$, where c_1, c_2, and c are arbitrary constants. Substituting x back into the equation $y' = -x - 4y$ gives the condition $c = 0$. Therefore, the given system of equations has the general solution

$$x = -3c_1\,e^{-t} - c_2\,e^{-3t}, \qquad y = c_1\,e^{-t} + c_2\,e^{-3t},$$

with two arbitrary constants c_1 and c_2. □

It is convenient to introduce the derivative operator, commonly denoted by $\texttt{D}$. Therefore, $\texttt{D}y$ means y' or $\dot{y}$, the derivative of the function y. The composition of $\texttt{D}$ with itself, when it operates on functions, is naturally denoted by $\texttt{D}^2$, which gives the second derivative when it is applied to a function. This allows us to define a polynomial in variable $\texttt{D}$. For example, $(\texttt{D}^2 + 2\,\texttt{D} - 1)y = \ddot{y} + 2\,\dot{y} - y$. Note that we consider only constant coefficient polynomials.

Example 6.4.3: Solve the system of differential equations

$$\begin{aligned} x_1' &= x_1 + 2x_2, \\ x_2' &= 2x_1 - 2x_2, \end{aligned}$$

subject to the initial conditions

$$x_1(0) = 5, \quad x_2(0) = 0.$$

Solution. Using the notation $\texttt{D} \stackrel{\text{def}}{=} \mathrm{d}/dt$ for the derivative operator, we rewrite the given system in the form

$$\begin{aligned} (\texttt{D} - 1)\, x_1 \quad &- \quad 2\, x_2 = 0, \\ -2\, x_1 \quad &+ \quad (\texttt{D} + 2)\, x_2 = 0. \end{aligned}$$

To eliminate x_1 from these equations, we multiply the first equation $(\texttt{D} - 1)\, x_1 - 2\, x_2 = 0$ by 2, which leads to

$$2(\texttt{D} - 1)\, x_1 - 4\, x_2 = 0.$$

Then we operate on the equation $-2\, x_1 + (\texttt{D} + 2)\, x_2 = 0$ with $\texttt{D} - 1$ to obtain

$$-2(\texttt{D} - 1)\, x_1 + (\texttt{D} - 1)(\texttt{D} + 2)\, x_2 = 0.$$

Adding these equation eliminates x_1 and yields

$$-4\, x_2 + (\texttt{D} - 1)(\texttt{D} + 2)\, x_2 = 0 \quad \text{or} \quad (\texttt{D}^2 + \texttt{D} - 6)\, x_2 = 0.$$

This constant coefficient differential equation has the characteristic polynomial $\chi(\lambda) = \lambda^2 + \lambda - 6 = (\lambda + 3)(\lambda - 2)$ with roots $\lambda_1 = -3$ and $\lambda_2 = 2$. Therefore, its general solution is

$$x_2(t) = c_1\, e^{-3t} + c_2\, e^{2t}$$

for some arbitrary constants c_1 and c_2. The initial condition $x_2(0) = 0$ is not enough to determine these two arbitrary constants—there remains a relation $c_1 = -c_2$ between constants. From the equation $-2\, x_1 + (\texttt{D}+2)\, x_2 = 0$, it follows that

$$\begin{aligned} 2\, x_1 &= (\texttt{D} + 2)\, x_2 = (\texttt{D} + 2)\, \left[c_1\, e^{-3t} + c_2\, e^{2t}\right] \\ &= \texttt{D}\left[c_1\, e^{-3t} + c_2\, e^{2t}\right] + 2\left[c_1\, e^{-3t} + c_2\, e^{2t}\right] \end{aligned}$$

or

$$2\, x_1 = -3\, c_1\, e^{-3t} + 2\, c_2\, e^{2t} + 2\, c_1\, e^{-3t} + 2\, c_2\, e^{2t} = -c_1\, e^{-3t} + 4\, c_2\, e^{2t}.$$

Hence,

$$x_1(t) = -\frac{1}{2}\, c_1\, e^{-3t} + 2\, c_2\, e^{2t}.$$

Using the given initial conditions, we obtain the system of algebraic equations

$$-\frac{1}{2}\, c_1 + 2\, c_2 = 5, \qquad c_1 + c_2 = 0.$$

Consequently, $c_1 = -2$ and $c_2 = 2$. Therefore,

$$x_1 = e^{-3t} + 4\, e^{2t}, \quad x_2 = -2\, e^{-3t} + 2\, e^{2t}.$$

Example 6.4.4: Let us consider the problem of finding solutions $x(t)$ and $y(t)$ for the system of differential equations of second order

$$\begin{cases} x'' + 3\,x - y' = t, \\ 8\,x' + y'' - 3\,y = 3. \end{cases}$$

Since we don't know any method of solving a system of differential equations, it is reasonable to attack this system by eliminating one dependent variable in order to reduce it to a single equation in the other dependent variable. We denote by $\texttt{D}$ the operator of differentiation, that is, $\texttt{D}u(t) = du/dt = \dot{u}(t)$. This operator assigns the derivative to every smooth function $u(t)$. With this in hand, we rewrite the given system of differential equations in operator form:

$$\begin{cases} (\texttt{D}^2 + 3)x - \texttt{D}y = t, \\ 8\,\texttt{D}x + (\texttt{D}^2 - 3)y = 3. \end{cases}$$

To eliminate the dependent variable x, we apply $8\,\texttt{D}$ to the first equation, $\texttt{D}^2 + 3$ to the second equation, and then subtract the results to obtain

$$\left[-8\,\texttt{D}^2 - (\texttt{D}^2 - 3)(\texttt{D}^2 + 3)\right] y = 8\,\texttt{D}t - (\texttt{D}^2 + 3)3$$

or

$$[-\texttt{D}^4 - 8\,\texttt{D}^2 + 9]y(t) = -1 \qquad \Longrightarrow \qquad [\texttt{D}^4 + 8\,\texttt{D}^2 - 9]y(t) = 1$$

because $\texttt{D}^2 t = 0$, $\texttt{D}t = 1$, and $\texttt{D}\,3 = 0$. A particular solution can be found using the method of undetermined coefficients. In our case, we choose a constant function, $y = a$. Evaluation of $[\texttt{D}^4 + 8\,\texttt{D}^2 - 9](a) = 1$ yields $a = -1/9$.

The general solution of this equation is a sum of the general solution of the corresponding homogeneous equation and a particular solution of the nonhomogeneous equation. Since $\texttt{D}^4 + 8\,\texttt{D}^2 - 9 = (\texttt{D}^2 - 1)(\texttt{D}^2 + 9)$, we have

$$y(t) = -\frac{1}{9} + a_1\, e^t + a_2\, e^{-t} + a_3\, \sin 3t + a_4\, \cos 3t,$$

for some constants a_1, a_2, a_3, and a_4. The function $y(t)$ may be eliminated in a similar manner, which leads to the equation

$$\left[(\texttt{D}^2 + 3)(\texttt{D}^2 - 3) + 8\,\texttt{D}^2\right] x(t) = (\texttt{D}^2 - 3)t + \texttt{D}3$$

or

$$\left[\texttt{D}^4 + 8\,\texttt{D}^2 - 9\right] x(t) = -3t.$$

Therefore its general solution is

$$x(t) = \frac{t}{3} + b_1\, e^t + b_2\, e^{-t} + b_3\, \sin 3t + b_4\, \cos 3t\,.$$

The constants a_i and b_i $i = 1, 2, 3, 4$, in these expressions for $x(t)$ and $y(t)$ should be chosen to satisfy the given system of differential equations. We substitute these functions into the corresponding homogeneous system to obtain

$$(4a_1 - b_1)e^t + (4a_2 + b_2)e^{-t} - 3(2a_3 - b_4)\sin 3t - 3(2a_4 + b_3)\cos 3t = 0.$$

The linearly independence of the functions e^t , e^{-t} , $\sin 3t$, and $\cos 3t$ requires that

$$4a_1 - b_1 = 0, \quad 4a_2 + b_2 = 0, \quad 2a_3 - b_4, \quad 2a_4 + b_3 = 0.$$

This allows us to express the b's in terms of the a's and create a general solution that contains four arbitrary constants:

$$\begin{aligned} x(t) &= \frac{t}{3} + 4a_1\, e^t - 4a_2\, e^{-t} - 2a_3\, \sin 3t + 2a_3\, \cos 3t, \\ y(t) &= -\frac{1}{9} + a_1\, e^t + a_2\, e^{-t} + a_3\, \sin 3t + a_4\, \cos 3t\,. \end{aligned}$$

We can reduce the given system of equations to an equivalent normal system of differential equations by setting $p = x'$ and $q = y'$. Then we have

$$\frac{d\mathbf{x}(t)}{dt} = \mathbf{A}\mathbf{x} + \mathbf{f}, \quad \text{where } \mathbf{x} = \begin{bmatrix} x(t) \\ y(t) \\ p(t) \\ q(t) \end{bmatrix}, \quad \mathbf{f} = \begin{bmatrix} 0 \\ 0 \\ t \\ 3 \end{bmatrix}, \quad \mathbf{A} = \begin{bmatrix} 0 & 0 & 1 & 0 \\ 0 & 0 & 0 & 1 \\ -3 & 0 & 0 & 1 \\ 0 & 3 & 8 & 0 \end{bmatrix}. \qquad \square$$

Polynomial Equations

The above elimination procedure works for any linear system of differential equations with constant coefficients regardless of the order of equations and the number of unknowns. We demonstrate this approach in the case of two equations with two unknown variables (which we usually denote by x_1 and x_2 or x and y) of arbitrary order.

Let $\mathtt{D} \stackrel{\text{def}}{=} d/dt$ be the operator of differentiation and $L(\lambda)$ be a constant coefficient polynomial in λ. Substituting $\mathtt{D}$ instead of λ, we obtain a differential operator $L[\mathtt{D}]$ or simply L. This operator is a linear differential operator, meaning that the following properties hold:

- $L[\mathtt{D}](x+y) = L[\mathtt{D}]x + L[\mathtt{D}]y$ for any two smooth functions $x(t)$ and $y(t)$.
- $L[\mathtt{D}](\alpha x) = \alpha L[\mathtt{D}]x$ for any constant α and a smooth function $x(t)$.

Let $L_1(\lambda)$, $L_2(\lambda)$, $L_3(\lambda)$, and $L_4(\lambda)$ be polynomials with constant coefficients. With these four polynomials, we consider the following system of two equations with two unknowns:

$$\begin{cases} L_1[\mathtt{D}]x_1 + L_2[\mathtt{D}]x_2 = f_1(t), \\ L_3[\mathtt{D}]x_1 + L_4[\mathtt{D}]x_2 = f_2(t), \end{cases} \tag{6.4.1}$$

where $f_1(t)$ and $f_2(t)$ are given functions and the variables $x_1(t)$ and $x_2(t)$ are to be determined. Since these four operators $L_1[\mathtt{D}]$, $L_2[\mathtt{D}]$, $L_3[\mathtt{D}]$, and $L_4[\mathtt{D}]$ are constant coefficient linear differential operators, they commute (i.e., $L_1L_2 = L_2L_1$) and we can eliminate one of the two variables x_1 or x_2. To do this, we multiply the first equation by $L_4[\mathtt{D}]$ and the second equation by $L_2[\mathtt{D}]$ to obtain

$$L_4L_1x_1 + L_4L_2x_2 = L_4f_1, \qquad L_2L_3x_1 + L_2L_4x_2 = L_2f_2.$$

Here we dropped explicit dependence on the derivative operator $\mathtt{D}$ and write L instead of $L[\mathtt{D}]$. Subtracting these results, we get the equation for one unknown variable: $L_1L_4x_1 - L_2L_3x_1 \equiv (L_1L_4 - L_2L_3)x_1 = L_4f_1 - L_2f_2$. Similarly, we obtain the equation for x_2, yielding two separate differential equations:

$$(L_1L_4 - L_2L_3)\,x_1 = L_4f_1 - L_2f_2, \qquad (L_3L_2 - L_1L_4)\,x_2 = L_3f_1 - L_1f_2.$$

Obviously, these two single differential equations may have a solution only when $L_1L_4 - L_2L_3 \neq 0$. This forces us to introduce another definition.

Definition 6.9: The operator

$$W[\mathtt{D}] = \det\begin{pmatrix} L_1[\mathtt{D}] & L_2[\mathtt{D}] \\ L_3[\mathtt{D}] & L_4[\mathtt{D}] \end{pmatrix} = L_1L_4 - L_2L_3$$

is called the **Wronskian determinant** or operational determinant of differential operators L_1, L_2, L_3, L_4. If the Wronskian is identically zero, then the system (6.4.1) is said to be **degenerate**.

Since a degenerate system may have either no solution or infinitely many independent solutions, we assume that $L_4L_1 - L_2L_3 \neq 0$. The number of arbitrary constants in the general solution of the system (6.4.1) is equal to the degree of its Wronskian as a polynomial in $\mathtt{D}$.

Example 6.4.5: Let $L_1 = \mathtt{D} - 1 = L_2$, $L_3 = \mathtt{D} + 1 = L_4$, and we consider the degenerate system of the following equations:

$$(\mathtt{D}-1)x + (\mathtt{D}-1)y = e^t, \qquad (\mathtt{D}+1)x + (\mathtt{D}+1)y = e^{-2t}.$$

After multiplying the first equation by $\mathtt{D}+1$ and the second equation by $\mathtt{D}-1$, we find their difference

$$0 = 2e^t + 3\,e^{-2t} \qquad \text{or} \qquad 2\,e^{3t} = -3.$$

Since the latter equation has no solutions (the exponential function cannot be negative), we claim that the given system of equations has no solution.

Example 6.4.6: The two-loop circuit considered in Example 6.1.2, page 343, leads to the following system of equations:

$$\begin{cases} \frac{1}{C}\, q + R_1\, I_1 - R_1\, I_2 = V(t), \\ -R_1\, I_1 + (L\,\mathtt{D} + R_1 + R_2)\, I_2 = 0, \\ \mathtt{D}\, q = I_1, \end{cases}$$

where $\mathtt{D} = \mathrm{d}/dt$ is the derivative operator. Eliminating I_1, we get the system with two equations:

$$\begin{cases} \left(R_1\,\mathtt{D} + C^{-1}\right) q - R_2\, I_2 = V(t), \\ -R_1\,\mathtt{D}\, q + (L\,\mathtt{D} + R_1 + R_2)\, I_2 = 0. \end{cases}$$

The above system of equations is of the form (6.4.1) with $L_1(\lambda) = R_1\lambda + C^{-1}$, $L_2(\lambda) = -R_2$, $L_3(\lambda) = -R_1\lambda$, and $L_4(\lambda) = L\lambda + R_1 + R_2$. Since

$$L_1\, L_4 - L_2\, L_3 = R_1 L\, \lambda^2 + \lambda\left(LC^{-1} + R_1^2\right) + C^{-1}\left(R_1 + R_2\right),$$

the system of differential equations with two unknowns $q(t)$ and $I_2(t)$ is not degenerate.

Example 6.4.7: The degenerate system

$$\mathtt{D}x - 2\,\mathtt{D}y = e^{-t}, \qquad 3\,\mathtt{D}x - 6\,\mathtt{D}y = 3\,e^{-t}$$

has a solution for any choice of $y(t)$ because one of them is a multiple of the other. Actually, this system is equivalent to the single equation $\mathtt{D}x = 2\,\mathtt{D}y(t) + e^{-t}$, which has a solution $x(t) = \int [2y' + e^{-t}]\, \mathrm{d}t = 2\,y(t) - e^{-t} + c$. □

Sometimes a system of n differential equations of the first order can be reduced to a single differential equation of an order less than n.

Example 6.4.8: Reduce the system of differential equations

$$\begin{aligned} \dot{y}_1(t) &= y_2(t) + y_3(t), \\ \dot{y}_2(t) &= y_1(t) + y_3(t), \\ \dot{y}_3(t) &= y_1(t) + y_2(t) \end{aligned}$$

to a single equation and solve it.

Solution. Differentiating the first equation, we obtain

$$\ddot{y}_1 = \dot{y}_2 + \dot{y}_3 = y_1(t) + y_3(t) + y_1(t) + y_2(t) = 2y_1(t) + \dot{y}_1(t)$$

since $y_2 + y_3 = \dot{y}_1$. Hence, we get a second order equation

$$\ddot{y}_1 - \dot{y}_1 - 2y_1 = 0$$

for $y_1(t)$, which has the general solution $y_1 = A\,e^{-t} + B\,e^{2t}$, containing two arbitrary constants A and B. Similarly, we can obtain second order differential equations for $y_2(t)$ and $y_3(t)$. However, the general solution of the given system of equations contains three arbitrary constants. For example, $y_1 = \left(e^{2t} + 2\,e^{-t}\right) c_1 + \left(e^{2t} - e^{-t}\right)(c_2 + c_3)$. Constants c_2 and c_3 are arbitrary, but we cannot use their sum $c_2 + c_3$ as a new arbitrary constant because other functions y_2, y_3 utilize these constants differently. For instance, $y_2 = \left(e^{2t} - e^{-t}\right)(c_1 + c_3) + \left(e^{2t} + 2\,e^{-t}\right) c_2$.

Problems

In all problems, $\mathtt{D}$ stands for the derivative operator, while $\mathtt{D}^0$, the identity operator, is omitted. The derivatives with respect to t are denoted by dots.

1. Classify each of the systems as linear or nonlinear.
   (a) $\dot{x} + \dot{y} = \sin t, \quad t\,\dot{x} + 2\,\dot{y} = x + y$. **(b)** $\ddot{x} + \dot{y} = t, \quad t\,\ddot{y} + x = t^2$.
   (c) $\dot{x} = t + x^2 + y, \quad \dot{y} = t^2 + x + y$. **(d)** $\dot{x} = t^2 + y, \quad \dot{y} = \sin t + x$.
   (e) $\dot{x} = x^2 + y^2, \quad \dot{y} = y^2 - x^2$. **(f)** $\dot{x} = x + y, \quad \dot{y} = y - 2x$.
2. Let $L = t\,\mathtt{D}^2 + 2\,\mathtt{D}$, where $\mathtt{D} = \mathrm{d}/dt$ is the derivative with respect to t. Find
   (a) $L(1)$, **(b)** $L(t)$, **(c)** $L(t^2)$, **(d)** $L(t^{-1})$.

3. Let $L = t^2\,\mathtt{D} + 2\,\mathtt{D}^2$, where $\mathtt{D} = \mathrm{d}/\mathrm{d}t$ is the derivative with respect to t. Find
   (a) $L(1)$, **(b)** $L(t)$, **(c)** $L(t^2)$, **(d)** $L(t^{-1})$.

4. Reduce each system to normal form with a single first derivative.

   (a) $\dot{x} + \dot{y} = y, \quad \dot{x} - \dot{y} = x.$ **(b)** $\dot{x} + 2\,\dot{y} = t, \quad \dot{x} - \dot{y} = x + y.$

   (c) $\dot{x} - \dot{y} = x + y - t, \quad 2\,\dot{x} + 3\,\dot{x} = 2x + 6.$

   (d) $2\,\dot{x} - \dot{y} = t, \quad 3\,\dot{x} + 2\,\dot{y} = y.$ **(e)** $5\,\dot{x} - 3\,\dot{y} = x + y, \quad 3\,\dot{x} - \dot{y} = t.$

   (f) $\dot{x} - 4\,\dot{y} = 0, \quad 2\,\dot{x} - 3\,\dot{y} = t + y.$ **(g)** $3\,\dot{x} + 2\,\dot{y} = \sin t, \quad \dot{x} - 2\,\dot{y} = x + t + y.$

5. Use elimination by operator multiplication to get rid of one of the dependent variables in the following systems of differential equations.

   (a) $\begin{cases} (\mathtt{D}-4)x + 3y = t, \\ -6x + (\mathtt{D}+7)y = 0; \end{cases}$ (b) $\begin{cases} (\mathtt{D}^2+5\,\mathtt{D}+6)x + \mathtt{D}(\mathtt{D}+1)y = t, \\ (\mathtt{D}+1)x + (\mathtt{D}+2)y = 0; \end{cases}$

   (c) $\begin{cases} (\mathtt{D}+2)x + \mathtt{D}y = \sin t, \\ (\mathtt{D}-3)x + (\mathtt{D}-2)y = 0; \end{cases}$ (d) $\begin{cases} (\mathtt{D}^2+2\,\mathtt{D}+1)x + (\mathtt{D}-1)y = 0, \\ (\mathtt{D}+2)x - (\mathtt{D}^2+4\,\mathtt{D}+3)y = t; \end{cases}$

   (e) $\begin{cases} (\mathtt{D}^2-4)x - 90\,y = 4t, \\ \mathtt{D}x + (\mathtt{D}+2)y = 1; \end{cases}$ (f) $\begin{cases} (\mathtt{D}^2+1)x + \mathtt{D}^2 y = 2\,e^{-t}, \\ (\mathtt{D}^2-1)x + \mathtt{D}y = 0; \end{cases}$

   (g) $\begin{cases} (\mathtt{D}^2-1)x + (\mathtt{D}+1)y = t, \\ (\mathtt{D}-1)x + \mathtt{D}y = e^{-t}; \end{cases}$ (h) $\begin{cases} (\mathtt{D}^2+1)x + (\mathtt{D}^2+2)y = 2\,e^{-t}, \\ (\mathtt{D}^2-1)x + \mathtt{D}^2 y = 0; \end{cases}$

   (i) $\begin{cases} \mathtt{D}^2-3\,\mathtt{D}-2)x + (\mathtt{D}^2-\mathtt{D}-2)y = 0, \\ (2\,\mathtt{D}^2-9\,\mathtt{D}-5)x + (3\,\mathtt{D}^2-2\,\mathtt{D}-1)y = 0. \end{cases}$

6. Show that each of the following systems has no solution.

   (a) $\begin{cases} (\mathtt{D}-2)x_1 + 2\,\mathtt{D}x_2 = t, \\ (2\,\mathtt{D}-4)x_1 + 4\,\mathtt{D}x_2 = t; \end{cases}$ (b) $\begin{cases} (\mathtt{D}-2)x_1 + 2\,\mathtt{D}x_2 = t, \\ (\mathtt{D}^2-2\,\mathtt{D})x_1 + 2\,\mathtt{D}^2 x_2 = 0; \end{cases}$

   (c) $\begin{cases} \mathtt{D}x_1 + (\mathtt{D}-5)x_2 = e^t, \\ 2\,\mathtt{D}x_1 + (2\,\mathtt{D}-10)x_2 = e^t; \end{cases}$ (d) $\begin{cases} (\mathtt{D}-1)x_1 + (\mathtt{D}+1)x_2 = e^t, \\ (2\,\mathtt{D}-2)x_1 + (2\,\mathtt{D}+2)x_2 = t^2. \end{cases}$

7. Find the general solution of the given system of differential equations by transforming it into a single equation for one unknown variable.

   (a) $\begin{cases} u' = 4u - v, \\ v' = -4u + 4v; \end{cases}$ (c) $\begin{cases} \dot{x} + x - y = t, \\ x + \dot{y} + 3y = \sin t; \end{cases}$ (e) $\begin{cases} x' = 4x + 2y, \\ y' = -x + y; \end{cases}$

   (b) $\begin{cases} w' = w - y - z, \\ y' = y + 3z, \\ z' = 3y + z; \end{cases}$ (d) $\begin{cases} \dot{x} = 3x + 4y, \\ \dot{y} = -2x - 3y; \end{cases}$ (f) $\begin{cases} x' = 2x - 3y, \\ y' = x - 2y. \end{cases}$

8. Solve the system of ordinary differential equations, where D denotes the derivative.

   (a) $\begin{cases} (\mathtt{D}^2-1)x_1 + x_2 = 0, \\ 4(\mathtt{D}-1)x_1 + (\mathtt{D}+1)x_2 = 0; \end{cases}$ (b) $\begin{cases} (\mathtt{D}-1)x + (\mathtt{D}+2)y = 0, \\ \mathtt{D}x + (\mathtt{D}-1)y = 0; \end{cases}$

   (c) $\begin{cases} (2\,\mathtt{D}+8)x + (\mathtt{D}-1)y = 0, \\ (\mathtt{D}+9)x + \mathtt{D}y = 9; \end{cases}$ (d) $\begin{cases} (\mathtt{D}^2-4)x - 8y = 4t, \\ (\mathtt{D}-2)x - (\mathtt{D}+2)y = 1. \end{cases}$

9. Solve the initial value problem

$$(\mathtt{D}-4)x + (\mathtt{D}-1)y = 3e^t, \quad (4-\mathtt{D})x + (\mathtt{D}+2)y = e^t, \qquad x(0) = y(0) = 1.$$

10. Transfer the following systems of equations (don't solve them!) into equivalent systems of first order differential equations in normal form:

    (a) $(\mathtt{D}-2)x + \mathtt{D}y = 0, \quad (\mathtt{D}^2-2\mathtt{D})x + y = 4;$ **(b)** $\mathtt{D}x - (\mathtt{D}^2-1)y = 0, \quad x + (\mathtt{D}+1)y = \sin t.$

6.5 Existence and Uniqueness

Since higher order not degenerate systems of differential equations can be converted into equivalent first order systems, we do not lose any generality by restricting our attention to the first order case throughout. "Equivalent" means that each solution to the higher order equation uniquely corresponds to a solution to the first order system and vice versa. Recall that $|a, b|$ denotes any interval (open, closed, or semi-closed) with endpoints a, b.

A first order system of ordinary differential equations has the general form

$$\frac{du_1}{dt} = f_1(t, u_1, \ldots, u_n), \quad \cdots \quad , \frac{du_n}{dt} = f_1(t, u_1, \ldots, u_n). \tag{6.5.1}$$

The unknowns $u_1(t)$, ..., $u_n(t)$ are scalar functions of a real variable t, which usually represents time. The right-hand side functions $f_1(t, u_1, \ldots, u_n)$, ..., $f_n(t, u_1, \ldots, u_n)$ are given functions of $n+1$ variables. It is customary to denote the derivative with respect to the time variable by a dot, that is, $\frac{du}{dt} = \dot{u}$. By introducing column vectors $\mathbf{u} = \langle u_1, u_2, \ldots, u_n \rangle^T$, $\mathbf{f} = \langle f_1, f_2, \ldots, f_n \rangle^T$, we rewrite Eq. (6.5.1) in a vector form:

$$\frac{d\mathbf{u}}{dt} = \mathbf{f}(t, \mathbf{u}) \qquad \text{or} \qquad \dot{\mathbf{u}} = \mathbf{f}(t, \mathbf{u}). \tag{6.5.2}$$

Definition 6.10: A **solution** to a system of differential equations (6.5.1) on an interval $|a, b|$ is a vector function $\mathbf{u}(t)$ with n components that are continuously differentiable on the interval $|a, b|$; moreover, $\mathbf{u}(t)$ satisfies the given vector equation on its interval of definition. Each solution $\mathbf{u}(t)$ serves to parameterize a curve in n-dimensional space, also known as a trajectory, streamline, or orbit of the system.

When we seek a particular solution that starts at the specified point, we impose the initial conditions

$$u_1(t_0) = u_{10}, \; u_2(t_0) = u_{20}, \quad \ldots \quad , u_n(t_0) = u_{n0} \qquad \text{or} \qquad \mathbf{u}(t_0) = \mathbf{u}_0. \tag{6.5.3}$$

Here t_0 is a prescribed initial time, while the column vector $\mathbf{u}_0 = \langle u_{10}, u_{20}, \ldots, u_{n0} \rangle^T$ fixes the initial position of the desired solution. In favorable situations, to be formulated shortly, the initial conditions serve to uniquely specify a solution to the differential system of equations—at least for nearby times. A system of equations (6.5.2) together with the initial conditions (6.5.3) form the **initial value problem** or the Cauchy problem.

Definition 6.11: A system of differential equations is called **autonomous** if the right-hand side does not explicitly depend upon the time t and so takes the form

$$\frac{d\mathbf{u}}{dt} = \mathbf{f}(\mathbf{u}) \qquad \text{or} \qquad \dot{\mathbf{u}} = \mathbf{f}(\mathbf{u}). \tag{6.5.4}$$

One important class of autonomous first order systems is steady state fluid flows, where $\mathbf{v} = \mathbf{f}(\mathbf{u})$ represents the fluid velocity at the position $\mathbf{u}$. The solution $\mathbf{u}(t)$ to the autonomous equation (6.5.4) subject to the initial condition $\mathbf{u}(t_0) = \mathbf{a}$ describes the motion of a fluid particle that starts at position $\mathbf{a}$ at time t_0. The vector differential equation (6.5.4) tells us that the fluid velocity at each point on the particle's trajectory matches the prescribed vector field generated by $\mathbf{f}$.

Theorem 6.3: Let $\mathbf{f}(t, \mathbf{u})$ be a continuous function. Then the initial value problem

$$\dot{\mathbf{u}} = \mathbf{f}(t, \mathbf{u}), \qquad \mathbf{u}(t_0) = \mathbf{u}_0 \tag{6.5.5}$$

admits a solution $\mathbf{u}(t)$ that is, at least, defined for nearby times, i.e., when $|t - t_0| < \delta$ for some positive δ.

Theorem 6.3 guarantees that the solution to the initial value problem exists in some neighborhood of the initial position. However, the interval of existence of the solution might be much larger. It is called the **validity interval**, and it is barred by singularities that the solution may have. So the interval of existence can be unbounded, possibly infinite, $-\infty < t < \infty$. Note that the existence theorem 6.3 can be readily adapted to any higher order system of ordinary differential equations by introducing additional variables and converting it into an equivalent first order system. The next statement is simply a reformulation of Picard's existence-uniqueness theorem 1.3 (page 23) in the vector case.

Theorem 6.4: Let $\mathbf{f}(t, \mathbf{u})$ be a vector-valued function with n components. If $\mathbf{f}$ is continuous in some domain and satisfies the Lipschitz condition

$$\| \mathbf{f}(t, \mathbf{u}_1) - \mathbf{f}(t, \mathbf{u}_2) \| \leqslant L \cdot \| \mathbf{x}_1 - \mathbf{x}_2 \|, \qquad L \text{ is a positive constant,}$$

then the initial value problem

$$\dot{\mathbf{u}} = \mathbf{f}(t, \mathbf{u}), \quad \mathbf{u}(t_0) = \mathbf{u}_0 \tag{6.5.6}$$

has a unique solution on some open interval containing the point t_0. Here $\|\mathbf{x}\| = \sqrt{x_1^2 + x_2^2 + \cdots + x_n^2}$ is the norm (length) of the column vector $\mathbf{x} = \langle x_1, x_2, \ldots, x_n \rangle^T$.

PROOF: It turns out that with the aid of vector notation, the proof of this theorem can be established using Picard's iteration method. Rewriting the given initial value problem in an equivalent integral form

$$\mathbf{u}(t) = \mathbf{u}_0 + \int_{t_0}^{t} \mathbf{f}(s, \mathbf{u}(s))\, \mathrm{d}s, \tag{6.5.7}$$

we find the required solution as the limit of the sequence of functions

$$\mathbf{u}(t) = \lim_{n \to \infty} \phi_n(t),$$

where the sequence $\{\phi_n(t)\}_{n \geqslant 0}$ is defined recursively by

$$\phi_0(t) = \mathbf{u}_0, \quad \phi_n(t) = \mathbf{u}_0 + \int_{t_0}^{t} \mathbf{f}(s, \phi_{n-1}(s))\, \mathrm{d}s, \qquad n = 1, 2, 3, \ldots. \tag{6.5.8}$$

Theorem 6.5: If $\mathbf{f}(t, \mathbf{u})$ is a holomorphic function (that can be represented by a convergent power series), then all solutions $\mathbf{u}(t)$ of Eq. (6.5.2) are holomorphic.

Corollary 6.1: If each of the components $f_1(t, \mathbf{u}), \ldots, f_n(t, \mathbf{u})$ of the vector function $\mathbf{f}(t, \mathbf{u}) = \langle f_1, f_2, \ldots, f_n \rangle^T$ and the partial derivatives $\partial f_1/\partial u_1, \ldots, \partial f_1/\partial u_n, \partial f_2/\partial u_1, \ldots, \partial f_n/\partial u_n$ are continuous in an $(n+1)$ region containing the initial point (6.5.3), then the initial value problem (6.5.6) has a unique solution in some neighborhood of the initial point. ■

As a first consequence, we find that the solutions of the autonomous system (6.5.4) are uniquely determined by its initial data. So the solution trajectories do not vary over time: the functions $\mathbf{u}(t)$ and $\mathbf{u}(t-a)$ parametrize the same curve in the n-dimensional space. All solutions passing through the point $\mathbf{u}_0$ follow the same trajectory, irrespective of the time they arrive there; therefore, orbits cannot touch or cross each other. For a linear system of differential equations we have a stronger result.

Theorem 6.6: Let the $n \times n$ matrix-valued function $\mathbf{P}(t)$ and the vector-valued function $\mathbf{g}(t)$ be continuous on the (bounded or unbounded) open interval (a, b) containing the point t_0. Then the initial value problem

$$\dot{\mathbf{x}} = \mathbf{P}(t)\mathbf{x} + \mathbf{g}(t), \quad \mathbf{x}(t_0) = \mathbf{x}_0, \quad t_0 \in (a, b),$$

has a continuous vector-valued solution $\mathbf{x}(t)$ on the interval (a, b).

Example 6.5.1: Consider the following initial value problem:

$$\dot{x}(t) = -y^3(t), \quad \dot{y}(t) = x^3(t), \qquad x(0) = 1, \quad y(0) = 0.$$

Since the corresponding system is autonomous, all conditions of Theorem 6.4 are satisfied. We denote its unique solution by $x(t) = \mathrm{cq}(t)$ and $y(t) = \mathrm{sq}(t)$, called the *cosquine* and *squine* functions, respectively. It is not hard to verify that $\mathrm{cq}^4(t) + \mathrm{sq}^4(t) \equiv 1$ (for all t).

Example 6.5.2: (Jacobi Elliptic Functions) Consider the initial value problem

$$\dot{x} = y\,z,\ \dot{y} = -x\,z,\ \dot{z} = -k^2 x\,y, \qquad x(0) = 0,\ y(0) = 1,\ z(0) = 1,$$

where dots stand for derivatives with respect to t, and k denotes a positive constant $\leqslant 1$. The parameter k is known as the *modulus*; its *complementary modulus* is $\kappa = \sqrt{1-k^2}$. Using recurrence (6.5.8), which takes the form

$$x_{(k+1)}(t) = \int_0^t y_{(k)}(s)\,z_{(k)}(s)\,\mathrm{d}s,$$

$$y_{(k+1)}(t) = 1 - \int_0^t x_{(k)}(s)\,z_{(k)}(s)\,\mathrm{d}s,$$

$$z_{(k+1)}(t) = 1 - k^2 \int_0^t x_{(k)}(s)\,y_{(k)}(s)\,\mathrm{d}s,$$

we find some first approximations:

$$\begin{cases} x_{(1)} &= t, \\ y_{(1)} &= 1, \\ z_{(1)} &= 1; \end{cases} \qquad \begin{cases} x_{(2)} &= t, \\ y_{(2)} &= 1 - \frac{t^2}{2!}, \\ z_{(2)} &= 1 - k^2\,\frac{t^2}{2!}; \end{cases} \qquad \begin{cases} x_{(3)} &= t - (1+k^2)\,\frac{t^3}{3!} + 6k^2\,\frac{t^5}{5!}, \\ y_{(3)} &= 1 - \frac{t^2}{2!} + 3k^2\,\frac{t^4}{4!}, \\ z_{(3)} &= 1 - k^2\,\frac{t^2}{2!} + 3k^2\,\frac{t^4}{4!}. \end{cases}$$

If we proceed successively, we find that the coefficients of the various powers of t ultimately become stable, i.e., they remain unchanged in successive iterations. This leads to representation of solutions as convergent power series, which are usually denoted by the symbols $\mathrm{sn}\,t$ (or $\mathrm{sn}\,(t,k)$ or $\mathrm{sn}\,(t,m)$ for $m = k^2$), $\mathrm{cn}\,t$, and $\mathrm{dn}\,t$, and are called *sine amplitude*, *cosine amplitude*, and *delta amplitude*, respectively. There are some famous relations: $\mathrm{cn}^2 t + \mathrm{sn}^2 t = 1$ and $\mathrm{dn}^2 = 1 - k^2\mathrm{sn}^2 t$. These functions were introduced by Carl Jacobi[67] in 1829.

Example 6.5.3: Consider the initial value problem for the first Painlevé[68] equation:

$$\ddot{y} = 6y^2 + t \qquad y(0) = 0, \quad y(0) = 1/2.$$

First, we convert the given Painlevé equation to the system of first order equations and write corresponding Picard's iterations:

$$\frac{\mathrm{d}}{\mathrm{d}t}\begin{bmatrix} y \\ v \end{bmatrix} = \begin{bmatrix} v \\ 6y^2 + t \end{bmatrix}, \quad \Longrightarrow \quad y_{k+1}(t) = \int_0^t v_k(s)\,\mathrm{d}s, \quad v_{k+1}(t) = \frac{1+t^2}{2} + 6\int_0^t y_k^2(s)\,\mathrm{d}s,$$

$k = 0, 1, 2, \ldots$. Integrating, we find a few first Picard approximations:

$$y_1 = \frac{t}{2}, \quad y_2 = \frac{t}{2} + \frac{t^3}{6}, \quad y_3 = \frac{t}{2} + \frac{t^3}{6} + \frac{t^4}{8}.$$

Problems

1. For what values of parameter p does the following initial value problem have a unique solution?

$$\dot{x}(t) = -y^{p-1}(t), \quad \dot{y}(t) = x^{p-1}(t), \qquad x(0) = 1, \quad y(0) = 1.$$

2. Verify that the following two initial value problems define trigonometric functions. What are they?

$$\begin{cases} \dot{x}(t) = -y(t), \quad \dot{y}(t) = x(t), \\ x(0) = 1, \quad y(0) = 0; \end{cases} \qquad \begin{cases} \dot{u}(t) = v^2(t), \quad \dot{v}(t) = u(t)\,v(t), \\ u(0) = 0, \quad v(0) = 1. \end{cases}$$

3. Verify that Jacobi elliptic functions $x = \mathrm{sn}\,(t,k)$, $y = \mathrm{cn}\,(t,k)$, and $z = \mathrm{dn}\,(t,k)$ satisfy the first order equations

$$\dot{x}^2 = \left(1 - x^2\right)\left(1 - k^2 x^2\right), \quad \dot{y}^2 = \left(1 - y^2\right)\left(\kappa^2 + k^2 y^2\right), \quad \dot{z}^2 = \left(1 - z^2\right)\left(z^2 - \kappa^2\right).$$

4. For the following second order differential equations subject to the initial conditions $y(0) = 1$, $y'(0) = 0$, find the first four Picard's iterations.
   (a) $y'' = 2x\,y' - 10y$; **(b)** $y'' + x\,y' = x^2 y$;
   (c) $y'' = 3x^2\,y' - y$; **(d)** $y'' = (1+x^2)\,y' - 6xy$.

---

[67] Carl Gustav Jakob Jacobi (1804–1851) was a German mathematician who made fundamental contributions to elliptic functions, dynamics, differential equations, and number theory. Jacobi was the first Jewish mathematician to be appointed professor at a German university.

[68] Paul Painlevé (1863–1933) was a French mathematician and politician. He served twice as Prime Minister of the Third French Republic.

Summary for Chapter 6

1. A **matrix** is a rectangular array of objects or entries, written in rows and columns. The numbers of rows and columns of a matrix are called its **dimensions**.
2. The matrix which is obtained from $m \times n$ matrix $\mathbf{A} = [a_{ij}]$ by interchanging rows and columns is called the **transpose** of $\mathbf{A}$ and is usually denoted by $\mathbf{A}^T$ or $\mathbf{A}^t$ or even $\mathbf{A}$'. Thus $\mathbf{A}^T = [a_{ji}]$. A matrix is called **symmetric** if $\mathbf{A}^T = \mathbf{A}$; that is, $a_{ij} = a_{ji}$.
3. The complex conjugate of the matrix $\mathbf{A} = [a_{ij}]$, denoted by $\overline{\mathbf{A}}$, is the matrix obtained from $\mathbf{A}$ by replacing each element a_{ij} by its conjugate $\overline{a}_{ij}$. The **adjoint** of the $m \times n$ matrix $\mathbf{A}$ is the transpose of its conjugate matrix and is denoted by $\mathbf{A}^*$, that is, $\mathbf{A}^* = \overline{\mathbf{A}}^T$. A matrix is called **self-adjoint** or **Hermitian** if $\mathbf{A}^* = \mathbf{A}$, that is, $\overline{a}_{ij} = a_{ji}$.
4. The **trace** of an $n \times n$ matrix $\mathbf{A} = [\,a_{ij}\,]$, denoted by $\operatorname{tr}(\mathbf{A})$ or $\operatorname{tr}\mathbf{A}$, is the sum of its diagonal elements, that is, $\operatorname{tr}(\mathbf{A}) = a_{11} + a_{22} + \cdots + a_{nn}$.
5. The first order system of linear differential equations in **normal form**:

$$\frac{d\mathbf{x}}{dt} = \mathbf{A}\,\mathbf{x} + \mathbf{f} \qquad \text{or} \qquad \dot{\mathbf{x}} = \mathbf{A}\,\mathbf{x} + \mathbf{f}.$$

6. The number of equations is called the **dimension** of the system. The system is said to be **homogeneous** or **undriven** if $f_1(t) = f_2(t) = \ldots = f_n(t) \equiv 0$. Otherwise the system is **nonhomogeneous** or **driven** and the vector function $\mathbf{f}(t)$ is referred to as a **nonhomogeneous term** or **forcing function** or **input**.
7. A second or higher order differential equation can be reduced to the equivalent system of equations in normal form.
8. The system

$$\begin{cases} L_1[\mathtt{D}]\,x_1 + L_2[\mathtt{D}]\,x_2 = f_1(t), \\ L_3[\mathtt{D}]\,x_1 + L_4[\mathtt{D}]\,x_2 = f_2(t), \end{cases}$$

where $\mathtt{D} = d/dt$ and $L_1(\lambda)$, $L_2(\lambda)$, $L_3(\lambda)$, $L_4(\lambda)$ are polynomials with constant coefficients, can be reduced to an equation with one unknown variable. We assume that $L_4L_1 - L_2L_3 \neq 0$, otherwise the method fails and the system is said to be **degenerate**. The operator $W = L_4L_1 - L_2L_3$ is called the **Wronskian determinant** of the above system.

9. The existence and uniqueness-existence theorems for a single differential equation are valid for the system of differential equations in normal form.

Review Questions for Chapter 6

Section 6.1 of Chapter 6 (Review)

1. Derive the system of ordinary differential equations for the currents I_1, I_2, and I_3 for each loop in the circuit presented at the right.

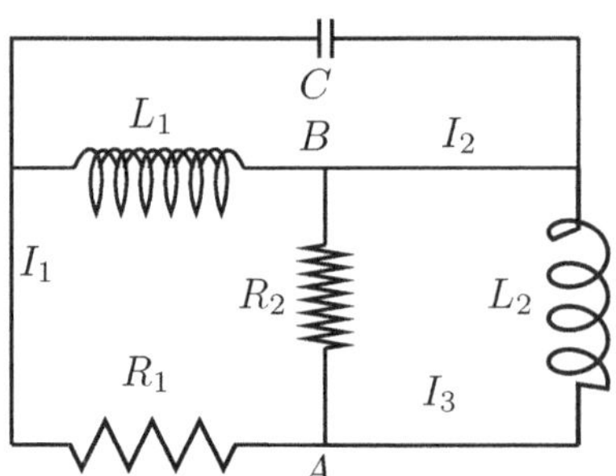

2. A popular child's toy consists of a small rubber ball of mass m attached to a wooden paddle by a rubber band of length ℓ cm. When the ball is launched vertically upward by a paddle with an initial speed v_0, the rubber band is observed to stretch up to L cm when the ball reached its highest point. Assume that the rubber band behaves like a spring obeying Hooke's law for the amount of stretching. Let $y(t)$ be the height of the ball with respect to the paddle, and t_ℓ and t_L represent the times at which the height of the ball is $y(t_\ell) = \ell$ and $y(t_L) = L$. Upon introducing the velocity $v(t) = \dot{y}$, transfer the given model system to the standard system of first order:

$$\begin{aligned} m\ddot{y} &= -mg, \quad 0 < t < t_\ell, \quad y(0) = 0, \ \dot{y}(0) = v_0, \\ m\ddot{y} &= -mg - k\,(y - \ell)\,, \quad t_\ell < t < t_L, \quad y(t_L) = L, \ \dot{y}(t_L) = 0. \end{aligned}$$

Section 6.2 of Chapter 6 (Review)

1. Show that the following 2×2 matrices do not commute.

(a) $\begin{bmatrix} 2 & 3 \\ 1 & 0 \end{bmatrix}$ and $\begin{bmatrix} 3 & -2 \\ 2 & 2 \end{bmatrix}$; (b) $\begin{bmatrix} 2 & 3 \\ 3 & 2 \end{bmatrix}$ and $\begin{bmatrix} 5 & -2 \\ -2 & 2 \end{bmatrix}$;

(c) $\begin{bmatrix} 3 & 4 \\ 5 & 6 \end{bmatrix}$ and $\begin{bmatrix} 3 & 5 \\ 4 & 6 \end{bmatrix}$; (d) $\begin{bmatrix} 1 & 4 \\ 1 & 6 \end{bmatrix}$ and $\begin{bmatrix} 3 & 1 \\ 1 & 1 \end{bmatrix}$;

(e) $\begin{bmatrix} 0 & 4 \\ 4 & 0 \end{bmatrix}$ and $\begin{bmatrix} 0 & 1 \\ 2 & 0 \end{bmatrix}$; (f) $\begin{bmatrix} 2 & 3 \\ 0 & 0 \end{bmatrix}$ and $\begin{bmatrix} 0 & 1 \\ 0 & 2 \end{bmatrix}$.

Section 6.3 of Chapter 6 (Review)

1. Reduce the given differential equations to the first order system of equations. If the equation is linear, identify the matrix $\mathbf{P}(t)$ and the driving term $\mathbf{f}$ (if any) in Eq. (6.3.3) on page 363.

(a) $y'' = y^2 + t^2$; (b) $(t^2+1)\, y'' + 2t\, y' + y = \sin t$;
(c) $ty'' + t^2 y' + t^3 y = 2$; (d) $y'' + y' \ln t - y = e^t$;
(e) $y''/t - t^2\, y' + y \sin t = 0$; (f) $t\, y''' + (t^2+4)\, y'' + y = \cos t$;
(g) $y''' + t\, y'' + t^2\, y' + t^3 = 0$; (h) $y''' + t^2\, y'' + y' \ln t + y = 5$;

2. In each problem, reduce the given linear differential equation with constant coefficients to the vector form, $\dot{\mathbf{x}} = \mathbf{A}\mathbf{x} + \mathbf{f}$, and identify the matrix $\mathbf{A}$ and the forcing vector $\mathbf{f}$ (if any).

(a) $y'' + 6y' + 9y = 2$; (b) $y'' - 4y' + 13y = \sin t$; (c) $y'' + 6y' + 13y = 0$;
(d) $y''' - 6y'' + 12y' - 8y = 0$; (e) $y'' + 2y' + 15y = t$; (f) $y''' - 2y'' - 5y' + 6y = 0$.

3. By setting $v = \dot{x}$, express the following second order nonlinear differential equations as first order systems of dimension 2.

(a) $\ddot{x} + x\,\dot{x} + x = 0$, (c) $\ddot{x} = \cot x + \sin x$, (e) $\ddot{x} + x^2 = 0$,
(b) $\ddot{x} + \dot{x}^3 + \sin x = 0$, (d) $\ddot{x} = x^2 + \dot{x}^2$, (f) $\ddot{x} + \cos x = 0$.

4. A **spherical pendulum** is one that pivots freely about a fixed point in 3-dimensional space (see §6.1.4). To describe its motion, we consider a spherical coordinate system (ℓ, θ, ϕ), where ℓ is the length of the pendulum, $\theta(t)$ is the azimuth angle of the pendulum that it makes with the downward vertical direction, and $\phi(t)$ is a polar angle measured from a fixed zenith direction (usually chosen as x), at time t. With this system, the rectangular coordinates of the mass are then

$$x = \ell \sin\theta\, \cos\phi, \quad y = \ell \sin\theta\, \sin\phi, \quad z = -\ell\cos\theta.$$

(a) Newton's second law applied to the pendulum of mass 1 leads to $\ddot{x} = 0$, $\ddot{y} = 0$, $\ddot{z} = -g$, where g is acceleration due to gravity. By differentiating $x(\theta,\phi)$, $y(\theta,\phi)$, and $z(\theta,\phi)$ with respect to time variable t, show that identities

$$\frac{\ddot{x}}{\ell}\cos\phi + \frac{\ddot{y}}{\ell}\sin\phi = 0, \quad \frac{\ddot{y}}{\ell}\cos\phi - \frac{\ddot{x}}{\ell}\sin\phi = 0$$

lead to the equations

$$\ddot{\theta}\cos\theta = \sin\theta\left(\dot{\theta}^2 + \dot{\phi}^2\right),$$
$$0 = 2\cos\theta\,\dot{\theta}\dot{\phi} + \sin\theta\,\ddot{\phi},$$

respectively.

(b) Show that these equations yield the following system of differential equations:

$$\ddot{\theta} = \dot{\phi}^2 \sin\theta\, \cos\theta - \frac{g}{\ell}\sin\theta,$$
$$\ddot{\phi} = -2\dot{\theta}\dot{\phi}\cot\theta, \qquad \theta \ne n\pi, \ n \text{ is an integer.}$$

Section 6.4 of Chapter 6 (Review)

1. Solve the following systems of homogeneous differential equations.

(a) $\begin{cases} (\mathtt{D}^2 - 8)y - 28\,\mathtt{D}x = 0, \\ \mathtt{D}^2 y + 4x = 0; \end{cases}$ (b) $\begin{cases} (5\,\mathtt{D} + 6)x + (2\,\mathtt{D} + 1)y = 0, \\ (4\,\mathtt{D} + 5)x + (3\,\mathtt{D} + 2)y = 0; \end{cases}$

(c) $\begin{cases} (\mathtt{D}^2 - 4)y - 5x = 0, \\ (\mathtt{D}^2 - 4)x - 5y = 0; \end{cases}$ (d) $\begin{cases} (\mathtt{D}^2 - 3\,\mathtt{D} + 2)x + (\mathtt{D}^2 - 4\,\mathtt{D} + 3)y = 0, \\ (3\,\mathtt{D} - 6)x + (\mathtt{D} - 3)y = 0. \end{cases}$

2. Use elimination by operator multiplication to get rid of one of the dependent variable in the following systems of nonhomogeneous differential equations. Then find the general solution.

(a) $\begin{cases} (3\,\mathtt{D}^2+1)x+(\mathtt{D}^2+3)y=0, \\ (2\,\mathtt{D}^2+1)x+(\mathtt{D}^2+2)y=0; \end{cases}$ (b) $\begin{cases} (\mathtt{D}+2)x+2\,\mathtt{D}y=\sin t, \\ (\mathtt{D}-3)x+(\mathtt{D}-3)y=0; \end{cases}$

(c) $\begin{cases} (\mathtt{D}^2-4)y-x=0, \\ 3\,\mathtt{D}^2y+x=4\,e^{3t}; \end{cases}$ (d) $\begin{cases} (\mathtt{D}-3)x-y=12\,e^{5t}, \\ (\mathtt{D}-2)y-2x=0; \end{cases}$

(e) $\begin{cases} \mathtt{D}^2x+(\mathtt{D}-2)y=13\,\sin t, \\ (\mathtt{D}+1)x-2y=0; \end{cases}$ (f) $\begin{cases} \mathtt{D}^2x-4y=85\,\cos 3t, \\ \mathtt{D}^2y+x=0. \end{cases}$

3. The following systems of equations illustrate the exceptional cases. Solve them by the method of elimination.

(a) $\begin{cases} (\mathtt{D}-1)^2x+(\mathtt{D}+3)^2y=0, \\ (\mathtt{D}+4)^2x+(\mathtt{D}+2)^2y=0; \end{cases}$ (b) $\begin{cases} (2\,\mathtt{D}+2)x+(\mathtt{D}+1)y=\sin t, \\ (\mathtt{D}+1)x-(\mathtt{D}+1)y=0; \end{cases}$

(c) $\begin{cases} (\mathtt{D}-4)x+(2\,\mathtt{D}-2)y=e^{2t}, \\ (\mathtt{D}-3)x+(\mathtt{D}-1)y=0; \end{cases}$ (d) $\begin{cases} (4\,\mathtt{D}-6)x+(2\,\mathtt{D}-3)y=e^{t}, \\ (2\,\mathtt{D}-3)x-(\mathtt{D}-1)y=0; \end{cases}$

(e) $\begin{cases} (3\,\mathtt{D}-6)x-(\mathtt{D}-2)y=16\,e^{2t}, \\ (\mathtt{D}-2)x+(5\,\mathtt{D}-10)y=0; \end{cases}$ (f) $\begin{cases} (3\,\mathtt{D}-3)x+(\mathtt{D}-2)y=4\,e^{t}, \\ (\mathtt{D}-1)x-(\mathtt{D}-2)y=0. \end{cases}$

4. For the following systems of differential equations, determine whether they are degenerate or not.

(a) $\begin{aligned} \dot{x} &= x+2y+z, \\ \dot{y} &= 2y, \\ \dot{z} &= -x-3y-z; \end{aligned}$ (b) $\begin{aligned} \dot{x} &= x+2y+2z, \\ \dot{y} &= 2x+3y+2z, \\ \dot{z} &= -2x-3y-2z; \end{aligned}$ (c) $\begin{aligned} \ddot{x} &= \dot{x}-\dot{y}+z, \\ \ddot{y} &= -\dot{x}+3y-\dot{z}, \\ \ddot{z} &= 2x+7y+\dot{z}; \end{aligned}$

5. When a drug is taken orally, it passes through two primary stages, first the digestive system and then the blood circulatory system. Assuming that on the boundary of these two stages the rate of change of drug concentration is proportional to the concentration present in that stage, we arrive at the following pair of differential equations

$$\begin{aligned} \dot{x} &= -ax+f(t), \qquad x(0)=0, \\ \dot{y} &= bx-cy, \qquad y(0)=0, \end{aligned}$$

where $x(t)$ is the drug concentration in the digestive system and $y(t)$ is the drug concentration in the circulatory system. The function $f(t)$ represents the rate at which the drug concentration is increased in the digestive system by external dosage. Assuming that parameters a, b, and c are constants, solve the equation for $x(t)$ and then substitute its solution into the equation for $y(t)$. Since the resulting equation in x is linear, solve it.

Section 6.5 of Chapter 6 (Review)

1. In each exercise, the initial conditions are assumed homogeneous, that is, $y(0)=\dot{y}(0)=0$.

   (a) Rewrite the given second order scalar initial value problem in vector form (6.5.1) by defining $\mathbf{u}=\langle u_1,u_2\rangle^T$, where $u_1=y$, $u_2=\dot{y}$.

   (b) Compute the four partial derivatives $\partial f_k(t,u_1,u_2)/\partial u_j$, $k,j=1,2$.

   (c) For the system obtained in part (a), determine where in 3-dimensional $t\mathbf{u}$-space the hypotheses of Corollary 6.1 are *not* satisfied.

   (a) $\ddot{y}+(\dot{y})^{1/2}+y^3=t$; **(b)** $\ddot{y}+(2+3y+4\dot{y})^{-1}=\sin t$;
   (c) $\ddot{y}+\sin(t\dot{y})+\cos y=0$; **(d)** $\ddot{y}+y^3/(\dot{y}-1)^{-1}=e^t$.

2. For the following second order differential equations subject to the initial conditions $y(0)=0$, $y'(0)=1$, find the first four Picard iterations.

   (a) $y''+2t\,y=0$; **(b)** $y''+3t^2\,y=0$; **(c)** $y''=t^2\,y'-y$; **(d)** $y''=y'+ty$;
   (e) $y''=t^2+y^2$; **(f)** $y''+t\,y^2=0$; **(g)** $y''=y'-t\,y$; **(h)** $y''=t\,y'+2y$.

Chapter 7

Topics from Linear Algebra

This chapter is devoted to some important topics from linear algebra that play an essential role in the study of systems of differential equations. Our main objective is to define a function of a square matrix. To achieve it, we present in detail four methods: diagonalization, Sylvester's, resolvent, and the spectral decomposition procedure, of which the last three do not require any knowledge about eigenvectors. Note that while we are focusing on these four, there are many other approaches that can be used to define a function of a matrix [22, 35]. In the next chapter, we apply our four techniques to solve vector differential equations of the first order, either homogeneous $\dot{\mathbf{y}} = \mathbf{A}\,\mathbf{y}$ or nonhomogeneous $\dot{\mathbf{y}} = \mathbf{A}\,\mathbf{y} + \mathbf{f}$, and equations of the second order $\ddot{\mathbf{y}} + \mathbf{A}\,\mathbf{y} = \mathbf{0}$, with a square matrix $\mathbf{A}$. In particular, we focus on constructing the fundamental matrix functions: $e^{\mathbf{A}t}$, needed for solution of the first order equations and $e^{\sqrt{\mathbf{A}}t}$, $\mathbf{A}^{-1/2}\sin\left(\mathbf{A}^{1/2}t\right)$, and $\cos\left(\mathbf{A}^{1/2}t\right)$, used in second order equations. This chapter contains many examples of matrix functions. The first four sections give an introduction to linear algebra and lay the foundation needed for future applications.

7.1 The Calculus of Matrix Functions

In this section, we use square matrices whose entries are functions of a real variable t, namely,

$$\mathbf{A}(t) = [a_{ij}(t)] = \begin{bmatrix} a_{11}(t) & a_{12}(t) & \cdots & a_{1n}(t) \\ a_{21}(t) & a_{22}(t) & \cdots & a_{2n}(t) \\ \vdots & \vdots & \ddots & \vdots \\ a_{m1}(t) & a_{m2}(t) & \cdots & a_{mn}(t) \end{bmatrix}.$$

The calculus of matrix functions is based on the definition of the limit:

$$\lim_{t\to t_0} \mathbf{A}(t) = \lim_{t\to t_0} [a_{ij}(t)] = \left[\lim_{t\to t_0} a_{ij}(t)\right] = \mathbf{B} = [b_{ij}].$$

This means that for every component of the $m \times n$ matrix $\mathbf{A}(t)$,

$$\lim_{t\to t_0} a_{ij}(t) = b_{ij}, \qquad 1 \leqslant i \leqslant m,\ 1 \leqslant j \leqslant n.$$

If the limit does not exist for at least one matrix entry, then the limit of the matrix does not exist. For example, $\lim_{t\to 0}\begin{bmatrix} t & t^{-1} \\ t^2 & t^3 \end{bmatrix}$ does not exist because one component t^{-1} fails to have a limit.

A matrix $\mathbf{A}(t)$ is said to be *continuous* on an interval $|\alpha,\beta|$ if each entry of $\mathbf{A}(t)$ is a continuous function on the given interval: $\lim_{t\to t_0} \mathbf{A}(t) = \mathbf{A}(t_0)$ for every $t_0 \in |\alpha,\beta|$. With matrix functions we can operate in a similar way as with functions. For example, we define the definite integral of a matrix function as

$$\int_\alpha^\beta \mathbf{A}(t)\,\mathrm{d}t = \left[\int_\alpha^\beta a_{ij}(t)\,\mathrm{d}t\right].$$

The derivative $\mathrm{d}\mathbf{A}(t)/\mathrm{d}t$ of a matrix $\mathbf{A}(t)$, also denoted by a dot, $\dot{\mathbf{A}}$, is defined as

$$\frac{\mathrm{d}\mathbf{A}}{\mathrm{d}t} = \lim_{h\to 0} \frac{1}{h}\left[\mathbf{A}(t+h) - \mathbf{A}(t)\right] = \left[\lim_{h\to 0} \frac{1}{h}\left(a_{ij}(t+h) - a_{ij}(t)\right)\right] = \left[\frac{\mathrm{d}a_{ij}(t)}{\mathrm{d}t}\right].$$

Many properties and formulas from calculus are extended to matrix functions; in particular,

$$\frac{\mathrm{d}}{\mathrm{d}t}(\mathbf{AB}) = \mathbf{A}\frac{\mathrm{d}\mathbf{B}}{\mathrm{d}t} + \frac{\mathrm{d}\mathbf{A}}{\mathrm{d}t}\mathbf{B},$$

$$\frac{\mathrm{d}}{\mathrm{d}t}(\mathbf{A}+\mathbf{B}) = \frac{\mathrm{d}\mathbf{A}}{\mathrm{d}t} + \frac{\mathrm{d}\mathbf{B}}{\mathrm{d}t}, \qquad \int (\mathbf{A}(t) + \mathbf{B}(t))\,\mathrm{d}t = \int \mathbf{A}(t)\,\mathrm{d}t + \int \mathbf{B}(t)\,\mathrm{d}t,$$

$$\frac{\mathrm{d}}{\mathrm{d}t}(\mathbf{C}\,\mathbf{A}(t)) = \mathbf{C}\,\frac{\mathrm{d}\mathbf{A}}{\mathrm{d}t}, \qquad \int (\mathbf{C}\,\mathbf{A}(t))\,\mathrm{d}t = \mathbf{C}\int \mathbf{A}(t)\,\mathrm{d}t,$$

where $\mathbf{C}$ is a constant matrix.

Example 7.1.1: Let

$$\mathbf{A}(t) = \begin{bmatrix} \cos t & e^t \\ \sin t & t^2 \end{bmatrix}, \qquad \mathbf{B}(t) = \begin{bmatrix} e^{2t} & 1 \\ 2 & t \end{bmatrix}, \qquad \mathbf{C} = \begin{bmatrix} 1 & 2 \\ 0 & 3 \end{bmatrix}.$$

Then

$$\mathbf{AB} = \begin{bmatrix} e^{2t}\cos t + 2\,e^t & \cos t + t\,e^t \\ e^{2t}\sin t + 2\,t^2 & \sin t + t^2 \end{bmatrix}.$$

Differentiation gives

$$\frac{\mathrm{d}\mathbf{A}}{\mathrm{d}t} = \begin{bmatrix} -\sin t & e^t \\ \cos t & 2t \end{bmatrix}, \qquad \frac{\mathrm{d}\mathbf{B}}{\mathrm{d}t} = \begin{bmatrix} 2\,e^{2t} & 0 \\ 0 & 1 \end{bmatrix},$$

$$\frac{\mathrm{d}}{\mathrm{d}t}(\mathbf{AB}) = \begin{bmatrix} 2\,e^{2t}\cos t - e^{2t}\sin t + 2\,e^t & t\,e^t + e^t - \sin t \\ 2\,e^{2t}\sin t + e^{2t}\cos t + 4t & \cos t + 3t^2 \end{bmatrix}.$$

Multiplication of the matrices yields

$$\mathbf{A}\frac{\mathrm{d}\mathbf{B}}{\mathrm{d}t} = \begin{bmatrix} 2\,e^{2t}\cos t & e^t \\ 2\,e^{2t}\sin t & t^2 \end{bmatrix},$$

$$\frac{\mathrm{d}\mathbf{A}}{\mathrm{d}t}\mathbf{B} = \begin{bmatrix} e^t - e^{2t}\sin t & t\,e^t - \sin t \\ e^{2t}\cos t + 2t & \cos t + 2t^2 \end{bmatrix}.$$

Hence

$$\frac{\mathrm{d}}{\mathrm{d}t}(\mathbf{AB}) = \mathbf{A}\frac{\mathrm{d}\mathbf{B}}{\mathrm{d}t} + \frac{\mathrm{d}\mathbf{A}}{\mathrm{d}t}\mathbf{B}. \tag{7.1.1}$$

Similarly,

$$\mathbf{CA} = \begin{bmatrix} \cos t + 2\sin t & e^t + 2t \\ 3\cos t & 6t \end{bmatrix},$$

$$\frac{\mathrm{d}}{\mathrm{d}t}(\mathbf{CA}) = \begin{bmatrix} -\sin t + 2\cos t & e^t + 4t \\ 3\cos t & 6t \end{bmatrix} = \mathbf{C}\frac{\mathrm{d}\mathbf{A}}{\mathrm{d}t}.$$

□

Note that the relation

$$\frac{\mathrm{d}\mathbf{A}^2(t)}{\mathrm{d}t} = 2\mathbf{A}\frac{\mathrm{d}\mathbf{A}(t)}{\mathrm{d}t}$$

is valid if and only if matrices $\mathbf{A}$ and $\frac{\mathrm{d}\mathbf{A}(t)}{\mathrm{d}t}$ commute, that is,

$$\mathbf{A}\frac{\mathrm{d}\mathbf{A}(t)}{\mathrm{d}t} = \frac{\mathrm{d}\mathbf{A}(t)}{\mathrm{d}t}\mathbf{A}.$$

In general, this is not true, as the following example shows.

Example 7.1.2: Let us consider the symmetric matrix $\mathbf{A}(t) = \begin{bmatrix} t & \sin t \\ \sin t & 1 \end{bmatrix}$. Then

$$\frac{\mathrm{d}\mathbf{A}(t)}{\mathrm{d}t} = \begin{bmatrix} 1 & \cos t \\ \cos t & 0 \end{bmatrix} \quad \text{and} \quad \mathbf{A}^2(t) = \begin{bmatrix} t^2 + \sin^2 t & t\sin t + \sin t \\ t\sin t + \sin t & \sin^2 t + 1 \end{bmatrix}.$$

Calculations show that

$$\mathbf{A}\,\frac{\mathrm{d}\mathbf{A}(t)}{\mathrm{d}t} = \begin{bmatrix} t + \sin t\cos t & t\cos t \\ \sin t + \cos t & \sin t\cos t \end{bmatrix}, \quad \frac{\mathrm{d}\mathbf{A}(t)}{\mathrm{d}t}\,\mathbf{A} = \begin{bmatrix} t + \sin t\cos t & \sin t + \cos t \\ t\cos t & \sin t\cos t \end{bmatrix}.$$

On the other hand,

$$\frac{\mathrm{d}\mathbf{A}^2(t)}{\mathrm{d}t} = \begin{bmatrix} 2t + 2\sin t\cos t & \sin t + t\cos t + \cos t \\ \sin t + t\cos t + \cos t & 2\sin t\cos t \end{bmatrix}. \qquad \square$$

Generally speaking, the derivative of $\mathbf{A}^2(t)$ is not equal to $2\mathbf{A}(t)\,\dot{\mathbf{A}}(t)$; however,

$$\frac{\mathrm{d}\mathbf{A}^2(t)}{\mathrm{d}t} = \mathbf{A}\,\frac{\mathrm{d}\mathbf{A}(t)}{\mathrm{d}t} + \frac{\mathrm{d}\mathbf{A}(t)}{\mathrm{d}t}\,\mathbf{A}. \tag{7.1.2}$$

Example 7.1.3: Let $\mathbf{A}(t) = \begin{bmatrix} t^2 - 1 & 0 \\ t^2 - 2 & 3 - t^2 \end{bmatrix}$. Then

$$\frac{\mathrm{d}\mathbf{A}(t)}{\mathrm{d}t} = \begin{bmatrix} 2t & 0 \\ 2t & -2t \end{bmatrix} = t\begin{bmatrix} 2 & 0 \\ 2 & -2 \end{bmatrix}.$$

Therefore,

$$\mathbf{A}\,\frac{\mathrm{d}\mathbf{A}(t)}{\mathrm{d}t} = \frac{\mathrm{d}\mathbf{A}(t)}{\mathrm{d}t}\,\mathbf{A} = 2t\begin{bmatrix} t^2 - 1 & 0 \\ 1 & -3 + t^2 \end{bmatrix}.$$

On the other hand, $\mathbf{A}^2 = \begin{bmatrix} (t^2-1)^2 & 0 \\ 2(t^2-2) & -(3-t^2)^2 \end{bmatrix}$ and

$$\frac{\mathrm{d}\mathbf{A}^2(t)}{\mathrm{d}t} = 4t\begin{bmatrix} t^2 - 1 & 0 \\ 1 & -3 + t^2 \end{bmatrix} = 2\,\mathbf{A}\,\frac{\mathrm{d}\mathbf{A}(t)}{\mathrm{d}t}.$$

Problems

1. For each of the given 2×2 matrix functions, show that $\frac{\mathrm{d}\mathbf{A}^2}{\mathrm{d}t} \ne 2\mathbf{A}\,\frac{\mathrm{d}\mathbf{A}}{\mathrm{d}t}$.

   (a) $\begin{bmatrix} 1 & t \\ t & t^2 \end{bmatrix}$; (b) $\begin{bmatrix} 1 & t^2 \\ t^3 & t^3 \end{bmatrix}$; (c) $\begin{bmatrix} e^t & e^{2t} \\ \sinh t & \sinh 2t \end{bmatrix}$; (d) $\begin{bmatrix} t & 2 \\ 3t & 4t \end{bmatrix}$;

   (e) $\begin{bmatrix} t & 2t+1 \\ 3t & 4t \end{bmatrix}$; (f) $\begin{bmatrix} 2 & 1 \\ 3t & 5t \end{bmatrix}$; (g) $\begin{bmatrix} t^2+1 & t+1 \\ t^2-1 & t-12 \end{bmatrix}$; (h) $\begin{bmatrix} 1 & 2 \\ t^2 & t^3 \end{bmatrix}$.

2. For each of the given 2×2 matrix functions, show that $\frac{\mathrm{d}\mathbf{A}^2}{\mathrm{d}t} = 2\mathbf{A}\,\frac{\mathrm{d}\mathbf{A}}{\mathrm{d}t}$.

   (a) $\begin{bmatrix} t & t \\ t & t \end{bmatrix}$; (b) $\begin{bmatrix} t & \sin t \\ \sin t & t \end{bmatrix}$; (c) $\begin{bmatrix} t & 1 \\ 1 & t \end{bmatrix}$; (d) $\begin{bmatrix} e^t & t \\ t & e^t \end{bmatrix}$;

   (e) $\begin{bmatrix} 2t & 3t \\ 3t & 4t \end{bmatrix}$; (f) $\begin{bmatrix} 1 & e^t \\ e^t & 1 \end{bmatrix}$; (g) $\begin{bmatrix} 3t & t \\ t & 2t \end{bmatrix}$; (h) $\begin{bmatrix} 1/t & 2/t \\ -1/t & -2/t \end{bmatrix}$.

3. Find a formula for the derivative of $\mathbf{P}^3(t)$, where $\mathbf{P}(t)$ is a square matrix.

4. Show that if $\mathbf{A}(t)$ is a differentiable and invertible (Definition 7.3 on page 382) square matrix function, then $\mathbf{A}^{-1}(t)$ is differentiable and $\left(\mathbf{A}^{-1}\right)' = -\mathbf{A}^{-1}\,\mathbf{A}'\,\mathbf{A}^{-1}$. *Hint:* differentiate the identity $\mathbf{A}^{-1}\,\mathbf{A} = \mathbf{I}$.

5. Find $\lim_{t\to 0}\mathbf{A}(t)$ or state why the limit does not exist.

   (a) $\mathbf{A}(t) = \begin{bmatrix} \frac{t}{\sin t} & \sec t \\ \cos t & \tan t \end{bmatrix}$; (b) $\mathbf{A}(t) = \begin{bmatrix} \cot t & \sin t \\ e^{\sin t} & \sinh t \end{bmatrix}$; (c) $\mathbf{A}(t) = \begin{bmatrix} e^{-t} & \tanh t \\ \cos t & \cosh t \end{bmatrix}$.

7.2 Inverses and Determinants

Starting with this section, we will deal with square matrices only. To each square matrix, it is possible to assign a number called the **determinant**[69]. Its definition is difficult, non-intuitive, and generally speaking hard to evaluate numerically when the dimension of the matrix exceeds 10×10. Fortunately, algorithms are known for calculating determinants without actual application of the definition. We begin with 2×2 matrices

$$\mathbf{A} = \begin{bmatrix} a & b \\ c & d \end{bmatrix}.$$

The number $ad - bc$ is called the **determinant** of the 2×2 matrix $\mathbf{A}$, denoted by

$$\det \mathbf{A} = \det \begin{bmatrix} a & b \\ c & d \end{bmatrix} = \begin{vmatrix} a & b \\ c & d \end{vmatrix} = ad - bc.$$

Note that the vertical bars distinguish a determinant from a matrix.

Example 7.2.1: Let

$$\mathbf{A} = \begin{bmatrix} 2 & 3 \\ 5 & 8 \end{bmatrix}.$$

Then its determinant is

$$\det \mathbf{A} = 2 \cdot 8 - 3 \cdot 5 = 1.$$

□

Permutations

Recall that a permutation of a set of the first n integers $\{1, 2, \ldots, n\}$ is a reordering of these integers. For each permutation σ, sign(σ) is $+1$ if σ is even and -1 if σ is odd. Evenness or oddness can be defined as follows: the permutation is even (odd) if the new sequence can be obtained by an even number (odd, respectively) of switches of numbers starting with the initial ordering $\sigma = (1, 2, \ldots, n)$, which has sign$(\sigma) = +1$ (zero switches). For $n = 3$, switching the positions of 2 and 3 yields $(1, 3, 2)$, with sign$(1, 3, 2) = -1$. Switching once more yields $(3, 1, 2)$, with sign$(3, 1, 2) = +1$ again. Finally, after a total of three switches (an odd number), the resulting permutation becomes $(3, 2, 1)$, with sign$(3, 2, 1) = -1$. Therefore $(3, 2, 1)$ is an odd permutation. Similarly, the permutation $(2, 3, 1)$ is even: $(1, 2, 3) \mapsto (2, 1, 3) \mapsto (2, 3, 1)$, with an even (2) number of switches. In general, there are $n!$ permutations of $\{1, 2, \ldots, n\}$.

In general, the **determinant** of a square $n \times n$ matrix $\mathbf{A} = [a_{ij}]$ is the sum of $n!$ terms. Each term is the product of n matrix entries, one element from each row and one element from each column. Furthermore, each product is assigned a plus or a minus sign:

$$\det(\mathbf{A}) = \sum \text{sign}(\sigma)\, a_{1i_1} a_{2i_2} \cdots a_{ni_n},$$

where the summation is over all $n!$ permutations $(i_1, i_2, \ldots, i_n)$ of the integers $1, 2, \ldots, n$ and sign(σ) is ± 1, which is determined by the parity of the permutation. Therefore, half of all the products in this sum are positive and half are negative.

Recursive definition of determinant

Next, we present the recursive definition of the determinant that reduces its evaluation of an $n \times n$ matrix to calculation of determinants of $(n-1) \times (n-1)$ matrices.

Definition 7.1: Let $\mathbf{A}$ be an $n \times n$ matrix. The **minor** of the kj^{th} entry of $\mathbf{A}$ is the determinant of the $(n-1) \times (n-1)$ submatrix obtained from $\mathbf{A}$ by deleting row k and column j:

$$\text{Minor}\,(a_{kj}) = \det \begin{bmatrix} a_{11} & \cdots & a_{1j} & \cdots & a_{1n} \\ \vdots & & & & \vdots \\ a_{k1} & & a_{kj} & & a_{kn} \\ \vdots & & & & \vdots \\ a_{n1} & \cdots & a_{nj} & \cdots & a_{nn} \end{bmatrix}$$

[69]The word determinant was first coined by Cauchy in 1812.

The **cofactor** of the entry a_{kj} is

$$\text{Cof}\,(a_{kj}) = (-1)^{k+j}\text{Minor}\,(a_{kj}).$$

Theorem 7.1: Let $\mathbf{A} = [a_{ij}]$ be a square $n \times n$ matrix. For any row k or any column j we define the **determinant** of the matrix $\mathbf{A}$ by

$$\det(\mathbf{A}) = \sum_{m=1}^{n} a_{mj}\text{Cof}\,(a_{mj}) = \sum_{m=1}^{n} a_{km}\text{Cof}\,(a_{km}). \tag{7.2.1}$$

The appropriate sign in the cofactor is easy to remember since it alternates in the following manner:

$$\begin{vmatrix} + & - & + & - & + & \cdots \\ - & + & - & + & - & \cdots \\ + & - & + & - & + & \cdots \\ \vdots & \vdots & \vdots & \vdots & \vdots & \ddots \end{vmatrix}.$$

Definition 7.2: A matrix $\mathbf{A}$ is called **singular** if $\det \mathbf{A} = 0$ and **nonsingular** if $\det \mathbf{A} \ne 0$.

Example 7.2.2: The determinant of a 3×3 matrix

$$\mathbf{A} = \begin{bmatrix} a_{11} & a_{12} & a_{13} \\ a_{21} & a_{22} & a_{23} \\ a_{31} & a_{32} & a_{33} \end{bmatrix}$$

is

$$\det(\mathbf{A}) = a_{11}a_{22}a_{33} + a_{12}a_{23}a_{31} + a_{21}a_{32}a_{13} - a_{31}a_{22}a_{13} - a_{21}a_{12}a_{33} - a_{11}a_{32}a_{23}.$$

Theorem 7.2: If **A** and **B** are square matrices of the same size, then

- $\det(\mathbf{AB}) = \det(\mathbf{A})\det(\mathbf{B}) = \det(\mathbf{BA})$;
- $\det(\mathbf{A}) = \det\left(\mathbf{A}^T\right)$;
- $\det(\alpha\mathbf{A}) = \alpha^n \det(\mathbf{A})$;
- the determinant of a triangular matrix is the product of its main diagonal elements.

Theorem 7.3: If $\mathbf{A}(t) = [a_{ij}(t)]$ is an $n \times n$ function matrix, then

$$\frac{\text{d}}{\text{d}t}\left(\det \mathbf{A}(t)\right) = \begin{vmatrix} \frac{\text{d}a_{11}(t)}{\text{d}t} & \frac{\text{d}a_{12}(t)}{\text{d}t} & \cdots & \frac{\text{d}a_{1n}(t)}{\text{d}t} \\ \cdots & \cdots & \cdots & \cdots \\ a_{n1}(t) & a_{n2}(t) & \cdots & a_{nn}(t) \end{vmatrix} + \cdots + \begin{vmatrix} a_{11}(t) & a_{12}(t) & \cdots & a_{1n}(t) \\ \cdots & \cdots & \cdots & \cdots \\ \frac{\text{d}a_{n1}(t)}{\text{d}t} & \frac{\text{d}a_{n2}(t)}{\text{d}t} & \cdots & \frac{\text{d}a_{nn}(t)}{\text{d}t} \end{vmatrix},$$

where vertical bars are used to identify the determinants. ■

Recall that the $n \times n$ identity matrix (see Definition 6.6 on page 359), denoted by $\mathbf{I}_n$ or simply by $\mathbf{I}$, has entries of zero in every position in the matrix except for those on the main diagonal. These diagonal entries will always equal one. For example,

$$\mathbf{I}_3 = \begin{bmatrix} 1 & 0 & 0 \\ 0 & 1 & 0 \\ 0 & 0 & 1 \end{bmatrix}.$$

Definition 7.3: If for a square matrix $\mathbf{A}$ there exists a unique matrix $\mathbf{B}$ such that

$$\mathbf{AB} = \mathbf{BA} = \mathbf{I},$$

where $\mathbf{I}$ is the identity matrix, then $\mathbf{A}$ is said to be **invertible** with **inverse B**. We denote this relation as $\mathbf{A}^{-1} = \mathbf{B}$ and vice versa, $\mathbf{B}^{-1} = \mathbf{A}$.

Theorem 7.4: If $\mathbf{A}$ is a nonsingular $n \times n$ matrix, then

$$\mathbf{A}^{-1} = \frac{1}{\det \mathbf{A}} \left[\mathrm{Cof}\,(a_{ji})\right]_{ij} \qquad \text{(transpose of the cofactor matrix).} \tag{7.2.2}$$

Example 7.2.3: Let us find the inverse[70] of the following matrix:

$$\mathbf{A} = \begin{bmatrix} -1 & 2 & 1 \\ 3 & 1 & 4 \\ -2 & 0 & -3 \end{bmatrix}.$$

First, we calculate its determinant: $\det \mathbf{A} = 7$. Hence, the matrix is not singular, and its inverse matrix will be

$$\mathbf{A}^{-1} = \frac{1}{7} \begin{bmatrix} \mathrm{Cof}\,(a_{11}) & \mathrm{Cof}\,(a_{21}) & \mathrm{Cof}\,(a_{31}) \\ \mathrm{Cof}\,(a_{12}) & \mathrm{Cof}\,(a_{22}) & \mathrm{Cof}\,(a_{32}) \\ \mathrm{Cof}\,(a_{13}) & \mathrm{Cof}\,(a_{23}) & \mathrm{Cof}\,(a_{33}) \end{bmatrix},$$

where entries are cofactors of the matrix $\mathbf{A}$. To determine the cofactors, we eliminate the corresponding row and column from the original matrix $\mathbf{A}$ to obtain

$$\begin{aligned}
\mathrm{Cof}\,(a_{11}) &= \mathrm{Cof}(-1) = \det \begin{bmatrix} 1 & 4 \\ 0 & -3 \end{bmatrix} = -3, \\
\mathrm{Cof}\,(a_{12}) &= -\mathrm{Cof}(2) = -\det \begin{bmatrix} 3 & 4 \\ -2 & -3 \end{bmatrix} = 1, \\
\mathrm{Cof}\,(a_{13}) &= \mathrm{Cof}(1) = \det \begin{bmatrix} 3 & 1 \\ -2 & 0 \end{bmatrix} = 2, \\
\mathrm{Cof}\,(a_{21}) &= -\mathrm{Cof}(3) = \det \begin{bmatrix} 2 & 1 \\ 0 & -3 \end{bmatrix} = 6, \\
\mathrm{Cof}\,(a_{22}) &= \mathrm{Cof}(1) = \det \begin{bmatrix} -1 & 1 \\ -2 & -3 \end{bmatrix} = 5,
\end{aligned}$$

and so on. Therefore, the inverse matrix is

$$\mathbf{A}^{-1} = \frac{1}{7} \begin{bmatrix} -3 & 6 & 7 \\ 1 & 5 & 7 \\ 2 & -4 & -7 \end{bmatrix}.$$

Theorem 7.5: If $\mathbf{A}$ and $\mathbf{B}$ are invertible matrices, then

- $\left(\mathbf{A}^{-1}\right)^{-1} = \mathbf{A}$;
- $\left(\mathbf{AB}\right)^{-1} = \mathbf{B}^{-1}\mathbf{A}^{-1}$;
- $\left(\alpha\mathbf{A}\right)^{-1} = \alpha^{-1}\mathbf{A}^{-1}$ for any nonzero constant α;
- $\left(\mathbf{A}^{T}\right)^{-1} = \left(\mathbf{A}^{-1}\right)^{T}$ and $\left(\mathbf{A}^{*}\right)^{-1} = \left(\mathbf{A}^{-1}\right)^{*}$;
- $\det\left(\mathbf{A}^{-1}\right) = \left(\det \mathbf{A}\right)^{-1}$.

[70]The concept of the inverse of a square matrix was first introduced into mathematics in 1855 by the English mathematician Arthur Cayley (1821–1895), a closed friend of James Sylvester.

Definition 7.4: The **resolvent** of a square matrix $\mathbf{A}$ is the matrix $\mathbf{R}_\lambda(\mathbf{A})$ defined by

$$\mathbf{R}_\lambda(\mathbf{A}) = (\lambda\mathbf{I} - \mathbf{A})^{-1},$$

where $\mathbf{I}$ is the identity matrix.

Example 7.2.4: Find the resolvent of the matrix

$$\mathbf{A} = \begin{bmatrix} 0 & 1 & 1 \\ 1 & 0 & 1 \\ 1 & 1 & 0 \end{bmatrix}. \tag{7.2.3}$$

Solution. Throughout the text, we will use the notation $\chi(\lambda) \stackrel{\text{def}}{=} \det(\lambda\mathbf{I} - \mathbf{A})$. For the matrix $\mathbf{A}$, we have $\lambda\mathbf{I} - \mathbf{A} = \begin{bmatrix} \lambda & -1 & -1 \\ -1 & \lambda & -1 \\ -1 & -1 & \lambda \end{bmatrix}$. Hence, the determinant of the latter matrix is $\chi(\lambda) = \det(\lambda\mathbf{I} - \mathbf{A}) = \lambda^3 - 3\lambda - 2 = (\lambda+1)^2(\lambda-2)$.

The inverse of $(\lambda\mathbf{I} - \mathbf{A})$ gives us the resolvent: $\mathbf{R}_\lambda(\mathbf{A}) = \dfrac{1}{(\lambda+1)(\lambda-2)} \begin{bmatrix} \lambda-1 & 1 & 1 \\ 1 & \lambda-1 & 1 \\ 1 & 1 & \lambda-1 \end{bmatrix}$.

Example 7.2.5: In MATLAB®, the command inv($\mathbf{A}$) gives the inverse matrix $\mathbf{A}^{-1}$. With det($\mathbf{A}$), MATLAB, MuPad provide the determinant of the matrix $\mathbf{A}$. The transposition of a real matrix in MATLAB can be found by typing $\mathbf{A}$'.

Example 7.2.6: *Maple*™ has two packages to handle matrices, **linalg** and **LinearAlgebra**. The former is deprecated, but still widely used for instance in MuPad, a CAS from MATLAB; it calls for commands `inverse(A)` and `det(A)`, respectively. The latter utilizes `MatrixInverse` and `Determinant` instead. Note that the **LinearAlgebra** package uses commands started with upper case letters similar to *Mathematica* while **linalg** package commands are all typed in lower case letters. To find the resolvent of the matrix $\mathbf{A}$, we type:

```
with(linalg): M:=Matrix(3,3,shape=identity); R := inverse(lambda*M-A);
```

Example 7.2.7: In *Mathematica*®, we define the matrix, its inverse, and the determinant with the following commands:

```
A:={{0,1,1},{1,0,1},{1,1,0}}
Inverse[A] // MatrixForm
Det[A]
```

and then hold "Shift" and press "Enter." The resolvent can be defined as `Inverse[lambda*IdentityMatrix[3] - A]` To solve the system of algebraic equations $\mathbf{Ax} = \mathbf{b}$, *Mathematica* has a special command: `LinearSolve[A,b]`, which gives the vector $\mathbf{x}$.

Example 7.2.8: *Maxima* allows us to find the inverse and the determinant of the matrix $\mathbf{A}$ by typing

```
B: A^^-1;  /* or */ invert(A);   /* or */ A^^-1, detout;
determinant(A);
```

In *wxMaxima*, you can also click "Algebra" and then "Determinant." Upon typing `minor(A,i,j)` it will return the matrix with row i and column j removed from $\mathbf{A}$. *Sage* has two dedicated commands: `M.determinant()` and `M.det()`, where `M` is defined previously matrix. SymPy uses the latter.

7.2.1 Solving Linear Equations

Now we turn our attention to the set of solutions of a vector algebraic equation: $\mathbf{Ax} = \mathbf{b}$ when $\mathbf{A}$ is a square $n \times n$ singular matrix, and $\mathbf{x}$ and $\mathbf{b}$ are n-vectors. This equation provides us another point of view on square matrices: they are transformations in a finite dimensional space because matrices map a vector $\mathbf{x}$ into another vector $\mathbf{b}$. Moreover, a square matrix can be considered as an example of a linear operator (Definition 4.1, page 190).

The term *linear operator* means a linear transformation from a vector space to itself. For some choice of basis, a linear operator corresponds to a square matrix, and vice versa. Let vectors $\mathbf{b}_1$, $\mathbf{b}_2$, $\ldots$, $\mathbf{b}_n$ form a basis in an

n-dimensional vector space. Then an arbitrary vector $\mathbf{v}$ can be written as a linear combination of these vectors, say $\mathbf{v} = v_1\mathbf{b}_1 + v_2\mathbf{b}_2 + \cdots + v_n\mathbf{b}_n$. We can identify $\mathbf{v}$ with the n-column vector $\langle v_1, v_2, \ldots, v_n\rangle^T$.

For a given linear operator L, let $\mathbf{u} = L[\mathbf{v}]$ (which we will denote as $L\mathbf{v}$ for short). In the chosen basis, the vector $\mathbf{u}$ has coordinates $\mathbf{u} = \langle u_1, u_2, \ldots, u_n\rangle^T$, meaning that $\mathbf{u} = u_1\mathbf{b}_1 + u_2\mathbf{b}_2 + \cdots + u_n\mathbf{b}_n$. Since L is a linear operator, we have

$$L\mathbf{v} = L\left[v_1\mathbf{b}_1 + v_2\mathbf{b}_2 + \cdots + v_n\mathbf{b}_n\right] = v_1 L\mathbf{b}_1 + v_2 L\mathbf{b}_2 + + \cdots + v_n L\mathbf{b}_n.$$

By expanding each vector $L\mathbf{b}_j = a_{1j}\mathbf{b}_1 + a_{2j}\mathbf{b}_2 + \cdots + a_{nj}\mathbf{b}_n$ $(j = 1, 2, \ldots, n)$ as a linear combination of the basis vectors, we obtain a square matrix $\mathbf{A} = [a_{ij}]$, called the standard matrix for the linear operator L with respect to the given basis $\{\mathbf{b}_k\}$.

Theorem 7.6: If a square matrix $\mathbf{A}$ is not singular, then the solution of the algebraic equation $\mathbf{A}\,\mathbf{x} = \mathbf{b}$ is $\mathbf{x} = \mathbf{A}^{-1}\mathbf{b}$. If $\det \mathbf{A} = 0$, then the vector equation $\mathbf{A}\,\mathbf{x} = \mathbf{b}$ either has no solution, or it has infinitely many solutions. In the latter case, the vector $\mathbf{b}$ must be orthogonal to every solution of $\mathbf{A}^*\mathbf{y} = \mathbf{0}$, namely, $\mathbf{b} \cdot \mathbf{y} = 0$. Here $\mathbf{A}^* = \overline{\mathbf{A}}^T$.

Recall that the rank of the matrix $\mathbf{A}$ is the maximum number of linearly independent column or row vectors of $\mathbf{A}$. If the rank of the matrix $\mathbf{A}$ is $r < n$, then the set of all solutions of the vector equation $\mathbf{Ax} = \mathbf{0}$, together with $\mathbf{x} = \mathbf{0}$, forms a vector space, called the **kernel** or **null space** (also **nullspace**). The dimension $n - r$ of the null space of the matrix $\mathbf{A}$ is called the **nullity** of $\mathbf{A}$. It is related to the rank of $\mathbf{A}$ by the equation

$$\text{rank}(\mathbf{A}) + \text{nullity}(\mathbf{A}) = n.$$

A nonhomogeneous algebraic equation $\mathbf{Ax} = \mathbf{b}$ has a solution only if $\mathbf{b}$ belongs to the column space of the matrix (sometimes called the range of a matrix), which is the set of all possible linear combinations of its column vectors. When $\mathbf{A}$ is a singular matrix, a system of equations $\mathbf{Ax} = \mathbf{b}$ has a solution only if $\mathbf{b}$ is orthogonal to every solution $\mathbf{y}$ of the adjoint homogeneous equation $\mathbf{A}^*\mathbf{y} = \mathbf{0}$ or $\mathbf{y}^*\mathbf{A} = \mathbf{0}$. The null space of $\mathbf{A}^*$ is called the **cokernel**, and it can be viewed as the space of constraints that must be satisfied if the equation $\mathbf{Ax} = \mathbf{b}$ is to have a solution.

Example 7.2.9: Let us consider a singular matrix of rank 2:

$$\mathbf{A} = \begin{bmatrix} 1 & 2 & 3 \\ 3 & 2 & 1 \\ 1 & 1 & 1 \end{bmatrix}.$$

This matrix has rank 2 because its first two rows $\langle 1, 2, 3\rangle$ and $\langle 3, 2, 1\rangle$ are linearly independent, but the last row $\langle 1, 1, 1\rangle$ is their linear combination: $\langle 1, 1, 1\rangle = \frac{1}{4}\langle 1, 2, 3\rangle + \frac{1}{4}\langle 3, 2, 1\rangle$. The vector equation $\mathbf{Ax} = \mathbf{0}$ is equivalent to the three equations

$$x_1 + 2x_2 + 3x_3 = 0, \quad 3x_1 + 2x_2 + x_3 = 0, \quad x_1 + x_2 + x_3 = 0,$$

where $\mathbf{x} = \langle x_1, x_2, x_3\rangle^T$ is a 3-column vector. From the latter equation, we get $x_3 = -x_1 - x_2$. Substituting x_3 into the first two equations, we obtain

$$\begin{aligned} x_1 + 2x_2 - 3\,(x_1 + x_2) = 0 \quad &\Longleftrightarrow \quad -2x_1 - x_2 = 0, \\ 3x_1 + 2x_2 - (x_1 + x_2) = 0 \quad &\Longleftrightarrow \quad 2x_1 + x_2 = 0. \end{aligned}$$

We can determine only one variable from these equations, for example, $x_2 = -2x_1$. Then $x_3 = -x_1 - x_2 = -x_1 + 2x_1 = x_1$. Substituting these values of x_3 and x_2 into $\mathbf{x} = \langle x_1, x_2, x_3\rangle^T$, we get the solution of $\mathbf{Ax} = \mathbf{0}$ to be $\langle x_1, -2x_1, x_1\rangle^T = x_1\,\langle 1, -2, 1\rangle^T$. Hence, the null space is spanned by the vector

$$\langle 1, -2, 1\rangle^T,$$

resulting in a one-dimensional vector space (a line).

Example 7.2.10: Let us return to a nonhomogeneous vector equation $\mathbf{Ax} = \mathbf{b}$, with the singular matrix $\mathbf{A}$ from the previous example:

$$\begin{bmatrix} 1 & 2 & 3 \\ 3 & 2 & 1 \\ 1 & 1 & 1 \end{bmatrix} \begin{bmatrix} x_1 \\ x_2 \\ x_3 \end{bmatrix} = \begin{bmatrix} b_1 \\ b_2 \\ b_3 \end{bmatrix}, \qquad \text{where} \quad \mathbf{x} = \begin{bmatrix} x_1 \\ x_2 \\ x_3 \end{bmatrix}, \quad \mathbf{b} = \begin{bmatrix} b_1 \\ b_2 \\ b_3 \end{bmatrix}.$$

First, we consider the adjoint homogeneous equation:

$$\mathbf{A}^T\mathbf{y} = \mathbf{0} \qquad \text{or} \qquad \begin{bmatrix} 1 & 3 & 1 \\ 2 & 2 & 1 \\ 3 & 1 & 1 \end{bmatrix} \begin{bmatrix} y_1 \\ y_2 \\ y_3 \end{bmatrix} = \begin{bmatrix} 0 \\ 0 \\ 0 \end{bmatrix},$$

which can be expressed by three equations

$$y_1 + 3y_2 + y_3 = 0, \qquad 2y_1 + 2y_2 + y_3 = 0, \qquad 3y_1 + y_2 + y_3 = 0.$$

Solving this system of equations, we get $y_1 = y_2$ and $y_3 = -4y_1$. Hence,

$$\mathbf{y} = \begin{bmatrix} y_1 \\ y_1 \\ -4y_1 \end{bmatrix} = y_1 \begin{bmatrix} 1 \\ 1 \\ -4 \end{bmatrix}.$$

The solution space of the equation $\mathbf{A}^*\mathbf{y} = \mathbf{0}$ is spanned by the vector $\langle 1, 1, -4\rangle^T$. So the cokernel of $\mathbf{A}$ is a one-dimensional space. Now we are ready to determine the column space of $\mathbf{A}$. This space consists of all vectors that are orthogonal to $\langle 1, 1, -4\rangle^T$. This leads to

$$\langle 1, 1, -4\rangle^T \cdot \langle b_1, b_2, b_3\rangle = 0 \qquad \text{or} \qquad b_1 + b_2 - 4b_3 = 0.$$

Since we have only one constraint, $b_1 = -b_2 + 4b_2$, this vector space is two dimensional:

$$\begin{bmatrix} b_1 \\ b_2 \\ b_3 \end{bmatrix} = \begin{bmatrix} -b_2 + 4b_2 \\ b_2 \\ b_3 \end{bmatrix} = b_2 \begin{bmatrix} -1 \\ 1 \\ 0 \end{bmatrix} + b_3 \begin{bmatrix} 4 \\ 0 \\ 1 \end{bmatrix}.$$

Therefore, the algebraic equation $\mathbf{A}\mathbf{x} = \mathbf{b}$ has a nontrivial solution if and only if the vector $\mathbf{b}$ belongs to the column space of $\mathbf{A}$ spanned by the two vectors $\langle -1, 1, 0\rangle^T$ and $\langle 4, 0, 1\rangle^T$.

Problems

1. Determine all minors and cofactors of each of the given matrices.

$$\text{(a)} \begin{bmatrix} 0 & 0 & -4 \\ 0 & -4 & 0 \\ -4 & 0 & 15 \end{bmatrix}; \qquad \text{(b)} \begin{bmatrix} 1 & 2 & -1 \\ 2 & 1 & 1 \\ -1 & 1 & -1 \end{bmatrix}; \qquad \text{(c)} \begin{bmatrix} 0 & 2 & -1 \\ 4 & 3 & 5 \\ 2 & 0 & -4 \end{bmatrix};$$

$$\text{(d)} \begin{bmatrix} 1 & 2 & 3 \\ 3 & 2 & 1 \\ 1 & 1 & 3 \end{bmatrix}; \qquad \text{(e)} \begin{bmatrix} 1 & 1 & 4 \\ 0 & 6 & -1 \\ 2 & 0 & 10 \end{bmatrix}; \qquad \text{(f)} \begin{bmatrix} 1 & 2 & 3 \\ 2 & 3 & 4 \\ 3 & 4 & 4 \end{bmatrix}.$$

2. Find the determinants of the following matrices:

$$\text{(a)}\ e^t \begin{bmatrix} \sin 2t & \cos 2t \\ \sin 2t + 2\cos 2t & \cos 2t - \sin 2t \end{bmatrix}; \qquad \text{(b)} \begin{bmatrix} t & t^2 \\ 1 - t & 1 - 3t^2 \end{bmatrix}.$$

3. For each of the following 2×2 matrices, find its inverse using Cayley–Hamilton formula $\mathbf{A}^{-1} = \frac{1}{\det \mathbf{A}}\left[(\mathrm{tr}\mathbf{A})\,\mathbf{I} - \mathbf{A}\right]$.

$$\text{(a)} \begin{bmatrix} 0 & 1 \\ -5 & 2 \end{bmatrix}; \quad \text{(b)} \begin{bmatrix} 1 & 2 \\ -1 & 3 \end{bmatrix}; \quad \text{(c)} \begin{bmatrix} 2 & 1 \\ -1 & 2 \end{bmatrix}; \quad \text{(d)} \begin{bmatrix} 2 & -1 \\ 4 & -2 \end{bmatrix}.$$

4. For each of the following 3×3 matrices, find its inverse.

$$\text{(a)} \begin{bmatrix} 1 & 2 & -1 \\ 2 & 1 & 1 \\ -1 & 1 & 0 \end{bmatrix}; \qquad \text{(b)} \begin{bmatrix} 0 & 2 & 2 \\ 2 & 0 & 2 \\ 2 & 2 & 0 \end{bmatrix}; \qquad \text{(c)} \begin{bmatrix} 1 & 1 & 1 \\ 1 & 0 & 1 \\ 1 & 1 & 0 \end{bmatrix};$$

$$\text{(d)} \begin{bmatrix} 1 & 3 & 0 \\ 0 & 0 & 2 \\ 0 & -1 & 5 \end{bmatrix}; \qquad \text{(e)} \begin{bmatrix} 1 & 2 & 3 \\ 2 & 3 & 5 \\ 0 & 1 & 2 \end{bmatrix}; \qquad \text{(f)} \begin{bmatrix} 1 & 1 & 1 \\ 1 & 2 & 3 \\ 2 & 3 & 5 \end{bmatrix}.$$

7.3 Eigenvalues and Eigenvectors

For a square $n \times n$ matrix $\mathbf{A} = (a_{ij})$, we consider an associated set of simultaneous linear algebraic equations

$$\begin{cases} a_{11}x_1 + a_{12}x_2 + \cdots + a_{1n}x_n = b_1, \\ a_{21}x_1 + a_{22}x_2 + \cdots + a_{2n}x_n = b_2, \\ \quad\ddots \\ a_{n1}x_1 + a_{n2}x_2 + \cdots + a_{nn}x_n = b_n, \end{cases}$$

which can be written in the matrix form $\mathbf{A}\,\mathbf{x} = \mathbf{b}$, where

$$\mathbf{x} = \begin{bmatrix} x_1 \\ x_2 \\ \vdots \\ x_n \end{bmatrix}, \quad \mathbf{b} = \begin{bmatrix} b_1 \\ b_2 \\ \vdots \\ b_n \end{bmatrix}, \quad \text{and} \quad \mathbf{A} = \begin{bmatrix} a_{11} & a_{12} & \cdots & a_{1n} \\ a_{21} & a_{22} & \cdots & a_{2n} \\ \vdots & \vdots & \ddots & \vdots \\ a_{n1} & a_{n2} & \cdots & a_{nn} \end{bmatrix}.$$

The equation $\mathbf{A}\,\mathbf{x} = \mathbf{b}$ establishes a relationship between column vectors $\mathbf{x}$ and $\mathbf{b}$. This means that $\mathbf{A}$ is a linear transformation in the vector space of all n-vectors. This transformation is one-to-one if and only if the matrix $\mathbf{A}$ is nonsingular (i.e., $\det \mathbf{A} \ne 0$). In addition, if we require that $\mathbf{b} = \lambda\mathbf{x}$ for some scalar λ, so that the matrix $\mathbf{A}$ transforms $\mathbf{x}$ into a parallel vector, we are led to a consideration of the equation

$$\mathbf{A}\,\mathbf{x} = \lambda\mathbf{x} \qquad \text{or} \qquad (\lambda\mathbf{I} - \mathbf{A})\,\mathbf{x} = \mathbf{0},$$

which has a nontrivial solution if and only if the matrix $\lambda\mathbf{I} - \mathbf{A}$ is a singular matrix. That is, λ is a root of the so-called **characteristic equation**:

$$\det(\lambda\mathbf{I} - \mathbf{A}) = \det \begin{bmatrix} \lambda - a_{11} & -a_{12} & \cdots & -a_{1n} \\ -a_{21} & \lambda - a_{22} & \cdots & -a_{2n} \\ \vdots & \vdots & \ddots & \vdots \\ -a_{n1} & -a_{n2} & \cdots & \lambda - a_{nn} \end{bmatrix} = 0.$$

The determinant of the matrix $\mathbf{A} - \lambda\mathbf{I}$, where $\mathbf{I}$ is the identity matrix, is clearly a polynomial in λ of degree n, with the leading term $(-1)^n\lambda^n$. It is more convenient to have the leading coefficient be 1 instead of $(-1)^n$, yielding the following definitions.

Definition 7.5: The **characteristic polynomial** of a square matrix $\mathbf{A}$, denoted $\chi_{\mathbf{A}}(\lambda)$ or simply $\chi(\lambda)$, is the determinant of the matrix $\lambda\mathbf{I} - \mathbf{A}$,

$$\chi(\lambda) \stackrel{\text{def}}{=} \det(\lambda\mathbf{I} - \mathbf{A}). \tag{7.3.1}$$

Obviously, $\chi(\lambda)$ has the leading term λ^n. Any solution of the characteristic equation $\chi(\lambda) = 0$ is said to be an **eigenvalue** of the matrix $\mathbf{A}$. The set of all eigenvalues is called the **spectrum** of the matrix $\mathbf{A}$, denoted by $\sigma(\mathbf{A})$.

Definition 7.6: A nonzero n-vector $\mathbf{x}$ such that

$$\mathbf{A}\mathbf{x} = \lambda\mathbf{x} \tag{7.3.2}$$

is called an **eigenvector** of a square matrix $\mathbf{A}$ corresponding to the **eigenvalue** λ.

For $\lambda = 0$, we have the relation $\mathbf{A}\mathbf{x} = \mathbf{0}$ instead of Eq. (7.3.2). All solutions of this equation form a vector space, called the **null space** or **kernel** of $\mathbf{A}$. Note that an eigenvector corresponding to a given eigenvalue is not unique; any its nonzero constant multiple is again an eigenvector. Moreover, any linear combination of eigenvectors corresponding to a fixed eigenvalue[71] is again an eigenvector.

[71]The prefix "eigen" is adopted from the old Dutch and German, meaning "self" or "proper."

Definition 7.7: Let N_λ be the collection of all eigenvectors corresponding to the eigenvalue λ. Since, by our definition, $\mathbf{0}$ is not an eigenvector, N_λ does not contain $\mathbf{0}$. If, however, we enlarge N_λ by adjoining the origin to it, then N_λ becomes a subspace, usually called the **eigenspace** or *proper space*. We define the **geometric multiplicity** of the eigenvalue λ as the dimension of the subspace N_λ. If the eigenvalue λ has multiplicity 1, it is said to be a *simple* eigenvalue.

Definition 7.8: Let λ be an eigenvalue of a matrix $\mathbf{A}$. The **algebraic multiplicity** of λ is called the multiplicity of λ as a root of the characteristic equation: $\det(\lambda\mathbf{I} - \mathbf{A}) = 0$.

These two concepts of multiplicity do not coincide (see Examples 7.3.1–7.3.3 on page 388). It is quite easy to see that the geometric multiplicity of λ is never greater than its algebraic multiplicity. Indeed, if $\mathbf{T}$ is any linear transformation with eigenvalue λ, then N_λ is invariant under $\mathbf{T}$. If $\mathbf{T}_0$ is the linear transformation of $\mathbf{T}$ restricted to N_λ only, then clearly $\det(\lambda\mathbf{I} - \mathbf{T}_0)$ is a factor of $\det(\lambda\mathbf{I} - \mathbf{T})$.

Theorem 7.7: Nonzero eigenvectors corresponding to distinct eigenvalues of a square matrix are linearly independent.

PROOF: Let $\mathbf{v}_1, \mathbf{v}_2, \ldots, \mathbf{v}_m$ be nonzero eigenvectors of a square matrix $\mathbf{T}$ corresponding to distinct eigenvalues $\lambda_1, \lambda_2, \ldots, \lambda_m$. Suppose $a_1, a_2, \ldots, a_m$ are complex numbers such that

$$a_1\mathbf{v}_1 + a_2\mathbf{v}_2 + \cdots + a_m\mathbf{v}_m = \mathbf{0}.$$

Applying the linear operator $(\lambda_2\mathbf{I} - \mathbf{T})(\lambda_3\mathbf{I} - \mathbf{T})\cdots(\lambda_m\mathbf{I} - \mathbf{T})$ to both sides, we get

$$a_1(\lambda_2 - \lambda_1)(\lambda_3 - \lambda_1)\cdots(\lambda_m - \lambda_1)\mathbf{v}_1 = \mathbf{0}.$$

Thus, $a_1 = 0$. In a similar fashion, $a_j = 0$ for each j, as desired. ■

Definition 7.9: The eigenvalue λ of a square matrix is called **defective** if its algebraic multiplicity is greater than its geometric one. The difference (which is always nonnegative) between the algebraic multiplicity and geometric multiplicity is called the **defect** of the eigenvalue λ.

Definition 7.10: For a square matrix $\mathbf{A}$, let λ be an eigenvalue of defect 1. If there exist two column vectors $\boldsymbol{\xi}$ and $\boldsymbol{\eta}$ such that

$$\mathbf{A}\boldsymbol{\xi} = \lambda\boldsymbol{\xi} \qquad \text{and} \qquad (\lambda\mathbf{I} - \mathbf{A})\,\boldsymbol{\eta} = \boldsymbol{\xi},$$

then the vector $\boldsymbol{\eta}$ is called the **generalized eigenvector** corresponding to the eigenvalue λ. In other words, for the vector $\boldsymbol{\eta}$ we have $(\lambda\mathbf{I} - \mathbf{A})^2\boldsymbol{\eta} = \mathbf{0}$, but $(\lambda\mathbf{I} - \mathbf{A})\,\boldsymbol{\eta} \ne \mathbf{0}$. If the eigenvalue λ has defect 2, the vectors $\boldsymbol{\eta}$ and $\boldsymbol{\zeta}$ that satisfy the equations $\mathbf{A}\boldsymbol{\xi} = \lambda\boldsymbol{\xi}$, $(\lambda\mathbf{I} - \mathbf{A})\,\boldsymbol{\eta} = \boldsymbol{\xi}$, and $(\lambda\mathbf{I} - \mathbf{A})\,\boldsymbol{\zeta} = \boldsymbol{\eta}$ are called the generalized eigenvectors. Therefore, for these vectors, we have $(\lambda\mathbf{I} - \mathbf{A})\,\boldsymbol{\xi} = \mathbf{0}$, $(\lambda\mathbf{I} - \mathbf{A})^2\boldsymbol{\eta} = \mathbf{0}$, and $(\lambda\mathbf{I} - \mathbf{A})^3\boldsymbol{\zeta} = \mathbf{0}$. In general, a vector $\boldsymbol{\eta}$ is a generalized eigenvector of order m associated with the eigenvalue λ if $(\lambda\mathbf{I} - \mathbf{A})^m\boldsymbol{\eta} = \mathbf{0}$ and $(\lambda\mathbf{I} - \mathbf{A})^{m-1}\boldsymbol{\eta} \ne \mathbf{0}$.

The set of generalized eigenvectors of the $n \times n$ square matrix $\mathbf{A}$ corresponding to an eigenvalue λ is a subspace of the n-th dimensional vector space. Problem 8 on page 427 asks you to prove that the set of generalized eigenvectors is mapped to $\mathbf{0}$ by $(\lambda\mathbf{I} - \mathbf{A})^n$.

The eigenspace for every eigenvalue λ is the kernel of the matrix $\lambda\mathbf{I} - \mathbf{A}$: $N_\lambda = \ker(\lambda\mathbf{I} - \mathbf{A})$, where $\mathbf{I}$ is the identity matrix. The dimension of this space N_λ is the geometric multiplicity of λ. If the matrix $\mathbf{A}$ is not defective, then the dimensions of all the kernels of $(\lambda\mathbf{I} - \mathbf{A})^k$ remain the same for any positive integer $k = 1, 2, \ldots$. However, if a matrix is defective, then $\ker(\lambda\mathbf{I} - \mathbf{A}) \subset \ker(\lambda\mathbf{I} - \mathbf{A})^2$. If λ has defect 1, then $\ker(\lambda\mathbf{I} - \mathbf{A})^2 = \ker(\lambda\mathbf{I} - \mathbf{A})^3$. For an eigenvalue λ with defect 3, we have $\ker(\lambda\mathbf{I} - \mathbf{A}) \subset \ker(\lambda\mathbf{I} - \mathbf{A})^2 \subset \ker(\lambda\mathbf{I} - \mathbf{A})^3 = \ker(\lambda\mathbf{I} - \mathbf{A})^4$.

Let $\lambda_1, \lambda_2, \ldots, \lambda_r$ be the distinct eigenvalues of a square matrix $\mathbf{A}$ with algebraic multiplicities $m_1, m_2, \ldots, m_r$, respectively; then

$$\det\mathbf{A} = \prod_{j=1}^{r} \lambda_j^{m_j}, \quad \operatorname{tr}\mathbf{A} = \sum_{j=1}^{r} m_j\lambda_j, \tag{7.3.3}$$

where the expression tr $\mathbf{A}$ is called the **trace** of $\mathbf{A}$. Recall (Definition 6.8 on page 360) that the trace of a square matrix is the sum of its diagonal elements.

The characteristic polynomial for any $n \times n$ matrix $\mathbf{A}$ can be written as

$$\chi(\lambda) = \lambda^n - c_{n-1}\lambda^{n-1} + c_{n-2}\lambda^{n-2} - \cdots + (-1)^n c_0,$$

where $c_{n-1} = \operatorname{tr}(\mathbf{A})$ and $c_0 = \det \mathbf{A}$. Actually all coefficients c_j can be expressed via eigenvalues, for example, $c_{n-2} = \sum_{\lambda_i < \lambda_j} \lambda_i \lambda_j$ (for real eigenvalues).

Example 7.3.1: Let us consider a three-dimensional space and a linear operator $\mathbf{T}$ acting on this space. Suppose this linear transformation is determined by the matrix

$$\mathbf{T} = \begin{bmatrix} 1 & 0 & 3 \\ 2 & 1 & 2 \\ 0 & 0 & 2 \end{bmatrix} \tag{7.3.4}$$

for a given basis. The spectrum of $\mathbf{T}$ consists of all roots of the characteristic equation $\chi(\lambda) = \det(\lambda\mathbf{I} - \mathbf{T}) = 0$. A short computation gives $\chi(\lambda) = (\lambda - 1)^2(\lambda - 2) = \lambda^3 - 4\lambda^2 + 5\lambda - 2$. Hence, the spectrum $\sigma(\mathbf{T})$ of the operator $\mathbf{T}$ consists of two real numbers $\lambda = 1$ and $\lambda = 2$. The geometric multiplicity and the algebraic multiplicity of the simple eigenvalue $\lambda = 2$ are the same.

The algebraic multiplicity of the number $\lambda = 1$ equals two, but the geometric multiplicity of this number equals one, so it has defect 1. Indeed, solving the system of equations $\mathbf{Tx} = \mathbf{x}$, where $\mathbf{x} = \langle x_1, x_2, x_3 \rangle^T$, we see that the relevant eigen subspace N_1 consists of column vectors of the form $\langle 0, x, 0 \rangle^T$, where x is any nonzero number and the index "T" denotes the transpose. To find the generalized eigenvector, we need to solve two systems of equations

$$\begin{bmatrix} 1 & 0 & 3 \\ 2 & 1 & 2 \\ 0 & 0 & 2 \end{bmatrix} \begin{bmatrix} \xi_1 \\ \xi_2 \\ \xi_3 \end{bmatrix} = \begin{bmatrix} \xi_1 \\ \xi_2 \\ \xi_3 \end{bmatrix}, \qquad \begin{bmatrix} 1 & 0 & 3 \\ 2 & 1 & 2 \\ 0 & 0 & 2 \end{bmatrix} \begin{bmatrix} \eta_1 \\ \eta_2 \\ \eta_3 \end{bmatrix} = \begin{bmatrix} \eta_1 \\ \eta_2 \\ \eta_3 \end{bmatrix} + \begin{bmatrix} \xi_1 \\ \xi_2 \\ \xi_3 \end{bmatrix}$$

with respect to $\boldsymbol{\xi} = \langle \xi_1, \xi_2, \xi_3 \rangle^T$ and $\boldsymbol{\eta} = \langle \eta_1, \eta_2, \eta_3 \rangle^T$. Since $\boldsymbol{\xi}$ is the eigenvector, we get $\xi_1 = \xi_3 = 0$. Then the latter system of algebraic equations becomes

$$\begin{cases} \eta_1 + 3\eta_3 = \eta_1, \\ 2\eta_1 + \eta_2 + 2\eta_3 = \eta_2 + \xi_2, \\ \eta_3 = \eta_3. \end{cases}$$

From the first equation, it follows that $\eta_3 = 0$. Since the last equation does not impose any condition, we obtain

$$2\eta_1 = \xi_2.$$

By choosing some numerical value for ξ_2, say $\xi_2 = 2$, we obtain two (linearly independent) eigenvectors corresponding to $\lambda = 1$: $\xi = \langle 0, 2, 0 \rangle^T$ and $\eta = \langle 1, 0, 0, \rangle^T$. Together with the eigenvector $\mathbf{x} = \langle 3, 8, 1 \rangle^T$ corresponding to the eigenvalue $\lambda = 2$ they form a basis for the three-dimensional space.

To analyze the kernels of the powers of $(\mathbf{I} - \mathbf{T})^n$, we introduce the matrix

$$\mathbf{K} = \mathbf{I} - \mathbf{T} = \begin{bmatrix} 0 & 0 & -3 \\ -2 & 0 & -2 \\ 0 & 0 & -1 \end{bmatrix} \quad \Longrightarrow \quad \mathbf{K}^2 = \begin{bmatrix} 0 & 0 & 3 \\ 0 & 0 & 8 \\ 0 & 0 & 1 \end{bmatrix}, \quad \mathbf{K}^3 = \begin{bmatrix} 0 & 0 & -3 \\ 0 & 0 & -8 \\ 0 & 0 & -1 \end{bmatrix} = -\mathbf{K}^2.$$

While the kernel of the matrix $\mathbf{K}$ is one dimensional and spanned on the vector $\langle 0, 1, 0 \rangle^T$, the kernel of the matrix $\mathbf{K}^2$ is two dimensional and spanned on two vectors, $\langle 0, 1, 0 \rangle^T$ and $\langle 1, 0, 0 \rangle^T$. Since $(\mathbf{I} - \mathbf{T})^n = (-1)^n \mathbf{K}$, $n = 2, 3, \ldots$, we have $\ker(\lambda\mathbf{I} - \mathbf{T}) \subset \ker(\lambda\mathbf{I} - \mathbf{T})^2 = \ker(\lambda\mathbf{I} - \mathbf{T})^3 = \ker(\lambda\mathbf{I} - \mathbf{T})^4 = \cdots$.

Example 7.3.2: (Example 7.2.4 revisited) The matrix (7.2.3) on page 383 has the characteristic polynomial $\chi(\lambda) = (\lambda - 2)(\lambda + 1)^2$. Consequently, the spectrum $\sigma(\mathbf{A})$ of the operator $\mathbf{A}$ includes two numbers $\lambda = 2$ (of multiplicity 1, with eigenvector $\langle 1, 1, 1 \rangle^T$) and $\lambda = -1$ (of multiplicity 2). Moreover, the geometric multiplicity of $\lambda = -1$ coincides with its algebraic multiplicity. Indeed, the eigen subspace N_{-1} consists of column vectors $\langle x_1, x_2, -x_1 - x_2 \rangle^T$, where x_1, x_2 are arbitrary numbers. For example, the two eigenvectors $\langle 1, 0, -1 \rangle^T$ and $\langle 0, 1, -1 \rangle^T$ are linearly independent.

Example 7.3.3: Let us consider linear operators defined by the following matrices:

$$\mathbf{T}_1 = \begin{bmatrix} 2 & 0 & 0 \\ 0 & 2 & 0 \\ 0 & 0 & 2 \end{bmatrix}, \quad \mathbf{T}_2 = \begin{bmatrix} 2 & 1 & 0 \\ 0 & 2 & 0 \\ 0 & 0 & 2 \end{bmatrix}, \quad \mathbf{T}_3 = \begin{bmatrix} 2 & 1 & a \\ 0 & 2 & 1 \\ 0 & 0 & 2 \end{bmatrix}, \tag{7.3.5}$$

where a is any number. The polynomial $\chi(\lambda) = (\lambda - 2)^3$ is the characteristic polynomial of all three matrices (7.3.5). Therefore, the spectrum, $\sigma(\mathbf{T}_j)$ $(j = 1, 2, 3)$, consists of one number, $\lambda = 2$, with algebraic multiplicity 3. However, the geometric multiplicities of $\lambda = 2$ for each of these operators differ. The eigen subspace N_2 for the operator $\mathbf{T}_1$ coincides with $\mathbb{R}^3$. The eigen subspace N_2 for the operator $\mathbf{T}_2$ consists of all the column vectors of the form $\langle x_1, 0, x_3\rangle^T$, and the eigen subspace N_2 for the operator $\mathbf{T}_3$ coincides with all the column vectors of the form $\langle x_1, 0, 0\rangle^T$. Here, x_1, x_3 are any (real) numbers.

Theorem 7.8: The eigenvalues and eigenvectors of any $n \times n$ self-adjoint matrix ($\mathbf{A} = \mathbf{A}^H = \mathbf{A}^*$, that is, $a_{ij} = \overline{a}_{ji}$), possess the following properties.

1. All eigenvalues are real numbers.
2. There exists a basis of n linearly independent eigenvectors.
3. Eigenvectors that correspond to different eigenvalues are orthogonal, that is, $(\mathbf{x}, \mathbf{y}) = \overline{\mathbf{x}} \cdot \mathbf{y}^T = 0$.

PROOF: A more general result is proved in Theorem 10.1, page 555.

Corollary 7.1: For a square matrix $\mathbf{A}$, the products of $\mathbf{A}$ with its adjoint, $\mathbf{A}\mathbf{A}^*$ and $\mathbf{A}^*\mathbf{A}$, have the same real eigenvalues.

Example 7.3.4: Find the eigenvalues and eigenvectors of each of the following matrices.

1. The matrix $\mathbf{A} = \begin{bmatrix} 2 & -2 & 3 \\ 10 & -4 & 5 \\ 5 & -4 & 6 \end{bmatrix}$ has the following eigenvalues and eigenvectors:

$$\lambda_1 = \lambda_2 = 1, \ \mathbf{x}_1 = \begin{pmatrix} 1 \\ 5 \\ 3 \end{pmatrix}, \quad \text{and} \qquad \lambda_3 = 2, \ \mathbf{x}_1 = \begin{pmatrix} 4/5 \\ 3 \\ 2 \end{pmatrix}.$$

   Therefore, the eigenvalue $\lambda = 1$ is defective.

2. The matrix $\mathbf{A} = \begin{bmatrix} -2 & 2 & -3 \\ 2 & 1 & -6 \\ -1 & -2 & 0 \end{bmatrix}$ has eigenvalues $\lambda_1 = 5$, $\lambda_2 = \lambda_3 = -3$ because its characteristic polynomial is $\chi(\lambda) = \lambda^3 + \lambda^2 - 21\lambda - 45 = (\lambda - 5)(\lambda + 3)^2$. An eigenvector corresponding to $\lambda_1 = 5$ is $\langle 1, 2, -1\rangle^T$, and the two linearly independent eigenvectors that correspond to $\lambda_2 = \lambda_3 = -3$ are $\langle -2, 1, 0\rangle^T$ and $\langle 3, 0, 1\rangle^T$. Therefore, the matrix is not defective.

3. The matrix $\mathbf{A} = \begin{bmatrix} 3 & 2 & -3 \\ 0 & 2 & 1 \\ 0 & 0 & 1 \end{bmatrix}$ has the characteristic polynomial $\chi(\lambda) = (\lambda - 3)(\lambda - 2)(\lambda - 1)$ with simple eigenvalues $\lambda_1 = 1$, $\lambda_2 = 2$, and $\lambda_3 = 3$. An eigenvector corresponding to $\lambda_1 = 1$ is $\langle 5, -2, 2\rangle^T$, an eigenvector corresponding to $\lambda_2 = 2$ is $\langle -2, 1, 0\rangle^T$, and an eigenvector corresponding $\lambda_3 = 3$ is $\langle 1, 0, 0\rangle^T$.

Example 7.3.5: In MATLAB, eigenvalues of a matrix $\mathbf{A}$ can be found by entering `eig(A)`. The output of this operation contains a column vector of eigenvalues. To determine eigenvalues and corresponding eigenvectors one should enter `[V,D] = eig( A )`. This produces a diagonal matrix $\mathbf{D}$ of eigenvalues and a full matrix $\mathbf{V}$ with columns of eigenvectors so that $\mathbf{A} * \mathbf{V} = \mathbf{V} * \mathbf{D}$ (the asterisk $*$ represents multiplication of matrices).

Maple has two packages—**linalg** and **LinearAlgebra**—to deal with matrices. The command `eigenvectors(A)` provides eigenvalues, their multiplicities, and eigenvectors within deprecated **linalg** package (where all commands are typed in lower case letters); when the command `eigenvalues(A)` is typed, it provides eigenvalues of the matrix **A**. On the other hand, **LinearAlgebra** packages calls for `Eigenvalues(A)` and `Eigenvectors(A)`. Eigenvalues also can be found from the resolvent. First, we find the matrix $\lambda\mathbf{I} - \mathbf{A}$ within `linalg` package:
`RR:= matadd(array(identity, 1..3, 1..3), -A, lambda, 1));`
and then factor it: `factor(RR)` or calculate the resolvent: `R:=inverse(RR)`.

Mathematica has two dedicated commands: `Eigenvectors[A]` and `Eigenvalues[A]` These two commands can be united into one: `Eigensystem[A]`, which gives a list of the eigenvalues and eigenvectors of the square matrix `A`. *Sage* utilizes the similar commands.

When using *Maxima*, its package `eigen` contains several functions devoted to the symbolic computation of eigenvalues and eigenvectors: `eigenvalues(A)` and `eigenvectors(A)` that can be united into one: `spectral_rep(A);` which displays eigenvectors and eigenvalues. The command `charpoly(A,lambda)` returns the determinant of $\det(\mathbf{A} - \lambda\mathbf{I}) = (-1)^n\,\chi(\lambda)$. An alternative to *Maxima*'s `charpoly` is the function `ncharpoly`, which requires the load of this package: `load ("nchrpl")`. SymPy uses special commands: `M.eigenvals()` and `M.eigenvects()`, while *Sage* utilizes similar ones: `M.eigenvalues()` and `M.eigenvectors()`.

Problems

1. For each of the following 2×2 matrices, find its eigenvalues and corresponding eigenvectors.

$$\text{(a)}\ \begin{bmatrix} 2 & -1 \\ 3 & -2 \end{bmatrix};\quad \text{(b)}\ \begin{bmatrix} -4 & -6 \\ 3 & 5 \end{bmatrix};\quad \text{(c)}\ \begin{bmatrix} 1 & 2 \\ 1 & 2 \end{bmatrix};\quad \text{(d)}\ \begin{bmatrix} 1 & 1 \\ 4 & -2 \end{bmatrix};\quad \text{(e)}\ \begin{bmatrix} 3 & 1 \\ 4 & 3 \end{bmatrix}.$$

2. For each of the following 2×2 matrices, find its eigenvalues and corresponding eigenvectors, including generalized eigenvectors.

$$\text{(a)}\ \begin{bmatrix} 0 & -1 \\ 1 & -2 \end{bmatrix};\quad \text{(b)}\ \begin{bmatrix} 2 & 1 \\ -1 & 4 \end{bmatrix};\quad \text{(c)}\ \begin{bmatrix} 1 & -1 \\ 9 & -5 \end{bmatrix};\quad \text{(d)}\ \begin{bmatrix} 1 & -1 \\ 1 & 3 \end{bmatrix};\quad \text{(e)}\ \begin{bmatrix} 0 & -3 \\ 3 & 6 \end{bmatrix};$$

$$\text{(f)}\ \begin{bmatrix} 15 & -9 \\ 16 & -9 \end{bmatrix};\quad \text{(g)}\ \begin{bmatrix} -4 & 3 \\ -12 & 8 \end{bmatrix};\quad \text{(h)}\ \begin{bmatrix} 10 & 3 \\ -3 & 4 \end{bmatrix};\quad \text{(i)}\ \begin{bmatrix} 8 & -4 \\ 1 & 4 \end{bmatrix};\quad \text{(j)}\ \begin{bmatrix} 9 & -4 \\ 1 & 5 \end{bmatrix}.$$

3. For each of the following 2×2 matrices, find its complex eigenvalues and corresponding eigenvectors.

$$\text{(a)}\ \begin{bmatrix} 1 & -1 \\ 4 & 1 \end{bmatrix};\quad \text{(b)}\ \begin{bmatrix} 3 & 4 \\ -4 & 3 \end{bmatrix};\quad \text{(c)}\ \begin{bmatrix} 5 & -5 \\ 2 & -1 \end{bmatrix};\quad \text{(d)}\ \begin{bmatrix} 3 & 5 \\ -2 & 1 \end{bmatrix};\quad \text{(e)}\ \begin{bmatrix} -1 & -2 \\ 1 & -3 \end{bmatrix};$$

$$\text{(f)}\ \begin{bmatrix} -5 & -2 \\ 20 & 7 \end{bmatrix};\quad \text{(g)}\ \begin{bmatrix} -4 & 2 \\ -10 & 4 \end{bmatrix};\quad \text{(h)}\ \begin{bmatrix} 2 & -5 \\ 4 & -2 \end{bmatrix};\quad \text{(i)}\ \begin{bmatrix} 1 & -5 \\ 1 & -1 \end{bmatrix};\quad \text{(j)}\ \begin{bmatrix} 5 & -9 \\ 2 & -1 \end{bmatrix}.$$

4. For each of the following 3×3 matrices, find its eigenvalues and corresponding eigenvectors.

$$\text{(a)}\ \begin{bmatrix} -1 & 2 & 3 \\ 0 & 1 & 6 \\ 0 & 0 & -2 \end{bmatrix};\qquad \text{(b)}\ \begin{bmatrix} 3 & -1 & -1 \\ -2 & 3 & 2 \\ 4 & -1 & -2 \end{bmatrix};\qquad \text{(c)}\ \begin{bmatrix} 3 & 5 & 8 \\ 1 & -1 & -2 \\ -1 & -1 & -1 \end{bmatrix};$$

$$\text{(d)}\ \begin{bmatrix} 1 & -1 & -2 \\ 1 & -2 & -3 \\ -4 & 1 & -1 \end{bmatrix};\qquad \text{(e)}\ \begin{bmatrix} 1 & -1 & -2 \\ 12 & -4 & 10 \\ -6 & 1 & -7 \end{bmatrix};\qquad \text{(f)}\ \begin{bmatrix} 4 & -1 & -4 \\ 4 & -3 & -2 \\ 1 & -1 & -1 \end{bmatrix}.$$

5. For each of the following 3×3 not defective matrices, find its eigenvalues and corresponding eigenvectors.

$$\text{(a)}\ \begin{bmatrix} -3 & 2 & 2 \\ 2 & -3 & 2 \\ 2 & 2 & -3 \end{bmatrix};\qquad \text{(b)}\ \begin{bmatrix} 3 & 2 & 4 \\ 2 & 0 & 2 \\ 4 & 2 & 3 \end{bmatrix};\qquad \text{(c)}\ \begin{bmatrix} 2 & -2 & 1 \\ -1 & 3 & -1 \\ 2 & -4 & 3 \end{bmatrix};$$

$$\text{(d)}\ \begin{bmatrix} 3 & -3 & -4 \\ 6 & -8 & -12 \\ -4 & 6 & 9 \end{bmatrix};\qquad \text{(e)}\ \begin{bmatrix} 1 & 2 & -4 \\ 1 & 0 & 4 \\ 1 & -2 & 6 \end{bmatrix};\qquad \text{(f)}\ \begin{bmatrix} 3 & 1 & -1 \\ 1 & 3 & -1 \\ 3 & 3 & -1 \end{bmatrix}.$$

7.4 Diagonalization

This section shows the reduction of matrices into their diagonal form. This approach, called diagonalization, allows us to define a function of a square matrix. We present the diagonalization procedure only for matrices with nondefective eigenvalues, that is, when the algebraic and geometric multiplicities are equal for every eigenvalue. The algorithm based on diagonalization can be extended for an arbitrary matrix; however, all calculations become more tedious as a result and require more work.

Definition 7.11: Two matrices $\mathbf{A}$ and $\mathbf{B}$ are said to be **similar** if there exists a nonsingular matrix $\mathbf{S}$ such that $\mathbf{S}^{-1}\mathbf{A}\mathbf{S} = \mathbf{B}$. We denote this relation as $\mathbf{A} \sim \mathbf{B}$.

Note that the matrix $\mathbf{S}$ is not unique. It is obvious that if $\mathbf{A} \sim \mathbf{B}$, then $\mathbf{B} \sim \mathbf{A}$ because $\mathbf{S}\mathbf{B}\mathbf{S}^{-1} = \mathbf{A}$. Also if $\mathbf{A} \sim \mathbf{B}$ and $\mathbf{B} \sim \mathbf{C}$, then $\mathbf{A} \sim \mathbf{C}$. Indeed, suppose that there exist two matrices $\mathbf{S}$ and $\mathbf{T}$ such that

$$\mathbf{S}^{-1}\mathbf{A}\mathbf{S} = \mathbf{B}, \quad \mathbf{T}^{-1}\mathbf{B}\mathbf{T} = \mathbf{C}.$$

We multiply both sides of the former equation by $\mathbf{T}^{-1}$ from the left and by $\mathbf{T}$ from the right. This yields

$$\mathbf{T}^{-1}\mathbf{S}^{-1}\mathbf{A}\mathbf{S}\mathbf{T} = \mathbf{T}^{-1}\mathbf{B}\mathbf{T} = \mathbf{C}$$

since $\mathbf{T}^{-1}\mathbf{B}\mathbf{T} = \mathbf{C}$. If we denote $\mathbf{R} \stackrel{\text{def}}{=} \mathbf{S}\mathbf{T}$, then $\mathbf{R}^{-1} = \mathbf{T}^{-1}\mathbf{S}^{-1}$, and we have

$$\mathbf{R}^{-1}\mathbf{A}\mathbf{R} = \mathbf{C}.$$

Theorem 7.9: For similar square matrices $\mathbf{A}$ and $\mathbf{B}$, we have

$$\det(\mathbf{A}) = \det(\mathbf{B}) \quad \text{and} \quad \operatorname{tr}(\mathbf{A}) = \operatorname{tr}(\mathbf{B}).$$

Example 7.4.1: Let us consider two nonsingular matrices:

$$\mathbf{S} = \begin{bmatrix} 0 & 1 & 1 \\ 1 & 0 & 1 \\ 1 & 1 & 0 \end{bmatrix} \quad \text{and} \quad \mathbf{A} = \begin{bmatrix} 1 & 3 & 0 \\ 0 & 0 & 2 \\ 0 & -1 & 5 \end{bmatrix}.$$

We use the matrix $\mathbf{S}$ to build a similar matrix to $\mathbf{A}$, which we denote by $\mathbf{B}$:

$$\begin{aligned} \mathbf{B} = \mathbf{S}^{-1}\,\mathbf{A}\,\mathbf{S} &= \frac{1}{2}\begin{bmatrix} -1 & 1 & 1 \\ 1 & -1 & 1 \\ 1 & 1 & -1 \end{bmatrix}\begin{bmatrix} 1 & 3 & 0 \\ 0 & 0 & 2 \\ 0 & -1 & 5 \end{bmatrix}\begin{bmatrix} 0 & 1 & 1 \\ 1 & 0 & 1 \\ 1 & 1 & 0 \end{bmatrix} \\ &= \frac{1}{2}\begin{bmatrix} 3 & 6 & -5 \\ 5 & 4 & 3 \\ 1 & -2 & 5 \end{bmatrix}. \end{aligned}$$

These two matrices have the same determinants and traces: $\det(\mathbf{A}) = \det(\mathbf{B}) = 2$ and $\operatorname{tr}(\mathbf{A}) = 6 = \operatorname{tr}(\mathbf{B})$.

Similarly, we can use the matrix $\mathbf{A}$ to determine $\mathbf{T}$, a similar matrix to $\mathbf{S}$:

$$\begin{aligned} \mathbf{T} = \mathbf{A}^{-1}\,\mathbf{S}\,\mathbf{A} &= \frac{1}{2}\begin{bmatrix} 2 & -15 & 6 \\ 0 & 5 & -2 \\ 0 & 1 & 0 \end{bmatrix}\begin{bmatrix} 0 & 1 & 1 \\ 1 & 0 & 1 \\ 1 & 1 & 0 \end{bmatrix}\begin{bmatrix} 1 & 3 & 0 \\ 0 & 0 & 2 \\ 0 & -1 & 5 \end{bmatrix} \\ &= \frac{1}{2}\begin{bmatrix} -9 & -14 & -49 \\ 3 & 4 & 21 \\ 1 & 2 & 5 \end{bmatrix}, \qquad \text{and } \operatorname{tr}\mathbf{T} = \operatorname{tr}\mathbf{S} = 0. \end{aligned}$$

Definition 7.12: An $n \times n$ matrix is called a **diagonal** matrix if its entries are 0 everywhere except possibly along the main diagonal (which will therefore contain all eigenvalues).

A diagonal matrix is completely determined by its main diagonal elements. The product and sum of any two diagonal matrices is again a diagonal matrix. Since a diagonal matrix is uniquely identified by a vector of its diagonal elements, there exists a one-to-one correspondence between diagonal $n \times n$ matrices and n-vectors.

Mathematica and *Maple* (needs the `LinearAlgebra` subroutine) share a dedicated command to define a diagonal matrix:

```
DiagonalMatrix[{1,2,3}] // MatrixForm
```

Maxima and MATLAB also have a similar command: `A : diag([a,b]);` for a diagonal 2×2 matrix.

Definition 7.13: An $n \times n$ matrix $\mathbf{A}$ is said to be **diagonalizable** if it is similar to a diagonal matrix, that is, there exists a diagonal matrix $\mathbf{\Lambda}$ and a nonsingular matrix $\mathbf{S}$ such that $\mathbf{\Lambda} = \mathbf{S}^{-1}\mathbf{A}\mathbf{S}$.

Definition 7.14: A square matrix is called **positive definite**, or **positive semidefinite**, if its eigenvalues are all positive numbers or nonnegative numbers, respectively.

A matrix is positive definite if and only if $(\mathbf{u}, \mathbf{A}\mathbf{u}) > 0$ for any nonzero vector $\mathbf{u}$, where $(\, , \,)$ is the inner product (Definition 6.1, page 356).

Theorem 7.10: A square matrix $\mathbf{A}$ is diagonalizable if and only if it is nondefective, that is, its eigenvalues have the same geometric and algebraic multiplicities. In other words, if a square $n \times n$ matrix $\mathbf{A}$ has n linearly independent eigenvectors, then $\mathbf{A}$ is diagonalizable. ■

We present some sufficient conditions that guarantee this property.

Theorem 7.11: A square matrix $\mathbf{A}$ is diagonalizable if it

1. is a self-adjoint matrix, that is, $\mathbf{A} = \mathbf{A}^*$ $\quad (\mathbf{A}^* = \overline{\mathbf{A}}^T$ is the adjoint matrix);
2. is normal, namely, $\mathbf{A}\mathbf{A}^* = \mathbf{A}^*\mathbf{A}$;
3. has distinct eigenvalues. ■

Other than Theorem 7.10 and Theorem 7.13 on page 400, necessary and sufficient conditions for a matrix to be diagonalizable are still unknown. When the entries of an $n \times n$ matrix $\mathbf{A}$ are integers, then the matrix $\mathbf{A}$ has the Smith[72] canonical form $\mathbf{U}\mathbf{A}\mathbf{V} = \mathbf{\Lambda}$, where $\mathbf{U}$ and $\mathbf{V}$ are unimodular matrices (which means that they have the determinant equal to ± 1), and $\mathbf{\Lambda}$ is a diagonal matrix with integer values, called the invariant factors of $\mathbf{A}$. However, this canonical form is rarely used in practical calculations.

We illustrate the product of matrices by considering two 2×2 matrices:

$$\mathbf{A} = \begin{bmatrix} a_{11} & a_{12} \\ a_{21} & a_{22} \end{bmatrix} \quad \text{and} \quad \mathbf{B} = \begin{bmatrix} b_{11} & b_{12} \\ b_{21} & b_{22} \end{bmatrix}.$$

Their product is

$$\mathbf{A}\mathbf{B} = \begin{bmatrix} a_{11}b_{11} + a_{12}b_{21} & a_{11}b_{12} + a_{12}b_{22} \\ a_{21}b_{11} + a_{22}b_{21} & a_{21}b_{12} + a_{22}b_{22} \end{bmatrix}.$$

It is convenient to rewrite the matrix $\mathbf{B}$ in column form as

$$\mathbf{B} = [\mathbf{b}_1, \mathbf{b}_2], \quad \text{where} \quad \mathbf{b}_1 = \begin{bmatrix} b_{11} \\ b_{21} \end{bmatrix}, \quad \mathbf{b}_2 = \begin{bmatrix} b_{12} \\ b_{22} \end{bmatrix}.$$

Then

$$\mathbf{A}\mathbf{B} = [\mathbf{A}\,\mathbf{b}_1, \ \mathbf{A}\,\mathbf{b}_2].$$

Therefore, in general,

$$\mathbf{A}\mathbf{B} = [\mathbf{A}\,\mathbf{b}_1, \ \mathbf{A}\,\mathbf{b}_2, \ldots, \mathbf{A}\,\mathbf{b}_n],$$

[72]The British mathematician Henry John Stephen Smith (1826–1883) discovered this property in 1861.

where $\mathbf{b}_j$ $(j = 1, 2, \ldots, n)$ is the j-th column vector of the $n \times n$ matrix $\mathbf{B} = [\mathbf{b}_1, \mathbf{b}_2, \ldots, \mathbf{b}_n]$.

Suppose that a square $n \times n$ matrix $\mathbf{A}$ has n linearly independent eigenvectors

$$\mathbf{x}_1,\ \mathbf{x}_2, \ldots, \mathbf{x}_n,$$

where $\mathbf{x}_i = \langle x_{1i}, x_{2i}, \ldots, x_{ni}\rangle^T$, $i = 1, 2, \ldots n$. Let λ_1, λ_2, …, λ_n be the corresponding eigenvalues. We construct a nonsingular matrix $\mathbf{S}$ from the eigenvectors, that is,

$$\mathbf{S} = \begin{bmatrix} x_{11} & x_{12} & \cdots & x_{1n} \\ x_{21} & x_{22} & \cdots & x_{2n} \\ \vdots & \vdots & \ddots & \vdots \\ x_{n1} & x_{n2} & \cdots & x_{nn} \end{bmatrix}, \qquad \det(\mathbf{S}) \neq 0,$$

where every column is an eigenvector. Recall that the condition $\det(\mathbf{S}) \neq 0$ is necessary and sufficient for vectors $\mathbf{x}_i$, $i = 1, 2, \ldots n$, to be linearly independent. Since $\mathbf{A}\mathbf{x}_i = \lambda_i \mathbf{x}_i$, we have

$$\mathbf{AS} = [\mathbf{A}\mathbf{x}_1, \mathbf{A}\mathbf{x}_2, \ldots, \mathbf{A}\mathbf{x}_n] = \begin{bmatrix} \lambda_1 x_{11} & \lambda_2 x_{12} & \cdots & \lambda_n x_{1n} \\ \lambda_1 x_{21} & \lambda_2 x_{22} & \cdots & \lambda_n x_{2n} \\ \vdots & \vdots & \ddots & \vdots \\ \lambda_1 x_{n1} & \lambda_2 x_{n2} & \cdots & \lambda_n x_{nn} \end{bmatrix}.$$

On the other hand, let $\mathbf{\Lambda}$ be the diagonal matrix having λ_1, λ_2, …, λ_n on its diagonal:

$$\mathbf{\Lambda} = \begin{bmatrix} \lambda_1 & 0 & \cdots & 0 \\ 0 & \lambda_2 & \cdots & 0 \\ \vdots & \vdots & \ddots & \vdots \\ 0 & 0 & \cdots & \lambda_n \end{bmatrix}. \tag{7.4.1}$$

Then

$$\mathbf{S}\,\mathbf{\Lambda} = \begin{bmatrix} x_{11}\lambda_1 & x_{12}\lambda_2 & \cdots & x_{1n}\lambda_n \\ x_{21}\lambda_1 & x_{22}\lambda_2 & \cdots & x_{2n}\lambda_n \\ \vdots & \vdots & \ddots & \vdots \\ x_{n1}\lambda_1 & x_{n2}\lambda_2 & \cdots & x_{nn}\lambda_n \end{bmatrix}.$$

From the above equation, it follows that $\mathbf{S}\,\mathbf{\Lambda} = \mathbf{AS}$ so $\mathbf{A} \sim \mathbf{\Lambda}$. Therefore, a matrix $\mathbf{S}$ can be constructed from the eigenvectors of the given matrix $\mathbf{A}$.

Example 7.4.2: Let $\mathbf{j}$ be the imaginary unit vector in the positive vertical direction on the complex plane $\mathbb{C}$ $(\mathbf{j}^2 = -1)$, and let

$$\mathbf{A}_1 = \begin{bmatrix} 1 & \mathbf{j} \\ \mathbf{j} & 3 \end{bmatrix}$$

be a symmetric matrix, but nonself-adjoint. It has one eigenvalue $\lambda = 2$ of algebraic multiplicity two. To this eigenvalue corresponds a one-dimensional space of eigenvectors that is spanned by the complex vector

$$\begin{bmatrix} \mathbf{j} \\ 1 \end{bmatrix}.$$

Hence, the geometric multiplicity of $\lambda = 2$ is 1 and, as a result, this symmetric matrix with complex entries is not diagonalizable. Note that a real symmetric matrix is always diagonalizable (Theorem 7.11, page 392). The self-adjoint matrix

$$\mathbf{A}_2 = \begin{bmatrix} 1 & \mathbf{j} \\ -\mathbf{j} & 3 \end{bmatrix}$$

has two distinct real positive eigenvalues

$$\lambda_1 = 2 + \sqrt{2} \approx 3.41421 \qquad \text{and} \qquad \lambda_2 = 2 - \sqrt{2} \approx 0.585786$$

with complex eigenvectors

$$\mathbf{x}_1 = \begin{bmatrix} \mathbf{j}(\sqrt{2}-1) \\ 1 \end{bmatrix} \quad \text{and} \quad \mathbf{x}_2 = \begin{bmatrix} \mathbf{j}(\sqrt{2}+1) \\ 1 \end{bmatrix} = \overline{\mathbf{x}}_1,$$

respectively. Hence, the matrix $\mathbf{A}_2$ is diagonalizable; all its eigenvalues are positive real numbers, so $\mathbf{A}_2$ is a positive matrix. The matrix $\mathbf{S}$ of these column eigenvectors is

$$\mathbf{S} = \begin{bmatrix} \mathbf{j}(\sqrt{2}-1) & -\mathbf{j}(\sqrt{2}+1) \\ 1 & 1 \end{bmatrix} \qquad \text{and} \qquad \det(\mathbf{S}) = 2\mathbf{j}\sqrt{2}.$$

Since $\mathbf{S}^{-1} = \frac{1}{2\mathbf{j}\sqrt{2}} \begin{bmatrix} 1 & \mathbf{j}(\sqrt{2}+1) \\ -1 & \mathbf{j}(\sqrt{2}-1) \end{bmatrix}$, we have

$$\mathbf{S}^{-1}\mathbf{A}_2\mathbf{S} = \begin{bmatrix} 2+\sqrt{2} & 0 \\ 0 & 2-\sqrt{2} \end{bmatrix}.$$

Example 7.4.3: (Normal matrix) The matrix $\begin{bmatrix} \mathbf{j} & 0 & -\mathbf{j} \\ 0 & \mathbf{j} & 0 \\ \mathbf{j} & 0 & \mathbf{j} \end{bmatrix}$ is normal, but not self-adjoint. It has three distinct (complex) eigenvalues $\mathbf{j}$ and $\pm 1 + \mathbf{j}$; therefore, the matrix is diagonalizable.

Another nonsymmetric matrix $\mathbf{M} = \begin{bmatrix} 4 & 0 & 2 \\ 2 & 4 & 0 \\ 0 & 2 & 4 \end{bmatrix}$ is normal, which means that

$$\mathbf{M}\,\mathbf{M}^* = \begin{bmatrix} 4 & 0 & 2 \\ 2 & 4 & 0 \\ 0 & 2 & 4 \end{bmatrix} \begin{bmatrix} 4 & 2 & 0 \\ 0 & 4 & 2 \\ 2 & 0 & 4 \end{bmatrix} = \mathbf{M}^*\,\mathbf{M} = \begin{bmatrix} 20 & 8 & 8 \\ 8 & 20 & 8 \\ 8 & 8 & 20 \end{bmatrix}.$$

The matrix $\mathbf{M}$ has three simple eigenvalues 3 and $3 \pm \mathbf{j}\sqrt{3}$, so it is diagonalizable.

Example 7.4.4: Let us consider a 3×2-matrix

$$\mathbf{B} = \begin{bmatrix} 1 & 2 \\ 0 & 1 \\ 1 & -1 \end{bmatrix}, \quad \text{with} \quad \mathbf{B}^T = \begin{bmatrix} 1 & 0 & 1 \\ 2 & 1 & -1 \end{bmatrix}.$$

We can construct two (diagonalizable) matrices:

$$\mathbf{B}_1 = \mathbf{B}\,\mathbf{B}^T = \begin{bmatrix} 5 & 2 & -1 \\ 2 & 1 & -1 \\ -1 & -1 & 2 \end{bmatrix} \quad \text{and} \quad \mathbf{B}_2 = \mathbf{B}^T\,\mathbf{B} = \begin{bmatrix} 2 & 1 \\ 1 & 6 \end{bmatrix}.$$

The singular matrix $\mathbf{B}_1$ has three real eigenvalues

$$\lambda_1 = 4 + \sqrt{5} \approx 6.23607, \quad \lambda_2 = 4 - \sqrt{5} \approx 1.76393, \quad \lambda_3 = 0.$$

The matrix $\mathbf{B}_2$ has two eigenvalues $\lambda_1 = 4 + \sqrt{5}$ and $\lambda_2 = 4 - \sqrt{5}$. It is not a coincidence that matrices $\mathbf{B}_1$ and $\mathbf{B}_2$ have common eigenvalues. □

Now we are in the position to define a function of a diagonal matrix (7.4.1) as

$$f(\mathbf{\Lambda}) = \begin{bmatrix} f(\lambda_1) & 0 & \cdots & 0 \\ 0 & f(\lambda_2) & \cdots & 0 \\ \vdots & \vdots & \ddots & \vdots \\ 0 & 0 & \cdots & f(\lambda_n) \end{bmatrix}. \tag{7.4.2}$$

The powers of diagonalizable matrices can be obtained as follows. Let $\mathbf{\Lambda} = \mathbf{S}^{-1}\mathbf{A}\mathbf{S}$ be a diagonal matrix similar to $\mathbf{A}$. Then

$$\mathbf{A}^2 = \mathbf{A}\mathbf{A} = \mathbf{S}\mathbf{\Lambda}\mathbf{S}^{-1}\mathbf{S}\mathbf{\Lambda}\mathbf{S}^{-1} = \mathbf{S}\mathbf{\Lambda}^2\mathbf{S}^{-1}.$$

Similarly, for any positive integer m, we have

$$\mathbf{A}^m = \mathbf{S}\mathbf{\Lambda}\mathbf{S}^{-1}\cdots\mathbf{S}\mathbf{\Lambda}\mathbf{S}^{-1} = \mathbf{S}\mathbf{\Lambda}^m\mathbf{S}^{-1}.$$

That is why for an arbitrary polynomial $f(\lambda) = a_0\lambda^m + a_{m-1}\lambda^{m-1} + \cdots + a_m$, we have

$$f(\mathbf{A}) = \mathbf{S}f(\mathbf{\Lambda})\mathbf{S}^{-1}.$$

With this in hand, we define a function of an arbitrary diagonalizable matrix $\mathbf{A}$:

$$f(\mathbf{A}) = \mathbf{S}\mathbf{S}^{-1}f(\mathbf{A})\mathbf{S}\mathbf{S}^{-1} = \mathbf{S}f\left(\mathbf{S}^{-1}\mathbf{A}\mathbf{S}\right)\mathbf{S}^{-1} = \mathbf{S}f\left(\mathbf{\Lambda}\right)\mathbf{S}^{-1}.$$

Thus,

$$f(\mathbf{A}) = \mathbf{S}f\left(\mathbf{\Lambda}\right)\mathbf{S}^{-1} = \mathbf{S}\begin{bmatrix} f(\lambda_1) & 0 & \cdots & 0 \\ 0 & f(\lambda_2) & \cdots & 0 \\ \vdots & \vdots & \ddots & \vdots \\ 0 & 0 & \cdots & f(\lambda_n) \end{bmatrix}\mathbf{S}^{-1}. \tag{7.4.3}$$

Remark. We cannot use this formula (7.4.3) if the function $f(\lambda)$ is undefined for some eigenvalue of a matrix $\mathbf{A}$. The function $f(\lambda)$ must be bounded on the spectrum of the matrix $\mathbf{A}$, otherwise $f(\mathbf{A})$ does not exist. For example, the exponential function $e^{\mathbf{A}t}$ exists for an arbitrary square matrix $\mathbf{A}$ because $e^{\lambda t}$ is defined for every λ (complex or real). However, the function $f(\lambda) = \lambda^{-1}$ can be applied only to nonsingular matrices.

Example 7.4.5: Let us consider the 2×2 matrix $\mathbf{A} = \begin{bmatrix} 9 & 8 \\ 7 & 8 \end{bmatrix}$, which has two simple eigenvalues $\lambda_1 = 1$ and $\lambda_2 = 16$. The corresponding eigenvectors are as follows: $\mathbf{x}_1 = \langle 1, -1\rangle^T$ and $\mathbf{x}_2 = \langle 8, 7\rangle^T$. Taking the diagonal matrix of its eigenvalues $\mathbf{\Lambda}$, and building the matrix of eigenvectors $\mathbf{S}$, we obtain

$$\mathbf{S} = \begin{bmatrix} 1 & 8 \\ -1 & 7 \end{bmatrix}, \quad \mathbf{S}^{-1} = \frac{1}{15}\begin{bmatrix} 7 & -8 \\ 1 & 1 \end{bmatrix}, \quad \text{and} \quad \mathbf{\Lambda} = \mathbf{S}^{-1}\mathbf{A}\mathbf{S} = \begin{bmatrix} 1 & 0 \\ 0 & 16 \end{bmatrix}.$$

According to Eq. (7.4.3), for the function $f(\lambda) = \frac{\lambda-1}{\lambda+1}$, we have that

$$\begin{aligned} \frac{\mathbf{A}-\mathbf{I}}{\mathbf{A}+\mathbf{I}} &= \mathbf{S}\begin{bmatrix} f(\lambda_1) & 0 \\ 0 & f(\lambda_2) \end{bmatrix}\mathbf{S}^{-1} = \mathbf{S}\begin{bmatrix} 0 & 0 \\ 0 & 15/17 \end{bmatrix}\mathbf{S}^{-1} \\ &= \begin{bmatrix} 8 & 8 \\ 7 & 7 \end{bmatrix}\frac{1}{34}\begin{bmatrix} 9 & -8 \\ -7 & 10 \end{bmatrix} = \frac{1}{17}\begin{bmatrix} 8 & 8 \\ 7 & 7 \end{bmatrix} = \frac{1}{17}(\mathbf{A}-\mathbf{I}). \end{aligned}$$

However, the fractions

$$\frac{\mathbf{A}+\mathbf{I}}{\mathbf{A}-\mathbf{I}} \quad \text{and} \quad \frac{\mathbf{A}+\mathbf{I}}{\mathbf{A}-16\,\mathbf{I}}$$

do not exist because the corresponding functions $(\lambda+1)(\lambda-1)^{-1}$ and $(\lambda+1)(\lambda-16)^{-1}$ are undefined for $\lambda = 1$ and $\lambda = 16$ (the spectrum of the matrix $\mathbf{A}$). We can determine $f(\mathbf{A})$ for functions that are defined on the spectrum of $\mathbf{A}$, for instance, for $f(\lambda) = \lambda^{-1}$ or $f(\lambda) = e^{\lambda t}$, as the following calculations show.

$$\mathbf{A}^{-1} = \mathbf{S}\begin{bmatrix} 1 & 0 \\ 0 & 1/16 \end{bmatrix}\mathbf{S}^{-1} = \frac{1}{16}\begin{bmatrix} 8 & -8 \\ -7 & 9 \end{bmatrix}.$$

$$e^{\mathbf{A}t} = \begin{bmatrix} 1 & 8 \\ -1 & 7 \end{bmatrix}\cdot\begin{bmatrix} e^t & 0 \\ 0 & e^{16t} \end{bmatrix}\cdot\frac{1}{15}\begin{bmatrix} 7 & -8 \\ 1 & 1 \end{bmatrix} = \frac{1}{15}\begin{bmatrix} 8\,e^{16t}+7\,e^t & 8\,e^{16t}-8\,e^t \\ 7\,e^{16t}-7\,e^t & 7\,e^{16t}+8\,e^t \end{bmatrix}.$$

Now we turn our attention to the definition of a root of a matrix. Since the square root, $f(\lambda) = \sqrt{\lambda}$, is not a function (that assigns a unique output to every input), but rather the analytic function of a complex variable: every input has two output values depending on the branch chosen. Using the square roots of $\lambda = 1, 16$, we may define four possible roots of the diagonal matrix:

$$\mathbf{\Lambda}_1 = \begin{bmatrix} 1 & 0 \\ 0 & 4 \end{bmatrix}, \quad \mathbf{\Lambda}_2 = \begin{bmatrix} -1 & 0 \\ 0 & -4 \end{bmatrix}, \quad \mathbf{\Lambda}_3 = \begin{bmatrix} -1 & 0 \\ 0 & 4 \end{bmatrix}, \quad \mathbf{\Lambda}_4 = \begin{bmatrix} 1 & 0 \\ 0 & -4 \end{bmatrix}.$$

According to Eq. (7.4.3), the corresponding roots of the matrix $\mathbf{A}$ are as follows:

$$\mathbf{R}_1 \stackrel{\text{def}}{=} \mathbf{S}\mathbf{\Lambda}_1\mathbf{S}^{-1} = \frac{1}{5}\begin{bmatrix} 13 & 8 \\ 7 & 12 \end{bmatrix}, \quad \mathbf{R}_2 \stackrel{\text{def}}{=} \mathbf{S}\mathbf{\Lambda}_2\mathbf{S}^{-1} = -\frac{1}{5}\begin{bmatrix} 13 & 8 \\ 7 & 12 \end{bmatrix} = -\mathbf{R}_1,$$

$$\mathbf{R}_3 \stackrel{\text{def}}{=} \mathbf{S}\mathbf{\Lambda}_3\mathbf{S}^{-1} = \frac{1}{3}\begin{bmatrix} 5 & 8 \\ 7 & 4 \end{bmatrix}, \quad \mathbf{R}_4 \stackrel{\text{def}}{=} \mathbf{S}\mathbf{\Lambda}_4\mathbf{S}^{-1} = -\frac{1}{3}\begin{bmatrix} 5 & 8 \\ 7 & 4 \end{bmatrix} = -\mathbf{R}_3.$$

Each of the matrices $\mathbf{R}_k$, $k = 1, 2, 3, 4$, is a root of the square matrix $\mathbf{A}$ because $\mathbf{R}_k^2 = \mathbf{A}$. These matrices have distinct eigenvalues, but they share the same eigenvectors, as the matrix $\mathbf{A}$. For each root, we define four exponential functions according to the formula $e^{\mathbf{R}_k t} = \mathbf{S}e^{\mathbf{\Lambda}_k t}\mathbf{S}^{-1}$ to obtain

$$e^{\mathbf{R}_1 t} = \frac{1}{15}\begin{bmatrix} 8\,e^{4t} + 7\,e^{t} & 8\,e^{4t} - 8\,e^{t} \\ 7\,e^{4t} - 7\,e^{t} & 7\,e^{4t} + 8\,e^{t} \end{bmatrix}, \qquad e^{\mathbf{R}_2 t} = \frac{1}{15}\begin{bmatrix} 8\,e^{-4t} + 7\,e^{-t} & 8\,e^{-4t} - 8\,e^{-t} \\ 7\,e^{-4t} - 7\,e^{-t} & 7\,e^{-4t} + 8\,e^{-t} \end{bmatrix},$$

$$e^{\mathbf{R}_3 t} = \frac{1}{15}\begin{bmatrix} 8\,e^{4t} + 7\,e^{-t} & 8\,e^{4t} - 8\,e^{-t} \\ 7\,e^{4t} - 7\,e^{-t} & 7\,e^{4t} + 8\,e^{-t} \end{bmatrix}, \qquad e^{\mathbf{R}_4 t} = \frac{1}{15}\begin{bmatrix} 8\,e^{-4t} + 7\,e^{t} & 8\,e^{-4t} - 8\,e^{t} \\ 7\,e^{-4t} - 7\,e^{t} & 7\,e^{-4t} + 8\,e^{t} \end{bmatrix}.$$

Now we consider two functions that involve the root

$$\frac{\sin(\sqrt{\lambda}t)}{\sqrt{\lambda}} = \sum_{k\geqslant 0}(-1)^k\lambda^k\frac{t^{2k+1}}{(2k+1)!}, \qquad \cos(\sqrt{\lambda}t) = \sum_{k\geqslant 0}(-1)^k\lambda^k\frac{t^{2k}}{(2k)!}.$$

However, as their Maclaurin series show, these functions are actually entire functions because their series converge for all of the parameter values. Therefore, we expect that Eq. (7.4.3) will give the same result for each root $\mathbf{R}_k$, $k = 1, 2, 3, 4$. Indeed,

$$\begin{aligned}
\frac{\sin(\sqrt{\mathbf{A}}t)}{\sqrt{\mathbf{A}}} &= \frac{\sin(\mathbf{R}_k t)}{\mathbf{R}_k} = \mathbf{S}\,\frac{\sin(\mathbf{\Lambda}_k t)}{\mathbf{\Lambda}_k}\,\mathbf{S}^{-1} = \begin{bmatrix} 1 & 8 \\ -1 & 7 \end{bmatrix}\begin{bmatrix} \sin t & 0 \\ 0 & \frac{\sin 4t}{4} \end{bmatrix}\frac{1}{15}\begin{bmatrix} 7 & -8 \\ 1 & 1 \end{bmatrix} \\
&= \frac{1}{15}\begin{bmatrix} 2\,\sin 4t + 7\,\sin t & 2\,\sin 4t - 8\,\sin t \\ \frac{7}{4}\,\sin 4t - 7\,\sin t & \frac{7}{4}\,\sin 4t + 8\,\sin t \end{bmatrix} \qquad (k = 1, 2, 3, 4), \\
\cos(\sqrt{\mathbf{A}}t) &= \cos(\mathbf{R}_k t) = \mathbf{S}\,\cos(\mathbf{\Lambda}_k t)\,\mathbf{S}^{-1} = \begin{bmatrix} 1 & 8 \\ -1 & 7 \end{bmatrix}\begin{bmatrix} \cos t & 0 \\ 0 & \cos 4t \end{bmatrix}\frac{1}{15}\begin{bmatrix} 7 & -8 \\ 1 & 1 \end{bmatrix} \\
&= \frac{1}{15}\begin{bmatrix} 8\,\cos 4t + 7\,\cos t & 8\,\cos 4t - 8\,\cos t \\ 7\,\cos 4t - 7\,\cos t & 7\,\cos 4t + 8\,\cos t \end{bmatrix} \qquad (k = 1, 2, 3, 4).
\end{aligned}$$

Maple can help you to find, say, $\cos(\mathbf{R}_1 t)$, with the following commands:

```
with(linalg):  s:=matrix(2,2,[1,8,-1,7]);
si:= inverse(s);  cc:=matrix(2,2,[cos(t),0,0,cos(4*t)]);
cosR1:=evalm(s&*cc&*si);
```

The trigonometric functions, $\cos(\sqrt{\mathbf{A}}t)$ and $\sin(\sqrt{\mathbf{A}}t)$, can also be obtained from the exponential function $e^{\sqrt{\mathbf{A}}\mathbf{j}t}$ by extracting the real part and imaginary part, respectively. The definition of the matrix function $\cos\left(\sqrt{\mathbf{A}}t\right)$ does not depend on what root $\sqrt{\mathbf{A}}$ is chosen, but the function $\sin\left(\sqrt{\mathbf{A}}t\right)$ is sensitive to such a choice because it depends on $\mathbf{R}_k$, $k = 1, 2, 3, 4$. Choosing, for instance, the root $\mathbf{R}_1$, we ask *Maple* to do this job:

```
d1:=matrix(2,2,[1,0,0,4]);  R1:=evalm(s&*d1&*si); exponential(R1,I*t);
```

Maple has a dedicated command to determine the function of a matrix (see Example 6.2.7, page 360).

Example 7.4.6: Let

$$\mathbf{A} = \begin{bmatrix} -3 & 2 & 2 \\ -6 & 5 & 2 \\ -7 & 4 & 4 \end{bmatrix} \quad \text{and} \quad \mathbf{\Lambda} = \begin{bmatrix} 1 & 0 & 0 \\ 0 & 2 & 0 \\ 0 & 0 & 3 \end{bmatrix}$$

be the 3×3 matrix and corresponding diagonal matrix of its eigenvalues. Then $\mathbf{\Lambda} = \mathbf{S}^{-1}\mathbf{A}\mathbf{S}$, where

$$\mathbf{S} = \begin{bmatrix} 1 & 2 & 3 \\ 1 & 2 & 4 \\ 1 & 3 & 5 \end{bmatrix} \quad \text{and} \quad \mathbf{S}^{-1} = \begin{bmatrix} 2 & 1 & -2 \\ 1 & -2 & 1 \\ -1 & 1 & 0 \end{bmatrix},$$

with $\det(\mathbf{S}) = -1$. Recall that each column vector of the matrix $\mathbf{S}$ is an eigenvector of the matrix $\mathbf{A}$. For the function $f(\lambda) = \frac{\sin(\lambda t)}{\lambda}$, we have

$$f(\mathbf{A}) = \frac{\sin(\mathbf{A}t)}{\mathbf{A}} = \mathbf{S}\,\frac{\sin(\mathbf{\Lambda}t)}{\mathbf{\Lambda}}\,\mathbf{S}^{-1} = \begin{bmatrix} 1 & 2 & 3 \\ 1 & 2 & 4 \\ 1 & 3 & 5 \end{bmatrix} \cdot \begin{bmatrix} \sin(t) & 0 & 0 \\ 0 & \frac{\sin(2t)}{2} & 0 \\ 0 & 0 & \frac{\sin(3t)}{3} \end{bmatrix} \cdot \begin{bmatrix} 2 & 1 & -2 \\ 1 & -2 & 1 \\ -1 & 1 & 0 \end{bmatrix}$$

$$= \sin t \begin{bmatrix} 3 + 2\cos t - 4\cos^2 t & -4\cos t + 4\cos^2 t & -1 + 2\cos t \\ \frac{10}{3} + 2\cos t - \frac{16}{3}\cos^2 t & -\frac{1}{3} - 4\cos t + \frac{16}{3}\cos^2 t & -1 + 2\cos t \\ \frac{11}{3} + 3\cos t - \frac{20}{3}\cos^2 t & -\frac{2}{3} - 6\cos t + \frac{20}{3}\cos^2 t & -1 + 3\cos t \end{bmatrix}.$$

Similarly, we have

$$g(\mathbf{A}) \stackrel{\text{def}}{=} \cos(\mathbf{A}t) = \mathbf{S}\,\cos(\mathbf{\Lambda}t)\,\mathbf{S}^{-1} = \begin{bmatrix} 1 & 2 & 3 \\ 1 & 2 & 4 \\ 1 & 3 & 5 \end{bmatrix} \cdot \begin{bmatrix} \sin(t) & 0 & 0 \\ 0 & \cos(2t) & 0 \\ 0 & 0 & \cos(3t) \end{bmatrix} \cdot \begin{bmatrix} 2 & 1 & -2 \\ 1 & -2 & 1 \\ -1 & 1 & 0 \end{bmatrix}$$

$$= \begin{bmatrix} 2\cos t + 2\cos(2t) - 3\cos(3t) & \cos t - 4\cos(2t) + 3\cos(3t) & 2\cos(2t) - \cos t \\ 2\cos t + 2\cos(2t) - 4\cos(3t) & \cos t - 4\cos(2t) + 4\cos(3t) & 2\cos(2t) - \cos t \\ 2\cos t + 3\cos(2t) - 5\cos(3t) & \cos t - 6\cos(2t) + 5\cos(3t) & 3\cos(2t) - \cos t \end{bmatrix}.$$

Both $f(\mathbf{A})$ and $g(\mathbf{A})$ are solutions of the following matrix differential equation $\ddot{\mathbf{\Phi}} + \mathbf{A}^2\mathbf{\Phi}(t) = \mathbf{0}$, namely,

$$\frac{d^2}{dt^2}\left(\frac{\sin(\mathbf{A}t)}{\mathbf{A}}\right) + \mathbf{A}^2\left(\frac{\sin(\mathbf{A}t)}{\mathbf{A}}\right) = \mathbf{0}, \qquad \frac{d^2}{dt^2}\left(\cos(\mathbf{A}t)\right) + \mathbf{A}^2\cos(\mathbf{A}t) = \mathbf{0}.$$

Problems

1. Are the following matrices diagonalizable?

$$\text{(a)} \begin{bmatrix} 2 & \mathbf{j}/2 \\ \mathbf{j}/2 & 3 \end{bmatrix}; \qquad \text{(b)} \begin{bmatrix} 2 & \mathbf{j}/2 \\ -\mathbf{j}/2 & 3 \end{bmatrix}; \qquad \text{(c)} \begin{bmatrix} 2 & -1 \\ 2 & 0 \end{bmatrix}.$$

2. For the following pairs of matrices (a, b, c, and d are real constants),

$$\text{(a)}\ \mathbf{A} = \begin{bmatrix} 0 & 1 \\ 1 & 0 \end{bmatrix} \text{ and } \mathbf{B} = \begin{bmatrix} -1 & 0 \\ 0 & 1 \end{bmatrix}; \qquad \text{(b)}\ \mathbf{A} = \begin{bmatrix} a & b \\ c & d \end{bmatrix} \text{ and } \mathbf{B} = \begin{bmatrix} a & -b \\ -c & d \end{bmatrix};$$

   (a) prove that they are similar;

   (b) show that matrices **A** and **B** have distinct eigenvectors.

3. Using the diagonalization procedure, find the functions $f(\mathbf{A})$ and $g(\mathbf{A})$, where $f(\lambda) = e^{\lambda t}$ and $g(\lambda) = \dfrac{\lambda^2 - 2}{\lambda^2 + 4}$, of the following 2×2 matrices:

$$\text{(a)} \begin{bmatrix} 0.8 & 0.3 \\ 0.2 & 0.7 \end{bmatrix}; \quad \text{(b)} \begin{bmatrix} -1 & 2 \\ -3 & 4 \end{bmatrix}; \quad \text{(c)} \begin{bmatrix} 1 & 3 \\ 4 & 2 \end{bmatrix}; \quad \text{(d)} \begin{bmatrix} 1 & 1 \\ 0 & -1 \end{bmatrix};$$

$$\text{(e)} \begin{bmatrix} -1 & 6 \\ 3 & 2 \end{bmatrix}; \quad \text{(f)} \begin{bmatrix} -4 & 6 \\ 3 & 3 \end{bmatrix}; \quad \text{(g)} \begin{bmatrix} 1 & 1 \\ 4 & -2 \end{bmatrix}; \quad \text{(h)} \begin{bmatrix} -3 & 4 \\ 6 & -5 \end{bmatrix}.$$

4. Using the diagonalization procedure, find the exponential matrix, $e^{\mathbf{A}t}$, for each of the 2×2 matrices in Problem 3 on page 390, Section 7.3.

5. Using the diagonalization procedure, find the exponential matrix, $e^{\mathbf{A}t}$, for each of the following 2×2 matrices:

(a) $\begin{bmatrix} 4 & 13 \\ 8 & 9 \end{bmatrix}$; (b) $\begin{bmatrix} 14 & 9 \\ -5 & 8 \end{bmatrix}$; (c) $\begin{bmatrix} 9 & -15 \\ 6 & 27 \end{bmatrix}$; (d) $\begin{bmatrix} 4 & 13 \\ 9 & 40 \end{bmatrix}$;

(e) $\begin{bmatrix} 4 & 1 \\ 3 & 2 \end{bmatrix}$; (f) $\begin{bmatrix} 7 & -2 \\ 12 & -3 \end{bmatrix}$; (g) $\begin{bmatrix} 17 & -10 \\ 4 & 3 \end{bmatrix}$; (h) $\begin{bmatrix} 17 & -7 \\ 8 & 2 \end{bmatrix}$.

6. Using the diagonalization procedure, find the functions $f(\mathbf{A})$ and $g(\mathbf{A})$, where $f(\lambda) = e^{\lambda t}$ and $g(\lambda) = \dfrac{\lambda^2 - 2}{\lambda^2 + 4}$, of the following 3×3 matrices:

(a) $\begin{bmatrix} 12 & -2 & 4 \\ 8 & -1 & 4 \\ -13 & 4 & -4 \end{bmatrix}$; (b) $\begin{bmatrix} 7 & -2 & 2 \\ 25 & -1 & 4 \\ -25 & 4 & -1 \end{bmatrix}$; (c) $\begin{bmatrix} 2 & -2 & 1 \\ -1 & 3 & -1 \\ 2 & -4 & 3 \end{bmatrix}$;

(d) $\begin{bmatrix} 1 & 12 & -8 \\ -1 & 9 & -4 \\ -18 & 7 & -18 \end{bmatrix}$; (e) $\begin{bmatrix} 4 & -50 & 50 \\ 10 & -2 & 3 \\ 1 & 0 & 1 \end{bmatrix}$; (f) $\begin{bmatrix} 3 & 2 & 4 \\ 12 & 5 & 12 \\ 4 & 2 & 3 \end{bmatrix}$;

(g) $\begin{bmatrix} 1 & 2 & 1 \\ -2 & 3 & 5 \\ 2 & 1 & -1 \end{bmatrix}$; (h) $\begin{bmatrix} 1 & 12 & -8 \\ -1 & 9 & -4 \\ -4 & 8 & -4 \end{bmatrix}$; (i) $\begin{bmatrix} 3 & 2 & 4 \\ 2 & 4 & 2 \\ 4 & 3 & 3 \end{bmatrix}$.

7. Using the diagonalization procedure, find the exponential matrix, $e^{\mathbf{A}t}$, for each of the 3×3 matrices in Problem 4 on page 390, §7.3.

8. For each of the following 2×2 matrices, find all square roots.

(a) $\begin{bmatrix} 20 & 4 \\ 5 & 21 \end{bmatrix}$; (b) $\begin{bmatrix} 2 & 1 \\ 2 & 3 \end{bmatrix}$; (c) $\begin{bmatrix} 5 & 4 \\ 11 & 12 \end{bmatrix}$; (d) $\begin{bmatrix} 8 & 4 \\ 1 & 5 \end{bmatrix}$;

(e) $\begin{bmatrix} 9 & 8 \\ 10 & 20 \end{bmatrix}$; (f) $\begin{bmatrix} 10 & 6 \\ 6 & 10 \end{bmatrix}$; (g) $\begin{bmatrix} 3 & -1 \\ -2 & 2 \end{bmatrix}$; (h) $\begin{bmatrix} 5 & -4 \\ -1 & 8 \end{bmatrix}$;

(i) $\begin{bmatrix} 31 & 5 \\ -18 & 10 \end{bmatrix}$; (j) $\begin{bmatrix} 1 & 3 \\ 1 & 3 \end{bmatrix}$; (k) $\begin{bmatrix} 3 & -3 \\ 2 & 10 \end{bmatrix}$; (l) $\begin{bmatrix} 7 & 3 \\ -6 & -2 \end{bmatrix}$;

(m) $\begin{bmatrix} 1 & 0 \\ 4 & 9 \end{bmatrix}$; (n) $\begin{bmatrix} -4 & -20 \\ 2 & 9 \end{bmatrix}$; (o) $\begin{bmatrix} 9 & 7 \\ 0 & 4 \end{bmatrix}$; (p) $\begin{bmatrix} -1 & 5 \\ -4 & 11 \end{bmatrix}$.

9. For each of the matrices in the previous Problem, determine $\frac{\sin(\sqrt{\mathbf{A}}\,t)}{\sqrt{\mathbf{A}}}$ and $\cos\left(\sqrt{\mathbf{A}}\,t\right)$, and show that the result does not depend on the choice of the root.

10. For each of the matrices in Problem 8, find $\mathbf{R}(t) \stackrel{\text{def}}{=} e^{\sqrt{\mathbf{A}}\,\mathrm{j}t}$ ($\mathrm{j}^2 = -1$) for each root of the matrix $\mathbf{A}$; then extract its real part (denote it by $\cos\left(\sqrt{\mathbf{A}}\,t\right)$) and imaginary part (denote it by $\sin\left(\sqrt{\mathbf{A}}\,t\right)$), and show that they satisfy the matrix equations:

$$\cos^2\left(\sqrt{\mathbf{A}}\,t\right) + \sin^2\left(\sqrt{\mathbf{A}}\,t\right) = \mathbf{I}, \qquad \frac{d^2}{dt^2}\left[\cos\left(\sqrt{\mathbf{A}}\,t\right)\right] + \mathbf{A}\cos\left(\sqrt{\mathbf{A}}\,t\right) = \mathbf{0}.$$

11. Determine which of the following 3×3 matrices is diagonalizable.

(a) $\begin{bmatrix} 1 & 0 & 1 \\ 1 & 1 & 0 \\ 0 & 0 & 1 \end{bmatrix}$; (b) $\begin{bmatrix} 7 & 2 & -6 \\ 3 & 29 & 3 \\ 6 & -5 & -5 \end{bmatrix}$; (c) $\begin{bmatrix} 3 & 0 & 7 \\ 15 & 1 & -5 \\ 0 & 0 & 3 \end{bmatrix}$;

(d) $\begin{bmatrix} 25 & -8 & 30 \\ 24 & -4 & 30 \\ -12 & 4 & -14 \end{bmatrix}$; (e) $\begin{bmatrix} 3 & 43 & 43 \\ -1 & 1 & -1 \\ 2 & 1 & 2 \end{bmatrix}$; (f) $\begin{bmatrix} 5 & -1 & 0 \\ -1 & 1 & 2 \\ 5 & 1 & 1 \end{bmatrix}$;

(g) $\begin{bmatrix} 91 & 72 & -18 \\ -135 & -107 & 27 \\ -105 & -84 & 22 \end{bmatrix}$; (h) $\begin{bmatrix} -35 & -99 & 99 \\ 51 & 151 & -153 \\ 39 & 117 & -119 \end{bmatrix}$; (i) $\begin{bmatrix} 3 & -2 & 2 \\ 0 & -1 & 0 \\ 4 & -2 & -3 \end{bmatrix}$;

(j) $\begin{bmatrix} -11 & 25 & 13 \\ 6 & -9 & -6 \\ -32 & 61 & 34 \end{bmatrix}$; (k) $\begin{bmatrix} 3 & 5 & 1 \\ 2 & -3 & -2 \\ -4 & 17 & 8 \end{bmatrix}$; (l) $\begin{bmatrix} 6 & 1 & -1 \\ 1 & 0 & -1 \\ 1 & 7 & 4 \end{bmatrix}$.

12. For each of the following nonnegative matrices, determine $\frac{\sin(\sqrt{\mathbf{A}}\,t)}{\sqrt{\mathbf{A}}}$ and $\cos\left(\sqrt{\mathbf{A}}\,t\right)$, and show that the result does not depend on the choice of the root.

(a) $\begin{bmatrix} 24 & -24 & -4 \\ 6 & -8 & -12 \\ -4 & 6 & 9 \end{bmatrix}$; (b) $\begin{bmatrix} 1 & 6 & 6 \\ 3 & 10 & -6 \\ 1 & 6 & 2 \end{bmatrix}$; (c) $\begin{bmatrix} 1 & 2 & 6 \\ 1 & 10 & -2 \\ 1 & 6 & 2 \end{bmatrix}$;

(d) $\begin{bmatrix} 1 & 10 & 6 \\ 5 & 10 & -10 \\ 1 & 6 & 2 \end{bmatrix}$; (e) $\begin{bmatrix} 1 & -1 & 1 \\ -1 & 3 & 1 \\ 1 & 1 & 3 \end{bmatrix}$; (f) $\begin{bmatrix} -1 & -1 & 1 \\ 0 & 1 & 1 \\ -2 & -1 & 3 \end{bmatrix}$;

(g) $\begin{bmatrix} 2 & 0 & -3 \\ 5 & 2 & -3 \\ 0 & 0 & 3 \end{bmatrix}$; (h) $\begin{bmatrix} -3 & 2 & 2 \\ 2 & -3 & 2 \\ 2 & 2 & -3 \end{bmatrix}$; (i) $\begin{bmatrix} 4 & 0 & 4 \\ 1 & 7 & 1 \\ 0 & 4 & 0 \end{bmatrix}$;

(j) $\begin{bmatrix} -4 & -2 & 6 \\ 6 & 3 & -6 \\ -2 & -1 & 4 \end{bmatrix}$; (k) $\begin{bmatrix} 6 & -2 & 3 \\ -2 & 3 & -2 \\ -4 & 1 & -1 \end{bmatrix}$; (l) $\begin{bmatrix} 21 & 10 & -2 \\ -22 & -11 & 50 \\ 110 & 50 & -11 \end{bmatrix}$;

(m) $\begin{bmatrix} 39 & 5 & -16 \\ -36 & -5 & 16 \\ 72 & 9 & -29 \end{bmatrix}$; (n) $\begin{bmatrix} 3 & 7 & 7 \\ -1 & 1 & -1 \\ 1 & 1 & 3 \end{bmatrix}$; (o) $\begin{bmatrix} 6 & -2 & -1 \\ -2 & 6 & 1 \\ -1 & 7 & 1 \end{bmatrix}$;

(p) $\begin{bmatrix} -2 & 3 & 3 \\ 0 & 4 & 0 \\ -6 & 3 & 7 \end{bmatrix}$; (q) $\begin{bmatrix} 9 & -15 & -5 \\ -5 & 19 & 5 \\ 15 & -45 & -11 \end{bmatrix}$; (r) $\begin{bmatrix} 15 & 17 & 1 \\ 8 & -15 & -8 \\ -10 & 65 & 26 \end{bmatrix}$.

13. For each of the following nonnegative matrices with complex entries, determine $\frac{\sin(\sqrt{\mathbf{A}}\,t)}{\sqrt{\mathbf{A}}}$ and $\cos\left(\sqrt{\mathbf{A}}\,t\right)$, and show that the result does not depend on the choice of the root.

(a) $\begin{bmatrix} 3+\mathbf{j} & 2+\mathbf{j} \\ 6-\mathbf{j} & 7-\mathbf{j} \end{bmatrix}$; (b) $\begin{bmatrix} 1+2\mathbf{j} & 2 \\ 2+3\mathbf{j} & 4-2\mathbf{j} \end{bmatrix}$; (c) $\begin{bmatrix} 9+2\mathbf{j} & 2-5\mathbf{j} \\ 2 & 10-2\mathbf{j} \end{bmatrix}$;

(d) $\begin{bmatrix} 2+3\mathbf{j} & 1-2\mathbf{j} \\ 1+5\mathbf{j} & 3-3\mathbf{j} \end{bmatrix}$; (e) $\begin{bmatrix} 6+\mathbf{j} & -3-\mathbf{j} \\ 1-2\mathbf{j} & 11-\mathbf{j} \end{bmatrix}$; (f) $\begin{bmatrix} 10+\mathbf{j} & 6-\mathbf{j} \\ 9+\mathbf{j} & 7-\mathbf{j} \end{bmatrix}$;

(g) $\begin{bmatrix} 9+\mathbf{j} & 1-\mathbf{j} \\ 3-2\mathbf{j} & 4-\mathbf{j} \end{bmatrix}$; (h) $\begin{bmatrix} 8+3\mathbf{j} & -7+\mathbf{j} \\ -2+\mathbf{j} & 5-3\mathbf{j} \end{bmatrix}$; (i) $\begin{bmatrix} 7+2\mathbf{j} & 2+2\mathbf{j} \\ 2-3\mathbf{j} & 6-2\mathbf{j} \end{bmatrix}$.

14. Show that if a square matrix $\mathbf{A}$ is nilpotent, then all its eigenvalues are zeroes. In particular, a nilpotent matrix is always singular ($\det \mathbf{A} = 0$) and its trace is zero. Recall that a matrix $\mathbf{A}$ is called nilpotent if $\mathbf{A}^p = \mathbf{0}$ for some positive integer p.

15. For each of the following 3×3 matrices, find all square roots.

(a) $\begin{bmatrix} -35 & 15 & 21 \\ 234 & -86 & -126 \\ -234 & 90 & 130 \end{bmatrix}$; (b) $\begin{bmatrix} -51 & 20 & 30 \\ -30 & 19 & 15 \\ -90 & 30 & 54 \end{bmatrix}$; (c) $\begin{bmatrix} -87 & -2 & -3 \\ -24 & 9 & 13 \\ 13 & -5 & -7 \end{bmatrix}$;

(d) $\begin{bmatrix} 97 & -32 & -48 \\ 98 & -15 & -24 \\ 144 & -48 & -71 \end{bmatrix}$; (e) $\begin{bmatrix} 177 & -48 & -126 \\ -88 & 25 & 64 \\ 264 & -72 & -191 \end{bmatrix}$; (f) $\begin{bmatrix} 16 & -3 & -9 \\ -84 & 25 & 63 \\ 48 & -12 & 32 \end{bmatrix}$.

16. For each of the following 3×3 positive matrices, determine $\frac{\sin(\sqrt{\mathbf{A}}\,t)}{\sqrt{\mathbf{A}}}$ and $\cos\left(\sqrt{\mathbf{A}}\,t\right)$, and show that the result does not depend on the choice of the root.

(a) $\begin{bmatrix} 3 & 3 & -1 \\ 0 & 6 & -5 \\ -3 & -3 & -1 \end{bmatrix}$; (b) $\begin{bmatrix} 6 & -2 & 0 \\ -5 & 5 & -2 \\ 0 & -2 & 3 \end{bmatrix}$; (c) $\begin{bmatrix} 11 & 1 & -7 \\ 1 & 2 & 4 \\ 1 & -2 & 8 \end{bmatrix}$.

17. Using the diagonalization procedure, find the exponential matrix, $e^{\mathbf{A}t}$, for each of the following 3×3 matrices:

(a) $\begin{bmatrix} 7 & 1 & 2 \\ 5 & 4 & -7 \\ 5 & 0 & 6 \end{bmatrix}$; (b) $\begin{bmatrix} 0 & -1 & -2 \\ 5 & 3 & 7 \\ -1 & 1 & 1 \end{bmatrix}$; (c) $\begin{bmatrix} 2 & -3 & 1 \\ 4 & 5 & -2 \\ 29 & -7 & 0 \end{bmatrix}$;

(d) $\begin{bmatrix} -3 & 7 & -6 \\ 7 & -1 & 7 \\ 1 & -1 & 4 \end{bmatrix}$; (e) $\begin{bmatrix} -2 & 7 & 5 \\ 2 & -1 & -2 \\ -6 & 7 & 9 \end{bmatrix}$; (f) $\begin{bmatrix} -15 & -7 & 4 \\ 34 & 16 & -4 \\ 17 & 7 & 5 \end{bmatrix}$.

7.5 Sylvester's Formula

This section presents probably the most effective algorithm to define a function of a square matrix—the Sylvester method. We discuss only a simple version of this algorithm that can be applied to diagonalizable matrices. When Sylvester's method is extended to the general case, it loses its simplicity and beauty.

Because a square matrix **A** can be viewed as a linear operator in n-dimensional vector space, there is a smallest positive integer m such that

$$\mathbf{I} = \mathbf{A}^0,\ \mathbf{A},\ \mathbf{A}^2,\ \ldots,\ \mathbf{A}^m$$

are not linearly independent. Thus, there exist (generally speaking complex) numbers q_0, q_1, ..., q_{m-1} such that

$$q(\mathbf{A}) = q_0\mathbf{I} + q_1\mathbf{A} + q_2\mathbf{A}^2 + \cdots + q_{m-1}\mathbf{A}^{m-1} + \mathbf{A}^m = \mathbf{0}.$$

Definition 7.15: A scalar polynomial $q(\lambda)$ is called an **annulled polynomial** (or *annihilating polynomial*) of the square matrix **A**, if $q(\mathbf{A}) = \mathbf{0}$, with the understanding that $\mathbf{A}^0 = \mathbf{I}$ replaces $\lambda^0 = 1$ in the substitution.

The annihilating polynomial, $\psi(\lambda)$, of least degree with leading coefficient 1 is called the **minimal polynomial** of **A**.

The degree of the minimal polynomial of an $n \times n$ matrix **A** is less than or equal to n; moreover, we can rewrite the product of its factors in compact form using the notation $\prod$, which is similar to Σ for summation, as follows:

$$\psi(\lambda) = (\lambda - \lambda_1)^{m_1}(\lambda - \lambda_2)^{m_2}\cdots(\lambda - \lambda_s)^{m_s} = \prod_{j=1}^{s}(\lambda - \lambda_j)^{m_j}, \tag{7.5.1}$$

where each number m_j, $j = 1, 2, \ldots, s$, is less than or equal to the algebraic multiplicity of the eigenvalue λ_j. So $m_1 + m_2 + \cdots + m_s \leqslant n$. The minimal polynomials can be determined from the resolvent of the matrix. From Eq. (7.2.2), page 382, it follows that the resolvent of a square matrix **A** has the form

$$\mathbf{R}_\lambda(\mathbf{A}) = (\lambda\mathbf{I} - \mathbf{A})^{-1} = \frac{1}{\chi(\lambda)}\,\mathbf{P}(\lambda),$$

where $\chi(\lambda) = \det(\lambda\mathbf{I} - \mathbf{A})$ and $\mathbf{P}(\lambda)$ is the $n \times n$ matrix of cofactors, which are polynomials in λ of degree less than n. If every entry of the matrix $\mathbf{P}(\lambda)$ has a common multiple $(\lambda - \lambda_j)$, where λ_j is an eigenvalue of **A**, then it can be canceled out with a corresponding factor of $\chi(\lambda)$. After removing all common factors, the resolvent can be written as

$$\mathbf{R}_\lambda(\mathbf{A}) = (\lambda\mathbf{I} - \mathbf{A})^{-1} = \frac{1}{\psi(\lambda)}\,\mathbf{Q}(\lambda),$$

where the minimal polynomial $\psi(\lambda)$ of the square matrix **A** and the polynomial $n \times n$ matrix $\mathbf{Q}(\lambda)$ have no common factors of positive degree. The following celebrated theorem, named after the British mathematician and lawyer Arthur Cayley (1821–1895) and an Irish physicist, astronomer, and mathematician William Rowan Hamilton (1805–1865) is crucial for understanding the material.

Theorem 7.12: [Cayley–Hamilton] Every matrix **A** is annulled by its characteristic polynomial, that is, $\chi(\mathbf{A}) = 0$, where $\chi(\lambda) = \det(\lambda\mathbf{I} - \mathbf{A})$.

Theorem 7.13: A square matrix with distinct eigenvalues μ_1, μ_2, ..., μ_s, is diagonalizable if and only if its minimal polynomial is $\psi(\lambda) = (\lambda - \mu_1)(\lambda - \mu_2)\cdots(\lambda - \mu_s)$.

Example 7.5.1: Let us consider two matrices

$$\mathbf{A}_1 = \begin{bmatrix} 1 & 1 & 0 \\ 0 & 1 & 0 \\ 0 & 0 & 1 \end{bmatrix} \quad \text{and} \quad \mathbf{A}_2 = \begin{bmatrix} 3 & -3 & 2 \\ -1 & 5 & -2 \\ -1 & 3 & 0 \end{bmatrix}.$$

Their characteristic polynomials are $\chi_1(\lambda) = (\lambda - 1)^3$ and $\chi_2(\lambda) = (\lambda - 2)^2(\lambda - 4)$, respectively, but the minimal polynomials for the matrices $\mathbf{A}_1$ and $\mathbf{A}_2$ are of the second degree: $\psi_1(\lambda) = (\lambda - 1)^2$ and $\psi_2(\lambda) = (\lambda - 2)(\lambda - 4) =$

$\lambda^2 - 6\lambda + 8$. Therefore, the second powers of these matrices can be expressed as a linear combination of $\mathbf{I}$, the identity matrix, and the matrix itself: $\mathbf{A}_1^2 - 2\mathbf{A}_1 + \mathbf{I} = \mathbf{0}$ and $\mathbf{A}_2^2 - 6\mathbf{A}_2 + 8\mathbf{I} = \mathbf{0}$. Thus, for non-diagonalizable matrix,

$$\mathbf{A}_1^2 = 2\mathbf{A}_1 - \mathbf{I} = \begin{bmatrix} 2 & 2 & 0 \\ 0 & 2 & 0 \\ 0 & 0 & 2 \end{bmatrix} - \begin{bmatrix} 1 & 0 & 0 \\ 0 & 1 & 0 \\ 0 & 0 & 1 \end{bmatrix} = \begin{bmatrix} 1 & 2 & 0 \\ 1 & 1 & 0 \\ 0 & 0 & 1 \end{bmatrix},$$

and for the diagonalizable matrix $\mathbf{A}_2$ we have

$$\mathbf{A}_2^2 = 6\mathbf{A}_2 - 8\mathbf{I} = \begin{bmatrix} 18 & -18 & 12 \\ -6 & 30 & -12 \\ -6 & 18 & 0 \end{bmatrix} - \begin{bmatrix} 8 & 0 & 0 \\ 0 & 8 & 0 \\ 0 & 0 & 8 \end{bmatrix} = \begin{bmatrix} 10 & -18 & 12 \\ -6 & 22 & -12 \\ -6 & 18 & -8 \end{bmatrix}. \qquad \square$$

Let $\mathbf{A}$ be an $n \times n$ diagonalizable matrix and let $\psi(\lambda)$ be its minimal polynomial that is also the product of linear factors:

$$\psi(\lambda) = (\lambda - \lambda_1)(\lambda - \lambda_2)\cdots(\lambda - \lambda_s) = \prod_{k=1}^{s} (\lambda - \lambda_k), \tag{7.5.2}$$

where λ_k, $k = 1, 2, \ldots s \leqslant n$, are distinct eigenvalues of the matrix $\mathbf{A}$. If $\psi(\lambda)$ coincides with the characteristic polynomial $\chi(\lambda)$, then s equals n, the dimension of the matrix $\mathbf{A}$.

Suppose that a function $f(\lambda)$ is defined on the spectrum $\sigma(\mathbf{A}) = \{\lambda_1, \lambda_2, \ldots, \lambda_s\}$ of the matrix $\mathbf{A}$, that is, the values $f(\lambda_k)$, $k = 1, 2, \ldots s$, are finite for every eigenvalue λ_k. Then we define $f(\mathbf{A})$ as

$$f(\mathbf{A}) = \sum_{k=1}^{s} f(\lambda_k)\, \mathbf{Z}_k(\mathbf{A}), \tag{7.5.3}$$

where

$$\mathbf{Z}_k(\mathbf{A}) = \frac{(\mathbf{A} - \lambda_1)\cdots(\mathbf{A} - \lambda_{k-1})(\mathbf{A} - \lambda_{k+1})\cdots(\mathbf{A} - \lambda_s)}{(\lambda_k - \lambda_1)\cdots(\lambda_k - \lambda_{k-1})(\lambda_k - \lambda_{k+1})\cdots(\lambda_k - \lambda_s)}, \quad k = 1, 2, \ldots s. \tag{7.5.4}$$

The matrices $\mathbf{Z}_k(\mathbf{A})$ are usually referred to as **Sylvester's**[73] **auxiliary matrices** (one can recognize in $\mathbf{Z}_k(\mathbf{A})$ the Lagrange interpolation polynomial). Actually, $\mathbf{Z}_k(\mathbf{A})$ is the projector operator on the eigenspace corresponding λ_k. Note that the notation $\mathbf{A} - \lambda$ means $\mathbf{A} - \lambda\mathbf{I}$, and we will learn later (§7.6) that $\mathbf{Z}_k(\mathbf{A}) = \mathrm{Res}_{\lambda_k}\, \mathbf{R}_\lambda(\mathbf{A})$.

Example 7.5.2: Let us consider a diagonalizable matrix

$$\mathbf{A} = \begin{bmatrix} 1 & 3 & 3 \\ -3 & -5 & -3 \\ 3 & 3 & 1 \end{bmatrix},$$

with the characteristic polynomial $\chi(\lambda) = \det(\lambda\mathbf{I} - \mathbf{A}) = (\lambda - 1)(\lambda + 2)^2$. Since the resolvent is

$$\mathbf{R}_\lambda(\mathbf{A}) = \frac{1}{(\lambda - 1)(\lambda + 2)} \begin{bmatrix} \lambda - 4 & -6 & -3(\lambda + 2) \\ 6 & \lambda + 8 & 3(\lambda + 2) \\ -3(\lambda + 2) & -3(\lambda + 2) & -(\lambda + 2)^2 \end{bmatrix},$$

its minimal polynomial is the polynomial in the denominator, $\psi(\lambda) = (\lambda + 2)(\lambda - 1)$, which has degree 2. So the matrix $\mathbf{A}$ has one simple eigenvalue $\lambda = 1$ with eigenspace spanned by the eigenvector $\langle 1, -1, 1\rangle^T$ and one double eigenvalue $\lambda = -2$ with two-dimensional eigenspace spanned by vectors $\langle 1, -1, 0\rangle^T$ and $\langle 1, 0, -1\rangle^T$. The corresponding auxiliary matrices are

$$\mathbf{Z}_{(1)} = \frac{\mathbf{A} + 2\mathbf{I}}{1 + 2} = \begin{bmatrix} 1 & 1 & 1 \\ -1 & -1 & -1 \\ 1 & 1 & 1 \end{bmatrix} \quad \text{and} \quad \mathbf{Z}_{(-2)} = \frac{\mathbf{A} - \mathbf{I}}{-2 - 1} = \begin{bmatrix} 0 & -1 & -1 \\ 1 & 2 & 1 \\ -1 & -1 & 0 \end{bmatrix}.$$

For the function $f(\lambda) = e^{\lambda t}$, we have

$$e^{\mathbf{A}t} = e^{-2t}\, \mathbf{Z}_{(-2)} + e^t\, \mathbf{Z}_{(1)} = e^{-2t} \begin{bmatrix} 0 & -1 & -1 \\ 1 & 2 & 1 \\ -1 & -1 & 0 \end{bmatrix} + e^t \begin{bmatrix} 1 & 1 & 1 \\ -1 & -1 & -1 \\ 1 & 1 & 1 \end{bmatrix}.$$

[73] James Joseph Sylvester (1814–1897) was a man of many talents who took music lessons from Gounoud and was prouder of his "high C" than his matrix achievements. Florence Nightingale was one of his students.

Once the auxiliary matrices are known, other functions of the matrix $\mathbf{A}$ can be determined without a problem. For instance, $\cos(\mathbf{A}t) = \cos t\, \mathbf{Z}_{(1)} + \cos 2t\, \mathbf{Z}_{(-2)}$.

Example 7.5.3: Let $\mathbf{A}$ be the matrix from Example 7.4.6 on page 396, that is,

$$\mathbf{A} = \begin{bmatrix} -3 & 2 & 2 \\ -6 & 5 & 2 \\ -7 & 4 & 4 \end{bmatrix}.$$

Since the eigenvalues of the matrix $\mathbf{A}$ are distinct, the minimal polynomial is also its characteristic polynomial, that is,

$$\psi(\lambda) = (\lambda - 1)(\lambda - 2)(\lambda - 3)$$

and Sylvester's auxiliary polynomials become

$$\mathbf{Z}_{(1)}(\mathbf{A}) = \frac{(\mathbf{A} - 2\mathbf{I})(\mathbf{A} - 3\mathbf{I})}{(1-2)(1-3)} = \frac{1}{2} \begin{bmatrix} -5 & 2 & 2 \\ -6 & 3 & 2 \\ -7 & 4 & 2 \end{bmatrix} \cdot \begin{bmatrix} -6 & 2 & 2 \\ -6 & 2 & 2 \\ -7 & 4 & 1 \end{bmatrix},$$

$$\mathbf{Z}_{(2)}(\mathbf{A}) = \frac{(\mathbf{A} - \mathbf{I})(\mathbf{A} - 3\mathbf{I})}{(2-1)(2-3)} = - \begin{bmatrix} -4 & 2 & 2 \\ -6 & 4 & 2 \\ -7 & 4 & 3 \end{bmatrix} \cdot \begin{bmatrix} -6 & 2 & 2 \\ -6 & 2 & 2 \\ -7 & 4 & 1 \end{bmatrix},$$

$$\mathbf{Z}_{(3)}(\mathbf{A}) = \frac{(\mathbf{A} - \mathbf{I})(\mathbf{A} - 2\mathbf{I})}{(3-1)(3-2)} = \frac{1}{2} \begin{bmatrix} -4 & 2 & 2 \\ -6 & 4 & 2 \\ -7 & 4 & 3 \end{bmatrix} \cdot \begin{bmatrix} -5 & 2 & 2 \\ -6 & 3 & 2 \\ -7 & 4 & 2 \end{bmatrix}.$$

Multiplying the matrices $\mathbf{A} - 2\mathbf{I}$ and $\mathbf{A} - 3\mathbf{I}$ (the order of multiplication does not matter), we get

$$\mathbf{Z}_{(1)}(\mathbf{A}) = \frac{1}{2} \begin{bmatrix} -5 & 2 & 2 \\ -6 & 3 & 2 \\ -7 & 4 & 2 \end{bmatrix} \cdot \begin{bmatrix} -6 & 2 & 2 \\ -6 & 2 & 2 \\ -7 & 4 & 1 \end{bmatrix} = \frac{1}{2} \begin{bmatrix} 6 & 2 & 2 \\ -6 & 2 & 2 \\ -7 & 4 & 1 \end{bmatrix} \cdot \begin{bmatrix} -5 & 2 & 2 \\ -6 & 3 & 2 \\ -7 & 4 & 2 \end{bmatrix} = \begin{bmatrix} 2 & 1 & -2 \\ 2 & 1 & -2 \\ 2 & 1 & -2 \end{bmatrix}.$$

Similarly,

$$\mathbf{Z}_{(2)}(\mathbf{A}) = \begin{bmatrix} 2 & -4 & 2 \\ 2 & -4 & 2 \\ 3 & -6 & 3 \end{bmatrix}, \qquad \mathbf{Z}_{(3)}(\mathbf{A}) = \begin{bmatrix} -3 & 3 & 0 \\ -4 & 4 & 0 \\ -5 & 5 & 0 \end{bmatrix}.$$

For the following functions of one variable $\frac{\lambda+1}{\lambda+2}$, $e^{\lambda t}$, and $\frac{\sin(\lambda t)}{\lambda}$, we define the corresponding matrix functions:

$$\begin{aligned}
\frac{\mathbf{A} + \mathbf{I}}{\mathbf{A} + 2\mathbf{I}} &= \frac{1+1}{1+2} \mathbf{Z}_{(1)}(\mathbf{A}) + \frac{2+1}{2+2} \mathbf{Z}_{(2)}(\mathbf{A}) + \frac{3+1}{3+2} \mathbf{Z}_{(3)}(\mathbf{A}) \\
&= \frac{2}{3} \begin{bmatrix} 2 & 1 & -2 \\ 2 & 1 & -2 \\ 2 & 1 & -2 \end{bmatrix} + \frac{3}{4} \begin{bmatrix} 2 & -4 & 2 \\ 2 & -4 & 2 \\ 3 & -6 & 3 \end{bmatrix} + \frac{4}{5} \begin{bmatrix} -3 & 3 & 0 \\ -4 & 4 & 0 \\ -5 & 5 & 0 \end{bmatrix}; \\
e^{\mathbf{A}t} &= e^t\, \mathbf{Z}_{(1)}(\mathbf{A}) + e^{2t}\, \mathbf{Z}_{(2)}(\mathbf{A}) + e^{3t}\, \mathbf{Z}_{(3)}(\mathbf{A}) \\
&= e^t \begin{bmatrix} 2 & 1 & -2 \\ 2 & 1 & -2 \\ 2 & 1 & -2 \end{bmatrix} + e^{2t} \begin{bmatrix} 2 & -4 & 2 \\ 2 & -4 & 2 \\ 3 & -6 & 3 \end{bmatrix} + e^{3t} \begin{bmatrix} -3 & 3 & 0 \\ -4 & 4 & 0 \\ -5 & 5 & 0 \end{bmatrix}; \\
\frac{\sin(\mathbf{A}t)}{\mathbf{A}} &= \sin t\, \mathbf{Z}_{(1)}(\mathbf{A}) + \frac{\sin 2t}{2} \mathbf{Z}_{(2)}(\mathbf{A}) + \frac{\sin 3t}{3} \mathbf{Z}_{(3)}(\mathbf{A}).
\end{aligned}$$

Example 7.5.4: Let $\mathbf{A}$ be the following singular matrix:

$$\mathbf{A} = \begin{bmatrix} 0 & 0 & 1 \\ 0 & 0 & -1 \\ 0 & 1 & 0 \end{bmatrix}.$$

The characteristic polynomial $\chi(\lambda) = \lambda(\lambda^2+1)$ has one real null $\lambda_1 = 0$ and two complex conjugate nulls $\lambda_{2,3} = \pm\mathbf{j}$. Since $\lambda = 0$ is an eigenvalue, $\mathbf{A}$ is a singular matrix. The minimal polynomial coincides with the characteristic one, namely, $\psi(\lambda) = \lambda(\lambda - \mathbf{j})(\lambda + \mathbf{j})$ and therefore the corresponding Sylvester auxiliary matrices are

$$\mathbf{Z}_{(0)} = \mathbf{A}^2 + \mathbf{I} = \begin{bmatrix} 0 & 1 & 0 \\ 0 & -1 & 0 \\ 0 & 0 & -1 \end{bmatrix} + \begin{bmatrix} 1 & 0 & 0 \\ 0 & 1 & 0 \\ 0 & 0 & 1 \end{bmatrix} = \begin{bmatrix} 1 & 1 & 0 \\ 0 & 0 & 0 \\ 0 & 0 & 0 \end{bmatrix},$$

$$\mathbf{Z}_{(\mathbf{j})} = \frac{\mathbf{A}(\mathbf{A}+\mathbf{j}\mathbf{I})}{\mathbf{j}(\mathbf{j}+\mathbf{j})} = -\frac{1}{2}\begin{bmatrix} 0 & 0 & 1 \\ 0 & 0 & -1 \\ 0 & 1 & 0 \end{bmatrix} \cdot \begin{bmatrix} \mathbf{j} & 0 & 1 \\ 0 & \mathbf{j} & -1 \\ 0 & 1 & \mathbf{j} \end{bmatrix} = \frac{1}{2}\begin{bmatrix} 0 & -1 & -\mathbf{j} \\ 0 & 1 & \mathbf{j} \\ 0 & -\mathbf{j} & 1 \end{bmatrix},$$

$$\mathbf{Z}_{(-\mathbf{j})} = \frac{\mathbf{A}(\mathbf{A}-\mathbf{j}\mathbf{I})}{-\mathbf{j}(-\mathbf{j}-\mathbf{j})} = \overline{\mathbf{Z}_{(\mathbf{j})}} = \frac{1}{2}\begin{bmatrix} 0 & -1 & \mathbf{j} \\ 0 & 1 & -\mathbf{j} \\ 0 & \mathbf{j} & 1 \end{bmatrix}.$$

For the functions $f(\lambda) = (\lambda+1)/(\lambda+2)$ and $g(\lambda) = e^{\lambda t}$, we have

$$\begin{aligned} \frac{\mathbf{A}+\mathbf{I}}{\mathbf{A}+2\mathbf{I}} &= \frac{0+1}{0+2}\mathbf{Z}_{(0)}(\mathbf{A}) + \frac{\mathbf{j}+1}{\mathbf{j}+2}\mathbf{Z}_{(\mathbf{j})}(\mathbf{A}) + \frac{-\mathbf{j}+1}{-\mathbf{j}+2}\mathbf{Z}_{(-\mathbf{j})}(\mathbf{A}) \\ &= \frac{1}{2}\mathbf{Z}_{(0)}(\mathbf{A}) + \frac{3+\mathbf{j}}{5}\mathbf{Z}_{(\mathbf{j})}(\mathbf{A}) + \frac{3-\mathbf{j}}{5}\mathbf{Z}_{(-\mathbf{j})}(\mathbf{A}) = \begin{bmatrix} 1/2 & -1/10 & -1/5 \\ 0 & 3/5 & 1/5 \\ 0 & -1/5 & 3/5 \end{bmatrix} \end{aligned}$$

and

$$\begin{aligned} e^{\mathbf{A}t} &= \mathbf{Z}_{(0)}(\mathbf{A}) + e^{\mathbf{j}t}\,\mathbf{Z}_{(\mathbf{j})} + e^{-\mathbf{j}t}\,\mathbf{Z}_{(-\mathbf{j})} = \mathbf{Z}_0(\mathbf{A}) + 2\Re\, e^{\mathbf{j}t}\,\mathbf{Z}_{(\mathbf{j})} \\ &= \begin{bmatrix} 1 & 1 & 0 \\ 0 & 0 & 0 \\ 0 & 0 & 0 \end{bmatrix} + e^{\mathbf{j}t}\frac{1}{2}\begin{bmatrix} 0 & -1 & -\mathbf{j} \\ 0 & 1 & \mathbf{j} \\ 0 & -\mathbf{j} & 1 \end{bmatrix} + e^{-\mathbf{j}t}\frac{1}{2}\begin{bmatrix} 0 & -1 & \mathbf{j} \\ 0 & 1 & -\mathbf{j} \\ 0 & \mathbf{j} & 1 \end{bmatrix}. \end{aligned}$$

We can break matrices $\mathbf{Z}_{(\mathbf{j})}$ and its complex conjugate $\mathbf{Z}_{(-\mathbf{j})}$ into sums

$$\mathbf{Z}_{(\mathbf{j})} = \mathbf{B} + \mathbf{j}\mathbf{C}, \quad \mathbf{Z}_{(-\mathbf{j})} = \mathbf{B} - \mathbf{j}\mathbf{C},$$

where

$$\mathbf{B} = \frac{1}{2}\begin{bmatrix} 0 & -1 & 0 \\ 0 & 1 & 0 \\ 0 & 0 & 1 \end{bmatrix}, \quad \mathbf{C} = \frac{1}{2}\begin{bmatrix} 0 & 0 & -1 \\ 0 & 0 & 1 \\ 0 & -1 & 0 \end{bmatrix}.$$

With this in hand, we rewrite $e^{\mathbf{A}t}$ as

$$\begin{aligned} e^{\mathbf{A}t} &= \mathbf{Z}_{(0)}(\mathbf{A}) + e^{\mathbf{j}t}\,(\mathbf{B}+\mathbf{j}\,\mathbf{C}) + e^{-\mathbf{j}t}\,(\mathbf{B}-\mathbf{j}\,\mathbf{C}) \\ &= \mathbf{Z}_{(0)}(\mathbf{A}) + \mathbf{B}\,\left(e^{\mathbf{j}t} + e^{-\mathbf{j}t}\right) + \mathbf{j}\,\mathbf{C}\,\left(e^{\mathbf{j}t} - e^{-\mathbf{j}t}\right) \\ &= \mathbf{Z}_{(0)}(\mathbf{A}) + 2\,\mathbf{B}\,\cos t - 2\,\mathbf{C}\,\sin t. \end{aligned}$$

Therefore,

$$e^{\mathbf{A}t} = \begin{bmatrix} 1 & 1 & 0 \\ 0 & 0 & 0 \\ 0 & 0 & 0 \end{bmatrix} + \cos t \begin{bmatrix} 0 & -1 & 0 \\ 0 & 1 & 0 \\ 0 & 0 & 1 \end{bmatrix} - \sin t \begin{bmatrix} 0 & 0 & -1 \\ 0 & 0 & 1 \\ 0 & -1 & 0 \end{bmatrix} = \begin{bmatrix} 1 & 1-\cos t & -\sin t \\ 0 & \cos t & \sin t \\ 0 & -\sin t & \cos t \end{bmatrix}.$$

Problems

1. For each of the given 2×2 matrices, find Sylvester's auxiliary matrices.

(a) $\begin{bmatrix} -2 & 6 \\ 3 & 5 \end{bmatrix}$; (b) $\begin{bmatrix} 1 & 1 \\ 4 & -2 \end{bmatrix}$; (c) $\begin{bmatrix} 1 & 3 \\ 4 & 2 \end{bmatrix}$; (d) $\begin{bmatrix} -3 & 6 \\ -1 & 4 \end{bmatrix}$;

(e) $\begin{bmatrix} 7 & 4 \\ 3 & 6 \end{bmatrix}$; (f) $\begin{bmatrix} 5 & -2 \\ 4 & 1 \end{bmatrix}$; (g) $\begin{bmatrix} -2 & 3 \\ 4 & 2 \end{bmatrix}$; (h) $\begin{bmatrix} -2 & 5 \\ 8 & 4 \end{bmatrix}$.

2. For each of the given 3×3 matrices, find Sylvester's auxiliary matrices.

(a) $\begin{bmatrix} -15 & -7 & 4 \\ 34 & 16 & -18 \\ 17 & 7 & 5 \end{bmatrix}$; (b) $\begin{bmatrix} 3 & 2 & 2 \\ -5 & -4 & -2 \\ 5 & 5 & 3 \end{bmatrix}$; (c) $\begin{bmatrix} 1 & -3 & 1 \\ 1 & 2 & 1 \\ -1 & 1 & 2 \end{bmatrix}$;

(d) $\begin{bmatrix} 4 & -3 & 4 \\ 1 & -1 & 3 \\ -3 & 3 & -3 \end{bmatrix}$; (e) $\begin{bmatrix} 5 & -4 & 5 \\ 1 & -1 & 4 \\ -4 & 4 & -4 \end{bmatrix}$; (f) $\begin{bmatrix} 4 & -3 & 6 \\ 1 & -1 & 3 \\ -1 & 3 & -3 \end{bmatrix}$.

3. For each of the given 3×3 matrices with multiple eigenvalues, find Sylvester's auxiliary matrices.

(a) $\begin{bmatrix} 1 & 0 & 6 \\ 2 & 1 & 2 \\ 0 & 0 & 2 \end{bmatrix}$; (b) $\begin{bmatrix} 2 & -2 & 1 \\ -1 & 3 & -1 \\ 2 & -4 & 3 \end{bmatrix}$; (c) $\begin{bmatrix} 2 & 1 & 1 \\ 1 & 2 & 1 \\ 1 & 1 & 2 \end{bmatrix}$;

(d) $\begin{bmatrix} 1 & 1 & 1 \\ 2 & 2 & 2 \\ 3 & 3 & 3 \end{bmatrix}$; (e) $\begin{bmatrix} 1 & 0 & 1 \\ 0 & 1 & 1 \\ 0 & 0 & -1 \end{bmatrix}$; (f) $\begin{bmatrix} -3 & 2 & 2 \\ -2 & 1 & 2 \\ -2 & 2 & 1 \end{bmatrix}$;

(g) $\begin{bmatrix} 23 & -6 & -16 \\ -11 & 4 & 8 \\ 33 & -9 & -23 \end{bmatrix}$; (h) $\begin{bmatrix} -2 & 1 & 3 \\ 28 & -5 & -21 \\ -16 & 4 & 14 \end{bmatrix}$; (i) $\begin{bmatrix} -19 & -15 & 9 \\ -42 & -28 & 18 \\ -126 & -90 & 56 \end{bmatrix}$.

4. For each of the given 3×3 matrices, find $e^{\mathbf{A}t}$, $\frac{\sin(\mathbf{A}t)}{\mathbf{A}}$, and $\cos(\mathbf{A}t)$.

(a) $\begin{bmatrix} 4 & 2 & 4 \\ 0 & 1 & 0 \\ -1 & -3 & -1 \end{bmatrix}$; (b) $\begin{bmatrix} 3 & -3 & -7 \\ 6 & -8 & -12 \\ -4 & 6 & 9 \end{bmatrix}$; (c) $\begin{bmatrix} -5 & -1 & 6 \\ -5 & 1 & 1 \\ -2 & -1 & 3 \end{bmatrix}$;

(d) $\begin{bmatrix} 6 & -2 & 10 \\ -1 & -1 & -2 \\ -1 & 1 & -1 \end{bmatrix}$; (e) $\begin{bmatrix} -3 & 7 & -12 \\ 3 & -1 & 3 \\ 2 & -2 & 4 \end{bmatrix}$; (f) $\begin{bmatrix} 7 & 0 & -4 \\ -2 & 4 & 5 \\ 1 & 0 & 2 \end{bmatrix}$;

(g) $\begin{bmatrix} -12 & -10 & 6 \\ -28 & -18 & 12 \\ -84 & -60 & 38 \end{bmatrix}$; (h) $\begin{bmatrix} 17 & 10 & -6 \\ 28 & 23 & -12 \\ 84 & 60 & -33 \end{bmatrix}$; (i) $\begin{bmatrix} 8 & 5 & -3 \\ 14 & 11 & -6 \\ 42 & 30 & -17 \end{bmatrix}$.

5. For each of the given 3×3 matrices, find square roots

(a) $\begin{bmatrix} 2 & -1 & -1 \\ 1 & 0 & -1 \\ 1 & -1 & 0 \end{bmatrix}$; (b) $\begin{bmatrix} -20 & -15 & 9 \\ -42 & -29 & 18 \\ -126 & -90 & 55 \end{bmatrix}$; (c) $\begin{bmatrix} 32 & 20 & -12 \\ 56 & 44 & -24 \\ 168 & 120 & -68 \end{bmatrix}$;

(d) $\begin{bmatrix} -31 & -25 & 15 \\ -70 & -46 & 30 \\ -210 & -150 & 94 \end{bmatrix}$; (e) $\begin{bmatrix} -167 & -120 & 72 \\ -336 & -239 & 144 \\ -1008 & -720 & 433 \end{bmatrix}$; (f) $\begin{bmatrix} 121 & 96 & -48 \\ -48 & -23 & 24 \\ 144 & 144 & -47 \end{bmatrix}$;

(g) $\begin{bmatrix} 265 & -432 & 336 \\ 240 & -407 & 336 \\ 120 & -216 & 193 \end{bmatrix}$; (h) $\begin{bmatrix} 89 & -144 & 112 \\ 80 & -135 & 112 \\ 40 & -72 & 65 \end{bmatrix}$; (i) $\begin{bmatrix} 64 & -100 & 80 \\ 45 & -71 & 60 \\ 15 & -25 & 24 \end{bmatrix}$;

(j) $\begin{bmatrix} -191 & -528 & 336 \\ 256 & 705 & -448 \\ 288 & 792 & -503 \end{bmatrix}$; (k) $\begin{bmatrix} 5 & 1 & -3 \\ 2 & -1 & 8 \\ 1 & -1 & 5 \end{bmatrix}$; (l) $\begin{bmatrix} 19 & 45 & -30 \\ 6 & 22 & -12 \\ 18 & 54 & -32 \end{bmatrix}$;

(m) $\begin{bmatrix} 88 & -693 & 441 \\ 21 & -206 & 147 \\ 21 & -231 & 172 \end{bmatrix}$; (n) $\begin{bmatrix} 16 & 12 & 72 \\ 3 & 7 & 18 \\ -3 & -3 & -14 \end{bmatrix}$; (o) $\begin{bmatrix} 94 & -220 & 50 \\ 45 & -106 & 25 \\ 45 & -110 & 29 \end{bmatrix}$.

7.6 The Resolvent Method

The goal of this section is to present another method for defining the function $f(\mathbf{A})$ for any square matrix $\mathbf{A}$ and an "arbitrary" function f. Although the resolvent method was used for a while in operator theory, its application to functions of matrices is new and was developed by the author of this textbook. In this section, we consider functions that are defined in a neighborhood of every eigenvalue from the spectrum $\sigma(\mathbf{A})$ of the matrix $\mathbf{A}$. This means that $f(\lambda)$ is $m_k - 1$ times differentiable at every eigenvalue λ_k of multiplicity m_k.

Recall from Definition 7.4 on page 383 that the **resolvent** of a square matrix $\mathbf{A}$,

$$\mathbf{R}_\lambda(\mathbf{A}) = (\lambda\mathbf{I} - \mathbf{A})^{-1}, \tag{7.6.1}$$

is a matrix function depending on a parameter λ.

Example 7.6.1: Let us consider three 2×2 matrices:

$$\mathbf{B}_1 = \begin{bmatrix} 1 & 3 \\ 1 & -1 \end{bmatrix}, \quad \mathbf{B}_2 = \begin{bmatrix} 3 & -2 \\ 5 & -3 \end{bmatrix}, \quad \mathbf{B}_3 = \begin{bmatrix} 1 & 2 \\ -2 & 5 \end{bmatrix}.$$

The resolvents of these matrices along with their characteristic polynomials are

$$\mathbf{R}_\lambda(\mathbf{B}_1) = \frac{1}{\lambda^2 - 4}\begin{bmatrix} \lambda+1 & 3 \\ 1 & \lambda-1 \end{bmatrix} \quad \text{with} \quad \chi(\lambda) = \lambda^2 - 4;$$

$$\mathbf{R}_\lambda(\mathbf{B}_2) = \frac{1}{\lambda^2 + 1}\begin{bmatrix} \lambda+3 & -2 \\ 5 & \lambda-3 \end{bmatrix} \quad \text{with} \quad \chi(\lambda) = \lambda^2 + 1;$$

$$\mathbf{R}_\lambda(\mathbf{B}_3) = \frac{1}{(\lambda - 3)^2}\begin{bmatrix} \lambda-5 & 2 \\ -2 & \lambda-1 \end{bmatrix} \quad \text{with} \quad \chi(\lambda) = (\lambda-3)^2.$$

Example 7.6.2: For symmetric matrices (nonself-adjoint)

$$\mathbf{A} = \begin{bmatrix} 1 & \mathbf{j} \\ \mathbf{j} & 3 \end{bmatrix} \quad \text{and} \quad \mathbf{B} = \begin{bmatrix} 1 & \mathbf{j} \\ \mathbf{j} & 1 \end{bmatrix},$$

we have

$$\mathbf{R}_\lambda(\mathbf{A}) = \frac{1}{\chi_A(\lambda)}\begin{bmatrix} \lambda-3 & \mathbf{j} \\ \mathbf{j} & \lambda-1 \end{bmatrix}, \qquad \mathbf{R}_\lambda(\mathbf{B}) = \frac{1}{\chi_B(\lambda)}\begin{bmatrix} \lambda-1 & \mathbf{j} \\ \mathbf{j} & \lambda-1 \end{bmatrix},$$

with corresponding characteristic polynomials $\chi_A(\lambda) = (\lambda-2)^2$, $\chi_B(\lambda) = (\lambda - 1 - \mathbf{j})(\lambda - 1 + \mathbf{j}) = (\lambda-1)^2 + 1 = \lambda^2 - 2\lambda + 2$. As we know from Example 7.4.2, page 393, the matrix $\mathbf{A}$ is defective.

Example 7.6.3: Let us reconsider matrices $\mathbf{T}$, $\mathbf{A}$, $\mathbf{T}_1, \mathbf{T}_2, \mathbf{T}_3$ from Examples 7.3.1 – 7.3.3, page 388. The minimal annulled polynomials of matrices (7.3.4) – (7.3.5) are as follows.

1. $\psi(\lambda) = (\lambda-1)^2(\lambda-2)$ for the matrix (7.3.4), page 388, because

$$\mathbf{R}_\lambda(\mathbf{T}) = \left(\lambda\begin{bmatrix} 1 & 0 & 0 \\ 0 & 1 & 0 \\ 0 & 0 & 1 \end{bmatrix} - \begin{bmatrix} 1 & 0 & 3 \\ 2 & 1 & 2 \\ 0 & 0 & 2 \end{bmatrix}\right)^{-1} = \begin{bmatrix} \frac{1}{\lambda-1} & 0 & \frac{3}{(\lambda-1)(\lambda-2)} \\ \frac{2}{(\lambda-1)^2} & \frac{1}{\lambda-1} & \frac{2(\lambda+1)}{(\lambda-1)^2(\lambda-2)} \\ 0 & 0 & \frac{1}{\lambda-2} \end{bmatrix}.$$

2. $\psi(\lambda) = (\lambda+1)(\lambda-2)$ for the matrix (7.2.3), page 383, because

$$\mathbf{R}_\lambda(\mathbf{A}) = \left(\lambda\begin{bmatrix} 1 & 0 & 0 \\ 0 & 1 & 0 \\ 0 & 0 & 1 \end{bmatrix} - \begin{bmatrix} 0 & 1 & 1 \\ 1 & 0 & 1 \\ 1 & 1 & 0 \end{bmatrix}\right)^{-1} = \begin{bmatrix} \lambda & -1 & -1 \\ -1 & \lambda & -1 \\ -1 & -1 & \lambda \end{bmatrix}^{-1}$$

$$= \frac{1}{(\lambda-2)(\lambda+1)}\begin{bmatrix} \lambda-1 & 1 & 0 \\ 1 & \lambda-1 & 1 \\ 1 & 1 & \lambda-1 \end{bmatrix}.$$

3. $\psi(\lambda) = \lambda - 2$ for the matrix (7.3.5)(a) because

$$\mathbf{R}_\lambda(\mathbf{T}_1) = \left(\begin{bmatrix} \lambda & 0 & 0 \\ 0 & \lambda & 0 \\ 0 & 0 & \lambda \end{bmatrix} - \begin{bmatrix} 2 & 0 & 0 \\ 0 & 2 & 0 \\ 0 & 0 & 2 \end{bmatrix} \right)^{-1} = \frac{1}{\lambda - 2} \begin{bmatrix} 1 & 0 & 0 \\ 0 & 1 & 0 \\ 0 & 0 & 1 \end{bmatrix} = \frac{1}{\lambda - 2}\,\mathbf{I}.$$

4. $\psi(\lambda) = (\lambda - 2)^2$ for the matrix (7.3.5)(b) because

$$\mathbf{R}_\lambda(\mathbf{T}_2) = \left(\begin{bmatrix} \lambda & 0 & 0 \\ 0 & \lambda & 0 \\ 0 & 0 & \lambda \end{bmatrix} - \begin{bmatrix} 2 & 1 & 0 \\ 0 & 2 & 0 \\ 0 & 0 & 2 \end{bmatrix} \right)^{-1} = \frac{1}{(\lambda - 2)^2} \begin{bmatrix} \lambda - 2 & 1 & 0 \\ 0 & \lambda - 2 & 0 \\ 0 & 0 & \lambda - 2 \end{bmatrix}.$$

5. $\psi(\lambda) = (\lambda - 2)^3$ for the matrix (7.3.5)(c) because

$$\begin{aligned} \mathbf{R}_\lambda(\mathbf{T}_3) &= \left(\begin{bmatrix} \lambda & 0 & 0 \\ 0 & \lambda & 0 \\ 0 & 0 & \lambda \end{bmatrix} - \begin{bmatrix} 2 & 1 & a \\ 0 & 2 & 1 \\ 0 & 0 & 2 \end{bmatrix} \right)^{-1} \\ &= \frac{1}{(\lambda - 2)^3} \begin{bmatrix} (\lambda - 2)^2 & \lambda - 2 & 1 + \lambda a - 2a \\ 0 & (\lambda - 2)^2 & \lambda - 2 \\ 0 & 0 & (\lambda - 2)^2 \end{bmatrix}. \end{aligned}$$ □

If the characteristic polynomial $\chi(\lambda)$ of a real symmetric matrix **A** has a multiple eigenvalue, then its minimal polynomial is of lesser degree than $\chi(\lambda)$.

Example 7.6.4: Let **A** be a symmetric diagonalizable matrix:

$$\mathbf{A} = \begin{bmatrix} 1 & 1 & 1 \\ 1 & 1 & -1 \\ 1 & -1 & 1 \end{bmatrix} \quad \Longrightarrow \quad \mathbf{R}_\lambda(\mathbf{A}) \equiv (\lambda\mathbf{I} - \mathbf{A})^{-1} = \frac{1}{\lambda^2 - \lambda - 2} \begin{bmatrix} \lambda & 1 & 1 \\ 1 & \lambda & -1 \\ 1 & -1 & \lambda \end{bmatrix}.$$

So its characteristic polynomial is $\chi(\lambda) = (\lambda - 2)^2(\lambda + 1)$, but the minimal polynomial, $\psi(\lambda) = (\lambda - 2)(\lambda + 1) = \lambda^2 - \lambda - 2$, is of the second degree. Therefore, $\mathbf{A}^2 - \mathbf{A} - 2\mathbf{I} = 0$ and $\mathbf{A}^2 = \mathbf{A} + 2\mathbf{I}$. Thus,

$$\mathbf{A}^2 = \begin{bmatrix} 1 & 1 & 1 \\ 1 & 1 & -1 \\ 1 & -1 & 1 \end{bmatrix} + \begin{bmatrix} 2 & 0 & 0 \\ 0 & 2 & 0 \\ 0 & 0 & 2 \end{bmatrix} = \begin{bmatrix} 3 & 1 & 1 \\ 1 & 3 & -1 \\ 1 & -1 & 3 \end{bmatrix}.$$

Definition 7.16: Let $f(\lambda) = P(\lambda)/Q(\lambda)$ be the ratio of two polynomials (or in general, of two entire functions) without common multiples. The point $\lambda = \lambda_0$ is said to be a **singular point** of $f(\lambda)$ if the function is not defined at this point. We say that this singular point is of multiplicity m if the nonzero limit

$$\lim_{\lambda \to \lambda_0} \frac{P(\lambda)(\lambda - \lambda_0)^m}{Q(\lambda)}$$

exists. Such a singular point is called a **pole**.

Definition 7.17: The **residue** of the function $f(\lambda) = P(\lambda)/Q(\lambda)$ at the pole λ_0 of multiplicity m is defined by

$$\operatorname*{Res}_{\lambda_0} \frac{P(\lambda)}{Q(\lambda)} = \frac{1}{(m-1)!} \frac{\mathrm{d}^{m-1}}{\mathrm{d}\lambda^{m-1}} \left. \frac{P(\lambda)(\lambda - \lambda_0)^m}{Q(\lambda)} \right|_{\lambda = \lambda_0}. \tag{7.6.2}$$

In particular, for $m = 1$ we have

$$\operatorname*{Res}_{\lambda_0} \frac{P(\lambda)}{Q(\lambda)} = \frac{P(\lambda_0)}{Q'(\lambda_0)}, \tag{7.6.3}$$

for $m = 2$:

$$\operatorname{Res}_{\lambda_0} \frac{P(\lambda)}{Q(\lambda)} = \frac{\mathrm{d}}{\mathrm{d}\lambda} \left. \frac{P(\lambda)(\lambda-\lambda_0)^2}{Q(\lambda)} \right|_{\lambda=\lambda_0}, \tag{7.6.4}$$

and for $m = 3$:

$$\operatorname{Res}_{\lambda_0} \frac{P(\lambda)}{Q(\lambda)} = \frac{1}{2} \frac{\mathrm{d}^2}{\mathrm{d}\lambda^2} \left. \frac{P(\lambda)(\lambda-\lambda_0)^3}{Q(\lambda)} \right|_{\lambda=\lambda_0}. \tag{7.6.5}$$

If a real-valued function $f(\lambda) = P(\lambda)/Q(\lambda)$ has a pair of complex conjugate poles $a \pm b\mathbf{j}$, then

$$\operatorname{Res}_{a+b\mathbf{j}} f(\lambda) + \operatorname{Res}_{a-b\mathbf{j}} f(\lambda) = 2\,\Re \operatorname{Res}_{a+b\mathbf{j}} f(\lambda) = 2\,\Re \operatorname{Res}_{a-b\mathbf{j}} f(\lambda), \tag{7.6.6}$$

where $\Re$ stands for the real part of a complex number (so $\Re(A + B\mathbf{j}) = A$).

Definition 7.18: Let $f(\lambda)$ be a function defined on the spectrum $\sigma(\mathbf{A})$ of a square matrix $\mathbf{A}$. Then

$$f(\mathbf{A}) = \sum_{\lambda_k \in \sigma(\mathbf{A})} \operatorname{Res}_{\lambda_k} f(\lambda)\, \mathbf{R}_\lambda(\mathbf{A}). \tag{7.6.7}$$

Example 7.6.5: (Example 7.4.5 revisited) The matrix $\mathbf{A} = \begin{bmatrix} 9 & 8 \\ 7 & 8 \end{bmatrix}$ has the resolvent

$$\mathbf{R}_\lambda(\mathbf{A}) = \begin{bmatrix} \lambda-9 & -8 \\ -7 & \lambda-8 \end{bmatrix}^{-1} = \frac{1}{(\lambda-1)(\lambda-16)} \begin{bmatrix} \lambda-8 & 8 \\ 7 & \lambda-9 \end{bmatrix}.$$

Using Definition 7.18, we define the exponential of the matrix $\mathbf{A}$ as

$$e^{\mathbf{A}t} = \operatorname{Res}_{\lambda=1} e^{\lambda t}\, \mathbf{R}_\lambda(\mathbf{A}) + \operatorname{Res}_{\lambda=16} e^{\lambda t}\, \mathbf{R}_\lambda(\mathbf{A}).$$

Since the eigenvalues of the matrix $\mathbf{A}$ are distinct and simple, we use the formula (7.6.3) to obtain

$$\operatorname{Res}_{\lambda=1} e^{\lambda t}\, \mathbf{R}_\lambda(\mathbf{A}) = \frac{e^{\lambda t}}{(\lambda-16)} \left. \begin{bmatrix} \lambda-8 & 8 \\ 7 & \lambda-9 \end{bmatrix} \right|_{\lambda=1} = \frac{e^t}{-15} \begin{bmatrix} -7 & 8 \\ 7 & -8 \end{bmatrix},$$

$$\operatorname{Res}_{\lambda=16} e^{\lambda t}\, \mathbf{R}_\lambda(\mathbf{A}) = \frac{e^{\lambda t}}{(\lambda-1)} \left. \begin{bmatrix} \lambda-8 & 8 \\ 7 & \lambda-9 \end{bmatrix} \right|_{\lambda=16} = \frac{e^{16t}}{15} \begin{bmatrix} 8 & 8 \\ 7 & 7 \end{bmatrix}.$$

Using these matrices, we get the exponential matrix from Eq. (7.6.7):

$$e^{\mathbf{A}t} = \frac{1}{15} \begin{bmatrix} 8\,e^{16t} + 7\,e^t & 8\,e^{16t} - 8\,e^t \\ 7\,e^{16t} - 7\,e^t & 7\,e^{16t} + 8\,e^t \end{bmatrix}.$$

Example 7.6.6: For the function $f(\lambda) = e^{\lambda t}$ and matrices from Example 7.6.2, page 405,

$$\mathbf{A} = \begin{bmatrix} 1 & \mathbf{j}/2 \\ \mathbf{j}/2 & 2 \end{bmatrix} \quad \text{and} \quad \mathbf{B} = \begin{bmatrix} 1 & \mathbf{j} \\ \mathbf{j} & 1 \end{bmatrix} \qquad (\mathbf{j}^2 = -1),$$

we can determine the exponential matrices using the residue method in the following way. In Example 7.6.2, we found the resolvents of the matrices $\mathbf{A}$ and $\mathbf{B}$. Therefore,

$$\begin{aligned} e^{\mathbf{A}t} &= \operatorname{Res}_{\lambda=3/2} e^{\lambda t}\, \mathbf{R}_\lambda(\mathbf{A}) = \frac{\mathrm{d}}{\mathrm{d}\lambda} e^{\lambda t} \left. \begin{bmatrix} \lambda-2 & \mathbf{j}/2 \\ \mathbf{j}/2 & \lambda-1 \end{bmatrix} \right|_{\lambda=3/2} \\ &= t\,e^{\lambda t} \left. \begin{bmatrix} \lambda-2 & \mathbf{j}/2 \\ \mathbf{j}/2 & \lambda-1 \end{bmatrix} \right|_{\lambda=3/2} + e^{\lambda t} \left. \begin{bmatrix} 1 & 0 \\ 0 & 1 \end{bmatrix} \right|_{\lambda=3/2} \\ &= \frac{t}{2}\, e^{3t/2} \begin{bmatrix} -1 & \mathbf{j} \\ \mathbf{j} & 1 \end{bmatrix} + e^{3t/2} \begin{bmatrix} 1 & 0 \\ 0 & 1 \end{bmatrix}, \end{aligned}$$

and

$$
\begin{aligned}
e^{\mathbf{B}t} &= \operatorname*{Res}_{1+\mathbf{j}} e^{\lambda t}\,\mathbf{R}_\lambda(\mathbf{B}) + \operatorname*{Res}_{1-\mathbf{j}} e^{\lambda t}\,\mathbf{R}_\lambda(\mathbf{B}) \\
&= \operatorname*{Res}_{1+\mathbf{j}} \frac{e^{\lambda t}}{\chi(\lambda)} \begin{bmatrix} \lambda-1 & \mathbf{j} \\ \mathbf{j} & \lambda-1 \end{bmatrix} + \operatorname*{Res}_{1-\mathbf{j}} \frac{e^{\lambda t}}{\chi(\lambda)} \begin{bmatrix} \lambda-1 & \mathbf{j} \\ \mathbf{j} & \lambda-1 \end{bmatrix} \\
&= \frac{e^{\lambda_1 t}}{\chi'(\lambda_1)} \begin{bmatrix} \lambda_1-1 & \mathbf{j} \\ \mathbf{j} & \lambda_1-1 \end{bmatrix} + \frac{e^{\lambda_2 t}}{\chi'(\lambda_2)} \begin{bmatrix} \lambda_2-1 & \mathbf{j} \\ \mathbf{j} & \lambda_2-1 \end{bmatrix}.
\end{aligned}
$$

Since $\lambda_1 = 1+\mathbf{j}$ and $\lambda_2 = 1-\mathbf{j}$ are simple nulls of the characteristic polynomial $\chi(\lambda) = \lambda^2 - 2\lambda + 2$, we can use Eq. (7.6.3). Calculations show that

$$
\chi'(\lambda_1) = 2(\lambda_1 - 1) = 2\mathbf{j} \quad \text{and} \quad \chi'(\lambda_2) = 2(\lambda_2 - 1) = -2\mathbf{j}.
$$

Hence,

$$
\begin{aligned}
e^{\mathbf{B}t} &= \frac{e^{(1+\mathbf{j})t}}{2\mathbf{j}} \begin{bmatrix} \mathbf{j} & \mathbf{j} \\ \mathbf{j} & \mathbf{j} \end{bmatrix} + \frac{e^{(1-\mathbf{j})t}}{-2\mathbf{j}} \begin{bmatrix} -\mathbf{j} & \mathbf{j} \\ \mathbf{j} & -\mathbf{j} \end{bmatrix} \\
&= e^{(1+\mathbf{j})t} \begin{bmatrix} 1/2 & 1/2 \\ 1/2 & 1/2 \end{bmatrix} + e^{(1-\mathbf{j})t} \begin{bmatrix} 1/2 & -1/2 \\ -1/2 & 1/2 \end{bmatrix} \\
&= e^t\,\frac{1}{2} \begin{bmatrix} e^{\mathbf{j}t} + e^{-\mathbf{j}t} & e^{\mathbf{j}t} - e^{-\mathbf{j}t} \\ e^{\mathbf{j}t} - e^{-\mathbf{j}t} & e^{\mathbf{j}t} + e^{-\mathbf{j}t} \end{bmatrix} = e^t \begin{bmatrix} \cos t & \mathbf{j}\sin t \\ \mathbf{j}\sin t & \cos t \end{bmatrix}
\end{aligned}
$$

because $\sin t = \frac{1}{2\mathbf{j}}\,e^{\mathbf{j}t} - \frac{1}{2\mathbf{j}}\,e^{-\mathbf{j}t}$ and $\cos t = \frac{1}{2}\,e^{\mathbf{j}t} + \frac{1}{2}\,e^{-\mathbf{j}t}$.

Example 7.6.7: (Example 7.6.1 revisited) For the matrix $\mathbf{B}_1$, we have

$$
\begin{aligned}
e^{\mathbf{B}_1 t} &= \operatorname*{Res}_{2} e^{\lambda t}\,\mathbf{R}_\lambda(\mathbf{B}_1) + \operatorname*{Res}_{-2} e^{\lambda t}\,\mathbf{R}_\lambda(\mathbf{B}_1) \\
&= e^{2t}\,\frac{1}{4} \begin{bmatrix} 3 & 3 \\ 1 & 1 \end{bmatrix} - e^{-2t}\,\frac{1}{4} \begin{bmatrix} -1 & 3 \\ 1 & -3 \end{bmatrix}.
\end{aligned}
$$

Similarly,

$$
\begin{aligned}
e^{\mathbf{B}_2 t} &= \operatorname*{Res}_{\mathbf{j}} e^{\lambda t}\,\mathbf{R}_\lambda(\mathbf{B}_2) + \operatorname*{Res}_{-\mathbf{j}} e^{\lambda t}\,\mathbf{R}_\lambda(\mathbf{B}_2) \\
&= e^{\mathbf{j}t}\,\frac{1}{2\mathbf{j}} \begin{bmatrix} 3+\mathbf{j} & -2 \\ 5 & \mathbf{j}-3 \end{bmatrix} - e^{-\mathbf{j}t}\,\frac{1}{2\mathbf{j}} \begin{bmatrix} 3-\mathbf{j} & -2 \\ 5 & -\mathbf{j}-3 \end{bmatrix} \\
&= \begin{bmatrix} \cos t + 3\sin t & -2\sin t \\ 5\sin t & \cos t - 3\sin t \end{bmatrix}.
\end{aligned}
$$

Actually, we don't need to evaluate the residue at $\lambda = -\mathbf{j}$ because we know the answer: it is a complex conjugate of the residue at $\lambda = \mathbf{j}$. Therefore, we can find the residue at one of the complex singular points, say at $\lambda = \mathbf{j}$, extract the real part (denoted by $\Re$), and double the result:

$$
e^{\mathbf{B}_2 t} = 2\Re \operatorname*{Res}_{\lambda=\mathbf{j}} e^{\lambda t}\,\mathbf{R}_\lambda(\mathbf{B}_2).
$$

For the matrix $\mathbf{B}_3$, we have

$$
\begin{aligned}
e^{\mathbf{B}_3 t} &= \operatorname*{Res}_{3} e^{\lambda t}\,\mathbf{R}_\lambda(\mathbf{B}_3) = \operatorname*{Res}_{3} \frac{e^{\lambda t}}{(\lambda-3)^2} \begin{bmatrix} \lambda-5 & 2 \\ -2 & \lambda-1 \end{bmatrix} \\
&= \left.\frac{\mathrm{d}}{\mathrm{d}\lambda} \left[e^{\lambda t} \begin{bmatrix} \lambda-5 & 2 \\ -2 & \lambda-1 \end{bmatrix} \right]\right|_{\lambda=3} \\
&= \left. t e^{\lambda t} \begin{bmatrix} \lambda-5 & 2 \\ -2 & \lambda-1 \end{bmatrix} \right|_{\lambda=3} + \left. e^{\lambda t} \begin{bmatrix} 1 & 0 \\ 0 & 1 \end{bmatrix} \right|_{\lambda=3} \\
&= t e^{3t} \begin{bmatrix} -2 & 2 \\ -2 & 2 \end{bmatrix} + e^{3t} \begin{bmatrix} 1 & 0 \\ 0 & 1 \end{bmatrix} = e^{3t} \begin{bmatrix} 1-2t & 2t \\ -2t & 1+2t \end{bmatrix}.
\end{aligned}
$$

Although the function $f(\lambda) = \cos\lambda t$ can be expressed as the sum of two exponential functions, $\cos\lambda t = \frac{1}{2}\,e^{\mathbf{j}\lambda t} + \frac{1}{2}\,e^{-\mathbf{j}\lambda t}$, it is instructive to follow the procedure. Therefore, we have

$$\begin{aligned}\cos(\mathbf{B}_1 t) &= \operatorname*{Res}_{2} \cos(\lambda t)\,\mathbf{R}_\lambda(\mathbf{B}_1) + \operatorname*{Res}_{-2} \cos(\lambda t)\,\mathbf{R}_\lambda(\mathbf{B}_1) \\ &= \operatorname*{Res}_{2} \frac{\cos(\lambda t)}{\lambda^2 - 4}\begin{bmatrix}\lambda+1 & 3 \\ 1 & \lambda-1\end{bmatrix} + \operatorname*{Res}_{-2} \frac{\cos(\lambda t)}{\lambda^2 - 4}\begin{bmatrix}\lambda+1 & 3 \\ 1 & \lambda-1\end{bmatrix} \\ &= \frac{\cos 2t}{4}\begin{bmatrix}3 & 3 \\ 1 & 1\end{bmatrix} + \frac{\cos 2t}{-4}\begin{bmatrix}-1 & 3 \\ 1 & -3\end{bmatrix} = \cos 2t\begin{bmatrix}1 & 0 \\ 0 & 1\end{bmatrix}.\end{aligned}$$

For the matrix $\mathbf{B}_2$, we calculate only one residue, say at the point $\lambda = \mathbf{j}$:

$$\begin{aligned}\operatorname*{Res}_{\mathbf{j}} \cos(\lambda t)\,\mathbf{R}_\lambda(\mathbf{B}_2) &= \operatorname*{Res}_{\mathbf{j}} \frac{\cos(\lambda t)}{\lambda^2 + 1}\begin{bmatrix}\lambda+3 & -2 \\ 5 & \lambda-3\end{bmatrix} = \frac{\cos(\lambda t)}{2\lambda}\begin{bmatrix}\lambda+3 & -2 \\ 5 & \lambda-3\end{bmatrix}\Bigg|_{\lambda=\mathbf{j}} \\ &= \frac{\cos(\mathbf{j}t)}{2\mathbf{j}}\begin{bmatrix}3+\mathbf{j} & -2 \\ 5 & \mathbf{j}-3\end{bmatrix}.\end{aligned}$$

Since $\cos(\mathbf{j}t) = \cosh t = \frac{1}{2}\,e^{t} + \frac{1}{2}\,e^{-t}$, we get

$$\operatorname*{Res}_{\mathbf{j}} \cos(\lambda t)\,\mathbf{R}_\lambda(\mathbf{B}_2) = \mathbf{j}\,\cosh t\begin{bmatrix}-3/2 & 1 \\ -5/2 & 3/2\end{bmatrix} + \frac{\cosh t}{2}\begin{bmatrix}1 & 0 \\ 0 & 1\end{bmatrix}.$$

Extracting the real part and multiplying it by 2, we obtain

$$\cos(\mathbf{B}_2 t) = (\cosh t)\,\mathbf{I},$$

with $\mathbf{I}$ being the identity matrix. For matrix $\mathbf{B}_3$, we use formula (7.6.4):

$$\begin{aligned}\cos(\mathbf{B}_3 t) &= \operatorname*{Res}_{3} \cos(\lambda t)\,\mathbf{R}_\lambda(\mathbf{B}_3) = \frac{\mathrm{d}}{\mathrm{d}\lambda}\,\cos\lambda t\begin{bmatrix}\lambda-5 & 2 \\ -2\lambda-1 & \end{bmatrix}\Bigg|_{\lambda=3} \\ &= \cos 3t\begin{bmatrix}1 & 0 \\ 0 & 1\end{bmatrix} - t\,\sin 3t\begin{bmatrix}-2 & 2 \\ -2 & 2\end{bmatrix}.\end{aligned}$$

Example 7.6.8: Let us consider the non-diagonalizable matrix

$$\mathbf{A} = \begin{bmatrix}5 & -1 \\ 1 & 3\end{bmatrix}.$$

Since the resolvent of the given matrix is

$$\mathbf{R}_\lambda(\mathbf{A}) = \frac{1}{(\lambda-4)^2}\begin{bmatrix}\lambda-3 & -1 \\ 1 & \lambda-5\end{bmatrix},$$

we find a root of $\mathbf{A}$ by calculating the residue at $\lambda = 4$:

$$\begin{aligned}\sqrt{\mathbf{A}} &= \operatorname*{Res}_{\lambda=4} \frac{\sqrt{\lambda}}{(\lambda-4)^2}\begin{bmatrix}\lambda-3 & -1 \\ 1 & \lambda-5\end{bmatrix} = \frac{\mathrm{d}}{\mathrm{d}\lambda}\,\sqrt{\lambda}\begin{bmatrix}\lambda-3 & -1 \\ 1 & \lambda-5\end{bmatrix}\Bigg|_{\lambda=4} \\ &= \frac{1}{4}\begin{bmatrix}1 & -1 \\ 1 & -1\end{bmatrix} + 2\begin{bmatrix}1 & 0 \\ 0 & 1\end{bmatrix} = \frac{1}{4}\begin{bmatrix}9 & -1 \\ 1 & 7\end{bmatrix},\end{aligned}$$

which is indeed the root of $\mathbf{A}$. Actually, the given matrix has two square roots; one is just the negative of the other.

Example 7.6.9: Let us consider the following 3×3 defective matrix:

$$\mathbf{A} = \begin{bmatrix}3 & 3 & -4 \\ -4 & -5 & 8 \\ -2 & -3 & 5\end{bmatrix}, \qquad \text{with} \quad \chi(\lambda) = \det(\lambda\mathbf{I} - \mathbf{A}) = (\lambda-1)^3.$$

Its minimal polynomial $\psi(\lambda) = (\lambda-1)^2$ is not equal to the characteristic polynomial $\chi(\lambda)$. The eigenvectors corresponding to the eigenvalue $\lambda = 1$ are as follows:

$$\mathbf{x}_1 = \langle 2, 0, 1\rangle^t, \qquad \mathbf{x}_2 = \langle -3, 2, 0\rangle^t.$$

The matrix $\mathbf{A}$ is neither diagonalizable nor normal, that is, $\mathbf{AA}^* \ne \mathbf{A}^*\mathbf{A}$ since

$$\mathbf{AA}^* = \begin{bmatrix} 34 & -59 & -35 \\ -59 & 105 & 63 \\ -35 & 63 & 38 \end{bmatrix}, \qquad \mathbf{A}^*\mathbf{A} = \begin{bmatrix} 29 & 35 & -54 \\ 35 & 43 & -67 \\ -54 & -67 & 105 \end{bmatrix}.$$

The resolvent of $\mathbf{A}$ is

$$\mathbf{R}_\lambda(\mathbf{A}) = \frac{1}{(\lambda-1)^2} \begin{bmatrix} \lambda+1 & 3 & -4 \\ -4 & \lambda-7 & 8 \\ -2 & -3 & \lambda+3 \end{bmatrix}.$$

Here are some functions of the matrix:

$$e^{\mathbf{A}t} = \operatorname*{Res}_{1} e^{\lambda t}\,\mathbf{R}_\lambda(\mathbf{A}) = \frac{\mathrm{d}}{\mathrm{d}\lambda}\, e^{\lambda t} \begin{bmatrix} \lambda+1 & 3 & -4 \\ -4 & \lambda-7 & 8 \\ -2 & -3 & \lambda+3 \end{bmatrix}_{\lambda=1}$$

$$= e^t \begin{bmatrix} 1+2t & 3t & -4t \\ -4t & 1-6t & 8t \\ -2t & -3t & 1+4t \end{bmatrix},$$

$$\frac{\mathbf{A}-\mathbf{I}}{\mathbf{A}+\mathbf{I}} = \operatorname*{Res}_{1} \frac{\lambda-1}{\lambda+1}\,\mathbf{R}_\lambda(\mathbf{A}) = \frac{\mathrm{d}}{\mathrm{d}\lambda}\,\frac{\lambda-1}{\lambda+1} \begin{bmatrix} \lambda+1 & 3 & -4 \\ -4 & \lambda-7 & 8 \\ -2 & -3 & \lambda+3 \end{bmatrix}_{\lambda=1}$$

$$= \begin{bmatrix} 1 & 3/2 & -2 \\ -2 & -3 & 4 \\ -1 & -3/2 & 2 \end{bmatrix}.$$

Problems

1. Use the resolvent method to compute $\mathbf{A}^2$, $\mathbf{A}^3$, $\mathbf{A}^4$, and $\mathbf{A}^5$ if $\mathbf{A}$ is one of the following 2×2 matrices:

(a) $\begin{bmatrix} 1 & 1 \\ 2 & 0 \end{bmatrix}$; (b) $\begin{bmatrix} 1 & 3 \\ 3 & 9 \end{bmatrix}$; (c) $\begin{bmatrix} 1 & 4 \\ -1 & 5 \end{bmatrix}$; (d) $\begin{bmatrix} 2 & -1 \\ 3 & -2 \end{bmatrix}$.

2. Find the resolvent of the each of the following 2×2 matrices and calculate $e^{\mathbf{A}t}$.

(a) $\begin{bmatrix} 1 & 2 \\ 0 & 3 \end{bmatrix}$; (b) $\begin{bmatrix} 1 & 2 \\ 1 & 2 \end{bmatrix}$; (c) $\begin{bmatrix} 4 & -1 \\ 5 & -2 \end{bmatrix}$; (d) $\begin{bmatrix} 9 & -3 \\ -1 & 11 \end{bmatrix}$;

(e) $\begin{bmatrix} 7 & -5 \\ 2 & 5 \end{bmatrix}$; (f) $\begin{bmatrix} 11 & -4 \\ 26 & -9 \end{bmatrix}$; (g) $\begin{bmatrix} 1 & 2 \\ -4 & 5 \end{bmatrix}$; (h) $\begin{bmatrix} 7 & 15 \\ -3 & 1 \end{bmatrix}$.

3. For each of the following 2×2 defective matrices, find a square root.

(a) $\begin{bmatrix} 10 & -1 \\ 1 & 8 \end{bmatrix}$; (b) $\begin{bmatrix} 7 & -9 \\ 4 & -5 \end{bmatrix}$; (c) $\begin{bmatrix} 11 & -5 \\ 5 & 21 \end{bmatrix}$; (d) $\begin{bmatrix} 23 & 4 \\ -1 & 27 \end{bmatrix}$;

(e) $\begin{bmatrix} 10 & 1 \\ -1 & 8 \end{bmatrix}$; (f) $\begin{bmatrix} 15 & -20 \\ 5 & 35 \end{bmatrix}$; (g) $\begin{bmatrix} 2 & -7 \\ 7 & 16 \end{bmatrix}$; (h) $\begin{bmatrix} 5 & -8 \\ 2 & 13 \end{bmatrix}$;

(i) $\begin{bmatrix} 5 & -1 \\ 1 & 3 \end{bmatrix}$; (j) $\begin{bmatrix} 18 & 4 \\ -1 & 14 \end{bmatrix}$; (k) $\begin{bmatrix} 2 & 2 \\ -2 & 6 \end{bmatrix}$; (l) $\begin{bmatrix} 35 & -7 \\ 28 & 63 \end{bmatrix}$.

4. For each of the matrices in the previous exercise, determine $\frac{\sin(\sqrt{\mathbf{A}}\,t)}{\sqrt{\mathbf{A}}}$ and $\cos\left(\sqrt{\mathbf{A}}\,t\right)$.

5. For each of the following 3×3 defective matrices, find a square root.

(a) $\begin{bmatrix} 3 & 0 & 7 \\ 15 & 1 & 4 \\ 0 & 0 & 3 \end{bmatrix}$; (b) $\begin{bmatrix} 3 & -2 & 1 \\ 2 & -1 & 1 \\ -4 & 4 & 1 \end{bmatrix}$; (c) $\begin{bmatrix} 3 & -8 & -10 \\ -3 & 7 & 9 \\ 3 & -6 & -8 \end{bmatrix}$;

(d) $\begin{bmatrix} 6 & 0 & 4 \\ 3 & 4 & 5 \\ 1 & 0 & 6 \end{bmatrix}$; (e) $\begin{bmatrix} 7 & 2 & -6 \\ 0 & 1 & 0 \\ 6 & -5 & -5 \end{bmatrix}$; (f) $\begin{bmatrix} 6 & 0 & -4 \\ 3 & 4 & 1 \\ -5 & 0 & 14 \end{bmatrix}$;

(g) $\begin{bmatrix} 1 & -3 & 2 \\ 1 & 3 & 2 \\ 2 & 1 & 4 \end{bmatrix}$; (h) $\begin{bmatrix} 1 & 2 & -1 \\ -1 & 1 & 2 \\ -1 & 1 & 2 \end{bmatrix}$; (i) $\begin{bmatrix} 1 & -2 & 1 \\ 1 & 3 & -1 \\ 2 & 4 & -1 \end{bmatrix}$.

6. For each of the matrices in the previous Problem, determine $\frac{\sin(\sqrt{\mathbf{A}}\,t)}{\sqrt{\mathbf{A}}}$ and $\cos\left(\sqrt{\mathbf{A}}\,t\right)$.

7. Use the resolvent method to compute $\mathbf{A}^2$, $\mathbf{A}^3$, $\mathbf{A}^4$, and $\mathbf{A}^5$ if $\mathbf{A}$ is one of the following 3×3 matrices:

(a) $\begin{bmatrix} 3 & -1 & -1 \\ -2 & 3 & 2 \\ 4 & -1 & -2 \end{bmatrix}$; (b) $\begin{bmatrix} 3 & 0 & 4 \\ 1 & 1 & 1 \\ -5 & 0 & -5 \end{bmatrix}$; (c) $\begin{bmatrix} 4 & 2 & -2 \\ -5 & 3 & 2 \\ -2 & 4 & 1 \end{bmatrix}$;

(d) $\begin{bmatrix} 3 & 2 & 4 \\ 2 & 0 & 2 \\ 4 & 2 & 3 \end{bmatrix}$; (e) $\begin{bmatrix} 4 & 2 & -6 \\ -6 & -3 & 8 \\ 2 & 1 & -3 \end{bmatrix}$; (f) $\begin{bmatrix} -3 & 7 & -8 \\ 4 & -1 & 4 \\ 2 & -2 & 5 \end{bmatrix}$.

8. Find the resolvent of the each of the following 3×3 matrices and calculate $e^{\mathbf{A}t}$.

(a) $\begin{bmatrix} 2 & 0 & 3 \\ 0 & 3 & 1 \\ 0 & -1 & 1 \end{bmatrix}$; (b) $\begin{bmatrix} 5 & 2 & -1 \\ -3 & 2 & 2 \\ 1 & 3 & 2 \end{bmatrix}$; (c) $\begin{bmatrix} 1 & 15 & -15 \\ -6 & 18 & -22 \\ -3 & 11 & -15 \end{bmatrix}$;

(d) $\begin{bmatrix} -15 & -7 & 4 \\ 34 & 16 & -18 \\ 17 & 7 & 5 \end{bmatrix}$; (e) $\begin{bmatrix} 1 & 8 & 3 \\ -4 & -1 & 2 \\ 4 & 5 & 2 \end{bmatrix}$; (f) $\begin{bmatrix} 3 & 5 & -3 \\ 3 & 2 & 1 \\ -1 & 2 & 7 \end{bmatrix}$;

(g) $\begin{bmatrix} -27 & 32 & -4 \\ -14 & 17 & -2 \\ 70 & -80 & 11 \end{bmatrix}$; (h) $\begin{bmatrix} 1 & 2 & -1 \\ 1 & 1 & -2 \\ 3 & -4 & -1 \end{bmatrix}$; (i) $\begin{bmatrix} 1 & 2 & -1 \\ 2 & 1 & -1 \\ 8 & -3 & -1 \end{bmatrix}$.

9. Find a cube root of the matrices

(a) $\begin{bmatrix} 11 & -9 \\ 1 & 5 \end{bmatrix}$; (b) $\begin{bmatrix} 29 & 21 \\ -2 & 6 \end{bmatrix}$; (c) $\begin{bmatrix} 24 & -3 \\ 3 & 30 \end{bmatrix}$; (d) $\begin{bmatrix} 61 & 9 \\ -1 & 67 \end{bmatrix}$;

(e) $\begin{bmatrix} 0 & 1 \\ -1 & 2 \end{bmatrix}$; (f) $\begin{bmatrix} -1 & 3 \\ -6 & 10 \end{bmatrix}$; (g) $\begin{bmatrix} 33 & 16 \\ 62 & 32 \end{bmatrix}$; (h) $\begin{bmatrix} 15 & 14 \\ 12 & 13 \end{bmatrix}$.

10. Find a cube root with real entries for each of the following matrices

(a) $\begin{bmatrix} 9 & -1 & -1 \\ 1 & 1 & -1 \\ 1 & -1 & 7 \end{bmatrix}$; (b) $\begin{bmatrix} -2 & 1 & 1 \\ -1 & 0 & 1 \\ -1 & 1 & 8 \end{bmatrix}$; (c) $\begin{bmatrix} 5 & 1 & -1 \\ -20 & 8 & 19 \\ -4 & 11 & 16 \end{bmatrix}$.

11. Show that the following 3×3 matrices have no square root. Nevertheless, find the matrix functions $\mathbf{\Psi}(t) = \cos\left(\sqrt{\mathbf{A}t}\right)$ and $\mathbf{\Phi}(t) = \mathbf{A}^{-1/2}\sin\left(\mathbf{A}^{1/2}t\right)$, and show that they satisfy the second order matrix differential equation $\ddot{\mathbf{X}} + \mathbf{A}\,\mathbf{X} = 0$.

(a) $\begin{bmatrix} -1 & 1 & 1 \\ -1 & 0 & 1 \\ -1 & 1 & 1 \end{bmatrix}$; (b) $\begin{bmatrix} -1 & 2 & -1 \\ 1 & -2 & 1 \\ 1 & -8 & 3 \end{bmatrix}$; (c) $\begin{bmatrix} -1 & -1 & 6 \\ -1 & -1 & -2 \\ -1 & -1 & 2 \end{bmatrix}$; (d) $\begin{bmatrix} -1 & 1 & -2 \\ -3 & -3 & 3 \\ 2 & -2 & 4 \end{bmatrix}$.

12. The following pairs of matrices have the same eigenvalues. Use this information to construct $e^{\mathbf{A}t}$ for each of the matrices.

(a) $\begin{bmatrix} -1 & 3 & -1 \\ 2 & -2 & 1 \\ 1 & -9 & 3 \end{bmatrix}$, $\begin{bmatrix} -1 & 3 & -1 \\ 1 & -2 & 2 \\ 1 & -3 & 3 \end{bmatrix}$; (b) $\begin{bmatrix} 1 & -3 & 6 \\ 2 & 3 & -1 \\ 2 & 4 & -2 \end{bmatrix}$, $\begin{bmatrix} 59 & -100 & 80 \\ 45 & -76 & 60 \\ 15 & -25 & 19 \end{bmatrix}$;

(c) $\begin{bmatrix} 1 & 3 & -1 \\ 3 & 1 & -1 \\ 8 & -2 & -1 \end{bmatrix}$, $\begin{bmatrix} 1 & 3 & -1 \\ 3 & 1 & -1 \\ 9 & -3 & -1 \end{bmatrix}$; (d) $\begin{bmatrix} 1 & 2 & -1 \\ 2 & 1 & -1 \\ 7 & -3 & -1 \end{bmatrix}$, $\begin{bmatrix} -15 & 8 & 4 \\ -24 & 13 & 6 \\ -8 & 4 & 3 \end{bmatrix}$.

7.7 The Spectral Decomposition Method

The matrix spectral decomposition method is a particular case of a more general method that is used in functional analysis and quantum mechanics, namely, spectral operator decomposition [42]. In this section, we will discuss in detail computation of the matrix exponential for an arbitrary square matrix, as well as some trigonometric functions of a matrix variable. Such functions play a central role in constructing solutions of vector linear differential equations.

The exponential function e^{λ} of the complex number λ may be defined by means of the corresponding Maclaurin's series

$$e^{\lambda} = 1 + \frac{\lambda}{1!} + \frac{\lambda^2}{2!} + \frac{\lambda^3}{3!} + \cdots + \frac{\lambda^n}{n!} + \cdots .$$

This gives us the key step to define the **exponential matrix** $e^{\mathbf{A}}$ of an $n \times n$ matrix $\mathbf{A}$ as the $n \times n$ matrix defined by the series

$$e^{\mathbf{A}} = \mathbf{A}^0 + \mathbf{A} + \frac{\mathbf{A}^2}{2!} + \frac{\mathbf{A}^3}{3!} + \cdots + \frac{\mathbf{A}^n}{n!} + \cdots ,$$

where $\mathbf{A}^0 = \mathbf{I}$ (the $n \times n$ identity matrix). Actually, we are interested in the general exponential function

$$e^{\mathbf{A}t} = \mathbf{I} + \mathbf{A}t + \frac{\mathbf{A}^2}{2!}\,t^2 + \frac{\mathbf{A}^3}{3!}\,t^3 + \cdots + \frac{\mathbf{A}^n}{n!}\,t^n + \cdots , \tag{7.7.1}$$

for some parameter t (usually associated with time). To emphasize that $e^{\mathbf{A}t}$ is a fundamental matrix (see §8.1), we also denote it by $\mathbf{\Phi}(t)$. The meaning of the infinite series on the right-hand side in Eq. (7.7.1) is given by

$$\mathbf{\Phi}(t) \stackrel{\text{def}}{=} e^{\mathbf{A}t} = \lim_{k\to\infty} \left(\sum_{m=0}^{k} \frac{\mathbf{A}^m\, t^m}{m!} \right), \tag{7.7.2}$$

where the limit is taken with respect to a matrix[74] norm. Actually, the matrix exponential (7.7.2) is the solution of the following matrix initial value problem, which is equivalent to the companion integral equation:

$$\dot{\mathbf{\Phi}} = \mathbf{A}\mathbf{\Phi}, \qquad \mathbf{\Phi}(0) = \mathbf{I} \qquad \Longleftrightarrow \qquad \mathbf{\Phi}(t) = \mathbf{\Phi}(0) + \int_0^t \mathbf{A}\mathbf{\Phi}(\tau)\,\mathrm{d}\tau.$$

To solve the above integral matrix equation, we apply the Picard method (see §2.3) starting with the initial approximation $\mathbf{\Phi}_0 = \mathbf{I}$, where $\mathbf{I}$ is the identity matrix. This leads to the sequence of matrix-valued functions

$$\mathbf{\Phi}_{n+1}(t) = \mathbf{I} + \mathbf{A}\int_0^t \mathbf{\Phi}_n(\tau)\,\mathrm{d}\tau, \qquad n = 0, 1, 2, \ldots .$$

In particular,

$$\begin{aligned} \mathbf{\Phi}_1(t) &= \mathbf{I} + \mathbf{A}\int_0^t \mathrm{d}\tau = \mathbf{I} + \mathbf{A}t, \\ \mathbf{\Phi}_2(t) &= \mathbf{I} + \mathbf{A}\int_0^t \mathbf{\Phi}_1(\tau)\,\mathrm{d}\tau = \mathbf{I} + \mathbf{A}t + \frac{1}{2}\,\mathbf{A}^2 t^2, \end{aligned}$$

and so on, which leads to Eq. (7.7.1) because $\mathbf{\Phi}_n(t) = \mathbf{I} + \mathbf{A}t + \frac{\mathbf{A}^2}{2!}\,t^2 + \frac{\mathbf{A}^3}{3!}\,t^3 + \cdots + \frac{\mathbf{A}^n}{n!}\,t^n$.

Theorem 7.14: For square matrices $\mathbf{A}$ and $\mathbf{B}$ of the same dimensions, the following relations hold:

- $\frac{\mathrm{d}}{\mathrm{d}t}\,e^{\mathbf{A}t} = \mathbf{A}\,e^{\mathbf{A}t} = e^{\mathbf{A}t}\,\mathbf{A}$;
- $\left(e^{\mathbf{A}t}\right)^{-1} = e^{-\mathbf{A}t}$;
- $e^{(\mathbf{A}+\mathbf{B})t} = e^{\mathbf{A}t}\,e^{\mathbf{B}t} = e^{\mathbf{B}t}\,e^{\mathbf{A}t}$ if and only if $\mathbf{A}\mathbf{B} = \mathbf{B}\mathbf{A}$;
- $\det e^{\mathbf{A}t} \ne 0$ for any t.

[74] There are many definitions for the norm, denoted by $\|\mathbf{A}\|$, of a square matrix $\mathbf{A}$. The norm of a square matrix is a generalization of the length of a vector.

The proof of these results requires a precise definition of convergence of an infinite series of matrices and it involves the properties of the norm of a matrix. This would take us far astray from the main goal of this book. ■

Example 7.7.1: Consider a defective 3×3 matrix

$$\mathbf{A} = \begin{bmatrix} 1 & 1 & 0 \\ 0 & 1 & 0 \\ 0 & 0 & 1 \end{bmatrix} = \mathbf{I} + \mathbf{E}, \qquad \text{where} \quad \mathbf{E} = \begin{bmatrix} 0 & 1 & 0 \\ 0 & 0 & 0 \\ 0 & 0 & 0 \end{bmatrix}$$

is a nilpotent matrix (so $\mathbf{E}^2 = \mathbf{0}$). The matrix $\mathbf{A}$ has the triple eigenvalue $\lambda = 1$ with only two linearly independent eigenvectors $\langle 1, 0, 0\rangle^T$ and $\langle 0, 0, 1\rangle^T$. Therefore, this matrix is not diagonalizable. Raising $\mathbf{A}$ to the second power, we get

$$\mathbf{A}^2 = (\mathbf{I} + \mathbf{E})^2 = \mathbf{I} + 2\,\mathbf{E} + \mathbf{E}^2 = \mathbf{I} + 2\,\mathbf{E}$$

since $\mathbf{E}^2 = \mathbf{0}$. Repetition of this process leads to

$$\mathbf{A}^n = \mathbf{I} + n\,\mathbf{E}, \quad n = 1, 2, \ldots .$$

This formula allows us to find the exponential matrix:

$$\begin{aligned} e^{\mathbf{A}t} &= \mathbf{I} + (\mathbf{I} + \mathbf{E})\,t + \frac{1}{2!}\,(\mathbf{I} + 2\,\mathbf{E})\,t^2 + \frac{1}{3!}\,(\mathbf{I} + 3\,\mathbf{E})\,t^3 + \cdots \\ &= \mathbf{I} + \mathbf{I}\,t + \frac{t^2}{2!}\,\mathbf{I} + \frac{t^3}{3!}\,\mathbf{I} + \frac{t^4}{4!}\,\mathbf{I} + \cdots \\ &\quad + \mathbf{E}\,t + \frac{2\,t^2}{2!}\,\mathbf{E} + \frac{3\,t^3}{3!}\,\mathbf{E} + \frac{4\,t^4}{4!}\,\mathbf{E} + \cdots \\ &= \mathbf{I}\left[1 + \frac{t}{1!} + \frac{t^2}{2!} + \frac{t^3}{3!} + \frac{t^4}{4!} + \cdots\right] + \mathbf{E}\,t\left[1 + t + \frac{t^2}{2!} + \frac{t^3}{3!} + \cdots\right] \\ &= \mathbf{I}\,e^t + \mathbf{E}\,t\,e^t = \mathbf{I}\,e^t\,(1 - t) + \mathbf{A}\,t\,e^t . \end{aligned}$$

Hence, the exponential matrix $e^{\mathbf{A}t}$ is a linear combination of two powers of the given matrix $\mathbf{I} = \mathbf{A}^0$ and $\mathbf{A}$. □

Eq. (7.7.2) forces us to compute a power $\mathbf{A}^m$, where m is an arbitrary positive integer and $\mathbf{A}$ is a square $n \times n$ matrix. We define the powers of a square matrix $\mathbf{A}$ inductively,

$$\mathbf{A}^{m+1} = \mathbf{A}\mathbf{A}^m, \quad \mathbf{A}^0 = \mathbf{I}, \quad m = 0, 1, \ldots .$$

It is known that sums and products of square matrices are again square matrices of the same dimensions. We can also multiply a matrix by a constant. Therefore, we can compute linear combinations of nonnegative integral powers of the matrix. For any polynomial $q(\lambda) = q_0 + q_1\lambda + q_2\lambda^2 + \cdots + q_m\lambda^m$, we can unambiguously define the function $q(\mathbf{A})$ for a matrix $\mathbf{A}$ as

$$q(\mathbf{A}) = q_0\mathbf{I} + q_1\,\mathbf{A} + q_2\,\mathbf{A}^2 + \cdots + q_m\mathbf{A}^m .$$

Having defined polynomial functions, we may consider other functions of an arbitrary matrix $\mathbf{A}$. If a function $f(\lambda)$ can be expressed as a convergent power series

$$f(\lambda) = \sum_{m=0}^{\infty} a_m\,\lambda^m ,$$

then we can define a matrix function $f(\mathbf{A})$ by

$$f(\mathbf{A}) = \lim_{N\to\infty} \sum_{m=0}^{N} a_m\,\mathbf{A}^m$$

for those matrices for which the indicated limit exists. The spectral decomposition method allows us to define a function of the square matrix without actually applying the limit.

If $\mathbf{v}$ is an eigenvector for the matrix operator $\mathbf{A}$ that corresponds to the eigenvalue λ, then $\mathbf{A}^2\mathbf{v} = \mathbf{A}(\mathbf{A}\mathbf{v}) = \mathbf{A}(\lambda\mathbf{v}) = \lambda\mathbf{A}\mathbf{v} = \lambda^2\mathbf{v}$.

In the general case, we have $q(\mathbf{A})\mathbf{v} = q(\lambda)\mathbf{v}$ for an arbitrary polynomial or analytical function $q(\lambda)$. Hence, every eigenvector of $\mathbf{A}$ belonging to the eigenvalue λ is an eigenvector of $q(\mathbf{A})$ corresponding to the eigenvalue $q(\lambda)$.

Example 7.7.2: Let $\mathbf{A} = \begin{bmatrix} 1 & 3 \\ 0 & 2 \end{bmatrix}$ and $q(\lambda) = (\lambda+1)^2 = \lambda^2 + 2\lambda + 1$. This matrix has two real eigenvalues, $\lambda_1 = 1$ and $\lambda_2 = 2$, with eigenvectors $\mathbf{u}_1 = \begin{bmatrix} 1 \\ 0 \end{bmatrix}$ and $\mathbf{u}_2 = \begin{bmatrix} 3 \\ 1 \end{bmatrix}$, respectively. Indeed, $\mathbf{A}\mathbf{u}_1 = \mathbf{u}_1$ and $\mathbf{A}\mathbf{u}_2 = 2\mathbf{u}_2$.

We can define $q(\mathbf{A})$ in two ways. For the first method, we calculate $\mathbf{B} \stackrel{\text{def}}{=} \mathbf{A} + \mathbf{I} = \begin{bmatrix} 2 & 3 \\ 0 & 4 \end{bmatrix}$, and raise $\mathbf{B}$ to the second power: $\mathbf{B}^2 = \begin{bmatrix} 4 & 15 \\ 0 & 9 \end{bmatrix}$. The second method consists of calculating the second power of $\mathbf{A}$: $\mathbf{A}^2 = \begin{bmatrix} 1 & 9 \\ 0 & 4 \end{bmatrix}$ and using $q(\lambda)$:

$$q(\mathbf{A}) = \mathbf{A}^2 + 2\mathbf{A} + \mathbf{I} = \begin{bmatrix} 1 & 9 \\ 0 & 4 \end{bmatrix} + 2 \begin{bmatrix} 1 & 3 \\ 0 & 2 \end{bmatrix} + \begin{bmatrix} 1 & 0 \\ 0 & 1 \end{bmatrix} = \begin{bmatrix} 4 & 15 \\ 0 & 9 \end{bmatrix}.$$

The matrix $q(\mathbf{A})$ has eigenvalues $\mu_1 = 4$ and $\mu_2 = 9$, corresponding to the same eigenvectors, $\mathbf{u}_1 = \langle 1, 0 \rangle^T$ and $\mathbf{u}_2 = \langle 3, 1 \rangle^T$.

Example 7.7.3: We reconsider Example 7.6.4 on page 406. Since the minimal polynomial for the given symmetric matrix is $\psi(\lambda) = (\lambda - 2)(\lambda + 1) = \lambda^2 - \lambda - 2$, we have the relation between the first two powers of $\mathbf{A}$: $\mathbf{A}^2 = \mathbf{A} + 2\mathbf{I}$. The next powers of $\mathbf{A}$ are

$$\begin{aligned} \mathbf{A}^3 &= \mathbf{A}\mathbf{A}^2 = \mathbf{A}(2\mathbf{I} + \mathbf{A}) = 2\,\mathbf{A} + \mathbf{A}^2 = 2\,\mathbf{A} + (2\,\mathbf{I} + \mathbf{A}) = 2\,\mathbf{I} + 3\,\mathbf{A}, \\ \mathbf{A}^4 &= 2\,\mathbf{A} + 3\,\mathbf{A}^2 = 2\,\mathbf{A} + 3\,(2\,\mathbf{I} + \mathbf{A}) = 6\,\mathbf{I} + 5\,\mathbf{A}, \\ \mathbf{A}^5 &= 6\,\mathbf{A} + 5\,\mathbf{A}^2 = 6\,\mathbf{A} + 5\,(2\,\mathbf{I} + \mathbf{A}) = 10\,\mathbf{I} + 11\,\mathbf{A}, \\ \mathbf{A}^6 &= 10\,\mathbf{A} + 11\,\mathbf{A}^2 = 10\,\mathbf{A} + 11\,(2\mathbf{I} + \mathbf{A}) = 22\,\mathbf{I} + 21\,\mathbf{A}, \end{aligned}$$

and so on. So all powers of $\mathbf{A}$ are expressed as a linear combination of the matrix $\mathbf{A}$ and the identity matrix $\mathbf{I}$. □

Let $\psi(\lambda)$ be the m-th degree minimal polynomial of an $n \times n$ matrix $\mathbf{A}$, that is,

$$\psi(\lambda) = c_0 + c_1\lambda + c_2\lambda^2 + \cdots + c_{m-1}\lambda^{m-1} + \lambda^m.$$

Then

$$\psi\,(\mathbf{A}) = c_0\mathbf{I} + c_1\mathbf{A} + c_2\mathbf{A}^2 + \cdots + c_{m-1}\mathbf{A}^{m-1} + \mathbf{A}^m = \mathbf{0},$$

where $\mathbf{I}$ is the identity matrix, and $\mathbf{0}$ is the $n \times n$ matrix of zeroes. We can express $\mathbf{A}^m$ as a linear combination of lower powers of the matrix $\mathbf{A}$:

$$\mathbf{A}^m = -\left[c_0\mathbf{A}^0 + c_1\mathbf{A} + c_2\mathbf{A}^2 + \cdots + c_{m-1}\mathbf{A}^{m-1}\right]. \tag{7.7.3}$$

Multiplication of this equality by $\mathbf{A}$ yields

$$\mathbf{A}\mathbf{A}^m = \mathbf{A}^{m+1} = -\left[c_0\mathbf{A} + c_1\mathbf{A}^2 + c_2\mathbf{A}^3 + \cdots + c_{m-1}\mathbf{A}^m\right].$$

We can substitute expression (7.7.3) for $\mathbf{A}^m$ to obtain a polynomial of degree $m-1$. Hence, the m-th and $(m+1)$-th powers of the matrix $\mathbf{A}$ are again linear combinations of powers $\mathbf{A}^j$, $j = 0, 1, 2, \ldots m-1$. In a similar manner, we can show that any power $\mathbf{A}^k$, $k \geqslant m$, is a linear combination of powers $\mathbf{A}^j$, $j = 0, 1, \ldots m-1$, that is,

$$\mathbf{A}^k = a_0(k)\mathbf{I} + a_1(k)\mathbf{A} + a_2(k)\mathbf{A}^2 + \cdots + a_{m-1}(k)\mathbf{A}^{m-1}, \quad k = m, m+1, \ldots. \tag{7.7.4}$$

We can extend this equation (7.7.4) for all positive k by setting

$$a_j(k) = \begin{cases} 1, & \text{if } j = k < m, \\ 0, & \text{if } j \neq k < m. \end{cases}$$

For example,

$$a_j(m) = -c_j, \quad j = 0, 1, 2, \ldots m-1.$$

The Cayley–Hamilton theorem (page 400) assures us that any power of a square $n \times n$ matrix $\mathbf{A}$ can be expressed as a linear combination of the first n powers of the matrix: $\mathbf{A}^0 = \mathbf{I}$, $\mathbf{A}$, $\mathbf{A}^2$, ..., $\mathbf{A}^{n-1}$. Therefore, any series in $\mathbf{A}$ can be expressed as a polynomial of degree at most $n-1$ in $\mathbf{A}$, regardless of the value of n. If the minimum polynomial is known to be of degree m ($m \leqslant n$), only m powers of the matrix are needed to define all other powers of $\mathbf{A}$. This allows us to make an important observation about any analytical function of a square matrix: it can be expressed as a linear combination of a finite number of the first m powers of the given matrix.

Now we substitute Eq. (7.7.4) into the expansion (7.7.1) for $e^{\mathbf{A}t}$:

$$\begin{aligned} e^{\mathbf{A}t} &= \sum_{k=0}^{\infty} \frac{t^k}{k!}\,\mathbf{A}^k = \sum_{k=0}^{\infty} \frac{t^k}{k!} \left(\sum_{j=0}^{m-1} a_j(k)\,\mathbf{A}^j \right) \\ &= \sum_{j=0}^{m-1} \left(\sum_{k=0}^{\infty} \frac{t^k}{k!}\, a_j(k) \right) \mathbf{A}^j . \end{aligned}$$

Therefore,

$$e^{\mathbf{A}t} = \sum_{j=0}^{m-1} b_j(t)\,\mathbf{A}^j , \tag{7.7.5}$$

where

$$b_j(t) = \sum_{k=0}^{\infty} \frac{t^k}{k!}\, a_j(k).$$

Thus, the infinite sum (7.7.1) that defines $e^{\mathbf{A}t}$ reduces to a finite sum. Of course, its coefficients $b_j(t)$ are represented as infinite series. Fortunately, coefficients $b_j(t)$, $j = 0, 1, \ldots, m-1$, in expansion (7.7.5) can be determined without the tedious computations of series. This simplification is stated in the following theorem.

Theorem 7.15: Let $\psi(\lambda)$ be the minimal polynomial of degree m for a square matrix $\mathbf{A}$. The coefficient functions $b_j(t)$, $j = 0, 1, \ldots, m-1$, in the exponential representation (7.7.5) satisfy the following equations:

$$e^{\lambda_k t} = b_0(t) + b_1(t)\lambda_k + \cdots + b_{m-1}(t)\lambda_k^{m-1}, \quad k = 1, 2, \ldots, s, \tag{7.7.6}$$

where λ_k, $k = 1, 2, \ldots, s$, are distinct eigenvalues of the square matrix $\mathbf{A}$.

If the expansion (7.5.1), page 400, of the minimal polynomial $\psi(\lambda)$ contains the multiple $(\lambda - \lambda_k)^{m_k}$, $m_k > 1$, we include in the system to be solved $m_k - 1$ additional equations

$$t^p\, e^{\lambda_k t} = \sum_{j=p}^{m_k-1} \frac{j!}{(j-p)!}\, b_j(t)\, \lambda_k^{j-p}, \quad p = 1, 2, \ldots, m_k - 1. \tag{7.7.7}$$

Remark 1. Equation (7.7.7) is equivalent to the following equation ($p = 1, 2, \ldots, m_k - 1$):

$$\left. \frac{\mathrm{d}^p\, e^{\lambda t}}{\mathrm{d}\lambda^p} \right|_{\lambda=\lambda_k} = \left. \frac{\mathrm{d}^p}{\mathrm{d}\lambda^p} \left[b_0(t) + b_1(t)\lambda + \cdots + b_{m-1}(t)\lambda^{m-1} \right] \right|_{\lambda=\lambda_k} . \tag{7.7.8}$$

Remark 2. Theorem 7.15 holds for any annulled polynomial, including the characteristic polynomial, instead of the minimal polynomial. The following example shows that using a characteristic polynomial to construct a function of a matrix is not optimal when the minimal polynomial has a lesser degree than the characteristic polynomial.

Example 7.7.4: (Example 7.5.2 revisited) Let

$$\mathbf{A} = \begin{bmatrix} 1 & 3 & 3 \\ -3 & -5 & -3 \\ 3 & 3 & 1 \end{bmatrix} .$$

Its characteristic polynomial, $\chi(\lambda) = (\lambda+2)^2(\lambda-1)$, is not equal to the minimal polynomial $\psi(\lambda) = (\lambda+2)(\lambda-1)$. So we use $\psi(\lambda)$ to define $e^{\mathbf{A}t}$. From Eq. (7.7.6), it follows that

$$\begin{aligned} e^t &= b_0(t) + b_1(t), \\ e^{-2t} &= b_0(t) - 2b_1(t). \end{aligned}$$

We subtract the first equation from the last one to obtain

$$-3b_1(t) = e^{-2t} - e^t \quad \text{or} \quad b_1(t) = \frac{1}{3}\,e^t - \frac{1}{3}\,e^{-2t}.$$

Then

$$b_0(t) = e^t - b_1(t) = \frac{2}{3}\,e^t + \frac{1}{3}\,e^{-2t}.$$

Plugging these coefficient functions $b_0(t)$ and $b_1(t)$ into Eq. (7.7.5) yields

$$\begin{aligned} e^{\mathbf{A}t} &= \left(\frac{2}{3}\,e^t + \frac{1}{3}\,e^{-2t}\right)\mathbf{I} + \left(\frac{1}{3}\,e^t - \frac{1}{3}\,e^{-2t}\right)\mathbf{A} \\ &= \begin{bmatrix} e^t & e^t - e^{-2t} & e^t - e^{-2t} \\ -e^{-t} + e^{-2t} & -e^t + 2\,e^{-2t} & -e^t + e^{-2t} \\ e^t - e^{-2t} & e^t - e^{-2t} & e^t \end{bmatrix}. \end{aligned}$$

Now suppose that we decide to use the characteristic polynomial $\chi(\lambda) = (\lambda + 2)^2(\lambda - 1)$ instead of $\psi(\lambda)$. Then

$$e^{\mathbf{A}t} = b_0(t)\mathbf{I} + b_1(t)\mathbf{A} + b_2(t)\mathbf{A}^2,$$

where the coefficient functions $b_0(t)$, $b_1(t)$, and $b_2(t)$ satisfy the following system of equations:

$$\begin{aligned} e^t &= b_0(t) + b_1(t) + b_2(t), \\ e^{-2t} &= b_0(t) - 2b_1(t) + 4b_2(t), \\ t\,e^{-2t} &= b_1(t) - 4b_2(t). \end{aligned}$$

We rewrite the above system of algebraic equations in vector form $\mathbf{X}\,\mathbf{b} = \mathbf{v}(t)$, where

$$\mathbf{X} = \begin{bmatrix} 1 & 1 & 1 \\ 1 & -2 & 4 \\ 0 & 1 & -4 \end{bmatrix}, \quad \mathbf{b} = \begin{bmatrix} b_0(t) \\ b_1(t) \\ b_2(t) \end{bmatrix}, \quad \mathbf{v}(t) = \begin{bmatrix} e^t \\ e^{-2t} \\ t\,e^{-2t} \end{bmatrix}.$$

Since the determinant of the matrix $\mathbf{X}$ is not zero, $\det \mathbf{X} = 9$, it is invertible, and we find the vector $\mathbf{b}$ directly:

$$\mathbf{b} = \mathbf{X}^{-1}\mathbf{v} = \frac{1}{9}\begin{bmatrix} 4 & 5 & 2 \\ 4 & -4 & -1 \\ 1 & -1 & -3 \end{bmatrix}\begin{bmatrix} e^t \\ e^{-2t} \\ t\,e^{-2t} \end{bmatrix} = \frac{1}{9}\begin{bmatrix} 4\,e^t + 5\,e^{-2t} + 6t\,e^{-2} \\ 4\,e^t - 4\,e^{-2t} - 3t\,e^{-2t} \\ e^t - e^{-2t} - 3t\,e^{-2t} \end{bmatrix}.$$

With this in hand, we calculate the exponential function

$$\begin{aligned} e^{\mathbf{A}t} &= \left(\frac{4}{9}\,e^t + \frac{5}{9}\,e^{-2t} + \frac{2}{3}\,t\,e^{-2t}\right)\begin{bmatrix} 1 & 0 & 0 \\ 0 & 1 & 0 \\ 0 & 0 & 1 \end{bmatrix} \\ &\quad + \left(\frac{4}{9}\,e^t - \frac{4}{9}\,e^{-2t} - \frac{1}{3}\,t\,e^{-2t}\right)\begin{bmatrix} 1 & 3 & 3 \\ -3 & -5 & -3 \\ 3 & 3 & 1 \end{bmatrix} \\ &\quad + \left(\frac{1}{9}\,e^t - \frac{1}{9}\,e^{-2t} - \frac{1}{3}\,t\,e^{-2t}\right)\begin{bmatrix} 1 & -3 & -3 \\ 3 & 7 & 3 \\ -3 & -3 & 1 \end{bmatrix} \\ &= \begin{bmatrix} e^t & e^t - e^{-2t} & e^t - e^{-2t} \\ -e^{-t} + e^{-2t} & -e^t + 2\,e^{-2t} & -e^t + e^{-2t} \\ e^t - e^{-2t} & e^t - e^{-2t} & e^t \end{bmatrix}. \end{aligned}$$

Spectral decomposition method for "arbitrary" functions

We can extend the definition of the exponential of a square matrix to an "arbitrary" function. Let $f(\lambda)$ be an analytic function in a neighborhood of the origin. In this case, it has a Maclaurin series representation:

$$f(\lambda) = f_0 + f_1\lambda + f_2\lambda^2 + \cdots + f_k\lambda^k + \cdots = \sum_{k=0}^{\infty} f_k\lambda^k.$$

Then $f(\mathbf{A})$ can be defined via the following series:

$$f(\mathbf{A}) = f_0\mathbf{I} + f_1\mathbf{A} + f_2\mathbf{A}^2 + \cdots f_k\mathbf{A}^k + \cdots = \sum_{k=0}^{\infty} f_k\mathbf{A}^k,$$

subject that this series converges. From Eq. (7.7.4), it follows that $f(\mathbf{A})$ is actually the sum of the m powers of $\mathbf{A}$, namely,

$$f(\mathbf{A}) = \sum_{k=0}^{\infty} f_k\mathbf{A}^k = \sum_{k=0}^{\infty} f_k\left(\sum_{j=0}^{m-1} a_j(k)\mathbf{A}^k\right) = \sum_{j=0}^{m-1} \mathbf{A}^k\left(\sum_{k=0}^{\infty} f_k a_j(k)\right).$$

Therefore,

$$f(\mathbf{A}) = \sum_{j=0}^{m-1} b_j\,\mathbf{A}^j, \tag{7.7.9}$$

where the coefficients

$$b_j = \sum_{k=0}^{\infty} f_k\,a_j(k), \quad j = 0, 1, \ldots, m-1,$$

should satisfy the following equations:

$$f(\lambda_k) = b_0 + b_1\lambda_k + \cdots + b_{m-1}\lambda_k^{m-1} \tag{7.7.10}$$

for each distinct eigenvalue λ_k. If the eigenvalue λ_k is of multiplicity m_k, then we need to add $m_k - 1$ auxiliary equations similar to Eq. (7.7.8):

$$\left.\frac{\mathrm{d}^p f(\lambda)}{\mathrm{d}\lambda^p}\right|_{\lambda=\lambda_k} = \left.\frac{\mathrm{d}^p}{\mathrm{d}\lambda^p}\left[b_0 + b_1\lambda + \cdots + b_{s-1}\lambda^{s-1}\right]\right|_{\lambda=\lambda_k}, \quad p = 1, 2, \ldots, m_k - 1. \tag{7.7.11}$$

The above formula shows that the function $f(\lambda)$ must have $m_k - 1$ derivatives at each eigenvalue λ_k.

If the matrix $\mathbf{A}$ is invertible ($\det\mathbf{A} \ne 0$), the above definition can be extended for functions containing negative powers. Since a unique inverse of $\mathbf{A}$ exists, the relation

$$\mathbf{A}^{-n} = \left(\mathbf{A}^{-1}\right)^n, \quad n = 1, 2, \cdots$$

returns us to positive powers of $\mathbf{A}^{-1}$. ■

Remark 3. The coefficients $b_0, b_1, \ldots, b_{m-1}$ in the expansion (7.7.9) are completely determined by the eigenvalues of the given square matrix $\mathbf{A}$. Therefore two different matrices with the same eigenvalues have the same coefficient functions b_j in Eq. (7.7.9).

Remark 4. Generally speaking, we are allowed only to use this approach for holomorphic (analytic) functions (which are sums of convergent power series) in a neighborhood of a spectrum. For example, we cannot determine $\sqrt{\mathbf{A}}$ for a singular square matrix $\mathbf{A}$ with the aid of the spectral decomposition method because the corresponding function $f(\lambda) = \lambda^{1/2}$ does not have a Maclaurin representation at $\lambda = 0$. However, if $f(\lambda)$ is a smooth function in a domain including the spectrum of $\mathbf{A}$, spectral decomposition is applicable to define the corresponding square matrix $f(\mathbf{A})$. ■

Let us show that for 2×2 matrices (as well as for any matrix having a minimal polynomial of the first degree), a square root can be obtained with the aid of the spectral decomposition method only when the matrix coefficients satisfy a special condition. Suppose we try to find a root of a square 2×2 matrix $\mathbf{A}$ in the form

$$\sqrt{\mathbf{A}} = b_0\mathbf{I} + b_1\mathbf{A}. \tag{7.7.12}$$

Squaring both sides, we obtain

$$\mathbf{A} = \left(\sqrt{\mathbf{A}}\right)^2 = (b_0\mathbf{I} + b_1\mathbf{A})^2 = b_0^2\mathbf{I} + 2b_0b_1\mathbf{A} + b_1^2\mathbf{A}^2$$

Suppose we know the minimal polynomial $\psi(\lambda) = \lambda^2 + c_1\lambda + c_0$, where $c_1 = -\mathrm{tr}\,(\mathbf{A})$ and $c_0 = \det\mathbf{A}$. Then $\mathbf{A}^2 = -c_0\mathbf{I} - c_1\mathbf{A}$. Using this equation, we find

$$\mathbf{A} = b_0^2\mathbf{I} + 2b_0b_1\mathbf{A} - b_1^2\left[c_0\mathbf{I} + c_1\mathbf{A}\right] = \left(b_0^2 - c_0b_1^2\right)\mathbf{I} + \left(2b_0b_1 - c_1b_1^2\right)\mathbf{A}.$$

From the latter, we obtain two nonlinear equations to be solved for b_0 and b_1:

$$b_0^2 = b_1^2\det\mathbf{A}, \qquad 1 = 2b_0b_1 + b_1^2\mathrm{tr}\,(\mathbf{A}). \tag{7.7.13}$$

If this system of equations has a solution (not necessarily unique), the spectral decomposition method is applicable to find a square root. When $\det\mathbf{A} = 0$ and $\mathrm{tr}\,\mathbf{A} = 0$, Eq. (7.7.13) has no solution.

Example 7.7.5: The nilpotent matrix $\mathbf{A} = \begin{bmatrix} 1 & -1 \\ 1 & -1 \end{bmatrix}$ has a double eigenvalue $\lambda = 0$ and $\mathbf{A}^2 = \mathbf{0}$. For this matrix, $\mathrm{tr}\,\mathbf{A} = 0$ and $\det\mathbf{A} = 0$, so the conditions (7.7.13) are violated. Suppose we want to find a square root of $\mathbf{A}$. Let us see what happens if we ignore this warning. Then we seek a root as a linear combination of matrices $\mathbf{I} = \mathbf{A}^0$ and $\mathbf{A}$:

$$\sqrt{\mathbf{A}} = b_0\mathbf{I} + b_1\mathbf{A},$$

where coefficients b_0 and b_1 should satisfy the equations

$$\sqrt{0} = b_0, \qquad \text{and} \qquad 1 = b_1\,2\sqrt{0}.$$

Since this system of equations with respect to b_0, b_1 has no solution, the given matrix has no square root. But matrix-functions $\cos\left(\mathbf{A}^{1/2}t\right) = \mathbf{I} - \frac{t^2}{2}\,\mathbf{A}$ and $\mathbf{A}^{-1/2}\sin\left(\mathbf{A}^{1/2}t\right) = t\,\mathbf{I} - \frac{t^3}{6}\,\mathbf{A}$ exist.

Let us look at the 3×3 matrices $\mathbf{B} = \begin{bmatrix} 4 & 2 & 0 \\ 0 & 0 & 0 \\ 0 & 0 & 0 \end{bmatrix}$, $\mathbf{B}^k = \begin{bmatrix} 4^k & 2 \times 4^{k-1} & 0 \\ 0 & 0 & 0 \\ 0 & 0 & 0 \end{bmatrix}$, and $\mathbf{R} = \begin{bmatrix} 2 & 1 & x \\ 0 & 0 & -2x \\ 0 & 0 & 0 \end{bmatrix}$. If we try to apply the spectral decomposition method to find a square root of $\mathbf{B}$, we would assume that

$$\sqrt{\mathbf{B}} = b_0\mathbf{I} + b_1\mathbf{B} + b_2\mathbf{B}^2.$$

Squaring both sides yields

$$\left(\sqrt{\mathbf{B}}\right)^2 = \mathbf{B} = b_0^2\mathbf{I} + b_1^2\mathbf{B}^2 + 16b_2^2\mathbf{B}^2 + 2b_0b_1\mathbf{B} + 2b_0b_2\mathbf{B}^2 + 8b_1b_2\mathbf{B}^2$$

because $\mathbf{B}^3 = 4\mathbf{B}^2$ and $\mathbf{B}^4 = 16\mathbf{B}^2$. Equating coefficients of $\mathbf{I}$, $\mathbf{B}$, and $\mathbf{B}^2$, we obtain the system of nonlinear equations

$$b_0^2 = 0, \quad 1 = 2b_0b_1, \quad 0 = b_1^2 + 16b_2^2 + 2b_0b_2 + 8b_1b_2,$$

which has no solution. Therefore, the given matrix $\mathbf{B}$ has no square root expressed through powers of $\mathbf{B}$. On the other hand, this matrix has infinite many roots because $\mathbf{R}^2 = \mathbf{B}$ for any x. The matrix-function $\cos\left(\sqrt{\mathbf{B}}t\right)$ exists:

$$\cos\left(\sqrt{\mathbf{B}}t\right) = \begin{bmatrix} \cos 2t & \frac{1}{4}\cos 2t - \frac{1}{4} & \frac{1}{16}\cos 2t + \frac{t^2}{8} - \frac{1}{16} \\ 0 & 1 & -\frac{t^2}{2} \\ 0 & 0 & 1 \end{bmatrix},$$ which is the solution of $\ddot{\mathbf{\Phi}} + \mathbf{B}\,\mathbf{\Phi} = \mathbf{0}$, $\mathbf{\Phi}(0) = \mathbf{I}$, $\dot{\mathbf{\Phi}}(0) = \mathbf{0}$.

Example 7.7.6: It is not hard to verify that the following diagonal matrix $\mathbf{A}$ has two square roots $\pm\mathbf{R}$, where

$$\mathbf{A} = \begin{bmatrix} 4 & 0 & 0 \\ 0 & 0 & 0 \\ 0 & 0 & 0 \end{bmatrix} \qquad \text{and} \qquad \mathbf{R} = \frac{1}{2}\,\mathbf{A} = \begin{bmatrix} 2 & 0 & 0 \\ 0 & 0 & 0 \\ 0 & 0 & 0 \end{bmatrix}.$$

Its minimal polynomial is $\psi(\lambda) = \lambda(\lambda - 4) = \lambda^2 - 4\lambda$, while its characteristic polynomial is $\chi(\lambda) = \lambda^2(\lambda - 4)$. Hence, $\mathbf{A}^2 = 4\mathbf{A}$. We use $\psi(\lambda)$ to construct a root of $\mathbf{A}$:

$$\sqrt{\mathbf{A}} = b_0\mathbf{I} + b_1\mathbf{A}$$

because it is assumed that any function of the matrix $\mathbf{A}$ is expressed through two powers of $\mathbf{A}$: $\mathbf{I} = \mathbf{A}^0$ and $\mathbf{A}$. Squaring both sides of Eq. (7.7.12) and substituting $\mathbf{A}^2 = 4\mathbf{A}$, we obtain

$$\mathbf{A} = (b_0\mathbf{I} + b_1\mathbf{A})^2 = b_0^2\mathbf{I} + (2b_0b_1 + 4b_1^2)\mathbf{A}.$$

Equating the same powers of $\mathbf{A}$, we see that coefficients b_0 and b_1 should have the following values: $b_0 = 0$ and $4b_1^2 = 1$. This leads to two roots: $\pm\mathbf{R}$.

Now suppose we would like to use the characteristic polynomial instead of the minimal polynomial. According to the Cayley–Hamilton theorem, page 400, we have

$$\chi(\mathbf{A}) = 0 \qquad \text{or} \qquad \mathbf{A}^3 = 4\mathbf{A}^2.$$

Hence, any power of $\mathbf{A}$ greater than 3 is expressed through its first three powers: $\mathbf{I} = \mathbf{A}^0$, $\mathbf{A}$, and $\mathbf{A}^2$. Now we assume that the square root of $\mathbf{A}$ is also represented as

$$\sqrt{\mathbf{A}} = b_0\mathbf{I} + b_1\mathbf{A} + b_2\mathbf{A}^2.$$

Squaring both sides, we get

$$\mathbf{A} = b_0^2\mathbf{I} + 2b_0b_1\mathbf{A} + 2b_0b_2\mathbf{A}^2 + 2b_1b_2\mathbf{A}^3 + b_1^2\mathbf{A}^2 + b_2^2\mathbf{A}^4.$$

Substituting $\mathbf{A}^3 = 4\mathbf{A}^2$ and $\mathbf{A}^4 = 16\mathbf{A}^2$ and comparing like powers of $\mathbf{A}$, we obtain the system of nonlinear equations

$$b_0^2 = 0, \quad 2b_0b_1 = 1, \quad b_1^2 + 2b_0b_2 + 8b_1b_2 + 16b_2^2 = 0,$$

which has no solution. Therefore, the characteristic polynomial cannot be used for the calculation of $\sqrt{\mathbf{A}}$. This conclusion is true not only for a square root function, but for any function having a derivative which is undefined at $\lambda = 0$. However, if the function has continuous derivatives in the neighborhoods of eigenvalues, it does not matter whether a minimal or characteristic polynomial is utilized.

Example 7.7.7: Consider two 2×2 matrices

$$\mathbf{A} = \begin{bmatrix} 3 & -4 \\ 1 & -2 \end{bmatrix} \qquad \text{and} \qquad \mathbf{B} = \begin{bmatrix} -4 & 9 \\ -2 & 5 \end{bmatrix}$$

having the same eigenvalues: $\lambda_1 = -1$ and $\lambda_2 = 2$. To find the exponential functions of these matrices, we need to determine the function $b_0(t)$ and $b_1(t)$ from two simultaneous algebraic equations:

$$e^{-t} = b_0(t) - b_1(t), \qquad e^{2t} = b_0(t) + 2b_1(t),$$

which yields $b_0(t) = \frac{1}{3}\,e^{2t} + \frac{2}{3}\,e^{-t}$, $b_1(t) = \frac{1}{3}\,e^{2t} - \frac{1}{3}\,e^{-t}$. We then construct both exponential functions using the same functions $b_0(t)$ and $b_1(t)$:

$$e^{\mathbf{A}\,t} = b_0(t)\,\mathbf{I} + b_1(t)\mathbf{A} = \frac{1}{3}\,e^{2t}\begin{bmatrix} 4 & -4 \\ 1 & -1 \end{bmatrix} + \frac{1}{3}\,e^{-t}\begin{bmatrix} -1 & 4 \\ -1 & 4 \end{bmatrix},$$

$$e^{\mathbf{B}\,t} = b_0(t)\,\mathbf{I} + b_1(t)\mathbf{B} = e^{2t}\begin{bmatrix} -1 & 3 \\ -2/3 & 2 \end{bmatrix} + e^{-t}\begin{bmatrix} 2 & -3 \\ 2/3 & -1 \end{bmatrix}. \qquad \square$$

If, for a square matrix $\mathbf{A}$, the minimal polynomial is $\psi(\lambda) = (\lambda - \lambda_0)^2$, then any (holomorphic) function of this matrix is

$$f\,(\mathbf{A}) = b_0\mathbf{I} + b_1\mathbf{A} = \left(f(\lambda_0) - \lambda_0 f'(\lambda_0)\right)\mathbf{I} + f'(\lambda_0)\mathbf{A}$$

because $b_0 = f(\lambda_0) - \lambda_0 f'(\lambda_0)$ and $b_1 = f'(\lambda_0)$. In particular,

$$e^{\mathbf{A}\,t} = (1 - \lambda_0 t)\,\mathbf{I}\,e^{\lambda_0 t} + t\,e^{\lambda_0 t}\mathbf{A}. \tag{7.7.14}$$

Example 7.7.8: The matrix $\mathbf{B} = \begin{bmatrix} 2 & 4 & 3 \\ -4 & -6 & -3 \\ 3 & 3 & 1 \end{bmatrix}$ has the characteristic polynomial $\chi(\lambda) = (\lambda+2)^2(\lambda-1)$, which coincides with the characteristic polynomial of the matrix **A** from Example 7.7.4. Since the determination of coefficients in the expansion (7.7.5) depends only on the characteristic/minimal polynomial, we can use them for any matrix having the same characteristic polynomial. Therefore,

$$\begin{aligned} e^{\mathbf{B}t} &= b_0(t)\mathbf{I} + b_1(t)\mathbf{B} + b_2(t)\mathbf{B}^2 \\ &= \left(\frac{4}{9}e^t + \frac{5}{9}e^{-2t} + \frac{2}{3}t\,e^{-2t}\right)\begin{bmatrix} 1 & 0 & 0 \\ 0 & 1 & 0 \\ 0 & 0 & 1 \end{bmatrix} \\ &+ \left(\frac{4}{9}e^t - \frac{4}{9}e^{-2t} - \frac{1}{3}t\,e^{-2t}\right)\begin{bmatrix} 2 & 4 & 3 \\ -4 & -6 & -3 \\ 3 & 3 & 1 \end{bmatrix} \\ &+ \left(\frac{1}{9}e^t - \frac{1}{9}e^{-2t} - \frac{1}{3}t\,e^{-2t}\right)\begin{bmatrix} -3 & -7 & -3 \\ 7 & 11 & 3 \\ -3 & -3 & 1 \end{bmatrix} \\ &= e^t\begin{bmatrix} 1 & 1 & 1 \\ -1 & -1 & -1 \\ 1 & 1 & 1 \end{bmatrix} + e^{-2t}\begin{bmatrix} 0 & -1 & -1 \\ 1 & 2 & 1 \\ -1 & -1 & 0 \end{bmatrix} + t\,e^{-2t}\begin{bmatrix} 1 & 1 & 0 \\ -1 & -1 & 0 \\ 0 & 0 & 0 \end{bmatrix}. \end{aligned}$$

We can use *Maple* to perform all of the operations and check our answer. First, invoke the `linalg` package and define the matrix of coefficients needed for determination of b_0, b_1, and b_2:

```
B:=matrix([[1,1,1],[1,-2,4],[0,1,-4]])
```

Next we find its inverse to calculate the values of coefficients b_j $(j = 1, 2, 3)$:

```
Binv := inverse(B)
y:=matrix([[exp(t)], [exp(2, t))], [exp(3, t))]])
BB := multiply(Binv, y);   II := Matrix(3, shape = identity);
expAt:=evalm(II*(BB[1, 1] + BB[2, 1]*A + BB[3, 1]*A*A)
```

If the coefficients b_j $(j = 1, 2, 3)$ are known, we can calculate $e^{\mathbf{A}t}$ by executing one command:

```
expAt:=evalm(scalarmul(array(identity,1..3,1..3),b0) + scalarmul(A,b1)
+ scalarmul(multiply(A,A),b2)))
```

Example 7.7.9: Let us define some functions for the matrix $\mathbf{B} = \begin{bmatrix} 1 & 4 \\ 2 & -1 \end{bmatrix}$. We start with the function $f(\lambda) = \frac{\lambda-1}{\lambda+1}$. The given matrix has the characteristic polynomial $\chi(\lambda) = \lambda^2 - 9 = (\lambda-3)(\lambda+3)$ with simple eigenvalues $\lambda = \pm 3$. We seek the ratio in the form

$$\frac{\mathbf{B}-\mathbf{I}}{\mathbf{B}+\mathbf{I}} = b_0\mathbf{I} + b_1\mathbf{B} \quad \text{and} \quad \frac{\lambda-2}{\lambda+1} = b_0 + b_1\lambda.$$

Setting $\lambda = 3$ and $\lambda = -3$ in the latter equation gives

$$\frac{3-1}{3+1} = b_0 + 3\,b_1, \quad \text{and} \quad \frac{-3-1}{-3+1} = b_0 - 3\,b_1.$$

Solving this system of algebraic equations with respect to b_0 and b_1, we obtain

$$b_0 = \frac{5}{4}, \quad b_1 = -\frac{1}{4}.$$

Hence,

$$\frac{\mathbf{B}-\mathbf{I}}{\mathbf{B}+\mathbf{I}} = \frac{5}{4}\mathbf{I} - \frac{1}{4}\mathbf{B} = \frac{5}{4}\begin{bmatrix} 1 & 0 \\ 0 & 1 \end{bmatrix} - \frac{1}{4}\begin{bmatrix} 1 & 4 \\ 2 & -1 \end{bmatrix} = \begin{bmatrix} 1 & -1 \\ -1/2 & 3/2 \end{bmatrix}.$$

To define $\cos(\mathbf{B}t)$, we need to find the function $b_0(t)$ and $b_1(t)$ such that

$$\cos(\mathbf{B}t) = b_0(t)\mathbf{I} + b_1(t)\mathbf{B}.$$

For these functions, we have two equations:

$$\cos 3t = b_0(t) \pm 3\, b_1(t),$$

from which $b_1 \equiv 0$ and $b_0 = \cos 3t$, and we get $\cos(\mathbf{B}t) = \cos 3t\mathbf{I} = \begin{bmatrix} \cos 3t & 0 \\ 0 & \cos 3t \end{bmatrix}$. Since $\mathbf{B}^2 = 9\mathbf{I}$, the matrix-function $\cos(\mathbf{B}t)$ satisfies the matrix differential equation

$$\frac{\mathrm{d}^2 \cos(\mathbf{B}t)}{\mathrm{d}t^2} + \mathbf{B}^2 \cos(\mathbf{B}t) = \mathbf{0} \qquad \text{or} \qquad \frac{\mathrm{d}^2 \cos(\mathbf{B}t)}{\mathrm{d}t^2} + 9\cos(\mathbf{B}t) = \mathbf{0}.$$

Problems

1. Let $\mathbf{B}$ be a square matrix similar to $\mathbf{A}$, so that $\mathbf{B} = \mathbf{S}^{-1}\mathbf{A}\mathbf{S}$, with $\det \mathbf{S} \ne 0$. For an entire function $f(\lambda)$, show that $f(\mathbf{B}) = \mathbf{S}^{-1} f(\mathbf{A})\,\mathbf{S}$. In particular, show that their resolvents are similar matrices: $\mathbf{R}_\lambda(\mathbf{B}) = (\lambda\mathbf{I} - \mathbf{B})^{-1} = \mathbf{S}^{-1}\mathbf{R}_\lambda(\mathbf{A})\,\mathbf{S}$.

2. (a) Compute $b_0(t)$ and $b_1(t)$ in the expansion (7.7.5) for matrices

$$\mathbf{A} = \begin{bmatrix} 1 & 0 \\ \alpha & 2 \end{bmatrix} \quad \text{and} \quad \mathbf{B} = \begin{bmatrix} 2 & 0 \\ \alpha & 1 \end{bmatrix}.$$

   (b) Compute the exponential matrices $e^{\mathbf{A}t}$ and $e^{\mathbf{B}t}$.

   (c) Do the matrices $e^{\mathbf{A}t}$ and $e^{\mathbf{B}t}$ commute with each other?

3. (a) Compute $b_0(t)$, $b_1(t)$, and $b_2(t)$ in the expansion (7.7.5) for the defective matrices

$$\mathbf{A} = \begin{bmatrix} -1 & 1 & 4 \\ 3 & 1 & -4 \\ -1 & 0 & 3 \end{bmatrix} \quad \text{and} \quad \mathbf{B} = \begin{bmatrix} 1 & 1 & 1 \\ 2 & 1 & -1 \\ 0 & -1 & 1 \end{bmatrix}.$$

   (b) Compute the exponential matrices $e^{\mathbf{A}t}$ and $e^{\mathbf{B}t}$.

   (c) Do the matrices $e^{\mathbf{A}t}$ and $e^{\mathbf{B}t}$ commute?

4. For each of the 2×2 matrices from Problems 1 – 3 in §7.3, page 390, compute $e^{\mathbf{A}t}$, $\dfrac{\sin(\mathbf{A}t)}{\mathbf{A}}$, and $\cos(\mathbf{A}t)$.

5. Show that the following matrices have no square roots:

   (a) $\begin{bmatrix} 0 & 1 \\ 0 & 0 \end{bmatrix}$; **(b)** $\begin{bmatrix} 1 & 3 \\ -1/3 & -1 \end{bmatrix}$; **(c)** $\begin{bmatrix} 1 & 1 \\ -1 & -1 \end{bmatrix}$; **(d)** $\begin{bmatrix} -2 & 1 \\ -4 & 2 \end{bmatrix}$;

   (e) $\begin{bmatrix} -4 & 2 \\ -8 & 4 \end{bmatrix}$; **(f)** $\begin{bmatrix} 2 & 0 & 0 \\ 0 & 0 & 4 \\ 0 & 0 & 0 \end{bmatrix}$; **(g)** $\begin{bmatrix} -3 & 1 \\ -9 & 3 \end{bmatrix}$; **(h)** $\begin{bmatrix} 0 & 0 & 0 \\ 1 & 0 & 0 \\ 0 & 0 & 1 \end{bmatrix}$.

6. For the given 2×2 matrices, compute $e^{\mathbf{A}t}$, $\cos(\sqrt{\mathbf{A}}t) = \Re\, e^{\mathrm{j}\sqrt{\mathbf{A}}t}$, and $\dfrac{\sin(\sqrt{\mathbf{A}}t)}{\sqrt{\mathbf{A}}}$.

   (a) $\begin{bmatrix} 11 & 5 \\ -5 & 21 \end{bmatrix}$; **(b)** $\begin{bmatrix} 15 & -1 \\ 1 & 17 \end{bmatrix}$; **(c)** $\begin{bmatrix} 31 & -9 \\ 4 & 19 \end{bmatrix}$; **(d)** $\begin{bmatrix} -2 & 3 \\ -3 & 4 \end{bmatrix}$;

   (e) $\begin{bmatrix} 10 & 1 \\ -1 & 8 \end{bmatrix}$; **(f)** $\begin{bmatrix} 47 & 4 \\ 23 & 3 \end{bmatrix}$; **(g)** $\begin{bmatrix} 1 & -9 \\ 1 & 7 \end{bmatrix}$; **(h)** $\begin{bmatrix} 20 & 2 \\ 40 & 9 \end{bmatrix}$;

   (i) $\begin{bmatrix} 18 & 18 \\ 17 & 19 \end{bmatrix}$; **(j)** $\begin{bmatrix} 30 & 6 \\ 5 & 31 \end{bmatrix}$; **(k)** $\begin{bmatrix} 100 & 51 \\ -51 & -2 \end{bmatrix}$; **(l)** $\begin{bmatrix} 20 & 8 \\ -2 & 12 \end{bmatrix}$.

7. For the given 3×3 matrices, compute $e^{\mathbf{A}t}$, $\cos(\sqrt{\mathbf{A}}t) = \Re\, e^{\mathrm{j}\sqrt{\mathbf{A}}t}$, and $\dfrac{\sin(\sqrt{\mathbf{A}}t)}{\sqrt{\mathbf{A}}}$.

(a) $\begin{bmatrix} 4 & 3 & -4 \\ -4 & -5 & 8 \\ -2 & -3 & 5 \end{bmatrix}$; (b) $\begin{bmatrix} 1 & 3 & -4 \\ 3 & 1 & 4 \\ 1 & -2 & 6 \end{bmatrix}$; (c) $\begin{bmatrix} -1 & -1 & 2 \\ -1 & 9 & 1 \\ -2 & -1 & 3 \end{bmatrix}$;

(d) $\begin{bmatrix} 3 & -1 & -1 \\ -1 & 3 & -1 \\ -4 & 8 & -2 \end{bmatrix}$; (e) $\begin{bmatrix} -4 & -2 & 6 \\ 6 & 3 & -8 \\ -2 & -1 & 3 \end{bmatrix}$; (f) $\begin{bmatrix} 6 & 0 & -4 \\ 3 & 4 & 7 \\ 1 & 0 & 2 \end{bmatrix}$;

(g) $\begin{bmatrix} 4 & 0 & -1 \\ 0 & 4 & -1 \\ -1 & 1 & 4 \end{bmatrix}$; (h) $\begin{bmatrix} 3 & 3 & -4 \\ -4 & -5 & 8 \\ -2 & -3 & 5 \end{bmatrix}$; (i) $\begin{bmatrix} 3 & -8 & -10 \\ -2 & 7 & 8 \\ 2 & -6 & -7 \end{bmatrix}$;

(j) $\begin{bmatrix} 1 & 2 & 0 \\ 1 & 1 & 2 \\ 0 & -1 & 1 \end{bmatrix}$; (k) $\begin{bmatrix} 1 & 1 & 0 \\ -2 & 1 & -2 \\ 0 & -1 & 1 \end{bmatrix}$; (l) $\begin{bmatrix} 3 & -8 & -10 \\ -3 & 7 & 9 \\ 3 & -6 & -8 \end{bmatrix}$.

8. For each of the following 3×3 matrices, compute $e^{\mathbf{A}t}$, determine $\frac{\sin(\sqrt{\mathbf{A}}\,t)}{\sqrt{\mathbf{A}}}$ and $\cos\left(\sqrt{\mathbf{A}}\,t\right)$, and then show that the result does not depend on the choice of the root.

(a) $\begin{bmatrix} -5 & 1 & 0 \\ 1 & -1 & -2 \\ -5 & -1 & -1 \end{bmatrix}$; (b) $\begin{bmatrix} 7 & 4 & -4 \\ 1 & 0 & -1 \\ 10 & 5 & -6 \end{bmatrix}$; (c) $\begin{bmatrix} -4 & 8 & 4 \\ -3 & 4 & -3 \\ -1 & 4 & 9 \end{bmatrix}$;

(d) $\begin{bmatrix} 2 & -2 & -1 \\ 2 & -2 & -1 \\ 3 & -3 & -4 \end{bmatrix}$; (e) $\begin{bmatrix} 5 & 1 & -11 \\ 7 & -1 & -13 \\ 4 & 0 & -8 \end{bmatrix}$; (f) $\begin{bmatrix} -3 & 1 & 1 \\ 1 & -3 & 1 \\ 4 & -8 & 2 \end{bmatrix}$;

(g) $\begin{bmatrix} -5 & 1 & -1 \\ 1 & -9 & 3 \\ 2 & -2 & -4 \end{bmatrix}$; (h) $\begin{bmatrix} 1 & 12 & -8 \\ -1 & 9 & -4 \\ -1 & 6 & -1 \end{bmatrix}$; (i) $\begin{bmatrix} 1 & 1 & 0 \\ -2 & 2 & -1 \\ 2 & -2 & 2 \end{bmatrix}$.

9. The following pairs of matrices have the same eigenvalues. Use this information to construct $e^{\mathbf{A}t}$ for each of the matrices.

(a) $\begin{bmatrix} 2 & -2 & 3 \\ -4 & 4 & 3 \\ 2 & 1 & 0 \end{bmatrix}$, $\begin{bmatrix} 2 & 4 & -7 \\ 1 & 5 & -5 \\ -4 & 4 & -1 \end{bmatrix}$; (b) $\begin{bmatrix} 3 & 0 & 4 \\ 1 & 1 & 1 \\ -1 & 0 & -1 \end{bmatrix}$, $\begin{bmatrix} -1 & 0 & 1 \\ 0 & -1 & 1 \\ 1 & -1 & -1 \end{bmatrix}$;

(c) $\begin{bmatrix} 3 & 1 & 0 \\ -8 & -6 & 2 \\ -9 & -9 & 4 \end{bmatrix}$, $\begin{bmatrix} -9 & -3 & -7 \\ 2 & 1 & 2 \\ 11 & 3 & 9 \end{bmatrix}$; (d) $\begin{bmatrix} 4 & 0 & 10 \\ 10 & 3 & 1 \\ -1 & 0 & 2 \end{bmatrix}$, $\begin{bmatrix} 5 & -1 & 1 \\ 10 & 3 & 0 \\ -3 & 2 & 1 \end{bmatrix}$;

(e) $\begin{bmatrix} 1 & 2 & -1 \\ 2 & 1 & -1 \\ 9 & -5 & -1 \end{bmatrix}$, $\begin{bmatrix} 39 & -46 & 6 \\ 38 & -45 & 6 \\ 38 & -46 & 7 \end{bmatrix}$; (f) $\begin{bmatrix} 1 & 2 & -3 \\ 2 & 1 & -1 \\ 2 & -2 & -1 \end{bmatrix}$, $\begin{bmatrix} 1 & 2 & -5 \\ 2 & 1 & -1 \\ 1 & -1 & -1 \end{bmatrix}$.

10. Calculate the powers $\mathbf{A}^n$ (n is an arbitrary positive integer) of the Fibonacci matrix $\mathbf{A} = \begin{bmatrix} 1 & 1 \\ 1 & 0 \end{bmatrix}$. Express your answer in terms of the golden ratio $\phi = \left(1 + \sqrt{5}\right)/2$.

11. Show that the matrix $\mathbf{A} = \begin{bmatrix} 0 & 0 \\ 2 & 0 \end{bmatrix}$, has no square root. Find two matrix-functions $\cos\left(\sqrt{\mathbf{A}}t\right)$ and $\mathbf{A}^{-1/2}\sin\left(\mathbf{A}^{1/2}t\right)$.

12. Repeat the previous question for the matrix $\mathbf{A} = \begin{bmatrix} 2 & -4 \\ 1 & -2 \end{bmatrix}$.

13. Show that a nilpotent matrix has no square root. Recall that a square matrix $\mathbf{A}$ is said to be nilpotent if $\mathbf{A}^p = \mathbf{0}$ for some positive integer p.

14. For each of the following 3×3 positive matrices, determine the matrix functions $\mathbf{\Psi}(t) = \cos\left(\sqrt{\mathbf{A}}t\right)$ and $\mathbf{\Phi}(t) = \mathbf{A}^{-1/2}\sin\left(\mathbf{A}^{1/2}t\right)$, and show that they satisfy the second order matrix differential equation $\ddot{\mathbf{X}} + \mathbf{A}\,\mathbf{X} = 0$.

(a) $\begin{bmatrix} 8 & 0 & -4 \\ 3 & 4 & 1 \\ -8 & 0 & 12 \end{bmatrix}$; (b) $\begin{bmatrix} 5 & -7 & 5 \\ -4 & 8 & -1 \\ -1 & -1 & 5 \end{bmatrix}$; (c) $\begin{bmatrix} -53 & 127 & 31 \\ -22 & 53 & 12 \\ -9 & 20 & 9 \end{bmatrix}$; (d) $\begin{bmatrix} -25 & -130 & 64 \\ -1 & 4 & 4 \\ -11 & -45 & 30 \end{bmatrix}$.

Summary for Chapter 7

1. A square $n \times n$ matrix $\mathbf{A}$ is called an **unitary** (or **isometric**) matrix if $\|\mathbf{Ax}\| = \|\mathbf{x}\|$ for all n-vectors $\mathbf{x}$. A square $n \times n$ matrix $\mathbf{A}$ is called **normal** if $\mathbf{AA}^* = \mathbf{A}^*\mathbf{A}$. Self-adjoint matrices and unitary matrices are normal.
2. A matrix $\mathbf{A}(t)$ is said to be continuous on an interval (α, β) if each element of $\mathbf{A}$ is a continuous function on the given interval. With matrix functions we can operate in a similar way as with functions.
3. The **determinant** of a square $n \times n$ matrix $\mathbf{A} = [\, a_{ij}\,]$ is the sum of $n!$ terms. Each term is the product of n matrix entries, one element from each row and one element from each column. Furthermore, each product is assigned a plus or a minus sign:
$$\det(\mathbf{A}) = \sum (-1)^{\sigma}\, a_{1i_1} a_{2i_2} \cdots a_{ni_n},$$
where the summation is over all permutations $(i_1, i_2, \ldots, i_n)$ of the integers $1, 2, \ldots, n$ and σ is the integer that is determined by the evenness of the permutation. Therefore, half of all products in this sum comes with a plus sign and the other half comes with a minus sign.
4. A matrix $\mathbf{A}$ is called **singular** (or degenerate) if $\det \mathbf{A} = 0$ and **nonsingular** (or invertible) if its determinant is not zero, that is, $\det \mathbf{A} \neq 0$.
5. Let $\mathbf{A}$ be an $n \times n$ matrix. The **minor** of the kj^{th} entry of $\mathbf{A}$ is the determinant of the $(n-1) \times (n-1)$ submatrix obtained from $\mathbf{A}$ by deleting row k and column j. The **cofactor** of the entry a_{kj} is Cof $(a_{kj}) = (-1)^{k+j}$ Minor (a_{kj}).
6. If $\mathbf{A}$ is a nonsingular $n \times n$ matrix, then $\mathbf{A}^{-1} = \dfrac{1}{\det \mathbf{A}} \left[\mathrm{Cof}(a_{ji})\right]_{ij}$.
7. The **resolvent** of a square matrix $\mathbf{A}$ is the matrix $\mathbf{R}_\lambda(\mathbf{A})$ defined as $\mathbf{R}_\lambda(\mathbf{A}) = (\lambda\mathbf{I} - \mathbf{A})^{-1}$, where $\mathbf{I}$ is the identity matrix.
8. A set of vectors $\mathbf{x}_1, \mathbf{x}_2, \ldots, \mathbf{x}_n$ is said to **span** the vector space $\mathbf{V}$ if every element $\mathbf{V}$ can be expressed as a linear combination $c_1\mathbf{x}_1 + c_2\mathbf{x}_2 + \cdots + c_n\mathbf{x}_n$ of these vectors.
9. A set of vectors $\mathbf{x}_1, \mathbf{x}_2, \ldots, \mathbf{x}_n$ in $\mathbf{V}$ is said to be **linearly dependent** if one of these vectors is a linear combination of the others.

   If these vectors are not linearly dependent then they are said to be **linearly independent**.
10. The **dimension** of a vector space $\mathbf{V}$, denoted by $\dim \mathbf{V}$, is the fewest number of linearly independent vectors that span $\mathbf{V}$. A vector space is said to be a finite dimensional space if its dimension is finite. Otherwise, we say that $\mathbf{V}$ is an infinite dimensional space if no set of finitely many elements spans $\mathbf{V}$.
11. Let $\mathbf{V}$ be a vector space. A subset X is said to be a **basis** for $\mathbf{V}$ if it has the following two properties:

    (a) any finite subset of X is linearly independent;

    (b) every vector in $\mathbf{V}$ is a linear combination of finitely many elements of X.
12. The **characteristic polynomial** $\chi(\lambda)$ is the determinant of the matrix $\lambda\mathbf{I} - \mathbf{A}$, that is, $\chi(\lambda) = \det(\lambda\mathbf{I} - \mathbf{A})$. Obviously $\chi(\lambda)$ has leading term λ^n. Any solution of the characteristic equation $\chi(\lambda) = 0$ is said to be an **eigenvalue**. The set of all eigenvalues is called the **spectrum** of the matrix $\mathbf{A}$, denoted $\sigma(\mathbf{A})$.
13. A nonzero vector $\mathbf{x}$ satisfying $\mathbf{Ax} = \lambda\mathbf{x}$ is called an **eigenvector** of a square matrix $\mathbf{A}$ corresponding to the **eigenvalue** λ.
14. Let N_λ be the collection of all vectors $\mathbf{x} \in X$ such that $\mathbf{Ax} = \lambda\mathbf{x}$. Since, by our definition, $\mathbf{0}$ is not an eigenvector, N_λ does not contain $\mathbf{0}$. If, however, we enlarge N_λ by adjoining the origin to it, then N_λ becomes a subspace, usually called the **eigenspace** or *proper space*. We define the **geometric multiplicity** of the eigenvalue λ as the dimension of the subspace N_λ. If the eigenvalue λ has multiplicity 1, it is said to be a *simple* eigenvalue.
15. Let λ be an eigenvalue of a matrix $\mathbf{A}$. The **algebraic multiplicity** of λ is called the multiplicity of λ as a root of the characteristic equation of $\mathbf{A}$, i.e., a root of $\chi(\lambda) = 0$.
16. The eigenvalue λ of a square matrix is called **defective** if its algebraic multiplicity is greater than the geometrical one. The difference (which is always nonnegative) between the algebraic multiplicity and geometrical multiplicity is called the **defect** of the eigenvalue λ.
17. An $n \times n$ matrix $\mathbf{A}$ is said to be **diagonalizable** if it is similar to a diagonal matrix, that is, there exists a diagonal matrix $\mathbf{D}$ and nonsingular matrix $\mathbf{S}$ such that $\mathbf{D} = \mathbf{S}^{-1}\mathbf{AS}$.
18. An $n \times n$ matrix $\mathbf{A}$ is diagonalizable if and only if its eigenvalues have the same geometrical and algebraic multiplicities.
19. If $\mathbf{A}$ is a diagonalizable matrix, namely, $\mathbf{A} \sim \mathbf{D}$, where $\mathbf{D}$ is a diagonal matrix of eigenvalues, then $\mathbf{A} = \mathbf{SDS}^{-1}$. A function of the matrix $\mathbf{A}$ is defined as $f(\mathbf{A}) = \mathbf{S}f(\mathbf{D})\mathbf{S}^{-1}$.

20. To implement the **diagonalization procedure**, do the following steps:

    (a) Determine the geometric multiplicities of eigenvalues. This is equivalent to constructing eigenvectors $\mathbf{x}_1$, $\mathbf{x}_2$, $\cdots$, $\mathbf{x}_m$. If $m = n$, the dimension of the matrix $\mathbf{A}$, then these eigenvectors span the n-dimensional vector space, and the matrix is diagonalizable.

    (b) Define a nonsingular matrix $\mathbf{S}$ that reduces the given matrix to a diagonal matrix. You may build the matrix $\mathbf{S}$ from eigenvectors, writing them in sequence as row vectors, namely, $\mathbf{S} = [\mathbf{x}_1\, \mathbf{x}_2\, \cdots\, \mathbf{x}_n]$. The determinant of the matrix $\mathbf{S}$ is not zero because $\mathbf{S}$ consists of column vectors of eigenvectors, which are linearly independent.

    (c) Calculate $\mathbf{S}^{-1}$.

    (d) Define the function of the matrix according to the formula

$$f(\mathbf{A}) = \mathbf{S}f(\mathbf{D})\,\mathbf{S}^{-1} = \mathbf{S}\begin{bmatrix} f(\lambda_1) & 0 & \cdots & 0 \\ 0 & f(\lambda_2) & \cdots & 0 \\ \vdots & \vdots & \ddots & \vdots \\ 0 & 0 & \cdots & f(\lambda_n) \end{bmatrix}\mathbf{S}^{-1}. \tag{7.4.3}$$

21. A scalar polynomial $q(\lambda)$ is called an **annulled polynomial** (or *annihilating polynomial*) of the square matrix $\mathbf{A}$, if $q(\mathbf{A}) = \mathbf{0}$, with the understanding that $\mathbf{A}^0 = \mathbf{I}$ replaces $\lambda^0 = 1$ in the substitution.

22. The annihilating polynomial, $\psi(\lambda)$, of least degree with leading coefficient 1 is called the **minimal polynomial** of $\mathbf{A}$.

23. The **Sylvester formula:** If $\mathbf{A}$ is a square diagonalizable matrix and

$$\psi(\lambda) = (\lambda - \lambda_1)(\lambda - \lambda_2)\cdots(\lambda - \lambda_s) = \prod_{k=1}^{s}(\lambda - \lambda_k),$$

    is its minimal polynomial, then for a function $f(\lambda)$ we define $f(\mathbf{A})$ as

$$f(\mathbf{A}) = \sum_{k=1}^{s} f(\lambda_k)\,\mathbf{Z}_k(\mathbf{A}), \tag{7.5.3}$$

    where

$$\mathbf{Z}_k(\mathbf{A}) = \operatorname*{Res}_{\lambda_k} \mathbf{R}_\lambda(\mathbf{A}) = \frac{(\mathbf{A} - \lambda_1)\cdots(\mathbf{A} - \lambda_{k-1})(\mathbf{A} - \lambda_{k+1})\cdots(\mathbf{A} - \lambda_s)}{(\lambda_k - \lambda_1)\cdots(\lambda_k - \lambda_{k-1})(\lambda_k - \lambda_{k+1})\cdots(\lambda_k - \lambda_s)}, \quad k = 1, 2, \ldots s,$$

    are known as Sylvester's auxiliary matrices. Note that $\mathbf{Z}_k(\mathbf{A})$, also called the Lagrange interpolation polynomial, is the projection operator onto the eigenspace of λ_k. If $\psi(\lambda)$ coincides with the characteristic polynomial $\chi(\lambda)$, then s equals n, the dimension of the matrix $\mathbf{A}$. In the above formula, λ_k, $k = 1, 2, \ldots s$, are distinct eigenvalues of the matrix $\mathbf{A}$.

24. **Resolvent Method:** For a given square $n \times n$ matrix $\mathbf{A}$ (defective or not), use the following procedure to define a function $f(\mathbf{A})$.

    (a) Find the characteristic polynomial $\chi(\lambda) = \det(\lambda\mathbf{I} - \mathbf{A})$.

    (b) By equating $\chi(\lambda) = 0$, determine eigenvalues λ_1, λ_2, $\ldots$, λ_s of a matrix $\mathbf{A}$ and their algebraic multiplicities.

    (c) Find the resolvent of the given matrix $\mathbf{A}$:

$$\mathbf{R}_\lambda(\mathbf{A}) = (\lambda\mathbf{I} - \mathbf{A})^{-1}.$$

    (d) Then for a function $f(\lambda)$ defined on the spectrum $\sigma(\mathbf{A})$ of square matrix $\mathbf{A}$ we set

$$f(\mathbf{A}) = \sum_{\lambda_k \in \sigma(\mathbf{A})} \operatorname*{Res}_{\lambda_k} f(\lambda)\,\mathbf{R}_\lambda(\mathbf{A}).$$

    (e) The residue of a ratio of two polynomials (or entire functions), $\operatorname{Res}_{\lambda_0} \frac{P(\lambda)}{Q(\lambda)}$, is defined according to the multiplicity, m, of the singular point λ_0 as follows:

$$\operatorname*{Res}_{\lambda_0} \frac{P(\lambda)}{Q(\lambda)} = \frac{1}{(m-1)!}\,\frac{\mathrm{d}^{m-1}}{\mathrm{d}\lambda^{m-1}}\,\frac{P(\lambda)(\lambda - \lambda_0)^m}{Q(\lambda)}\bigg|_{\lambda=\lambda_0}. \tag{7.6.2}$$

    In particular, for $m = 1$, we have

$$\operatorname*{Res}_{\lambda_0} \frac{P(\lambda)}{Q(\lambda)} = \frac{P(\lambda_0)}{Q'(\lambda_0)}. \tag{7.6.3}$$

25. **Spectral Decomposition Method:** For a square matrix $\mathbf{A}$, let

$$\psi(\lambda) = (\lambda - \lambda_1)^{m_1} (\lambda - \lambda_2)^{m_2} \cdots (\lambda - \lambda_s)^{m_s} \tag{7.5.1}$$

be its minimal polynomial of degree $m = m_1 + \cdots + m_s$, where λ_k, $k = 1, 2, \ldots, s$, are distinct eigenvalues. The exponential matrix can be defined as

$$e^{\mathbf{A}t} = b_0(t)\,\mathbf{I} + b_1(t)\,\mathbf{A} + b_2(t)\,\mathbf{A}^2 + \cdots + b_{m-1}(t)\,\mathbf{A}^{m-1},$$

where the coefficient functions $b_j(t)$, $j = 0, 1, \ldots, m-1$ satisfy the following equations:

$$e^{\lambda_k t} = b_0(t) + b_1(t)\lambda_k + \cdots + b_{m-1}(t)\lambda_k^{m-1}, \; k = 1, 2, \ldots, s. \tag{7.7.6}$$

If the expansion (7.5.1) of the minimal polynomial $\psi(\lambda)$ contains the multiple $(\lambda - \lambda_k)^{m_k}$, $m_k > 1$, we include in the system to be solved $m_k - 1$ additional equations

$$t^p\, e^{\lambda_k t} = \sum_{j=p}^{m_k - 1} \frac{j!}{(j-p)!}\, b_j(t)\, \lambda_k^{j-p}, \quad p = 1, 2, \ldots, m_k - 1. \tag{7.7.7}$$

A similar procedure is used to define $f(\mathbf{A}) = \sum_{j=0}^{m-1} b_j A^j$, where the coefficients b_j can be found from the equations

$$f(\lambda_k) = b_0 + b_1 \lambda_k + \cdots + b_{m-1}\lambda_k^{m-1}, \qquad k = 1, 2, \ldots, m.$$

If there is a multiple root in $\psi(\lambda)$ of multiplicity s, then you have to differentiate the corresponding equation $s - 1$ times, so that the number of equations is equal to the number of unknown coefficients b_j.

Review Questions for Chapter 7

Section 7.1

1. Suppose that $\mathbf{P}(t)$ is a differentiable square matrix, find a formula for the derivative of $\mathbf{P}^n(t)$, where n is any positive integer.
2. Find $\lim_{t\to 0} \mathbf{A}(t)$ or state why the limit does not exist.

$$\textbf{(a)}\;\; \mathbf{A}(t) = \begin{bmatrix} t^{-2}(\cos t - 1) & t \csc t \\ (t^2+1)^{-2} & (t^2 - 1) \end{bmatrix}, \qquad \textbf{(b)}\;\; \mathbf{A}(t) = \begin{bmatrix} t^{-1} e^t & e^{\cos t} \\ 1/\sqrt{1-t} & t^2 \end{bmatrix}.$$

3. In each problem, determine $\mathbf{A}(t)$ where

$$\textbf{(a)}\;\; \mathbf{A}'(t) = \begin{bmatrix} \sin t & \sec^2 t \\ 2t & \tan t \end{bmatrix} \quad \text{and} \quad \mathbf{A}(0) = \begin{bmatrix} 1 & 2 \\ 3 & 4 \end{bmatrix};$$

$$\textbf{(b)}\;\; \mathbf{A}'(t) = \begin{bmatrix} \cos t & (t+2)^{-1} \\ t^2 & \sin 2t \end{bmatrix} \quad \text{and} \quad \mathbf{A}(0) = \begin{bmatrix} 1 & 0 \\ 0 & 2 \end{bmatrix};$$

$$\textbf{(c)}\;\; \mathbf{A}'(t) = \begin{bmatrix} \cos 2t & (t+1)^{-2} \\ 4t^3 & \cot(t+1) \end{bmatrix} \quad \text{and} \quad \mathbf{A}(0) = \begin{bmatrix} 0 & 1 \\ 1 & 0 \end{bmatrix}.$$

Section 7.2 of Chapter 7 (Review)

1. Find the determinants of the following matrices:

$$\textbf{(a)}\;\; e^{-t}\begin{bmatrix} \cos t & -\sin t \\ \sin 2t & \cos 2t \end{bmatrix}; \qquad \textbf{(b)}\;\; \begin{bmatrix} t-1 & 2t+1 \\ 1-t & 1+t^2 \end{bmatrix}.$$

2. For each of the following 2×2 matrices, find its inverse.

$$\textbf{(a)}\;\; \begin{bmatrix} 4 & -20 \\ 1 & -4 \end{bmatrix}; \qquad \textbf{(b)}\;\; \begin{bmatrix} -4 & 10 \\ -1 & 2 \end{bmatrix}; \qquad \textbf{(c)}\;\; \begin{bmatrix} -9 & 5 \\ -10 & 6 \end{bmatrix}; \qquad \textbf{(d)}\;\; \begin{bmatrix} 8 & 1 \\ 4 & 8 \end{bmatrix}.$$

3. For each of the following 3×3 matrices, find its inverse.

$$\textbf{(a)}\;\; \begin{bmatrix} 3 & 4 & 2 \\ 2 & 1 & -2 \\ -2 & -4 & -1 \end{bmatrix}; \qquad \textbf{(b)}\;\; \begin{bmatrix} 4 & -8 & -10 \\ -1 & 6 & 5 \\ 1 & -8 & -7 \end{bmatrix}; \qquad \textbf{(c)}\;\; \begin{bmatrix} 1 & -2 & 2 \\ -4 & 3 & 2 \\ 4 & -2 & -1 \end{bmatrix};$$

$$\textbf{(d)}\;\; \begin{bmatrix} 2 & 1 & 2 \\ 4 & 5 & -2 \\ 5 & 1 & -1 \end{bmatrix}; \qquad \textbf{(e)}\;\; \begin{bmatrix} 4 & -2 & -1 \\ -1 & 3 & -1 \\ 1 & -2 & 2 \end{bmatrix}; \qquad \textbf{(f)}\;\; \begin{bmatrix} -17 & 4 & 2 \\ 5 & -25 & 1 \\ 3 & 12 & -12 \end{bmatrix}.$$

4. Find the element in the third row and second column of $\mathbf{A}^{-1}$ if

$$\mathbf{A} = \begin{bmatrix} 0 & -1 & -3 & 0 \\ 1 & 2 & -1 & 1 \\ 2 & 0 & 1 & 2 \\ 3 & 1 & 2 & 1 \end{bmatrix}.$$

5. Show that $\mathbf{A}^2 = \mathbf{0}$ for a 2×2 matrix $\mathbf{A}$ if $\operatorname{tr}\mathbf{A} = \det\mathbf{A} = 0$ (such matrix is called nilpotent).

6. Suppose that 2×2 matrices $\mathbf{A}$ and $\mathbf{B}$ satisfy $\mathbf{A}^2 + \mathbf{B}^2 = 2\mathbf{AB}$.

   (a) Prove that $\operatorname{tr}\mathbf{A} = \operatorname{tr}\mathbf{B}$.

   (b) Prove that $\mathbf{AB} = \mathbf{BA}$.

Section 7.3 of Chapter 7 (Review)

1. For each of the following 2×2 matrices, find its eigenvalues and corresponding eigenvectors.

$$\textbf{(a)}\ \begin{bmatrix} 2 & 8 \\ 2 & 2 \end{bmatrix}; \quad \textbf{(b)}\ \begin{bmatrix} 1 & 3 \\ -2 & 6 \end{bmatrix}; \quad \textbf{(c)}\ \begin{bmatrix} -3 & 2 \\ -3 & 4 \end{bmatrix}; \quad \textbf{(d)}\ \begin{bmatrix} 3 & 8 \\ 5 & -3 \end{bmatrix}; \quad \textbf{(e)}\ \begin{bmatrix} 3 & -1 \\ 5 & -3 \end{bmatrix}.$$

2. For each of the following 2×2 matrices, find its eigenvalues and corresponding eigenvectors, including generalized eigenvectors.

$$\textbf{(a)}\ \begin{bmatrix} 1 & -9 \\ 1 & -5 \end{bmatrix}; \quad \textbf{(b)}\ \begin{bmatrix} 1 & -2 \\ 2 & 5 \end{bmatrix}; \quad \textbf{(c)}\ \begin{bmatrix} 3 & -1 \\ 1 & 5 \end{bmatrix}; \quad \textbf{(d)}\ \begin{bmatrix} 4 & -1 \\ 1 & 6 \end{bmatrix}; \quad \textbf{(e)}\ \begin{bmatrix} 2 & -4 \\ 1 & 6 \end{bmatrix}.$$

3. For each of the following 2×2 matrices, find its complex eigenvalues and corresponding eigenvectors.

$$\textbf{(a)}\ \begin{bmatrix} 1 & -1 \\ 5 & -1 \end{bmatrix}; \quad \textbf{(b)}\ \begin{bmatrix} 3 & -10 \\ 5 & -7 \end{bmatrix}; \quad \textbf{(c)}\ \begin{bmatrix} 6 & -1 \\ 5 & 8 \end{bmatrix}; \quad \textbf{(d)}\ \begin{bmatrix} 6 & -1 \\ 5 & 10 \end{bmatrix}; \quad \textbf{(e)}\ \begin{bmatrix} 6 & -4 \\ 5 & 10 \end{bmatrix}.$$

4. For each of the following 3×3 matrices, find its eigenvalues and corresponding eigenvectors.

$$\textbf{(a)}\ \begin{bmatrix} 1 & 1 & 0 \\ 1 & 0 & 1 \\ 0 & 1 & 1 \end{bmatrix}; \quad \textbf{(b)}\ \begin{bmatrix} 0 & 4 & 1 \\ 2 & 2 & 1 \\ 4 & 0 & 4 \end{bmatrix}; \quad \textbf{(c)}\ \begin{bmatrix} 6 & -3 & -8 \\ 2 & 1 & -2 \\ 3 & -3 & -5 \end{bmatrix};$$

$$\textbf{(d)}\ \begin{bmatrix} 1 & 2 & 1 \\ 6 & -1 & 0 \\ -1 & -2 & -1 \end{bmatrix}; \quad \textbf{(e)}\ \begin{bmatrix} 5 & 0 & -6 \\ 2 & -1 & -2 \\ 4 & -2 & -4 \end{bmatrix}; \quad \textbf{(f)}\ \begin{bmatrix} 3 & 2 & 2 \\ -5 & -4 & -2 \\ 5 & 5 & 3 \end{bmatrix}.$$

5. For each of the following 3×3 matrices, find its eigenvalues and corresponding eigenvectors.

$$\textbf{(a)}\ \begin{bmatrix} 1 & 1 & -1 \\ 0 & 2 & 0 \\ -1 & 1 & 1 \end{bmatrix}; \quad \textbf{(b)}\ \begin{bmatrix} 1 & 0 & 3 \\ 2 & -2 & 2 \\ 3 & 0 & 1 \end{bmatrix}; \quad \textbf{(c)}\ \begin{bmatrix} 1 & 0 & -1 \\ -2 & 0 & 2 \\ 3 & 0 & -3 \end{bmatrix}.$$

6. For each of the following 3×3 matrices, find its eigenvalues and corresponding eigenvectors, including generalized eigenvectors.

$$\textbf{(a)}\ \begin{bmatrix} 1 & 2 & 0 \\ -1 & -1 & 1 \\ 0 & 1 & 1 \end{bmatrix}; \quad \textbf{(b)}\ \begin{bmatrix} 4 & 0 & 1 \\ 2 & 3 & 2 \\ -1 & 0 & 2 \end{bmatrix}; \quad \textbf{(c)}\ \begin{bmatrix} 0 & 1 & 0 \\ 0 & 0 & 1 \\ 2 & -5 & 4 \end{bmatrix};$$

$$\textbf{(d)}\ \begin{bmatrix} -2 & 0 & 2 \\ -5 & -1 & 1 \\ 1 & 3 & 3 \end{bmatrix}; \quad \textbf{(e)}\ \begin{bmatrix} 5 & -1 & 1 \\ 1 & 3 & 0 \\ -3 & 2 & 1 \end{bmatrix}; \quad \textbf{(f)}\ \begin{bmatrix} 1 & 1 & 0 \\ 1 & 0 & 1 \\ 1 & -2 & 3 \end{bmatrix};$$

$$\textbf{(g)}\ \begin{bmatrix} 1 & 1 & 0 \\ 1 & 4 & 1 \\ 1 & -2 & 2 \end{bmatrix}; \quad \textbf{(h)}\ \begin{bmatrix} 2 & -1 & 3 \\ 1 & -1 & 1 \\ -5 & 5 & -5 \end{bmatrix}; \quad \textbf{(i)}\ \begin{bmatrix} 2 & -9 & 3 \\ 1 & -1 & 1 \\ -9 & 9 & -9 \end{bmatrix};$$

$$\textbf{(j)}\ \begin{bmatrix} 4 & -2 & -2 \\ -2 & 3 & -1 \\ 2 & -1 & 3 \end{bmatrix}; \quad \textbf{(k)}\ \begin{bmatrix} 3 & 4 & -10 \\ 2 & 1 & -2 \\ 2 & 2 & -5 \end{bmatrix}; \quad \textbf{(l)}\ \begin{bmatrix} 1 & 1 & 0 \\ -2 & 1 & -2 \\ 0 & -1 & 1 \end{bmatrix}.$$

7. For each of the following 3×3 matrices, find its complex eigenvalues and corresponding eigenvectors.

(a) $\begin{bmatrix} 1 & 1 & 2 \\ 1 & 0 & -1 \\ -1 & -2 & -1 \end{bmatrix}$; (b) $\begin{bmatrix} 3 & -4 & -2 \\ -5 & 7 & -8 \\ -10 & 13 & -8 \end{bmatrix}$; (c) $\begin{bmatrix} 6 & 0 & -3 \\ -3 & 3 & 3 \\ 1 & -2 & 6 \end{bmatrix}$;

(d) $\begin{bmatrix} 4 & -4 & 4 \\ -10 & 3 & 15 \\ 2 & -3 & 1 \end{bmatrix}$; (e) $\begin{bmatrix} 4 & 4 & 0 \\ 8 & 10 & -20 \\ 2 & 3 & -2 \end{bmatrix}$; (f) $\begin{bmatrix} 1 & -1 & -2 \\ 1 & 3 & 2 \\ 1 & -1 & 2 \end{bmatrix}$.

8. Prove that the set of generalized eigenvectors of a square $n \times n$ matrix $\mathbf{T}$ corresponding to an eigenvalue λ equals $\ker(\lambda\mathbf{I} - \mathbf{T})^n$.
????

9. For each of the following 3×3 matrices, find its eigenvalues and corresponding eigenvectors, including generalized eigenvectors.

(a) $\begin{bmatrix} 2 & 12 & -10 \\ -2 & 24 & -11 \\ -2 & 24 & -8 \end{bmatrix}$; (b) $\begin{bmatrix} 1 & 1 & 1 \\ 1 & 3 & -1 \\ 0 & 2 & 2 \end{bmatrix}$; (c) $\begin{bmatrix} -1 & 2 & 3 \\ 1 & -2 & -3 \\ 1 & -1 & -2 \end{bmatrix}$;

(d) $\begin{bmatrix} 1 & 1 & -3 \\ -4 & -4 & 3 \\ -2 & 1 & 0 \end{bmatrix}$; (e) $\begin{bmatrix} 6 & -5 & 3 \\ 2 & -1 & 3 \\ 2 & 1 & 1 \end{bmatrix}$; (f) $\begin{bmatrix} -1 & 1 & -1 \\ -2 & 0 & 2 \\ -1 & 3 & -1 \end{bmatrix}$.

10. For each of the following 3×3 matrices, find its complex eigenvalues and corresponding eigenvectors.

(a) $\begin{bmatrix} -5 & 5 & 4 \\ -8 & 7 & 6 \\ 1 & 0 & 0 \end{bmatrix}$; (b) $\begin{bmatrix} 1 & -1 & -2 \\ 1 & 3 & 2 \\ 1 & -1 & 2 \end{bmatrix}$; (c) $\begin{bmatrix} -3 & 1 & -3 \\ 4 & -1 & 2 \\ 4 & -2 & 3 \end{bmatrix}$;

(d) $\begin{bmatrix} 3 & -3 & 1 \\ 0 & 2 & 2 \\ 5 & 1 & 1 \end{bmatrix}$; (e) $\begin{bmatrix} 4 & 2 & 4 \\ 6 & 0 & 12 \\ -8 & 0 & -6 \end{bmatrix}$; (f) $\begin{bmatrix} 0 & 2 & 2 \\ -5 & 2 & 5 \\ 2 & 1 & 0 \end{bmatrix}$.

Section 7.4 of Chapter 7 (Review)

1. Using the diagonalization procedure, find the functions $f(\mathbf{A})$ and $g(\mathbf{A})$, where $f(\lambda) = e^{\lambda t}$ and $g(\lambda) = \dfrac{\lambda^2 - 1}{\lambda^2 + 1}$, of the following 2×2 matrices:

(a) $\begin{bmatrix} -11 & 7 \\ -12 & 8 \end{bmatrix}$; (b) $\begin{bmatrix} -12 & 7 \\ -13 & 8 \end{bmatrix}$; (c) $\begin{bmatrix} 18 & -11 \\ 3 & 4 \end{bmatrix}$; (d) $\begin{bmatrix} 8 & -1 \\ 3 & 4 \end{bmatrix}$;

(e) $\begin{bmatrix} 14 & -3 \\ 3 & 4 \end{bmatrix}$; (f) $\begin{bmatrix} 11 & 3 \\ -4 & 4 \end{bmatrix}$; (g) $\begin{bmatrix} 20 & 3 \\ -5 & 4 \end{bmatrix}$; (h) $\begin{bmatrix} 15 & 7 \\ -4 & 4 \end{bmatrix}$.

2. Using the diagonalization procedure, find the functions $f(\mathbf{A})$ and $g(\mathbf{A})$, where $f(\lambda) = e^{\lambda t}$ and $g(\lambda) = \dfrac{\lambda^2 - 1}{\lambda^2 + 1}$, of the following 3×3 matrices:

(a) $\begin{bmatrix} 3 & -2 & 0 \\ -1 & 3 & -2 \\ 0 & -1 & 3 \end{bmatrix}$; (b) $\begin{bmatrix} -8 & -11 & -2 \\ 6 & 9 & 2 \\ -6 & -6 & 1 \end{bmatrix}$; (c) $\begin{bmatrix} 2 & -3 & 1 \\ 4 & 5 & -2 \\ 29 & -7 & 0 \end{bmatrix}$;

(d) $\begin{bmatrix} 2 & 1 & 1 \\ 1 & 3 & 1 \\ -2 & 3 & -1 \end{bmatrix}$; (e) $\begin{bmatrix} 0 & 1 & 2 \\ -5 & 4 & 7 \\ 1 & -3 & 11 \end{bmatrix}$; (f) $\begin{bmatrix} 2 & 9 & 3 \\ 1 & 4 & -3 \\ 1 & 3 & 0 \end{bmatrix}$.

3. For each of the following 2×2 matrices, find all square roots.

(a) $\begin{bmatrix} 30 & 21 \\ 19 & 28 \end{bmatrix}$; (b) $\begin{bmatrix} 19 & 18 \\ 25 & 54 \end{bmatrix}$; (c) $\begin{bmatrix} 41 & 32 \\ 40 & 49 \end{bmatrix}$; (d) $\begin{bmatrix} 15 & -6 \\ -49 & 22 \end{bmatrix}$;

(e) $\begin{bmatrix} 25 & -6 \\ -96 & 25 \end{bmatrix}$; (f) $\begin{bmatrix} 7 & 6 \\ 9 & 10 \end{bmatrix}$; (g) $\begin{bmatrix} 16 & -6 \\ -40 & 24 \end{bmatrix}$; (h) $\begin{bmatrix} 15 & -11 \\ -10 & 14 \end{bmatrix}$;

(i) $\begin{bmatrix} 16 & -9 \\ -12 & 13 \end{bmatrix}$; (j) $\begin{bmatrix} 20 & 19 \\ 61 & 62 \end{bmatrix}$; (k) $\begin{bmatrix} 21 & -17 \\ -15 & 19 \end{bmatrix}$; (l) $\begin{bmatrix} 8 & -4 \\ -41 & 45 \end{bmatrix}$;

(m) $\begin{bmatrix} 7 & 12 \\ 9 & 19 \end{bmatrix}$; (n) $\begin{bmatrix} 12 & -4 \\ -3 & 13 \end{bmatrix}$; (o) $\begin{bmatrix} 8 & -8 \\ -28 & 60 \end{bmatrix}$; (p) $\begin{bmatrix} 40 & -12 \\ -78 & 25 \end{bmatrix}$.

Section 7.5 of Chapter 7 (Review)

1. For each of the given 3 × 3 matrices with multiple eigenvalues, find Sylvester's auxiliary matrices.

(a) $\begin{bmatrix} 25 & -8 & 30 \\ 24 & -7 & 30 \\ -12 & 4 & -14 \end{bmatrix}$; (b) $\begin{bmatrix} 7 & -2 & 2 \\ 8 & -1 & 4 \\ -8 & 4 & -1 \end{bmatrix}$; (c) $\begin{bmatrix} 3 & 1 & 1 \\ 2 & 4 & 2 \\ -1 & -1 & 1 \end{bmatrix}$;

(d) $\begin{bmatrix} -19 & 12 & 84 \\ 0 & 5 & 0 \\ -8 & 4 & 33 \end{bmatrix}$; (e) $\begin{bmatrix} 3 & 2 & 4 \\ 6 & 2 & 6 \\ 4 & 2 & 3 \end{bmatrix}$; (f) $\begin{bmatrix} 3 & 2 & 4 \\ 10 & 4 & 10 \\ 4 & 2 & 3 \end{bmatrix}$.

2. For each of the given 2 × 2 matrices, find $e^{\mathbf{A}t}$, $\frac{\sin(\mathbf{A}t)}{\mathbf{A}}$, and $\cos(\mathbf{A}t)$.

(a) $\begin{bmatrix} 1 & 10 \\ 5 & -4 \end{bmatrix}$; (b) $\begin{bmatrix} -2 & 6 \\ 7 & -13 \end{bmatrix}$; (c) $\begin{bmatrix} 1 & 4 \\ 1 & 1 \end{bmatrix}$; (d) $\begin{bmatrix} 1 & 5 \\ 3 & -13 \end{bmatrix}$;

(e) $\begin{bmatrix} 4 & 9 \\ 1 & -4 \end{bmatrix}$; (f) $\begin{bmatrix} 1 & 3 \\ 5 & -1 \end{bmatrix}$; (g) $\begin{bmatrix} 4 & 3 \\ 4 & -7 \end{bmatrix}$; (h) $\begin{bmatrix} 1 & 3 \\ 11 & -7 \end{bmatrix}$.

3. For each of the given 2 × 2 positive matrices, find all roots $\sqrt{\mathbf{A}}$ and determine $e^{\sqrt{\mathbf{A}}t}$.

(a) $\begin{bmatrix} 8 & 4 \\ 1 & 5 \end{bmatrix}$; (b) $\begin{bmatrix} 7 & 3 \\ 2 & 6 \end{bmatrix}$; (c) $\begin{bmatrix} 13 & 3 \\ 4 & 12 \end{bmatrix}$; (d) $\begin{bmatrix} 15 & 3 \\ 2 & 10 \end{bmatrix}$;

(e) $\begin{bmatrix} 0 & 2 \\ -2 & 5 \end{bmatrix}$; (f) $\begin{bmatrix} 3 & 3 \\ 4 & 7 \end{bmatrix}$; (g) $\begin{bmatrix} 6 & 2 \\ 3 & 7 \end{bmatrix}$; (h) $\begin{bmatrix} 2 & 4 \\ 1 & 2 \end{bmatrix}$.

4. For each of the given 3 × 3 matrices with multiple eigenvalues, find $e^{\mathbf{A}t}$ using Sylvester's method.

(a) $\begin{bmatrix} 3 & 2 & 4 \\ 2 & 0 & 2 \\ 4 & 2 & 3 \end{bmatrix}$; (b) $\begin{bmatrix} 3 & 1 & -1 \\ 3 & 5 & 1 \\ -6 & 2 & 4 \end{bmatrix}$; (c) $\begin{bmatrix} 3 & 2 & -2 \\ -2 & 7 & -2 \\ -10 & 10 & -5 \end{bmatrix}$;

(d) $\begin{bmatrix} 3 & 2 & 4 \\ 14 & 6 & 14 \\ 4 & 2 & 3 \end{bmatrix}$; (e) $\begin{bmatrix} 3 & 1 & -1 \\ 1 & 3 & -1 \\ 3 & 3 & -1 \end{bmatrix}$; (f) $\begin{bmatrix} -2 & -2 & 4 \\ -2 & 1 & 2 \\ -4 & -2 & 6 \end{bmatrix}$;

(g) $\begin{bmatrix} -10 & 20 & -16 \\ -9 & 17 & -12 \\ -3 & 5 & -2 \end{bmatrix}$; (h) $\begin{bmatrix} 13 & -18 & 14 \\ 10 & -15 & 14 \\ 5 & -9 & 10 \end{bmatrix}$; (i) $\begin{bmatrix} 14 & -20 & 16 \\ 9 & -13 & 12 \\ 3 & -5 & 6 \end{bmatrix}$.

5. Find $\dfrac{\sin\left(t\sqrt{\mathbf{A}}\right)}{\sqrt{\mathbf{A}}}$ and $\cos\left(t\sqrt{\mathbf{A}}\right)$ for each of the following 3 × 3 matrices. Then show that these two functions are solutions to the matrix differential equation $\ddot{\mathbf{\Phi}} + \mathbf{A}\,\mathbf{\Phi}(t) = \mathbf{0}$.

(a) $\begin{bmatrix} 9 & -40 & 25 \\ -25 & 204 & -125 \\ -40 & 320 & -196 \end{bmatrix}$; (b) $\begin{bmatrix} -116 & -330 & 210 \\ 160 & 444 & -280 \\ 180 & 495 & -311 \end{bmatrix}$; (c) $\begin{bmatrix} -28 & 16 & -8 \\ -48 & 28 & 12 \\ -16 & 8 & 8 \end{bmatrix}$;

(d) $\begin{bmatrix} -47 & -48 & 192 \\ -96 & -95 & 384 \\ -48 & -48 & 193 \end{bmatrix}$; (e) $\begin{bmatrix} -41 & -45 & 180 \\ -90 & -86 & 360 \\ -45 & -45 & 184 \end{bmatrix}$; (f) $\begin{bmatrix} -8 & -12 & 48 \\ -24 & -20 & 96 \\ -12 & -12 & 52 \end{bmatrix}$;

(g) $\begin{bmatrix} 2 & 9 & -4 \\ -14 & 43 & -40 \\ -7 & 9 & 5 \end{bmatrix}$; (h) $\begin{bmatrix} 17 & 8 & -32 \\ 16 & 25 & -64 \\ 8 & 8 & -23 \end{bmatrix}$; (i) $\begin{bmatrix} -7 & -21 & 74 \\ -32 & -38 & 158 \\ -16 & -21 & 83 \end{bmatrix}$;

(j) $\begin{bmatrix} -55 & 32 & 16 \\ -96 & 57 & 24 \\ -32 & 16 & 17 \end{bmatrix}$; (k) $\begin{bmatrix} -20 & 12 & 6 \\ -36 & 22 & 9 \\ -12 & 6 & 7 \end{bmatrix}$; (l) $\begin{bmatrix} 44 & -20 & -10 \\ 60 & -26 & -15 \\ 20 & -10 & -1 \end{bmatrix}$;

(m) $\begin{bmatrix} 289 & -96 & -144 \\ 144 & -47 & -72 \\ 432 & -144 & -215 \end{bmatrix}$; (n) $\begin{bmatrix} 16 & -96 & 60 \\ -60 & 484 & -300 \\ -96 & 768 & -476 \end{bmatrix}$; (o) $\begin{bmatrix} 1 & 24 & -15 \\ 15 & -116 & 75 \\ 24 & -192 & 124 \end{bmatrix}$.

Section 7.6 of Chapter 7 (Review)

1. For each of the following 3 × 3 defective matrices, find $e^{\mathbf{A}t}$ using the resolvent method.

(a) $\begin{bmatrix} 1 & 1 & 1 \\ 1 & 3 & -1 \\ 0 & 2 & 2 \end{bmatrix}$; (b) $\begin{bmatrix} 5 & 1 & -11 \\ 7 & -1 & -13 \\ 4 & 0 & -8 \end{bmatrix}$; (c) $\begin{bmatrix} 5 & -1 & 1 \\ -1 & 9 & -3 \\ -2 & 2 & 4 \end{bmatrix}$;

(d) $\begin{bmatrix} -2 & 17 & 4 \\ -1 & 6 & 1 \\ 0 & 1 & 2 \end{bmatrix}$; (e) $\begin{bmatrix} -3 & 5 & -5 \\ 3 & -1 & 3 \\ 8 & -8 & 10 \end{bmatrix}$; (f) $\begin{bmatrix} -15 & -7 & 4 \\ 34 & 16 & -11 \\ 17 & 7 & 5 \end{bmatrix}$;

(g) $\begin{bmatrix} 4 & -1 & 1 \\ 10 & -2 & 3 \\ 1 & 0 & 1 \end{bmatrix}$; (h) $\begin{bmatrix} 5 & -1 & 1 \\ 14 & -3 & 6 \\ 5 & -2 & 5 \end{bmatrix}$; (i) $\begin{bmatrix} 3 & -8 & -10 \\ -2 & 7 & 8 \\ 2 & -6 & -7 \end{bmatrix}$.

2. For each of the following 3 × 3 matrices, express $e^{\mathbf{A}t}$ through real-valued functions using the resolvent method.

(a) $\begin{bmatrix} 1 & 2 & -2 \\ 2 & 5 & -2 \\ 4 & 12 & -5 \end{bmatrix}$; (b) $\begin{bmatrix} 2 & 2 & 9 \\ 1 & -1 & 3 \\ -1 & -1 & -4 \end{bmatrix}$; (c) $\begin{bmatrix} 3 & 1 & -2 \\ 3 & 2 & 1 \\ -2 & 3 & 1 \end{bmatrix}$;

(d) $\begin{bmatrix} 3 & 2 & -1 \\ -2 & 4 & -3 \\ -5 & 2 & 7 \end{bmatrix}$; (e) $\begin{bmatrix} 3 & 1 & -1 \\ 2 & 3 & -1 \\ 4 & 4 & -1 \end{bmatrix}$; (f) $\begin{bmatrix} 5 & -2 & 1 \\ 4 & -2 & 0 \\ 8 & 4 & 1 \end{bmatrix}$;

(g) $\begin{bmatrix} 8 & 12 & -4 \\ -9 & 12 & 4 \\ 37 & 12 & 20 \end{bmatrix}$; (h) $\begin{bmatrix} 4 & -3 & 3 \\ 1 & 4 & -3 \\ 2 & 5 & -1 \end{bmatrix}$; (i) $\begin{bmatrix} 5 & 2 & -1 \\ -3 & 2 & 2 \\ 1 & 3 & 2 \end{bmatrix}$;

(j) $\begin{bmatrix} 3 & -9 & -1 \\ 2 & -1 & 3 \\ 1 & 2 & 10 \end{bmatrix}$; (k) $\begin{bmatrix} 1 & -2 & -1 \\ 1 & 3 & -1 \\ 2 & 4 & -1 \end{bmatrix}$; (l) $\begin{bmatrix} 1 & -3 & -1 \\ 2 & 6 & -1 \\ 2 & 4 & -2 \end{bmatrix}$.

3. For each of the following 3 × 3 defective matrices, find a square root and determine $e^{\sqrt{\mathbf{A}}t}$.

(a) $\begin{bmatrix} 10 & 6 & -3 \\ -12 & -11 & 12 \\ -6 & -6 & 7 \end{bmatrix}$; (b) $\begin{bmatrix} -3 & -4 & 12 \\ 4 & 5 & -12 \\ 0 & 0 & 1 \end{bmatrix}$; (c) $\begin{bmatrix} 12 & 8 & -17 \\ 1 & 2 & 4 \\ 1 & -2 & 8 \end{bmatrix}$;

(d) $\begin{bmatrix} -6 & -14 & -8 \\ 10 & 18 & 8 \\ -38 & -46 & 24 \end{bmatrix}$; (e) $\begin{bmatrix} 17 & 8 & -8 \\ -2 & -1 & 2 \\ 14 & 6 & -5 \end{bmatrix}$; (f) $\begin{bmatrix} 12 & 6 & -15 \\ 1 & 2 & 4 \\ 1 & -2 & 8 \end{bmatrix}$;

(g) $\begin{bmatrix} 6 & 0 & -4 \\ -2 & 4 & 5 \\ 1 & 0 & 2 \end{bmatrix}$; (h) $\begin{bmatrix} -2 & -14 & -8 \\ 6 & 18 & 8 \\ -54 & -30 & 28 \end{bmatrix}$; (i) $\begin{bmatrix} 1 & 16 & -12 \\ 6 & -3 & 24 \\ 5 & -10 & 24 \end{bmatrix}$;

(j) $\begin{bmatrix} -4 & 8 & 4 \\ -3 & 4 & -3 \\ -1 & 4 & 9 \end{bmatrix}$; (k) $\begin{bmatrix} 2 & -1 & -1 \\ 1 & 1 & -1 \\ 1 & -1 & 0 \end{bmatrix}$; (l) $\begin{bmatrix} 5 & -1 & -1 \\ 1 & 1 & -1 \\ 1 & -1 & 3 \end{bmatrix}$.

4. Show that the following 3 × 3 matrices have no square root. Nevertheless, find the matrix functions $\mathbf{\Psi}(t) = \cos\left(\sqrt{\mathbf{A}}t\right)$ and $\mathbf{\Phi}(t) = \mathbf{A}^{-1/2}\sin\left(\mathbf{A}^{1/2}t\right)$, and show that they satisfy the second order matrix differential equation $\ddot{\mathbf{X}} + \mathbf{A}\,\mathbf{X} = 0$.

(a) $\begin{bmatrix} -2 & 1 & 2 \\ -1 & 0 & 1 \\ -2 & 1 & 2 \end{bmatrix}$; (b) $\begin{bmatrix} -1 & 2 & -7 \\ 1 & -2 & 1 \\ 1 & -2 & 3 \end{bmatrix}$; (c) $\begin{bmatrix} -1 & -1 & 1 \\ -1 & -1 & 2 \\ -1 & -1 & 2 \end{bmatrix}$; (d) $\begin{bmatrix} -1 & 1 & -1 \\ 1 & -3 & 2 \\ 2 & -6 & 4 \end{bmatrix}$.

5. For each of the following 3 × 3 positive matrices, determine $\frac{\sin(\sqrt{\mathbf{A}}\,t)}{\sqrt{\mathbf{A}}}$ and $\cos\left(\sqrt{\mathbf{A}}\,t\right)$, and show that the result does not depend on the choice of the root.

(a) $\begin{bmatrix} -34 & 13 & 19 \\ -63 & 25 & 34 \\ -20 & 7 & 12 \end{bmatrix}$; (b) $\begin{bmatrix} 2 & 8 & -1 \\ -1 & 7 & 0 \\ -1 & 5 & 3 \end{bmatrix}$; (c) $\begin{bmatrix} -17 & 46 & 13 \\ -10 & 26 & 6 \\ 3 & -7 & 3 \end{bmatrix}$; (d) $\begin{bmatrix} 2 & -40 & -8 \\ -1 & 4 & 4 \\ -2 & -15 & 6 \end{bmatrix}$.

Section 7.7 of Chapter 7 (Review)

1. Some of the following 3×3 matrices have the same spectrum. Therefore, the coefficients $b_0(t)$, $b_1(t)$, and $b_2(t)$ in expansion (7.7.5), page 415, are the same for these matrices. Find these coefficients and construct the propagator matrix $e^{\mathbf{A}t}$ for each given matrix.

(a) $\begin{bmatrix} 1 & 10 & -12 \\ 2 & 2 & 3 \\ 2 & -1 & 6 \end{bmatrix}$; (b) $\begin{bmatrix} 1 & 12 & -8 \\ -1 & 9 & -4 \\ -1 & 6 & -1 \end{bmatrix}$; (c) $\begin{bmatrix} 4 & 0 & 1 \\ 1 & 3 & 1 \\ -1 & 0 & 2 \end{bmatrix}$;

(d) $\begin{bmatrix} 2 & 1 & 0 \\ -3 & 2 & -3 \\ 0 & -1 & 2 \end{bmatrix}$; (e) $\begin{bmatrix} -3 & 5 & -5 \\ 3 & -1 & 3 \\ 8 & -8 & 10 \end{bmatrix}$; (f) $\begin{bmatrix} -2 & 17 & 4 \\ -1 & 6 & 1 \\ 0 & 1 & 2 \end{bmatrix}$;

(g) $\begin{bmatrix} 1 & 8 & -12 \\ 2 & 2 & 3 \\ 2 & -1 & 6 \end{bmatrix}$; (h) $\begin{bmatrix} 4 & 0 & 5 \\ 5 & 3 & 1 \\ -1 & 0 & 2 \end{bmatrix}$; (i) $\begin{bmatrix} 1 & 12 & -8 \\ -1 & 9 & -4 \\ -1 & 7 & -1 \end{bmatrix}$;

(j) $\begin{bmatrix} 6 & -5 & 10 \\ -1 & 2 & -2 \\ -1 & 3 & -1 \end{bmatrix}$; (k) $\begin{bmatrix} 6 & -1 & 10 \\ -1 & 2 & -2 \\ -1 & 1 & -1 \end{bmatrix}$; (l) $\begin{bmatrix} 6 & -9 & 18 \\ -1 & 2 & -2 \\ -1 & 1 & -1 \end{bmatrix}$.

2. For each of the following 3×3 defective matrices, find $e^{\mathbf{A}t}$ using the *spectral decomposition* method.

(a) $\begin{bmatrix} -1 & -1 & 2 \\ -1 & 3 & 1 \\ -2 & -1 & 3 \end{bmatrix}$; (b) $\begin{bmatrix} -2 & -1 & 3 \\ -2 & 1 & 1 \\ -2 & -1 & 3 \end{bmatrix}$; (c) $\begin{bmatrix} 6 & -5 & 10 \\ -1 & 6 & -6 \\ -1 & 1 & -1 \end{bmatrix}$;

(d) $\begin{bmatrix} 7 & -5 & 12 \\ -1 & 3 & -6 \\ -1 & 1 & -1 \end{bmatrix}$; (e) $\begin{bmatrix} 2 & 1 & -2 \\ 1 & 2 & -2 \\ -1 & 1 & 1 \end{bmatrix}$; (f) $\begin{bmatrix} 4 & -1 & 1 \\ 10 & -2 & 3 \\ 1 & 0 & 1 \end{bmatrix}$;

(g) $\begin{bmatrix} 4 & -5 & 6 \\ -1 & 0 & -3 \\ -1 & 1 & -1 \end{bmatrix}$; (h) $\begin{bmatrix} 6 & -5 & 10 \\ -1 & 2 & -6 \\ -1 & 1 & -1 \end{bmatrix}$; (i) $\begin{bmatrix} 1 & 12 & -8 \\ -1 & 9 & -4 \\ -1 & 6 & -1 \end{bmatrix}$;

(j) $\begin{bmatrix} 5 & -1 & 1 \\ 1 & 3 & 0 \\ -3 & 2 & 1 \end{bmatrix}$; (k) $\begin{bmatrix} -15 & -7 & 4 \\ 34 & 16 & -11 \\ 17 & 7 & 5 \end{bmatrix}$; (l) $\begin{bmatrix} 3 & 3 & -4 \\ -4 & -5 & 8 \\ -2 & -3 & 5 \end{bmatrix}$.

3. For each of the following 2×2 positive matrices, determine $\frac{\sin(\sqrt{\mathbf{A}}\,t)}{\sqrt{\mathbf{A}}}$ and $\cos\left(\sqrt{\mathbf{A}}\,t\right)$, and show that the result does not depend on the choice of the root.

(a) $\begin{bmatrix} 19 & -3 \\ 3 & 13 \end{bmatrix}$; (b) $\begin{bmatrix} 13 & 4 \\ 3 & 12 \end{bmatrix}$; (c) $\begin{bmatrix} 14 & 22 \\ 5 & 15 \end{bmatrix}$; (d) $\begin{bmatrix} 33 & -9 \\ 1 & 39 \end{bmatrix}$;

(e) $\begin{bmatrix} 51 & -4 \\ 1 & 47 \end{bmatrix}$; (f) $\begin{bmatrix} 21 & 119 \\ 4 & 32 \end{bmatrix}$; (g) $\begin{bmatrix} 41 & 10 \\ 4 & 44 \end{bmatrix}$; (h) $\begin{bmatrix} 30 & 133 \\ 3 & 28 \end{bmatrix}$;

(i) $\begin{bmatrix} 90 & -3 \\ 123 & 40 \end{bmatrix}$; (j) $\begin{bmatrix} 40 & 18 \\ 20 & 49 \end{bmatrix}$; (k) $\begin{bmatrix} 34 & 25 \\ 30 & 39 \end{bmatrix}$; (l) $\begin{bmatrix} 36 & 16 \\ 26 & 17 \end{bmatrix}$.

4. For each of the following 3×3 positive matrices, determine $\frac{\sin(\sqrt{\mathbf{A}}\,t)}{\sqrt{\mathbf{A}}}$ and $\cos\left(\sqrt{\mathbf{A}}\,t\right)$, and show that the result does not depend on the choice of the root.

(a) $\begin{bmatrix} 3 & 0 & 4 \\ 1 & 1 & 1 \\ -1 & 0 & -1 \end{bmatrix}$; (b) $\begin{bmatrix} -1 & 0 & 1 \\ 0 & -1 & 1 \\ 1 & -1 & -1 \end{bmatrix}$; (c) $\begin{bmatrix} 5 & -5 & 8 \\ -1 & 1 & -2 \\ -1 & 1 & -1 \end{bmatrix}$;

(d) $\begin{bmatrix} 5 & 2 & 5 \\ 2 & 3 & 2 \\ -5 & -3 & -5 \end{bmatrix}$; (e) $\begin{bmatrix} 5 & 1 & -1 \\ 1 & 5 & -1 \\ 5 & 5 & -1 \end{bmatrix}$; (f) $\begin{bmatrix} 1 & 2 & 1 \\ 0 & 9 & 0 \\ -1 & -3 & -1 \end{bmatrix}$;

(g) $\begin{bmatrix} -1 & -1 & 2 \\ -1 & 9 & 1 \\ -2 & -1 & 3 \end{bmatrix}$; (h) $\begin{bmatrix} -2 & 18 & 4 \\ -1 & 6 & 1 \\ 1 & 1 & 2 \end{bmatrix}$; (i) $\begin{bmatrix} -1 & -1 & 2 \\ -5 & 1 & 5 \\ -2 & -1 & 3 \end{bmatrix}$.

Chapter 8

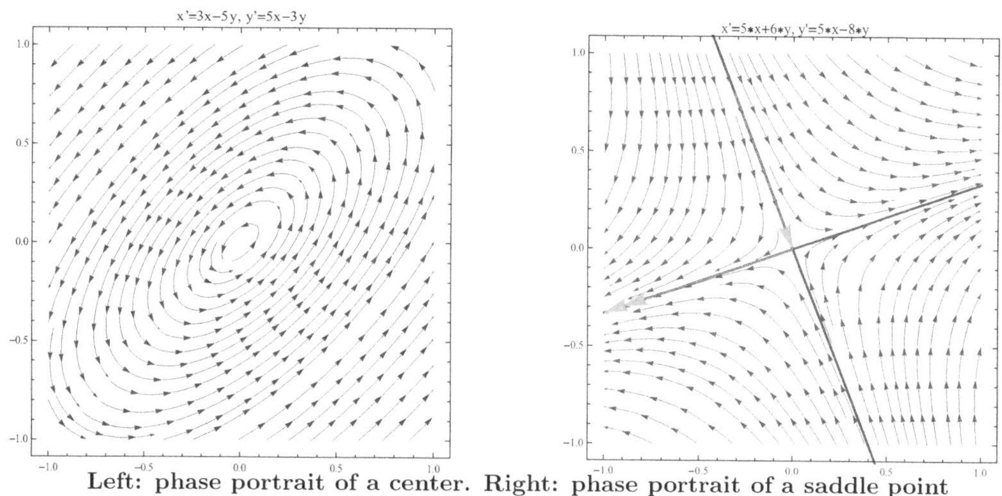

Left: phase portrait of a center. Right: phase portrait of a saddle point

Systems of First Order Linear Differential Equations

As is well known [14], many practical problems lead to systems of differential equations. In the present chapter, we concentrate our attention only on systems of linear first order differential equations in normal form when the number of dependent variables (unknown functions) is the same as the number of equations. The theory of such systems parallels very closely the theory of higher order linear differential equations (see Chapter 4); however, there are some features that are not observed in the theory of single differential equations.

The opening sections present general properties of linear vector differential equations with variable and constant coefficients. In the next sections, we discuss solutions of the initial value problems for nonhomogeneous systems using the variation of parameters, method of undetermined coefficients, and the Laplace transformation. While a second order system of equations can be reduced to one of first order, we present a direct solution of such systems in the concluding section.

8.1 Systems of Linear Differential Equations

In this section, we discuss the properties of the solutions for systems of linear nonhomogeneous differential equations with variable coefficients in the normal form:

$$\frac{\mathrm{d}\mathbf{x}(t)}{\mathrm{d}t} = \mathbf{P}(t)\,\mathbf{x}(t) + \mathbf{f}(t) \qquad \text{or} \qquad \dot{\mathbf{x}} = \mathbf{P}(t)\,\mathbf{x}(t) + \mathbf{f}(t), \tag{8.1.1}$$

where the dot represents the derivative with respect to t: $\dot{\mathbf{x}} \stackrel{\text{def}}{=} \mathrm{d}\mathbf{x}/\mathrm{d}t$, $\mathbf{P}(t) = [p_{ij}(t)]$ is a square $n \times n$ matrix with continuous entries $p_{ij}(t)$ on some interval, called the **coefficient matrix**, $\mathbf{f}(t) = \langle f_1(t), f_2(t), \ldots, f_n(t)\rangle^T$ is a given continuous column vector function on the same interval, and $\mathbf{x}(t) = \langle x_1(t), x_2(t), \ldots, x_n(t)\rangle^T$ is an n-vector of unknown functions. A column vector $\mathbf{x}(t)$, which is to be determined, is usually called the state of the system at time t.

A system of linear differential equations in normal form (8.1.1) is called a **vector differential equation**. If the column vector $\mathbf{f}(t)$, called the **nonhomogeneous term**, the **forcing** or **driving function**, or the input vector, is not identically zero, then Eq. (8.1.1) is called *nonhomogeneous* or *inhomogeneous* or *driven*. Otherwise we have a *homogeneous* vector equation

$$\frac{\mathrm{d}\mathbf{x}(t)}{dt} = \mathbf{P}(t)\,\mathbf{x}(t) \qquad \text{or} \qquad \dot{\mathbf{x}} = \mathbf{P}(t)\,\mathbf{x}(t). \tag{8.1.2}$$

Equation (8.1.2) is called the **complementary equation** to nonhomogeneous equation (8.1.1), and its general solution is called the **complementary function**, which contains n arbitrary constants. The homogeneous equation (8.1.2) obviously has identically zero solution $\mathbf{x}(t) \equiv 0$, which is referred to as the trivial solution. The first observation about solutions of the homogeneous vector equation (8.1.2) follows from the linearity of the problem.

Theorem 8.1: [Superposition Principle for Homogeneous Equations] Let $\mathbf{x}_1(t)$, $\mathbf{x}_2(t)$, ..., $\mathbf{x}_m(t)$ be a set of solution vectors of the homogeneous system of differential equations (8.1.2) on an interval $|a, b|$. Then their linear combination,

$$\mathbf{x}(t) = c_1\,\mathbf{x}_1(t) + c_2\,\mathbf{x}_2(t) + \cdots + c_m\,\mathbf{x}_m(t),$$

where c_i, $i = 1, 2, \ldots, m$, are arbitrary constants, is also a solution to Eq. (8.1.2) on the same interval.

Definition 8.1: A set of n vector functions $\mathbf{x}_1(t)$, $\mathbf{x}_2(t)$, ..., $\mathbf{x}_n(t)$ is said to be **linearly dependent** on an interval $|a, b|$ if there exists a set of numbers c_1, c_2, ..., c_n, with at least one nonzero, such that

$$c_1\,\mathbf{x}_1(t) + c_2\,\mathbf{x}_2(t) + \cdots + c_n\,\mathbf{x}_n(t) \equiv 0 \quad \text{for all } t \in |a, b|.$$

Otherwise, these vector functions are called **linearly independent**.

Two vectors are linearly dependent if and only if each of them is a constant multiple of another. The following example shows that the functions can be linearly independent on some interval, but also can be linearly dependent vectors at every point on this interval. As we will see shortly (Theorem 8.2), this situation is never observed when vector functions are solutions of a linear homogeneous system of equations (8.1.2).

Example 8.1.1: The vector functions

$$\mathbf{x}(t) = \begin{bmatrix} e^t \\ t\,e^t \end{bmatrix} = e^t \begin{bmatrix} 1 \\ t \end{bmatrix} \qquad \text{and} \qquad \mathbf{y}(t) = \begin{bmatrix} 1 \\ t \end{bmatrix}$$

are linearly independent on $(-\infty, \infty)$ since $\mathbf{x}(t) = e^t\,\mathbf{y}(t)$ and there is no constant C such that $\mathbf{x}(t) = C\,\mathbf{y}(t)$.

Theorem 8.2: Let $\mathbf{x}_1(t)$, ..., $\mathbf{x}_m(t)$ be solutions of the homogeneous vector equation (8.1.2) on an interval $|a, b|$, and let t_0 be any point in this interval. Then the set of vector functions $\{\mathbf{x}_1(t), \ldots, \mathbf{x}_m(t)\}$ is linearly dependent if and only if the set of vectors $\{\mathbf{x}_1(t_0), \ldots, \mathbf{x}_m(t_0)\}$ is linearly dependent.

PROOF: It is obvious that if the set of vector functions is linearly dependent, then at any point the corresponding vectors are linearly dependent. Now, we suppose the opposite, and assume that constant vectors $\{\mathbf{x}_1(t_0), \ldots, \mathbf{x}_m(t_0)\}$ are linearly dependent. Then there exist constants c_1, ..., c_m, which are not all zeroes, such that

$$c_1\mathbf{x}_1(t_0) + c_2\mathbf{x}_2(t_0) + \cdots + c_m\mathbf{x}_m(t_0) = 0.$$

Since each vector function $\mathbf{x}_j(t)$, $j = 1, 2, \ldots, m$, is a solution of the homogeneous vector equation $\dot{\mathbf{x}} = \mathbf{P}(t)\,\mathbf{x}$, the vector

$$\mathbf{x}(t) = c_1\mathbf{x}_1(t) + c_2\mathbf{x}_2(t) + \cdots + c_m\mathbf{x}_m(t)$$

is a solution of the initial value problem

$$\dot{\mathbf{x}} = \mathbf{P}(t)\,\mathbf{x}(t), \qquad \mathbf{x}(t_0) = \mathbf{0}.$$

From Theorem 6.6, page 372, it follows that this initial value problem has only trivial solution $\mathbf{x}(t) \equiv \mathbf{0}$ for all t in the given interval. Hence, the set of functions $\{\mathbf{x}_1(t), \ldots, \mathbf{x}_m(t)\}$ is linearly dependent. ∎

Next, we are going to show that the dimension of the solution space of the vector equation (8.1.2) is exactly n. To prove this, we need the following corollary from the previous theorem.

Corollary 8.1: Let $\mathbf{x}_k(t)$, $k = 1, 2, \ldots, n$, be solutions to the initial value problems

$$\frac{\mathrm{d}\mathbf{x}_k}{\mathrm{d}t} = \mathbf{P}(t)\mathbf{x}_k(t), \qquad \mathbf{x}_k(t_0) = \mathbf{e}_k, \quad k = 1, 2, \ldots, n,$$

where

$$\mathbf{e}_1 = \begin{pmatrix} 1 \\ 0 \\ \vdots \\ 0 \end{pmatrix}, \quad \mathbf{e}_2 = \begin{pmatrix} 0 \\ 1 \\ \vdots \\ 0 \end{pmatrix}, \quad \ldots, \quad \mathbf{e}_n = \begin{pmatrix} 0 \\ 0 \\ \vdots \\ 1 \end{pmatrix}.$$

Then $\mathbf{x}_k(t)$, $k = 1, 2, \ldots, n$, are linearly independent solutions of the system $\dot{\mathbf{x}} = \mathbf{P}(t)\mathbf{x}$. ∎

Corollary 8.1 affirms that there exist at least n linearly independent solutions of the homogeneous system of differential equations (8.1.2). Now suppose that the dimension of its solution set exceeds n and there are $n+1$ linearly independent solutions $\mathbf{u}_1(t), \ldots, \mathbf{u}_n(t), \mathbf{u}_{n+1}(t)$. Theorem 8.2 would imply that the vectors $\mathbf{u}_1(t_0), \ldots, \mathbf{u}_{n+1}(t_0)$ are linearly independent in n dimensional space $\mathbb{R}^n$, which is impossible. Therefore, we proved the following statement:

Theorem 8.3: The dimension of the solution space of the $n \times n$ system of differential equations $\dot{\mathbf{x}}(t) = \mathbf{P}(t)\,\mathbf{x}(t)$ is n.

Definition 8.2: Let $\mathbf{P}(t)$ be an $n \times n$ matrix that is continuous on an interval $|a, b|$ $(a < b)$. Any set of n solutions $\mathbf{x}_1(t), \mathbf{x}_2(t), \ldots, \mathbf{x}_n(t)$ of the homogeneous vector equation $\dot{\mathbf{x}}(t) = \mathbf{P}(t)\,\mathbf{x}(t)$ that is linearly independent on the interval $|a, b|$ is called a **fundamental set of solutions** (or fundamental solution set). In this case, the $n \times n$ nonsingular matrix, written in column form

$$\mathbf{X}(t) = [\,\mathbf{x}_1(t), \mathbf{x}_2(t), \ldots, \mathbf{x}_n(t)\,] = \begin{bmatrix} | & | & & | \\ \mathbf{x}_1(t) & \mathbf{x}_2(t) & \cdots & \mathbf{x}_n(t) \\ | & | & & | \end{bmatrix},$$

where each column vector is a solution of the homogeneous vector equation $\dot{\mathbf{x}}(t) = \mathbf{P}(t)\,\mathbf{x}(t)$, is called a **fundamental matrix** for the system of differential equations.

In other words, a square matrix whose columns form a linearly independent set of solutions of the homogeneous system of differential equations (8.1.2) is called a fundamental matrix for the system. A product of a fundamental matrix and a nonsingular constant matrix is a fundamental matrix. This leads to the conclusion that a fundamental matrix is not unique. Because the column vectors of the fundamental matrix $\mathbf{X}(t)$ satisfy the vector differential equation $\mathbf{x}'(t) = \mathbf{P}(t)\,\mathbf{x}(t)$, the matrix-function itself is a solution of the matrix differential equation:

$$\frac{\mathrm{d}\mathbf{X}(t)}{\mathrm{d}t} = \mathbf{P}(t)\,\mathbf{X}(t) \qquad \text{or} \qquad \dot{\mathbf{X}}(t) = \mathbf{P}(t)\,\mathbf{X}(t) \tag{8.1.3}$$

that contains n^2 differential equations for each entry of the $n \times n$ matrix $\mathbf{X} = [x_{ij}(t)]$. For example, consider a two-dimensional case:

$$\mathbf{X}(t) = \begin{bmatrix} x_{11}(t) & x_{12}(t) \\ x_{21}(t) & x_{22}(t) \end{bmatrix} \qquad \text{and} \qquad \mathbf{P}(t) = \begin{bmatrix} p_{11}(t) & p_{12}(t) \\ p_{21}(t) & p_{22}(t) \end{bmatrix}.$$

Then the matrix differential equation (8.1.3) can be written as

$$\begin{bmatrix} \dot{x}_{11}(t) & \dot{x}_{12}(t) \\ \dot{x}_{21}(t) & \dot{x}_{22}(t) \end{bmatrix} = \begin{bmatrix} p_{11}(t) & p_{12}(t) \\ p_{21}(t) & p_{22}(t) \end{bmatrix} \begin{bmatrix} x_{11}(t) & x_{12}(t) \\ x_{21}(t) & x_{22}(t) \end{bmatrix}$$
$$= \begin{bmatrix} p_{11}x_{11} + p_{12}x_{21} & p_{11}x_{12} + p_{12}x_{22} \\ p_{21}x_{11} + p_{22}x_{21} & p_{21}x_{12} + p_{22}x_{22} \end{bmatrix}.$$

This matrix equation is equivalent to two separate vector equations (for each column):

$$\begin{cases} \dot{x}_{11} = p_{11}x_{11} + p_{12}x_{21}, \\ \dot{x}_{21} = p_{21}x_{11} + p_{22}x_{21}, \end{cases} \quad \text{and} \quad \begin{cases} \dot{x}_{12} = p_{11}x_{12} + p_{12}x_{22}, \\ \dot{x}_{22} = p_{21}x_{11} + p_{22}x_{21}. \end{cases}$$

The established relation between the matrix differential equation (8.1.3) and the vector equation (8.1.2) leads to the following sequence of statements.

Theorem 8.4: If $\mathbf{X}(t)$ is a solution of the $n \times n$ matrix differential equation (8.1.3), then for any constant column vector $\mathbf{c} = \langle c_1, c_2, \ldots, c_n \rangle^T$, the n-vector $\mathbf{u} = \mathbf{X}(t)\,\mathbf{c}$ is a solution of the vector equation (8.1.2).

Theorem 8.5: If an $n \times n$ matrix $\mathbf{P}(t)$ has continuous entries on an open interval, then the vector differential equation $\dot{\mathbf{x}} = \mathbf{P}(t)\,\mathbf{x}(t)$ has an $n \times n$ fundamental matrix $\mathbf{X}(t) = [\,\mathbf{x}_1(t), \mathbf{x}_2(t), \ldots, \mathbf{x}_n(t)\,]$ on the same interval. Every solution $\mathbf{x}(t)$ to this system can be written as a linear combination of the column vectors of the fundamental matrix in a unique way:

$$\mathbf{x}(t) = c_1\mathbf{x}_1(t) + c_2\mathbf{x}_2(t) + \cdots + c_n\mathbf{x}_n(t) \quad \text{or in vector form} \quad \mathbf{x}(t) = \mathbf{X}(t)\mathbf{c} \tag{8.1.4}$$

for appropriate constants $c_1, c_2, \ldots, c_n$, where $\mathbf{c} = \langle c_1, c_2, \ldots, c_n \rangle^T$ is a column vector of these constants. ■

Throughout the text, we will refer to (8.1.4) as the **general solution** to the homogeneous vector differential equation (8.1.2).

Theorem 8.6: The general solution of a nonhomogeneous linear vector equation (8.1.1) is the sum of the general solution of the complement homogeneous equation (8.1.2) and a particular solution of the inhomogeneous equation (8.1.1). That is, every solution to Eq. (8.1.1) is of the form

$$\mathbf{x}(t) = c_1\,\mathbf{x}_1(t) + c_2\,\mathbf{x}_2(t) + \cdots + c_n\,\mathbf{x}_n(t) + \mathbf{x}_p(t) \tag{8.1.5}$$

for some constants $c_1, c_2, \ldots, c_n$, where $\mathbf{x}_h(t) = c_1\mathbf{x}_1(t) + c_2\mathbf{x}_2(t) + \cdots + c_n\mathbf{x}_n(t)$ is the general solution of the homogeneous linear equation (8.1.2) and $\mathbf{x}_p(t)$ is a particular solution of the nonhomogeneous equation (8.1.1).

Theorem 8.7: [Superposition Principle for Inhomogeneous Equations] Let $\mathbf{P}(t)$ be an $n \times n$ matrix function that is continuous on an interval $|a, b|$, and let $\mathbf{x}_1(t)$ and $\mathbf{x}_2(t)$ be two vector solutions of the nonhomogeneous equations

$$\dot{\mathbf{x}}_1(t) = \mathbf{P}(t)\,\mathbf{x}_1 + \mathbf{f}_1(t), \qquad \dot{\mathbf{x}}_2(t) = \mathbf{P}(t)\,\mathbf{x}_2 + \mathbf{f}_2(t), \quad t \in |a, b|,$$

respectively. Then, for arbitrary constants α and β, their linear combination $\mathbf{x}(t) = \alpha\mathbf{x}_1(t) + \beta\mathbf{x}_2(t)$ is a solution of the following nonhomogeneous equation

$$\dot{\mathbf{x}}(t) = \mathbf{P}(t)\,\mathbf{x} + \alpha\mathbf{f}_1(t) + \beta\mathbf{f}_2(t), \qquad t \in |a, b|.$$

Corollary 8.2: The difference between any two solutions of the nonhomogeneous vector equation $\dot{\mathbf{x}} = \mathbf{P}(t)\,\mathbf{x} + \mathbf{f}$ is a solution of the complementary homogeneous equation $\dot{\mathbf{x}} = \mathbf{P}(t)\,\mathbf{x}$.

Example 8.1.2: It is not hard to verify that the vector functions

$$\mathbf{x}_1(t) = \begin{bmatrix} t^2 \\ t \end{bmatrix}, \quad \mathbf{x}_2(t) = \begin{bmatrix} e^t \\ 0 \end{bmatrix}$$

are two linearly independent solutions to the following homogeneous vector differential equation:

$$\dot{\mathbf{x}}(t) = \mathbf{P}(t)\,\mathbf{x}(t), \qquad \mathbf{P}(t) = \begin{bmatrix} 1 & 2-t \\ 0 & 1/t \end{bmatrix} \quad (t \neq 0).$$

Therefore, the corresponding fundamental matrix is

$$\mathbf{X}(t) = \begin{bmatrix} t^2 & e^t \\ t & 0 \end{bmatrix}, \qquad \det \mathbf{X}(t) = -t\, e^t.$$

Definition 8.3: The determinant, $W(t) = \det \mathbf{X}(t)$, of a square matrix $\mathbf{X}(t) = [\mathbf{x}_1(t),\ \mathbf{x}_2(t),\ \ldots,\ \mathbf{x}_n(t)]$ formed from the set of n vector functions $\mathbf{x}_1(t),\ \mathbf{x}_2(t),\ \ldots,\ \mathbf{x}_n(t)$ is called the **Wronskian** of these column vectors $\{\mathbf{x}_1(t), \ldots, \mathbf{x}_n(t)\}$.

Theorem 8.8: [N. Abel] Let $\mathbf{P}(t)$ be an $n \times n$ matrix with entries $p_{ij}(t)$ $(i, j = 1, 2, \ldots, n)$ that are continuous functions on some interval. Let $\mathbf{x}_k(t)$, $k = 1, 2, \ldots, n$, be n solutions to the homogeneous vector differential equation $\dot{\mathbf{x}} = \mathbf{P}(t)\, \mathbf{x}(t)$. Then the Wronskian of the set of vector solutions is

$$W(t) = W(t_0)\, \exp\left(\int_{t_0}^{t} \operatorname{tr} \mathbf{P}(t)\, \mathrm{d}t \right), \tag{8.1.6}$$

with t_0 being a point within an interval where the trace $\operatorname{tr} \mathbf{P}(t) = p_{11}(t) + p_{22}(t) + \cdots + p_{nn}(t)$ is continuous. Here $W(t) = \det \mathbf{X}(t)$, where $\mathbf{X}(t) = [\mathbf{x}_1(t),\ \mathbf{x}_2(t),\ \ldots,\ \mathbf{x}_n(t)]$ is the matrix formed from the set of column vectors $\{\mathbf{x}_1(t), \mathbf{x}_2(t), \ldots, \mathbf{x}_n(t)\}$.

PROOF: Differentiating the Wronskian $W(t) = \det \mathbf{X}(t)$ using the product rule, we get

$$\frac{\mathrm{d}}{\mathrm{d}t} W(t) = |\mathbf{x}_1'(t),\ \mathbf{x}_2(t),\ \ldots,\ \mathbf{x}_n| + |\mathbf{x}_1(t),\ \mathbf{x}_2'(t),\ \ldots,\ \mathbf{x}_n| + \cdots + |\mathbf{x}_1(t),\ \mathbf{x}_2(t),\ \ldots,\ \mathbf{x}_n'(t)|.$$

Looking at a typical term in this sum, we use the vector differential equation (8.1.2) to obtain

$$\left|\mathbf{x}_1(t),\ \ldots,\ \mathbf{x}_j'(t),\ \ldots,\ \mathbf{x}_n\right| = |\mathbf{x}_1(t),\ \ldots,\ \mathbf{P}(t)\, \mathbf{x}_j(t),\ \ldots,\ \mathbf{x}_n(t)|.$$

Now we write out the product $\mathbf{P}(t)\, \mathbf{x}_j(t)$ in its k-th entry form:

$$|\mathbf{P}\, \mathbf{x}_j|_k = \sum_{s=1}^{n} P_{ks} [\mathbf{x}_j]_s.$$

Since each k-th column of the determinant is a linear combination of the other columns except the k-th one, we sum all these determinants to obtain

$$\frac{\mathrm{d}}{\mathrm{d}t} W(t) = (\operatorname{tr}\mathbf{P})\, W(t), \tag{8.1.7}$$

which is a first order separable equation. Integration yields **Abel's formula** displayed in Eq. (8.1.6). Problem 1 on page 438 asks you to show all the details for a two-dimensional case.

Corollary 8.3: Let $\mathbf{x}_1(t),\ \mathbf{x}_2(t),\ \ldots,\ \mathbf{x}_n(t)$ be column solutions of the homogeneous vector equation $\dot{\mathbf{x}} = \mathbf{P}(t)\mathbf{x}$ on some interval $|a, b|$, where $n \times n$ matrix $\mathbf{P}(t)$ is continuous. Then the corresponding matrix $\mathbf{X}(t) = [\mathbf{x}_1(t),\ \mathbf{x}_2(t),\ \ldots,\ \mathbf{x}_n(t)]$ of these column vectors is either a singular matrix for all $t \in |a, b|$ or else nonsingular. In other words, $\det \mathbf{X}(t)$ is either identically zero or it never vanishes on the interval $|a, b|$.

Corollary 8.4: Let $\mathbf{P}(t)$ be an $n \times n$ matrix function that is continuous on an interval $|a, b|$. If $\{\mathbf{x}_1(t),\ \mathbf{x}_2(t),\ \ldots,\ \mathbf{x}_n(t)\}$ is a linearly independent set of solutions to the homogeneous differential equation $\dot{\mathbf{x}} = \mathbf{P}(t)\mathbf{x}$ on $|a, b|$, then the Wronskian

$$W(t) = \det[\mathbf{x}_1(t),\ \mathbf{x}_2(t),\ \ldots,\ \mathbf{x}_n(t)]$$

is not zero at every point t in $|a, b|$.

Example 8.1.3: (Example 8.1.2 revisited) The matrix

$$\mathbf{P}(t) = \begin{bmatrix} 1 & 2 & t \\ 0 & & 1/t \end{bmatrix} \qquad (0 < t)$$

has the trace $\mathrm{tr}\mathbf{P} = 1 + 1/t$. From Abel's theorem, it follows that the Wronskian

$$\begin{aligned} W(t) &= C\,e^{\int \mathrm{tr}\mathbf{P}(t)\,\mathrm{d}t} \\ &= C\,e^{\int (1+1/t)\,\mathrm{d}t} = C\,e^{t+\ln t} = C\,t\,e^{t}. \end{aligned}$$

On the other hand, direct calculations show that the Wronskian of the functions $\mathbf{x}_1(t)$ and $\mathbf{x}_2(t)$ is

$$W(t) = \det \begin{bmatrix} t^2 & e^t \\ t & 0 \end{bmatrix} = -t\,e^t \neq 0 \quad \text{for } t \neq 0.$$

□

Now let us consider the initial value problem

$$\frac{\mathrm{d}\mathbf{x}}{\mathrm{d}t} = \mathbf{P}(t)\,\mathbf{x}(t), \qquad \mathbf{x}(t_0) = \mathbf{x}_0. \tag{8.1.8}$$

From Theorem 8.5, page 434, it follows that the general solution of the vector differential equation (8.1.2) is

$$\mathbf{x}(t) = \mathbf{X}(t)\,\mathbf{c},$$

where $\mathbf{c} = \langle c_1, c_2, \ldots, c_n \rangle^T$ is the column vector of arbitrary constants. To satisfy the initial condition, we set

$$\mathbf{X}(t_0)\,\mathbf{c} = \mathbf{x}_0 \qquad \text{or} \qquad \mathbf{c} = \mathbf{X}^{-1}(t_0)\,\mathbf{x}_0.$$

Therefore, the solution to the initial value problem (8.1.8) becomes

$$\mathbf{x}(t) = \mathbf{\Phi}(t, t_0)\mathbf{x}_0 = \mathbf{X}(t)\mathbf{X}^{-1}(t_0)\,\mathbf{x}_0. \tag{8.1.9}$$

The square matrix $\mathbf{\Phi}(t, s) = \mathbf{X}(t)\mathbf{X}^{-1}(s)$ is usually referred to as a **propagator matrix**. Thus, the following statement is proved:

Theorem 8.9: Let $\mathbf{X}(t)$ be a fundamental matrix for the homogeneous linear system $\mathbf{x}' = \mathbf{P}(t)\mathbf{x}$, meaning that $\mathbf{X}(t)$ is a solution of the matrix differential equation (8.1.3) and $\det \mathbf{X}(t) \neq 0$. Then the unique solution of the vector initial value problem (8.1.8) is given by Eq. (8.1.9).

Corollary 8.5: For a fundamental matrix $\mathbf{X}(t)$, the propagator matrix-function $\mathbf{\Phi}(t, t_0) = \mathbf{X}(t)\mathbf{X}^{-1}(t_0)$ is the unique solution of the following matrix initial value problem:

$$\frac{\mathrm{d}\mathbf{\Phi}(t, t_0)}{\mathrm{d}t} = \mathbf{P}(t)\mathbf{\Phi}(t, t_0), \quad \mathbf{\Phi}(t_0, t_0) = \mathbf{I},$$

where $\mathbf{I}$ is the identity matrix. Hence, $\mathbf{\Phi}(t, t_0)$ is a fundamental matrix of Eq. (8.1.2).

Corollary 8.6: Let $\mathbf{X}(t)$ and $\mathbf{Y}(t)$ be two fundamental matrices of the homogeneous vector equation (8.1.2). Then there exists a nonsingular constant square matrix $\mathbf{C}$ such that $\mathbf{X}(t) = \mathbf{Y}(t)\mathbf{C}$, $\det \mathbf{C} \neq 0$. This means that the solution space of the matrix differential equation (8.1.3) is 1.

Example 8.1.4: Solve the initial value problem

$$\frac{\mathrm{d}\mathbf{x}}{\mathrm{d}t} = \mathbf{P}(t)\mathbf{x}, \quad \mathbf{x}(1) = \mathbf{x}_0, \qquad \text{where} \quad \mathbf{P}(t) = \frac{1}{2t^2} \begin{bmatrix} 3t & -t^2 \\ -1 & t \end{bmatrix} \ (t \neq 0), \quad \mathbf{x}_0 = \begin{bmatrix} 1 \\ 2 \end{bmatrix},$$

given that vectors

$$\mathbf{x}_1(t) = \begin{bmatrix} t \\ 1 \end{bmatrix} \quad \text{and} \quad \mathbf{x}_2(t) = \begin{bmatrix} -t^2 \\ t \end{bmatrix} = t \begin{bmatrix} -t \\ 1 \end{bmatrix}$$

constitute the fundamental set of solutions.

Solution. A fundamental matrix is

$$\mathbf{X}(t) = \left[\ \mathbf{x}_1(t),\ \mathbf{x}_2(t)\ \right] = \begin{bmatrix} t & -t^2 \\ 1 & t \end{bmatrix}, \quad \text{with} \quad \mathbf{X}^{-1}(t) = \frac{1}{2t^2}\begin{bmatrix} t & t^2 \\ -1 & t \end{bmatrix}.$$

Then

$$\mathbf{X}^{-1}(1) = \frac{1}{2}\begin{bmatrix} 1 & 1 \\ -1 & 1 \end{bmatrix} \quad \Longrightarrow \quad \mathbf{X}^{-1}(1)\,\mathbf{x}_0 = \frac{1}{2}\begin{bmatrix} 3 \\ 1 \end{bmatrix}.$$

Therefore, the solution of the given initial value problem becomes

$$\mathbf{x}(t) = \mathbf{\Phi}(t,1)\mathbf{x}_0 = \mathbf{X}(t)\mathbf{X}^{-1}(1)\,\mathbf{x}_0 = \begin{bmatrix} t & -t^2 \\ 1 & t \end{bmatrix}\cdot\frac{1}{2}\begin{bmatrix} 3 \\ 1 \end{bmatrix} = \frac{1}{2}\begin{bmatrix} 3t-t^2 \\ 3+t \end{bmatrix}.$$

□

A linear inhomogeneous equation of order n

$$y^{(n)} + a_{n-1}(t)\,y^{(n-1)} + \cdots + a_1(t)\,y' + a_0(t)\,y = f(t) \tag{8.1.10}$$

can be replaced with an equivalent first order system of dimension n. To do this, we introduce new variables

$$x_1 = y, \quad x_2 = y', \quad \ldots, \quad x_n = y^{(n-1)}. \tag{8.1.11}$$

It turns out that these new variables satisfy the system of first order equations

$$\begin{aligned} x_1' &= x_2, \\ &\vdots \\ x_{n-1}' &= x_n, \\ x_n' &= -a_{n-1}(t)x_n - \cdots - a_1(t)x_2 - a_0(t)x_1 + f(t). \end{aligned} \tag{8.1.12}$$

The system (8.1.12) is a system of n first order linear differential equations involving the n unknown functions $x_1(t)$, $x_2(t)$, …, $x_n(t)$. By introducing n-column vector functions $\mathbf{x}(t) = \langle x_1(t), \ldots, x_n(t)\rangle^T$ and $\mathbf{f}(t) = \langle 0, 0, \ldots, 0, f(t)\rangle^T$, we can rewrite the system of equations (8.1.12) in vector form (8.1.1), page 431, where

$$\mathbf{P}(t) = \begin{bmatrix} 0 & 1 & 0 & \cdots & 0 \\ 0 & 0 & 1 & \cdots & 0 \\ \vdots & \vdots & \vdots & \ddots & \vdots \\ -a_0(t) & -a_1(t) & -a_2(t) & \cdots & -a_{n-1}(t) \end{bmatrix}. \tag{8.1.13}$$

Therefore, every result for a single differential equation of order larger than 1 can be translated into a similar result for a first order system of differential equations.

8.1.1 The Euler Vector Equations

The Euler system of equations

$$t\,\dot{\mathbf{x}} = \mathbf{A}\,\mathbf{x}(t), \qquad t > 0, \tag{8.1.14}$$

with a constant square matrix $\mathbf{A}$, is a vector version of the scalar n-th order Euler equation (see §4.6.2)

$$a_n t^n y^{(n)} + a_{n-1}t^{n-1}y^{(n-1)} + \cdots + a_1 t\, y' + a_0 y = 0, \tag{8.1.15}$$

where a_k, $(k = 0, 1, \ldots, n)$ are constant coefficients. While Eq. (8.1.14) can be reduced to a constant coefficient vector equation by substitution $x = \ln t$, we seek its solution in the form $\mathbf{x}(t) = t^{\lambda}\boldsymbol{\xi}$, where $\boldsymbol{\xi}$ is a column vector independent of t. Since $\dot{\mathbf{x}} = \lambda\, t^{\lambda-1}\boldsymbol{\xi}$, we get from Eq. (8.1.14) that λ must be a root of the algebraic equation

$$\lambda\boldsymbol{\xi} = \mathbf{A}\,\boldsymbol{\xi}.$$

This means that λ must be an eigenvalue of the matrix $\mathbf{A}$, and $\boldsymbol{\xi}$ must be its eigenvector. If the given matrix $\mathbf{A}$ is diagonalizable, then we get n eigenvalues λ_1, …, λ_n (not necessarily different) with corresponding linearly independent eigenvectors $\boldsymbol{\xi}_1, \ldots, \boldsymbol{\xi}_n$. The general solution of Eq. (8.1.14) is a linear combination of these n column vectors

$$\mathbf{x} = c_1 t^{\lambda_1} \boldsymbol{\xi}_1 + c_2 t^{\lambda_2} \boldsymbol{\xi}_2 + \cdots + c_n t^{\lambda_n} \boldsymbol{\xi}_n, \tag{8.1.16}$$

where $c_1, \ldots, c_n$ are arbitrary constants. When the matrix $\mathbf{A}$ is defective, we do not enjoy having n linearly independent eigenvectors, but we get only m of them, where $m < n$. We do not pursue this case because it requires the use of more advanced material—generalized eigenvectors and functions of a complex variable.

Example 8.1.5: Consider the system of equations with a singular matrix:

$$t\,\dot{\mathbf{x}}(t) = \mathbf{A}\,\mathbf{x}(t), \qquad \text{where} \quad \mathbf{A} = \begin{bmatrix} 2 & 2 \\ -1 & -1 \end{bmatrix}.$$

Since the eigenvalues and corresponding eigenvectors of the matrix $\mathbf{A}$ are $\lambda = 0$, $\boldsymbol{\xi}_0 = \langle -1, 1 \rangle^T$ and $\lambda = 1$, $\boldsymbol{\xi}_1 = \langle 2, -1 \rangle^T$, the general solution of the given equation becomes

$$\mathbf{x}(t) = c_1 \begin{bmatrix} -1 \\ 1 \end{bmatrix} + c_2\, t \begin{bmatrix} 2 \\ -1 \end{bmatrix}.$$

Example 8.1.6: Consider the system of equations with a diagonalizable matrix:

$$t\,\dot{\mathbf{x}}(t) = \mathbf{A}\,\mathbf{x}(t), \qquad \text{where} \quad \mathbf{A} = \begin{bmatrix} 5 & -6 \\ 3 & -4 \end{bmatrix}.$$

Using eigenvalues and corresponding eigenvectors $\lambda = 2$, $\boldsymbol{\xi}_2 = \langle 2, 1 \rangle^T$ and $\lambda = -1$, $\boldsymbol{\xi}_{-1} = \langle 1, 1 \rangle^T$ of the matrix $\mathbf{A}$, we construct the general solution

$$\mathbf{x}(t) = c_1 t^2 \begin{bmatrix} 2 \\ 1 \end{bmatrix} + c_2\, t^{-1} \begin{bmatrix} 1 \\ 1 \end{bmatrix}.$$

Example 8.1.7: Consider the vector differential equation with a defective matrix:

$$t\,\dot{\mathbf{x}}(t) = \mathbf{A}\,\mathbf{x}(t), \qquad \text{where} \quad \mathbf{A} = \begin{bmatrix} -3 & 2 \\ -2 & 1 \end{bmatrix}. \tag{8.1.17}$$

The matrix $\mathbf{A}$ has a double eigenvalue $\lambda = -1$ of geometric multiplicity 1. This means that the eigenspace is spanned on the vector $\boldsymbol{\xi} = \langle 2, 2 \rangle^T$. Thus, one solution of the system (8.1.17) is $\mathbf{x}_1(t) = t^{-1}\boldsymbol{\xi}$, but another linearly independent solution has a different form.

Let $\boldsymbol{\eta} = \langle 0, 1 \rangle^T$ be a generalized eigenvector, so $(\mathbf{A} + \mathbf{I})\,\boldsymbol{\eta} = \boldsymbol{\xi}$. Based on the reduction of order procedure presented in §4.6.1, we attempt to find a second linearly independent solution of the system (8.1.17) in the form

$$\mathbf{x}_2(t) = t^{-1} \ln|t|\, \boldsymbol{\xi} + t^{-1}\boldsymbol{\eta}.$$

Upon differentiation, we obtain

$$\dot{\mathbf{x}}_2 = -t^{-2} \ln|t| \boldsymbol{\xi} + t^{-2}\boldsymbol{\xi} - t^{-2}\boldsymbol{\eta} = -t^{-2} \ln|t| \boldsymbol{\xi} + t^{-2} \left(\boldsymbol{\xi} - \boldsymbol{\eta} \right).$$

Since $\mathbf{A}\boldsymbol{\xi} = -\boldsymbol{\xi}$ and $\mathbf{A}\boldsymbol{\eta} = \boldsymbol{\xi} - \boldsymbol{\eta}$, the vector $\mathbf{x}_2(t)$ is indeed a solution of the system (8.1.17) and linearly independent from $\mathbf{x}_1(t)$.

Problems

1. Prove Abel's theorem 8.8 on page 435 for a 2×2 matrix $\mathbf{P}(t)$ using the identity

$$\frac{\mathrm{d}W}{\mathrm{d}t} = \begin{vmatrix} \frac{\mathrm{d}x_1}{\mathrm{d}t} & y_1(t) \\ \frac{\mathrm{d}x_2}{\mathrm{d}t} & y_2(t) \end{vmatrix} + \begin{vmatrix} x_1(t) & \frac{\mathrm{d}y_1}{\mathrm{d}t} \\ x_2(t) & \frac{\mathrm{d}y_2}{\mathrm{d}t} \end{vmatrix}, \qquad \text{where} \quad W(t) = \det \begin{bmatrix} x_1(t) & y_1(t) \\ x_2(t) & y_2(t) \end{bmatrix}.$$

2. Prove Corollary 8.6, page 436.

3. Let us consider the homogeneous system of equations

$$\frac{\mathrm{d}\mathbf{x}}{\mathrm{d}t} = \mathbf{P}(t)\mathbf{x}, \qquad \mathbf{P}(t) = \begin{bmatrix} \frac{3-t}{t} & t \\ \frac{2-t}{t^2} & 1 \end{bmatrix} \quad (t \neq 0). \tag{8.1.18}$$

(a) Compute the Wronskian of two solutions of Eq. (8.1.18): $\mathbf{x}_1 = \langle t^3, t^2\rangle^T$ and $\mathbf{x}_2 = \langle t^2, t-1\rangle^T$.

(b) On what intervals are $\mathbf{x}_1(t)$ and $\mathbf{x}_2(t)$ linearly independent?

(c) Construct a fundamental matrix $\mathbf{X}(t)$ and compute the propagator matrix $\mathbf{X}(t)\mathbf{X}^{-1}(2)$.

4. In each exercise, two vector functions $\mathbf{v}_1(t)$ and $\mathbf{v}_2(t)$, defined on the interval $-\infty < t < \infty$, are given.

(a) Evaluate the determinant of the 2×2 matrix formed by these column vectors $\mathbf{V}(t) = [\mathbf{v}_1(t),\ \mathbf{v}_2(t)]$.

(b) At $t = 0$, the determinant, $\det \mathbf{V}(0)$, calculated in part (a) is zero. Therefore, there exists a nonzero number k such that $\mathbf{v}_1(0) = k\mathbf{v}_2(0)$. Does this fact prove that the given vector functions are linearly dependent on $-\infty < t < \infty$?

(c) At $t = 2$, the determinant $\det \mathbf{V}(2) \ne 0$. Does this fact prove that the given vector functions are linearly independent on $-\infty < t < \infty$?

$$\text{(a)}\quad \mathbf{v}_1(t) = \begin{bmatrix} t \\ 1 \end{bmatrix},\quad \mathbf{v}_2(t) = \begin{bmatrix} t^3 \\ 2 \end{bmatrix}; \qquad \text{(b)}\quad \mathbf{v}_1(t) = \begin{bmatrix} 2 \\ e^t \end{bmatrix},\quad \mathbf{v}_2(t) = \begin{bmatrix} \sin \pi t \\ t \end{bmatrix}.$$

5. In each exercise, verify that two given matrix functions, $\mathbf{X}(t)$ and $\mathbf{Y}(t)$, are fundamental matrices for the vector differential equation $\dot{\mathbf{x}} = \mathbf{P}(t)\mathbf{x}$, with specified matrix $\mathbf{P}(t)$. Find a constant nonsingular matrix $\mathbf{C}$ such that $\mathbf{X}(t) = \mathbf{Y}(t)\mathbf{C}$.

$$\text{(a)}\ \mathbf{P}(t) = \text{sech}2t \begin{bmatrix} 1+\sinh 2t & -1 \\ 2 & \sinh 2t - 1 \end{bmatrix},\quad \mathbf{X} = \begin{bmatrix} e^t/2 & e^{-t}/2 \\ \sinh t & \cosh t \end{bmatrix},$$

$$\mathbf{Y} = \begin{bmatrix} \cosh t - 5\sinh t & \cosh t + 3\sinh t \\ 6\cosh t - 4\sinh t & 4\sinh t - 2\cosh t \end{bmatrix};$$

$$\text{(b)}\ \mathbf{P} = \frac{e^{-3t}}{2\cosh 3t}\begin{bmatrix} 2e^{6t}+1 & -e^{4t} \\ 5e^{2t} & 3e^{6t}-2 \end{bmatrix},\quad \mathbf{X} = \begin{bmatrix} e^t & e^{2t} \\ e^{3t} & -e^{2t} \end{bmatrix},\quad \mathbf{Y} = \begin{bmatrix} e^t + e^{2t} & 3e^t \\ e^{3t} - e^{-2t} & 3e^{3t} \end{bmatrix};$$

$$\text{(c)}\ \mathbf{P}(t) = \begin{bmatrix} \frac{5}{3t} & -\frac{4}{3} \\ -\frac{1}{6t^2} & \frac{1}{3t} \end{bmatrix},\quad \mathbf{X} = \begin{bmatrix} 2t & 4t^2 \\ 1 & -t \end{bmatrix},\quad \mathbf{Y} = \frac{1}{3}\begin{bmatrix} 4t+4t^2 & 4t^2-2t \\ 2-t & -1-t \end{bmatrix}.$$

6. Consider the system of equations and two of its solutions

$$\frac{d\mathbf{x}}{dt} = \mathbf{P}(t)\mathbf{x},\qquad \mathbf{P}(t) = \begin{bmatrix} \frac{t-2}{t(t-1)} & \frac{t}{t-1} \\ -\frac{2}{t^2(t-1)} & \frac{t+1}{t-1} \end{bmatrix},\qquad \mathbf{x}_1(t) = \begin{bmatrix} t \\ 1/t \end{bmatrix},\quad \mathbf{x}_2(t) = \begin{bmatrix} t^2 \\ t \end{bmatrix}.$$

(a) Show that the column vectors $\mathbf{x}_1(t)$ and $\mathbf{x}_2(t)$ are solutions to the given vector equation $\dot{\mathbf{x}} = \mathbf{P}(t)\mathbf{x}$.

(b) Compute the Wronskian of $\mathbf{x}_1$ and $\mathbf{x}_2$.

(c) On what intervals are $\mathbf{x}_1(t)$ and $\mathbf{x}_2(t)$ linearly independent?

(d) Construct a fundamental matrix $\mathbf{X}(t)$ and compute $\mathbf{X}(t)\mathbf{X}^{-1}(2)$.

7. In each exercise (a) through (h), a nonsingular matrix $\mathbf{X}(t)$ is given. Find $\mathbf{P}(t)$ so that the matrix $\mathbf{X}(t)$ is a solution of the matrix equation $\dot{\mathbf{X}} = \mathbf{P}(t)\mathbf{X}$. On what interval would $\mathbf{X}(t)$ be the fundamental matrix? *Hint:* $\mathbf{P}(t) = \dot{\mathbf{X}}(t)\mathbf{X}^{-1}(t)$.

$$\text{(a)}\ \begin{bmatrix} \sin t & 1 \\ \cos t & t \end{bmatrix};\quad \text{(b)}\ \begin{bmatrix} t & t^2 \\ \ln t & t^{-1} \end{bmatrix};\quad \text{(c)}\ \begin{bmatrix} \tan t & 1 \\ \cot t & t \end{bmatrix};\quad \text{(d)}\ \begin{bmatrix} e^t & e^{2t} \\ t^{-1} & t \end{bmatrix};$$

$$\text{(e)}\ \begin{bmatrix} \sinh t & t \\ \cosh t & t^2 \end{bmatrix};\quad \text{(f)}\ \begin{bmatrix} \ln t & 1 \\ \ln t & t^2 \end{bmatrix};\quad \text{(g)}\ \begin{bmatrix} \tanh t & t \\ \coth t & 1 \end{bmatrix};\quad \text{(h)}\ \begin{bmatrix} t & t^{-1} \\ t^2 & t^{-2} \end{bmatrix}.$$

8.2 Constant Coefficient Homogeneous Systems

This section provides a qualitative analysis of autonomous vector linear differential equations of the form

$$\dot{\mathbf{y}}(t) = \mathbf{A}\,\mathbf{y}(t), \tag{8.2.1}$$

where $\mathbf{A}$ is an $n\times n$ constant matrix and $\mathbf{y}(t)$ is an n-column vector (which is $n\times 1$ matrix) of n unknown functions. Here we use a dot to represent the derivative with respect to t: $\dot{\mathbf{y}}(t) = d\mathbf{y}/dt$. A solution of Eq. (8.2.1) is a curve in n-dimensional space. It is called an **integral curve**, a **trajectory**, a **streamline**, or an **orbit** of the system. When the variable t is associated with time, we can call a solution $\mathbf{y}(t)$ the **state of the system** at time t. Since a constant matrix $\mathbf{A}$ is continuous on any interval, Theorem 6.3 (page 371) assures us that all solutions of Eq. (8.2.1) are determined on $(-\infty, \infty)$. Therefore, when we speak of solutions to the vector equation $\dot{\mathbf{y}} = \mathbf{A}\,\mathbf{y}$, we consider solutions on the whole real axis.

We refer to a constant solution $\mathbf{y}(t) = \mathbf{y}^*$ of a system as an **equilibrium** if $\mathrm{d}\mathbf{y}(t)/\mathrm{d}t = \mathbf{0}$. Such a constant solution is also called a **critical** or **stationary** point of the system. An equilibrium solution is isolated if there is a neighborhood to the critical point that does not contain any other critical point. If matrix $\mathbf{A}$ is not singular ($\det \mathbf{A} \ne 0$), then $\mathbf{0}$ is the only critical point of the system (8.2.1). Otherwise, the system (8.2.1) has a subspace of equilibrium solutions.

Example 8.2.1: Consider a linear system of differential equations with a singular matrix:

$$\frac{\mathrm{d}\mathbf{y}}{\mathrm{d}t} = \mathbf{A}\,\mathbf{y}, \qquad \text{where} \quad \mathbf{y} = \begin{bmatrix} x(t) \\ y(t) \end{bmatrix}, \quad \mathbf{A} = \begin{bmatrix} 1 & -2 \\ 2 & -4 \end{bmatrix}.$$

The characteristic polynomial $\chi(\lambda) = \lambda(\lambda + 3)$ of the given matrix $\mathbf{A}$ has two distinct real nulls $\lambda = 0$ and $\lambda = -3$. Since the eigenspace corresponding to the eigenvalue $\lambda = 0$ is spanned on the vector $\langle 2, 1\rangle^T$, every element from this vector space is a critical point of the given system of differential equations. Indeed, the general solution of the given vector differential equation is

$$\mathbf{y}(t) = e^{\mathbf{A}t}\mathbf{c}, \quad \text{where} \quad e^{\mathbf{A}t} = \frac{1}{3}\begin{bmatrix} 4 - e^{-3t} & 2\,e^{-3t} - 2 \\ 2 - 2\,e^{-3t} & 4\,e^{-3t} - 1 \end{bmatrix}$$

and $\mathbf{c} = \langle c_1, c_2\rangle^T$ is a column vector of arbitrary constants. As t approaches infinity, the general solution tends to

$$\mathbf{y}(t) \mapsto \frac{1}{3}\begin{bmatrix} 4c_1 - 2c_2 \\ 2c_1 - c_2 \end{bmatrix} = \frac{c_1}{3}\begin{bmatrix} 4 \\ 2 \end{bmatrix} + \frac{c_2}{3}\begin{bmatrix} 2 \\ 1 \end{bmatrix} \qquad \text{as} \quad t \to \infty.$$

To plot a direction field and then some solutions, we type the following script in *Mathematica:*

```
VectorPlot[{x - 2*y, 2*x - 4 y}, {x, -3, 3}, {y, -3, 3},
 AxesLabel -> {x, y}, Axes -> True, VectorPoints -> 20,
 VectorScale -> {Tiny, Automatic, None}]
StreamPlot[{x - 2*y, 2*x - 4 y}, {x, -3, 3}, {y, -3, 3}, AxesLabel -> {x, y}, Axes -> True,
 StreamScale -> {Tiny, Automatic, None},
 StreamPoints -> {{.1, 2}, {0, 0}, {1, 2}, {-1, -1}, {1, 1}}, StreamStyle -> {Black, "Line"}]
```

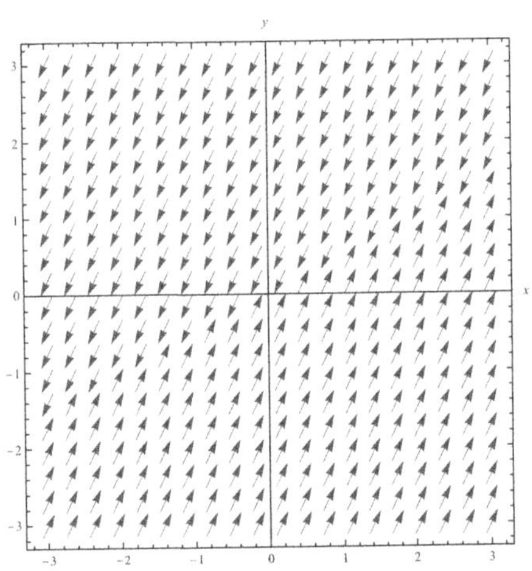

Figure 8.1: Example 8.2.1, direction field, plotted with `VectorPlot` (*Mathematica*).

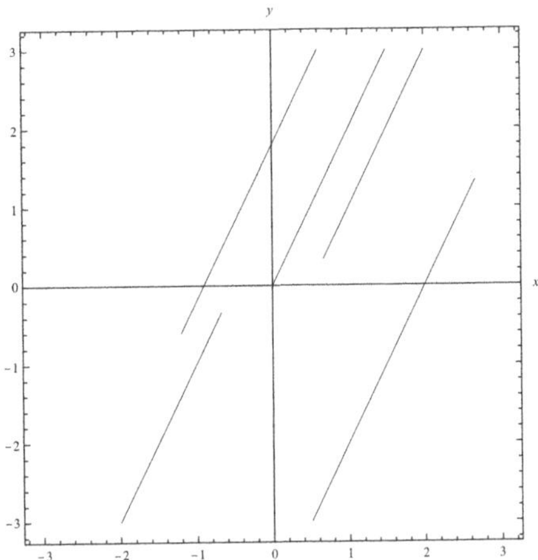

Figure 8.2: Example 8.2.1, some solutions, plotted with *Mathematica*.

As we can see, every solution is the straight line $y = 2x + c$, for some constant c (see Fig. 8.2), because

$$\frac{\mathrm{d}y}{\mathrm{d}x} = \frac{\mathrm{d}y/\mathrm{d}t}{\mathrm{d}x/\mathrm{d}t} = \frac{2x - 4y}{x - 2y} = 2.$$

Let us consider a similar vector equation $\dot{y} = \mathbf{B}\mathbf{y}$ with the matrix $\mathbf{B} = \begin{bmatrix} -3 & 0 \\ 0 & 0 \end{bmatrix}$ that has the same characteristic polynomial $\chi(\lambda) = \lambda(\lambda + 3)$. Since the corresponding system of equations $\dot{x} = -3x$, $\dot{y} = 0$, is equivalent to one equation $\frac{\mathrm{d}y}{\mathrm{d}x} = \frac{\dot{y}}{\dot{x}} = \frac{0}{-3x} = 0$, the abscissa $y = 0$ is a stationary line. The orbits will be horizontal lines, pointed toward the stationary line (see Fig. 8.3). The general solution of the equation $\dot{\mathbf{y}} = \mathbf{B}\mathbf{y}$ is

$$\mathbf{y}(t) = e^{\mathbf{B}t}\mathbf{c} = \begin{bmatrix} e^{-3t} & 0 \\ 0 & 1 \end{bmatrix} \begin{bmatrix} c_1 \\ c_2 \end{bmatrix} = c_1\, e^{-3t} \begin{bmatrix} 1 \\ 0 \end{bmatrix} + c_2 \begin{bmatrix} 0 \\ 1 \end{bmatrix},$$

where c_1, c_2 are arbitrary constants. Therefore, $\mathbf{y}(t) \mapsto c_2\langle 0, 1\rangle^T$ as $t \to \infty$.

Example 8.2.2: Consider another vector differential equation

$$\frac{\mathrm{d}\mathbf{y}}{\mathrm{d}t} = \mathbf{A}\,\mathbf{y}, \qquad \text{where} \quad \mathbf{y} = \begin{bmatrix} x(t) \\ y(t) \end{bmatrix}, \quad \mathbf{A} = \begin{bmatrix} -2 & 1 \\ -4 & 2 \end{bmatrix}.$$

The nilpotent matrix $\mathbf{A}$ has a double (deficient) eigenvalue $\lambda = 0$ with the eigenspace spanning on the vector $\langle 1, 2\rangle^T$ because its characteristic polynomial is $\chi(\lambda) = \det(\lambda\mathbf{I} - \mathbf{A}) = \lambda^2$. Solutions of the corresponding linear vector differential equation $\dot{\mathbf{y}} = \mathbf{A}\,\mathbf{y}$ are also straight lines, similar to Example 8.2.1. They are pointed in different directions and are separated by the stationary line $y = 2x$ (see Fig. 8.4). □

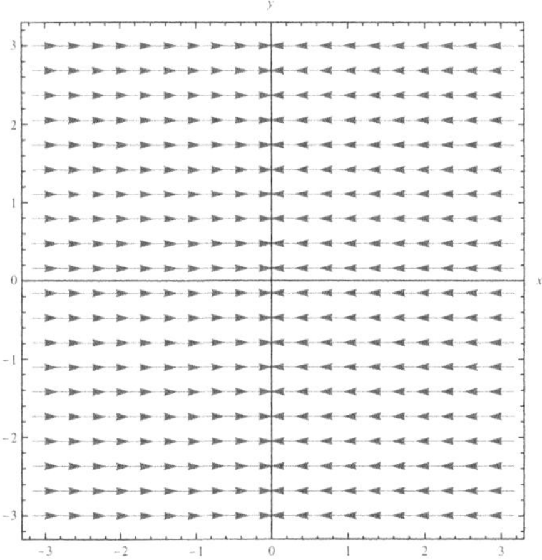

Figure 8.3: Example 8.2.1, direction field for the system with matrix **B**.

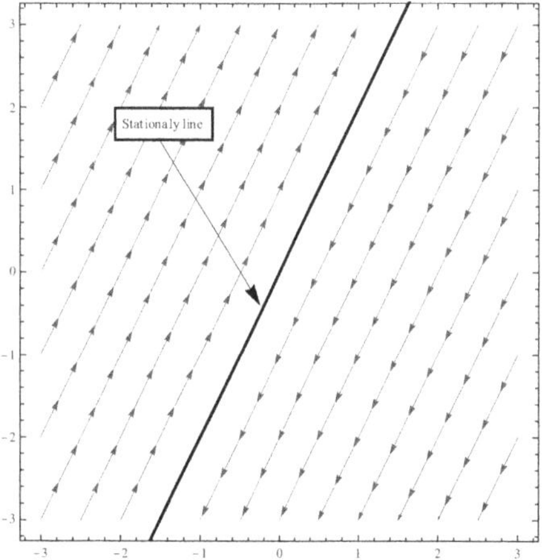

Figure 8.4: Example 8.2.2, direction field, plotted with *Mathematica*.

From Theorem 7.14, page 412, we know that the fundamental matrix $\mathbf{\Phi}(t, t_0) = e^{\mathbf{A}(t-t_0)}$ is a solution to the matrix equation Eq. (8.1.3), page 433, for any t_0. Since for a constant coefficient vector equation (8.2.1) the propagator matrix depends always on the difference $t - t_0$, it will be denoted as $\mathbf{\Phi}(t - t_0)$. From the definition of the exponential matrix (7.7.1), page 412, it follows that $\mathbf{\Phi}(0) = \mathbf{I}$, the identity matrix. Multiplication of the propagator $\mathbf{\Phi}(t) = e^{\mathbf{A}t}$ by a constant nonsingular matrix $\mathbf{C}$ leads to another fundamental matrix $\mathbf{X}(t) = e^{\mathbf{A}t}\mathbf{C}$ that satisfies the matrix equation (8.1.3) subject to the initial condition $\mathbf{X}(0) = \mathbf{C}$.

The following statements give us the main properties of the exponential matrix.

Theorem 8.10: Let $\mathbf{A}$ be an $n \times n$ matrix with constant real entries. The propagator $\mathbf{\Phi}(t) \stackrel{\text{def}}{=} e^{\mathbf{A}t}$ is a fundamental matrix for the system of differential equations (8.2.1). In other words, the column vectors of the exponential matrix $e^{\mathbf{A}t}$ are linearly independent solutions of the vector equation $\dot{\mathbf{y}}(t) = \mathbf{A}\,\mathbf{y}(t)$.

Corollary 8.7: Let $\mathbf{A}$ be an $n \times n$ matrix with constant entries. Then the exponential matrix $\mathbf{\Phi}(t) = e^{\mathbf{A}t}$ is the unique solution of the matrix differential equation subject to the initial condition

$$\frac{\mathrm{d}\,\mathbf{\Phi}}{\mathrm{d}t} = \mathbf{A}\,\mathbf{\Phi}(t), \qquad \mathbf{\Phi}(0) = \mathbf{I}, \tag{8.2.2}$$

where $\mathbf{I}$ is the $n \times n$ identity matrix.

Theorem 8.11: Let $\mathbf{Y}(t) = [\mathbf{y}_1(t), \mathbf{y}_2(t), \ldots, \mathbf{y}_n(t)]$ be a fundamental matrix for the vector differential equation (8.2.1) with a constant square matrix $\mathbf{A}$. Then

$$e^{\mathbf{A}(t-t_0)} = \mathbf{Y}(t)\mathbf{Y}^{-1}(t_0) = \mathbf{\Phi}(t-t_0).$$

Corollary 8.8: The inverse matrix to $\mathbf{\Phi}(t) = e^{\mathbf{A}t}$ is $\mathbf{\Phi}^{-1}(t) = e^{-\mathbf{A}t}$.

Theorem 8.12: For any constant $n \times n$ matrix $\mathbf{A}$, the column vector function

$$\mathbf{y}(t) = e^{\mathbf{A}t}\,\mathbf{c} = c_1\,\mathbf{y}_1(t) + c_2\,\mathbf{y}_2(t) + \cdots + c_n\,\mathbf{y}_n(t) \tag{8.2.3}$$

is the general solution of the linear vector differential equation $\dot{\mathbf{y}} = \mathbf{A}\,\mathbf{y}(t)$. Here $\mathbf{y}_k(t)$ is k-th column ($k = 1, 2, \ldots, n$) of the exponential matrix $\mathbf{\Phi}(t) = e^{\mathbf{A}t}$ and $\mathbf{c} = \langle c_1, c_2, \ldots, c_n\rangle^T$ is a column vector of arbitrary constants. Moreover, the column vector

$$\mathbf{y}(t) = \mathbf{\Phi}(t-t_0)\,\mathbf{y}(t_0) = e^{\mathbf{A}(t-t_0)}\,\mathbf{y}_0 \tag{8.2.4}$$

is the unique solution of the initial value problem:

$$\dot{\mathbf{y}}(t) = \mathbf{A}\,\mathbf{y}(t), \quad \mathbf{y}(t_0) = \mathbf{y}_0. \tag{8.2.5}$$

If $\mathbf{y}(t)$ is a solution of a constant coefficient system $\dot{\mathbf{y}}(t) = \mathbf{A}\,\mathbf{y}(t)$ and if t_0 is a fixed value of t, then $\mathbf{y}(t \pm t_0)$ is also a solution. However, these solutions determine the same trajectory because the corresponding initial value problem (8.2.5) has a unique solution expressed explicitly through the propagator matrix $\mathbf{y}(t) = \mathbf{\Phi}(t)\mathbf{y}_0 = e^{\mathbf{A}t}\mathbf{y}_0$.

So if two solutions of the same linear system of equations (8.2.1) with constant coefficients coincide at one point, then they are identical at all points. Thus, an integral curve of the vector differential equation $\dot{\mathbf{y}} = \mathbf{A}\mathbf{y}$ is a trajectory of infinitely many solutions. Therefore, distinct integral curves of Eq. (8.2.1) do not touch each other, which means that the vector equation $\dot{\mathbf{y}}(t) = \mathbf{A}\,\mathbf{y}(t)$ has no singular solution.

Example 8.2.3: If $\mathbf{A}$ is a square constant matrix, then software packages can be used to calculate the fundamental matrix $\mathbf{\Phi}(t) = e^{\mathbf{A}t}$ for the system $\dot{\mathbf{y}} = \mathbf{A}\mathbf{y}$. After the matrix

$$\mathbf{A} = \begin{bmatrix} -13 & -10 \\ 21 & 16 \end{bmatrix}$$

has been entered, either the *Maxima* command

```
load(linearalgebra)$
matrixexp(A,t);
```

the *Maple*™ command

```
with(LinearAlgebra):  MatrixExponential(A,t)
```

or `MatrixFunction(A,F,lambda)` with $F(\lambda) = e^{\lambda t}$, or

```
with(linalg):  exponential(A,t)
```

the *Mathematica*® command

```
MatrixExp[A t]
```

or the MATLAB® command

```
syms t, expm(A*t)
```

yields the matrix exponential (propagator)

$$e^{\mathbf{A}t} = \begin{bmatrix} 15\,e^{t} - 14\,e^{2t} & 10\,e^{t} - 10\,e^{2t} \\ 21\,e^{2t} - 21\,e^{t} & 15\,e^{2t} - 14\,e^{t} \end{bmatrix}. \qquad \square$$

If we seek a solution of Eq. (8.2.1) in the form $\mathbf{y}(t) = \mathbf{v}\,e^{\lambda t}$, then by substitution for $\mathbf{y}$ into Eq. (8.2.1), we obtain

$$(\lambda\mathbf{I} - \mathbf{A})\,\mathbf{v} = \mathbf{0}.$$

Therefore, λ is an eigenvalue and $\mathbf{v}$ is a corresponding eigenvector of the coefficient matrix $\mathbf{A}$. So if an $n \times n$ matrix $\mathbf{A}$ has m ($m \leqslant n$) distinct eigenvalues λ_k, $k = 1, 2, \ldots, m$, then the vector differential equation $\dot{\mathbf{y}}(t) = \mathbf{A}\,\mathbf{y}(t)$ has at least m linearly independent exponential solutions $\mathbf{v}_k\, e^{\lambda_k t}$ because eigenvectors $\mathbf{v}_k$ corresponding to different eigenvalues λ_k are linearly independent (Theorem 7.7, page 387).

If the matrix $\mathbf{A}$ is diagonalizable, then we have exactly n linearly independent solutions of the form $\mathbf{v}_k\, e^{\lambda_k t}$. The following theorem states that we have at least m linearly independent exponential solutions to the vector differential equation $\dot{\mathbf{y}} = \mathbf{A}\mathbf{y}$, where m is the number of distinct eigenvalues of the constant matrix $\mathbf{A}$.

Theorem 8.13: Suppose that an $n \times n$ constant matrix $\mathbf{A}$ has m ($m \leqslant n$) distinct eigenvalues $\lambda_1, \lambda_2, \ldots, \lambda_m$ with corresponding m eigenvectors $\mathbf{v}_1, \mathbf{v}_2, \ldots, \mathbf{v}_m$. Then the column functions

$$\mathbf{y}_1(t) = \mathbf{v}_1\, e^{\lambda_1 t}, \quad \mathbf{y}_2(t) = \mathbf{v}_2\, e^{\lambda_2 t}, \quad \ldots, \quad \mathbf{y}_m(t) = \mathbf{v}_m\, e^{\lambda_m t}$$

are linearly independent solutions of the vector equation (8.2.1).

Theorem 8.14: Suppose that an $n \times n$ constant diagonalizable matrix $\mathbf{A}$ has n real or complex (not necessarily distinct) eigenvalues $\lambda_1, \lambda_2, \ldots, \lambda_n$ with corresponding n linearly independent eigenvectors $\mathbf{v}_1, \mathbf{v}_2, \ldots, \mathbf{v}_n$. Then the general solution of the homogeneous system of differential equations $\dot{\mathbf{y}}(t) = \mathbf{A}\,\mathbf{y}(t)$ is

$$\mathbf{y}(t) = c_1\, \mathbf{v}_1\, e^{\lambda_1 t} + c_2\, \mathbf{v}_2\, e^{\lambda_2 t} + \cdots + c_n\, \mathbf{v}_n\, e^{\lambda_n t}, \tag{8.2.6}$$

where $c_1, c_2, \ldots, c_n$ are arbitrary constants. ■

It should be noted that the general solution in the form (8.2.6) is not convenient for solving the initial value problem since it leads to determination of arbitrary constants from an algebraic system of equations. For a nondefective square matrix $\mathbf{A}$, let us form a fundamental matrix from the column vectors specified in Theorem 8.14:

$$\mathbf{Y}(t) = \left[e^{\lambda_1 t}\mathbf{v}_1, e^{\lambda_2 t}\mathbf{v}_2, \ldots, e^{\lambda_n t}\mathbf{v}_n \right].$$

Therefore, there exists a fundamental matrix $\mathbf{Y}(t)$ for a system of homogeneous equations with a nondefective matrix $\mathbf{A}$ that can be expressed through the exponential matrix and the matrix generated by its eigenvectors $\mathbf{v}_k$ ($k = 1, 2, \ldots, n$):

$$\mathbf{Y}(t) = \left[e^{\lambda_1 t}\mathbf{v}_1, e^{\lambda_2 t}\mathbf{v}_2, \ldots, e^{\lambda_n t}\mathbf{v}_n \right] = e^{\mathbf{A}t} \left[\mathbf{v}_1,\ \mathbf{v}_2, \ldots, \mathbf{v}_n \right]. \tag{8.2.7}$$

Example 8.2.4: Let us consider $\mathbf{A} = \begin{bmatrix} 1 & 2 \\ 0 & 3 \end{bmatrix}$, then the propagator

$$\mathbf{\Phi}(t) = e^{\mathbf{A}t} = \begin{bmatrix} e^t & e^{3t} - e^t \\ 0 & e^{3t} \end{bmatrix}$$

is an example of a fundamental matrix because its columns are linearly independent solutions of $\mathrm{d}\mathbf{x}/\mathrm{d}t = \mathbf{A}\mathbf{x}$. Substituting $\mathbf{x} = e^{\lambda t}\,\mathbf{v}$ into the latter equation, we obtain two linearly independent solutions $\mathbf{x}_1 = e^{\lambda_1 t}\,\mathbf{v}_1$ and $\mathbf{x}_2 = e^{\lambda_2 t}\,\mathbf{v}_2$, where $\mathbf{v}_1 = \langle 1, 0 \rangle^T$ and $\mathbf{v}_2 = \langle 1, 1 \rangle^T$ are eigenvectors corresponding to two eigenvalues $\lambda_1 = 1$ and $\lambda_2 = 3$, respectively. Therefore, another fundamental matrix for the same system will be

$$\mathbf{Y}(t) = \begin{bmatrix} e^t & e^{3t} \\ 0 & e^{3t} \end{bmatrix} = \left[\mathbf{v}_1 e^t, \mathbf{v}_2 e^{3t} \right] = \left(e^t \begin{bmatrix} 1 \\ 0 \end{bmatrix}, e^{3t} \begin{bmatrix} 1 \\ 1 \end{bmatrix} \right).$$

These two fundamental matrices, $\mathbf{Y}(t)$ and $e^{\mathbf{A}t}$, are related by

$$\mathbf{Y}(t) = e^{\mathbf{A}t} \begin{bmatrix} 1 & 1 \\ 0 & 1 \end{bmatrix} \qquad \text{or} \qquad \begin{bmatrix} e^t & e^{3t} \\ 0 & e^{3t} \end{bmatrix} = \begin{bmatrix} e^t & e^{3t} - e^t \\ 0 & e^{3t} \end{bmatrix} \begin{bmatrix} 1 & 1 \\ 0 & 1 \end{bmatrix}.$$

On the other hand,

$$\mathbf{Y}^{-1}(s) = -\frac{1}{e^{4s}} \begin{bmatrix} 0 & -e^s \\ -e^{3s} & e^{3s} - e^s \end{bmatrix} = \begin{bmatrix} 0 & -e^{-3s} \\ -e^{-s} & e^{-s} - e^{-3s} \end{bmatrix} = \begin{bmatrix} 1 & -1 \\ 0 & 1 \end{bmatrix} e^{-\mathbf{A}s}$$

and

$$\begin{aligned} \mathbf{\Phi}(t, s) = \mathbf{Y}(t)\mathbf{Y}^{-1}(s) &= \begin{bmatrix} e^{3t} - e^t & e^t \\ e^{3t} & 0 \end{bmatrix} \cdot \begin{bmatrix} 0 & -e^{-3s} \\ -e^{-s} & e^{-s} - e^{-3s} \end{bmatrix} \\ &= \begin{bmatrix} e^{t-s} & e^{3(t-s)} - e^{t-s} \\ 0 & e^{3(t-s)} \end{bmatrix} = e^{\mathbf{A}(t-s)} = \mathbf{\Phi}(t - s). \end{aligned}$$

Example 8.2.5: Suppose that two square $n\times n$ matrices $\mathbf{A}$ and $\mathbf{B}$ satisfy the anticommutative relation $\mathbf{AB} = -\mathbf{BA}$. Then $\mathbf{A}^k\mathbf{B} = (-1)^k\,\mathbf{BA}^k$ for any nonnegative integer k. By expanding the exponential matrix $e^{\mathbf{A}t}$ into the power series, it then follows that $\mathbf{B}\,e^{\mathbf{A}t} = e^{-\mathbf{A}t}\,\mathbf{B}$. Let $\mathbf{y}(t)$ be the solution to the initial value problem

$$\frac{\mathrm{d}\mathbf{y}}{\mathrm{d}t} = \mathbf{B}\,e^{2\mathbf{A}t}\,\mathbf{y}, \qquad \mathbf{y}(0) = \mathbf{y}_0. \tag{8.2.8}$$

Calculations show that

$$\begin{aligned}\frac{\mathrm{d}\left(e^{\mathbf{A}t}\,\mathbf{y}\right)}{\mathrm{d}t} &= e^{\mathbf{A}t}\left(e^{-2\mathbf{A}t}\,\mathbf{B}\,\mathbf{y}\right) + \mathbf{A}\,e^{\mathbf{A}t}\,\mathbf{y} = e^{-\mathbf{A}t}\,\mathbf{B}\,\mathbf{y} + \mathbf{A}\,e^{\mathbf{A}t}\,\mathbf{y}\\ &= \mathbf{B}\,e^{\mathbf{A}t}\,\mathbf{y} + \mathbf{A}\,e^{\mathbf{A}t}\,\mathbf{y} = (\mathbf{A}+\mathbf{B})\,e^{\mathbf{A}t}\,\mathbf{y}.\end{aligned}$$

Since $e^{\mathbf{A}t}\,\mathbf{y} = \mathbf{y}_0$ when $t=0$, and $e^{\mathbf{A}t}\,\mathbf{y} = e^{(\mathbf{A}+\mathbf{B})t}\,\mathbf{y}_0$, we obtain the desired solution of the IVP (8.2.8) to be

$$\mathbf{y}(t) = e^{-\mathbf{A}t}\,e^{(\mathbf{A}+\mathbf{B})t}\,\mathbf{y}_0.$$

8.2.1 Simple Real Eigenvalues

For a nondefective (see Definition 7.9 on page 387) constant matrix $\mathbf{A}$, every solution of the vector equation $\dot{y} = \mathbf{A}\,\mathbf{y}$ can be rewritten as the sum of exponential terms (Theorem 8.14 on page 443)

$$\mathbf{y}(t) = e^{\mathbf{A}t}\,\mathbf{c} = \boldsymbol{\xi}_1\,e^{\lambda_1 t} + \boldsymbol{\xi}_2\,e^{\lambda_2 t} + \cdots + \boldsymbol{\xi}_n\,e^{\lambda_n t}, \tag{8.2.9}$$

where $\boldsymbol{\xi}_k = c_k\,\mathbf{v}_k$ and $\mathbf{v}_k$ are eigenvectors corresponding to eigenvalues λ_k, and c_k are arbitrary constants, $k = 1, 2, \ldots, n$. In what follows, we consider only vector equations $\dot{\mathbf{y}} = \mathbf{A}\mathbf{y}$ with nonsingular matrices. In this case, the system has only one critical point—the origin. The objective of the following material in this section is to provide a qualitative description of solutions to the vector equation $\dot{\mathbf{y}} = \mathbf{A}\mathbf{y}$ in a neighborhood of the isolated critical point.

If all eigenvalues are real and negative, then exponentials in Eq. (8.2.9) decrease very fast as t increases. This means that the general solution approaches the origin when t is large. We say in this case that the origin is the **attractor**. Every solution approaches the origin as $t\to\infty$, hence it is asymptotically stable. If all eigenvalues are real and positive, then every solution moves away from the origin (except the origin itself to which corresponds $\mathbf{c} = \mathbf{0}$), it would be natural to call the origin a **repeller** and to refer to this critical point as **unstable**.

If some of the eigenvalues are real and positive and some are real and negative, then we cannot call the origin a repeller or attractor. Such a case deserves a special name—we call it a **saddle point**. For example, let $\lambda_1 > 0$ while all other eigenvalues are real and negative. Then, the solution $\mathbf{y}_1 = \boldsymbol{\xi}_1\,e^{\lambda_1 t}$ approaches infinity as $t\to\infty$. As t increases, all other linearly independent solutions $\mathbf{y}_k(t) = \boldsymbol{\xi}_k\,e^{\lambda_k t}$, $k = 2, 3, \ldots, n$, approach zero. This means that the solution (8.2.9) is asymptotic, as $t\to\infty$, to the line spanned on $\boldsymbol{\xi}_1 = c_1\mathbf{v}_1$ (unless $c_1 = 0$). The presence of solutions near the origin that move away from it would lead us to call the origin unstable.

Now we turn our attention to the two-dimensional case assuming that $\mathbf{A}$ in Eq. (8.2.1) is a 2×2 constant matrix and $\mathbf{y}(t)$ is a 2-column vector function. Then the system (8.2.1) is called a **planar system**, and we can exhibit qualitatively the behavior of solutions by sketching its **phase portrait**—trajectories with arrows indicating the direction in which the integral curve is traversed. Visualization of planar systems not only facilitates understanding of geometrical properties of solutions, but also helps in examination of higher-dimensional systems.

Suppose that the characteristic polynomial $\chi(\lambda) = \det(\lambda\mathbf{I} - \mathbf{A})$ of a 2×2 matrix $\mathbf{A}$ has two distinct real nulls, that is, $\chi(\lambda) = (\lambda-\lambda_1)(\lambda-\lambda_2)$, $\lambda_1 \ne \lambda_2$, and λ_1, λ_2 are real numbers. Then any solution of Eq. (8.2.1) is

$$\mathbf{y}(t) = c_1\mathbf{v}_1\,e^{\lambda_1 t} + c_2\mathbf{v}_2\,e^{\lambda_2 t} = \boldsymbol{\xi}_1\,e^{\lambda_1 t} + \boldsymbol{\xi}_2\,e^{\lambda_2 t}, \tag{8.2.10}$$

where $\mathbf{v}_1$, $\boldsymbol{\xi}_1 = c_1\mathbf{v}_1$ and $\mathbf{v}_2$, $\boldsymbol{\xi}_2 = c_2\mathbf{v}_2$ are linearly independent eigenvectors corresponding to eigenvalues λ_1 and λ_2, respectively; here c_1, c_2 are arbitrary constants.

Let L_1 and L_2 denote lines through the origin parallel to $\mathbf{v}_1$ and $\mathbf{v}_2$, respectively. A **half-line** of L_1 (or L_2) is the ray obtained by removing the original along with the remaining part of the line from L_1 (or L_2). Let λ_2 be the largest eigenvalue of the matrix $\mathbf{A}$. To emphasize it, we associate a double arrow with vector $\mathbf{v}_2$, see Fig. 8.5.

Letting $c_1 = 0$ in Eq. (8.2.10) yields $\mathbf{y}(t) = c_2\mathbf{v}_2\,e^{\lambda_2 t}$. If $c_2 \ne 0$, the streamline defined by this formula is half-line of L_2. The direction of motion is away from the origin if $\lambda_2 > 0$, toward the origin if $\lambda_2 < 0$. Similarly, the trajectory of $\mathbf{y}(t) = c_1\mathbf{v}_1\,e^{\lambda_1 t}$ with $c_1 \ne 0$ is a half-line of L_1.

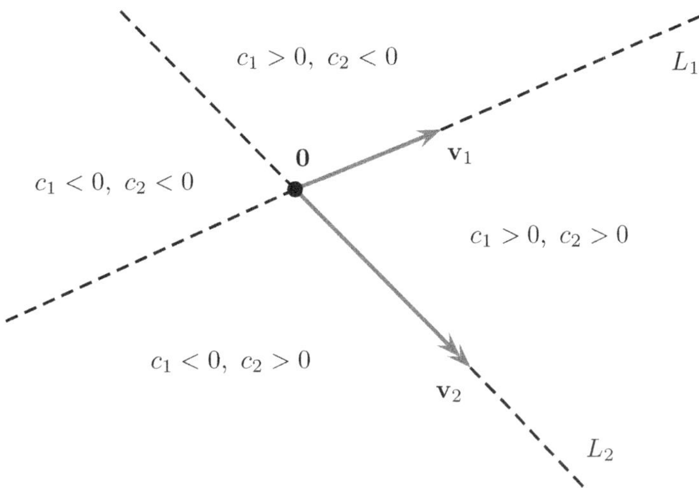

Figure 8.5: Open sectors bounded by L_1 and L_2.

Henceforth we assume that c_1 and c_2 in (8.2.10) are both nonzero. In this case, the solution curve cannot touch or cross L_1 or L_2 since every point on these lines belongs to the trajectory of a solution for which either $c_1 = 0$ or $c_2 = 0$. Hence, the streamline of (8.2.10) must lie entirely in one of the four open sectors bounded by L_1 and L_2, but should not contain any point from these lines. The position of trajectory is totally determined by the initial point $\mathbf{y}(0) = c_1\mathbf{v}_1 + c_2\mathbf{v}_2$. Therefore, the signs of c_1, c_2 determine which sector contains the solution curve.

Assuming $\lambda_2 > \lambda_1$, we factor the exponential term $e^{\lambda_2 t}$ to obtain

$$\mathbf{y}(t) = e^{\lambda_2 t}\left[\boldsymbol{\xi}_2 + \boldsymbol{\xi}_1\, e^{(\lambda_1-\lambda_2)t}\right].$$

The term $\boldsymbol{\xi}_1\, e^{(\lambda_1-\lambda_2)t}$ is negligible compared to $\boldsymbol{\xi}_2$ for t sufficiently large since $\lambda_1 - \lambda_2 < 0$. Therefore, the trajectory is asymptotically parallel to L_2 as $t \to \infty$. The shape and direction of traversal of the streamline depend upon the signs of eigenvalues. We are going to analyze these cases separately.

Suppose that λ_1 and λ_2 are both negative. Let, for example, $\lambda_1 < \lambda_2 < 0$. The solution moves toward the origin tangent to $\boldsymbol{\xi}_2 = c_2\mathbf{v}_2$ as $t \to +\infty$ and we have an asymptotically stable critical point $\mathbf{x} = \mathbf{0}$. We say in this case that the critical point is a **node** or a **nodal sink**.

If eigenvalues λ_1 and λ_2 are both positive, say $0 < \lambda_1 < \lambda_2$, then the solution $\mathbf{y}(t)$ moves away from the origin and we call the critical point $\mathbf{y} = \mathbf{0}$ a **nodal source** (unstable).

Now assume that the given diagonalizable matrix has two real eigenvalues of different sign, $\lambda_1 < 0 < \lambda_2$. Then the general solution is comprised by a linear combination of the exponential terms

$$\mathbf{y}(t) = c_1\, e^{\lambda_1 t}\,\mathbf{v}_1 + c_2\, e^{\lambda_2 t}\,\mathbf{v}_2,$$

where $\mathbf{v}_1$ is the eigenvector corresponding to the negative eigenvalue λ_1 and $\mathbf{v}_2$ is the eigenvector corresponding to the positive eigenvalue λ_2. Since one of the solutions, $c_1\, e^{\lambda_1 t}$, tends to zero as $t \to \infty$, while the other one, $c_2\, e^{\lambda_2 t}$, grows boundlessly when $c_2 \neq 0$, the origin is called a **saddle point**, and it is unstable. The lines through the origin along the eigenvectors separate the solution curves into distinct classes (see Fig. 8.8, page 447), and for this reason each line is referred to as a **separatrix**.

Example 8.2.6: (Nodal source, repeller) The matrix

$$\mathbf{A} = \begin{bmatrix} 0 & 2 \\ -1 & 3 \end{bmatrix}$$

has two positive eigenvalues $\lambda_1 = 1$ and $\lambda_2 = 2$ with corresponding eigenvectors

$$\lambda_1 = 1,\ \mathbf{v}_1 = \begin{bmatrix} 2 \\ 1 \end{bmatrix}, \qquad \lambda_2 = 2,\ \mathbf{v}_2 = \begin{bmatrix} 1 \\ 1 \end{bmatrix}.$$

Then the general solution of Eq. (8.2.1) with this 2×2 matrix $\mathbf{A}$ is

$$\mathbf{y}(t) = c_1\, e^{t}\,\mathbf{v}_1 + c_2\, e^{2t}\,\mathbf{v}_2$$

with two arbitrary constants c_1 and c_2. The corresponding direction field is presented in Fig. 8.6, where the dominating vector is indicated with double arrows.

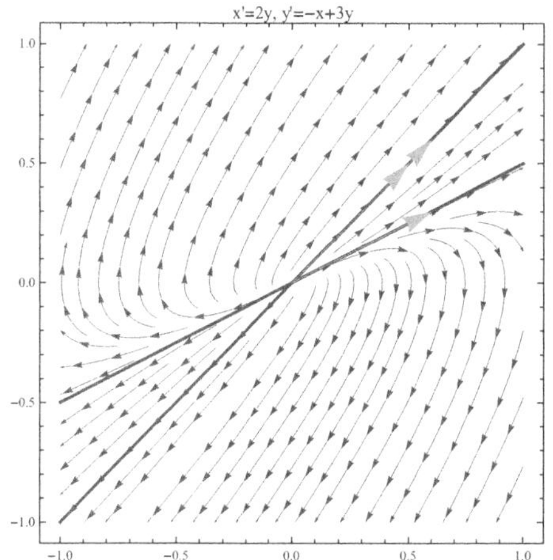

Figure 8.6: Example 8.2.6, repeller, plotted with *Mathematica*.

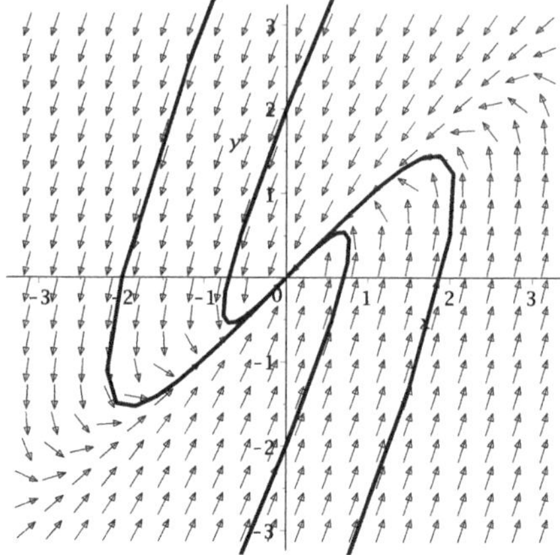

Figure 8.7: Example 8.2.7, attractor, plotted with *Maple*.

Example 8.2.7: (Nodal sink, attractor) The matrix

$$\mathbf{A} = \begin{bmatrix} 1 & -2 \\ 4 & -5 \end{bmatrix}$$

has two negative eigenvalues $\lambda_1 = -1$ and $\lambda_2 = -3$ with corresponding eigenvectors

$$\lambda_1 = -1, \ \mathbf{v}_1 = \begin{bmatrix} 1 \\ 1 \end{bmatrix}; \quad \lambda_2 = -3, \ \mathbf{v}_2 = \begin{bmatrix} 1 \\ 2 \end{bmatrix}.$$

Using the exponential matrix

$$e^{\mathbf{A}t} = \begin{bmatrix} 2\,e^{-t} - e^{-3t} & e^{-3t} - e^{-t} \\ 2\,e^{-t} - 2\,e^{-3t} & 2\,e^{-3t} - e^{-t} \end{bmatrix},$$

we obtain the general solution of $\dot{\mathbf{y}} = \mathbf{A}\mathbf{y}$:

$$\mathbf{y}(t) = c_1\,e^{-t}\,\mathbf{v}_1 + c_2\,e^{-3t}\,\mathbf{v}_2 = e^{\mathbf{A}t}\,[\,\mathbf{v}_1, \mathbf{v}_2\,]\,\mathbf{c}$$

with an arbitrary constant vector $\mathbf{c} = \langle c_1, c_2 \rangle^T$. Here $[\,\mathbf{v}_1, \mathbf{v}_2\,]$ is the square matrix of column eigenvectors $\mathbf{v}_1$ and $\mathbf{v}_2$. The phase portrait of the corresponding system is given in Fig. 8.7.

Example 8.2.8: (Saddle point) The matrix

$$\mathbf{A} = \begin{bmatrix} 0 & 2 \\ 1 & 1 \end{bmatrix}$$

has one negative eigenvalue $\lambda_1 = -1$ and one positive eigenvalue $\lambda_2 = 2$ with corresponding eigenvectors

$$\lambda_1 = -1, \ \mathbf{v}_1 = \begin{bmatrix} 2 \\ -1 \end{bmatrix}, \quad \lambda_2 = 2, \ \mathbf{v}_2 = \begin{bmatrix} 1 \\ 1 \end{bmatrix}.$$

The general solution of Eq. (8.2.1) with this 2×2 matrix $\mathbf{A}$ contains two arbitrary constants c_1 and c_2:

$$\mathbf{y}(t) = c_1\,e^{-t}\,\mathbf{v}_1 + c_2\,e^{2t}\,\mathbf{v}_2 = e^{\mathbf{A}t}\,[\,\mathbf{v}_1, \mathbf{v}_2\,]\,\mathbf{c},$$

where

$$e^{\mathbf{A}t} = \frac{1}{3}\begin{bmatrix} 2\,e^{-t} + e^{2t} & 2\,e^{2t} - 2\,e^{-t} \\ e^{2t} - e^{-t} & 2\,e^{2t} + e^{-t} \end{bmatrix}, \quad [\,\mathbf{v}_1, \mathbf{v}_2\,] = \begin{bmatrix} 2 & 1 \\ -1 & 1 \end{bmatrix}, \quad \mathbf{c} = \begin{bmatrix} c_1 \\ c_2 \end{bmatrix}.$$

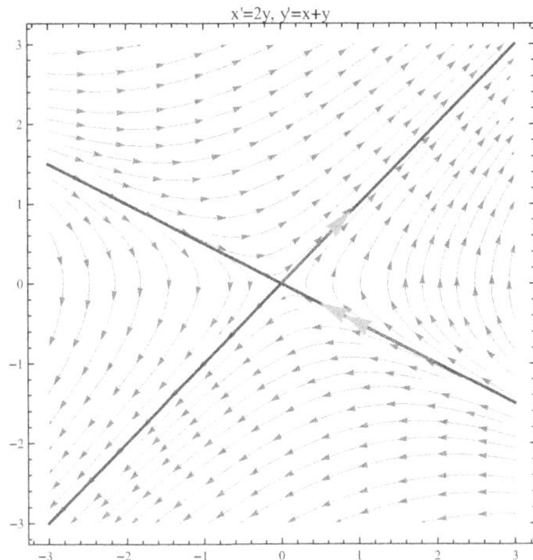

Figure 8.8: Example 8.2.8, saddle point, plotted with *Mathematica*.

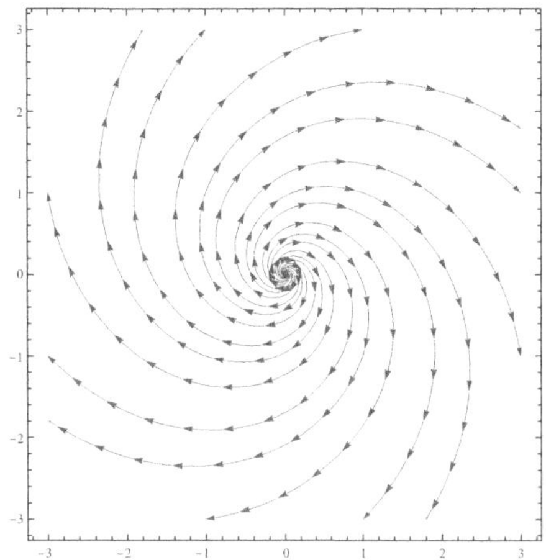

Figure 8.9: Example 8.2.9, spiral source, plotted with *Mathematica*.

8.2.2 Complex Eigenvalues

Suppose that a real-valued square matrix $\mathbf{A}$ of the system of differential equations $\dot{\mathbf{y}}(t) = \mathbf{A}\,\mathbf{y}(t)$ has a complex eigenvalue $\lambda = \alpha + \mathbf{j}\beta$ with $\mathbf{w} = \mathbf{u} + \mathbf{j}\mathbf{v}$ being its associated eigenvector. Here $\mathbf{j}$ is the unit vector in the positive vertical direction on the complex plane so that $\mathbf{j}^2 = -1$. Since all entries of the matrix $\mathbf{A}$ are real numbers, $\overline{\lambda} = \alpha - \mathbf{j}\beta$ is also an eigenvalue of $\mathbf{A}$ with associated eigenvector $\overline{\mathbf{w}} = \mathbf{u} - \mathbf{j}\mathbf{v}$. According to Theorem 8.14, page 443, the general solution contains the term

$$\mathbf{y}(t) = c_1\,\mathbf{w}\,e^{(\alpha+\mathbf{j}\beta)t} + c_2\,\overline{\mathbf{w}}\,e^{(\alpha-\mathbf{j}\beta)t},$$

where c_1 and c_2 are arbitrary (complex) numbers. To make $\mathbf{y}(t)$ real, we have to assume that $c_2 = \overline{c_1} = a - \mathbf{j}b$ is the complex conjugate of $c_1 = a + \mathbf{j}b$, in which a and b are some real constants. This means that $(\alpha + \mathbf{j}\beta)(\mathbf{u} + \mathbf{j}\mathbf{v}) = \mathbf{A}\,(\mathbf{u} + \mathbf{j}\mathbf{v})$, which leads (after separating real and imaginary parts) to two simultaneous vector equations:

$$\mathbf{A}\mathbf{u} = \alpha\mathbf{u} - \beta\mathbf{v} \qquad \text{and} \qquad \mathbf{A}\mathbf{v} = \alpha\mathbf{v} + \beta\mathbf{u}.$$

Using Euler's formula, $e^{\mathbf{j}\theta} = \cos\theta + \mathbf{j}\sin\theta$, we transform $\mathbf{y}(t)$ into a real-valued form:

$$\begin{aligned}
\mathbf{y}(t) &= c_1\,(\mathbf{u} + \mathbf{j}\mathbf{v})\,e^{(\alpha+\mathbf{j}\beta)t} + c_2\,(\mathbf{u} - \mathbf{j}\mathbf{v})\,e^{(\alpha-\mathbf{j}\beta)t} \\
&= (a + \mathbf{j}b)\,(\mathbf{u} + \mathbf{j}\mathbf{v})\,e^{(\alpha+\mathbf{j}\beta)t} + (a - \mathbf{j}b)\,(\mathbf{u} - \mathbf{j}\mathbf{v})\,e^{(\alpha-\mathbf{j}\beta)t} \\
&= e^{\alpha t}\left[(a + \mathbf{j}b)\,(\mathbf{u} + \mathbf{j}\mathbf{v})\,(\cos\beta t + \mathbf{j}\sin\beta t) + (a - \mathbf{j}b)\,(\mathbf{u} - \mathbf{j}\mathbf{v})\,(\cos\beta t - \mathbf{j}\sin\beta t)\right] \\
&= 2\,e^{\alpha t}\left[(a\mathbf{u} - b\mathbf{v})\cos\beta t - (b\mathbf{u} + a\mathbf{v})\sin\beta t\right].
\end{aligned}$$

If we denote $\boldsymbol{\xi}_1 = 2(a\mathbf{u} - b\mathbf{v})$ and $\boldsymbol{\xi}_2 = -2(b\mathbf{u} + a\mathbf{v})$, then $\mathbf{y}(t)$ has the form

$$\mathbf{y}(t) = e^{\alpha t}\left[\boldsymbol{\xi}_1\,\cos\beta t + \boldsymbol{\xi}_2\,\sin\beta t\right], \tag{8.2.11}$$

where $\boldsymbol{\xi}_1$ and $\boldsymbol{\xi}_2$ are real-valued vector solutions of the following system of algebraic equations:

$$\mathbf{A}\boldsymbol{\xi}_1 = \alpha\boldsymbol{\xi}_1 + \beta\boldsymbol{\xi}_2, \qquad \mathbf{A}\boldsymbol{\xi}_2 = \alpha\boldsymbol{\xi}_2 - \beta\boldsymbol{\xi}_1.$$

The trigonometric functions $\cos\beta t$ and $\sin\beta t$ are both periodic with period $2\pi/|\beta|$ and frequency $|\beta|/(2\pi)$, measured in hertz ($|\beta|$ is called the angular frequency). Consequently, the vector-valued function $e^{-\alpha t}\,\mathbf{y}(t)$ exhibits an oscillating behavior.

If a 2×2 matrix $\mathbf{A}$ of the system $\dot{\mathbf{y}} = \mathbf{A}\mathbf{y}$ has complex conjugate eigenvalues $\lambda = \alpha \pm \mathbf{j}\beta$, we refer the origin as a **spiral point**. If its real part is negative, $\alpha < 0$, the point is asymptotically stable because all solutions approach $\mathbf{0}$, and the point is called an attractor. If α is positive, all solutions leave the origin, and the critical point $\mathbf{0}$ is an unstable spiral point (repeller). When the real part is zero, $\text{Re}\lambda = \Re\lambda = \alpha = 0$, all solutions oscillate around the origin. We refer to this last case as a **center** (stable but not asymptotically stable).

Example 8.2.9: (Spiral source) Let us consider the initial value problem for the system of ordinary differential equations

$$\begin{cases} \dot{x} = x(t) + 2y(t), \\ \dot{y} = -2x(t) + y(t), \end{cases} \qquad x(0) = 1, \quad y(0) = 2,$$

with the corresponding matrix

$$\mathbf{A} = \begin{bmatrix} 1 & 2 \\ -2 & 1 \end{bmatrix}.$$

Eigenvalues of the matrix $\mathbf{A}$ are $\lambda_1 = 1+2\mathbf{j}$ and $\lambda_2 = \overline{\lambda_1} = 1-2\mathbf{j}$ because they annihilate its characteristic polynomial $\chi(\lambda) = \det(\lambda\mathbf{I} - \mathbf{A}) = (\lambda - \lambda_1)(\lambda - \lambda_2) = (\lambda - 1)^2 + 4$.

To find a function of the matrix $\mathbf{A}$ applying Sylvester's method, we first have to determine the auxiliary matrices

$$\begin{aligned} \mathbf{Z}_{(\lambda_1)}(\mathbf{A}) &= \frac{\mathbf{A} - \lambda_2}{\lambda_1 - \lambda_2} = \frac{\mathbf{A} - 1 + 2\mathbf{j}}{1 + 2\mathbf{j} - 1 + 2\mathbf{j}} \\ &= \frac{1}{4\mathbf{j}} \begin{bmatrix} 2\mathbf{j} & 2 \\ -2 & 2\mathbf{j} \end{bmatrix} = \begin{bmatrix} \frac{1}{2} & \frac{1}{2\mathbf{j}} \\ -\frac{1}{2\mathbf{j}} & \frac{1}{2} \end{bmatrix}, \\ \mathbf{Z}_{(\lambda_2)}(\mathbf{A}) &= \frac{\mathbf{A} - \lambda_1}{\lambda_2 - \lambda_1} = \frac{\mathbf{A} - 1 - 2\mathbf{j}}{1 - 2\mathbf{j} - 1 - 2\mathbf{j}} \\ &= -\frac{1}{4\mathbf{j}} \begin{bmatrix} -2\mathbf{j} & 2 \\ -2 & -2\mathbf{j} \end{bmatrix} = \begin{bmatrix} \frac{1}{2} & -\frac{1}{2\mathbf{j}} \\ \frac{1}{2\mathbf{j}} & \frac{1}{2} \end{bmatrix} = \overline{\mathbf{Z}}_{(\lambda_1)}(\mathbf{A}). \end{aligned}$$

Therefore, using Euler's formula, $e^{\mathbf{j}\theta} = \cos\theta + \mathbf{j}\theta$, we get the fundamental exponential matrix

$$e^{\mathbf{A}t} = e^{\lambda_1 t}\,\mathbf{Z}_{(\lambda_1)}(\mathbf{A}) + e^{\lambda_2 t}\,\mathbf{Z}_{(\lambda_2)}(\mathbf{A}) = e^t \begin{bmatrix} \cos 2t & \sin 2t \\ -\sin 2t & \cos 2t \end{bmatrix}.$$

Since the general solution $\mathbf{y}(t) = e^{\mathbf{A}t}\mathbf{c} = e^t \begin{bmatrix} \cos 2t & \sin 2t \\ -\sin 2t & \cos 2t \end{bmatrix} \begin{bmatrix} c_1 \\ c_2 \end{bmatrix}$ contains the exponential multiple e^t, the origin is a spiral source (unstable). Substituting $c_1 = 1$ and $c_2 = 2$, we obtain the required solution to the given initial value problem.

Sometimes it is convenient to use polar coordinates $x = r\cos\theta$, $y = r\sin\theta$, where $r = \sqrt{x^2 + y^2}$ and $\theta = \arctan(y/x)$. Then the given system of differential equations can be rewritten as

$$\dot{r} = \frac{\partial r}{\partial x}\,\dot{x} + \frac{\partial r}{\partial y}\,\dot{y} = \frac{1}{r}\,(x\dot{x} + y\dot{y}) = r, \qquad \dot{\theta} = \frac{x\dot{y} - y\dot{x}}{r^2} = -2.$$

Since these ordinary differential equations are decoupled, we solve them separately to obtain

$$r(t) = \sqrt{5}\,e^t, \qquad \theta(t) = -2\theta + \arctan(2).$$

Hence, trajectories spiral clockwise away from the origin as t increases (see Fig. 8.9 on page 447).

Example 8.2.10: (Center) Consider another system of ordinary differential equations

$$\frac{\mathrm{d}}{\mathrm{d}t} \begin{bmatrix} y_1(t) \\ y_2(t) \end{bmatrix} = \begin{bmatrix} 1 & 2 \\ -\frac{5}{2} & -1 \end{bmatrix} \begin{bmatrix} y_1 \\ y_2 \end{bmatrix}, \qquad \text{with} \quad \mathbf{A} = \begin{bmatrix} 1 & 2 \\ -\frac{5}{2} & -1 \end{bmatrix}.$$

The matrix $\mathbf{A}$ has two pure imaginary conjugate eigenvalues $\lambda = \pm 2\mathbf{j}$; its fundamental matrix is

$$e^{\mathbf{A}\,t} = \begin{bmatrix} \cos 2t + \frac{1}{2}\sin 2t & \sin 2t \\ -\frac{5}{4}\sin 2t & \cos 2t - \frac{1}{2}\sin 2t \end{bmatrix}.$$

Therefore, the origin is the center (stable but not asymptotically stable), see Fig. 8.10 on page 449.

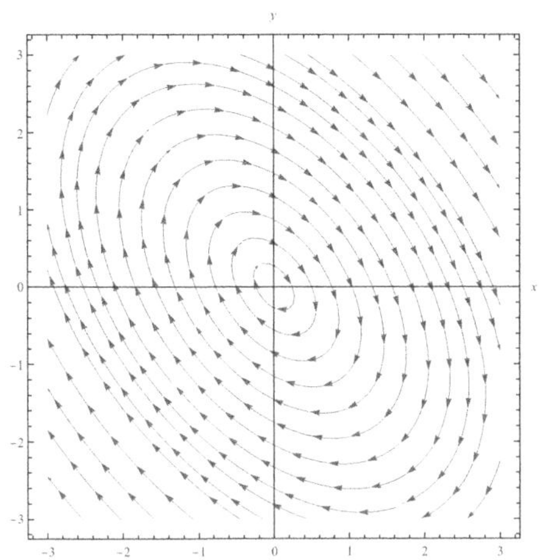

Figure 8.10: Example 8.2.10, center, plotted with *Mathematica.*

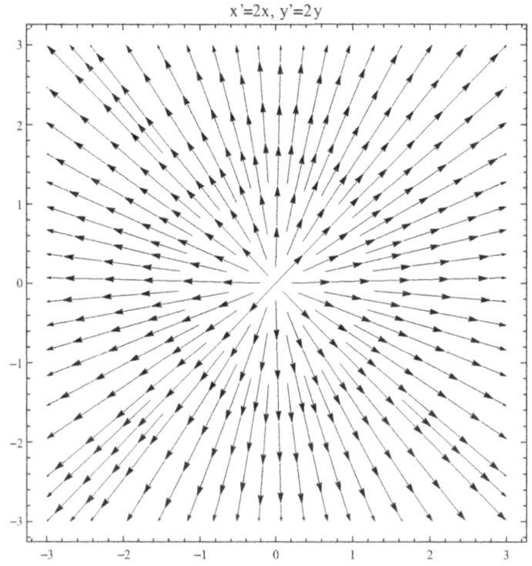

Figure 8.11: Example 8.2.11, proper unstable node, plotted with *Mathematica.*

8.2.3 Repeated Eigenvalues

Let us look more closely at the structure of the exponential matrix function $e^{\mathbf{A}\,t}$. According to the resolvent formula (see §7.6), the exponential matrix function is equal to the sum of residues over all of the eigenvalues of the square matrix $\mathbf{A}$:

$$e^{\mathbf{A}\,t} = \sum_{\lambda_k \in \sigma(\mathbf{A})} \operatorname*{Res}_{\lambda_k} e^{\lambda\,t}\,\mathbf{R}_\lambda(\mathbf{A}), \tag{8.2.12}$$

where $\sigma(\mathbf{A})$ is the set of all eigenvalues (also called the spectrum) of the $n \times n$ matrix $\mathbf{A}$, and $\mathbf{R}_\lambda(\mathbf{A}) = (\lambda\mathbf{I} - \mathbf{A})^{-1}$ is the resolvent of the matrix. Since the evaluation of the residue of a function may contain, at most, an $(n-1)$-th derivative with respect to λ, such a derivative, when applied to the product $g(\lambda)\,e^{\lambda\,t}$, gives

$$\frac{\mathrm{d}^{n-1}}{\mathrm{d}\lambda^{n-1}}\,g(\lambda)\,e^{\lambda\,t} = \sum_{k=0}^{n-1} \binom{n-1}{k} g^{(n-k-1)}\,\frac{\mathrm{d}^k}{\mathrm{d}\lambda^k}\,e^{\lambda\,t} = \sum_{k=0}^{n-1} \binom{n-1}{k} g^{(n-k-1)}\,t^k\,e^{\lambda\,t},$$

where $\binom{n}{k} = n!/k!/(n-k)!$ is the binomial coefficient. Therefore, if the minimal polynomial $\psi(\lambda)$ of an $n \times n$ matrix $\mathbf{A}$ contains a multiple $(\lambda - \lambda_0)^m$, then the residue at the point $\lambda = \lambda_0$ produces a polynomial in t of degree $m-1$ times the exponential term, $e^{\lambda_0\,t}$. In general, the exponential matrix $e^{\mathbf{A}\,t}$ is equal to the sum of exponential terms $e^{\lambda_j\,t}$, times polynomials in t of degree one less than the corresponding multiplicity of its minimal polynomial expansion.

Since the exponential function $e^{\lambda_j\,t}$ grows or decreases faster than any polynomial in t, the behavior of the terms from the exponential matrix $e^{\mathbf{A}\,t}$ that correspond to the multiple eigenvalue λ_j completely depends on the sign of the real part of λ_j. If the real part $\Re\,\lambda_j > 0$, the critical point is unstable because solutions containing the exponential term $e^{\lambda_j\,t} = e^{\Re\lambda_j t}\,[\cos(\Im\lambda_j t) + \mathbf{j}\,\sin(\Im\lambda_j t)]$ approach infinity. If $\Re\,\lambda_j < 0$, the corresponding terms containing $e^{\lambda_j\,t}$ die out. When singular matrix has eigenvalue $\lambda = 0$ of multiplicity m, its residue at this point turns into a polynomial in t of order $m-1$.

Suppose that a matrix has an eigenvalue λ^* of multiplicity $m > 1$. If its geometric multiplicity is equal to m, we have m linearly independent eigenvectors corresponding to λ^*, and the origin is called a **star** or **proper node**. If the geometric multiplicity is less than m, we call it defective and the origin is referred to as a **deficient** or **degenerate** or **improper node**. Its stability is determined by the sign of $\Re\,\lambda^*$, the real part of λ^*.

Example 8.2.11: (Proper node) Consider the differential equation

$$\dot{\mathbf{y}} = \mathbf{A}\,\mathbf{y}, \qquad \text{where } \mathbf{A} = \begin{bmatrix} 2 & 0 \\ 0 & 2 \end{bmatrix},$$

which actually consists of two uncoupled equations. The matrix $\mathbf{A}$ has double eigenvalue $\lambda = 2$ with two linearly independent eigenvectors: $\mathbf{v}_1 = \langle 1, 0\rangle^T$ and $\mathbf{v}_2 = \langle 0, 1\rangle^T$. Therefore, the origin is a proper node (or star), which is unstable (Fig. 8.11).

Example 8.2.12: (Degenerate node) The matrix

$$\mathbf{A} = \begin{bmatrix} 2 & 1 \\ 1 & 4 \end{bmatrix}$$

is defective because its characteristic polynomial $\chi(\lambda) = (\lambda - 3)^2$ has one double root $\lambda = 3$ to which corresponds only one eigenvector $\langle 1, -1 \rangle^T$. The exponential matrix is

$$e^{\mathbf{A}t} = e^{3t} \begin{bmatrix} 1-t & -t \\ t & 1+t \end{bmatrix},$$

so the general solution of the vector equation $\dot{\mathbf{y}} = \mathbf{A}\,\mathbf{y}$ becomes

$$\mathbf{y}(t) = \begin{bmatrix} y_1(t) \\ y_2(t) \end{bmatrix} = e^{\mathbf{A}t}\mathbf{c} = e^{3t} \begin{bmatrix} (1-t)c_1 - tc_2 \\ tc_1 + (1+t)c_2 \end{bmatrix},$$

where $\mathbf{c} = \langle c_1, c_2 \rangle^T$ is a column vector of arbitrary constants. As $t \to +\infty$, the solution blows up, so the origin is an unstable deficient node.

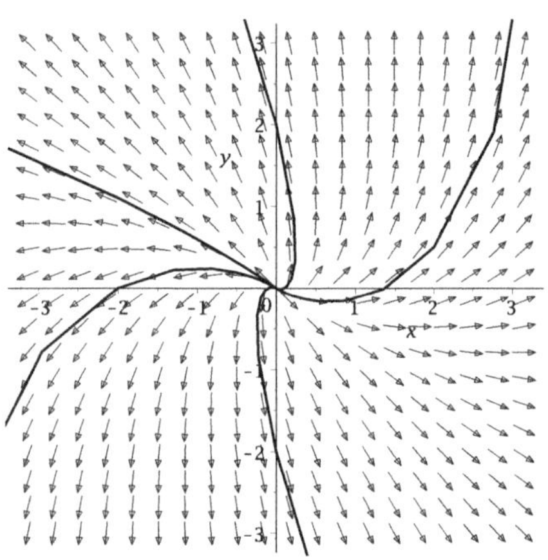

Figure 8.12: Example 8.2.12, degenerate unstable node, plotted with *Maple*.

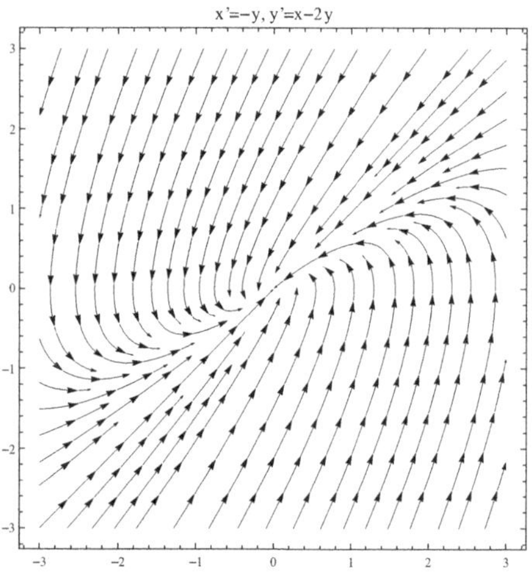

Figure 8.13: Example 8.2.13, degenerate stable node, plotted with *Mathematica*.

Example 8.2.13: (Degenerate node) The matrix

$$\mathbf{A} = \begin{bmatrix} 0 & -1 \\ 1 & -2 \end{bmatrix}$$

is defective since its characteristic polynomial $\chi(\lambda) = (\lambda + 1)^2$ has one double root $\lambda = -1$ with an eigenvector $\langle 1, 1 \rangle^T$. The exponential matrix is

$$e^{\mathbf{A}t} = e^{-t} \begin{bmatrix} 1+t & -t \\ t & 1-t \end{bmatrix},$$

so the general solution of the vector equation $\dot{\mathbf{y}} = \mathbf{A}\,\mathbf{y}$ becomes

$$\mathbf{y}(t) = \begin{bmatrix} x(t) \\ y(t) \end{bmatrix} = e^{\mathbf{A}t}\mathbf{c} = e^{-t} \begin{bmatrix} (1+t)c_1 & -tc_2 \\ tc_1 & (1-t)c_2 \end{bmatrix} = e^{-t}c_1 \begin{bmatrix} 1+t \\ -t \end{bmatrix} + e^{-t}c_2 \begin{bmatrix} -t \\ 1-t \end{bmatrix},$$

where $\mathbf{c} = \langle c_1, c_2 \rangle^T$ is a column vector of arbitrary constants. As $t \to +\infty$, the solution approaches zero, so the origin is an asymptotically stable deficient node.

8.2.4 Qualitative Analysis of Linear Systems

This subsection is an introduction to vector autonomous differential equations. We summarize the observations made previously and give a relatively simple classification of constant coefficient homogeneous vector equations with solutions of a specified kind. Treating the independent variable t as time, we discuss the behavior of the solutions as $t \to \infty$ based on an analysis of the corresponding matrix. In particular, we consider the system of differential equations

$$\dot{\mathbf{y}} \stackrel{\text{def}}{=} \mathrm{d}\mathbf{y}/\mathrm{d}t = \mathbf{A}\mathbf{y}, \tag{8.2.1}$$

where $\mathbf{y} = \mathbf{y}(t) = \langle y_1(t), y_2(t), \ldots, y_n(t)\rangle^T$ is an n-column vector of unknown functions ("T" stands for transposition), and $\mathbf{A}$ is an $n \times n$ square nonsingular matrix that does not depend on time t. Its general solution is known to be

$$\mathbf{y} = e^{\mathbf{A}\,t}\,\mathbf{c}, \tag{8.2.3}$$

where $\mathbf{c} = \langle c_1(t), c_2(t), \ldots, c_n(t)\rangle^T$ is an n-column vector of arbitrary constants. The exponential matrix $e^{\mathbf{A}\,t}$ may have a rather complicated form. Nevertheless, we know that each of its entries contains a polynomial of a degree less than or equal to $n-1$ times the exponential term: $e^{\lambda t}$, where λ is an eigenvalue of the matrix $\mathbf{A}$. This observation gives us a clue to its long-term behavior. When the matrix $\mathbf{A}$ is not deficient, the general solution of Eq. (8.2.1) is the sum (Theorem 8.14, page 443) of exponential terms

$$\mathbf{y}(t) = c_1\,\mathbf{v}_1\,e^{\lambda_1 t} + c_2\,\mathbf{v}_2\,e^{\lambda_2 t} + \cdots + c_n\,\mathbf{v}_n\,e^{\lambda_n t}, \tag{8.2.6}$$

where $\mathbf{v}_1$, $\mathbf{v}_2$, ..., $\mathbf{v}_n$ are linearly independent eigenvectors corresponding to eigenvalues λ_1, λ_2, ..., λ_n (some of them can be equal).

If all eigenvalues are real and negative, then exponentials in Eq. (8.2.6) decrease as t increases. This means that the general solution approaches the origin for a large t. We say in this case that the origin is the **attractor** or **sink**. Every solution approaches the origin as $t \to \infty$; hence, it is asymptotically stable. If all eigenvalues are real and positive, then every solution moves away from the origin (except the origin itself, which corresponds to $\mathbf{c} = \mathbf{0}$); we can call the origin a **repeller** or **source** and describe the critical point as being **unstable**.

If some of the eigenvalues are real and positive and some are real and negative, then we cannot call the origin a repeller or an attractor. Such cases deserve a special name—we call the origin a **saddle point**. For example, let $\lambda_1 > 0$ and all the other eigenvalues be real and negative. Then the solution $\mathbf{y}_1(t) = \boldsymbol{\xi}_1\,e^{\lambda_1 t}$ $(\boldsymbol{\xi}_1 = c_1\mathbf{v}_1)$ approaches infinity as $t \to \infty$ unless $\boldsymbol{\xi}_1 = \mathbf{0}$. As t increases, all other linearly independent solutions $\mathbf{y}_k(t) = \boldsymbol{\xi}_k\,e^{\lambda_k t}$, $k = 2, 3, \ldots, n$, approach zero. This means that the solution (8.2.6) is asymptotic to the line determined[75] by $\boldsymbol{\xi}_1$ as $t \to \infty$. The presence of solutions that move away suggests that the origin is unstable. As shown in §8.2.3, the existence of multiple eigenvalues does not effect the stability of the critical point unless $\Re\,\lambda = 0$.

When the characteristic polynomial $\chi(\lambda) = \det(\lambda\mathbf{I} - \mathbf{A})$ has a complex null $\lambda = \alpha + \mathbf{j}\beta$, its complex conjugate is also an eigenvalue. Since we only consider systems of equations with real-valued matrices, all complex eigenvalues appear in pairs with their complex conjugates. To the pair of complex roots of $\chi(\lambda) = 0$ of multiplicity m corresponds $2m$ real-valued solutions of the system (8.2.1) that have the multiple $e^{\alpha t}$ times polynomial in t of degree $m-1$ times a trigonometric function $\sin\beta t$ or $\cos\beta t$. Since an exponential function grows/decreases faster than any polynomial, the behavior of this solution is determined by the real part α of the eigenvalue $\lambda = \alpha \pm \mathbf{j}\beta$. If $\alpha > 0$, we have a repeller, if $\alpha < 0$, we get an attractor. The presence of trigonometric functions leads to oscillating behavior of solutions. If this occurs, then the origin is called the **spiral point** (or focus).

If pure imaginary eigenvalues $\lambda = \pm\mathbf{j}\beta$ are simple roots of the characteristic equation $\chi(\lambda) = 0$, the corresponding solution is stable, and the stationary point is called the **center**. However, if a pure imaginary eigenvalue is defective, then the solution is always unstable.

These observations illustrate the following stability result.

Theorem 8.15: Let $\mathbf{A}$ be a real invertible ($\det\mathbf{A} \neq 0$) square matrix. Then the linear vector equation $\dot{\mathbf{y}}(t) = \mathbf{A}\mathbf{y}(t)$ has the only one equilibrium point—the origin. This critical point is

1. asymptotically stable if all eigenvalues of $\mathbf{A}$ have a negative real part;
2. stable but not asymptotically stable if all eigenvalues of $\mathbf{A}$ are simple pure imaginary numbers;

[75] The line, which is determined by a vector $\boldsymbol{\xi}$, consists of all points $\mathbf{x} = c\boldsymbol{\xi}$, c is an arbitrary constant. So this line spans $\boldsymbol{\xi}$ and includes the origin.

3. neutrally stable if all eigenvalues have negative real parts but at least one is a pure imaginary of multiplicity 1;

4. unstable if all eigenvalues of defective matrix **A** are pure imaginary but at least one of them has multiplicity larger than 1;

5. unstable if at least one eigenvalue of **A** has a positive real part.

Now we turn our attention to the planar case, assuming that **A** is a 2×2 matrix and **y** is a 2-column vector in Eq. (8.2.1). The two-dimensional case facilitates the visualization of solution curves of the system. When $n = 2$, the system (8.2.1) is a **plane system**, and we can qualitatively observe the behavior of solutions by sketching **phase portraits**—trajectories with arrows indicating the direction in which the integral curve is traversed. Consider the following system of linear differential equations:

$$\begin{aligned} \dot{x} &= a\,x(t) + b\,y(t), \\ \dot{y} &= c\,x(t) + d\,y(t), \end{aligned} \qquad \text{or in vector form:} \qquad \frac{\mathrm{d}\,\mathbf{y}(t)}{\mathrm{d}t} = \mathbf{A}\,\mathbf{y}(t),$$

where the matrix $\mathbf{A} = \begin{bmatrix} a & b \\ c & d \end{bmatrix}$ of the system is assumed to be nonsingular (with $\det \mathbf{A} \neq 0$) and $\mathbf{y}(t) = \langle x(t), y(t) \rangle^T$. This system can be solved by the elimination method (see §6.4):

$$\frac{\mathrm{d}y}{\mathrm{d}x} = \frac{\dot{y}}{\dot{x}} = \frac{cx + dy}{ax + by} = \frac{a + dy/x}{a + by/x}.$$

The behavior of the solutions near the origin (which is the only critical point) depends on the nature of the eigenvalues λ_1 and λ_2 of the 2×2 matrix **A**:

$$\lambda_{1,2} = \frac{\operatorname{tr} \mathbf{A}}{2} \pm \frac{1}{2}\sqrt{(\operatorname{tr} \mathbf{A})^2 - 4 \det \mathbf{A}}, \tag{8.2.13}$$

where $\operatorname{tr} \mathbf{A} = a + d$ is the trace and $\det \mathbf{A} = ad - bc \neq 0$ is the determinant of the nonsingular matrix **A**. There are four kinds of stability for equilibrium solutions:

Critical Point	Stability	Conditions
Center	stable	$\operatorname{tr}\mathbf{A} = 0$, $\det\mathbf{A} > 0$
Sink	asymptotically stable	$\operatorname{tr}\mathbf{A} < 0$ and $\det\mathbf{A} > 0$
Source	unstable: all trajectories recede	$\operatorname{tr}\mathbf{A} > 0$ and $\det\mathbf{A} > 0$
Saddle point	unstable, but some solutions may approach the critical point	$\det\mathbf{A} < 0$

Further classification of critical points depends on how trajectories approach or recede from them. An equilibrium solution $\mathbf{y}^*$ is called a **node** if every trajectory approaches it or if every trajectory recedes from it, and these orbits do not reverse their directions in the neighborhood of $\mathbf{y}^*$. This means that every trajectory is tangent to a line through the critical point.

A critical point is a **proper node** or **star** if solution curves approach it or recede from it in all directions. A critical point is an **improper** or **degenerate node** if all trajectories approach or emanate from it in at most two directions. A node is called **deficient** if the orbits only approach it or recede from it in one direction.

An equilibrium solution is a **spiral point** if trajectories wind around the critical point as they approach it or recede from it. If solutions near an isolated critical point neither approach it nor recede from it, we call such an equilibrium solution a **center**.

Therefore, there are five types of critical points:

- a proper node (stable or unstable);
- an improper or degenerate node (stable or unstable);
- spiral point (stable or unstable);
- center (always stable);

- saddle point (always unstable).

The behavior of solutions near a critical point depends on eigenvalues of the corresponding matrix; therefore, we consider the following cases:

- eigenvalues are real and distinct of the same sign;
- real eigenvalues of opposite sign;
- equal eigenvalues;
- complex conjugate eigenvalues with a nonzero real part;
- pure imaginary eigenvalues.

For planar systems, all cases are summarized in the following table:

Eigenvalues	**Type of Critical Point**	**Stability**
$\lambda_1 > \lambda_2 > 0$	Nodal source (node)	Unstable
$\lambda_1 < \lambda_2 < 0$	Nodal sink (node)	Asymptotically stable
$\lambda_1 < 0 < \lambda_2$	Saddle point	Unstable
$\lambda_1 = \lambda_2 > 0$, independent eigenvectors	Proper node/star point	Unstable
$\lambda_1 = \lambda_2 < 0$, independent eigenvectors	Proper node/star point	Asymptotically stable
$\lambda_1 = \lambda_2 > 0$, missing eigenvector	Improper/degenerate node	Unstable
$\lambda_1 = \lambda_2 < 0$, missing eigenvector	Improper/degenerate node	Asymptotically stable
$\lambda = \alpha \pm \mathbf{j}\beta$, $\alpha > 0$	Spiral point	Unstable
$\lambda = \alpha \pm \mathbf{j}\beta$, $\alpha < 0$	Spiral point	Asymptotically stable
$\lambda = \pm\beta\mathbf{j}$	Center	Stable

Example 8.2.14: Let $\mathbf{A}$ be the singular matrix

$$\mathbf{A} = \begin{bmatrix} 1 & -7 & 3 \\ -1 & -1 & 1 \\ 4 & -4 & 0 \end{bmatrix}.$$

Its characteristic polynomial $\Delta(\lambda) = \det(\lambda\mathbf{I} - \mathbf{A}) = \lambda(\lambda^2 - 16)$ has a multiple λ, so to the eigenvalue $\lambda = 0$ corresponds the eigenvector $\mathbf{y}_0 = \langle 1, 1, 2\rangle^T$. Therefore, the origin is not an isolated critical point for the differential equation

$$\dot{\mathbf{y}}(t) = \mathbf{A}\,\mathbf{y}(t), \qquad \text{where } \mathbf{y} = \langle y_1, y_2, y_3\rangle^T.$$

This means that the equilibrium solutions are spanned on the eigenvector $\mathbf{y}_0$.

Example 8.2.15: Consider the vector differential equation

$$\dot{y} = \mathbf{A}\,\mathbf{y}, \qquad \text{with } \mathbf{A} = \begin{bmatrix} 3 & 2 & -3 \\ -6 & -3 & 8 \\ 2 & 1 & -4 \end{bmatrix},$$

where $\mathbf{y} = \langle y_1(t), y_2(t), y_3(t)\rangle^T$ is a 3D vector function to be determined. The matrix $\mathbf{A}$ of the equation has two pure imaginary eigenvalues $\lambda_{1,2} = \pm\mathbf{j}$ and one negative eigenvalue $\lambda_3 = -4$. The fundamental matrix is

$$e^{\mathbf{A}t} = \frac{1}{17}\begin{bmatrix} 25\cos t + 19\sin t & 3\cos t + 22\sin t & -19\cos t + 25\sin t \\ -16\cos t - 38\sin t & 11\cos t - 27\sin t & 38\cos t - 16\sin t \\ 8\cos t + 2\sin t & 3\cos t + 5\sin t & -2\cos t + 8\sin t \end{bmatrix}$$
$$+ \frac{1}{17}e^{-4t}\begin{bmatrix} -8 & -3 & 19 \\ 16 & 6 & -38 \\ -8 & -3 & 19 \end{bmatrix}.$$

Because the exponential multiple e^{-4t} tends to zero as $t \mapsto +\infty$, the general solution oscillates. Therefore, the origin is a neutrally stable critical point.

Problems

1. In each exercise (a) through (d), verify that the matrix-function written in vector form $\mathbf{Y}(t) = \left[e^{\lambda_1 t}\mathbf{v}_1, e^{\lambda_2 t}\mathbf{v}_2\right]$ is the fundamental matrix for the given vector differential equation $\dot{\mathbf{y}} = \mathbf{A}\mathbf{y}$, with specified constant diagonalizable matrix $\mathbf{A}$. Then find a constant nonsingular matrix $\mathbf{C}$ such that $\mathbf{Y}(t) = e^{\mathbf{A}t}\mathbf{C}$.

(a) $\mathbf{A} = \begin{bmatrix} 3 & 2 \\ 4 & 5 \end{bmatrix}$, $\qquad \mathbf{Y}(t) = \left[e^{7t}\begin{bmatrix} 1 \\ 2 \end{bmatrix}, \ e^{t}\begin{bmatrix} 1 \\ -1 \end{bmatrix}\right]$;

(b) $\mathbf{A} = \begin{bmatrix} 5 & -3 \\ 3 & -5 \end{bmatrix}$, $\qquad \mathbf{Y}(t) = \left[e^{4t}\begin{bmatrix} 3 \\ 1 \end{bmatrix}, \ e^{-4t}\begin{bmatrix} 1 \\ 3 \end{bmatrix}\right]$;

(c) $\mathbf{A} = \begin{bmatrix} -4 & 1 \\ 3 & -2 \end{bmatrix}$, $\qquad \mathbf{Y}(t) = \left[e^{-5t}\begin{bmatrix} -1 \\ 1 \end{bmatrix}, \ e^{-t}\begin{bmatrix} 1 \\ 3 \end{bmatrix}\right]$;

(d) $\mathbf{A} = \begin{bmatrix} 1 & -8 \\ 2 & 1 \end{bmatrix}$, $\qquad \mathbf{Y}(t) = e^{t}\left[\begin{bmatrix} 2\cos 4t \\ \sin 4t \end{bmatrix}, \ \begin{bmatrix} -2\sin 4t \\ \cos 4t \end{bmatrix}\right]$.

2. Compute the propagator matrix $e^{\mathbf{A}t}$ for each system $\dot{\mathbf{y}} = \mathbf{A}\,\mathbf{y}$ given in exercises (a) through (d).

(a) $\dot{x} = 6x - 6y, \ \dot{y} = 4x - 4y$;

(b) $\dot{x} = 11x - 15y, \ \dot{y} = 6x - 8y$;

(c) $\dot{x} = 9x + 2y, \ \dot{y} = 2x + 6y$;

(d) $\dot{x} = 9x - 8y, \ \dot{y} = 6x - 5y$.

3. In each of exercises (a) through (h):
   - Find the general solution of the given system of homogeneous equations $\dot{\mathbf{y}} = \mathbf{A}\,\mathbf{y}$ and describe the behavior of its solutions as $t \to +\infty$.
   - Draw a direction field and plot a few trajectories of the system.

(a) $\begin{bmatrix} 2 & 3 \\ 1 & 0 \end{bmatrix}$; (b) $\begin{bmatrix} 3 & 2 \\ -1 & 0 \end{bmatrix}$; (c) $\begin{bmatrix} -3 & 4 \\ \frac{1}{2} & -4 \end{bmatrix}$; (d) $\begin{bmatrix} 5 & 6 \\ -3 & -4 \end{bmatrix}$;

(e) $\begin{bmatrix} 7 & 6 \\ 2 & 6 \end{bmatrix}$; (f) $\begin{bmatrix} 13 & 4 \\ 4 & 7 \end{bmatrix}$; (g) $\begin{bmatrix} 6 & -7 \\ 1 & -2 \end{bmatrix}$; (h) $\begin{bmatrix} 4 & 2 \\ 3 & -1 \end{bmatrix}$.

4. For each of the following matrices in the homogeneous equation $\dot{\mathbf{y}} = \mathbf{B}\,\mathbf{y}$, classify the critical point $(0,0)$ as to type, and determine whether it is stable, asymptotically stable, or unstable. In the case of centers and spirals you are asked to determine the direction of rotation. Also sketch the phase portrait, which should show all special trajectories and a few generic trajectories. At each trajectory, the direction of motion should be indicated by an arrow.
   - In the case of centers, sketch a few closed trajectories with the right direction of rotation. For spirals, one generic trajectory is sufficient.
   - In the case of saddles or nodes, the sketch should include all half-line trajectories (corresponding to eigenvectors) and a generic trajectory in each of the four regions separated by the half-line trajectories. The half-line trajectories should be sketched correctly, that is, you have to compute eigenvalues as well as eigenvectors.
   - In the case of nodes you should also distinguish between fast (indicated by a double arrow, which corresponds to the largest eigenvalue) and slow (with a single arrow) motions.

$\mathbf{B}_1 = \begin{bmatrix} 13 & 4 \\ 4 & 7 \end{bmatrix}$; $\mathbf{B}_2 = \begin{bmatrix} 0.2 & 1 \\ -1 & 0.2 \end{bmatrix}$; $\mathbf{B}_3 = \begin{bmatrix} 3 & 1 \\ 2 & 2 \end{bmatrix}$; $\mathbf{B}_4 = \begin{bmatrix} 1 & 8 \\ 2 & 7 \end{bmatrix}$; $\mathbf{B}_5 = \begin{bmatrix} 4 & 3 \\ 1 & 2 \end{bmatrix}$;

$\mathbf{B}_6 = \begin{bmatrix} 2 & -5 \\ 1 & -2 \end{bmatrix}$; $\mathbf{B}_7 = \begin{bmatrix} 2 & -2.5 \\ 1.8 & -1 \end{bmatrix}$; $\mathbf{B}_8 = \begin{bmatrix} 4 & -10 \\ 2 & -4 \end{bmatrix}$; $\mathbf{B}_9 = \begin{bmatrix} 1 & 4 \\ -1 & 1 \end{bmatrix}$; $\mathbf{B}_{10} = \begin{bmatrix} 2 & 3 \\ -3 & -4 \end{bmatrix}$;

$\mathbf{B}_{11} = \begin{bmatrix} 2 & 3 \\ -3 & 8 \end{bmatrix}$; $\mathbf{B}_{12} = \begin{bmatrix} 2 & 1.5 \\ -1.5 & -1 \end{bmatrix}$; $\mathbf{B}_{13} = \begin{bmatrix} 2 & 5 \\ -3 & -6 \end{bmatrix}$; $\mathbf{B}_{14} = \begin{bmatrix} 2 & 6 \\ 3 & 5 \end{bmatrix}$; $\mathbf{B}_{15} = \begin{bmatrix} 2 & 5 \\ -4 & -2 \end{bmatrix}$.

5. Find the general solution of the homogeneous system of differential equations $\dot{\mathbf{y}} = \mathbf{A}\,\mathbf{y}$ for the given square matrix $\mathbf{A}$ and determine the stability or instability of the origin.

(a) $\begin{bmatrix} 5 & 6 \\ -5 & -8 \end{bmatrix}$; (b) $\begin{bmatrix} 3 & 2 \\ -1 & 0 \end{bmatrix}$; (c) $\begin{bmatrix} -1 & -4 \\ \frac{1}{2} & -4 \end{bmatrix}$; (d) $\begin{bmatrix} 5 & 6 \\ -3 & -4 \end{bmatrix}$;

(e) $\begin{bmatrix} 4 & 1 \\ -1 & 2 \end{bmatrix}$; (f) $\begin{bmatrix} 3 & 1 \\ -1 & 1 \end{bmatrix}$; (g) $\begin{bmatrix} -6 & 3 \\ 1 & -4 \end{bmatrix}$; (h) $\begin{bmatrix} 2 & 13 \\ -1 & -2 \end{bmatrix}$.

8.3 Variation of Parameters

Suppose we need to find a particular solution $\mathbf{x}_p(t)$ of the inhomogeneous linear system of differential equations with variable or constant coefficients

$$\dot{\mathbf{x}}(t) = \mathbf{P}(t)\,\mathbf{x} + \mathbf{f}(t), \tag{8.3.1}$$

where $\mathbf{P}(t) = [p_{ij}(t)]$ is a given $n \times n$ matrix with continuous coefficients $p_{ij}(t)$ $(i, j = 1, 2, \ldots, n)$, $\mathbf{f}(t)$ is a known n-column vector ($(n\times 1)$-matrix) of integrable functions, and a dot denotes the derivative with respect to t: $\dot{\mathbf{x}} \stackrel{\text{def}}{=} \mathrm{d}\mathbf{x}/\mathrm{d}t$. Here $\mathbf{x}(t) = \langle x_1, x_2, \ldots, x_n\rangle^T$ is the column vector of n unknown functions that are to be determined. Suppose that we know the fundamental matrix $\mathbf{X}(t)$ for the associated homogeneous system $\dot{\mathbf{x}}(t) = \mathbf{P}(t)\,\mathbf{x}$. This matrix is convenient to write as a collection of linearly independent column vectors

$$\mathbf{X}(t) = [\,\mathbf{x}_1(t),\, \mathbf{x}_2(t),\, \ldots,\, \mathbf{x}_n(t)\,], \qquad \det \mathbf{X}(t) \neq 0,$$

where each n-column $\mathbf{x}_k(t)$, $k = 1, 2, \ldots, n$, is a solution of the homogeneous equation:

$$\dot{\mathbf{x}}(t) = \mathbf{P}(t)\,\mathbf{x}(t). \tag{8.3.2}$$

From Theorem 8.5, page 434, the general solution of the homogeneous equation (8.3.2) is known to be

$$\mathbf{x}(t) = \mathbf{X}(t)\,\mathbf{c}$$

for an arbitrary constant column vector $\mathbf{c}$. The **variation of parameters method** (also sometimes referred to as Lagrange's method) calls to replace the constant vector $\mathbf{c}$ with a variable vector $\mathbf{u}(t)$, and seek a particular solution of Eq. (8.3.1) in the form

$$\mathbf{x}_p(t) = \mathbf{X}(t)\,\mathbf{u}(t).$$

According to the product rule, its derivative is

$$\dot{\mathbf{x}}_p(t) = \dot{\mathbf{X}}(t)\,\mathbf{u}(t) + \mathbf{X}(t)\,\dot{\mathbf{u}}(t).$$

Its substitution into Eq. (8.3.1) yields

$$\dot{\mathbf{X}}(t)\,\mathbf{u}(t) + \mathbf{X}(t)\,\dot{\mathbf{u}}(t) = \mathbf{P}(t)\mathbf{X}(t)\,\mathbf{u}(t) + \mathbf{f}(t).$$

Since $\dot{\mathbf{X}}(t) = \mathbf{P}(t)\mathbf{X}(t)$, we have

$$\mathbf{X}(t)\dot{\mathbf{u}}(t) = \mathbf{f}(t) \qquad \text{or} \qquad \dot{\mathbf{u}}(t) = \mathbf{X}^{-1}(t)\,\mathbf{f}(t).$$

Thus, if $\mathbf{u}(t)$ is the solution of the latter equation, namely,

$$\mathbf{u}(t) = \int \mathbf{X}^{-1}(t)\,\mathbf{f}(t)\,\mathrm{d}t + \mathbf{c},$$

where $\mathbf{c}$ is an arbitrary constant vector of integration, then a particular solution becomes

$$\mathbf{x}_p(t) = \mathbf{X}(t)\mathbf{u}(t) = \mathbf{X}(t)\int \mathbf{X}^{-1}(t)\,\mathbf{f}(t)\,\mathrm{d}t + \mathbf{X}(t)\,\mathbf{c}.$$

The latter term $\mathbf{X}(t)\,\mathbf{c}$ is the general solution of the complementary equation $\dot{\mathbf{x}}(t) = \mathbf{P}(t)\,\mathbf{x}$ while the former gives a particular solution of the nonhomogeneous equation (8.3.1). Therefore, we have proved the following statement.

Theorem 8.16: If $\mathbf{X}(t)$ is a fundamental matrix for the homogeneous system $\dot{\mathbf{x}} = \mathbf{P}(t)\,\mathbf{x}(t)$ on some interval where the square matrix-function $\mathbf{P}(t)$ is a continuous and column vector $\mathbf{f}(t)$ is integrable, then a particular solution of the inhomogeneous system of equations (8.3.1) is given by

$$\mathbf{x}_p(t) = \mathbf{X}(t)\int \mathbf{X}^{-1}(t)\,\mathbf{f}(t)\,\mathrm{d}t = \mathbf{X}(t)\int_{t_0}^{t} \mathbf{X}^{-1}(\tau)\,\mathbf{f}(\tau)\,\mathrm{d}\tau. \tag{8.3.3}$$

Corollary 8.9: If $\mathbf{X}(t)$ is a fundamental matrix for the homogeneous system $\dot{\mathbf{x}} = \mathbf{P}(t)\,\mathbf{x}(t)$ on some interval $|a,b|$, where $\mathbf{P}(t)$ and $\mathbf{f}(t)$ are continuous, then the initial value problem

$$\dot{\mathbf{x}} = \mathbf{P}(t)\mathbf{x}(t) + \mathbf{f}(t), \qquad \mathbf{x}(t_0) = \mathbf{x}_0 \quad (t_0 \in |a,b|),$$

has the unique solution:

$$\mathbf{x}(t) = \mathbf{X}(t)\mathbf{X}^{-1}(t_0)\,\mathbf{x}_0 + \mathbf{X}(t)\int_{t_0}^{t} \mathbf{X}^{-1}(\tau)\,\mathbf{f}(\tau)\,\mathrm{d}\tau.$$

Example 8.3.1: Find a particular solution of

$$\dot{\mathbf{x}}(t) = \mathbf{P}(t)\,\mathbf{x}(t) + \mathbf{f}(t), \qquad \text{where} \quad \mathbf{P}(t) = \begin{bmatrix} 1 & e^{-2t} \\ e^{2t} & 3 \end{bmatrix}, \quad \mathbf{f}(t) = \begin{bmatrix} 1 \\ e^{2t} \end{bmatrix},$$

given that

$$\mathbf{X}(t) = \begin{bmatrix} e^{2t} & -1 \\ e^{4t} & e^{2t} \end{bmatrix}$$

is a fundamental matrix for the complementary system $\dot{\mathbf{x}} = \mathbf{P}(t)\mathbf{x}$.

Solution. We seek a particular solution $\mathbf{x}_p(t)$ of the given nonhomogeneous vector differential equation in the form

$$\mathbf{x}_p(t) = \mathbf{X}(t)\mathbf{u}(t),$$

where $\mathbf{u}(t) = \langle u_1(t), u_2(t)\rangle^T$ is an unknown 2-vector of functions $u_1(t)$, $u_2(t)$ to be determined. Substituting $\mathbf{x}_p(t)$ into the given equation leads to

$$\mathbf{X}(t)\dot{\mathbf{u}}(t) = \mathbf{f}(t) \qquad \text{or} \qquad \dot{\mathbf{u}}(t) = \mathbf{X}^{-1}(t)\mathbf{f}(t).$$

The inverse of the fundamental matrix is

$$\mathbf{X}^{-1}(t) = \frac{1}{2\,e^{4t}}\begin{bmatrix} e^{2t} & 1 \\ -e^{4t} & e^{2t} \end{bmatrix} = \frac{1}{2}\begin{bmatrix} e^{-2t} & e^{-4t} \\ -1 & e^{-2t} \end{bmatrix}.$$

Therefore,

$$\mathbf{X}^{-1}(t)\mathbf{f}(t) = \frac{1}{2}\begin{bmatrix} e^{-2t} & e^{-4t} \\ -1 & e^{-2t} \end{bmatrix} \cdot \begin{bmatrix} 1 \\ e^{2t} \end{bmatrix} = \begin{bmatrix} e^{-2t} \\ 0 \end{bmatrix}.$$

By integrating and eliminating the constants of integration, we get

$$\mathbf{u}(t) = \int \begin{bmatrix} e^{-2t} \\ 0 \end{bmatrix} \mathrm{d}t = -\frac{1}{2}\begin{bmatrix} e^{-2t} \\ 0 \end{bmatrix}.$$

Hence, a particular solution becomes

$$\mathbf{x}_p(t) = \mathbf{X}(t)\mathbf{u}(t) = -\frac{1}{2}\begin{bmatrix} 1 \\ e^{2t} \end{bmatrix}.$$

Example 8.3.2: Find the general solution of $\dot{\mathbf{x}}(t) = \mathbf{P}(t)\mathbf{x}(t) + \mathbf{f}(t)$, where

$$\mathbf{P}(t) = \begin{bmatrix} 3 & e^{t} & e^{2t} \\ e^{-t} & 2 & e^{t} \\ e^{-2t} & e^{-t} & 1 \end{bmatrix}, \qquad \mathbf{f}(t) = \begin{bmatrix} e^{2t} \\ e^{t} \\ -2 \end{bmatrix},$$

given that

$$\mathbf{X}(t) = \begin{bmatrix} e^{5t} & e^{2t} & 0 \\ e^{4t} & 0 & e^{t} \\ e^{3t} & -1 & -1 \end{bmatrix}$$

is a fundamental matrix for the complementary system $\dot{\mathbf{x}} = \mathbf{P}(t)\mathbf{x}$.

Solution. The determinant of $\mathbf{X}(t)$ is the Wronskian

$$\det \mathbf{X}(t) = 3\,e^{6t} \neq 0.$$

Therefore, the inverse matrix exists everywhere and

$$\mathbf{X}^{-1}(t) = \frac{1}{3\,e^{6t}} \begin{bmatrix} e^t & e^{2t} & e^{3t} \\ 2\,e^{4t} & -e^{5t} & -e^{6t} \\ -e^{4t} & 2\,e^{5t} & -e^{6t} \end{bmatrix} = \frac{1}{3} \begin{bmatrix} e^{-5t} & e^{-4t} & e^{-3t} \\ 2\,e^{-2t} & -e^{-t} & -1 \\ -e^{-2t} & 2\,e^{-t} & -1 \end{bmatrix}.$$

We seek a particular solution in the form $\mathbf{x}_p(t) = \mathbf{X}(t)\mathbf{u}(t)$, where $\dot{\mathbf{u}} = \mathbf{X}^{-1}(t)\mathbf{f}(t)$. Integration yields

$$\begin{aligned} \mathbf{u} &= \int \mathbf{X}^{-1}(t)\mathbf{f}(t)\,\mathrm{d}t + \mathbf{c} = \frac{1}{3}\int \begin{bmatrix} e^t & e^{2t} & e^{3t} \\ 2\,e^{4t} & -e^{5t} & -e^{6t} \\ -e^{4t} & 2\,e^{5t} & -e^{6t} \end{bmatrix} \begin{bmatrix} e^{2t} \\ e^t \\ -2 \end{bmatrix} \mathrm{d}t + \begin{bmatrix} c_1 \\ c_2 \\ c_3 \end{bmatrix} \\ &= \frac{1}{3}\int \begin{bmatrix} 0 \\ 3 \\ 3 \end{bmatrix} \mathrm{d}t + \mathbf{c} = \begin{bmatrix} 0 \\ t \\ t \end{bmatrix} + \mathbf{c}, \end{aligned}$$

where $\mathbf{c} = \langle c_1, c_2, c_3 \rangle^T$ is an arbitrary constant column vector. Substituting this result into the solution form $\mathbf{x}(t) = \mathbf{X}(t)\mathbf{u}(t)$, we get the general solution

$$\mathbf{x}(t) = \mathbf{X}(t) \begin{bmatrix} 0 \\ t \\ t \end{bmatrix} + \mathbf{X}(t)\mathbf{c} = t \begin{bmatrix} e^{2t} \\ e^t \\ -2 \end{bmatrix} + \begin{bmatrix} c_1 e^{5t} + c_2\,e^{2t} \\ c_1\,e^{4t} + c_3\,e^t \\ c_1\,e^{3t} - c_2 - c_3 \end{bmatrix}.$$

8.3.1 Equations with Constant Coefficients

In this subsection, we will reconsider the driven vector differential equation (8.3.1) when the corresponding matrix has constant entries:

$$\dot{\mathbf{y}}(t) = \mathbf{A}\,\mathbf{y}(t) + \mathbf{f}(t), \qquad t \in |a, b|. \tag{8.3.4}$$

From Theorem 8.16, page 455, it follows

Corollary 8.10: If $\mathbf{A}$ is a constant square matrix, then the initial value problem

$$\dot{\mathbf{y}} = \mathbf{A}\mathbf{y}(t) + \mathbf{f}(t), \qquad \mathbf{y}(t_0) = \mathbf{y}_0$$

has the unique solution:

$$\mathbf{y}(t) = e^{\mathbf{A}(t-t_0)}\,\mathbf{y}_0 + \int_{t_0}^{t} e^{\mathbf{A}(t-\tau)}\,\mathbf{f}(\tau)\,\mathrm{d}\tau = e^{\mathbf{A}(t-t_0)}\,\mathbf{y}_0 + e^{\mathbf{A}t}\int_{t_0}^{t} e^{-\mathbf{A}\tau}\,\mathbf{f}(\tau)\,\mathrm{d}\tau. \tag{8.3.5}$$

The general solution of the nonhomogeneous vector equation (8.3.4) is the sum

$$\mathbf{y}(t) = \mathbf{y}_h(t) + \mathbf{y}_p(t),$$

where $\mathbf{y}_h(t) = e^{\mathbf{A}\,t}\,\mathbf{c}$, with arbitrary constant vector $\mathbf{c}$, is the general solution (which is usually referred to as the complementary function) of the corresponding homogeneous equation $\dot{\mathbf{y}} = \mathbf{A}\,\mathbf{y}$, and $\mathbf{y}_p(t)$ is a particular solution of the nonhomogeneous equation (8.3.4):

$$\mathbf{y}_p = \int_{t_0}^{t} e^{\mathbf{A}(t-\tau)}\,\mathbf{f}(\tau)\,\mathrm{d}\tau = e^{\mathbf{A}t}\int_{t_0}^{t} e^{-\mathbf{A}\tau}\,\mathbf{f}(\tau)\,\mathrm{d}\tau, \tag{8.3.6}$$

where t_0 is an initial point. We clarify the construction of the general solution with the following elaborative examples.

Example 8.3.3: Consider the following system of inhomogeneous constant coefficients differential equations $\dot{\mathbf{y}} = \mathbf{A}\mathbf{y} + \mathbf{f}$ with

$$\mathbf{A} = \begin{bmatrix} 1 & 1 \\ 4 & -2 \end{bmatrix} \qquad \text{and} \qquad \mathbf{f}(t) = \begin{bmatrix} e^t \\ e^{-t} \end{bmatrix}.$$

Since the given matrix $\mathbf{A}$ has two distinct real eigenvalues $\lambda_1 = -3$ and $\lambda_2 = 2$, we get the fundamental matrix (propagator)

$$\mathbf{\Phi}(t) \stackrel{\text{def}}{=} e^{\mathbf{A}t} = \frac{1}{5}\begin{bmatrix} 4\,e^{2t} + e^{-3t} & e^{2t} - e^{-3t} \\ 4\,e^{2t} - 4\,e^{-3t} & e^{2t} + 4\,e^{-3t} \end{bmatrix}.$$

Then we multiply $e^{\mathbf{A}(t-\tau)}$ by $\mathbf{f}(\tau)$ to obtain

$$\begin{aligned} e^{\mathbf{A}(t-\tau)}\,\mathbf{f}(\tau) &= \frac{1}{5}\begin{bmatrix} 4\,e^{2(t-\tau)} + e^{-3(t-\tau)} & e^{2(t-\tau)} - e^{-3(t-\tau)} \\ 4\,e^{2(t-\tau)} - 4\,e^{-3(t-\tau)} & e^{2(t-\tau)} + 4\,e^{-3(t-\tau)} \end{bmatrix} \begin{bmatrix} e^{\tau} \\ e^{-\tau} \end{bmatrix} \\ &= \frac{1}{5}\begin{bmatrix} 4\,e^{2t-\tau} + e^{-3t+4\tau} + e^{2t-3\tau} - e^{-3t+2\tau} \\ 4\,e^{2t-\tau} - 4\,e^{-3t+4\tau} + e^{2t-3\tau} + 4\,e^{-3t+2\tau} \end{bmatrix} \\ &= \frac{4}{5}\,e^{2t}\begin{bmatrix} e^{-\tau} \\ e^{-\tau} \end{bmatrix} + \frac{1}{5}\,e^{-3t}\begin{bmatrix} e^{4\tau} \\ -4\,e^{4\tau} \end{bmatrix} + \frac{1}{5}\,e^{2t}\begin{bmatrix} e^{-3\tau} \\ e^{-3\tau} \end{bmatrix} + \frac{1}{5}\,e^{-3t}\begin{bmatrix} -e^{2\tau} \\ 4\,e^{2\tau} \end{bmatrix}. \end{aligned}$$

Choosing the origin as the initial point when $t = 0$ and using antiderivatives of the four functions

$$a\int_0^t e^{-a\tau}\,\mathrm{d}\tau = -e^{-at} + 1, \quad b\int_0^t e^{b\tau}\,\mathrm{d}\tau = e^{bt} - 1, \quad a = 1, 3, \quad b = 4, 2,$$

we get

$$\begin{aligned} \int_0^t e^{\mathbf{A}(t-\tau)}\,\mathbf{f}(\tau)\,\mathrm{d}\tau &= \frac{4}{5}\,e^{2t}\left(1 - e^{-t}\right)\begin{bmatrix} 1 \\ 1 \end{bmatrix} + \frac{1}{5}\,e^{-3t}\begin{bmatrix} 1 \\ -4 \end{bmatrix}\frac{1}{4}\left(e^{4t} - 1\right) \\ &\quad + \frac{1}{5}\,e^{2t}\begin{bmatrix} 1 \\ 1 \end{bmatrix}\frac{1}{3}\left(1 - e^{-3t}\right) + \frac{1}{5}\,e^{-3t}\begin{bmatrix} -1 \\ 4 \end{bmatrix}\frac{1}{2}\left(e^{2t} - 1\right). \end{aligned}$$

Upon simplification, we obtain

$$\begin{aligned} \int_0^t e^{\mathbf{A}(t-\tau)}\,\mathbf{f}(\tau)\,\mathrm{d}\tau &= e^t\left(\frac{1}{20}\begin{bmatrix} 1 \\ -4 \end{bmatrix} - \frac{4}{5}\begin{bmatrix} 1 \\ 1 \end{bmatrix}\right) + e^{-t}\left(\frac{1}{10}\begin{bmatrix} -1 \\ 4 \end{bmatrix} - \frac{1}{15}\begin{bmatrix} 1 \\ 1 \end{bmatrix}\right) \\ &\quad + e^{2t}\begin{bmatrix} 1 \\ 1 \end{bmatrix}\left(\frac{4}{5} + \frac{1}{15}\right) + e^{-3t}\begin{bmatrix} 1 \\ -4 \end{bmatrix}\left(\frac{1}{10} - \frac{1}{20}\right). \end{aligned}$$

The last two terms can be dropped because they can be included in the complementary function $e^{\mathbf{A}\,t}\,\mathbf{c}$ with appropriate constant vector $\mathbf{c} = \langle c_1, c_2 \rangle^T$. Substituting this integral into Eq. (8.3.6), we obtain a particular solution

$$\mathbf{y}_p(t) = -\frac{1}{4}\,e^t\begin{bmatrix} 3 \\ 4 \end{bmatrix} + \frac{1}{6}\,e^{-t}\begin{bmatrix} -1 \\ 2 \end{bmatrix}.$$

Using *Maxima*, we can check the results:

```
load(linearalgebra)$
A: matrix([1,1],[4,-2]);
x0:  matrix([a],[b]); /* arbitrary constants */
ev(matrixexp(A,t).x0); /* solution of the homogeneous equation */
f(t):= matrix([exp(t)],[exp(-t)]);
integrate(matrixexp(A,t-s).f(s),s,0,t);
```

Example 8.3.4: Find a particular solution of the driven system

$$\dot{\mathbf{y}}(t) = \mathbf{A}\mathbf{y}(t) + \mathbf{f}(t), \quad \mathbf{A} = \begin{bmatrix} 11 & -4 \\ 25 & -9 \end{bmatrix}, \quad \mathbf{f}(t) = \begin{bmatrix} 1 + 2\,e^t \\ 1 + t \end{bmatrix}.$$

Solution. The general solution of the complementary system $\dot{\mathbf{y}} = \mathbf{A}\mathbf{y}$ is

$$\mathbf{y}_h(t) = e^{\mathbf{A}t}\mathbf{c},$$

where

$$e^{\mathbf{A}t} = e^t \begin{bmatrix} 1+10t & -4t \\ 25t & 1-10t \end{bmatrix} = e^t \begin{bmatrix} 1 & 0 \\ 0 & 1 \end{bmatrix} + t\,e^t \begin{bmatrix} 10 & -4 \\ 25 & -10 \end{bmatrix} \quad \text{and} \quad \mathbf{c} = \begin{bmatrix} c_1 \\ c_2 \end{bmatrix}$$

is a column vector of arbitrary constants. Therefore, the general solution of the homogeneous equation $\dot{\mathbf{y}} = \mathbf{A}\mathbf{y}$ becomes

$$\mathbf{y}_h(t) = e^{\mathbf{A}t}\mathbf{c} = e^t \begin{bmatrix} c_1 \\ c_2 \end{bmatrix} + t\,e^t \begin{bmatrix} 10c_1 - 4c_2 \\ 25c_1 - 10c_2 \end{bmatrix}.$$

We seek a particular solution $\mathbf{y}_p(t)$ of the given inhomogeneous vector differential equation in the form

$$\mathbf{y}_p(t) = e^{\mathbf{A}t}\mathbf{u}(t) = e^t \begin{bmatrix} 1+10t & -4t \\ 25t & 1-10t \end{bmatrix} \begin{bmatrix} u_1(t) \\ u_2(t) \end{bmatrix},$$

where $\dot{\mathbf{u}}(t) = e^{-\mathbf{A}t}\mathbf{f}(t)$. From Eq. (8.3.5), it follows that

$$\begin{aligned}
\mathbf{y}_p(t) &= \int_0^t e^{\mathbf{A}(t-\tau)}\,\mathbf{f}(\tau)\,\mathrm{d}\tau \\
&= \int_0^t \mathrm{d}\tau\, e^{t-\tau} \begin{bmatrix} 1+10(t-\tau) & -4\,(t-\tau) \\ 25\,(t-\tau) & 1-10(t-\tau) \end{bmatrix} \begin{bmatrix} 1+2\,e^{\tau} \\ 1+\tau \end{bmatrix} \\
&= \int_0^t \mathrm{d}\tau\, e^{t-\tau} \begin{bmatrix} (1+10t-10\tau)\,(1+2\,e^{\tau}) - 4(t-\tau)(1+\tau) \\ 1+5t\,(3+10\,e^{\tau}-2\tau) + 2\tau\,(5\tau-7-25\,e^{\tau}) \end{bmatrix} \\
&= \begin{bmatrix} e^t\left(3+4t+10t^2\right) - 3 - 4t \\ e^t\left(7+5t+25t^2\right) - 7 - 11t \end{bmatrix}.
\end{aligned}$$

To check our calculations, we differentiate $\mathbf{y}_p(t)$ to obtain

$$\dot{\mathbf{y}}_p(t) = e^t \begin{bmatrix} 7+24t+10t^2 \\ 12+55t+25t^2 \end{bmatrix} - \begin{bmatrix} 4 \\ 11 \end{bmatrix}.$$

On the other hand,

$$\begin{aligned}
\mathbf{A}\mathbf{y}_p &= \begin{bmatrix} 11 & -4 \\ 25 & -9 \end{bmatrix} \left(e^t \begin{bmatrix} 3+4t+10t^2 \\ 7+5t+25t^2 \end{bmatrix} - \begin{bmatrix} 3+4t \\ 7+11t \end{bmatrix} \right) \\
&= \begin{bmatrix} 5+24t+10t^2 \\ 12+55t+25t^2 \end{bmatrix} e^t - \begin{bmatrix} 5 \\ 12+t \end{bmatrix},
\end{aligned}$$

so

$$\mathbf{A}\mathbf{y} + \mathbf{f} = \begin{bmatrix} 5+24t+10t^2 \\ 12+55t+25t^2 \end{bmatrix} e^t - \begin{bmatrix} 5 \\ 12+t \end{bmatrix} + \begin{bmatrix} 1+2\,e^t \\ 1+t \end{bmatrix} = \dot{\mathbf{y}}_p(t).$$

Mathematica is helpful to verify the particular solution obtained:

```
A = {{11, -4}, {25, -9}}
f[t_] = {{1 + 2*E^t}, {1 + t}}
Integrate[MatrixExp[A*(t - s)].f[s], {s, 0, t}]
```

Example 8.3.5: Consider the nonhomogeneous equation

$$\frac{\mathrm{d}\mathbf{y}(t)}{\mathrm{d}t} = \mathbf{A}\mathbf{y}(t) + \mathbf{f}(t), \quad \mathbf{A} = \begin{bmatrix} 2 & -5 \\ 1 & -2 \end{bmatrix}, \quad \mathbf{f}(t) = \begin{bmatrix} 0 \\ \cos t \end{bmatrix}.$$

The characteristic polynomial $\chi(\lambda) = \det(\lambda\mathbf{I} - \mathbf{A})$ has two complex nulls (eigenvalues of the matrix $\mathbf{A}$) $\lambda_{1,2} = \pm\mathbf{j}$. The fundamental matrix is

$$e^{\mathbf{A}t} = \begin{bmatrix} \cos t + 2\sin t & -5\sin t \\ \sin t & \cos t - 2\sin t \end{bmatrix}.$$

Therefore, the general solution $\mathbf{y}(t)$ is the sum of a particular solution,

$$\mathbf{y}_p(t) = \int_0^t e^{\mathbf{A}(t-\tau)}\,\mathbf{f}(\tau)\,\mathrm{d}\tau = \int_0^t \mathrm{d}\tau \begin{bmatrix} -5\cos(\tau)\,\sin(t-\tau) \\ \cos(\tau)\,(\cos(t-\tau) - 2\sin(t-\tau)) \end{bmatrix}$$
$$= \frac{1}{2}\begin{bmatrix} -5t\,\sin t \\ t\,\cos t - 2t\,\sin t + \sin t \end{bmatrix},$$

and the general solution $\mathbf{y}_h(t)$ of the associated homogeneous equation

$$\mathbf{y}_h(t) = e^{\mathbf{A}t}\,\mathbf{c} = c_1 \begin{bmatrix} \cos t + 2\sin t \\ \sin t \end{bmatrix} + c_2 \begin{bmatrix} -5\sin t \\ \cos t - 2\sin t \end{bmatrix},$$

where $\mathbf{c} = \langle c_1, c_2\rangle^T$ is a column vector of two arbitrary constants c_1 and c_2. The following *Maple* commands were used to check the calculations.

```
with(LinearAlgebra):
A := Matrix(2, 2, [2, -5, 1, -2])
funct := t-> <0, cos(t)>
u := MatrixExponential(A, t-s).funct(s)
simplify(map(int, u, s = 0 ..  t))
```

Problems

1. In each of the exercises, (a) through (h), find the general solution of the nonhomogeneous planar system of equations $\dot{\mathbf{x}}(t) = \mathbf{P}(t)\,\mathbf{x}(t) + \mathbf{f}(t)$, given that $\mathbf{X}(t)$ is a fundamental 2×2 matrix for the complementary system.

   (a) $\mathbf{P}(t) = \frac{1}{t}\begin{bmatrix} 2 & t \\ 0 & 2 \end{bmatrix}$, $\mathbf{f} = \begin{bmatrix} t^2 \\ 1 \end{bmatrix}$, $\mathbf{X}(t) = \begin{bmatrix} t^3 & t^2 \\ t^2 & 0 \end{bmatrix}$;

   (b) $\mathbf{P}(t) = \begin{bmatrix} -\tan 2t & \sec 2t \\ \sec 2t & -\tan 2t \end{bmatrix}$, $\mathbf{f} = \begin{bmatrix} \sin 2t \\ 1 \end{bmatrix}$, $\mathbf{X}(t) = \begin{bmatrix} \sin t & \cos t \\ \cos t & \sin t \end{bmatrix}$;

   (c) $\mathbf{P}(t) = \begin{bmatrix} 1 - e^{-2t} & e^{-2t} \\ -e^{-2t} & 1 + e^{-2t} \end{bmatrix}$, $\mathbf{f} = \begin{bmatrix} e^{2t} \\ 1 \end{bmatrix}$, $\mathbf{X}(t) = \begin{bmatrix} e^t & \cosh t \\ e^t & -\sinh t \end{bmatrix}$;

   (d) $\mathbf{P}(t) = \frac{1}{t^2}\begin{bmatrix} 0 & t^2 \\ -18 & 8t \end{bmatrix}$, $\mathbf{f} = \begin{bmatrix} 1 \\ t^2 \end{bmatrix}$, $\mathbf{X}(t) = \begin{bmatrix} t^3 & t^6 \\ 3t^2 & 6t^5 \end{bmatrix}$;

   (e) $\mathbf{P}(t) = \frac{1}{t}\begin{bmatrix} 1 & 2t \\ -2t & 1 \end{bmatrix}$, $\mathbf{f} = \begin{bmatrix} 2t \\ 4t^2 \end{bmatrix}$, $\mathbf{X}(t) = \begin{bmatrix} t\,\cos 2t & -t\,\sin 2t \\ t\,\sin 2t & t\,\cos 2t \end{bmatrix}$;

   (f) $\mathbf{P}(t) = \frac{1}{2}\begin{bmatrix} \sec^2 2t\,(2+\cos 4t) & -\sec^2 2t\,\cos 4t \\ -3\csc^2 2t\,\cos 4t & -2\csc 4t\,(2+\cos 4t) \end{bmatrix}$, $\mathbf{f} = \begin{bmatrix} 8 \\ 24 \end{bmatrix}$,
   $\mathbf{X}(t) = \begin{bmatrix} \sec 2t & -\sin 2t \\ 3\csc 2t & \cos 2t \end{bmatrix}$;

   (g) $\mathbf{P}(t) = \begin{bmatrix} 1 & 0 \\ 0 & 1 + t^{-1} \end{bmatrix}$, $\mathbf{f} = t\begin{bmatrix} 1 \\ -1 \end{bmatrix}$, $\mathbf{X}(t) = \begin{bmatrix} e^t & -e^t \\ t\,e^t & t\,e^t \end{bmatrix}$;

   (h) $\mathbf{P}(t) = \frac{1}{t}\begin{bmatrix} 3 & -2 \\ 1 & 0 \end{bmatrix}$, $\mathbf{f} = t^2\begin{bmatrix} 1 \\ 2 \end{bmatrix}$, $\mathbf{X}(t) = t\begin{bmatrix} 2 & 2t \\ 2 & t \end{bmatrix}$.

2. In each of the exercises, (a) through (f), find a particular solution of the nonhomogeneous system of equations $\dot{\mathbf{x}}(t) = \mathbf{P}(t)\mathbf{x}(t) + \mathbf{f}$, given that $\mathbf{X}(t)$ is a fundamental 3×3 matrix for the complementary system.

   (a)
$$\mathbf{P} = \frac{1}{2}\begin{bmatrix} 3 & e^{-t} & -1 \\ 2\,e^t & 4 & -2\,e^t \\ 1 & -e^{-t} & 1 \end{bmatrix}, \qquad \mathbf{X}(t) = \begin{bmatrix} e^t & 0 & e^{2t} \\ 0 & e^t & e^{3t} \\ e^t & 1 & 0 \end{bmatrix}, \qquad \mathbf{f} = \begin{bmatrix} 2\,e^{2t} \\ e^{-3t} \\ -e^{-t} \end{bmatrix};$$

   (b)
$$\mathbf{P} = \frac{1}{2\,\sinh t}\begin{bmatrix} 2\,e^t - e^{-t} & e^{4t} + e^{2t} + 2\,e^{-2t} - 4 & -e^{3t} \\ 0 & 0 & 0 \\ -e^{-3t} & -8\,e^t\,\sinh^3 t & e^t - 2\,e^{-t} \end{bmatrix},$$
$$\mathbf{X}(t) = \begin{bmatrix} e^t & e^{-t} & e^{2t} \\ 0 & 1 & 0 \\ e^{-t} & e^t & e^{-2t} \end{bmatrix}, \qquad \mathbf{f} = \begin{bmatrix} 1 \\ e^t \\ e^t \end{bmatrix};$$

(c)

$$\mathbf{P} = \frac{1}{1+2t^3}\begin{bmatrix} 6t^2 + \frac{1}{t} & -8t^4 & 8t^2 \\ \frac{1}{t^3} - \frac{1}{2t^2} & -\frac{2}{t} - 2t^2 & -2 \\ 0 & 0 & 0 \end{bmatrix},$$

$$\mathbf{X}(t) = \begin{bmatrix} 2t & 0 & 4t^3 \\ 1/t & 1/t^2 & 0 \\ 0 & 1 & 1 \end{bmatrix}, \quad \mathbf{f} = \begin{bmatrix} 0 & 4\,t \\ 4\,t^3 & \end{bmatrix};$$

(d)

$$\mathbf{P}(t) = \frac{1}{t}\begin{bmatrix} 1 & 2t & 0 \\ -t & 1 & t \\ 0 & -2t & 1 \end{bmatrix}, \qquad \mathbf{X}(t) = t\begin{bmatrix} 1 & \cos 2t & \sin 2t \\ 0 & -\sin 2t & \cos 2t \\ 1 & -\cos 2t & -\sin 2t \end{bmatrix}, \quad \mathbf{f} = 2t\begin{bmatrix} 1 \\ 1 \\ 1 \end{bmatrix};$$

(e)

$$\mathbf{P}(t) = \begin{bmatrix} 6 - 1/t & e^{-2t}(6t-1) - 8 & e^{-2t}(1-6t) \\ 4 & -6 - 1/t - 4t\,e^{-2t} & -4\,e^{-2t} \\ 4 & 4t\,e^{-2t} - 8 & 2 - 1/t - 4t\,e^{-2t} \end{bmatrix},$$

$$\mathbf{X}(t) = \frac{1}{t}\begin{bmatrix} 2\,e^{2t} & -e^{-2t} & 2t \\ e^{2t} & -e^{-2t} & 0 \\ e^{2t} & -e^{-2t} & 2\,e^{2t} \end{bmatrix}, \quad \mathbf{f} = t\begin{bmatrix} 3 \\ 2 \\ 2 \end{bmatrix};$$

(f)

$$\mathbf{P}(t) = \begin{bmatrix} \frac{1}{t} - 6 & 3 + \frac{1}{t} & \frac{3}{2t} \\ -12 - \frac{2}{t} & 6 + \frac{4}{t} & \frac{3}{t} \\ 12 & -6 & 0 \end{bmatrix}, \qquad \mathbf{X}(t) = \begin{bmatrix} t^3 & t^2 & 1 \\ 2t^3 & t^2 & -1 \\ 0 & 2t^3 & 2 \end{bmatrix}, \quad \mathbf{f} = 3\begin{bmatrix} t^2 \\ t \\ 0 \end{bmatrix}.$$

3. In exercises (a) through (h), use the method of *variation of parameters* to find the general solution to the vector equation $\dot{\mathbf{y}} = \mathbf{A}\mathbf{y} + \mathbf{f}$ with a constant 2×2 matrix $\mathbf{A}$.

(a) $\begin{bmatrix} -4 & 9 \\ -5 & 2 \end{bmatrix}$, $\mathbf{f} = \begin{bmatrix} 12\,e^{-t} \\ 0 \end{bmatrix}$;

(b) $\begin{bmatrix} -7 & 6 \\ -12 & 5 \end{bmatrix}$, $\mathbf{f} = \begin{bmatrix} 6\,e^{-t} \\ 37 \end{bmatrix}$;

(c) $\begin{bmatrix} -7 & 10 \\ -10 & 9 \end{bmatrix}$, $\mathbf{f} = \begin{bmatrix} 18\,e^{t} \\ 37 \end{bmatrix}$;

(d) $\begin{bmatrix} -14 & 39 \\ -6 & 16 \end{bmatrix}$, $\mathbf{f} = \begin{bmatrix} 78 \sinh t \\ 6 \cosh t \end{bmatrix}$;

(e) $\begin{bmatrix} -7 & 4 \\ -5 & 1 \end{bmatrix}$, $\mathbf{f} = e^{3t}\begin{bmatrix} 10 \\ 5 \end{bmatrix}$;

(f) $\begin{bmatrix} 7 & -4 \\ 4 & -1 \end{bmatrix}$, $\mathbf{f} = e^{-3t}\begin{bmatrix} 4 \\ -2 \end{bmatrix}$;

(g) $\begin{bmatrix} -6 & 4 \\ -5 & 6 \end{bmatrix}$, $\mathbf{f} = e^{-2t}\begin{bmatrix} 4 \\ 8 \end{bmatrix}$;

(h) $\begin{bmatrix} 8 & 13 \\ 3 & -2 \end{bmatrix}$, $\mathbf{f} = \begin{bmatrix} 60\,e^{5t} \\ 132\,e^{-11t} \end{bmatrix}$.

4. In exercises (a) through (d), use the method of *variation of parameters* to find the general solution to the vector equation $\dot{\mathbf{y}}(t) = \mathbf{A}\mathbf{y}(t) + \mathbf{f}(t)$ with a constant 3×3 matrix $\mathbf{A}$.

(a) $\begin{bmatrix} 2 & 4 & -2 \\ 4 & 2 & -2 \\ -1 & 3 & 1 \end{bmatrix}$, $\mathbf{f} = \begin{bmatrix} -2 \sinh t \\ 10 \cosh t \\ 5 \end{bmatrix}$;

(b) $\begin{bmatrix} 2 & 6 & -2 \\ 6 & 2 & -2 \\ -1 & 6 & 1 \end{bmatrix}$, $\mathbf{f} = \begin{bmatrix} 50\,e^{t} \\ 21\,e^{-t} \\ 9 \end{bmatrix}$;

(c) $\begin{bmatrix} -2 & -2 & 4 \\ -2 & 1 & 2 \\ -4 & -2 & 6 \end{bmatrix}$, $\mathbf{f} = \begin{bmatrix} 0 \\ 0 \\ e^{2t} \end{bmatrix}$;

(d) $\begin{bmatrix} 3 & -2 & 3 \\ 1 & -1 & 2 \\ -2 & 2 & -2 \end{bmatrix}$, $\mathbf{f} = \begin{bmatrix} 0 \\ 2\,e^{-t} \\ 0 \end{bmatrix}$.

5. In exercises (a) through (d), use the method of *variation of parameters* to find a particular solution to the vector equation $\dot{\mathbf{y}}(t) = \mathbf{A}\mathbf{y}(t) + \mathbf{f}(t)$ with a constant 2×2 matrix $\mathbf{A}$ that satisfies the initial conditions $\mathbf{y}(0) = \langle 1, -1 \rangle^T$.

(a) $\begin{bmatrix} 7 & 1 \\ -4 & 3 \end{bmatrix}$, $\mathbf{f} = \begin{bmatrix} -6\,e^{t} - 1 \\ 4\,e^{t} - 3 \end{bmatrix}$;

(b) $\begin{bmatrix} 3 & -2 \\ 9 & -3 \end{bmatrix}$, $\mathbf{f} = \begin{bmatrix} 24 \sin t \\ 12 \cos t \end{bmatrix}$;

(c) $\begin{bmatrix} 7 & -4 \\ 3 & 14 \end{bmatrix}$, $\mathbf{f} = \begin{bmatrix} 10\,e^{t} \\ 6\,e^{2t} \end{bmatrix}$;

(d) $\begin{bmatrix} -7 & 4 \\ -5 & 2 \end{bmatrix}$, $\mathbf{f} = \begin{bmatrix} 6\,e^{3t} \\ 6\,e^{2t} \end{bmatrix}$.

8.4 Method of Undetermined Coefficients

In this section, we present the technique applicable to nonhomogeneous constant coefficient differential equations when the forcing functions are of a special form (compare with §4.7.2). Namely, we consider the inhomogeneous vector linear differential equation

$$\frac{\mathrm{d}\mathbf{y}}{\mathrm{d}t} = \mathbf{A}\,\mathbf{y}(t) + \mathbf{f}(t), \qquad t \in (a, b), \tag{8.4.1}$$

with a constant square matrix $\mathbf{A}$ and the nonhomogeneous term $\mathbf{f}(t)$, also called the driving term or forcing term, to be of a special form: it is assumed that $\mathbf{f}(t)$ is a solution of some undriven constant coefficient system of differential equations. In other words, the forcing term $\mathbf{f}(t)$ in Eq. (8.1.7) is a linear combination (with constant vector coefficients) of products of polynomials, exponential functions, and sines and cosines because only in this case $\dot{\mathbf{f}}(t) = \mathbf{M}\mathbf{f}(t)$ for some constant square matrix $\mathbf{M}$. The main idea of the method of undetermined coefficients is essentially the same as for a single linear differential equation when we make an intelligent guess about the general form of a particular solution. Namely, the method considers the following two cases:

- **The forcing vector function $\mathbf{f}(t)$ is not a solution of the complementary equation $\dot{\mathbf{y}}(t) = \mathbf{A}\,\mathbf{y}(t)$.** Then a particular solution has the same form as the driving function $\mathbf{f}(t)$.
- **The input term $\mathbf{f}(t)$ is a solution of the complementary equation $\dot{\mathbf{y}}(t) = \mathbf{A}\,\mathbf{y}(t)$.** Then a particular solution has the same form as the driven function $\mathbf{f}(t)$ multiplied by a polynomial in t of a degree equal to the multiplicity of the corresponding eigenvalue.

To understand the method of undetermined coefficients better, we recommend the reader go over the following examples; however, before doing this, we must first make some observations. It is convenient to introduce the derivative operator $\mathtt{D}$ (where $\mathtt{D}g = \mathrm{d}g/\mathrm{d}t = \dot{g}$ for any differentiable function g) and rewrite the given vector equation (8.4.1) in operator form:

$$(\mathtt{D}\mathbf{I} - \mathbf{A})\,\mathbf{y}(t) = \mathbf{f}(t),$$

where $\mathbf{I}$ is the identity matrix. If we choose its solution as a power function $\mathbf{y}(t) = \mathbf{a}t^p$ with some constant column vector $\mathbf{a}$, then

$$(\mathtt{D}\mathbf{I} - \mathbf{A})\,\mathbf{a}t^p = \mathbf{a}p\,t^{p-1} - \mathbf{A}\mathbf{a}t^p.$$

The right-hand side is always a polynomial in t of degree p unless $\mathbf{A}\mathbf{a} = \mathbf{0}$, that is, $\mathbf{a}$ belongs to the kernel of the matrix $\mathbf{A}$ (consult §7.2.1). So, if the matrix in Eq. (8.4.1) is not singular ($\det \mathbf{A} \ne 0$) and $\mathbf{y}(t)$ is a polynomial in t, then $(\mathtt{D}\mathbf{I} - \mathbf{A})\,\mathbf{y}$ is a polynomial in t of the same degree as the vector $\mathbf{y}(t)$.

Example 8.4.1: Using the method of undetermined coefficients, solve the system of differential equations

$$x' = x + 2y + t, \qquad y' = 2x + y - t.$$

Solution. This system can be rewritten in a vector form:

$$\dot{\mathbf{y}} \overset{\text{def}}{=} \frac{\mathrm{d}\mathbf{y}(t)}{\mathrm{d}t} = \mathbf{A}\,\mathbf{y} + \mathbf{f}, \qquad \text{where} \quad \mathbf{A} = \begin{bmatrix} 1 & 2 \\ 2 & 1 \end{bmatrix}, \quad \mathbf{f}(t) = t\begin{bmatrix} 1 \\ -1 \end{bmatrix}.$$

Since the eigenvalues of the matrix $\mathbf{A}$ are $\lambda_1 = -1$ and $\lambda_2 = 3$, and the control number (consult §4.7.2) of the column vector $\mathbf{f}$ is 0, we are looking for a particular solution as a linear function:

$$\mathbf{y}_p(t) = \mathbf{a} + t\mathbf{b},$$

where $\mathbf{a}$ and $\mathbf{b}$ are some constant column vectors to be determined. Substitution into the given system of equations leads to

$$\dot{\mathbf{y}} = \mathbf{b} = \mathbf{A}\mathbf{a} + t\mathbf{A}\mathbf{b} + \mathbf{f}(t).$$

Equating coefficients of like powers of t, we obtain

$$\mathbf{b} = \mathbf{A}\mathbf{a} \qquad \text{and} \qquad \mathbf{A}\mathbf{b} + \mathbf{f}/t = \mathbf{0}.$$

So

$$\mathbf{b} = -\mathbf{A}^{-1}\mathbf{f}/t = -\frac{1}{-3}\begin{bmatrix} -1 & 2 \\ 2 & -1 \end{bmatrix}\begin{bmatrix} 1 \\ -1 \end{bmatrix} = \begin{bmatrix} 1 \\ -1 \end{bmatrix}$$

and

$$\mathbf{a} = \mathbf{A}^{-1}\mathbf{b} = \frac{1}{-3}\begin{bmatrix} -1 & 2 \\ 2 & -1 \end{bmatrix}\begin{bmatrix} 1 \\ -1 \end{bmatrix} = \begin{bmatrix} -1 \\ 1 \end{bmatrix}.$$

Therefore, a particular solution becomes

$$\mathbf{y}_p(t) = \begin{bmatrix} -1 \\ 1 \end{bmatrix} + t\begin{bmatrix} 1 \\ -1 \end{bmatrix}.$$

Now we solve the given vector equation using the variation of parameters method. First, we determine the exponential matrix using Sylvester's method:

$$e^{\mathbf{A}t} = e^{-t}\,\mathbf{Z}_{(-1)} + e^{3t}\,\mathbf{Z}_{(3)}, \quad \mathbf{Z}_{(-1)} = \frac{\mathbf{A}-3}{-1-3} = \frac{1}{2}\begin{bmatrix} 1 & -1 \\ -1 & 1 \end{bmatrix}, \quad \mathbf{Z}_{(3)} = \frac{1}{2}\begin{bmatrix} 1 & 1 \\ 1 & 1 \end{bmatrix}.$$

A particular solution of this system of differential equations is

$$\mathbf{y}_p(t) = e^{\mathbf{A}t}\mathbf{u}(t), \quad \text{where } \mathbf{u}(t) = \int e^{-\mathbf{A}t}\mathbf{f}(t)\,\mathrm{d}t = \begin{bmatrix} 1 \\ -1 \end{bmatrix}\int te^{t}\,\mathrm{d}t = \begin{bmatrix} 1 \\ -1 \end{bmatrix} e^{t}(t-1),$$

because $\mathbf{Z}_{(3)}\mathbf{f}(t) \equiv \mathbf{0}$. Therefore,

$$\mathbf{y}_p(t) = (t-1)\frac{1}{2}\begin{bmatrix} 1 & -1 \\ -1 & 1 \end{bmatrix}\begin{bmatrix} 1 \\ -1 \end{bmatrix} + (t-1)\,e^{4t}\frac{1}{2}\begin{bmatrix} 1 & 1 \\ 1 & 1 \end{bmatrix}\begin{bmatrix} 1 \\ -1 \end{bmatrix} = (t-1)\begin{bmatrix} 1 \\ -1 \end{bmatrix}.$$

Example 8.4.2: We consider a problem of finding a particular solution to the nonhomogeneous vector equation

$$\dot{\mathbf{y}}(t) = \mathbf{A}\mathbf{y}(t) + \mathbf{f}(t), \quad \mathbf{A} = \begin{bmatrix} 1 & 4 \\ 3 & 2 \end{bmatrix}, \quad \mathbf{f}(t) = 7\begin{bmatrix} -5 + 7\,e^{-2t} \\ 10 + 10t \end{bmatrix}.$$

Solution. We break $\mathbf{f}(t)$ into the sum of a polynomial (having a control number $\sigma = 0$) and the exponential function (with control number $\sigma = -2$):

$$\mathbf{f}(t) = 7\begin{bmatrix} -5 \\ 10 \end{bmatrix} + 7\begin{bmatrix} 0 \\ 10 \end{bmatrix} t + 7\begin{bmatrix} 7 \\ 0 \end{bmatrix} e^{-2t}.$$

Since the eigenvalues of the matrix $\mathbf{A}$ are $\lambda_1 = 5$ and $\lambda_2 = -2$, the second eigenvalue matches the control number $\sigma = -2$ of the exponential term. Therefore, we seek a particular solution $\mathbf{y}_p(t)$ of the given nonhomogeneous vector differential equation in a similar form as the forcing function $\mathbf{f}(t)$ and add an additional term to incorporate this match:

$$\mathbf{y}_p(t) = \mathbf{a} + \mathbf{b}t + \mathbf{c}\,e^{-2t} + \mathbf{d}t\,e^{-2t},$$

where $\mathbf{a}$, $\mathbf{b}$, $\mathbf{c}$, and $\mathbf{d}$ are some vectors to be determined later. Differentiation yields

$$\dot{\mathbf{y}}_p(t) = \mathbf{b} - 2\mathbf{c}\,e^{-2t} + \mathbf{d}\,e^{-2t} - 2\mathbf{d}t\,e^{-2t} = \mathbf{b} + (\mathbf{d} - 2\mathbf{c})\;e^{-2t} - 2\mathbf{d}t\,e^{-2t}.$$

We substitute the assumed solution into the given inhomogeneous vector differential equation and get

$$\mathbf{b} + (\mathbf{d} - 2\mathbf{c})\;e^{-2t} - 2\mathbf{d}t\,e^{-2t} = \mathbf{A}\mathbf{a} + \mathbf{A}\mathbf{b}t + \mathbf{A}\mathbf{c}\,e^{-2t} + \mathbf{A}\mathbf{d}t\,e^{-2t} + \mathbf{f}(t).$$

By collecting similar terms, we obtain the following algebraic equations for $\mathbf{a}$, $\mathbf{b}$, $\mathbf{c}$, and $\mathbf{d}$:

$$\mathbf{b} = \mathbf{A}\mathbf{a} + \begin{bmatrix} -35 \\ 70 \end{bmatrix}; \tag{8.4.2}$$

$$\mathbf{0} = \mathbf{A}\mathbf{b} + \begin{bmatrix} 0 \\ 70 \end{bmatrix}; \tag{8.4.3}$$

$$\mathbf{d} - 2\mathbf{c} = \mathbf{A}\mathbf{c} + \begin{bmatrix} 49 \\ 0 \end{bmatrix}; \tag{8.4.4}$$

$$-2\mathbf{d} = \mathbf{A}\mathbf{d}. \tag{8.4.5}$$

From Eq. (8.4.3), we find the vector $\mathbf{b}$:

$$\mathbf{b} = -\mathbf{A}^{-1}\begin{bmatrix} 0 \\ 70 \end{bmatrix} = 7\begin{bmatrix} 2 & -4 \\ -3 & 1 \end{bmatrix}\begin{bmatrix} 0 \\ 1 \end{bmatrix} = 7\begin{bmatrix} -4 \\ 1 \end{bmatrix} = \begin{bmatrix} -28 \\ 7 \end{bmatrix}$$

because

$$\mathbf{A}^{-1} = -\frac{1}{10}\begin{bmatrix} 2 & -4 \\ -3 & 1 \end{bmatrix} = \frac{1}{10}\begin{bmatrix} -2 & 4 \\ 3 & -1 \end{bmatrix}.$$

Solving Eq. (8.4.2), we obtain

$$\mathbf{a} = \mathbf{A}^{-1}\left(\mathbf{b} - \begin{bmatrix} -35 \\ 70 \end{bmatrix}\right) = -\frac{7}{10}\begin{bmatrix} 2 & -4 \\ -3 & 1 \end{bmatrix}\begin{bmatrix} 1 \\ -9 \end{bmatrix} = -\frac{7}{10}\begin{bmatrix} 38 \\ -12 \end{bmatrix} = \begin{bmatrix} -26.6 \\ 8.4 \end{bmatrix}.$$

From Eq. (8.4.5), it follows that $\mathbf{d}$ is an eigenvector of the matrix $\mathbf{A}$ corresponding to the eigenvalue $\lambda = -2$. Therefore $\mathbf{d} = \alpha\langle 4, -3\rangle^T$, where α is any nonzero constant. From Eq. (8.4.4), we obtain

$$(\mathbf{A} + 2\mathbf{I})\,\mathbf{c} = \mathbf{d} - \begin{bmatrix} 49 \\ 0 \end{bmatrix} = \alpha\begin{bmatrix} 4 \\ -3 \end{bmatrix} - \begin{bmatrix} 49 \\ 0 \end{bmatrix}.$$

This system of algebraic equations can be rewritten in the form

$$\begin{bmatrix} 3 & 4 \\ 3 & 4 \end{bmatrix}\mathbf{c} = \begin{bmatrix} 4\alpha - 49 \\ -3\alpha \end{bmatrix}. \tag{8.4.6}$$

Since the matrix $\mathbf{B} = \mathbf{A} + 2\mathbf{I} = \begin{bmatrix} 3 & 4 \\ 3 & 4 \end{bmatrix}$ is singular, the algebraic equation (8.4.6) has a nontrivial solution if and only if the right-hand side vector $\langle 4\alpha - 49, -3\alpha\rangle^T$ is orthogonal to solutions of the adjoint problem:

$$\mathbf{B}^T\mathbf{z} = \mathbf{0} \qquad \text{or} \qquad \begin{bmatrix} 3 & 3 \\ 4 & 4 \end{bmatrix}\mathbf{z} = \begin{bmatrix} 0 \\ 0 \end{bmatrix}.$$

Hence, $\mathbf{z} = \langle z_1, -z_1\rangle^T = z_1\langle 1, -1\rangle^T$ is the general solution of $\mathbf{B}^T\mathbf{z} = \mathbf{0}$ with arbitrary constant z_1, and the vector $\langle 4\alpha - 49, -3\alpha\rangle^T$ must be orthogonal to the vector $\langle 1, -1\rangle^T$:

$$\langle 4\alpha - 49, -3\alpha\rangle \perp \langle 1, -1\rangle \qquad \Longleftrightarrow \qquad 4\alpha - 49 + 3\alpha = 0.$$

The latter defines the value of α to be $\alpha = 7$, which finally identifies the vectors $\mathbf{d} = 7\langle 4, -3\rangle^T = \langle 28, -21\rangle^T$ and

$$\mathbf{c} = \begin{bmatrix} -7 \\ 0 \end{bmatrix} + c_2\begin{bmatrix} -4 \\ 3 \end{bmatrix},$$

where c_2 is an arbitrary constant. Note that the vector $c_2\langle -4, 3\rangle^T$ is an eigenvector of the matrix $\mathbf{A}$ corresponding to the eigenvalue $\lambda = -2$. Therefore, this vector can be included in the general solution of the homogeneous equation $\dot{\mathbf{y}}(t) = \mathbf{A}\mathbf{y}(t)$. Finally, we get the general solution:

$$\mathbf{y}(t) = \begin{bmatrix} -26.6 \\ 8.4 \end{bmatrix} + 7\begin{bmatrix} -4 \\ 1 \end{bmatrix}t + \begin{bmatrix} -7 \\ 0 \end{bmatrix}e^{-2t} + 7\begin{bmatrix} 4 \\ -3 \end{bmatrix}t\,e^{-2t} + c_1\begin{bmatrix} 1 \\ 1 \end{bmatrix}e^{5t} + c_2\begin{bmatrix} -4 \\ 3 \end{bmatrix}e^{-2t}.$$

Problems

1. In exercises (a) through (d), apply the technique of *undetermined coefficients* to find a particular solution of the nonhomogeneous system of equations $\dot{\mathbf{y}}(t) = \mathbf{A}\mathbf{y}(t) + \mathbf{f}(t)$, given the constant 2×2 matrix $\mathbf{A}$ and the forcing vector function $\mathbf{f}(t)$.

   (a) $\begin{bmatrix} 1 & 5 \\ 19 & -13 \end{bmatrix}$, $\mathbf{f} = \begin{bmatrix} 10\sinh t \\ 24\sinh t \end{bmatrix}$;

   (b) $\begin{bmatrix} 9 & -3 \\ -1 & 11 \end{bmatrix}$, $\mathbf{f} = \begin{bmatrix} -6\,t \\ 10\,t \end{bmatrix}$;

   (c) $\begin{bmatrix} 10 & -9 \\ 4 & -2 \end{bmatrix}$, $\mathbf{f} = \begin{bmatrix} 9\,e^t \\ 2\,e^{4t} \end{bmatrix}$;

   (d) $\begin{bmatrix} 7 & -4 \\ 2 & 3 \end{bmatrix}$, $\mathbf{f} = \begin{bmatrix} 9\,e^t \\ 25^{-t} \end{bmatrix}$.

8.5 The Laplace Transformation

Consider a nonhomogeneous system of ordinary differential equations with **constant coefficients** subject to the initial condition:

$$\dot{\mathbf{y}}(t) = \mathbf{A}\mathbf{y}(t) + \mathbf{f}(t), \qquad \mathbf{y}(0) = \mathbf{y}_0, \tag{8.5.1}$$

where $\dot{\mathbf{y}} \stackrel{\text{def}}{=} \mathrm{d}\mathbf{y}/\mathrm{d}t$ is the derivative with respect to t, and $\mathbf{f}(t)$ is a given vector function. For some matrices $\mathbf{A}$ and forcing terms $\mathbf{f}(t)$, the system (8.5.1) may have a unique constant or periodic solution with the property that all solutions of the system approach this unique solution as $t \mapsto +\infty$. This constant or periodic solution is called the **steady state solution**. We analyze when it happens using the Laplace transform. Note that the initial conditions are irrelevant in the long run because eventually any solution will look like a steady state trajectory.

Recall that the Laplace transform of a function $\mathbf{y}(t)$ is denoted as

$$\mathcal{L}[\mathbf{y}](\lambda) = \mathbf{y}^L(\lambda) \stackrel{\text{def}}{=} \int_0^\infty e^{-\lambda t}\mathbf{y}(t)\,\mathrm{d}t. \tag{8.5.2}$$

Since the Laplace transform of the derivative is $\mathcal{L}[\dot{\mathbf{y}}] = \lambda\mathbf{y}^L - \mathbf{y}(0)$, we reduce the given initial value problem (8.5.1) to the algebraic vector problem

$$\lambda\mathbf{y}^L - \mathbf{y}(0) = \mathbf{A}\mathbf{y}^L + \mathbf{f}^L \quad \text{or} \quad (\lambda\mathbf{I} - \mathbf{A})\,\mathbf{y}^L = \mathbf{f}^L + \mathbf{y}(0),$$

where $\mathbf{I}$ is the identity square matrix. Solving the above system of algebraic equations, we get

$$\mathbf{y}^L = (\lambda\mathbf{I} - \mathbf{A})^{-1}\left(\mathbf{y}(0) + \mathbf{f}^L\right) = \mathbf{y}_h^L + \mathbf{y}_p^L, \tag{8.5.3}$$

where $\mathbf{R}_\lambda(\mathbf{A}) = (\lambda\mathbf{I} - \mathbf{A})^{-1}$ is the resolvent to the matrix $\mathbf{A}$, and

$$\mathbf{y}_h^L = (\lambda\mathbf{I} - \mathbf{A})^{-1}\mathbf{y}(0), \qquad \mathbf{y}_p^L = (\lambda\mathbf{I} - \mathbf{A})^{-1}\mathbf{f}^L.$$

Application of the inverse Laplace transform to both sides of Eq. (8.5.3) yields

$$\mathbf{y}(t) = \mathcal{L}^{-1}\left[\mathbf{R}_\lambda(\mathbf{A})\right]\mathbf{y}(0) + \mathcal{L}^{-1}\left[\mathbf{R}_\lambda(\mathbf{A})\mathbf{f}^L\right] = \mathbf{y}_h(t) + \mathbf{y}_p(t). \tag{8.5.4}$$

In accordance with Theorem 8.6, page 434, the vector function $\mathbf{y}_h(t) = \mathcal{L}^{-1}\left[\mathbf{R}_\lambda(\mathbf{A})\right]\mathbf{y}(0)$ is the solution of the corresponding initial value problem for the homogeneous equation

$$\dot{\mathbf{y}}_h(t) = \mathbf{A}\mathbf{y}_h(t), \qquad \mathbf{y}_h(0) = \mathbf{y}_0, \tag{8.5.5}$$

and the function $\mathbf{y}_p(t)$ is the solution of the driven equation subject to the homogeneous initial condition

$$\dot{\mathbf{y}}_p(t) = \mathbf{A}\mathbf{y}_p(t) + \mathbf{f}(t), \qquad \mathbf{y}_p(0) = \mathbf{0}. \tag{8.5.6}$$

The formula (8.2.4), page 442, gives us the explicit solution of the IVP (8.5.5) :

$$\mathbf{y}_h(t) = e^{\mathbf{A}t}\,\mathbf{y}_0,$$

which is equal to $\mathcal{L}^{-1}\left[\mathbf{R}_\lambda(\mathbf{A})\right]\mathbf{y}(0)$. Therefore, we establish the relation between the resolvent $\mathbf{R}_\lambda(\mathbf{A})$ of a constant square matrix $\mathbf{A}$ and the exponential matrix:

$$e^{\mathbf{A}t} = \mathcal{L}^{-1}\left[\mathbf{R}_\lambda(\mathbf{A})\right] \qquad \text{or} \qquad \mathbf{R}_\lambda(\mathbf{A}) = (\lambda\mathbf{I} - \mathbf{A})^{-1} = \mathcal{L}\left[e^{\mathbf{A}t}\right]. \tag{8.5.7}$$

Finally, the solution (8.5.4) of the initial value problem (8.5.1) is the sum of two functions:

$$\mathbf{y}_h(t) = \mathcal{L}^{-1}\left[\mathbf{R}_\lambda(\mathbf{A})\right]\mathbf{y}(0) = e^{\mathbf{A}t}\,\mathbf{y}(0), \tag{8.5.8}$$

$$\mathbf{y}_p(t) = \mathcal{L}^{-1}\left[\mathbf{R}_\lambda(\mathbf{A})\,\mathbf{f}^L\right]. \tag{8.5.9}$$

These formulas allow us to establish the existence of a steady state solution.

Theorem 8.17: If all eigenvalues of the constant matrix $\mathbf{A}$ have negative real parts, and the smooth forcing vector function $\mathbf{f}(t)$ is periodic with period T, then the system (8.5.1) has a unique periodic steady state solution and T is its period.

Example 8.5.1: Let us consider the initial value problem (8.5.1) with the following matrices:

$$\mathbf{A} = \begin{bmatrix} 21 & 10 & -2 \\ -22 & -11 & 2 \\ 110 & 9 & -11 \end{bmatrix} \quad \text{and} \quad \mathbf{f}(t) = \begin{bmatrix} 0 \\ \sin t \\ 0 \end{bmatrix}, \quad \mathbf{y}_0 = \begin{bmatrix} 0 \\ 0 \\ 0 \end{bmatrix}.$$

Since the resolvent of the matrix $\mathbf{A}$ is

$$\mathbf{R}_\lambda(\mathbf{A}) = \frac{1}{(\lambda+1)(\lambda^2+9^2)} \begin{bmatrix} \lambda^2+22\lambda+103 & 10\lambda+92 & -2(\lambda+1) \\ -22(\lambda+1) & \lambda^2-10\lambda-11 & 2(\lambda+1) \\ 110\,\lambda+1012 & 9\lambda+911 & \lambda^2-10\lambda-11 \end{bmatrix},$$

we get from Eq. (8.5.9) that

$$\mathbf{y}(t) = \mathcal{L}^{-1}\left[\mathbf{R}_\lambda(\mathbf{A})\,\mathbf{f}^L\right] = \mathcal{L}^{-1}\left[\frac{\langle 10\lambda+92, \lambda^2-10\lambda-11, 9\lambda+911\rangle^T}{(\lambda+1)(\lambda^2+9^2)(\lambda^2+1)}\right].$$

Careful evaluation of the inverse Laplace transforms of each component of $\mathbf{y}(t) = \langle y_1(t), y_2(t),\, y_3(t)\rangle^T$ leads to

$$\begin{aligned} y_1(t) &= \frac{1}{2}\,e^{-t} + \frac{1}{80}\,(51\,\sin t - 41\,\cos t) + \frac{1}{720}\,(9\,\cos 9t - 11\,\sin 9t)\,, \\ y_2(t) &= \frac{1}{720}\,(9\,\cos t - 99\,\sin t - 9\,\cos 9t + 11\,\sin 9t)\,, \\ y_3(t) &= \frac{11}{2}\,e^{-t} + \frac{1}{80}\,(460\,\sin t - 451\,\cos t) + \frac{1}{720}\,(99\,\cos 9t - 20\,\sin 9t)\,. \end{aligned}$$

Obviously, the term $\frac{1}{2}\,e^{-t}\,\langle 1, 0, 11\rangle^T$ dies out, leaving the steady state solution, which has two periods: 2π caused by the driving term $\mathbf{f}(t)$ and $2\pi/9$ inherited from the eigenvalues $\pm 9\mathbf{j}$ of the matrix $\mathbf{A}$.

Example 8.5.2: When the Laplace transformation is applied to a system of linear differential equations, it does not matter whether the given system is in normal form or not. Let us consider the following initial value problem:

$$\begin{aligned} \dot{x} - \dot{y} &= 4y - 3x + \sin(t), \\ 2\dot{x} + \dot{y} &= 2x - 4y - \cos(t), \end{aligned} \qquad x(0) = 1, \quad y(0) = -8.$$

Application of the Laplace transform to both sides of the differential equations gives

$$\begin{aligned} \lambda\, x^L - 1 - \lambda\, y^L + 8 &= 4y^L - 3x^L + \left(1+\lambda^2\right)^{-1}, \\ 2\lambda\, x^L - 2 + \lambda\, y^L - 8 &= 2x^L - 4y^L - \lambda\left(1+\lambda^2\right)^{-1}, \end{aligned}$$

where x^L and y^L are Laplace transforms of the unknown functions $x(t)$ and $y(t)$, respectively; here $\mathcal{L}\left[\sin(t)\right] = (1+\lambda^2)^{-1}$ and $\mathcal{L}\left[\cos(t)\right] = \lambda\left(1+\lambda^2\right)^{-1}$ are Laplace transforms of the driving terms. Solving this system of algebraic equations with respect to x^L and y^L, we get

$$x^L = \frac{4-\lambda+3\lambda^2}{(1+3\lambda)(1+\lambda^2)}, \qquad y^L = -\frac{18+19\lambda+15\lambda^2+24\lambda^3}{(4+\lambda)(1+3\lambda)(1+\lambda^2)}.$$

Application of the inverse Laplace transform provides the answer:

$$\begin{aligned} x(t) &= \frac{1}{5}\left[7\,e^{-t/3} - 2\,\cos t - \sin t\right] H(t), \\ y(t) &= \left[\frac{1354}{187}\,e^{-4t} + \frac{56}{55}\,e^{-t/3} - \frac{22}{85}\,\cos t - \frac{31}{85}\,\sin t\right] H(t), \end{aligned}$$

where $H(t)$ is the Heaviside function, Eq. (5.1.5) on page 274. One can easily identify the transient part (exponential terms) and steady state part (trigonometric functions).

Example 8.5.3: (Discontinuous forcing input) Consider the following initial value problem with an intermittent forcing function:

$$\dot{\mathbf{y}}(t) = \mathbf{A}\mathbf{y}(t) + \mathbf{f}_1(t) + \mathbf{f}_2(t), \qquad \mathbf{y}(0) = \mathbf{0}, \qquad \mathbf{A} = \begin{bmatrix} 2 & -5 \\ 1 & -2 \end{bmatrix},$$

where

$$\mathbf{f}_1(t) = \begin{bmatrix} H(t) - H(t-1) \\ 0 \end{bmatrix}, \quad \mathbf{f}_2(t) = \begin{bmatrix} 0 \\ \sin t\,[H(t) - H(t-\pi)] \end{bmatrix}.$$

Here $H(t)$ is the Heaviside function, Eq. (5.1.5) on page 274. Application of the Laplace transform to the given initial value problem yields the algebraic equation

$$(\lambda\mathbf{I} - \mathbf{A})\,\mathbf{y}^L = \mathbf{f}_1^L + \mathbf{f}_2^L,$$

where $\mathbf{y}^L$ is the Laplace transform (8.5.2) of the unknown function $\mathbf{y}(t)$, and

$$\mathbf{f}_1^L = \frac{1}{\lambda}\begin{bmatrix} 1 - e^{-\lambda} \\ 0 \end{bmatrix}, \qquad \mathbf{f}_2^L = \frac{1}{1+\lambda^2}\begin{bmatrix} 0 \\ 1 - e^{-\lambda\pi} \end{bmatrix}.$$

Calculations show that the resolvent $\mathbf{R}_\lambda(\mathbf{A}) = (\lambda\mathbf{I} - \mathbf{A})^{-1}$ of the matrix $\mathbf{A}$ is

$$\mathbf{R}_\lambda(\mathbf{A}) = \mathcal{L}\left[e^{\mathbf{A}t}\right](\lambda) = \int_0^\infty e^{-\lambda t}\, e^{\mathbf{A}t}\,\mathrm{d}t = \frac{1}{\lambda^2+1}\begin{bmatrix} \lambda+2 & -5 \\ 1 & \lambda-2 \end{bmatrix}.$$

Therefore,

$$\begin{aligned}\mathbf{y}^L &= \mathbf{R}_\lambda(\mathbf{A})\left(\mathbf{f}_1^L + \mathbf{f}_2^L\right) \\ &= \frac{1}{\lambda(\lambda^2+1)}\begin{bmatrix} \lambda+2 \\ 1 \end{bmatrix}\left(1 - e^{-\lambda}\right) + \frac{1}{(\lambda^2+1)^2}\begin{bmatrix} -5 \\ \lambda-2 \end{bmatrix}\left(1 - e^{-\lambda\pi}\right).\end{aligned}$$

Let us introduce the auxiliary functions that are defined as the inverse Laplace transforms of the following expressions:

$$\begin{aligned}
g_1(t) &= \mathcal{L}^{-1}\left[\frac{\lambda+2}{\lambda(\lambda^2+1)}\right] = (2 + \sin t - 2\cos t)\,H(t),\\
g_2(t) &= \mathcal{L}^{-1}\left[\frac{1}{\lambda(\lambda^2+1)}\right] = (1 - \cos t)\,H(t),\\
g_3(t) &= \mathcal{L}^{-1}\left[\frac{-5}{(\lambda^2+1)^2}\right] = 5\left(\cos t - 1 + \frac{t}{2}\,\sin t\right)H(t),\\
g_4(t) &= \mathcal{L}^{-1}\left[\frac{\lambda-2}{(\lambda^2+1)^2}\right] = \left(2\cos t + t\sin t - 2 + \frac{1}{2}\,\sin t - \frac{t}{2}\,\cos t\right)H(t).
\end{aligned}$$

Then we express the required solution through these functions:

$$\mathbf{y}(t) = \mathcal{L}^{-1}\left[\mathbf{x}_p^L\right](t) = \begin{bmatrix} g_1(t) - g_1(t-1) \\ g_2(t) - g_2(t-1) \end{bmatrix} + \begin{bmatrix} g_3(t) - g_3(t-\pi) \\ g_4(t) - g_4(t-\pi) \end{bmatrix}.$$

Problems

1. In each exercise (a) through (h), use the *Laplace transform* to solve the vector differential equation $\dot{\mathbf{y}}(t) = \mathbf{A}\mathbf{y}(t) + \mathbf{f}$ subject to the initial condition $\mathbf{y}(0) = \langle 1, 1\rangle^T$, where a constant 2×2 matrix $\mathbf{A}$ and the driving term $\mathbf{f}$ are specified.

(a) $\begin{bmatrix} 2 & -1 \\ 7 & 10 \end{bmatrix}$, $\mathbf{f} = \begin{bmatrix} e^t - 36\,e^{3t} \\ -9\,e^t \end{bmatrix}$;

(b) $\begin{bmatrix} 7 & -5 \\ 1 & 3 \end{bmatrix}$, $\mathbf{f} = e^{2t}\begin{bmatrix} 5 \\ 1 \end{bmatrix}$;

(c) $\begin{bmatrix} 7 & -4 \\ 1 & 11 \end{bmatrix}$, $\mathbf{f} = e^{3t}\begin{bmatrix} 4 \\ 1 \end{bmatrix}$;

(d) $\begin{bmatrix} 5 & -1 \\ 20 & -3 \end{bmatrix}$, $\mathbf{f} = e^{t}\begin{bmatrix} 1 \\ 4 \end{bmatrix}$;

(e) $\begin{bmatrix} 6 & 3 \\ -2 & 1 \end{bmatrix}$, $\mathbf{f} = \begin{bmatrix} 10\,e^{-t} \\ 6\,e^t \end{bmatrix}$;

(f) $\begin{bmatrix} 1 & 5 \\ -2 & 3 \end{bmatrix}$, $\mathbf{f} = e^{-2t}\begin{bmatrix} 5 \\ 5 \end{bmatrix}$;

(g) $\begin{bmatrix} -4 & 5 \\ -13 & 4 \end{bmatrix}$, $\mathbf{f} = \cos t\begin{bmatrix} 48 \\ 48 \end{bmatrix}$;

(h) $\begin{bmatrix} 5 & -1 \\ 20 & 1 \end{bmatrix}$, $\mathbf{f} = e^{-3t}\begin{bmatrix} 8 \\ 20 \end{bmatrix}$.

2. Use the method of *Laplace transform* to solve the given initial value problems. Here $\dot{x}$ and $\dot{y}$ denote derivatives with respect to t.

(a) $$\begin{aligned} \dot{x} + \dot{y} - 2y &= e^{2t} + e^{=2t}, & x(0) &= 1, \\ 2\dot{y} - 3\dot{x} - 4x + 8y &= e^{2t} - e^{-2t}; & y(0) &= 0. \end{aligned}$$

(b) $$\begin{aligned} 3\dot{x} - \dot{y} + 3y &= 9e^{2t} + 1, & x(0) &= 0, \\ \dot{y} + 2\dot{x} - 3y - x &= 7t - t^2; & y(0) &= 3. \end{aligned}$$

(c) $$\begin{aligned} \dot{x} + \dot{y} - x + y &= 3\sin t + \cos t, & x(0) &= 0, \\ \dot{y} - 2\dot{x} - x - 2y &= \sin t - 2\cos t; & y(0) &= -1. \end{aligned}$$

(d) $$\begin{aligned} \dot{x} + 2\dot{y} - y &= 12t - 3, & x(0) &= 0, \\ 2\dot{y} - \dot{x} + y + 6x &= 6t^3 + 3; & y(0) &= 1. \end{aligned}$$

(e) $$\begin{aligned} 3\ddot{x} + \dot{y} - 6x &= -6t^2, & x(0) &= 1, \quad \dot{x}(0) = 2, \\ \ddot{y} + 3\dot{x} + y &= -9e^{2t}; & y(0) &= -3, \quad \dot{y}(0) = -12. \end{aligned}$$

(f) $$\begin{aligned} \ddot{x} + 3y - 3x &= 9t^2, & x(0) &= -1, \quad \dot{x}(0) = 2, \\ \ddot{y} + 3y - 3x &= 3 + 9t^2; & y(0) &= -2, \quad \dot{y}(0) = 0. \end{aligned}$$

(g) $$\begin{aligned} \ddot{x} - 3\dot{y} + 2x &= -2, & x(0) &= -1, \quad \dot{x}(0) = 0, \\ \ddot{y} + \dot{x} - 6y &= 0; & y(0) &= 0, \quad \dot{y}(0) = 2. \end{aligned}$$

(h) $$\begin{aligned} \ddot{x} + \dot{y} + 2x &= 5, & x(0) &= 3, \quad \dot{x}(0) = -1, \\ \ddot{y} + 6\dot{x} + y &= t; & y(0) &= 3, \quad \dot{y}(0) = -3. \end{aligned}$$

3. In each exercise (a) through (h), use the *Laplace transform* to solve the vector differential equation $\dot{\mathbf{y}}(t) = \mathbf{A}\mathbf{y}(t) + \mathbf{f}$ subject to the initial condition $\mathbf{y}(0) = \langle 0, 1, 0\rangle^T$, where a constant 3×3 matrix $\mathbf{A}$ and the driving term $\mathbf{f}$ are specified.

(a) $\begin{bmatrix} 3 & 2 & 4 \\ 45 & 4 & 45 \\ 4 & 2 & 3 \end{bmatrix}, \ \mathbf{f} = \begin{bmatrix} 27\,e^t \\ 34\,e^{2t} \\ 76 \end{bmatrix};$ (b) $\begin{bmatrix} 3 & 4 & -10 \\ 10 & 1 & -10 \\ 2 & 2 & -5 \end{bmatrix}, \ \mathbf{f} = \begin{bmatrix} 15t \\ 45t \\ 0 \end{bmatrix};$

(c) $\begin{bmatrix} 1 & -9 & 0 \\ -1 & 3 & -3 \\ 0 & 2 & 1 \end{bmatrix}, \ \mathbf{f} = \begin{bmatrix} 32t \\ 16t \\ -16t \end{bmatrix};$ (d) $\begin{bmatrix} -4 & 8 & 4 \\ 3 & 1 & 3 \\ -1 & 1 & -9 \end{bmatrix}, \ \mathbf{f} = \begin{bmatrix} 8\,e^t \\ 0 \\ e^t \end{bmatrix};$

(e) $\begin{bmatrix} 3 & 2 & 4 \\ 10 & 4 & 10 \\ 4 & 2 & 3 \end{bmatrix}, \ \mathbf{f} = e^t \begin{bmatrix} 22 \\ 21 \\ 6 \end{bmatrix};$ (f) $\begin{bmatrix} 3 & 4 & -6 \\ 6 & 5 & -9 \\ 2 & 1 & -1 \end{bmatrix}, \ \mathbf{f} = e^t \begin{bmatrix} 10 \\ 8 \\ 3 \end{bmatrix};$

(g) $\begin{bmatrix} 1 & -8 & 1 \\ -1 & 3 & -3 \\ 1 & 2 & 1 \end{bmatrix}, \ \mathbf{f} = \begin{bmatrix} 52\cos t \\ 26\sin t \\ -52\sin t \end{bmatrix};$ (h) $\begin{bmatrix} -4 & 8 & 4 \\ 7 & 1 & 7 \\ -1 & 5 & -9 \end{bmatrix}, \ \mathbf{f} = e^t \begin{bmatrix} 117 \\ 56 \\ 17 \end{bmatrix}.$

4. For each of the following matrices and given vector function $\mathbf{f}$, solve the inhomogeneous system of equations of second order $\ddot{\mathbf{x}}(t) + \mathbf{A}\mathbf{x}(t) = \mathbf{f}(t)$ using Laplace transform. Initial conditions can be chosen homogeneous.

(a) $\begin{bmatrix} 10 & 1 \\ 6 & 15 \end{bmatrix}, \quad \mathbf{f} = \begin{bmatrix} \sin 3t \\ \cos 4t \end{bmatrix};$ (b) $\begin{bmatrix} 19 & 10 \\ 6 & 15 \end{bmatrix}, \quad \mathbf{f} = \begin{bmatrix} \sin 3t \\ t \end{bmatrix};$

(c) $\begin{bmatrix} 30 & 21 \\ 6 & 15 \end{bmatrix}, \quad \mathbf{f} = \begin{bmatrix} \sin 3t \\ \cos 6t \end{bmatrix};$ (d) $\begin{bmatrix} 16 & -3 \\ -36 & 13 \end{bmatrix}, \quad \mathbf{f} = \begin{bmatrix} \sin 4t \\ \cos 5t \end{bmatrix};$

(e) $\begin{bmatrix} 16 & -18 \\ -6 & 13 \end{bmatrix}, \quad \mathbf{f} = \begin{bmatrix} \sin 4t \\ \cos 5t \end{bmatrix};$ (f) $\begin{bmatrix} 17 & -26 \\ -4 & 12 \end{bmatrix}, \quad \mathbf{f} = \begin{bmatrix} \sin 4t \\ \cos 5t \end{bmatrix};$

(g) $\begin{bmatrix} 22 & -21 \\ -27 & 28 \end{bmatrix}, \quad \mathbf{f} = \begin{bmatrix} \sin t \\ \cos 2t \end{bmatrix};$ (h) $\begin{bmatrix} 19 & -34 \\ -5 & 26 \end{bmatrix}, \quad \mathbf{f} = \begin{bmatrix} \sin 3t \\ \cos 2t \end{bmatrix}.$

8.6 Second Order Linear Systems

For a constant $n \times n$ matrix $\mathbf{A}$, let us consider a second order vector equation

$$\frac{\mathrm{d}^2\mathbf{x}}{\mathrm{d}t^2} + \mathbf{A}\mathbf{x} = \mathbf{0} \qquad \text{or} \qquad \ddot{\mathbf{x}} + \mathbf{A}\mathbf{x} = \mathbf{0}, \tag{8.6.1}$$

where $\mathbf{x}(t)$ is a vector function with n components to be determined. Recall that all vectors are assumed to be column vectors. Any nonsingular matrix $\mathbf{\Phi}(t)$ that satisfies the matrix differential equation

$$\ddot{\mathbf{\Phi}} + \mathbf{A}\mathbf{\Phi} = \mathbf{0} \tag{8.6.2}$$

is called the **fundamental matrix** for the vector equation (8.6.1). When a fundamental matrix is known, the general solution of the vector equation (8.6.1) can be expressed through this matrix.

Theorem 8.18: If $\mathbf{\Phi}(t)$ is a fundamental matrix for Eq. (8.6.1), then for any two constant column vectors $\mathbf{a} = \langle a_1, a_2, \ldots, a_n\rangle^T$ and $\mathbf{b} = \langle b_1, b_2, \ldots, b_n\rangle^T$, the n-vector $\mathbf{u} = \mathbf{\Phi}(t)\mathbf{a} + \dot{\mathbf{\Phi}}(t)\mathbf{b}$ is the general solution of the vector equation (8.6.1).

Fortunately, a fundamental matrix can be constructed explicitly when $\mathbf{A}$ is a constant matrix in the system (8.6.1).

Theorem 8.19: Any solution of the vector equation (8.6.1) can be represented as

$$\mathbf{x}(t) = \cos\left(t\sqrt{\mathbf{A}}\right)\mathbf{x}(0) + \frac{\sin\left(t\sqrt{\mathbf{A}}\right)}{\sqrt{\mathbf{A}}}\dot{\mathbf{x}}(0), \tag{8.6.3}$$

where $\mathbf{x}(0)$ and $\dot{\mathbf{x}}(0)$ are specified initial column vectors.

PROOF: Applying the Laplace transform to the initial value problem

$$\ddot{\mathbf{x}} + \mathbf{A}\,\mathbf{x} = \mathbf{0}, \qquad \mathbf{x}(0) = \mathbf{x}_0, \quad \dot{\mathbf{x}}(0) = \mathbf{v}_0,$$

we obtain

$$\lambda^2\mathbf{x}^L - \dot{\mathbf{x}}(0) - \lambda\mathbf{x}(0) + \mathbf{A}\,\mathbf{x}^L = \mathbf{0},$$

where $\mathbf{x}^L(\lambda)$ is the Laplace transform of the unknown vector function $\mathbf{x}(t)$. The above algebraic system of equations is not hard to solve:

$$\mathbf{x}^L = \frac{\lambda}{\lambda^2\mathbf{I} + \mathbf{A}}\,\mathbf{x}(0) + \frac{1}{\lambda^2\mathbf{I} + \mathbf{A}}\,\dot{\mathbf{x}}(0).$$

Application of the inverse Laplace transform yields Eq. (8.6.3) because

$$\begin{aligned}
\frac{\sin\left(t\sqrt{\mathbf{A}}\right)}{\sqrt{\mathbf{A}}} &= \mathcal{L}^{-1}\left[\frac{1}{\lambda^2\mathbf{I} + \mathbf{A}}\right] = \frac{1}{2\mathbf{j}\sqrt{\mathbf{A}}}\,\mathcal{L}^{-1}\left[\frac{1}{\lambda\mathbf{I} - \mathbf{j}\sqrt{\mathbf{A}}} - \frac{1}{\lambda\mathbf{I} + \mathbf{j}\sqrt{\mathbf{A}}}\right], \\
\cos\left(t\sqrt{\mathbf{A}}\right) &= \mathcal{L}^{-1}\left[\frac{\lambda}{\lambda^2\mathbf{I} + \mathbf{A}}\right] = \frac{1}{2}\,\mathcal{L}^{-1}\left[\frac{1}{\lambda\mathbf{I} - \mathbf{j}\sqrt{\mathbf{A}}} + \frac{1}{\lambda\mathbf{I} + \mathbf{j}\sqrt{\mathbf{A}}}\right]. \qquad \blacksquare
\end{aligned}$$

Here we used factorization $\lambda^2\mathbf{I} + \mathbf{A} = \left(\lambda\mathbf{I} - \mathbf{j}\sqrt{\mathbf{A}}\right)\left(\lambda\mathbf{I} + \mathbf{j}\sqrt{\mathbf{A}}\right)$, assuming that a root $\sqrt{\mathbf{A}}$ exists. As usual, $\mathbf{j}$ denotes the unit vector in the positive vertical direction in the complex plane, so $\mathbf{j}^2 = -1$. The matrix equation (8.6.2) has two linearly independent fundamental matrices

$$\mathbf{\Phi}_1(t) = \left(\sqrt{\mathbf{A}}\right)^{-1}\sin\left(t\sqrt{\mathbf{A}}\right), \qquad \mathbf{\Phi}_2(t) = \cos\left(t\sqrt{\mathbf{A}}\right), \tag{8.6.4}$$

each of which is a derivative of another: $\mathbf{\Phi}_2(t) = \dot{\mathbf{\Phi}}_1$ and $\mathbf{\Phi}_1(t) = -\mathbf{A}\dot{\mathbf{\Phi}}_2$. Both matrices do not depend which root $\sqrt{\mathbf{A}}$ has been chosen and they are solutions of the second order matrix differential equation (8.6.2), subject to the initial conditions

$$\mathbf{\Phi}_2(0) = \mathbf{I}, \quad \dot{\mathbf{\Phi}}_2(0) = \mathbf{0} \qquad \text{and} \qquad \mathbf{\Phi}_1(0) = \mathbf{0}, \quad \dot{\mathbf{\Phi}}_1(0) = \mathbf{I},$$

where $\mathbf{I}$ is the identity $n \times n$ matrix and $\mathbf{0}$ is the zero square matrix. Using these two fundamental matrices (8.6.4), we solve the initial value problem

$$\ddot{\mathbf{x}} + \mathbf{A}\mathbf{x} = 0, \qquad \mathbf{x}(0) = \mathbf{x}_0, \quad \dot{\mathbf{x}}(0) = \mathbf{v}_0$$

explicitly:

$$\mathbf{x}(t) = \mathbf{\Phi}_2(t)\,\mathbf{x}_0 + \mathbf{\Phi}_1(t)\,\mathbf{v}_0. \tag{8.6.5}$$

Now we turn our attention to the initial value problem for a driven vector equation

$$\ddot{\mathbf{x}} + \mathbf{A}\mathbf{x} = \mathbf{f}, \qquad \mathbf{x}(0) = \mathbf{x}_0, \quad \dot{\mathbf{x}}(0) = \mathbf{v}_0, \tag{8.6.6}$$

where $\mathbf{x}_0$ and $\mathbf{v}_0$ are specified vectors. Its solution is the sum of two vector functions: $\mathbf{x}(t) = \mathbf{x}_h(t) + \mathbf{x}_p(t)$, where $\mathbf{x}_h(t)$ is the complementary function, which is the solution to the IVP $\ddot{\mathbf{x}} + \mathbf{A}\mathbf{x} = \mathbf{0}$, $\mathbf{x}(0) = \mathbf{x}_0$, $\dot{\mathbf{x}}(0) = \mathbf{v}_0$. The second term, $\mathbf{x}_p(t)$, is a solution of the IVP (8.6.6) with $\mathbf{x}_0 = \mathbf{0}$ and $\mathbf{v}_0 = \mathbf{0}$. Since the complementary function is given explicitly by Eq. (8.6.3), our main concern is a particular solution $\mathbf{x}_p(t)$.

If the forcing vector function $\mathbf{f}(t)$ is an intermittent one, the most appropriate method to determine $\mathbf{x}_p(t)$ is the Laplace transformation. Using it, we get

$$\mathbf{x}_p(t) = \mathcal{L}^{-1}\left[\frac{1}{\lambda^2\mathbf{I} + \mathbf{A}}\,\mathbf{f}^L\right]. \tag{8.6.7}$$

If the driving function is smooth, it is convenient to apply the variation of parameter method. Note that the Lagrange method works for an arbitrary input, but practical application of the variation of parameters method becomes cumbersome when the forcing function $\mathbf{f}(t)$ is piecewise continuous.

According to the variation of parameters method, we seek a particular solution of the nonhomogeneous equation in the form

$$\mathbf{x}_p(t) = \mathbf{\Phi}_1(t)\,\mathbf{u}_1(t) + \mathbf{\Phi}_2(t)\,\mathbf{u}_2(t), \tag{8.6.8}$$

where the fundamental matrix functions $\mathbf{\Phi}_1(t)$ and $\mathbf{\Phi}_2(t)$ are given in Eq. (8.6.4), and vector functions $\mathbf{u}_1(t)$ and $\mathbf{u}_2(t)$ are to be determined. Next we proceed in a way similar to the scalar case (see §4.8). This leads to a system of algebraic equations for first derivatives of $\mathbf{u}_1(t)$ and $\mathbf{u}_2(t)$:

$$\begin{aligned}
\mathbf{\Phi}_1(t)\,\dot{\mathbf{u}}_1(t) + \mathbf{\Phi}_2(t)\,\dot{\mathbf{u}}_2(t) &= \mathbf{0},\\
\dot{\mathbf{\Phi}}_1(t)\,\dot{\mathbf{u}}_1(t) + \dot{\mathbf{\Phi}}_2(t)\,\dot{\mathbf{u}}_2(t) &= \mathbf{f}(t).
\end{aligned}$$

Its solution is

$$\dot{\mathbf{u}}_1(t) = \mathbf{\Phi}_2(t)\,\mathbf{f}(t), \qquad \dot{\mathbf{u}}_2(t) = -\mathbf{\Phi}_1(t)\,\mathbf{f}(t)$$

because their Wronskian is $\dot{\mathbf{\Phi}}_1(t)\mathbf{\Phi}_2(t) - \mathbf{\Phi}_1(t)\dot{\mathbf{\Phi}}_2(t) = \mathbf{I}$. Integrating, we get

$$\mathbf{u}_1(t) = \int \cos\left(t\sqrt{\mathbf{A}}\right)\mathbf{f}(t)\,\mathrm{d}t, \quad \mathbf{u}_2(t) = -\int \frac{\sin\left(t\sqrt{\mathbf{A}}\right)}{\sqrt{\mathbf{A}}}\,\mathbf{f}(t)\,\mathrm{d}t. \tag{8.6.9}$$

Substituting these formulas into Eq. (8.6.8), we obtain

$$\mathbf{x}_p(t) = \int_0^t \frac{\sin\left((t-\tau)\sqrt{\mathbf{A}}\right)}{\sqrt{\mathbf{A}}}\,\mathbf{f}(\tau)\,\mathrm{d}\tau = \frac{1}{\sqrt{\mathbf{A}}}\int_0^t \sin\left((t-\tau)\sqrt{\mathbf{A}}\right)\mathbf{f}(\tau)\,\mathrm{d}\tau. \tag{8.6.10}$$

As an example, consider the mechanical system of two bodies and three springs discussed in §6.1.2. It was shown (see page 345) that this system is modeled by the vector differential equation:

$$\begin{aligned}
m_1\,\frac{\mathrm{d}^2x_1}{\mathrm{d}t^2} + (k_1+k_2)x_1 - k_2x_2 &= 0,\\
m_2\,\frac{\mathrm{d}^2x_2}{\mathrm{d}t^2} + (k_2+k_3)x_2 - k_2x_1 &= 0,
\end{aligned}$$

where k_1, k_2, and k_3 are spring constants, and x_1, x_2 are displacements (from their equilibrium positions) of two bodies with masses m_1 and m_2, respectively.

The given system of differential equations can be written either as the second order vector equation

$$\frac{\mathrm{d}^2\mathbf{x}}{\mathrm{d}t^2} + \mathbf{A}\mathbf{x} = \mathbf{0}, \quad \text{with} \quad \mathbf{A} = \begin{bmatrix} \frac{k_1+k_2}{m_1} & -\frac{k_2}{m_1} \\ -\frac{k_2}{m_2} & \frac{k_2+k_3}{m_2} \end{bmatrix}, \quad \mathbf{x} = \begin{bmatrix} x_1 \\ x_2 \end{bmatrix}, \tag{8.6.11}$$

or as the first order vector equation

$$\frac{\mathrm{d}\mathbf{y}}{\mathrm{d}t} = \mathbf{B}\mathbf{y}, \quad \text{with} \quad \mathbf{B} = \begin{bmatrix} 0 & 0 & 1 & 0 \\ 0 & 0 & 0 & 1 \\ -\frac{k_1+k_2}{m_1} & \frac{k_2}{m_1} & 0 & 0 \\ \frac{k_2}{m_2} & -\frac{k_2+k_3}{m_2} & 0 & 0 \end{bmatrix}, \quad \mathbf{y} = \begin{bmatrix} y_1 \\ y_2 \\ y_3 \\ y_4 \end{bmatrix}, \tag{8.6.12}$$

where $y_1 = x_1$, $y_2 = x_2$ are displacements of two bodies, and $y_3 = \dot{x}_1$, $y_4 = \dot{x}_2$ are their velocities. In this particular physical problem, the obtained matrix **A** is positive, which means that all its eigenvalues are positive. For a positive matrix, the square roots of its eigenvalues are called (circular) **natural frequencies**. Let λ_1, λ_2 be two distinct positive eigenvalues of the matrix **A** given in Eq. (8.6.11), then the 4×4 matrix **B** has four distinct pure imaginary eigenvalues: $\pm\mathbf{j}\omega_1$ and $\pm\mathbf{j}\omega_2$, where $\omega_1 = \sqrt{\lambda_1}$ and $\omega_2 = \sqrt{\lambda_2}$ are natural frequencies, and the solution of Eq. (8.6.12) becomes

$$\mathbf{y}(t) = e^{\mathbf{B}t}\,\mathbf{y}_0, \tag{8.6.13}$$

where 4-column vector $\mathbf{y}_0$ is comprised from the initial displacement $\mathbf{x}_0$ and the initial velocity $\mathbf{v}_0$:

$$\mathbf{y}_0^T = \langle \mathbf{x}_0^T, \mathbf{v}_0^T \rangle.$$

Example 8.6.1: (A Multiple Spring-Mass System) Suppose that the spring constants for this system have the following values: $k_1 = 8$, $k_2 = 2$, $k_3 = 8$, and the masses are $m_1 = 2/3$, $m_2 = 1$. Then the system of equations (8.6.11) becomes

$$\frac{\mathrm{d}^2\mathbf{x}}{\mathrm{d}t^2} + \mathbf{A}\mathbf{x} = \mathbf{0}, \quad \text{where} \quad \mathbf{A} = \begin{bmatrix} 15 & -3 \\ -2 & 10 \end{bmatrix}, \quad \mathbf{x}(t) = \begin{bmatrix} x_1(t) \\ x_2(t) \end{bmatrix}. \tag{8.6.14}$$

The matrix of this system is positive because its eigenvalues are $\lambda_1 = 16$ and $\lambda_2 = 9$. The general solution is expressed through the formula (8.6.3), where $\mathbf{x}(0) = \mathbf{x}_0 = \langle a_1, a_2 \rangle^T$ is the initial displacement at $t = 0$ and $\dot{\mathbf{x}}(0) = \mathbf{v}_0 = \langle b_1, b_2 \rangle^T$ is the initial velocity. Using the Sylvester auxiliary matrices

$$\mathbf{Z}_{(9)} = \frac{\mathbf{A} - 16}{9 - 16} = \frac{1}{7}\begin{bmatrix} 1 & 3 \\ 2 & 6 \end{bmatrix}, \qquad \mathbf{Z}_{(16)} = \frac{\mathbf{A} - 9}{16 - 9} = \frac{1}{7}\begin{bmatrix} 6 & -3 \\ -2 & 1 \end{bmatrix},$$

we construct two fundamental matrices:

$$\begin{aligned} \mathbf{\Phi}_1(t) &= \frac{\sin\left(t\sqrt{\mathbf{A}}\right)}{\sqrt{\mathbf{A}}} = \frac{\sin(3t)}{3}\,\mathbf{Z}_{(9)} + \frac{\sin(4t)}{4}\,\mathbf{Z}_{(16)}, \\ \mathbf{\Phi}_2(t) &= \cos\left(t\sqrt{\mathbf{A}}\right) = \cos(3t)\;\mathbf{Z}_{(9)} + \cos(4t)\;\mathbf{Z}_{(16)}. \end{aligned}$$

According Eq. (8.6.5), we express the solution of the initial value problem for the vector equation (8.6.14) explicitly:

$$\mathbf{x}(t) = \frac{1}{7}\begin{bmatrix} 1 & 3 \\ 2 & 6 \end{bmatrix}\left(\mathbf{x}_0\cos 3t + \mathbf{v}_0\,\frac{\sin 3t}{3}\right) + \frac{1}{7}\begin{bmatrix} 6 & -3 \\ -2 & 1 \end{bmatrix}\left(\mathbf{x}_0\cos 4t + \mathbf{v}_0\,\frac{\sin 4t}{4}\right).$$

The given mechanical system has two natural frequencies $\omega_1 = 3$ and $\omega_2 = 4$. Therefore, the physical system is comprised of two natural modes of oscillations. For instance, if the initial displacement is $\mathbf{x}(0) = \mathbf{x}_0 = \langle 7, 0 \rangle^T$ and the initial velocity is $\dot{\mathbf{x}}(0) = \mathbf{v}_0 = \langle 0, 28 \rangle^T$, the vector $\mathbf{x}(t)$ becomes

$$\begin{aligned} \mathbf{x}(t) &= \begin{bmatrix} 1 \\ 2 \end{bmatrix}\cos 3t + \begin{bmatrix} 4 \\ 8 \end{bmatrix}\sin 3t + \begin{bmatrix} 6 \\ -2 \end{bmatrix}\cos 4t + \begin{bmatrix} -3 \\ 1 \end{bmatrix}\sin 4t \\ &= \begin{cases} \sqrt{17}\cos(3t - \alpha) + 3\sqrt{5}\sin(4t - \beta), \\ 2\sqrt{17}\cos(3t - \alpha) - \sqrt{5}\sin(4t - \beta), \end{cases} \end{aligned}$$

where $\alpha = \arccos\left(1/\sqrt{17}\right) \approx 1.32582$ and $\beta = \arccos\left(1/\sqrt{5}\right) \approx 1.10715$. If we transfer the given problem (8.6.14) to the first order vector equation (8.6.12), its solution is expressed through the exponential (4×4)-matrix $\mathbf{B}$:

$$
\begin{aligned}
e^{\mathbf{B}t} &= \frac{\cos 3t}{7}\begin{bmatrix} 1 & 3 & 0 & 0 \\ 2 & 6 & 0 & 0 \\ 0 & 0 & 1 & 3 \\ 0 & 0 & 2 & 6 \end{bmatrix} + \frac{\cos 4t}{7}\begin{bmatrix} 6 & -3 & 0 & 0 \\ -2 & 1 & 0 & 0 \\ 0 & 0 & 6 & -3 \\ 0 & 0 & -2 & 1 \end{bmatrix} \\
&+ \frac{\sin 3t}{7}\begin{bmatrix} 0 & 0 & 1/3 & 1 \\ 0 & 0 & 2/3 & 2 \\ -3 & -3 & 0 & 0 \\ -6 & -18 & 0 & 0 \end{bmatrix} + \frac{\sin 4t}{7}\begin{bmatrix} 0 & 0 & 3/2 & -3/4 \\ 0 & 0 & -1/2 & 1/4 \\ -24 & -12 & 0 & 0 \\ -8 & -4 & 0 & 0 \end{bmatrix}.
\end{aligned}
$$

The next application of $e^{\mathbf{B}t}$ to the initial vector $\mathbf{y}_0 = \langle 7, 0, 0, 28 \rangle^T$ yields the same solution.

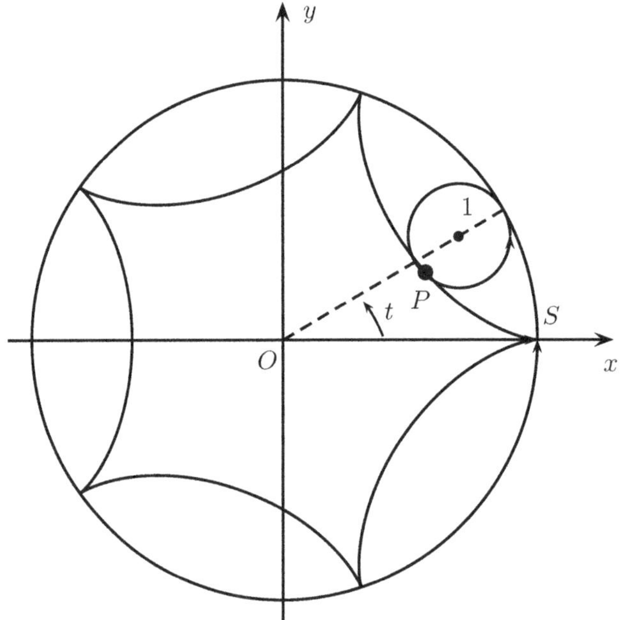

Example 8.6.2: (Hypocycloid) Consider a plane curve generated by the trace of a fixed point on a small circle of radius $r = 1$ that rolls within a larger circle of radius $R = 5$. Such a curve is called **hypocycloid**. The coordinates of this curve are expressed as

$$
\begin{aligned}
x(t) &= 4\,\cos(t) + \cos(4t), \\
y(t) &= 4\,\sin(t) - \sin(4t),
\end{aligned}
$$

where t is the angle of inclination of the center of the small circle. Indeed, an equation of a circle of radius r in polar coordinates is

$$
x = x_0 + r\,\cos\theta, \qquad y = y_0 + r\,\sin\theta,
$$

where (x_0, y_0) is its center. Since the small circle has radius $r = 1$ and it is rolled inside the large circle, its center is moved along the circumference of radius 4. Therefore, $x_0 = 4\,\cos(t)$ and $y_0 = 4\,\sin(t)$. Now we relate the angle θ with inclination t of the line that connects the center of the small circle with the origin. A pivot point on the small circle starts at point S and moves along the hypocycloid curve by rotating in a negative direction with respect to its own center (x_0, y_0). In polar coordinates, the length of any arc is proportional to the angle it ascribes. Since the center of the small circle moves in a positive direction with respect to the origin, and the small circle moves in the opposite direction, we get $\theta = -(5t - t) = -4t$.

Now we show that the coordinates (x, y) of the hypocycloid are solutions of the following initial value problem:

$$
\ddot{x} - 3\dot{y} + 4x = 0, \qquad \ddot{y} + 3\dot{x} + 4y = 0, \qquad x(0) = 5, \quad y(0) = \dot{y}(0) = \dot{x}(0) = 0.
$$

Upon application of the Laplace transform, we get

$$
\lambda^2 x^L - 5\lambda - 3\lambda\, y^L + 4x^L = 0, \qquad \lambda^2 y^L + 3\lambda\, x^L - 15 + 4y^L = 0.
$$

Solving this system of algebraic equations with respect to x^L and y^L and applying the inverse Laplace transform, we obtain the coordinates of the hypocycloid:

$$x(t) = \mathcal{L}^{-1}\left[\frac{5\lambda\,(\lambda+13)}{\lambda^4+17\lambda^2+16}\right] = 4\,\cos t + \cos(4t), \quad y(t) = \mathcal{L}^{-1}\left[\frac{60}{\lambda^4+17\lambda^2+16}\right] = 4\,\sin t - \sin(4t).$$

Example 8.6.3: (Forced Oscillations) Consider a spring-mass system that consists of two masses m_1 and m_2, connected by two springs to a wall, and there is an external periodic force, $f(t) = \sin(\omega t)$, acting on the mass, m_1 (see Fig. 8.14 on text page).

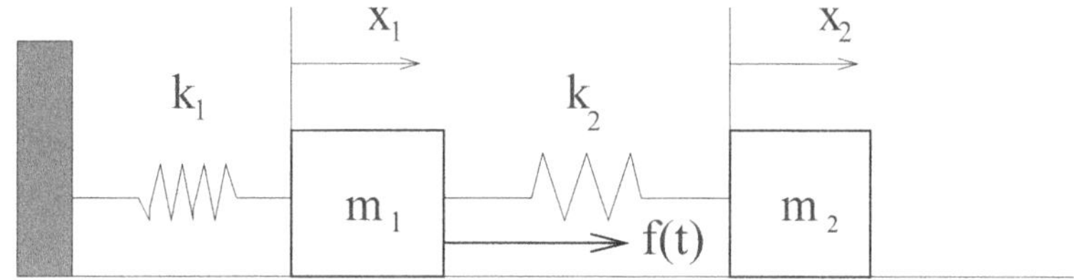

Figure 8.14: Spring-mass system with external force.

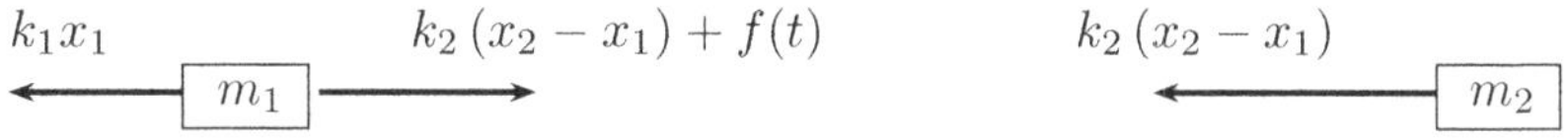

$k_1 x_1$ m_1 $k_2\,(x_2 - x_1) + f(t)$ $k_2\,(x_2 - x_1)$ m_2

Figure 8.15: The free-body diagram.

Using Newton's second law and referring to Fig. 8.15, we derive the system of linear second order equations that simulate the motion of the given spring-mass system:

$$m_1\,\ddot{x} = -k_1 x_1 + k_2\,(x_2 - x_1) + f(t), \qquad m_2\,\ddot{x} = -k_2\,(x_2 - x_1).$$

In the system, assume that $k_1 = 8$, $k_2 = 4$, $m_1 = 1$, and $\omega = 2$. However, we are going to choose the value of the second mass m_2 in such a way that in the resulting steady periodic motion the mass m_1 will oscillate due to its own natural frequencies. If we denote the ratio $k_2/m_2 = 4/m_2$ by α, the given system can be written in vector form:

$$\frac{\mathrm{d}^2}{\mathrm{d}t^2}\begin{bmatrix} x \\ y \end{bmatrix} + \begin{bmatrix} 12 & -4 \\ -\alpha & \alpha \end{bmatrix}\begin{bmatrix} x \\ y \end{bmatrix} = \begin{bmatrix} \sin\omega t \\ 0 \end{bmatrix},$$

where the initial conditions can be chosen homogeneous. Application of the Laplace transform reduces the given vector differential equation to the algebraic system of equations:

$$\lambda^2 x^L + 12x^L - 4y^L = \frac{\omega}{\lambda^2+\omega^2}, \qquad \lambda^2 y^L - \alpha x^L + \alpha y^L = 0,$$

where x^L and y^L are Laplace transforms of unknown functions $x(t)$ and $y(t)$, respectively. The above system of algebraic equations can be solved without a problem, which gives

$$x^L = \frac{\omega}{\lambda^2+\omega^2}\cdot\frac{\alpha+\lambda^2}{8\alpha+(\alpha+12)\lambda^2+\lambda^4}, \qquad y^L = \frac{\omega}{\lambda^2+\omega^2}\cdot\frac{\alpha}{8\alpha+(\alpha+12)\lambda^2+\lambda^4}.$$

To find the inverse Laplace transforms of $x^L(\lambda)$ and $y^L(\lambda)$, we need to determine the points where their common denominator is zero. This denominator is a product of two terms, $\lambda^2+\omega^2$, which is due to the oscillating external force $f(t) = \sin\omega t$, and $8\alpha+(\alpha+12)\lambda^2+\lambda^4$, which defines natural frequencies of oscillations of the mechanical system. Taking the inverse Laplace transform, we obtain

$$x(t) = x_\omega(t) + x_n(t), \qquad y(t) = y_\omega(t) + y_n(t),$$

where $x_\omega(t)$ and $y_\omega(t)$ are the steady periodic terms caused by the external force $f(t) = \sin\omega t$:

$$x_\omega(t) = 2\,\Re \operatorname*{Res}_{\lambda=\mathbf{j}\omega} x^L(\lambda)\,e^{\lambda t} = -\frac{(\omega^2-\alpha)\sin(\omega t)}{\omega\,(8\alpha-(12+\alpha)\omega^2+\omega^4)},$$
$$y_\omega(t) = 2\,\Re \operatorname*{Res}_{\lambda=\mathbf{j}\omega} y^L(\lambda)\,e^{\lambda t} = \frac{\sin(\omega t)}{\omega\,(8\alpha-(12+\alpha)\omega^2+\omega^4)},$$

and the terms $x_n(t)$ and $y_n(t)$ represent natural oscillations because they are sums of residues over all four pure imaginary roots $\pm\mathbf{j}\omega_k$ $(k=1,2)$ of the equation $8\alpha+(\alpha+12)\lambda^2+\lambda^4=0$:

$$x_n(t) = 2\,\Re\sum_{k=1}^{2}\operatorname*{Res}_{\mathbf{j}\omega_k} x^L(\lambda)\,e^{\lambda t}, \qquad y_n(t) = 2\,\Re\sum_{k=1}^{2}\operatorname*{Res}_{\mathbf{j}\omega_k} y^L(\lambda)\,e^{\lambda t}.$$

When $\alpha=\omega^2$ (so $m_2=1$), the first mass will not contain the oscillating part with frequency ω and it will oscillate due to its own natural frequencies only, which are approximately 1.53073 and 3.69552; therefore, in this mechanical system, mass m_2 will neutralize the effect of the periodic force on the first mass.

This example can be used to model the effect of an earthquake on a multi-story building. Typically, the structural elements in large buildings are made of steel, a highly elastic material. We suppose that the k-th floor of a building has mass m_k and that successive floors are connected by an elastic connector whose effect resembles that of a spring. This example also has an electrical analogy that cable companies use to prevent unsubscribed users from seeing some TV channels.

Problems

1. Consider a molecule of carbon dioxide (CO_2) that consists of three atoms. Numbering the atoms from 1 to 3, starting from the left, we have three position coordinates x_1, x_2, and x_3. We also assume that the atoms in the molecule act as three particles that are connected by springs. The atoms are in the positions, with equilibrium distances between them denoted as d_{12} and d_{23}, called the equilibrium bond lengths. For small displacements, the force is proportional to the amount that the distances between the atoms differ from the equilibrium distance. Show that the equations of motion for the three atoms of masses m_1, m_2, and m_3 are

$$\begin{aligned} m_1\ddot{x}_1 &= k_{12}\,(x_2-x_1-d_{12}),\\ m_2\ddot{x}_2 &= k_{23}\,(x_3-x_2-d_{23}) - k_{12}\,(x_2-x_1-d_{12}),\\ m_3\ddot{x}_3 &= -k_{23}\,(x_3-x_2-d_{23}). \end{aligned}$$

Show that this system of equations can be reduced to a homogeneous system

$$\begin{aligned} \ddot{y}_1 &= \frac{k_{12}}{m_1}\,(y_2-y_1),\\ \ddot{y}_2 &= \frac{k_{23}}{m_2}\,(y_3-y_2) - \frac{k_{12}}{m_2}\,(y_2-y_1),\\ \ddot{y}_3 &= -\frac{k_{23}}{m_3}\,(y_3-y_2), \end{aligned}$$

where $y_1=x_1$, $y_2=x_2-d_{12}$, and $y_3=x_3-d_{12}-d_{23}$. Rewrite this system in the vector form $\ddot{\mathbf{y}}(t)+\mathbf{A}\,\mathbf{y}(t)=\mathbf{0}$, where $\mathbf{y}(t)=\langle y_1,y_2,y_3\rangle^T$. In reality, the numerical values of coefficients of the matrix $\mathbf{A}$ are always truncated rational numbers. For this particular example, choose $m_1=3$, $m_2=15$, $m_3=45/2$, $k_{12}=20$, $k_{23}=45$, $d_{12}=3$, and $d_{23}=1$. Find the general solution of the corresponding vector equation $\ddot{\mathbf{y}}(t)+\mathbf{A}\mathbf{y}(t)=\mathbf{0}$.

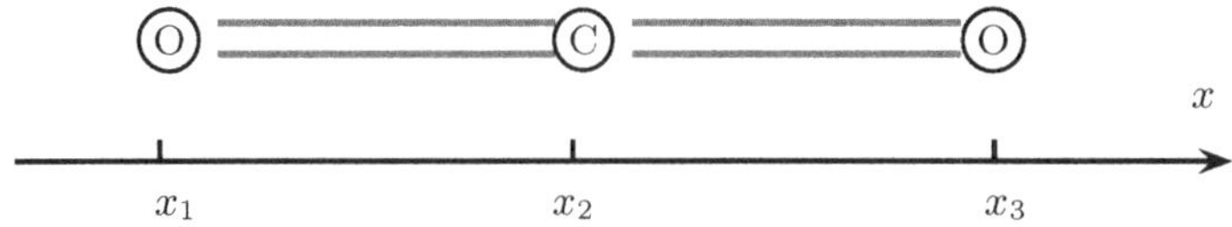

A molecule of carbon dioxide

2. Construct two fundamental matrices for the given system of differential equations of the second order $\ddot{\mathbf{x}} + \mathbf{A}\mathbf{x} = \mathbf{0}$, where the square matrix $\mathbf{A}$ is specified.

(a) $\begin{bmatrix} 50 & 14 \\ 14 & 50 \end{bmatrix}$; **(b)** $\begin{bmatrix} 37 & 1 \\ 12 & 48 \end{bmatrix}$; **(c)** $\begin{bmatrix} 8 & 4 \\ 8 & 12 \end{bmatrix}$; **(d)** $\begin{bmatrix} 14 & 10 \\ 11 & 15 \end{bmatrix}$;

(e) $\begin{bmatrix} 50 & 14 \\ 31 & 67 \end{bmatrix}$; **(f)** $\begin{bmatrix} 50 & 1 \\ 14 & 63 \end{bmatrix}$; **(g)** $\begin{bmatrix} 60 & 11 \\ 21 & 70 \end{bmatrix}$; **(h)** $\begin{bmatrix} 67 & 3 \\ 14 & 78 \end{bmatrix}$.

3. In each exercise, (a) through (d), solve the initial value problem.

(a) $\ddot{x} - 2y = 0, \quad \ddot{y} + 8x + 17y = 0, \quad x(0) = 0,\ \dot{x}(0) = 4 \quad y(0) = 15,\ \dot{y}(0) = -32.$
(b) $\ddot{x} + 11x + 10y = 0, \quad \ddot{y} + 14x + 15y = 0, \quad x(0) = -1,\ \dot{x}(0) = 0 \quad y(0) = 1,\ \dot{y}(0) = 6.$
(c) $\ddot{x} + 6x - 3y = 0, \quad \ddot{y} - 5x + 4y = 0, \quad x(0) = 3,\ \dot{x}(0) = 24, \quad y(0) = 5,\ \dot{y}(0) = 0.$
(d) $\ddot{x} + 15x - 21y = 0, \quad \ddot{y} - 14x + 22y = 0, \quad x(0) = 5,\ \dot{x}(0) = 3, \quad y(0) = 0,\ \dot{y}(0) = 2.$

4. Consider a mechanical system of two masses m_1 and m_2 connected by three elastic springs with spring constants k_1, k_2, and k_3, respectively (see Fig. 6.4, page 344). Find the natural frequencies of the mass-and-spring system and determine the general solution of the corresponding system of differential equations

$$m_1\,\ddot{x} + k_1 x - k_2(y - x) = 0, \qquad m_2\,\ddot{y} + k_3 y + k_2(y - x) = 0.$$

(a) $m_1 = 5,\ m_2 = 1,\ k_1 = 230,\ k_2 = 50,\ k_3 = 11.$
(b) $m_1 = 13,\ m_2 = 11,\ k_1 = 13,\ k_2 = 143,\ k_3 = 11.$
(c) $m_1 = 25,\ m_2 = 23,\ k_1 = 25,\ k_2 = 575,\ k_3 = 23.$
(d) $m_1 = 9,\ m_2 = 20,\ k_1 = 306,\ k_2 = 180,\ k_3 = 740.$
(e) $m_1 = 31,\ m_2 = 32,\ k_1 = 31,\ k_2 = 992,\ k_3 = 32.$
(f) $m_1 = 1,\ m_2 = 1,\ k_1 = 1,\ k_2 = 40,\ k_3 = 1.$
(g) $m_1 = 21,\ m_2 = 19,\ k_1 = 21,\ k_2 = 798,\ k_3 = 19.$
(h) $m_1 = 1,\ m_2 = 1,\ k_1 = 6,\ k_2 = 2,\ k_3 = 3.$

5. For each of the following matrices and given vector function $\mathbf{f}$, solve the inhomogeneous system of equations $\ddot{\mathbf{x}} + \mathbf{A}\mathbf{x} = \mathbf{f}$. Initial conditions can be chosen homogeneous.

(a) $\begin{bmatrix} 29 & 13 \\ 20 & 36 \end{bmatrix}, \quad \mathbf{f}(t) = \begin{bmatrix} \sin 3t \\ \cos 4t \end{bmatrix}$; **(b)** $\begin{bmatrix} 20 & 4 \\ 16 & 32 \end{bmatrix}, \quad \mathbf{f}(t) = \begin{bmatrix} \sin 4t \\ t \end{bmatrix}$;

(c) $\begin{bmatrix} 19 & 3 \\ 6 & 22 \end{bmatrix}, \quad \mathbf{f}(t) = \begin{bmatrix} \sin 4t \\ \cos 5t \end{bmatrix}$; **(d)** $\begin{bmatrix} 55 & 39 \\ 26 & 42 \end{bmatrix}, \quad \mathbf{f}(t) = \begin{bmatrix} \sin 9t \\ 1 \end{bmatrix}$;

(e) $\begin{bmatrix} 34 & 18 \\ 30 & 46 \end{bmatrix}, \quad \mathbf{f}(t) = \begin{bmatrix} \sin 4t \\ t \end{bmatrix}$; **(f)** $\begin{bmatrix} 18 & -7 \\ -14 & 11 \end{bmatrix}, \quad \mathbf{f}(t) = \begin{bmatrix} \sin 4t \\ \cos 5t \end{bmatrix}$.

6. The Fermi–Pasta–Ulam (FPU) problem bears the name of the three scientists who were looking for a theoretical physics problem suitable for an investigation with one of the very first computers. The original idea, proposed by Enrico Fermi, was to simulate the one-dimensional analogue of atoms in a crystal: a long chain of particles linked by springs that obey Hooke's law (a linear interaction), but with a weak nonlinear correction

$$\ddot{u}_n = (u_{n+1} + u_{n-1} - 2u_n) + \alpha\left[(u_{n+1} - u_n)^2 - (u_n - u_{n-1})^2\right], \tag{8.6.15}$$

where α is a parameter. Solve the linear three-dimensional version of the FPU-system assuming that $\alpha = 0$.

7. Prove Theorem 8.18, page 469.

Summary for Chapter 8

1. The system of linear differential equations in normal form

$$\frac{\mathrm{d}\mathbf{x}(t)}{\mathrm{d}t} = \mathbf{P}(t)\,\mathbf{x}(t) + \mathbf{f}(t) \qquad \text{or} \qquad \dot{\mathbf{x}}(t) = \mathbf{P}(t)\,\mathbf{x}(t) + \mathbf{f}(t), \tag{8.1.1}$$

where $\mathbf{P}(t)$ is $n \times n$ matrix, is called a nonhomogeneous system of equations. The homogeneous system

$$\dot{\mathbf{x}}(t) = \mathbf{P}(t)\,\mathbf{x}(t) \tag{8.1.2}$$

is called the complementary system for Eq. (8.1.1), and its solution is referred to as a complementary vector-function (which depends on n arbitrary constants). The column vector $\mathbf{f}(t)$ is usually called the **nonhomogeneous** or **driving term** or the **forcing** or input **function**.

2. Let $\mathbf{x}_1(t)$, $\mathbf{x}_2(t)$, ..., $\mathbf{x}_k(t)$ be a set of solution vectors of the homogeneous system (8.1.2) on an interval $|a, b|$. Their linear combination
$$\mathbf{x}(t) = c_1\,\mathbf{x}_1(t) + c_2\,\mathbf{x}_2(t) + \cdots + c_k\,\mathbf{x}_k(t),$$
where c_i, $i = 1, 2, \ldots, k$, are arbitrary constants, is also a solution to Eq. (8.1.2) on the same interval.

3. Let $\mathbf{P}(t)$ be an $n \times n$ matrix that is continuous on an open interval (a, b). Any set of n solutions $\mathbf{x}_1(t)$, $\mathbf{x}_2(t)$, ..., $\mathbf{x}_n(t)$ to the equation $\dot{\mathbf{x}}(t) = \mathbf{P}(t)\,\mathbf{x}(t)$ that is linearly independent on the interval (a, b) is called a **fundamental set of solutions** (or fundamental solution set). The $n \times n$ matrix
$$\mathbf{X}(t) = [\,\mathbf{x}_1(t), \mathbf{x}_2(t), \ldots, \mathbf{x}_n(t)\,] = \begin{bmatrix} | & | & & | \\ \mathbf{x}_1(t) & \mathbf{x}_2(t) & \cdots & \mathbf{x}_n(t) \\ | & | & & | \end{bmatrix},$$
where its column vectors are solution vectors of the vector equation $\dot{\mathbf{x}}(t) = \mathbf{P}(t)\,\mathbf{x}(t)$, is called a **fundamental matrix** if $\det \mathbf{X}(t) \neq 0$.

4. Let $\mathbf{P}(t)$ be an $n \times n$ matrix with entries $p_{ij}(t)$ that are continuous functions on some interval. Let $\mathbf{x}_k(t)$, $k = 1, 2, \ldots, n$, be n column solutions to the vector differential equation $\dot{\mathbf{x}} = \mathbf{P}(t)\mathbf{x}$. Then for an $n \times n$ matrix formed from these column solutions
$$\mathbf{X}(t) = [\,\mathbf{x}_1(t), \mathbf{x}_2(t), \ldots, \mathbf{x}_n(t)\,],$$
we define its Wronskian as the determinant
$$W(t) = \det \mathbf{X}(t) = C\,\exp\left(\int \mathrm{tr}\mathbf{P}(t)\,dt\right), \qquad \mathrm{tr}\mathbf{P}(t) = p_{11} + p_{22} + \cdots + p_{nn},$$
where C is some constant. Therefore $\det \mathbf{X}(t)$ is either never zero, if $C \neq 0$, or else identically zero, if $C = 0$. The above formula is named after Abel.

5. Let $\mathbf{X}(t)$ be a fundamental matrix for the homogeneous linear system $\dot{\mathbf{x}} = \mathbf{P}(t)\mathbf{x}$. Then the unique solution of the initial value problem
$$\dot{\mathbf{x}} = \mathbf{P}(t)\,\mathbf{x}(t), \qquad \mathbf{x}(t_0) = \mathbf{x}_0 \tag{8.1.8}$$
is explicitly expressed through the propagator
$$\mathbf{x}(t) = \mathbf{\Phi}(t, t_0)\,\mathbf{x}_0 = \mathbf{X}(t)\mathbf{X}^{-1}(t_0)\,\mathbf{x}_0. \tag{8.1.9}$$

6. Let $\mathbf{A}$ be an $n \times n$ matrix with constant entries and let $\mathbf{y}_k(t)$ be the k-th column of the exponential matrix $e^{\mathbf{A}t}$. Then the vector functions $\mathbf{y}_1(t)$, $\mathbf{y}_2(t)$, ..., $\mathbf{y}_n(t)$ are linearly independent.

7. Let $\mathbf{X}(t) = [\,\mathbf{y}_1(t), \mathbf{y}_2(t), \ldots, \mathbf{y}_n(t)\,]$ be a fundamental solution set for the vector differential equation $\dot{\mathbf{y}}(t) = \mathbf{A}\,\mathbf{y}(t)$ with constant square matrix $\mathbf{A}$. Then
$$e^{\mathbf{A}(t-t_0)} = \mathbf{X}(t)\mathbf{X}^{-1}(t_0).$$

8. For any constant $n \times n$ matrix $\mathbf{A}$, the vector
$$\mathbf{y}(t) = e^{\mathbf{A}t}\,\mathbf{c} = c_1\,\mathbf{y}_1 + c_2\,\mathbf{y}_2 + \cdots + c_n\,\mathbf{y}_n$$
is the general solution of the linear vector differential equation $\dot{\mathbf{y}} = \mathbf{A}\,\mathbf{y}(t)$. Here $\mathbf{y}_k(t)$ is the k-th column of the exponential matrix $\mathbf{\Phi}(t) = e^{\mathbf{A}t}$ and $\mathbf{c} = \langle c_1, c_2, \ldots, c_n\rangle^T$ is a constant column vector with entries c_k, $k = 1, 2, \ldots, n$. Moreover, the column vector
$$\mathbf{y}(t) = e^{\mathbf{A}t}\,\mathbf{x}_0$$
satisfies the initial condition $\mathbf{y}(0) = \mathbf{x}_0$.

9. If all solutions of the homogeneous linear vector differential equations with constant coefficients
$$\dot{\mathbf{y}}(t) = \mathbf{A}\,\mathbf{y}(t), \tag{8.2.1}$$
where $\mathbf{A}$ is an $n \times n$ constant matrix, that starts in a small neighborhood of the origin approach zero when t is large, we call the origin the **attractor** or **sink**. If opposite, all solutions leave the origin, we call it a **repeller** or **source** and refer to the origin as **unstable**.

10. A critical point $\mathbf{x}^*$ of an autonomous vector equation $\dot{\mathbf{x}} = \mathbf{f}(\mathbf{x})$ is stable if all solutions that start sufficiently close to $\mathbf{x}^*$ remain close to it. A critical point is said to be asymptotically stable if all solutions originated in a neighborhood of $\mathbf{x}^*$ approach it. There are four kinds of stability:
    - A center is stable, but not asymptotically stable.
    - A sink is asymptotically stable.

- A source is unstable and all trajectories recede from the critical point.
- A saddle point is unstable, although some trajectories are drawn to the critical point and other trajectories recede.

11. There are several types of critical points:
    - Proper node (stable or unstable)
    - Improper node (stable or unstable)
    - Spiral (stable or unstable)
    - Center (always stable)
    - Saddle point (always unstable)
12. A particular solution of the nonhomogeneous linear vector equation with variable coefficients
$$\dot{\mathbf{y}}(t) = \mathbf{P}(t)\,\mathbf{y} + \mathbf{f}(t), \tag{8.3.1}$$
can be obtained explicitly
$$\mathbf{y}_p(t) = \mathbf{X}(t)\int \mathbf{X}^{-1}(t)\,\mathbf{f}(t)\,\mathrm{d}t = \mathbf{X}(t)\int_{t_0}^{t} \mathbf{X}^{-1}(\tau)\,\mathbf{f}(\tau)\,\mathrm{d}\tau,$$
provided that the fundamental matrix $\mathbf{X}(t)$ is known.
13. If $\mathbf{A}$ is a constant square matrix, a particular solution to the vector equation $\dot{\mathbf{y}}(t) = \mathbf{A}\,\mathbf{y}(t) + \mathbf{f}(t)$ is
$$\mathbf{y}_p(t) = \int_{t_0}^{t} e^{\mathbf{A}(t-\tau)}\,\mathbf{f}(\tau)\,\mathrm{d}\tau = e^{\mathbf{A}t}\int_{t_0}^{t} e^{-\mathbf{A}\tau}\,\mathbf{f}(\tau)\,\mathrm{d}\tau.$$
14. The method of undetermined coefficients is a special technique used to find a particular solution of the driven vector equation $\mathbf{x}'(t) = \mathbf{A}\,\mathbf{x} + \mathbf{f}(t)$ with constant matrix $\mathbf{A}$ when the driving term $\mathbf{f}(t)$ is a linear combination (with constant vector coefficients) of products of polynomials, exponential functions, and sines and cosines.
15. The resolvent of a constant square matrix $\mathbf{A}$ is the Laplace transform of the exponential matrix: $\mathbf{R}_\lambda(\mathbf{A}) = (\lambda\mathbf{I} - \mathbf{A})^{-1} = \mathcal{L}\left[e^{\mathbf{A}t}\right]$.
16. The second order constant coefficient vector differential equation $\ddot{\mathbf{x}} + \mathbf{A}\mathbf{x} = \mathbf{0}$ has two fundamental matrices: $\mathbf{A}^{-1/2}\sin\left(t\mathbf{A}^{-1/2}\right)$ and $\cos\left(t\mathbf{A}^{-1/2}\right)$. Their definitions do not depend on the chosen root.

Review Questions for Chapter 8

Section 8.1

1. In each exercise, verify that two given matrix functions, $\mathbf{X}(t)$ and $\mathbf{Y}(t)$, are fundamental matrices for the given vector differential equation $\dot{\mathbf{x}} = \mathbf{P}(t)\mathbf{x}$, with specified square matrix $\mathbf{P}(t)$. Find a constant nonsingular matrix $\mathbf{C}$ such that $\mathbf{X}(t) = \mathbf{Y}(t)\mathbf{C}$.
$$\text{(a) } \mathbf{P} = \begin{bmatrix} 1+t^{-1} & -t^{-1} \\ \frac{t-2}{t-1} & \frac{1}{t-1} \end{bmatrix}, \quad \mathbf{X} = \begin{bmatrix} t & e^t \\ t^2 & e^t \end{bmatrix}, \quad \mathbf{Y} = \frac{1}{2}\begin{bmatrix} 3e^t - 4t & 2t - e^t \\ 3e^t - 4t^2 & 2t^2 - e^t \end{bmatrix};$$
$$\text{(b) } \mathbf{P} = \begin{bmatrix} 1 & t \\ 0 & -t^{-1} \end{bmatrix}, \quad \mathbf{X} = \begin{bmatrix} -1 & e^t \\ t^{-1} & 0 \end{bmatrix}, \quad \mathbf{Y} = \begin{bmatrix} -1 & 1 + e^t/6 \\ 1/t & -1/t \end{bmatrix};$$
$$\text{(c) } \mathbf{P} = \frac{1}{t^3}\begin{bmatrix} t^2+t & 1 \\ -t^2 & 2t^2 - t \end{bmatrix}, \quad \mathbf{X} = \begin{bmatrix} -t & t-1 \\ t^2 & t \end{bmatrix}, \quad \mathbf{Y} = \begin{bmatrix} 1-3t & 5t/2 - 1 \\ 2t^2 - t & t - 3t^2/2 \end{bmatrix}.$$
2. Prove that if $\mathbf{x}(t) = \mathbf{u}(t) + \mathbf{j}\mathbf{v}(t)$ is a complex-valued solution of the vector equation $\mathbf{x}'(t) = \mathbf{P}(t)\,\mathbf{x}(t)$ ($\det\mathbf{P} \ne 0$,) for a given real-valued square matrix $\mathbf{P}(t)$ with continuous coefficients, then its real part $\mathbf{u}(t) = \mathrm{Re}\,\mathbf{x} = \Re\mathbf{x}(t)$ and its imaginary part $\mathbf{v}(t) = \mathrm{Im}\,\mathbf{x} = \Im\mathbf{x}(t)$ are real-valued solutions to this equation.
3. Show that $\mathbf{X}(t)$ is a fundamental matrix for the linear vector system $\dot{\mathbf{x}} = \mathbf{P}(t)\,\mathbf{x}$ if and only if $\mathbf{X}^{-1}(t)$ is a fundamental matrix for the system $\dot{\mathbf{x}}^T = -\mathbf{x}^T\mathbf{P}(t)$.
4. Suppose that $\mathbf{x}(t)$ is a solution of the $n \times n$ system $\dot{\mathbf{x}} = \mathbf{P}(t)\,\mathbf{x}$ on some interval $|a, b|$, and that the $n \times n$ matrix $\mathbf{A}(t)$ is not singular ($\det\mathbf{A} \ne 0$) and differentiable on $|a, b|$. Find a matrix $\mathbf{B}$ such that the function $\mathbf{y} = \mathbf{A}\mathbf{x}$ is a solution of $\dot{\mathbf{y}} = \mathbf{B}\mathbf{y}$ on $|a, b|$.

5. Rewrite the second order Euler equation $at^2\ddot{y} + bt\,\dot{y} + cy = 0$ in a vector form $t\dot{\mathbf{x}} = \mathbf{A}\mathbf{x}$, with a constant matrix $\mathbf{A}$.
6. In exercises (a) through (h), find a general solution to the given Euler vector equation (8.1.14) for $t > 0$ with the given 2×2 matrix $\mathbf{A}$.

(a) $\begin{bmatrix} 2 & -1 \\ 7 & 10 \end{bmatrix}$; (b) $\begin{bmatrix} 1 & -1 \\ -3 & 3 \end{bmatrix}$; (c) $\begin{bmatrix} 1 & 3 \\ 4 & 2 \end{bmatrix}$; (d) $\begin{bmatrix} 5 & 3 \\ 2 & 4 \end{bmatrix}$;
(e) $\begin{bmatrix} 3 & 2 \\ 1 & 2 \end{bmatrix}$; (f) $\begin{bmatrix} 1 & 2 \\ 2 & 4 \end{bmatrix}$; (g) $\begin{bmatrix} -1 & 7 \\ 0 & 4 \end{bmatrix}$; (h) $\begin{bmatrix} 1 & 6 \\ 0 & 6 \end{bmatrix}$.

7. Let $y_1(t), \ldots, y_n(t)$ be solutions of the homogeneous equation corresponding to (8.1.10), and $\mathbf{x}_1 = \langle y_1, y_1', \ldots, y_1^{(n-1)} \rangle^T$, $\ldots$, $\mathbf{x}_n = \langle y_n, y_n', \ldots, y_n^{(n-1)} \rangle^T$ be solutions of the equivalent vector equation $\dot{\mathbf{x}} = \mathbf{P}(t)\,\mathbf{x}$, where the square matrix $\mathbf{P}(t)$ is given in Eq. (8.1.13). Show that the Wronskian of the set of functions $\{y_1, \ldots, y_n\}$ and the Wronskian of the column vectors $\mathbf{x}_1, \ldots, \mathbf{x}_n$ are the same.

Section 8.2 of Chapter 8 (Review)

1. In each exercise (a) through (d), verify that the matrix function written in vector form $\mathbf{Y}(t) = \left[e^{\lambda_1 t}\mathbf{v}_1, e^{\lambda_2 t}\mathbf{v}_2\right]$ is the fundamental matrix for the given vector differential equation $\dot{\mathbf{y}} = \mathbf{A}\mathbf{y}$, with specified constant diagonalizable matrix $\mathbf{A}$. Then find a constant nonsingular matrix $\mathbf{C}$ such that $\mathbf{Y}(t) = e^{\mathbf{A}t}\mathbf{C}$.

(a) $\mathbf{A} = \begin{bmatrix} -3 & -15 \\ 3 & 3 \end{bmatrix}$, $\quad \mathbf{Y}(t) = \left(\begin{bmatrix} \cos 6t - 8 \sin 6t \\ 2 \cos 6t - \sin 6t \end{bmatrix}, \begin{bmatrix} 3 \cos 6t + 2 \sin 6t \\ \sin 6t \end{bmatrix} \right)$;

(b) $\mathbf{A} = \begin{bmatrix} 4 & 5 \\ 2 & 1 \end{bmatrix}$, $\quad \mathbf{Y}(t) = \left(e^{6t} \begin{bmatrix} 5 \\ 2 \end{bmatrix}, e^{-t} \begin{bmatrix} 1 \\ -1 \end{bmatrix} \right)$;

(c) $\mathbf{A} = \begin{bmatrix} 11 & -9 \\ 1 & 5 \end{bmatrix}$, $\quad \mathbf{Y}(t) = e^{8t} \left(\begin{bmatrix} 3 \\ 1 \end{bmatrix}, \begin{bmatrix} 4 \\ 1 \end{bmatrix} + t \begin{bmatrix} 3 \\ 1 \end{bmatrix} \right)$;

(d) $\mathbf{A} = \begin{bmatrix} 1 & -8 \\ 1 & -3 \end{bmatrix}$, $\quad \mathbf{Y}(t) = e^{-t} \left(\begin{bmatrix} 4 \cos 2t \\ \sin 2t + \cos 2t \end{bmatrix}, \begin{bmatrix} 2 \cos 2t + 2 \sin 2t \\ \sin 2t \end{bmatrix} \right)$.

2. Compute the propagator matrix $e^{\mathbf{A}t}$ for each system $\dot{\mathbf{y}}(t) = \mathbf{A}\,\mathbf{y}(t)$ given in problems (a) through (d).

(a) $\dot{x} = x - 3y$, $\dot{y} = 4x - 12y$;
(b) $\dot{x} = x - 2y$, $\dot{y} = 2x - 3y$;
(c) $\dot{x} = x + y$, $\dot{y} = -4x + 6y$;
(d) $\dot{x} = 3x - 2y$, $\dot{y} = 5x - 3y$.

3. In each of exercises (a) through (h):
   - Find the general solution of the given system of homogeneous equations and describe the behavior of its solutions as $t \to +\infty$.
   - Draw a direction field and plot a few trajectories of the planar system.

(a) $\begin{bmatrix} 8 & -1 \\ 1 & 10 \end{bmatrix}$; (b) $\begin{bmatrix} -1 & 1 \\ -5 & 3 \end{bmatrix}$; (c) $\begin{bmatrix} 1 & 2 \\ 2 & -2 \end{bmatrix}$; (d) $\begin{bmatrix} 3 & -5 \\ 5 & -3 \end{bmatrix}$;
(e) $\begin{bmatrix} 4 & -2 \\ 1 & 1 \end{bmatrix}$; (f) $\begin{bmatrix} 0 & 3 \\ 1 & 2 \end{bmatrix}$; (g) $\begin{bmatrix} -3 & -1 \\ 13 & 3 \end{bmatrix}$; (h) $\begin{bmatrix} 3 & -5 \\ 5 & -5 \end{bmatrix}$.

4. Find the general solution of the homogeneous system of differential equations $\dot{\mathbf{x}} = \mathbf{A}\,\mathbf{x}$ for the given square matrix $\mathbf{A}$ and determine the stability or instability of the origin.

(a) $\begin{bmatrix} 1 & 2 \\ 5 & -2 \end{bmatrix}$; (b) $\begin{bmatrix} -3 & -1 \\ 29 & 1 \end{bmatrix}$; (c) $\begin{bmatrix} -3 & -2 \\ 9 & 3 \end{bmatrix}$; (d) $\begin{bmatrix} -5 & 2 \\ -18 & 7 \end{bmatrix}$;
(e) $\begin{bmatrix} 11 & -22 \\ 5 & -10 \end{bmatrix}$; (f) $\begin{bmatrix} 7 & -3 \\ 16 & -7 \end{bmatrix}$; (g) $\begin{bmatrix} 4 & -3 \\ 4 & -4 \end{bmatrix}$; (h) $\begin{bmatrix} 24 & -7 \\ 7 & 38 \end{bmatrix}$.

5. In each of exercises (a) – (f), solve the initial value problem $\dot{y} = \mathbf{A}\mathbf{y}$, $\mathbf{y}(0) = \mathbf{y}_0$ for the given 3×3 matrix $\mathbf{A}$ and 3×1 initial vector $\mathbf{y}_0$.

(a) $\begin{bmatrix} 2 & 1 & 0 \\ 0 & 2 & 0 \\ 0 & 0 & 1 \end{bmatrix}$, $\mathbf{y}(0) = \begin{bmatrix} 1 \\ 2 \\ 3 \end{bmatrix}$; (b) $\begin{bmatrix} 14 & 66 & -42 \\ 4 & 24 & -14 \\ 10 & 55 & -33 \end{bmatrix}$, $\mathbf{y}(0) = \begin{bmatrix} 1 \\ -1 \\ -1 \end{bmatrix}$;

(c) $\begin{bmatrix} 0 & 1 & 0 \\ 4 & 3 & -4 \\ 1 & 2 & -1 \end{bmatrix}$, $\mathbf{y}(0) = \begin{bmatrix} 2 \\ 2 \\ -1 \end{bmatrix}$; (d) $\begin{bmatrix} -2 & 0 & 3 \\ 0 & 4 & 0 \\ -6 & 0 & 7 \end{bmatrix}$, $\mathbf{y}(0) = \begin{bmatrix} 1 \\ 2 \\ -1 \end{bmatrix}$;

(e) $\begin{bmatrix} 3 & 0 & 2 \\ 0 & 3 & -2 \\ 2 & -2 & 1 \end{bmatrix}$, $\mathbf{y}(0) = \begin{bmatrix} 1 \\ 1 \\ 1 \end{bmatrix}$; (f) $\begin{bmatrix} 1 & 2 & 0 \\ 0 & 1 & 0 \\ -3 & 3 & 5 \end{bmatrix}$, $\mathbf{y}(0) = \begin{bmatrix} 2 \\ -1 \\ 2 \end{bmatrix}$.

6. Find the values of the real constant α for which $\mathbf{x} = \mathbf{0}$ is the steady state of $\dot{\mathbf{x}} = \mathbf{A}\,\mathbf{x}$.

(a) $\begin{bmatrix} 1 & -\alpha \\ 3 & -4 \end{bmatrix}$; (b) $\begin{bmatrix} 2 & -\alpha \\ 5 & 7 \end{bmatrix}$; (c) $\begin{bmatrix} -8 & 13 \\ -\alpha & 4 \end{bmatrix}$.

7. A matrix $\mathbf{A}$ is said to be nilpotent if there exists some positive integer p such that $\mathbf{A}^p = \mathbf{0}$. Show that the following 2×2 matrices are nilpotent and plot the phase portraits for the corresponding vector equation $\dot{\mathbf{y}}(t) = \mathbf{A}\,\mathbf{y}(t)$.

(a) $\mathbf{A} = \begin{bmatrix} 1 & -1 \\ 1 & -1 \end{bmatrix}$; (b) $\mathbf{A} = \begin{bmatrix} 3 & -1 \\ 9 & -3 \end{bmatrix}$; (c) $\mathbf{A} = \begin{bmatrix} 2 & -1 \\ 4 & -2 \end{bmatrix}$; (d) $\mathbf{A} = \begin{bmatrix} 4 & -2 \\ 8 & -4 \end{bmatrix}$;

(e) $\mathbf{A} = \begin{bmatrix} 5 & -1 \\ 25 & -5 \end{bmatrix}$; (f) $\mathbf{A} = \begin{bmatrix} 6 & -9 \\ 4 & -6 \end{bmatrix}$; (g) $\mathbf{A} = \begin{bmatrix} 7 & -1 \\ 49 & -7 \end{bmatrix}$; (h) $\mathbf{A} = \begin{bmatrix} 9 & -3 \\ 27 & -9 \end{bmatrix}$.

8. Show that the matrix $\mathbf{A}$ from Example 8.2.14 is not nilpotent. Evaluate the trace and determinant of this matrix.

9. Find the general solution of the homogeneous system of differential equations $\dot{\mathbf{y}} = \mathbf{A}\,\mathbf{y}$ for the given square matrix $\mathbf{A}$ and determine the stability or instability of the origin.

(a) $\begin{bmatrix} 2 & 1 & 2 \\ -1 & 0 & -2 \\ 0 & 0 & 1 \end{bmatrix}$; (b) $\begin{bmatrix} 1 & 2 & 2 \\ -1 & -1 & 0 \\ 0 & 0 & 1 \end{bmatrix}$; (c) $\begin{bmatrix} -5 & 5 & 2 \\ -10 & 5 & 4 \\ -20 & 10 & 9 \end{bmatrix}$;

(d) $\begin{bmatrix} 1 & 0 & 1 \\ 1 & 0 & 1 \\ 0 & 1 & 1 \end{bmatrix}$, (e) $\begin{bmatrix} -19 & 12 & 13 \\ 1 & 0 & -1 \\ -5 & 4 & 3 \end{bmatrix}$; (f) $\begin{bmatrix} 2 & 5 & 1 \\ 4 & 1 & -1 \\ -4 & -5 & -3 \end{bmatrix}$;

(g) $\begin{bmatrix} 2 & 7 & 1 \\ -1 & 4 & 1 \\ 3 & 1 & 0 \end{bmatrix}$; (h) $\begin{bmatrix} 1 & 4 & 2 \\ 2 & 1 & 4 \\ 3 & 2 & 6 \end{bmatrix}$; (i) $\begin{bmatrix} 1 & -3 & 2 \\ 1 & 3 & 2 \\ 2 & 1 & 4 \end{bmatrix}$;

(j) $\begin{bmatrix} 4 & -3 & 1 \\ 1 & 2 & 3 \\ 4 & -2 & 2 \end{bmatrix}$; (k) $\begin{bmatrix} 5 & -2 & 3 \\ 1 & 2 & 3 \\ 4 & -2 & -2 \end{bmatrix}$; (l) $\begin{bmatrix} 4 & -2 & 3 \\ 3 & 1 & 2 \\ 4 & -2 & 3 \end{bmatrix}$.

10. Draw a phase portrait for each of the following linear systems of linear differential equations $\dot{\mathbf{y}}(t) = \mathbf{A}\,\mathbf{y}(t)$, where

(a) $\mathbf{A} = \begin{bmatrix} 1 & 1 \\ 1 & 1 \end{bmatrix}$; (b) $\mathbf{A} = \begin{bmatrix} 1 & 1 \\ -1 & -1 \end{bmatrix}$; (c) $\mathbf{A} = \begin{bmatrix} 2 & -4 \\ 1 & -2 \end{bmatrix}$.

Then solve these systems of equations. Is there a line $y = kx$ that separates solutions into different categories of behavior?

Section 8.3 of Chapter 8 (Review)

1. In exercises (a) through (d), use the method of *variation of parameters* to find a particular solution to the vector equation $\dot{\mathbf{y}}(t) = \mathbf{A}\mathbf{y}(t) + \mathbf{f}(t)$ with a constant 3×3 matrix $\mathbf{A}$ that satisfies the initial conditions $\mathbf{y}(0) = \langle 1, 2, 3 \rangle^T$.

(a) $\begin{bmatrix} 3 & -3 & 1 \\ 0 & 2 & 2 \\ 5 & 1 & 1 \end{bmatrix}$, $\mathbf{f} = \begin{bmatrix} 0 \\ 29\,e^{-t} \\ 39\,e^{t} \end{bmatrix}$; (b) $\begin{bmatrix} 2 & 1 & -1 \\ 0 & 1 & 1 \\ 1 & 0 & 1 \end{bmatrix}$, $\mathbf{f} = \begin{bmatrix} 5 \sin t \\ -10 \cos t \\ 2 \end{bmatrix}$;

(c) $\begin{bmatrix} -3 & 3 & 1 \\ 1 & -5 & -3 \\ -3 & 7 & 3 \end{bmatrix}$, $\mathbf{f} = \begin{bmatrix} 5 \sin 2t \\ 5 \cos 2t \\ 23\,e^{t} \end{bmatrix}$; (d) $\begin{bmatrix} -3 & 1 & -3 \\ 4 & -1 & 2 \\ 4 & -2 & 3 \end{bmatrix}$, $\mathbf{f} = e^{t} \begin{bmatrix} 2 \\ 4 \\ 4 \end{bmatrix}$.

2. In exercises (a) through (f), use the method of *variation of parameters* to find a particular solution to the vector equation $\dot{\mathbf{y}}(t) = \mathbf{A}\,\mathbf{y}(t) + \mathbf{f}(t)$ with a constant matrix $\mathbf{A}$.

(a) $\begin{bmatrix} -16 & 9 & 4 \\ -14 & 7 & 4 \\ -38 & 18 & 11 \end{bmatrix}$, $\mathbf{f} = \begin{bmatrix} 3\,e^{t} \\ 4\,e^{-3t} \\ 4\,e^{-t} \end{bmatrix}$; (b) $\begin{bmatrix} -1 & -4 & 0 \\ -1 & 1 & -1 \\ -2 & 4 & -3 \end{bmatrix}$, $\mathbf{f} = \begin{bmatrix} 12\,e^{3t} \\ e^{t} \\ 4\,e^{-t} \end{bmatrix}$;

(c) $\begin{bmatrix} -2 & 16 & 4 \\ -1 & 6 & 1 \\ 6 & 1 & 2 \end{bmatrix}$, $\mathbf{f} = \begin{bmatrix} 53\cos 2t \\ 13 \\ 5\,e^{2t} \end{bmatrix}$; (d) $\begin{bmatrix} -3 & 8 & -5 \\ 3 & -1 & 3 \\ 8 & -8 & 10 \end{bmatrix}$, $\mathbf{f} = \begin{bmatrix} 3\,e^{5t} \\ 6\,e^{2t} \\ 9\,e^{-t} \end{bmatrix}$;

(e) $\begin{bmatrix} 2 & -1 & 4 \\ 1 & -1 & 1 \\ -1 & 1 & -1 \end{bmatrix}$, $\mathbf{f} = \begin{bmatrix} 2\sin t \\ 2\cos t \\ 1 \end{bmatrix}$; (f) $\begin{bmatrix} 4 & -4 & -4 \\ 6 & -8 & -12 \\ -4 & 6 & 9 \end{bmatrix}$, $\mathbf{f} = \begin{bmatrix} 5\sin 2t \\ 5\cos 2t \\ 1 \end{bmatrix}$.

3. In exercises (a) through (f), find a particular solution to the Euler vector equation $t\,\dot{\mathbf{y}}(t) = \mathbf{A}\,\mathbf{y}(t) + \mathbf{f}(t)$. Before using the method of *variation of parameters*, find a complementary solution of the homogeneous equation $t\,\dot{\mathbf{y}}(t) = \mathbf{A}\mathbf{y}(t)$.

(a) $\begin{bmatrix} 7 & 1 \\ 3 & 9 \end{bmatrix}$, $\mathbf{f} = t^{8}\begin{bmatrix} 2 \\ -3 \end{bmatrix}$; (d) $\begin{bmatrix} 1 & 5 \\ 3 & -13 \end{bmatrix}$, $\mathbf{f} = t^{-2}\begin{bmatrix} -5 \\ 12 \end{bmatrix}$;

(b) $\begin{bmatrix} 7 & -5 \\ 1 & 13 \end{bmatrix}$, $\mathbf{f} = t^{8}\begin{bmatrix} 2 \\ -1 \end{bmatrix}$; (e) $\begin{bmatrix} -2 & 2 \\ 2 & 1 \end{bmatrix}$, $\mathbf{f} = \begin{bmatrix} 5\,t \\ 5/t \end{bmatrix}$;

(c) $\begin{bmatrix} -6 & 4 \\ -5 & 3 \end{bmatrix}$, $\mathbf{f} = t^{-1}\begin{bmatrix} 4 \\ 3 \end{bmatrix}$; (f) $\begin{bmatrix} 2 & 9 \\ 5 & -2 \end{bmatrix}$, $\mathbf{f} = \begin{bmatrix} 14\,t \\ 14/t^2 \end{bmatrix}$.

4. In exercises (a) through (d), find the solution to the Euler vector equation $t\,\dot{\mathbf{y}}(t) = \mathbf{A}\,\mathbf{y}(t) + \mathbf{f}(t)$. Before using the method of *variation of parameters*, find a complementary solution of the homogeneous equation $t\,\dot{\mathbf{y}}(t) = \mathbf{A}\mathbf{y}(t)$.

(a) $\begin{bmatrix} 1 & 5 & 0 \\ 1 & 0 & 1 \\ 1 & -2 & 2 \end{bmatrix}$, $\mathbf{f} = \begin{bmatrix} 4 \\ 4 \\ 4 \end{bmatrix}$; (c) $\begin{bmatrix} -2 & 3 & 1 \\ -8 & 13 & 5 \\ 11 & -17 & -6 \end{bmatrix}$, $\mathbf{f} = \begin{bmatrix} 12\,t^{3} \\ 9\,t^{2} \\ 6 \end{bmatrix}$;

(b) $\begin{bmatrix} 1 & 10 & 0 \\ 1 & 0 & 1 \\ 1 & -2 & 2 \end{bmatrix}$, $\mathbf{f} = \begin{bmatrix} 9 \\ t \\ t^{2} \end{bmatrix}$; (d) $\begin{bmatrix} 3 & -8 & -10 \\ -2 & 7 & 9 \\ 2 & -6 & -8 \end{bmatrix}$, $\mathbf{f} = \begin{bmatrix} 9t \\ 2t^{2} \\ 30t^{3} \end{bmatrix}$.

Section 8.4 of Chapter 8 (Review)

1. In exercises (a) through (f), use the *method of undetermined coefficients* to determine only the form of a particular for the system $\dot{\mathbf{x}}(t) = \mathbf{A}\mathbf{x}(t) + \mathbf{f}(t)$ with a given constant 3×3 matrix $\mathbf{A}$, and specified vector $\mathbf{f}$.

(a) $\begin{bmatrix} 1 & 12 & -8 \\ -1 & 9 & -4 \\ -2 & 9 & -2 \end{bmatrix}$, $\mathbf{f} = e^{-3t}\begin{bmatrix} 8 \\ 8 \\ 1 \end{bmatrix}$; (b) $\begin{bmatrix} 1 & 12 & -8 \\ -1 & 9 & -4 \\ -7 & 6 & -7 \end{bmatrix}$, $\mathbf{f} = e^{2t}\begin{bmatrix} 78 \\ 43 \\ 77 \end{bmatrix}$;

(c) $\begin{bmatrix} 1 & -1 & -2 \\ 1 & -2 & -3 \\ -4 & 1 & -1 \end{bmatrix}$, $\mathbf{f} = \begin{bmatrix} 24\,e^{3t} \\ 4\,e^{t} \\ 4\,e^{-2t} \end{bmatrix}$; (d) $\begin{bmatrix} -20 & 11 & 13 \\ 16 & 0 & -8 \\ -48 & 21 & 31 \end{bmatrix}$, $\mathbf{f} = e^{t}\begin{bmatrix} 29 \\ 96 \\ 42 \end{bmatrix}$;

(e) $\begin{bmatrix} 3 & -3 & -4 \\ 2 & -8 & -12 \\ -4 & 6 & 9 \end{bmatrix}$, $\mathbf{f} = e^{t}\begin{bmatrix} 12 \\ 4 \\ -3 \end{bmatrix}$; (f) $\begin{bmatrix} -3 & 8 & -5 \\ 3 & -1 & 3 \\ 8 & -8 & 10 \end{bmatrix}$, $\mathbf{f} = 3\,e^{2t}\begin{bmatrix} 1 \\ 1 \\ 1 \end{bmatrix}$.

2. Using matrix algebra technique and the method of *undetermined coefficients*, solve the nonhomogeneous vector equations subject to the initial condition $\mathbf{y}(0) = \langle 1, 0 \rangle^{T}$.

(a) $\dot{\mathbf{y}} = \begin{bmatrix} 8 & 13 \\ 4 & -1 \end{bmatrix}\mathbf{y} - \begin{bmatrix} 8\sin t + 12\cos t \\ 3\sin t - \cos t \end{bmatrix}$; (d) $\dot{\mathbf{y}} = \begin{bmatrix} 7 & -5 \\ 1 & 5 \end{bmatrix}\mathbf{y} + \begin{bmatrix} 5t - 7 \\ -5t \end{bmatrix}$;

(b) $\dot{\mathbf{y}} = \begin{bmatrix} 9 & 7 \\ 4 & 12 \end{bmatrix}\mathbf{y} - \begin{bmatrix} 10 - 2t \\ 4 + 7t \end{bmatrix}$; (e) $\dot{\mathbf{y}} = \begin{bmatrix} 8 & -9 \\ 9 & 8 \end{bmatrix}\mathbf{y} - \begin{bmatrix} 7 - 10t \\ 7 + 25t \end{bmatrix}$;

(c) $\dot{\mathbf{y}} = \begin{bmatrix} 7 & -4 \\ 1 & 3 \end{bmatrix}\mathbf{y} + e^{5t}\begin{bmatrix} 2 \\ 1 \end{bmatrix}$; (f) $\dot{\mathbf{y}} = \begin{bmatrix} -7 & 4 \\ -5 & 5 \end{bmatrix}\mathbf{y} + \begin{bmatrix} 64\,e^{-5t} \\ 4\,e^{3t} \end{bmatrix}$.

3. In exercises (a) through (d), apply the technique of *undetermined coefficients* to find a particular solution of the nonhomogeneous system of equations $\dot{\mathbf{y}}(t) = \mathbf{A}\mathbf{y}(t) + \mathbf{f}(t)$, given the constant 2×2 matrix $\mathbf{A}$ and the forcing vector function $\mathbf{f}(t)$.

(a) $\begin{bmatrix} 8 & 13 \\ 6 & -15 \end{bmatrix}$, $\mathbf{f} = \begin{bmatrix} 69\,e^{5t} \\ 52\,e^{-2t} \end{bmatrix}$;

(b) $\begin{bmatrix} 4 & -3 \\ 2 & 11 \end{bmatrix}$, $\mathbf{f} = \begin{bmatrix} 75\,e^{-5t} \\ 28\,e^{-2t} \end{bmatrix}$;

(c) $\begin{bmatrix} 7 & -1 \\ 18 & -4 \end{bmatrix}$, $\mathbf{f} = \begin{bmatrix} 2\,e^{2t} \\ 30\,e^{-5t} \end{bmatrix}$;

(d) $\begin{bmatrix} 4 & -3 \\ 6 & -2 \end{bmatrix}$, $\mathbf{f} = e^{-t} \begin{bmatrix} 5 \\ 6 \end{bmatrix}$.

4. Apply the technique of *undetermined coefficients* to find a particular solution of the nonhomogeneous system of equations $\dot{\mathbf{y}}(t) = \mathbf{A}\mathbf{y}(t) + \mathbf{f}(t)$, given the constant 3×3 matrix $\mathbf{A}$ and the forcing vector function $\mathbf{f}(t)$.

(a) $\begin{bmatrix} -4 & -2 & 1 \\ 4 & 2 & -2 \\ 8 & 4 & 0 \end{bmatrix}$, $\mathbf{f} = e^{-t} \begin{bmatrix} 1 \\ 2 \\ -21 \end{bmatrix}$;

(b) $\begin{bmatrix} 1 & 4 & 0 \\ 1 & -1 & 1 \\ 2 & -4 & 3 \end{bmatrix}$, $\mathbf{f} = e^{t} \begin{bmatrix} 4 \\ 2 \\ 4 \end{bmatrix}$;

(c) $\begin{bmatrix} 1 & 4 & 1 \\ -3 & -6 & -1 \\ 3 & 4 & -3 \end{bmatrix}$, $\mathbf{f} = 37\,e^{3t} \begin{bmatrix} 1 \\ 0 \\ 1 \end{bmatrix}$;

(d) $\begin{bmatrix} -16 & 9 & 4 \\ -14 & 7 & 4 \\ -38 & 18 & 11 \end{bmatrix}$, $\mathbf{f} = e^{-t} \begin{bmatrix} 3 \\ 2 \\ -2 \end{bmatrix}$;

(e) $\begin{bmatrix} -19 & 12 & 84 \\ 2 & 5 & 2 \\ -8 & 4 & 33 \end{bmatrix}$, $\mathbf{f} = e^{t} \begin{bmatrix} 3 \\ 1 \\ 1 \end{bmatrix}$;

(f) $\begin{bmatrix} 10 & -2 & 2 \\ -2 & 1 & 2 \\ 2 & 2 & 10 \end{bmatrix}$, $\mathbf{f} = \begin{bmatrix} -16 \\ 2 \\ -8 \end{bmatrix}$.

Section 8.5 of Chapter 8 (Review)

1. Solve the initial value problems with intermittent driving terms, containing the Heaviside function $H(t)$ (see Definition 5.3 on page 274).

(a) $\dot{x} = -2x + y + H(t) - H(t-1)$, $\quad x(0) = 3$,
$\dot{y} = 6x + 3y + 3\,H(t) - 3\,H(t-2)$; $\quad y(0) = 1$.

(b) $\dot{x} = 3x - 2y + 17\,\cos 2t\,[H(t) - H(t-\pi)]$, $\quad x(0) = 2$,
$\dot{y} = 4x - y + 17\,\sin 2t\,[H(t) - H(t-\pi)]$; $\quad y(0) = 1$.

(c) $\dot{x} = 3x - 2y + 2\,t\,[H(t) - H(t-1)]$, $\quad x(0) = 2$,
$\dot{y} = 5x - 3y + H(t) - H(t-1)$; $\quad y(0) = 0$.

(d) $\dot{x} = 3x - 2y + 2\,\sin t\,[H(t) - H(t-\pi)]$, $\quad x(0) = 1$,
$\dot{y} = 2x - y + 4\,\cos t\,[H(t) - H(t-\pi)]$; $\quad y(0) = 1$.

Section 8.6 of Chapter 8 (Review)

1. Construct two fundamental matrices (8.6.4) for the given system of differential equations of the second order $\ddot{\mathbf{x}} + \mathbf{A}\mathbf{x} = \mathbf{0}$, where the square matrix $\mathbf{A}$ is specified.

(a) $\begin{bmatrix} 25 & 21 \\ 11 & 15 \end{bmatrix}$; (b) $\begin{bmatrix} 19 & 18 \\ 25 & 34 \end{bmatrix}$; (c) $\begin{bmatrix} 32 & 28 \\ 32 & 36 \end{bmatrix}$; (d) $\begin{bmatrix} 32 & 28 \\ 49 & 53 \end{bmatrix}$;

(e) $\begin{bmatrix} 30 & 5 \\ 19 & 44 \end{bmatrix}$; (f) $\begin{bmatrix} 30 & 5 \\ 34 & 59 \end{bmatrix}$; (g) $\begin{bmatrix} 55 & 30 \\ 26 & 51 \end{bmatrix}$; (h) $\begin{bmatrix} 30 & 5 \\ 6 & 31 \end{bmatrix}$.

2. Construct two fundamental matrices (8.6.4) for the given system of differential equations of the second order $\ddot{\mathbf{x}} + \mathbf{A}\mathbf{x} = \mathbf{0}$, where the square matrix $\mathbf{A}$ is specified.

(a) $\begin{bmatrix} 1 & 0 & -1 \\ 0 & 1 & -1 \\ -1 & 1 & 4 \end{bmatrix}$; (b) $\begin{bmatrix} 9 & 0 & -9 \\ 0 & 9 & -9 \\ -1 & 1 & 1 \end{bmatrix}$; (c) $\begin{bmatrix} 1 & 1 & 0 \\ -2 & 4 & -2 \\ 0 & -1 & 1 \end{bmatrix}$.

3. In each of exercises (a) through (d), solve the initial value problem.

(a) $\ddot{x} + 15x - 7y = 0$, $\ddot{y} - 42x + 22y = 0$, $x(0) = 0$, $\dot{x}(0) = 1$, $y(0) = -2$, $\dot{y}(0) = 2$.

(b) $\ddot{x} + 15x - 2y = 0$, $\ddot{y} - 147x + 22y = 0$, $x(0) = 2$, $\dot{x}(0) = 0$, $y(0) = -2$, $\dot{y}(0) = 3$.

(c) $\ddot{x} + 3x - y = 0$, $\ddot{y} - 2x + 2y = 0$, $x(0) = 1$, $\dot{x}(0) = 2$, $y(0) = -3$, $\dot{y}(0) = 0$.

(d) $\ddot{x} + 5x - 2y = 0$, $\ddot{y} - 8x + 5y = 0$, $x(0) = 1$, $\dot{x}(0) = 2$, $y(0) = 2$, $\dot{y}(0) = 0$.

4. Consider a plane curve, called hypocycloid, generated by the trace of a fixed point on a small circle of radius r that rolls within a larger circle of radius $R > r$. Show that the coordinates of this curve are solutions of the initial value problem

$$\begin{aligned} &\ddot{x} - (R-2r)\,\dot{y}/r + (R-r)\,x/r = 0, \\ &\ddot{y} + (R-2r)\,\dot{x}/r + (R-r)\,y/r = 0, \end{aligned} \qquad x(0) = R,\ \dot{x}(0) = y(0) = \dot{y}(0) = 0.$$

5. Consider a mechanical system of two masses m_1 and m_2 connected by three elastic springs with spring constants k_1, k_2, and k_3, respectively (see Fig. 6.4, page 344). Find the natural frequencies of the mass-and-spring system and determine the general solution of the corresponding system of differential equations

$$m_1\,\ddot{x} + k_1 x - k_2(y-x) = 0, \qquad m_2\,\ddot{y} + k_3 y + k_2(y-x) = 0.$$

(a) $m_1 = 1,\ m_2 = 3,\ k_1 = 1,\ k_2 = 6,\ k_3 = 3.$
(b) $m_1 = 7,\ m_2 = 8,\ k_1 = 7,\ k_2 = 56,\ k_3 = 8.$
(c) $m_1 = 7,\ m_2 = 2,\ k_1 = 427,\ k_2 = 14,\ k_3 = 86.$
(d) $m_1 = 4,\ m_2 = 11,\ k_1 = 4,\ k_2 = 44,\ k_3 = 11.$
(e) $m_1 = 1,\ m_2 = 1,\ k_1 = 34,\ k_2 = 6,\ k_3 = 39.$
(f) $m_1 = 1,\ m_2 = 4,\ k_1 = 57,\ k_2 = 8,\ k_3 = 312.$
(g) $m_1 = 11,\ m_2 = 5,\ k_1 = 539,\ k_2 = 110,\ k_3 = 245.$
(h) $m_1 = 34,\ m_2 = 29,\ k_1 = 34,\ k_2 = 986,\ k_3 = 29.$

Chapter 9

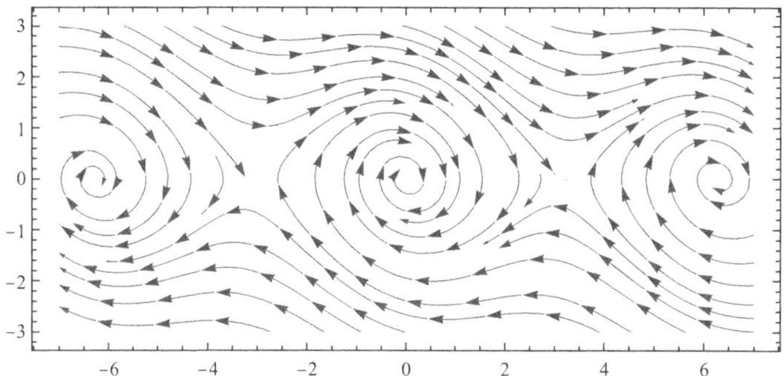

Phase portrait of a damped pendulum

Qualitative Theory of Differential Equations

Nonlinear differential equations and systems of simultaneous nonlinear differential equations have been encountered in many applications. It is often difficult, if not impossible, to solve a given nonlinear differential equation or system of equations. This is not simply because ingenuity fails, but because the repertory of standard functions in terms of which solutions may be expressed is too limited to accommodate the variety of solutions to differential equations encountered in practice. Even if a solution can be found, its expression is often too complicated to display clearly the principal features of the solution.

This chapter gives an introduction to qualitative theory of ordinary differential equations when properties of solutions can be determined without actually solving equations explicitly or implicitly. Its theory originated from the independent work of two mathematicians at the turn of 20th century, A. M. Lyapunov and H. Poincaré. We are often primarily interested in certain properties of the solutions as, for example, growing without bound as $t \to \infty$, or approaching a finite limit, or having a periodic solution, and so on. Also, some attention is given to the influence of coefficients on the solutions of the systems.

9.1 Autonomous Systems

Some solution graphs and phase plots for nonautonomous vector equations $\dot{\mathbf{x}}(t) = \mathbf{f}(t, \mathbf{x})$ are so irregular that they display very little apparent order. Since it is hard to visualize solutions of nonautonomous equations even in the two-dimensional case, we concentrate our attention on autonomous vector differential equations in normal form:

$$\dot{\mathbf{x}} = \mathbf{f}(\mathbf{x}), \tag{9.1.1}$$

where

$$\mathbf{x}(t) = \begin{bmatrix} x_1(t) \\ x_2(t) \\ \vdots \\ x_n(t) \end{bmatrix} \quad \text{and} \quad \mathbf{f}(\mathbf{x}) = \begin{bmatrix} f_1(x_1, x_2, \ldots, x_n) \\ f_2(x_1, x_2, \ldots, x_n) \\ \vdots \\ f_n(x_1, x_2, \ldots, x_n) \end{bmatrix}$$

are n-column vectors, and a dot stands for the derivative, $\dot{\mathbf{x}} \stackrel{\text{def}}{=} d\mathbf{x}/dt$, with respect to variable t, or time. The term **autonomous** means self-governing, justified by the absence of the time variable t in the vector function $\mathbf{f}(\mathbf{x})$. It is assumed that components of the vector function $\mathbf{f}(\mathbf{x})$ are continuously differentiable (or at least Lipschitz continuous) in some region of n-dimensional space. Then, according to the existence and uniqueness theorem 6.4, page 372, there exists a unique solution $\mathbf{x}(t)$ of the initial value problem

$$\dot{\mathbf{x}}(t) = \mathbf{f}(\mathbf{x}), \qquad \mathbf{x}(t_0) = \mathbf{x}_0, \tag{9.1.2}$$

that is defined in some open interval containing t_0. A maximum interval in which the solution exists is called the **validity interval**. Solutions to autonomous systems have a "time-shift immunity" in the sense that the function $\mathbf{x}(t - c)$ is a solution of the given system for an arbitrary c provided that $\mathbf{x}(t)$ is a solution.

Definition 9.1: A point $\mathbf{x}^*$ where all components of the rate vector function $\mathbf{f}$ are zeroes, $\mathbf{f}(\mathbf{x}^*) = 0$, is called a **critical point**, or **equilibrium point**, of the autonomous system (9.1.1). The corresponding constant solution $\mathbf{x}(t) \equiv \mathbf{x}^*$ is called an **equilibrium** or **stationary solution**. The set of all critical points is called the critical point set.

Definition 9.2: A critical point $\mathbf{x}^*$ of the autonomous system $d\mathbf{x}/dt = \mathbf{f}(\mathbf{x})$ is called an **isolated equilibrium point** if there are no other stationary points arbitrarily closed to it.

The concept of equilibrium plays a central role in various applied sciences, such as physics (especially mechanics), economics, engineering, transportation, sociology, chemistry, biology, and other fields. If one can formulate a problem as a mathematical model, its equilibrium solutions can be used for forecasting the future behavior of very complex systems and also for correcting the current state of the system under control. There is no supporting theory to find equilibria for all possible vector functions $\mathbf{f}(\mathbf{x})$ in Eq. (9.1.1). However, there is a rich library of special numerical methods for solving systems of nonlinear algebraic equations, including celebrated numerical methods such as Newton's or Chebyshev's methods, which are usually applied in combination with the bisection method. Nevertheless, sometimes numerical algorithms fail to determine all critical points due to a machine's inability to perform exact calculations involving irrational numbers. As an alternative, computer algebra systems offer convenient codes to solve the equations, when possible, including symbolic solutions.

Critical points of the autonomous equation $\dot{\mathbf{x}} = \mathbf{f}(\mathbf{x})$ are simultaneous solutions of the vector equation $\mathbf{f}(\mathbf{x}) = \mathbf{0}$. If $\mathbf{f}(\mathbf{x}) = \langle f_1(\mathbf{x}), f_2(\mathbf{x}), \ldots, f_n(\mathbf{x})\rangle^T$ and $\mathbf{x} = \langle x_1, x_2, \ldots, x_n\rangle^T$, then equilibrium points are solutions of the system of n algebraic equations

$$\begin{cases} f_1(x_1, x_2, \ldots, x_n) = 0, \\ f_2(x_1, x_2, \ldots, x_n) = 0, \\ \quad\vdots \\ f_n(x_1, x_2, \ldots, x_n) = 0, \end{cases} \tag{9.1.3}$$

with n unknowns x_1, x_2, ..., x_n. The solution set of each equation $f_k(x_1, x_2, \ldots, x_n) = 0$, $k = 1, 2, \ldots, n$, is referred to as the k**-th nullcline**. Therefore, equilibrium or stationary points are intersections of all k-nullclines. Each k-th nullcline separates solutions into disjoint subsets: in one of them the k-th component of the vector solution always increase, while in the other subsets the k-th component decreases.

The equilibrium points can be classified by the behavior of solutions in their neighborhoods. Imagine a circle with radius ε around a critical point $\mathbf{x}^*$—it is the set of points $\mathbf{x}$ such that $\|\mathbf{x}-\mathbf{x}^*\| < \varepsilon$, where $\|\mathbf{x}\| = \sqrt{x_1^2 + x_2^2 + \cdots + x_n^2}$. Next imagine a second smaller circle with radius δ. Let us take a point $\mathbf{x}_0$ in the δ-circle. If the trajectory starting at $\mathbf{x}_0$ leaves the ε-circle for every $\varepsilon > 0$, then the critical point is **unstable**. If, however, for every ε-circle we can find a δ-circle such that the orbit starting at an arbitrary point $\mathbf{x}_0$ in the δ-circle remains within the ε-circle, the critical point is **stable**. Now we give definitions of such points, attributed to A. Lyapunov.

Definition 9.3: Let $\mathbf{x}^*$ be an isolated equilibrium point for the autonomous vector equation (9.1.1), that is, $f(\mathbf{x}^*) = 0$. Then the critical point $\mathbf{x}^*$ is said to be **stable** if, given any $\varepsilon > 0$, there exists a $\delta > 0$ such that whenever the initial condition $\mathbf{x}(0)$ is inside a neighborhood of $\mathbf{x}^*$ of distance δ, that is,

$$\| \mathbf{x}(0) - \mathbf{x}^* \| < \delta, \tag{9.1.4}$$

Example 9.1.1: The equilibrium points of the nonlinear system

$$\dot{x} = 4x - y^2,$$
$$\dot{y} = x + y,$$

are found by solving the algebraic system

$$\begin{cases} 4x - y^2 = 0, \\ x + y = 0. \end{cases}$$

The x-nullcline is a parabola $4x = y^2$, and the y-nullcline is the straight line $y = -x$. Their intersection consists of two stationary points: the origin $(0, 0)$ and $(4, -4)$. As seen from Fig. 9.1, these nullclines separate the plane into subdomains where one of the components increases or declines. □

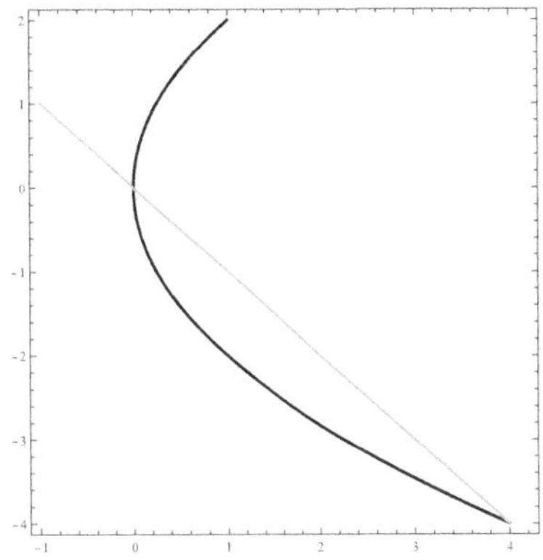

Figure 9.1: Example 9.1.1.

then every solution $\mathbf{x}(t)$ of the initial value problem

$$\dot{\mathbf{x}} = f(\mathbf{x}), \quad \mathbf{x}(0) = \mathbf{x}_0 \tag{9.1.5}$$

satisfies $\| \mathbf{x}(t) - \mathbf{x}^* \| < \varepsilon$ for all $t \geqslant 0$.

Definition 9.4: A critical point $\mathbf{x}^*$ is called **asymptotically stable** or an attractor if, in addition to being stable, there exists a $\delta > 0$ such that whenever the initial position satisfies Eq. (9.1.4), we have for every solution $\mathbf{x}(t)$ of the initial value problem (9.1.2):

$$\lim_{t\to+\infty} \| \mathbf{x}(t) - \mathbf{x}^* \| = 0 \quad \Longleftrightarrow \quad \lim_{t\to\infty} \mathbf{x}(t) = \mathbf{x}^*.$$

If all solutions approach the equilibrium point, then the critical point is **globally asymptotically stable**. If instead there are solutions starting arbitrary close to $\mathbf{x}^*$ that are infinitely often a fixed distance away from it, then the critical point is said to be **unstable** or a repeller.

For an asymptotically stable point, trajectories approach the stationary point only when they happen to be in close proximity of the equilibrium. On the other hand, if a solution curve stays around the critical point, then it is stable, but not asymptotically stable.

Definition 9.5: A set of states $S \subset \mathbb{R}^n$ is called an **invariant set** if for all $\mathbf{x}_0 \in S$, the solution of the initial value problem (9.1.2) belongs to S for all $t \geqslant t_0$.

Therefore, if an invariant set S contains the initial value, it must contain the entire solution trajectory from that point on.

Definition 9.6: For every asymptotically stable critical point, its **basin of attraction** or the **region of asymptotic stability** consists of all initial points leading to long-time behavior that approaches that stationary point. A trajectory that bounds a basin of attraction is called a **separatrix**.

9.1.1 Two-Dimensional Autonomous Equations

In this subsection, we consider only autonomous systems, given by a pair of autonomous differential equations,

$$\dot{x}(t) = f(x(t), y(t)), \qquad \dot{y}(t) = g(x(t), y(t)), \tag{9.1.6}$$

for unknown functions $x(t)$ and $y(t)$. This system of dimension 2 is called a **planar autonomous system**. By introducing 2×1 vector functions

$$\mathbf{x}(t) = \begin{bmatrix} x(t) \\ y(t) \end{bmatrix}, \qquad \mathbf{f}(\mathbf{x}) = \begin{bmatrix} f(x,y) \\ g(x,y) \end{bmatrix},$$

we can rewrite Eq. (9.1.6) in the vector form (9.1.1), page 483. A solution of such a system is a pair of points $\langle x(t), y(t)\rangle$ on the xy-plane, called a **phase plane**, for every t. As t increases, the point $\langle x(t), y(t)\rangle$ will trace out some path, which is called a **trajectory** of the system. Planetary orbits and particle paths in fluid flows have historically been prime examples of solution curves, so the terms **orbit**, **path**, or **streamline** are often used instead of trajectory.

Let us consider a curve in a three-dimensional space parameterized by $t \mapsto \langle t, x(t), y(t)\rangle$. At each point of this curve, there is a tangent vector, which is obtained by differentiating with respect to t:

$$\langle 1, \dot{x}(t), \dot{y}(t)\rangle \equiv \langle 1, f(x(t), y(t)), g(x(t), y(t))\rangle.$$

This tangent vector can be computed at any point (t, x, y) in $\mathbb{R}^3$ without knowing the solution to the system (9.1.6). Therefore, the set of such tangent vectors is called the **direction field** for the system (9.1.6). It is completely analogous to the definition of the direction field for a single equation. The solution curve to Eq. (9.1.6) must be tangent to the direction vector $\langle 1, f(x(t_0), y(t_0)), g(x(t_0), y(t_0))\rangle$ at any point $(t_0, x(t_0), y(t_0))$.

A picture that shows the critical points together with the collection of typical solution curves in the xy-plane is called a **phase portrait** or **phase diagram**. Usually, these trajectories are plotted with small arrows indicating the direction of traversal of a point on the solution curve. When it is not possible to attach arrows to streamlines, the phase portrait is plotted along with a corresponding direction field. Such pictures provide a great illumination of solution behaviors. See, for instance, the phase portrait of the damped pendulum on the opening page 483.

While a solution of the planar system (9.1.6) is a pair of functions $\langle x(t), y(t)\rangle$, solutions to the autonomous systems can be depicted on one graph. Another way of visualizing the solutions to the autonomous system is to construct a slope or direction field in the xy-plane by drawing typical line segments having slope

$$\frac{\mathrm{d}y}{\mathrm{d}x} = \frac{\mathrm{d}y/\mathrm{d}t}{\mathrm{d}x/\mathrm{d}t} = \frac{g(x,y)}{f(x,y)}. \tag{9.1.7}$$

However, the trajectory of a solution to Eq. (9.1.7) in the xy-plane contains less information than the original graphs because t-dependence has been suppressed. To restore this information, we indicate the direction of time increase with arrowheads on the curve, which show the evolution of a point traveling along the trajectory.

Equation (9.1.7) may have singular points where $f(x,y) = 0$ but $g(x,y) \ne 0$. Points where $f(x,y)$ and $g(x,y)$ are both zero

$$f(x,y) = 0, \qquad g(x,y) = 0 \tag{9.1.8}$$

are called **points of equilibrium**, or **fixed points**, or **critical points** of a system $\dot{\mathbf{x}} = \mathbf{f}(\mathbf{x})$. If (x^*, y^*) is a solution of Eq. (9.1.8), then

$$x(t) = x^*, \qquad y(t) = y^*$$

are constant solutions of Eq. (9.1.6). In applied literature, it may be called a **stationary point**, **steady state**, or **rest point**. Assuming uniqueness of the initial value problem for the system (9.1.6), no other trajectory $\mathbf{x}(t) = \langle x(t), y(t)\rangle$ in the phase plane can touch or cross an equilibrium point (x^*, y^*) or each other.

The points where the nullclines cross are precisely the critical points. In some cases the complete picture of the solutions of Eq. (9.1.6) can be established just by considering the nullclines, steady states, and how the sign of $\mathrm{d}y/\mathrm{d}x = \dot{y}/\dot{x}$ changes as we go between regions demarcated by nullclines. However, this level of detail may be insufficient, and we must study more carefully how solutions behave near a stationary point. For example, solutions may approach a steady state directly or may show an oscillatory behavior by spiraling into the equilibrium solution.

A second order differential equation

$$\ddot{x} = f(x, \dot{x}, t)$$

can be interpreted as an equation of motion for a mechanical system, in which $x(t)$ represents displacement of a particle of unit mass at time t, $\dot{x}$ its velocity, $\ddot{x}$ its acceleration, and f the applied force. After introducing a new variable $y = \dot{x}$, the given second order equation $\ddot{x} = f(x, \dot{x}, t)$ becomes equivalent to the system of first order equations

$$\dot{x} = y, \qquad \dot{y} = f(x, y, t).$$

A mechanical system is in equilibrium if its state does not change with time. This implies that $\dot{x} = 0$ at any constant solution. Such constant solutions are therefore the constant solutions (if any) of the equation

$$f(x, 0, t) = 0.$$

A typical nonautonomous equation models the damped linear oscillator with a harmonic forcing term

$$\ddot{x} + k\dot{x} + \omega_0^2 x = F\,\cos\omega t,$$

in which $f(x, \dot{x}, t) = -k\dot{x} - \omega_0^2 x + F\,\cos\omega t$, where k, ω_0, ω, and F are positive constants. We convert this equation to the system of first order equations:

$$\dot{x} = y, \qquad \dot{y} = -ky - \omega_0^2 x + F\,\cos\omega t. \tag{9.1.9}$$

To visualize, we ask *Maple* to plot the corresponding phase portrait for some particular values of coefficients $k = 0.1$, $\omega_0^2 = \omega^2 = 4$, $F = 3$:

```
with(DEtools):
DE1:=diff(x(t),t)=y(t); DE2:=diff(y(t),t)=-.1*y(t)-4*x(t)+3*cos(4*t);
phaseportrait([DE1, DE2],[x, y],t =-3..3, [[x(0)=1,y(0)=0],
[x(0)=0,y(0)=2]],x=-2.2..2,color=blue,linecolor=black,stepsize=0.05)
DEplot3d({DE1,DE2},{x(t),y(t)},t = 0 .. 11, [[x(0)=1,y(0)=0]],
 scene = [t, x, y], linecolor = black, stepsize = 0.005)
```

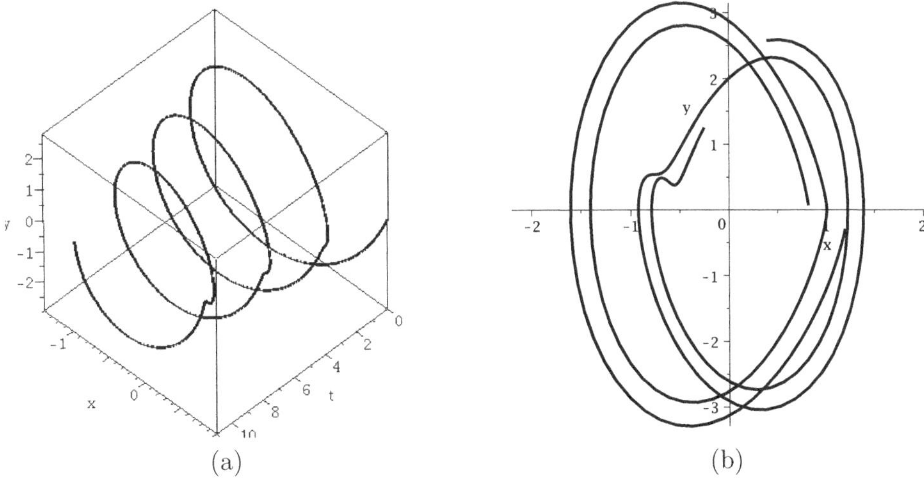

Figure 9.2: Phase portrait of a damped harmonic oscillator, Eq. (9.1.9), plotted with *Maple:* (a) 3D plot and (b) its projection on the xy-plane.

As we see from Fig. 9.2(b), projections of solutions on the xy-plane can cross each other, so the notion of direction field is no longer meaningful. Of course, we could make the equation autonomous by inserting a third variable, but we can't plot three-dimensional phase portraits with this function.

There are no equilibrium states for the system (9.1.9). Stationary points are not usually associated with nonautonomous equations, although they can occur, for instance, in the Mathieu equation

$$\ddot{x} + (\alpha + \beta\,\cos t)\,x = 0,$$

in which the origin is an equilibrium state.

Example 9.1.2: (Pendulum) We visualize the stability concepts and basin of attraction with an example of an oscillating pendulum, which consists of a bob of mass m that is attached to one end of a rigid, but weightless, rod of length ℓ. The other end of the rod is supported at a point O, which we choose as the origin of the coordinate system associated with our problem. The rod is free to rotate with respect to the pivot O. If pendulum oscillations occur in one plane, the position of the bob is uniquely determined by the angle θ of inclination of the rod and the downward vertical direction, with the counterclockwise direction taken as positive (see Fig. 6.5 on page 348).

The motion of the bob is determined by two forces acting on it: the gravitational force mg, which acts downward, and the resistance force acting in a direction always contrary to the direction of motion. Assuming that the damping force is proportional to the velocity, the equation of motion for the pendulum can be obtained by equating the sum of the moments about pivot O, $-\ell\left(k\,\dot\theta + mg\,\sin\theta\right)$, to the product of the pendulum's moment of inertia ($m\ell^2$) and angular acceleration (see details in [14]). The resulting equation becomes

$$m\ell^2\ddot\theta = -\ell\left(k\,\dot\theta + mg\,\sin\theta\right) \quad \text{or} \quad \ddot\theta + \gamma\,\dot\theta + \omega^2\sin\theta = 0, \tag{9.1.10}$$

where $\gamma = k/(m\ell)$, $\omega^2 = g/\ell$, g is the acceleration due to gravity, and k is the positive coefficient of proportionality used to model the damped force.

We convert the pendulum equation into a system of two first order autonomous equations by introducing new variables: $x = \theta$ and $y = \dot\theta$. Then

$$\dot x = y, \qquad \dot y = -\omega^2\sin x - \gamma y. \tag{9.1.11}$$

The critical points of Eq. (9.1.11) are found by solving the equations

$$y = 0, \qquad -\omega^2\sin x - \gamma y = 0 \quad \Longleftrightarrow \quad \sin x = 0.$$

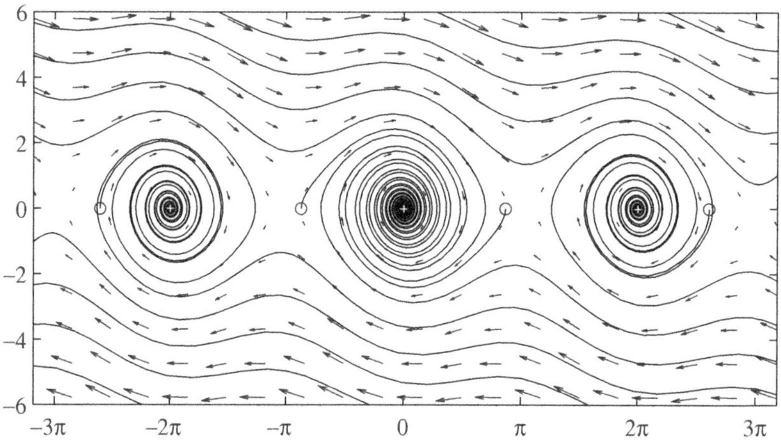

Figure 9.3: Example 9.1.2: The basin of attraction for a damped pendulum, plotted with MATLAB®.

Therefore, we have countably many critical points $(n\pi, 0)$, $n = 0, \pm1, \pm2, \ldots$, but they correspond to only two distinct physical equilibrium positions of the bob: either downward ($x = \theta = 0$) or upward ($x = \theta = \pi$). If the mass is slightly displaced from the lower equilibrium position, the bob will oscillate back and forth with gradually decreasing amplitude due to dissipation of energy caused by the resistance force. Since the bob will eventually stop in the downward position, this type of motion illustrates asymptotic stability.

On the other hand, if the mass is slightly displaced from the upper equilibrium position $\theta = \pi$, it will rapidly fall, under the influence of gravity, and will ultimately converge to the downward position. This type of motion illustrates the instability of the critical point $x = \theta = \pi$. By plotting the trajectories starting at various initial points in the phase plane, we obtain the phase portrait shown in the opening figure on page 483. A half of stationary points $(2k\pi, 0)$, where $k = 0 \pm 1, \pm2, \ldots$, correspond to the downward stable position of the pendulum, and the other half, $((2k+1)\pi, 0)$, where $k = 0 \pm 1, \pm2, \ldots$, correspond to the upward unstable position. Near each asymptotically stable critical point, the orbits are clockwise spirals that represent a decaying oscillation. The basins of attraction for asymptotically stable points are shown in Fig. 9.3. It is bounded by the trajectories that enter the two adjacent saddle points $((2k+1)\pi, 0)$. The bounding streamlines are separatrices. Each asymptotically stable equilibrium point has its own region of asymptotic stability, which is bounded by the separatrices entering the two neighboring saddle points.

Now we consider the case when the resistance force is proportional to the square of the velocity, $-k\dot\theta|\dot\theta|$ instead of $-k\dot\theta$. Then the modeling system of differential equations becomes

$$\dot x = y, \qquad \dot y = -\omega^2\sin x - \gamma y\,|y|.$$

This system has the same set of critical points $(n\pi, 0)$, $n = 0, \pm 1, \pm 2, \ldots$, half of them (for n even) stable, and the other half (odd n) unstable. The phase portrait presented in Fig. 9.4 confirms this.

Finally, consider the ideal pendulum when the resistance force is neglected, which corresponds to the case when $k = \gamma = 0$. If the bob is displaced slightly from its lower equilibrium position, it will oscillate indefinitely with constant amplitude about the equilibrium position $x = \theta = 0$. Since the damping force is absent, there is no dissipation of energy, and the mass will remain near the equilibrium position but will not approach it asymptotically. This type of motion is stable but not asymptotically stable (see Fig. 9.21 on page 516).

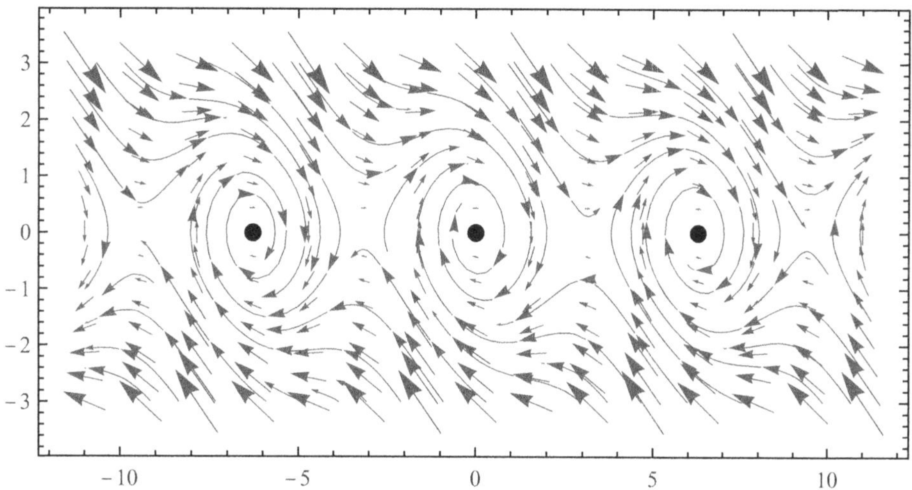

Figure 9.4: Example 9.1.2: Phase portrait for a damped pendulum when the resistance force is proportional to the velocity squared, plotted with *Mathematica*.

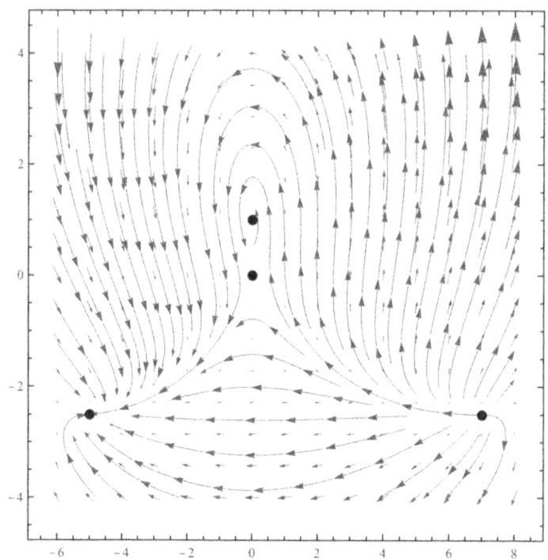

Figure 9.5: Example 9.1.3, phase portrait of the nonlinear system, plotted with *Mathematica*.

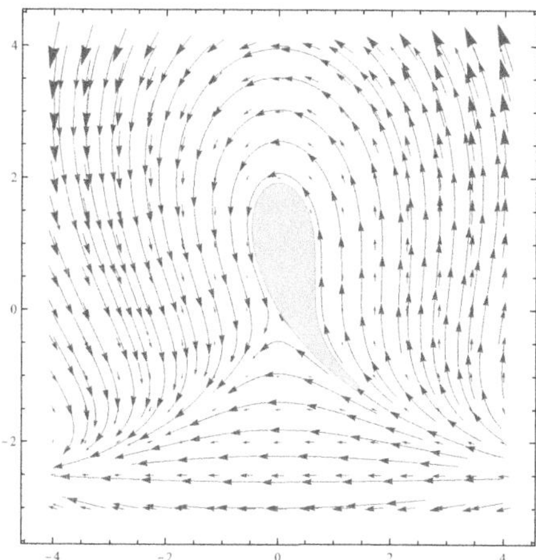

Figure 9.6: Basin of attraction for the spiral point $(0, 1)$ is shaded, plotted with *Mathematica*.

Example 9.1.3: Locate the equilibrium points, and sketch the phase paths and the basin of attraction to the autonomous system of equations

$$\dot{x} = (2y - x)\,(1 - y - x/2)\,, \qquad \dot{y} = x\,(5 + 2y)\,.$$

Solution. The stationary points occur at the simultaneous solutions of

$$(2y - x)\,(1 - y - x/2) = 0, \qquad x\,(5 + 2y) = 0.$$

These are four solution pairs of the above algebraic system of equations: $(0,0)$, $(-5,-2.5)$, $(0,1)$, $(7,-2.5)$. By plotting a direction field and critical points, we see from Fig. 9.5 that $(0,1)$ is a spiral point, $(-5,-2.5)$ is an attractor (asymptotically stable node), the origin is a saddle point, and $(7,-2.5)$ is an unstable node.

Problems In all problems, dot stands for the derivative with respect to t.
Locate all the equilibrium points for each of the systems in Problems 1 through 8.

1. $\begin{cases} \dot{x} = y, \\ \dot{y} = x + y^2 - 2; \end{cases}$
2. $\begin{cases} \dot{x} = -y, \\ \dot{y} = 2x + x^2; \end{cases}$
3. $\begin{cases} \dot{x} = 4x - xy^2, \\ \dot{y} = y^3 - x; \end{cases}$
4. $\begin{cases} \dot{x} = x - y + 2, \\ \dot{y} = y^2 - x^2; \end{cases}$
5. $\begin{cases} \dot{x} = y, \\ \dot{y} = x^2 + y^2 - 4; \end{cases}$
6. $\begin{cases} \dot{x} = x(2-y), \\ \dot{y} = y(2+3x); \end{cases}$
7. $\begin{cases} \dot{x} = 3 + 4y, \\ \dot{y} = 1 - 4x^2; \end{cases}$
8. $\begin{cases} \dot{x} = 5x - 2x^2 - 3xy, \\ \dot{y} = 6y - 4y^2 - xy; \end{cases}$
9. $\begin{cases} \dot{x} = 2x - x^2 - 3xy, \\ \dot{y} = 3y - 4y^2 - 2xy; \end{cases}$
10. $\begin{cases} \dot{x} = y\left(1 - x^2\right), \\ \dot{y} = x\left(4 - x^2\right); \end{cases}$
11. $\begin{cases} \dot{x} = 4x - x^2 - xy, \\ \dot{y} = 6y - x^2 y - y^2; \end{cases}$
12. $\begin{cases} \dot{x} = x(x-3), \\ \dot{y} = y\left(x^2 - 9y\right); \end{cases}$
13. $\begin{cases} \dot{x} = x\left(x^2 + y^2 - 16\right), \\ \dot{y} = y\left(xy - 8\right); \end{cases}$
14. $\begin{cases} \dot{x} = 2x - x^2 - xy, \\ \dot{y} = 2y - y^2 - x^2 y; \end{cases}$
15. $\begin{cases} \dot{x} = 4\,x - x^2 - x\,y, \\ \dot{y} = y(y - 3x); \end{cases}$
16. $\begin{cases} \dot{x} = x\left(x^2 + 2\,y - 1\right), \\ \dot{y} = y\left(1 - y - 3\,x\right). \end{cases}$

Each of the systems in Problems 17–19 has infinitely many equilibrium points. Find all of them.

17. $\begin{cases} \dot{x} = y^2 - 1, \\ \dot{y} = \sin(x); \end{cases}$
18. $\begin{cases} \dot{x} = \cos(y), \\ \dot{y} = \sin(x); \end{cases}$
19. $\begin{cases} \dot{x} = \sin(x) + \cos(y), \\ \dot{y} = \sin(x) + \sin(y). \end{cases}$

In Problems 20 through 25, rewrite the given scalar differential equation as a first order system and find the equilibrium points.

20. $\ddot{x} - x^2 + \dot{x} = 0.$
21. $\ddot{x} + 5x + 3\dot{x} = 0.$
22. $\ddot{x} - 2\dot{x} + 20x = 0.$
23. $\ddot{x} - 4x\dot{x} + x^3 = 0.$
24. $\ddot{x} + (\cos x)\dot{x} + \sin x = 1.$
25. $\ddot{x} + \frac{4x}{1+\dot{x}^2} - 2x^2 = 0.$

26. Compare the phase diagrams of the systems

$$\begin{aligned} \dot{x} &= y, \\ \dot{y} &= -x, \end{aligned} \qquad \text{and} \qquad \begin{aligned} \dot{x} &= xy, \\ \dot{y} &= -x^2. \end{aligned}$$

In Problems 27 through 30, use the information provided to determine the unspecified constants.

27. The system
$$\dot{x} = ax + 2xy, \qquad \dot{y} = by - 3xy$$
has equilibrium points at $(x,y) = (0,0)$ and $(1,2)$.
28. The system
$$\dot{x} = ax + bxy + 3, \qquad \dot{y} = cx + dy^2 - 4$$
has critical points at $(x,y) = (3,0)$ and $(1,1)$.
29. Consider the system
$$\dot{x} = x + ay^3, \qquad \dot{y} = bx^2 + y - y^2.$$
The slopes of the phase-plane orbits passing through the points $(x,y) = (1,2)$ and $(-1,1)$ are 9 and 3, respectively.

30. Consider the system

$$\dot{x} = ax^2 + by - 2, \qquad \dot{y} = cx + y + y^2.$$

The slopes of the phase-plane orbits passing through the points $(x, y) = (1, 2)$ and $(-2, 3)$ are 1 and 14/11, respectively.

31. Consider a ball of radius R that can float on the water surface, so its density ρ is less than 1.

   (a) Suppose that the center of the ball is beyond k units of the water's level when the ball is in equilibrium position. Find the algebraic equation of degree 3 for k using Archimedes' law of buoyancy stating that the upward force acting upon an object at the given instant is the weight of the water displaced at that instant.

   (b) Suppose that the ball is disturbed from equilibrium at some instant. Its position will be identified by the distance $z(t)$ of the center from its equilibrium at time t. Apply Newton's second law of motion to show that the function $z(t)$ is a solution of the differential equation (9.1.12), where g represents the acceleration due to gravity. Your

$$\ddot{z} + \frac{g\,z}{4\rho\,R}\left(3 - \frac{3k^2}{R^2} - \frac{3kz}{r^2} - \frac{z^2}{R^2}\right) = 0, \tag{9.1.12}$$

   derivation should be based on equating the acting force $m\ddot{z}$, where m is the mass of the ball, to the net downward force (which is the ball weight minus the upward buoyant force).

   (c) Rewrite differential equation (9.1.12) as an equivalent two-dimensional first order system.

In each of the systems in Problems 32 through 43:

(a) Locate all critical points.

(b) Use a computer to draw a direction field and phase portrait for the system.

(c) Determine whether each critical point is asymptotically stable, stable, or unstable.

(d) Describe the basin of attraction for each asymptotically stable equilibrium solution.

32. $\dot{x} = x(1-y), \quad \dot{y} = y(4+x)$.
33. $\dot{x} = (5+y)(2x+y), \quad \dot{y} = y(3-x)$.
34. $\dot{x} = 3x - 4x^2 - xy, \quad \dot{y} = 4y - y^2 - 5xy$.
35. $\dot{x} = 3 + y, \quad \dot{y} = 1 - 4x^2$.
36. $\dot{x} = y(3 - 2x - 2y), \quad \dot{y} = -x - 2y - xy$.
37. $\dot{x} = 5xy - x, \quad \dot{y} = 10y - y^2 - x^2$.
38. $\dot{x} = (3+x)(y-3x), \quad \dot{y} = y(8 + 2x - x^2)$.
39. $\dot{x} = y + 4x, \quad \dot{y} = x - \frac{1}{5}x^3 - \frac{1}{5}y$.
40. $\dot{x} = (4+x)(3y-x), \quad \dot{y} = (4-x)(y+5x)$.
41. $\dot{x} = (3-x)(3y-x), \quad \dot{y} = y(2 - x - x^2)$.
42. $\dot{x} = x(1 - 2y + 3x), \quad \dot{y} = 2y + xy - x$.
43. $\dot{x} = x(1 - 2y + 3x), \quad \dot{y} = (1-y)(3+x)$.
44. $\dot{x} = 1 + 2y - x^2, \quad \dot{y} = y - a$.
45. $\dot{x} = 2y - 3x^2 + a, \quad \dot{y} = 3y + 3x^2 - a$.
46. $\dot{x} = 1 + 2y - ax^3, \quad \dot{y} = y - 3x$.
47. $\dot{x} = a + 2y - x^2, \quad \dot{y} = y + 4x^2$.

48. The differential equation $\ddot{x} + x - 4x + x^3 = 0$ has three distinct constant solutions. What are they?

49. Let $\mathbf{f}(\mathbf{x})$ be a continuous vector field in Eq. (9.1.1), page 483. Prove that if either of the limits $\lim_{t\to+\infty} \mathbf{x}(t)$ or $\lim_{t\to-\infty} \mathbf{x}(t)$ exists, then the limit vector is an equilibrium solution for the system (9.1.1).

50. Consider the harmonic oscillator equation $\ddot{y} + \text{sign}(\dot{y}) + y = 0$ modified by the Coulomb friction. Here $\text{sign}(v)$ gives -1, 0, or 1 depending on whether v is negative, zero, or positive. Rewrite this differential equation as a system of first order differential equations. By plotting phase portraits show that the motion stops completely in finite time regardless of the initial conditions.

51. Every nonautonomous system of differential equations can be transferred to "equivalent" autonomous system of equations by introducing additional dependent variable. For example, consider the forced Duffing equation (or Duffing oscillator, named after Georg Wilhelm Christian Caspar Duffing (1861–1944))

$$\ddot{x} + a\,\dot{x} - x + x^3 = A\,\cos\omega t,$$

where a is the friction coefficient, A is the strength of the driving force which oscillates at a frequency ω, t represents time and $\dot{x} = dx/dt$. Convert Duffing's equation into autonomous system of first order differential equation upon introducing new variables $v = \dot{x}$ and $\theta = \omega\,t$.

9.2 Linearization

We will restrict ourselves by considering isolated equilibrium points, that is, equilibrium solutions that do not have other critical points arbitrarily close to it. All functions are assumed to have as many derivatives as needed.

Suppose $f(x_1, x_2, \ldots, x_n)$ is a real-valued function of n variables $x_1, x_2, \ldots, x_n$ that takes values in $\mathbb{R}$, i.e., $f : \mathbb{R}^n \mapsto \mathbb{R}$. To illustrate the Taylor series expansion in the multi-dimensional case, we consider first f as a function of two variables. The Taylor series of second order about the point (a, b) is

$$\begin{aligned} f(x,y) \approx\ & f(a,b) + f_x(a,b)\,(x-a) + f_y(a,b)\,(y-a) \\ & + \frac{1}{2}\left[f_{xx}(a,b)\,(x-a)^2 + 2\,f_{xy}(a,b)\,(x-a)(y-b) + f_{yy}(a,b)\,(y-b)^2\right], \end{aligned}$$

where $f_x(a, b)$ denotes the partial derivative of f with respect to x evaluated at (a, b), etc. Now, let $\mathbf{a} = (a, b)$ and $\mathbf{x} = (x, y)$ be the 2-vectors in $\mathbb{R}^2$. Then the above expansion can be written in compact form as

$$f(\mathbf{x}) = f(\mathbf{a}) + \nabla f(\mathbf{a})\,(\mathbf{x}-\mathbf{a}) + \frac{1}{2}\,(\mathbf{x}-\mathbf{a})^T\,\mathtt{D}^2 f(\mathbf{a})\,(\mathbf{x}-\mathbf{a}) + \cdots, \tag{9.2.1}$$

where $\nabla f(\mathbf{a}) = \langle f_x(a,b), f_y(a,b)\rangle$ is the gradient of f evaluated at $\mathbf{a}$, and $\mathtt{D}^2 f(\mathbf{a}) = \left[\mathtt{D}_i\mathtt{D}_j f(\mathbf{x})\right]_{\mathbf{x}=\mathbf{a}}$ is the Hessian matrix[76] of second partial derivatives evaluated at the point $\mathbf{a}$ ($\mathtt{D}_i = \partial/\partial x_i$). Once the vector notation is in use, the formula (9.2.1) becomes valid for arbitrary n-dimensional space. The last term in Eq. (9.2.1) is

$$\frac{1}{2}\,(\mathbf{x}-\mathbf{a})^T\,\mathtt{D}^2 f(\mathbf{a})\,(\mathbf{x}-\mathbf{a}) = O\left(\|\mathbf{x}-\mathbf{a}\|^2\right),$$

where the $O\left(\|\mathbf{x}\|^2\right)$ means the Landau symbol:

$$g = O\left(\|\mathbf{x}\|^2\right) \qquad \Longleftrightarrow \qquad |g(\mathbf{x})| \leqslant C\,\|\mathbf{x}\|^2$$

for some positive constant C. Recall that $\|\mathbf{x}\|$ is the norm of $\mathbf{x}$, i.e., the distance of $\mathbf{x}$ from the origin: $\|\mathbf{x}\|^2 = x_1^2 + x_2^2 + \cdots + x_n^2$. The second order term in Eq. (9.2.1) is bounded by a constant times the square of the distance of $\mathbf{x}$ from the point $\mathbf{a}$. Actually, the formula (9.2.1) follows from the one-dimensional Taylor series if one will consider $g\left(t(\mathbf{x}-\mathbf{a})\right)$ and expand it into the Maclaurin series in t.

Example 9.2.1: Consider a function of two variables $f(x, y) = \cos(x)\,\sin(y)$. Using Maclaurin's series for trigonometric functions, we obtain

$$f(x,y) = \left[\sum_{n\geqslant 0} \frac{(-1)^n\,x^{2n}}{(2n)!}\right]\left[\sum_{k\geqslant 0} \frac{(-1)^k\,y^{2k+1}}{(2k+1)!}\right] = y - \frac{1}{2}\,x^2 y - \frac{1}{6}\,y^3 + \cdots.$$

Suppose we want to find the Taylor series about the point $\left(\pi, \frac{\pi}{2}\right)$. Evaluating the gradient $\nabla f = \langle \sin(x)\,\sin(y), \cos(x)\,\cos(y)\rangle$ at that point, we get its Taylor series expansion at $\left(\pi, \frac{\pi}{2}\right)$ to be

$$f(x,y) = -1 + \frac{1}{2}\,(x-\pi)^2 + \frac{1}{2}\left(y - \frac{\pi}{2}\right)^2 + \cdots. \qquad \square$$

Let us consider the nonlinear vector differential equation $\dot{\mathbf{x}} = \mathbf{f}(\mathbf{x})$, where $\mathbf{f}(\mathbf{x})$ is a vector-valued function. To derive the Taylor series for a vector function $\mathbf{f}(\mathbf{x}) = \langle f_1, \ldots, f_n\rangle^T$ in a neighborhood of a point $\mathbf{x}^*$, we simply expand each of the functions $f_j(\mathbf{x})$, $j = 1, 2, \ldots, n$, in a Taylor series using formula (9.2.1). The higher order terms are again more complicated but we disregard them and keep only terms of the first order:

$$\mathbf{f}(\mathbf{x}) = \mathbf{f}(\mathbf{x}^*) + \mathtt{D}\,\mathbf{f}(\mathbf{x}^*)\,(\mathbf{x}-\mathbf{x}^*) + O\left(\|\mathbf{x}-\mathbf{x}^*\|^2\right).$$

Here $\mathtt{D}\,\mathbf{f}(\mathbf{x}^*)$ is the **Jacobian matrix** evaluated at the point $\mathbf{x}^*$:

$$\left[\mathtt{D}\,\mathbf{f}(\mathbf{x}^*)\right]_{ij} = \left.\frac{\partial f_i}{\partial x_j}\right|_{\mathbf{x}=\mathbf{x}^*}, \qquad i,j = 1,2,\ldots n. \tag{9.2.2}$$

[76] Named after a German mathematician, Ludwig Otto Hesse (1811–1874).

Thus, for $\mathbf{x}$ near $\mathbf{x}^*$,

$$\mathbf{f}(\mathbf{x}) \approx \mathbf{f}(\mathbf{x}^*) + \mathtt{D}\,\mathbf{f}(\mathbf{x}^*)\,(\mathbf{x} - \mathbf{x}^*)\,.$$

This is the linearization of a vector-valued function of several variables. Notice that upon neglecting terms of higher than first order, the linear approximation will only potentially give a good indication of the vector function $\mathbf{f}(\mathbf{x})$ while $\|\mathbf{x} - \mathbf{x}^*\|$ remains small.

Consider the autonomous equation for a smooth vector function $\mathbf{f}(\mathbf{x})$:

$$\dot{\mathbf{x}} = \mathbf{f}(\mathbf{x}). \tag{9.2.3}$$

Suppose $\mathbf{x}^*$ is an equilibrium point, meaning that $\mathbf{f}(\mathbf{x}^*) = \mathbf{0}$. Therefore, if the solution starts at $\mathbf{x}^*$, it stays there forever. The question we want to answer is: What happens if we start near $\mathbf{x}^*$? Will it approach the equilibrium? Will it behave in some regular way? To answer these questions, we expand the function $\mathbf{f}(\mathbf{x})$ into the Taylor series

$$\dot{\mathbf{x}} = \underbrace{\mathbf{f}(\mathbf{x}^*)}_{=0} + \mathtt{D}\,\mathbf{f}(\mathbf{x}^*)\,(\mathbf{x} - \mathbf{x}^*) + \cdots\,.$$

If we disregard the "$\cdots$" terms, then

$$\dot{\mathbf{x}} \approx \mathtt{D}\,\mathbf{f}(\mathbf{x}^*)\,(\mathbf{x} - \mathbf{x}^*)\,.$$

This is the same as

$$\frac{\mathrm{d}}{dt}\,(\mathbf{x} - \mathbf{x}^*) \approx \mathtt{D}\,\mathbf{f}(\mathbf{x}^*)\,(\mathbf{x} - \mathbf{x}^*)\,.$$

Upon introducing a new dependent variable $\mathbf{y} = \mathbf{x}(t) - \mathbf{x}^*$, the solutions of Eq. (9.2.3) may be approximated near the equilibrium point by the solutions of the linear system

$$\dot{\mathbf{y}} = \mathbf{J}\,\mathbf{y}, \qquad \mathbf{y}(t) = \mathbf{x}(t) - \mathbf{x}^*, \tag{9.2.4}$$

where $\mathbf{J} = \mathtt{D}\,\mathbf{f}(\mathbf{x}^*)$ is the Jacobian matrix (9.2.2) evaluated at the equilibrium point. When will this be true? The answer gives us the **linearization theorem**, also known[77] as the **Grobman–Hartman Theorem**. It says that as long as $\mathtt{D}\,\mathbf{f}(\mathbf{x}^*)$ is **hyperbolic**, meaning that none of its eigenvalues are purely imaginary, then the solutions of Eq. (9.2.3) may be mapped to solutions of Eq. (9.2.4) by a 1-to-1 and continuous function. In other words, the behavior of a dynamical system near a hyperbolic equilibrium point is qualitatively the same as the behavior of its linearization near this equilibrium point, so solutions of nonlinear vector differential equation (9.2.3) may be approximated by solutions of linear system (9.2.4) near $\mathbf{x}^*$, and the approximation is better the closer you get to $\mathbf{x}^*$. Equation (9.2.4) is called the linearization of the system (9.2.3) at the point $\mathbf{x}^*$.

Note that the linearization procedure cannot determine the stability of an autonomous nonlinear system but only asymptotic stability or instability. A heuristic argument that the stability properties for a linearized system around a critical point should be the same as the stability properties of the original nonlinear system becomes more clear when we turn our attention to the plane case.

9.2.1 Two-Dimensional Autonomous Equations

The two-dimensional case is singled out first for detailed attention because of its relevance to the systems associated with the widely applicable second order autonomous equation $\ddot{x} = f(x, \dot{x})$. However, the most extreme instabilities occur for systems of dimensions more than one.

Definition 9.7: Let the origin $(0, 0)$ be a critical point of the autonomous system

$$\begin{cases} \dot{x} = ax + by + F(x, y), \\ \dot{y} = cx + dy + G(x, y), \end{cases} \tag{9.2.5}$$

[77]This theorem was first proved in 1959 by the Russian mathematician David Matveevich Grobman (born in 1922) from Moscow University, student of Nemytskii. The next year, Philip Hartman (born in 1915) at Johns Hopkins University (USA) independently confirmed this result.

where a, b, c, d are constants and $F(x, y)$, $G(x, y)$ are continuous functions in a neighborhood about the origin. Assume that $ad \neq bc$ so that the origin is an isolated critical point for the corresponding linear system, which we obtain by setting $F \equiv G \equiv 0$. The system (9.2.5) is said to be **almost linear** near the origin if

$$\frac{F(x,y)}{\sqrt{x^2+y^2}} \to 0 \quad \text{and} \quad \frac{G(x,y)}{\sqrt{x^2+y^2}} \to 0 \qquad \text{as} \quad \sqrt{x^2+y^2} \to 0.$$

We now state a classical result discovered by Poincaré[78] that relates the stability of an almost linear planar system to the stability at the origin of the corresponding linear system.

Theorem 9.1: Let λ_1, λ_2 be the roots of the characteristic equation

$$\lambda^2 - (a+d)\lambda + (ad - bc) = 0$$

for the linear system

$$\frac{\mathrm{d}}{\mathrm{d}t}\begin{bmatrix} x(t) \\ y(t) \end{bmatrix} = \begin{bmatrix} a & b \\ c & d \end{bmatrix}\begin{bmatrix} x(t) \\ y(t) \end{bmatrix} \tag{9.2.6}$$

corresponding to the almost linear system (9.2.5). Then the stability properties of the critical point at the origin for the almost linear system are the same as the stability properties of the origin for the corresponding linear system with one exception: When $(a-d)^2 + 4bc = 0$, the roots of the characteristic equation are purely imaginary, and the stability properties for the almost linear system cannot be deduced from the linear system.

Corollary 9.1: Let $\mathbf{J}^* = \mathbf{J}(\mathbf{x}^*) = \mathtt{Df}(\mathbf{x}^*)$ be the Jacobian matrix (9.2.2) evaluated at the equilibrium point $\mathbf{x}^*$. Its characteristic polynomial reads as

$$\lambda^2 - (\operatorname{tr}\mathbf{J}^*)\,\lambda + \det\mathbf{J}^* = 0.$$

1. If $\det\mathbf{J}^* < 0$, then $\mathbf{J}^*$ has eigenvalues of opposite sign and $\mathbf{x}^*$ is a saddle point.
2. If $\det\mathbf{J}^* > 0$ and the trace of the matrix $\mathbf{J}^*$ is positive, then the real parts of the eigenvalues of $\mathbf{J}^*$ are positive and $\mathbf{x}^*$ is unstable.
3. If $\det\mathbf{J}^* > 0$ and the trace of the matrix $\mathbf{J}^*$ is negative, then the real parts of the eigenvalues of $\mathbf{J}^*$ are negative and $\mathbf{x}^*$ is locally stable.
4. If $\det\mathbf{J}^* > 0$ and $(\operatorname{tr}\mathbf{J}^*)^2 \geqslant 4\det\mathbf{J}^*$, then $\mathbf{x}^*$ is a node; and if $(\operatorname{tr}\mathbf{J}^*)^2 < 4\det\mathbf{J}^*$, then $\mathbf{x}^*$ is a spiral.
5. If for each $(x,y) \in \mathbb{R}^2$, $\det\mathbf{J}(x,y) > 0$ and $(\operatorname{tr}\mathbf{J})(x,y) < 0$, then $\mathbf{x}^*$ is a global attractor.

Example 9.2.2: Consider the system

$$\dot{x} = 2x + y^2, \qquad \dot{y} = -2y + 4x^2.$$

Linearization of this system around $(0,0)$ yields

$$\dot{\mathbf{x}}(t) = \mathbf{J}\mathbf{x}(t), \qquad \text{where} \quad \mathbf{J}(0,0) = \begin{bmatrix} 2 & 0 \\ 0 & -2 \end{bmatrix}, \quad \mathbf{x} = \begin{bmatrix} x \\ y \end{bmatrix}.$$

Phase portraits for the original nonlinear system and its linearization are presented in Figures 9.7 and 9.8, respectively. Since the eigenvalues of the matrix $\mathbf{J}(0,0)$ are real numbers of opposite signs, the origin is an unstable saddle point.

The given system of differential equations has another critical point

$$(x^*, y^*) = \left(-2^{-1/3}, 2^{1/3}\right) \approx (-0.793701, 1.25992).$$

[78] Jules Henri Poincaré (1854–1912) was a French mathematician who made many original fundamental contributions to pure and applied mathematics, mathematical physics, and celestial mechanics.

The Jacobian at this point

$$\mathbf{J}(x^*, y^*) = \begin{bmatrix} 2 & 2y \\ 8x & -2 \end{bmatrix}_{x=x^*, y=y^*} = \begin{bmatrix} 2 & 2^{4/3} \\ -2^{8/3} & -2 \end{bmatrix}$$

has two purely imaginary eigenvalues $\pm 2\mathbf{j}\sqrt{3}$. Therefore, this point is not hyperbolic, and linearization is inconclusive.

Example 9.2.3: (Electric Circuit) Consider an electric circuit consisting of a capacitor, a resistor, and an inductor, in series. Suppose that these elements are connected in a closed loop. The effect of each component of the circuit is measured in terms of the relationship between current and voltage. An ideal model gives the following relations:

$$\begin{aligned} v_R &= f(i_R) && \text{(resistor)}, \\ L\,\dot{i}_L &= v_L && \text{(inductor)}, \\ C\dot{v}_C &= i_C && \text{(capacitor)}, \end{aligned}$$

where v_R represents the voltage across the resistor, i_R represents the current through the resistor, and so on. The function $f(x)$ is called the *v-i characteristic* of the resistor. For a passive resistor, the function $f(x)$ has the same sign as x; however, in active resistor, $f(x)$ and x have opposite signs. In the classical linear model of the RLC-circuit, it is assumed that $f(x) = Rx$, where $R > 0$ is the resistance.

According to Kirchhoff's current law, the sum of the currents flowing into a node equal the sum of the currents flowing out:

$$i_R = i_L = i_C.$$

Kirchhoff's voltage law states that the sum of voltage drops along a closed loop must add up to zero:

$$v_R + v_L + v_C = 0.$$

Assuming that $f(x) = x^3 + Rx$, we introduce two new variables: $x = i_R = i_L = i_C$ and $y = v_C$. Since v_L is known to be $v_L = L\,\dot{i}_L = L\,\dot{x}$, we find $v_R = -v_C - v_L = -y - L\,\dot{x}$. This allows us to model such a circuit as the system of first order differential equations:

$$\begin{cases} L\dot{x} = -ax^3 - Rx - y, \\ C\dot{y} = x. \end{cases}$$

Since this system has the only one critical point, the origin, we linearize the system around this point, with the corresponding Jacobian

$$\begin{cases} L\dot{x} = -Rx - y, \\ C\dot{y} = x; \end{cases} \qquad \Longrightarrow \qquad \mathbf{J} = \begin{bmatrix} -R/L & -1/L \\ 1/C & 0 \end{bmatrix}.$$

Since the Jacobian matrix $\mathbf{J}$ has two real eigenvalues of different signs, $\lambda_{1,2} = -\frac{R}{2L} \pm \frac{1}{2L}\sqrt{R^2 + 4/C}$, the origin is an unstable saddle point.

Example 9.2.4: (FitzHugh–Nagumo model) Neurons are cells in the body that transmit information to the brain by amplifying an incoming stimulus (electric charge input) and transmitting it to neighboring neurons, then turning off to be ready for the next stimulus. Neurons have fast and slow mechanisms to open ion channels in response to electrical charges. Neurons use changes of sodium and potassium ions across the cell membrane to amplify and transmit information. Voltage-gated channels exist for each kind of ion. They are closed in a resting neuron, but may be open in response to voltage differences. When a burst of positive charge enters the cell, making the potential less negative, the voltage-gated sodium channels open. Since there is an excess of sodium ions outside the cell, more sodium ions enter, increasing the potential until it eventually becomes positive. Next, a slow mechanism acts to open voltage-gated potassium channels. Both of these diminish the buildup of positive charge by blocking sodium ions from entering and allowing excess potassium ions to leave. When the potential decreases to or below the resting potential, these slow mechanisms turn off, and then the process can start over. If the electrical excitation reaches a sufficiently high level, called an action potential, the neuron fires and transmits the excitations to other neurons.

The most successful and widely used model[79] of neurons has been developed from Hodgkin and Huxley's 1952 work. Using data from the giant squid axon, they applied a Markov kinetic approach to derive a realistic and biophysically sound four-dimensional model that bears their names. Their ideas have been extended and applied to

[79] The British neuroscientists Alan Hodgkin (1914–1998) and Andrew Huxley (1917–2012) were awarded the Nobel Prize in 1963.

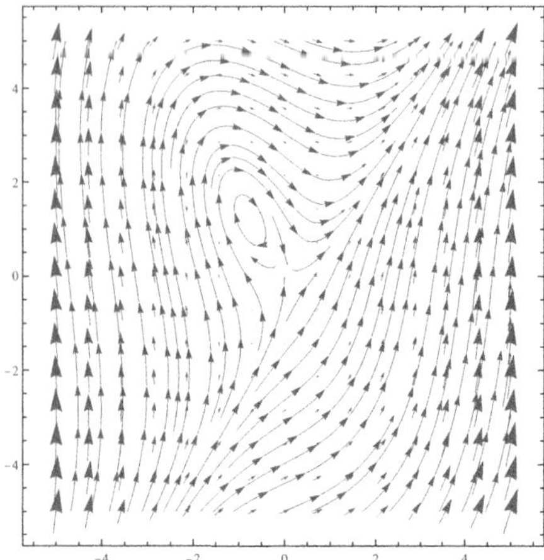

Figure 9.7: Example 9.2.2, phase portrait of the nonlinear system, plotted with *Mathematica*.

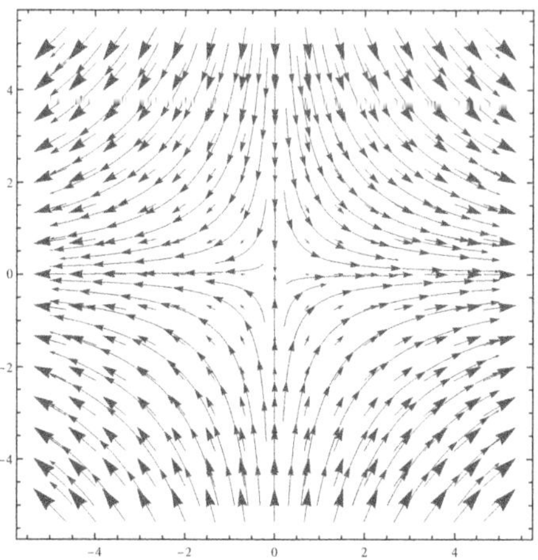

Figure 9.8: Example 9.2.2, phase portrait of the linear approximation at the origin, plotted with *Mathematica*.

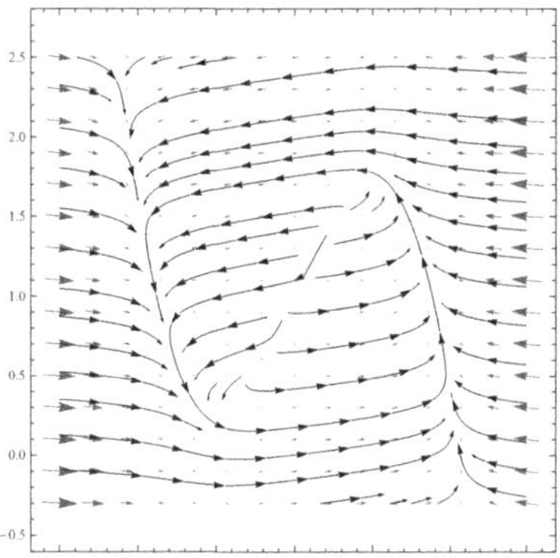

Figure 9.9: Phase portrait in Example 9.2.4, plotted with *Mathematica*.

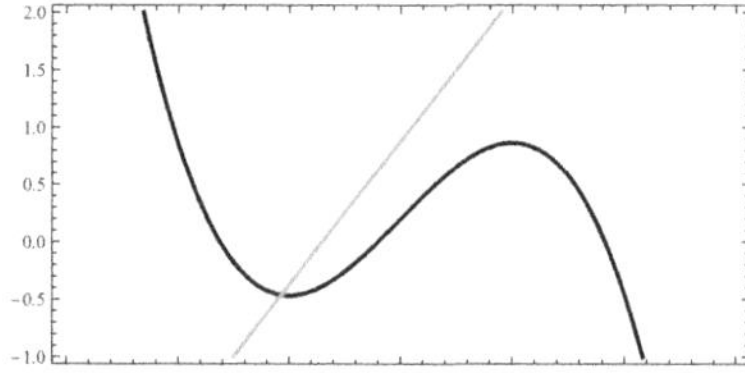

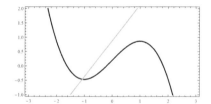

Figure 9.10: Nullclines in the FitzHugh–Nagumo model, plotted with *Mathematica*.

a wide variety of excitable cells. Sweeping simplifications to Hodgkin–Huxley were introduced[80] by FitzHugh and Nagumo in 1961 and 1962:

$$\dot{V} = V - V^3/3 - W + I_{\text{ext}}, \qquad \dot{W} = a\,(V + b - cW)\,,$$

where experimentally estimated dimensionless positive parameters are $a = 0.08$, $b = 0.7$, $c = 0.8$, and I_{ext} is an external stimulus current. Here V is the membrane potential and W denotes the strength of the blocking mechanism with $W = 0$ (turned off) when $V = 0$.

To plot the phase portrait of the FitzHugh–Nagumo (FN for short) model, we first identify critical points by plotting V-nullcline, which is the N-shaped cubic curve obtained from the condition $\dot{V} = 0$, and W-nullcline, which is a straight line obtained from the condition $\dot{W} = 0$. As seen from Fig. 9.10, the equilibrium point (V^*, W^*) is not at the origin (for instance, if $I_{\text{ext}} = 0.2$, the critical point is at $(-1.06939, -0.46174)$). We can check the stability of the equilibria by linearizing around the critical point and computing the eigenvalues of the linear system

$$\frac{\text{d}}{\text{d}t}\begin{bmatrix} v \\ w \end{bmatrix} = \begin{bmatrix} 1-(V^*)^2 & -1 \\ a & -ac \end{bmatrix}\begin{bmatrix} v \\ w \end{bmatrix}.$$

The eigenvalues can be computed easily, obtaining two complex conjugate values $\lambda_{1,2} \approx -0.1 \pm 0.28\mathbf{j}$. Therefore, the origin is an asymptotically stable spiral point and the system will oscillate before reaching it.

[80] Richard FitzHugh (1922–2007) from Johns Hopkins University created one of the most influential models of excitable dynamics. Jin-Ichi Nagumo (1926–1999) from the University of Tokyo, Japan, made fundamental contributions in the fields of nonlinear circuit theory, bioengineering, and mathematical biology.

It is convenient to scale variables, which leads to the following homogeneous system of equations:

$$\dot{v} = -v(v-\alpha)(v-1) - w, \qquad \dot{w} = \epsilon(v - \xi w), \tag{9.2.7}$$

where the parameter $0 < \alpha < 1$. The only equilibrium of the scaled FitzHugh–Nagumo system (9.2.7) is the origin.

To analyze the response of the FN model on an instantaneous current pulse J, we replace the system (9.2.7) with

$$\dot{v} = -v(v-\alpha)(v-1) - w + J, \qquad \dot{w} = \epsilon(v - \xi w). \tag{9.2.8}$$

When J increases, the equilibrium point $(0, 0)$ moves into the first quadrant of the (v, w)-plane. This equilibrium is asymptotically stable for small values of J but becomes unstable for larger values of J. This happens when the real part of the eigenvalues changes sign at two locations $V_{\pm}^{*} = \pm\sqrt{1-ac}$. Once the real part becomes zero and then positive, even infinitesimally small perturbations will become amplified and diverge away from the equilibrium. In this case, the model exhibits periodic (tonic spiking) activity (see Fig 9.9).

9.2.2 Scalar Equations

In this subsection, we illustrate linearization stability analysis for some second order nonlinear autonomous scalar differential equations $\ddot{x} = f(x, \dot{x})$. For any its solution $x(t)$, the vector-valued function $\langle x(t), \dot{x}(t)\rangle$ describes a path in the $(x, \dot{x})-$plane, called the Poincaré phase plane.

Example 9.2.5: (Pendulum) The pendulum equation $\ddot{\theta} + \gamma\,\dot{\theta} + \omega^2 \sin\theta = 0$ can be rewritten as an autonomous (nonlinear) system of differential equations:

$$\dot{x} = y, \qquad \dot{y} = -\gamma\, y - \omega^2 \sin x.$$

For small values of $x = \theta$, we may replace $\sin x$ by x to obtain a linear system of equations:

$$\dot{x} = y, \qquad \dot{y} = -\gamma\, y - \omega^2\, x,$$

which can be written in vector form:

$$\frac{\mathrm{d}}{\mathrm{d}t}\begin{bmatrix} x \\ y \end{bmatrix} = \begin{bmatrix} 0 & 1 \\ -\omega^2 & -\gamma \end{bmatrix}\begin{bmatrix} x \\ y \end{bmatrix}.$$

When $\gamma = 0$, we obtain the ideal pendulum (linear) equation $\ddot{\theta} + \omega^2\,\theta = 0$, which can be converted to a system of equations of first order by substitution: $y_1 = \omega\theta$, $y_2 = \dot{y}_1 = \omega\dot{\theta}$. Therefore,

$$\frac{\mathrm{d}}{\mathrm{d}t}\begin{bmatrix} y_1 \\ y_2 \end{bmatrix} = \begin{bmatrix} 0 & \omega \\ -\omega & 0 \end{bmatrix}\begin{bmatrix} y_1 \\ y_2 \end{bmatrix}.$$

Example 9.2.6: (Duffing Equation) When $\sin\theta$ is replaced by its two-term Taylor approximation, the ideal pendulum equation $\ddot{\theta} + \omega^2 \sin\theta = 0$ becomes

$$\ddot{\theta} + \omega^2\theta - \omega^2\,\theta^3/6 = 0. \tag{9.2.9}$$

A standard mass-spring harmonic equation $\ddot{x} + \omega^2 x = 0$ is derived by applying Hooke's law: the restoring force exerted by a spring under tension or compression is proportional to the displacement. This assumption is valid only when displacements are small. For larger ones, a restoring force can be modeled by $f_R(x) = -\dfrac{2k\delta}{\pi}\,\tan\left(\dfrac{\pi x}{2\delta}\right)$, where the value δ represents the maximum amount of x the spring can be stretched or compressed. This leads to the nonlinear differential equation

$$m\,\ddot{x} + \frac{2k\delta}{\pi}\,\tan\left(\frac{\pi x}{2\delta}\right) = 0 \qquad (-\delta < x < \delta),$$

where m is the mass of a particle. Assuming that the ratio $|\pi x/(2\delta)|$ is small, we can replace the tangent function by its two-term Maclaurin approximation:

$$m\,\ddot{x} + \frac{2k\delta}{\pi}\left(\frac{\pi x}{2\delta} + \frac{1}{3}\left(\frac{\pi x}{2\delta}\right)^3\right) = 0. \tag{9.2.10}$$

Equations (9.2.9) and (9.2.10) can be united into one equation, called[81] the **Duffing equation:**

$$\ddot{x} + \omega^2 x + \beta x^3 = 0. \tag{9.2.11}$$

It can be rewritten as a system:

$$\frac{\mathrm{d}}{\mathrm{d}t}\begin{bmatrix} x \\ y \end{bmatrix} = \begin{bmatrix} y \\ -\omega^2 x - \beta x^3 \end{bmatrix}.$$

The term βx^3 could be thought of as a small perturbation of the standard mass-spring harmonic oscillator equation.
If $\beta < 0$, then we have a soft spring and the corresponding solution is no longer bounded.
When $\beta = 0$, we get an ideal spring–mass harmonic oscillator.
For $\beta > 0$, it is a hard spring, so its slope is always negative:

$$\frac{\mathrm{d}y}{\mathrm{d}x} = \frac{\mathrm{d}y/\mathrm{d}t}{\mathrm{d}x/\mathrm{d}t} = -\frac{\omega^2 x + \beta x^3}{y}.$$

The general solution of the latter is $2(y^2 + \omega^2 x^2) + \beta x^4 = C$.

Problems

1. Convert the following single equation of the third order into a system of differential equations of the first order; then find all its critical points and classify them.
   (a) $\theta''' - \theta'\theta'' + 2\theta^2 = 8$; **(b)** $\theta''' - (\theta')^2 + \theta'' + \theta^3 = 8$; **(c)** $\theta''' - (\theta')^3 + \theta^3 = 1$.
2. Convert the following single equation of the second order into a system of differential equations of the first order; then find all its critical points and classify them.
   (a) $\ddot{\theta} + \dot{\theta} + \theta^4 = 1$; **(d)** $\ddot{\theta} + \dot{\theta}^2 - 4\theta = 0$; **(g)** $\ddot{\theta} + \dot{\theta}\,\theta + \theta^3 = 8$;
   (b) $\ddot{\theta} + 2\dot{\theta}/\theta + \theta = 2$; **(e)** $\ddot{\theta} + \dot{\theta}\,\theta - 4\theta = 0$; **(h)** $\ddot{\theta} = \theta^2 - \theta$;
   (c) $\ddot{\theta} - \dot{\theta} + \cos\theta = 0$; **(f)** $\ddot{\theta} - 2\dot{\theta} - \sin\theta = 0$; **(i)** $\ddot{\theta} + \theta - \theta^4 = 0$.
3. Find linearization of the following systems at the origin.
   (a) $\begin{cases} \dot{x} = x + x^2 + xy, \\ \dot{y} = y + y^{3/2}. \end{cases}$ **(d)** $\begin{cases} \dot{x} = x\cos y, \\ \dot{y} = y\,(\cos x - 1). \end{cases}$
   (b) $\begin{cases} \dot{x} = x^3, \\ \dot{y} = y + y\sin x. \end{cases}$ **(e)** $\begin{cases} \dot{x} = x^2, \\ \dot{y} = -y. \end{cases}$
   (c) $\begin{cases} \dot{x} = x^2 e^y, \\ \dot{y} = y\,(e^x - 1). \end{cases}$ **(f)** $\begin{cases} \dot{x} = \frac{x}{2} - y - \frac{1}{2}\left(x^3 + xy^2\right), \\ \dot{y} = x + \frac{y}{2} - \frac{1}{2}\left(y^3 + x^2 y\right). \end{cases}$
4. In each system of nonlinear differential equations, determine all critical points, and then apply the linearization theorem to classify them.
   (a) $\begin{cases} \dot{x} = x\,(-1 - x + y), \\ \dot{y} = y\,(3 - x - y). \end{cases}$ **(d)** $\begin{cases} \dot{x} = x^2 y - 3x + 3, \\ \dot{y} = y\,(3 + xy). \end{cases}$
   (b) $\begin{cases} \dot{x} = x(1 - y), \\ \dot{y} = y(2x - 1). \end{cases}$ **(e)** $\begin{cases} \dot{x} = x\,(3 - y - 2xy), \\ \dot{y} = y\left(2 - 3xy + x^2\right). \end{cases}$
   (c) $\begin{cases} \dot{x} = xy - 4, \\ \dot{y} = x\,(x - y). \end{cases}$ **(f)** $\begin{cases} \dot{x} = x\left(2 - xy + y^2\right), \\ \dot{y} = y\left(3 - 2x - x^2\right). \end{cases}$
5. Each system of linear differential equations has a single stationary point. Apply Theorem 9.1 to classify this critical point by type and stability.
   (a) $\begin{cases} \dot{x} = -5x + 8y + 1, \\ \dot{y} = -4x + 7y - 1. \end{cases}$ **(b)** $\begin{cases} \dot{x} = -3x + 6y, \\ \dot{y} = -2x + 5y - 1. \end{cases}$

[81] The Duffing equation is named after the German electrical engineer Georg Duffing (1861–1944).

(c) $\begin{cases} \dot{x} = 4x + y - 2, \\ \dot{y} = 3x + 2y + 1. \end{cases}$

(d) $\begin{cases} \dot{x} = 4x - 10y + 2, \\ \dot{y} = x - 2y + 1. \end{cases}$

(e) $\begin{cases} \dot{x} = x - 5y + 1, \\ \dot{y} = x + 3y + 1. \end{cases}$

(f) $\begin{cases} \dot{x} = 7x + 6y + 7, \\ \dot{y} = 2x - 4y + 2. \end{cases}$

(g) $\begin{cases} \dot{x} = 5x - 2y - 2, \\ \dot{y} = x + 3y + 3. \end{cases}$

(h) $\begin{cases} \dot{x} = 10x - 24y - 4, \\ \dot{y} = 4x - 10y + 2. \end{cases}$

6. Consider the linear system of differential equations containing a parameter ϵ:

$$\dot{x} = \epsilon x - 4y, \qquad \dot{y} = 9x + \epsilon y.$$

Show that the critical point $(0,0)$ is **(a)** a stable spiral point if $\epsilon < 0$; **(b)** a center if $\epsilon = 0$; **(c)** an unstable spiral point if $\epsilon > 0$. The value of parameter $\epsilon = 0$ is called bifurcation.

7. Building upon the FitzHugh–Nagumo model, Hindmarsh and Rose proposed in 1984 a model of neuronal activity described by three coupled first order differential equations:

$$\dot{x} = y + 3x^2 - x^3 - z + I, \quad \dot{y} = 1 - 5x^2 - y, \quad \dot{z} = r\left(4\left(x + \frac{8}{5}\right) - z\right),$$

where $r^2 = x^2 + y^2 + z^2$ and $r \approx 10^{-2}$. Find eigenvalues of the linearized system at the critical point when $I = 1$.

8. Show that the critical point x^* of a single differential equation $\dot{x} = f(x)$ is asymptotically stable if $f'(x^*) < 0$ and unstable if $f'(x^*) > 0$.

9. Consider a pendulum of length ℓ revolving about a vertical axis at constant speed ω_0 (`radians/sec`) and swinging horizontally in the plane perpendicular to the rod. Denoting its lumped mass by m, we get the equation of motion (see details in [14])

$$\ddot{\theta} = \omega_0^2 \sin\theta\, \cos\theta - \omega^2 \sin\theta - \gamma\,\dot{\theta},$$

where θ is the angular displacement of the pendulum in the vertical direction, $\omega^2 = g/\ell$, g is the acceleration due to gravity, $\gamma = \kappa/(m\ell)$, the damping force is assumed to be approximately proportional to the angular velocity with the coefficient κ. Find all critical points depending on the bifurcation parameter ω_0 and consider two cases when $\omega_0 < \omega$ and when $\omega_0 \geqslant \omega$.

10. Consider the linear system of differential equations containing a parameter ϵ

$$\dot{x} = \epsilon y - x, \qquad \dot{y} = x - 3y.$$

Find all bifurcation points for parameter ϵ and analyze stability depending on the values of this parameter.

11. Show that for arbitrary positive ε, the origin is a global attractor for the system

$$\dot{x} = (x - y)^3 - \varepsilon x, \qquad \dot{y} = (x - y)^3 - \varepsilon y.$$

12. Show that the two systems

$$\dot{x} = x\left(x^2 + y^2\right) - y, \qquad \dot{y} = y\left(x^2 + y^2\right) + x$$

and

$$\dot{x} = -x\left(x^2 + y^2\right) - y, \qquad \dot{y} = x - y\left(x^2 + y^2\right)$$

both have the same linearizations at the origin, but that their phase portraits are qualitatively different.

13. Show that the origin is not a hyperbolic critical point of the system $\dot{x} = y,\ \dot{y} = -x^3$. By plotting the phase portrait, verify that the origin is a stable stationary point but not asymptotically stable.

14. Consider the three systems

$$\text{(a)} \begin{cases} \dot{x} = x^2 + y, \\ \dot{y} = x - y^2; \end{cases} \qquad \text{(b)} \begin{cases} \dot{x} = y - x^3, \\ \dot{y} = x + y^2; \end{cases} \qquad \text{(c)} \begin{cases} \dot{x} = x^2 - y, \\ \dot{y} = y^2 - x. \end{cases}$$

All three have an equilibrium point at $(0,0)$. Which two systems have a phase portrait with the same "local picture" near the origin?

15. Consider the system of two differential equations

$$\dot{x} = x^2, \qquad \dot{y} = y - x.$$

(a) Show that the origin is the only one equilibrium solution of the given system.

(b) Find eigenvalues of the linearized system around the origin—equlibrium solution.

(c) Find the general solution to the given system of differential equations.

9.3 Population Models

About 2,500 years ago, an ancient Greek philosopher, Heraclitus, stated that nature is always in a state of flux. Today, rapid change in nature is an idea we all accept, if not welcome. Building a successful mathematical model that can predict species' population levels remains a great challenge. The models must be adjusted to each specific population—dynamic systems cannot be used blindly. A mathematical model must predict behavior that does not contradict valid observations, else it is flawed.

A remarkable variety of population models are known and used to describe specific interactions between species. Their derivation is usually based on suppressing or ignoring other factors that do not play a significant role. We need to remember Einstein's warning that "everything should be made as simple as possible, but not simpler."

In this section, we present some continuous models from population biology that are part of a larger class of dynamic models, called Kolmogorov's systems. Such models take the form $\dot{x}_i = x_i f_i(x_1, \ldots, x_n)$ for $i = 1, \ldots, n$, where n is the number of species and the smooth functions f_i describe the per capita growth rate for the ith species. In the planar case, the orbits of Kolmogorov's systems when starting on the axes stay on the axes, and interior trajectories cannot reach the axes in finite time.

Kolmogorov's systems provide examples of autonomous systems that model multi-species populations like several species of trout, who compete for food from the same resource pool, as well as foxes, wolves, and rabbits, who interact in a predator-prey environment. These models originated in the first part of the twentieth century through works[82] by Alfred Lotka (1925), Vito Volterra (1920s), Georgii Gause (1934), and Andrew Kolmogorov (1936). Later these systems were extended, generalized, and adapted to ecological models and areas not related to biology (for instance, economics and criminology).

9.3.1 Competing Species

In the absence of competitors, it is reasonable to model their population growth via a logistic equation (developed by Belgian mathematician Pierre Verhulst in 1838):

$$\dot{P} = \mathrm{d}P/\mathrm{d}t = rP - aP^2.$$

Here $P = P(t)$ is the population size at time t, r is the intrinsic growth rate, and $a \ll r$ is a measure of the strength of resource limitations. The smaller a is, the more room there is to grow. However, Verhulst's model does not take into account many other factors that affect population growth. Species do not exist in isolation of one another. The simple models of exponential and logistic growth fail to capture the fact that species can assist or emulate members of the same population and fight, exclude, or kill species from another population.

The competition between two or more species for some limited resource is called **interspecific competition**. This limited resource can be food or nutrients, space, mates, nesting sites—anything for which demand is greater than supply. When one species is a better competitor, interspecific competition negatively influences the other species by reducing its population size, which in turn affects the growth rate of the competitor. To be more specific, let us start with the following example of competition between hardwood and softwood trees, which one can observe in any unmanaged piece of forest area.

Example 9.3.1: Hardwood trees grow slowly, but are more durable, more resistant to disease, and produce more valuable timber. Softwood trees compete with the hardwoods by growing rapidly and consuming the available water and soil nutrients.

Competition is caused by resource limitations. The presence of softwood trees limits the amount of sunlight, water, land, etc., available for the hardwood, and vice versa. The loss in growth rate due to competition depends on the size of both populations. A simple assumption is that this loss is proportional to the product of the two. Given these assumptions about population growth and competition, we would like to know whether one species will die out over time, or whether there exist equilibrium populations.

[82]Alfred James Lotka was born 1880 in Lemberg, Austria-Hungary (now Lvov, Ukraine), and died in 1949 in New York. Lotka's parents were US nationals and he moved to the USA in 1902, where he pursued his carrier as mathematician, physical chemist, biophysicist, and statistician. Vito Volterra (1860–1940) was a distinguished Italian mathematician and physicist, famous for his contributions to mathematical biology and integral equations. Even though he was born in a very poor family, Vito became a professor of mechanics at the University of Turin in 1892 and then, in 1900, professor of mathematical physics at the University of Rome. Georgii Frantsevich Gause (1910–1986), was a Russian biologist. Later, Gause devoted most of his life to the research of antibiotics. Andrey Nikolaevich Kolmogorov (1903–1987) was one of the most famous Russian mathematicians of the 20th century. His accomplishments included solving two of Hilbert's problems, founding the axioms of probability theory, and many others.

Let $x_1(t)$ denote the population of hardwood trees at time t and $x_2(t)$ be the population of softwood trees. Assuming that state variables $x_1 \geqslant 0$, $x_2 \geqslant 0$, we consider the following logistic model:

$$\begin{aligned} \dot{x}_1 &= r_1 x_1 - a_1 x_1^2 - b_1 x_1 x_2, \\ \dot{x}_2 &= r_2 x_2 - a_2 x_2^2 - b_2 x_1 x_2. \end{aligned} \tag{9.3.1}$$

In the growth rate $r_k x_k - a_k x_k x_k - b_k x_k x_j$ $(k = 1, 2)$, the first term, $r_k x_k$, represents unrestricted growth, the second term, $a_k x_k^2$, represents the effect of competition within a population, and the third term, $b_k x_k x_j$, models competition between populations of different species. When the model is in the equilibrium point, we say that the system is in steady state because it remains there forever. At this point, all of the rates of change are equal to zero, and all of the forces acting on the system are in balance. Our first step is to locate equilibrium solutions by solving the algebraic system of equations:

$$\begin{cases} r_1 x_1 - a_1 x_1^2 - b_1 x_1 x_2 = 0, \\ r_2 x_2 - a_2 x_2^2 - b_2 x_1 x_2 = 0. \end{cases} \tag{9.3.2}$$

Factoring out x_1 from the first equation and x_2 from the second, we find three obvious (extinguishing) solutions:

$$\begin{aligned} &(0, 0), \\ &(0, r_2/a_2), \\ &(r_1/a_1, 0), \end{aligned}$$

and the fourth at the intersection of these two lines:

$$\begin{cases} r_1 = a_1 x_1 + b_1 x_2, \\ r_2 = b_2 x_1 + a_2 x_2. \end{cases} \tag{9.3.3}$$

If these two lines do not cross inside the first quadrant $(x_1, x_2 \geqslant 0)$, then there are only three equilibria. In this case, the two species cannot coexist in peaceful equilibrium, and at least one of them will die out. This case is referred to as the **competitive exclusion principle**, Gause's law of competitive exclusion, or just Gause's law because it was originally formulated by Gause in 1934 on the basis of experimental evidence. It states that two species competing for the same resources cannot coexist if other ecological factors are constant. For instance, when gray squirrels were introduced to Britain in about 30 sites between 1876 and 1929, they dominated native red squirrels, which led to their extinction. Another example gives a competition in the 20th century between the USA and USSR, which resulted in disintegration of the latter. However, there is considerable doubt about the universality of the Gause law because it is a consequence of linearity of per capita growth rates, and it is not a biological principle.

Only nonnegative population sizes are meaningful. We are interested in knowing the conditions under which $x_1, x_2 > 0$ and whether there exist equilibrium populations. It is reasonable to assume that $a_k > b_k$ for $k = 1, 2$ since the effect of competition between members of the same species should prevail over the competition between distinct species. Therefore, $a_1 a_2 - b_1 b_2 > 0$, and the solution of Eq. (9.3.3) is

$$x_1^* = \frac{a_2 r_1 - b_1 r_2}{a_1 a_2 - b_1 b_2}, \qquad x_2^* = \frac{a_1 r_2 - b_2 r_1}{a_1 a_2 - b_1 b_2}. \tag{9.3.4}$$

The conditions for coexistence become

$$\begin{aligned} a_2 r_1 - b_1 r_2 > 0, \\ a_1 r_2 - b_2 r_1 > 0, \end{aligned}$$

or, in other words,

$$\frac{r_2}{a_2} < \frac{r_1}{b_1} \qquad \text{and} \qquad \frac{r_1}{a_1} < \frac{r_2}{b_2}. \tag{9.3.5}$$

The ratios $K = r_1/a_1$ and $M = r_2/a_2$ are saturation levels or carrying capacities in the absence of competition between species: the population stops growing of its own accord. However, these definitions become confusing for organisms whose population dynamics are determined by the balance of reproduction and mortality processes (e.g., most insect populations). In this case, these ratios have no clear biological meaning. Similarly, if we neglect the factor of competition within a population, the net growth becomes

$$r_k x_k - b_k x_k x_j = x_k \left(r_k - b_k x_j\right), \qquad k, j = 1, 2.$$

In this case, the ratio r_k/b_j represents the level of population j necessary to put an end to the growth of population k. Therefore, peaceful coexistence is observed when each population reaches the point where it limits its own saturation level before it reaches the point where it limits the competitor's growth. □

Now we analyze stability of the system (9.3.1) using a linearization technique (see §9.2). We consider each equilibrium solution separately based on the properties of the corresponding Jacobian matrix:

$$\mathbf{J}(x_1, x_2) = \begin{bmatrix} r_1 - 2a_1x_1 - b_1x_2 & -b_1x_1 \\ -b_2x_2 & r_2 - 2a_2x_2 - b_2x_1 \end{bmatrix}. \tag{9.3.6}$$

I. The origin $(0, 0)$, with the Jacobian matrix (which sometimes is also called the community matrix)

$$\mathbf{J}(0,0) = \begin{bmatrix} r_1 & 0 \\ 0 & r_2 \end{bmatrix},$$

has to be an unstable node because it has two positive eigenvalues.

II. $(K, 0)$, with $K = r_1/a_1$. The Jacobian matrix is

$$\mathbf{J}(K,0) = \begin{bmatrix} -r_1 & -b_1r_1/a_1 \\ 0 & (a_1r_2 - b_2r_1)/a_1 \end{bmatrix},$$

with one negative eigenvalue $\lambda_1 = -r_1$ and another one $\lambda_2 = (a_1r_2 - b_2r_1)/a_1$.

III. $(0, M)$, with $M = r_2/a_2$. The Jacobian matrix is

$$\mathbf{J}(0,M) = \begin{bmatrix} (r_1a_2 - b_1r_2)/a_2 & 0 \\ -b_2r_2/a_2 & -r_2 \end{bmatrix},$$

with real eigenvalues $\lambda_1 = (r_1a_2 - b_1r_2)/a_2$ and $\lambda_2 = -r_2 < 0$.

IV. (x_1^*, x_2^*), with the Jacobian matrix

$$\mathbf{J}(x_1^*, x_2^*) = \begin{bmatrix} r_1 - 2a_1x_1^* - b_1x_2^* & -b_1x_1^* \\ -b_2x_2^* & r_2 - 2a_2x_2^* - b_2x_1^* \end{bmatrix}. \tag{9.3.7}$$

We will distinguish four cases, corresponding to the four possible sign combinations for numerators in Eq. (9.3.4), $a_2r_1 - b_1r_2$ and $a_1r_2 - b_2r_1$, but ignoring the possibilities of $x_1^* = 0$ and $x_2^* = 0$. As Problem 1, page 511, shows, it is not possible for the system (9.3.1) to have a spiral point or a center.

Case 1: $\dfrac{b_2}{a_1} < \dfrac{r_2}{r_1} < \dfrac{a_2}{b_1}$ and $a_1a_2 - b_1b_2 > 0$. The asymptotically stable critical point (x_1^*, x_2^*) is in the first quadrant of the phase plane. Since the Jacobian matrices for the equilibria $(K, 0) = (r_1/a_2, 0)$ and $(0, M) = (0, r_2/a_1)$ both have negative determinants, these stationary points are saddle points (see phase portrait in Fig. 9.11).

If interspecific competition is not too strong, the two populations can cohabit, but at lower sizes than their respective saturation levels. While the species may coexist, the price that they pay for competing with each other is that they do not reach their carrying capacities when the other species are absent.

Case 2: $\dfrac{a_2}{b_1} < \dfrac{r_2}{r_1} < \dfrac{b_2}{a_1}$ and $a_1a_2 - b_1b_2 < 0$. The critical point (9.3.4) is in the first quadrant of the phase plane, but because the determinant of the Jacobian matrix is negative (so the discriminant is positive and greater than its trace), this stationary point is a saddle point—see Eq. (8.2.13) on page 452. The equilibria $(K, 0)$ and $(0, M)$ are both asymptotically stable nodes. The steady state $(0, 0)$ is unstable. The phase portrait in Fig. 9.12 shows the separatrices (in black) going through (x_1^*, x_2^*). The separatrix splits the phase plane into two regions; interior trajectories above the separatrix go to the steady state $(0, r_2/a_2)$ and below the separatrix they approach the stationary point $(r_1/a_1, 0)$.

From an ecological point of view, interspecific competition is aggressive and ultimately one population wins, while the other is driven to extinction. The winner depends upon which has the starting advantage.

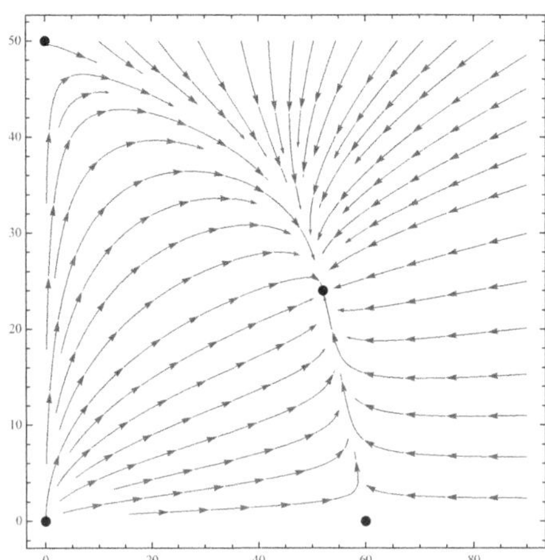

Figure 9.11: Phase portrait in Example 9.3.2, plotted with *Mathematica*.

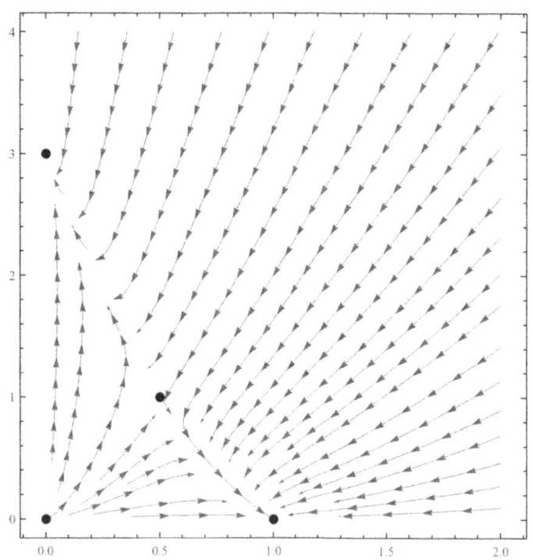

Figure 9.12: Phase portrait in Example 9.3.3, plotted with *Mathematica*.

Case 3: $\frac{r_2}{r_1} < \frac{a_2}{b_1}$ and $\frac{r_2}{r_1} < \frac{b_2}{a_1}$. There is no equilibrium in the interior of the first quadrant. The stationary point $(K, 0)$ is asymptotically stable, while the critical point $(0, M)$ is a saddle point (Fig. 9.13). All orbits tend to $(K, 0)$ as $t \to \infty$, corresponding to extinction of the x_2 species and survival of the x_1 species for all initial population sizes.

Case 4: $\frac{a_2}{b_1} < \frac{r_2}{r_1}$ and $\frac{b_2}{a_1} < \frac{r_2}{r_1}$. This case is similar to the previous one—no equilibrium solutions inside of the first quadrant. Now $(K, 0)$ is a saddle point while $(0, M)$ is an asymptotically stable node (Fig. 9.14).

Interspecific competition of one species dominates the other and, since the stable node in each case is globally stable, the species with the strongest competition always drives the other to extinction. ■

Conditions in cases 2 through 4 rule out the possibility of coexistence of two species. The experimental evidence is somewhat equivocal and there are known models of Kolmogorov's type for which the Gause principle does not work.

Example 9.3.2: (*Case 1*) Determine the outcome of a competition modeled by the system

$$\dot{x} = x\,(180 - 3x - y)\,, \qquad \dot{y} = y\,(100 - x - 2y)\,.$$

Solution. A coexisting equilibrium is found by solving the system of algebraic equations

$$180 - 3x - y = 0, \qquad 100 - x - 2y = 0.$$

By eliminating one variable, we obtain the critical point $(52, 24)$. The Jacobian matrix at this point becomes

$$\mathbf{J}(52, 24) = \begin{bmatrix} -156 & -52 \\ -24 & -48 \end{bmatrix},$$

with two negative eigenvalues $\lambda = -102 \pm 2\sqrt{1041}$. The other equilibrium points $(0, 0)$, $(0, 50)$, and $(60, 0)$ are unstable stationary points.

Example 9.3.3: (*Case 2*) Determine the outcome of a competition modeled by the system

$$\dot{x} = x\,(2 - 2x - y)\,, \qquad \dot{y} = y\,(3 - 4x - y)\,.$$

Solution. The critical points are obtained from the system

$$2 - 2x - y = 0, \qquad 3 - 4x - y = 0.$$

This system has one stationary point $\left(\frac{1}{2}, 1\right)$ in the first quadrant, which is a saddle point. The equilibria $(1, 0)$ and $(0, 3)$ are asymptotically stable nodes, which correspond to the situation of extinction for one species. The origin is always an unstable node (see Fig. 9.12).

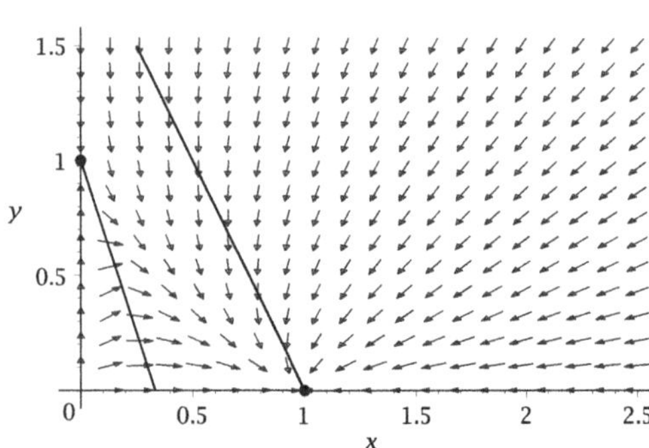

Figure 9.13: Phase portrait in Example 9.3.4, plotted with *Maple*.

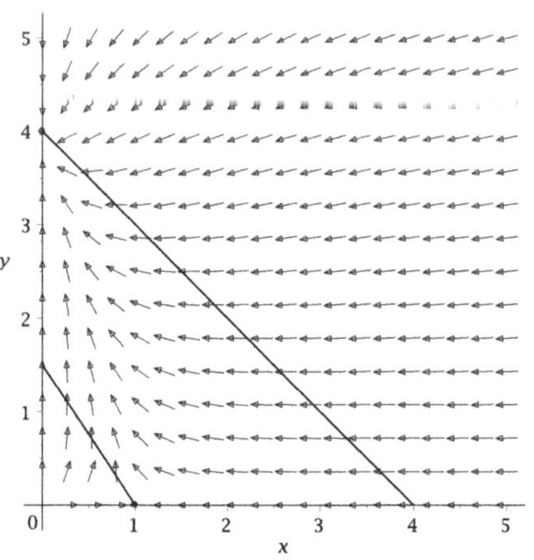

Figure 9.14: Phase portrait in Example 9.3.5, plotted with *Maple*.

Example 9.3.4: (*Case 3*) Determine the outcome of a competition modeled by the system

$$\dot{x} = x\,(2 - 2x - y)\,, \qquad \dot{y} = y\,(1 - 3x - y)\,.$$

Solution. This system has only three critical points inside the first quadrant: $(1, 0)$, $(0, 1)$, and the origin. According to the Grobman–Hartman theorem (see §9.2), we linearize the given system in neighborhoods of every point. The point $(1, 0)$ is a proper node because the community matrix $\begin{bmatrix} -2 & -1 \\ 0 & -2 \end{bmatrix}$ has a double negative eigenvalue. Therefore, this critical point is asymptotically stable, and all trajectories approach the carrying capacity. The Jacobian matrix $\begin{bmatrix} 1 & 0 \\ -3 & -1 \end{bmatrix}$ at another stationary point $(0, 1)$ has two real eigenvalues of distinct signs, so it is a saddle point.

Example 9.3.5: (*Case 4*) Determine the outcome of a competition modeled by the system

$$\dot{x} = x\,(3 - 3x - 2y)\,, \qquad \dot{y} = y\,(4 - x - y)\,.$$

Solution. This system has only three critical points inside the first quadrant: $(1, 0)$, $(0, 4)$, and the origin (which is unstable). The former is a saddle point (unstable) because the corresponding community matrix $\begin{bmatrix} -3 & -2 \\ 0 & 3 \end{bmatrix}$ has positive and negative eigenvalues. Another stationary point $(0, 4)$ is asymptotically stable because the Jacobian at this point is $\begin{bmatrix} -5 & 0 \\ -4 & -4 \end{bmatrix}$. □

Only stable equilibria are significant because they represent the population sizes for cohabitation. Unstable stationary solutions cannot be observed in practice. A point in the phase space that is not an equilibrium point corresponds to population sizes that change with time. These points on the phase plane present snapshots of population sizes that are subject to flux. In this case, biologists expect population sizes of two species to undergo change until they reach approximately the observable values.

Competitive interactions between organisms can have a great deal of influence on species evolution, the structuring of communities (which species coexist, which don't, relative abundances, etc.), and the distributions of species (where they occur). Modeling these interactions provides a useful framework for predicting outcomes.

9.3.2 Predator-Prey Equations

This subsection is concerned with the functional dependence of one species on another, where the first species depends on the second for its food. Such a situation occurs when a predator lives off its prey or a parasite lives off its host, harming it and possibly causing death. A standard example is a population of robins and worms that cohabit an ecosystem. The robins eat the worms, which are their only source of food. A few examples of parasites are tapeworms, fleas, and barnacles. Tapeworms are segmented flatworms that attach themselves to the insides of the intestines of humans and animals such as cows, pigs.

In 1925, during a conversation with Vito Volterra, a young zoologist by the name[83] of Umberto D'Ancona, Volterra's future son-in-law, asked Vito to explain his observation that the proportion of predator fish caught in the Upper Adriatic sea was up from before, whereas the proportion of prey fish was down. This phenomenon was later predicted by one of Volterra's models. In the same year, A. Lotka published a book titled "Elements of Physical Biology" where he utilized the same model, which now is known as the **Lotka–Volterra model**. It is interesting that the predator-prey model was initially proposed by Alfred Lotka in the theory of autocatalytic chemical reactions in 1910.

We denote by $x(t)$ and $y(t)$ the populations (or biomass or density) of the prey and predator, respectively, at time t. In 1926, Volterra came up with a model to describe the evolution of predator and prey based on the following assumptions:

1. in the absence of predators the per capita prey growth rate is a constant, but linearly falls as a function of the predator population when predation is present;

2. in the absence of prey, the per capita growth rate of the predator is a negative constant, and increases linearly with the prey population when prey is present.

This leads to the following system of differential equations:

$$\frac{1}{x}\frac{\mathrm{d}x}{\mathrm{d}t} = r - by(t), \qquad \frac{1}{y}\frac{\mathrm{d}y}{\mathrm{d}t} = -\mu + \beta x(t).$$

The positive constants r, b, μ, β represent the following: r is the per capita growth rate of the prey population when predators are not present and μ is the per capita death rate of predators when there is no food. The constants b and β represent the effect of interaction between two species. The above system can be rewritten as

$$\dot{x} = x\,(r - by)\,, \qquad \dot{y} = y\,(-\mu + \beta x)\,. \tag{9.3.8}$$

Since each of the equations in (9.3.8) is separable, we can solve the system explicitly:

$$\frac{\mathrm{d}y}{\mathrm{d}x} = \frac{\dot{y}}{\dot{x}} = \frac{y\,(-\mu + \beta x)}{x\,(r - by)} \qquad \Longleftrightarrow \qquad \frac{r - by}{y}\,\mathrm{d}y = \frac{-\mu + \beta x}{x}\,\mathrm{d}x.$$

However, we put this approach on the back burner (see Example 9.4.4, page 519) and pursue our main technique—linearization. First, solving the nullcline equations

$$x\,(r - by) = 0, \qquad y\,(-\mu + \beta x) = 0,$$

we obtain two critical points:

$$(0, 0) \qquad \text{and} \qquad \left(\frac{\mu}{\beta}, \frac{r}{b}\right).$$

The Jacobian matrix at the origin,

$$\mathbf{J}(0, 0) = \begin{bmatrix} r & 0 \\ 0 & -\mu \end{bmatrix},$$

has two real eigenvalues of opposite sign. Therefore, it is a saddle point. At another critical point, we have

$$\mathbf{J}\left(\frac{\mu}{\beta}, \frac{r}{b}\right) = \begin{bmatrix} r - by & -bx \\ \beta y & -\mu + \beta x \end{bmatrix}_{\substack{x=\mu/\beta \\ y=r/b}} = \begin{bmatrix} 0 & -b\mu/\beta \\ r\beta/b & 0 \end{bmatrix}.$$

Since its characteristic equation $\lambda^2 + r\mu = 0$ has two pure imaginary roots, $\pm\mathbf{j}\sqrt{r\mu}$, the stationary point $\left(\frac{\mu}{\beta}, \frac{r}{b}\right)$ is not hyperbolic, and the Grobman–Hartman theorem (see §9.2) is not applicable. We analyze this case in the next two sections, and now just plot the phase portrait. From Fig. 9.15, it follows that trajectories are closed curves in

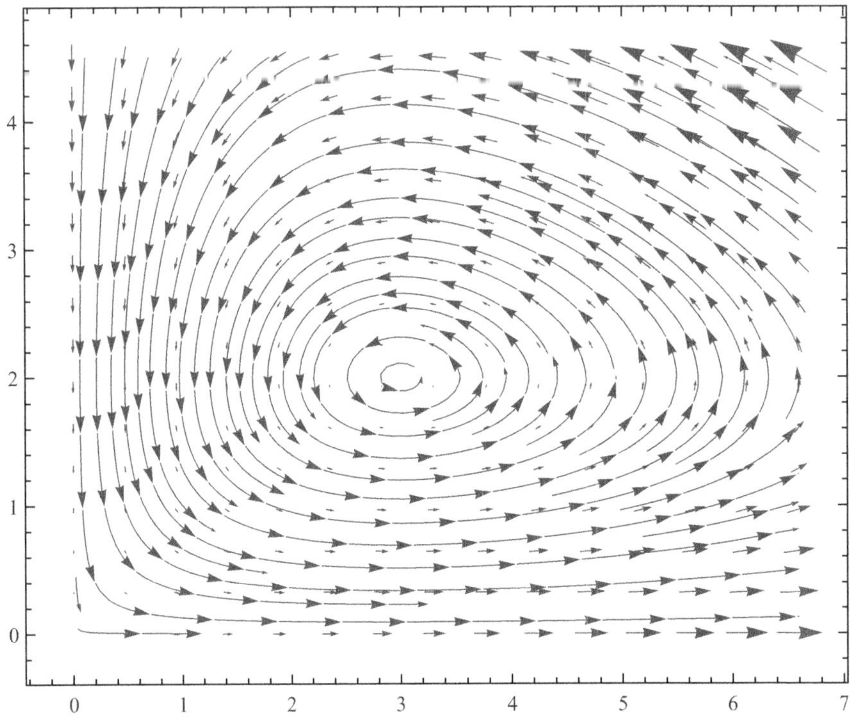

Figure 9.15: Phase portrait of the Lotka–Volterra model (9.3.8), plotted with *Mathematica*.

the first quadrant that spiral around the critical point $\left(\frac{\mu}{\beta}, \frac{r}{b}\right)$. Since no orbit can cross a coordinate axis, every solution starting in the first quadrant remains there for all time.

By plotting two populations of prey $x(t)$ and predators $y(t)$ with respect to time t, we see from Fig. 9.16 that the oscillation of the predator population lags behind that of the prey. Starting from a state in which both populations are small, the population of prey increases first because there is little predation. Later, the predators increase in size because of abundant food. This causes heavier predation, and the prey population shrinks. Finally, with a diminishing food supply, the predator population also decreases, and the system returns to the original state.

Example 9.3.6: Consider the Lotka–Volterra system of differential equations

$$\dot{x} = x\,(0.3 - 0.024\,y)\,, \qquad \dot{y} = y\,(0.02\,x - 0.4)$$

for $x(t)$ and $y(t)$ positive when $t \geqslant 0$. The critical points of this sytem are the solutions of the simultaneous algebraic equations

$$x\,(0.3 - 0.024\,y) = 0, \qquad y\,(0.02\,x - 0.4) = 0,$$

namely, the points $(0,0)$ and $(20, 12.5)$. The Jacobian at the origin, $\mathbf{J}(0,0) = \begin{bmatrix} 0.3 & 0 \\ 0 & -0.4 \end{bmatrix}$ has two eigenvalues $\lambda = 0.3$ and $\lambda = -0.4$; therefore this point is a saddle point, unstable. The Jacobian at another stationary point $\mathbf{J}(20, 12.5) = \begin{bmatrix} 0 & -0.48 \\ 0.25 & 0 \end{bmatrix}$ has two pure imaginary eigenvalues, so linearization does not provide enough information about its stability. The phase portrait on Fig. 9.15 confirms that the trajectories are closed curves, and this point is a center (stable, but not asymptotically stable). □

The Lotka–Volterra model of interspecific competition, also known as the predator-prey model, is frequently used to describe the dynamics of biological systems in which two species interact, one as a predator and the other as prey. Even though it is a very simple model, it explains cyclic variations in populations of two species observed in reality. However, its simplicity leads to some obvious flaws:

[83] Umberto D'Ancona (1896–1964) wrote more than 300 scientific articles and textbooks. His field of interest was extremely vast and ranged from physiology, to embryology, hydrobiology, oceanography, and evolutionary theory. Umberto interrupted his study at the University of Rome during World War I to go and fight as an artillery officer, where he was wounded and decorated. In 1926 he married Luisa Volterra.

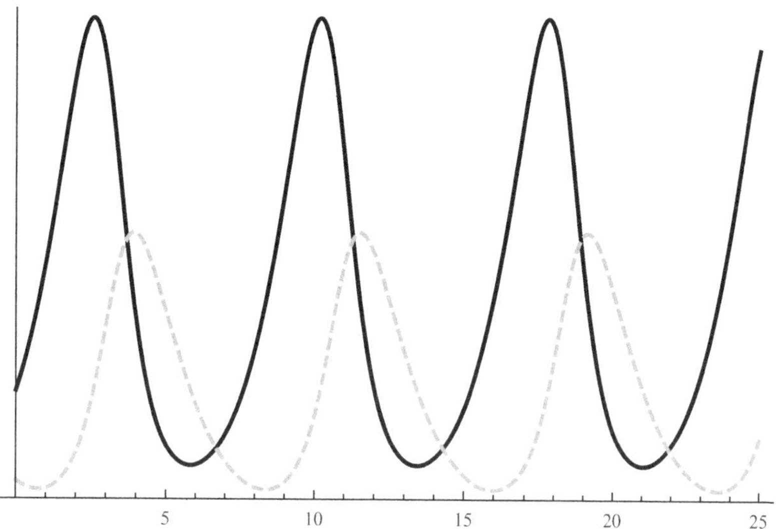

Figure 9.16: Variations of the prey (in black) and predator (in dashed blue) populations with time for the Lotka–Volterra system (9.3.8).

1. There is no possibility of either population being driven to extinction.
2. Changing the birth r and death μ rates does nothing but change the period of the oscillation, i.e., none can dominate.
3. A prey population in the absence of predators would grow exponentially toward infinity.
4. Population orbits always go in a counterclockwise direction while there are known quantitative data for which the direction should be clockwise.
5. The population size of prey usually oscillates even in the absence of predators, most likely due to climatic variations and to epidemics.

There are many known refinements to the Lotka–Volterra (LV for short) model that we cannot analyze due to size constraints. Therefore, we consider only one more model by introducing a self-limiting term for the growth of the prey and predator, reducing the equations to a logistic model:

$$\begin{aligned} \dot{x} &= x\,(r - ax - by)\,, \\ \dot{y} &= y\,(-\mu - \alpha y + \beta x)\,, \end{aligned} \tag{9.3.9}$$

where, as usual, $x(t)$ denotes the population of prey, and $y(t)$ stands for the population of the predator. Let us consider the nullclines, which are solutions to

$$\begin{aligned} \dot{x} = 0: &\quad x = 0 \quad \text{or} \quad r - ax - by = 0, \\ \dot{y} = 0: &\quad y = 0 \quad \text{or} \quad -\mu - \alpha y + \beta x = 0. \end{aligned}$$

This system of algebraic equations has four critical points:

$$(0,0)\,, \quad \left(0, -\frac{\mu}{\alpha}\right), \quad \left(\frac{r}{a}, 0\right), \quad \left(\frac{r\alpha + b\mu}{a\alpha + b\beta}, \frac{r\beta - a\mu}{a\alpha + b\beta}\right).$$

Since all coefficients are positive, we have to disregard $(0, -\mu/\alpha)$ as being biologically meaningless. Therefore, this system has at most three nonnegative solutions, but only one

$$x^* = \frac{r\alpha + b\mu}{a\alpha + b\beta}, \qquad y^* = \frac{r\beta - a\mu}{a\alpha + b\beta} \tag{9.3.10}$$

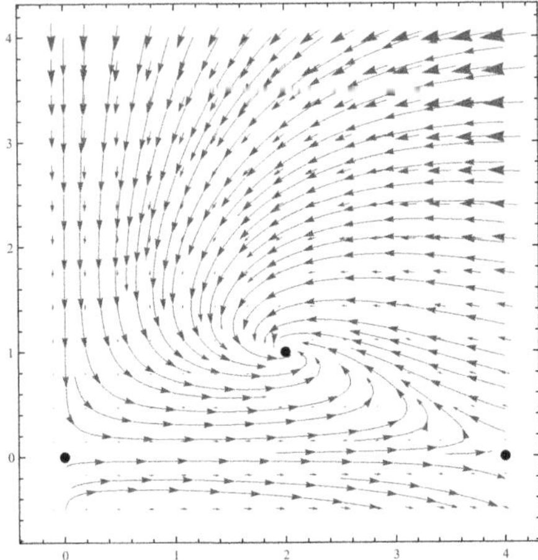

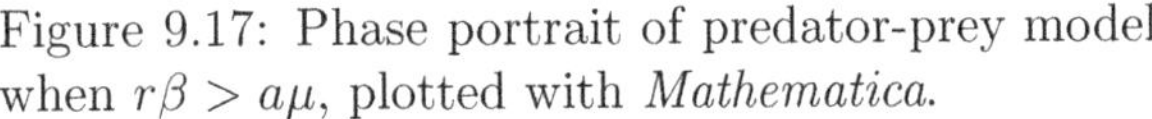

Figure 9.17: Phase portrait of predator-prey model when $r\beta > a\mu$, plotted with *Mathematica.*

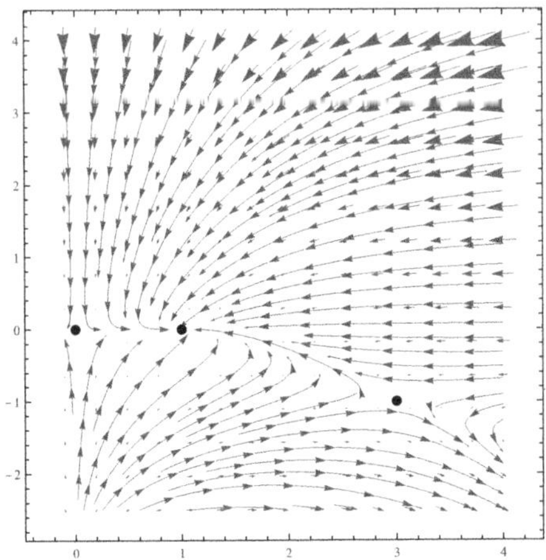

Figure 9.18: Phase portrait of predator-prey model when $r\beta < a\mu$, plotted with *Mathematica.*

will be in the first quadrant when $r\beta - a\mu > 0$. When $r\beta < a\mu$, we have two critical points on the boundary of the first quadrant. Phase portraits of these two possible cases when $r\beta > a\mu$ and $r\beta < a\mu$ are plotted in Figures 9.17 and 9.18, respectively. On the boundary $y = 0$ (which corresponds to extinction of predators), we have one critical point $x = r/a$, which is unstable when $r\beta > a\mu$ and asymptotically stable when $r\beta < a\mu$. There is no (positive) critical point on another boundary $x = 0$. The origin $(0, 0)$ is always an unstable saddle point.

It is not obvious whether the trajectories of the LV model are closed paths or spirals (or something else?) in a neighborhood of the stationary points. To complete the phase plots, we need to determine the correct behavior of the trajectories near steady states, i.e., perform the linear stability analysis. For the Jacobian matrix, we obtain

$$\mathbf{J}(x, y) = \begin{bmatrix} r - 2ax - by & -bx \\ \beta y & -\mu - 2\alpha y + \beta x \end{bmatrix}.$$

Hence, at $(0, 0)$, we have

$$\mathbf{J}(0, 0) = \begin{bmatrix} r & 0 \\ 0 & -\mu \end{bmatrix},$$

so that the eigenvalues r and $-\mu$ are of opposite sign, showing that the origin is a saddle point. At $(r/a, 0)$, we have

$$\mathbf{J}\left(\frac{r}{a}, 0\right) = \begin{bmatrix} -r & -rb/a \\ 0 & -\mu + \beta r/a \end{bmatrix}.$$

The eigenvalues of $\mathbf{J}(r/a, 0)$ are thus $\lambda_1 = -r < 0$ and $\lambda_2 = -\mu + \beta r/a$. In the case when $\beta r > a\mu$, the eigenvalue λ_2 is positive, so that there is no interior steady state, and the critical point $(r/a, 0)$ is unstable (saddle). When $\beta r < a\mu$, both eigenvalues $\lambda_{1,2}$ are negative, and the stationary point $(r/a, 0)$ is a stable node, so the interior steady state exists.

Finally, we consider the linear stability of (x^*, y^*). We have

$$\mathbf{J}(x^*, y^*) = \begin{bmatrix} -a\,\dfrac{r\alpha + b\mu}{a\alpha + b\beta} & -b\,x^* \\ \beta\, y^* & \dfrac{3a\alpha\mu + 2b\beta\mu - r\alpha\beta}{a\alpha + b\beta} \end{bmatrix}.$$

When $r\beta > a\mu$, the matrix $\mathbf{J}(x^*, y^*)$ has two complex conjugate eigenvalues with a negative real part. Therefore, the stationary point (x^*, y^*) is an asymptotically stable spiral point (see Fig. 9.17).

Example 9.3.7: Consider a simple example of interactions of parasites on hosts modeled by the following system of differential equations:

$$\dot{h} = h(a - bp), \qquad \dot{p} = p\left(ch^2 - d - sp\right),$$

where $h(t)$ is the population density of the hosts, $p(t)$ is the mean number of parasites per host, and a, b, c, d, s are some positive constants. The parasite-host system has two critical points in the first quadrant: the origin and $(x^*, y^*) = \left(\sqrt{\frac{bd+as}{cb}}, \frac{a}{b} \right)$. The Jacobian matrices $\mathbf{J}(x,y) = \begin{bmatrix} a-by & -bx \\ 2cxy & cx^2-d-2sy \end{bmatrix}$ at these points are

$$\mathbf{J}(0,0) = \begin{bmatrix} a & 0 \\ 0 & -d \end{bmatrix} \quad \text{and} \quad \mathbf{J}(x^*, y^*) = \begin{bmatrix} 0 & -bx^* \\ \frac{2acx^*}{b} & -\frac{as}{b} \end{bmatrix}.$$

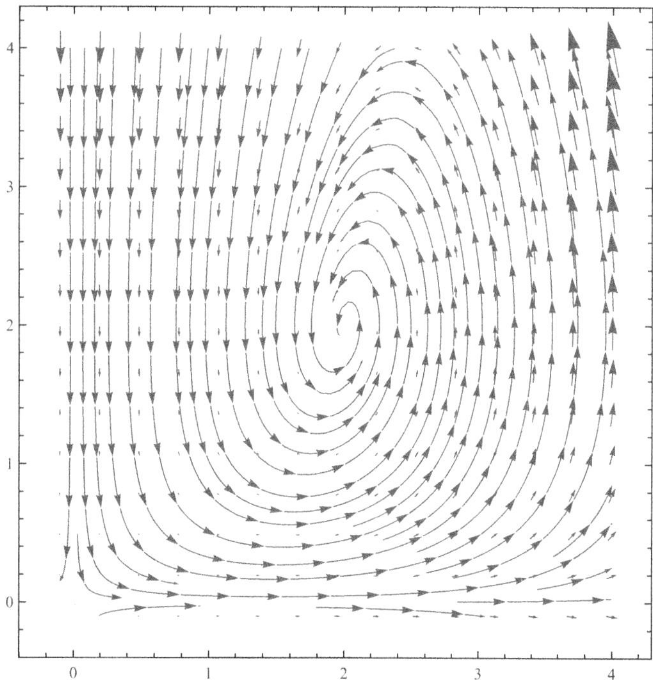

Figure 9.19: Example 9.3.7: phase portrait of the model describing interactions of parasites on hosts, plotted with *Mathematica*.

Since the eigenvalues of the matrix $\mathbf{J}(0,0)$ are $\lambda_1 = a > 0$ and $\lambda_2 = -d < 0$, the origin is an unstable saddle point. At the stationary point (x^*, y^*), the trace of the Jacobian matrix is $\operatorname{tr}\mathbf{J}(x^*, y^*) = -\frac{as}{b} < 0$, and its determinant is $\det \mathbf{J}(x^*, y^*) = \frac{2a}{b}(bd + as) > 0$. Therefore, (x^*, y^*) is always a stable equilibrium solution for any positive values of coefficients: it is a spiral sink when $(\operatorname{tr}\mathbf{J})^2 < 4 \det \mathbf{J}$ and it is an attractor when the latter inequality fails. □

Cycle variations of predator and prey as predicted by the Lotka–Volterra model (9.3.8) have been observed in nature. A classical example of interacting populations in which oscillations have been observed is the data collected by the Hudson's Bay Company in Canada during the period 1821 – 1940 on furs of the snowshoe hare (prey) and Canadian lynx (predator). Even though the data may not accurately describe the total population sizes, the plotting graphs of the hare and lynx unambiguously indicate that their trajectories are going clockwise while models (9.3.8) and (9.3.9) predict the opposite direction. Various suggestions have been made to explain the anomaly. One possibility is that the predator-prey models are too sensitive to actual perturbation in populations: relationships among species are often complex and subtle, especially when the number of distinct species exceeds three.

Nevertheless, mathematical models of biological systems have a long history of use not only in population dynamics but also in other disciplines such as economic theory. Frequently, a system of the Lotka–Volterra type is used to give some indication of the type of behavior one might expect in multidimensional cases. These models predict observable cohabitation of distinct species or industries. Biological experiments suggest that initial population sizes close to the equilibrium values cause populations to stay near the initial sizes, even though the populations oscillate periodically. Observations by biologists of large population variations seem to verify that individual populations oscillate periodically around the ideal cohabitation sizes.

9.3.3 Other Population Models

There are situations in which the interaction of two species is mutually beneficial, for example, plant-pollinator systems. The interaction may be *facultative*, meaning that the two species could survive separately, or *obligatory*, meaning that each species will become extinct without the assistance of the other. A mutualistic system can be modeled by a pair of differential equations with linear per capita growth rates:

$$\dot{x} = x\,(r - ax + by)\,, \qquad \dot{y} = y\,(\mu + cx - dy)\,. \tag{9.3.11}$$

In the above system, the mutualism of the interaction is modeled by the positive nature of the interaction terms cx and by. In a facultative interaction where two species can survive separately, the constants r and μ are positive, while in an obligatory relation these constants are negative. In each type of interaction there are two possibilities, depending on the relation between the slope a/b of the x-nullcline and the slope c/d of the y-nullcline. Since their analysis is very similar to those described in §9.3.1, we restrict ourselves to one example.

Example 9.3.8: Consider the following model:

$$\dot{x} = x\,(1 - x + 3y)\,, \qquad \dot{y} = y\,(3 + x - 5y)\,.$$

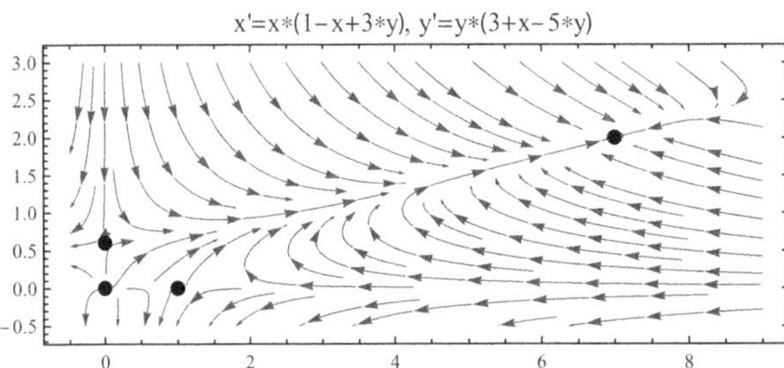

Figure 9.20: Example 9.3.8: phase portrait of the model describing mutually beneficial interaction of species, plotted with *Mathematica*.

The above system has four critical points:

$$(0,0)\,, \quad (0,0.6)\,, \quad (1,0)\,, \quad (7,2)\,,$$

but only one of them, $(7,2)$, is biologically meaningful because it leads to mutual coexistence. The Jacobian at this point is

$$\mathbf{J}\,(7,2) = \begin{bmatrix} -7 & 21 \\ 2 & -10 \end{bmatrix}, \quad \text{with} \quad \operatorname{tr}\mathbf{J} = -17, \quad \det\mathbf{J} = 28.$$

Since this matrix has two negative real eigenvalues $\lambda_1 = -(17+\sqrt{177})/2 \approx -15.1521$ and $\lambda_2 = -(17-\sqrt{177})/2 \approx -1.84793$, the critical point $(7,2)$ is an attractor. □

Now we go in another direction and briefly discuss the impact of humans on population dynamics—a very important and rapidly developing subject. In studying models for competition between two species (§9.3.1), we begin with the system (9.3.1). We will consider only the harvesting of one of the two species, say x-species. With constant-yield harvesting, the model is

$$\dot{x} = x\,(r_1 - a_1 x - b_1 y) - H, \qquad \dot{y} = y\,(r_2 - a_2 x - b_2 y)\,.$$

The x-nullcline, instead of being the pair of lines $x = 0$, $r_1 = a_1 x + b_1 y$, is now the curve $x\,(r_1 - a_1 x - b_1 y) = H$, which is a hyperbola having the lines $x = 0$ and $a_1 x + b_1 y = r_1$ as asymptotes and which moves away from these asymptotes as H increases.

Example 9.3.9: (Example 9.3.2 revisited) Determine the response of the system

$$\dot{x} = x\,(180 - 3x - y)\,, \qquad \dot{y} = y\,(100 - x - 2y)$$

to constant-yield harvesting of the x-species.

Solution. With no harvesting, there is an asymptotically stable equilibrium at $(52, 24)$. A harvesting system has the form

$$\dot{x} = x\,(180 - 3x - y) - H, \qquad \dot{y} = y\,(100 - x - 2y)\,.$$

Equilibria are given by the pair of equations

$$x\,(180 - 3x - y) = H, \qquad x + 2y = 100.$$

Replacing x by $100 - 2y$ in the first of these equations, we obtain a quadratic equation for y:

$$(100 - 2y)\,(5y - 120) = H \qquad \Longrightarrow \qquad y^2 - 74y + (120 + H/10) = 0.$$

From the quadratic formula, we have

$$y = 37 \pm \sqrt{169 - H/10}.$$

For $H = 0$, these roots are $y = 50$ (which gives $x = 0$) and $y = 24$ (which gives $x = 52$). There is an asymptotically stable stationary point at $(52, 24)$ and a saddle point at $(0, 50)$. As H increases, these equilibria move along the line $x + 2y = 100$ until they coalesce at $(26, 74)$ for $H = 1690$.

Problems

Systems of equations in Problems 1 and 2 can be interpreted as describing the interaction of two species with populations $x(t)$ and $y(t)$. In each of these problems, perform the following steps.

(a) Find the critical points.

(b) Find the corresponding linearization for each stationary point. Then find the eigenvalues of the linear system; classify each critical point as to type, and determine whether it is asymptotically stable, stable, or unstable.

(c) Plot the phase portrait of the given nonlinear system.

(d) Determine the limiting behavior of $x(t)$ and $y(t)$ as $t \to \infty$, and interpret the results in terms of the populations of the two species.

1. Equations modeling competitions of species:

(a) $\begin{cases} \dot{x} = x\,(4 - x - y)\,, \\ \dot{y} = y\,(6 - x - 3y)\,. \end{cases}$

(b) $\begin{cases} \dot{x} = x\,(3 - 2x - y)\,, \\ \dot{y} = y\,(3 - x - 2y)\,. \end{cases}$

(c) $\begin{cases} \dot{x} = x\,(100 - 4x - y)\,, \\ \dot{y} = y\,(60 - 2x - 3y)\,. \end{cases}$

(d) $\begin{cases} \dot{x} = x\,(80 - 2x - y)\,, \\ \dot{y} = y\,(120 - x - 3y)\,. \end{cases}$

(e) $\begin{cases} \dot{x} = x\,(60 - 2x - y)\,, \\ \dot{y} = y\,(75 - x - 3y)\,. \end{cases}$

(f) $\begin{cases} \dot{x} = x\,(40 - x - 2y)\,, \\ \dot{y} = y\,(90 - 3x - y)\,. \end{cases}$

(g) $\begin{cases} \dot{x} \;= x\,(80 - x - 2y)\,, \\ \dot{y} \;= y\,(90 - 3x - y)\,. \end{cases}$

(h) $\begin{cases} \dot{x} = x\,(40 - 3x - 2y)\,, \\ \dot{y} = y\,(40 - x - y)\,. \end{cases}$

(i) $\begin{cases} \dot{x} = x\left(1 - \frac{2}{3}\,x - y/9\right), \\ \dot{y} = y\left(\frac{1}{2} - \frac{1}{4}\,x - \frac{2}{3}\,y\right). \end{cases}$

(j) $\begin{cases} \dot{x} = x\,(1 - 0.3\,x - 0.3\,y)\,, \\ \dot{y} = y\,(1 - 0.25\,x - y)\,. \end{cases}$

(k) $\begin{cases} \dot{x} = x\left(\frac{2}{3} - \frac{2}{3}\,x - \frac{y}{4}\right), \\ \dot{y} = y\left(\frac{7}{9} - \frac{3}{4}\,x - \frac{1}{3}\,y\right). \end{cases}$

(l) $\begin{cases} \dot{x} = x\left(\frac{3}{2} - x - y/2\right), \\ \dot{y} = y\left(2 - \frac{3}{4}\,x - y\right). \end{cases}$

(m) $\begin{cases} \dot{x} = x\left(\frac{3}{2} - x/2 - y\right), \\ \dot{y} = y\left(3 - \frac{5}{4}\,x - y\right). \end{cases}$

(n) $\begin{cases} \dot{x} = x\left(\frac{5}{2} - x/4 - \frac{3}{2}\,y\right), \\ \dot{y} = y\left(5 - \frac{3}{4}\,x - y\right). \end{cases}$

2. Predator-prey vector equations:

(a) $\begin{cases} \dot{x} = x(6 - 2y), \\ \dot{y} = y(4x - 16); \end{cases}$

(b) $\begin{cases} \dot{x} = \frac{x}{4}\,(3 - y)\,, \\ \dot{y} = y\left(\frac{x}{2} - 1\right); \end{cases}$

(c) $\begin{cases} \dot{x} = x\left(\frac{y}{2} - 1\right), \\ \dot{y} = 3y\left(1 - \frac{x}{4}\right); \end{cases}$

(d) $\begin{cases} \dot{x} = x\left(\frac{1}{3} - \frac{y}{9}\right), \\ \dot{y} = y\left(\frac{x}{2} - \frac{1}{6}\right). \end{cases}$

3. What is the outcome of a competition of two species modeled by the system

$$\dot{x} = x\left(12 - x - y - x^2\right), \qquad \dot{y} = y\left(16 - 3x - y - x^2\right)?$$

4. Determine the qualitative behavior of a predator-prey interaction modeled by the Holling system.

(a) $\begin{cases} \dot{x} = x\left(1 - \frac{x}{30} - \frac{y}{x+5}\right), \\ \dot{y} = y\left(\frac{x}{x+5} - \frac{4}{5}\right). \end{cases}$

(b) $\begin{cases} \dot{x} = x\left(3 - \frac{x}{20} - \frac{y}{x+5}\right), \\ \dot{y} = y\left(\frac{x}{x+5} - \frac{4}{5}\right). \end{cases}$

(c) $\begin{cases} \dot{x} = x\left(3 - \frac{x}{40} - \frac{2y}{x+25}\right), \\ \dot{y} = y\left(\frac{2x}{x+25} - \frac{3}{2}\right). \end{cases}$

(d) $\begin{cases} \dot{x} = x\left(3 - \frac{3x}{10} - \frac{2y}{x+5}\right), \\ \dot{y} = y\left(\frac{x}{x+5} - \frac{1}{6}\right). \end{cases}$

(e) $\begin{cases} \dot{x} = x\left(1 - \frac{3x}{40} - \frac{17y}{x+2}\right), \\ \dot{y} = y\left(\frac{x}{x+2} - \frac{1}{2}\right). \end{cases}$

(f) $\begin{cases} \dot{x} = x\left(1 - \frac{3x}{20} - \frac{11y}{x+3}\right), \\ \dot{y} = y\left(\frac{x}{x+3} - \frac{1}{2}\right). \end{cases}$

(g) $\begin{cases} \dot{x} = x\left(5 - \frac{x}{8} - \frac{y}{x+7}\right), \\ \dot{y} = y\left(\frac{x}{x+7} - \frac{1}{8}\right). \end{cases}$

(h) $\begin{cases} \dot{x} = x\left(16 - \frac{x}{10} - \frac{y}{x+9}\right), \\ \dot{y} = y\left(\frac{x}{x+9} - \frac{1}{10}\right). \end{cases}$

5. Show that the stationary point (x^*, y^*) in the first quadrant ($x^* > 0$, $y^* > 0$) of the predator-prey system modeled by

$$\dot{x} = rx\left(1 - \frac{x}{K} - \frac{ay}{x+A}\right), \qquad \dot{y} = ry\left(\frac{ax}{x+A} - \frac{aM}{M+A}\right)$$

is unstable if $K > A + 2M$, and asymptotically stable if $M < K < A + 2M$.

6. Let (x^*, y^*) be a critical point (9.3.4) of the system (9.3.1). Show that the discriminant of the Jacobian matrix (9.3.7) does not contain quadratic terms:

$$\det \mathbf{J}(x_1^*, x_2^*) = r_1 r_2 - b_2 r_1 x_1^* - b_1 r_2 x_2^*.$$

7. The populations $x(t)$ and $y(t)$ of two species satisfy the system of equations

$$\dot{x} = x\left(7 - 3x - 2y\right), \qquad \dot{y} = y\left(4 - 2x - y\right)$$

(after scaling). Find the stationary points of the system. What happens to the species in the cases with initial conditions?

(a) $x = 1,\ y = 3$; **(b)** $x = 3,\ y = 1$.

8. Suppose that you need to model populations of blue whales and fin whales that inhabit some part of Pacific ocean. Since they both rely on the same source of food, they can be thought of as competitors and their populations can be modeled by the system of equations (9.3.1), page 501, where $x_1(t)$ represents the population of blue whales and $x_2(t)$ stands for the population of fin whales. Units for the population sizes might be in thousands of species. The intrinsic growth rate of each species is estimated at 3% per year for the blue whales and 5% per year for the fin whale. The environmental carrying capacity is estimated at 15 thousand blues whales and 35 thousand fin whales. The extent to which the whales compete is unknown, and you simplify your model by choosing $b_1 = b_2 = b$ in Eq. (9.3.1).
   (a) Estimate an interval of values of b for which coexistence of both types of whales is possible.
   (b) Find equilibrium solutions for $b = 0.02$ and determine the type and stability of each critical point. Describe what happens to the two populations over time.
   (c) Answer the previous question for $b = 0.05$.

9. One of the favorite foods of the blue whale is called krill. These tiny shrimp-like creatures are devoured in massive amounts to provide the principal food source for the huge whales. If $x(t)$ denotes the biomass of krills and $y(t)$ denotes the population of whales, then their interaction can be modeled by the following Lotka–Volterra equations:

$$\dot{x} = x\left(0.3 - 0.0075y\right), \qquad \dot{y} = y\left(-0.2 + 0.0025x\right).$$

Determine the behavior of the two populations over time.

10. Redo the previous problem based on the logistic equation

$$\dot{x} = x\left(0.3 - \frac{x}{1200} - 0.0075y\right), \qquad \dot{y} = y\left(-0.2 + 0.0025x\right).$$

The maximum sustainable population for krill is 1200 tons/hectare. In the absence of predators, the krill population grows at a rate of 30% per year.

11. Suppose your pond contains two kinds of freshwater fish: green sunfish and bluegill. You estimate their carrying capacities as 800 species of each population, and their unrestricted rate to be 20%. The corresponding competition model becomes
$$\dot{x} = 0.2\,x\left(1 - \frac{x}{800} - 0.016\,y\right), \qquad \dot{y} = 0.2\,y\left(1 - \frac{y}{800} - 0.008\,x\right),$$
where $x(t)$ is the population of green sunfish and $y(t)$ is the population of bluegill. Which species appears to be the stronger competitor? What outcome would you predict if we started with 100 of each species?

12. You have both green sunfish and bluegill available for stocking your pond. You model their interaction with the following system of equations:
$$\dot{x} = 0.2\,x\left(1 - \frac{x}{1200} - 0.06\,y\right), \qquad \dot{y} = 0.12\,y\left(1 - \frac{y}{800} - 0.08\,x\right).$$
Which species appears to be the stronger competitor?

13. The robin and worm populations at time t years are denoted by $r(t)$ and $w(t)$, respectively. The equations governing the growth of the two populations are
$$\dot{r}(t) = 3w(t) - 30,$$
$$\dot{w}(t) = 15 - r(t).$$
If initially 8 robins and 10 worms occupy the ecosystem, determine the behavior of the two populations over time.

14. For competition described by the Holling–Tanner system of equations with $r = s = 1$, $K = 10$, and $h = 5$:
$$\dot{x} = xr\left(1 - \frac{x}{K}\right) - yx^2, \qquad \dot{y} = ys\left(1 - \frac{hy}{x}\right),$$
which of two species outcompetes the other?

15. In an ecosystem, let $x(t)$ stand for the population of preys, and $y(t)$ and $z(t)$ denote populations of two distinct predator species that compete for the same source of food—prey $x(t)$. Suppose that their interaction is modeled by the following system of equations:
$$\dot{x} = x\,(12 - 2y - 3z)\,,$$
$$\dot{y} = y\,(-2 + 2x - 3z)\,,$$
$$\dot{z} = z\,(-1 + x - y)\,.$$
Find all critical points in the first quadrant of the above system and classify them as stable, asymptotically stable, or unstable.

16. In an ecosystem, let $x(t)$ and $y(t)$ stand for two populations of competing preys, and $z(t)$ denote populations of predator species that prey on both preys, $x(t)$ and $y(t)$. Suppose that their interaction is modeled by the following system of equations:
$$\dot{x} = x\left(\frac{1}{2} - x - \frac{1}{4}\,y - z\right),$$
$$\dot{y} = y\left(\frac{3}{2} - x - y - 2z\right),$$
$$\dot{z} = z\,(4x + 3y - 4)\,.$$
Find all critical points in the first quadrant of the above system and classify them as stable, asymptotically stable, or unstable.

17. Termite assassin bugs use their long rostrum to inject a lethal saliva that liquefies the insides of the prey, which are then sucked out. We may be interested in quantifying the number of termites that can be consumed in an hour by the assassin bug. Determine the equilibrium behavior of the corresponding predator-prey system modeled by
$$\dot{x} = x\left(1 - \frac{x}{30} - \frac{y}{x+1}\right), \qquad \dot{y} = y\left(\frac{x}{x+10} - \frac{1}{3}\right).$$

18. Consider the predator-prey system
$$\dot{x} = 5x - xy + \epsilon x\,(3 - x)\,, \qquad \dot{y} = xy - 3y,$$
containing a parameter ϵ. For this system of equations, a bifurcation occurs at the value $\epsilon = 0$.
    (a) Find all critical points for $\epsilon = 1, -1$, and 0.
    (b) Find the Jacobian matrices at all these points.
    (c) Determine stability at each equilibrium point. Identify cases where linearization does not work.

19. In each exercise from Problem 1, determine the response of the system to constant-yield harvesting of the x-species.

20. In each exercise from Problem 1, determine the response of the system to constant-yield harvesting of the y-species.

9.4 Conservative Systems

Many differential equations arise from problems in mechanics, electrical engineering, quantum mechanics, and other areas where conservation laws may be applied. In particular, a conservation law states that some physical quantity, which is usually energy, remains constant. In reality, a physical system is never conservative. However, mathematical models often neglect effects such as friction, electrical resistance, or temperature fluctuation if they are small enough. Therefore, we operate with idealized mathematical models that may obey conservative laws. We will see later that in many cases mathematical expressions that have no physical meaning behave conservatively. In this section we analyze mathematical models using systems of autonomous differential equations for which conservation laws can be applied.

Consider a mechanical system that is governed by Newton's second law,

$$F = m\,\ddot{y}, \qquad \ddot{y} = \mathrm{d}^2y/\mathrm{d}t^2,$$

where the force $F = F(y)$ depends only on displacement y. By dividing the initial equation $F = m\,\ddot{y}$ throughout by the bothersome mass m, we can rewrite it as

$$\ddot{y} + f(y) = 0, \tag{9.4.1}$$

where $f(y) = -F(y)/m$. Now we show that Eq. (9.4.1) possesses a conservative law. We first multiply the equation by $\dot{y}$, obtaining

$$\dot{y}\,\ddot{y} + f(y)\,\dot{y} = 0.$$

Recalling the chain rule of calculus, we see that the first term on the left-hand side is

$$\dot{y}\,\ddot{y} = \frac{\mathrm{d}}{\mathrm{d}t}\left[\frac{1}{2}\,(\dot{y})^2\right].$$

Likewise, if $\Pi(y)$ denotes an antiderivative of $f(y)$, we express the second term as

$$f(y)\,\dot{y} = \frac{\mathrm{d}}{\mathrm{d}t}\,\Pi(y), \qquad \text{with} \quad \frac{\mathrm{d}}{\mathrm{d}y}\,\Pi(y) = f(y).$$

Using these derivative expressions, we form the differential equation

$$\frac{\mathrm{d}}{\mathrm{d}t}\left[\frac{1}{2}\,(\dot{y})^2 + \Pi(y)\right] = 0.$$

Recognizing in the expression $(1/2)(\dot{y})^2$, the kinetic energy, and in $\Pi(y)$, the potential energy of the system, we see that the total energy is a constant (denoted by K):

$$E(y,\dot{y}) \stackrel{\text{def}}{=} \frac{1}{2}\,(\dot{y})^2 + \Pi(y) = K. \tag{9.4.2}$$

Equation (9.4.2) is the underlying conservative law. For our mechanical system, if $y(t)$ represents a displacement, then the term $(1/2)(\dot{y})^2$ is kinetic energy per unit mass, and $\Pi(y)$ is potential energy per unit mass.

Differential equation (9.4.1) can be recast as the first order autonomous system

$$\dot{x}_1 = x_2,$$
$$\dot{x}_2 = -f(x_1),$$

where $x_1(t) = y(t)$ and $x_2 = \dot{y}(t)$. Thus, the conservative law (9.4.2) takes the form

$$E(x_1,x_2) \stackrel{\text{def}}{=} \frac{1}{2}\,x_2^2 + \Pi(x_1) = K, \qquad \text{where} \quad \frac{\mathrm{d}}{\mathrm{d}x_1}\,\Pi(x_1) = f(x_1).$$

The family of curves obtained by graphing $E(x_1,x_2) = K$ for different energy levels K is a set of phase-plane trajectories describing the motion. It is called the Poincaré phase plane of the differential equation (9.4.1). Solving Eq. (9.4.2) with respect to the velocity $v = \dot{y} = x_2$, we obtain

$$v = x_2 = \pm\sqrt{2}\,(K - \Pi(y))^{1/2}, \qquad K \text{ is a constant.}$$

Hence, the (real-valued) velocity exists only when $K - \Pi(y) \geqslant 0$. This leads to the following three types of trajectory behavior.

1. The potential function $\Pi(y)$ has a local minimum at $y_{\min}$. Then the level curves $E(x_1, x_2) = K$, where K is slightly greater than $\Pi(y_{\min})$, are closed trajectories encircling the critical point $(y_{\min}, 0)$. Therefore, the equilibrium solution $y = y_{\min}$ is a center.

2. The potential function $\Pi(y)$ has a local maximum at $y_{\max}$. Then the level curves $E(x_1, x_2) = K$ in a neighborhood of the critical point $(y_{\max}, 0)$ on the phase plane $(x_1, x_2) = (y, \dot{y})$ are not closed trajectories. For $K > \Pi(y_{\max})$, there are trajectories moving away from the critical point because their velocities ($v = \pm\sqrt{2} \times (K - \Pi(y))^{1/2}$) are either positive or negative . However, for $K < \Pi(y_{\max})$, there is an interval around it where we do not observe any solution (the root $(K - \Pi(y))^{1/2}$ is purely imaginary). Therefore, the equilibrium solution $y = y_{\max}$ is a saddle point.

3. Away from critical points of $\Pi(y)$, the level curves may be part of a closed trajectory or may be unbounded.

Example 9.4.1: (Pendulum, Example 2.6.18 revisited) The differential equation $\ddot{\theta} + \omega^2 \sin\theta = 0$ used to model the motion of the ideal pendulum in the plane does not account for any resistive force, obeying the conservation law. It can be rewritten in the Poincaré phase plane form:

$$\begin{aligned} \dot{x} &= y, \\ \dot{y} &= -\omega^2 \sin x, \end{aligned}$$

where $x = \theta$ is the angle of the bob's inclination. Now we express the dependence of the velocity $y = \dot{\theta}$ with respect to inclination x. To find it, we write

$$\frac{\mathrm{d}y}{\mathrm{d}x} = \frac{\dot{y}}{\dot{x}} = \frac{\mathrm{d}y/\mathrm{d}t}{\mathrm{d}x/\mathrm{d}t} = -\frac{\omega^2 \sin x}{y}$$

and separate variables

$$y\,\mathrm{d}y = -\omega^2 \sin x\,\mathrm{d}x.$$

Then integrate both sides from $x = 0$ to $x = x$ on the right-hand side and from $y = 0$ to $y = y$ on the left-hand side. This yields

$$\frac{1}{2}\,y^2 + \omega^2\,(1 - \cos x) = K \tag{9.4.3}$$

for some constant K. The term $\omega^2\,(1 - \cos\theta) = \omega^2\,(1 - \cos x)$ represents the potential energy of the pendulum bob, measured relative to a zero reference at the horizontal line of the downward position. Since the first term $y^2/2 = \dot{x}^2/2$ corresponds to the kinetic energy, Eq. (9.4.3) describes the conservation law of mechanical energy for the undamped pendulum. Using the trigonometric identity $1 - \cos x = 2\,\sin^2(x/2)$, we solve Eq. (9.4.3) and express $y = \dot{x}$ in terms of x explicitly:

$$y = \dot{x} = \pm\left(2K - 4\omega^2 \sin^2(x/2)\right)^{1/2}. \tag{9.4.4}$$

We set up a frame of Cartesian axes x, y, called the **phase plane**, and plot the one-parameter family of curves obtained from Eq. (9.4.4) by using different values of energy level K. This leads to the picture (see Fig. 9.21), called a **phase diagram** or a **phase portrait** for the problem, and the solution curves are called the phase paths or trajectories. Various types of phase paths can be identified in terms of K. The critical value of the energy $K = 2\omega^2$ corresponds to the paths joining $(-\pi, 0)$ and $(\pi, 0)$. These two curves

$$y = \pm 2\omega\sqrt{1 - \sin^2(x/2)} = \pm 2\omega\left|\cos\frac{x}{2}\right| \qquad (K = 2\omega^2)$$

separate the family of solutions into disjoint sets where $K > \omega^2$ and where $K < \omega^2$, with different properties. Therefore, each of these curves (plotted in blue in Fig. 9.21) is called a **separatrix** because it separates phase curves representing two distinct behaviors: one is periodic, called oscillation, and plotted within shaded areas in Fig. 9.21 where the pendulum swings back and forth, and the other is aperiodic, called rotation, where the pendulum swings over the top. Each separatrix approaches an equilibrium point without reaching it: they are unstable saddle points $((2k+1)\pi, 0)$, $k = 0, \pm 1, \pm 2, \ldots$, that correspond to the upward position of the pendulum. The basins of attraction for stable stationary points $(2k\pi, 0)$ are shaded regions in Fig. 9.21 bounded by separatricies. Each of these points has its own region of asymptotic stability, which is bounded by separatricies entering the two neighboring unstable points.

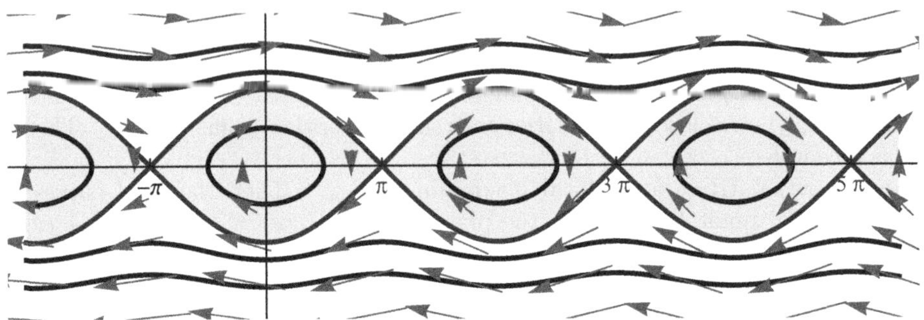

Figure 9.21: Example 9.4.1: The level curves for an undamped pendulum, plotted with *Mathematica*.

A given pair of values (x, y), or $(x, \dot{x})$, represented by a point $P(x, y)$ on the diagram is called a state of the system. A state gives the angular velocity $y = \dot{x}$ at a particular inclination $x = \theta$. A given state $(x, \dot{x})$ serves also as a pair of initial conditions for the original pendulum equation; therefore, a given state determines all subsequent states, which are obtained by following the trajectory that passes through the point P.

The direction of the point moving along these trajectories coincides with the direction field arrows that are tangent to the phase paths. When $y = \dot{x} > 0$, x must increase as t increases. This leads to the conclusion that the required directions are always from right to left in the upper half-plane. Similarly, the directions are always from left to right in the lower half-plane.

9.4.1 Hamiltonian Systems

We now discuss a class of autonomous first order systems that satisfy a conservative law. A system is called a **Hamiltonian system**[84] if there exists a real-valued function that is constant along any solution of the system. They are completely described by a scalar function $H(\mathbf{q}, \mathbf{p})$, where both $\mathbf{q}$ and $\mathbf{p}$ are vectors with the same dimension, called generalized coordinates and momentum, respectively. Therefore, Hamiltonian systems necessarily have even dimensions. For simplicity, we consider only two-dimensional systems.

Let $H(x, y)$ be a smooth function of two independent variables. If we replace x and y in $H(x, y)$ by functions $x(t)$ and $y(t)$, the composition $H(x(t), y(t))$ will become a function of the variable t. According to the chain rule, its derivative becomes

$$\dot{H} = \frac{\mathrm{d}}{\mathrm{d}t} H(x(t), y(t)) = \frac{\partial H}{\partial x}\frac{\mathrm{d}x}{\mathrm{d}t} + \frac{\partial H}{\partial y}\frac{\mathrm{d}y}{\mathrm{d}t}.$$

If these two functions $x = x(t)$ and $y = y(t)$ are solutions of the two-dimensional autonomous system

$$\begin{aligned} \dot{x} &= f(x, y), \\ \dot{y} &= g(x, y), \end{aligned} \tag{9.4.5}$$

then this system is called a **Hamiltonian system** if

$$\dot{x} = f(x, y) = \frac{\partial H}{\partial y}, \qquad \dot{y} = g(x, y) = -\frac{\partial H}{\partial x}. \tag{9.4.6}$$

The function $H(x, y)$ is called the **Hamiltonian function** or simply the **Hamiltonian** of the system (9.4.5). Since the composition $H(x(t), y(t))$ is a conservative quantity of the system, we get the general solution of Eq. (9.4.5):

$$H(x(t), y(t)) = c,$$

for some constant c. The following theorem gives the necessary and sufficient conditions for a system (9.4.5) to be conservative.

[84]Sir William Rowan Hamilton (1805–1865) was an Irish physicist, astronomer, and mathematician. Shortly before this death, he was elected the first foreign member of the United States National Academy of Science.

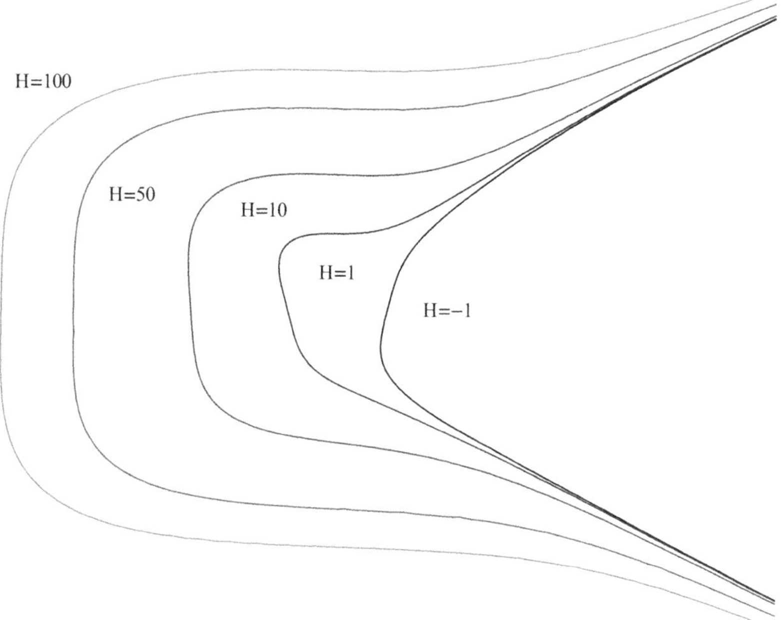

Figure 9.22: Example 9.4.2: The level curves, plotted with *Mathematica*.

Theorem 9.2: Consider the two-dimensional autonomous system (9.4.5). Assume that $f(x, y)$ and $g(x, y)$ along with their first partial derivatives are continuous in the xy-plane. Then the system is a Hamiltonian system if and only if

$$\frac{\partial f}{\partial x} = -\frac{\partial g}{\partial y} \quad \text{for all } (x, y). \tag{9.4.7}$$

Example 9.4.2: Consider the autonomous system

$$\dot{x} = 4y^3 + \sin x,$$
$$\dot{y} = 3x^2 + 1 - y\,\cos x.$$

Setting $f(x, y) = 4y^3 + \sin x$ and $g(x, y) = 3x^2 + 1 - y\,\cos x$, we apply the Hamiltonian test (9.4.7):

$$\frac{\partial f}{\partial x} = \cos x = -\frac{\partial g}{\partial y}.$$

Since our system is Hamiltonian, there exists a function $H(x, y)$ such that

$$\frac{\partial H}{\partial x} = -g(x, y) = -3x^2 - 1 + y\,\cos x,$$
$$\frac{\partial H}{\partial y} = f(x, y) = 4y^3 + \sin x.$$

For integration, we choose one of these equations, say the latter, and compute an anti-partial-derivative, obtaining

$$H(x, y) = y^4 + y\,\sin x + k(x),$$

where $k(x)$ is an arbitrary differentiable function of x. Using the former equation, we get

$$\frac{\partial H}{\partial x} = y\,\cos x + k'(x) = -g(x, y) = -3x^2 - 1 + y\,\cos x.$$

Therefore, the derivative of $k(x)$ is

$$k'(x) = -3x^2 - 1 \qquad \Longrightarrow \qquad k(x) = -x^3 - x,$$

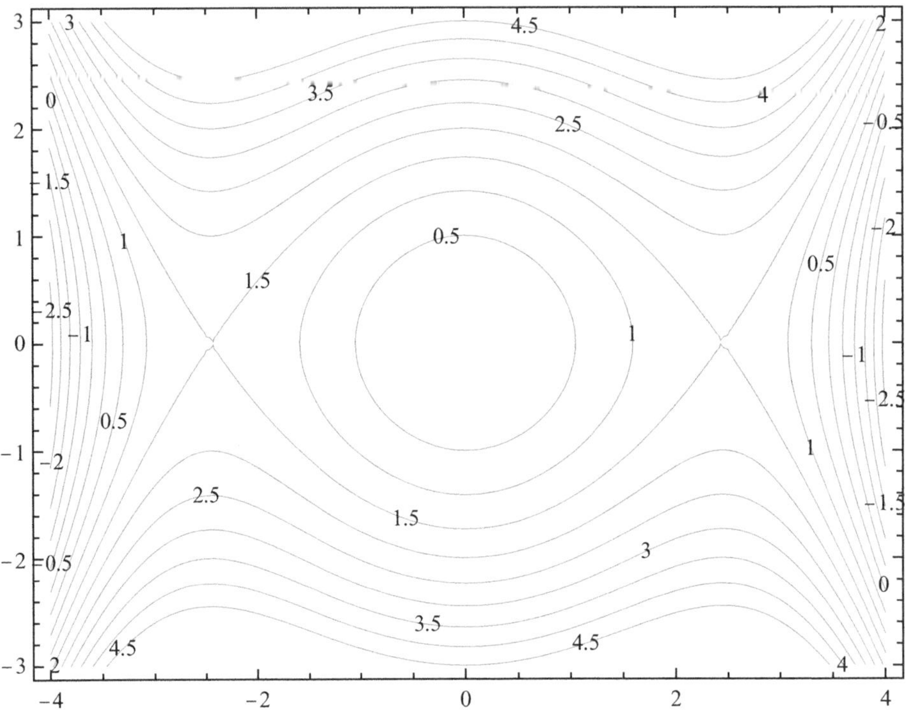

Figure 9.23: Example 9.4.3: The level curves, plotted with *Mathematica*.

where we dropped an arbitrary constant of integration. Hence, the Hamiltonian becomes

$$H(x, y) = y^4 + y\,\sin x - x^3 - x.$$

Example 9.4.3: (Pendulum) If we approximate $\sin\theta$ by $\theta - \theta^3/3!$ in the pendulum equation $\ddot{\theta} + \omega^2 \sin\theta = 0$, we will arrive at the following nonlinear system:

$$\dot{x} = y, \qquad \dot{y} = -x + x^3/6.$$

The level curves of the trajectories can be obtained from the differential equation

$$\frac{\mathrm{d}y}{\mathrm{d}x} = \frac{-x + x^3/6}{y},$$

which, after separation of variables, gives

$$y\,\mathrm{d}y = \left(\frac{1}{6}\,x^3 - x\right)\mathrm{d}x \qquad \Longrightarrow \qquad \frac{1}{2}\,y^2 = \frac{1}{24}\,x^4 - \frac{1}{2}\,x^2 + c.$$

Therefore, the Hamiltonian of this system reads

$$H(x, y) = \frac{1}{2}\,y^2 + \frac{1}{2}\,x^2 - \frac{1}{24}\,x^4.$$

To plot these level curves (Fig. 9.23), we use the following *Mathematica* script:

```
ContourPlot[(1/2)*y*y + (1/2)*x*x - (1/24)*x*x*x*x, {x, -4,
  4}, {y, -3, 3}, ContourShading -> None, ContourLabels -> True,
 PlotRange -> {-5, 5}, Contours -> 19, AspectRatio -> Automatic]
```

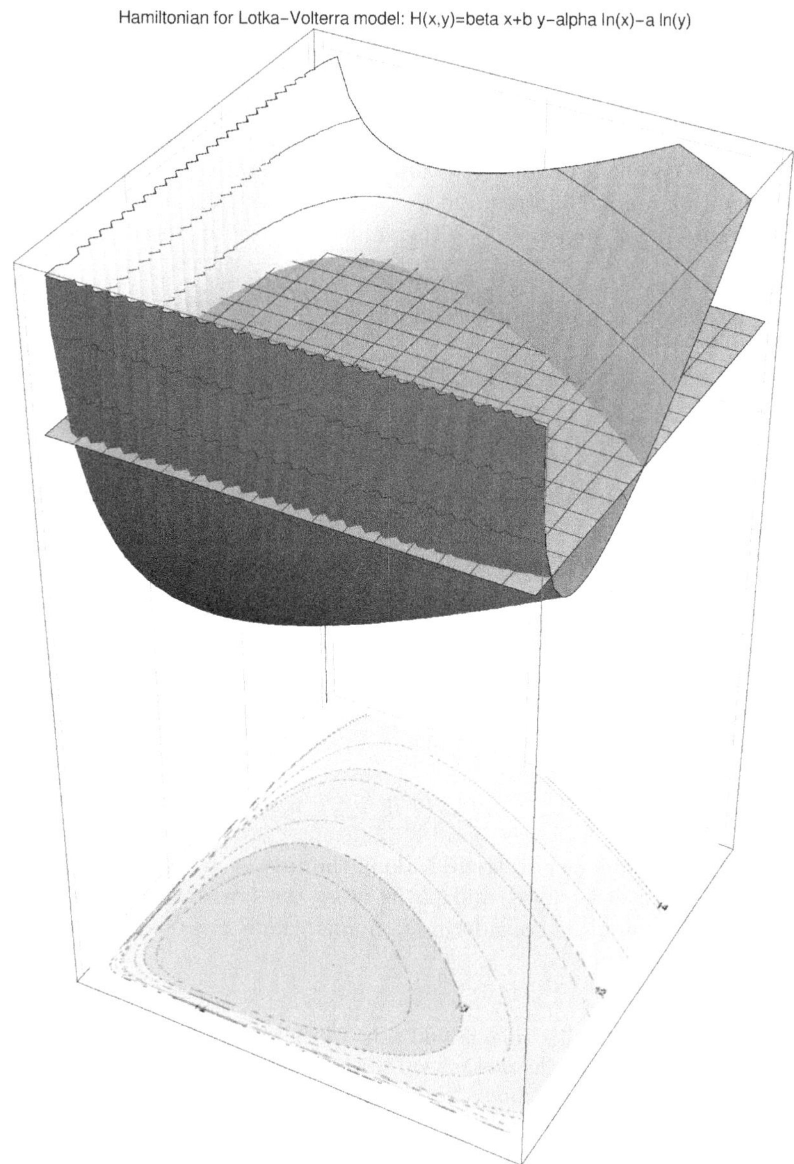

Hamiltonian for Lotka–Volterra model: H(x,y)=beta x+b y–alpha ln(x)–a ln(y)

Figure 9.24: Example 9.4.4: Lotka–Volterra level curves, plotted with *Mathematica*.

Example 9.4.4: (Lotka–Volterra model) Consider the predator-prey model (9.3.8):

$$\dot{x} = x\,(r - by)\,, \qquad \dot{y} = y\,(\beta x - \mu)\,.$$

Upon multiplying the former equation by $(\beta x - \mu)\,/x$ and the latter by $(r - by)\,/y$ and subtracting the results, we obtain the exact equation

$$\frac{-\mu + \beta x}{x}\,\frac{\mathrm{d}x}{\mathrm{d}t} - \frac{r - by}{y}\,\frac{\mathrm{d}y}{\mathrm{d}t} = 0$$

or

$$\frac{\mathrm{d}}{\mathrm{d}t}\left(\beta x + by - \mu\,\ln x - r\,\ln y\right) = 0.$$

Therefore, the Hamiltonian of the predator-prey equations is $H(x, y) = \beta x + by - \mu\,\ln x - r\,\ln y$. All orbits of the Lotka–Volterra model evolve so that they keep the Hamiltonian constant: $H(x(t), y(t)) = H(x(0), y(0))$. Now we claim that $H(x, y)$ is a concave function, which follows from the equations

$$H_{xx} = \alpha/x^2, \quad H_{yy} = a/y^2, \quad H_{xy} = H_{yx} = 0.$$

The stationary point is the point where $\nabla H = 0$, that is, $(x^*, y^*) = \left(\frac{\mu}{\beta}, \frac{r}{b}\right)$ is a critical point. Since at this point $H_{xx} H_{yy} - H_{xy}^2 - r\mu/(xy)^2 > 0$ and $H_{xx} > 0$, $H_{yy} > 0$, the Hamiltonian attains a minimum at (x^*, y^*): $H(x^*, y^*) = \mu - b - \mu \ln \frac{\mu}{\beta} - r \ln \frac{r}{b}$.

Therefore, all trajectories of the predator-prey equations are the projections of the level curves $H(x, y) = H_0$ of the Hamiltonian. Notice that this corresponds to the unique steady state of the system (9.3.8), page 505. Since H is a strictly convex function with a unique minimum in the positive quadrant, every trajectory must be a closed curve. Thus, the orbits are a one-parameter (depending on the value of H) set of closed curves starting at the steady state.

Example 9.4.5: Solve the system of autonomous equations

$$\dot{x} = y\left(1 - x^2\right), \qquad \dot{y} = x\left(y^2 - 4\right).$$

Solution. The stationary points occur at the simultaneous solutions of

$$y\left(1 - x^2\right) = 0, \qquad x\left(y^2 - 4\right) = 0.$$

These are five solution pairs of the above algebraic system of equations: $(0, 0)$, $(1, 2)$, $(1, -2)$, $(-1, 2)$, $(-1, -2)$. The phase paths satisfy the differential equation

$$\frac{\mathrm{d}y}{\mathrm{d}x} = \frac{x\left(y^2 - 4\right)}{y\left(1 - x^2\right)},$$

which is a first order separable equation. Upon separation of variables and integration, we obtain the general solution:

$$\left(y^2 - 4\right)\left(1 - x^2\right) = c, \qquad \text{a constant.}$$

Notice that there are special solutions (separatricies) along the lines $x = \pm 1$ and $y = \pm 2$ where $c = 0$. Streamlines cross the vertical axis $x = 0$ with zero slope, and paths cross the horizontal axis $y = 0$ with infinite slope. The directions of streamlines may be found by considering domains where $\dot{x} > 0$ or $\dot{x} < 0$, and similarly for $\dot{y}$.

Problems

1. For computing the motion of two bodies (planet and sun) which attract each other, we choose one of the bodies (sun) as the center of our coordinate system; the motion will then stay in a plane and we can use two-dimensional coordinates $q = (q_1, q_2)$ for the position of the second body. Newton's laws, with a suitable normalization, then yield the following differential equations:

$$\ddot{q}_1 = -\frac{q_1}{\left(q_1^2 + q_2^2\right)^{3/2}}, \qquad \ddot{q}_2 = -\frac{q_2}{\left(q_1^2 + q_2^2\right)^{3/2}}.$$

   Find a Hamiltonian for this system of equations.

2. Find the potential energy function $\Pi(y)$ and the energy function $E(x, y)$ for the given equations.

   (a) $\ddot{x} + 2x - 5x^4 = 0$; **(c)** $\ddot{x} + 2e^{2x} - 1 = 0$; **(e)** $\ddot{x} + \cos x = 0$;

   (b) $\ddot{x} + 3x^2 - 6x + 1 = 0$; **(d)** $\ddot{x} + 2x/(1 + x^2) = 0$; **(f)** $\ddot{x} + x \sin x = 0$.

3. Find Hamiltonians of the following systems.

   (a) $\begin{cases} \dot{x} = 2x^2 - 3xy^2 + 4y^3 - x^3, \\ \dot{y} = y^3 + 3x^2 y - 2xy; \end{cases}$ (c) $\begin{cases} \dot{x} = 2xy \sin x - x^2 \sin y, \\ \dot{y} = -y^2 \sin x - 2x \cos y - xy^2 \cos x; \end{cases}$

   (b) $\begin{cases} \dot{x} = 2y\left(1 + x^2 + y^2\right) e^{x^2 + y^2}, \\ \dot{y} = 2x\left(1 + x^2 + y^2\right) e^{x^2 + y^2}; \end{cases}$ (d) $\begin{cases} \dot{x} = -2y \sin\left(x^2 + y^2\right), \\ \dot{y} = 2x \sin\left(x^2 + y^2\right). \end{cases}$

4. For the following systems, determine whether they are Hamiltonian. If so, find the Hamiltonian function.

   (a) $\begin{cases} \dot{x} = x + y^2, \\ \dot{y} = y^2 - x; \end{cases}$ (b) $\begin{cases} \dot{x} = 2y - x \sin(y), \\ \dot{y} = -\cos(y). \end{cases}$

5. **(Hénon–Heiles problem, 1964)** The polynomial Hamiltonian[85] in four degrees of freedom
$$H(\mathbf{q},\mathbf{p}) = \frac{1}{2}\left(q_1^2 + q_2^2 + p_1^2 + p_2^2\right) + q_1^2 q_2 - \frac{1}{3}\,q_2^3, \quad \mathbf{q} = \langle q_1, q_2\rangle, \ \mathbf{p} = \langle p_1, p_2\rangle,$$
defines a Hamiltonian differential equation that can have chaotic solutions. Write the corresponding system of differential equations of first order and make a computational experiment by plotting some solutions.

6. Find a Hamiltonian for the system $\dot{x} = y$, $\dot{y} = x - x^2$, and plot the level curves. What do you notice?

7. Consider a solar system with N planets. Their motion can be modeled by a system of ordinary differential equations with a Hamiltonian
$$H(\mathbf{q},\mathbf{p}) = \frac{1}{2}\sum_{i=0}^{N} \frac{1}{m_i}\,\mathbf{p}_i^2 - g\sum_{i=1}^{N}\sum_{j=0}^{i-1} \frac{m_i m_j}{\|\mathbf{q}_i - \mathbf{q}_j\|},$$
where $m_0 = 1$ is the mass of the sun, and m_i are masses relative to m_0 of the planets, $g = 2.95\times 10^{-4}$ is the gravitational constant. Write this system of equations explicitly, and make a computer experiment with $N = 2$ planets by solving numerically the system and plotting their solutions.

8. Consider a second order linear system
$$\ddot{x} = -ax - by, \qquad \ddot{y} = -bx - cy,$$
where a, b, and c are constants.
   (a) Show that the system is equivalent to a first order Hamiltonian system, with Hamiltonian $H(x,y,u,v) = \frac{1}{2}\left(ax^2 + 2bxy + cy^2 + u^2 + v^2\right)$, where $\dot{x} = u$, $\dot{y} = v$.
   (b) Show that the system has a stable equilibrium at the origin if $a > 0$ and $ac > b^2$.

9. Let $\mathbf{q} = \langle q_1, q_2, \ldots, q_n\rangle$, $\mathbf{p} = \langle p_1, p_2, \ldots, p_n\rangle$. Consider a Hamiltonian
$$H(\mathbf{q},\mathbf{p}) = \frac{1}{2}\sum_{i=1}^{N} \frac{1}{m_i}\,p_i^2 + \sum_{i=2}^{N}\sum_{j=1}^{i-1} V_{ij}\left(\|q_i - q_j\|\right)$$
that can be used to model the interaction between N pairs of neutral atoms or molecules with Lennard-Jones potential[86]
$$V_{ij}(r) = 4\varepsilon_{ij}\left[\left(\frac{\sigma}{r}\right)^{12} - \left(\frac{\sigma}{r}\right)^{6}\right] = \varepsilon_{ij}\left[\left(\frac{r_m}{r}\right)^{12} - 2\left(\frac{r_m}{r}\right)^{6}\right],$$
where ε is the depth of the potential well, σ is the finite distance at which the inter-particle potential is zero, r is the distance between the particles, and r_m is the distance at which the potential reaches its minimum. At r_m, the potential function has the value $-\varepsilon$. The distances are related as $r_m = 2^{1/6}\sigma$. Here q_i and p_i are position and momenta of the i-th atom of mass m_i.
Write a system of differential equations with the corresponding Hamiltonian.

10. Using the energy function $E(x,y,z) = x^2 + 4y^2 + 9z^2$, show that $E(x,y,z)$ is a constant along the motion of the system
$$\dot{x} = 9xz - 4yz, \quad \dot{y} = 9yz + xz, \quad \dot{z} = -x^2 - 4y^2.$$
Then show that the equilibrium point $(0,0,-a)$ $(a > 0)$ is neutrally stable.

11. When expressed in plane polar coordinates (r,θ) given $x = r\cos\theta$, $y = r\sin\theta$, show that the Hamiltonian equations (9.4.6) become
$$\dot{r} = \frac{1}{r}\,\frac{\partial H}{\partial \theta}, \qquad \dot{\theta} = -\frac{1}{r}\,\frac{\partial H}{\partial r}.$$

12. Sketch a phase portrait for each of the following Hamiltonian systems written in polar coordinates (r,θ). Add the equation $\dot{\theta} = 1$.

(a) $\dot{r} = \begin{cases} r\,(r-4), & \text{if } r \leqslant 4, \\ 0, & \text{otherwise;} \end{cases}$ **(b)** $\dot{r} = \begin{cases} 0, & \text{if } r \leqslant 4, \\ r\,(r-4), & \text{otherwise.} \end{cases}$

[85] The corresponding system was published in a 1964 paper by French mathematician Michel Hénon (1931–2013) and American astrophysicist Carl Heiles (born in 1939).

[86] Sir John Edward Lennard-Jones (1894–1954) was a British mathematician who was a professor of theoretical physics at Bristol University and then of theoretical science at Cambridge University. He may be regarded as the founder of modern computational chemistry.

9.5 Lyapunov's Second Method

The linearization procedure, discussed in §9.2, can be used for determination of the stability of a critical point only if a corresponding linear system has the leading matrix without purely imaginary eigenvalues. This approach is of limited use because it provides only local information: the results are valid only for solutions that have initial values in a neighborhood of a critical point. There is no information about solutions that start far away from equilibrium points.

In this section we discuss a global method that provides information about stability, asymptotic stability, and instability of any autonomous system of equations, not necessarily almost linear systems. This method[87] is referred to as **Lyapunov's second method** in honor of Alexander Lyapunov, who derived it in 1892.

The Lyapunov method is also called the **direct method** because it can be applied to differential equations without any knowledge of the solutions. Furthermore, the method gives us a way of estimating the regions of asymptotic stability. This is something the linear approximation would never be able to do.

Definition 9.8: Let $\mathbf{x}^*$ be an isolated critical point for the autonomous vector differential equation

$$\dot{\mathbf{x}} = \mathbf{f}(\mathbf{x}), \qquad \mathbf{x} \in \mathbb{R}^n, \tag{9.5.1}$$

so $\mathbf{f}(\mathbf{x}^*) = \mathbf{0}$. A continuously differentiable function $V : U \mapsto \mathbb{R}$, where $U \subseteq \mathbb{R}^n$ is an open set with $\mathbf{x}^* \in U$, is called a **Lyapunov function** (also called a weak Lyapunov function) for the differential equation (9.5.1) at $\mathbf{x}^*$ provided that

1. $V(\mathbf{x}^*) = 0$,
2. $V(\mathbf{x}) > 0$ for $\mathbf{x} \in U \setminus \{\mathbf{x}^*\}$,
3. the function $\mathbf{x} \mapsto \nabla V(\mathbf{x})$ is continuous for $\mathbf{x} \in U \setminus \{\mathbf{x}^*\}$, and, on this set, $\dot{V}(\mathbf{x}) = \nabla V(\mathbf{x}) \cdot \mathbf{f}(\mathbf{x}) \leqslant 0$.

If, in addition,

4. $\dot{V}(\mathbf{x}) < 0$ for $\mathbf{x} \in U \setminus \{\mathbf{x}^*\}$, then V is called a **strong** (strict) **Lyapunov function**.

Theorem 9.3: [**Lyapunov**] If $\mathbf{x}^*$ is an isolated critical point for the differential equation (9.5.1) and there exists a Lyapunov function V for the system at $\mathbf{x}^*$, then $\mathbf{x}^*$ is stable. If, in addition, V is a strong Lyapunov function, then $\mathbf{x}^*$ is asymptotically stable.

PROOF: We outline the proof only for a two-dimensional case. The idea of Lyapunov's method is very simple. Let $\phi(t) = \langle x(t), y(t)\rangle$ denote a trajectory of the vector equation

$$\dot{x} = f(x, y), \quad \dot{y} = g(x, y). \tag{9.5.2}$$

It is reasonable to assume that the level set $S = \{(x, y) \in \mathbb{R}^2 : V(x, y) = c\}$ of the Lyapunov function V is a closed curve in the xy-plane and contains the only critical point of the system (9.5.2). The gradient ∇V is an outer normal for the level curve $S = \{\mathbf{x} \in \mathbb{R}^n : V(\mathbf{x}) = c\}$ because it points in the direction where the function $V(x, y)$ increases fastest. If $\phi(t) = \langle x(t), y(x)\rangle$ is a trajectory of the system (9.5.2), then $\mathbf{T}(t) = \dot{x}(t)\,\mathbf{i} + \dot{y}(t)\,\mathbf{j}$ is a tangent vector to it at time t. The derivative of V with respect to the given system (9.5.2) is a dot product of two vectors:

$$\dot{V}(t) = V_x(x, y)\,\dot{x}(t) + V_y(x, y)\,\dot{y}(t) = \nabla V(x, y) \cdot \mathbf{T}(t) = |\nabla V| \cdot |\mathbf{T}| \cos\theta,$$

where θ is the angle between ∇V and $\mathbf{T}$ at time t. Recall that $\cos\theta < 0$ for $\pi/2 < \theta < 3\pi/2$; therefore, $\dot{V}$ is negative when the trajectory passing through the level curve $V(x, y) = c$ is going inward. At points where $\dot{V} = 0$, the orbit is tangent to the level curve. Thus, V is not increasing on the curve $t \mapsto \langle x(t), y(t)\rangle$ at $t = 0$, and, as a result, the image of this curve either lies in the level surface S, or the set $\{\langle x(t), y(t)\rangle : t > 0\}$ is a subset of the set in the plane with outer boundary S. The same result is true for every point on S. Therefore, a solution starting on S is trapped: it either stays in S, or it stays in the set $\{\mathbf{x} \in \mathbb{R}^2 : V(\mathbf{x}) < c\}$. So the region where $\dot{V}(\mathbf{x}) < 0$ is

[87] Aleksandr Mikhailovich Lyapunov (1857–1918) was a famous Russian mathematician, mechanician, and physicist who made great contributions to probability theory, mathematical physics, and the theory of dynamic systems. In 1917 he moved to Odessa because of his wife's frail health. Shortly after her death Lyapunov committed suicide.

contained in the basin of attraction of $\mathbf{x}^*$. The stability of the critical point follows easily from this result. If V is a strict Lyapunov function, then the solution curve definitely crosses the level surface S and remains inside the set $\{\mathbf{x} \in \mathbb{R}^2 : V(\mathbf{x}) < c\}$ for all $t > 0$. Because the same property holds at all level sets "inside" S, the equilibrium point $\mathbf{x}^*$ is asymptotically stable. ∎

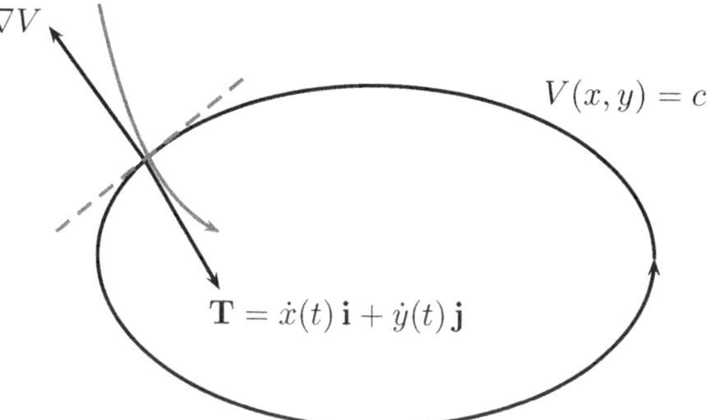

A Lyapunov function V can be thought of as a generalized energy function for a system. A function $V(\mathbf{x})$ that satisfies the first two conditions in Definition 9.8 is usually called **positive definite** on the domain U. If the inequality in the second condition $V(\mathbf{x}) > 0$ is replaced by $V(\mathbf{x}) \geqslant 0$, then the function $V(\mathbf{x})$ is referred to as **positive semidefinite** on U. By reversing inequalities in these definitions, we obtain negative definite ($V(\mathbf{x}) < 0$ for $\mathbf{x} \in U \setminus \{\mathbf{x}^*\}$) and negative semidefinite ($V(\mathbf{x}) \leqslant 0$ for $\mathbf{x} \in U \setminus \{\mathbf{x}^*\}$). Obviously, if $V(\mathbf{x})$ is positive definite/semidefinite, then $-V(\mathbf{x})$ is negative definite/semidefinite, and vice versa. Some simple positive definite and positive semidefinite polynomial functions are as follows:

$$a\,x^2 + b\,y^2, \quad a\,x^2 + b\,y^4, \quad a\,x^4 + b\,y^2, \quad \text{and} \quad a\,x^2, \quad b\,y^4, \quad (x-y)^2,$$

respectively, where a and b are positive constants. Also the quadratic function

$$V(x,y) = a\,x^2 + b\,xy + c\,y^2 \tag{9.5.3}$$

is positive definite if and only if $a > 0$ and $b^2 < 4ac$ (a, b, and c are constants).

Example 9.5.1: In the two-dimensional case, the function $V = 1 - \cos\left(x^2 + 4y^2\right)$ of two variables (x,y) is positive definite on the domain $U : -\pi < x^2 + 4y^2 < \pi$ since $V(0,0) = 0$ and $V(x,y) > 0$ for all other points in U.

However, the same function is positive semidefinite on $R : -\pi < x^2 + 4y^2 < 3\pi$ because there exists the ellipse of points, $x^2 + 4y^2 = 2\pi$, where the function $V(x,y)$ is zero.

Theorem 9.4: **[Lyapunov]** Let $\mathbf{x}^*$ be an isolated critical point of the autonomous system (9.5.1). Suppose that there exists a continuously differentiable function $V(\mathbf{x})$ such that $V(\mathbf{x}^*) = 0$ and that in every neighborhood of $\mathbf{x}^*$ there is at least one point at which V is positive. If there exists a region $R \ni \mathbf{x}^*$ such that the derivative of V with respect to the system (9.5.1), $\dot{V}(\mathbf{x}) = \nabla V(\mathbf{x}) \cdot \mathbf{f}(\mathbf{x})$, is positive definite on R, then $\mathbf{x} = \mathbf{x}^*$ is an unstable equilibrium solution.

There are many versions of Lyapunov stability theorems that make different regularity assumptions. Their power comes from simplicity because one does not need to know any solutions. However, finding an appropriate Lyapunov function is generally considered as an art to come up with good candidates.

Example 9.5.2: Lyapunov's function for the system of equations

$$\dot{x} = \alpha x - \beta y + y^2, \quad \dot{y} = \beta x + \alpha y - xy$$

can be chosen as $V(x,y) = ax^2 + by^2$, with positive coefficients a, b. Then the derivative of V with respect to the given system is

$$\dot{V} = \frac{\mathrm{d}V}{\mathrm{d}t} = \frac{\partial V}{\partial x}\frac{\mathrm{d}x}{\mathrm{d}t} + \frac{\partial V}{\partial y}\frac{\mathrm{d}y}{\mathrm{d}t} = 2\alpha\left[ax^2 + by^2\right] + 2xy^2(a-b) + 2\beta xy(b-a).$$

If we choose $a = b > 0$, then

$$\dot{V} = 2\alpha\left[ax^2 + by^2\right] = 2\alpha a\left(x^2 + y^2\right)$$

and the origin is stable if $\alpha < 0$ and unstable if $\alpha > 0$.

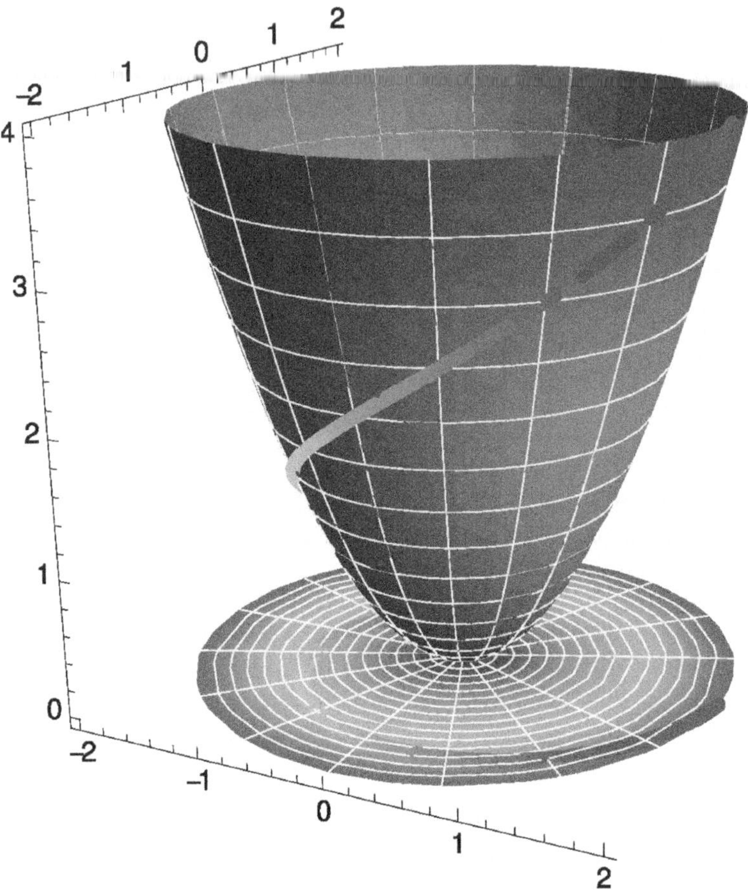

Figure 9.25: Example of a Lyapunov function, plotted with *Mathematica*.

Example 9.5.3: (van der Pol Equation) Consider a modification of the harmonic oscillator equation $\ddot{x} = -x$ by adding a term $-\epsilon\left(x^2-1\right)\dot{x}$ that has a damping effect when $|x| > 1$ and an amplifying effect when $|x| < 1$:

$$\ddot{x} + \epsilon\left(x^2-1\right)\dot{x} + x = 0, \tag{9.5.4}$$

where ϵ is a positive parameter. This (unforced) differential equation was introduced by Lord Rayleigh in 1883. The Dutch electrical engineer and physicist Balthasar van der Pol (1889–1959) investigated this oscillator more extensively when he studied the equation in 1926 as a model of the voltage in a triode circuit. Since its introduction, this differential equation has been suggested as a model for many different physical phenomena (for instance, as a model for the human heartbeat). Rewriting it as an equivalent system of equations

$$\dot{x} = y,$$
$$\dot{y} = -x + \epsilon\left(1-x^2\right)y,$$

we see that it has an equilibrium solution at the origin. Now we choose a Lyapunov function in the form $V(x,y) = ax^2 + by^2$, which is positive definite when constants a, b are positive. The derivative of V with respect to solutions of the van der Pol equation is

$$\dot{V}(x,y) = (2ax, 2by)\cdot\left(y, -x+\epsilon\left(1-x^2\right)y\right) = 2axy - 2bxy + 2b\epsilon y^2\left(1-x^2\right).$$

By taking $a = b$, we eliminate the first two terms in $\dot{V}$. Then, assuming that $\epsilon > 0$, we get $\dot{V}(x,y) = 2b\epsilon y^2\left(1-x^2\right) \geqslant 0$ for arbitrary $(x,y) \neq (0,0)$ and for any $b > 0$, whenever $|x| \leqslant 1$. Evidently $\dot{V}(x,y) \geqslant 0$ inside the unit circle $x^2+y^2 \leqslant 1$, so a solution trajectory starting in or on a circle of radius 1 about the origin must leave the stationary point. Also $\dot{V}(x,y) > 0$ when $x^2 < 1$ and $y \neq 0$. The Lyapunov theorem guarantees that every trajectory starting inside the unit circle not only departs the origin but tends to a closed loop (see Fig. 9.33 on page 535) because $\dot{V}(x,y) \leqslant 0$ for $x^2 > 1$. Therefore, the origin is an unstable stationary point for small x. The actual region of asymptotic stability includes this unit circle.

On the other hand, when x is large, the term x^2 becomes dominant and the damping becomes positive. In this case, $\dot{V}(x,y) \leqslant 0$. Therefore, the dynamics of the system is expected to be restricted in some area around the fixed point. Actually, the van der Pol equation satisfies the Levinson–Smith theorem 9.7, page 529, ensuring that there is a stable limit cycle in the phase space. The van der Pol system is therefore a Liénard equation (see §9.6).

Example 9.5.4: Determine the stability of the critical point at the origin for the system

$$\dot{x} = -y^5, \qquad \dot{y} = -x^3.$$

Solution. The origin is not an isolated equilibrium solution for the corresponding linear system. Therefore, we cannot apply the linearization technique from §9.2. To apply Lyapunov's instability theorem 9.4, we try the function $V(x,y) = -xy$. It is continuous, $V(0,0) = 0$, and it is positive at all points in the second and fourth quadrants. Moreover, its derivative with respect to the solutions of the given system is

$$\dot{V} = \frac{\partial V}{\partial x}\,\dot{x} + \frac{\partial V}{\partial y}\,\dot{y} = x^4 + y^6,$$

which is positive definite in any domain containing the origin. Hence, the origin is an unstable critical point.

Example 9.5.5: (SIRS Epidemic Model) Consider a population of individuals infected with a nonfatal disease. A simple SIRS model with disease-induced death and bilinear incidence that describes the spread of the disease is as follows:

$$\begin{aligned} \dot{S} &= A - \mu S - \beta I\, S + \delta R, \\ \dot{I} &= \beta I\, S - (\gamma + \mu + \alpha)\, I, \\ \dot{R} &= \gamma I - (\mu + \delta)\, R, \end{aligned} \tag{9.5.5}$$

where the dot stands for the derivative with respect to time t, and β is the contact rate, μ is the average death rate, $1/\gamma$ is the average infection period, and δ is the average temporary immunity period. Let $N(t) = S + I + R$, with $N(0) > 0$, be the total population, where S is the number of susceptible people, R is the number of recovered individuals with immunity, and I is the number of infected people. Then

$$\dot{N} = A - \mu N - \alpha I \qquad \Longrightarrow \qquad \limsup_{t\mapsto\infty} N(t) \leqslant A/\mu.$$

The region

$$\Omega = \left\{ (S,I,R) \in \mathbb{R}^3_+ \ : \ S + I + R \leqslant A/\mu \right\}$$

is a positively invariant set to Eq. (9.5.5). Let $R_0 = \beta A/\mu/(\mu + \alpha + \gamma)$ be the basic reproductive number. We summarize the stability properties of the SIRS system in the following statement, adapted from [33].

Theorem 9.5: When $R_0 \leqslant 1$, the disease-free equilibrium $P_0(A/\mu, 0, 0)$ is globally stable in Ω. If $R_0 > 1$, the disease-free equilibrium P_0 is unstable, and there is a unique endemic equilibrium $P^*(S^*, I^*, R^*)$ that is globally stable in the interior of Ω, where

$$S^* = \frac{A}{\mu\, R_0}, \quad I^* = \frac{(\delta+\mu)A}{(\alpha+\mu)(\delta+\mu)+\gamma\mu}\left(1 - \frac{1}{R_0}\right),$$

$$R^* = \frac{\gamma A}{(\alpha+\mu)(\delta+\mu)+\gamma\mu}\left(1 - \frac{1}{R_0}\right).$$

PROOF: Straightforward calculations can show the existence of the disease-free equilibrium P_0 and the endemic equilibrium P^*. To prove the global stability of P_0 in Ω for $R_0 \leqslant 1$, we choose the Lyapunov function I in Ω. Then

$$\begin{aligned} \frac{\mathrm{d}I}{\mathrm{d}t} &= \left[\beta S - (\gamma + \mu + \alpha)\right] I \\ &\leqslant \left[\beta\, \frac{A}{\mu} - (\gamma + \mu + \alpha)\right] I = (\gamma + \mu + \alpha)(R_0 - 1)\, I \leqslant 0. \end{aligned}$$

Therefore, the disease-free equilibrium P_0 is globally stable in Ω for $R_0 \leqslant 1$.

To prove the global stability of P^* in the interior of Ω, we consider the following equivalent system of (9.5.5):

$$\begin{aligned} \dot{I} &= I\left[\beta(N-I-R)\right] - (\gamma+\mu+\alpha), \\ \dot{R} &= \gamma I - (\mu+\delta)\,R, \\ \dot{N} &= A - \mu N - \alpha I. \end{aligned} \tag{9.5.6}$$

The positively invariant set of (9.5.6), corresponding to the positively invariant set Ω of (9.5.5), is $\Omega' = \{(I, R, N) \in \mathbb{R}^3_+ \,:\, I + R < N \leqslant A/\mu\}$, and, when $R_0 > 1$, Eq. (9.5.6) has a unique endemic equilibrium, $\tilde{P}^*(I^*, R^*, N^*)$, corresponding to $P^*(S^*, I^*, R^*)$, where

$$N^* = S^* + I^* + R^* = \frac{A}{\mu}\left[1 - \frac{\alpha(\delta+\mu)}{(\alpha+\mu)(\delta+\mu)+\gamma\mu}\left(1 - \frac{1}{R_0}\right)\right].$$

Hence, when $R_0 > 1$, the system of equations (9.5.6) can be rewritten as

$$\begin{aligned} \dot{I} &= \beta I\left[(N-N^*) - (I-I^*) - (R-R^*)\right], \\ \dot{R} &= \gamma(I-I^*) - (\mu+\delta)\,(R-R^*), \\ \dot{N} &= -\mu(N-N^*) - \alpha(I-I^*). \end{aligned} \tag{9.5.7}$$

Consider the Lyapunov function

$$V(I,R,N) = \alpha\gamma\left(I - I^* - I^*\ln\frac{I}{I^*}\right) + \frac{\alpha\beta}{2}\,(R-R^*)^2 + \frac{\beta\gamma}{2}\,(N-N^*)^2,$$

which is a positive definite function in region Ω'. The total derivative of $V(I,R,N)$ along the solutions of (9.5.7) is given by

$$\frac{\mathrm{d}V}{\mathrm{d}t} = -\alpha\beta\gamma\,(I-I^*)^2 - \alpha\beta(\mu+\delta)\,(R-R^*)^2 - \beta\gamma\mu\,(N-N^*)^2.$$

Since $\dot{V}$ is negative definite, it follows from the Lyapunov theorem that the endemic equilibrium $\tilde{P}^*$ of (9.5.6) is globally stable in the interior Ω'; that is, the unique endemic equilibrium P^* of (9.5.5) is globally stable in the interior of Ω. □

The previous examples show that finding an appropriate Lyapunov function can be a challenging problem. In systems coming from physics, often the total energy can be chosen as a Lyapunov function. More precisely, there are two important classes of systems for which Lyapunov functions are almost ready to be used. One of these classes we studied before in §9.4—conservative systems, and in particular, Hamiltonian ones. If the Hamiltonian $H(\mathbf{x})$ is known (here $\mathbf{x}$ is a phase point of even dimension), then its gradient ∇H is perpendicular to the vector field of the Hamiltonian system. Therefore, we can choose $H(\mathbf{x}) - H(\mathbf{x}^*)$ as a Lyapunov function because the trajectories of the system lie on level surfaces of H.

Every second order system with n degrees of freedom is equivalent to a $2n$-dimensional first order system and therefore meets the even-dimensionality requirement for Hamiltonian systems. If the given system is conservative, it can be written as $\ddot{\mathbf{x}} = -\nabla\Pi(\mathbf{x})$, where $\Pi(\mathbf{x})$ is the potential energy. Then $H(\mathbf{x}, \dot{\mathbf{x}}) = \frac{1}{2}\,\dot{\mathbf{x}}^2 + \Pi(\mathbf{x})$ is its Hamiltonian.

Apart from conservative systems, there is another kind of system of the form $\dot{\mathbf{x}} = k\nabla G(\mathbf{x})$, where $G : \mathbb{R}^n \mapsto \mathbb{R}$ is a continuously differentiable function and k is a constant. Such a system is called a **gradient system**. Suppose that the function $G(\mathbf{x})$ has an isolated local minimum/maximum value at the point $\mathbf{x} = \mathbf{x}^*$. Then this point will be a critical point of the given system. Its orbits follow the path of steepest descent/increase of G depending on the sign of k. In this case, G itself will be a Lyapunov function for the system at $\mathbf{x}^*$. Indeed, the total derivative of G becomes

$$\dot{G}\,(\mathbf{x}) = \nabla G\,(\mathbf{x}) \cdot (k\nabla G\,(\mathbf{x})) = k\,|\nabla G\,(\mathbf{x})|^2.$$

If $k < 0$, then $G\,(\mathbf{x}) > G\,(\mathbf{x}^*)$ except at $\mathbf{x} = \mathbf{x}^*$, and $\mathbf{x} = \mathbf{x}^*$ will be an asymptotically stable equilibrium solution. If $k > 0$ and $\mathbf{x} = \mathbf{x}^*$ is a local maximum, then $-G(\mathbf{x})$ is a Lyapunov function. Note that the linearized system at any equilibrium has only real eigenvalues. Therefore, a gradient system can have no spiral sources, spiral sinks, or center: nondegenerate critical points of a planar analytic gradient system are either a saddle or a node. A two-dimensional system (9.5.2) is a gradient system if and only if

$$\frac{\partial f}{\partial y} = \frac{\partial g}{\partial x}. \tag{9.5.8}$$

Example 9.5.6: (Example 9.4.1 revisited) The total energy for an ideal pendulum

$$\frac{1}{2}m\ell^2\dot{\theta}^2 + mg\ell\left(1-\omega_0^2\cos\theta\right) = E(\theta,\dot{\theta})$$

is a natural candidate for a Lyapunov function. Taking out $m\ell^2$, we get $V(x,y) = E(x,y)/(m\ell^2)$, where $x=\theta$, $\dot{x}=y$, a Lyapunov function: it is positive in the open domain $\Omega = \{(x,y)\,:\, -\pi/2 < x < \pi/2, -\infty < y < \infty\}$ except at the origin, where it is zero. Since its derivative with respect to the pendulum equation is identically zero, $\dot{V}$ is negative semidefinite. By Theorem 9.3, the origin is a stable critical point.

The total energy of the undamped system becomes a Lyapunov function if damping is added. Indeed, for the damped pendulum equation

$$\begin{aligned}\dot{x} &= y,\\ \dot{y} &= -\gamma y - \omega^2\sin x,\end{aligned} \tag{9.5.9}$$

the total energy $V(x,\dot{x}) = \frac{1}{2}\,\dot{x}^2 + \omega^2\left(1-\cos x\right)$ satisfies the inequality

$$\dot{V} = \dot{x}\,\omega^2\sin x + \dot{x}\,\ddot{x} = -\dot{x}^2\gamma \leqslant 0.$$

This shows that $V(x,y)$ is a Lyapunov function; however, we expect the original to be an asymptotically stable point. So we need to find an appropriate strong Lyapunov function instead (see Problem 2).

Example 9.5.7: (Example 9.2.6 revisited) The dynamics of the unforced system can be modeled by the Duffing equation

$$\ddot{x} + \delta\dot{x} + \beta x + \alpha x^3 = 0,$$

where the damping constant obeys $\delta \geqslant 0$. When there is no damping ($\delta = 0$), the equation can be integrated as

$$E(t) \stackrel{\text{def}}{=} \frac{1}{2}\left(\dot{x}\right)^2 + \frac{1}{2}\,\beta x^2 + \frac{1}{4}\,\alpha x^4 = \text{constant}.$$

Therefore, in this case, the Duffing equation is a Hamiltonian system.

When $\delta > 0$, $E(t)$ satisfies

$$\frac{\mathrm{d}}{\mathrm{d}t}\,E(t) = -\delta\left(\dot{x}\right)^2 \leqslant 0;$$

therefore, the trajectory $x(t)$ moves on the surface of $E(t)$ so that $E(t)$ decreases until $x(t)$ converges to one of the equilibria where $\dot{x} = x$. For positive α, β, and γ, $E(t)$ is a Lyapunov function and $\mathbf{x}^* = (0,0)$ is globally asymptotically stable in this case.

Problems

1. Construct a suitable Lyapunov function to determine the stability of the origin for the following systems of equations.

(a) $\begin{cases}\dot{x}_1 = -x_1^3 + 3x_2^4,\\ \dot{x}_2 = -x_1x_2^3.\end{cases}$

(b) $\begin{cases}\dot{x}_1 = -x_1^3 - x_1x_2^2,\\ \dot{x}_2 = -2x_1^2x_2 - x_2^3.\end{cases}$

(c) $\begin{cases}\dot{x}_1 = x_1^3 + x_1x_2^2,\\ \dot{x}_2 = x_1^2x_2 + x_2.\end{cases}$

(d) $\begin{cases}\dot{x}_1 = 2x_1^3 + x_1x_2^2,\\ \dot{x}_2 = -2x_1^2x_2 + x_2^3.\end{cases}$

(e) $\begin{cases}\dot{x}_1 = -2x_1 - 3x_1x_2^2,\\ \dot{x}_2 = -3x_2 - 2x_1^2x_2.\end{cases}$

(f) $\begin{cases}\dot{x}_1 = -3x_1^3 + x_1x_2^2,\\ \dot{x}_2 = -x_1^2x_2 - x_2^3.\end{cases}$

(g) $\begin{cases}\dot{x} = y - x^3,\\ \dot{y} = -x - y.\end{cases}$

(h) $\begin{cases}\dot{x} = 2xy - x^3,\\ \dot{y} = -x^2 - y^5.\end{cases}$

2. With some choice of a constant a, show that $V(x,y) = \frac{1}{2}\,y^2 + \frac{1}{2}\left(ax+\gamma y\right)^2 - \gamma\omega^2\cos x + \gamma\omega^2$ is a strong Lyapunov function for the damped pendulum equation (9.5.9), with $\gamma \ne 1$. *Hint:* Use Maclaurin series approximation for $\sin x$ and polar coordinates.

3. Show that the system $\dot{x} = y$, $\dot{y} = -x^3$ has a stable critical point at the origin which is not asymptotically stable.

4. The following SIRS epidemic model with disease-induced death and standard incidence

$$\begin{aligned}\dot{S} &= \Lambda - \mu S - \beta I\, S/N + \delta R,\\ \dot{I} &= \beta I\, S/N - (\gamma + \mu + \alpha)\, I,\\ \dot{R} &= \gamma I - (\mu + \delta)\, R,\end{aligned}$$

is similar to model (9.5.5); the region $\Omega = \{(S, I, R) \in \mathbb{R}^{e}_{+} \,:\, S + I + R \leqslant A/\mu\}$ is a positively invariant set for the above system of equations. Show that the disease-free equilibrium $P_0(A/\mu, 0, 0)$ is globally stable in Ω when $\beta < \mu + \alpha + \gamma$.

5. One of the milestones for the current renaissance in the field of neural networks was the associative model proposed by the American physicist John Hopfield in 1982. Hopfield considered a simplified model in which each neuron is represented by a linear circuit consisting of a resistor and capacitor, and is connected to the other neurons via nonlinear sigmoidal activation functions. An example of such a model gives the following system of ordinary differential equations:

$$\dot{x} = z_2(y) - x, \qquad \dot{y} = z_1(x) - y,$$

where

$$z_1(x) = k\, \arctan(ax), \qquad z_2(y) = k\, \arctan(ay),$$

with some positive constants k and a. To analyze the stability of this system, Hopfield suggested using a Lyapunov function of the form

$$V(x, y) = -z_1(x)\, z_2(y) - 2\left[\ln\left(\cos z_1(x)\right) + \ln\left(\cos z_1(x)\right)\right].$$

By plotting $V(x, y)$, with $k = 1.4$ and $a = 2$, verify that the Lyapunov function is positive definite in a neighborhood of the origin.

6. For the system of equations

$$\dot{x} = x - xy - 2\,x^2, \qquad \dot{y} = \frac{3}{4}\,y - \frac{1}{2}\,xy - y^2,$$

modeling competition of species, find all equilibrium points and investigate their stabilities using the appropriate Lyapunov function of the form $V(x, y) = a\,x^2 + b\,xy + c\,y^2$.

7. Use Lyapunov's direct method to determine the stability of the zero solution for the given equations of second order.

$$\textbf{(a)}\ \ddot{x} + 3x^2\,\dot{x} + 4\,x^5 = 0; \qquad \textbf{(b)}\ \ddot{x} + (\cos x)\,\dot{x} + \sin x = 0.$$

8. Show that a system $\dot{\mathbf{x}} = f(\mathbf{x})$ is at the same time a Hamiltonian system and a gradient system if and only if the Hamiltonian H is a harmonic function (which satisfies the Laplace equation: $\nabla^2 H = 0$).

9. Show that the one-dimensional gradient system with $G(x) = x^8\, \sin(1/x)$ has $x = 0$ as a stable equilibrium, but $x = 0$ is not a local minimum of $G(x)$.

10. Find a Hamiltonian for the system

$$\dot{x}_1 = y_1, \quad \dot{x}_2 = y_2, \quad \dot{y}_1 = -x_1 - 2x_1x_2, \quad \dot{y}_2 = x_2^2 - x_2 - x_1^2.$$

11. Find all critical points and determine their stability for the system with the Hamiltonian $H(x, y) = \frac{1}{2}\,y^2 + \frac{1}{2}\,x^2 - \frac{1}{4}\,x^4$.

12. Find a Hamiltonian for the following systems of equations.

(a) $\begin{cases}\dot{x} = -2y + 3x^2,\\ \dot{y} = -2x - 6xy.\end{cases}$

(b) $\begin{cases}\dot{x} = 5\,\sin x\,\cos y,\\ \dot{y} = -5\,\cos x\,\sin y.\end{cases}$

(c) $\begin{cases}\dot{x} = 10x + y + 3x^2 - x + 4y^3,\\ \dot{y} = 2x - 10y - 3x^2 - 6xy + y.\end{cases}$

(d) $\begin{cases}\dot{x} = -x + 4y + 3y^2,\\ \dot{y} = y - 2x - 2y^2.\end{cases}$

13. Show that the function $V(x, y) = 3x^2 + 7y^2$ is a strong Lyapunov function for the linear system

$$\dot{x} = -x + 7y, \qquad \dot{y} = -3x - y.$$

Also show that V cannot generate a gradient system.

9.6 Periodic Solutions

The linearization procedure discussed in §9.2 tells us nothing about the solution to an autonomous vector equation except in a neighborhood of certain types of isolated singular points. Local properties, that is, properties holding in the neighborhood of points, can be analyzed by stability analysis or the power series method. However, in many practical problems, one needs information about the global behavior of solutions. These solutions may occur in mechanical, electrical, or other nonconservative systems in which some external source of energy compensates for the energy dissipated.

The periodic phenomena occur in many applications; therefore, their modeling and analysis is an important task. We start with a conservative system modeled by the initial value problem

$$\ddot{y} + f(y) = 0, \qquad y(0) = y_0, \quad \dot{y}(0) = v_0. \tag{9.6.1}$$

If the function $f(y)$ is similar to an odd function, then the following theorem (see Problem 5 for proof) guarantees the existence of a periodic solution to the IVP (9.6.1).

Theorem 9.6: Suppose that the function $f(y)$ in Eq. (9.6.1) is continuous and $y\,f(y)$ is positive for $|y| \leqslant a$, where a is some positive number. For initial points (y_0, v_0) sufficiently close to the origin, the solutions to the IVP (9.6.1) are nonintersecting closed loops traced clockwise around $(0,0)$.

Example 9.6.1: (Example 9.2.6 revisited) The Duffing equation (9.2.9) on page 497 gives an example of conservative system (9.6.1) with $f(y) = \omega^2\left(y - y^3/6\right)$. Taking $a = \sqrt{6}$, we see that conditions of Theorem 9.6 are fulfilled, and the Duffing equation has a periodic solution, which is confirmed by Fig. 9.23 on page 518. □

While important, Eq. (9.6.1) is of limited application. We consider a more general second order differential equation or its equivalent system named after the French physicist **Alfred-Marie Liénard**[88]

$$\ddot{x} + f(x)\,\dot{x} + g(x) = 0 \qquad \text{or} \qquad \begin{cases} \dot{x} = y - F(x), \\ \dot{y} = -g(x), \end{cases} \tag{9.6.2}$$

where $F(x)$ is an antiderivative of $f(x)$ (so $F'(x) = f(x)$). The following theorem [31], proved by N. Levinson and O.K. Smith in 1942, guarantees an existence of a unique periodic solution to the equation (9.6.2).

Theorem 9.7: [Levinson–Smith] Let $F(x)$ and $G(x)$ be antiderivatives of functions $f(x)$ and $g(x)$, respectively, in the Liénard equation (9.6.2) subject to $F(0) = G(0) = 0$. The Liénard equation has a unique periodic solution whenever all of the following conditions hold:

1. $f(x)$ is even and continuous, and $g(x)$ is an odd continuous function with $g(x) > 0$ for $x > 0$.
2. $F(x) < 0$ when $0 < x < a$ and $F(x) > 0$ when $x > a$, for some $a > 0$.
3. $F(x) \mapsto +\infty$ as $x \mapsto +\infty$, monotonically for $x > a$.
4. $G(x) \mapsto +\infty$ as $x \mapsto \pm\infty$.

Next, we discuss the existence of periodic solutions of planar autonomous vector equations (9.1.6)

$$\dot{\mathbf{x}} = \mathbf{f}(\mathbf{x})$$

that satisfy the condition $\mathbf{x}(t) = \mathbf{x}(t+T)$ for some positive constant T, called the period. Any equilibrium solution is a periodic function with arbitrary period. In general, it is difficult to discover whether a given system of differential equations does or does not have periodic solutions—it is still an active area of mathematical research. Even if one can prove the existence of such a periodic solution, it is almost impossible, except in some exceptional cases, to find that solution explicitly.

[88]Liénard (1869–1958) was a professor at the École des Mines de Saint-Étienne and during 1908–1911 he was professor of electrical engineering at the École des Mines de Paris. In World War I he served in the French Army. He is most well known for his invention of the Liénard–Wiechert electromagnetic potentials.

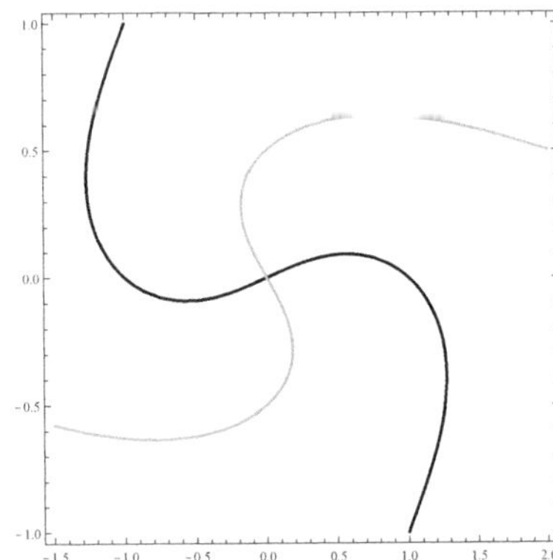

Figure 9.26: Nullclines in Example 9.6.2.

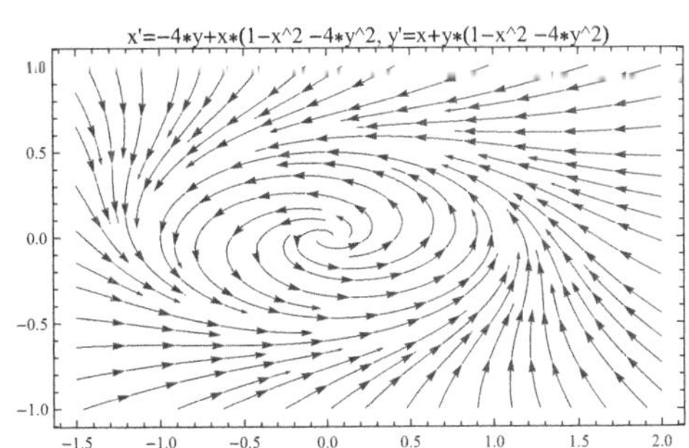

Figure 9.27: Phase portrait in Example 9.6.2, plotted with *Mathematica*.

> **Definition 9.9:** **A limit cycle** C is an isolated closed trajectory having the property that all other orbits in its neighborhood are spirals winding themselves onto C for $t \to +\infty$ (a stable limit cycle) or $t \to -\infty$ (an unstable limit cycle). All trajectories approach the limit cycle independent of the choice of initial conditions.

Semistable limit cycles[89] displaying a combination of both stable and unstable behaviors can also occur. We start with an illustrative example.

Example 9.6.2: Consider a family of systems of ordinary differential equations

$$\dot{x} = -4ky + x\left(1 - x^2 - 4y^2\right), \qquad \dot{y} = kx + y\left(1 - x^2 - 4y^2\right), \tag{9.6.3}$$

depending on a positive parameter k. By plotting x- and y-nullclines $4ky = x\left(1 - x^2 - 4y^2\right)$ and $kx+y\left(1 - x^2 - 4y^2\right) = 0$ (see Fig. 9.26), it is clear that the system has only one critical point—the origin (unstable). Equation (9.6.3) can be solved explicitly by introducing polar coordinates: $x = 2r\,\cos\theta$, $y = r\,\sin\theta$. Since $\tan\theta = 2y/x$, we get

$$\theta = \arctan\left(\frac{2y}{x}\right) \qquad \Longrightarrow \qquad \dot{\theta} = \frac{\mathrm{d}}{\mathrm{d}t}\left(\arctan\left(\frac{2y}{x}\right)\right) = 2\,\frac{x\dot{y} - y\dot{x}}{x^2 + 4y^2}. \tag{9.6.4}$$

Then substituting the expressions given in Eq. (9.6.3) for $\dot{x}$ and $\dot{y}$, we obtain

$$\dot{\theta} = \frac{\mathrm{d}\theta}{\mathrm{d}t} = 2\,\frac{k\left(x^2 + 4y^2\right)}{x^2 + 4y^2} = 2k.$$

Solving this separable equation, we get

$$\theta = 2kt + \theta_0, \qquad \text{where } \theta_0 = \theta(0).$$

This formula tells us that trajectories revolve clockwise around the origin if $k < 0$ and counterclockwise if $k > 0$. Since $4r^2 = x^2 + 4y^2$, we differentiate both sides to obtain

$$4r\,\dot{r} = x\,\dot{x} + 4y\,\dot{y} \qquad \Longleftrightarrow \qquad \dot{r} = r\left(1 - 4r^2\right).$$

Separating variables and integrating, we get

$$r(t) = \pm\frac{r(0)\,e^t}{\sqrt{1 + 4\,r^2(0)\left(e^{2t} - 1\right)}}.$$

Thus, the typical solutions of Eq. (9.6.3) may be expressed in the form

$$x(t) = 2r(t)\,\cos\left(kt + \theta_0\right), \qquad y(t) = r(t)\,\sin\left(kt + \theta_0\right).$$

[89]This term was introduced by Henri Poincaré.

A family of equations (9.6.3) gives an unusual example of systems that can be solved explicitly. Consider the modified system

$$\dot{x} = -4y\left(1 - x^2 - 4y^2\right)^2 + x\left(1 - x^2 - 4y^2\right)^3 - 4y^3,$$
$$\dot{y} = x\left(1 - x^2 - 4y^2\right)^2 + y\left(1 - x^2 - 4y^2\right)^3 + xy^2$$

with slopes obtained from the original ones upon multiplication by $\left(1 - x^2 - 4y^2\right)^2$ and addition of the vector $y^2\langle -4y, x\rangle$. By plotting a phase portrait for the modified system, we cannot see any geometric difference from Fig. 9.27. However, in addition to a common stationary point at the origin, the modified system has equilibrium points at $(1, 0)$ and $(-1, 0)$ that aren't shared by the simpler system. Since these additional critical points lie on the ellipse $x^2 + 4y^2 = 1$, this ellipse is not a limit cycle as in Eq. (9.6.3). □

It is often useful to be able to rule out the existence of a limit cycle in a region of the plane. We present a theorem, referred to as Bendixson's first theorem or more commonly as Bendixson's negative criterion. This theorem as well as its generalization, proved[90] by Dulac, sometimes enables us to establish the nonexistence of limit cycles for the basic system of autonomous equations

$$\dot{x} = f(x, y), \qquad \dot{y} = g(x, y). \tag{9.6.5}$$

Theorem 9.8: [Bendixson's negative criterion] If the expression

$$\frac{\partial f}{\partial x} + \frac{\partial g}{\partial y} \neq 0 \tag{9.6.6}$$

does not change its sign within a simply connected domain D of the phase plane, no periodic motions of Eq. (9.6.5) can exist in that domain.

This negative criterion is a straightforward consequence of Green's theorem in the plane. A simply connected domain is a region without holes. In other words, it is a region having the property that any closed curve or surface lying in the region can be shrunk continuously to a point without going outside of the region. Note that if $f_x + g_y$ changes sign in the domain, then no conclusion can be made.

Example 9.6.3: Consider a system (9.6.5) with

$$f(x, y) = y + x\left(x^2 + 4y^2 - 1\right), \qquad g(x, y) = -x + y\left(x^2 + 4y^2 - 1\right),$$

so

$$\frac{\partial f}{\partial x} + \frac{\partial g}{\partial y} = 4\left(x^2 + 4y^2 - \frac{1}{2}\right).$$

Since this expression does not change sign for $r < 1/\sqrt{2}$, where $r = \sqrt{x^2 + 4y^2}$, no closed trajectory can exist inside the ellipse $x^2 + 4y^2 = 1$. The quantity $\frac{\partial f}{\partial x} + \frac{\partial g}{\partial y}$ also does not change sign outside the radial distance $r = 1/\sqrt{2}$. However, we cannot apply Bendixson's negative criterion because the domain is no longer simply connected—the region $r < 1/\sqrt{2}$ must be excluded to keep the sign of $f_x + g_y$ unchanged. It can be verified that there exists a limit cycle at $r = 1$ but Bendixson's theorem should not be applied.

Example 9.6.4: (Pendulum) Consider the nonhomogeneous pendulum differential equation

$$\ddot{\theta} + \gamma\dot{\theta} + \omega^2 \sin\theta = \mu,$$

where $\gamma > 0$ and μ are constants, and θ is an angular variable. This differential equation is a model for an unbalanced rotor or pendulum with viscous damping $\gamma\dot{\theta}$ and external torque μ. The equivalent system of the first order equations

$$\dot{x} = \dot{\theta} = y, \qquad \dot{y} = \mu - \omega^2 \sin\theta - \gamma y \tag{9.6.7}$$

has no periodic solutions (see Fig. 9.29), and has no rest points. Since the expression $f_x + g_y = -\gamma < 0$, where $f = y$ and $g = \mu - \gamma y - \omega^2 \sin\theta$, does not change its sign, the system has no periodic solution according to Bendixson's theorem.

[90]Henri Dulac (1870–1955) was a French mathematician. He proved Theorem 9.9 in 1933.

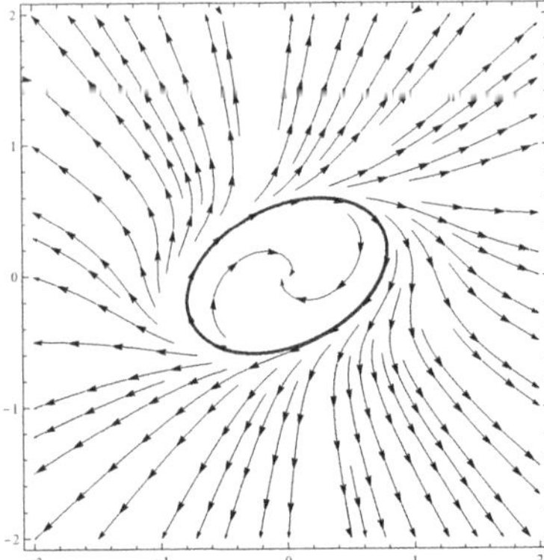

Figure 9.28: Phase portrait and unstable limit cycle in Example 9.6.3.

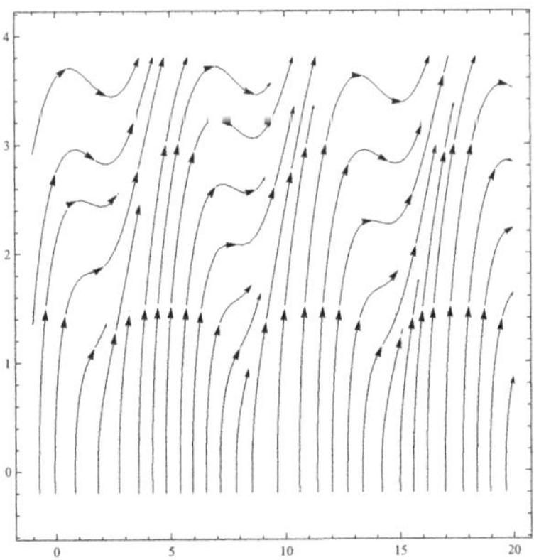

Figure 9.29: Phase portrait in Example 9.6.4, plotted with *Mathematica*.

Theorem 9.9: [Dulac's Criterion] Let Ω be a simply connected region of the phase plane. If there exists a continuously differentiable function $\phi(x,y)$ such that

$$\operatorname{div}(\phi\mathbf{f}) = \frac{\partial}{\partial x}\left[\phi(x,y)\, f(x,y)\right] + \frac{\partial}{\partial y}\left[\phi(x,y)\, g(x,y)\right], \quad \mathbf{f} = \langle f, g\rangle,$$

is of constant sign in Ω and is not identically zero on any subregion of Ω, then the dynamical system (9.6.5) has no closed orbits wholly contained in Ω.

Dulac's criterion suffers the same problem as when finding Lyapunov functions in that it is often difficult to find a suitable function $\phi(x,y)$. The case $\phi(x,y) = 1$ represents the negative criterion of Bendixson. A function $\phi(x,y)$ that satisfies the condition of the above theorem is called the **Dulac function** to the system. The existence of the Dulac function can be used to estimate the number of limit cycles of system (9.6.5) in some regions. Namely, system (9.6.5) has at most $p-1$ limit cycles in a p-connected region Ω if a Dulac function exists.

Example 9.6.5: Show that the system

$$\dot{x} = y, \qquad \dot{y} = -x - y + x^2 + y^2$$

has no closed orbits anywhere.

Solution. First, we apply Bendixson's negative criterion:

$$\frac{\partial y}{\partial x} + \frac{\partial}{\partial y}\left(-x - y + x^2 + y^2\right) = -1 + 2y.$$

Although this shows that there is no closed orbit contained in either half-plane $y < \frac{1}{2}$, or $y > \frac{1}{2}$, it does not rule out the existence of a closed orbit in the whole plane since there may be such an orbit which crosses the line $y = \frac{1}{2}$.

Now let us try the Dulac function $\phi(x,y) = e^{ax}$.

$$\frac{\partial}{\partial x}\left[e^{ax}\right] + \frac{\partial}{\partial y}\left[e^{ax}\left(-x - y + x^2 + y^2\right)\right] = e^{ax}\left[(a+2)y - 1\right].$$

Choosing $a = -2$ reduces the expression to $-e^{-2x}$, which is negative everywhere. Hence, there are no closed orbits. □

Another theorem due to Poincaré and Bendixson[91] gives the necessary and sufficient conditions for the existence of a limit cycle. Unfortunately, the theorem is often difficult to apply because it requires a preliminary knowledge

[91] Ivar Otto Bendixson (1861–1935) was a Swedish mathematician. Being the Professor of Pure Mathematics at the Royal Institute of Technology, he was intrigued by the complicated behavior of the integral curves in the neighborhood of singular points. Bendixson substantially improved an earlier result of Poincaré in 1901.

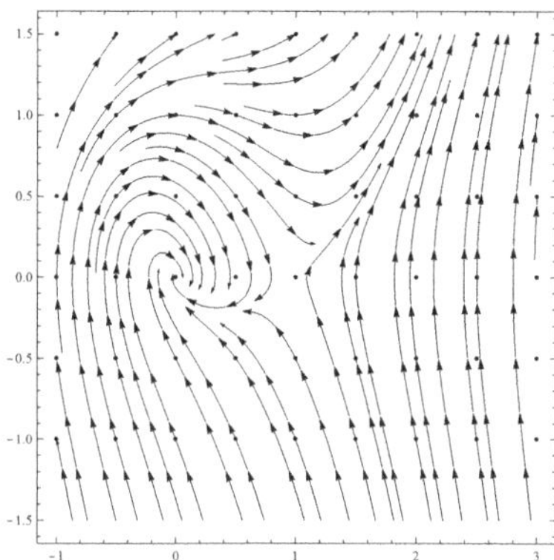

Figure 9.30: Phase portrait in Example 9.6.5, plotted with *Mathematica*.

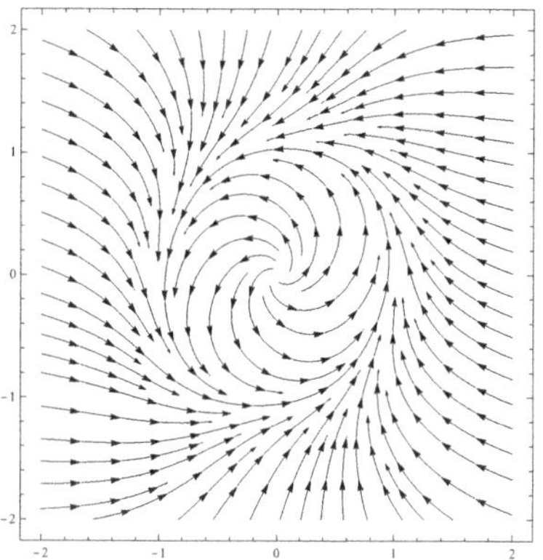

Figure 9.31: Phase portrait in Example 9.6.7, plotted with *Mathematica*.

of the nature of orbits. The Poincaré–Bendixson theorem involves the concept of a half-trajectory, which is the set of points traced by an orbit starting at a particular position, usually identified by $t = 0$. This representative point divides the orbit into two half-trajectories (for $t > 0$ and for $t < 0$).

Theorem 9.10: [Poincaré–Bendixson] Let the rate functions $f(x, y)$ and $g(x, y)$ in Eq. (9.6.5) be continuously differentiable, and let $\langle x(t), y(t)\rangle$ be the parametric equation of a half-trajectory Γ which remains inside the finite domain Ω for $t \mapsto +\infty$ without approaching any singularity. Then only two cases are possible: either Γ is itself a closed trajectory or Γ approaches such a trajectory.

Example 9.6.6: (Example 9.6.2 revisited) Recall that the system (9.6.3) can be rewritten in polar form as

$$\dot{r} = r\left(1 - 4r^2\right), \qquad \dot{\theta} = 2k > 0.$$

Then $\mathrm{d}r/\mathrm{d}\theta = r\left(1 - 4r^2\right)/(2k)$ with $\mathrm{d}r/\mathrm{d}\theta > 0$ for $r < 1/2$ and $\mathrm{d}r/\mathrm{d}\theta > 0$ for $r > 1/2$. If we choose an annular domain (donut shaped) R of inner radius $r = 1/4$ and outer radius $R = 1$, the trajectories must cross the inner ellipse from the region $r < 1/4$ to the region $r > 1/4$ because $\mathrm{d}r/\mathrm{d}\theta > 0$. On the other hand, the orbits must cross the outer ellipse $r = 1$ towards the region $r < 1$ since $\mathrm{d}r/\mathrm{d}\theta < 0$.

Example 9.6.7: Show that the system of equations

$$\dot{x} = x - y - xy^2 - x^5 - x^3y^2, \qquad \dot{y} = x + y + x^3 - x^2y - y^3 \tag{9.6.8}$$

has a limit cycle.

Solution. We are going to find the Lyapunov function $V(x, y)$ and a positive function $f(x, y) > 0$ such that the given system (9.6.8) has the form:

$$\dot{x} = -V_y - f\,V_x, \qquad \dot{y} = V_x - f\,V_y. \tag{9.6.9}$$

We look for functions of the form $f = f(r^2)$ and $V = g(r^2) + h(x)$, where $r^2 = x^2 + y^2$ and f, g, and h are polynomials. Substituting these functions into Eq. (9.6.8), we get

$$-y + x(1 - r^2) + x^3(1 - r^2) = -2y\,g' - 2xf\,g' - fg',$$
$$x + x^3 + y(1 - r^2) = 2x\,g' + h' - 2yfg'.$$

Since y cannot be written as a polynomial in r^2 and x, the terms multiplying y above must match. Thus: $g' = 1/2$ and $f = r^2 - 1$. The equations then become

$$(x + x^3)(1 - r^2) = (x + h')(1 - r^2) \qquad \text{and} \qquad x^3 = h'.$$

This yields the final answer:

$$f = r^2 - 1, \quad V = \frac{1}{2}r^2 + \frac{1}{4}x^4 \qquad (r^2 = x^2 + y^2). \tag{9.6.10}$$

Now, as a consequence of the form of the equations in (9.6.9), the derivative of V along the solutions is

$$\dot{V} = \frac{\mathrm{d}}{\mathrm{d}t}\,V(x,y) = \dot{x}\,V_x + \dot{y}\,V_y = -\left(V_x^2 + V_y^2\right).$$

Clearly, in any region where $V > 0$, the function V is a Lyapunov function, and the level curves of V provide trapping boundaries: the solutions cross these curves in only one direction (V decreasing). So the conditions of Theorem 9.10 are fulfilled.

Next we find and classify the critical points for Eq. (9.6.8). From Eq. (9.6.9), we find that at a critical point

$$\begin{bmatrix} V_y & V_x \\ -V_x & V_y \end{bmatrix} \begin{bmatrix} 1 \\ f \end{bmatrix} = 0.$$

Hence the determinant of the coefficient matrix must vanish, and $V_x = V_y = 0$. From Eq. (9.6.10), we conclude that there is only one critical point: the origin, which is an unstable spiral.

Note that the level curves of V are a set of nested closed curves, all containing the origin. The level curve $V(x,y) = V^* > 0$ contains the disk of radius $r_*^2 = \sqrt{1+V^*} - 1$ centered at the origin. Thus, the function f is positive on any level curve where $V^* > 0$.

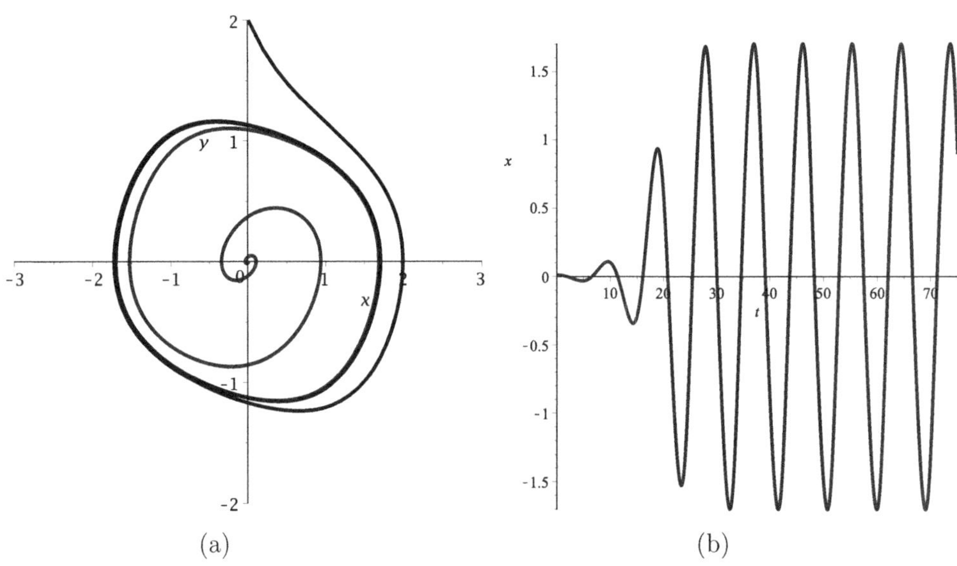

Figure 9.32: Example 9.6.8 (a) phase trajectories of Rayleigh's equation, both approaching the limit cycle; (b) a solution of Rayleigh's equation, plotted with *Maple*.

Example 9.6.8: The British mathematical physicist Lord Rayleigh (John William Strutt, 1842–1919) introduced in 1877 an equation of the form

$$m\ddot{x} + kx = a\dot{x} - b(\dot{x})^3$$

(with nonlinear velocity damping) to model the oscillations of a clarinet reed. With $y = \dot{x}$, we get the autonomous system

$$y = \dot{x}, \qquad \dot{y} = \left(-kx + ay - by^3\right)/m.$$

Next, we show the existence of a limit cycle by plotting trajectories using the following *Maple* script:

```
with(DEtools): with(plots): m:=2: k:=1: a:=1: b:=1:
deq1:= diff(x(t),t)=y;   deq2:= m*diff(y(t),t)=-k*x+k*y-b*y^3 ;
DEplot([deq1,deq2],[x(t),y(t)],t=0..75,x=-3..3,y=-3..3,
 [[x(0)=0.01, y(0)=0]], stepsize=0.1,linecolor=blue, arrows=none);
```

Example 9.6.9: (Example 9.5.3 revisited) Consider the van der Pol equation

$$\ddot{y} + \epsilon\left(y^2 - 1\right)\dot{y} + y = 0. \tag{9.6.11}$$

Multiplying it by the Dulac function[92] $\phi(x, y) = (x^2 + y^2 - 1)^{-1/2}$, we obtain

$$\frac{\partial}{\partial x}\, y\,\phi(x,y) + \frac{\partial}{\partial y}\left[\epsilon(1 - x^2)y - x\right]\phi(x,y) = -\frac{\epsilon\left(x^2 - 1\right)^2}{\left(x^2 + y^2 - 1\right)^{3/2}}.$$

Since this expression does not change sign in any domain not containing the unit circle $x^2 + y^2 = 1$, the van der Pol equation has at most one limit cycle (similar to Example 9.6.6).

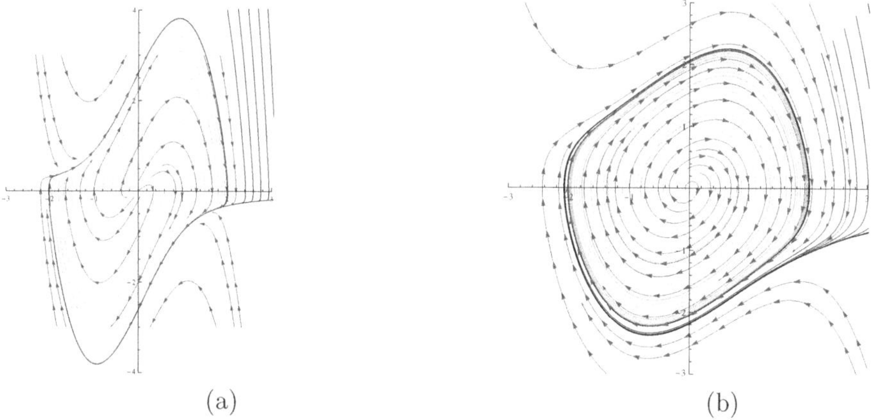

Figure 9.33: Example 9.6.9, phase portraits of the van der Pol equation for (a) $\epsilon = 2$ and (b) for $\epsilon = 0.5$, plotted with *Mathematica*.

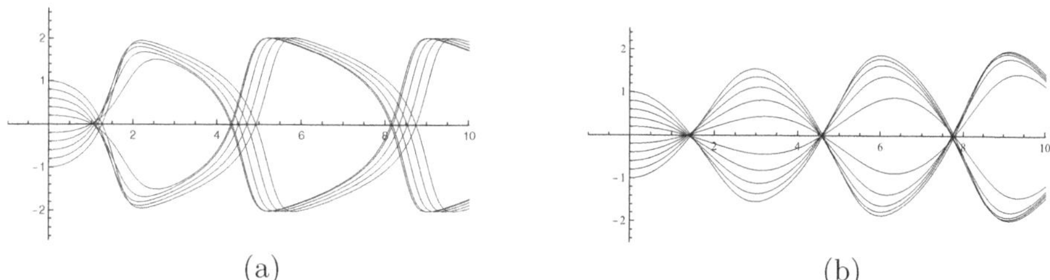

Figure 9.34: Example 9.6.9, solutions of the van der Pol equation for (a) $\epsilon = 2$ and (b) for $\epsilon = 0.5$, plotted with *Mathematica*.

9.6.1 Equations with Periodic Coefficients

A fundamental engineering problem is to determine the response of a physical system to an applied force. Consider a Duffing oscillator depending on the amplitude F chosen for the forcing periodic term

$$\ddot{x} + 2\gamma\dot{x} + \alpha x + \beta x^3 = F\,\cos(\omega t),$$

with damping coefficient γ and the driving frequency ω. In a typical introductory classical mechanics or electrical circuit course, students first solve a linear approximation (when $\beta = 0$) by assuming that the transient solution has died away and seek the steady state periodic solutions vibrating at the same frequency ω as that of the oscillatory driving force. A nonlinear system may respond in many additional ways that are not possible for the linear system. For instance, a solution may "blow up" or exhibit chaotic behavior. Since we are looking for a periodic response, it can be shown that if a Duffing oscillator has a periodic solution, its period is an integer multiple of ω.

[92]This function was discovered by Leonid Cherkas (1937–2011), a famous Russian mathematician.

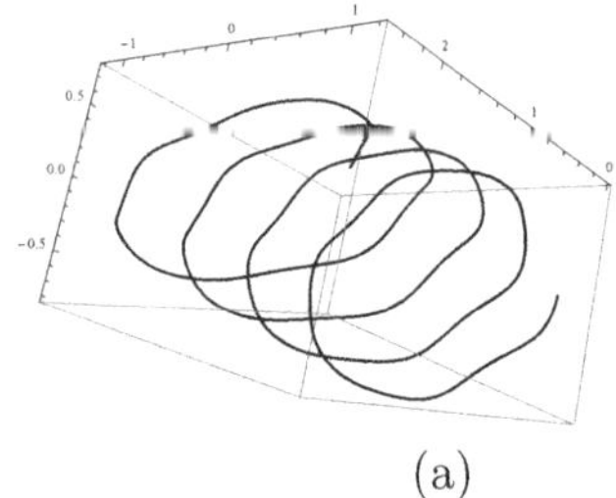

(a)

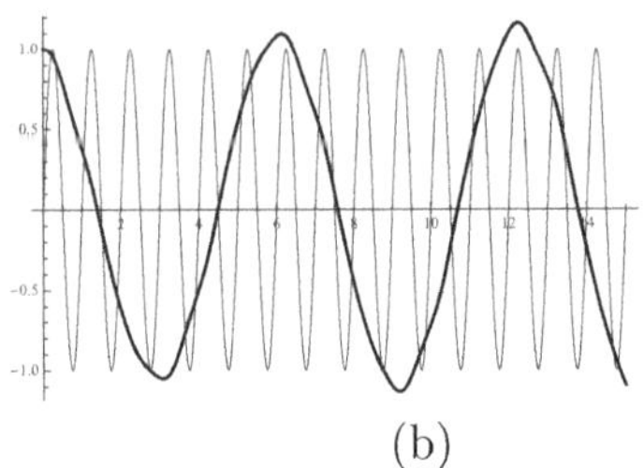

(b)

Figure 9.35: Example 9.6.10 (a) a solution trajectory, and (b) input (in blue) and output, plotted with *Mathematica*.

In 1950, Jose Luis Massera[93] showed an example of a linear system of differential equations

$$\begin{aligned}\dot{x} &= f(t)\,x + [f(t)-\omega]\,y - f(t)\,z + f(t),\\ \dot{y} &= [g(t)+\omega]\,x + g(t)\,y - g(t)\,z + g(t),\\ \dot{z} &= [h(t)+\omega]\,x + [h(t)-\omega]\,y - h(t)\,z + h(t),\end{aligned}$$

where f, g, and h are continuous T-periodic functions, that admits a periodic solution

$$x = a\,\cos(\omega t + b), \quad y = a\,\sin(\omega t + b), \quad z = a\left[\cos(\omega t + b) + \sin(\omega t + b) + 1\right]$$

(here a and b are arbitrary constants) with period $T_1 = 2\pi/\omega$, which, generally speaking, is not a rational multiple of T. This example illustrates a counter intuitive phenomenon that multi-dimensional systems subject to a periodic input may have a periodic response with different periods. This is something we never observe in one- or two-dimensional ordinary differential equations: if there is a bounded solution of the differential equation and the solutions are continuous for all following times, then there is a T-periodic solution. Systems of equations that admit solutions, called exceptional solutions, with periods that are not rational multiples of the periodic input are called **strongly irregular**, and oscillations modeled by this system are called **asynchronous**. In particular, all constant solutions are exceptional solutions.

It was observed in practice that there exist strongly irregular mechanical and electrical systems. We present one such example.

Example 9.6.10: Consider two ideal pendula of equal masses m and the same length ℓ connected by an elastic string (see figure on the opening page 341 of Chapter 6 and Example 6.1.4 on page 349). Upon linearization of Eq. (6.1.8), page 349, for small angles of inclinations θ_1 and θ_2, the two pendula system can be modeled by two coupled equations:

$$\ddot{\theta}_k + \nu^2\theta_k - a^2 m^{-1}\ell^{-2}k(t)\,(\theta_{3-k} - \theta_k) = 0 \qquad (\nu^2 = g/\ell;\ k = 1, 2),$$

where g is acceleration due to gravity and a is the distance from the pivot to the point where the spring is attached. Here the elastic coefficient $k(t)$ is assumed to be a T-periodic function. By plotting solutions, Fig. 9.35, we see that the system has a periodic response with an asynchronous period compared with $k(t)$.

Problems

1. When solving planar differential equations, it is helpful to use polar coordinates $x = a\,r\,\cos\theta$, $y = b\,r\,\sin\theta$, where a and b are some positive numbers. Using the chain rule, show that
$$r\,\dot{r} = \frac{x}{a^2}\,\dot{x} + \frac{y}{b^2}\,\dot{y}, \qquad abr^2\dot{\theta} = x\,\dot{y} - y\,\dot{x}.$$
2. As an example of an analytically derived limit cycle, consider the set of coupled nonlinear ordinary differential equations
$$\dot{x} = -4y + \frac{x}{\sqrt{x^2+4y^2}}\left(1 - x^2 - 4y^2\right), \quad \dot{y} = x + \frac{y}{\sqrt{x^2+4y^2}}\left(1 - x^2 - 4y^2\right). \tag{9.6.12}$$
To solve this system, introduce the plane polar coordinates $x = 2r\cos\theta$ and $y = r\,\sin\theta$. Then, after multiplying the first and the second equations by x and $4y$, respectively, and adding, we obtain $\frac{1}{2}\frac{\text{d}}{\text{d}t}\left(x^2+4y^2\right) = \left(x^2+4y^2\right)^{1/2}\left(1-x^2-4y^2\right)$, which is an autonomous equation with respect to r: $2\,\dot{r} = 1 - 4r^2$. Find a solution of Eq. (9.6.12).

[93]José Luis Massera (1915–2002) was an Uruguayan mathematician who researched the stability of differential equations.

3. When parameter ϵ in the van der Pol equation $\ddot{x} - \epsilon\left(1 - x^2\right)\dot{x} + x = 0$ is large ($\epsilon \gg 1$), it is convenient to use Liénard's transformation $y = x - x^3/3 - \dot{x}/\epsilon$. This allows us to rewrite it as the system of first order equations

$$\dot{x} = \epsilon\left(x - \frac{x^3}{3} - y\right), \qquad \dot{y} = \frac{x}{\epsilon}, \tag{9.6.13}$$

which can be regarded as a special case of the FitzHugh–Nagumo model (also known as Bonhoeffer–van der Pol model). By plotting phase portraits for different values of the parameter ϵ, observe that the system has a stable limit cycle.

4. Show that the following nonlinear systems have no periodic solutions by applying the Bendixson negative criterion 9.8 (page 531).

   (a) $\dot{x} = 2y^2 + y^3 - 3x$, $\dot{y} = 5x^2 - 4y$; **(b)** $\dot{x} = 4y - 2x$, $\dot{y} = 3x - 4y^3$;

   (c) $\dot{x} = y^2 - x\,e^{2x^2+2y^2}$, $\dot{y} = x^3 - y\,e^{3x^2+3y^2}$; **(d)** $\dot{x} = 4y - 2x^3$, $\dot{y} = -5x - 3y^3$;

   (e) $\dot{x} = 2x^3 - 3x^2y + y^3$, $\dot{y} = 2y^3 - 3xy^2 + x^3$; **(f)** $\dot{x} = y^3 - x$, $\dot{y} = x^3 - 2y^3$.

5. Prove Theorem 9.6 on page 529.

6. **The pursuit problem of A.S. Hathaway** (1920). A playful dog initially at the center O (origin) of a large pond swims straight for a duck, which is swimming at constant speed in a circle of radius a centered on O and taunting the dog. The dog is swimming k times as fast as the duck.

   (a) Show that the path of the pursuing dog is determined by the coupled nonlinear equations

   $$\mathrm{d}\phi/\mathrm{d}\theta = (a\,\cos\phi)/\rho - 1, \qquad \mathrm{d}\rho/\mathrm{d}\theta = a\,\sin\phi - ka,$$

   where θ is the polar angle of the duck's position, ρ is the distance between the instantaneous positions of the dog and the duck, and ϕ is the angle formed by tangent lines to the paths of the duck (circle) and the dog (pursuit).

   (b) The duck is safe for any $k < 1$! For $k = 3/4$, numerically show that the dog never reaches the duck, but instead traces out a path which asymptotically approaches a circle of radius $3a/4$ about the origin. By choosing other starting positions for the dog, show by plotting the pursuit trajectories that the circle is a stable limit cycle. Take $a = 1$ and start the duck out at $x = 1$, $y = 0$, swimming counterclockwise. The value of k dictates the size of the limit cycle.

7. **Multiple nested limit cycles.** Given the pair of equations depending on a positive parameter k:

$$\dot{x} = -k^2 y + x\,f\left(\sqrt{x^2+k^2y^2}\right), \quad \dot{y} = x + y\,f\left(\sqrt{x^2+k^2y^2}\right),$$

with $f\left(\sqrt{x^2+k^2y^2}\right) = kr\,\sin(2/kr)$, where $k^2r^2 = x^2 + k^2y^2$. Show analytically that all of the ellipses $x^2 + k^2y^2 = 1/(n\pi)^2$, $n = 1, 2, 3, \ldots$, are stable limit cycles.

8. By carrying out an exact analytic solution, show that the system of equations

$$\dot{x} = -9y\left(x^2 + 9y^2 + 1\right) + x\left(x^2 + 9y^2 - 1\right), \quad \dot{y} = x\left(x^2 + 9y^2 + 1\right) + y\left(x^2 + 9y^2 - 1\right)$$

has a limit cycle and determine whether it is stable, unstable, or semistable. Confirm your analysis by plotting representative trajectories in the phase plane.

9. **Wasley Krogdahl** (1919–2009) from the University of Kentucky proposed in 1955 a model to explain the pulsation of variable stars of the Cepheid type. He modified van der Pol's equation by adding some extra terms:

$$\ddot{x} - b\left(1 - x^2\right)\dot{x} - a\left(1 - \frac{3a}{2}\right)\dot{x}^2 + x - ax^2 + \frac{7}{6}\,a^2x^3 = 0,$$

where a and b are positive constants. Taking $a = 1/12$, $b = 1$, plot the phase portrait and show that a stable limit cycle exists. Also examine how the shape of the limit cycle changes with deferring values of a and b.

10. Consider the Duffing equation with variable damping

$$\ddot{x} - \left(1.69 - x^2\right)\dot{x} - 2.25\,x + x^3 = 0.$$

   (a) Apply Bendixson's negative criterion to this problem and state what conclusion you would come to regarding the existence of a limit cycle.

   (b) Find and identify the stationary points of this system. Confirm the nature of two of these stationary points by producing a phase plane portrait for the initial values $x = 0.75$, $\dot{x} = 0$ and $x = -0.75$, $\dot{x} = 0$. Take interval $t = 0..30$.

(c) With the same time interval as in (b), let the initial phase plane coordinates be $x = 0.1$, $\dot{x} = 0$. What conclusion can you draw from this numerical result? How do you reconcile your answer with (a)?

11. Find the equilibrium points and the limit cycles of the system $\dot{r} = r\,\cos(r/2)$, $\dot{\theta} = 1$, written in polar coordinates.

12. By plotting the phase portrait, show that the system

$$\dot{x} = 2x - y - x^3, \qquad \dot{x} = x + 2.5\,y - y^3$$

has a unique globally attracting limit cycle. Also, prove that the origin is an unstable spiral point.

13. **Zhukovskii's model of a glider**. Imagine a glider operating in a vertical plane. Let v be the speed of glider and θ be the angle flight path makes with the horizontal. In the absence of drag (friction), the dimensionless equations of motion are

$$\dot{\theta} = v - (\cos\theta)/v, \qquad \dot{v} = -\sin\theta.$$

By plotting the phase portrait for this system, do you observe periodic solutions? How many different families of periodic solutions exist? Find an equation for the separatrix.

14. The **Rössler system** of three nonlinear ordinary differential equations,

$$\dot{x} = -y - z, \quad \dot{y} = x + ay, \quad \dot{z} = b + z(x - c),$$

was originally studied by a German biochemist Otto Rössler (born in 1940) in the late 1970s. Consider the Rössler system with $a = b = 0.2$. Confirm that three-dimensional periodic solutions exist for $c = 2.5$, $c = 3.5$, and $c = 4.5$. Identify the periodicity in each case. Are these periodic solutions limit cycles?

15. The following model due to Rössler produces a three-dimensional limit cycle for $\epsilon = 0.1$ and $a = 1/4$, $b = 1/16$, which is a distortion in the z direction of the limit cycle, which is the ellipse $a\,x^2 + b\,y^2 = 1$ rotated by $\pi/4$.

$$\dot{x} = x\left(1 - a\,x^2 - b\,y^2\right) - y, \quad \dot{y} = y\left(1 - a\,x^2 - b\,y^2\right) + x, \quad \dot{z} = (1 - z^2)(x + z - 1) - \epsilon z.$$

Confirm that this model yields a three-dimensional limit cycle of the indicated shape.

16. Prove that the system $\dot{x} = x(a + bx + cy)$, $\dot{y} = y(\alpha + \beta x + \gamma y)$ has no limit cycles. *Hint:* Find a Dulac function of the form $x^r y^s$. This statement is attributed to the Russian mathematician Nikolai N. Bautin (1908–1993).

17. Prove that the following Dirichlet boundary value problem has a solution.

$$y'' = 4 - y^2, \qquad y(0) = 0, \quad y(4) = 0$$

18. **(A. K. Demenchuk)** Verify that the equation

$$\dddot{x} + (\sin t - 1)\,\ddot{x} + k^2\dot{x} + k^2\,(\sin t - 1)\,x = 0, \qquad k \ne 0 \text{ and } k \ne 1,$$

has a two-parametric family of asynchronous periodic solutions $x(t) = c_1\,\cos kt + c_2\,\sin kt$.

19. **(A. K. Demenchuk)** Verify that the system

$$\dot{x} = y\,\cos t + 2z - \cos t + 2,\ \dot{y} = \sin t\,(y - 1)\,,\ \dot{z} = \sin 2t\,(y - 1) - x,$$

has a two-parametric family of asynchronous periodic solutions $x(t) = c_1\sqrt{2}\,\sin\sqrt{2}t$ $+\, c_2\sqrt{2}\,\cos\sqrt{2}t$, $y(t) = 1$, $z = c_1\cos\sqrt{2}t - c_2\sin\sqrt{2}t - 1$.

Summary for Chapter 9

1. A solution $\mathbf{x}(t)$ of a system of differential equations $\dot{\mathbf{x}} = \mathbf{f}\,(t, \mathbf{x})$ can be interpreted as a curve in n-dimensional space. It is called an **integral curve**, a **trajectory**, or an **orbit** of the system. When the variable t is associated with time, we can call a solution $\mathbf{x}(t)$ the **state of the system** at time t.

2. We refer to a constant solution $\mathbf{x}(t) = \mathbf{x}^*$ of a system

$$\dot{\mathbf{x}}(t) = \mathbf{A}\,\mathbf{x}(t), \quad \mathbf{A} \text{ is an } n \times n \text{ matrix},$$

as an **equilibrium** if $\dot{\mathbf{x}} = \mathrm{d}\mathbf{x}(t)/\mathrm{d}t = \mathbf{0}$. Such a constant solution is also called a **critical** or **stationary** point of the system.

3. Critical points can be stable, asymptotically stable or unstable (see Definitions 9.3 and 9.4)

4. A picture that shows its critical points together with collection of typical solution curves in the xy-plane is called a **phase portrait** or **phase diagram**.
5. A linearization of an autonomous vector equation $\dot{\mathbf{x}} = \mathbf{f}(\mathbf{x})$ near an isolated critical point $\mathbf{x} = \mathbf{x}^*$ is a linear system $\dot{\mathbf{x}} = \mathbf{J}\,\mathbf{x}$, with the Jacobian matrix

$$\mathbf{J} = [\mathtt{D}\,\mathbf{f}(\mathbf{x}^*)]_{ij} = \left.\frac{\partial f_i}{\partial x_j}\right|_{\mathbf{x}=\mathbf{x}^*}, \qquad i,j = 1,2,\ldots n.$$

6. The Grobman–Hartman theorem states that as long as $\mathtt{D}\,\mathbf{f}(\mathbf{x}^*)$ is **hyperbolic**, meaning that none of its eigenvalues are purely imaginary, the solutions of $\dot{\mathbf{x}} = \mathbf{f}(\mathbf{x})$ may be mapped to solutions of $\dot{\mathbf{x}} = \mathbf{J}\,\mathbf{x}$ by a 1-to-1 and continuous function.
7. There are only two situations in which the long-term behavior of solutions near a critical point of the nonlinear system and its linearization can differ. One occurs when the equilibrium solution of the linearized system is a center. The other is when the linearized system has zero eigenvalue.
8. A remarkable variety of population models are known and used to describe specific interactions between species. We mention the two most popular models: the competition one

$$\begin{cases} \dot{x}_1 = r_1 x_1 - a_1 x_1^2 - b_1 x_1 x_2, \\ \dot{x}_2 = r_2 x_2 - a_2 x_2^2 - b_2 x_1 x_2, \end{cases} \tag{9.3.1}$$

and predator-prey equations (or Lotka–Volterra model)

$$\dot{x} = x\,(r - by)\,, \qquad \dot{y} = y\,(-\mu + \beta x)\,. \tag{9.3.8}$$

9. A system of equations $\dot{\mathbf{x}} = \mathbf{f}(\mathbf{x})$ is called the **Hamiltonian** if there exists a real-valued function that is constant along any solution of the system.
10. Theorem 9.3 **(Lyapunov)** If $\mathbf{x}^*$ is an isolated critical point for the differential equation $\dot{\mathbf{x}} = \mathbf{f}(\mathbf{x})$ and there exists a Lyapunov function V for the system at $\mathbf{x}^*$, then $\mathbf{x}^*$ is stable. If, in addition, V is a strong Lyapunov function, then $\mathbf{x}^*$ is asymptotically stable.
11. Theorem 9.4 **(Lyapunov)** Let $\mathbf{x}^*$ be an isolated critical point of the autonomous system $\dot{\mathbf{x}} = \mathbf{f}(\mathbf{x})$. Suppose that there exists a continuously differentiable function $V(\mathbf{x})$ such that $V(\mathbf{x}^*) = 0$ and that in every neighborhood of $\mathbf{x}^*$ there is at least one point at which V is positive. If there exists a region $R \ni \mathbf{x}^*$ such that the derivative of V with respect to the system $\dot{V}(\mathbf{x}) = \nabla V(\mathbf{x}) \cdot \mathbf{f}(\mathbf{x})$ is positive definite on R, then $\mathbf{x} = \mathbf{x}^*$ is an unstable equilibrium solution.
12. A **limit cycle** C is an isolated closed trajectory having the property that all other orbits in its neighborhood are spirals winding themselves onto C for $t \to +\infty$ (a stable limit cycle) or $t \to -\infty$ (an unstable limit cycle).
13. Theorem 9.8 **(Bendixson's negative criterion)** If the expression

$$\frac{\partial f}{\partial x} + \frac{\partial g}{\partial y} \neq 0 \tag{9.6.6}$$

does not change its sign within a simply connected domain Ω of the phase plane, no periodic motions can exist in Ω.

14. Theorem 9.10 **(Poincaré–Bendixson)** Let the rate functions $f(x,y)$ and $g(x,y)$ in Eq. (9.6.5) be continuously differentiable, and let $\phi(t) = \langle x(t),\, y(t)\rangle$ be the parametric equation of a half-trajectory Γ which remains inside the finite domain Ω for $t \mapsto +\infty$ without approaching any singularity. Then only two cases are possible: either Γ is itself a closed trajectory or Γ approaches such a trajectory.

Review Questions for Chapter 9

Section 9.2 of Chapter 9 (Review)

1. Show that the system $\dot{x} = e^{x+y} - y, \quad \dot{y} = xy - x$ has only one critical point $(-1,1)$. Find the linearization of the system near this point.
2. Each system of nonlinear differential equations has a single stationary point. Apply Theorem 9.1 to classify this critical point as to type and stability.

(a) $\begin{cases} \dot{x} = 81 - xy, \\ \dot{y} = x - 16y^3. \end{cases}$ (b) $\begin{cases} \dot{x} = x^2 - x, \\ \dot{y} = x + y^2 - 4y + 3. \end{cases}$

(c) $\begin{cases} \dot{x} = x^2 - y^2 - x, \\ \dot{y} = x^2 - 4y. \end{cases}$

(d) $\begin{cases} \dot{x} = 4x + 2y - 2xy + y^2, \\ \dot{y} = y^2 + 2y + xy. \end{cases}$

(e) $\begin{cases} \dot{x} = 3x + 6y + x^2 + 3xy, \\ \dot{y} = 3y + y^2 + xy + y^2. \end{cases}$

(f) $\begin{cases} \dot{x} = x + y, \\ \dot{y} = 5y - 3xy + y^2. \end{cases}$

(g) $\begin{cases} \dot{x} = \sin(y - x), \\ \dot{y} = 3x^2 + y^2 - 1. \end{cases}$

(h) $\begin{cases} \dot{x} = x - 2y, \\ \dot{y} = x - y + y^2 - 2. \end{cases}$

3. Consider the three systems

(a) $\begin{cases} \dot{x} = 2\sin 2x + 3y, \\ \dot{y} = 4x + y^2; \end{cases}$ (b) $\begin{cases} \dot{x} = 4x + 3\cos y - 3, \\ \dot{y} = 4\sin x + y^2; \end{cases}$ (c) $\begin{cases} \dot{x} = 4x - 3y, \\ \dot{y} = 2\sin 2x. \end{cases}$

All three have an equilibrium point at $(0, 0)$. Which two systems have phase portrait with the same "local picture" near the origin?

In Problems 4 through 6, each system depends on the parameter ϵ. In each exercise,

(a) find all critical points;

(b) determine all values of ϵ at which a bifurcation occurs.

4. $\dot{x} = x(\epsilon - y), \qquad \dot{y} = y(2 + 3x).$

5. $\dot{x} = \epsilon x - y^2, \qquad \dot{y} = 1 + x - 2y.$

6. The van der Pol equation $\ddot{x} + \epsilon\left(x^2 - 1\right)\dot{x} + x = 0.$

Section 9.3 of Chapter 9 (Review)

Systems of equations in Problems 1 and 2 can be interpreted as describing the interaction of two species with populations $x(t)$ and $y(t)$. In each of these problems, perform the following steps.

(a) Find the critical points.

(b) For each stationary point find the corresponding linearization. Then find the eigenvalues of the linear system; classify each critical point as to type, and determine whether it is asymptotically stable or unstable.

(c) Plot the phase portrait of the nonlinear system.

(d) Determine the limiting behavior of $x(t)$ and $y(t)$ as $t \to \infty$, and interpret the results in terms of the populations of the two species.

1. Equations modeling competitions of species:

(a) $\begin{cases} \dot{x} = x\left(1 - \frac{x}{2} - \frac{y}{4}\right), \\ \dot{y} = y\left(1 - \frac{x}{6} - \frac{3y}{4}\right); \end{cases}$

(b) $\begin{cases} \dot{x} = x\left(1 - \frac{x}{3} - \frac{y}{2}\right), \\ \dot{y} = y\left(1 - \frac{x}{6} - \frac{3y}{4}\right); \end{cases}$

(c) $\begin{cases} \dot{x} = x\left(1 - \frac{x}{2} - \frac{y}{2}\right), \\ \dot{y} = y\left(1 - \frac{x}{3} - \frac{2y}{3}\right); \end{cases}$

(d) $\begin{cases} \dot{x} = x\left(1 - \frac{x}{2} - \frac{y}{3}\right), \\ \dot{y} = y\left(1 - \frac{x}{4} - \frac{2y}{3}\right); \end{cases}$

(e) $\begin{cases} \dot{x} = x\left(1 - \frac{x}{2} - \frac{y}{3}\right), \\ \dot{y} = y\left(1 - \frac{2x}{5} - \frac{2y}{3}\right); \end{cases}$

(f) $\begin{cases} \dot{x} = x\left(1 - \frac{x}{4} - \frac{y}{4}\right), \\ \dot{y} = y\left(1 - \frac{x}{6} - \frac{y}{3}\right). \end{cases}$

2. Predator-prey vector equations:

(a) $\begin{cases} \dot{x} = x(1 - \frac{1}{6}\,y), \\ \dot{y} = 4y\,(x - 3)\,; \end{cases}$

(b) $\begin{cases} \dot{x} = \frac{x}{3}\,(3 - y)\,, \\ \dot{y} = \frac{y}{2}\,(x - 2)\,. \end{cases}$

3. In each exercise from Problem 1, determine the response of the system to constant-yield harvesting of the x-species.

4. Bees and flowering plants are famous for benefitting each other. In the following model

$$\dot{x} = x\left(1 + \frac{y}{3} - \frac{x}{2}\right), \qquad \dot{y} = y\left(2 - \frac{y}{3} + \frac{x}{4}\right),$$

determine all equilibrium solutions.

Section 9.4 of Chapter 9 (Review)

1. Show that the Hamiltonian system $\dot{x} = 2y\,e^{x^2+y^2}$, $\dot{y} = -2x\,e^{x^2+y^2}$ has a stable equilibrium at the origin. Then show that a linearization technique from §9.2 fails to lead to this conclusion.

2. In each system, find the potential function $\Pi(x)$ and the energy function $E(x, v)$. Select E so that $E(0, 0) = 0$.

(a) $\ddot{x} + 3x^2 - 6x + 5 = 0$;

(b) $\ddot{x} + x\cos x = 0$;

(c) $\ddot{x} + x^2/(1+x)^2 = 0$;

(d) $\ddot{x} + 2x - x^3 + x^5 = 0$;

(e) $\ddot{x} + 2x/(1+x^2)^2 = 0$;

(f) $\ddot{x} + 2x\,e^{x^2} - 1 = 0$.

3. Show that a Hamiltonian system has no attractors and no repellers.
4. Find Hamiltonians of the following systems.

(a) $\begin{cases} \dot{x} = x\,(y+1)\,, \\ \dot{y} = -y\,(1+y/2)\,; \end{cases}$

(b) $\begin{cases} \dot{x} = x\,(x\,e^y - \cos y)\,, \\ \dot{y} = \sin y - 2x\,e^y; \end{cases}$

(c) $\begin{cases} \dot{x} = \sec x \ (|x| < \pi/2), \\ \dot{y} = -y\,\sec x\,\tan x; \end{cases}$

(d) $\begin{cases} \dot{x} = 2xy - 3x^3, \\ \dot{y} = 9x^2y - y^2 - 4x^3. \end{cases}$

Section 9.5 of Chapter 9 (Review)

1. Construct a suitable Lyapunov function of the form $V(x,y) = ax^2 + by^2$, where a and b are positive constants to be determined. Then identify the stability of the critical point at the origin.

(a) $\dot{x} = y - x^3$, $\dot{y} = -5x^3 - y^5$;

(b) $\dot{x} = 6xy^2 - x^3$, $\dot{y} = -9y^3$;

(c) $\dot{x} = 3y^3 - x^3$, $\dot{y} = -3xy^2$;

(d) $\dot{x} = 2xy^2 - x^3$, $\dot{y} = -y^3$;

(e) $\dot{x} = -2x^3$, $\dot{y} = 2x^2y - y^3$;

(f) $\begin{cases} \dot{x} = x^3 + 2x^2y^3 - 3x^5, \\ \dot{y} = -3x^4 - y^5; \end{cases}$

(g) $\begin{cases} \dot{x} = 3y - 4x^3 - 10xy^2, \\ \dot{y} = -3x - 9y^3 - 2x^2y; \end{cases}$

(h) $\dot{x} = x^3 + 2x^2y^3 - 3x^5$, $\dot{y} = -3x^4 - y^5$;

(i) $\dot{x} = 5y^3 - x^3$, $\dot{y} = -5xy^2$;

(j) $\dot{x} = 5xy^2 - 3x^3$, $\dot{y} = -4x^2y - 7y^3$;

(k) $\dot{x} = x^3 - y^3$, $\dot{y} = xy^2 + 4x^2y + 2y^3$;

(l) $\dot{x} = 3x^3 - 2y^3$, $\dot{y} = 2xy^2 + 6x^3y$;

(m) $\begin{cases} \dot{x} = 3y^2 + 3xy^2 - 4x^3, \\ \dot{y} = 5x^2y - 3xy - 4y^3; \end{cases}$

(n) $\begin{cases} \dot{x} = x^3y + x^2y^3 - x^5, \\ \dot{y} = -2x^4 - 6x^3y^2 - 2y^5. \end{cases}$

2. Consider the Hamiltonian system $\dot{x} = H_y(x,y)$, $\dot{y} = -H_x(x,y)$, where $H(x,y) = x^{\mu}y^{r}\,e^{-\beta x - by}$, and constants μ, β, r, and b are positive.

   (a) Show that the Hamiltonian system has the same solution trajectories as the Lotka–Volterra system from Example 9.4.4, page 519.

   (b) Show that $H(x,y)$ has a strict maximum at the equilibrium $(\mu/\beta, r/b)$, making it unsuitable as a Lyapunov function for the system.

3. Consider the nonlinear system of equations

$$\dot{x} = y + \alpha x\left(4x^2 + y^2\right), \qquad \dot{y} = -x + \alpha y\left(4x^2 + y^2\right),$$

   containing a real parameter α.

   (a) Show that the only equilibrium point is the origin regardless of the value of α.

   (b) Show that a Lyapunov function can be chosen in the form $V(x,y) = ax^2 + by^2$, with some positive constants a and b. Then show that the origin is asymptotically stable if $\alpha < 0$.

   (c) Use $V(x,y)$ to show that the origin is unstable if $\alpha > 0$.

   (d) What can you say about stability if $\alpha = 0$?

4. Construct a suitable Lyapunov function to determine the stability of the origin for the following systems of equations.

(a) $\begin{cases} \dot{x} = -7y^3, \\ \dot{y} = 3x - 4y^3; \end{cases}$

(b) $\begin{cases} \dot{x} = y, \\ \dot{y} = -3x - y + 6x^2; \end{cases}$

(c) $\begin{cases} \dot{x} = y - 2x, \\ \dot{y} = 2x - y - x^3; \end{cases}$

(d) $\begin{cases} \dot{x} = x - y - x^3, \\ \dot{y} = x + y - y^3; \end{cases}$

(e) $\begin{cases} \dot{x} = 3y - x^5, \\ \dot{y} = -x^3 - 2y^3; \end{cases}$

(f) $\begin{cases} \dot{x} = 6y^3 - 7x^3, \\ \dot{y} = -3x^3 - 5y^3. \end{cases}$

5. Use Lyapunov's direct method to prove instability of the origin for the following systems of equations.

(a) $\begin{cases} \dot{x} = 3x^3, \\ \dot{y} = 4x^2y - 2y^3; \end{cases}$ (b) $\begin{cases} \dot{x} = y - 3x + y^5, \\ \dot{y} = x + 5y + x^3. \end{cases}$

6. Let $f(x,y)$ be a continuously differentiable function in some domain Ω containing the origin. Show that the system
$$\dot{x} = y + x\,f(x,y), \qquad \dot{y} = -x + y\,f(x,y)$$
is asymptotically stable when $f(x,y) < 0$ in Ω and unstable when $f(x,y) > 0$ in Ω.

7. Find a gradient system $\dot{\mathbf{x}} = \nabla G$, where the function G is given. What is a corresponding Lyapunov function for this gradient system?

(a) $G(x,y) = x^3 - xy^2$; (b) $G(x,y) = x^2 + y^2 - \left(x^4 + y^4\right) - 5x^2y^2$.

8. Show that the nonlinear equations $\dot{x} = 7x^3 - 5xy^2$, $\dot{y} = 8y^5 - 5x^2y$, form a gradient system.

9. Consider the Hamiltonian system $\dot{x} = H_y$, $\dot{y} = -H_x$ and the gradient system $\dot{x} = H_x$, $\dot{y} = H_y$, where $H(x,y)$ is the same function in each case. What can you say about the relationship between the two phase portraits of these two systems?

10. Using the appropriate Lyapunov function, show that the origin is asymptotically stable for the van der Pol equation $\ddot{x} + x + \epsilon\left(\dot{x} - b^2\dot{x}^3\right) = 0$.

11. Prove that $V(x,y) = x^2 + x^2y^2 + y^4$, $(x,y) \in \mathbb{R}^2$, is a strong Lyapunov function for the system
$$\dot{x} = 1 - 3x + 3x^2 + 2y^2 - x^3 - 2xy^2,$$
$$\dot{y} = y - 2xy + x^2y - y^3$$
at the critical point $(1,0)$.

12. Consider the Liénard equation $\ddot{y} + f(y)\,\dot{y} + g(y) = 0$ with $f(u) > 0$ and $u\,g(u) > 0$ for $u \ne 0$, where $f(u)$ and $g(u)$ are continuous functions. Rewrite the Liénard equation as a system of first order differential equations and show that the origin is a stable critical point. *Hint:* Use a Lyapunov function of the form $V(x,y) = 2\int_0^x g(s)\,\mathrm{d}s + y^2$.

13. Consider the damped pendulum system $\dot{x} = y$, $\dot{y} = -\omega^2 \sin x - \gamma y$.

(a) Use the Lyapunov function $V = 4\omega^2\left(1 - \cos(x - 2n\pi)\right) + 2y^2 + 2\gamma(x - 2n\pi)y + \gamma^2\,(x - 2n\pi)^2$ to show that this system is asymptotically stable at every critical point $(2\pi n, 0)$, $n = 0, \pm 1, \ldots$.

(b) Use the Lyapunov function $V = \gamma\left[1 - \cos\left(x - 2k\pi - \pi\right)\right] + y \sin\left(x - 2k\pi - \pi\right)$ to show that this system is unstable at every critical point $(2k\pi + \pi, 0)$, $k = 0, \pm 1, \ldots$.

14. Determine whether the following equations are gradient systems and, in case they are, find the gradient function $G(x,y)$.

(a) $\begin{cases} \dot{x} = 2x - 3x^2 + 3y^2, \\ \dot{y} = x - 2y + x^2 - 3y^2. \end{cases}$ (c) $\begin{cases} \dot{r} = 8r\theta - \theta\sec^2 r, \\ \dot{\theta} = 4t^2 - \tan r. \end{cases}$

(b) $\begin{cases} \dot{x} = \cos x\,\sin y, \\ \dot{y} = \sin x\,\cos y. \end{cases}$ (d) $\begin{cases} \dot{x} = 4x^3 - 2y, \\ \dot{y} = 4y^3 - 2x. \end{cases}$

15. Show that the origin is an unstable stationary point of $\ddot{x} - \dot{x}^2 \text{sign}(\dot{x}) + x = 0$.

16. Using a Lyapunov function of the form $V(x,y,z) = ax^2 + by^2 + cz^2$, show that $(0,0,0)$ is an asymptotically stable equilibrium point of the systems given.

(a) $\begin{aligned} \dot{x} &= y + z^3 - x^3, \\ \dot{y} &= -x - x^2y + z^2 - y^3, \\ \dot{z} &= -yz - y^2z - xz^2 - z^5; \end{aligned}$ (b) $\begin{aligned} \dot{x} &= -2y + yz - x^3, \\ \dot{y} &= x - xz - y^3, \\ \dot{z} &= xy - z^3. \end{aligned}$

17. Consider the differential equation (9.4.1) on page 514, $\ddot{y} + f(y) = 0$, where $f(0) = 0$, $f(y) > 0$ for $0 < y < k$, and $f(y) < 0$ for $-k < y < 0$. Introducing the velocity variable $v = \dot{y}$, we rewrite Eq. (9.4.1) as a system of two differential equations for which the origin is a critical point.

Show that the total energy function $V(y,v) = \dfrac{1}{2}\,v^2 + \displaystyle\int_0^y f(s)\,\mathrm{d}s$ is positive definite, so $V(y,v)$ is a Lyapunov function. Use this conclusion to determine stability of the critical point $(0,0)$.

Section 9.6 of Chapter 9 (Review)

1. Prove that the equation $\ddot{y} + y\,\sin t/\left(3 + \sin t\right) = 0$ does not have a fundamental set of periodic solutions. Does it have a nonzero periodic solution?

2. In each of the following problems, an autonomous system is expressed in polar coordinates. Determine all periodic solutions, all limit cycles, and the stability characteristic of all periodic solutions.

(a) $\dot{r} = r\,(1-r)\,, \quad \dot{\theta} = r;$

(b) $\dot{r} = r\,(1-r)^2\,, \quad \dot{\theta} = r;$

(c) $\dot{r} = r\,(1-r)^3\,, \quad \dot{\theta} = r;$

(d) $\dot{r} = r\,(1-r)\,(4-r)\,, \quad \dot{\theta} = 1;$

(e) $\dot{r} = r\,(1-r)^2\,, \quad \dot{\theta} = -1;$

(f) $\dot{r} = \sin(\pi r), \quad \dot{\theta} = -1.$

3. Find periodic solutions to the system and determine their stability types:

$$\dot{x} = 2x - y - 2x\left(x^2+y^2\right), \qquad \dot{y} = x + 2y - 2y\left(x^2+y^2\right), \qquad \dot{z} = -kz,\ k > 0.$$

4. Living cells obtain energy by breaking down sugar, a process known[94] as glycolysis. In yeast cells, this process proceeds in an oscillatory manner with a period of a few minutes. A model proposed by Selkov in 1968 to describe the oscillations is

$$\dot{x} = -x + \alpha y + x^2 y, \qquad \dot{y} = \beta - \alpha y - x^2 y.$$

Here x and y are the normalized concentrations of adenosine diphosphate (ADS) and fructose-6-phosphate (F6P), and α and β are positive constants.

(a) Show that the nonlinear system has an equilibrium solution $x = \beta$, $y = \beta/(\alpha+\beta^2)$. Show that the stationary point is an unstable focal or nodal point if $(\beta^2+\alpha)^2 < \beta^2 - \alpha$.

(b) To apply the Poincaré–Bendixson theorem 9.10, page 533, choose a domain Ω in the form of a slanted rectangle on the upper right corner having a slope of -1. By calculating the dot product $\vec{v}\cdot\vec{n}$ on each boundary of the domain, where $\vec{v} = \langle \dot{x}, \dot{y}\rangle$ is the direction of the tangent line to a trajectory, determine a domain of the indicated shape, such that all trajectories cross the boundaries from the outside to the inside. Is there a limit cycle inside Ω?

(c) Confirm that the system with $\alpha = 0.05$, $\beta = 0.5$ has a limit cycle by plotting the phase portrait.

5. Find conditions on a smooth function $f(x)$ so that a differential equation $\ddot{x} = f(x) - \epsilon f'(x)\,\dot{x}$, where ϵ is a parameter, has a limit cycle. *Hint:* Use a Liénard transformation $y = \epsilon f(x) + \dot{x}$, $\dot{y} = f(x)$.

6. By plotting a phase portrait, show that the system

$$\dot{x} = 3y + x\left(1 - 2x^2 - y^2\right), \qquad \dot{y} = -2x$$

has a stable limit cycle.

7. Draw the phase portrait of the system to observe a stable limit cycle.

$$\dot{x} = 5y + 3x\left(1 - x^2 - 2y^2\right), \qquad \dot{y} = -2x.$$

8. Determine $\lim_{t\mapsto\infty} x(t)$, where $x(t)$ denotes the solution of the initial value problem

$$\ddot{x} + \dot{x} + 2x + 2x^5 = 0, \qquad x(0) = 1, \quad \dot{x}(0) = 0.$$

9. Show that the differential equation $\ddot{x} + 2\left(\dot{x}^2 + x^2 - 1\right)\dot{x} + x = 0$ has a unique stable limit cycle.

10. By plotting the phase portrait, show that the system

$$\dot{x} = -y + \frac{xy}{2}, \qquad \dot{y} = x + \frac{1}{4}\left(x^2 - y^2\right)$$

has periodic solutions, but no limit cycles.

11. Find several values of parameter k such that the initial value problem has a periodic solution.

$$\dot{x} = 0.5 - x + y^3, \qquad \dot{y} = 0.5\,y + kxy + y^2, \qquad x(0) = 0.9,\ y(0) = 0.5.$$

12. Draw the phase portrait for the system $\ddot{x} = x^2/2 - x^3$. Is the solution with the initial conditions $x(0) = 3/4$ and $\dot{x} = 0$ periodic?

13. Draw the phase portrait of the Hamiltonian system $\ddot{x} + 1.02\,x - x^2 = 0$. Give an explicit formula for the Hamiltonian and use it to justify the features of the phase portrait.

14. For the system

$$\dot{x} = y + x^2 - 2y^2, \qquad \dot{y} = -2x - 2xy,$$

find a Hamiltonian and prove that the system has a limit cycle or periodic solution.

[94]The model was first proposed by the famous Russian scientist Evgenii Selkov in the paper "On the Mechanism of Single-Frequency Self-Oscillations in Glycolysis. I. A Simple Kinetic Model," *Eur. J. Biochem.* **4**(1), 79–86, 1968.

15. For positive constants a and b, show that the system
$$\dot{x} = bx - ay + bx\, f(r)/r, \qquad \dot{y} = bx + ay + ay\, f(r)/r,$$
where $x = ar\cos\theta$, $y = br\sin\theta$, $a^2b^2r^2 = b^2x^2 + a^2y^2$, and $\tan\theta = (ay)/(bx)$, has periodic solutions corresponding to the values of r such that $r + f(r) = 0$. Also determine the direction of rotation for the orbits of this system.
16. Show that the given system has only constant periodic solutions in some neighborhood of the origin.
(a) $\begin{cases} \dot{x} = 4x - 5y + 3x^2y, \\ \dot{y} = 3y^3 + x^2y - x; \end{cases}$ (c) $\begin{cases} \dot{x} = 6x + 5y - xy^2, \\ \dot{y} = x3 - 7y - x^2y; \end{cases}$
(b) $\begin{cases} \dot{x} = 2y + x^5 - 3x, \\ \dot{y} = 7x - 2y + y^5; \end{cases}$ (d) $\begin{cases} \dot{x} = 5x - 3y - 2xy^2, \\ \dot{y} = 6x + 3y - 3x^2y. \end{cases}$
17. Show that the given system has a nonconstant periodic solution.
(a) $\begin{cases} \dot{x} = y + x^2 - y^2 + 4x^2y^3, \\ \dot{y} = -2x - 2xy - 2x\,y^4; \end{cases}$ (c) $\begin{cases} \dot{x} = 2\,y + x^2 - 2\,y^2 + x^2y^3, \\ \dot{y} = -2x - 2xy + xy^4 + x^3y^2; \end{cases}$
(b) $\begin{cases} \dot{x} = y, \\ \dot{y} = -x - \left(x^4 + y^4 - 1\right) y; \end{cases}$ (d) $\begin{cases} \dot{x} = 2y, \\ \dot{y} = -2x^3 + 3y\ln\left(x^2 + y^2 + 1/4\right). \end{cases}$
18. Show that each of the given Liénard equations has a unique nonconstant periodic solution.
(a) $\ddot{x} + x^2\left(x^2 - 9\right)\dot{x} + 4x^3 = 0$; **(c)** $\ddot{x} + \left(3x^2 - 4\right)\dot{x} + 2x = 0$;
(b) $\ddot{x} + \left(3x^2 - \cos x\right)\dot{x} + x + \sin x = 0$; **(d)** $\ddot{x} + \left(5x^4 - 16\right)\dot{x} + 6x^5 = 0$.
19. Find a continuous periodic solution of the perturbed harmonic oscillator in each of the following systems:
(a) nonlinear weakly damped van der Pol equation (c is a positive constant)
$$\ddot{x} + \epsilon\left(x^2 - 1\right)\dot{x} + \omega^2 x - c^2x^3 = 0;$$
(b) modified van der Pol equation (ϵ is a positive parameter)
$$\ddot{x} + \epsilon\left(x^2 + \dot{x}^2 - 1\right)\dot{x} + x = 0.$$
20. Find numerical values of ϵ and ω^2 for which the weakly damped van der Pol equation $\ddot{x} + \epsilon\left(x^2 - 1\right)\dot{x} + \omega^2 x = 0$ has periodic solutions.
21. The **Morse** potential, named after American physicist Philip M. Morse (1903–1985), is a convenient model for the potential energy of a diatomic molecule. The corresponding model reads $\ddot{y} = D_e\left(e^{-2a(y-r_e)} - e^{-a(y-r_e)}\right)$, where D_e is the dissociation energy, r_e is the equilibrium bond distance, y is the distance between the atoms, and the parameter a controls the "width" of the potential. By plotting a phase portrait, show that the equation has periodic solutions.
22. The **Rosen–Morse** (1932) differential equation[95]
$$y'' + \left[\frac{\alpha}{\cosh^2(ax)} + \beta\,\tanh(ax) + \gamma\right] y = 0,$$
where α, β, γ, and a are positive parameters, has been used in molecular physics and quantum chemistry. By plotting a phase portrait for some numerical values of parameters, observe that it has periodic solutions.
23. **(N. P. Erugin)** Verify that the system
$$\dot{x} = y + \left(x^2 + y^2 - 1\right)\sin\omega t, \qquad \dot{y} = -x,$$
has an asynchronous solution $x = -\cos t$, $y = \sin t$.
24. It is convenient to reformulate the SIRS model in terms $x = S/N$, $y = I/N$, and $z = R/N$, which are the fractions of the susceptibles, infectives, and removeds, respectively:
$$\dot{y} = \left[\beta - (\gamma + \alpha + \beta) - (\beta - \alpha)\,y - \beta z + \frac{rN}{K}\right] y,$$
$$\dot{z} = \gamma y - (\delta + b)\,z + \frac{rNz}{K} + \alpha yz, \qquad \dot{N} = \left[r\left(1 - \frac{N}{K}\right) - \alpha y\right] N.$$
Show that the above system has a unique positive equilibrium $P_4(y_4, z_4, N_4)$ in the interior of Ω if and only if
$$R_0 = \frac{\beta}{\gamma + \alpha + b - r} > 1, \quad \text{and} \quad R_2 = \frac{r\beta}{\alpha(\beta - \gamma - \alpha - \mu)}\left(1 + \frac{\gamma}{\delta + \mu}\right) > 1,$$
where
$$y_4 = \left(1 - \frac{1}{R_0}\right)\frac{\delta + \mu}{\gamma + \delta + \mu}, \quad z_4 = \left(1 - \frac{1}{R_0}\right)\frac{\gamma}{\gamma + \delta + \mu}, \quad N_4 = K\left(1 - \frac{1}{R_4}\right).$$

[95] Nathan Rosen (1909–2005) was an American-Israeli physicist who worked with Albert Einstein and Boris Podolsky.

Chapter 10

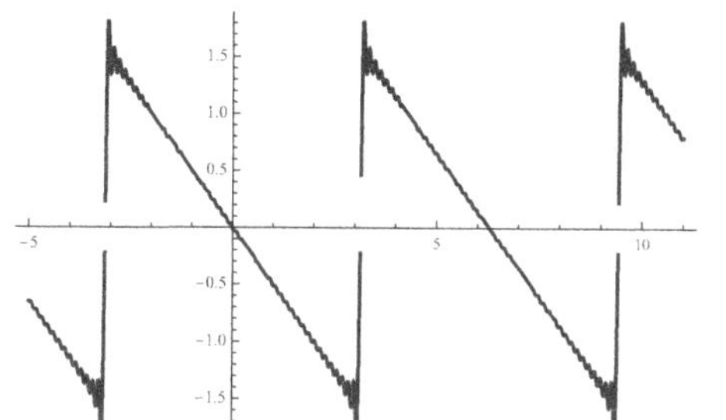

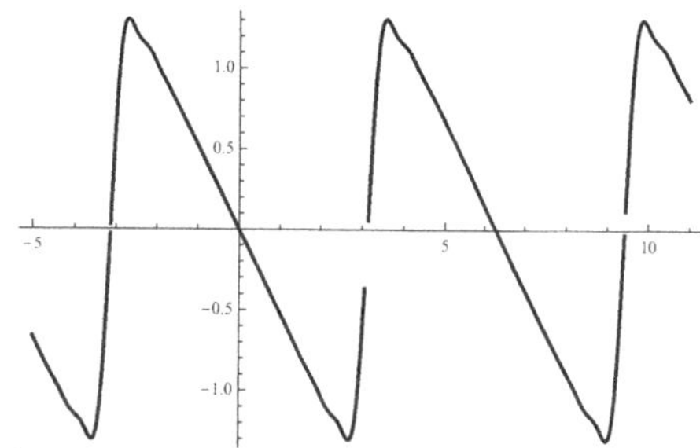

Fourier partial sum approximation (left) of $-x/2$ with 45 terms versus Cesàro partial sum approximation with the same number of terms.

Orthogonal Expansions

The objective of this chapter is to present the topic of orthogonal expansions of functions with respect to a set of orthogonal functions. This topic is an essential part of the method of separation of variables, used to solve some linear partial differential equations (Chapter 11). The set of orthogonal functions is usually determined by solving the so called Sturm–Liouville problem that deals with ordinary (or partial) differential equations containing a parameter subject to some auxiliary conditions. As an application, we give an introduction to the fascinating example of such an orthogonal expansion—Fourier series.

10.1 Sturm–Liouville Problems

Many physical applications lead to linear differential equations that are subject to some auxiliary conditions that could be either boundary or boundedness conditions rather than initial conditions. To every such differential equation can be assigned a linear operator acting in some vector space of functions. Moreover, there are many instances where a differential equation contains a parameter, which is typically denoted by λ.

Definition 10.1: Let $L : V \to V$ be a linear operator acting in a vector space V with domain $W \subseteq V$. A real or complex number λ is called the **eigenvalue** of the operator L if there exists a nonzero (also called nontrivial) element y from W such that

$$L\,y = \lambda y. \tag{10.1.1}$$

The corresponding solution y of the equation (10.1.1) is called the eigenfunction (or eigenvector) of the operator L. The problem of determination of eigenvalues and eigenfunctions is called the **Sturm–Liouville problem**. A linear operator is called **positive/nonnegative** if all its eigenvalues are positive/nonnegative.

In other words, a Sturm–Liouville problem requires finding such values of parameter λ for which the given problem (10.1.1) has a nontrivial (not identically zero) solution, and then find these solutions. Our objective is to solve some Sturm–Liouville problems for linear differential operators acting in a space of functions. What makes this problem special is the presence of a (real or complex) parameter λ. The theory of such equations originated with the pioneer work of Sturm from 1829 to 1836 and was then followed by the short but significant joint paper of Sturm and Liouville in 1837 on second order linear ordinary differential equations with an eigenvalue[96] parameter.

Note that Eq. (10.1.1) resembles an eigenvalue problem $\mathbf{A}\mathbf{x} = \lambda\mathbf{x}$ for a square matrix $\mathbf{A}$ and a column vector $\mathbf{x}$ that we discussed in §7.3. While a Sturm–Liouville problem embraces a matrix equation, it is usually referred to as an infinitely dimensional vector space of functions.

Let us start with a motivated example involving a nonnegative differential operator $L\,y = -y''$ of the second order:

$$-y''(x) = \lambda y(x), \qquad -\infty < x < \infty.$$

This homogeneous differential equation always has the solution $y(x) \equiv 0$, which is referred to as the trivial solution. The identically zero solution is rarely of interest. However, the given equation has a bounded nontrivial solution for any positive λ:

$$y(x) = C_1 \cos\left(x\sqrt{\lambda}\right) + C_2 \sin\left(x\sqrt{\lambda}\right),$$

where C_1 and C_2 are arbitrary constants. When $\lambda = 0$, the equation $y'' = 0$ has a constant nontrivial solution. For negative λ, the given equation has only unbounded exponential solutions that are disregarded. Since the eigenfunction $y(x)$ exists for every $\lambda \geqslant 0$, the corresponding differential operator is nonnegative. Clearly, an eigenfunction is not unique and can be multiplied by an arbitrary nonzero constant. The set of all eigenvalues is usually referred to as a **spectrum** (plural spectra). In our simple case, we say that the differential operator L has a continuous nonnegative spectrum.

Now we consider the same differential equation on the finite interval:

$$y''(x) + \lambda y(x) = 0, \qquad 0 < x < \ell, \tag{10.1.2}$$

where ℓ is some positive real number and λ is a (real or complex) parameter. Among many possible boundary conditions, we begin our journey with simple homogeneous conditions of the Dirichlet type:

$$y(0) = 0, \qquad y(\ell) = 0. \tag{10.1.3}$$

A linear operator corresponding to the given problem (10.1.2), (10.1.3) is $L[\mathtt{D}] = -\mathtt{D}^2 = -\mathrm{d}^2/\mathrm{d}x^2$ acting in the space of functions defined on the finite interval $[0, \ell]$ that vanish at the end points $x = 0$ and $x = \ell$.

To solve the Sturm–Liouville problem (10.1.2), (10.1.3), we need to consider separately three cases depending on the sign of λ because the form of the solution of Eq. (10.1.2) is different in each of these cases. It will be shown in the next section that the Sturm–Liouville problem for a self-adjoint operator ($L\,y = -y''$ is one of them) does not have a complex eigenvalue.

1. If $\lambda < 0$, then the general solution of the differential equation (10.1.2) is

   $$y(x) = C_1\, e^{x\sqrt{-\lambda}} + C_2\, e^{-x\sqrt{-\lambda}},$$

   for some constants C_1, C_2. Satisfying the boundary conditions (10.1.3), we get

   $$C_1 + C_2 = 0, \qquad C_1\, e^{\ell\sqrt{-\lambda}} + C_2\, e^{-\ell\sqrt{-\lambda}} = 0.$$

   Since the determinant of the corresponding system of algebraic equations

   $$\det \begin{bmatrix} 1 & 1 \\ e^{\ell\sqrt{-\lambda}} & e^{-\ell\sqrt{-\lambda}} \end{bmatrix} = e^{-\ell\sqrt{-\lambda}} - e^{\ell\sqrt{-\lambda}} = -2\,\sinh\left(\ell\sqrt{-\lambda}\right) \ne 0,$$

   the given problem has only a trivial (identically zero) solution.

2. If $\lambda = 0$, the general solution of the differential equation (10.1.2) becomes $y(x) = C_1 + C_2\,x$. From the boundary conditions (10.1.3), it follows that

   $$C_1 + C_2 \cdot 0 = C_1 = 0, \qquad C_1 + C_2\,\ell = C_2\,\ell = 0.$$

   Hence, we don't have a nontrivial solution.

[96] Jacques Charles François Sturm (1803–1855), a French mathematician of German ancestry, was known for his work in differential equations, projective geometry, optics, and mechanics. He made the first accurate measurements of the speed of sound in water in 1826. Joseph Liouville (1809–1882) was a French mathematician who, besides his academic achievements, was very talented in organizational matters. The definition of positiveness was introduced by the German-American mathematician Kurt Otto Friedrichs (1901–1982). He was the co-founder of the Courant Institute at New York University and recipient of the National Medal of Science.

3. If $\lambda > 0$, then the general solution of the differential equation (10.1.2) is

$$y(x) = C_1 \cos\left(x\sqrt{\lambda}\right) + C_2 \sin\left(x\sqrt{\lambda}\right).$$

Satisfying the boundary conditions (10.1.3), we get

$$C_1 + C_2 \cdot 0 = C_1 = 0, \qquad C_2 \sin\left(\ell\sqrt{\lambda}\right) = 0.$$

Assuming that $C_2 \neq 0$ (otherwise, we would have a trivial solution), we obtain the transcendent equation

$$\sin\left(\ell\sqrt{\lambda}\right) = 0,$$

which has infinite many discrete solutions (called eigenvalues)

$$\lambda_n = \left(\frac{n\pi}{\ell}\right)^2, \qquad n = 1, 2, 3, \ldots.$$

To these eigenvalues correspond eigenfunctions (nontrivial solutions):

$$y_n(x) = \sin\left(\frac{n\pi x}{\ell}\right), \qquad n = 1, 2, 3, \ldots.$$

Positive and negative values of n which are equal in magnitude correspond to the same eigenfunctions up to a multiplicative constant. Thus we choose only positive indices for n to label eigenfunctions and eigenvalues. Other numbering schemes such as $n = -1, 2, -3, 4, \ldots$ also work. Therefore, the Sturm–Liouville problem (10.1.2), (10.1.3) has a positive discrete spectrum $\{\lambda_n\}$ $(n = 1, 2, \ldots)$ to which correspond eigenfunctions $\sin\left(\frac{n\pi x}{\ell}\right)$ up to an arbitrary multiplicative constant. ■

Now we turn our attention to Neumann boundary conditions

$$y'(0) = 0, \qquad y'(\ell) = 0. \tag{10.1.4}$$

To solve the corresponding Sturm–Liouville problem (10.1.2), (10.1.4), we need again to consider three cases depending on the sign of λ. When $\lambda < 0$, we have only a trivial solution. For $\lambda = 0$, we substitute the general solution $y(x) = C_1 + C_2\, x$ into the Neumann conditions to obtain

$$C_2 = 0$$

because its derivative $y' = C_2$ must vanish in order to satisfy the boundary conditions (10.1.4). Therefore, $\lambda = 0$ is an eigenvalue to which corresponds a constant eigenfunction $y \equiv C_1$; it is convenient to choose $C_1 = 1$.

For $\lambda > 0$, the general solution of Eq. (10.1.2) is a linear combination of periodic functions $y(x) = C_1 \cos\left(x\sqrt{\lambda}\right) + C_2 \sin\left(x\sqrt{\lambda}\right)$, which upon substitution into the Neumann boundary conditions (10.1.4) yields

$$y'(0) = C_2\sqrt{\lambda} = 0, \qquad y'(\ell) = C_1\sqrt{\lambda}\, \sin\left(\ell\sqrt{\lambda}\right) = 0.$$

From the former, we get $C_2 = 0$ because $\lambda > 0$. Solving the transcendent equation $\sin\left(x\sqrt{\lambda}\right) = 0$, we obtain a discrete set of eigenvalues

$$\lambda_n = \left(\frac{n\pi}{\ell}\right)^2, \qquad n = 0, 1, 2, 3, \ldots.$$

Note that we include $n = 0$ to embrace the case $\lambda = 0$. To these eigenvalues correspond the eigenfunctions $y_n(x) = \cos\left(\frac{n\pi x}{\ell}\right)$. ■

If we have a Dirichlet condition at one end and a Neumann condition at the other end,

$$y(0) = 0, \qquad y'(\ell) = 0, \tag{10.1.5}$$

then these mixed boundary conditions are called of the third kind. This leads to the Sturm–Liouville problem that consists of the differential equation (10.1.2) with a parameter λ subject to the boundary conditions of the third kind

(10.1.5). Similarly to the previous discussion, it can be shown that the Sturm–Liouville problem (10.1.2), (10.1.5) has only a trivial solution for $\lambda \leqslant 0$. Assuming $\lambda > 0$, we substitute the general solution $y(x) = C_1 \cos\left(x\sqrt{\lambda}\right) + C_2 \sin\left(x\sqrt{\lambda}\right)$ into the boundary conditions (10.1.5) to obtain

$$y(0) = C_1 = 0, \qquad y'(\ell) = C_2\sqrt{\lambda}\cos\left(\ell\sqrt{\lambda}\right) = 0.$$

To avoid a trivial solution, we set $C_2 \ne 0$ and get the eigenvalues from the equation

$$\cos\left(\ell\sqrt{\lambda}\right) = 0 \qquad \Longrightarrow \qquad \lambda_n = \left(\frac{\pi(1+2n)}{2\ell}\right)^2, \qquad n = 0, 1, 2, \ldots.$$

The corresponding eigenfunctions become

$$y_n(x) = \sin\frac{\pi(1+2n)x}{2\ell}, \quad n = 0, 1, 2, \ldots.$$

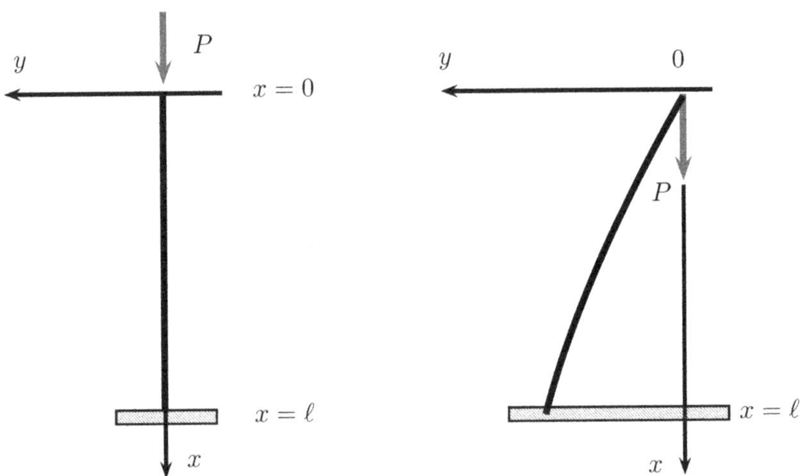

Figure 10.1: Example 10.1.1.

Example 10.1.1: Let us consider a rod of length ℓ; one end of it $x = \ell$ is fixed, while another one $x = 0$ is free. A stretching force P (with units in `newtons`) is applied to the free end $x = 0$ along the axis of the rod. It is known that when the load P is small, the form of the rod is stable; however, there exists a critical value P_0 (known as the Euler load) of the force that the form of the rod becomes unstable when $P > P_0$, and the rod bends (or buckles).

We consider the beginning of this bending; in other words, we assume that it is slightly different from its equilibrium position along a straight line. Then the equation (credited to Euler) of the bent axis of the rod $y = y(x)$ becomes

$$P\,y = -EI\,y'' \qquad (0 < x < \ell),$$

where I is the area moment of inertia of the cross-section about an axis through the centroid perpendicular to the xy-plane (it has dimensions of length to the fourth power), and E is Young's modulus (or elastic modulus, which measures the stiffness of an elastic material and has dimensions of force per length squared). If the rod is homogeneous of constant cross-section, then EI is a constant with units `N·m`^{2}. Setting $\lambda = P/(EI)$, we obtain the boundary value problem

$$-y'' = \lambda y, \qquad y'(0) = 0, \quad y(\ell) = 0.$$

Similar Sturm–Liouville problem has been considered previously. The critical loads $P_n = EI\,\lambda_n$ correspond to the eigenvalues $\lambda_n = \left(\frac{2n-1}{2\ell}\right)^2$, and the eigenfunctions $y_n(x) = \cos\frac{\pi(2n-1)x}{2\ell}$, $n = 1, 2, \ldots$, define the equilibrium positions of the rod.

If the given rod is nonhomogeneous, then EI is a function of x. If we set $\rho(x) = (EI)^{-1}$, we get the Sturm–Liouville problem

$$-y'' = P\rho(x)\,y, \qquad y'(0) = 0, \quad y(\ell) = 0,$$

which cannot be solved using elementary functions in general.

Example 10.1.2: Consider an elastic column of length ℓ; one of its end is clamped, but the other one is simply supported. Let $y(x)$ be the deflection of the column at point x from its equilibrium position.

The bending moment is $M = Py - Hx = -EI\,y''$, where H is the horizontal force of reaction, and P is a load. After differentiation twice, we get

$$(EI\,y'')'' = -Py'', \qquad y(0) = y''(0) = 0, \quad y(\ell) = y'(\ell) = 0.$$

If flexural rigidity EI is a constant, we set $k^2 = P/(EI)$ (with units $\mathtt{m}^{-2}$) and get the Sturm–Liouville problem

$$y^{(IV)} + k^2\,y'' = 0, \qquad y(0) = y''(0) = 0, \quad y(\ell) = y'(\ell) = 0.$$

The general solution of the differential equation $y^{(IV)} + k^2\,y'' = 0$ is

$$y = a + bx + c\,\cos kx + d\,\sin kx,$$

with some constants a, b, c, and d. The boundary conditions dictate that $a = c = 0$, and the eigenvalues k_n $(n = 1, 2, \ldots)$ are roots of the transcendent equation

$$\sin(k\ell) = k\ell\,\cos(k\ell).$$

The eigenfunctions $y_n(x) = \sin k_n x - x k_n\,\cos k_n \ell$ correspond to these roots.

When $k = 0$, the general solution is $y = a + bx + cx^2 + dx^3$. The boundary conditions are satisfied only when $a = b = c = d = 0$. Hence, $\lambda = 0$ is not an eigenvalue. □

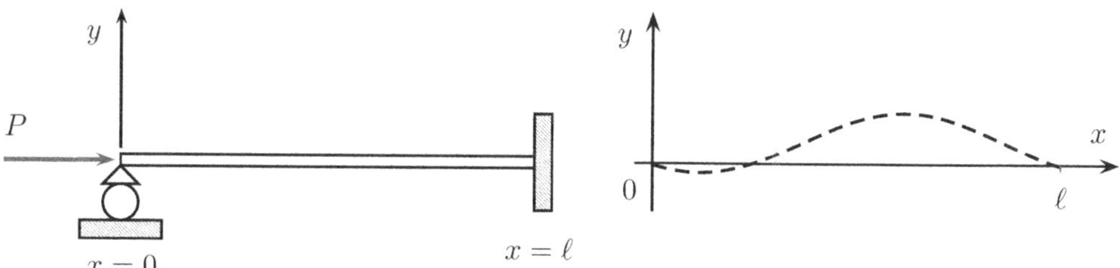

Figure 10.2: Example 10.1.2.

The conditions (10.1.5) constitute a particular case of the boundary conditions of the third kind:

$$\alpha_0 y(0) - \beta_0 y'(0) = 0, \qquad \alpha_1 y(\ell) + \beta_1 y'(\ell) = 0 \tag{10.1.6}$$

with some specified values α_0, α_1, β_0, and β_1. The periodic endpoint condition $y(0) = y(\ell)$ may be imposed in some Sturm–Liouville problems instead of traditional boundary conditions. Equation (10.1.2) can be generalized to

$$\frac{\mathrm{d}}{\mathrm{d}x}\left[p(x)\,\frac{\mathrm{d}y}{\mathrm{d}x}\right] - q(x)\,y(x) + \lambda\rho(x)\,y(x) = 0, \qquad 0 < x < \ell, \tag{10.1.7}$$

where $p(x) > 0$ has a continuous derivative, and $q(x)$, and $\rho(x) > 0$ are continuous functions on the finite interval $[0, \ell]$. The function $\rho(x)$ is called the "weight" or "density" function. By introducing the derivative operator $\mathtt{D} = \mathrm{d}/\mathrm{d}x$, Eq. (10.1.7) can be rewritten in the operator form $L[y] = \lambda\rho\,y$, where

$$L = L[x, \mathtt{D}] = -\mathtt{D}\,(p(x)\,\mathtt{D}) + q(x), \qquad \mathtt{D} = \mathrm{d}/\mathrm{d}x, \tag{10.1.8}$$

is the linear self-adjoint differential operator. Such differential equations are typical in many applications and are usually subject to boundary conditions of the third kind (10.1.6). The corresponding Sturm–Liouville problem (10.1.7), (10.1.6) is much harder to solve and analyze. We illustrate it in the following example.

Example 10.1.3: Solve the Sturm–Liouville problem

$$y'' + \lambda y = 0 \qquad (0 < x < 2), \tag{10.1.9}$$

$$y(0) - y'(0) = 0, \qquad y(2) = 0. \tag{10.1.10}$$

We disregard negative values of λ because the problem (10.1.9), (10.1.10) has only a trivial solution when $\lambda < 0$. For $\lambda = 0$, the general solution of Eq. (10.1.9) is a linear function

$$y(x) = C_1 + C_2\, x \qquad \Longrightarrow \qquad y'(x) = C_2.$$

The two boundary conditions require that

$$y(0) - y'(0) = 0 = C_1 - C_2, \qquad y(2) = 0 = C_2.$$

This leads to $C_1 = C_2 = 0$, and we get the identically zero (trivial) solution.

If $\lambda > 0$, the general solution of Eq. (10.1.9) is a linear combination of the trigonometric functions

$$y(x) = C_1 \sin\left(x\sqrt{\lambda}\right) + C_2 \cos\left(x\sqrt{\lambda}\right),$$

where, as usual, we choose a positive root $\sqrt{\lambda} > 0$. The boundary condition at $x = 0$ requires $C_2 - \sqrt{\lambda}C_1 = 0$ or $C_2 = C_1\sqrt{\lambda}$. From another boundary condition at $x = 2$, we get

$$C_1 \left[\sin\left(2\sqrt{\lambda}\right) + \sqrt{\lambda}\cos\left(2\sqrt{\lambda}\right)\right] = 0.$$

For the Sturm–Liouville problem (10.1.9), (10.1.10) to have a nontrivial solution, we must have $C_1 \ne 0$, and $\mu = \sqrt{\lambda}$ must be a positive root of the transcendent equation

$$\sin(2\mu) + \mu \cos(2\mu) = 0 \qquad (\mu = \sqrt{\lambda}).$$

Since the sine and cosine functions cannot annihilate the same point simultaneously, we may assume that both of them are not zero: $\sin(2\mu) \ne 0$ and $\cos(2\mu) \ne 0$. Dividing the previous equation by $\cos(2\mu)$, we get

$$\tan t = -t/2, \qquad \text{where} \quad t = 2\sqrt{\lambda}. \tag{10.1.11}$$

This equation does not have an analytic solution expressed through elementary functions; however, it can be solved numerically. The roots of Eq. (10.1.11) can also be found approximately by sketching the graphs $f(t) = \tan t$ and $g(t) = -t/2$ for $t > 0$. From Fig. 10.3, it follows that the straight line $g(t) = -t/2$ intersects the graph of the tangent at infinitely many discrete points t_n, $n = 1, 2, \ldots$. The first three positive solutions of the equation $\tan t + t/2 = 0$ are $t_1 \approx 2.28893$, $t_2 \approx 5.08699$, and $t_3 \approx 8.09616$, to which correspond eigenvalues $\lambda_1 = (t_1/2)^2 \approx 1.3098$, $\lambda_2 = (t_2/2)^2 \approx 6.46935$, and $\lambda_3 = (t_3/2)^2 \approx 16.387$, respectively. The other roots are given with reasonable accuracy by

$$t_n \approx \frac{\pi}{2} + (n-1)\pi, \qquad n = 3, 4, \ldots.$$

For instance, $7\pi/2 \approx 10.9956$ gives a good approximation to $t_4 \approx 11.1727$ with correct 2 decimal places. Hence, the eigenvalues are $\lambda_n = (t_n/2)^2 \approx \left(\frac{\pi(2n-1)}{4}\right)^2$ $(n = 3, 4, \ldots)$, where precision of this estimation becomes better as n grows. Say for $n = 15$, we have $t_{15} \approx 45.5969$ and $29\pi/2 \approx 45.5531$. □

Finally, it should be noted that, generally speaking, a Sturm–Liouville problem may have complex eigenvalues.

Example 10.1.4: Consider the following boundary value problem from Example 10.1.2:

$$y^{(IV)} + k^2\, y'' = 0, \qquad y(0) = y'(0) = y''(0) = 0, \quad y(\ell) = 0.$$

As previously, it can be shown that $k = 0$ is not an eigenvalue. Nontrivial solutions exist when k is a root of the transcendent equation $\sin k\ell = k\ell$, which has only complex roots.

Problems

In each of Problems 1 through 6, either solve the given boundary value problem or else show that it has no solution.

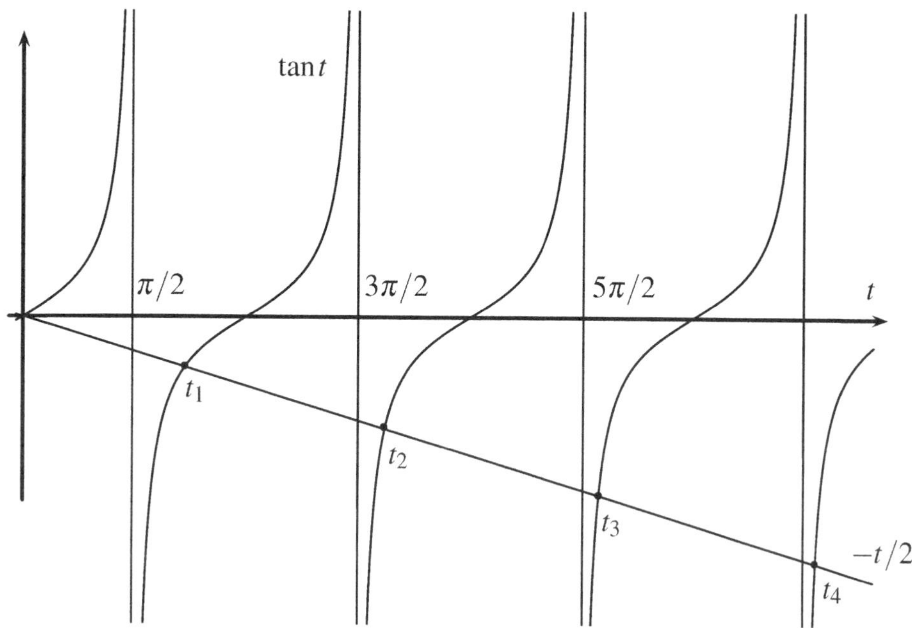

Figure 10.3: Graphical solution of $\tan t = -t/2$.

1. $y'' + 4y = 4x^2, \quad y(0) = 0, \; y'(\pi) = 0.$
2. $y'' + 9y = 8\cos x, \quad y'(0) = 0, \; y(\pi) = 0.$
3. $y'' + y = 3\sin 2x, \quad y(0) = 0, \; y(\pi) = 0.$
4. $y'' + y = 2\sin x, \quad y(0) = 0, \; y(\pi) = 0.$
5. $y'' + 4y = 4x, \quad y'(0) = 0, \; y(\pi) = 0.$
6. $y'' + 2y' + y = \cos 3x, \; y'(0) = 0, \; y'(1) = 0.$

In each of Problems 7 through 10, find the eigenvalues and eigenfunctions of the given boundary value problem. Assume that all eigenvalues are real.

7. $y'' - 4y' + \lambda y = 0, \quad y(0) = 0, \quad y(\ell) = 0, \quad \ell > 1.$
8. $y'' + 2y' + \lambda y = 0, \quad y(0) = 0, \quad y(\ell) = 0, \quad \ell > 1.$
9. $y'' + 4\lambda y = 0, \quad y(-1) = 0, \quad y(1) = 0.$
10. $y'' + 2y' + \lambda y = 0, \quad y'(0) = 0, \quad y(2) = 0$

In each of the following two problems, determine the real eigenvalues and corresponding eigenfunctions (if any).

11. $y'' + y' + 2\lambda\,(y' + y) = 0 \quad (0 < x < 1), \quad y(0) = 0, \quad y'(1) = 0.$
12. $x^2 y'' - \lambda\,(x\,y' - y) = 0 \quad (1 < x < 4), \quad y(1) = 0, \quad y(4) + y'(4) = 0.$

In each of Problems 13 through 15, assume that all eigenvalues are real.

(a) Determine the form of the eigenfunctions and find the equation for nonzero eigenvalues.

(b) Determine whether $\lambda = 0$ is an eigenvalue.

(c) Find approximate values for λ_1 and λ_2, the nonzero eigenvalues of smallest absolute value.

(d) Estimate λ_n for large values of n.

13. $y'' + \lambda y = 0, \quad y(0) = 0, \quad y(1) + 2y'(1) = 0.$
14. $y'' + \lambda y = 0, \quad y(0) - 2y'(0) = 0, \quad y(1) = 0.$
15. $y'' + \lambda y = 0, \quad y(0) - 2y'(0) = 0, \quad y(1) + 2y'(1) = 0.$
16. Solve the Sturm–Liouville problem $y'' - 4y' + \lambda y = 0, \quad y'(0) = 0, \; y'(1) = 0.$
17. Solve the Sturm–Liouville problem $y'' + \lambda y = 0, \quad y(0) - hy'(0) = 0, \; y(\ell) + hy'(\ell) = 0$, where $h, \ell > 0$.
18. Solve the Sturm–Liouville problem
$$y'' - 6y' + (9 + \lambda)y = 0, \qquad y'(0) = 0, \quad y(1) = 0.$$
19. Solve the Sturm–Liouville problem
$$y'' + 2y' + (1 + \lambda)y = 0, \qquad y(0) = 0, \quad y'(1) = 0.$$

20. A quantum particle of mass m in a one-dimensional box of length ℓ is modeled by the Schrodinger equation subject to the boundary conditions ($\hbar = 1.054571800\ldots \times 10^{-34}\text{m}^2$ kg/s is the reduced Planck constant)

$$-\frac{\hbar^2}{2m}\,\psi''(x) = E\,\psi(x), \qquad \psi(0) = 0, \quad \psi(\ell) = 0.$$

Find the eigenvalues and eigenfunctions of this problem.

21. Consider the Sturm–Liouville problem with a positive parameter α

$$y'' + \lambda^2 y = 0, \qquad y'(0) - \alpha y(0) = 0, \quad y'(1) = 0.$$

(a) Show that for all values of α there is an infinite sequence of eigenvalues and corresponding even eigenfunctions.

(b) Show that independently of α, the eigenvalues $\lambda_n \mapsto \pi(1+2n)/2$ as $n \mapsto \infty$.

(c) Show that $\lambda = 0$ is an eigenvalue only if $\alpha = 0$.

22. For each of the following boundary conditions, find the smallest real eigenvalue and determine the corresponding eigenfunction of the Sturm–Liouville problem for the buckled column equation $y^{(4)} + k^2 y'' = 0 \quad (0 < x < \ell)$, subject to the boundary conditions

(a) $y(0) = y'(0) = 0, \quad y'(\ell) = y''(\ell) = 0;$

(b) $y(0) = y'''(0) = 0, \quad y(\ell) = y'(\ell) = 0;$

(c) $y(0) = y'(0) = 0, \quad y(\ell) = y'(\ell) = 0.$

23. Consider the Sturm–Liouville problem assuming that all eigenvalues are real

$$y'' - 4y' + (4+\lambda)y = 0, \qquad y(0) = 0, \quad y(\pi) = 0.$$

(a) By making the Bernoulli substitution $y = u\,v$, determine the function $u(x)$ from the condition that the differential equation for $v(x)$ has no v' term.

(b) Solve the boundary value problem for v and thereby find the eigenvalues and eigenfunctions of the original problem.

24. Consider the Sturm–Liouville problem:

$$y'' + \lambda y = 0, \qquad y(0) = 0, \quad 3y(1) - y'(1) = 0.$$

(a) Find the determinantal equation satisfied by the positive eigenvalues.

(b) Show that there is an infinite sequence of such eigenvalues.

(c) Find the first two eigenvalues, and then show that $\lambda_n \approx [(2n+1)\pi/2]^2$ for large n.

(d) Find the determinantal equation satisfied by the negative eigenvalues.

(e) Show that there is exactly one negative eigenvalue and find its value.

25. Determine the real eigenvalues and the corresponding eigenfunctions in the boundary problem

$$x^2 y'' - \lambda\,(xy' - y) = 0, \qquad y(1) = 0, \quad y(4) = y'(4).$$

26. Consider the Sturm–Liouville problem with a positive parameter α

$$y'' + 2\lambda y = 0, \qquad y'(0) = 0, \quad \alpha y(1) + y'(1) = 0.$$

(a) Show that for all values of $\alpha > 0$ there is an infinite sequence of positive eigenvalues.

(b) Show that all (real) eigenvalues are positive.

(c) Show that $\lambda = 0$ is an eigenvalue only if $\alpha = 0$.

(d) Show that the eigenvalues $\lambda_n = \mu_n^2/2$, where $\mu_n \approx n\pi - \alpha/(n\pi)$ for large n.

In each of Problems 27 through 30, convert the given problem into a corresponding boundary value problem for the Prüfer variables. Assume that $R(x)$ is not zero at the end points.

27. $y' + 3y + \lambda y = 0, \; y(0) = y(1) = 0.$

28. $y' - 4y + 5\lambda y = 0, \; y(0) = y(1) = 0.$

29. $y' - 7y + 8\lambda y = 0, \; y(0) = y(2) = 0.$

30. $y' - 2xy + (3+2\lambda)y = 0, \; y(0) = y'(1) = 0.$

10.2 Orthogonal Expansions

In this section we continue analyzing the Sturm–Liouville problems and their applications. However, before presenting this material, we must first develop certain properties of a set of functions; therefore, we start with some definitions.

Definition 10.2: Let $f(x)$ and $g(x)$ be two real-valued or complex-valued functions defined over an interval $|a, b|$ and $\rho(x)$ be a positive function over the same interval. The **inner product** or **scalar product** of these functions with weight ρ, denoted by $\langle f \mid g \rangle$ or simply by (f, g), is a complex number

$$\langle f \mid g \rangle = (f, g) = \int_a^b \overline{f}(x)\, g(x)\, \rho(x)\, \mathrm{d}x,$$

where $\overline{f} = u - \mathbf{j}v$ is the complex conjugate of $f = u + \mathbf{j}v$. If f and g are real-valued functions, then their scalar product is a real number

$$(f, g) = \int_a^b f(x)\, g(x)\, \rho(x)\, \mathrm{d}x.$$

If $\rho(x) = 1$, then

$$(f, g) = \int_a^b f(x)\, g(x)\, \mathrm{d}x.$$

The bra-ket notation $\langle f \mid g \rangle$ for the inner product was introduced in quantum mechanics in 1939 by Paul Dirac (1902–1984). This definition is a natural generalization of the finite dimensional dot product of two n-vectors $(\mathbf{u}, \mathbf{v}) = \overline{\mathbf{u}} \cdot \mathbf{v} = \sum_{k=1}^{n} \overline{u}_k\, v_k$, where $\mathbf{u} = \langle u_1, \ldots, u_n \rangle$, $\mathbf{v} = \langle v_1, \ldots, v_n \rangle$, as the dimension n of the vector space, which is the number of components involved, becomes infinitely large. From the definition of the inner product, it follows that

$$(f, g) = \overline{(g, f)}. \tag{10.2.1}$$

Definition 10.3: A linear operator L is called **self-adjoint** if for every pair of elements u and v from the domain of L, we have

$$(Lu, v) = (u, Lv).$$

Definition 10.4: The positive square root of the definite integral (if it exists)

$$\int_a^b |f(x)|^2\, \rho(x)\, \mathrm{d}x \quad \text{in the complex case,} \qquad \int_a^b f^2(x)\, \rho(x)\, \mathrm{d}x \quad \text{in the real case,}$$

is called the **norm** of the function $f(x)$ (with weight $\rho > 0$). It is denoted by $\|f\|_2$ or simply $\|f\|$, that is,

$$\|f\|^2 = (f, f) = \int_a^b |f(x)|^2\, \rho(x)\, \mathrm{d}x. \tag{10.2.2}$$

A function f that has unit norm $\|f\| = 1$ is said to be **normalized**.

If the norm of f is zero, then the integral of the nonnegative function $|f(x)|^2$ over the interval $|a, b|$ must vanish. This means that $f(x)$ is almost everywhere zero and it may differ from zero over any range of zero length. It is convenient to speak of a such function as a *trivial function*. In particular, if f is continuous everywhere in an interval $|a, b|$, and has a zero norm over that interval, then f must vanish everywhere in $|a, b|$. Functions with finite norm (10.2.2) are called **square integrable**.

Definition 10.5: For a real-valued function $f(x)$ defined on interval $|a,b|$, its **root mean square** value (r.m.s.) is

$$\overline{\overline{f}} = \text{r.m.s.}\, f = \sqrt{\frac{1}{b-a}\int_a^b f^2(x)\,\mathrm{d}x}.$$

Squaring both sides, we obtain the **mean square** value $\left(\overline{\overline{f}}\right)^2 = (b-a)^{-1}\|f\|^2$.

For oscillating quantities, caused by alternating currents and electromotive forces, we are most interested in their averages. For instance, a current $I(t)$ amperes, produced by an electromotive force of $E(t)$ volts across a resister of R ohms, generates heat at a rate

$$0.24\, I^2 R = 0.24\, E^2/R\ (\texttt{cal/sec}).$$

The number of calories generated during any time interval $a < t < b$ may be expressed in terms of the root mean square values for that interval as

$$0.24\left(\overline{\overline{I}}\right)^2 R(b-a) = 0.24\,\|I(t)\|^2 R = 0.24\left(\overline{\overline{E}}\right)^2 (b-a)/R = 0.24\,\|E(t)\|^2/R.$$

Definition 10.6: Let $\rho(x)$ be a positive function over the interval $|a,b|$. We say that two functions $f(x)$ and $g(x)$ are **orthogonal** on the interval $|a,b|$ (with weight ρ) if

$$(f,g) = \int_a^b \rho(x)\, f(x) g(x)\,\mathrm{d}x = 0.$$

The next definition uses the **Kronecker delta**[97] notation.

Definition 10.7: Let $\phi_1(x)$, $\phi_2(x)$, $\ldots, \phi_n(x)$, $\ldots$ be some set of functions over the interval $|a,b|$. We call such a set of functions an **orthogonal** set if

$$(\phi_i, \phi_j) = \int_a^b \overline{\phi}_i(x)\,\phi_j(x)\,\rho(x)\,\mathrm{d}x = 0, \quad i \ne j.$$

This set is called an **orthonormal** set if

$$(\phi_i, \phi_j) = \int_a^b \overline{\phi}_i(x)\,\phi_j(x)\,\rho(x)\,\mathrm{d}x = \delta_{ij} \equiv \begin{cases} 0 & \text{if } i \ne j, \\ 1 & \text{if } i = j, \end{cases}$$

where δ_{ij} is the Kronecker delta.

Example 10.2.1: The set of linearly independent functions $\phi_n(x) = x^n$ with odd or even powers n is an orthogonal set on any symmetrical interval $(-\ell, \ell)$ with the weight $\rho(x) = x$. Calculations show that

$$(\phi_k, \phi_n) = \int_{-\ell}^{\ell} x^k\, x^n\, x\,\mathrm{d}x = \int_{-\ell}^{\ell} x^{k+n+1}\,\mathrm{d}x = \frac{1}{k+n+2}\, x^{k+n+2}\Big|_{x=-\ell}^{x=\ell} = 0$$

since $k+n+2$ is an even integer. In general, an infinitely differentiable function can be extended in a Taylor series about a point x_0 as $f(x) = \sum_{k=0}^{\infty} \frac{f^{(k)}(x_0)}{k!}\,(x-x_0)^k$, which may contain all powers of $(x-x_0)$. □

[97] Leopold Kronecker (1823–1891), a German mathematician (of Jewish descent) at Berlin University, made important contributions to algebra, group theory, and number theory.

Consider a self-adjoint differential operator (10.1.8), page 549, which we denote for simplicity by L. Upon multiplication of $L[u](x)$ by a function $v(x)$ and integration by parts within an interval $[0,\ell]$, we get

$$\begin{aligned}\int_0^\ell L[u]\,v\,\mathrm{d}x &= \int_0^\ell \left(-(pu')'\,v + quv\right)\mathrm{d}x \\ &= -(pu')'\,v\Big|_{x=0}^{x=\ell} + p(x)u(x)\,v'\Big|_{x=0}^{x=\ell} + \int_0^\ell \left(-u(pv')' + quv\right)\mathrm{d}x \\ &= -p(x)\left[u'(x)\,v(x) - u(x)\,v(x)\right]\Big|_{x=0}^{x=\ell} + \int_0^\ell u\,L[v]\,\mathrm{d}x.\end{aligned}$$

Thus, we obtain the so called Lagrange identity (for brevity, the independent variable is dropped in the left-hand side)

$$\int_0^\ell \left\{L[u]\,v - u\,L[v]\right\}\mathrm{d}x = -p(x)\left[u'(x)\,v(x) - u(x)\,v(x)\right]\Big|_{x=0}^{x=\ell}. \tag{10.2.3}$$

Theorem 10.1: All eigenvalues of the self-adjoint operator are real numbers.

PROOF: Let λ be an eigenvalue of the self-adjoint operator L, and y be the corresponding eigenfunction. Then from the identity $L\,y = \lambda y$, it follows that $(Ly,y) = (\lambda y,y) = \lambda(y,y)$. Since $(y,y) = \|y\|^2 > 0$, and $(Ly,y) = \overline{(y,Ly)} = \overline{(Ly,y)}$ is a real number (because (Ly,y) is equal to its complex conjugate), then

$$\lambda = \frac{(Ly,y)}{(y,y)}$$

is also a real number.

Theorem 10.2: Eigenfunctions $y_n(x)$ and $y_m(x)$ of the Sturm–Liouville problem:

$$(p\,y')' - qy + \lambda\rho y = 0, \qquad \alpha_0 y(0) - \beta_0 y'(0) = 0, \quad \alpha_1 y(\ell) + \beta_1 y'(\ell) = 0, \tag{10.2.4}$$

corresponding to distinct eigenvalues λ_n and λ_m, are orthogonal:

$$\int_0^\ell \rho(x) y_n(x) y_m(x)\,\mathrm{d}x = 0.$$

PROOF: Let $L = L[x,\mathtt{D}] = -\mathtt{D}\left(p(x)\mathtt{D}\right) + q(x)\mathbf{I}$ be the differential operator of the given equation, where $\mathtt{D} = \mathrm{d}/\mathrm{d}x$ and $\mathbf{I}$ is the identity operator. For two eigenfunctions $y_n(x)$ and $y_m(x)$, we have $L[y_n] = \lambda_n\rho y_n$ and $L[y_m] = \lambda_m \rho y_m$. Using Lagrange's identity (10.2.3), we get

$$\begin{aligned}\int_0^\ell \left(L[y_n]y_m - L[y_m]y_n\right)\mathrm{d}x &= (\lambda_n - \lambda_m)\int_0^\ell \rho\, y_n y_m\,\mathrm{d}x \\ &= -p(x)\left[y_n'(x)\,y_m(x) - y_n(x)\,y_m(x)\right]\Big|_{x=0}^{x=\ell}.\end{aligned}$$

From the boundary conditions (10.1.6), it follows that the latter is zero and we have

$$(\lambda_n - \lambda_m)\int_0^\ell \rho(x)\,y_n(x)\,y_m(x)\,\mathrm{d}x = 0.$$

Since $\lambda_n \ne \lambda_m$, we conclude that the eigenfunctions y_n and y_m are orthogonal.

Example 10.2.2: Let us consider a set of linearly independent complex-valued functions:

$$\phi_n(x) = e^{\mathbf{j}n\omega x} = \left(e^{\mathbf{j}\omega x}\right)^n, \qquad n = 0,,\pm 1, \pm 2, \ldots,$$

where ω is a real positive number, x is an arbitrary real number from a finite interval, and $\mathbf{j}$ is the unit vector in the positive vertical direction on the complex plane so that $\mathbf{j}^2 = -1$. Note that the complex variable $z = e^{\mathbf{j}\omega x}$ has unit length independently of x and ω. Each of these complex-valued functions $\phi_n(x)$ of a real variable x repeats itself after passing an interval of length $T = 2\pi/\omega$. It is convenient to denote $T = 2\ell$ and choose the basic interval $[-\ell, \ell]$.

Now we show that these functions are orthogonal on the interval $[-\ell, \ell]$, where $\ell = \frac{\pi}{\omega}$:

$$(\phi_k, \phi_n) = \int_{-\ell}^{\ell} \overline{e^{\mathbf{j}k\omega x}}\, e^{\mathbf{j}n\omega x}\,\mathrm{d}x = \int_{-\ell}^{\ell} e^{\mathbf{j}(n-k)\omega x}\,\mathrm{d}x = \begin{cases} 0, & \text{if } k \ne n, \\ 2\ell, & \text{if } k = n. \end{cases} \tag{10.2.5}$$

Indeed, taking the antiderivative, we get

$$\int_{-\ell}^{\ell} e^{\mathbf{j}(n-k)\omega x}\,\mathrm{d}x = \frac{1}{\mathbf{j}(n-k)\omega}\, e^{\mathbf{j}(n-k)\omega x}\Big|_{x=-\ell}^{x=\ell} = \frac{1}{\mathbf{j}(n-k)\omega}\left(e^{\mathbf{j}(n-k)\omega\ell} - e^{-\mathbf{j}(n-k)\omega\ell}\right).$$

Since $\omega\ell = \pi$, the right-hand side is zero when $n \ne k$. For $k = n$, the identity is obviously true.

Using Euler's formulas

$$e^{\mathbf{j}\theta} = \cos\theta + \mathbf{j}\sin\theta, \quad \cos\theta = \frac{1}{2}\left(e^{\mathbf{j}\theta} + e^{-\mathbf{j}\theta}\right), \quad \sin\theta = \frac{1}{2\mathbf{j}}\left(e^{\mathbf{j}\theta} - e^{-\mathbf{j}\theta}\right), \tag{10.2.6}$$

we obtain from Eq. (10.2.5) the orthogonality relationships for trigonometric functions

$$\int_{-\pi}^{\pi} \cos(k\theta)\,\sin(n\theta)\,\mathrm{d}\theta = \int_{0}^{2\pi} \cos(k\theta)\,\sin(n\theta)\,\mathrm{d}\theta = 0; \tag{10.2.7}$$

$$\int_{-\pi}^{\pi} \cos(k\theta)\,\cos(n\theta)\,\mathrm{d}\theta = \int_{0}^{2\pi} \cos(k\theta)\,\cos(n\theta)\,\mathrm{d}\theta = \begin{cases} 2\pi, & \text{if } k = n = 0, \\ \pi, & \text{if } k = n > 0, \\ 0, & \text{otherwise;} \end{cases} \tag{10.2.8}$$

$$\int_{-\pi}^{\pi} \sin(k\theta)\,\sin(n\theta)\,\mathrm{d}\theta = \int_{0}^{2\pi} \sin(k\theta)\,\sin(n\theta)\,\mathrm{d}\theta = \begin{cases} \pi, & \text{if } k = n > 0, \\ 0, & \text{otherwise.} \end{cases} \tag{10.2.9}$$

The above formulas tell us that the average of the product of two sines, of two cosines, or of a sine and a cosine, of commensurable but numerically unequal frequencies, taken over any interval of length 2π is zero. □

Let $\{\,\phi_k(x)\,\}$, $k = 1, 2, \ldots$, be a set of linearly independent *orthogonal* functions on an interval $|a, b|$, which means that none of the functions $\{\,\phi_k(x)\,\}$ can be expressed as a linear combination of the other functions. In many applications, these functions are eigenfunctions of some linear differential operator. For any square integrable (with weight ρ) function $f(x)$ defined in the interval $|a, b|$, we may calculate the scalar product of $f(x)$ with each function $\phi_k(x)$:

$$c_k = \frac{(f, \phi_k)}{\|\phi_k\|^2} = \frac{1}{\|\phi_k\|^2}\int_a^b f(x)\,\phi_k(x)\,\rho(x)\,\mathrm{d}x, \quad k = 1, 2, \ldots. \tag{10.2.10}$$

These coefficients are called the **Fourier constants** (or coefficients) of $f(x)$ with respect to the set of orthogonal functions $\{\phi_k(x)\}_{k\geqslant 1}$ and the weight function $\rho(x)$. With this picturesque terminology, the representation of a square integrable function $f(x)$ as

$$f(x) = \sum_{k\geqslant 1} c_k\,\phi_k(x) \tag{10.2.11}$$

can be interpreted as an expansion of the given function in terms of the set of orthogonal functions.

The set of functions $\{\phi_k(x)\}_{k\geqslant 1}$ is analogous to a set of n mutually orthogonal vectors in n-vector space, and we may think of the numbers c_1, c_2, …, c_k, … as the scalar components of $f(x)$ relative to this basis. We can identify $f(x)$ with the infinite vector $\mathbf{c} = \langle c_1, c_2, \ldots\rangle$. This correspondence along with uniqueness of the Fourier series representation establishes a one-to-one mapping between a certain set of functions (which usually includes square integrable functions) and a discrete but infinite set of sequences.

We may try to find a finite approximation of the given real-valued square integrable function $f(x)$ as a linear combination of functions $\{\phi_k(x)\}_{k\geqslant 1}$:

$$f(x) \approx \sum_{k=1}^{n} a_k\,\phi_k(x) \qquad (a < x < b). \tag{10.2.12}$$

It is convenient to assume that the set of functions is normalized, that is, $\|\phi_k(x)\| = 1$, $k = 1, 2, \ldots$. We want to determine the coefficients a_k in such a way that the norm of the difference between the function $f(x)$ and its partial sum (10.2.12) over the interval (a, b) is as small as possible:

$$\Delta \stackrel{\text{def}}{=} \left\| f(x) - \sum_{k=1}^{n} a_k\,\phi_k(x) \right\|^2 = \int_a^b \left[f(x) - \sum_{k=1}^{n} a_k\,\phi_k(x) \right]^2 \rho(x)\,\mathrm{d}x.$$

The approximation to be obtained, over the interval (a, b), is thus the best possible in the "least squares" sense. Now, we reduce brackets to obtain

$$\begin{aligned}
\Delta &= \int_a^b f^2(x)\,\rho(x)\,\mathrm{d}x - 2\sum_{k=1}^{n} a_k \int_a^b f(x)\,\phi_k(x)\,\rho(x)\,\mathrm{d}x + \int_a^b \left[\sum_{k=1}^{n} a_k\,\phi_k(x)\right]^2 \rho(x)\,\mathrm{d}x \\
&= \int_a^b f^2(x)\,\rho(x)\,\mathrm{d}x - 2\sum_{k=1}^{n} a_k\,c_k \|\phi_k\|^2 + \sum_{k,j=1}^{n} a_k\,a_j \int_a^b \phi_k(x)\,\phi_j(x)\,\rho(x)\,\mathrm{d}x \\
&= \|f\|^2 - 2\sum_{k=1}^{n} a_k\,c_k + \sum_{k=1}^{n} a_k^2 = \|f\|^2 - \sum_{k=1}^{n} c_k^2 + \sum_{k=1}^{n} (c_k - a_k)^2 \qquad (\|\phi_k\| = 1).
\end{aligned}$$

It is clear that, since f and its Fourier constant c_k are fixed, Δ takes a minimum value when the coefficients a_k are chosen such that

$$a_k = c_k \qquad (k = 1, 2, \ldots).$$

Therefore, the best approximation (10.2.12) in the mean square sense is obtained when a_k is taken as the Fourier constant of $f(x)$ relative to $\phi_k(x)$ over the interval (a, b).

Since Δ is a positive number, we have the relation

$$\|f\|^2 = \int_a^b f^2(x)\,\rho(x)\,\mathrm{d}x \geqslant \sum_{k=1}^{n} c_k^2 \|\phi_k\|^2, \tag{10.2.13}$$

known as **Bessel's inequality**. The geometric meaning of inequality (10.2.13) is that the orthogonal projection of a function f on the linear span of the elements, $\{\phi_k(x)\}$, $k = 1, 2, \ldots, n$, has a norm which does not exceed the norm of f.

Suppose now that a dimension n of the finite orthogonal set $\{\phi_k(x)\}$, $k = 1, 2, \ldots, n$, is increased without a limit. The positive series in the right-hand side of Eq. (10.2.13) must increase with n because it is the sum of positive numbers. Since the series cannot become greater than the fixed number $\|f\|^2$, we conclude that the series $\sum c_k^2$ always converges to some positive number less than or equal to $\int_a^b f^2\,\rho\,\mathrm{d}x = \|f\|^2$.

However, there is no assurance that the limit to which this series converges will actually coincide with this integral. If this is the case for every square integrated function f, we say that the set[98] of functions $\phi_1(x)$, $\phi_2(x)$, $\ldots$, $\phi_n(x)$, $\ldots$, is **complete** (with respect to mean square convergence). Hence, for a complete set of functions, we have

$$\lim_{n\to\infty} \int_a^b \left[f(x) - \sum_{k=1}^{n} c_k\,\phi_k(x) \right]^2 \mathrm{d}x = 0. \tag{10.2.14}$$

Generally speaking, we cannot guarantee that the **Fourier series**

$$\lim_{n\to\infty} \sum_{k=1}^{n} c_k\,\phi_k(x) = \sum_{k=1}^{\infty} c_k\,\phi_k(x)$$

converges pointwise to $f(x)$ for every $x \in (a, b)$. We know only that the mean square error in (a, b) goes to zero, and we say accordingly that if Eq. (10.2.14) is valid, then the series *converges in the mean square* sense to $f(x)$. This is an essentially different type of convergence compared to a pointwise convergence because an infinite series may converge at every point but diverge in the mean square sense. And vice versa, a series may converge in the

[98]The term "complete" was introduced in 1910 by the Russian mathematician Vladimir Andreevich Steklov (1864–1926), a student of Aleksander Lyapunov.

mean square sense to a function that differs from a pointwise limit in a discrete number of points. However, if $f(x)$ and all functions $\phi_k(x)$ are *continuous* over the interval (a, b) and the series $\sum_k c_k\, \phi_k(x)$ converges uniformly in the interval, then the Fourier series (10.2.14) pointwise converges everywhere in (a, b).

Theorem 10.3: Every function from the domain of a self-adjoint differential operator can be expanded into uniformly convergent Fourier series (10.2.14) over the set of its eigenfunctions.

The proof of this theorem is based on reduction of the problem to an integral equation. Then using the Hilbert–Schmidt theorem, a function is expanded into series over eigenfunctions. See details in a course on integral equations, for instance, [40]. ■

Recall that the domain of a linear differential operator of order n consists of all functions that have continuous derivatives up to the order n and satisfy the boundary conditions that generate the linear operator L. Therefore, Theorem 10.3 refers to such functions. In particular, if L is an operator (10.1.7) of the second order and f is a function having two continuous derivatives subject to the corresponding boundary conditions (10.1.6), then the Fourier series converges uniformly to $f(x)$.

Finally, we turn our attention to the Sturm–Liouville problems (10.1.7), (10.1.6) generated by the self-adjoint differential operator $L[x, \mathtt{D}] = -\mathtt{D}\,(p(x)\,\mathtt{D}) + q(x)$ on a finite interval $[0, \ell]$ subject to traditional boundary conditions (10.1.6). This problem is usually referred to as a regular Sturm–Liouville problem when $p(x) > 0$ and $\rho(x) > 0$.

Theorem 10.4: Any regular Sturm–Liouville problem has an infinite sequence of real eigenvalues $\lambda_0 < \lambda_1 < \lambda_2 < \cdots$ with $\lim_{n\to\infty} \lambda_n = \infty$. The eigenfunctions $\phi_n(x)$ are uniquely determined up to a constant factor.

Next statement assures us that any square integrable function can be expanded into Fourier series over eigenfunctions generated by a regular Sturm–Liouville problem and this series converges in mean square sense.

Theorem 10.5: The set of eigenfunctions of any regular Sturm–Liouville problem is complete in the space of square integrable continuous functions, on the interval $0 \leqslant x \leqslant \ell$ relative to the wight function $\rho(x)$.

Problems

1. This problem consists of two parts. In the first one, you are asked to show that pointwise convergence does not apply to a mean square convergence. In the second part, you are asked to show that a sequence may converge pointwise, but diverge in the mean square sense.
   (a) Let $f_n(x) = x^n$ be a sequence of functions on the unit interval $[0, 1]$, which converges pointwise to $f(x) \equiv 0$ on the semi-open interval $[0, 1)$ and to 1 at $x = 1$. Show that the sequence $f_n(x)$ converges in the mean square sense to 0 on $[0, 1]$.
   (b) Consider a sequence of functions $s_n(x) = n\, e^{-nx}$ on the unit interval $0 \leqslant\leqslant 1$. Show that $s_n(x) \to 0$ as $n \to \infty$ for each x in $(0, 1]$. On the other hand, show that
$$\text{mean error} = \int_0^1 s_n^2(x)\,\mathrm{d}x = \frac{n}{2}\left(1 - e^{-2n}\right) \to \infty.$$
2. Find the square norm for each of the following functions on the interval $-2 < x < 2$:
$$\textbf{(a)}\ 4; \qquad \textbf{(b)}\ \cos\frac{x}{2}; \qquad \textbf{(c)}\ \cosh x; \qquad \textbf{(d)}\ 2x; \qquad \textbf{(e)}\ x^2.$$
3. Find the square norm for each function of the previous problem on the intervals $0 < x < 2$ and $-2 < x < 0$.
4. Find the norm and the root mean square value for each of the following functions on the interval indicated
$$\textbf{(a)}\ \cos t; \quad 0 < t < \pi/2; \qquad \textbf{(b)}\ 1/t, \quad 2 < t < 6; \qquad \textbf{(c)}\ t^4, \quad 0 < t < 1.$$
5. Find the norm and the root mean square value of the product $\cos x\, \sin x$ on the interval $0 < x < \pi/2$ and $-\pi/2 < x < \pi/2$.
6. What multiple of $\sin x$ is closest in the least square sense to $\sin^3 x$ on the interval $(0, \pi)$?
7. Find the norm and the root mean square value of the function $f(x) = \begin{cases} 2x, & \text{if } 0 < x < 1, \\ 3 - x, & \text{if } 1 < x < 3; \end{cases}$ over the interval $0 < x < 3$.

10.3 Fourier Series

How can a string vibrate with a number of different frequencies at the same time? The answer for this question was given by Jean Baptiste Joseph Fourier[99] (1768–1830), who made a claim in 1807 that a periodic wave can be decomposed into a (usually infinite) sum of sines and cosines.

10.3.1 Music as Motivation

People have enjoyed music since their appearance on the earth—it is one of the oldest pleasurable human activities. Reproducing music and transferring it from generation to generation and from one nation to another requires a special universal language. This language is not our objective; however, we try to explain music using mathematics. The partnership between mathematics and music traces back at least 2500 years when the ancient Greeks established a connection between the two. The Pythagoreans tried to explain the pleasing harmonics of some sounds and the distracting effects of others—you can observe this phenomena by running a long chalk bar over a black board. We use music as a motivating example for clarifying the topic in this section.

The sounds we hear arise from vibrations in air pressure. Humans don't hear all sounds. For example, we can't hear the sound a dog whistle makes, but dogs can hear that sound. Marine animals can often hear sounds in much larger frequency range than humans. What sound vibrations we can hear depends on the frequency and intensity of the air oscillations and one's individual hearing sensitivity. Frequency is the rate of repetition of a regular event. The number of cycles of a wave per second is expressed in units of hertz[100] (`Hz`). Intensity is the average amount of sound power (sound energy per unit time) transmitted through a unit area in a specified direction; therefore, the unit of intensity is watts per square meter. The sound intensity that scientists measure is not the same as loudness. Loudness describes how people perceive sound. Humans can hear sounds at frequencies from about 20 `Hz` to 20,000 `Hz`, though we hear sounds best at around 3,000 to 4,000 `Hz`, where human speech is centered. Scientists often specify sound intensity as a ratio, in decibels[101] (written as `dB`), which is defined as 10 times the logarithm of the ratio of the intensity of a sound wave to a reference intensity.

When we detect sounds, or noise, our body is changing the energy in sound waves into nerve impulses which the brain then interprets. Such transmission occurs in human ears where sound waves cause the eardrum to vibrate. The vibrations pass through 3 connected bones in the middle ear, which causes fluid motion in the inner ear. Moving fluid bends thousands of delicate hair-like cells which convert the vibrations into nerve impulses that are carried to the brain by the auditory nerve; however, how the brain converts these electrical signals into what we "hear" is still unknown.

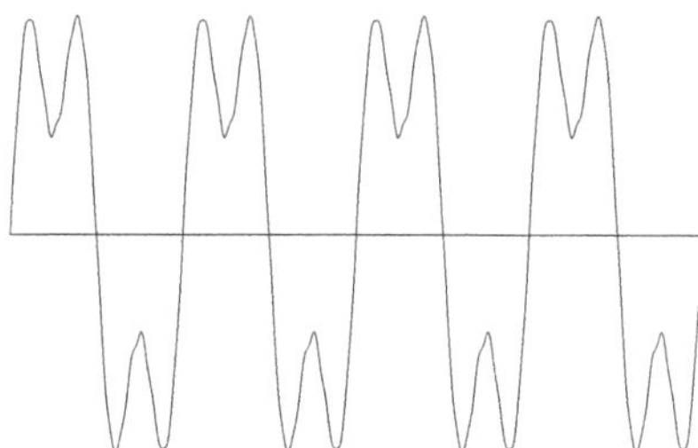

Figure 10.4: The sound of a clarinet.

Sounds consist of vibrations of the air caused by its compression and decompression. Air is a gas containing atoms or molecules that can move freely in an arbitrary direction. The average velocity of air molecules at room temperature, under normal conditions, is around 450–500 meters per second. The mean free paths of air molecules before they collide with other air molecules is about 6×10^{-8} meters. So air consists of a large number of molecules in close proximity that collide with each other on average 10^{10} times per second, which is perceived as air pressure.

[99] Joseph Fourier was born in Auxerre (France), the ninth child of a master tailor. The name "Fourier" is a variant on the word *fourrier*, which means in a military sense a quartermaster and in a figurative connotation a precursor or anticipator of ideas. More often than not during his life Fourier was to have his name spelled in that way. He took active role in French revolution, and more than once his life was in danger. In 1798, Fourier accompanied Napoleon on his expedition to Egypt, and was Prefect (Governor) of Grenoble twice.

[100] The word "hertz" is named for a German physicist Heinrich Rudolf Hertz (1857–1894), who was the first to conclusively prove the existence of electromagnetic waves.

[101] The unit for intensity is the `bel`, named in honor of the eminent British scientist Alexander Graham Bell (1847–1922), the inventor of the telephone.

When an object vibrates, it causes waves of increased and decreased pressure in the air that travel at about 340 meters per second. Figure 10.4 represents variation of air pressure on the vertical scale produced by a clarinet with time along the horizontal axis. The greater the variation in the vertical direction, the louder the sound.

A musical tone (pl. *tonoi*) is a steady periodic sound, which is often used in conjunction with pitch. A simple tone, or pure tone, has a sinusoidal waveform. Such a tone is identified by its frequency and magnitude, while its corresponding pitch is a subjective psychoacoustical attribute of sound; however, tone is commonly used by musicians as a synonym for pitch. From a mathematical point of view, a tone is a sine (or cosine) function $\sin(2\pi\nu t)$, where ν is the frequency of the pitch. In theory, it is often called the mode[102] (from the Latin *modus*, which means measure, standard, or size).

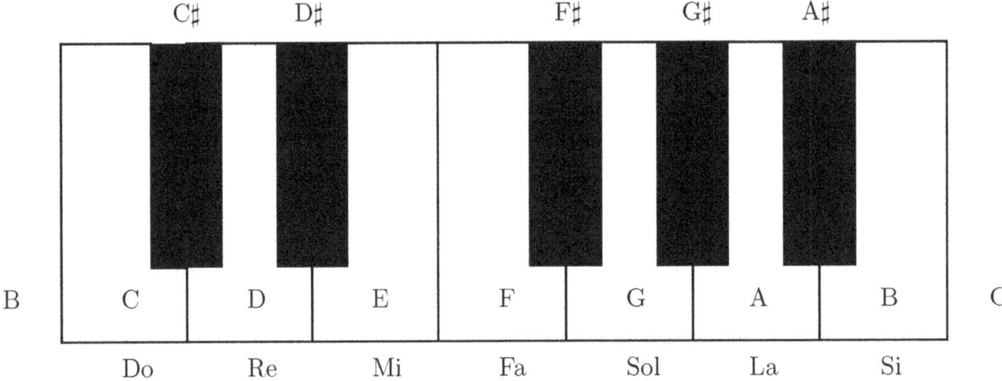

To represent a pitch class and duration of a musical sound, musicians use special symbols, called notes. Many years ago, it was discovered that doubling the frequency of a note results in a higher note that, in some sense, people perceive to be the same as the original. In musical terminology, the note with double the frequency of another note lies one octave higher and is called an overtone. Likewise, descending by an octave—halving the frequency—produces a lower note that is one octave below the original note. All such notes are called the pitch class.

One can easily recognize notes of the same pitch class on a classical 88-key piano that has seven octaves accompanied by a minor third. A 12-key pattern with seven white keys and five black keys repeats up and down the piano. The lowest note marked A_0 corresponds to a frequency of 27.5 Hz. The highest note is C_8 with a frequency of about 4186.01 Hz. The interval between two successive piano keys is called a semitone. Therefore, an octave includes 12 semitones. A sharp ♯ raises a note by a semitone or half-step, and a flat ♭ lowers it by the same amount. In modern tuning a half-step has a frequency ratio of $\sqrt[12]{2} = 2^{1/12}$, approximately 1.059. The accidentals are written after the note name: so, for example, F♯ represents F-sharp, which is the same as G♭, G-flat. In European terminology, they are called Fa-diesis or Sol-bemolle.

In traditional music theory within the English-speaking world, pitch classes are typically represented by the first seven letters of the Latin alphabet (A, B, C, D, E, F, and G). Many countries in Europe and most in Latin America, however, use the naming convention Do-Re-Mi-Fa-Sol-La-Si-Do. The eighth note, or octave, is given the same name as the first, but has double its frequency. These names follow the original names reputedly given[103] by Guido d'Arezzo.

"Concert pitch" is the universal pitch to which all instruments in a concert setting are tuned, so that they all produce the same frequency corresponding to the middle C or Do (although technically, the A or La above middle C is used as the compass, and has a frequency of 440 Hz). Middle C's frequency is $220 \times 2^{1/4} \approx 261.626$ Hz. The middle C, which is designated C_4 in scientific pitch notation, has key number $n = 40$. On a piano keyboard, the frequency of the n-th note is $2^{(n-49)/12} \times 440$ Hz.

Overtones are often referred to as harmonics and are higher in frequency. Therefore, they can be heard distinctly from other tones played at the same time. Tones of the lowest frequency of periodic waveforms, referred to as fundamental frequencies or tones, generate harmonics. Pythagoreans also discovered that there exist other pairs of notes (called chords) that sound pleasant to human ears. One of them, called the *fifth*, is a musical interval encompassing five staff positions, and the *perfect fifth* (often abbreviated P5) is a fifth spanning seven semitones, for instance, C and G (or D and A). If the first note in the fifth has frequency f, then the next note has the frequency approximately of $3f/2$. This note together with the note of frequency $2f$ (one octave above) constitute the fourth;

[102] A Roman philosopher Anicius Boethius (480–524) used the term *modus* to translate the Greek *tonos*, or key, which is the source of our word "tone."

[103] Guido of Arezzo (also Guido Aretinus, Guido da Arezzo, Guido Monaco, or Guido d'Arezzo) (991/992–after 1033) was a music theorist of the Medieval era.

they are also consonant. For example, G and the C located an octave up is a fourth. Its frequency ratio is 4/3 because $\frac{3}{2} \cdot \frac{4}{3} = 2$. Therefore, the octave is divided into a fifth comprised by the fourth.

Not every chord results in as pleasant a harmony as an octave, which has a simple frequency ratio of 2, or a fifth (with a ratio about 3/2). Some note pairs make us cringe. As an example, let us look at a chord to find out why these notes sound good together:

C	261.626 Hz
E	329.628 Hz
G	391.995 Hz

The ratio of E to C is about 5/4. This means that every 5th wave of E matches up with every 4th wave of C. The ratio of G to E is about 5/4 as well. The ratio of G to C is about 3/2. Since every note's frequency matches up well with every other note's frequencies (at regular intervals), they all sound good together!

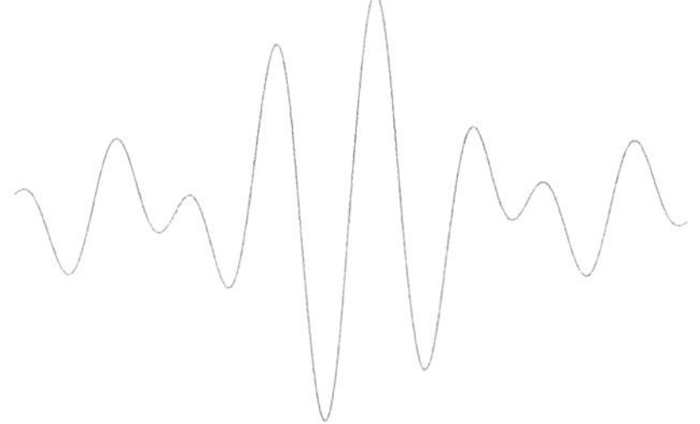

Figure 10.5: The chord C-E-G.

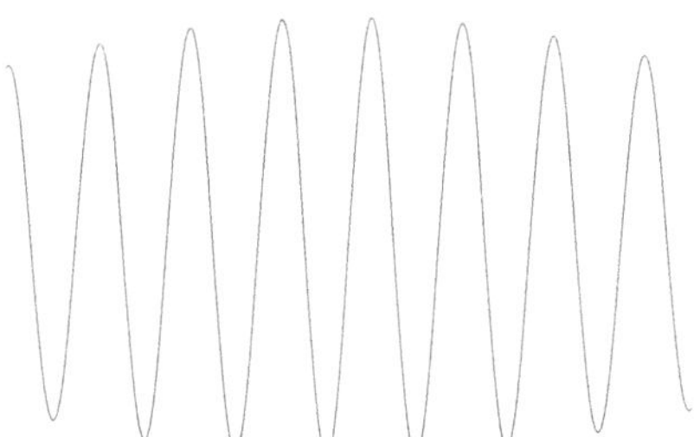

Figure 10.6: The chord C-C♯.

Now consider two chromatically adjacent notes C and C♯ (of frequency 277.183 Hz), which make up the smallest musical interval, known as a half-tone interval. Played alone, they produce an unpleasant, clashing sound (see Fig. 10.6).

These examples show that musical sounds or notes can be interpreted as a combination of pure tones of sine wavefronts. In reality, any actual sound is not exactly periodic because periodic functions have no starting point and no end point. Nevertheless, within some time interval, almost every musical sound exhibits periodic movements. The next question to address is whether any sound can be decomposed into a linear combination of pure tones? The affirmative answer was given by Joseph Fourier, and is presented in §10.3.3.

10.3.2 Sturm–Liouville Periodic Problem

To understand sound waves and their wavefronts, we have to create mental pictures of them because they cannot be seen. As we saw previously, a sound can be modeled by a composition of sine and cosine functions. What makes these trigonometric functions so special? Could we decompose sound waves using other families of periodic functions? The answer lies in the differential equation for a simple harmonic motion

$$m\,\ddot{y} = -k\,y,$$

when a particle of mass m is subject to an elastic force moving it toward its equilibrium position $y = 0$. Here $\ddot{y} = \mathrm{d}^2y/\mathrm{d}t^2$ is the acceleration of the particle and k is the constant of proportionality.

Recall that a function defined for all real values of its argument is said to be **periodic** if there exists a positive number $T > 0$ such that

$$f(t+T) = f(t) \quad \text{for every value of } t. \tag{10.3.1}$$

It is obvious that if, for some T, Eq. (10.3.1) is valid, then it is true for $2T$, $3T$, and so on. The smallest positive value of T for which Eq. (10.3.1) holds is called the **fundamental period** or simply the **period**.

In the case of the human ear, the harmonic motion equation $m\,\ddot{y} + k\,y = 0$ may be taken as a close approximation to the equation of motion of a particular point on the basilar membrane, or anywhere else along the chain of transmission between the outside air and the cochlea, which separates out sound into various frequency components. Of course, harmonic motion is an approximation that does not take into account damping, nonlinearity of the restoring force, or the membrane equation, which is actually a partial differential equation. Nevertheless, since harmonic motion gives a reasonable approximation to actual phenomena, we can start with it.

Setting $\lambda = k/m$, we consider the Sturm–Liouville problem with the following periodic conditions:

$$\ddot{y} + \lambda y = 0, \qquad y(t) = y(t + 2\ell) = y(t + 2n\ell),$$

where $T = 2\ell$ is the period and n is any integer. Obviously, $\lambda = 0$ is an eigenvalue, to which corresponds the constant eigenfunction. When $\lambda < 0$, the general solution of the equation $\ddot{y} + \lambda y = 0$ is a linear combination of exponential functions that are not periodic. Therefore, negative λ cannot be an eigenvalue. Assuming that $\lambda > 0$, we find the general solution of the equation $y'' + \lambda y = 0$ to be

$$y(t) = a\,\cos\left(t\sqrt{\lambda}\right) + b\,\sin\left(t\sqrt{\lambda}\right) = c\,\sin\left(t\,\sqrt{\lambda} + \varphi\right),$$

where a, b or c, φ are arbitrary constants ($c = \sqrt{a^2 + b^2}$ is called the peak amplitude and the φ phase). This function is periodic with period 2ℓ if $\sqrt{\lambda} = 2\pi n/(2\ell)$, which leads to the sequence of eigenvalues $\lambda_n = (n\pi/\ell)^2$ and corresponding eigenfunctions

$$y_n(t) = a_n \cos\left(\frac{n\pi t}{\ell}\right) + b_n \sin\left(\frac{n\pi t}{\ell}\right), \qquad n = 0, 1, 2, \ldots.$$

If n is a positive integer, $\omega = (2\pi)/(2\ell) = \pi/\ell$ is a positive number, and a_n, b_n are real constants, the eigenfunction

$$a_n \cos(n\omega t) + b_n \sin(n\omega t) = \sqrt{a_n^2 + b_n^2}\,\sin\left(n\omega t + \varphi_n\right)$$

represents an oscillation of frequency $\nu = n\omega/(2\pi)$. The period of each oscillation is the reciprocal of the frequency, $\nu^{-1} = 2\pi/(n\omega)$. Hence, in a time interval of length $\tau = 2\pi/\omega$, the function $y_n(t)$ completes n oscillations.

10.3.3 Fourier Series

In 1822, the French mathematician/physicist/engineer/politician Joseph Fourier published a book ("Théorie analytique de la chaleur," which is translated as "The analytical theory of heat") in which he summarized his research, including an astonishing discovery that virtually every periodic function could be represented by an infinite series of elementary trigonometric functions: sines and cosines. Fourier's claim was so remarkable and counter intuitive that most of the leading mathematicians of that time did not believe him. Now we enjoy applications of his discovery: Fourier analysis lies at the heart of signal processing, including audio, speech, images, videos, seismic data, radio transmissions, and so on.

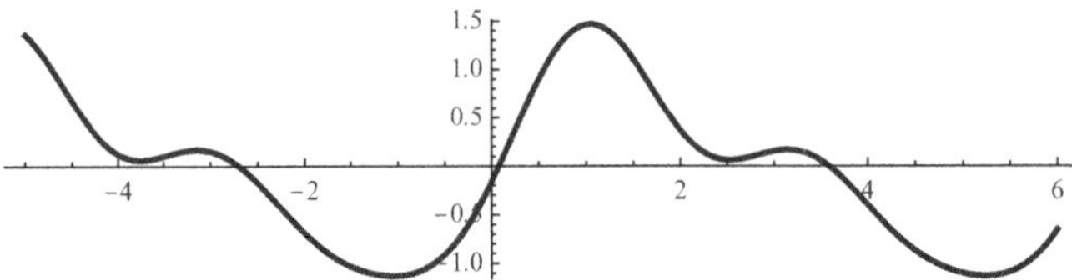

Figure 10.7: The graph of the sum of three eigenfunctions.

It is obvious that the sum of two sinusoidal functions with the same frequency, $a_n \cos(n\omega x) + b_n \sin(n\omega x)$ and $A_n \cos(n\omega x) + B_n \sin(n\omega x)$, is again an eigenfunction with the same frequency. On the other hand, the sum of two or more sinusoidal eigenfunctions with different frequencies is a periodic function, but it is not an eigenfunction. For example, the sum of three eigenfunctions

$$\sin x + \frac{1}{2}\sin 2x - \frac{1}{6}\cos 3x$$

is a periodic function, which graph has little resemblance to a sinusoidal function (see Fig. 10.7). It is natural to ask an opposite question: is it possible to represent an arbitrary periodic function as a linear combination of a finite or infinite number of such oscillations? For a wide set of functions, an affirmative answer for this question will be given in the next section. Now we start with a finite sum of eigenfunctions:

$$S_N(x) = \sum_{n=0}^{N} \left(a_n \cos(n\omega x) + b_n \sin(n\omega x)\right), \qquad \omega = \frac{\pi}{\ell},$$

that defines a **trigonometric polynomial** (of order N) that repeats itself after each interval of length $\tau = 2\pi/\omega = 2\ell$, that is, $S_N(x+\tau) = S_N(x)$. Therefore, τ is the fundamental period for $S_N(x)$ independent of the number of terms, N. Sometimes, it is convenient to introduce a new variable $t = \omega x = \pi x/\ell$, so the function $S_N(x) = S_N(t\ell/\pi)$ becomes a finite sum of simple harmonics with the standard period 2π in variable t.

Suppose that we are given a periodic function $f(x)$ of period $T = 2\ell$, which we approximate with a trigonometric polynomial $S_N(x)$. As the number N of terms in the finite sum $S_N(x)$ increases without bounds, the limit function, if it exists, defines the infinite series. But it may be neither differentiable nor continuous. The limit is called the **Fourier series** of $f(x)$:

$$f(x) \sim \frac{a_0}{2} + \sum_{k=1}^{\infty} a_k \cos\left(\frac{k\pi x}{\ell}\right) + b_k \sin\left(\frac{k\pi x}{\ell}\right). \tag{10.3.2}$$

The first coefficient $a_0/2$ in this sum is written in this specific form for convenience that will be explained shortly. When we do not know whether the infinite sum (10.3.2) converges or diverges, or we are not sure whether it is equal to the given function at all points or not, we prefer to use the symbol "$\sim$" instead of "=." This series (10.3.2) can be simplified further for a function $F(t) = f(t\ell/\pi)$ with the fundamental period 2π:

$$F(t) \sim \frac{a_0}{2} + \sum_{k=1}^{\infty} (a_k \cos kt + b_k \sin kt). \tag{10.3.3}$$

To find the values of coefficients in these series (10.3.2) and (10.3.3), we first assume that they converge uniformly to the given functions. Then the symbol "$\sim$" in Eqs. (10.3.2) and (10.3.3) should be replaced with an equal sign. Multiplying both sides of the relation (10.3.3) by $\sin mt$ for some integer m, and integrating the results over an interval $(-\pi, \pi)$, we get

$$\int_{-\pi}^{\pi} F(t) \sin mt\,\mathrm{d}t = \frac{a_0}{2}\int_{-\pi}^{\pi} \sin mt\,\mathrm{d}t + \sum_{k=1}^{\infty}\int_{-\pi}^{\pi} (a_k \cos kt + b_k \sin kt)\sin mt\,\mathrm{d}t.$$

Here we interchanged the order of integration and summation, which is justified for uniformly convergent series. All integrals in the right-hand side are zeroes because of orthogonality formulas (10.2.7)–(10.2.9) except $k = m$. Repeating this procedure for $\cos mt$ (involving multiplication by $\cos mt$ and integration), we finally obtain

$$\begin{aligned} a_k &= \frac{1}{\pi}\int_{-\pi}^{\pi} F(t)\cos(kt)\,\mathrm{d}t, \quad k = 0, 1, 2, \ldots, \\ b_k &= \frac{1}{\pi}\int_{-\pi}^{\pi} F(t)\sin(kt)\,\mathrm{d}t, \quad k = 1, 2, 3, \ldots. \end{aligned} \tag{10.3.4}$$

The coefficients a_k, b_k $(k = 1, 2, \ldots)$ are twice averages of $F(t)$ times the trigonometric functions over the interval $[-\pi, \pi]$. Similarly, for a periodic function $f(x)$ of period $T = 2\ell$, we have

$$\begin{aligned} a_k &= \frac{2}{T}\int_{-T/2}^{T/2} f(x)\cos\left(\frac{2k\pi x}{T}\right)\mathrm{d}x, \quad k = 0, 1, 2, \ldots, \\ b_k &= \frac{2}{T}\int_{0}^{T} f(x)\sin\left(\frac{2k\pi x}{T}\right)\mathrm{d}x, \quad k = 1, 2, 3, \ldots. \end{aligned} \tag{10.3.5}$$

The formulas (10.3.4) and (10.3.5) were discovered by L. Euler in the second part of the eighteenth century, and independently by J. Fourier at the beginning of nineteenth century. Therefore, they are usually referred to as the **Euler–Fourier formulas**. The numbers $a_0, a_1, \ldots, b_1, b_2, \ldots$ are the **Fourier coefficients** of $f(x)$ on $[-\ell, \ell]$ with respect to the set of trigonometric functions $\left\{\cos\left(\frac{k\pi x}{\ell}\right), \sin\left(\frac{k\pi x}{\ell}\right)\right\}_{k\geqslant 0}$. Now it is clear why we have chosen the free coefficient ($a_0/2$ is the average value of the corresponding function) in such a form—it simplifies and unifies the formulas (10.3.4) and (10.3.5). It is possible to show (see Problem 11) that for a periodic function $f(x)$, the Fourier coefficients can be evaluated over any interval of length $T = 2\ell$.

Since the trigonometric functions in expansion (10.3.2) are periodic, the series (10.3.2) thereby defines a periodic function with period $T = 2\ell$. If originally we are given a function $f(x)$ on an interval of length T, it is convenient to expand $f(x)$ from this interval by making it periodic with the period $T = 2\ell$, that is, we set $f(x+T) = f(x)$. With this approach, we expect that the trigonometric series (10.3.2) represents extended periodically function for all real x. There is nothing special that initially a function $f(x)$ is required to be defined in the symmetrical interval $(-\ell, \ell)$ of length $T = 2\ell$. An arbitrary interval (a, b) of length $T = b - a$ can be used instead. This will affect the limits of integrations in Eq. (10.3.5) but not the values of coefficients. The Fourier series (10.3.2) defines a periodic extension of the given function from the finite interval. The following examples clarify the topic.

Example 10.3.1: Let us consider the function $f(t) = -t/2$ on the interval $(-\pi, \pi)$. Expanding it periodically with the period 2π (so $\ell = \pi$), we obtain a sawtooth function (see Fig. 10.8). Other sawtooth functions are discussed in Exercise 2, page 568. From the Euler–Fourier formulas (10.3.4), it follows

$$
\begin{aligned}
a_0 &= \frac{1}{\pi}\int_{-\pi}^{\pi} \frac{-t}{2}\,\mathrm{d}t = -\left.\frac{t^2}{4\pi}\right|_{t=\pi}^{\pi} = 0, \\
a_k &= \frac{1}{\pi}\int_{-\pi}^{\pi} \frac{-t}{2}\cos kt\,\mathrm{d}t = -\frac{1}{2\pi k^2}\left.(tk\,\sin kt + \cos kt)\right|_{t=-\pi}^{t=\pi} = 0.
\end{aligned}
$$

Similarly,

$$
\begin{aligned}
b_k &= \frac{1}{\pi}\int_{-\pi}^{\pi} \frac{-t}{2}\sin kt\,\mathrm{d}t = \frac{1}{2\pi k^2}\left.(kt\,\cos kt - \sin kt)\right|_{t=-\pi}^{t=\pi} \\
&= \frac{k\pi\,\cos k\pi - \sin k\pi}{k^2\pi} = \frac{(-1)^k}{k}.
\end{aligned}
$$

Therefore,

$$
-\frac{t}{2} = \sum_{k=1}^{\infty} \frac{(-1)^k}{k}\sin kt, \qquad -\pi < t < \pi.
$$

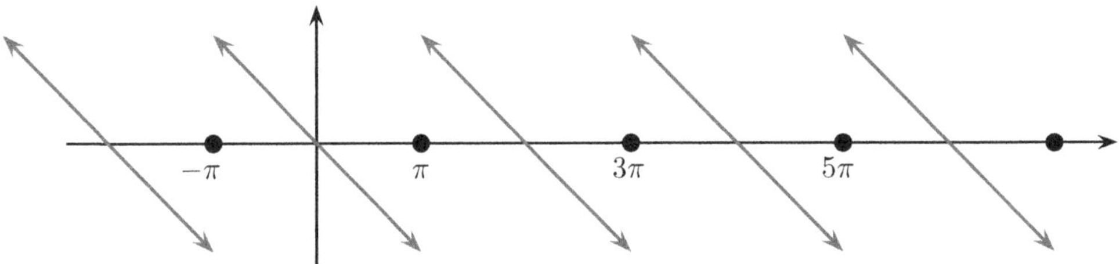

Figure 10.8: Example 10.3.1, graph of the function $f(t) = -t/2$ extended periodically from the interval $(-\pi, \pi)$.

The series converges to 0 at the points of discontinuity ($t = 2n\pi + \pi$, $n = 0, \pm 1, \pm 2, \ldots$). The graph of its partial sum, $\sum_{k=1}^{N} \frac{(-1)^k}{k}\sin kt$ with $N = 45$ terms, is presented in the figure on the front page 545 of Chapter 10.

If we consider the same function $f(t) = -t/2$ on the interval $[0, 2\pi]$, we will get the following Fourier series:

$$
-\frac{t}{2} = -\frac{\pi}{2} + \sum_{k=1}^{\infty} \frac{1}{k}\sin kt, \qquad 0 < t < 2\pi.
$$

To estimate the quality of partial sum approximation (see Fig. 10.9), we calculate the mean square errors with $N = 15$ and $N = 50$ terms:

$$
\Delta_N = \int_0^{2\pi} \left(\frac{\pi - t}{2} - \sum_{k=1}^{N} \frac{1}{k}\sin kt\right)^2 \mathrm{d}t,
$$

which gives $\Delta_{15} \approx 0.202613$ and $\Delta_{50} \approx 0.0622077$.

Example 10.3.2: Now we find the Fourier series for the piecewise continuous function on the interval $|-2, 2|$:

$$
g(x) = \begin{cases} -x, & -2 < x < 0, \\ 1, & 0 < x < 2. \end{cases}
$$

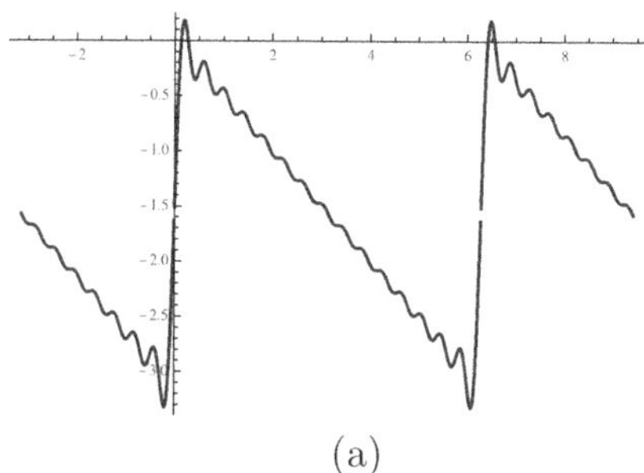

(a)

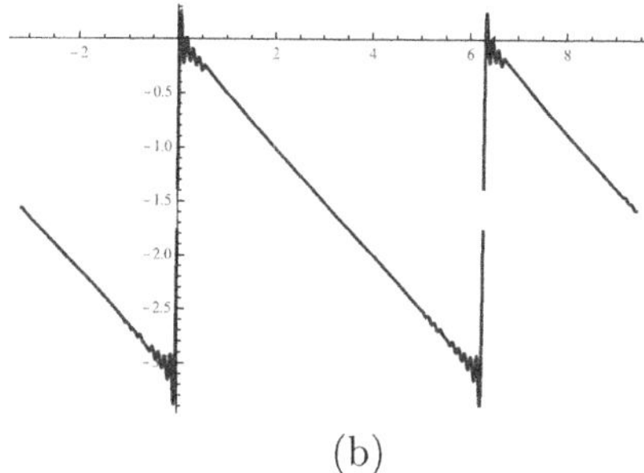

(b)

Figure 10.9: Example 10.3.1. (a) Fourier approximation of the periodic extension of the function $f(t) = -t/2$ defined initially on the interval $(0, 2\pi)$ with $N = 15$ terms and (b) the same approximation with $N = 50$ terms.

Using the Euler–Fourier formulas (10.3.5) with $\ell = 2$, we obtain

$$
\begin{aligned}
a_0 &= \frac{1}{2}\int_0^2 \mathrm{d}x - \frac{1}{2}\int_{-2}^0 x\,\mathrm{d}x = 2,\\
a_k &= \frac{1}{2}\int_0^2 \cos\left(\frac{k\pi x}{2}\right)\mathrm{d}x - \frac{1}{2}\int_{-2}^0 x\cos\left(\frac{k\pi x}{2}\right)\mathrm{d}x = \frac{2}{k^2\pi^2}\left(\cos(k\pi) - 1\right),\\
b_k &= \frac{1}{2}\int_0^2 \sin\left(\frac{k\pi x}{2}\right)\mathrm{d}x - \frac{1}{2}\int_{-2}^0 x\sin\left(\frac{k\pi x}{2}\right)\mathrm{d}x = \frac{1}{k\pi}\left(\cos(k\pi) + 1\right).
\end{aligned}
$$

Recall that $\cos(k\pi) = (-1)^k$, for integer values of k, so $\cos(k\pi) - 1 = (-1)^k - 1 = -2$ for odd k, and 0 for all other values of k. Similarly, $\cos(k\pi) + 1 = (-1)^k + 1 = 2$ for even k, and 0 otherwise. Thus, we can simplify the expressions for the Fourier coefficients:

$$
a_k = \begin{cases} 0, & \text{if } k \text{ is even},\\ -\frac{4}{k^2\pi^2}, & \text{if } k = 2n-1 \text{ is odd}; \end{cases} \qquad b_k = \begin{cases} \frac{2}{k\pi}, & \text{if } k = 2n \text{ is even},\\ 0, & \text{if } k \text{ is odd}. \end{cases}
$$

Substituting these coefficients into the Fourier series and separating even and odd indices, we get

$$
g(x) = 1 + \sum_{n\geqslant 1} \frac{1}{n\pi}\sin(n\pi x) - \frac{4}{\pi^2}\sum_{n\geqslant 1}\frac{1}{(2n-1)^2}\cos\left(\frac{(2n-1)\pi x}{2}\right).
$$

These sums can be evaluated explicitly:

$$
g_1(x) \stackrel{\text{def}}{=} 1 + \sum_{n\geqslant 1}\frac{1}{n\pi}\sin(n\pi x) = \begin{cases} \frac{1-x}{2}, & -2 < x < 0,\\ 1 + \frac{1-x}{2}, & 0 < x < 2; \end{cases}
$$

$$
g_2(x) \stackrel{\text{def}}{=} -\frac{4}{\pi^2}\sum_{n\geqslant 1}\frac{1}{(2n-1)^2}\cos\left(\frac{(2n-1)\pi x}{2}\right) = \begin{cases} -\frac{1+x}{2}, & -2 < x < 0,\\ \frac{x-1}{2}, & 0 < x < 2. \end{cases}
$$

This allows us to represent $g(x)$ as $g(x) = g_1(x) + g_2(x)$. The partial sums $1 + \sum_{n=1}^{N}\frac{1}{n\pi}\sin(n\pi x)$ converge to $g_1(x)$ at every point where the function is continuous (except points $2k\pi$, $k = 0, \pm 1, \pm 2, \ldots$). The series for $g_2(x)$ defines a continuous function on the whole line. The partial sums can be plotted using the following *Maple* commands:

```
N:=15:  fN:= x→ 1+sum((n*Pi)^(-1)*sin(n*Pi*x),n=1..N) -
4*Pi^(-2)*sum((2*n-1)^(-2)*cos((2*n-1)*Pi*x/2),n=1..N)
plot(fN(x),x=-2..2, color=blue)
```

We choose $N = 15$, the number of terms in the partial sums, for simplicity. The graph of the corresponding truncated Fourier series for the function $g(x)$ is presented in Fig. 10.10(a).

Example 10.3.3: Let us consider a continuous function on the interval $(-1, 3)$:

$$
h(x) = \begin{cases} x, & -1 < x \leqslant 1,\\ 1, & 1 \leqslant x \leqslant 2,\\ -x+3 & 2 \leqslant x < 3. \end{cases}
$$

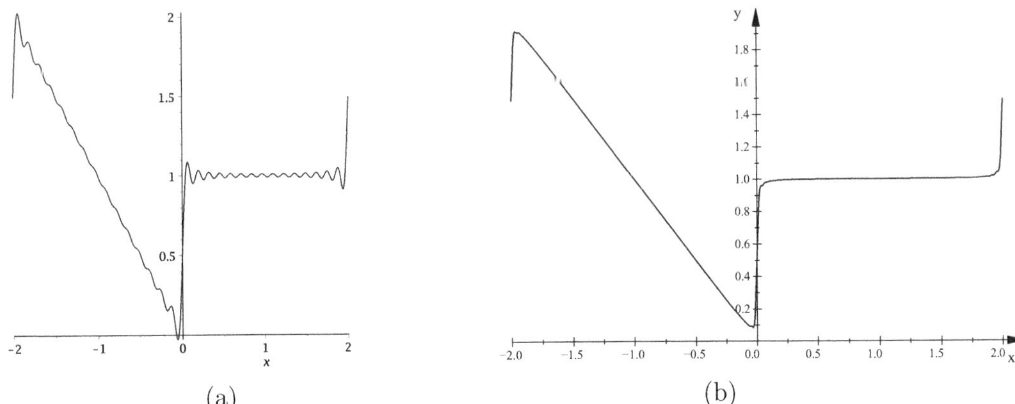

Figure 10.10: Example 10.3.2. (a) Fourier approximation of the periodic extension of the function $f(x)$ and (b) the corresponding Cesàro approximation, plotted with *Maple*.

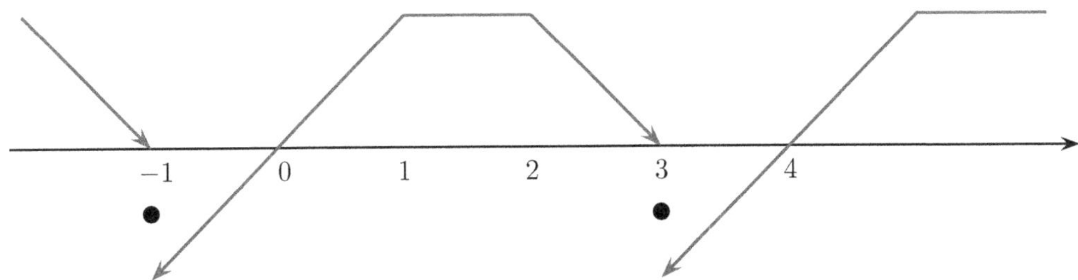

Figure 10.11: Example 10.3.3, graph of the function $h(x)$.

If we extend it periodically $h(x) = h(x+4)$, then its extension becomes a discontinuous function on the whole line $(-\infty, \infty)$, see Fig. 10.11. Since the given interval has length $T = 4 = 2\ell$, we use the Euler–Fourier formulas (10.3.5) with $\ell = 2$:

$$\begin{aligned}
a_0 &= \frac{1}{2}\int_{-1}^{1} x\,\mathrm{d}x + \frac{1}{2}\int_{1}^{2} 1\,\mathrm{d}x + \frac{1}{2}\int_{2}^{3} (3-x)\,\mathrm{d}x = 0 + \frac{1}{2} + \frac{1}{4} = \frac{3}{4}, \\
a_n &= \frac{1}{2}\int_{-1}^{1} x\,\cos\frac{n\pi x}{2}\,\mathrm{d}x + \frac{1}{2}\int_{1}^{2} 1\,\cos\frac{n\pi x}{2}\,\mathrm{d}x + \frac{1}{2}\int_{2}^{3} (3-x)\,\cos\frac{n\pi x}{2}\,\mathrm{d}x \\
&= \frac{1}{n^2\pi^2}\left(2\,\cos n\pi - n\pi\,\sin\frac{n\pi}{2} - 2\,\cos\frac{3n\pi}{2}\right), \\
b_n &= \frac{1}{2}\int_{-1}^{1} x\,\sin\frac{n\pi x}{2}\,\mathrm{d}x + \frac{1}{2}\int_{1}^{2} 1\,\sin\frac{n\pi x}{2}\,\mathrm{d}x + \frac{1}{2}\int_{2}^{3} (3-x)\,\sin\frac{n\pi x}{2}\,\mathrm{d}x \\
&= \frac{1}{n^2\pi^2}\left(4\,\sin\frac{n\pi}{2} - 2\,\sin\frac{3n\pi}{2} - n\pi\,\cos\frac{n\pi}{2}\right).
\end{aligned}$$

Substituting these values into Eq. (10.3.2), we obtain the Fourier series for the given function:

$$\begin{aligned}
h(x) &= \frac{3}{8} + \sum_{n\geqslant 1} \frac{1}{n^2\pi^2}\left(2\,(-1)^n - n\pi\,\sin\frac{n\pi}{2} - 2\,\cos\frac{3n\pi}{2}\right) \cos\left(\frac{n\pi}{2}\,x\right) \\
&\quad + \sum_{n\geqslant 1} \frac{1}{n^2\pi^2}\left(4\,\sin\frac{n\pi}{2} - 2\,\sin\frac{3n\pi}{2} - n\pi\,\cos\frac{n\pi}{2}\right) \sin\left(\frac{n\pi}{2}\,x\right).
\end{aligned}$$

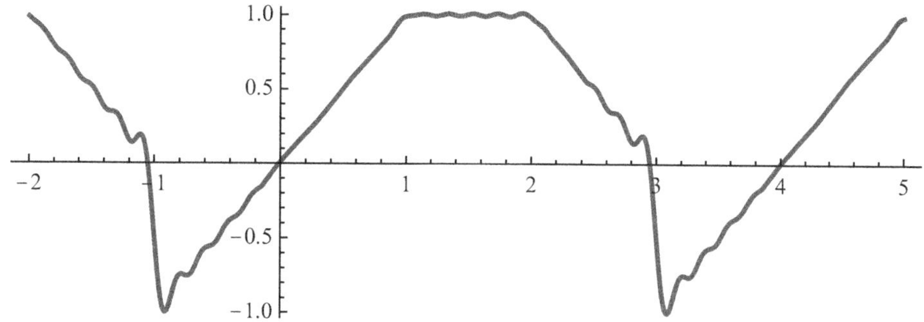

Figure 10.12: Example 10.3.3; Fourier approximation with $N = 20$ terms.

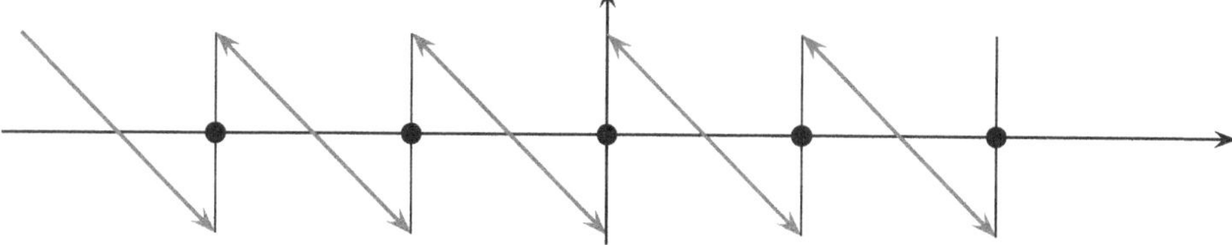

Figure 10.13: Problem 2(a), graph of the sawtooth function.

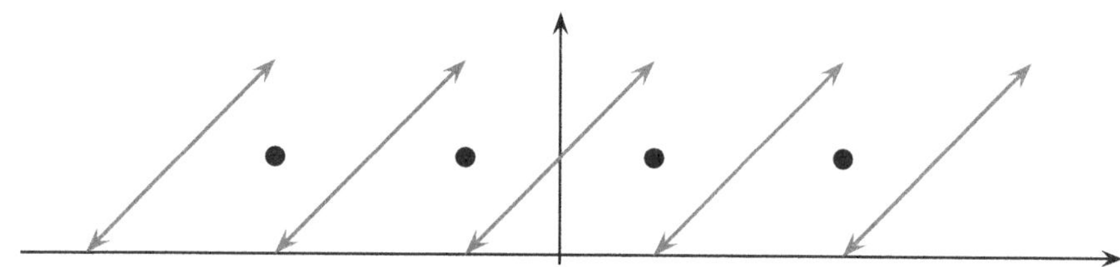

Figure 10.14: Problem 2(b), graph of the sawtooth function.

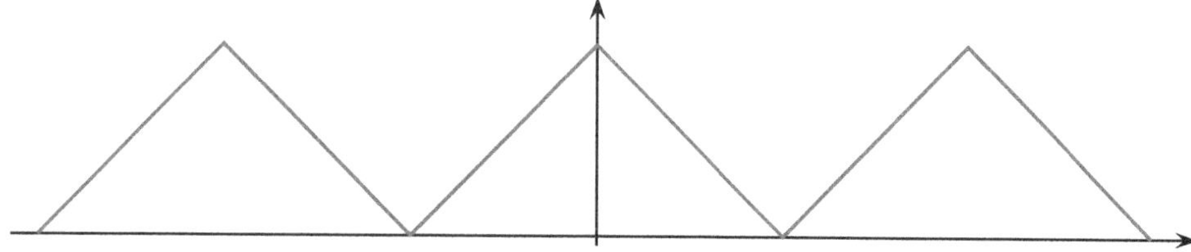

Figure 10.15: Problem 3, graph of the function $f(x)$.

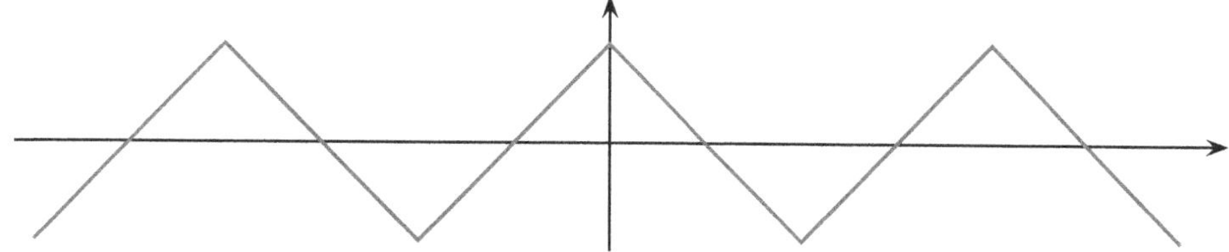

Figure 10.16: Problem 3, graph of the function $g(x)$, plotted with `pstricks`.

Problems

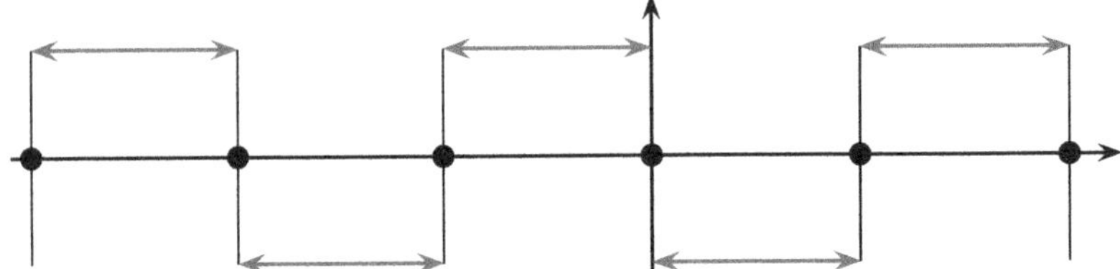

Figure 10.17: Problem 5, graph of the square wave, plotted with `pstricks`.

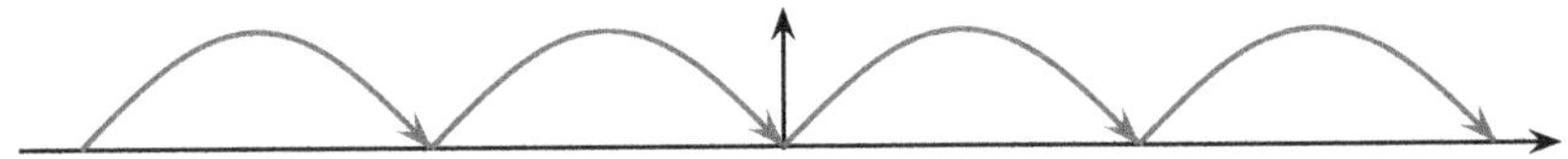

Figure 10.18: Problem 6, graph of $|\sin x|$, plotted with `pstricks`.

1. Using the geometric series sum $\sum_{k=1}^{n} q^k = \frac{q-q^{n+1}}{1-q}$ and Euler's formula (10.2.6), show the following relations:

$$\sum_{k=1}^{n} \sin(2k\theta) = \frac{\cos\theta - \cos(2n+1)\theta}{2\sin\theta} = \frac{\sin(n+1)\theta\,\sin n\theta}{\sin\theta}, \tag{10.3.6}$$

$$\sum_{k=1}^{n} \cos(2k\theta) = \frac{\sin(2n+1)\theta - \sin\theta}{2\sin\theta} = \frac{\cos(n+1)\theta\,\sin n\theta}{\sin\theta}, \tag{10.3.7}$$

$$\sum_{k=1}^{n} (-1)^k \sin(2k\theta) = \frac{1}{\cos\theta}\,\cos\frac{n\pi + 2(n+1)\theta)}{2}\,\sin\frac{n\pi + 2n\theta}{2}, \tag{10.3.8}$$

$$\sum_{k=1}^{n} (-1)^k \cos(2k\theta) = \frac{1}{\cos\theta}\,\cos\frac{(n+1)(\pi + 2\theta)}{2}\,\sin\frac{n\pi + 2n\theta}{2}. \tag{10.3.9}$$

2. Prove the following Fourier series for sawtooth functions:
   (a) Euler (1755):
   $$\frac{\pi - t}{2} = \sum_{k=1}^{\infty} \frac{1}{k}\sin kt, \qquad 0 < t < 2\pi;$$
   (b)
   $$1 + t = 1 - 2\sum_{k=1}^{\infty} \frac{(-1)^k}{k}\sin kt, \qquad -\pi < t < \pi;$$
   (c) Euler (1755):
   $$x = \frac{2a}{\pi}\sum_{k\geqslant 1} \frac{(-1)^k}{k}\sin\left(\frac{k\pi x}{a}\right), \qquad -a < x < a;$$
   (d)
   $$\frac{2x+a}{4} = \frac{a}{4} - \frac{a}{2\pi}\sum_{k=1}^{\infty} \frac{(-1)^k}{k}\sin\frac{2k\pi x}{a}, \qquad -a/2 < x < a/2.$$

3. Consider two continuous ramp functions
   $$f(x) = \begin{cases} 2+x, & -2 < x \leqslant 0, \\ 2-x, & 0 \leqslant x < 2; \end{cases} \qquad \text{and} \qquad g(x) = \begin{cases} 1+x, & -2 < x \leqslant 0, \\ 1-x, & 0 \leqslant x < 2. \end{cases}$$
   By extending these functions periodically: $f(x) = f(x+4)$ and $g(x) = g(x+4)$, find their Fourier series.
4. Compute the Fourier series for another ramp function: $f(x) = |x|$, $-a < x < a$.
5. By differentiating the function in the previous problem, find the Fourier series for the square wave function
   $$w(x) = \begin{cases} 1, & -\ell < x \leqslant 0, \\ -1, & 0 \leqslant x < \ell. \end{cases}$$
   In particular, choose $\ell = 2$.

6. Find the Fourier series for the periodic function $f(t) = |\sin t|$ (the absolute value of $\sin t$).
7. Find the Fourier series on the interval $[-\pi, \pi]$ of each of the following functions

$$\textbf{(a)}\ \sin^3 t - \cos^3 t; \qquad \textbf{(b)}\ \cos^4 t - \sin^4 t; \qquad \textbf{(c)}\ \sin 2t - \cos^2 t.$$

8. Find the Fourier series of the following intermittent function:

$$f(x) = \begin{cases} \ell, & \text{if } -\ell < x < 0, \\ 2x, & \text{if } 0 < x < \ell. \end{cases}$$

9. Find the Fourier series for a periodic function of period 6, which is zero in the interval $-3 < x < 0$, and in the interval $0 < x < 3$ is equal to

$$\textbf{(a)}\ x; \qquad \textbf{(b)}\ x^2; \qquad \textbf{(c)}\ \sin x; \qquad \textbf{(d)}\ e^x \quad (0 < x < 3).$$

10. Find the Fourier series which represent the function of period 2π, defined in the interval $-\pi < t < \pi$ as equal to

$$\textbf{(a)}\ e^t; \qquad \textbf{(b)}\ t \sin t; \qquad \textbf{(c)}\ t \cos t.$$

11. Suppose that $f(x)$ is an integrable periodic function with period T. Show that for any $0 \leqslant a \leqslant T$,

$$\int_0^T f(x)\,\mathrm{d}x = \int_a^{a+T} f(x)\,\mathrm{d}x.$$

12. A function is periodic, of period 8. Find the Fourier series which represent it, if it is defined in the interval $0 < x < 8$ by

$$f(x) = \begin{cases} 5, & \text{if } 0 < x < 2; \\ 2, & \text{if } 2 < x < 4; \\ -5, & \text{if } 4 < x < 6; \\ -2, & \text{if } 6 < x < 8. \end{cases}$$

13. Expand the function into Fourier series assuming that it is periodic with period 4:

$$\textbf{(a)}\ f(x) = \begin{cases} 1, & \text{if } -2 < x < 0, \\ x^2, & \text{if } 0 < x < 2; \end{cases} \qquad \textbf{(b)}\ \begin{cases} 0, & \text{if } -2 < x < 0, \\ x^2, & \text{if } 0 < x < 2; \end{cases} \qquad \textbf{(c)}\ \begin{cases} -x, & \text{if } -2 < x < 0, \\ x^2, & \text{if } 0 < x < 2. \end{cases}$$

14. Expand the function into Fourier series assuming that it is periodic with period 6:

$$f(x) = \begin{cases} 0, & \text{if } -3 < x < -1, \\ 3, & \text{if } -1 < x < 1, \\ 0, & \text{if } 1 < x < 3. \end{cases}$$

15. Expand the function into Fourier series assuming that it is periodic with period 4:

$$f(x) = \begin{cases} x + 3, & \text{if } -2 < x < 0, \\ 3 - 2x, & \text{if } 0 < x < 2. \end{cases}$$

16. Expand the function $f(x) = x^3/3$ into Fourier series assuming that it is periodic with period 4: $f(x+4) = f(x)$.

In Problems 17 through 20, compute the Fourier series for the given function on the interval $[-2, 2]$.

17. $f(x) = \begin{cases} -x^3, & -2 < x \leqslant 0, \\ x^2, & 0 \leqslant x < 2; \end{cases}$

18. $f(x) = \begin{cases} x + 2, & -2 < x < 0, \\ x, & 0 < x < 2; \end{cases}$

19. $f(x) = \begin{cases} e^{-x}, & -2 < x \leqslant 0, \\ e^x, & 0 \leqslant x < 2; \end{cases}$

20. $f(x) = \begin{cases} -x^2, & -2 < x \leqslant 0, \\ x^2, & 0 \leqslant x < 2. \end{cases}$

21. Find the Fourier series of the 3ℓ-periodic function $f(x) = \begin{cases} \frac{x}{2\ell}, & 0 < x < 2\ell, \\ 3 - \frac{x}{\ell}, & 2\ell < x < 3\ell. \end{cases}$

22. Find the Fourier series for the function

$$f(x) = \begin{cases} -1, & -\ell < x \leqslant 0, \\ x^2, & 0 \leqslant x < \ell. \end{cases}$$

In particular, choose $\ell = 2$.

10.4 Convergence of Fourier Series

The classical subject of Fourier series is about approximating periodic functions by sines and cosines. This topic became a part of harmonic analysis, which includes representations of functions as infinite series of eigenfunctions corresponding to a Sturm–Liouville problem. The sines and cosines are the "basic" periodic functions in terms of which all other functions are expressed. In chemical terminology, the sines and cosines are the atoms; the other functions are the molecules. However, this language is not strictly speaking appropriate because it does not reflect the existence of other atoms, other eigenfunctions, that can serve as the "basic" functions.

The process of understanding is always facilitated if more complicated structures are known to be synthesized from simpler ones. In mathematics, we don't usually get a full decomposition into the simpler things, but an approximation. For example, real numbers are approximated by rationals. Another example gives the Taylor series expansion of a function, which provides an approximation by a polynomial. Of course, for a function to have a Taylor series, it must (among other things) be infinitely differentiable in some interval, and this is a very restrictive condition. Sines and cosines serve as much more versatile "prime elements" than powers of x because they can be used to approximate nondifferentiable functions.

In any realistic situation, of course, we can only compute expansions involving a finite number of terms anyway. We need some sort of assurance that the partial Fourier sums

$$S_N(x) = \frac{a_0}{2} + \sum_{n=1}^{N} \left(a_n \cos \frac{n\pi x}{\ell} + b_n \sin \frac{n\pi x}{\ell} \right) \tag{10.4.1}$$

converge to the given function as $N \to \infty$ in some sense. Using completeness of the set of trigonometric functions $\left\{ \cos \frac{n\pi x}{\ell},\ \sin \frac{n\pi x}{\ell} \right\}_{n \geqslant 0}$ in the space of square integrable functions, we obtain our first result.

Theorem 10.6: Let $f(x)$ be a square integrable function on the interval $[-\ell, \ell]$, that is, $\|f\|^2 = \int_{-\ell}^{\ell} |f(x)|^2 \, dx < \infty$. Then the Fourier series (10.3.2), page 563, converges to $f(x)$ in the mean square sense.

However, the mean square convergence tells us nothing about pointwise convergence. Its study is both deep and subtle. Since the Fourier coefficients (10.3.5) are obtained by averaging the given function with trigonometric functions, we cannot recover the function from its Fourier coefficients without some further smoothness conditions. The most obvious problem is that if two functions differ at a discrete number of points, then their Fourier coefficients will be identical.

Definition 10.8: An absolutely integrable function $f(x)$ is said to be **piecewise monotone** or satisfy Dirichlet's conditions on the finite interval $[a, b]$ if:

1. $f(x)$ has at most finitely many points of discontinuity in the close interval $[a, b]$, which can be broken into finitely many subintervals such that $f(x)$ is either nondecreasing or nonincreasing on each subinterval;
2. the right limits of the function $f(x)$ exist at every point x_0 from $[a, b)$ (which we denote as $f(x_0 + 0)$ or simply $f(x_0^+)) : f(x_0 + 0) = \lim_{x \to x_0 + 0} f(x)$ if $a \leqslant x_0 < b$;
3. the left limits exist at every point x_0 from the interval $(a, b] : f(x_0 - 0) = f(x_0^-) = \lim_{x \to x_0 - 0} f(x)$ if $a < x_0 \leqslant b$.

Note that a piecewise monotone function is bounded everywhere on the interval because it has left and right limits at every point: the function is either continuous or has finite jumps (in discrete number of points within the closed interval). So $\tan(x)$ is not a piecewise continuous function because it has infinite jumps at points $k\pi + \pi/2$, $k = 0, \pm 1, \pm 2, \ldots$. Also the function $x \sin(1/x)$ does not satisfy Dirichlet's conditions because it has infinite many local maximum and minimum points around $x = 0$.

Other examples of piecewise monotone functions are given in the previous section. For instance, every sawtooth function (see Example 10.3.1 and Problem 2) is a piecewise monotone function but not continuous. Their derivatives are identically 1 except a discrete number of points where they are undefined.

Sufficient conditions for the pointwise convergence of the Fourier series were discovered by P. Dirichlet[104] in 1829. He proved that the Fourier series of a piecewise smooth integrable function converges at each point x to its average of the limiting values.

[104] Peter Gustav Lejeune Dirichlet (1805–1859) was a German mathematician of French origin, a student of Poisson and Fourier.

Theorem 10.7: [Dirichlet] Suppose that $f(x)$ is a piecewise monotone (satisfies Dirichlet's conditions) periodic real-valued function with the period 2ℓ. Then the function f is represented by a convergent Fourier series

$$f(x) \sim \frac{a_0}{2} + \sum_{n=1}^{\infty} \left(a_n \cos \frac{n\pi x}{\ell} + b_n \sin \frac{n\pi x}{\ell} \right), \tag{10.4.2}$$

whose coefficients are given by

$$a_n = \frac{1}{\ell} \int_{-\ell}^{\ell} f(t) \cos \frac{n\pi t}{\ell} \, dt, \qquad n = 0, 1, 2, \ldots; \tag{10.4.3}$$

$$b_n = \frac{1}{\ell} \int_{-\ell}^{\ell} f(t) \sin \frac{n\pi t}{\ell} \, dt, \qquad n = 1, 2, \ldots. \tag{10.4.4}$$

The Fourier series (10.4.2) converges to $f(x)$ at all points where f is continuous and to

$$\frac{1}{2} \left[f(x+0) + f(x-0) \right] = \lim_{\epsilon \to 0} \frac{1}{2} \left[f(x+\epsilon) + f(x-\epsilon) \right] \tag{10.4.5}$$

at all points where f is discontinuous. This convergence is uniform on any closed interval that does not contain a discontinuity of f.

PROOF: We are going to outline the main idea of the Dirichlet theorem, while the complete proof can be found elsewhere, for example, in [27].

Equation (10.4.2) means that at every point x in the interval $[-\ell, \ell]$, the function $f(x)$ is the limit of the partial sums

$$f(x) = \lim_{N \to \infty} S_N(x), \tag{10.4.6}$$

where the trigonometric polynomial $S_N(x)$ is given in Eq. (10.4.1). We have used t for the dummy variable of integration in Eqs. (10.4.3), (10.4.4) to avoid confusion with the fixed x for which we are considering the sum of the series equals $f(x)$. Since x is a constant for the variable of integration, we substitute the expressions (10.4.3), (10.4.4) of the Fourier coefficients into the general term of the sum (10.4.2):

$$\begin{aligned} a_n \cos \frac{n\pi x}{\ell} + b_n \sin \frac{n\pi x}{\ell} &= \frac{1}{\ell} \int_{-\ell}^{\ell} f(t) \left(\cos \frac{n\pi x}{\ell} \cos \frac{n\pi t}{\ell} + \sin \frac{n\pi x}{\ell} \sin \frac{n\pi t}{\ell} \right) dt \\ &= \frac{1}{\ell} \int_{-\ell}^{\ell} f(t) \cos \frac{n\pi (x-t)}{\ell} \, dt. \end{aligned}$$

Then the partial sum (10.4.1) can be expressed as

$$S_N(x) = \frac{1}{\ell} \int_{-\ell}^{\ell} \left[\frac{1}{2} + \sum_{n=1}^{N} \cos \frac{n\pi (x-t)}{\ell} \right] f(t) \, dt.$$

We now introduce a new variable of integration $t = x + u$ and get

$$S_N(x) = \frac{1}{\ell} \int_{-\ell}^{\ell} \left[\frac{1}{2} + \sum_{n=1}^{N} \cos \frac{n\pi u}{\ell} \right] f(x+u) \, dt = \frac{1}{\ell} \int_{-\ell}^{\ell} s_N(u) \, f(x+u) \, dt.$$

To find a simpler expression for the finite sum $s_N(u) = \dfrac{1}{2} + \displaystyle\sum_{n=1}^{N} \cos \frac{n\pi u}{\ell}$, we multiply both sides by $2 \sin \frac{u\pi}{2\ell}$. This yields

$$\begin{aligned} 2 \sin \frac{u\pi}{2\ell} \, s_N(u) &= \sin \frac{u\pi}{2\ell} + \sum_{n=1}^{N} 2 \sin \frac{u\pi}{2\ell} \cos \frac{n\pi u}{\ell} \\ &= \sin \frac{u\pi}{2\ell} + \sum_{n=1}^{N} \left[\sin \left(n + \frac{1}{2} \right) \frac{u\pi}{\ell} - \sin \left(n - \frac{1}{2} \right) \frac{u\pi}{\ell} \right]. \end{aligned}$$

Since the sum is telescopic, all the terms except the one before the last appear twice, with opposite signs, and so cancel out. This leads to

$$2\sin\frac{u\pi}{2\ell}\, s_N(u) = \sin\left[\left(N+\frac{1}{2}\right)\frac{u\pi}{\ell}\right] \quad \Longrightarrow \quad S_N(x) = \frac{1}{\ell}\int_{-\ell}^{\ell} \frac{\sin\left[\left(N+\frac{1}{2}\right)\frac{u\pi}{\ell}\right]}{2\,\sin\frac{u\pi}{2\ell}}\, f(x+u)\,\mathrm{d}u.$$

The next step involves representation of the partial sum $S_N(x)$ in a symmetric form:

$$S_N(x) = \frac{1}{2\ell}\int_0^{\ell} \frac{\sin\left[\left(N+\frac{1}{2}\right)\frac{u\pi}{\ell}\right]}{\sin\frac{u\pi}{2\ell}}\,[f(x+u)+f(x-u)]\,\mathrm{d}u \tag{10.4.7}$$

and finding the limit as $N\to\infty$. To achieve this, the interval of integration $[0,\ell]$ can be broken into two subintervals, one of length $\delta>0$, $[0,\delta]$, and the rest $[\delta,\ell]$, where $0<\sin\frac{\delta\pi}{2\ell}\leqslant\sin\frac{u\pi}{2\ell}$. Since the limit of the integral over the interval $[\delta,\ell]$ is zero, we get

$$S_N(x) = \frac{1}{2\ell}\int_0^{\delta} \frac{\sin\left[\left(N+\frac{1}{2}\right)\frac{u\pi}{\ell}\right]}{\sin\frac{u\pi}{2\ell}}\,[f(x+u)+f(x-u)]\,\mathrm{d}u + o(1),$$

where $o(1)$ approaches zero as $N\to\infty$. Using the monotonic property of the function $f(x)$ on $[0,\delta]$, we apply the mean value theorem to obtain the desired formula (10.4.5). ■

Note that the Fourier series (10.4.2) converges to the function that provides a periodic expansion with period $T=2\ell$ of the given function $f(x)$ defined initially on the interval of length T. How many terms are needed to achieve a good approximation depends on the order of convergence of the general term to zero. The faster the general term decreases, the more accurate approximation partial sums provide. If a periodic function $f(x)$ is continuous, with a piecewise continuous first derivative, integration by parts and periodicity yield

$$|a_n|,\ |b_n| \leqslant \frac{1}{n\pi}\int_{-\ell}^{\ell} |f'(x)|\,\mathrm{d}x.$$

If $f(x)$ has finite jumps, then its Fourier coefficients will decrease as $1/n$ as $n\to\infty$. When $f(x)$ has more continuous derivatives, we may iterate this procedure and obtain

$$|a_n|,\ |b_n| \leqslant \frac{\ell^{k-1}}{n^k\pi^k}\int_{-\ell}^{\ell} |f^{(k)}|(x)\,\mathrm{d}x, \quad k=1,2,\ldots. \tag{10.4.8}$$

The estimate (10.4.8) can be used in a comparison test to show that the Fourier series is absolutely dominated by a p-series of the form $\sum_{n\geqslant 1} n^{-p}$, which converges for $p>1$. A condition that f' is a piecewise continuous function is sufficient: it is known that for any discrete number of points, there exists a continuous function for which its Fourier series diverges at this set of points. In 1903, Lebesgue[105] proved that the Fourier coefficients a_n and b_n of a Lebesgue- or Riemann-integrable function approach 0 as $n\to\infty$. Andrew Kolmogorov gave an example in 1929 of an integrable function whose Fourier series diverges at every point. For uniform convergence of the Fourier series, we need to impose an additional condition on the function.

Theorem 10.8: Let $f(x)$ be a continuous real-valued function on $(-\infty,\infty)$ and periodic with a period of 2ℓ. If f is piecewise monotonic on $[-\ell,\ell]$, then the Fourier series (10.4.2) converges uniformly to $f(x)$ on $[-\ell,\ell]$ and hence on any closed interval. That is, for every $\varepsilon>0$, there exists an integer $N_0=N_0(\varepsilon)$ that depends on ε such that

$$|S_N(x)-f(x)| = \left|\frac{a_0}{2}+\sum_{n=1}^{N}\left(a_n\cos\frac{n\pi x}{\ell}+b_n\sin\frac{n\pi x}{\ell}\right)-f(x)\right| < \varepsilon,$$

for all $N\geqslant N_0$, and all $x\in(-\infty,\infty)$.

Theorem 10.9: If a function $f(x)$ satisfies the Dirichlet conditions on the interval $[-\ell,\ell]$, then its Fourier series converges to $f(x)$ at every point x where $f(x)$ has a derivative.

[105]Henri Léon Lebesgue (1875–1941) was a French mathematician most famous for his theory of integration.

In many practical questions, we deal with piecewise continuous functions that are not differentiable. For example, music noise, which resembles Brownian motion, is an example of everywhere continuous functions that are not differentiable. Sawtooth functions and square waves are typical examples of functions that occur in music synthesis. In 1899, Lipót Fejér[106] proved that any continuous function on a closed interval is the Cesàro[107] limit of its Fourier partial sums (10.4.1) on that interval:

$$f(x) = \lim_{N\to\infty} C_N(x), \quad \text{where} \quad C_N(x) = \frac{S_0(x) + S_1(x) + \cdots + S_N(x)}{N+1}. \tag{10.4.9}$$

Cesàro's approximation is highly recommended because it speeds convergence and reduces problems caused by the Gibbs phenomenon (see §10.4.2). There is known another method that accelerates convergence and eliminates most of unwanted Gibbs oscillations (see §10.4.2) in partial sums. The **method of σ-factors** represents a function as $f(x) = \lim_{N\to\infty} s_N(x)$, where

$$s_N(x) = \frac{a_0}{2} + \sum_{n=1}^{N-1} \left(a_n \cos\frac{n\pi x}{\ell} + b_n \sin\frac{n\pi x}{\ell}\right) \operatorname{sinc}\left(\frac{n}{N}\right) = \frac{N}{2\ell}\int_{-\ell/N}^{\ell/N} S_N(x+t)\,\mathrm{d}t, \tag{10.4.10}$$

where $S_N(x)$ is the Fourier N-th partial sum (10.4.2) and the multiple $\operatorname{sinc}(n/N) = \frac{N}{n\pi}\sin\left(\frac{n\pi}{N}\right)$ is called the Lanczos[108] σ-factor. However, the question of convergence of the Fourier series is not the same as the question of whether the function $f(x)$ can be reconstructed from its Fourier coefficients (10.4.3) and (10.4.4). This question is of great practical interest when a sound must be recovered from its Fourier coefficients transmitted through the Internet or by cellular phones.

Example 10.4.1: (Example 10.3.1 revisited) The function $f(x) = -x/2$, expanded periodically from the interval $(-\pi, \pi)$, is piecewise smooth but not continuous (see its graph in Fig. 10.8 on page 564). Therefore, the Dirichlet theorem is applicable for this function, but Theorem 10.8 is not. Indeed, the Fourier series

$$S_n(x) = \sum_{k=1}^{n} \frac{(-1)^k}{k}\sin kx \quad \underset{n\to\infty}{\longrightarrow} \quad S(x) = \sum_{k\geqslant 1} \frac{(-1)^k}{k}\sin kx$$

converges at every point, but it does not converge uniformly. According to the Weierstrass M-test, a series $\sum_{k\geqslant 1} f_k(x)$ converges uniformly if the general term can be estimated by a constant $|f_k(x)| \leqslant M_n$ for all x in the domain and the series $\sum_{k\geqslant 1} M_n$ converges. By estimating the general term $\left|\frac{(-1)^k}{k}\sin kx\right| \leqslant \frac{1}{k}$, we see that the M-test does not work because it leads to the harmonic series $\sum_{k\geqslant 1} k^{-1}$, which diverges. However, the series $S(x)$ converges uniformly (but very slowly) in every finite closed interval not containing a discrete set of points $n\pi$ ($n = 0, \pm 1, \pm 2, \ldots$) of discontinuity. The infinite sum $S(x)$ defines a periodic extension of the function $f(x) = -x/2$ outside the interval $(-\pi, \pi)$, where $S(x) = -x/2$.

Now we calculate the Cesàro partial sum

$$C_N(x) = \frac{1}{N}\sum_{n=1}^{N} S_n(x) = \frac{1}{N}\sum_{n=1}^{N}\sum_{k=1}^{n} \frac{(-1)^k}{k}\sin kx = \sum_{k=1}^{N}\left(1 - \frac{k-1}{N}\right)\frac{(-1)^k}{k}\sin kx,$$

and σ-factor partial sum

$$\begin{aligned} s_N(x) &= \frac{N}{2\pi}\int_{-\pi/N}^{\pi/N}\sum_{k=1}^{N}\frac{(-1)^k}{k}\sin k(x+t)\,\mathrm{d}t = \frac{N}{2\pi}\sum_{k=1}^{N}\frac{(-1)^k}{k}\int_{-\pi/N}^{\pi/N}\sin k(x+t)\,\mathrm{d}t \\ &= \sum_{k=1}^{N}\frac{(-1)^k}{k}\,\frac{N}{k\pi}\sin\left(\frac{k\pi}{N}\right)\sin kx. \end{aligned}$$

---

[106] Lipót Fejér (1880–1959) was a Hungarian mathematician who was born as Leopold Weiss in Jewish family (Weiss in German means "white" while the Hungarian for white is "feher"). As the chair of mathematics at the University of Budapest, Fejér led a highly successful Hungarian school of analysis being the thesis adviser of great mathematicians such as John von Neumann, Paul Erdős, George Pólya, Marcel Riesz, Gábor Szegő, and Pál Turán.

[107] Ernesto Cesàro (1859–1906) was an Italian mathematician.

[108] Cornelius Lanczos (1893–1974) was a Hungarian mathematician and physicist.

By plotting two finite sums, the Cesàro $C_{10}(x)$ with 10 terms and Fourier partial sum $S_{45}(x)$, we see that $C_{10}(x)$ gives a smoother approximation. The graphs of these partial sums are presented on the opening page 545 of the chapter. Their mean square errors are

$$\Delta_{S_N} = \int_{-\pi}^{\pi} \left| \frac{x}{2} + S_N(x) \right|^2 dx \quad \Longrightarrow \quad \Delta_5 \approx 0.569643, \quad \Delta_{45} \approx 0.0690432;$$

$$\Delta_{C_N} = \int_{-\pi}^{\pi} \left| \frac{x}{2} + C_N(x) \right|^2 dx \quad \Longrightarrow \quad \Delta_5 \approx 0.80802, \quad \Delta_{10} \approx 0.127738.$$

$$\Delta_{s_N} = \int_{-\pi}^{\pi} \left| \frac{x}{2} + s_N(x) \right|^2 dx \quad \Longrightarrow \quad \Delta_5 \approx 0.955757, \quad \Delta_{10} \approx 0.476326.$$

Example 10.4.2: Consider the following continuous function defined on the interval $[0, 3\ell]$ by the equation

$$f(x) = \begin{cases} \frac{x}{\ell}, & 0 \leqslant x \leqslant \ell, \\ \frac{3}{2} - \frac{x}{2\ell}, & \ell \leqslant x \leqslant 3\ell. \end{cases}$$

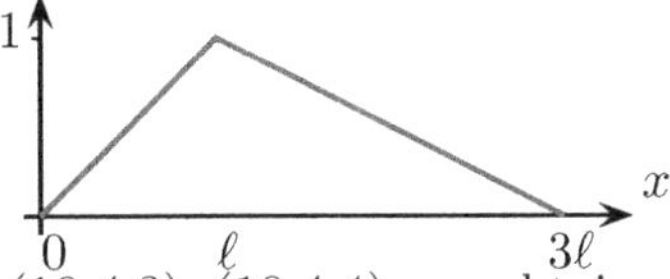

Calculating the Fourier coefficients according to Eqs. (10.4.3), (10.4.4), we obtain

$$a_0 = \frac{2}{3\ell} \int_0^\ell \frac{t}{\ell}\, dt + \frac{2}{3\ell} \int_\ell^{3\ell} \left(\frac{3}{2} - \frac{t}{2\ell} \right) dt = 1,$$

$$a_n = \frac{2}{3\ell} \int_0^\ell \frac{t}{\ell} \cos \frac{2n\pi t}{3\ell}\, dt + \frac{2}{3\ell} \int_\ell^{3\ell} \left(\frac{3}{2} - \frac{t}{2\ell} \right) \cos \frac{2n\pi t}{3\ell}\, dt = \frac{9}{4n^2\pi^2} \left(\cos \frac{2n\pi}{3} - 1 \right),$$

$$b_n = \frac{2}{3\ell} \int_0^\ell \frac{t}{\ell} \sin \frac{2n\pi t}{3\ell}\, dt + \frac{2}{3\ell} \int_\ell^{3\ell} \left(\frac{3}{2} - \frac{t}{2\ell} \right) \sin \frac{2n\pi t}{3\ell}\, dt = \frac{3}{n^2\pi^2} \sin^3 \frac{2n\pi}{3}.$$

The Fourier series

$$\frac{1}{2} + \sum_{n \geqslant 1} \frac{9}{4n^2\pi^2} \left(\cos \frac{2n\pi}{3} - 1 \right) \cos \frac{2n\pi x}{3\ell} + \sum_{n \geqslant 1} \frac{3}{n^2\pi^2} \sin^3 \frac{2n\pi}{3} \sin \frac{2n\pi x}{3\ell}$$

converges uniformly to $f(x)$ on the interval $[0, 3\ell]$ because its general term decreases as $O\left(n^{-2}\right)$. The sum defines a continuous extension of $f(x)$ on the whole line.

Example 10.4.3: Consider a continuously differentiable function on the interval $(-\pi/4, \pi/4)$ that does not satisfy the Dirichlet conditions: $f(x) = x^3 \sin(1/x)$ because it has infinitely many local maxima and minima in a neighborhood of the origin. The graph of $f(x)$ along with its continuous derivative $f'(x) = 3x^2 \sin(1/x) - x \cos(1/x)$ is presented in Fig. 10.19. Looking at Fig. 10.20, we are not convinced that its partial sum Fourier approximation

$$f_N(x) = \frac{a_0}{2} + \sum_{k=1}^{N} a_k \cos(4kx),$$

where

$$a_k = \frac{4}{\pi} \int_{-\pi/4}^{\pi/4} x^3 \sin(1/x) \cos(4kx)\, dx, \qquad k = 0, 1, 2, \ldots,$$

converges pointwise (the first 10 values of coefficients a_k are given in Fig. 10.20). Nevertheless, the Fourier series for the function $f(x)$ converges pointwise despite it does not satisfy (sufficient) conditions of the Dirichlet theorem. Also, the mean square error of its approximation tends to zero because $f(x)$ is a square integrable function (Theorem 10.6, page 570). For instance, with $N = 30$ terms, the error is about $\Delta_{30} \approx 5.7 \times 10^{-7}$.

10.4.1 Complex Fourier Series

Sometimes it is convenient to express the Fourier series in complex form using the complex exponentials

$$e^{\mathbf{j}k\pi x/\ell} = \cos\left(\frac{k\pi x}{\ell} \right) + \mathbf{j} \sin\left(\frac{k\pi x}{\ell} \right), \qquad k = 0, \pm 1, \pm 2, \ldots. \tag{10.4.11}$$

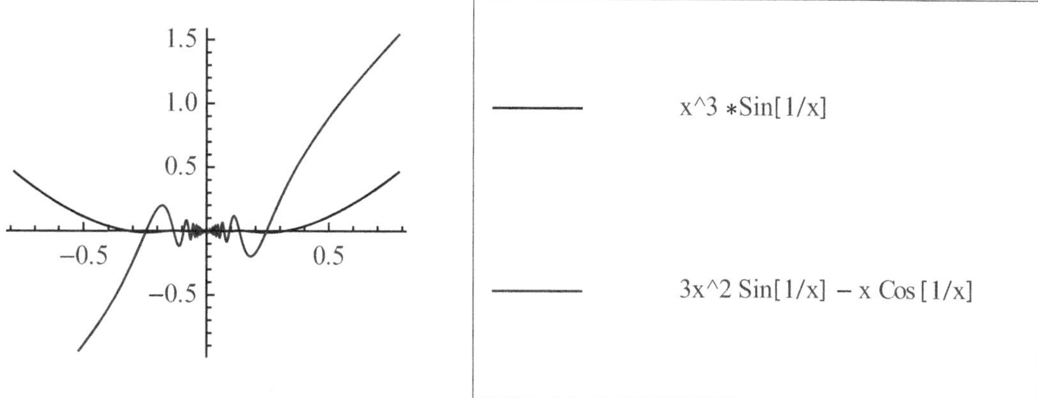

Figure 10.19: Example 10.4.3, graphs of the function $f(x) = x^3 \sin(1/x)$ along with its derivative, plotted with *Mathematica*.

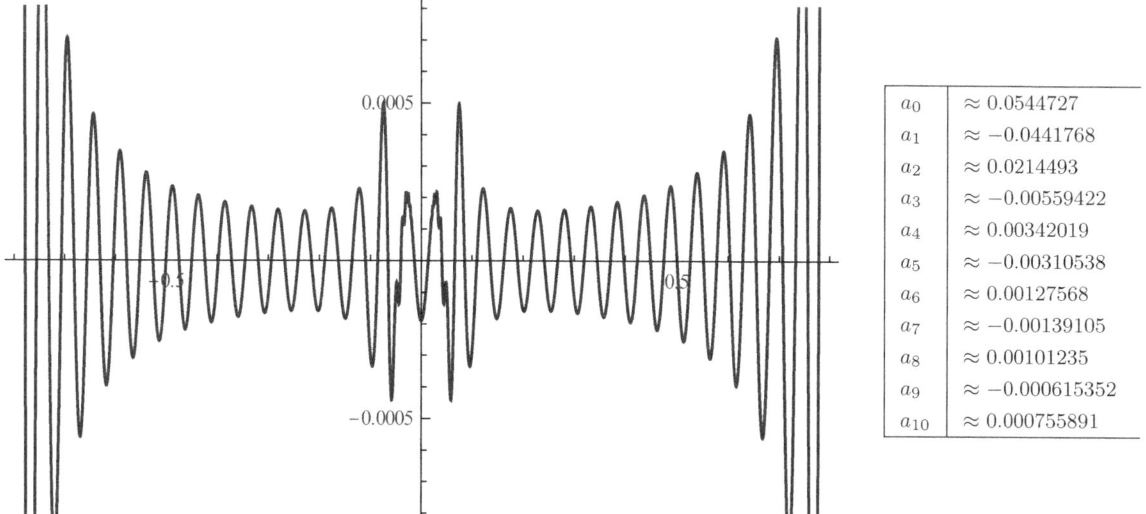

a_0	≈ 0.0544727
a_1	≈ -0.0441768
a_2	≈ 0.0214493
a_3	≈ -0.00559422
a_4	≈ 0.00342019
a_5	≈ -0.00310538
a_6	≈ 0.00127568
a_7	≈ -0.00139105
a_8	≈ 0.00101235
a_9	≈ -0.000615352
a_{10}	≈ 0.000755891

Figure 10.20: Example 10.4.3, the graph of the difference between the function $f(x) = x^3 \sin(1/x)$ and its Fourier partial sum approximation with $N = 30$ terms.

Here $\mathbf{j}$ is the unit vector in the positive vertical direction on the complex plane, so that $\mathbf{j}^2 = -1$. From the formula (10.4.11), called the **Euler formula**, it follows that

$$\cos\left(\frac{k\pi x}{\ell}\right) = \frac{1}{2}\left(e^{\mathbf{j}k\pi x/\ell} + e^{-\mathbf{j}k\pi x/\ell}\right) = \Re\, e^{\mathbf{j}k\pi x/\ell}, \quad \sin\left(\frac{k\pi x}{\ell}\right) = \frac{1}{2\mathbf{j}}\left(e^{\mathbf{j}k\pi x/\ell} - e^{-\mathbf{j}k\pi x/\ell}\right) = \Im\, e^{\mathbf{j}k\pi x/\ell}.$$

The general term in the Fourier series (10.3.2), page 563, which we denote by

$$\begin{aligned} f_k &\stackrel{\text{def}}{=} a_k \cos\left(\frac{k\pi x}{\ell}\right) + b_k \sin\left(\frac{k\pi x}{\ell}\right) \\ &= \frac{a_k}{2}\left(e^{\mathbf{j}k\pi x/\ell} + e^{-\mathbf{j}k\pi x/\ell}\right) + \frac{b_k}{2\mathbf{j}}\left(e^{\mathbf{j}k\pi x/\ell} - e^{-\mathbf{j}k\pi x/\ell}\right) \end{aligned}$$

can be rewritten in the following form:

$$f_k = \frac{a_k - \mathbf{j}b_k}{2}\, e^{\mathbf{j}k\pi x/\ell} + \frac{a_k + \mathbf{j}b_k}{2}\, e^{-\mathbf{j}k\pi x/\ell} = \alpha_k\, e^{\mathbf{j}k\pi x/\ell} + \alpha_{-k}\, e^{-\mathbf{j}k\pi x/\ell},$$

where

$$\alpha_k \stackrel{\text{def}}{=} \frac{a_k - \mathbf{j}b_k}{2} \qquad \text{and} \qquad \alpha_{-k} \stackrel{\text{def}}{=} \frac{a_k + \mathbf{j}b_k}{2} = \overline{\alpha}_k, \quad \text{for } k \geqslant 1.$$

If we write the constant term as $f_0 = \alpha_0 = a_0/2$, we obtain an elegant expansion,

$$f(x) \sim \sum_{k=-\infty}^{\infty} \alpha_k \, e^{\mathbf{j}k\pi x/\ell}, \tag{10.4.12}$$

called the **complex Fourier series**. The coefficients in the complex series (10.4.12) are expressed through a compact formula:

$$\alpha_k = \frac{1}{2\ell} \int_{-\ell}^{\ell} f(x)\, e^{-\mathbf{j}k\pi x/\ell} \, dx, \qquad k = 0, \pm 1, \pm 2, \ldots \tag{10.4.13}$$

because

$$\begin{aligned} \alpha_k &= \frac{a_k - \mathbf{j}b_k}{2} = \frac{1}{2\ell} \int_{-\ell}^{\ell} f(x) \left[\cos\left(\frac{k\pi x}{\ell}\right) - \mathbf{j}\, \sin\left(\frac{k\pi x}{\ell}\right) \right] dx \\ &= \frac{1}{2\ell} \int_{-\ell}^{\ell} f(x)\, e^{-\mathbf{j}k\pi x/\ell} \, dx. \end{aligned}$$

If $f(x)$ is a real-valued function, then the coefficients of the Fourier series (10.3.2) are expressed through the coefficients of the complex Fourier series (10.4.12) as

$$a_k = \alpha_k + \overline{\alpha_k} = 2\,\Re\, \alpha_k \qquad \text{and} \qquad b_k = \mathbf{j}(\alpha_k - \overline{\alpha_k}) = -2\,\Im\, \alpha_k,$$

where $\mathrm{Re}\, z = \Re\, z = \mathrm{Re}\,(a+\mathbf{j}b) = a$ is the real part of the complex number $z = a+\mathbf{j}b$ and $\mathrm{Im}\, z = \Im\, z = \mathrm{Im}\,(a+\mathbf{j}b) = b$ is its imaginary part. Since the expression $\omega = e^{\mathbf{j}\pi x/\ell}$ is a complex number of unit length, the complex Fourier series (10.4.12) can be considered as a Laurent series[109] on the unit circle:

$$f(x) \sim \sum_{k=-\infty}^{\infty} \alpha_k \left(e^{\mathbf{j}\pi x/\ell}\right)^k = \sum_{k=-\infty}^{\infty} \omega^k, \qquad \omega = e^{\mathbf{j}\pi x/\ell}. \tag{10.4.14}$$

The Dirichlet theorem (page 571) shows that the sequence of trigonometric functions $\left\{\cos \frac{n\pi x}{\ell}, \sin \frac{n\pi x}{\ell}\right\}_{n\geqslant 0}$ is complete in the space of piecewise smooth functions, and for such a function we have Parseval's identity[110]

$$\|f\|^2 = \int_{-\ell}^{\ell} |f(x)|^2 \, dx = 2\ell \sum_{k=-\infty}^{\infty} |\alpha_k|^2 = \ell \left(\frac{a_0^2}{2} + \sum_{n=1}^{\infty} a_n^2 + \sum_{n=1}^{\infty} b_n^2 \right), \tag{10.4.15}$$

where Fourier coefficients α_k and a_k, b_k are defined by equations (10.4.13) and (10.3.5) on page 563, respectively.

Example 10.4.4: The function $f(x) = 1 + x$ becomes a sawtooth function when extended periodically from the interval $(-1, 1)$. Its complex Fourier coefficients are

$$\alpha_k = \frac{1}{2} \int_{-1}^{1} (1+x)\, e^{-\mathbf{j}k\pi x} \, dx = -\frac{\mathbf{j}\, \sin(k\pi)}{k^2\pi^2} + \frac{\mathbf{j}}{k\pi}\, e^{-\mathbf{j}k\pi}, \qquad k = \pm 1, \pm 2, \ldots.$$

Since $\sin(k\pi) = 0$ for any integer k, the coefficients are simplified to

$$\alpha_k = \frac{\mathbf{j}}{k\pi}\, e^{-\mathbf{j}k\pi} = \frac{\mathbf{j}}{k\pi} \left(\cos k\pi - \mathbf{j}\, \sin k\pi\right) = \frac{\mathbf{j}}{k\pi} \,(-1)^k.$$

For $k = 0$, we have

$$\alpha_0 = \frac{a_0}{2} = \frac{1}{2} \int_{-1}^{1} (1+x)\, dx = 1.$$

Hence, the complex Fourier series for the function $f(x) = 1 + x$ on the interval $(-1, 1)$ becomes (compare with the series in Problem 2(b), page 568, from §10.3)

$$\begin{aligned} 1 + x &= 1 + \sum_{k=1}^{\infty} \frac{\mathbf{j}}{k\pi} \,(-1)^k\, e^{\mathbf{j}k\pi x} + \sum_{k=-\infty}^{-1} \frac{\mathbf{j}}{k\pi}\,(-1)^k\, e^{\mathbf{j}k\pi x} \\ &= 1 + \sum_{k=1}^{\infty} \frac{\mathbf{j}}{k\pi}\,(-1)^k \left[e^{\mathbf{j}k\pi x} - e^{-\mathbf{j}k\pi x}\right] = 1 - 2\sum_{k=1}^{\infty} \frac{1}{k\pi}\,(-1)^k \sin(k\pi x). \end{aligned}$$

[109]Pierre Alphonse Laurent (1813–1854) was a French mathematician best known as the discoverer of the Laurent series.

[110]It is named after French mathematician Marc-Antoine Parseval (1755–1836), who discovered it in 1799.

Since the given function $f(x) = 1 + x$ is odd, its complex Fourier series becomes a sine Fourier series (see §10.5). One can plot partial sums, using, for instance, *Maple:*

```
N:=10:  fN:=x-> 1 - 2*sum((-1)^k*(k*Pi)^(-1)*sin(k*Pi*x),k=1..N);
plot(fN(x),x=-1..1);
```

Example 10.4.5: (Example 10.3.2 revisited) Using formulas (10.4.13), we obtain complex Fourier coefficients of the function from Example 10.3.2:

$$\begin{aligned}\alpha_k &= \frac{1}{4}\int_{-2}^{0}(-x)\,e^{-\mathbf{j}k\pi x/2}\,\mathrm{d}x + \frac{1}{4}\int_{0}^{2} e^{-\mathbf{j}k\pi x/2}\,\mathrm{d}x \\ &= \frac{1-\mathbf{j}k\pi}{k^2\pi^2}\,e^{\mathbf{j}k\pi} - \frac{1}{k^2\pi^2} - \frac{\mathbf{j}}{2k\pi} + \frac{\mathbf{j}}{2k\pi}\,e^{-\mathbf{j}k\pi} \\ &= \frac{2-\mathbf{j}k\pi}{2k^2\pi^2}\,(-1)^k - \frac{1}{k^2\pi^2} - \frac{\mathbf{j}}{2k\pi} = -\frac{\mathbf{j}}{2k\pi}\left(1+(-1)^k\right) \qquad (k = \pm 1, \pm 2, \ldots)\end{aligned}$$

because

$$e^{\pm \mathbf{j}k\pi} = \cos(k\pi) \pm \mathbf{j}\,\sin(k\pi) = (-1)^k, \quad k = 0, \pm 1, \pm 2, \ldots.$$

Note that $\alpha_0 = 1$.

Example 10.4.6: (Pulse streams) One of the practical applications in analogue synthesis where Fourier series play an important role is a stream of square pulses, which is defined by the function

$$f(t) = \begin{cases} 1, & \text{when } 0 \leqslant t < p/2, \\ 0, & \text{when } p/2 < t < T - p/2, \\ 1, & \text{when } T - p/2 < t < T. \end{cases}$$

Here p is some number between 0 and T, and $f(t+T) = f(t)$. The Fourier coefficients in complex series (10.4.12)

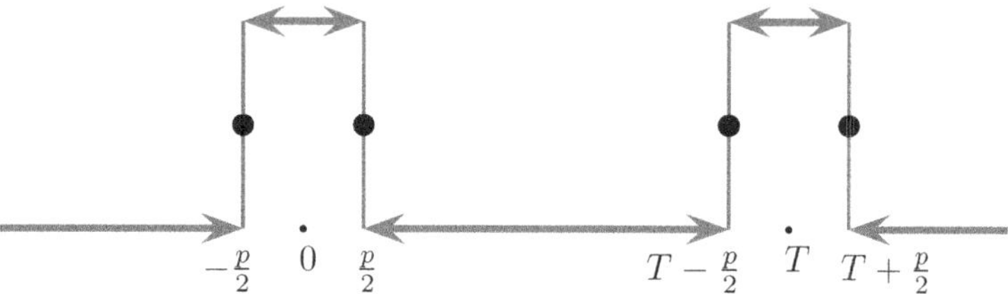

Figure 10.21: Example 10.4.6: Pulse stream.

become real numbers, given by

$$\alpha_0 = \frac{p}{T}, \quad \alpha_k = \frac{1}{T}\int_{-p/2}^{p/2} e^{-2k\pi\mathbf{j}/T}\,\mathrm{d}t = \frac{1}{k\pi}\,\sin\left(\frac{k\pi p}{T}\right), \qquad k = \pm 1, \pm 2, \ldots.$$

By increasing T while keeping p constant, the shape of the spectrum stays the same, but it is vertically scaled down in proportion, so as to keep the energy density along the horizontal axis constant.

10.4.2 The Gibbs Phenomenon

Many functions encountered in electrical engineering and in the theory of synthesized sound are intermittent but not continuous. These functions include waveforms such as the square wave or sawtooth function. Other examples are given in §10.3.3. A piecewise continuous function cannot be represented by a Taylor series globally because it is comprised by distinct smooth functions. Each of these branches may have a Taylor series expansion, but they cannot be united into a single series. Therefore, the sequence of polynomials $\{x^n\}_{n\geqslant 0}$ is not an appropriate set for expansions in the space of not smooth functions.

On the other hand, we saw previously that many intermittent functions can be expanded into a single Fourier series. Since basic trigonometric functions (sine and cosine) are infinitely differentiable, one may expect a problem representing a discontinuous function by a Fourier series. The American mathematician Josiah Willard Gibbs (1839–1903) observed in 1898 that near points of discontinuity, the N-th partial sums (10.4.1) of the Fourier series may overshoot/undershoot the jump by approximately 9%, regardless of the number of terms, N. This observation is referred to as the **Gibbs phenomenon**, which was first noticed and analyzed by the English mathematician Henry Wilbraham (1825–1883) in 1848. The term "Gibbs phenomenon" was introduced by the American mathematician Maxime Bôcher in 1906. The history of its discovery can be found in [21].

To understand the Gibbs phenomenon, we consider an example, say, the sawtooth function $\phi(\theta) = \theta/2$ on the interval $(-\pi, \pi)$. It has the following Fourier expansion (see Example 10.3.1, page 564):

$$\phi(\theta) = \frac{\theta}{2} = \sum_{k\geqslant 1} \frac{(-1)^{k+1}}{k} \sin(k\theta), \qquad \phi(\theta + 2\pi) = \phi(\theta). \tag{10.4.16}$$

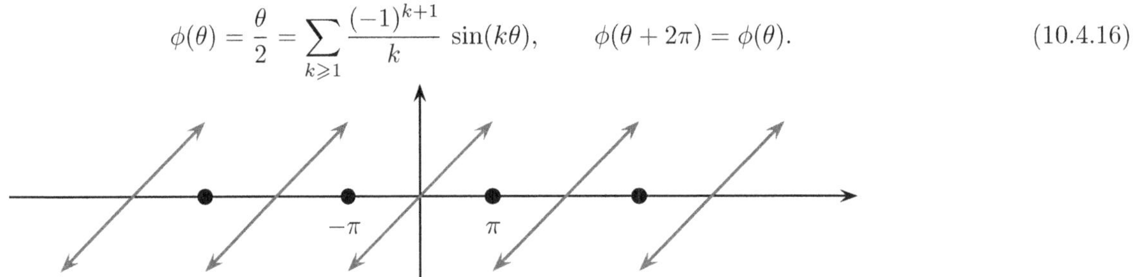

Figure 10.22: The graph of the periodically extended function $\frac{\theta}{2}$ from the interval $(-\pi, \pi)$.

This function satisfies conditions of Dirichlet's theorem 10.7, page 571; therefore, its Fourier series converges to $\phi(\theta)$ at every point $\theta \in (-\pi, \pi)$. Let

$$S_n(\theta) = \sum_{k=1}^{n} \frac{(-1)^{k+1}}{k} \sin k\theta$$

be the partial sum of the corresponding Fourier series. Although this function converges pointwise to $\phi(\theta)$ at every point, the convergence is not uniform. This means that for a given $\varepsilon > 0$, there exists N such that

$$|\phi(\theta) - S_n(\theta)| < \varepsilon$$

for all $n > N$. If this number N can be chosen independently of the point θ, we get a uniform convergence, which leads to a continuous limiting function. In our case, the convergence is not uniform and our limiting function is discontinuous. At the points of discontinuity $\theta = (2n+1)\pi$, $n = 0, \pm1, \pm2, \ldots$, the series (10.4.16) converges to zero because all of the terms are zero. The peak of the overshoot/undershoot gets closer and closer to the discontinuity while convergence holds for every particular value of θ.

To demonstrate such an overshoot/undershoot, we differentiate $S_n(\theta)$ to find its local maxima and minima. We consider only the interval $-\pi < \theta < \pi$ because $S_n(2\pi - \theta) = -S_n(\theta)$. Using the geometric series sum and Euler's formula (10.4.11), we have (according to Eq. (10.3.9), page 568)

$$S_n'(\theta) = \sum_{k=1}^{n} (-1)^k \cos k\theta = \Re \sum_{k=1}^{n} e^{\mathrm{j}k(\theta+\pi)} = \frac{\cos\frac{(n+1)(\pi+\theta)}{2} \sin\frac{n\pi + n\theta}{2}}{\cos(\theta/2)}.$$

Since the function in the denominator, $\cos(\theta/2)$, is positive throughout the interval $-\pi < \theta < \pi$, we need to consider only functions in the numerator. Hence, the zeroes of $S_n'(\theta)$ occur at $\theta = \frac{(2j+1)\pi}{n+1} - \pi$ where $\cos\frac{(n+1)(\pi+\theta)}{2} = 0$ and at $\theta = \frac{2j\pi}{n} - \pi$ where $\sin\frac{n\pi+n\theta}{2} = 0$, $j = 0 \pm 1, \pm2, \ldots$. To apply the second derivative test, we evaluate $S_n''(\theta)$ at these points:

$$S_n''\left(\frac{(2j+1)\pi}{n+1} - \pi\right) = (-1)^j \frac{1+n}{2} \csc\left(\frac{(2j+1)\pi}{n+1}\right) \sin\left(\frac{(2j+1)n\pi}{n+1}\right)$$

$$S_n''\left(\frac{2j\pi}{n} - \pi\right) = -\frac{n}{2} \cot\left(\frac{j\pi}{n}\right), \quad j = 0, \pm1, \pm2, \ldots.$$

At these points, the partial sum $S_n(\theta)$ attains either a local minimum or local maximum. These maxima and minima alternate. For instance, the first local minimum value of $S_n(\theta)$ near the point $\theta = -\pi$ happens at $\theta = \frac{\pi}{n+1} - \pi$

because $S_n''\left(\frac{\pi}{n+1} - \pi\right) > 0$. However, at the next point $\theta = \frac{3\pi}{n+1} - \pi$ the function $S_n(\theta)$ attains a maximum because $S_n''\left(\frac{3\pi}{n+1} - \pi\right) < 0$.

The value of $S_n(\theta)$ at the first minimum is

$$S_n\left(\frac{\pi}{n+1} - \pi\right) = \sum_{k=1}^{n} \frac{(-1)^{k+1}}{k} \sin\left(\frac{k\pi}{n+1} - k\pi\right) = -\frac{\pi}{n+1} \sum_{k=1}^{n} \frac{\sin \frac{k\pi}{n+1}}{\frac{k\pi}{n+1}}$$

because $\sin(\theta - k\pi) = \sin\theta \cos k\pi - \cos\theta \sin k\pi = (-1)^k \sin\theta$ based on identities $\cos k\pi = (-1)^k$, $\sin k\pi = 0$. The expression in the right-hand side is the Riemann sum for the negative of the integral (denoted by `SinIntegral` in *Mathematica*)

$$\text{Si}(\pi) = \int_0^{\pi} \frac{\sin t}{t}\, dt \approx 1.851937051982468,$$

where $\text{Si}(x) = \int_0^x \frac{\sin t}{t}\, dt$ is the sine integral. Since the exact value $\phi(-\pi) = -\pi/2 \approx -1.570796327\ldots$, the Fourier sum undershoots it by a factor of 1.1789797. Of course, the size of the discontinuity is not $\frac{\pi}{2}$ but π; hence, as a proportion of the size of the discontinuity, it is half of $1.851937\ldots/1.570796327\ldots \approx 1.1789797\ldots$ or about 8.9490%. After the series undershoots, it returns to overshoot, then undershoot again, and so on, each time with a smaller value than before. As n increases, more terms are added to the partial sum and the ripples increase in frequency and decrease in amplitude at every point. Correspondingly, both the highest peak (the overshoot) and the lowest peak (the undershoot) narrow and move to $\theta = \pi$, the point of discontinuity.

In general, if a function $f(x)$ has a finite jump discontinuity at the point x_0, the vertical span extending from the top of the overshoot to the bottom of the undershoot has length

$$\frac{2}{\pi}\,\text{Si}(\pi)\,|f(x_0+0) - f(x_0-0)| \approx 1.1789797444721672\,|f(x_0+0) - f(x_0-0)|\,. \tag{10.4.17}$$

For the sawtooth function $\phi(\theta) = \theta/2$, we see that $\phi(\pi - 0) - \phi(\pi + 0) = \pi$. Therefore, Eq. (10.4.17) predicts an overshoot having approximate height 1.85. This prediction is confirmed in the figure on front page 545, where the graph shows that partial sums have a negative peak value of about -1.85 and a positive peak value of about 1.85 near the point $x = 0$.

The same function, considered on the interval $(0, 2\pi)$, has the different Fourier series

$$\phi(\theta) = \frac{\theta}{2} = \frac{\pi}{2} - \sum_{k\geqslant 1} \frac{1}{k} \sin kx.$$

The figure at the right clearly shows the Gibbs phenomenon at the point $\theta = 0$, where partial sums with $N = 50$ terms undershoot $\phi(0+) = 0$ by approximately 0.25 and overshoot $\phi(0-) = \pi$ by about the same value. When $N \to \infty$, these undershoots/overshoots increase in magnitude and approach approximately $-0.281141/3.42273$, respectively.

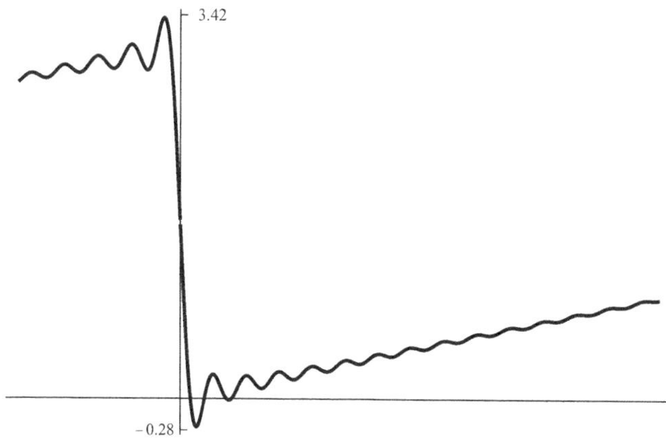

Therefore, formula (10.4.17) allows one to predict the maximum/minimum error in the Fourier pointwise approximation.

The Gibbs phenomenon is a good example to illustrate the distinction between pointwise convergence and uniform convergence. For pointwise convergence of a sequence of functions $S_n(\theta)$ to a function $\phi(\theta)$, it is required that for each value of θ, the values $S_n(\theta)$ must converge to $\phi(\theta)$. For uniform convergence, it is required that the distance between $S_n(\theta)$ and $\phi(\theta)$ is bounded by a quantity which depends on n and not on θ, and which approaches zero as n tends to infinity.

Problems

1. Prove the formula
$$D_m(\theta) = \sum_{n=-m}^{m} e^{\mathrm{j}n\theta} = 1 + 2\sum_{n=1}^{m} \cos n\theta = \frac{\sin\left(m+\frac{1}{2}\right)\theta}{\sin\left(\frac{1}{2}\,\theta\right)}.$$
The function $D_m(\theta)$ is called the Dirichlet kernel.
2. Prove the following properties of the Dirichlet kernel:
$$\text{(a)}\ D_m(0) = 2m+1; \quad \text{(b)} \int_0^{\pi} D_m(x)\,\mathrm{d}x = \pi; \quad \text{(c)} \int_{-\pi}^{\pi} D_m^2(x)\,\mathrm{d}x = \pi(2m+1).$$
3. Prove the formula
$$K_m(\theta) = \frac{1}{m+1}\sum_{n=0}^{m} D_n(\theta) = \sum_{n=-m}^{m} \frac{m+1-|n|}{m+1}\,e^{\mathrm{j}n\theta} = \frac{1}{m+1}\left(\frac{\sin\dfrac{m+1}{2}\,\theta}{\sin\left(\dfrac{1}{2}\,\theta\right)}\right)^2.$$
The function $K_m(\theta)$ is called the Fejér kernel.
4. Find the complex Fourier series for the function $\left(2-e^{\mathrm{j}x}\right)^{-1} + \left(2-e^{-\mathrm{j}x}\right)^{-1}$, $|x| < \pi$.
5. Find the complex Fourier series for the function $h(x) = \begin{cases} 1/h, & \text{for } -h/2 < x < h/2, \\ 0, & \text{elsewhere,} \end{cases}$
on the interval $(-3h, 3h)$, where $h > 0$.
6. Consider the 2-periodic function $f(x) = \begin{cases} 1, & -1 < x < 0, \\ x, & 0 < x < 1. \end{cases}$
   (a) Find its Fourier series. Observe the Gibbs phenomenon for the corresponding Fourier series: what is the highest peak you expect near the point of discontinuity $x = 0$?
   (b) Calculate Cesàro partial sum $C_{10}(x)$ and σ-factor partial sum $s_{10}(x)$ with 10 terms. By plotting these sums, do you observe the Gibbs phenomenon?
7. Consider the function of period 12 defined by the following equation on the interval $-6 \leqslant x \leqslant 6$:
$$f(x) = \begin{cases} 0, & -6 \leqslant x \leqslant -3, \\ x+3, & -3 < x \leqslant 0, \\ 3-x, & 0 \leqslant x \leqslant 3, \\ 0, & 3 \leqslant x \leqslant 6. \end{cases}$$
Find its Fourier series. By differentiating, determine the Fourier series for its first derivative $f'(x)$. Which of these two Fourier series for $f(x)$ and $f'(x)$ converges uniformly?
8. Does the function $f(x) = x^2 \sin(1/x)$ satisfy the conditions of the Dirichlet theorem?
9. **Acceleration of convergence.** Consider the series
$$S(x) = \sum_{k\geqslant 2} \frac{k}{k^2-1}\sin kx, \qquad 0 < x < \pi,$$
which converges rather slowly. Using the identity $\frac{k}{k^2-1} = \frac{1}{k} + \frac{1}{k(k^2-1)}$, we can represent $S(x)$ as the sum $S(x) = f(x) + g(x)$, where
$$f(x) = \sum_{k\geqslant 2} \frac{1}{k}\sin kx, \qquad g(x) = \sum_{k\geqslant 2} \frac{1}{k(k^2-1)}\sin kx.$$
Find the explicit formula for $f(x)$. Note that $g(x)$ is a continuous function because its Fourier series has the general term that decreases as k^{-3} when $k \to \infty$.

In each of Problems 10 through 13, assume that the given function is periodically extended outside the original interval $(0, \pi)$.

(a) Find the Fourier series for the extended function.
(b) Calculate the least square error for partial sums with $N = 10$ and $N = 20$ terms.
(c) What is the highest peak value predicted near $x = \pi$ for the partial Fourier sum?
(d) Sketch the graph of the function to which the series converges on the interval $(-\pi, 2\pi)$.

10. $f(x) = x^2 - \pi^2$;
11. $f(x) = \cos(x/2)$;
12. $f(x) = \cos^3(x)$;
13. $f(x) = \sinh x$.

14. Prove the formulas:

$$\sum_{k=1}^{n} e^{\mathrm{j}kt} = \frac{\sin(nt/2)}{\sin(t/2)}\, e^{\mathrm{j}(n+1)t/2} \quad (t \ne 2m\pi), \qquad \sum_{k=1}^{n} e^{\mathrm{j}(2k-1)t} = \frac{\sin(nt)}{\sin(t)}\, e^{\mathrm{j}nt} \quad (t \ne m\pi).$$

15. Express explicitly the Cesàro partial sum

$$\frac{S_0(x) + S_1(x) + \cdots + S_N(x)}{N+1}$$

similar to Eq. (10.4.7).

16. Find the complex Fourier series for the function $\left(3 - e^{\mathrm{j}x}\right)^{-1} + \left(3 - e^{-\mathrm{j}x}\right)^{-1}$, $|x| < \pi$.
17. Find the complex form of the Fourier series for $f(x) = e^x$ on the interval $(-\pi, \pi)$.
18. Consider the 2-periodic function $f(x) = \begin{cases} 0, & -1 < x < 0, \\ 1 - x, & 0 < x < 1. \end{cases}$
    Find its Fourier series. Observe the Gibbs phenomenon for the corresponding Fourier series: what is the highest peak you expect near the point of discontinuity $x = 0$?
19. Find the Fourier series of a square pulse function on the interval $(0, 2\pi)$:

$$s(x) = 1 \quad \text{for} \quad |x| < \pi, \qquad s(x) = 0 \quad \text{for} \quad \pi < |x| < 2\pi.$$

20. Assuming that the function $f(x) = x^2$ is extended periodically from the interval $(0, 2)$, expand it into the Fourier series. Does it converge uniformly?
21. Assuming that the function $g(x) = x^2$ is extended periodically from the interval $(-2, 2)$, expand it into the Fourier series. Does it converge uniformly?
22. Find the Cesàro partial sums for each of the functions $f(x)$ and $g(x)$ in two previous exercises. Which of these partial sums gives better approximation compared to the Fourier partial sums?
23. Consider the 4-periodic function $f(x) = \begin{cases} x^2, & -2 < x < 0, \\ 1, & 0 < x < 2. \end{cases}$
    Find its Fourier series. Observe the Gibbs phenomenon for the corresponding Fourier series: what are the highest peaks you expect near the points of discontinuity $x = 0$ and $x = 2$? Then find and plot the Cesàro partial sums; do you observe the Gibbs phenomenon?
24. Find the Fourier series for the function and determine the maximum peak its partial sums are approached,
    $f(x) = \begin{cases} x + 1, & -2 < x < 0, \\ -x, & 0 < x < 2; \end{cases} \qquad f(x+4) = f(x).$
25. **Acceleration of Convergence.** In this problem, we show how it is possible to improve the speed of convergence of a Fourier series. Consider the series

$$S(x) = \sum_{k \geqslant 1} \frac{k}{k^2 + 1} \cos\left(\frac{k\pi}{2}\right) \sin kx, \qquad 0 < x < \pi,$$

which converges rather slowly. Using the identity $\frac{k}{k^2+1} = \frac{1}{k} - \frac{1}{k(k^2+1)}$, we can represent $S(x)$ as the sum $S(x) = f(x) - g(x)$, where

$$f(x) = \sum_{k \geqslant 1} \frac{1}{k} \cos\left(\frac{k\pi}{2}\right) \sin kx, \qquad g(x) = \sum_{k \geqslant 1} \frac{1}{k(k^2+1)} \cos\left(\frac{k\pi}{2}\right) \sin kx.$$

Find the explicit formula for $f(x)$. Note that $g(x)$ is a continuous function.

In each exercise, expand the given function into complex Fourier series (10.4.12), page 576, on the interval $|-\pi, \pi|$.

26. $f(t) = e^{at}$, $a \ne 0$;
27. $f(t) = \cosh at$, $a \ne 0$;
28. $f(t) = (2 - \cos t)^{-1}$;
29. $f(t) = 4 + t$.

10.5 Even and Odd Functions

Recall that f is called an **even function** if its domain contains the point $-x$ whenever it contains the point x, and if

$$f(-x) = f(x)$$

for each x in the domain of f. Similarly, f is said to be an **odd function** if its domain contains the point $-x$ whenever it contains the point x, and if

$$f(-x) = -f(x)$$

for each x in the domain of f. Any function $f(x)$ can be decomposed uniquely as a sum of its *even part* and its *odd part:*

$$f(x) = \frac{f(x) + f(-x)}{2} + \frac{f(x) - f(-x)}{2}.$$

Geometrically, an even function is characterized by the property that its graph to the left of the ordinate is a mirror image of that to the right of the vertical axis. The graph of the odd function to the left of the ordinate is obtained from that on the right by a single rotation of π radians about an axis through the origin perpendicular to the coordinate plane.

Any function that is a linear combination of monomials x^p with even (odd) powers p is an even (odd) function. Since the Maclaurin series for cosine function $\cos x = \sum_{k\geqslant 0}(-1)^k \frac{x^{2k}}{(2k)!}$ contains only even powers, it is an even function. Similarly, the sine function $\sin x = \sum_{k\geqslant 0}(-1)^k \frac{x^{2k+1}}{(2k+1)!}$ is an example of an odd function. A sum or difference of two or more even functions is an even function; for instance, $x^2 + 1 - \cos x$ is an even function. On the other hand, a product of two even or odd functions is an even function, while the product of an even and an odd function is an odd function. For instance, $\sin^2 x$ and $\cos^2 x$ are both even functions, while $2 \sin x \cos x = \sin(2x)$ is an odd function.

If we average an even function over a symmetrical interval $-a < x < a$, we get the same result that we would obtain for the interval $-a < x < 0$ or for the interval $0 < x < a$:

$$\int_{-a}^{0} g(x)\,\mathrm{d}x = -\int_{a}^{0} g(x)\,\mathrm{d}x = \int_{0}^{a} g(x)\,\mathrm{d}x,$$

for an even function $g(x) = g(-x)$. If we average an odd function over the interval $-a < x < a$, we get the negative of the average for the interval $0 < x < a$, and the average for the interval $-a < x < 0$. For any odd function $f(x)$ and any positive number a, we have

$$\int_{-a}^{0} f(x)\,\mathrm{d}x = -\int_{0}^{a} f(x)\,\mathrm{d}x \qquad \Longrightarrow \qquad \int_{-a}^{a} f(x)\,\mathrm{d}x = 0.$$

For example, if $g(x)$ is even and periodic with period 2ℓ, then the product of $\sin(k\pi x/\ell)$ and $g(x)$ is odd for any integer k. Thus, the Fourier coefficients b_k in Eq. (10.4.4), page 571, are all zeroes. Similarly, if $f(x)$ is odd and periodic with period 2ℓ, then $\cos(k\pi x/\ell)\, f(x)$ is odd, and so the Fourier coefficients a_k in Eq. (10.4.3), page 571, are all zeroes.

Therefore, when an odd function is represented by a Fourier series, its expansion will be a sine Fourier series, so all coefficients of cosine terms are zeroes: $a_0 = 0$ and $a_n = 0$ for all n; hence,

$$f(x) = \sum_{n\geqslant 1} b_k \sin\frac{k\pi x}{\ell}, \tag{10.5.1}$$

where

$$b_k = \frac{2}{\ell}\int_0^{\ell} f(x) \sin\frac{k\pi x}{\ell}\,\mathrm{d}x = \frac{1}{\ell}\int_{-\ell}^{\ell} f(x) \sin\frac{k\pi x}{\ell}\,\mathrm{d}x, \qquad k = 1, 2, \ldots. \tag{10.5.2}$$

We refer to this series (10.5.1) as the **Fourier sine series**. It can be considered a series for the function $f(x)$ with a domain of the interval $[0, \ell]$, which is extended in an odd manner to the interval $[-\ell, 0]$ (that is, $f(-x) = -f(x)$). Similarly, if a function $g(x)$ is an even function on an interval $[-\ell, \ell]$, then its Fourier series contains only cosine functions. Therefore, such a series is called the **Fourier cosine series** (all coefficients b_n in Eq. (10.4.4) are zeroes):

$$g(x) = \frac{a_0}{2} + \sum_{n\geqslant 1} a_n \cos\frac{n\pi x}{\ell}, \tag{10.5.3}$$

where

$$a_n = \frac{2}{\ell} \int_0^{\ell} g(x) \cos \frac{n\pi x}{\ell} \, dx, \qquad n = 0, 1, 2, \ldots . \tag{10.5.4}$$

A function on an interval $[0, \ell]$ can be extended in either an odd way or an even way on the interval $[-\ell, 0]$. As a result, the same function can have two different Fourier series representation on the interval $[0, \ell]$ with respect to cosine and sine functions, depending on its extension.

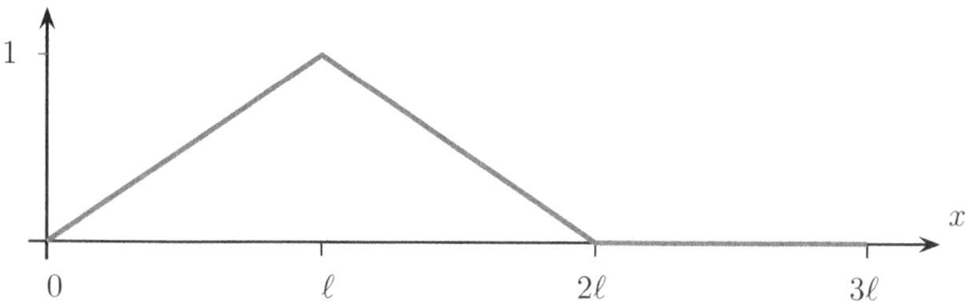

Figure 10.23: Graph of the function $f(x)$ from Example 10.5.1.

Example 10.5.1: Consider the function defined on the interval $[0, 3\ell]$, $\ell > 0$, by the equation

$$f(x) = \begin{cases} \frac{x}{\ell}, & 0 \leqslant x \leqslant \ell, \\ 2 - \frac{x}{\ell}, & \ell \leqslant x \leqslant 2\ell, \\ 0, & 2\ell \leqslant x \leqslant 3\ell. \end{cases}$$

First, we expand this function into Fourier series (10.4.2), which has period 3ℓ:

$$f(x) = \frac{1}{3} + \frac{6}{\pi^2} \sum_{n\geqslant 1} \frac{1}{n^2} \sin^2 \frac{n\pi}{3} \cos\left(\frac{2n\pi x}{3\ell} - \frac{2n\pi}{3} \right).$$

Figure 10.24: Partial Fourier sum approximation (10.4.1) of the function $f(x)$ with $N = 5$ terms.

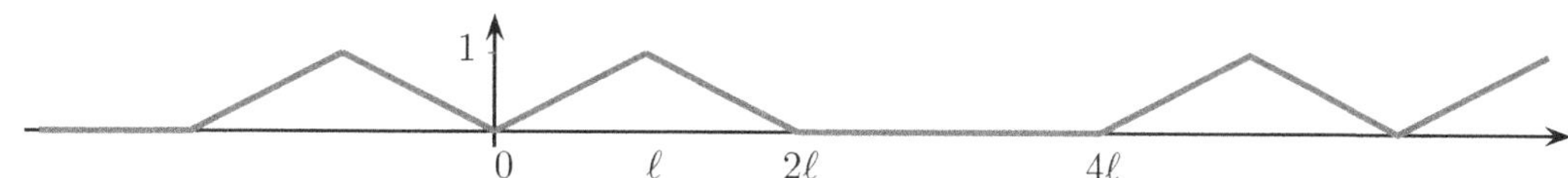

Figure 10.25: Graph of the even expansion of $f(x)$ from Example 10.5.1.

Its cosine Fourier series is

$$f(x) = \frac{1}{3} + \frac{24}{\pi^2} \sum_{n\geqslant 1} \frac{1}{n^2} \cos \frac{n\pi}{3} \sin^2 \frac{n\pi}{6} \cos\left(\frac{n\pi x}{3\ell} \right).$$

Similarly, the sine Fourier expansion becomes

$$f(x) = \frac{48}{\pi^2} \sum_{n\geqslant 1} \frac{1}{n^2} \cos \frac{n\pi}{6} \sin^3 \frac{n\pi}{6} \sin\left(\frac{n\pi x}{3\ell} \right).$$

Figure 10.26: Partial cosine Fourier sum approximation of the function $f(x)$ with $N = 5$ terms.

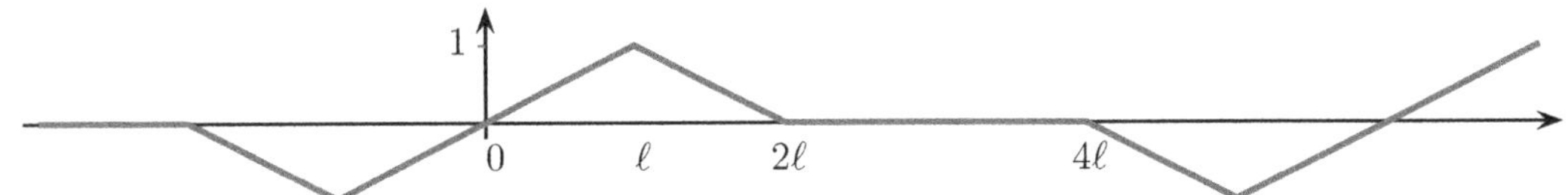

Figure 10.27: Graph of the odd expansion of $f(x)$ from Example 10.5.1.

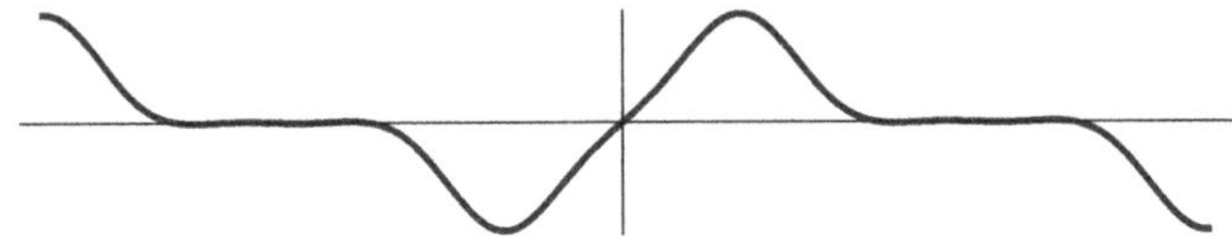

Figure 10.28: Example 10.5.1: Partial sine Fourier sum approximation of the function $f(x)$ with $N = 5$ terms.

Example 10.5.2: Consider the function $f(x) = x^2$ on the interval $[0, 2]$. First, we extend it in an even way (see Fig. 10.29(a)), which leads to the Fourier cosine series

$$x^2 = \frac{a_0}{2} + \sum_{n\geqslant 1} a_n \cos\frac{n\pi x}{2},$$

where

$$a_0 = \int_0^2 x^2\,\mathrm{d}x = \frac{8}{3},$$
$$a_n = \int_0^2 x^2 \cos\frac{n\pi x}{2}\,\mathrm{d}x = \frac{16}{n^2\pi^2}\cos(n\pi) = \frac{16}{n^2\pi^2}(-1)^n.$$

This yields the following cosine series:

$$x^2 = \frac{4}{3} + \frac{16}{\pi^2}\sum_{n\geqslant 1}\frac{(-1)^n}{n^2}\cos\left(\frac{n\pi x}{2}\right) \qquad (-2 < x < 2).$$

For $N > 0$, its partial sum approximations are

$$x^2 \sim C_N(x) = \frac{4}{3} + \frac{16}{\pi^2}\sum_{n=1}^{N}\frac{(-1)^n}{n^2}\cos\left(\frac{n\pi x}{2}\right). \tag{10.5.5}$$

Now we extend the function x^2 into a negative semi-axis in an odd way (see Fig. 10.29(b)), which leads to the sine Fourier series

$$x^2 = \sum_{n\geqslant 1} b_n \sin\left(\frac{n\pi x}{2}\right) \qquad (0 < x < 2),$$

where

$$b_n = \int_0^2 x^2 \sin\left(\frac{n\pi x}{2}\right)\mathrm{d}x = \frac{16}{n^3\pi^3}\left[(-1)^n - 1\right] - \frac{8}{n\pi}(-1)^n.$$

Therefore,

$$x^2 = -\frac{8}{\pi}\sum_{n\geqslant 1}\frac{(-1)^n}{n}\sin\frac{n\pi x}{2} - \frac{32}{\pi^3}\sum_{k\geqslant 0}\frac{1}{(2k+1)^3}\sin\frac{(2k+1)\pi x}{2},$$

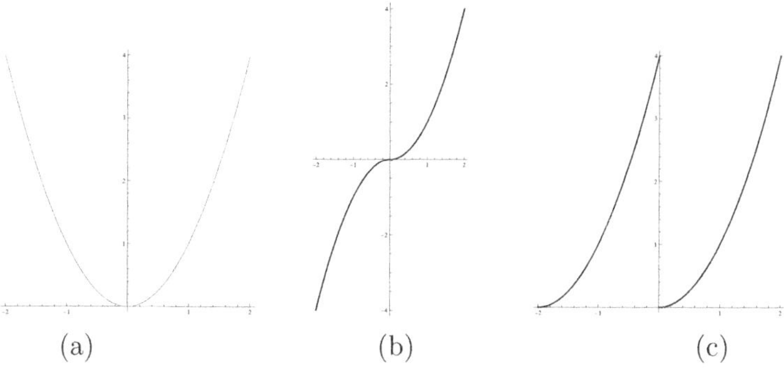

Figure 10.29: Extensions of the function x^2 to the negative semi-axis: (a) evenly, (b) oddly, and (c) in periodic way.

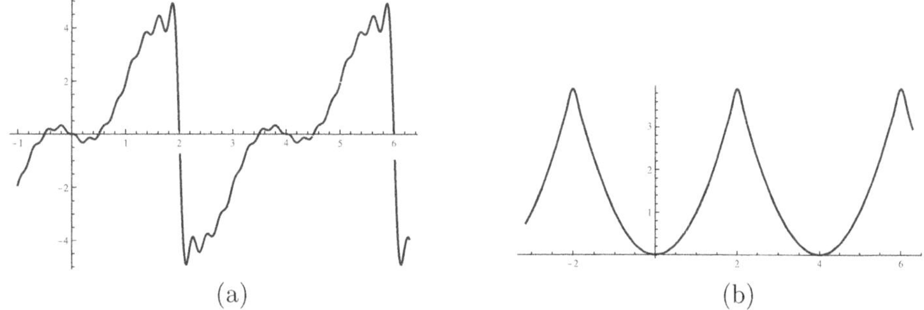

Figure 10.30: Example 10.5.2. Partial sum approximations with $N = 15$ terms of (a) $S_N(x)$ and (b) $C_N(x)$.

and its N-th partial sum becomes

$$x^2 \sim S_N(x) = -\frac{8}{\pi}\sum_{n=1}^{N}\frac{(-1)^n}{n}\sin\frac{n\pi x}{2} - \frac{32}{\pi^3}\sum_{k=0}^{(N-1)/2}\frac{1}{(2k+1)^3}\sin\frac{(2k+1)\pi x}{2},$$

where we use only odd indices: $n = 2k+1$ in the latter sum. This series is comprised of two series that converge at different speeds: the latter converges uniformly to a continuous function because its general term decays as $(2k+1)^{-3}$. However, the former converges to a piecewise continuous function. Therefore, the trigonometric sine polynomials $S_N(x)$ converge to x^2 on interval $(0, 2)$ much slower than $C_N(x)$, the cosine series. Such behavior of partial sums $S_N(x)$ and $C_N(x)$ is predicted: the even expansion of x^2 is a continuous function, while its odd expansion is discontinuous.

For the periodic extension (see Fig. 10.29(c)) with half period $\ell = 1$, we have the general Fourier series

$$x^2 = \frac{A_0}{2} + \sum_{n\geqslant 1}\left[A_n\,\cos(n\pi x) + B_n\sin(n\pi x)\right],$$

where

$$\begin{aligned}
A_0 &= \int_{-1}^{0}(x+2)^2\,\mathrm{d}x + \int_0^1 x^2\,\mathrm{d}x = \int_0^2 x^2\,\mathrm{d}x = \frac{8}{3},\\
A_n &= \int_{-1}^{0}(x+2)^2\,\cos(n\pi x)\,\mathrm{d}x + \int_0^1 x^2\,\cos(n\pi x)\,\mathrm{d}x = \frac{4}{n^2\pi^2},\\
B_n &= \int_{-1}^{0}(x+2)^2\,\sin(n\pi x)\,\mathrm{d}x + \int_0^1 x^2\,\sin(n\pi x)\,\mathrm{d}x = -\frac{4}{n\pi}.
\end{aligned}$$

This leads to

$$x^2 \sim F_N(x) \equiv \frac{4}{3} + \frac{4}{\pi^2}\sum_{n=1}^{N}\frac{1}{n^2}\cos\left(n\pi x\right) - \frac{4}{\pi}\sum_{n=1}^{N}\frac{1}{n}\sin(n\pi x) \quad (0 < x < 2).$$

To estimate the accuracy of these truncated series, let us summarize the calculations of the N-th partial sums for different values of N at $x = 2$ and the corresponding mean square errors, Δ_S, Δ_C, and Δ_F, for each approximation, in the following table:

N	sin series	Δ_S	cos series	Δ_C	Fourier series	Δ_F	True value
10	0	2.74809	3.84572	0.000753343	1.96143	0.154325	4
20	0	2.44911	3.92094	0.000101564	1.98023	0.0790706	4
100	0	0.0645216	3.98387	8.6×10^{-7}	1.99597	0.0161307	4
1000	0	0.0064813	3.99838	8.7×10^{-10}	1.99959	0.0016207	4

Here $\Delta_S(N) = \int_0^2 \left|x^2 - S_N(x)\right|^2 \mathrm{d}x$, $\Delta_C(N) = \int_0^2 \left|x^2 - C_N(x)\right|^2 \mathrm{d}x$, and $\Delta_F(N) = \int_0^2 \left|x^2 - F_N(x)\right|^2 \mathrm{d}x$ are mean square errors for each approximation. At the point $x = 1$ (where the given function x^2 is continuous), we have

N	sin series	cos series	Fourier series	True value
10	1.12562	0.993457	1.00183	1
20	0.936559	1.00183	1.00048	1
100	0.987269	1.00008	1.00002	1

The Fourier series partial sums (except the cosine series because an even extension of x^2 is a continuous function) demonstrate the Gibbs phenomenon near the points of discontinuity $x = 0$ and $x = 2$, which is clearly seen from Figures 10.30(a) and 10.31. □

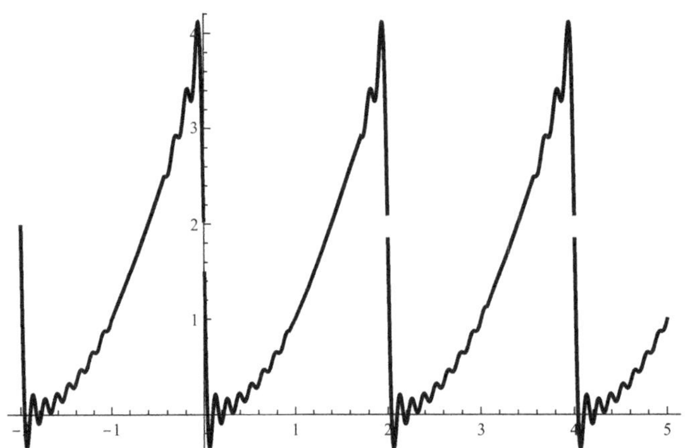

Figure 10.31: Partial Fourier sum approximation, with $N = 15$ terms, of the function x^2 extended periodically with period $T = 2$.

As previous examples show, a function defined on a finite interval $[0, \ell]$ may have three different Fourier series expansions depending on how the function is extended outside the given interval. Its selection is usually dictated by the reason of such series that may have different rates of convergence.

Sometimes we can make another observation when Fourier coefficients with even indices or odd indices are zeroes, so we need a definition.

Definition 10.9: A periodic function with period T that satisfies the relation

$$f\left(x + \frac{T}{2}\right) = -f(x)$$

is called an **odd-harmonic** function or *half-period antisymmetric*. Similarly, the function $g(x)$ that satisfies $g\left(x + \frac{T}{2}\right) = g(x)$ is referred to as an **even-harmonic** function or *half-period symmetric*.

For example, sine and cosine are odd-harmonic with $T = 2\pi$ because $\sin(x+\pi) = -\sin x$ and $\cos(x+\pi) = -\cos x$. Any function $\phi(x)$ can be represented as a sum of its half-period symmetric and antisymmetric parts:

$$\phi(x) = \frac{\phi(x) + \phi(x+T/2)}{2} + \frac{\phi(x) - \phi(x+T/2)}{2}.$$

Products and sums of half-period symmetric and antisymmetric functions have the same properties as corresponding properties of even and odd functions. Namely, a sum of two odd-harmonic functions is an odd-harmonic one. Similarly, the product of two odd-harmonic functions is an even-harmonic one, and so on. If $f(x)$ is half-period antisymmetric and $g(x)$ is half-period symmetric, then

$$\int_{T/2}^{T} f(x)\,dx = -\int_0^{T/2} f(x)\,dx \qquad \text{and} \qquad \int_{T/2}^{T} g(x)\,dx = \int_0^{T/2} g(x)\,dx \tag{10.5.6}$$

and so $\int_0^T f(x)\,dx = 0$ and $\int_0^T g(x)\,dx = 2\int_0^{T/2} g(x)\,dx$.

The trigonometric functions $\sin\frac{k\pi x}{\ell} = \sin\frac{2k\pi x}{T}$ and $\cos\frac{k\pi x}{\ell} = \sin\frac{2k\pi x}{T}$ are both half-period symmetric if k is even, and half-period antisymmetric if k is odd. Therefore, if $g(x)$ is half-period symmetric, $g(x+T/2) = g(x)$, then its Fourier coefficients with odd indices (a_{2k+1} and b_{2k+1}) are zeroes, while if $f(x)$ is antisymmetric, $f(x+T/2) = -f(x)$, then its Fourier coefficients with even indices (a_{2k} and b_{2k}) are both zeroes.

Theorem 10.10: If a periodic function $f(x)$ with period T is also an odd-harmonic function, then its Fourier series is

$$f(x) = \sum_{n\geqslant 0} a_{2n+1}\cos\frac{(2n+1)2\pi x}{T} + \sum_{n\geqslant 0} b_{2n+1}\sin\frac{(2n+1)2\pi x}{T}, \tag{10.5.7}$$

$$a_{2n+1} = \frac{4}{T}\int_0^{T/2} f(t)\cos\frac{(2n+1)2\pi t}{T}\,dt, \quad b_{2n+1} = \frac{4}{T}\int_0^{T/2} f(t)\sin\frac{(2n+1)2\pi t}{T}\,dt. \tag{10.5.8}$$

If a periodic function $g(x)$ with period T is also an even-harmonic function, then its Fourier series is

$$g(x) = \frac{a_0}{2} + \sum_{n\geqslant 1} a_{2n}\cos\frac{4n\pi x}{T} + \sum_{n\geqslant 1} b_{2n}\sin\frac{4n\pi x}{T}, \tag{10.5.9}$$

$$a_{2n} = \frac{4}{T}\int_0^{T/2} f(t)\cos\frac{4n\pi t}{T}\,dt, \quad b_{2n} = \frac{4}{T}\int_0^{T/2} f(t)\sin\frac{4n\pi t}{T}\,dt. \tag{10.5.10}$$

Example 10.5.3: The square wave sounds vaguely like the waveform produced by a clarinet:

$$s(x) = \begin{cases} 1, & \text{if } 0 < x < \pi, \\ -1, & \text{if } \pi < x < 2\pi; \end{cases} \qquad s(x+2\pi) = s(x).$$

When the function $s(x)$ is extended periodically to all values of x with period 2π, it becomes an odd function and its Fourier expansion contains only sine functions. Moreover, since it is also half-period antisymmetric, its Fourier coefficients can be calculated using either Eq. (10.4.4) or along half period interval according to Eq. (10.5.8):

$$\begin{aligned}
a_n &= \frac{1}{\pi}\int_0^{\pi}\cos(nt)\,dt - \frac{1}{\pi}\int_{\pi}^{2\pi}\cos(nt)\,dt = \frac{1}{\pi}\left.\frac{\sin(nt)}{n}\right|_{t=0}^{\pi} - \frac{1}{\pi}\left.\frac{\sin(nt)}{n}\right|_{t=\pi}^{2\pi} = 0, \\
b_n &= \frac{1}{\pi}\int_0^{\pi}\sin(nt)\,dt - \frac{1}{\pi}\int_{\pi}^{2\pi}\sin(nt)\,dt = -\frac{1}{\pi}\left.\frac{\cos(nt)}{n}\right|_{t=0}^{\pi} + \frac{1}{\pi}\left.\frac{\cos(nt)}{n}\right|_{t=\pi}^{2\pi} \\
&= \frac{1}{\pi}\left(-\frac{(-1)^n}{n} + \frac{1}{n} + \frac{1}{n} - \frac{(-1)^n}{n}\right) = \begin{cases} \frac{4}{n\pi}, & \text{if } n = 2k+1 \text{ is odd}, \\ 0, & \text{if } n \text{ is even}, \end{cases} \\
b_{2k+1} &= \frac{2}{\pi}\int_0^{\pi}\sin(2k+1)t\,dt = \frac{4}{\pi(2k+1)}, \qquad k = 0, 1, 2, \ldots.
\end{aligned}$$

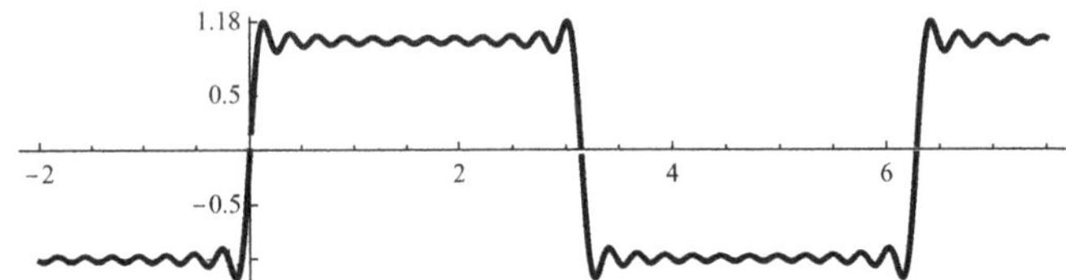

Figure 10.32: Example 10.5.3, graph of the partial sum $\frac{4}{\pi}\sum_{k=0}^{12}\frac{\sin(2k+1)x}{2k+1}$.

Therefore, the sine Fourier series for this square wave is

$$\frac{4}{\pi}\sum_{k\geqslant 0}\frac{\sin(2k+1)x}{2k+1} = \frac{4}{\pi}\left(\sin x + \frac{\sin 3x}{3} + \frac{\sin 5x}{5} + \cdots\right) = \begin{cases} 1, & \text{if } 0 < x < \pi, \\ -1, & \text{if } \pi < x < 2\pi. \end{cases}$$

The above series does not converge uniformly because its general term approaches zero as $(2k+1)^{-1}$. Let us consider its partial sums $S_n(x) = \sum_{k=0}^{n}\frac{\sin(2k+1)x}{2k+1}$. By plotting its graph (Fig. 10.32), we definitely observe the Gibbs phenomenon. The estimate (10.4.17) on page 579 predicts the height of an overshoot to be about 1.18.

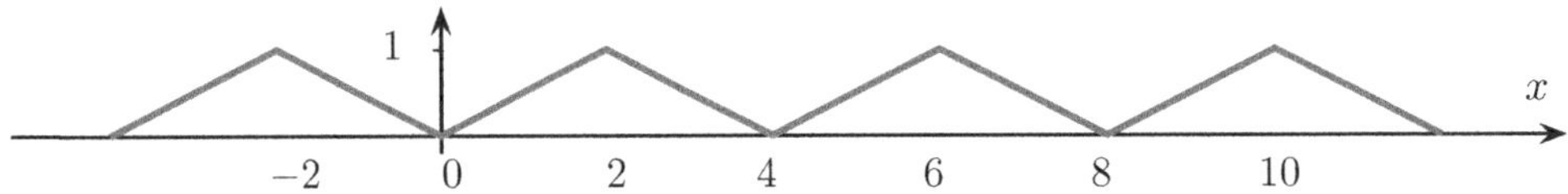

Figure 10.33: Graph of the periodic expansion of $g(x)$ from Example 10.5.4.

Example 10.5.4: Consider the continuous function of period $T = 8$ defined by the following equations in the interval $-4 \leqslant x \leqslant 4$:

$$g(x) = \begin{cases} (x+4)/2, & -4 \leqslant x \leqslant -2, \\ -x/2, & -2 \leqslant x \leqslant 0, \\ x/2, & 0 \leqslant x \leqslant 2, \\ (4-x)/2, & 2 \leqslant x \leqslant 4. \end{cases}$$

The graph of this function is shown in Fig. 10.33. This function cannot be represented by a single Taylor series, but a Fourier series is appropriate in this case. The given function is both even and half-period symmetric. Therefore, we can find coefficients in its Fourier series using either standard Euler–Fourier formulas (10.3.5) or half-period formulas (10.5.4) or Eq. (10.5.10) for half-period symmetric functions—they all yield the same Fourier cosine series:

$$g(x) = \frac{1}{2} + \sum_{n\geqslant 1} a_n \cos\frac{n\pi x}{4} = \frac{1}{2} + \sum_{n\geqslant 1} A_n \cos\frac{n\pi x}{2} = \frac{1}{2} - \frac{4}{\pi^2}\sum_{k\geqslant 0}\frac{1}{(1+2k)^2}\cos\frac{(1+2k)\pi x}{2},$$

where coefficients a_n and A_n can be calculated as follows

$$a_n = \frac{1}{4}\int_{-4}^{4} g(x)\cos\frac{n\pi x}{4}\,\mathrm{d}x = \frac{1}{2}\int_{0}^{4} g(x)\cos\frac{n\pi x}{4}\,\mathrm{d}x, \quad A_n = \int_0^2 g(x)\cos\frac{n\pi x}{2}\,\mathrm{d}x.$$

Problems

1. Evaluate $\int_0^{2\pi}\sin(\sin x)\,\sin(4x)\,\mathrm{d}x$.
2. Determine whether the given function is odd, even, or neither.

(a) $\sqrt{1+x^4}$; **(c)** $\cot 2x$; **(e)** $x^{1/3}+\sin x$;
(b) x^2+4x^5; **(d)** $x^{1/3}\cos x$; **(f)** $\csc x^2$.

3. If $f(x) = x^2 - x^3$, $0 \leqslant x \leqslant 1$, sketch the graph for $-3 < x < 3$ if
   (a) $f(x)$ is an odd function, and of period 2;
   (b) $f(x)$ is an even function, and of period 2;
   (c) $f(x)$ is of period 1;
   (d) $f(x)$ is an odd-harmonic function, and of period 2.
4. Which of the following are odd functions, which are even functions, and which are odd-harmonic functions?

   (a) $65\cos 11x + 5\cos 33x - \cos 99x$; **(b)** $5\sin 4x - 4\sin 12x + 3\sin 20x$.

5. Prove that if a function is both an odd function and an odd-harmonic function, in determining the coefficients of the odd sine terms, we may take the averages from 0 to $T/4$.
6. Prove that if a periodic function of period T is both an even function, and an odd-harmonic function, in determining the coefficients of the odd cosine terms, we may take the averages from 0 to $T/4$.
7. Find the sine and cosine Fourier series of the function defined on the interval $[0,3]$:

$$f(x) = \begin{cases} (x+2)^2, & \text{if } 0 < x < 1, \\ x^2, & \text{if } 1 < x < 3. \end{cases}$$

8. Show that the sine series of period $T = 2\ell$ for the constant function $f(x) = \pi/4$ on the interval $0 < x < \ell$ is

$$\frac{\pi}{4} = \sum_{k\geqslant 1} \frac{1}{2k-1} \sin\left(\frac{(2k-1)\pi x}{\ell}\right) = \sin\frac{\pi x}{\ell} + \frac{1}{3}\sin\frac{3\pi x}{\ell} + \frac{1}{5}\sin\frac{5\pi x}{\ell} + \cdots .$$

9. Prove the identity (established by L. Euler in 1772)

$$\frac{4}{\pi}\sum_{k\geqslant 0}(-1)^k \frac{\cos(2k+1)x}{2k+1} = \begin{cases} 1, & |x| < \frac{\pi}{2}, \\ -1, & -\frac{\pi}{2} < |x| < \pi. \end{cases}$$

10. Find the Fourier series of each of the following functions on the interval $|x| < \pi$:

    (a) $\sin^3 x$; **(b)** $\cos^4 x$; **(c)** $\sin^4 x$; **(d)** $\sin 2x \cos x$.

11. If a function is expanded in a sine series $S(x)$ of period 2ℓ, which represents it on the interval $[0,\ell[$, and a cosine series $C(x)$ of period 2ℓ, which represents it on the same interval, show that $\frac{1}{2}[S(x)+C(x)]$ is a Fourier series for the function, which is zero on the interval $-\ell < x < 0$, and equal to $f(x)$ on the interval $0 < x < \ell$.
12. Consider the function $f(x) = x^3$ on the interval $(0,2)$. Find the Fourier series, the sine Fourier series, and cosine Fourier series for this function on the given interval. Which of these series converges the fastest?

In each of Problems 13 through 20, assume that the given function is periodically extended outside the given interval $(0,\ell)$. If explicit formulas are not possible to obtain, use numerical approximations with at least 6 decimal figures.

(a) Sketch the graphs of the even extension and the odd extension of the given function of period 2ℓ over three periods.
(b) Find the Fourier cosine and sine series for the given function.
(c) Calculate the least square error for partial sums with $N = 10$ and $N = 20$ terms.

13. $f(x) = x(4-x)$, $0 < x < 4$;
14. $f(x) = \frac{\sin x}{x}$, $0 < x < \frac{\pi}{2}$;
15. $f(x) = \begin{cases} x^2, & 0 < x < 1, \\ 1, & 1 < x < 2, \\ 0, & 2 < x < 3; \end{cases}$
16. $f(x) = \begin{cases} x, & 0 < x < \pi, \\ \pi, & \pi < x < 2\pi, \\ x-\pi, & 2\pi < x < 3\pi; \end{cases}$
17. $f(x) = x^3 - x$, $0 < x < 1$;
18. $f(x) = \frac{\cos x - 1}{x}$, $0 < x < \frac{\pi}{2}$;

19. $f(x) = \begin{cases} 0, & 0 < x < 1, \\ 1, & 1 < x < 2, \\ (x-1)^2, & 2 < x < 3; \end{cases}$

20. $f(x) = \begin{cases} \sin x, & 0 < x < \pi, \\ 0, & \pi < x < 2\pi, \\ \cos x - 1, & 2\pi < x < 3\pi. \end{cases}$

Summary for Chapter 10

1. For a linear operator $L : V \to V$, a complex number λ is called its **eigenvalue** if there exists a nonzero (also called nontrivial) element y such that
$$Ly = \lambda y. \tag{10.1.1}$$
The corresponding solution y of Eq. (10.1.1) is called the eigenfunction (or eigenvector) of the operator L. The problem of determination of eigenvalues and eigenfunctions is called the **Sturm–Liouville problem**.

2. The **norm** of the function $f(x)$ (with weight $\rho > 0$) is
$$\|f\|^2 = (f, f) = \int_a^b |f(x)|^2 \rho(x)\,\mathrm{d}x. \tag{10.2.2}$$
Convergence with respect to norm (10.2.2) is referred to as L^2-convergence or mean-square convergence.

3. A square integrable function $f(x)$ can be expanded with respect to the complete set of eigenfunctions:
$$f(x) = \sum_{k\geqslant 1} c_k\, \phi_k(x), \tag{10.2.11}$$
where the coefficients
$$c_k = \frac{(f, \phi_k)}{\|\phi_k\|^2} = \frac{1}{\|\phi_k\|^2} \int_a^b f(x)\, \phi_k(x)\, \rho(x)\,\mathrm{d}x, \quad k = 1, 2, \ldots.$$
are called the **Fourier constants** (or coefficients) of $f(x)$ with respect to the set of orthogonal functions $\{\phi_k(x)\}_{k\geqslant 1}$ and the weight function $\rho(x)$.

4. A periodic function $f(x)$ of period $T = 2\ell$ can be expanded into the **Fourier series**
$$f(x) \sim \frac{a_0}{2} + \sum_{k=1}^{\infty} a_k \cos\left(\frac{k\pi x}{\ell}\right) + b_k \sin\left(\frac{k\pi x}{\ell}\right),$$
where
$$a_k = \frac{2}{T} \int_{-T/2}^{T/2} f(x) \cos\left(\frac{2k\pi x}{T}\right) \mathrm{d}x, \qquad b_k = \frac{2}{T} \int_0^T f(x) \sin\left(\frac{2k\pi x}{T}\right) \mathrm{d}x.$$

5. For a square integrable function on the interval $[-\ell, \ell]$, its Fourier series converges to $f(x)$ in the mean square sense.

6. Complex Fourier series: $f(x) \sim \sum_{k=-\infty}^{\infty} \alpha_k\, e^{\mathbf{j}k\pi x/\ell}$, where
$$\alpha_k = \frac{1}{2\ell} \int_{-\ell}^{\ell} f(x)\, e^{-\mathbf{j}k\pi x/\ell}\,\mathrm{d}x, \qquad k = 0, \pm 1, \pm 2, \ldots.$$

7. At the point of discontinuity, the Fourier series overshoots/undershoots the jump by about 8.9%. This observation is referred to as the **Gibbs phenomenon**.

8. An odd function can be expanded into the sine Fourier series
$$f(x) = \sum_{n\geqslant 1} b_k \sin\frac{k\pi x}{\ell},$$
where
$$b_k = \frac{2}{\ell} \int_0^{\ell} f(x) \sin\frac{k\pi x}{\ell}\,\mathrm{d}x = \frac{1}{\ell} \int_{-\ell}^{\ell} f(x) \sin\frac{k\pi x}{\ell}\,\mathrm{d}x, \qquad k = 1, 2, \ldots.$$

9. An even function can be expanded into the cosine Fourier series
$$g(x) = \frac{a_0}{2} + \sum_{n\geqslant 1} a_n \cos\frac{n\pi x}{\ell},$$
where
$$a_n = \frac{2}{\ell} \int_0^{\ell} g(x) \cos\frac{n\pi x}{\ell}\,\mathrm{d}x, \qquad n = 0, 1, 2, \ldots.$$

Review Questions for Chapter 10

Section 10.1

In each of Problems 1 through 6 find the eigenvalues and eigenfunctions of the given boundary value problem. Assume that all eigenvalues are real. *Hint:* Seek a solution to the Euler equation $ax^2y'' + by' + cy = 0$ in the form $y = x^m (c_1 \cos(k \ln x) + c_2 \sin(k \ln x))$.

1. $x^2y'' - 7xy' + \lambda y = 0 \quad (0 < x < 4), \qquad y(1) = 0, \quad y(4) = 0.$
2. $x^2y'' + xy' + \lambda y = 0 \quad (0 < x < 2), \qquad y(1) = 0, \quad y(2) = 0.$
3. $x^2y'' - 3xy' + \lambda y = 0 \quad (0 < x < 3), \qquad y'(1) = 0, \quad y(3) = 0.$
4. $x^2y'' + 5xy' + \lambda y = 0 \quad (0 < x < 4), \qquad y(1) = 0, \quad y'(4) = 0.$
5. $x^2y'' - 3xy' + \lambda y = 0, \quad (0 < x < \ell), \qquad y(1) = 0, \quad y(\ell) = 0, \quad \ell > 1.$
6. $x^2y'' + 5xy' + \lambda y = 0, \quad (0 < x < \ell), \qquad y(1) = 0, \quad y(\ell) = 0, \quad \ell > 1.$

In each of Problems 7 through 15 assume that all eigenvalues are real.

(a) Determine the form of the eigenfunctions and find the equation for nonzero eigenvalues.

(b) Determine whether $\lambda = 0$ is an eigenvalue.

(c) Find approximate values for λ_1 and λ_2, the nonzero eigenvalues of smallest absolute value.

(d) Estimate λ_n for large values of n.

7. $y'' + \lambda y = 0 \quad (0 < x < 1), \qquad y'(0) = 0, \quad y(1) + 4y'(1) = 0.$
8. $y'' + 4\lambda y = 0 \quad (0 < x < 1), \qquad y(0) - 3y'(0) = 0, \quad y'(1) = 0.$
9. $y'' + \lambda y = 0 \quad (0 < x < 1), \qquad 3y(0) + y'(0) = 0, \quad y(1) = 0.$
10. $y'' + 9\lambda^2 y = 0, \qquad y(-1/3) = 0, \quad y(1/3) = 0.$
11. $y'' + 4y' + (\lambda + 3)y = 0 \quad (0 < x < 1), \qquad y(0) = 0, \quad y(1) = 0.$
12. $y'' - 8y' + \lambda y = 0 \quad (0 < x < 3), \qquad y(0) = 0, \quad y'(3) = 0.$
13. $y'' + 6y' + \lambda y = 0 \quad (1 < x < 3), \qquad y'(1) = 0, \quad y(3) = 0.$
14. $y'' - 8y' + (16 + \lambda)y = 0 \quad (0 < x < 1), \qquad y'(0) = 0, \quad y(1) = 0.$
15. $y'' + 6y' + (8 + \lambda)y = 0 \quad (0 < x < 1), \qquad y'(0) = 0, \quad y'(1) = 0.$
16. Consider the Sturm–Liouville problem
$$x^2y'' = 3\lambda\left(xy' - y\right), \qquad y'(1) = 0, \quad y(2) = 0.$$
Show that its eigenvalues are complex numbers.
17. Determine the real eigenvalues and the corresponding eigenfunctions in the boundary value problem
$$y'' + 2y' - \lambda(2y + y') = 0, \qquad y'(0) = 0, \quad y'(3) = 0.$$
18. Solve the Sturm–Liouville problem for a higher order differential equation
$$y^{(4)} - \lambda^4 y = 0 \quad (0 < x < \ell), \qquad y(0) = y'(0) = 0, \quad y''(\ell) = y'''(\ell) = 0.$$
19. In some buckling problems the eigenvalue parameter appears in the boundary conditions as illustrated in the following one:
$$y^{(4)} + \lambda^2 y'' = 0 \quad (0 < x < \ell), \qquad y(0) = 0, \quad y'''(0) - 2\lambda\, y''(0) = 0, \quad y(\ell) = y'(\ell) = 0.$$
Solve the Sturm–Liouville problem and find the smallest eigenvalue.
20. A quantum particle freely moving on a circle is modeled by the Schrödinger equation and the periodic boundary conditions
$$-\frac{\hbar^2}{2m}\,\psi''(x) = E\,\psi(x), \qquad \psi(0) = \psi(\ell), \quad \psi'(0) = \psi'(\ell).$$
Solve the corresponding Sturm–Liouville problem.

The Prüfer[111] substitution is aimed at replacing unknown variables $y(x)$ and $y'(x)$ in the self-adjoint differential expression (10.1.7) with an equivalent pair of variables $R(x)$ and $\theta(x)$ according to the equations

$$p(x)y'(x) = R(x)\cos(\theta(x)), \qquad y(x) = R(x)\sin(\theta(x)).$$

The variables R and θ are polar coordinates in the Poincaré phase plane (py', y); they are referred to as the amplitude and phase variables, respectively.

[111] Ernst Paul Heinz Prüfer (1896–1934) was a German mathematician from the University of Münster.

21. Show that the Prüfer variables R and θ satisfy the polar equations
$$R^2 = (py')^2 + y^2, \qquad \tan\theta = \frac{y}{py'}.$$

22. Show that, in terms of the Prüfer variables, the self-adjoint differential operator (10.1.7) is transformed into a pair of first order differential equations
$$\begin{cases} d\theta/dx = (\lambda\rho - q)\sin^2\theta + \frac{1}{p}\cos^2\theta, \\ dR/dx = \left(p^{-1} - \lambda\rho + q\right) R\sin\theta\cos\theta. \end{cases}$$

Section 10.2 of Chapter 10 (Review)

1. Express the root mean square value of a function $f(x)$ for the interval $a < x < c$, denoted as $\overline{\overline{f}}_{ac}$, in terms of the average of the same function for the intervals $a < x < b$, denoted by $\overline{\overline{f}}_{ab}$, and $b < x < c$, denoted by $\overline{\overline{f}}_{bc}$, and the lengths of these intervals.
2. Find the norm for each of the following functions on the interval indicated
$$\textbf{(a)}\ e^x; \quad 0 < x < \pi/4; \quad \textbf{(b)}\ 1/x^2, \quad 2 < x < 5; \quad \textbf{(c)}\ 2x^2 - 4x + 3, \quad 0 < x < 2.$$
3. By taking the proper averages of the expressions
$$\textbf{(a)}\ \cos(x+a) = A + A_1\cos x + A_2\sin x, \quad \textbf{(b)}\ \sin(x+a) = B + B_1\cos x + B_2\sin x,$$
determine the values of coefficients A's and B's.
4. Find the norm of the function $f(x) = \begin{cases} x, & \text{if } 0 < x < 2, \\ 4 - x, & \text{if } 2 < x < 4; \end{cases}$ over the interval $0 < x < 4$.
5. Find the norm of the function $f(x) = \begin{cases} \cos x, & \text{if } 0 < x < \pi/2, \\ 0, & \text{if } \pi/2 < x < 2\pi; \end{cases}$ over the interval $0 < x < 2\pi$.
6. What constant function is closest in the least square sense to $\sin^2 x$ on the interval $(0, \pi)$?
7. What multiple of $\cos x$ is closest in the least square sense to $\cos^3 x$ on the interval $(0, 2\pi)$?
8. Let $Ly = y^{(4)}$. Suppose that the domain of L consists of all functions that have four continuous derivatives on the interval $[0, \ell]$ and satisfy
$$y(0) = y'(0) = 0 \qquad \text{and} \qquad y''(\ell) = y'''(\ell) = 0.$$
Show that L is a self-adjoint operator.
9. Show that the series $\phi_1(x) + \dfrac{\phi_2(x)}{\sqrt{2}} + \cdots + \dfrac{\phi_n(x)}{\sqrt{n}} + \cdots$ is not the eigenfunction series for any square integrable function, but the series $\phi_1(x) + \dfrac{\phi_2(x)}{\sqrt{2}\ln 2} + \cdots + \dfrac{\phi_n(x)}{\sqrt{n}\ln n} + \cdots$ is.
10. For what positive value of a will the functions x and $\sin x$ be orthogonal on the interval $[0, a]$?
11. Consider the Sturm–Liouville problem (10.2.4), page 555, where $p(x)$, $p'(x)$, $q(x)$, and $\rho(x)$ are continuous functions, and $p(x) > 0$, $\rho(x) > 0$.
    (a) Show that if λ is an eigenvalue and ϕ a corresponding eigenfunction, then
    $$\lambda\int_0^\ell \phi^2\rho\,dx = \int_0^\ell \left(p\phi'^2 + q\phi^2\right)dx + \frac{\alpha_1}{\beta_1}\,p(\ell)\phi^2(\ell) + \frac{\alpha_0}{\beta_0}\,p(0)\phi^2(0), \tag{10.5.1}$$
    provided that $\beta_0 \ne 0$ and $\beta_1 \ne 0$.
    (b) Modify the previous formula when $\beta_0 = 0$ or $\beta_1 = 0$.
    (c) Show that if $q(x) \geqslant 0$ and if α_0/β_0 and α_1/β_1 are nonnegative, then the eigenvalue λ is nonnegative.
    (d) Under the conditions of part (c), show that the eigenvalue λ is strictly positive unless $\alpha_0 = \alpha_1 = 0$ and $q(x) \equiv 0$ for each $x \in [0, \ell]$.
12. Let $\phi_1(x)$ and $\phi_2(x)$ be two eigenfunctions of the Sturm–Liouville problem (10.2.4) corresponding to the same eigenvalue λ. By computing the Wronskian $W[\phi_1, \phi_2]$, show that it is identically zero for all x. Using this, show that the eigenvalues of the boundary value problem (10.2.4) are all simple. *Hint:* use Abel's theorem 4.10, page 201.
13. Find ω so that the exponential functions $\phi_k(x) = e^{\mathbf{j}k\omega x}$ $(k = 0, \pm1, \pm2, \ldots)$ become orthogonal on an interval (a, b). Here $\mathbf{j}$ is the unit vector in the positive vertical direction on the complex plane such that $\mathbf{j}^2 = -1$.
14. Prove the law of cosines: $\|f + g\|^2 = \|f\|^2 + \|g\|^2 + 2(f, g)$.

In each given Sturm–Liouville problem, determine all real eigenvalues and eigenfunctions.

15. $y'' + y' + y = -\lambda y \quad (0 < x < 1),\ y(0) = y'(1) = 0;$

16. $y'' + 2y' + 2y = -\lambda y \quad (0 < x < 1),\ y'(0) = y(1) = 0;$

17. $y''+2y'+y = -\lambda y \quad (0 < x < 1),\ y(0) = y(1)+y'(1) = 0;$

18. $x^2 y'' + x y' = -\lambda y \quad (1 < x < 8),\ y(1) = y'(8) = 0.$

Section 10.3 of Chapter 10 (Review)

1. Find the Fourier series for the piecewise continuous function with period 3ℓ

$$f(x) = \begin{cases} \frac{x}{\ell}, & 0 < x \leqslant \ell, \\ 1, & \ell \leqslant x \leqslant 2\ell, \\ 3 - \frac{x}{\ell}, & 2\ell \leqslant x < 3\ell. \end{cases}$$

2. Find the Fourier series for the periodic function $f(x) = |\cos x|$ (the absolute value of $\cos x$).

3. Show that the Fourier series (10.3.3), page 563, may be written in **amplitude-phase** form

$$F(t) \sim \frac{a_0}{2} + \sum_{k\geqslant 1} d_k \sin(kt + \varphi_k).$$

The coefficient d_k is called the **amplitude** and φ_k the **phase** (or phase angle) of the kth component.

4. Let $F(t)$ be a square integrable function of the interval $[-\pi, \pi]$, and c be some constant. Given the Fourier series $F(t) = \frac{a_0}{2} + \sum_{k\geqslant 1} (a_k \cos kt + b_k \sin kt)$, what is the Fourier series for $F(ct)$?

5. A function is periodic, of period 40. Find the Fourier series which represent it, if it is defined in the interval $-20 < x < 20$ by

$$f(x) = \begin{cases} -10, & \text{if } -20 < x < 0; \\ 0, & \text{if } 0 < x < 10; \\ 20, & \text{if } 10 < x < 20. \end{cases}$$

6. Find the Fourier series for the function $t(\pi - t)$ on the interval $[0, \pi]$.

7. Find the Fourier series for the function $\pi^2 t - t^3$ on the interval $[-\pi, \pi]$.

8. Find the Fourier series for the function $\sin t + \cos 2t$ on the interval $[-\pi, \pi]$.

9. Find the Fourier series for the following piecewise continuous functions.

(a) $f(x) = \begin{cases} 0, & -\pi < x < -\pi/2, \\ x + \pi/2, & -\pi/2 < x < 0, \\ x, & 0 < x < \pi/2, \\ 0, & \pi/2 < x < \pi. \end{cases}$

(b) $f(x) = \begin{cases} 0, & -\pi < x \leqslant 0, \\ \sin x, & 0 \leqslant x < \pi. \end{cases}$

(c) $f(x) = \begin{cases} 0, & \frac{\pi}{2} < x \leqslant \frac{3\pi}{2}, \\ \cos x, & -\frac{\pi}{2} \leqslant x < \frac{\pi}{2}. \end{cases}$

(d) $f(x) = \begin{cases} 0, & -\pi < x < -\pi/2, \\ -1, & -\pi/2 < x < 0, \\ 1, & 0 < x < \pi/2, \\ 0, & \pi/2 < x < \pi. \end{cases}$

(e) $f(x) = \begin{cases} 0, & -\ell < x < 0, \\ x, & 0 < x < \ell. \end{cases}$

(f) $f(x) = \begin{cases} x^2, & -\pi < x \leqslant 0, \\ 0, & 0 \leqslant x < \pi. \end{cases}$

Section 10.4 of Chapter 10 (Review)

In each of Problems 1 through 4, assume that the given function is periodically extended outside the original interval $(0, \pi)$.

(a) Find the Fourier series for the extended function.

(b) Calculate the root mean square error for partial sums with $N = 10$ and $N = 20$ terms.

(c) What is the highest peak value predicted near $x = \pi$ for the partial Fourier sum?

(d) Sketch the graph of the function to which the series converges for three periods.

1. $f(x) = x^3$;

2. $f(x) = \cos 2x$;

3. $f(x) = \sin^2(x + \pi/2)$;

4. $f(x) = \cosh x$.

In each Exercise 5 through 10, determine the value of Gibb's overshoot/undershoot at each point of discontinuity when the given function is expanded into the Fourier series.

5. $f(x) = \begin{cases} 0, & -1 < x < 0, \\ x + \pi, & 0 < x < 1; \end{cases}$ $f(x+2) = f(x)$;

8. $f(x) = \begin{cases} x, & 0 < x < 1, \\ 2 - x, & 1 < x < 2; \end{cases}$ $f(x+2) = f(x)$;

6. $f(x) = \begin{cases} 2, & -1 < x < 0, \\ x, & 0 < x < 1; \end{cases}$ $f(x+2) = f(x)$;

9. $f(x) = \begin{cases} -1, & -1 < x < 0, \\ 2, & 0 < x < 1; \end{cases}$ $f(x+2) = f(x)$;

7. $f(x) = \begin{cases} 1, & 0 < x < 1, \\ x - 2, & 1 < x < 2; \end{cases}$ $f(x+2) = f(x)$;

10. $f(x) = \begin{cases} 2, & 0 < x < 1, \\ x - 1, & 1 < x < 2; \end{cases}$ $f(x+2) = f(x)$.

Section 10.5 of Chapter 10 (Review)

1. Which of the following functions is even, odd, or neither?

$$\textbf{(a)}\ \sin\frac{x}{2}; \qquad \textbf{(b)}\ 2x^2; \qquad \textbf{(c)}\ \cos(2x); \qquad \textbf{(d)}\ \sinh x; \qquad \text{(e)}\ e^{x^2}.$$

2. Decompose each of the following functions into the sum of an even and an odd function

$$\textbf{(a)}\ e^t; \qquad \textbf{(b)}\ t^3 - 2t^2 + 4t - 3; \qquad \textbf{(c)}\ t\,e^{-t}.$$

3. Which of the following are odd functions, which are even functions, and which are odd-harmonic functions?

$$\textbf{(a)}\ 5 - 6\cos\frac{x}{3} + 8\cos\frac{7x}{3}; \qquad \textbf{(b)}\ 5\sin\frac{x}{6} + 2\sin\frac{11x}{6}.$$

4. If a function $f(t)$ is periodic of period $T = 4\ell$, odd, and also an odd-harmonic function, show that its Fourier coefficients (10.4.3) and (10.4.4) are given by

$$a_0 = a_n = b_{2n} = 0, \qquad b_{2n+1} = \frac{2}{\ell}\int_0^{\ell} f(t)\,\sin\frac{2\pi(2n+1)t}{4\ell}\,dt \quad (n = 0, 1, \ldots).$$

5. If a function $f(t)$ is periodic of period $T = 4\ell$, even, and also an odd-harmonic function, show that its Fourier coefficients (10.4.3) and (10.4.4) are given by

$$a_0 = a_{2n} = b_n = 0, \qquad a_{2n+1} = \frac{2}{\ell}\int_0^{\ell} f(t)\,\cos\frac{2\pi(2n+1)t}{4\ell}\,dt \quad (n = 0, 1, \ldots).$$

6. Expand the function $f(x) = x$ defined on the interval $0 < x < \ell$ into Fourier sine and cosine series.

7. Prove the following expansions on the interval $(-\pi, \pi)$:

(a) $\displaystyle\sum_{n\geqslant 1}(-1)^n\frac{\sin(nx)}{n} = -\frac{x}{2}$;

(b) $\displaystyle\sum_{n\geqslant 1}(-1)^n\frac{\cos(nx)}{n^2} = \frac{3x^2-\pi^2}{12}$;

(c) $\displaystyle\sum_{k\geqslant 0}\frac{\sin(2k+1)x}{(2k+1)^2} = \operatorname{sign}(x)\left[1 - \frac{4}{\pi^2}\left(x\operatorname{sign}(x) - \frac{\pi}{2}\right)^2\right]$;

(d) $\displaystyle\sum_{k\geqslant 0}\frac{\cos(2k+1)x}{(2k+1)^2} = \frac{\pi^2}{8} - \frac{\pi}{4}|x|$;

(e) $\displaystyle\sum_{k\geqslant 1}(-1)^k\frac{\sin(2k-1)x}{(2k-1)^2} = -\frac{8x}{\pi^2}$;

(f) $\displaystyle\sum_{k\geqslant 1}(-1)^k\frac{\cos(2k-1)x}{(2k-1)^2} = \frac{4x^2}{\pi^2} - 1$ for $|x| \leqslant \frac{\pi}{2}$.

8. Prove the following expansions on the interval $(0, 2\pi)$:

(a) $\displaystyle\sum_{n\geqslant 1}\frac{\cos(nx)}{n} = -\ln\left|2\sin\frac{x}{2}\right|$;

(b) $\displaystyle\sum_{n\geqslant 1}\frac{\cos(nx)}{n^2} = \frac{3x^2 - 6\pi|x| + 2\pi^2}{12}$;

(c) $\displaystyle\sum_{n\geqslant 1}\frac{\sin(nx)}{n} = \frac{\pi - x}{2}$;

(d) $\displaystyle\sum_{n\geqslant 1}\frac{\cos(nx)}{n^3} = \frac{x\left(x^2 - 3\pi|x| + 2\pi^2\right)}{12}$;

(e) $\displaystyle\sum_{n\geqslant 1}(-1)^n\frac{\pi^2 n^2 - 6}{n^4}\cos(nx) = \frac{x^4}{8} - \frac{\pi^4}{40}$;

(f) $\displaystyle\sum_{n\geqslant 1}\frac{\sin(nx)}{n^3} = \frac{x^3 - 3\pi x^2 + 2\pi^2 x}{12}$;

(g) $\displaystyle\sum_{n\geqslant 1}\frac{\cos(nx)}{n^4} = \frac{1}{48}\left(\frac{8\pi^4}{15} - 4\pi^2x^2 + 4\pi x^3 - x^4\right)$.

9. Using expansions from Exercise 8, prove the identities

$$\text{(a)}\ \frac{\pi^2}{6} = \sum_{k\geqslant 1}\frac{1}{k^2}; \qquad \text{(b)}\ \frac{\pi^2}{8} = \sum_{k\geqslant 0}\frac{1}{(1+2k)^2}; \qquad \text{(c)}\ \frac{\pi^4}{96} = \sum_{k\geqslant 0}\frac{1}{(1+2k)^4}.$$

In each of Problems 10 through 19, assume that the given function is periodically extended outside the original interval $(0, \ell)$.

(a) Sketch the graphs of the even extension and the odd extension of the given function of period 2ℓ over three periods.

(b) Find the Fourier cosine and sine series for the given function.

(c) Calculate the root mean square error for partial sums with $N = 10$ and $N = 20$ terms.

10. $f(x) = \pi^2 - x^2,\ 0 < x < \pi.$

11. $f(x) = \begin{cases} 1, & 0 < x < 2, \\ x - 1, & 2 < x < 3. \end{cases}$

12. $f(x) = \begin{cases} 2x, & 0 < x < 2, \\ x^2, & 2 < x < 3; \end{cases}$

13. $f(x) = \begin{cases} 1, & 0 < x < \pi, \\ 0, & \pi < x < 2\pi, \\ 2, & 2\pi < x < 3\pi. \end{cases}$

14. $f(x) = (x-1)^2,\ 0 < x < 2.$

15. $f(x) = \begin{cases} x^2, & 0 < x < 1, \\ 1, & 1 < x < 3. \end{cases}$

16. $f(x) = \begin{cases} x^2, & 0 < x < 1, \\ x, & 1 < x < 2; \end{cases}$

17. $f(x) = \begin{cases} x, & 0 < x < 1, \\ 1, & 1 < x < 2, \\ 3 - x, & 2 < x < 3; \end{cases}$

18. $f(x) = \begin{cases} \frac{1}{4} - x, & 0 < x < \frac{1}{2}, \\ x - \frac{3}{4}, & \frac{1}{2} < x < 1; \end{cases}$

19. $f(x) = \begin{cases} 1 + x, & 0 < x < 1, \\ 3 - x, & 1 < x < 3. \end{cases}$

20. Show that

(a) $$\frac{4}{\ell} \sum_{k\geqslant 0} \frac{1}{2k+1} \sin \frac{(2k+1)\pi x}{\ell} = \begin{cases} 1, & 0 < x < \ell, \\ -1, & -\ell < x < 0; \end{cases}$$

(b) $$\sum_{k\geqslant 0} \frac{(-1)^k}{2k+1} \sin \frac{(2k+1)\pi x}{\ell} = -\frac{1}{2} \ln \left[\cot \left(\frac{\pi x}{2\ell} + \frac{\pi}{4} \right) \right], \qquad -\frac{\ell}{2} < x < \frac{\ell}{2};$$

(c) $$\frac{24}{\pi^2} \sum_{k\geqslant 0} \frac{(-1)^k}{(2k+1)^2} \sin \frac{(2k+1)\pi x}{6} = \begin{cases} -6 - x, & -6 < x < -3, \\ x, & -3 < x < 3, \\ 6 - x, & 3 < x < 6; \end{cases}$$

(d) $$\frac{1152}{\pi^3} \sum_{k\geqslant 0} \frac{1}{(2k+1)^3} \sin \frac{(2k+1)\pi x}{12} = 12x - x^2, \qquad 0 < x < 12;$$

(e) $$\frac{2\pi}{3} \sum_{k\geqslant 0} \frac{1}{(2k+1)} \cos \frac{(2k+1)\pi x}{\ell} = -\ln \left[\tan \left(\frac{\pi x}{2\ell} \right) \right], \qquad 0 < x < \ell;$$

(f) $$\frac{4}{\pi} \sum_{k\geqslant 0} \frac{(-1)^k}{2k+1} \cos \frac{(2k+1)\pi x}{\ell} = \begin{cases} 1, & 0 < x < \frac{\ell}{2}, \\ -1, & \frac{\ell}{2} < x < \ell; \end{cases}$$

(g) $$\frac{2\ell}{\pi^2} \sum_{k\geqslant 0} \frac{1}{(2k+1)^2} \cos \frac{(2k+1)\pi x}{\ell} = \frac{\ell}{2} - x, \qquad 0 < x < \ell.$$

21. For arbitrary positive number z, expand the function $\cos zx$ into even Fourier series on the interval $|x| < \pi$. By choosing particular values of x, prove the formulas

$$\frac{1}{\sin \pi z} = \frac{2z}{\pi} \left[\frac{1}{2z^2} + \sum_{k\geqslant 1} \frac{1}{k^2 - z^2} \right], \qquad \cot \pi z = \frac{1}{\pi} \left[\frac{1}{z} - \sum_{k\geqslant 1} \frac{2z}{k^2 - z^2} \right].$$

22. The acoustical wave form $w(t) = e^{-t^2} \cos(2\pi \cdot 200t)$ corresponds to a flute-like tone with a pitch of 200 Hz that sounds from $t \approx -2$ sec to $t \approx 2$ sec. Expand $w(t)$ into the Fourier cosine series.

23. Consider a continuous function $f(x) = x^2$ on the open interval $(0, 2)$.

(a) Find the Fourier series for $f(x)$ assuming that $f(x)$ is extended periodically with period $T = 2$.

(b) Determine points of discontinuity and find corresponding overshoots and undershoot values for the Fourier partial sums in the previous part.

(c) Expand the function $f(x)$ into the cosine Fourier series. Does the series converge uniformly?

(d) Expand the function $f(x)$ into the sine Fourier series. Does this series exhibit the Gibbs phenomenon? What are the values of overshoot and undershoot for the partial sine Fourier sums?

24. Evaluate the integral $\int_{-\pi}^{\pi} \left(t^5 \cos 5t - t^2 \sin 3t \right) dt.$ *Hint:* Don't spend more than 10 seconds.

Chapter 11

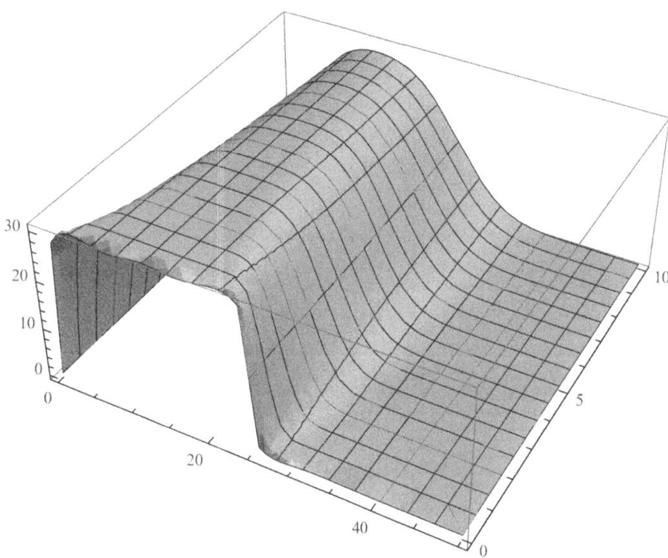

Partial Differential Equations

This chapter serves as an introduction to linear partial differential equations (PDEs, for short) that are used for modeling physical phenomena with more than one independent variable. Frequently, the independent variables are time t and one or more of the spacial variables, which are usually denoted by x, y, and z. For example, $u(x, y, z, t)$ might represent the temperature of the three-dimensional solid at spacial point (x, y, z) and time t. A partial differential equation is an equation involving partial derivatives of a dependent variable.

Traditionally, a course on partial differential equations includes three types of equations: parabolic, hyperbolic, and elliptic equations. Their prototypes originated from the heat transfer and diffusion equations ($u_t = \alpha\nabla^2 u$), wave propagation ($u_{tt} = c^2\nabla^2 u$), and time-independent or steady processes ($\nabla^2 u = 0$), respectively, where ∇ is the gradient operator. Their derivations and detailed physical interpretations can be found elsewhere [14]. Therefore, we consider three different types of equations that possess different properties and that can all be solved using the same method. This method is known as separation of variables, and it is often called the Fourier method. It was first derived and studied by J. d'Alambert and later used by L. Euler to solve spring problems.

11.1 Separation of Variables for the Heat Equation

We start this section by considering the temperature distribution in a uniform bar or wire of length ℓ with a perfectly insulated lateral surface and certain boundary and initial conditions. Let the x-axis be chosen to lie along the axis of the bar, and let $x = 0$ and $x = \ell$ denote the ends of the bar. We also assume that the cross-sectional dimensions are so small that the temperature can be considered constant within any cross-section but may vary from section to section. To describe the problem, let $u(x, t)$ represent the temperature of the point x of the bar at time t. Assuming the absence of internal sources of heat, the function $u(x, t)$ can be shown (see [14]) to satisfy the one-dimensional heat conduction (or transfer) equation:

$$\frac{\partial u}{\partial t} = \alpha\,\frac{\partial^2 u}{\partial x^2} \qquad \text{or in a shorthand notation} \quad u_t = \alpha\, u_{xx}. \tag{11.1.1}$$

The positive constant α in Eq. (11.1.1) is known as **thermal diffusivity**; it is a material property which describes the rate at which heat flows through a material, typically measured in $\texttt{mm}^2\texttt{/sec}$ or $\texttt{cm}^2\texttt{/sec}$. Usually, the thermal diffusivity is very sensitive to the variation in temperature and is chosen as a constant for simplicity; it can be calculated as

$$\alpha = \kappa/(\rho s),$$

where κ is the thermal conductivity (W/(m·°K)), ρ is the density ($\texttt{g/cm}^3$), and s is the specific heat of the material of the bar or wire (J/(kg·°K)). Common values of α are given in the following table:

Material	Thermal diffusivity $\texttt{mm}^2\texttt{/sec}$
Gold	127
Copper at 25°C	110.8
Aluminum	84.18
Water at 25°C	0.143

Material	Thermal diffusivity $\texttt{mm}^2\texttt{/sec}$
Silver, pure (99.9%)	165.63
Inconel 600 at 25°C	3.428
Steel, 1% carbon	11.72
Paraffin at 25°C	0.081

In addition, we assume that the ends of the bar are held at constant temperature 0°C and that initially at $t = 0$ the temperature distribution in the bar is known (or measured) to be $f(x)$. Then $u(x,t)$ satisfies the **boundary conditions**

$$u(0,t) = 0, \qquad u(\ell,t) = 0, \tag{11.1.2}$$

and the **initial condition**

$$u(x,0) = f(x) \qquad (0 < x < \ell). \tag{11.1.3}$$

Equation (11.1.1) together with the conditions (11.1.2) and (11.1.3) is called the **initial boundary value problem** (IBVP, for short). The conditions of (11.1.2) are called the Dirichlet boundary conditions or conditions of the first kind. The described physical problem imposes a **compatibility constraint** (or conditions) on the function $f(x)$: it must satisfy $f(0) = f(\ell) = 0$ to match the homogeneous boundary condition (11.1.2). However, the solution to the IBVP exists without any constraint on $f(x)$ at end points.

The heat transfer problem (11.1.1) – (11.1.3) is *linear* because unknown function $u(x,t)$ appears only to the first power throughout. Since the differential equation (11.1.1) and boundary conditions (11.1.2) are also *homogeneous*, we solve this problem using the method of separation of variables. While the method originated in the works by d'Alembert, Daniel Bernoulli, and Euler, its systematic application was carried out by Joseph Fourier. When boundary conditions are not homogeneous, direct application of the method is not possible. To bypass this obstacle, we present interesting and important variations on the heat problem in the following section.

To solve the IBVP (11.1.1) – (11.1.3), we apply the separation of variables method. It starts with disregarding the initial conditions (11.1.3) and seeking nontrivial (meaning not identically zero) solutions of the partial differential equation (11.1.1) subject to the **homogeneous** boundary conditions (11.1.2) and represented as a product of two functions:

$$u(x,t) = X(x)\,T(t), \tag{11.1.4}$$

where $X(x)$ is a function of x alone, and $T(t)$ is a function of t alone. Substitution of the above product into Eq. (11.1.1) yields

$$X(x)\,\dot{T}(t) = \alpha\,X''(x)\,T(t),$$

where $\dot{T}(t)$ is the first derivative of $T(t)$ with respect to time t and $X''(x)$ is the second derivative of the function $X(x)$ with respect to the spatial variable x. Dividing both sides of the latter equation by $\alpha X(x)\,T(t)$ (we will not worry about $X(x)\,T(t)$ being 0), we get

$$\frac{\dot{T}(t)}{\alpha\,T(t)} = \frac{X''(x)}{X(x)}.$$

On the left-hand side we have a function depending only on time variable t, and in the right-hand side there is a function depending only on the spacial variable x; therefore, we separate the variables! Since x and t are independent variables, the only way a function of t can equal a function of x is if both functions are constants. Consequently, there is a constant that we denote by $-\lambda$, such that

$$\frac{\dot{T}(t)}{\alpha\,T(t)} = -\lambda \qquad \text{and} \qquad \frac{X''(x)}{X(x)} = -\lambda,$$

or

$$\dot{T}(t) + \alpha\lambda T(t) = 0, \tag{11.1.5}$$

$$X''(x) + \lambda\, X(x) = 0. \tag{11.1.6}$$

Substituting $u(x,t) = X(x)\,T(t)$ into the boundary conditions (11.1.2), we get

$$X(0)\,T(t) = 0, \qquad X(\ell)\,T(t) = 0.$$

A product of two quantities is zero if at least one of them is zero. Since we are after a nontrivial solution, $T(t)$ cannot be identically zero, and we get the boundary conditions

$$X(0) = 0, \qquad X(\ell) = 0. \tag{11.1.7}$$

The homogeneous differential equation (11.1.6), which contains a parameter λ, subject to the homogeneous boundary conditions (11.1.7) has obviously a trivial solution $X(x) \equiv 0$. However, we seek nontrivial solutions; therefore, we need to find such values of parameter λ for which the problem has nontrivial solutions. Such values of λ are called **eigenvalues** and the corresponding nontrivial solutions are called **eigenfunctions**. The problem (11.1.6), (11.1.7) is usually referred to as the **Sturm–Liouville problem** (see §10.1).

Since we know from §10.1 that negative values of λ cannot be eigenvalues, we consider only the case when $\lambda > 0$. Equation (11.1.6) has the general solution

$$X(x) = A\,\cos\left(x\sqrt{\lambda}\right) + B\,\sin\left(x\sqrt{\lambda}\right),$$

where A and B are arbitrary constants. The boundary conditions demand

$$X(0) = A = 0, \qquad X(\ell) = B\,\sin\left(\ell\sqrt{\lambda}\right) = 0.$$

Since B cannot be zero (otherwise we have a trivial solution), we get the condition for determining λ:

$$\sin\left(\ell\sqrt{\lambda}\right) = 0 \qquad \Longrightarrow \qquad \sqrt{\lambda} = \frac{n\pi}{\ell}, \quad n = 1, 2, 3, \ldots.$$

Therefore, the Sturm–Liouville problem (11.1.6), (11.1.7) has a sequence of solutions, called eigenfunctions, that correspond to eigenvalues:

$$X_n(x) = \sin\frac{n\pi x}{\ell}, \qquad \lambda_n = \left(\frac{n\pi}{\ell}\right)^2, \quad n = 1, 2, 3, \ldots.$$

Turning now to Eq. (11.1.5) and substituting λ_n for λ, we get the first order differential equation

$$\dot{T} + \alpha\lambda_n T(t) = 0,$$

which has the general solution $T_n(t) = C_n\, e^{-\alpha\lambda_n t}$, where C_n is an arbitrary constant. Multiplying functions $X_n(x)$ and $T_n(t)$, we obtain a sequence

$$u_n(x,t) = X_n(x)\,T_n(t) = C_n\, e^{-\alpha\lambda_n t}\,\sin\frac{n\pi x}{\ell}, \quad n = 1, 2, \ldots,$$

of partial solutions of the differential equation (11.1.1) that satisfy the boundary conditions (11.1.2). Since both Eq. (11.1.1) and the boundary conditions (11.1.2) are homogeneous, we can use the linearity of the heat equation to conclude that any sum of partial nontrivial solutions $u_n(x,y)$ is a solution of Eq. (11.1.1), for which boundary conditions (11.1.2) hold. Therefore, the sum-function

$$u(x,t) = \sum_{n\geqslant 1} u_n(x,t) = \sum_{n\geqslant 1} C_n\, e^{-\alpha\lambda_n t}\,\sin\frac{n\pi x}{\ell} \tag{11.1.8}$$

is a solution of boundary problem (11.1.1), (11.1.2). This solution (11.1.8) is formal because we did not prove the convergence of the series (11.1.8) and its differentiability. Such a proof will require some lengthy mathematical arguments that would not improve our understanding. So we just assume that this is true. It remains only to satisfy the initial condition (11.1.3). Here is where the Fourier series makes its appearance. Assuming that the series

(11.1.8) converges uniformly (which means that we can interchange the signs of limit as $t \to 0$ and the summation), we have to satisfy

$$u(x,0) = \sum_{n\geqslant 1} u_n(x,0) = \sum_{n\geqslant 1} C_n \sin\frac{n\pi x}{\ell} = f(x). \tag{11.1.9}$$

Equation (11.1.9) looks familiar because it is a Fourier expansion of the function $f(x)$ with respect to sine functions (see §10.5). Therefore, its coefficients are

$$C_n = \frac{2}{\ell}\int_0^{\ell} f(x)\, \sin\frac{n\pi x}{\ell}\, dx \qquad (n = 1, 2, \ldots). \tag{11.1.10}$$

For a square integrable function $f(x)$, the coefficients C_n in Eq. (11.1.10) tend to zero as $n \mapsto \infty$ (see §10.4). If $t \geqslant \varepsilon > 0$, the series solution (11.1.8) converges uniformly with respect to independent variables x and t because it contains an exponentially decreasing multiple $e^{-\alpha\lambda_n t}$. Therefore, for practical evaluation of $u(x,t)$, the series (11.1.8) is truncated by keeping a few first terms. When t is small or zero, the convergence of the series $\sum_{n\geqslant 1} u_n(x,t)$ depends on the speed at which the Fourier coefficients C_n approach zero. From §10.4, it is known that the smoother the function $f(x)$ is, the faster its Fourier coefficients tend to zero, and hence, the fewer terms in the truncated series are needed to obtain an accurate approximation.

Example 11.1.1: (Zero temperature ends) Suppose that a copper rod of length 50 cm was placed into a reservoir with hot water at 50°C so that half of it is in the air at 20°C. At $t = 0$, the rod is taken out and its ends are kept at constant ambient temperature of 20°C. Let us denote the difference between the rod's temperature and the ambient temperature by $u(x,t)$, where x is the distance from the left end of the rod, $x = 0$. Then $u(x,t)$ is a solution of the following initial boundary value problem:

$$\dot{u}_t = \alpha u_{xx}, \qquad u(0,t) = u(50,t) = 0, \quad u(x,0) = f(x),$$

where $\alpha \approx 1.14$ and

$$f(x) = \begin{cases} 30, & \text{if } 0 < x < 25, \\ 0, & \text{if } 25 < x < 50. \end{cases}$$

For this problem, the solution, according to Eq. (11.1.8), is

$$u(x,t) = \sum_{n\geqslant 1} C_n\, e^{-\alpha\lambda_n t} \sin\frac{n\pi x}{50}, \qquad \lambda_n = \left(\frac{n\pi}{50}\right)^2,$$

where the coefficients C_n $(n = 1, 2, 3, \ldots)$ are calculated to be

$$C_n = \frac{2}{50}\int_0^{50} f(x)\, \sin\frac{n\pi x}{50}\, dx = \frac{30}{25}\int_0^{25} \sin\frac{n\pi x}{50}\, dx = \frac{120}{n\pi}\, \sin^2\left(\frac{n\pi}{4}\right).$$

Therefore, the solution-sum becomes

$$u(x,t) = \frac{120}{\pi}\sum_{n\geqslant 1} \frac{1}{n}\, e^{-\alpha\lambda_n t} \sin\left(\frac{n\pi x}{50}\right) \sin^2\left(\frac{n\pi}{4}\right) \qquad (\alpha \approx 1.14).$$

The M-th partial sum gives approximation to the exact solution:

$$S_M(x,t) = \frac{120}{\pi}\sum_{n=1}^{M} \frac{1}{n}\, e^{-\alpha\lambda_n t} \sin\left(\frac{n\pi x}{50}\right) \sin^2\left(\frac{n\pi}{4}\right) \qquad (\alpha \approx 1.14).$$

We can plot the M-th partial sum using *Mathematica* (see the graph of the partial sum with 100 terms on page 597).

```
SF[x_, t_, M_] := (120/Pi)* Sum[(1/n)*Exp[-(1.14*n*Pi/50)^2 *t]*
    Sin[n*Pi*x/50] *(Sin[n*Pi/4])^2, {n, 1, M}]
Plot3D[SF[x, t, 100], {x, 0, 50}, {t, 0, 10}]
```

□

Heat flux or thermal flux is the rate of heat energy transfer through a given surface per unit time. The SI unit of heat flux is watts per square meter. Heat rate is a scalar quantity, while heat flux is a vector quantity. According to Fourier's first law, the heat flux q (with units $\mathtt{W{\cdot}m}^{-2}$) is proportional to the gradient of the temperature u: $q = -\kappa \nabla u$, where κ is the thermal conductivity of the material and ∇ is the gradient operator. In a one-dimensional case, we have

$$q = -\kappa \frac{\partial u}{\partial x} = -\kappa \times \begin{cases} u_x, & \text{if the direction of flux is along } x\text{-axis}, \\ -u_x, & \text{if the direction of flux is opposite } x\text{-axis}. \end{cases}$$

Example 11.1.2: (Insulated ends) Consider a thin rod with insulated ends, which means that heat flux through the end points is zero. Since heat can neither enter nor leave the bar, all thermal energy initially present is "trapped" in the bar. The temperature $u(x, t)$ inside the rod of length ℓ at the spacial point x and at time t is a solution to the heat equation (11.1.1):

$$\dot{u}_t = \alpha\, u_{xx}, \qquad 0 < x < \ell, \quad 0 < t < t^* < \infty,$$

subject to boundary conditions of the second kind

$$u_x(0,t) = 0, \qquad u_x(\ell,t) = 0, \qquad 0 < t < t^* < \infty, \tag{11.1.11}$$

and the initial condition

$$u(x, 0) = f(x).$$

The boundary conditions (11.1.11) are usually referred to as the **Neumann**[112] boundary conditions. In general, boundary conditions of the second kind specify the normal derivative on the boundary of the spacial domain,

To solve this initial boundary value problem, we use separation of variables. So we seek partial nontrivial solutions of the heat equation subject to the Neumann boundary conditions and represented as the product of two functions

$$u(x, t) = X(x)\, T(t).$$

Substituting this form into the heat equation, we get two differential equations that we met before:

$$\dot{T} + \alpha\lambda T(t) = 0, \tag{11.1.5}$$

$$X''(x) + \lambda\, X(x) = 0. \tag{11.1.6}$$

From the boundary conditions (11.1.11), we have

$$X'(0) = 0, \qquad X'(\ell) = 0.$$

which, together with Eq. (11.1.6), constitute the Sturm–Liouville problem for $X(x)$. Note that $\lambda = 0$ is an eigenvalue, to which corresponds the eigenfunction $X_0 = 1$ (or any constant). Indeed, setting $\lambda = 0$ in Eq. (11.1.6), we have $X'' = 0$, so the general solution becomes $X(x) = C_1 + C_2 x$. Since $X' = C_2$, the boundary condition causes $C_2 = 0$ and the constant $X(x) = C_1$ is an eigenfunction, for arbitrary C_1.

Assuming that $\lambda > 0$, we obtain the general solution of Eq. (11.1.11):

$$X(x) = C_1 \cos\left(x\sqrt{\lambda}\right) + C_2 \sin\left(x\sqrt{\lambda}\right)$$

for constants C_1 and C_2. Since its derivative is $X'(x) = -C_1\sqrt{\lambda}\,\sin\left(x\sqrt{\lambda}\right) + C_2\sqrt{\lambda}\,\cos\left(x\sqrt{\lambda}\right)$, we get from the boundary conditions that

$$C_2\,\sqrt{\lambda} = 0 \qquad \text{and} \qquad C_1\sqrt{\lambda}\,\sin\left(\ell\sqrt{\lambda}\right) + C_2\sqrt{\lambda}\,\cos\left(\ell\sqrt{\lambda}\right) = 0.$$

Remember that we assume $\lambda > 0$, so $C_2 = 0$, and from the latter equation, it follows that

$$C_1\, \sin\left(\ell\sqrt{\lambda}\right) = 0.$$

[112]Carl (also Karl) Gottfried Neumann (1832–1925) was a German mathematician who can be considered as the initiator of the theory of integral equations. For most of his career, Neumann was a professor at the universities of Halle, Basel, Tübingen, and Leipzig.

If we choose $C_1 = 0$, then we get a trivial solution. Therefore, we reject such an assumption and from the equation $\sin\left(\ell\sqrt{\lambda}\right) = 0$ find eigenvalues and their eigenfunctions:

$$\lambda_n = \left(\frac{n\pi}{\ell}\right)^2, \qquad X_n(x) = \cos\frac{xn\pi}{\ell}, \quad n = 0, 1, 2, \ldots.$$

The case $n = 0$ incorporates the eigenvalue $\lambda = 0$. Substituting $\lambda = \lambda_n$ into the equation for $T(t)$: $\dot{T} + a^2\lambda T = 0$, and solving it, we find

$$T_n(t) = C_n\, e^{-\alpha\lambda_n t},$$

where C_n is a constant. Now we can form the solution of the given initial boundary value problem as the sum of all partial nontrivial solutions:

$$u(x,t) = \sum_{n\geqslant 0} X_n(x)\, T_n(t) = \sum_{n\geqslant 0} C_n\, e^{-\alpha\lambda_n t}\, \cos\frac{xn\pi}{\ell}.$$

To satisfy the initial condition $u(x,0) = f(x)$, we have to choose the coefficients C_n in such a way that

$$\sum_{n\geqslant 0} C_n\, X_n(x) = \sum_{n\geqslant 0} C_n\, \cos\frac{xn\pi}{\ell} = f(x).$$

Since this is a Fourier cosine series for $f(x)$, we get the coefficients (see Eq. (10.5.4) on page 583)

$$C_n = \frac{2}{\ell}\int_0^{\ell} f(x)\, \cos\frac{xn\pi}{\ell}\, \mathrm{d}x, \qquad n = 0, 1, 2, \ldots.$$

The temperature inside the bar tends toward a constant nonzero value:

$$\lim_{t\to\infty} u(x,t) = C_0 = \frac{2}{\ell}\int_0^{\ell} f(x)\, \mathrm{d}x$$

because all other terms in the sum have an exponential multiple $e^{-\alpha\lambda_n t}$, which goes to zero as time elapses.

Example 11.1.3: (One insulated end) Consider a thin rod of length ℓ with one insulated end $x = 0$ and other end $x = \ell$ is maintained at zero degrees. Then $u(x,t)$, the temperature in the rod at section x and time t, is a solution of the following initial boundary value problem:

$$\dot{u}_t = \alpha\, u_{xx}, \qquad u'_x(0,t) = 0, \quad u(\ell,t) = 0, \quad u(x,0) = f(x).$$

The compatibility conditions read as $f'(0) = f(\ell) = 0$. Our objective is to determine how the initial temperature distribution $f(x)$ within the bar changes as time progresses. Following the general procedure of Fourier's method, we seek partial nontrivial solutions in the form $u(x,t) = X(x)\,T(t)$, which yields the following Sturm–Liouville problem:

$$X''(x) + \lambda X(x) = 0, \qquad X'(0) = 0, \quad X(\ell) = 0.$$

Substituting the general solution $X(x) = C_1\,\cos\left(x\sqrt{\lambda}\right) + C_2\,\sin\left(x\sqrt{\lambda}\right)$ into the boundary conditions, we get

$$C_2 = 0, \qquad C_1\,\cos\left(\ell\sqrt{\lambda}\right) = 0.$$

Hence, λ must be a solution of the transcendental equation $\cos\left(\ell\sqrt{\lambda}\right) = 0$, so $\ell\sqrt{\lambda} = \frac{\pi}{2} + n\pi$, and we find the solution of the Sturm–Liouville problem:

$$\lambda_n = \left(\frac{\pi(1+2n)}{2\ell}\right)^2, \qquad X_n(x) = \cos\left(\frac{x\pi(1+2n)}{2\ell}\right), \qquad n = 0, 1, 2, \ldots.$$

This leads to a series solution:

$$u(x,t) = \sum_{n\geqslant 0} c_n\, e^{-\alpha\lambda_n t}\, \cos\left(\frac{x\pi(1+2n)}{2\ell}\right).$$

To satisfy the initial condition $u(x,0) = f(x)$, we should choose the coefficients so that

$$C_n = \frac{2}{\ell}\int_0^{\ell} f(x)\, \cos\left(\frac{x\pi(1+2n)}{2\ell}\right) \mathrm{d}x, \qquad n = 0, 1, 2, \ldots.$$

11.1.1 Two-Dimensional Heat Equation

Suppose that a thin solid object has a rectangular shape with insulated faces and of dimensions a and b. Denoting by $u(x,y,t)$ its temperature at the point (x,y) and time t, the two-dimensional heat equation becomes

$$u_t = \alpha \nabla^2 u \qquad \text{or} \qquad u_t = \alpha \left(u_{xx} + u_{yy} \right). \tag{11.1.12}$$

Here the constant α denotes the thermal diffusivity of the material, and ∇ is the gradient operator. We consider the Dirichlet boundary conditions when its edges are kept at zero temperature:

$$\begin{aligned} u(0,y,t) &= u(a,y,t) = 0, \quad 0 < y < b, \quad 0 < t, \\ u(x,0,t) &= u(x,b,t) = 0, \quad 0 < x < a, \quad 0 < t. \end{aligned} \tag{11.1.13}$$

Assuming that the initial temperature distribution $f(x,y)$ is known, we get the initial condition:

$$u(x,y,0) = f(x,y), \quad 0 < x < a, \quad 0 < y < b. \tag{11.1.14}$$

The solution of the problem is based on the separation of variables and follows a step-by-step procedure used to obtain a solution of one-dimensional heat equation. The details are outlined in Problem 30 (page 604). This gives the following "explicit" formula for the unique (formal) solution:

$$u(x,y,t) = \sum_{n=1}^{\infty} \sum_{k=1}^{\infty} A_{nk}\, e^{-\alpha \lambda_{nk} t} \sin \frac{k\pi x}{a} \sin \frac{n\pi y}{b}, \tag{11.1.15}$$

where

$$\lambda_{nk} = \pi^2 \left(\frac{k^2}{a^2} + \frac{n^2}{b^2} \right), \quad k, n = 1, 2, \ldots, \tag{11.1.16}$$

and

$$A_{nk} = \frac{4}{ab} \int_0^a \mathrm{d}x \int_0^b \mathrm{d}y\, f(x,y) \sin \frac{k\pi x}{a} \sin \frac{n\pi y}{b}. \tag{11.1.17}$$

Problems

In each of Problems 1 through 4, determine whether the method of separation of variables can be used to replace the given partial differential equation by a pair of ordinary differential equations with a parameter λ.

1. $u_t = x^2 u_{xx}$,
2. $u_t = x^2 u_{xx} + t^2 u_{xt}$,
3. $x\, u_t = t\, u_{xx}$,
4. $u_{tt} = u_{xx} + u_{xt}$.

5. Scaling is a common procedure in solving differential equations by introducing dimensionless variables.
   (a) Show that if the dimensionless variable $\xi = x/\ell$ is introduced, the heat equation (11.1.1) becomes
   $$\frac{\partial u}{\partial t} = \frac{\alpha}{\ell^2} \frac{\partial^2 u}{\partial \xi^2}, \qquad 0 < \xi < 1.$$
   (b) Show that if the dimensionless variable $\tau = \alpha t$ is introduced, the heat equation (11.1.1) becomes
   $$\frac{\partial u}{\partial \tau} = \frac{\partial^2 u}{\partial x^2}.$$

In Problems 6 through 11, find a formal solution to the given initial boundary value problem subject to the Dirichlet conditions

$$u_t = \alpha u_{xx}, \qquad u(0,t) = u(\ell,t) = 0, \quad u(x,0) = f(x),$$

when the coefficient of diffusivity α, the length of the rod ℓ, and the initial temperature $f(x)$ are specified.

6. $\alpha = 2$, $\ell = \pi$, and $f(x) = 7 \sin 3x$.
7. $\alpha = 1$, $\ell = 4$, and $f(x) = x^2$ (violates the compatibility constraint).
8. $\alpha = 2$, $\ell = 3$, and $f(x) = 5$ (violates the compatibility constraint).
9. $\alpha = 3$, $\ell = 2$, and $f(x) = x^2(4 - x^2)$.

10. $\alpha = 2$, $\ell = 5$, and $f(x) = \begin{cases} x, & \text{if } 0 \leqslant x \leqslant 2, \\ (10-2x)/3, & \text{if } 2 \leqslant x \leqslant 5. \end{cases}$

11. $\alpha = 1$, $\ell = 6$, and $f(x) = \begin{cases} 0, & \text{if } 0 \leqslant x \leqslant 1, \\ 1, & \text{if } 1 \leqslant x \leqslant 3, \\ 0, & \text{if } 3 \leqslant x \leqslant 6. \end{cases}$

In Problems 12 through 15, find a formal solution to the given initial boundary value problem subject to the Neumann conditions

$$u_t = \alpha u_{xx}, \qquad u_x(0,t) = u_x(\ell,t) = 0, \quad u(x,0) = f(x),$$

when the coefficient of diffusivity α, the length of the rod ℓ, and the initial temperature $f(x)$ are specified.

12. $\alpha = 3$, $\ell = 3$, and $f(x) = x^2(3-x)^2$.
13. $\alpha = 1$, $\ell = 1$, and $f(x) = e^x$ (violates the compatibility constraint).
14. $\alpha = 2$, $\ell = 1$, and $f(x) = 1 - \sin(3\pi x)$ (violates the compatibility constraint).
15. $\alpha = 2$, $\ell = 4$, and $f(x) = 1 - \cos(3\pi x)$.

In Problems 16 through 21, find a formal solution to the given initial boundary value problem with boundary conditions of the third kind

$$u_t = \alpha u_{xx}, \qquad u(0,t) = u_x(\ell,t) = 0, \quad u(x,0) = f(x),$$

when the coefficient of diffusivity α, the length of the rod ℓ, and the initial temperature $f(x)$ are specified.

16. $\alpha = 2$, $\ell = \pi/2$, and $f(x) = 3\sin 5x - 7\sin 9x$.
17. $\alpha = 1$, $\ell = 1$, and $f(x) = x^2\left(1 - x^2\right)$ (violates the compatibility constraint).
18. $\alpha = 2$, $\ell = \pi/2$, and $f(x) = x$ (violates the compatibility constraint).
19. $\alpha = 3$, $\ell = 3$, and $f(x) = x(3-x)^2$.
20. $\alpha = 4$, $\ell = 2$, and $f(x) = \sin(5\pi x/4)$.
21. $\alpha = 9$, $\ell = 3$, and $f(x) = 1 - \cos(5\pi x)$.

In Problems 22 through 27, find a formal solution to the given initial boundary value problem with boundary conditions of the third kind

$$u_t = \alpha u_{xx}, \qquad u_x(0,t) = u(\ell,t) = 0, \quad u(x,0) = f(x),$$

when the coefficient of diffusivity α, the length of the rod ℓ, and the initial temperature $f(x)$ are specified.

22. $\alpha = 2$, $\ell = \pi/2$, and $f(x) = 4\cos 3x - 6\cos 5x$.
23. $\alpha = 3$, $\ell = 3$, and $f(x) = x$ (violates the compatibility constraint).
24. $\alpha = 1$, $\ell = 2$, and $f(x) = \sin \pi x$ (violates the compatibility constraint).
25. $\alpha = 2$, $\ell = 2$, and $f(x) = x^2 - 4$.
26. $\alpha = 2$, $\ell = 4$, and $f(x) = 8\cos(7\pi x/8)$.
27. $\alpha = 9$, $\ell = 3$, and $f(x) = x^2 - 9$.
28. Consider the conduction of heat in a copper rod ($\alpha \approx 1.11\ \mathtt{cm}^2/\mathtt{sec}$) 50 cm in length whose one end $x = 0$ is maintained at 0°C while the other one $x = 50$ is insulated. At $t = 0$, the temperature profile is 0°C for $0 < x < 25$ and $x - 25$ for $25 < x < 50$.
    (a) Find the temperature distribution $u(x,t)$.
    (b) Plot u versus x for $t = 0.5$, $t = 1$, and $t = 5$.
    (c) Determine the steady state temperature in the rod when $t \mapsto +\infty$.
    (d) Draw a three-dimensional plot of u versus x and t.
29. In the previous problem, find the time that will elapse before the left end $x = 50$ cools to a temperature of 10°C if the bar is made of

    (a) cooper $\alpha \approx 1.11$; **(b)** molybdenum $\alpha \approx 0.54$; **(c)** silver $\alpha \approx 1.66$.
30. Prove the formulas (11.1.15) – (11.1.17) by performing the following steps:
    (a) Assuming that the heat equation (11.1.12) has a partial, nontrivial solution of the form $u(x,y,t) = v(x,y)\,T(t)$, derive the differential equations for functions $v(x,y)$ and $T(t)$:

    $$\dot{T} + \alpha\lambda T(t) = 0 \qquad \text{and} \qquad \nabla^2 v(x,y) + \lambda\, v(x,y) = 0,$$

    where λ can be any constant.

(b) Derive the boundary conditions for $v(x, y)$:

$$v(0, y, t) = v(a, y, t) = 0, \quad 0 < y < b,$$
$$v(x, 0, t) = v(x, b, t) = 0, \quad 0 < x < a.$$

(c) Assuming that a solution of the Sturm–Liouville problem for $v(x, y)$ has the form $v(x, y) = X(x)\, Y(y)$, derive the corresponding Sturm–Liouville problems for $X(x)$ and $Y(y)$:

$$X''(x) + \mu\, X(x) = 0, \quad X(0) = X(a) = 0;$$
$$Y''(y) + (\lambda - \mu)\, Y(y) = 0, \quad Y(0) = Y(b) = 0.$$

(d) Solve these Sturm–Liouville problems for $X(x)$ and $Y(y)$ to determine the eigenfunctions:

$$X_k(x) = \sin \frac{k\pi\, x}{a}, \qquad Y_n(y) = \sin \frac{n\pi\, y}{b}, \qquad \lambda_{nk} = \pi \sqrt{\frac{k^2}{a^2} + \frac{n^2}{b^2}}, \qquad k, n = 1, 2, \ldots.$$

(e) Substitute the eigenvalue λ_{nk} that depends on two parameters n and k into the differential equation for $T(t)$ and solve it:

$$T_{nk}(t) = A_{nk}\, e^{-\alpha \lambda_{nk}^2 t}, \qquad k, n = 1, 2, \ldots.$$

(f) Take a double infinite series of the product $T_{nk}(t) X_k(x) Y_n(y)$ as a formal solution of the given IBVP and find its coefficients (11.1.17).

31. In mathematical finance, the Black–Scholes[113] equation is a partial differential equation (PDE) governing the price evolution of a European call or European put under the Black–Scholes model. Broadly speaking, the term may refer to a similar PDE that can be derived for a variety of options, or more generally, derivatives.

For a European call or put on an underlying stock paying no dividends, the equation is:

$$\frac{\partial V}{\partial t} + \frac{1}{2} \sigma^2 x^2 \frac{\partial^2 V}{\partial x^2} + rx \frac{\partial V}{\partial x} - r\, V = 0,$$

where $V(x, t)$ denotes the value V of the option to buy or sell a particular security at price x and time t. The parameter σ^2 is a measure of the volatility of the security's return to the investor, and the constant r is the current interest rate on the risk free investment such as a government bond. The option to buy or sell is called a *derivative* of the underlying security. The formula led to a boom in options trading and legitimized scientifically the activities of the Chicago Board Options Exchange and other options markets around the world. The key financial insight behind the equation is that one can perfectly hedge the option by buying and selling the underlying asset in just the right way and consequently "eliminate risk."

Show that the Black–Scholes differential equation has solutions of the form $V(x, t) = C\, e^{-\alpha t} x^{\beta}$, assuming $x > 0$, and α, β, and C are positive constants. Find the equation that relates these constants.

In Problems 32 through 37, find a formal solution to the given initial boundary value problem subject to the Dirichlet conditions

$$u_t = \alpha\, u_{xx} \qquad u(0, t) = u(\ell, t) = 0, \quad u(x, 0) = f(x),$$

when the coefficient of diffusivity α, the length of the rod ℓ, and the initial temperature $f(x)$ are specified.

32. $\alpha = 5$, $\ell = 4$, and $f(x) = 6 \sin 2\pi x$.

33. $\alpha = 9$, $\ell = 3$, and $f(x) = x^3$ (violates the compatibility constraint).

34. $\alpha = 16$, $\ell = 2$, and $f(x) = 2 - x$ (violates the compatibility constraint).

35. $\alpha = 4$, $\ell = 2$, and $f(x) = x(2 - x)$.

36. $\alpha = 25$, $\ell = 10$, and $f(x) = \begin{cases} 3x, & \text{if } 0 \leqslant x \leqslant 4, \\ 20 - 2x, & \text{if } 4 \leqslant x \leqslant 10. \end{cases}$

37. $\alpha = 4$, $\ell = 4$, and $f(x) = \begin{cases} 0, & \text{if } 0 \leqslant x \leqslant 1, \\ 2, & \text{if } 1 \leqslant x \leqslant 2, \\ 0, & \text{if } 2 \leqslant x \leqslant 4. \end{cases}$

[113] The Black–Scholes model was first published by an American economist Fischer Black (1938–1995) and a Canadian-American financial economist Myron Scholes (born in 1941) in their 1973 paper, "The Pricing of Options and Corporate Liabilities," published in the *Journal of Political Economy*.

11.2 Other Heat Conduction Problems

Previously, we considered initial boundary value problems for heat transfer equations when both the equation and the boundary conditions were homogeneous. Now we consider problems when neither the equation nor the boundary conditions are homogeneous and show that it is not an obstacle for the separation of variables method. To demonstrate how the method works, we will do some examples.

We start with a heat conduction problem subject to "arbitrary" boundary conditions of the first kind (Dirichlet):

$$\dot{u}_t = \alpha\, u_{xx} + \Phi(x,t), \qquad u(0,t) = T_0(t), \quad u(\ell,t) = T_\ell(t), \quad u(x,0) = f(x),$$

where $f(x)$ provides the initial temperature distribution in the bar of length ℓ, $\Phi(x,t)$ is a given function representing a rate of change in the temperature provided by external sources, and $T_0 = T_0(t)$, $T_\ell = T_\ell(t)$ are known functions of time. Since the separation of variables method can be applied only when the boundary conditions are homogeneous, we shift the boundary data. In other words, we represent the required solution $u(x,t)$ as the sum of two functions

$$u(x,t) = v(x,t) + w(x,t),$$

where $v(x,t)$ is any function that satisfies the boundary conditions. For instance, we choose it as

$$v(x,t) = \frac{x}{\ell}\, T_\ell(t) + \frac{\ell - x}{\ell}\, T_0(t).$$

Since $v(x,t)$ is a linear function in x, we have $v_{xx} \equiv 0$. This yields the following problem for unknown function $w = u - v$:

$$\dot{w}_t = \alpha\, w_{xx} - \dot{v} + \Phi(x,t), \qquad w(0,t) = 0, \quad w(\ell,t) = 0, \qquad w(x,0) = f(x) - v(x,0).$$

If we set $F(x,t)$ to be $\Phi(x,t) - \dot{v}$ and $\varphi(x) = f(x) - v(x,0)$, we get the initial boundary value problem for determining $w(x,t)$:

$$\dot{w}_t = \alpha\, w_{xx} + F(x,t), \qquad w(0,t) = 0, \quad w(\ell,t) = 0, \qquad w(x,0) = \varphi(x), \tag{11.2.1}$$

which is similar to one for $u(x,t)$, but it has homogeneous boundary conditions. Our objective now is to solve a nonhomogeneous heat equation subject to homogeneous boundary conditions as specified in problem (11.2.1). Note that temperature $u(x,t)$ does not depend on our choice of the function $v(x,t)$; however, the functions $F(x,t)$ and $\varphi(x)$ do.

When the partial differential equation is not homogeneous, the separation of variables method asks us to find first eigenfunctions and eigenvalues of the Sturm–Liouville problem that corresponds to the similar problem for a homogeneous equation subject to the homogeneous boundary conditions:

$$\dot{w}_t = \alpha\, w_{xx}, \qquad w(0,t) = 0, \quad w(\ell,t) = 0, \qquad w(x,0) = \varphi(x).$$

Since its solution was obtained previously in §11.1, we seek a formal solution of the initial boundary value problem (11.2.1) in the form of an infinite series over eigenfunctions:

$$w(x,t) = \sum_{n\geqslant 1} C_n(t)\, X_n(x), \tag{11.2.2}$$

where the eigenfunctions $X_n(x) = \sin\left(\frac{nx\pi}{\ell}\right)$ were obtained in §11.1, but $C_n(t)$ should be determined. The next step consists of expanding the function $F(x,t)$ into a Fourier series with respect to the same system of eigenfunctions:

$$F(x,t) = \sum_{n\geqslant 1} f_n(t)\, X_n(x), \qquad f_n(t) = \frac{2}{\ell}\int_0^\ell F(x,t)\, \sin\frac{xn\pi}{\ell}\, dx.$$

Substitution of the Fourier series for $w(x,t)$ and $F(x,t)$ into the given heat equation yields

$$\dot{w}_t = \sum_{n\geqslant 1} \dot{C}_n(t)\, X_n(x) = a^2\, w_{xx} + F(x,t) = a^2 \sum_{n\geqslant 1} C_n(t)\, X_n''(x) + \sum_{n\geqslant 1} f_n(t)\, X_n(x).$$

Since $X_n'' = -\lambda_n\, X_n$, we get

$$\dot{w}_t = \alpha\, w_{xx} + F(x,t) = \sum_{n\geqslant 1} \dot{C}_n(t)\, X_n(x) = -\alpha \sum_{n\geqslant 1} C_n(t)\, \lambda_n\, X_n(x) + \sum_{n\geqslant 1} f_n(t)\, X_n(x).$$

The above equation can be reduced to one sum:

$$\sum_{n\geqslant 1}\left[\dot{C}_n(t)\,X_n(x)+\alpha\lambda_n\,X_n(x)-f_n(t)\,X_n(x)\right]=0,$$

or after factoring $X_n(x)$ out, to

$$\sum_{n\geqslant 1}\left[\dot{C}_n(t)+\alpha\lambda_n\,C_n(t)-f_n(t)\right]X_n(x)=0.$$

It is known from §10.2 that the set of eigenfunctions $\{X_n(x)\}_{n\geqslant 1}$ is complete in the space of square integrable functions. Therefore, the latter is valid only when all coefficients of this sum are zeroes:

$$\dot{C}_n(t)+\alpha\lambda_n\,C_n(t)-f_n(t)=0,\qquad n=1,2,\ldots.$$

Substituting Eq. (11.2.2) into the initial conditions $w(x,0)=\varphi(x)$, we get

$$\sum_{n\geqslant 1}C_n(0)\,X_n(x)=\varphi(x)=\sum_{n\geqslant 1}\varphi_n\,X_n(x),$$

where

$$\varphi_n=\frac{2}{\ell}\int_0^{\ell}\varphi(x)\,X_n(x)\,\mathrm{d}x$$

are Fourier coefficients of the known function $\varphi(x)$ over the system of eigenfunctions for the corresponding Sturm–Liouville problem. To satisfy the initial conditions, we demand that $C_n(0)=\varphi_n$. This leads to the following initial value problem for determining coefficients $C_n(t)$:

$$\dot{C}_n(t)+\alpha\lambda_n\,C_n(t)=f_n(t),\qquad C_n(0)=\varphi_n.$$

Since the differential equation for $C_n(t)$ is linear with constant coefficients, the reader may consult §2.5 and verify that its solution is

$$C_n(t)=e^{-\alpha\lambda_n t}\int_0^t f_n(\tau)\,e^{\alpha\lambda_n\tau}\,\mathrm{d}\tau+\varphi_n\,e^{-\alpha\lambda_n t},\quad n=1,2,\ldots. \tag{11.2.3}$$

Example 11.2.1: (Temperature at the ends is specified) We reconsider Example 11.1.1, but now with nonhomogeneous boundary conditions of the first type:

$$\dot{u}_t=\alpha\,u_{xx},\qquad u(0,t)=T_0,\quad u(\ell,t)=T_\ell,\quad u(x,0)=f(x),$$

where $f(x)$ is the initial temperature distribution in the bar of length ℓ, and T_0, T_ℓ are given constants. (They are chosen as constants for simplicity but could be functions of time t.) Since the separation of variables method can be applied only when the boundary conditions are homogeneous, we represent the required solution $u(x,t)$ as the sum of two functions:

$$u(x,t)=v(x)+w(x,t),$$

where $v(x)$ is a function that satisfies the given boundary conditions, for instance,

$$v(x)=\frac{x}{\ell}\,T_\ell+\frac{\ell-x}{\ell}\,T_0.$$

Since $v(x)$ does not depend on time t, it is a steady state temperature. Recall that a **steady state** temperature is one that does not depend on time. This choice for $v(x)$ is not unique—there exist infinitely many functions that satisfy the nonhomogeneous boundary conditions. For example, we can choose $v(x,t)$ as

$$v(x,t)=T_\ell(t)\,\sin\frac{\pi x}{2\ell}+T_0(t)\,\cos\frac{\pi x}{2\ell}.$$

It does not matter which function is chosen for $v(x,t)$ to satisfy the boundary conditions because the given IBVP has a unique solution. Now we return to the problem for w:

$$\dot{w}_t=\alpha\,w_{xx}+(\alpha v_{xx}-v_t),\qquad w(0,t)=0,\quad w(\ell,t)=0,\qquad w(x,0)=f(x)-v(x).$$

Since in our case $v_t = \alpha v_{xx}$, the heat equation for w becomes homogeneous, and we can use its series representation from Example 11.1.1:

$$w(x,t) = \sum_{n\geqslant 1} C_n\, e^{-\alpha\lambda_n t} \sin\frac{n\pi x}{\ell}, \qquad \lambda_n = \left(\frac{n\pi}{\ell}\right)^2,$$

where

$$\begin{aligned} C_n &= \frac{2}{\ell}\int_0^\ell f(x)\,\sin\frac{xn\pi}{\ell}\,dx - \frac{2}{\ell}\int_0^\ell \left[\frac{x}{\ell}\,T_\ell + \frac{\ell - x}{\ell}\,T_0\right] \sin\frac{n\pi x}{\ell}\,dx \\ &= \frac{2}{\ell}\int_0^\ell f(x)\,\sin\frac{xn\pi}{\ell}\,dx + \frac{2}{n\pi}\,T_\ell\,(-1)^n + \frac{2}{n\pi}\,T_0. \end{aligned}$$

Example 11.2.2: (Heated rod) Consider a thin rod of length ℓ when one end is kept at 0°C and the other end is perfectly insulated. Assume that at time $t = 0$ the insulated end of the rod is lifted out of ice and one third of it becomes heated with a constant uniform source for finite time period $(0, \tau)$, $\tau > 0$. The model IBVP is

$$\begin{array}{lll} \text{(PDE)} & u_t = \alpha u_{xx} + \Phi(x,t), & 0 < x < \ell,\ 0 < t \leqslant t^* < \infty, \\ \text{(BC)} & u(0,t) = 0, \quad u_x(\ell,t) = 0, & 0 < t \leqslant t^* < \infty, \\ \text{(IC)} & u(x,0) = 0, & \end{array} \tag{11.2.4}$$

where the external heat source $\Phi(x,t)$ is expressed through the Heaviside function $H(t)$ as

$$\Phi(x,t) = [H(t) - H(t-\tau)] \times \begin{cases} 0, & \text{if } 0 < x < 2\ell/3, \\ 1, & \text{if } 2\ell/3 < x < \ell. \end{cases}$$

To find the separated solutions of the homogeneous PDE that satisfy the given boundary conditions, we insert $u = X(x)\,T(t)$ into the PDE and BC, separate variables, and obtain the Sturm–Liouville problem

$$X''(x) + \lambda\, X(x) = 0, \qquad X(0) = X'(\ell) = 0.$$

Solving the eigenvalue problem, we see that $X_n(x) = \sin\frac{\pi(1+2n)x}{2\ell}$, $n = 0, 1, 2, \ldots$, are eigenfunctions corresponding to the eigenvalues $\lambda_n = \left[\frac{\pi(1+2n)}{2\ell}\right]^2$. So we seek a formal solution as a series over the eigenfunctions

$$u(x,t) = \sum_{n\geqslant 0} C_n(t)\, X_n(x) = \sum_{n\geqslant 0} C_n(t)\, \sin\frac{\pi\,(1+2n)\,x}{2\ell}. \tag{11.2.5}$$

Expanding the source function $\Phi(x,t)$ into a similar series, we get

$$\Phi(x,t) = [H(t) - H(t-\tau)] \sum_{n\geqslant 0} \frac{4}{\pi\,(1+2n)}\,\cos\frac{\pi\,(1+2n)}{3}\,\sin\frac{\pi\,(1+2n)\,x}{2\ell}.$$

Substituting these Fourier expansions into the given problem (11.2.4), we obtain the following initial value problem for the coefficients $C_n(t)$:

$$\begin{aligned} &\dot{C}_n + \alpha\lambda_n\, C_n(t) = [H(t) - H(t-\tau)]\,\frac{4}{\pi\,(1+2n)}\,\cos\frac{\pi\,(1+2n)}{3}, \\ &C_n(0) = 0, \qquad n = 0, 1, 2, \ldots. \end{aligned}$$

To solve this initial value problem, we apply the Laplace transform and obtain

$$C_n(t) = \frac{4}{\pi\,(1+2n)}\,\cos\left(\frac{\pi\,(1+2n)}{3}\right)[g_n(t) - g_n(t-\tau)]\,,$$

where

$$g_n(t) = \frac{1}{a_n}\left[1 - e^{-a_n t}\right] H(t), \qquad a_n = \left(\frac{c\pi\,(1+2n)}{2\ell}\right)^2.$$

Substitution of these values $C_n(t)$ into Eq. (11.2.5) gives the formal solution of the initial boundary value problem (11.2.4). Since the coefficients $C_n(t)$ decrease with n as n^{-2}, the series (11.2.5) converges uniformly and the function $u(x,t)$ is continuous. □

So far, we discussed the Dirichlet (if the rod's ends are maintained with specified temperature) and Neumann (when the rod's ends are insulated) boundary conditions and their variations. However, a more general type of boundary conditions occur when the rod's ends undergo convective heat transfer with the ambient temperature. In this case, we use Fourier's law of heat conduction

$$\left.\frac{\partial u}{\partial \mathbf{n}}\right|_{P\in\partial R} = (\nabla u\cdot\mathbf{n})|_{P\in\partial R} = k\left[U(t) - u(P,t)\right], \tag{11.2.6}$$

where $U(t)$ is the ambient temperature outside domain R, ∂R is the boundary of R, k is a positive coefficient, which must be in units of $\texttt{sec}^{-1}$, and is therefore sometimes expressed in terms of a characteristic time constant t_0 given by $k = 1/t_0 = -(\mathrm{d}u(t)/\mathrm{d}t)/\Delta u$, $\mathbf{n}$ is the outward unit vector normal to the boundary ∂R, and $\partial u/\partial\mathbf{n}$ is the directional derivative of u along the outward normal vector. The time constant t_0 is expressed as $t_0 = C/(hA)$, where $C = m\,s = \mathrm{d}Q/\mathrm{d}u$ is the total heat capacity of a system of mass m, s is the specific heat, h is the heat transfer coefficient (assumed independent of u) with units $\texttt{W/(m}^2\cdot{}^\circ\texttt{K)}$, and A is the surface area of the heat being transferred ($\texttt{m}^2$). The boundary conditions (11.2.6) corresponding to the convective heat transfer are referred to as boundary conditions[114] of the **third kind**.

In the one-dimensional case, the boundary condition (11.2.6) at the end $x = 0$ reads as

$$-u_x(0,t) = k\left[U(t) - u(0,t)\right], \quad t \geqslant 0.$$

At other end $x = \ell$, we have

$$u_x(\ell,t) = k\left[U(t) - u(\ell,t)\right], \quad t \geqslant 0.$$

When the method of separation of variables is applied to the heat equation subject to boundary conditions of the third kind, it leads to the Sturm–Liouville problems (see §10.1) that do not admit "explicit" solutions and can be solved only numerically.

Problems

In Problems 1 through 9, solve the initial boundary value problems.

1. $u_t = 9\,u_{xx} + e^{-2t}$ $(0 < x < 3)$, $u(0,t) = u_x(3,t) = 0$, $u(x,0) = 0$.
2. $u_t = 4\,u_{xx}$ $(0 < x < 2)$, $u(0,t) = t\,e^{-t}$, $u(2,t) = 0$, $u(x,0) = x(2-x)$.
3. $u_t = u_{xx}$ $(0 < x < \pi/2)$, $u_x(0,t) = 1$, $u(\pi/2,t) = \pi/2$, $u(x,0) = \sin(2x)\cos(4x)$.
4. $u_t = 25\,u_{xx}$ $(0 < x < 5)$, $u_x(0,t) = u_x(5,t) = 1$, $u(x,0) = \cos\frac{4\pi x}{5}$.
5. $u_t = 16\,u_{xx} + \cos\frac{7\pi x}{4}$ $(0 < x < 2)$, $u_x(0,t) = u(2,t) = 1$, $u(x,0) = x^2 - 4$.
6. $u_t = 4\,u_{xx} + H(t) - H(t-2)$ $(0 < x < 2)$, $u(0,t) = u_x(2,t) = 1$, $u(x,0) = x(x-2)^2$.
7. $u_t = 9\,u_{xx}$ $(0 < x < 3)$, $u(0,t) = t$, $u_x(3,t) = 0$, $u(x,0) = x$.
8. $u_t = u_{xx} + t$ $(0 < x < 1)$, $u_x(0,t) = 0$, $u_x(1,t) = 1$, $u(x,0) = 0$.
9. $u_t = 25\,u_{xx} + 1$ $(0 < x < 10)$, $u_x(0,t) = 1$, $u(10,t) = 0$, $u(x,0) = x$.
10. Suppose that an aluminum rod 50 cm long with an insulated lateral surface is heated to a uniform temperature of 20°C, and that at time $t = 0$ one end $x = 0$ is kept in ice at 0°C while the other end $x = 0.5$ is maintained with a constant temperature of 100°C. Find the formal series solution for the temperature $u(x,t)$ of the rod. Assume that the thermal diffusivity is a constant, which is equal to 8.4×10^{-5} $\texttt{m}^2\texttt{/sec}$ at 25°C. In reality, it is not a constant and varies with temperature in a wide range.
11. A stainless steel rod 1 m long with an insulated lateral surface has initial temperature $u(x,0) = 1 - x$, and at time $t = 0$ one of its ends $(x = 0)$ is insulated while the other end $(x = 1)$ is embedded in ice at 0°C. Find the formal series solution for the temperature $u(x,t)$ of the rod when it is heated with a constant rate q. The thermal diffusivity can be assumed to be constant at $4.2\times10^{-5}\,\texttt{m}^2\texttt{/sec}$.
12. Suppose that a wire of length ℓ loses heat to the surrounding medium at a rate proportional to the temperature $u(x,t)$. Then the function $u(x,t)$ is a solution of the following IBVP:

$$u_t = \alpha u_{xx} - hu, \quad u(0,t) = u_x(\ell,t) = 0, \quad u(x,0) = f(x).$$

By making substitution $u(x,t) = e^{-ht}\,v(x,t)$, find the formal series solution of the problem.

[114]Sometimes the boundary conditions of the third kind are mistakenly associated with (Victor) Gustave Robin (1855–1897), a professor of mathematical physics at the Sorbonne in Paris. Actually, Robin never used this boundary condition.

11.3 Wave Equation

Consider a string stretched from $x = 0$ to $x = \ell$. Let $u(x, t)$ describe the vertical displacement of the wire at position x and time t. If damping effects, such as air resistance, are neglected, the force of gravity is ignored, and oscillations of the string are sufficiently small so that the tension forces can be treated as elastic, then $u(x, t)$ satisfies the **wave equation**

$$\Box u \equiv \frac{\partial^2 u}{\partial t^2} - c^2 \frac{\partial^2 u}{\partial x^2} = 0, \qquad 0 < x < \ell,\ 0 < t \leqslant t^* < \infty. \tag{11.3.1}$$

where $\Box = \Box_c = \partial^2 u/\partial t^2 - c^2 \nabla^2 u = \partial^2 u/\partial t^2 - c^2 \partial^2 u/\partial x^2$ is the wave or d'Alembert operator, simply called d'Alembertian, and the positive constant c is the wave velocity, with units [length/time]. Under these conditions, the equilibrium position of the string $u(x, t) = 0$ corresponds to a straight line between two fixed end points.

D'Alembert[115] discovered around 1744–1746 a strikingly simple method for finding the general solution to the wave equation. Roughly speaking, his idea was to factorize the one-dimensional wave operator:

$$\Box = \frac{\partial^2}{\partial t^2} - c^2 \frac{\partial^2}{\partial x^2} = \left(\frac{\partial}{\partial t} - c\frac{\partial}{\partial x}\right)\left(\frac{\partial}{\partial t} + c\frac{\partial}{\partial x}\right).$$

This allows one to reduce the second order partial differential equation (11.3.1) to two first order equations

$$\frac{\partial v}{\partial t} - c\frac{\partial v}{\partial x} = 0 \qquad \text{and} \qquad \frac{\partial v}{\partial t} + c\frac{\partial v}{\partial x} = 0.$$

By introducing a new variable $\xi = x \pm ct$, each of the above equations is reduced to a simple ordinary differential equation

$$\frac{\mathrm{d}v}{\mathrm{d}\xi} = \frac{\partial v}{\partial x}\frac{\partial x}{\partial \xi} + \frac{\partial v}{\partial t}\frac{\partial t}{\partial \xi} = \frac{\partial v}{\partial x} \pm \frac{1}{c}\frac{\partial v}{\partial t} = 0,$$

which can be integrated directly. This leads to the conclusion that a solution of the wave equation (11.3.1) is the sum

$$u(x, t) = f(x + ct) + g(x - ct) \tag{11.3.2}$$

of two functions $f(\xi)$ and $g(\xi)$ of one variable. This formula represents a superposition of two waves, one traveling to the right and one traveling to the left, each with velocity c. However, in practice, traveling waves are excited by the initial disturbance

$$u(x, 0) = d(x), \qquad \left.\frac{\partial u}{\partial t}\right|_{t=0} = v(x), \tag{11.3.3}$$

where $d(x)$ is the initial displacement (initial configuration) and $v(x)$ is the initial velocity of the string. Upon substituting the general solution (11.3.2) into Eq. (11.3.3), we arrive at two equations

$$d(x) = f(x) + g(x),$$
$$v(x) = c\,f'(x) - c\,g'(x).$$

Integrating the latter, we get

$$\int_0^x v(x)\,\mathrm{d}x = c\,f(x) - c\,g(x).$$

This enables us to express the general solution in terms of the initial displacement and the initial velocity (called the d'Alembert's formula)

$$u(x, t) = \frac{d(x + ct) + d(x - ct)}{2} + \frac{1}{2c}\int_{x-ct}^{x+ct} v(\xi)\,\mathrm{d}\xi. \tag{11.3.4}$$

The above formula allows us to make an important observation. Although the wave equation only makes sense for functions with second order partial derivatives, the solution (11.3.4) exists for any continuous functions $d(\xi)$ and $v(\xi)$ of one variable ξ. (Discontinuous functions cannot represent displacement of an unbroken string!) Therefore, justification that the function $u(x, t)$, defined through the formula (11.3.4) for continuous but not differentiable functions, satisfies the wave equation (11.3.1) requires a more advanced mathematical technique, which is usually done with the aid of generalized functions (distributions).

[115] Jean-Baptiste le Rond d'Alembert (1717–1783) was a French mathematician, mechanician, physicist, philosopher, and music theorist.

Also, the formula (11.3.4) shows that the solution of the initial value problem for the wave equation (11.3.1), (11.3.3) is unique. Moreover, the formula presents $u(x,t)$ as the sum of two solutions of the wave equation: one with prescribed initial displacement $d(x)$ and zero velocity, and the other with zero initial displacement but specified initial velocity $v(x)$. ■

Assuming the string is fixed at both end points, the displacement function $u(x,t)$ will satisfy the initial boundary value problem that consists of the wave equation (11.3.1), the Dirichlet boundary conditions

$$u(0,t) = u(\ell,t) = 0, \tag{11.3.5}$$

and the initial conditions (11.3.3). For compatibility, we require that the initial displacement be such that $d(0) = d(\ell) = 0$.

First, we try to satisfy the first boundary condition $u(0,t) = 0$. From the general formula (11.3.2), it follows that

$$0 = f(ct) + g(-ct)$$

for all $t \geqslant 0$, so that $g(\xi) = -f(-\xi)$ for any value ξ. Thus, the solution of the wave equation that satisfies the boundary condition $u(0,t) = 0$ is expressed through one (unknown) function

$$u(x,t) = f(x+ct) - f(ct-x). \tag{11.3.6}$$

Physically, this means that the wave, traveling to the left, hits the end $x = 0$ of the string and returns inverted as a wave traveling to the right. This phenomenon is called the *Principle of Reflection* (see Fig. 11.1, page 614).

Substituting $u(x,t)$ from Eq. (11.3.6) into the other boundary condition $u(\ell,t) = 0$, we get $f(\ell + ct) = f(ct - \ell)$ for all t, so that

$$f(\xi) = f(\xi + 2\ell)$$

for all values of ξ. This means that the function $f(\xi)$ in Eq. (11.3.6) must be a periodic function with the fundamental period $T = 2\ell$. As a result, the function $f(\xi)$ admits a Fourier series representation. To find the corresponding coefficients, we use the separation of variables method.

In order to apply the separation of variables method, we need to find nontrivial solutions of the wave equation $u_{tt} = c^2 u_{xx}$, subject to the homogeneous boundary conditions $u(0,t) = u(\ell,t) = 0$, that is represented as the product of two functions, each depending on one independent variable:

$$u(x,t) = T(t)\,X(x).$$

Substituting this expression into the wave equation, we obtain

$$\ddot{T}(t)\,X(x) = c^2\,T(t)\,X''(x).$$

After division by $c^2\,T(t)\,X(x)$, we get

$$\frac{\ddot{T}(t)}{c^2\,T(t)} = \frac{X''(x)}{X(x)}, \qquad 0 < x < \ell, \quad 0 < t < t^* < \infty.$$

In the above equation, the prime denotes differentiation with respect to a spacial variable, and the dot represents differentiation with respect to the time variable. The only way that the above equation can hold is if both expressions equal a common constant, which we denote by $-\lambda$. This leads to two equations:

$$\ddot{T} + c^2\lambda\,T(t) = 0, \qquad 0 < t < t^* < \infty, \tag{11.3.7}$$

$$X''(x) + \lambda\,X(x) = 0, \qquad 0 < x < \ell. \tag{11.3.8}$$

Imposing the homogeneous boundary conditions on $u(x,t) = T(t)\,X(x)$ yields

$$X(0) = 0, \qquad X(\ell) = 0. \tag{11.3.9}$$

Equation (11.3.8) together with the boundary conditions (11.3.9) is recognized as a Sturm–Liouville problem, considered in §10.1. Hence, the eigenpairs are

$$\lambda_n = \left(\frac{n\pi}{\ell}\right)^2, \qquad X_n(x) = \sin\left(\frac{n\pi x}{\ell}\right), \quad n = 1, 2, 3, \ldots.$$

With $\lambda = \lambda_n = (n\pi/\ell)^2$, Eq. (11.3.7) becomes

$$\ddot{T} + \left(\frac{cn\pi}{\ell}\right)^2 T_n(t) = 0, \qquad 0 < t < t^* < \infty.$$

The general solution of $T_n(t)$ is

$$T_n(t) = A_n \cos\frac{cn\pi t}{\ell} + B_n \sin\frac{cn\pi t}{\ell}, \qquad n = 1, 2, 3, \ldots,$$

where A_n and B_n are arbitrary constants to be specified. Since we consider the homogeneous wave equation $\ddot{u} = c^2 u_{xx}$, any sum of partial nontrivial solutions $u_n(x,t) = T_n(t)\,X_n(x)$ will also be a solution of the wave equation. Further, this sum satisfies the homogeneous boundary conditions $u(0,t) = u(\ell,t) = 0$. Now we look for a solution of the given initial boundary value problem in the form of the infinite series

$$u(x,t) = \sum_{n\geqslant 1}\left[A_n \cos\left(\frac{cn\pi t}{\ell}\right) + B_n \sin\left(\frac{cn\pi t}{\ell}\right)\right]\sin\left(\frac{n\pi x}{\ell}\right). \tag{11.3.10}$$

Since its initial displacement is specified, we have

$$u(x,0) = d(x) = \sum_{n\geqslant 1} A_n \sin\left(\frac{n\pi x}{\ell}\right).$$

This equation is the Fourier sine expansion for the initial configuration of the string. Its coefficients are

$$A_n = \frac{2}{\ell}\int_0^\ell d(x)\sin\left(\frac{n\pi x}{\ell}\right)\mathrm{d}x, \qquad n = 1, 2, 3, \ldots. \tag{11.3.11}$$

Term-by-term differentiation of the series (11.3.10) (this is justified if the series converges uniformly) leads to

$$\dot{u} = \frac{\partial u}{\partial t} = \sum_{n\geqslant 1}\left[-\frac{cn\pi}{\ell} A_n \sin\left(\frac{cn\pi t}{\ell}\right) + \frac{cn\pi}{\ell} B_n \cos\left(\frac{cn\pi t}{\ell}\right)\right]\sin\left(\frac{n\pi x}{\ell}\right).$$

Then, setting $t = 0$, we obtain

$$\dot{u}(x,0) = \sum_{n\geqslant 1}\frac{cn\pi}{\ell} B_n \sin\left(\frac{n\pi x}{\ell}\right) = v(x).$$

Using Eq. (10.5.2) on page 582, we get

$$B_n = \frac{2}{cn\pi}\int_0^\ell v(x)\sin\left(\frac{n\pi x}{\ell}\right)\mathrm{d}x, \qquad n = 1, 2, 3, \ldots. \tag{11.3.12}$$

Series (11.3.10), with the coefficients (11.3.11) and (11.3.12), is the (formal) solution of the given initial boundary value problem.

Example 11.3.1: Solve the initial boundary value problem

$$\begin{cases} \ddot{u} = 25\,u_{xx}, & 0 < x < 20, \quad 0 < t < t^* < \infty; \\ u(0,t) = u(20,t) = 0, & 0 < t < t^* < \infty; \\ u(x,0) = \sin(\pi x), \quad \dot{u}(x,0) = \sin(2\pi x), & 0 < x < 20. \end{cases}$$

Solution. According to Eq. (11.3.10), the given problem has a series-solution:

$$u(x,t) = \sum_{n\geqslant 1}\left[A_n \cos\left(\frac{n\pi t}{4}\right) + B_n \sin\left(\frac{n\pi t}{4}\right)\right]\sin\left(\frac{n\pi x}{20}\right)$$

because the string has length $\ell = 20$. Imposing the initial conditions leads to the equations

$$u(x,0) = \sin(\pi x) = \sum_{n\geqslant 1} A_n \sin\left(\frac{n\pi x}{20}\right),$$

$$\dot{u}(x,0) = \sin(2\pi x) = \sum_{n\geqslant 1}\frac{n\pi}{4} B_n \sin\left(\frac{n\pi x}{20}\right).$$

While we could use formulas (10.5.2) on page 582, it is simpler to observe that the initial functions are actually eigenfunctions of the corresponding Sturm–Liouville problem. Since Fourier series expansion is unique for every "suitable" function, we conclude that these initial functions are Fourier series themselves. Therefore,

$$\begin{aligned} A_{20} &= 1, & A_n &= 0 \quad \text{for } n \ne 20, \\ \frac{40\pi}{4} B_{40} &= 1, & B_n &= 0 \quad \text{for } n \ne 40. \end{aligned}$$

Hence, $B_{40} = 1/(10\pi)$ and we get the required solution

$$u(x,t) = \cos(5\pi t)\,\sin(\pi x) + \frac{1}{10\pi}\,\sin(10\pi t)\,\sin(2\pi x).$$

Example 11.3.2: (Guitar) Suppose we pluck a string by pulling it upward and release it from rest (see Fig. 10.23 on page 583). If the point of the pluck is in the third of a string of length ℓ (which is usually the case when playing guitar), we can model the vibration of the string by solving the following initial boundary value problem:

$$\begin{aligned} &\ddot{u} = c^2\, u_{xx}, && 0 < x < \ell, \quad 0 < t < t^* < \infty; \\ &u(0,t) = u(\ell,t) = 0, && 0 < t < t^* < \infty; \\ &u(x,0) = f(x) = \begin{cases} \dfrac{3x}{\ell}, & 0 < x < \frac{\ell}{3}, \\ \dfrac{3}{2\ell}\,(\ell - x), & \frac{\ell}{3} < x < \ell, \end{cases} \\ &\dot{u}(x,0) = v(x) \equiv 0, && 0 < x < \ell. \end{aligned}$$

Solution. The formal Fourier series solution is given by Eq. (11.3.10). Since the initial velocity is zero, we have

$$u(x,t) = \sum_{n\geqslant 1} A_n\, \cos\left(\frac{cn\pi t}{\ell}\right)\, \sin\left(\frac{n\pi x}{\ell}\right).$$

The coefficients A_n in the above series are evaluated as

$$\begin{aligned} A_n &= \frac{2}{\ell}\int_0^{\ell} f(x)\, \sin\left(\frac{n\pi x}{\ell}\right) \mathrm{d}x \\ &= \frac{2}{\ell}\int_0^{\ell/3} \frac{3x}{\ell}\, \sin\left(\frac{n\pi x}{\ell}\right) \mathrm{d}x + \frac{2}{\ell}\int_{\ell/3}^{\ell} \frac{3}{2\ell}\,(\ell - x)\, \sin\left(\frac{n\pi x}{\ell}\right) \mathrm{d}x \\ &= \frac{2}{n^2\pi^2}\left(3\, \sin\frac{n\pi}{3} - n\pi\, \cos\frac{n\pi}{3}\right) + \frac{1}{n^2\pi^2}\left(2n\pi\, \cos\frac{n\pi}{3} + 3\, \sin\frac{n\pi}{3}\right) \\ &= \frac{9}{n^2\pi^2}\, \sin\frac{n\pi}{3}, \qquad n = 1, 2, \ldots. \end{aligned}$$

This gives the solution

$$u(x,t) = \sum_{n\geqslant 1} \frac{9}{n^2\pi^2}\, \sin\frac{n\pi}{3}\, \cos\left(\frac{cn\pi t}{\ell}\right)\, \sin\left(\frac{n\pi x}{\ell}\right). \tag{11.3.13}$$

The coefficients in Eq. (11.3.13) decrease with n as $1/n^2$, so the series converges uniformly, and its sum is a continuous function. By plotting $u(x,t)$ for a fixed value of t, we get the shape of the string at that time. Its partial sum approximations are presented in Fig.11.1 on page 614. □

According to Eq. (11.3.10), the solution of the vibrating string problem is an infinite sum of the normal modes (also called the standing waves or harmonics)

$$u_n(x,t) = \left[A_n\, \cos\left(\frac{cn\pi t}{\ell}\right) + B_n\, \sin\left(\frac{\pi t}{\ell}\right)\right] \sin\left(\frac{n\pi x}{\ell}\right), \qquad n = 1, 2, \ldots. \tag{11.3.14}$$

The first normal mode ($n = 1$) is called the **fundamental mode** or fundamental harmonic: all other modes are known as **overtones** (see §10.3.1). In music, the intensity of the sound produced by a given normal mode depends on the magnitude $\sqrt{A_n^2 + B_n^2}$, which is called the amplitude of the n-th normal mode. The circular or natural

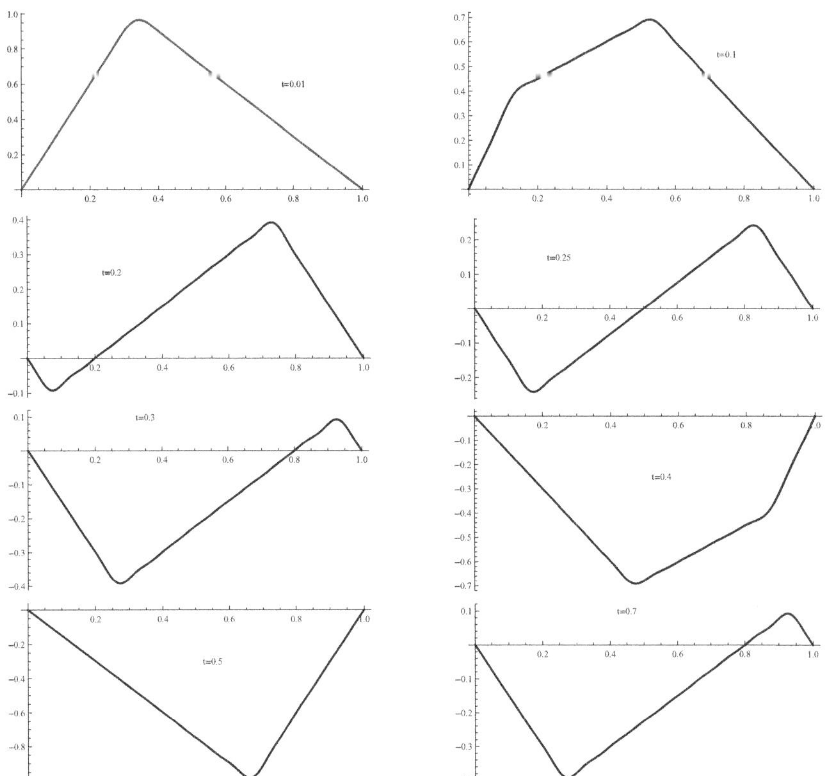

Figure 11.1: Several snapshots of the guitar string, Example 11.3.2.

frequency of the normal mode, which gives the number of oscillations in 2π units of time, is $\lambda_n = nc\pi/\ell$. The larger the natural frequency, the higher the pitch of the sound produced.

For a fixed integer n, the n-th standing wave $u_n(x,t)$ is periodic in time t with period $2\ell/(cn)$; it therefore represents a vibration of the string having this period with wavelength $2\ell/n$. The quantities $cn/(2\ell)$ are known as the **natural frequencies** of the string, that is, the frequencies at which the string will freely vibrate. The fundamental mode oscillates with frequency $c/(2\ell)$. The factor $A_n \cos\frac{cn\pi t}{\ell} + B_n \sin\frac{cn\pi t}{\ell}$ represents the displacement pattern occurring in the string at time t. When the string vibrates in normal mode, some points on the string are fixed at all times. These are solutions of the equation $\sin\frac{n\pi x}{\ell} = 0$.

Example 11.3.3: (Piano) Sounds from a piano, unlike the guitar, are put into effect by striking strings. When a player presses a key, it causes a hammer to strike the strings. The corresponding IBVP is

$$\begin{array}{lll} \text{(PDE)} & u_{tt} = c^2 u_{xx}, & 0 < x < \ell,\ 0 < t \leqslant t^* < \infty, \\ \text{(BC)} & u(0,t) = 0, \quad u(\ell,t) = 0, & 0 < t \leqslant t^* < \infty, \\ \text{(IC)} & u(x,0) = 0, \quad u_t(x,0) = v(x), & \end{array} \tag{11.3.15}$$

where the initial velocity is the step function

$$v(x) = \begin{cases} 1, & \text{when } s < x < s+h, \\ 0, & \text{otherwise.} \end{cases}$$

Here s is a position of the left hammer's end and h is the width of the hammer. It is assumed that both s and $s+h$ are within the string length ℓ.

Solution. Using series solution (11.3.10), we get

$$u(x,t) = \sum_{n\geqslant 1} B_n \sin\left(\frac{cn\pi t}{\ell}\right) \sin\frac{n\pi x}{\ell},$$

where the coefficients B_n are determined upon integration:

$$B_n = \frac{2}{cn\pi} \int_s^{s+h} \sin \frac{n\pi x}{\ell} \, dx = \frac{2\ell}{cn^2\pi^2} \left[\cos \frac{n\pi s}{\ell} - \cos \frac{n\pi(s+h)}{\ell} \right].$$

11.3.1 Transverse Vibrations of Beams

The free vertical vibrations of a uniform beam of length ℓ are modeled by the **beam equation**

$$u_{tt} + c^2 u_{xxxx} = 0 \qquad (0 < x < \ell), \tag{11.3.16}$$

where $c^2 = EI/(\rho A)$, E is Young's modulus (with units in `pascal`), I is the moment of inertia (units $\mathtt{kg}\cdot\mathtt{m}^2$) of a cross-section of the beam with respect to an axis through its center of mass and perpendicular to the (x, u)-plane, ρ is density (mass per unit volume), and A is the area of cross-section ($\mathtt{m}^2$). It is assumed that the beam is of uniform density throughout, which means that the cross-sections are constant and that in its equilibrium position the centers of mass of the cross-sections lie on the x-axis. The dependent variable $u(x, t)$ represents the displacement of the point on the beam corresponding to position x at time t.

The boundary conditions that accompany Eq. (11.3.16) depend on the end supports of the beam. Let us consider, for instance, a pin support of the left end $x = 0$ and a roller at the right end $x = \ell$. Both boundary conditions prevent vertical translation and both allow rotation. The only difference between a pin and a roller is that a pin prevents horizontal movement, whereas a roller does not. This case of boundary conditions is described by

$$u(0, t) = 0, \quad u(\ell, t) = 0, \quad u_{xx}(0, t) = 0, \quad u_{xx}(\ell, t) = 0. \tag{11.3.17}$$

To complete the description of the problem, we specify the initial conditions

$$u(x, 0) = f(x), \quad u_t(x, 0) = v(x), \qquad 0 < x < \ell. \tag{11.3.18}$$

To solve the IBVP (11.3.16) – (11.3.18), we use the method of separation of variables. Upon substitution of $u(x, t) = X(x)\,T(t)$ into the beam equation (11.3.16), we get

$$\frac{X^{(4)}(x)}{X(x)} = -\frac{\ddot{T}(t)}{c^2 T(t)} = \lambda^4.$$

Taking into consideration the boundary conditions (11.3.17), we obtain the Sturm–Liouville problem

$$X^{(4)}(x) - \lambda^4 X(x) = 0, \quad X(0) = X(\ell) = X''(0) = X''(\ell) = 0. \tag{11.3.19}$$

The general solution of the fourth order equation $X^{(4)}(x) - \lambda^4 X(x) = 0$ is

$$X(x) = A \cosh \lambda x + B \sinh \lambda x + C \cos \lambda x + D \sin \lambda x,$$

where A, B, C, and D are arbitrary constants. The boundary conditions at $x = 0$ dictate that $A = C = 0$; correspondingly, from conditions at $x = \ell$, we get

$$B \sinh \lambda\ell + D \sin \lambda\ell = 0, \qquad B \sinh \lambda\ell - D \sin \lambda\ell = 0.$$

These are equivalent to $B \sinh \lambda\ell = 0$ and $D \sin \lambda\ell = 0$, from which follows that $B = 0$ because $\sinh \lambda\ell \ne 0$ for $\lambda > 0$. This allows us to find the eigenfunctions and eigenvalues:

$$X_n(x) = \sin \frac{n\pi x}{\ell}, \quad \lambda_n = \frac{n\pi}{\ell}, \qquad n = 1, 2, 3, \ldots.$$

Going back to the equation $\ddot{T} + c^2 \lambda_n^4 T(t) = 0$, we obtain the corresponding solutions

$$T_n(t) = A_n \cos\left(c\lambda_n^2 t\right) + B_n \sin\left(c\lambda_n^2 t\right).$$

Forming the product solutions and superposing, we find the general form of the solution:

$$u(x, t) = \sum_{n\geqslant 1} \sin \frac{n\pi x}{\ell} \left(A_n \cos \frac{cn^2\pi^2 t}{\ell^2} + B_n \sin \frac{cn^2\pi^2 t}{\ell^2} \right). \tag{11.3.20}$$

Using the initial conditions (11.3.18), we are forced to choose the values of coefficients A_n and B_n so that the Fourier expansions

$$f(x) = \sum_{n\geqslant 1} A_n \sin\frac{n\pi x}{\ell}, \qquad v(x) = \sum_{n\geqslant 1} B_n\, c\lambda_n^2 \sin\frac{n\pi x}{\ell}$$

hold. This leads to the following formulas:

$$A_n = \frac{2}{\ell}\int_0^{\ell} f(x)\,\sin\frac{n\pi x}{\ell}\,\mathrm{d}x, \quad B_n = \frac{2\ell}{n^2\pi^2 c}\int_0^{\ell} v(x)\,\sin\frac{n\pi x}{\ell}\,\mathrm{d}x, \quad n = 1, 2, \ldots.$$

Problems

Consider the initial boundary value problem for the one-diemnsional wave equation

$$\begin{aligned} &u_{tt} = c^2 u_{xx}, \qquad 0 < x < \ell, \quad 0 < t < t^* < \infty; \\ &u(0,t) = u(\ell,t) = 0, \qquad 0 < t < \infty; \\ &u(x,0) = f(x), \qquad u_t(x,0) = v(x), \qquad 0 < x < \ell. \end{aligned}$$

In Problems 1 through 10, solve this problem for the given parameter values (c and ℓ) and the given initial conditions.

1. $c = 5$, $\ell = 3$, $u(x,0) = 0$, $u_t(x,0) = 5\,\sin(\pi x)$.
2. $c = 4$, $\ell = 1$, $u(x,0) = \sin(\pi x)$, $u_t(x,0) = 0$.
3. $c = 2$, $\ell = \pi$, $u(x,0) = \begin{cases} 3x/\pi, & \text{if } 0 < x < \pi/3, \\ 3(\pi - x)/(2\pi), & \text{if } \pi/3 < x < \pi, \end{cases} \quad u_t(x,0) = 0.$
4. $c = 3$, $\ell = 3$, $u(x,0) = \begin{cases} x, & \text{if } 0 < x < 1/3, \\ 1, & \text{if } 1 < x < 2, \\ 3 - x, & \text{if } 2 < x < 3, \end{cases} \quad u_t(x,0) = 0.$
5. $c = 2$, $\ell = 2$, $u(x,0) = x(2-x)^2$, $u_t(x,0) = \sin(2\pi x)$.
6. $c = 1$, $\ell = 10$, $u(x,0) = \sin \pi x$, $u_t(x,0) = \begin{cases} x/5, & \text{if } 0 < x < 5, \\ (10 - x)/5, & \text{if } 5 < x < 10. \end{cases}$
7. $c = 2$, $\ell = 20$, $u(x,0) = x^2(20 - x)$, $u_t(x,0) = \sin \pi x$.
8. $c = 1$, $\ell = \ell$, $u(x,0) = \begin{cases} \sin\frac{2\pi x}{\ell}, & \text{if } 0 < x < \ell/2, \\ 0, & \text{if } \ell/2 < x < \ell, \end{cases} \quad u_t(x,0) = 1.$
9. $c = 3$, $\ell = 10$, $u(x,0) = 0$, $u_t(x,0) = \begin{cases} 1, & \text{if } 0 < x < 5, \\ 0, & \text{if } 5 < x < 10. \end{cases}$
10. $c = 2$, $\ell = 4$, $u(x,0) = 0$, $u_t(x,0) = \begin{cases} 0, & \text{if } 0 < x < 1, \\ x - 1, & \text{if } 1 < x < 2, \\ 3 - x, & \text{if } 2 < x < 3, \\ 0, & \text{if } 3 < x < 4. \end{cases}$
11. The initial boundary value problem below models the vertical displacement $u(x,t)$ of a taut flexible string tied at both ends with homogeneous initial data and acted on by gravity (with constant acceleration $g \approx 9.81\,\mathtt{m/sec}^2$). Solve this problem.

$$\begin{aligned} &\text{(PDE)} \quad \ddot{u}_{tt} = c^2 u_{xx} + g, \quad 0 < x < \ell,\ 0 < t \leqslant t^* < \infty, \\ &\text{(BC)} \quad u(0,t) = 0, \quad u(\ell,t) = 0, \quad 0 < t \leqslant t^* < \infty, \\ &\text{(IC)} \quad u(x,0) = 0, \qquad \dot{u}(x,0) = 0. \end{aligned}$$

12. Using Eq. (11.3.10) on page 612, show that for all x and t, $u(x, t + \ell/c) = -u(\ell - x, t)$. What does this imply about the shape of the string at half a time period?
13. **(Damped vibrations of a string)** In the presence of resistance proportional to velocity, the one-dimensional wave equation becomes

$$u_{tt} + 2k\,u_t = c^2 u_{xx} \quad (0 < x < \ell,\ 0 < t < t^* < \infty),$$

where k is a positive constant. Assuming that $k\ell/(\pi c)$ is not an integer, find the general solution of this equation subject to the homogeneous Dirichlet conditions (11.3.5) and caused into motion by a displacement $d(x)$ without initial velocity.

11.4 Laplace Equation

A temperature distribution that we get when time elapses unboundedly ($t \mapsto +\infty$) is called the **steady state solution** (or distribution). Since it is independent of t, we must have $\partial u/\partial t = 0$. Substituting this into two-dimensional heat equation (11.1.12), page 603, we see that the steady state distribution satisfies the so called the Laplace equation[116] (in two spacial variables):

$$\nabla^2 u = 0 \qquad \text{or} \qquad u_{xx} + u_{yy} = 0, \tag{11.4.1}$$

where $\nabla = \langle \partial/\partial x, \partial/\partial y \rangle$ is the gradient operator, and u_{xx} and u_{yy} are shortcuts for partial derivatives, so $u_{xx} = \partial^2 u/\partial x^2$ and $u_{yy} = \partial^2 u/\partial y^2$. Laplace's equation also occurs in other branches of mathematical physics, including electrostatics, hydrodynamics, elasticity, and many others. A smooth function $u(x, y)$ that satisfies Laplace's equation is called a **harmonic function**.

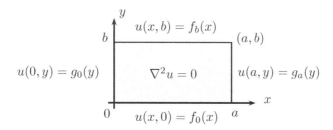

Figure 11.2: The Dirichlet conditions for a rectangular domain.

For simplicity, we consider a rectangular plate with specified temperature along the boundary. More specifically, we impose the Dirichlet boundary conditions

$$\begin{aligned} u(x,0) &= f_0(x), \quad u(x,b) = f_b(x), \quad 0 < x < a, \\ u(0,y) &= g_0(y), \quad u(a,y) = g_a(y), \quad 0 < y < b, \end{aligned} \tag{11.4.2}$$

as illustrated in Figure 11.2. A problem consisting of Laplace's equation on a region in the plane when the values of unknown function are specified on its boundary is called a **Dirichlet problem** or the first boundary value problem. Thus, Eq. (11.4.1) together with the boundary conditions (11.4.2) is a Dirichlet problem for the Laplace equation over a rectangle.

Rather than attacking this problem in its full generality, we split it into four auxiliary problems that are easier to handle. First we observe that this boundary value problem (11.4.1), (11.4.2) can be decomposed into four similar problems, each of which has only one nonzero boundary condition and $u(x, y)$ will be zero along three edges of the rectangle. Namely, we consider the following four auxiliary boundary conditions for the Laplace equation:

1. $u(x,0) = f_0(x), \quad u(x,b) = u(0,y) = u(a,y) = 0;$

2. $u(x,0) = 0, \quad u(x,b) = f_b(x), \quad u(0,y) = u(a,y) = 0;$

3. $u(x,0) = u(x,b) = 0, \quad u(0,y) = g_0(y), \quad u(a,y) = 0;$

4. $u(x,0) = u(x,b) = u(0,y) = 0, \quad u(a,y) = g_a(y).$

[116] Laplace's equation is named after French mathematician Pierre-Simon de Laplace (1749–1827), who studied it in 1782; however, the equation first appeared in 1752 in a paper by Leonhard Euler (1707–1783) on hydrodynamics.

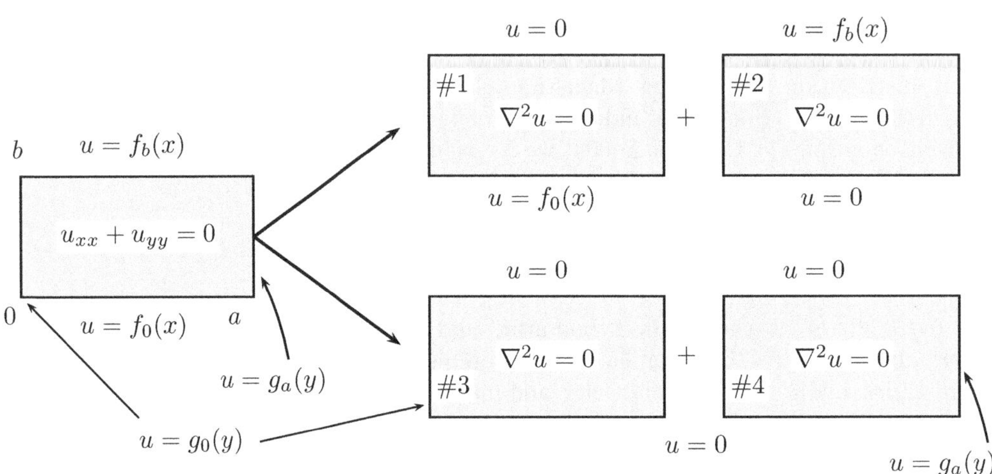

Since all these auxiliary boundary problems are similar, we consider only one of them. Namely, we solve the Laplace equation (11.4.1) in the rectangular domain $0 < x < a$, $0 < y < b$ subject to the boundary conditions

$$u(x,0) = f_0(x), \qquad u(x,b) = u(0,y) = u(a,y) = 0. \tag{11.4.3}$$

Substituting $u(x,y) = X(x)\,Y(y)$ into Laplace's equation gives $X''Y + X\,Y'' = 0$, so

$$\frac{X''}{X} = -\frac{Y''}{Y} = -\lambda$$

for some constant λ. Since the boundary conditions at $x = 0$ and $x = a$ are homogeneous, the function $X(x)$ must satisfy the familiar eigenvalue problem

$$X'' + \lambda X = 0, \qquad X(0) = X(a) = 0.$$

The eigenvalues and associated eigenfunctions were found previously, hence,

$$\lambda_n = \left(\frac{n\pi}{a}\right)^2, \quad X_n(x) = \sin\frac{n\pi x}{a}, \qquad n = 1, 2, 3, \ldots.$$

From the differential equation $Y'' - \lambda_n Y = 0$, we find

$$Y_n(y) = A_n \cosh\frac{n\pi y}{a} + B_n \sinh\frac{n\pi y}{a},$$

where $\cosh z = \frac{1}{2}\left(e^z + e^{-z}\right)$ is the hyperbolic cosine and $\sinh z = \frac{1}{2}\left(e^z - e^{-z}\right)$ is the hyperbolic sine function. Therefore, the formal solution of the auxiliary problem #1 is

$$u(x,y) = \sum_{n\geqslant 1}\left[A_n \cosh\frac{n\pi y}{a} + B_n \sinh\frac{n\pi y}{a}\right]\sin\frac{n\pi x}{a}.$$

To satisfy the boundary conditions (11.4.3), the coefficients A_n and B_n must be chosen such that

$$u(x,0) = f_0(x) = \sum_{n\geqslant 1} A_n \sin\frac{n\pi x}{a},$$

$$u(x,b) = 0 = \sum_{n\geqslant 1}\left[A_n \cosh\frac{n\pi b}{a} + B_n \sinh\frac{n\pi b}{a}\right]\sin\frac{n\pi x}{a}.$$

The first equation is recognized as the Fourier series expansion of $f_0(x)$; therefore,

$$A_n = \frac{2}{a}\int_0^a f_0(x)\,\sin\frac{n\pi x}{a}\,\mathrm{d}x, \quad n = 1, 2, \ldots. \tag{11.4.4}$$

The boundary condition at $y = b$ leads to

$$A_n \cosh\frac{n\pi b}{a} + B_n \sinh\frac{n\pi b}{a} = 0.$$

Hence, the coefficient B_n is expressed through the known value A_n:

$$B_n = -A_n \coth \frac{n\pi b}{a},$$

and the solution of the given auxiliary problem #1 becomes

$$u(x,y) = \sum_{n\geqslant 1} A_n \left[\cosh \frac{n\pi y}{a} - \coth \frac{n\pi b}{a} \sinh \frac{n\pi y}{a}\right] \sin \frac{n\pi x}{a}. \qquad \square$$

Other auxiliary boundary problems can be solved in a similar way, and they are left as exercises. A smooth solution of Laplace's equation is unique if it satisfies some boundary conditions. However, not every boundary condition leads to existence and uniqueness of the corresponding boundary value problem. Let $\mathbf{n}$ be the outward normal vector to the boundary ∂R of the region R. If the outward normal derivative of the harmonic function $u(x, y)$ is specified, $\left.\frac{\partial u}{\partial \mathbf{n}}\right|_{\partial R} = f$, we get the boundary conditions of the second kind, usually referred to as the Neumann condition. For the rectangular domain $(0 < x < a,\ 0 < y < b)$, the Neumann boundary conditions are read as

$$\frac{\partial u}{\partial x}(0,y) = g_0(y), \ \frac{\partial u}{\partial x}(a,y) = g_a(y), \ \frac{\partial u}{\partial y}(x,0) = f_0(x), \ \frac{\partial u}{\partial y}(x,b) = f_b(x). \tag{11.4.5}$$

The functions $g_0(y)$, $g_a(y)$, $f_0(x)$, $f_b(x)$ in the boundary conditions (11.4.5) cannot be chosen arbitrarily. Since we consider a steady state case, the total flux of heat across the boundary of R must be 0. This means that the total integral along the boundary must vanish:

$$\int_0^b g_0(y)\,\mathrm{d}y + \int_0^b g_a(y)\,\mathrm{d}y + \int_0^a f_0(x)\,\mathrm{d}x + \int_0^a f_b(x)\,\mathrm{d}x = 0.$$

This is a necessary condition for a Neumann problem to have a solution. Moreover, this solution, if it exists, is not unique—any constant can be added.

Note: Since the boundary of the rectangle has four corner points, we have to impose the **compatibility conditions** at these points. Usually, any irregularity at the boundary leads to a condition on the solution behavior in a neighborhood of that point.

11.4.1 Laplace Equation in Polar Coordinates

Previously, we solved several problems for partial differential equations using expansions with respect to eigenfunctions. The success of the method of separation of variables depended to a large extent on the fact that the domains under consideration were easily described in Cartesian coordinates. In this subsection, we address boundary value problems for a circular domain that are easily described in polar coordinates.

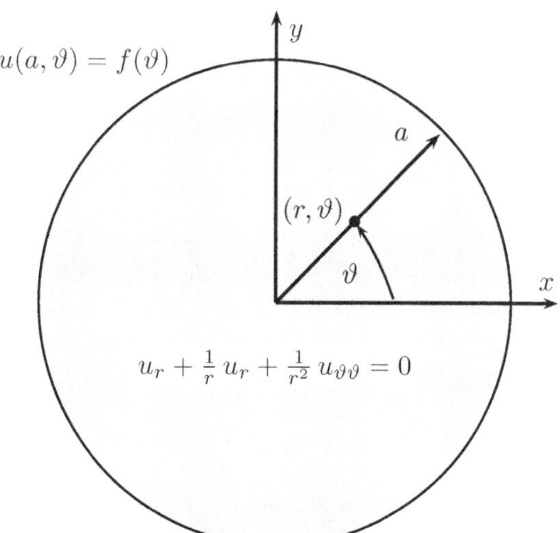

We consider the steady state (or time-independent) temperature distribution in a circular disk of radius a with insulated faces. To accommodate the geometry of the disk, we express the temperature, which is denoted by $u(r,\vartheta)$, in terms of polar coordinates r and ϑ, with $x = r\cos\vartheta$ and $y = r\sin\vartheta$. Then $u(r,\vartheta)$ satisfies the two-dimensional Laplace equation in polar coordinates:

$$\nabla^2 u \equiv \frac{\partial^2 u}{\partial r^2} + \frac{1}{r}\frac{\partial u}{\partial r} + \frac{1}{r^2}\frac{\partial^2 u}{\partial \vartheta^2} = 0. \tag{11.4.6}$$

If the function $u(r,\vartheta)$ is specified at the circumference of radius a, we get the boundary condition of the first kind or Dirichlet condition:

$$u(a,\vartheta) = f(\vartheta), \qquad 0 \leqslant \vartheta < 2\pi. \tag{11.4.7}$$

Equation (11.4.6) subject to the boundary condition (11.4.7) is called the **Dirichlet problem** for the Laplace equation. There are two known Dirichlet problems: the inner (interior) problem when the solution is defined inside the circle ($r < a$), and the outer (exterior) problem, when the solution is defined outside the circle ($r > a$).

In order for $u(r,\vartheta)$ to be a single-valued bounded function in the disk $0 \leqslant r < a$, we require the function to be 2π-periodic:

$$u(r,\vartheta) = u(r,\vartheta + 2\pi) \qquad \text{and} \qquad u(0,\vartheta) < \infty. \tag{11.4.8}$$

For an outer Dirichlet problem ($r > a$), the conditions (11.4.8) are replaced by

$$u(r,\vartheta) = u(r,\vartheta + 2\pi) \qquad \text{and} \qquad \lim_{r\mapsto+\infty} u(r,\vartheta) < \infty.$$

To use the method of separation of variables, we seek partial, nontrivial 2π-periodic solutions of the inner Laplace equation (11.4.6) in the form

$$u(r,\vartheta) = R(r)\,\Theta(\vartheta), \qquad 0 \leqslant r < a, \quad -\pi \leqslant \vartheta < \pi.$$

Substituting into Eq. (11.4.6) and separating variables gives

$$\frac{r^2 R''(r) + r\,R'(r)}{R(r)} = -\frac{\Theta''(\vartheta)}{\Theta(\vartheta)} = \lambda,$$

where λ is any constant. This leads to the two ordinary differential equations

$$r^2 R''(r) + r\,R'(r) - \lambda\,R(r) = 0, \tag{11.4.9}$$

$$\Theta''(\vartheta) + \lambda\,\Theta(\vartheta) = 0. \tag{11.4.10}$$

By adding the periodic condition

$$\Theta(\vartheta) = \Theta(\vartheta + 2\pi), \tag{11.4.11}$$

we get the eigenvalue problem (11.4.10), (11.4.11) for the function $\Theta(\vartheta)$. This Sturm–Liouville problem has nontrivial solutions only when λ is a nonnegative integer $\lambda = n$ ($n = 0, 1, 2, \ldots$) to which correspond eigenfunctions $\Theta_n(\vartheta) = a_n\cos n\vartheta + b_n\sin n\vartheta$, where a_n and b_n are arbitrary constants.

Equation (11.4.9) is an Euler equation (see §4.6.2) and has the general solution

$$R_n(r) = k_1 r^n + k_2 r^{-n} \quad \text{if } n > 0, \quad \text{and} \quad R_0(r) = k_1 + k_2\ln r \quad \text{if } n = 0,$$

with some constant k_1, k_2. The logarithmic term cannot be accepted because the function $u(r,\vartheta)$ is to remain bounded as either $r \mapsto 0$ or $r \mapsto +\infty$, which the logarithmic term does not satisfy. Thus, the eigenfunction corresponding to $\lambda = 0$ should be chosen as a constant. For $\lambda = n > 0$, one of the coefficients k_1 or k_2 must be zero to make the function $R_n(r)$ bounded either at $r = 0$ (inner problem) or at $r \mapsto +\infty$ (outer problem). Since we consider an inner Dirichlet problem, the partial nontrivial solutions will be

$$u_n(r,\vartheta) = R_n(r)\,\Theta(\vartheta) = r^n\left(a_n\cos n\vartheta + b_n\sin n\vartheta\right), \qquad n = 0, 1, 2, \ldots.$$

By forming an infinite series from these nontrivial solutions, we express $u(r,\vartheta)$ as their linear combination:

$$u(r,\vartheta) = \frac{a_0}{2} + \sum_{n\geqslant 1} r^n\left(a_n\cos n\vartheta + b_n\sin n\vartheta\right).$$

It is more convenient to write this series in the equivalent form

$$u(r,\vartheta)=\frac{a_0}{2}+\sum_{n\geqslant 1}\left(\frac{r}{a}\right)^n(a_n\cos n\vartheta+b_n\sin n\vartheta)\quad (r\leqslant a). \tag{11.4.12}$$

The boundary condition (11.4.7) then requires that

$$u(a,\vartheta)=f(\vartheta)=\frac{a_0}{2}+\sum_{n\geqslant 1}(a_n\cos n\vartheta+b_n\sin n\vartheta),$$

which is the Fourier expansion of the function $f(\vartheta)$. Its coefficients are obtained from the Euler–Fourier formulas (10.3.4), page 563.

For the outer Dirichlet problem, the series solution is

$$u(r,\vartheta)=\frac{a_0}{2}+\sum_{n\geqslant 1}\left(\frac{a}{r}\right)^n(a_n\cos n\vartheta+b_n\sin n\vartheta)\quad (r\geqslant a), \tag{11.4.13}$$

where coefficients a_n and b_n are determined by Eq. (10.3.4), page 563. ■

If the outward normal derivative is specified at the boundary such that

$$\left.\frac{\partial u(r,\vartheta)}{\partial r}\right|_{r=a}=g(\vartheta),\qquad 0\leqslant\vartheta<2\pi, \tag{11.4.14}$$

then we have the boundary condition of the second kind or Neumann boundary condition. The corresponding boundary value problem (11.4.1), (11.4.14) is called the **Neumann problem** or the second boundary value problem. There are also two known problems—the outer (for $r>a$) and the inner (for $r<a$) problems.

The procedure to solve the Neumann problem is exactly the same as before, and its solution is represented either by formula (11.4.12) for the inner problem or (11.4.13) for the outer problem. To satisfy the boundary condition (11.4.14), we have

$$\left.\frac{\partial u(r,\vartheta)}{\partial r}\right|_{r=a}=g(\vartheta)=\frac{1}{a}\sum_{n\geqslant 1}n\,(a_n\cos n\vartheta+b_n\sin n\vartheta)\quad\text{(inner problem)},$$

$$\left.\frac{\partial u(r,\vartheta)}{\partial r}\right|_{r=a}=g(\vartheta)=-\frac{1}{a}\sum_{n\geqslant 1}n\,(a_n\cos n\vartheta+b_n\sin n\vartheta)\quad\text{(outer problem)}.$$

Since the above Fourier series do not contain the free term a_0, the term must be zero:

$$a_0=\frac{1}{\pi}\int_{-\pi}^{\pi}g(\vartheta)\,\mathrm{d}\vartheta=0,$$

which is the compatibility constraint. For the inner problem, the Euler–Fourier coefficients are

$$a_n=\frac{a}{n\pi}\int_{-\pi}^{\pi}g(\vartheta)\cos(n\vartheta)\,\mathrm{d}\vartheta,\quad b_n=\frac{a}{n\pi}\int_{-\pi}^{\pi}g(\vartheta)\sin(n\vartheta)\,\mathrm{d}\vartheta.$$

For the outer Neumann problem, the coefficients have the same values but opposite signs.

Problems

1. Solve the second auxiliary interior Dirichlet auxiliary boundary problem in the rectangle $\nabla^2u=0$, $u(x,0)=0$, $u(x,b)=f_b(x)$, $u(0,y)=u(a,y)=0$.
2. Solve the third auxiliary interior Dirichlet auxiliary boundary problem in the rectangle $\nabla^2u=0$, $u(x,0)=u(x,b)=0$, $u(0,y)=g_0(y)$, $u(a,y)=0$.
3. Solve the fourth auxiliary interior Dirichlet auxiliary boundary problem in the rectangle $\nabla^2u=0$, $u(x,0)=u(x,b)=u(0,y)=0$, $u(a,y)=g_a(y)$.
4. Solve the first auxiliary Dirichlet boundary value problem in the rectangle $0<x<\pi$, $0<y<1$, given $f_0(x)=5\sin 3x-1$.
5. Solve the second auxiliary Dirichlet boundary value problem in the rectangle $0<x<3$, $0<y<2$, given $f_b(x)=x(3-x)$.

6. Solve the third auxiliary Dirichlet boundary value problem in the rectangle $0 < x < 1$, $0 < y < 1/2$, given $g_0(y) = 7\sin(4\pi y)$.
7. Solve the fourth auxiliary Dirichlet boundary value problem in the rectangle $0 < x < 4$, $0 < y < 1/4$, given $g_a(y) = 5\sin(8\pi y)$.
8. Solve the Neumann problem for Laplace's equation in the rectangle $0 < x < 2$, $0 < y < 1$ subject to the boundary conditions
$$\left.\frac{\partial u}{\partial y}\right|_{y=0} = \left.\frac{\partial u}{\partial y}\right|_{y=1} = 0, \quad \left.\frac{\partial u}{\partial x}\right|_{x=0} = 0, \quad \left.\frac{\partial u}{\partial x}\right|_{x=2} = 2y-1.$$
9. Solve the Neumann problem for Laplace's equation in the rectangle $0 < x < 2$, $0 < y < 4$ subject to the boundary conditions
$$\left.\frac{\partial u}{\partial x}\right|_{x=0} = \left.\frac{\partial u}{\partial x}\right|_{x=2} = 0, \quad \left.\frac{\partial u}{\partial y}\right|_{y=0} = 3x^2 - 8x, \quad \left.\frac{\partial u}{\partial x}\right|_{y=4} = 0.$$
10. Solve the third boundary problem for Laplace's equation in the rectangle $0 < x < \pi/2$, $0 < y < 1$ subject to the boundary conditions $u_x(0,y) = u(\pi/2, y) = u(x,1) = 0$, $u(x,0) = x^2$.
11. Solve the third boundary problem for Laplace's equation in the rectangle $0 < x < \pi$, $0 < y < 2\pi$ subject to the boundary conditions $u(0,y) = u_x(\pi,y) = u(x,0) = 0$, $u(x,2\pi) = \sin x$.
12. Solve the third boundary problem for Laplace's equation in the rectangle $0 < x < 4$, $0 < y < 1$ subject to the boundary conditions $u_x(0,y) = 1 - y$, $u(4,y) = u_y(x,0) = u(x,1) = 0$.
13. Find the solution $u(x,y)$ of Laplace's equation (11.4.1) in the semi-infinite strip $0 < x < \infty$, $0 < y < 1$ that satisfies the boundary conditions
$$u(x,0) = 0, \qquad u(x,1) = 0, \qquad u(0,y) = 1 - \cos^2(\pi y).$$
14. A **Newton potential** is a harmonic function of the form $v = v(r)$, where r is the distance to the origin, which is defined for all $r > 0$. Find all Newton potentials on $\mathbb{R}^2$ that are solutions of the differential equation $\frac{\mathrm{d}}{\mathrm{d}r}\left(r\,\frac{\mathrm{d}v}{\mathrm{d}r}\right) = 0$.
15. When a harmonic function in cylindrical coordinates does not depend on angle ϑ, it is called axially symmetric. Assuming that partial nontrivial solutions of Laplace's axially symmetric equation
$$u_{rr} + (1/r)\,u_r + u_{zz} = 0$$
can be written in the form $u(r,z) = R(r)\,Z(z)$, show that these functions $R(r)$ and $Z(z)$ are solutions of the following ordinary differential equations:
$$r\,R' + R' + \lambda^2 r\,R = 0, \qquad Z'' - \lambda^2 Z = 0.$$
The equation for $R(r)$ is the Bessel equation of order zero.

Consider Laplace's equation in the domain $a < r < b$ bounded by two circles $r = a$ and $r = b$ written in polar coordinates. Since the boundary of the annulus consists of two circles, the boundary conditions should be specified at each circle. This time, the origin is not in the problem domain. Therefore, there is no reason to discard the separation of variables solutions $\ln r$ and r^n or r^{-n} as we did when solving boundary value problems for circular domains. In Problems 16 through 20, solve Laplace's equation (11.4.6) in the annulus having given inner radius a, given outer radius b, and subject to the given boundary conditions.

16. $(1 < r < 3)$ $u(1,\vartheta) = \cos 3\vartheta$, $\quad u(3,\vartheta) = \sin 2\vartheta$.
17. $(2 < r < 5)$ $u_r(2,\vartheta) = \cos 2\vartheta$, $\quad u(5,\vartheta) = \sin 4\vartheta$.
18. $(2 < r < 4)$ $u(2,\vartheta) = \begin{cases} \sin 4\vartheta, & \text{if } 0 < \vartheta < \pi, \\ 0, & \text{if } \pi < \vartheta < 2\pi; \end{cases}$

    $u(4,\vartheta) = \begin{cases} 0, & \text{if } 0 < \vartheta < \pi, \\ \cos 5\vartheta, & \text{if } \pi < \vartheta < 2\pi. \end{cases}$
19. $(1 < r < 5)$ $u(1,\vartheta) = \cos 5\vartheta$, $\quad u_r(5,\vartheta) = \cos 3\vartheta$.
20. $(2 < r < 3)$ $u_r(2,\vartheta) = \cos 4\vartheta$, $\ u_r(3,\vartheta) = \sin 7\vartheta$.

In Problems 21 through 26, you are asked to solve Laplace's equation (11.4.6) in circular domains; therefore, it is appropriate to use polar coordinates. Using separation of variables, solve the corresponding outer or inner boundary value problems.

21. Find the harmonic solution $u(r,\vartheta)$ for $r > 2$ that satisfies the boundary condition $u(2,\vartheta) = \vartheta$. Note that this function is discontinuous at $\vartheta = 0$.

22. Determine the harmonic solution $u(r,\vartheta)$ for $r < 3$ that satisfies the boundary condition $u(3,\vartheta) = 2 - 2\cos 2\vartheta$.
23. Find the harmonic solution $u(r,\vartheta)$ for $r > 4$ that satisfies the boundary condition
$$u(4,\vartheta) = \begin{cases} 1, & \text{if } 0 < \vartheta < \pi/2, \\ 0, & \text{if } \pi/2 < \vartheta < 2\pi. \end{cases}$$
24. Find the harmonic solution $u(r,\vartheta)$ for $r < 1$ that satisfies the boundary condition $u(1,\vartheta) = 3\cos 3\vartheta$.
25. Find the harmonic solution $u(r,\vartheta)$ for $r < 2$ that satisfies the boundary condition $u_r(2,\vartheta) = \sin 3\vartheta - 4\cos 5\vartheta$.
26. Find the harmonic solution $u(r,\vartheta)$ for $r > 3$ that satisfies the boundary condition
$$u_r(3,\vartheta) = \begin{cases} 0, & \text{if } 0 < \vartheta < \pi, \\ \cos 3\vartheta, & \text{if } \pi < \vartheta < 2\pi. \end{cases}$$

Consider Laplace's equation in the circular sector $0 < r < a$, $0 < \vartheta < \alpha$. In Problems 27 through 29, solve Laplace's equation (11.4.6) in the sector subject to the given boundary conditions.

27. Find the harmonic function in the circular sector $0 < r < \sqrt{2}$, $0 < \vartheta << \pi/4$ subject to the boundary conditions $u(r,0) = 1$, $u(r,\pi/4) = 1 + r$, $u(\sqrt{2},\vartheta) = 1 + 2\sin\vartheta$.
28. Find the harmonic function in the circular sector $0 < r < 1$, $0 < \vartheta << \pi/2$ subject to the boundary conditions $u(r,0) = r^2$, $u(r,\pi/2) = -r^2 - r^3$, $u(1,\vartheta) = \cos 2\vartheta + \sin 3\vartheta$.
29. Find the harmonic function in the circular sector $0 < r < 1$, $0 < \vartheta << \pi/4$ subject to the boundary conditions $u(r,0) = 0$, $u(r,\pi/4) = 1 + r^4$, $u(1,\vartheta) = 1 - \cos 4\vartheta + \sin 3\vartheta$.

Summary for Chapter 11

1. The heat transfer (or conduction) equation $u_t = c^2\nabla^2 u$, subject to various boundary conditions and the initial condition $u(x,0) = f(x)$, has a unique solution that can be obtained by using the separation of variables method (also known as the Fourier method).
2. The separation of variables method is applicable only when the boundary conditions are homogeneous. Otherwise, it requires shifting the boundary data.
3. When the separation of variables method is applied to the one-dimensional heat equation $u_t = c^2 u_{xx}$ with the Dirichlet boundary conditions $u(0,t) = u(\ell,t) = 0$, it leads to the Sturm–Liouville problem $X'' + \lambda X = 0$, $X(0) = X(\ell) = 0$, where $u(x,y) = X(x)\,T(t)$. Finding the eigenfunctions $X_n(x) = \sin\frac{n\pi x}{\ell}$, $n = 1, 2, \ldots$, we build the general solution as a sum of all partial nontrivial solutions:
$$u(x,t) = \sum_{n\geqslant 1} C_n\, e^{-c^2\lambda_n^2 t}\,\sin\frac{n\pi x}{\ell},$$
where the coefficients C_n are determined by the Euler–Fourier formulas (11.1.10), page 600.
4. When boundary conditions are not of the first kind (Dirichlet), it leads to a Sturm–Liouville problem subject other boundary conditions. Then there exists a series solution in the form $\sum_{n\geqslant 1} T_n(t)\,X_n(x)$, where $X_n(x)$ are eigenfunctions of the corresponding Sturm–Liouville problem.
5. When the boundary conditions are not homogeneous, it requires a shift of these data to make them homogeneous and then application of the separation of variables method.
6. The wave equation $u_{tt} = c^2 u_{xx}$ can also be solved by separation of variables. The solution of the corresponding initial boundary value problem (11.3.1), (11.3.2), and (11.3.5) is given by formula (11.3.10), which is an infinite sum of the normal modes
$$u_n(x,t) = \left[A_n\cos\left(\frac{cn\pi t}{\ell}\right) + B_n\sin\left(\frac{cn\pi t}{\ell}\right)\right]\sin\left(\frac{n\pi x}{\ell}\right), \qquad n =, 1, 2, 3, \ldots.$$
The first normal mode is called the fundamental mode or fundamental standing wave; all other modes (standing waves) are known as overtones.
7. A steady state temperature distribution $u(x,y)$ inside a plane domain satisfies the Laplace equation $u_{xx} + u_{yy} = 0$ or simply $\nabla^2 u = 0$, where ∇ is the gradient operator. Any smooth solution of Laplace's equation is called the harmonic function.
8. For Laplace's equation in some planar domain R, it is common to consider three types of boundary conditions: the first kind or Dirichlet $u|_{\partial R} = f$, Neumann or second kind $\partial u/\partial\mathbf{n}|_{\partial R} = f$, and the third kind $u|_{\partial R} + k\partial u/\partial\mathbf{n}|_{\partial R} = f$, where f is a given function and $\partial u/\partial\mathbf{n}|_{\partial R}$ is the outward normal derivative to the boundary.
9. The following trigonometric identities are proved to be useful in many calculations:
$$2\cos^2\theta = 1 + \cos 2\theta, \qquad 2\sin^2\theta = 1 - \cos 2\theta.$$

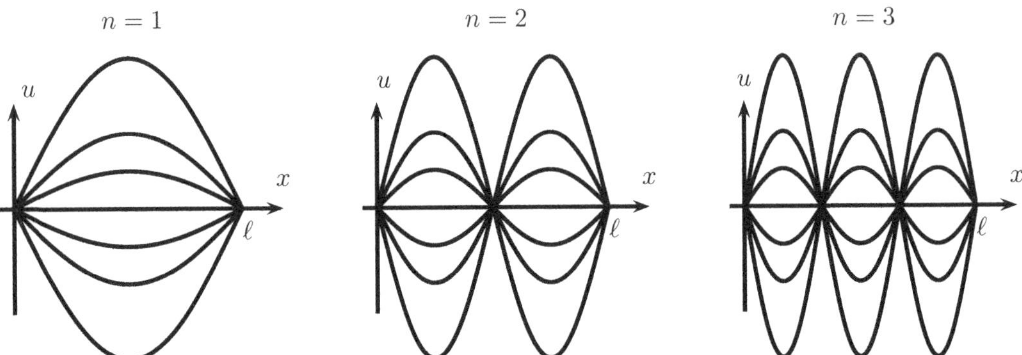

Figure 11.3: Standing waves.

Review Questions for Chapter 11

Section 11.1

In Problems 1 through 4, find a formal solution to the given initial boundary value problem subject to the Neumann boundary conditions

$$u_t = \alpha\, u_{xx} \qquad u_x(0,t) = u_x(\ell,t) = 0, \quad u(x,0) = f(x),$$

when the coefficient of diffusivity α, the length of the rod ℓ, and the initial temperature $f(x)$ are specified.

1. $\alpha = 9$, $\ell = 1$, and $f(x) = \cos(\pi x) - 10\cos(3\pi x)$.
2. $\alpha = 1$, $\ell = 4$, and $f(x) = 5x$ (violates the compatibility constraint).
3. $\alpha = 9$, $\ell = \pi$, and $f(x) = x^2$ (violates the compatibility constraint).
4. $\alpha = 1$, $\ell = 1/2$, and $f(x) = \cos^2(2\pi x)$.

In Problems 5 through 8, find a formal solution to the given initial boundary value problem with boundary conditions of the third kind

$$u_t = \alpha\, u_{xx} \qquad u(0,t) = u_x(\ell,t) = 0, \quad u(x,0) = f(x),$$

when the coefficient of diffusivity α, the length of the rod ℓ, and the initial temperature $f(x)$ are specified.

5. $\alpha = 4$, $\ell = \pi$, and $f(x) = x\,\sin x$.
6. $\alpha = 4$, $\ell = 1$, and $f(x) = x^2 - x$ (violates the compatibility constraint).
7. $\alpha = 1$, $\ell = \pi/2$, and $f(x) = x\,\cos x$ (violates the compatibility constraint).
8. $\alpha = 1$, $\ell = 1/2$, and $f(x) = x(1/2 - x)^2$.

In Problems 9 through 12, find a formal solution to the given initial boundary value problem with boundary conditions of the third kind

$$u_t = \alpha\, u_{xx} \qquad u_x(0,t) = u(\ell,t) = 0, \quad u(x,0) = f(x),$$

when the coefficient of diffusivity α, the length of the rod ℓ, and the initial temperature $f(x)$ are specified.

9. $\alpha = 16$, $\ell = 2\pi$, and $f(x) = 5\cos\frac{3x}{4}$.
10. $\alpha = 4$, $\ell = 1$, and $f(x) = x^2 - 2x$ (violates the compatibility constraint).
11. $\alpha = 4$, $\ell = \pi$, and $f(x) = \cos x$ (violates the compatibility constraint).
12. $\alpha = 9$, $\ell = 3$, and $f(x) = x^2\,(3 - x)$.
13. The ends of a thin, laterally insulated bar of length π are maintained at a temperature of zero degrees. At time $t = 0$, the temperature profile is $u(x,0) = 59\sin(x)$. At time $t = 1/2$, the temperature $u(x,t)$ at the center $x = \pi/2$ of the bar has decreased to a value of about 8°C; that is, $u(\pi/2, 1/2) \approx 8$. What is the thermal diffusivity of the bar? What is $u_x(\pi, 1)$?
14. The ends of a thin, laterally insulated bar of length 2 are maintained at a temperature of zero degrees. At time $t = 0$, the temperature profile is $u(x,0) = 100\,(1 - \cos\pi x)$. At time $t = 1$, the temperature at $x = 1$ has decreased to a value of about 49; that is, $u(1,1) \approx 49$. What is the thermal diffusivity of the bar? What is $u_x(0,1)$?

15. Consider a rod of length $\ell = 1$, one end of which is maintained at zero temperature, $u(0,t) = 0$, but another end $x = 1$ of the rod is in a solution of water fixed at the same temperature of zero. (Zero refers to some reference temperature.) Using Newton's law of cooling, the boundary conditions can be shown to be

$$u(0,t) = 0, \qquad u_x(1,t) + h\,u(1,t) = 0,$$

where $h > 0$ is a constant that depends on the rate of heat flow across the boundary. Find the temperature in such insulated rod.

Section 11.2 of Chapter 11 (Review)

1. Assume that a thin, laterally insulated bar of 10 units length has its two ends maintained at constant temperatures, $u(0,t) = 10$ and $u(40,t) = T_\ell$, which is to be determined. At time $t = 0$, the initial temperature in the bar is known to be $u(x,0) = x$. A probe inserted into the bar center measures the temperature and finds it to be approximately 27°C and 29°C at times $t = 1.45$ and $t = 2.5$. Determine the unknown end point temperature T_ℓ and the thermal diffusivity of the bar's material.
2. Find all solutions of the form $u(x,t) = X(x)\,T(t)$ for each of the equations.

$$\textbf{(a)}\ \ u_t = x^2 u_{xx} + x\,u_x, \qquad \textbf{(b)}\ \ u_t = u_{xx} + 2\,u_x.$$

In each Problem 3 through 7, use the separation of variables method to find a formal solution of the given initial boundary value problem.

3. $u_t = 4\,u_{xx}$ $(0 < x < 2)$, $u(0,t) = 0$, $u(2,t) = 2\,e^{-t}$, $u(x,0) = x$.
4. $u_t = 25\,u_{xx}$ $(0 < x < 5)$, $u(0,t) = 20$, $u(5,t) = 10$, $u(x,0) = 20 - 2x + \sin(2\pi x)$.
5. $u_t = 9\,u_{xx} + 2$ $(0 < x < 3)$, $u_x(0,t) = 0$, $u(3,t) = 90$, $u(x,0) = 10\,x^2$.
6. $u_t = u_{xx} + x(x-1)^2$ $(0 < x < 1)$, $u(0,t) = 0$, $u_x(1,t) = 0$, $u(x,0) = 0$.
7. $u_t = 4\,u_{xx} + \cos(\pi x)$ $(0 < x < 2)$, $u_x(0,t) = 0$, $u_x(2,t) = 0$, $u(x,0) = 2\cos^2(\pi x)$.

In Problems 8 through 13, use separation of variables to find the series solution $u(x,t) = \sum_{n\geqslant 1} X_n(x)\,T_n(t)$ of the given initial boundary value problems for the heat conduction equation with non-smooth temperature profiles. Then plot $u_{30}(x,t)$ for $t = 1,\ 5,\ 10$, where $u_M(x,t) = \sum_{n=1}^{M} X_n(x)\,T_n(t)$ is the truncated partial sum with M terms. Observe a striking characteristic of the heat equation expressed in the smoothing effect on the initial temperature distribution.

8. $u_t = u_{xx}$ $(0 < x < 5)$, $u(0,t) = 3$, $u(5,t) = 1$, $u(x,0) = \begin{cases} x + 3, & \text{if } 0 < x < 2, \\ 1 + \frac{4}{3}\,(5 - x), & \text{if } 2 < x < 5. \end{cases}$
9. $u_t = 3\,u_{xx} + 1$ $(0 < x < 3)$, $u_x(0,t) = 0$, $u_x(3,t) = 0$, $u(x,0) = \begin{cases} 1, & \text{if } 0 < x < 2, \\ 0, & \text{if } 2 < x < 3. \end{cases}$
10. $u_t = 8\,u_{xx}$ $(0 < x < 4)$, $u(0,t) = 2$, $u_x(4,t) = 0$, $u(x,0) = \begin{cases} 2, & \text{if } 0 < x < 3, \\ 0, & \text{if } 3 < x < 4. \end{cases}$
11. $u_t = 4\,u_{xx}$ $(0 < x < 2)$, $u(0,t) = 1$, $u(2,t) = 1$, $u(x,0) = |\cos \pi x|$.
12. $u_t = 9\,u_{xx}$ $(0 < x < 3)$, $u_x(0,t) = 0$, $u(3,t) = 1$, $u(x,0) = \begin{cases} \sin^2(\pi x), & \text{if } 0 < x < 1, \\ 1, & \text{if } 1 < x < 3. \end{cases}$
13. $u_t = u_{xx}$ $(0 < x < \pi)$, $u(0,t) = 0$, $u(\pi,t) = 1$, $u(x,0) = \begin{cases} 0, & \text{if } 0 < x < \pi/2, \\ \cos^2 x, & \text{if } \pi/2 < x < \pi. \end{cases}$
14. Consider a bar 50 cm long that is made of a material for which $\alpha = 4$ and whose ends ($x = 0$ and $x = 0.5$) are insulated. Suppose that the bar is heated by an external source with a rate which is proportional to $1 - x^2$. Derive the formal series solution assuming the bar had initial zero temperature.
15. A copper rod 5 cm long with an insulated lateral surface has initial temperature $u(x,0) = (1 - 2x)^2$, and at time $t = 0$ both of its ends ($x = 0$, $x = 0.05$) are maintained with a constant temperature of 40°C. Find the temperature distribution in the rod assuming that its thermal diffusivity is constant with the value $\alpha \approx 1.11 \times 10^{-4}\,\mathtt{m}^2/\mathtt{sec}$.
16. Suppose that a tin rod 20 cm long with an insulated lateral surface is heated to a uniform temperature of 20°C, and that at time $t = 0$ one end ($x = 0$) is insulated while the other end ($x = 0.2$) is maintained with a constant temperature of 40°C. Find the formal series solution for the temperature $u(x,t)$ of the rod. Assume that the thermal diffusivity is a constant, which is approximately equal to $4 \times 10^{-5}\ \mathtt{m}^2/\mathtt{sec}$ at 20°C.
17. Consider a silicon rod of length 30 cm whose initial temperature is given by $u(x,0) = x^2(30 - x)/15$. Suppose that the thermal diffusivity is approximately $\alpha \approx 0.9\,\mathtt{cm}^2/\mathtt{sec}$, and that one end ($x = 0$) is insulated while the other end ($x = 30$) is maintained with a constant temperature of 20°C. Find the formal series solution for the temperature $u(x,t)$ of the rod.

18. One face of the slab $0 \leqslant x \leqslant 2\ell$ is insulated while the other end $x = 2\ell$ is kept at a constant temperature 20°C. The initial temperature distribution is given by $u(x,0) = 0$ for $0 < x < \ell$, $u(x,0) = 20$ for $\ell < x < 2\ell$. Derive the formal series solution.

19. Let a niobium wire of length 1 m be initially at the uniform temperature of 20°C. Suppose that at time $t = 0$, the end $x = 0$ is cooled to 5°C while the end $x = 1$ is heated to 55°C, and both are thereafter maintained at those temperatures. Find the temperature distribution in the wire at any time assuming the thermal diffusivity to a constant $\alpha \approx 2.5 \times 10^{-5}\,\mathtt{m}^2/\mathtt{sec}$.

Section 11.3 of Chapter 11 (Review)

1. The derivation of the wave equation starts by obtaining an equation of the form $u_{tt} = c^2 u_{ss}$, where s is arc length. Upon introducing the more convenient coordinate, the spacial variable x, its derivative is expressed as $\mathrm{d}s/\mathrm{d}x = \sqrt{1 + u_x^2}$. Show that the wave equation becomes
$$u_{tt} = c^2 \left(1 + u_x^2\right)^{-2} u_{xx},$$
which leads to the linear equation (11.3.1) when u_x^2 is disregarded as being small.

2. A string has length π, and units are chosen so that $c = 1$ in the wave equation (11.3.1). Initially, the string is in its equilibrium position, with velocity $\frac{\partial u}{\partial t}(x,0) = \sin x - \sin(3x)$. Determine the displacement function $u(x,t)$.

3. Use the method of separation of variables to derive a formal solution to the **telegraph equation** $u_{tt} + u_t + u = c^2 u_{xx}$, when $4\pi^2 c^2 > 3\ell^2$, in the finite domain $0 < x < \ell$ subject to the boundary and initial conditions
$$u(0,t) = u(\ell,t) = 0, \qquad u(x,0) = f(x), \quad u_t(x,0) = 0.$$

4. Find all product solutions $u(x,t) = X(x)\,T(t)$ of the modified wave equation $u_{tt} + k\,u = c^2 u_{xx}$.

Consider the initial boundary value problem for the wave equation
$$\begin{aligned} &u_{tt} = c^2 u_{xx}, \qquad 0 < x < \ell, \quad 0 < t < t^* < \infty; \quad u(0,t) = u(\ell,t) = 0, \qquad 0 < t < \infty; \\ &u(x,0) = f(x), \qquad u_t(x,0) = v(x), \qquad 0 < x < \ell. \end{aligned}$$

In Problems 5 through 8, solve the problem for the given parameter values (c and ℓ) and the given initial conditions.

5. $c = 1$, $\ell = 2$, $f = 0$, $g = \cos^2(5\pi x)$.

6. $c = 2$, $\ell = 4$, $f = 0$,
$$v = \begin{cases} 1 - \cos(2\pi x), & 0 < x < 2, \\ 0, & 2 < x < 4; \end{cases}.$$

7. $c = 3$, $\ell = 6$, $f = \cos 7\pi x$, $g = 0$.

8. $c = \pi$, $\ell = 2\pi$, $f = 0$,
$$v = \begin{cases} x, & 0 < x < \pi, \\ 0, & \pi < x < 2\pi. \end{cases}$$

Consider an elastic rod of length ℓ. The rod is set in motion with an initial displacement $d(x)$ and initial velocity $v(x)$. In each Problem 9 through 14, carry out the following steps given longitudinal velocity c, the length ℓ, and the initial and boundary conditions.

(a) Solve the wave equation $u_{tt} = c^2 u_{xx}$ subject to the given initial and boundary conditions.

(b) Plot $u(x,t)$ versus $x \in [0,\ell]$ for several values of $t = 1$, 2, 10, and 20.

(c) Plot $u(x,t)$ versus t for $0 \leqslant t \leqslant 20$ and for several values of $x = 0$, $\ell/4$, $\ell/2$, $3\ell/4$, and ℓ.

(d) Draw a three-dimensional plot of u versus x and t.

9. $u_{tt} = u_{xx}$ $(0 < x < \pi/2)$, $u(0,t) = u(\pi/2,t) = 0$, $u(x,0) = x$, $u_t(x,0) = (2x - \pi)^2$.

10. $u_{tt} = 25\,u_{xx}$ $(0 < x < 10)$, $u_x(0,t) = u(10,t) = 0$, $u(x,0) = 0$, $u_t(x,0) = x^2(10 - x)$.

11. $u_{tt} = 64\,u_{xx}$ $(0 < x < 4)$, $u_x(0,t) = u(4,t) = 0$, $u(x,0) = x(4 - x)$, $u_t(x,0) = 1$.

12. $u_{tt} = 36 u_{xx}$ $(0 < x < 3)$, $u(0,t) = u_x(3,t) = 0$, $u(x,0) = 0$, $u_t(x,0) = 0$ for $0 < x < 1$ and $u_t(x,0) = x - 1$ for $1 < x < 3$.

13. $u_{tt} = 4\,u_{xx}$ $(0 < x < 1)$, $u(0,t) = u(1,t) = 0$, $u(x,0) = \sin \pi x$, $u_t(x,0) = 2\cos^2(3\pi x) - 1$.

14. $u_{tt} = 16\,u_{xx}$ $(0 < x < 2)$, $u(0,t) = u_x(2,t) = 0$, $u(x,0) = \sin(3\pi x/4)$, $u_t(x,0) = \cos(3\pi x/4)$.

Mechanical waves propagate through a material medium (solid, liquid, or gas) at a wave speed which depends on the inertial properties of that medium. There are two basic types of wave motion for mechanical waves: longitudinal waves and transverse waves. In a transverse wave, particles of the medium are displaced in a direction perpendicular to the direction of energy transport. In a longitudinal wave, particles of the medium are displaced in a direction parallel to energy transport. Longitudinal waves are observed, for instance, in elastic bars or rods when their vertical dimensions are small. By placing the x-axis along the bar's direction so that its left end coincides with the origin, we can assume that its vibrations are uniform over each cross-section. Then the longitudinal displacement $u(x,t)$ satisfies the one-dimensional wave equation (11.3.1). In Problems 15 through 20, find longitudinal displacements in rods under the given initial and boundary conditions.

15. $u_{tt} = u_{xx}$ $(0 < x < \pi)$, $u(0,t) = u_x(\pi,t) = 0$, $u(x,0) = x$, $u_t(x,0) = 0$.
16. $u_{tt} = 4\,u_{xx}$ $(0 < x < 2)$, $u_x(0,t) = u(2,t) = 0$, $u(x,0) = \cos(\pi x)$, $u_t(x,0) = \sin(2\pi x)$.
17. $u_{tt} = 9\,u_{xx}$ $(0 < x < 3)$, $u_x(0,t) = u_x(2,t) = 0$, $u(x,0) = x^2(3-x)^2$, $u_t(x,0) = \sin(2\pi x)$.
18. $u_{tt} = c^2 u_{xx} - 2h\,u_t$ $(0 < x < \ell)$, $u(0,t) = u(\ell,t) = 0$, $u(x,0) = f(x)$, $u_t(x,0) = 0$.
19. $u_{tt} = 25\,u_{xx}$ $(0 < x < 5)$, $u_x(0,t) = u(5,t) = 0$, $u(x,0) = \cos\frac{9\pi x}{10}$, $u_t(x,0) = x(x-5)$.
20. $u_{tt} = u_{xx}$ $(0 < x < 2)$, $u(0,t) = u_x(2,t) = 0$, $u(x,0) = 0$, $u_t(x,0) = \begin{cases} 0, & 0 < x < 1, \\ 1, & 1 < x < 2. \end{cases}$
21. Consider the wave equation with nonhomogeneous boundary conditions:

$$\begin{aligned} &\text{(PDE)} \quad u_{tt} = c^2 u_{xx}, \quad 0 < x < \ell,\ 0 < t \leqslant t^* < \infty, \\ &\text{(BC)} \quad u(0,t) = a, \quad u(\ell,t) = b, \quad 0 < t \leqslant t^* < \infty, \\ &\text{(IC)} \quad u(x,0) = d(x), \quad u_t(x,0) = v(x), \end{aligned}$$

where a and b are not both zero. This problem can be reduced to a similar problem with homogeneous boundary conditions by substitution $u(x,t) = u_1(x,t) + v(x,t)$, where $u_1(x,t) = a + \frac{b-a}{\ell}\,x$. Unlike the case of the heat equation, the solution $u(x,t)$ of the given wave equation does not have a limit in general as $t \to \infty$. Hence, $u_1(x,t)$ is not the time-asymptotic form of the solution. Show that the solution $u(x,t)$ wiggles about $u_1(x,t)$. For this reason, a physicist still call $u_1(x,t)$ the equlibrium solution.

22. Show the total energy

$$E(t) = \frac{c^2}{2}\int_0^\ell (u_x)^2 \mathrm{d}x + \frac{1}{2}\int_0^\ell (u_t)^2 \mathrm{d}x,$$

where $u(x,t)$ is a solution of the IBVP (11.3.1), (11.3.3), (11.3.5), is independent of time.

23. If a piano string of length 0.5 m has the fundamental frequency 261.626 Hz (it corresponds to "middle C" or "Do" in the European terminology), what is the numerical value of constant c in the wave equation?
24. A wire of length 50 cm is stretched between two pins. The velocity constant c in the corresponding wave equation (11.3.1) is approximately 560 m/sec. The wire is plucked at a point 10 centimeters from the end by displacing that point 0.1 cm. This caused the initial wire profile to be the union of two straight lines connecting end points with the plucked point. Determine the displacement function of the wire, assuming zero initial velocity.
25. To approximate the effect of an initial moment impulse P applied at the midpoint $x = \ell/2$ of a simply supported beam, solve the beam equation (11.3.16) subject to the boundary conditions (11.3.17) and the initial conditions (11.3.18), where $f(x) \equiv 0$ and the initial velocity is

$$v(x) = \begin{cases} \frac{P}{2\rho\varepsilon}, & \text{if } \frac{\ell}{2} - \varepsilon < x < \frac{\ell}{2} + \varepsilon, \\ 0, & \text{otherwise.} \end{cases}$$

Then find the limit as $\varepsilon \to 0$.

26. Solve simply supported beam equations (11.3.16), (11.3.17), which is put into vibration by constant initial velocity, so $f(x) \equiv 0$ and $g(x) = v_0$ in Eq. (11.3.18).

Section 11.4 of Chapter 11 (Review)

1. Show that the real and imaginary parts of $(x + \mathbf{j}y)^n$ are harmonic functions on the plane $\mathbb{R}^2$, where n is a positive integer and $\mathbf{j}^2 = -1$.
2. Find all product solutions $u(x,y) = X(x)\,Y(y)$ of the **Helmholtz equation** $u_{xx} + u_{yy} + k\,u = 0$, where $k > 0$.
3. Apply separation of variables to obtain a formal series solution to the **Poisson equation** $\nabla^2 u = f(x,y)$ inside rectangle $0 < x < 2$, $0 < y < \pi$, subject to the homogeneous boundary condition $u = 0$ on all four sides of the rectangle. Consider $f(x,y) = 1 + xy$ inside the rectangle.

In Problems 4 through 9, solve Laplace's equation $\nabla^2 u = 0$ in the rectangular domain $(0 < x < a,\ 0 < y < b)$ subject to the given boundary conditions.

4. $(0 < x < 1,\ 0 < y < 2)$ $u_x(0,y) = u(1,y) = 0$, $u(x,0) = 0$, $u_y(x,2) = \cos 2\pi x$.
5. $(0 < x < 2,\ 0 < y < 3)$ $u(0,y) = y^2$, $u(2,y) = 0$, $u(x,0) = 0$, $u_y(x,3) = 0$.
6. $(0 < x < 1,\ 0 < y < 1)$ $u_x(0,y) = 0$, $u_x(1,y) = 3\cos 3\pi y$, $u_y(x,0) = u_y(x,1) = 0$.
7. $(0 < x < 2\pi,\ 0 < y < 1)$ $u(0,y) = u(2\pi,y) = 0$, $u_y(x,0) = \cos x$, $u(x,1) = 0$.
8. $(0 < x < \pi,\ 0 < y < 2\pi)$ $u(0,y) = 0$, $u_x(\pi,y) = 0$, $u(x,0) = 0$, $u_y(x,2\pi) = x$.

9. $(0 < x < 4,\ 0 < y < 2)$ $u(0, y) = 0$, $u_x(4, y) = \sin y$, $u_y(x, 0) = u_y(x, 2) = 0$.

10. Consider the general Dirichlet problem (11.4.1), (11.4.2) for the Laplace equation in a rectangle $0 \leqslant x \leqslant a$, $0 \leqslant y \leqslant b$. Its solution can be represented via the series

$$u(x,y) = \sum_{n\geqslant 1} A_n \sin\left(\frac{n\pi x}{a}\right) \sinh \frac{n\pi}{b}(b-y) + \sum_{n\geqslant 1} B_n \sin\left(\frac{n\pi x}{a}\right) \sinh\frac{n\pi}{b} y$$
$$+ \sum_{n\geqslant 1} C_n \sinh\frac{n\pi}{b}(a-x) \sin\left(\frac{n\pi x}{b}\,y\right) + \sum_{n\geqslant 1} D_n \sinh\left(\frac{n\pi x}{b}\right) \sin\left(\frac{n\pi x}{b}\,y\right).$$

Find the values of coefficients A_n, B_n, C_n, and D_n.

In Problems 11 through 16, a circular disk of radius a is given, as well as a function defined on the boundary $r = a$ of the disk. Using separation of variables, solve the corresponding outer or inner boundary value problem.

11. $r < 2$, $u(2, \vartheta) = 8 \sin 3\vartheta$.
12. $r > 3$, $u_r(3, \vartheta) = 3\vartheta^2 - 3\pi\vartheta - \pi^2$.
13. $r > 1$, $u_r(1, \vartheta) = 3 \cos 3\vartheta$.
14. $r < 4$, $u(4, \vartheta) = \vartheta(2\pi - \vartheta)$.
15. $r < 1$, $u_r(1, \vartheta) = 8 \sin 4\vartheta$.
16. $r > 2$, $u_r(2, \vartheta) = \vartheta - \pi$.

Consider Laplace's equation in the domain $a < r < b$ bounded by two circles $r = a$ and $r = b$ written in polar coordinates. Since the boundary of the annulus consists of two circles, the boundary conditions should be specified at each circle. In Problems 17 through 19, solve Laplace's equation (11.4.6) in the annulus having given inner radius a, given outer radius b, and subject to the given boundary conditions.

17. $(3 < r < 4)$ $u(3, \vartheta) = \begin{cases} 0, & \text{if } 0 < \vartheta < \pi, \\ \cos 3\vartheta, & \text{if } \pi < \vartheta < 2\pi; \end{cases}$ $\quad u(4, \vartheta) = \begin{cases} \sin 2\vartheta, & \text{if } 0 < \vartheta < \pi, \\ 0, & \text{if } \pi < \vartheta < 2\pi. \end{cases}$

18. $(1 < r < 3)$ $u(1, \vartheta) = \begin{cases} \pi - 4\vartheta, & \text{if } 0 < \vartheta < \pi/2, \\ 0, & \text{if } \pi/2 < \vartheta < 2\pi; \end{cases}$ $\quad u(3, \vartheta) = \begin{cases} 0, & \text{if } 0 < \vartheta < 3\pi/2, \\ 7\pi - 4\vartheta, & \text{if } 3\pi/2 < \vartheta < 2\pi. \end{cases}$

19. $(2 < r < 5)$ $u(2, \vartheta) = \sin 3\vartheta$, $\quad u(5, \vartheta) = \cos 6\vartheta$.

20. $(1 < r < 3)$ $u(1, \vartheta) = \sin 3\vartheta$, $\quad u_r(3, \vartheta) = \sin 5\vartheta$.

Chapter 12

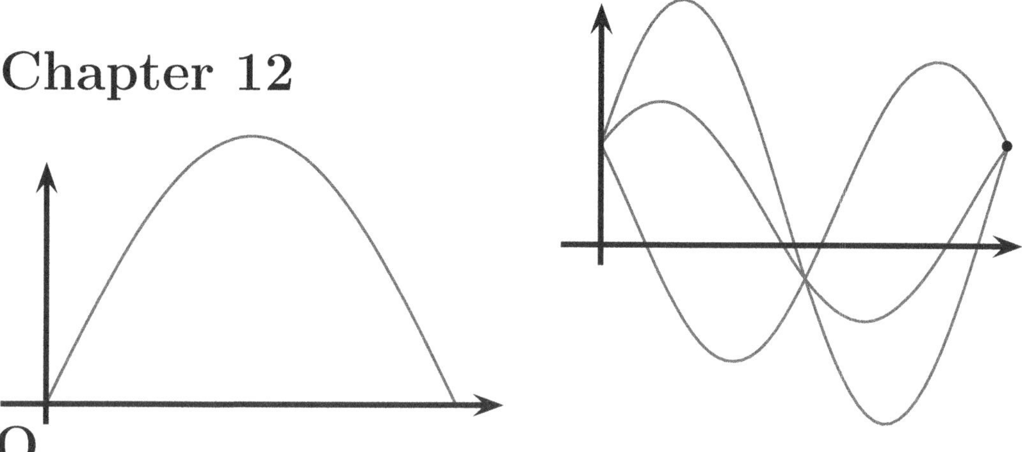

Left: unique solution. Right: multiple solutions

Boundary Value Problems

As a result of separating variables in a partial differential equation presented in §11.2, we may encounter the nonhomogeneous differential equation

$$-\frac{\mathrm{d}}{\mathrm{d}x}\left[p(x)\,\frac{\mathrm{d}u}{\mathrm{d}x}\right] + q(x)\,u = \lambda\rho(x)\,u + f(x), \tag{12.0.1}$$

subject to some boundary conditions. Unlike homogeneous equations, the boundary value problem for the driven differential equation (12.0.1) does not need to have any solution (even trivial) or may have infinitely many solutions (see figure on this page).

In this chapter, we will derive a procedure to find solutions of boundary value problems based on Green's functions and discuss some Sturm–Liouville boundary value problems that play an important role in physics applications. Note that we used Green's function previously for differential equations (§4.8) and initial value problems (§5.5).

12.1 Green's Functions

This section gives an introduction to Green's functions and their applications for self-adjoint differential operators of the second order. We start with a motivational example.

Example 12.1.1: Consider the cable or board of length ℓ shown in Figure 12.1 (page 630) that stays motionless along horizontal x-axis in its equilibrium position. It is pinned at $x = 0$ and $x = \ell$ and subject to a vertical load with a density of $w(x)$ units of force per unit length. The loading density includes the cable weight. Under the loading, the cable deforms at distance $y(x)$ from its equilibrium position.

To derive a corresponding mathematical problem, we cut off a differential cable segment and impose the conditions of static equilibrium because the cable is motionless. Newton's second law tells us that the sum of the forces acting upon the differential segment in both the horizontal and vertical directions is zero. Let $T(x)$ be the tension at the point x. This force is applied tangentially to the cable curve because it offers no resistance to bending. Then the state equilibrium constraints are

$$\begin{aligned} T(x+\mathrm{d}x)\,\cos\theta(x+\mathrm{d}x) - T(x)\,\cos\theta(x) &= 0 \quad \text{(projection on abscissa)},\\ T(x+\mathrm{d}x)\,\sin\theta(x+\mathrm{d}x) - T(x)\,\sin\theta(x) &= w(x)\,\mathrm{d}x \quad \text{(projection on ordinate)}. \end{aligned} \tag{12.1.1}$$

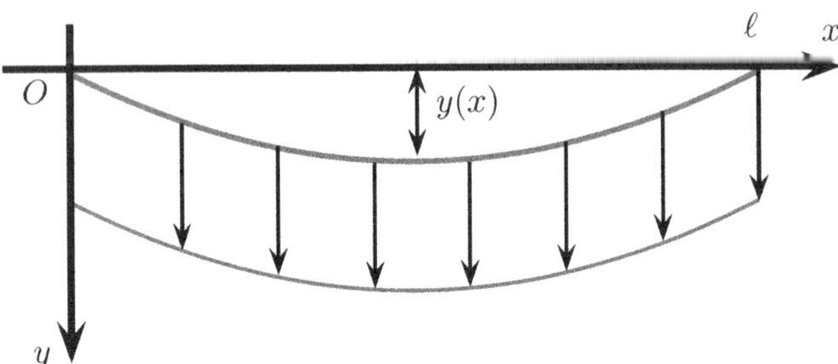

Figure 12.1: The cable is pinned at $x = 0$ and $x = \ell$.

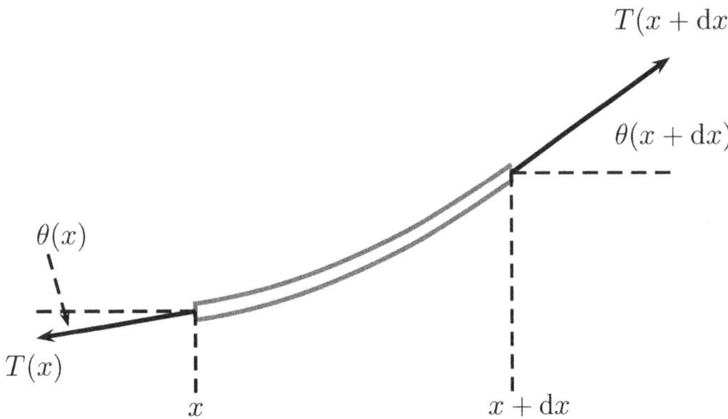

Figure 12.2: Forces acting on the differential cable element.

Dividing both equations in (12.1.1) by $\mathrm{d}x$ and letting $\mathrm{d}x \mapsto 0$, we obtain

$$\frac{\mathrm{d}}{\mathrm{d}x}\,T(x)\,\cos\theta(x) = 0, \qquad \frac{\mathrm{d}}{\mathrm{d}x}\,T(x)\,\sin\theta(x) = w(x).$$

Since there is no motion in the horizontal direction, the x-component $T(x)\,\cos\theta(x)$ of the tension must be a constant, which we denote by T_0. Then $T(x) = T_0/\cos\theta(x)$, and the latter equation leads to

$$\frac{\mathrm{d}}{\mathrm{d}x}\,T(x)\,\sin\theta(x) = \frac{\mathrm{d}}{\mathrm{d}x}\,\frac{T_0}{\cos\theta(x)}\,\sin\theta(x) = \frac{\mathrm{d}}{\mathrm{d}x}\,T_0\tan\theta(x) = w(x).$$

Using notation $\tan\theta(x) = -\mathrm{d}y/\mathrm{d}x$, where the minus sign arises from assumption that the downward position is positive, we get

$$\frac{\mathrm{d}}{\mathrm{d}x}\,\tan\theta(x) = -\frac{\mathrm{d}^2}{\mathrm{d}x^2}\,y(x) = \frac{w(x)}{T_0} \stackrel{\text{def}}{=} f(x).$$

Since the deformation is motionless at end points $x = 0$ and $x = \ell$, we obtain the following two-point boundary value problem for a positive differential operator $L[\mathtt{D}] = -\mathtt{D}^2 = -\mathrm{d}^2/\mathrm{d}x^2$:

$$-\frac{\mathrm{d}^2 y}{\mathrm{d}x^2} = f(x), \qquad y(0) = 0, \quad y(\ell) = 0. \tag{12.1.2}$$

By successive antidifferentiation of the equation, we get

$$y(x) = c_0 x + c_1\,(\ell - x) - \frac{1}{2}\int_0^x \mathrm{d}t \int_0^t \mathrm{d}s\, f(s) - \frac{1}{2}\int_x^\ell \mathrm{d}t \int_t^\ell \mathrm{d}s\, f(s),$$

where c_0 and c_1 are constants of integration. Upon changing the order of integration, we find

$$y(x) = c_0 x + c_1\,(\ell - x) - \frac{1}{2}\int_0^x (x - s)\,f(s)\,\mathrm{d}s - \frac{1}{2}\int_x^\ell (s - x)\,f(s)\,\mathrm{d}s.$$

Imposing the boundary conditions $y(0) = 0$ and $y(\ell) = 0$, we determine the values of constants c_0 and c_1:

$$c_1 \ell = \frac{1}{2} \int_0^\ell s\, f(s)\, \mathrm{d}s, \quad c_0 \ell = \frac{1}{2} \int_0^\ell (\ell - s)\, f(s)\, \mathrm{d}s.$$

Thus, the deformation function is given by

$$y(x) = \int_0^\ell G(x,t)\, f(t)\, \mathrm{d}t, \tag{12.1.3}$$

where

$$G(x,t) = \begin{cases} \dfrac{t(\ell - x)}{\ell}, & 0 \leqslant t \leqslant x, \\ \dfrac{x(\ell - t)}{\ell}, & x \leqslant t \leqslant \ell, \end{cases}$$

is called the **Green's function** for this boundary value problem (12.1.2). □

Now we turn our attention to the general case. Our starting point is the linear two-point boundary value problem for the self-adjoint (positive if $q(x) > 0$) differential operator $L[x, \mathtt{D}] = -\mathtt{D}p(x)\mathtt{D} + q(x)$, depending on the derivative operator $\mathtt{D} = \mathrm{d}/\mathrm{d}x$:

$$-\frac{\mathrm{d}}{\mathrm{d}x}\left[p(x)\,\frac{\mathrm{d}y}{\mathrm{d}x}\right] + q(x)\, y = f(x), \qquad \alpha_0 y(0) - \alpha_1 y'(0) = 0, \quad \beta_0 y(\ell) + \beta_1 y'(\ell) = 0, \tag{12.1.4}$$

on the interval $0 < x < \ell$. The functions $p(x)$, $p'(x)$, $q(x)$, and $f(x)$ are assumed to be continuous on the closed interval $[0, \ell]$, and

$$|\alpha_0| + |\alpha_1| > 0, \qquad |\beta_0| + |\beta_1| > 0.$$

There is nothing special for choosing the left end point of the interval $(0, \ell)$ to be zero: any interval can be scaled into this particular case. It is convenient to introduce two boundary operators

$$B_0[y] = \alpha_0 y(0) - \alpha_1 y'(0) \qquad \text{and} \qquad B_\ell[y] = \beta_0 y(\ell) + \beta_1 y'(\ell).$$

Then the boundary value problem can be written in compact form:

$$L[x, \mathtt{D}]y = f, \qquad B_0[y] = 0, \quad B_\ell[y] = 0, \tag{12.1.5}$$

where $L[x, \mathtt{D}] = -\mathtt{D}p(x)\mathtt{D} + q(x)$ with $\mathtt{D} = \mathrm{d}/\mathrm{d}x$. We consider self-adjoint differential operators of the second order for two reasons. First, all eigenvalues of a self-adjoint operator are real numbers and eigenfunctions corresponding to distinct eigenvalues are orthogonal. Second, a nonself-adjoint differential equation can be reduced (see §4.1.3) to a self-adjoint counterpart. Moreover, if the boundary value problem generated by the differential operator $L[x, \mathtt{D}]$ and the boundary conditions B_0, B_ℓ is nonself-adjoint, then the corresponding Green's function is not symmetric and its eigenvectors are not orthogonal.

Throughout this section we assume that the corresponding homogeneous problem ($L[x, \mathtt{D}]y = 0$, $B_0[y] = 0$, $B_\ell[y] = 0$) has only the trivial (identically zero) solution; this means that $\lambda = 0$ is not an eigenvalue of the boundary value problem. It then follows from the Fredholm alternative (see [40]) that the inhomogeneous problem (12.1.4) has a unique solution.

Our next task is to construct the Green's function explicitly provided that the fundamental set of solutions $\{\phi(x), \psi(x)\}$ to the equation $L[x, \mathtt{D}]y = 0$ is known. That is, we assume that $L[x, \mathtt{D}]\phi = 0$ and $L[x, \mathtt{D}]\psi = 0$ and the functions $\phi(x)$ and $\psi(x)$ are linearly independent; namely, the Wronskian $W[\phi, \psi](x) = \phi\,\psi' - \phi'\,\psi$ of these functions is not zero on the interval $[0, \ell]$. Moreover, $p(x)\,W[\phi, \psi](x)$ is a constant for a self-adjoint differential operator $L[x, \mathtt{D}] = -\mathtt{D}p(x)\mathtt{D} + q(x)$ because $\mathtt{D}\,p(x)\,\mathtt{D}\,W(x) = 0$ for $\mathtt{D} = \mathrm{d}/\mathrm{d}x$, which can be verified by substitution (see exercise 4 on page 634).

To apply the formulas from §4.8 (the variation of parameters method), we seek a particular solution of $L[x, \mathtt{D}]y = f$ in the form

$$y(x) = c_1(x)\,\phi(x) + c_2(x)\,\psi(x), \tag{12.1.6}$$

where

$$c_1'(x) = \frac{f(x)\,\psi(x)}{p(x)\,W[\phi, \psi]}, \qquad c_2'(x) = -\frac{f(x)\,\phi(x)}{p(x)\,W[\phi, \psi]}.$$

Integrating the above equations, we obtain explicit representations of $c_1(x)$ and $c_2(x)$ in the quadrature form:

$$c_1(x) = \int_x^{\ell} \frac{-f(t)\,\psi(t)}{p(t)\,W[\phi,\psi]}\,\mathrm{d}t + c_1, \qquad c_2(x) = \int_0^{x} \frac{-f(t)\,\phi(t)}{p(t)\,W[\phi,\psi]}\,\mathrm{d}t + c_2, \tag{12.1.7}$$

where c_1 and c_2 are constants of integration. Substituting these expressions into (12.1.6), we obtain the solution to the boundary value problem (12.1.5). This solution is the sum of a particular solution $y_p(x)$ of the nonhomogeneous equation $L[x,\mathtt{D}]y = f$ and the solution $y_h(x) = c_1\phi(x) + c_2\psi(x)$ of the homogeneous equation $L[x,\mathtt{D}]y = 0$. Using formulas (12.1.7), we can represent $y_p(x)$ in integral form (12.1.3), where the kernel

$$G(x,t) = \begin{cases} \dfrac{-\phi(t)\psi(x)}{p(t)\,W[\phi,\psi](t)}, & 0 \leqslant t \leqslant x, \\ \dfrac{-\phi(x)\psi(t)}{p(t)\,W[\phi,\psi](t)}, & x \leqslant t \leqslant \ell, \end{cases} \tag{12.1.8}$$

is the Green's function for the nonhomogeneous equation $L[x,\mathtt{D}]y = f$. Since $\phi(x)$ and $\psi(x)$ are arbitrary linearly independent solutions of the homogeneous equation $L[x,\mathtt{D}]y = 0$, the function $y_p(x)$ does not need to satisfy the boundary conditions. It is assumed that the coefficients c_1 and c_2 can be chosen in such a way that the sum $y = y_p + y_h$ will satisfy the homogeneous boundary conditions $B_0[y] = 0$, $B_\ell[y] = 0$. However, if we set $\phi(x) = y_1(x)$ to be a solution of $L[y] = 0$ satisfying the boundary condition $B_0[y] = 0$ at $x = 0$, and $\psi(x) = y_2(x)$ to be a solution of $L[y] = 0$ satisfying the boundary condition $B_\ell[y] = 0$ at $x = \ell$, then no determination of arbitrary constants is required because y_p, as given by integral formula (12.1.3) for the Green's function, automatically satisfies the boundary conditions since the Green's function (12.1.8) does.

Note that if we set $\mathbf{A} = \mathtt{D}p(x)\mathtt{D}$ and consider the boundary value problem (12.1.5) in the space of functions that satisfy the homogeneous boundary conditions $B_0[y] = 0$, $B_\ell[y] = 0$, then it will resemble the problem $(q\mathbf{I} - \mathbf{A})\,\mathbf{x} = \mathbf{b}$ that we discussed in §7.2.1 for matrices, and the Green's function becomes an analog of the resolvent. The next theorem summarizes some basic properties of the Green's functions.

Theorem 12.1: Let $G(x,t)$ be the Green's function for the boundary value problem (12.1.4) that is assumed to have a unique solution. Then

(a) $G(x,t)$ is continuous on closed square $[0,\ell] \times [0,\ell]$;

(b) $G(x,t)$ is a symmetric function, $G(x,t) = G(t,x)$;

(c) for each fixed t, the partial derivatives $\partial G/\partial x$ and $\partial^2 G/\partial x^2$ are continuous functions of x for $x \ne t$;

(d) at $x = t$, the first partial derivative $\partial G/\partial x$ has a jump discontinuity:

$$\lim_{x\mapsto t+0} \frac{\partial G}{\partial x}(x,t) - \lim_{x\mapsto t-0} \frac{\partial G}{\partial x}(x,t) = -\frac{1}{p(t)};$$

(e) for each fixed t, the function $G(x,t)$ is a solution of the corresponding homogeneous boundary value problem

$$L[x,\mathtt{D}]G(\cdot,t) = 0, \qquad B_0[G] = 0, \quad B_\ell[G] = 0;$$

(f) there is only one function satisfying the above properties.

Example 12.1.2: Consider the nonself-adjoint boundary value problem

$$y'' + y' - 2y = -f(t), \qquad y(0) = 0, \quad y'(1) = 0.$$

Since the characteristic equation $\lambda^2 + \lambda - 2 = 0$ has two real roots $\lambda = 1$ and $\lambda = -2$, the homogeneous differential equation $y'' + y' - 2y = 0$ has two linearly independent solutions $\phi(x) = e^x$ and $\psi(x) = e^{-2x}$. According to Eq. (12.1.3), a particular solution of the inhomogeneous equation can be represented in the quadrature form: $y_p(x) = \int_0^1 G_0(x,t)\,f(t)\,\mathrm{d}t$, where the kernel is

$$G_0(x,t) = \begin{cases} \frac{1}{3}\,e^{2(t-x)}, & 0 \leqslant t \leqslant x, \\ \frac{1}{3}\,e^{x-t}, & x \leqslant t \leqslant 1. \end{cases}$$

Then the solution of the given boundary value problem becomes

$$y(t) = \int_0^1 G_0(x,t)\, f(t)\, \mathrm{d}t + c_1\phi(x) + c_2\psi(x).$$

We choose constants c_1 and c_2 in such a way that the above function satisfies the given boundary conditions $y(0) = 0$ and $y'(1) = 0$. This yields

$$\begin{aligned} y(x) &= \frac{1}{3}\int_0^x f(t)\, e^{2(t-x)}\, \mathrm{d}t + \frac{1}{3}\int_x^1 f(t)\, e^{x-t}\, \mathrm{d}t - \frac{1}{3}e^{-2x}\int_0^1 f(t)\, e^{-t}\, \mathrm{d}t \\ &+ \frac{2}{3\,(2+e^3)}\left(e^{-2x} - e^x\right)\int_x^1 f(t)\left(e^{-t} - e^{2t}\right)\mathrm{d}t = \int_0^1 G(x,t)\, f(t)\, \mathrm{d}t, \end{aligned}$$

where $G(x,t)$ is the kernel of this integral representation. As we see from the above formula, the Green's function $G(x,t)$ is not symmetric with respect to x and t, but it satisfies the boundary conditions: $G(0,t) = 0$ and $G_x(1,t) = 0$. Its first derivative with respect to x has a jump of discontinuity at $x = t$:

$$\lim_{x\mapsto t+0}\frac{\partial G}{\partial x}(x,t) - \lim_{x\mapsto t-0}\frac{\partial G}{\partial x}(x,t) = -\frac{2}{3} - \frac{1}{3} = -1.$$

Example 12.1.3: Consider the self-adjoint boundary value problem for the operator $L[x,\mathtt{D}] = -\mathtt{D}x^2\mathtt{D} + 6$:

$$-\frac{\mathrm{d}}{\mathrm{d}x}\left(x^2\frac{\mathrm{d}}{\mathrm{d}x}y\right) + 6y(x) = f(x) \qquad (1 < x < 5), \qquad y'(1) = 0, \quad y(5) = 0.$$

The corresponding homogeneous equation $x^2y'' + 2xy' - 6y = 0$ is an example of the Euler equation (see §4.6.2), which has two linearly independent solutions $\phi = x^2$ and $\psi = x^{-3}$. The Wronskian determinant of ϕ and ψ is

$$W[\phi,\psi](x) = \det\begin{bmatrix} x^2 & x^{-3} \\ 2x & -3x^{-4}\end{bmatrix} = -5x^{-2}.$$

While we can use Eq. (4.8.8), page 240, to obtain the explicit expression (12.1.8), we proceed directly to show every step. Using variation of parameters (see §4.8), we represent the general solution of the nonhomogeneous equation as

$$y(x) = \frac{1}{5}x^2\int_x^5 t^{-3}f(t)\,\mathrm{d}t + \frac{1}{5}x^{-3}\int_1^x t^2 f(t)\,\mathrm{d}t + c_1x^2 + c_2x^{-3}.$$

Since its derivative is

$$y'(x) = \frac{2}{5}x\int_x^5 t^{-3}f(t)\,\mathrm{d}t - \frac{3}{5}x^{-4}\int_1^x t^2 f(t)\,\mathrm{d}t + 2c_1x - 3c_2x^{-4},$$

we get the conditions on constants c_1 and c_2 to satisfy the given boundary conditions:

$$y'(1) = \frac{2}{5}\int_1^5 t^{-3}f(t)\,\mathrm{d}t + 2c_1 - 3c_2 = 0, \quad y(5) = 5^{-4}\int_1^5 t^2 f(t)\,\mathrm{d}t + c_1 5^2 + c_2 5^{-3} = 0.$$

Solving this system of algebraic equations, we obtain

$$c_1 = \frac{-1}{46885}\int_1^5\left(3t^2 + 2t^{-3}\right) f(t)\,\mathrm{d}t, \quad c_2 = \frac{2}{9377}\int_1^5\left(5^4t^{-3} - 5^{-1}t^2\right) f(t)\,\mathrm{d}t.$$

This allows us to represent the solution in integral form:

$$\begin{aligned} y(x) &= \int_0^5 G(x,t)\, f(t)\,\mathrm{d}t = \frac{1}{5}x^2\int_x^5 t^{-3}f(t)\,\mathrm{d}t + \frac{1}{5}x^{-3}\int_1^x t^2 f(t)\,\mathrm{d}t \\ &+ \frac{1}{46885}\int_1^5\left[-x^2\left(3t^2 + 2t^{-3}\right) + 10x^{-3}\left(5^4t^{-3} - 5^{-1}t^2\right)\right] f(t)\,\mathrm{d}t, \end{aligned}$$

where the Green's function is

$$G(x,t) = \begin{cases} \frac{3\cdot 5^4}{9377}\,x^{-3}t^2 - \frac{x^2}{46885}\left(3t^2+2t^{-3}\right) + \frac{2\cdot 5^4}{9377}\,x^{-3}t^{-3}, & \text{for } 1 \leqslant t \leqslant x, \\ \frac{3\cdot 5^4}{9377}\,x^2t^{-3} - \frac{3}{46885}\,x^2t^2 + \frac{2\,x^{-3}}{9377}\left(5^4t^{-3} - 5^{-1}t^{-3}\right), & \text{for } x \leqslant t \leqslant 5. \end{cases}$$

Calculations show that the Green's function satisfies the boundary conditions, $G_x(1,t) = 0$ and $G(5,t) = 0$. Also

$$\lim_{x\mapsto t+0}\frac{\partial G}{\partial x}(x,t) - \lim_{x\mapsto t-0}\frac{\partial G}{\partial x}(x,t) = -\frac{1}{t^2}.$$

To see that the Green's function is symmetric, we choose $y_1(x) = 3x^2 + 2x^{-3}$ and $y_2(x) = x^2 - 5^5x^{-3}$ as two linearly independent solutions of the homogeneous differential equation $x^2y'' + 2xy' - 6y = 0$. Their Wronskian is $W[y_1,y_2](x) = 46885\,x^{-2} \neq 0$ on the interval $[1,5]$. Since $y_1'(1) = 0$ and $y_2(5) = 0$, we can use Eq. (12.1.8) to construct the Green's function:

$$G(x,t) = \frac{1}{46885} \times \begin{cases} \left(5^5x^{-3} - x^2\right)\left(3t^2 + 2t^{-3}\right), & 1 \leqslant t \leqslant x, \\ \left(5^5t^{-3} - t^2\right)\left(3x^2 + 2x^{-3}\right), & x \leqslant t \leqslant 5. \end{cases}$$

Problems

1. For each of the boundary value problems, determine whether there are infinitely many solutions, a unique solution, or there is no solution.

   (a) $y'' + y = 0, \quad y(0) = 0, \; y'(2\pi) = 1.$

   (b) $y'' + 4y = 0, \quad y'(0) = 2, \; y(\pi) = 0.$

   (c) $y'' + 9y = 0, \quad y'(0) = 0, \; y'(2\pi) = 0.$

   (d) $y'' + 9y = 0, \quad y'(0) = 0, \; y'(2\pi) = 1.$

2. Show that the homogeneous second order ODE $a_2(x)\,y''(x) + a_1(x)\,y'(x) + a_0(x)\,y(x) = 0$ can be put into self-adjoint form by multiplying by the integrating factor

$$\mu(x) = \frac{1}{a_2(x)}\,\exp\left\{\int \frac{a_1(x)}{a_2(x)}\,\mathrm{d}x\right\},$$

   and find the form of $p(x)$ and $q(x)$ in Eq. (12.1.4) in terms of the coefficients $a_2(x)$, $a_1(x)$, and $a_0(x)$.

3. In each exercise, find an integrating factor $\mu(t)$ needed to convert the given differential equation into self-adjoint form $\left(\mu(t)\,y'(t)\right)' + \mu(t)\,q(t)\,y = \mu(t)\,g(t)$.

   (a) $y'' + 4y' + 4y = \cos 2t;$

   (b) $e^t y'' - y' = e^{2t};$

   (c) $t^{1/2}y'' + t^{-1/2}y' + e^{2t}\,y = t;$

   (d) $(\sin t)y'' - (\cos t)\,y' + y = t^2.$

4. Show that the product of $p(x)$ and the Wronskian $W[\phi,\psi](x)$ of two linearly independent solutions ϕ, ψ to the second order differential equation $L[x,\mathtt{D}]\,y = 0$ generated by the self-adjoint differential operator $L[x,\mathtt{D}] = -\mathtt{D}p(x)\mathtt{D} + q(x)$ is a constant.

5. Derive the Green's function for the given two-point boundary value problems.

   (a) $-y'' = f(x), \quad y(0) - y'(0) = 0, \; y(1) = 0.$

   (b) $-y'' - y = f(x), \quad y(0) = 0, \; y'(\pi) = 0.$

   (c) $-y'' + y = f(x), \quad y'(0) = 0, \; y(1) = 0.$

   (d) $-y'' + y = f(x), \quad y(0) - y'(0) = 0, \; y(1) + y'(1) = 0.$

   (e) $x\,y'' + y' = f(x), \quad y(0) < \infty, \; y(1) = 0.$

   (f) $(1-x^2)\,y'' - 2x\,y' = f(x), \quad y'(0) = 0, \; y(b) = 0, \; b < 1.$

6. Find the Green's function for the given differential operator $L[x,\mathtt{D}]$, where $\mathtt{D} = \mathrm{d}/\mathrm{d}x$.

   (a) $\mathtt{D}\,e^{2x}\mathtt{D} + e^{2x};$

   (b) $\mathtt{D}\,x^3\mathtt{D} + 1/4;$

   (c) $\mathtt{D}\,x\,\mathtt{D} - 9/x;$

   (d) $\mathtt{D}\,x^2\mathtt{D}.$

12.2 Green's Functions for Linear Systems

In this section, we consider boundary value problems for scalar linear differential equations of the n-th order and their reformulations as first order systems. We start with the existence and uniqueness theorem of a two point boundary value problem for second order scalar differential equations [20].

Theorem 12.2: Let $p(x)$, $q(x)$, and $f(x)$ be functions continuous on $0 \leqslant x \leqslant \ell$, where $q(x) > 0$. Let α_0, α_1, β_0, β_1 be constants, where $|\alpha_0| + |\alpha_1| > 0$ and $|\beta_0| + |\beta_1| > 0$. In addition, suppose that

$$0 \leqslant \alpha_0\alpha_1, \qquad 0 \leqslant \beta_0\beta_1, \quad \text{and} \quad |\alpha_0| + |\beta_0| > 0.$$

Then for any values α and β, the boundary value problem

$$-y'' - p(x)\, y' + q(x)\, y = f(x), \qquad 0 < x < \ell,$$

$$\alpha_0 y(0) - \alpha_1 y'(0) = \alpha, \qquad \beta_0 y(\ell) + \beta_1 y(\ell) = \beta,$$

has a unique solution.

In Chapter 6, we saw that systems of first order differential equations form a conceptual framework that includes the theory of n-th order scalar differential equations. Now we show that two-point boundary value problems for scalar equations can be recast as problems for first order systems.

Let $\mathbf{P}(x)$ be an $(n \times n)$ matrix with continuous entries on the interval $[0, \ell]$. Let $\mathbf{f}(x) = \langle f_1(x), \ldots, f_n(x)\rangle^T$ be an $(n \times 1)$ column vector whose component functions $f_i(x)$, $i = 1, 2, \ldots, n$, are also continuous on $[0, \ell]$. In the language of Chapter 6, $\mathbf{P}(x)$ and $\mathbf{f}(x)$ are continuous matrix-valued functions defined on $[0, \ell]$. Let $\mathbf{B}^{[0]}$ and $\mathbf{B}^{[\ell]}$ be given constant $(n \times n)$ matrices and $\boldsymbol{\alpha}$ be a known constant n-column vector.

Consider the linear inhomogeneous first order system subject to the given boundary conditions

$$\mathbf{y}'(x) = \mathbf{P}(x)\mathbf{y}(x) + \mathbf{f}(x), \qquad \mathbf{B}^{[0]}\mathbf{y}(0) + \mathbf{B}^{[\ell]}\mathbf{y}(\ell) = \boldsymbol{\alpha}. \tag{12.2.1}$$

Without any loss of generality, the vector $\boldsymbol{\alpha}$ can be chosen as zero. Indeed, let $\mathbf{z}(x)$ be any column function that satisfies the given boundary conditions

$$\mathbf{B}^{[0]}\mathbf{z}(0) + \mathbf{B}^{[\ell]}\mathbf{z}(\ell) = \boldsymbol{\alpha}.$$

Then the n-column vector $\mathbf{u}(x) = \mathbf{y}(x) - \mathbf{z}(x)$ is a solution of a similar two-point boundary value problem with homogeneous boundary conditions:

$$\mathbf{u}'(x) = \mathbf{P}(x)\mathbf{u}(x) + \mathbf{P}(x)\mathbf{z}(x) - \mathbf{z}'(x) + \mathbf{f}(x), \qquad \mathbf{B}^{[0]}\mathbf{u}(0) + \mathbf{B}^{[\ell]}\mathbf{u}(\ell) = \mathbf{0}.$$

In the vector equation, $\mathbf{P}(x)\mathbf{z}(x) - \mathbf{z}'(x) + \mathbf{f}(x)$ can be considered as a known column vector $\mathbf{g}(x)$ and the above problem can be written in the vector form with homogeneous boundary conditions: $\mathbf{u}'(x) = \mathbf{P}(x)\mathbf{u}(x) + \mathbf{g}(x)$, $\mathbf{B}^{[0]}\mathbf{u}(0) + \mathbf{B}^{[\ell]}\mathbf{u}(\ell) = \mathbf{0}$. Now suppose that a fundamental matrix $\boldsymbol{\Phi}(x)$ for the homogeneous equation is known, so

$$\mathrm{d}\boldsymbol{\Phi}/\mathrm{d}x = \mathbf{P}(x)\boldsymbol{\Phi} \qquad \text{and} \qquad \det \boldsymbol{\Phi}(x) \neq 0 \quad \text{on } [0, \ell].$$

Using the variation of parameters formula (8.3.3), page 455, we find the general solution of the nonhomogeneous vector equation $\mathbf{y}'(x) = \mathbf{P}(x)\mathbf{y}(x) + \mathbf{f}(x)$ to be

$$\mathbf{y}(x) = \boldsymbol{\Phi}(x)\,\mathbf{c} + \frac{1}{2}\,\boldsymbol{\Phi}(x)\left[\int_0^x \boldsymbol{\Phi}^{-1}(t)\,\mathbf{f}(t)\,\mathrm{d}t - \int_x^\ell \boldsymbol{\Phi}^{-1}(t)\,\mathbf{f}(t)\,\mathrm{d}t\right],$$

where $\mathbf{c}$ is an $(n \times 1)$ vector of arbitrary constants. Now we choose this column vector $\mathbf{c}$ so that the boundary conditions in Eq. (12.2.1) are satisfied:

$$\mathbf{B}^{[0]}\boldsymbol{\Phi}(0)\mathbf{c} - \frac{1}{2}\,\mathbf{B}^{[0]}\boldsymbol{\Phi}(0)\int_0^\ell \boldsymbol{\Phi}^{-1}(t)\,\mathbf{f}(t)\,\mathrm{d}t + \mathbf{B}^{[\ell]}\boldsymbol{\Phi}(\ell)\mathbf{c} + \frac{1}{2}\,\mathbf{B}^{[\ell]}\boldsymbol{\Phi}(\ell)\int_0^\ell \boldsymbol{\Phi}^{-1}(t)\,\mathbf{f}(t)\,\mathrm{d}t = \boldsymbol{\alpha}.$$

This leads to the vector equation with respect to $\mathbf{c}$:

$$\left[\mathbf{B}^{[0]}\boldsymbol{\Phi}(0) + \mathbf{B}^{[\ell]}\boldsymbol{\Phi}(\ell)\right]\mathbf{c} = \frac{1}{2}\left[\mathbf{B}^{[0]}\boldsymbol{\Phi}(0) - \mathbf{B}^{[\ell]}\boldsymbol{\Phi}(\ell)\right]\int_0^\ell \boldsymbol{\Phi}^{-1}(t)\,\mathbf{f}(t)\,\mathrm{d}t + \boldsymbol{\alpha}.$$

We rewrite this system of equations in vector form:

$$\mathbf{B}\,\mathbf{c} = \mathbf{b}, \tag{12.2.2}$$

where

$$\begin{aligned} \mathbf{B} &= \mathbf{B}^{[0]}\boldsymbol{\Phi}(0) + \mathbf{B}^{[\ell]}\boldsymbol{\Phi}(\ell) && \text{is } n\times n \text{ matrix,} \\ \mathbf{b} &= \frac{1}{2}\left[\mathbf{B}^{[0]}\boldsymbol{\Phi}(0) - \mathbf{B}^{[\ell]}\boldsymbol{\Phi}(\ell)\right]\int_0^\ell \boldsymbol{\Phi}^{-1}(t)\,\mathbf{f}(t)\,\mathrm{d}t + \boldsymbol{\alpha} && \text{is } n\text{-colomn vector.} \end{aligned} \tag{12.2.3}$$

If the matrix $\mathbf{B}$ is invertible ($\det \mathbf{B} \ne 0$), then equation (12.2.2) has a unique solution $\mathbf{c} = \mathbf{B}^{-1}\mathbf{b}$ and the boundary value problem (12.2.1) has the unique solution

$$\begin{aligned} \mathbf{y}(x) &= \boldsymbol{\Phi}(x)\,\mathbf{B}^{-1}\mathbf{b} + \frac{1}{2}\,\boldsymbol{\Phi}(x)\left[\int_0^x \boldsymbol{\Phi}^{-1}(t)\,\mathbf{f}(t)\,\mathrm{d}t - \int_x^\ell \boldsymbol{\Phi}^{-1}(t)\,\mathbf{f}(t)\,\mathrm{d}t\right] \\ &= \int_0^\ell \mathbf{G}(x,t)\,\mathbf{f}(t)\,\mathrm{d}t, \end{aligned} \tag{12.2.4}$$

where the kernel $\mathbf{G}(x,t)$ is called the Green's matrix-function. If the matrix $\mathbf{B}$ is singular, then the vector equation (12.2.2) has either no solution or infinitely many solutions. The latter happens when vector $\mathbf{b}$ is orthogonal to any solution $\mathbf{z}$ of the adjoint homogeneous equation $\mathbf{B}^*\mathbf{z} = \mathbf{0}$ (see §7.2.1). For a singular matrix $\mathbf{B}$, the homogeneous boundary value problem $\mathbf{y}'(x) = \mathbf{P}(x)\mathbf{y}(x)$, $\mathbf{B}^{[0]}\mathbf{y}(0) + \mathbf{B}^{[\ell]}\mathbf{y}(\ell) = \mathbf{0}$ has nontrivial solutions.

If the matrix $\mathbf{B}$ is not singular ($\det \mathbf{B} \ne 0$), then the unique solution (12.2.4) does not depend upon any particular fundamental matrix. Indeed, according to Corollary 8.6, page 436, any two fundamental matrices $\boldsymbol{\Phi}(x)$ and $\boldsymbol{\Psi}(x)$ to the homogeneous vector equation $\mathbf{y}'(x) = \mathbf{P}(x)\mathbf{y}(x)$ differ by a constant multiple. Therefore, there exists a constant $n\times n$ matrix $\mathbf{C}$ such that $\boldsymbol{\Phi}(x) = \boldsymbol{\Psi}(x)\,\mathbf{C}$ and $\det \mathbf{C} \ne 0$. Then expressions

$$\boldsymbol{\Phi}(x)\mathbf{B}^{-1} \qquad \text{and} \qquad \boldsymbol{\Phi}(x)\boldsymbol{\Phi}^{-1}(t)$$

do not depend on particular choice of the fundamental matrix, and $\boldsymbol{\Phi}$ can be replaced by $\boldsymbol{\Phi}(x) = \boldsymbol{\Psi}(x)\,\mathbf{C}$. Indeed,

$$\mathbf{B} = \mathbf{B}^{[0]}\boldsymbol{\Phi}(0) + \mathbf{B}^{[\ell]}\boldsymbol{\Phi}(\ell) = \left[\mathbf{B}^{[0]}\boldsymbol{\Psi}(0) + \mathbf{B}^{[\ell]}\boldsymbol{\Psi}(\ell)\right]\mathbf{C}.$$

Then its inverse is

$$\mathbf{B}^{-1} = \mathbf{C}^{-1}\left[\mathbf{B}^{[0]}\boldsymbol{\Psi}(0) + \mathbf{B}^{[\ell]}\boldsymbol{\Psi}(\ell)\right]^{-1}.$$

Hence

$$\boldsymbol{\Phi}(x)\mathbf{B}^{-1} = \boldsymbol{\Phi}(x)\,\mathbf{C}^{-1}\left[\mathbf{B}^{[0]}\boldsymbol{\Psi}(0) + \mathbf{B}^{[\ell]}\boldsymbol{\Psi}(\ell)\right]^{-1} = \boldsymbol{\Psi}(x)\left[\mathbf{B}^{[0]}\boldsymbol{\Psi}(0) + \mathbf{B}^{[\ell]}\boldsymbol{\Psi}(\ell)\right]^{-1}.$$

Similarly, it can be shown that $\boldsymbol{\Phi}(x)\boldsymbol{\Phi}^{-1}(t)$ and $\mathbf{b}$ do not depend on what fundamental matrix has been chosen.

Example 12.2.1: Consider the boundary value problem (12.1.5), page 631, for the self-adjoint differential operator $L[x,\mathtt{D}] = -\mathtt{D}p(x)\mathtt{D} + q(x)$, where $\mathtt{D} = \mathrm{d}/\mathrm{d}x$. We reformulate this scalar problem in vector form by introducing a 2-column vector

$$\mathbf{y}(x) = \langle y_1(x), y_2(x)\rangle^T = \langle y(x), p(x)\,y'(x)\rangle^T.$$

Then the scalar equation $L[x,\mathtt{D}]y = f$ will be equivalent to the vector equation

$$\frac{\mathrm{d}}{\mathrm{d}x}\begin{bmatrix} y_1(x) \\ y_2(x) \end{bmatrix} = \begin{bmatrix} 0 & 1/p(x) \\ q(x) & 0 \end{bmatrix}\begin{bmatrix} y_1(x) \\ y_2(x) \end{bmatrix} - \begin{bmatrix} 0 \\ f(x) \end{bmatrix}.$$

The homogeneous boundary conditions $\alpha_0 y(0) - \alpha_1 y'(0) = 0$ and $\beta_0 y(\ell) + \beta_1 y'(\ell) = 0$ are incorporated as follows:

$$\begin{bmatrix} \alpha_0 & -\alpha_1/p(0) \\ 0 & 0 \end{bmatrix}\mathbf{y}(0) + \begin{bmatrix} 0 & 0 \\ \beta_0 & \beta_1/p(\ell) \end{bmatrix}\mathbf{y}(\ell) = \begin{bmatrix} 0 \\ 0 \end{bmatrix}.$$

Therefore, the corresponding matrices become

$$\mathbf{P}(x) = \begin{bmatrix} 0 & 1/p(x) \\ q(x) & 0 \end{bmatrix}, \qquad \mathbf{B}^{[0]} = \begin{bmatrix} \alpha_0 & -\alpha_1/p(0) \\ 0 & 0 \end{bmatrix}, \qquad \mathbf{B}^{[\ell]} = \begin{bmatrix} 0 & 0 \\ \beta_0 & \beta_1/p(\ell) \end{bmatrix}.$$

Example 12.2.2: (Example 12.1.2 revisited) Upon introducing the 2-column vector $\mathbf{y} = \langle y_1(x), y_2(x)\rangle^T = \langle y(x), y'(x)\rangle^T$, we rewrite the two-point boundary value problem in the vector form:

$$\frac{\mathrm{d}}{\mathrm{d}x}\begin{bmatrix} y_1(x) \\ y_2(x) \end{bmatrix} = \begin{bmatrix} 0 & 1 \\ 2 & -1 \end{bmatrix}\begin{bmatrix} y_1(x) \\ y_2(x) \end{bmatrix} + \begin{bmatrix} 0 \\ -f(x) \end{bmatrix}, \qquad 0 < x < 1;$$

$$\begin{bmatrix} 1 & 0 \\ 0 & 0 \end{bmatrix}\mathbf{y}(0) + \begin{bmatrix} 0 & 0 \\ 0 & 1 \end{bmatrix}\mathbf{y}(1) = \begin{bmatrix} 0 \\ 0 \end{bmatrix}.$$

The fundamental matrix for the homogeneous system of first order equations is

$$\mathbf{\Phi}(x) = e^{\mathbf{A}x} = \frac{1}{3}\,e^{-2x}\begin{bmatrix} 1 & -1 \\ -2 & 2 \end{bmatrix} + \frac{1}{3}\,e^{x}\begin{bmatrix} 2 & 1 \\ 2 & 1 \end{bmatrix}, \qquad \text{where } \mathbf{A} = \begin{bmatrix} 0 & 1 \\ 2 & -1 \end{bmatrix}.$$

Since $\mathbf{\Phi}(0) = \mathbf{I}$, we have from Eq. (12.2.3) that

$$\mathbf{B} = \mathbf{B}^{[0]} + \mathbf{B}^{[1]}\mathbf{\Phi}(1) = \begin{bmatrix} 1 & 0 \\ 0 & 0 \end{bmatrix} + \begin{bmatrix} 0 & 0 \\ 0 & 1 \end{bmatrix} e^{\mathbf{A}} = \frac{1}{3}\begin{bmatrix} 3 & 0 \\ 2\left(e - e^{-2}\right) & e + 2\,e^{-2} \end{bmatrix},$$

and

$$\mathbf{b} = \frac{1}{6}\begin{bmatrix} 3 & 0 \\ 2\left(e^{-2} - e\right) & -e - 2\,e^{-2} \end{bmatrix}\int_0^1 e^{-\mathbf{A}t}\mathbf{f}(t)\,\mathrm{d}t$$

Example 12.2.3: (Beam equation) The beam deflection equation, considered in §11.3.1, is

$$u_{tt} + c^2 u_{xxxx} = q(x) \qquad (0 < x < \ell),$$

where c is assumed to be a positive constant and $q(x)$ is the distributed load. Application of separation of variables method (or Laplace transformation) leads to an ordinary differential equation subject to cantilever boundary conditions:

$$\frac{\mathrm{d}^4 y}{\mathrm{d}x^4} - \mu^4 y = f(x) \quad (0 < x < \ell), \qquad \begin{array}{ll} y(0) = 0, & y'(0) = 0, \\ y''(\ell) = 0, & y'''(\ell) = 0. \end{array}$$

In this equation, μ is a positive constant depending upon the radian frequency of the periodic loading and the physical properties of the beam, while $f(x)$ represents the strength of the loading at point x along the beam. The left end ($x = 0$) of the beam is assumed to be anchored, and the right end ($x = \ell$) is free.

Now we rewrite this scalar two-point boundary value problem in vector form by introducing a 4-column vector of unknowns

$$\mathbf{y}(x) = \langle y_1(x), y_2(x), y_3(x), y_4(x)\rangle^T = \langle y(x), y'(x), y''(x), y'''(x)\rangle^T.$$

Then the scalar beam equation will be equivalent to the vector equation: $\mathbf{y}' = \mathbf{A}\,\mathbf{y} + \mathbf{f}$, where the square 4×4 matrix $\mathbf{A}$ and the known column-vector $\mathbf{f}(x)$ are identified from the equation:

$$\frac{\mathrm{d}}{\mathrm{d}x}\begin{bmatrix} y_1(x) \\ y_2(x) \\ y_3(x) \\ y_4(x) \end{bmatrix} = \begin{bmatrix} 0 & 1 & 0 & 0 \\ 0 & 0 & 1 & 0 \\ 0 & 0 & 0 & 1 \\ \mu^4 & 0 & 0 & 0 \end{bmatrix}\begin{bmatrix} y_1(x) \\ y_2(x) \\ y_3(x) \\ y_4(x) \end{bmatrix} + \begin{bmatrix} 0 \\ 0 \\ 0 \\ f(x) \end{bmatrix}, \qquad 0 < x < \ell.$$

Its general solution can be written (see §8.3.1) as

$$\mathbf{y}(x) = \frac{1}{2}\int_0^x e^{\mathbf{A}(x-\tau)}\,\mathbf{f}(\tau)\,\mathrm{d}\tau - \frac{1}{2}\int_x^\ell e^{\mathbf{A}(x-\tau)}\,\mathbf{f}(\tau)\,\mathrm{d}\tau + e^{\mathbf{A}x}\,\mathbf{c},$$

where $\mathbf{c} = \langle c_1, c_2, c_3, c_4\rangle^T$ is a column vector of arbitrary constants. Since the matrix $\mathbf{A}$ has four distinct eigenvalues $\lambda_{1,2} = \pm\mu$ and $\lambda_{3,4} = \pm\mathbf{j}\mu$, its fundamental exponential matrix becomes

$$e^{\mathbf{A}\,t} = \frac{1}{2}\begin{bmatrix} \cosh\mu t + \cos\mu t & \frac{1}{\mu}\sinh\mu t + \frac{1}{\mu}\sin\mu t & \frac{1}{\mu^2}\cosh\mu t - \frac{1}{\mu^2}\cos\mu t & \frac{1}{\mu^3}\sinh\mu t - \frac{1}{\mu^3}\sin\mu t \\ \mu\sinh\mu t - \mu\sin\mu t & \cosh\mu t + \cos\mu t & \frac{1}{\mu}\sinh\mu t + \frac{1}{\mu}\sin\mu t & \frac{1}{\mu^2}\cosh\mu t - \frac{1}{\mu^2}\cos\mu t \\ \mu^2\cosh\mu t - \mu^2\cos\mu t & \mu\sinh\mu t - \mu\sin\mu t & \cosh\mu t + \cos\mu t & \frac{1}{\mu}\sinh\mu t + \frac{1}{\mu}\sin\mu t \\ \mu^3\sinh\mu t + \mu^3\sin\mu t & \mu^2\cosh\mu t - \mu^2\cos\mu t & \mu\sinh\mu t - \mu\sin\mu t & \cosh\mu t + \cos\mu t \end{bmatrix}$$

Then

$$\mathbf{y}(x) = \frac{1}{4\,\mu^3}\int_0^x f(\tau)\begin{bmatrix} \sinh\mu(x-\tau) - \sin\mu(x-\tau) \\ \mu\,(\cosh\mu(x-\tau) - \cos\mu(x-\tau)) \\ \mu^2\,(\sinh\mu(x-\tau) + \sin\mu(x-\tau)) \\ \mu^3\,(\cosh\mu(x-\tau) + \cos\mu(x-\tau)) \end{bmatrix} \mathrm{d}\tau$$

$$- \frac{1}{4\,\mu^3}\int_x^\ell f(\tau)\begin{bmatrix} \sinh\mu(x-\tau) - \sin\mu(x-\tau) \\ \mu\,(\cosh\mu(x-\tau) - \cos\mu(x-\tau)) \\ \mu^2\,(\sinh\mu(x-\tau) + \sin\mu(x-\tau)) \\ \mu^3\,(\cosh\mu(x-\tau) + \cos\mu(x-\tau)) \end{bmatrix} \mathrm{d}\tau + e^{\mathbf{A}x}\,\mathbf{c}.$$

The boundary constraints arising from the cantilever connections become

$$c_1 = \frac{1}{4\,\mu^3}\int_0^\ell f(\tau)\,[-\sinh(\mu\tau) + \sin(\mu\tau)]\,\mathrm{d}\tau, \qquad c_2 = \frac{1}{\mu^2}\int_0^\ell f(\tau)\,[\cosh(\mu\tau) - \cos(\mu\tau)]\,\mathrm{d}\tau,$$

$$c_3 = \frac{1}{4\mu}\int_0^\ell f(\tau)\,[\sinh(\mu\tau) + \sin(\mu\tau)]\,\mathrm{d}\tau, \qquad c_4 = -\frac{1}{4}\int_0^\ell f(\tau)\,[\cosh(\mu\tau) + \cos(\mu\tau)]\,\mathrm{d}\tau.$$

Problems

1. Show that the Green's matrix $\mathbf{G}(x,t)$ for the two-point boundary value problem (12.2.1) with homogeneous boundary conditions ($\boldsymbol{\alpha} = 0$) is

$$\mathbf{G}(x,t) = \frac{1}{2}\,\boldsymbol{\Phi}(x)\left[\mathbf{B}^{[0]}\boldsymbol{\Phi}(0) + \mathbf{B}^{[\ell]}\boldsymbol{\Phi}(\ell)\right]^{-1}\left[\mathbf{B}^{[0]}\boldsymbol{\Phi}(0) - \mathbf{B}^{[\ell]}\boldsymbol{\Phi}(\ell)\right]\boldsymbol{\Phi}^{-1}(t)$$
$$\times\,\frac{1}{2}\begin{cases} \boldsymbol{\Phi}(x)\,\boldsymbol{\Phi}^{-1}(t), & 0 \leqslant t < x, \\ -\boldsymbol{\Phi}(x)\,\boldsymbol{\Phi}^{-1}(t), & x \leqslant t \leqslant \ell \end{cases}$$

2. Rewrite the given boundary value problem as an equivalent boundary value problem for a first order system (12.2.1).
   (a) $\left(x^2 y'\right)' - 6\,y = -f(x) \quad 1 < x < 2, \quad y'(1) = 0, \; y(2) = 0.$
   (b) $(x\,y')' - 9\,y/x = -f(x) \quad 1 < x < 3, \quad y(1) = 0, \; y'(3) = 0.$
   (c) $\left(e^{-2x}\,y'\right)' - 3\,e^{-2x}y = -f(x) \quad 0 < x < 1, \quad y(0) = 0, \; 2\,y(1) + y'(1) = 0.$
   (d) $\left(e^{3x}\,y'\right)' + 2\,e^{3x}y = -f(x) \quad 0 < x < 1, \quad y(0) - 2y'(0) = 0, \; y(1) = 0.$
3. In each exercise from the previous problem, determine the Green's matrix for the corresponding first order system.
4. In each exercise, you are given boundary conditions for the two-point boundary value problem (12.2.1), where

$$\mathbf{P} = \begin{bmatrix} 3 & -13 \\ 1 & -3 \end{bmatrix}, \quad \mathbf{y}(x) = \begin{bmatrix} y_1(x) \\ y_2(x) \end{bmatrix}, \quad \boldsymbol{\alpha} = \begin{bmatrix} \alpha_1 \\ \alpha_2 \end{bmatrix}.$$

   Note that the fundamental matrix for the corresponding homogeneous equation is given to be $e^{\mathbf{P}\,x}$. Form the matrix $\mathbf{B}$ from Eq. (12.2.3) and determine whether the boundary value problem has a unique solution for every $\mathbf{f}(x)$ and $\boldsymbol{\alpha}$.
   (a) $y_1(0) = \alpha_1, \; y_2(\pi) = \alpha_2;$
   (b) $y_1(0) = \alpha_1, \; y_1(\pi) = \alpha_2;$
   (c) $y_1(0) - y_2(0) = \alpha_1, \; y_1(\pi) + y_2(\pi) = \alpha_2;$
   (d) $y_1(0) + y_2(0) = \alpha_1, \; y_1(\pi) + y_2(\pi) = \alpha_2.$
5. Show that the two-point boundary value problem

$$\mathbf{y}' = \mathbf{A}\,\mathbf{y} \quad (0 < t < 1) \qquad y_1(0) = 1, \; y_2(1) = 2, \; y_2(1) = 3,$$

   where

$$\mathbf{y}(t) = \begin{pmatrix} y_1(t) \\ y_2(t) \\ y_3(t) \end{pmatrix}, \qquad \mathbf{A} = \begin{bmatrix} 1 & 2 & 0 \\ 0 & -1 & 0 \\ 3 & 0 & 1 \end{bmatrix},$$

   has a unique solution and find it.
6. Consider the two-point boundary value problem

$$\frac{\mathrm{d}}{\mathrm{d}x}\begin{bmatrix} y_1 \\ y_2 \end{bmatrix} = a(x)\begin{bmatrix} 1 & -1 \\ 1 & -1 \end{bmatrix}, \qquad 0 < x < \ell, \quad y_1(0) = \alpha, \quad y_2'(\ell) = \beta.$$

   Upon introducing a new variable $t = \int_0^t a(x)\,\mathrm{d}x$, reduce the given differential equation to the constant coefficient one and solve the corresponding boundary value problem.

12.3 Singular Sturm–Liouville Problems

In certain applications, we come across Sturm–Liouville boundary value problems that do not satisfy all conditions (see §10.1) to be qualified as "regular" problems. To be more precise, we consider the self-adjoint (positive if $q(x) > 0$) differential operator (10.1.8):

$$L = L[x, \texttt{D}] = -\texttt{D}\,(p(x)\,\texttt{D}) + q(x), \qquad \texttt{D} = \text{d}/\text{d}x, \tag{12.3.1}$$

where $p(x)$, $p'(x)$ and $q(x)$ are real-valued continuous functions on $[a, b]$. It is assumed that the operator (12.3.1) acts on smooth functions that are defined on the interval $[a, b]$ subject to the boundary conditions of the third kind

$$\alpha_0 y(a) - \beta_0 y'(a) = 0, \qquad \alpha_1 y(b) + \beta_1 y'(b) = 0. \tag{12.3.2}$$

For every such operator (12.3.1) and a positive smooth function $\rho(x)$, we consider the Sturm–Liouville problem that consists of the differential equation generated by the self-adjoint differential expression

$$L[x, \texttt{D}]y = \lambda\rho(x)\,y \quad \text{or} \quad -\left(p(x)\,y'\right)' + q(x)\,y = \lambda\rho(x)\,y, \qquad a < x < b, \tag{12.3.3}$$

together with the boundary conditions (12.3.2).

Definition 12.1: The Sturm–Liouville boundary value problem (12.3.3), (12.3.2) is said to be regular if the functions $p(x)$, $p'(x)$, $q(x)$, $1/p(x)$, and $\rho(x)$ are continuous, and $p(x) > 0$, $\rho(x) > 0$ on the closed interval $[a, b]$.

Now we consider a certain class of boundary value problems for the differential operator $L[x, \texttt{D}]$ in which coefficients may be zero at some endpoints. When one or both of the boundary points in a Sturm–Liouville problem goes to $\pm\infty$, or when the coefficient $p(x)$ in the differential operator (12.3.1) diverges at an endpoint, the problem becomes singular. Usually the boundary condition at the singular endpoint is forfeited, imposing only the finiteness condition instead. The corresponding system is called a **singular Sturm–Liouville problem** on the interval $[a, b]$ if one of the following conditions is fulfilled:

- $p(a) = 0$ and boundary condition at $x = a$ is dropped, but assumed $y(a) < \infty$;
- $p(b) = 0$ and boundary condition at $x = b$ is dropped, but assumed $y(b) < \infty$;
- $p(a) = p(b) = 0$ and no boundary conditions; $y(x)$ is assumed to be a square integrable function.

Accordingly, there are four different situations each arising from the zeroes of $p(x)$. The corresponding differential equations may contain two real parameters α and β along with a nonnegative integer n, used to identify the eigenvalue.

1. If the function $p(x)$ has two distinct zeroes at end points $x = a$ and $x = b$. Then by appropriate translation and scaling, the Jacobi differential equation is discovered:

$$\left(1 - x^2\right) y'' + \left((\beta - \alpha) - (2 + \alpha + \beta)x\right) y' + n\,(n + \alpha + \beta + 1)\,y = 0. \tag{12.3.4}$$

2. If $p(x)$ has a single zero, the generalized (or associated) Laguerre equation (discovered by N. Ya. Sonin in 1880) emerges:

$$x\,y'' + (\alpha + 1 - x)\,y' + ny = 0. \tag{12.3.5}$$

3. If there is no zero, but the interval is infinite, we find the Hermite equation:

$$y'' - 2x\,y' + 2n\,y = 0 \qquad (-\infty < x < \infty). \tag{12.3.6}$$

4. If $p(x)$ has a double zero, the Bessel equation is discovered:

$$x^2 y'' + (\alpha x + \beta)\,y' - n\,(n + \alpha - 1)\,y = 0. \tag{12.3.7}$$

In the following, we discuss these four cases along with some of their applications in partial differential equations. However, instead of focusing on the general case of the Jacobi equation, we consider its two important particular cases—the Chebyshev and Legendre equations, discussed in §12.4.

Example 12.3.1: The boundary value problem

$$\left(x^2\, y'\right)' + (\lambda^2 x^2 - n^2)\, y = 0, \quad 0 < x < \ell, \quad y(\ell) = 0,$$

is not a regular Sturm–Liouville problem but a singular one. For this equation, we have $p(x) = \rho(x) = x^2$ and $q(x) = n^2$. Since the functions $p(x)$ and $\rho(x)$ are not positive at $x = 0$, we have an example of a singular Sturm–Liouville problem.

12.3.1 Green's Function

Consider the equation

$$x^2 y'' + xP(x)y' + Q(x)y = 0, \tag{12.3.8}$$

where $P(x)$ and $Q(x)$ are real-valued analytic functions of a real variable about the origin:

$$P(x) = p_0 + p_1 x + p_2 x^2 + \cdots, \qquad Q(x) = q_0 + q_1 x + q_2 x^2 + \cdots, \tag{12.3.9}$$

with convergent power series expansions. For small values of x, the equation (12.3.8) resembles

$$x^2 u'' + xP(0)u' + Q(0)u = 0.$$

This is Euler's differential equation, which we discussed in §4.6.2. Recall that solutions of the Euler equation are obtained by looking for functions of the form $u(x) = x^m$. It is reasonable to assume that equation (12.3.8) has a solution close to x^m for small values of x. In other words, near the regular singular point $x = 0$, we seek a solution of Eq. (12.3.8) in the form

$$y(x) = x^m [c_0 + c_1 x + c_2 x^2 + \cdots] = \sum_{n=0}^{\infty} c_n x^{m+n},$$

where $c_0 \neq 0$. Substituting this formula into Eq. (12.3.8), we get the indicial equation $m(m-1)+P(0)m+Q(0) = 0$, which is assumed to have two real roots, $m_1 \geqslant m_2$. We make the transformation $y = x^{m_1} u$ and substitute it into Eq. (12.3.8); this yields

$$xu'' + [2m_1 + P(x)]u' + \frac{m_1(m_1-1) + m_1 P(x) + Q(x)}{x}\, u = 0.$$

Let $p(x) = 2m_1 + P(x)$ and $q(x) = [m_1(m_1-1) + m_1 P(x) + Q(x)]/x$. The limit

$$\begin{aligned}
\lim_{x\to 0+} q(x) &= \lim_{x\to 0+} \frac{m_1(m_1-1) + m_1 P(x) + Q(x) - [m_1(m_1-1) + m_1 P(0) + Q(0)]}{x} \\
&= m_1 \lim_{x\to 0+} \frac{P(x) - P(0)}{x} + \lim_{x\to 0+} \frac{Q(x) - Q(0)}{x} \\
&= m_1 P'(0+) + Q'(0+)
\end{aligned}$$

is finite. So, we actually need to consider the equation

$$xu'' + p(x)u' + q(x)u = 0, \tag{12.3.10}$$

where p and q are continuous in some interval containing the origin and $p(x)$ is differentiable in it. Also, since $1 - P(0) = m_1 + m_2$, we have $p_0 = p(0) = 2m_1 + P(0) = 1 + (m_1 - m_2) \geqslant 1$.

We multiply equation $tu'' + p(t)u' + q(t)u = 0$ by a function $G(t, x)$ and integrate with respect to t by parts from 0 to x. This leads to

$$tu'G(t,x)\big|_{t=0}^{t=x} - \int_0^x [(tG)' - p(t)G(t,x)]\, u'\, \mathrm{d}t + \int_0^x q(t)u(t)G(t,x)\, \mathrm{d}t = 0.$$

If the function $G(x,t)$ satisfies the following equation and the conditions:

$$\frac{\mathrm{d}}{\mathrm{d}x}\left[x\, G(x,t)\right] - p(x)G(x,t) = 1, \qquad G(t,t) = 0, \quad \lim_{x\to 0+} x\, G(x,t) = 0, \tag{12.3.11}$$

then it is called a Green's function, and

$$u(x) = u(0) + \int_0^x q(t)G(x,t)u(t)\,\mathrm{d}t. \tag{12.3.12}$$

To solve Eq. (12.3.11), we set $Y(t) = tG(t,x)$, then the equation for $Y(t)$ becomes $Y' - p(t)Y(t)/t = 1$, or, after multiplication by an integrating factor $\mu(t) = \exp\left\{-\int \frac{p(t)}{t}\,\mathrm{d}t\right\}$, we get

$$Y(t) = \frac{1}{\mu(t)}\left[\int_a^t \mu(t)\,\mathrm{d}t + C\right]$$

with $C = \int_x^a \mu(t)\,\mathrm{d}t$ to satisfy the condition $G(x,x) = 0$. Let $p_0 = p(0)$, then

$$\mu(t) = \exp\left\{-\int_1^t \frac{p(t)}{t}\,\mathrm{d}t\right\} = \exp\left\{-\int_1^t \frac{p_0}{t}\,\mathrm{d}t\right\}\exp\left\{-\int_1^t \frac{p(t)-p_0}{t}\,\mathrm{d}t\right\} = t^{-p_0}F(t),$$

where $F(t) = \exp\left\{-\int_1^t \frac{p(t)-p(0)}{t}\,\mathrm{d}t\right\}$. We can rearrange terms to obtain

$$G(x,t) = \frac{t^{p_0-1}}{F(t)}\int_x^t \frac{F(\xi)}{\xi^{p_0}}\,\mathrm{d}\xi, \tag{12.3.13}$$

where $F(t)$ is a continuous positive function in the neighborhood of the origin because $\lim_{t\to 0}\dfrac{p(t)-p_0}{t} = p'(0)$. Since $F(x)$ is bounded in some interval $[0,h]$, there exist positive constants A and B such that $0 < A \leqslant F(x) \leqslant B$ in $[0,h]$. Then for $p_0 > 1$ we have

$$|G(x,t)| \leqslant \frac{B}{A}\,x^{p_0-1}\int_t^x \frac{\mathrm{d}\xi}{\xi^{p_0}} = \frac{B}{A(p_0-1)}\left[1 - \left(\frac{t}{x}\right)^{p_0-1}\right]$$

for all $0 < t \leqslant x \leqslant h$. If $p_0 = 1$, then

$$|G(x,t)| \leqslant \frac{B}{A}\int_t^x \frac{\mathrm{d}\xi}{\xi} = \frac{B}{A}\ln\frac{x}{t}.$$

So the function $G(x,t)$ is integrable and, in any case, for every x such that $0 < x \leqslant h$ the limit of the product $x\,G(x,t)$ tends to zero: $\lim_{x\to 0+} x\,G(x,t) = 0$.

Now we have to prove that a solution of the integral equation (12.3.12) satisfies the differential equation (12.3.10). Assume that $u(x)$ is a solution of the integral equation (12.3.12) for $0 < x < h$. Then

$$\begin{aligned} u'(x) &= \int_0^x q(t)G_x(x,t)u(t)\,\mathrm{d}t = -\frac{F(x)}{x^{p_0}}\int_0^x \frac{q(t)u(t)t^{p_0-1}}{F(t)}\,\mathrm{d}t, \\ u''(x) &= -\frac{q(x)u(x)}{x} + \int_0^x q(t)u(t)G_{xx}(t,x)\,\mathrm{d}t = -\frac{q(x)u(x)}{x} - p(x)u'(x), \end{aligned}$$

because

$$G_x(x,t) = -\frac{t^{p_0-1}}{F(t)}\frac{F(x)}{x^{p_0}}, \qquad G_x(x,x) = -\frac{1}{x}, \qquad G_{xx}(t,x) = \frac{t^{p_0-1}}{F(t)}\frac{F(x)}{x^{p_0}}\,p(x).$$

Hence $xu'' + p(x)u'(x) + q(x)u = 0$.

12.3.2 Orthogonality of Bessel Functions

We start with a famous Bessel function (see §4.9). Other examples of singular Sturm–Liouville problems and their solutions are provided in the next section.

The Bessel functions $J_\nu(z)$ are used to express eigenfunctions $\phi_n(x) = J_\nu(\alpha_{\nu,n}x/\ell)$ of the singular Sturm–Liouville problem

$$\frac{\mathrm{d}}{\mathrm{d}x}\left(x\,\frac{\mathrm{d}y}{\mathrm{d}x}\right) + \left(\lambda x - \frac{\nu^2}{x}\right)y = 0, \quad 0 < x < \ell, \quad y(0) < \infty, \ y(\ell) = 0,$$

corresponding to the eigenvalues $\lambda_n = \alpha_{\nu,n}^2/\ell^2$, where $\alpha_{\nu,n}$ is the n-th positive root of $J_\nu(z) = 0$. Let us consider the parametric Bessel equation (4.9.8), page 249, containing parameters α and β;

$$x^2 y'' + x y' + (\alpha^2 x^2 - \nu^2) y = 0 \quad \text{and} \quad x^2 y'' + x y' + (\beta^2 x^2 - \nu^2) y = 0.$$

We know from §4.9 that the functions $u(x) = J_\nu(\alpha x)$ and $v(x) = J_\nu(\beta x)$ are their solutions. Using Lagrange's identity (10.2.3) on page 555, we obtain

$$(\beta^2 - \alpha^2) \int_0^\ell x\, J_\nu(\alpha x)\, J_\nu(\beta x)\, \mathrm{d}x = J_\nu'(\alpha\ell)\, J_\nu(\beta\ell) - J_\nu(\alpha\ell)\, J_\nu'(\beta\ell).$$

Upon choosing $\alpha\ell$ and $\beta\ell$ to be zeroes either of the Bessel function ($J_\nu(\alpha\ell) = 0$ and $J_\nu(\beta\ell) = 0$) or its derivative ($J_\nu'(\alpha\ell) = 0$ and $J_\nu'(\beta\ell) = 0$), the right-hand side of the above formula vanishes and we get the orthogonality relation:

$$\int_0^\ell x\, J_\nu(\alpha x)\, J_\nu(\beta x)\, \mathrm{d}x = 0, \qquad \text{if } \alpha \ne \beta. \tag{12.3.14}$$

The first few approximations to the positive roots of the equation $J_\nu(x) = 0$ are given in the following table.

$\nu = 0$	2.404825557696	5.520078110286	8.653727912911	11.79153443901
$\nu = 1$	3.831705970208	7.015586669816	10.17346813507	13.32369193631

For any positive $\alpha = \beta$, we have

$$\|J_\nu\|^2 = \int_0^\ell x\, J_\nu^2(\alpha x)\, \mathrm{d}x = \frac{1}{2\alpha^2} \left\{ \left[\alpha\ell\, J_\nu'(\alpha\ell)\right]^2 + \left(\alpha^2\ell^2 - \nu^2\right) J_\nu^2(\alpha\ell) \right\}. \tag{12.3.15}$$

The formula (12.3.15) is simplified when α is chosen to be a root of either $J_\nu(\alpha\ell) = 0$ or $J_\nu'(\alpha\ell) = 0$. The orthogonality property (12.3.14) allows one to expand an "arbitrary" function into the Fourier–Bessel series of order ν:

$$f(x) = \sum_{r\geqslant 1} A_r\, J_\nu\left(x\alpha_r\right), \qquad A_r = \frac{1}{\|J_\nu\|^2} \int_0^\ell f(x) J_\nu(x\alpha_r)\, x\, \mathrm{d}x, \tag{12.3.16}$$

where $\{\alpha_r\}_{r\geqslant 1}$ is a sequence of all roots of $J_\nu(\alpha\ell) = 0$ and $\|J_\nu\|^2 = \ell^2 J_{\nu+1}^2(\alpha)/2$. If $\{\alpha_r\}_{r\geqslant 1}$ is a sequence of all roots of $J_\nu'(\alpha\ell) = 0$, then expansion (12.3.16) is valid with the square norm $\|J_\nu\|^2 = \frac{1}{2}\left(\ell^2 - \frac{\nu^2}{\alpha^2}\right) J_\nu^2(\alpha\ell)$. At first glance, expansion (12.3.16) has only theoretical meaning with no chance to use it manually. However, a computer solver makes this task pretty manageable, as the following examples show. For instance, *Mathematica* and *Maple* share two dedicated built-in symbols, `BesselJ` and `BesselJZero`/`BesselJZeros`, to evaluate the Bessel function and find its roots, respectively. MATLAB has two similar commands—`besselj` and `besseljzero`. *Maxima* uses `bessel_j`, while *Sage* utilizes the nomenclatures `bessel_J`, `bessel_Y`, `bessel_I`, and `bessel_K`.

Example 12.3.2: Consider the function $f(x) = x(1-x)$ on the interval $[0,1]$. We expand it into the Fourier–Bessel series with respect to two Bessel functions: $J_0(x)$ and $J_1(x)$. Recall that $J_0(0) = 1$ and $J_1(0) = 0$. Since $f(0) = 0$, we expect that the series (12.3.16) with respect to $J_1(x)$ will give a better approximation to $f(x)$. First, we calculate the square norms of the Bessel functions, corresponding to first six roots of $J_\nu(\alpha) = 0$:

	$r = 1$	$r = 2$	$r = 3$	$r = 4$	$r = 5$	$r = 6$
$\|J_0\|^2 =$	0.134757	0.0578901	0.0368432	0.0270188	0.0213307	0.0176211
$\|J_1\|^2 =$	0.0811076	0.0450347	0.0311763	0.0238404	0.0192993	0.0162114

Then we calculate the values of coefficients A_r ($r = 1, 2, \ldots, 5$) in Eq. (12.3.16). Next we build N-th term finite sum approximation and plot this partial sum with $N = 5$ terms in Fig. 12.3. The mean square error of approximation with respect to the Bessel functions of order 0 is about $\Delta_5(J_0) \approx 0.000111967$, and with respect to the Bessel functions of order 1 it is much smaller, $\Delta_5(J_1) \approx 7.21913 \times 10^{-7}$. The fact that $J_0(0) = 1$ hinders the approximation at $x = 0$ with respect to Bessel functions of the zero order.

Example 12.3.3: Now we expand the function $g(x) = (1+x^2)^{-1}$ into the Fourier–Bessel series:

$$g(x) = \frac{1}{1+x^2} = \sum_{r\geqslant 1} A_r\, J_\nu\left(x\alpha_r\right), \qquad A_r = \frac{1}{\|J_\nu\|^2} \int_0^\ell \frac{J_\nu(x\alpha_r)}{1+x^2}\, x\, \mathrm{d}x.$$

Since $g(0) = 1/2 \ne 0$, we expect a better approximation of the Fourier–Bessel series with respect to the Bessel function of order 0. Indeed, the mean square error of 5-term approximation with respect to the Bessel functions of order 0 is about $\Delta_5(J_0) \approx 0.0122404$, and with respect to the Bessel functions of order 1 is much larger, $\Delta_5(J_1) \approx 0.0442942$.

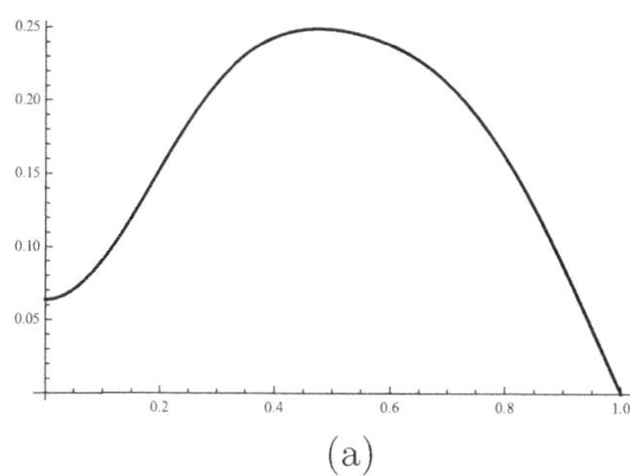

(a)

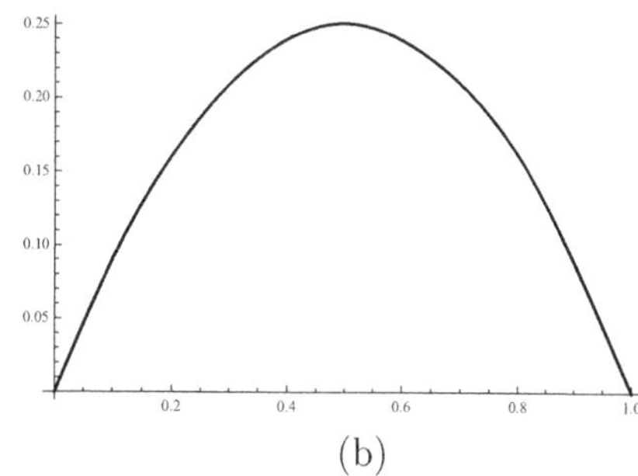

(b)

Figure 12.3: Example 12.3.1. The graph of the N-th partial sum approximations with $N = 5$ terms with respect to (a) $J_0(x)$ and (b) $J_1(x)$, plotted with *Mathematica*.

Example 12.3.4: Consider a thin elastic membrane that vibrates in accordance with the two-dimensional wave equation

$$u_{tt} = c^2 \left(u_{xx} + u_{yy}\right), \qquad x^2 + y^2 \leqslant a^2.$$

Here, a is the radius of a circular membrane; it is convenient to use wave equation in polar coordinates

$$\ddot{u} = c^2 \left(u_{rr} + \frac{1}{r}\, u_r + \frac{1}{r^2}\, u_{\theta\theta} \right), \qquad 0 \leqslant r < a, \quad 0 \leqslant \theta < 2\pi. \tag{12.3.17}$$

Assuming that the displacement $u(r, \theta, t)$ from equilibrium is zero at the rim, we get the following boundary conditions:

$$\left.\begin{aligned} &|u(0,\theta,t)| < \infty && \text{(bounded at the origin)}, \\ &u(a,\theta,t) = 0 && \text{(fixed edge)}, \\ &u(r,\theta,t) = u(r,\theta + 2\pi, t) && \text{(single-valued function)}. \end{aligned}\right\} \tag{12.3.18}$$

We seek a solution of the wave equation (12.3.17) when the initial displacement and velocity are specified

$$u(r,\theta,t=0) = f_1(r,\theta), \qquad u_t(r,\theta,t=0) = f_2(r,\theta). \tag{12.3.19}$$

In accordance with separation of variables, we assume that partial nontrivial solutions have the form

$$u(r,\theta,t) = v(r,\theta)\, T(t).$$

Then we obtain the equation for $T(t)$:

$$\ddot{T}(t) + c^2 \lambda\, T(t) = 0,$$

and the following eigenvalue problem for v:

$$\frac{1}{r}\frac{\partial}{\partial r}\left(r\, \frac{\partial v}{\partial r} \right) + \frac{1}{r^2}\frac{\partial^2 v}{\partial \theta^2} + \lambda v = 0 \qquad (0 \leqslant r < a),$$

subject to the boundary conditions (12.3.18).

Let us assume again that

$$v(r,\theta) = R(r)\,\Theta(\theta).$$

Substituting this form of the solution in our equation and dividing by $R(r)\,\Theta(\theta)$, we obtain

$$\frac{r}{R}\frac{\mathrm{d}}{\mathrm{d}r}\left(r\,\frac{\mathrm{d}R}{\mathrm{d}r}\right) + \frac{\Theta'}{\Theta} + \lambda r^2 = 0 \iff \frac{r}{R}\frac{\mathrm{d}}{\mathrm{d}r}\left(r\,\frac{\mathrm{d}R}{\mathrm{d}r}\right) + \lambda r^2 = -\frac{\Theta'}{\Theta} = \mu,$$

which leads to

$$\Theta'' + \mu\Theta = 0.$$

The function $v(r,\theta)$ must be a single-valued and differentiable function, as should $\Theta(\theta)$ be. Since $\Theta(\theta)$ is a periodic function with period 2π, i.e., $\Theta(\theta) = \Theta(\theta + 2\pi)$, we get $\mu = n^2$, a positive integer; otherwise, a nontrivial periodic solution does not exist. Therefore,

$$\Theta_n(\theta) = A_n \cos n\theta + B_n \sin n\theta, \qquad n = 0, 1, 2, \ldots,$$

where A_n and B_n are some constants. Using this value of $\mu = n^2$, we get the following singular Sturm–Liouville problem for $R(r)$:

$$\frac{1}{r}\frac{\mathrm{d}}{\mathrm{d}r}\left(r\,\frac{\mathrm{d}R}{\mathrm{d}r}\right)+\left(\lambda-\frac{n^2}{r^2}\right)R(r)=0 \quad (0\leqslant r<a),$$

subject to the corresponding boundary conditions

$$R(a)=0, \qquad |R(0)|<\infty.$$

Introducing a new variable $x=\sqrt{\lambda}r$ and setting $R(r)=R\left(\frac{x}{\sqrt{\lambda}}\right)=y(x)$, we obtain Bessel's equation of the n-th order:

$$\frac{\mathrm{d}^2y}{\mathrm{d}x^2}+\frac{1}{x}\frac{\mathrm{d}y}{\mathrm{d}x}+\left(1-\frac{n^2}{x^2}\right)y=0$$

with the boundary conditions

$$y\left(a\sqrt{\lambda}\right)=0, \qquad |y(0)|<\infty.$$

The only function that satisfies (up to an arbitrary constant multiple) the Bessel's equation, which is bounded at the origin, is $J_n(x)=J_n\left(r\sqrt{\lambda}\right)$. From the first boundary condition, it follows that $J_n\left(a\sqrt{\lambda}\right)=0$. If $\mu_m^{(n)}$ is the m-th root of the equation $J_n(\mu)=0$, then

$$\lambda_{n,m}=\left(\frac{\mu_m^{(n)}}{a}\right)^2, \qquad n=0,1,2,\ldots;\ m=1,2,\ldots, \tag{12.3.20}$$

is the sequence of eigenvalues. The eigenfunction corresponding to this eigenvalue is

$$R_{n,m}(r)=J_n\left(\frac{\mu_m^{(n)}}{a}\,r\right). \tag{12.3.21}$$

Using $R_{n,m}(r)$ and $\Theta_n(\theta)$, we form partial nontrivial solutions

$$v_{n,m}(r,\theta)=J_n\left(\frac{\mu_m^{(n)}}{a}\,r\right)\left(A_{n,m}\cos n\theta+B_{n,m}\sin n\theta\right)$$

that we have to multiply by $T_{n,m}(t)$, the general solution of $T''+c^2\lambda_{n,m}T=0$:

$$T_{n,m}(t)=a_{n.m}\cos\left(c\lambda_{n,m}t\right)+b_{n,m}\sin\left(c\lambda_{n,m}t\right).$$

Summing all nontrivial solution, we get the general solution of the wave equation (12.3.17) subject to the boundary conditions (12.3.18):

$$\begin{aligned}u(r,\theta,t)&=\sum_{n\geqslant 0}\cos n\theta\sum_{m\geqslant 1}\left[a_{n.m}\cos\left(c\lambda_{n,m}\right)t+b_{n,m}\sin\left(c\lambda_{n,m}t\right)\right]J_n\left(\frac{\mu_m^{(n)}}{a}\,r\right)\\&\quad+\sum_{n\geqslant 1}\sin n\theta\sum_{m\geqslant 1}\left[A_{n.m}\cos\left(c\lambda_{n,m}t\right)+B_{n,m}\sin\left(c\lambda_{n,m}t\right)\right]J_n\left(\frac{\mu_m^{(n)}}{a}\,r\right)\end{aligned}$$

that contains four arbitrary constants. For their determination, we use the initial conditions:

$$u(r,\theta,0)=\sum_{m\geqslant 1}\left[\sum_{n\geqslant 0}a_{n.m}\cos n\theta+\sum_{n\geqslant 1}A_{n.m}\sin n\theta\right]J_n\left(\frac{\mu_m^{(n)}}{a}\,r\right)=f_1(r,\theta),$$

$$u_t(r,\theta,0)=\sum_{m\geqslant 1}c\lambda_{n,m}\left[\sum_{n\geqslant 0}b_{n.m}\cos n\theta+\sum_{n\geqslant 1}B_{n.m}\sin n\theta\right]J_n\left(\frac{\mu_m^{(n)}}{a}\,r\right)=f_2(r,\theta).$$

The coefficients of the above Fourier-Bessel expansions can be obtained using orthogonal relations. For instance,

$$A_{n.m} = \frac{1}{\pi}\int_0^{2\pi} \sin n\theta \, \mathrm{d}\theta \, \frac{1}{\|J_n\|^2}\int_0^a r\mathrm{d}r \, f_1(r,\theta)\, J_n\left(\frac{\mu_m^{(n)}}{a}\, r\right).$$

Now suppose that the initial displacement is $f_1(r,\theta) = (r-a)^2 \sin 2\theta$ while the initial velocity is zero. These initial conditions dictate that $a_{n,m} = b_{n,m} = B_{n,m} = 0$, and $A_{n,m} = 0$ for $n \ne 2$. The coefficients $A_{2,m}$ must satisfy the relation

$$\begin{aligned}
(r-a)^2 &= \sum_{m\geqslant 1} A_{2,m} J_2\left(\frac{\mu_m^{(2)}}{a}\, r\right), \\
\|J_2\|^2 A_{2,m} &= \int_0^a r\mathrm{d}r \, (r-a)^2 J_2\left(\frac{\mu_m^{(2)}}{a}\, r\right) \\
&= \frac{2\,a^4}{\left(\mu_m^{(2)}\right)^2} + \frac{a^2}{\left(\mu_m^{(2)}\right)^3} J_1\left(\mu_m^{(2)}\right)\left[8 - 3\,\mu_m^{(2)}\,\pi\, H_0\left(\mu_m^{(2)}\right)\right] \\
&\quad + \frac{3\,a^2}{\left(\mu_m^{(2)}\right)^2} J_0\left(\mu_m^{(2)}\right)\left[\pi\, H_1\left(\mu_m^{(2)}\right) - 2\right],
\end{aligned}$$

where $H_s(z) = \sum_{k\geqslant 0} \frac{(-1)^k}{\Gamma\left(k+\frac{3}{2}\right)\Gamma\left(k+s+\frac{3}{2}\right)}\left(\frac{z}{2}\right)^{2k+s+1}$ is the s-th Struve function and $\mu_m^{(2)}$ $(m = 1, 2, \ldots)$ are roots of the equation $J_2(\mu) = 0$. The square norm (12.3.15) is simplified as $\|J_2\|^2 = J_3^2\left(\mu_m^{(2)}\right)/2$.

Problems

1. Expand the function $f(x) = x^2$ into Fourier–Bessel series (12.3.16) on the interval $[0, 1]$ with $N = 5$ terms with respect to the Bessel function of order 0 and order 1. Which of these two approximations gives better result?
2. Expand the function $g(x) = 1 - x^2$ into Fourier–Bessel series on the interval $[0, 1]$ with $N = 5$ terms with respect to the Bessel function of order 0 and order 1. Which of these two approximations gives better result?
3. Prove the Dini[117] series expansion:

$$f(x) = \sum_{r\geqslant 1} A_r\, J_n\left(x\alpha_r\right), \qquad A_r = \frac{2\alpha_r^2}{\left(\alpha_r^2\ell^2 - n^2 + h^2\ell^2\right)\left[J_n(\alpha_r\ell)\right]^2}\int_0^\ell f(x) J_n(x\alpha_r)\, x\, \mathrm{d}x,$$

where α_r is defined by the boundary condition $hJ_n\left(\alpha\ell\right) + \alpha\, J_n'\left(\alpha\ell\right) = 0$.

4. Prove the Fourier–Bessel series expansion:

$$f(x) = c_0 + \sum_{r\geqslant 1} c_r\, J_n\left(x\alpha_r\right), \qquad c_r = \frac{2\,\alpha_r^2}{\left(\alpha_r^2\ell^2 - n^2\right)\left[J_0(\alpha_r\ell)\right]^2}\int_0^\ell f(x) J_0(x\alpha_r)\, x\, \mathrm{d}x,$$

where α_r is defined by the boundary condition $J_0'\left(\alpha\ell\right) = 0$ and $c_0 = \frac{2}{\ell^2}\int_0^\ell f(x)\, x\, \mathrm{d}x$.

5. Consider the singular Sturm–Liouville problem

$$\begin{aligned}
-\left(x\, y'\right)' &= \lambda x\, y(x) \qquad (0 < x < \ell), \\
y(x),\; y'(x) \text{ bounded as } x \mapsto 0, &\qquad y'(\ell) = 0.
\end{aligned}$$

Show that $\lambda_0 = 0$ is an eigenvalue of this problem corresponding to the eigenfunction $y_0(x) = 1$. If $\lambda > 0$, show formally that the eigenfunctions are given by $\phi_n(x) = J_0\left(x\sqrt{\alpha_n}/\ell\right)$, where α_n is the n-th positive root (in increasing order) of the equation $J_0'(\sqrt{\alpha}) = 0$.

6. By evaluating at $x = 1, 2, 3$ and performing the appropriate numerical integration, give empirical evidence that the following formulas may indeed be correct. You may want to show that these integrals satisfy the appropriate differential equations subject to corresponding initial conditions.

[117] Ulisse Dini (1845–1918) was an Italian mathematician and politician born in Pisa.

(a) $J_0(x) = \dfrac{1}{\pi} \displaystyle\int_0^{\pi} \cos(x\,\cos\theta)\,\mathrm{d}\theta;$ (b) $J_0(x) = \dfrac{2}{\pi} \displaystyle\int_0^{1} \dfrac{\cos(xs)}{\sqrt{1-s^2}}\,\mathrm{d}s.$

7. Evaluating numerically at $x = 1, 2, 3$, give empirical evidence that each of the following formulas may indeed be correct.

(a) $J_0(x) + 2\sum_{k\geqslant 1} J_{2k}(x) = 1;$

(b) $J_0(x) + 2\sum_{k\geqslant 1} (-1)^k J_{2k}(x) = \cos x;$

(c) $2\sum_{k\geqslant 1} (-1)^k J_{2k+1}(x) = \sin x.$

For each of the functions in Exercises 8 through 22 on the specified interval $[0, \ell]$, obtain $\sum_{r=1}^{5} A_r J_0(x\alpha_r)$, the five-term partial sum of the Fourier–Bessel series and graph it along with the given function. Then repeat the calculations for the Fourier–Bessel finite sum approximation $\sum_{r=1}^{5} B_r J_1(x\beta_r)$ with respect to the Bessel function of the order 1. Which of these two approximations gives better result?

8. $f(x) = 1 - x^2/9$, $\ell = 3$;
9. $f(x) = \sqrt{x}$, $\ell = 4$;
10. $f(x) = e^{-x}$, $\ell = 2$;
11. $f(x) = e^{x}/(1+x^2)$, $\ell = 3/2$;
12. $f(x) = \cos^2 x$, $\ell = 2\pi$;
13. $f(x) = 1 - \sqrt{2x}$, $\ell = 1/2$;
14. $f(x) = x\,\sin x$, $\ell = \pi$;
15. $f(x) = e^{x}\,\sin x$, $\ell = \pi/2$; (new)
16. $f(x) = |x-1|$, $\ell = 2$;
17. $f(x) = |\sin x|$, $\ell = 2\pi$;
18. $f(x) = \begin{cases} x, & 0 \leqslant x < 1, \\ 1, & 1 \leqslant x < 2, \end{cases}$, $\ell = 2$;
19. $f(x) = \begin{cases} x, & 0 \leqslant x < 1, \\ 1-x, & 1 \leqslant x < 2, \end{cases}$, $\ell = 2$;
20. $f(x) = \begin{cases} e^{-x}, & 0 \leqslant x < 1, \\ 1, & 1 \leqslant x < 2, \end{cases}$, $\ell = 2$;
21. $f(x) = \begin{cases} \sin x, & 0 \leqslant x < \pi/2, \\ 1, & \pi/2 \leqslant x < \pi, \end{cases}$, $\ell = \pi$;
22. $f(x) = \begin{cases} x, & 0 \leqslant x < 1, \\ 1, & 1 \leqslant x < 2,, \\ 3-x, & 2 \leqslant x \leqslant 3 \end{cases}$ $\ell = 3$;

23. Using separation of variables, solve the initial value problem for the wave equation in polar coordinates

$$\frac{\partial^2 u}{\partial t^2} = c^2\left(\frac{\partial^2 u}{\partial r^2} + \frac{1}{r}\frac{\partial u}{\partial r} + \frac{1}{r^2}\frac{\partial^2 u}{\partial \theta^2}\right) \qquad (0 \leqslant r < a),$$
$$u(a,\theta,t) = 0, \quad u(r,\theta,0) = f(r.\theta), \quad u_t(r,\theta,0) = 0.$$

24. Consider the steady state temperature distribution $u(x, y)$ in semi-infinite strip $0 < x < \infty$, $0 < y < \pi$. Physically speaking, of course, this is an unrealistic idealization, perhaps concocted to simulate a rectangular configuration when one of the sides is much larger than another. Using sine-Fourier integral transform

$$f^S(\omega) = \int_0^{\infty} f(x)\,\sin(\omega x)\,\mathrm{d}x, \qquad f(x) = \frac{2}{\pi}\int_0^{\infty} f^S(\omega)\,\sin(\omega x)\,\mathrm{d}\omega,$$

solve the following Dirichlet problem:

$$u_{xx} + u_{yy} = 0, \qquad u(x,0) = 0, \quad u(x,\pi) = f(x), \quad u(0,y) = 0.$$

25. Expand the function $f(x) = x^2$ into Fourier–Bessel series (12.3.16) on the interval $[0, 1]$ with 5 terms with respect to the Bessel function of order 2 and order 3. Which of these two approximations gives better result?

12.4 Orthogonal Polynomials

Let us choose a set of linearly independent integer powers $q_n(x) = x^n$ $(n = 0, 1, 2, \ldots)$ and some interval (a, b), which can be infinite. For a fixed weight function $\rho(x)$, this set $\{q_n(x)\}_{n\geqslant 0}$ is not orthogonal. However, we can construct a linearly independent set of polynomials $p_n(x)$ (each of degree n, $n = 0, 1, 2, \ldots$) that are orthogonal on that interval with weight $\rho(x)$. Some famous orthogonal polynomials are summarized in the following table[118].

Interval	Weight	Notation	Author
$(-1,1)$	1	$P_n(x)$	Legendre
$(-1,1)$	$(1-x^2)^{-1/2}$	$T_n(x)$	Chebyshev
$(-\infty,\infty)$	e^{-x^2}	$H_n(x)$	Chebyshev–Hermite
$(0,\infty)$	$x^n e^{-x}$	$L_n(x)$	Chebyshev–Laguerre

These orthogonal polynomials (as well as many others) are eigenfunctions of the corresponding singular Sturm–Liouville problem, generated by a self-adjoint differential equation (12.3.3), page 639. These polynomials have numerous applications, and their properties have been intensively analyzed. In this section, we concentrate on only one very important property—orthogonality, followed by corresponding orthogonal expansions with respect to these polynomials.

12.4.1 Chebyshev's Polynomials

The differential equation on the interval $(-1, 1)$ with a positive parameter λ

$$\left(1-x^2\right) y'' - xy' + \lambda^2 y = 0 \quad \text{or} \quad \frac{\mathrm{d}}{\mathrm{d}x}\left(\sqrt{1-x^2}\, y'\right) + \frac{\lambda^2}{\sqrt{1-x^2}}\, y = 0 \tag{12.4.1}$$

is called **Chebyshev's[119] equation**. By imposing conditions at the end points for solution $y(x)$ to be finite, we obtain the singular Sturm–Liouville problem that has eigenvalues $\lambda = n$, integers. Any point from the interval $(-1, 1)$ is an ordinary point for Chebyshev's equation. Thus, any series solution will converge for $|x| < 1$. Assuming that

$$y(x) = \sum_{k=0}^{\infty} a_k\, x^k\,,$$

substitution into Chebyshev's equation (12.4.1) gives us

$$\sum_{k=0}^{\infty} a_{k+2}(k+2)(k+1)x^k - \sum_{k=2}^{\infty} a_k k(k-1)x^k - \sum_{k=1}^{\infty} a_k k x^k + \lambda^2 \sum_{k=0}^{\infty} a_k x^k = 0,$$

or

$$2a_2 + a_0\lambda^2 + \left[6a_3 - a_1 + \lambda^2 a_1\right] x + \sum_{k=2}^{\infty} \left[(k+2)(k+1)a_{k+2}(\lambda^2 - k^2)a_k\right] x^k = 0.$$

Equating coefficients of like powers of x to zero, we obtain the following relations, listed in the table.

Power of x	Coefficients		Recurrence
x^0	$2a_2 + \lambda^2 a_0 = 0$	or	$a_2 = -\lambda^2\, a_0/2$
x^1	$3 \cdot 2a_3 + (\lambda^2 - 1)a_1 = 0$	or	$a_3 = (1-\lambda^2)\, a_1/6$
x^2	$4 \cdot 3a_4 + (\lambda^2 - 2^2)a_2 = 0$	or	$a_4 = -(\lambda^2 - 2^2)\, a_2/12$
$\vdots$	$\vdots$	$\vdots$	$\vdots$
x^k	$(k+2)(k+1)a_{k+2} + (\lambda^2 - k^2)a_k = 0$	or	$a_{k+2} = -\frac{\lambda^2-k^2}{(k+2)(k+1)}\, a_k$

[118]Polynomials $P_n(x)$ were introduced in 1785 by the French mathematician, Adrien-Marie Legendre (1752–1833), who made important contributions to special functions, elliptic integrals, number theory, and the calculus of variations. Other polynomials, $T_n(x)$, $H_n(x)$, and $L_n(x)$ were defined by the Russian mathematician, Pafnuty Chebyshev (1821–1894), in 1859. A generalization of Laguerre polynomials were discovered by Russian professors Yu. V. Sokhotskii (1842–1927) and somewhat later Nikolay Sonin (1849–1915). In 1864, the French mathematician Charles Hermite (1822–1901) studied polynomials $H_n(x)$, and the French mathematician Edmond Laguerre (1834–1886) analyzed $L_n(x)$ in 1879. See `dlmf.nist.gov` for other special functions.

[119]Pafnuty L. Chebyshev (1821–1894), professor at St. Petersburg University, was the first to study solutions of this equation. The collected works of this eminent savant are available in Russian and French.

From the recurrence relation

$$a_{k+2} = -\frac{\lambda^2 - k^2}{(k+2)(k+1)}\, a_k,$$

we can determine the values of all coefficients by its direct substitution to be

$$a_{2m} = (-1)^m\, \frac{\lambda^2(\lambda^2-2^2)(\lambda^2-4^2)\cdots(\lambda^2-(2m-2)^2)}{(2m)!}\, a_0,$$

and

$$a_{2m+1} = (-1)^m\, \frac{(\lambda^2-1^2)(\lambda^2-3^2)\cdots(\lambda^2-(2m-1)^2)}{(2m+1)!}\, a_1.$$

When $\lambda = n$ is a positive integer, one of these recurrences will terminate depending on the parity of n. In this case, the coefficients $a_{n+2}, a_{n+4}, \ldots$ are all equal to zero. Then we obtain a polynomial of degree n as one of the solutions of Eq. (12.4.1). If we set the leading coefficient a_n of this polynomial to be 2^{n-1}, then a polynomial solution is called a **Chebyshev polynomial** of the first kind and denoted by $T_n(x)$. The letter T is used because of the alternative transliteration of the name Chebyshev as Tchebycheff (actually, Chebyshev himself used five distinct spellings with Latin letters and two Russian spellings).

It is customary to define Chebyshev's polynomials via either the following formula:

$$T_n(x) = \cos(n \arccos x) \qquad (|x| \leqslant 1) \tag{12.4.2}$$

or the recurrence relation (for arbitrary x)

$$T_n(x) = 2x\, T_{n-1}(x) - T_{n-2}(x), \qquad T_0(x) = 1,\ T_1(x) = x. \tag{12.4.3}$$

Let us list the first few Chebyshev polynomials:

$$T_2(x) = 2\,x^2 - 1, \quad T_3(x) = 4\,x^3 - 3x, \quad T_4(x) = 1 - 8x^2 + 8x^4, \quad T_5(x) = 16x^5 - 20x^3 + 5x.$$

Chebyshev's polynomials possess so many remarkable properties and can be defined in many ways that we simply cannot present them (see [11, 34]). The reader can find some of their properties in the exercises. We provide here the orthogonality property:

$$\int_{-1}^{1} T_n(x)\, T_m(x)\, \frac{\mathrm{d}x}{\sqrt{1-x^2}} = 0, \quad m \ne n, \qquad \int_{-1}^{1} T_n^2(x)\, \frac{\mathrm{d}x}{\sqrt{1-x^2}} = \begin{cases} \frac{\pi}{2}, & n \ne 0, \\ \pi, & n = 0. \end{cases}$$

Chebyshev's polynomial of the second kind, denoted by

$$U_n(x) = \frac{\sin[(n+1)\arccos x]}{\sqrt{1-x^2}} \qquad \text{for } |x| \leqslant 1,$$

are solutions of the following boundary value problem

$$(1-x^2)\, y'' - 3x\, y' + n(n+1)\, y = 0, \qquad U_n(1) = n+1 = (-1)^n U(-1). \tag{12.4.4}$$

The Chebyshev polynomials of first and second kind turn out to be the best choice for most applications, mainly due to the good convergence properties of the corresponding series:

$$f(x) = \frac{c_0}{2} + \sum_{n\geqslant 1} c_n\, T_n(x), \qquad c_n = \frac{2}{\pi}\int_{-1}^{1} f(x)\, T_n(x)\, \frac{\mathrm{d}x}{\sqrt{1-x^2}} = \frac{2}{\pi}\int_0^{\pi} f(\cos\theta)\, \cos(n\theta)\, \mathrm{d}\theta. \tag{12.4.5}$$

We isolate the coefficient c_0 because $\|T_n\|^2 = \pi/2$ if $n \ne 0$ and $\|T_0\|^2 = \pi$. For Chebyshev's polynomials of the second kind, we have an expansion similar to Eq. (12.4.5):

$$f(x) = \sum_{n\geqslant 0} a_n\, U_n(x), \qquad a_n = \frac{2}{\pi}\int_{-1}^{1} f(x)\, U_n(x)\, \sqrt{1-x^2}\, \mathrm{d}x. \tag{12.4.6}$$

To invoke the Chebyshev polynomial, the latest version of MATLAB uses the commands `orthpoly::chebyshev1(n, x)` and `orthpoly::chebyshev2(n, x)` for these polynomials of the first and second kind, respectively. *Mathematica* and *Maple* share almost the same name `ChebyshevT[n,x]` and `ChebyshevT(n, x)`, respectively. Also, *Maple* has a special command `chebyshev(f,x,eps)` (when the package `numapprox` is invoked) to expand the function f into series with respect to Chebyshev polynomials. *Maxima* uses `chebyshev_t(n, x)` and `chebyshev_u(n, x)`, while *Sage* utilizes `chebyshev_T(n, x)` and `chebyshev_U(n, x)`. SymPy uses `chebyt(n,x)` and `chebyu(n,x)`.

There is an open-source special package, called Chebfun, written in MATLAB, for numerical evaluation of Chebyshev expansions with functions to 15-digit accuracy. It was proposed in 2002 by the famous British mathematician Lloyd N. Trefethen and his student Zachary Battles. The mathematical basis of the system combines tools of Chebyshev expansions, fast Fourier transform, barycentric interpolation, recursive zerofinding, and automatic differentiation.

Example 12.4.1: We expand the function $g(x) = (1 + x^2)^{-1}$ into the Chebyshev series (12.4.5) on the interval $[-1, 1]$. To achieve it, we calculate the first few coefficients in the series (12.4.5):

$$c_0 = \frac{1}{\pi}\int_{-1}^{1} \frac{g(x)}{\sqrt{1-x^2}}\,\mathrm{d}x = \frac{1}{\sqrt{2}}, \quad c_2 = 4 - 3\sqrt{2}, \quad \frac{c_4}{2} = \frac{17}{\sqrt{2}} - 12, \quad \frac{c_6}{2} = 70 - \frac{99}{\sqrt{2}}.$$

Note that odd numbered coefficients are all zeroes. The Chebyshev truncated series with 4 terms (which is a polynomial of degree 6),

$$S_6(x) = \frac{1}{\sqrt{2}} + c_2\,T_2(x) + c_4\,T_4(x) + c_6\,T_6(x),$$

gives a very accurate approximation of $g(x)$: it has a mean square error of about 1.71413×10^{-6}. Next we use expansion (12.4.6):

$$u_6(x) = 2(\sqrt{2} - 1) + a_2\,U_2(x) + a_4\,U_4(x) + a_6\,U_6(x),$$

where

$$a_2 = 2\left(7 - 5\sqrt{2}\right), \quad a_4 = 2\left(29\sqrt{2} - 41\right), \quad a_6 = 2\left(239 - 169\sqrt{2}\right),$$

since $a_0 = \frac{2}{\pi}\int_{-1}^{1} \frac{\sqrt{1-x^2}}{1+x^2}\,\mathrm{d}x = 2(\sqrt{2} - 1)$. This finite sum $u_6(x)$ also gives a very good approximation to $g(x)$ because it has a mean square error of about 1.84542×10^{-6}.

12.4.2 Legendre's Equation

The closest relative to the Chebyshev polynomial of the second kind is the Legendre polynomial, conventionally denoted as $P_\nu(x)$, which is an eigenfunction of the singular Sturm–Liouville problem:

$$\frac{\mathrm{d}}{\mathrm{d}x}\left[(1-x^2)\,\frac{\mathrm{d}y}{\mathrm{d}x}\right] + \lambda\,y = 0 \quad (-1 < x < 1), \quad y(-1) < \infty, \; y(1) < \infty.$$

Setting λ to be equal to the eigenvalue $\lambda = \nu(\nu+1)$ ($\nu \geqslant 0$), we obtain

$$(1-x^2)\,y'' - 2x\,y' + \nu(\nu+1)\,y = 0, \tag{12.4.7}$$

where ν is a nonnegative real number. Despite the fact that this is a differential equation of order 2, it is known as **Legendre's equation of order** ν. Eq. (12.4.7) is frequently encountered in physics and other numerous problems, especially in those exhibiting spherical symmetry.

These polynomials have a default script in all software packages: MATLAB – `orthpoly :: legendre(n, x)`; *Mathematica* and *Maple* – `LegendreP[n, x]` and `LegendreP(n, x)`, respectively, and *Maxima* – `legendre_p(n,m,x)`. Note that *Maple* has short-cut `P(n,x)` when the package `orthopoly` is invoked. SymPy uses the same command as MATLAB to evaluate the Legendre polynomials. *Sage* utilizes `legendre_P(x,n)`.

Legendre's equation has regular singular points at $x = 1$, $x = -1$, and $x = \infty$, and ordinary points elsewhere. For simplicity, we consider only the case when $\nu = n$, a positive integer. A solution of Eq. (12.4.7) has a Maclaurin expansion that converges when $|x| < 1$, so we set

$$y(x) = \sum_{k=0}^{\infty} a_k\,x^k.$$

We substitute the series and its first two derivatives into Eq. (12.4.7) to obtain

$$(1-x^2)\sum_{k=2}^{\infty} k(k-1)\,a_k\,x^{k-2} - 2x\sum_{k=1}^{\infty} ka_k\,x^{k-1} + \nu(\nu+1)\sum_{k=0}^{\infty} a_k\,x^k = 0,$$

or

$$\sum_{k=2}^{\infty} k(k-1)\,a_k\,x^{k-2} - \sum_{k=2}^{\infty} k(k-1)\,a_k\,x^k - \sum_{k=1}^{\infty} 2ka_k\,x^k + \sum_{k=0}^{\infty} \nu(\nu+1)\,a_k\,x^k = 0.$$

By shifting the index of summation in the first series in this expression, we find that

$$\sum_{k=0}^{\infty} (k+2)(k+1)\,a_{k+2}\,x^k - \sum_{k=2}^{\infty} k(k-1)\,a_k\,x^k - \sum_{k=1}^{\infty} 2ka_k\,x^k + \sum_{k=0}^{\infty} \nu(\nu+1)\,a_k\,x^k = 0.$$

Next, we collect like terms to obtain

$$\begin{aligned}&1\cdot 2\,a_2 + \nu(\nu+1)\,a_0 + [2\cdot 3\,a_3 - 2\,a_1 + \nu(\nu+1)\,a_1]\;x\\ &+\sum_{k=2}^{\infty}\left\{(k+2)(k+1)\,a_{k+2} - k(k-1)\,a_k - 2ka_k + \nu(\nu+1)\,a_k\right\}\,x^k = 0.\end{aligned}$$

Setting the sums of the coefficients of like powers of x equal to zero gives[120]

$$\begin{aligned}1\cdot 2a_2 + \nu(\nu+1)\,a_0 &= 0,\\ 2\cdot 3\,a_3 + (\nu+2)(\nu-1)\,a_1 &= 0,\\ (k+2)(k+1)\,a_{k+2} + (\nu+k+1)(\nu-k)\,a_k &= 0, \quad k\geqslant 2.\end{aligned} \tag{12.4.8}$$

These relations show us that all coefficients a_k from $k\geqslant 2$ can be determined onward in terms of a_0 and a_1, which leads to

$$\begin{aligned}y(x) = a_0&\left[1 - \frac{\nu(\nu+1)}{2!}\,x^2 + \frac{(\nu+3)(\nu+1)\nu(\nu-2)}{4!}\,x^4 - \cdots\right]\\ &+ a_1\left[x - \frac{(\nu+2)(\nu-1)}{3!}\,x^3 + \frac{(\nu+4)(\nu+2)(\nu-1)(\nu-3)}{5!}\,x^5 - \cdots\right].\end{aligned}$$

Although the bracketed series in this solution are rather unwieldy, it could be shown that if ν is not an integer, then each of them converges when $|x|<1$ and diverges when $|x|>1$. The series also diverges when $x=\pm 1$, though this is not easy to prove. When ν is an integer, however, one of the series terminates (depending on the parity of ν) and, therefore, is a polynomial. This polynomial $P_\nu(x)$, called **Legendre's polynomial**, is uniquely defined by setting $P_\nu(1)=1$. This condition defines the coefficient a_0 to be 1 and the coefficient a_1 to be zero.

Legendre's polynomials diverge at the singular point $x=\infty$. The efficient way to define Legendre's polynomials gives[121] **Rodrigues's formula**:

$$P_n(x) = \frac{1}{2^n n!}\,\frac{\mathrm{d}^n}{\mathrm{d}x^n}\,(x^2-1)^n = \frac{1}{2^n}\sum_{k=0}^{\lfloor n/2\rfloor}(-1)^k\binom{n}{k}\binom{2n-2k}{n}\,x^{n-2k}. \tag{12.4.9}$$

The first six Legendre polynomials are seen to be

$$P_0(x)=1, \qquad P_1(x)=x, \qquad P_2(x)=\frac{1}{2}\left(3x^2-1\right), \qquad P_3(x)=\frac{1}{2}\left(5x^3-3x\right),$$

$$P_4(x)=\frac{1}{8}\left(35x^4-30x^2+3\right), \qquad P_5(x)=\frac{1}{8}\left(63x^5-70x^3+15x\right).$$

[120] We use the relations $\nu(\nu+1)-2=(\nu+2)(\nu-1)$ and $-k(k-1)-2k+\nu(\nu+1)=\nu(\nu+1)-k(k+1)=(\nu+k+1)(\nu-k)$.

[121] Benjamin Olinde Rodrigues (1795–1851), more commonly known as Olinde Rodrigues, was a French banker, mathematician, and social reformer; he derived this formula in 1816.

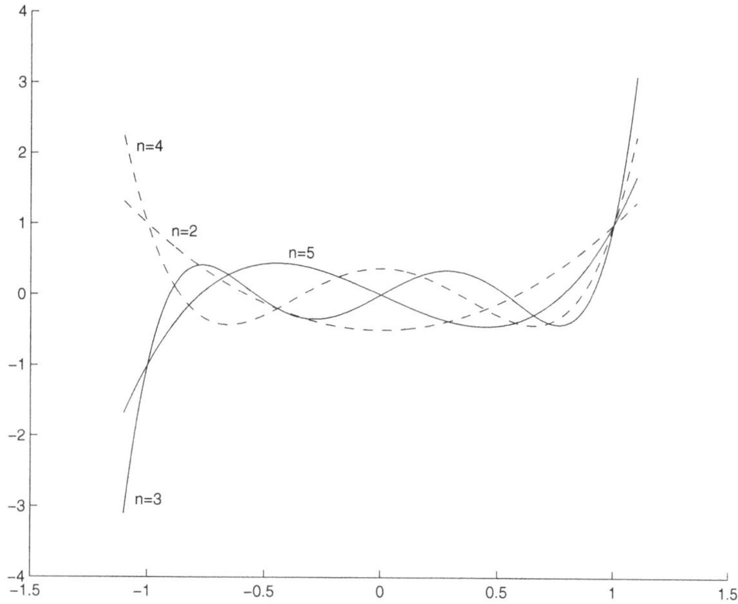

Figure 12.4: Graphs of the Legendre polynomials $P_n(x)$ for $n = 2, 3, 4, 5$, plotted with MATLAB.

The Legendre polynomials are interconnected by the following relations, known as **recurrence formulas**:

$$\frac{\mathrm{d}}{\mathrm{d}x}\left[x^n\, P_{n-1}(x)\right] = x^{n-1}\, P_n'(x), \tag{12.4.10}$$

$$P_n'(x) - x\, P_{n-1}'(x) - n\, P_{n-1}(x) = 0, \tag{12.4.11}$$

$$(n+1)P_{n+1}(x) = (2n+1)x\, P_n(x) - nP_{n-1}(x). \tag{12.4.12}$$

The latter difference equation is well suited for computations. Some important quadratures of Legendre polynomials can be evaluated with the aid of the following theorems (their proofs are left as exercises, but also can be found in [29]) .

Theorem 12.3:

$$\int_{-1}^{1} x^m\, P_n(x)\,\mathrm{d}x = \begin{cases} 0, & \text{if } m < n \\ \frac{2^{n+1}\,(n!)^2}{(2n+1)!}, & \text{if } n = m. \end{cases}$$

Theorem 12.4:

$$\int_{-1}^{1} P_m(x)P_n(x)\,\mathrm{d}x = \begin{cases} 0, & \text{if } m \ne n, \\ \frac{2}{2n+1}, & \text{if } n = m. \end{cases}$$

Corollary 12.1: Let $q(x)$ be a polynomial of degree less than n, where $n \geqslant 1$. Then

$$\int_{-1}^{1} P_n(x)\, q(x)\,\mathrm{d}x = 0.$$

Theorem 12.5: The Legendre polynomial $P_n(x)$ has n distinct zeros in the open interval $(-1, 1)$.

The orthogonality property of Legendre's polynomials (Theorem 12.4) leads to an orthogonal expansion:

$$\frac{f(x-0) + f(x+0)}{2} = \sum_{n\geqslant 0} a_n\, P_n(x), \qquad a_n = \frac{2n+1}{2}\int_{-1}^{1} f(x)\, P_n(x)\,\mathrm{d}x. \tag{12.4.13}$$

Example 12.4.2: (Example 12.4.1 revisited) Expanding the function $g(x) = (1+x^2)^{-1}$ into the Legendre series (12.4.13) and keeping only 6 terms, we obtain

$$l_6(x) = a_0 + a_2\, P_2(x) + a_4\, P_4(x) + a_6\, P_6(x),$$

where

$$a_0 = \frac{1}{2}\int_{-1}^{1} \frac{\mathrm{d}x}{1+x^2} = \frac{\pi}{4}, \quad a_2 = \frac{3-\pi}{2}, \quad a_4 = \frac{17\pi}{8} - \frac{20}{3}, \quad a_6 = \frac{161}{5} - \frac{41\pi}{4}.$$

Its mean square error is about 0.0333262.

Example 12.4.3: Imagine a solid ball of radius 1. We introduce the spherical coordinates (r, θ, ϕ) in this ball so that the origin coincides with the center of the ball. Let steady state temperature be rotationally invariant, so it does not depend on the azimuthal angle ϕ. If the temperature inside the ball is denoted by $u(r, \theta)$, then it will satisfy Laplace's equation

$$\frac{1}{r^2}\frac{\partial}{\partial r}\left(r^2 \frac{\partial u}{\partial r}\right) + \frac{1}{r^2 \sin\theta}\left(\sin\theta\, \frac{\partial u}{\partial \theta}\right) = 0. \tag{12.4.14}$$

We seek partial nontrivial solutions of this equation in the form $u(r,\theta) = R(r)\,\Theta(\theta)$. Substituting into (12.4.14) gives

$$\Theta \cdot \frac{\mathrm{d}}{\mathrm{d}r}\left(r^2 \frac{\mathrm{d}R(r)}{\mathrm{d}r}\right) + R(r)\cdot \frac{1}{\sin\theta}\cdot\frac{\mathrm{d}}{\mathrm{d}\theta}\left(\sin\theta\, \frac{\mathrm{d}\Theta}{\mathrm{d}\theta}\right) = 0.$$

Separating variables, we get

$$\frac{1}{R}\frac{\mathrm{d}}{\mathrm{d}r}\left(r^2 \frac{\mathrm{d}R(r)}{\mathrm{d}r}\right) = -\frac{1}{\Theta}\frac{1}{\sin\theta}\cdot\frac{\mathrm{d}}{\mathrm{d}\theta}\left(\sin\theta\, \frac{\mathrm{d}\Theta}{\mathrm{d}\theta}\right)$$

The right-hand side depends only on θ and the left-hand side depends only on r. We conclude that both sides are equal to a constant that we denote by λ. This leads to two separate equations, one for $R(r)$ and one for $\Theta(\theta)$. We start with the latter:

$$\frac{1}{\sin\theta}\cdot\frac{\mathrm{d}}{\mathrm{d}\theta}\left(\sin\theta\, \frac{\mathrm{d}\Theta}{\mathrm{d}\theta}\right) + \lambda\,\Theta(\theta) = 0.$$

We make the change of variables

$$x = \cos\theta, \qquad y(x) = \Theta(\theta).$$

With standard identities

$$\sin^2\theta = 1 - x^2 \qquad \text{and} \qquad \frac{\mathrm{d}}{\mathrm{d}x} = \frac{\mathrm{d}\theta}{\mathrm{d}x}\,\frac{\mathrm{d}}{\mathrm{d}\theta} = -\frac{1}{\sin\theta}\,\frac{\mathrm{d}}{\mathrm{d}\theta},$$

we convert our equation into

$$\frac{\mathrm{d}}{\mathrm{d}x}\left((1-x^2)\,\frac{\mathrm{d}y}{\mathrm{d}x}\right) + \lambda\, y = 0.$$

This is equivalent to Legendre's equation (12.4.7). Its eigenvalues are $\lambda = n(n+1)$, where n is nonnegative integer, and corresponding eigenfunctions are $y(x) = P_n(x) = P_n(\cos\theta)$.

Our next task is to solve the equation for $R(r)$:

$$\frac{\mathrm{d}}{\mathrm{d}r}\left(r^2 \frac{\mathrm{d}R}{\mathrm{d}r}\right) = n(n+1)\, R(r).$$

Since this equation is of Euler's form (see §8.1.1, page 437), it has a polynomial solution

$$R_n(r) = c_n r^n + d_n\, r^{-n-1} \qquad \text{for } n > 0.$$

Here, of course, c_n and d_n are some arbitrary constants. Since the function $R_n(r)$ must be bounded at $r = 0$, we must set d_n equal 0. When $n = 0$, the general solution becomes

$$R_0(r) = c_0 + d_0\, \ln r.$$

Again, we set $d_0 = 0$ because the logarithm function is unbounded at the origin, and obtain the sequence of eigenfunctions $R_n(r) = c_n r^n$, $n = 0, 1, 2, \ldots$. Putting this information together with our solution in θ, we find partial nontrivial solutions of Laplace's equation to be

$$u_n(r,\theta) = c_n r^n P_n(\cos\theta), \qquad n = 0, 1, 2, \ldots.$$

Now we invoke the familiar idea of the Fourier method, and write our general solution as the sum over all possible partial nontrivial solutions:

$$u(r,\theta) = \sum_{n\geqslant 0} c_n r^n P_n(\cos\theta).$$

For Dirichlet boundary condition, the function $u(1,\theta) = f(\theta)$ is specified, so we need to determine the values of coefficients c_n from the Fourier–Legendre expansion:

$$u(1,\theta) = f(\theta) = \sum_{n\geqslant 0} c_n\, P_n(\cos\theta).$$

Using Eq. (12.4.13), we can determine coefficients c_n accordingly. □

Legendre's differential equation has two linearly independent solutions. If $\nu = n$ is a positive integer, one solution is the Legendre polynomial (12.4.9). Another solution, having a logarithmic singularity at points ± 1, is called the **Legendre function of the second kind**. It may be defined recursively according to Eq. (12.4.12) with the following first few functions:

$$Q_0(x) = \frac{1}{2}\ln\frac{1+x}{1-x}, \quad Q_1(x) = \frac{x}{2}\ln\frac{1+x}{1-x} - 1, \quad Q_2(x) = \frac{3x^2-1}{4}\ln\frac{1+x}{1-x} - \frac{3x}{2}.$$

12.4.3 Hermite's Polynomials

The classical one dimensional harmonic oscillator in quantum mechanics is described by the Schrödinger equation

$$\psi'' + \frac{2m}{\hbar^2}\left(E - V(\xi)\right)\psi = 0,$$

where $\psi(\xi)$ is the state of a particle of mass m in the potential $V(\xi)$ with energy E. The constant $\hbar = h/(2\pi) \approx 6.626\times 10^{-34}\,\mathtt{m^2kg/s}$ is called the reduced Planck constant, or Dirac constant. We will suppose that ψ depends only on the position ξ, and that the potential V is defined by $V(\xi) = \frac{k}{2}\,\xi^2$, which corresponds to the elastic force $-k\xi$. Hence, we get the equation

$$\psi'' + \frac{2m}{\hbar^2}\left(E - \frac{k}{2}\,\xi^2\right)\psi = 0,$$

and the values of energy E must be determined from the condition that the solution is bounded in the whole line $-\infty < \xi < \infty$. Let us introduce new parameters:

$$\alpha^2 = \frac{mk}{\hbar^2}, \qquad \alpha\mu = \frac{2mE}{\hbar^2} \quad (\alpha > 0).$$

The constant α^2 is known, but μ is a parameter to be determined by solving a corresponding Sturm–Liouville problem. So we get

$$\psi'' + \left(\alpha\mu - \alpha^2\xi^2\right)\psi = 0.$$

By introducing a new independent variable $x = \xi\sqrt{\alpha}$, we reduce the above equation to the following one:

$$\psi'' + \left(\mu - x^2\right)\psi = 0. \tag{12.4.15}$$

This linear differential equation has an irregular singular point at infinity ($x = \infty$). If we seek a solution as a product

$$\psi(x) = e^{-x^2/2} y(x),$$

then $y(x)$ must satisfy the differential equation

$$y'' - 2xy' + 2\lambda y = 0\,, \qquad -\infty < x < \infty, \tag{12.4.16}$$

which is called **Hermite's equation**. Here, we set $2\lambda = \mu - 1$. Eq. (12.4.16) can be rewritten in self-adjoint form:

$$\frac{\mathrm{d}}{\mathrm{d}x}\left[e^{-x^2}\frac{\mathrm{d}y}{\mathrm{d}x}\right] + 2\lambda e^{-x^2}\,y = 0. \tag{12.4.17}$$

The **Chebyshev–Hermite polynomials**, or simply **Hermite polynomials**, are defined as the eigenfunctions of Eq. (12.4.17) on an infinitely straight line $-\infty < x < \infty$ that tends to infinity not faster than a polynomial as $x \to \infty$.

Since Eq. (12.4.16) has no singular points in the finite plane, $x = 0$ is an ordinary point of the equation. We shall look for a solution in the form of the power series

$$y(x) = \sum_{k=0}^{\infty} a_k \, x^k.$$

Substitution into Eq. (12.4.16) leads to

$$\sum_{k=0}^{\infty} \left[a_{k+2}(k+2)(k+1) - 2ka_k + 2\lambda a_k\right] x^k = 0.$$

Hence, we obtain the recurrence for its coefficients to be

$$a_{k+2} = \frac{2(k-\lambda)}{(k+2)(k+1)}\, a_k, \qquad k = 0, 1, 2, \ldots.$$

The coefficients a_0 and a_1 are arbitrary and all others can be found from this difference equation. Thus, we have the general solution:

$$\begin{aligned} y(x) &= a_0 \left[1 + \sum_{k=1}^{\infty} \frac{2^k(-\lambda)(2-\lambda)\cdots(2k-2-\lambda)}{(2k)!}\, x^{2k}\right] \\ &= a_1 \left[x + \sum_{k=1}^{\infty} \frac{2^k(1-\lambda)(1-\lambda+2)\cdots(1-\lambda+2k-2)}{(2k+1)!}\, x^{2k+1}\right], \end{aligned}$$

valid for all finite x. If $\lambda = n$, a positive integer, then this solution always has a polynomial solution. If n is an even integer, the multiple of a_0 contains a terminating series, each term for $k \geqslant (n+2)/2$ being zero. If n is an odd integer, the multiple of a_1 is a polynomial because each term is zero for $k \geqslant (n+2)/2$. So the result is

$$H_n(x) = \sum_{k=0}^{\lfloor \frac{n}{2} \rfloor} \frac{(-1)^k\, n!}{k!\,(n-2k)!}\,(2x)^{n-2k} = e^{x^2/2}\left(x - \frac{\mathrm{d}}{\mathrm{d}x}\right)^n e^{-x^2/2} = (-1)^n e^{x^2}\,\frac{\mathrm{d}^n}{\mathrm{d}x^n}\, e^{-x^2},$$

in which $\lfloor \frac{n}{2} \rfloor$ stands for the greatest integer $\leqslant \frac{n}{2}$, called the floor of $\frac{n}{2}$. The polynomial $H_n(x)$ is the **Hermite polynomial**. Using this relation, we evaluate the first few polynomials:

$$H_0(x) = 1, \quad H_1(x) = 2x, \quad H_2(x) = 4x^2 - 2, \quad H_3(x) = 8x^3 - 12x, \quad H_4(x) = 16x^4 - 48x^2 + 12.$$

It can be shown that the Hermite polynomials satisfy the differential equation:

$$H_n'(x) = 2n\, H_{n-1}(x) \quad \text{or} \quad H_n'(x) = 2x\, H_n(x) - H_{n+1}(x),$$

from which follows the recurrence relation:

$$H_{n+1}(x) - 2x\, H_n(x) + 2n\, H_{n-1}(x) = 0 \qquad n = 1, 2, \ldots.$$

Next, differentiation shows that $H_n(x)$ are eigenfunctions of the Sturm–Liouville problem (12.4.16) corresponding to eigenvalues $\lambda_n = n$. The functions $\psi_n(x) = H_n(x)\, e^{-x^2/2}$ $(n = 0, 1, 2, \ldots)$, called **Hermite functions**, are eigenfunctions to the differential operator $-\mathtt{D}^2 + x^2$, so that $-\psi_n''(x) + x^2\psi_n(x) = (2n+1)\,\psi_n(x)$.

The Hermite polynomials form an orthogonal system with weight $\rho(x) = e^{-x^2}$:

$$\int_{-\infty}^{\infty} \psi_m(x)\psi_n(x)\,\mathrm{d}x = \int_{-\infty}^{\infty} H_m(x)H_n(x)\, e^{-x^2}\,\mathrm{d}x = \begin{cases} 0, & \text{if } m \ne n, \\ 2^n n!\sqrt{\pi}, & \text{if } m = n. \end{cases}$$

This allows us to define the following expansion:

$$f(x) = \sum_{k\geqslant 0} c_n\, e^{-x^2/2}\, H_n(x), \qquad c_n = \frac{1}{2^n n!\sqrt{\pi}} \int_{-\infty}^{\infty} f(x)H_n(x)e^{-x^2/2}\,\mathrm{d}x. \tag{12.4.18}$$

In MATLAB, Hermite's polynomial can be defined by `orthpoly::hermite(n, x)`, while *Maxima* uses a similar command `hermite(n,x)`. *Maple* and *Mathematica* share the same script: `HermiteH(n, x)` and `HermiteH[n, x]`, respectively. Note that *Maple* has a short-cut for this polynomial: `H(n,x)` when the package `orthopoly` is invoked. *Sage* and SymPy utilize `hermite(n,x)`.

Example 12.4.4: (Example 12.4.1 revisited) The function $g(x) = (1+x^2)^{-1}$ can be expanded into the Hermite series (12.4.18), where coefficients c_n are evaluated only numerically:

$$c_n = \frac{1}{2^n n! \sqrt{\pi}} \int_{-\infty}^{\infty} \frac{H_n(x)}{1+x^2}\, e^{-x^2/2}\, \mathrm{d}x, \qquad n = 0, 1, 2, \ldots.$$

Since $c_{2k+1} = 0$ $(k = 0, 1, 2, \ldots)$ and other coefficients with even indices decrease pretty fast,

$$c_0 \approx 0.9272709, \quad c_2 \approx 0.0116536, \quad c_4 \approx 0.00674567, \quad c_6 \approx 0.0001393194718,$$

finite sums with relatively small number of terms give good approximations. For instance, the mean square error of such approximation with 11 terms (five of them are identically zero) is about 0.00608485677, while with 31 terms, it is about 0.0013093589.

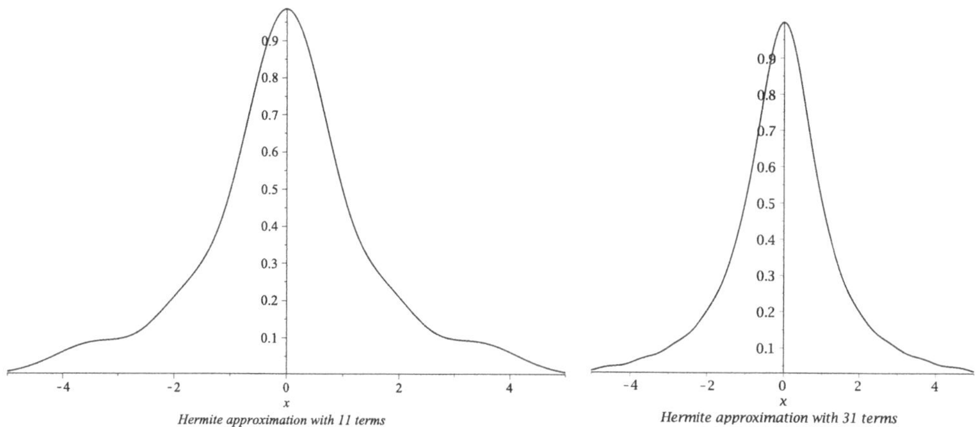

Hermite approximation with 11 terms

Hermite approximation with 31 terms

12.4.4 Laguerre's Polynomials

The differential equation

$$x\, y'' + (1-x)\, y' + \lambda\, y = 0, \qquad 0 < x < \infty, \tag{12.4.19}$$

is called **Laguerre's equation**. We rewrite this equation in a self-adjoint form:

$$\frac{\mathrm{d}}{\mathrm{d}x}\left[x\, e^{-x}\, \frac{\mathrm{d}y}{\mathrm{d}x}\right] + \lambda e^{-x}\, y = 0, \quad 0 < x < \infty.$$

The Laguerre polynomials are eigenfunctions of the latter equation subject to the following conditions: the solution should be bounded at $x = 0$ and tend to infinity not faster than a finite power of x as $x \to \infty$.

The point $x = 0$ is a regular singular point of Eq. (12.4.19) with indicial equation $\sigma^2 = 0$. Since it has a double root $\sigma_1 = \sigma_2 = 0$, it is natural to look for a solution as a Maclaurin series

$$y(x) = \sum_{k=0}^{\infty} a_k\, x^k.$$

Substituting this series into Eq. (12.4.19) yields

$$\sum_{k=0}^{\infty} \left[a_{k+1} k(k+1) + a_{k+1}(k+1) - k a_k + \lambda a_k\right] x^k = 0$$

or

$$\sum_{k=0}^{\infty} \left[a_{k+1}(k+1)^2 - a_k(k-\lambda) \right] x^k = 0.$$

We obtain the recurrence for its coefficients to be

$$a_{k+1} = \frac{k-\lambda}{(k+1)^2}\, a_k, \qquad k = 0, 1, 2, \ldots.$$

If $\lambda = n$ is a positive integer, Eq. (12.4.19) has a polynomial solution. Choosing a_0 in such a way that the coefficient of the highest power of x^k is equal to $(-1)^n$, we obtain the **Chebyshev–Laguerre**, or simply **Laguerre polynomials**:

$$L_n(x) = \sum_{k=0}^{n} \frac{(-1)^k n! x^k}{(k!)^2 (n-k)!}.$$

They can be defined in the form:

$$L_n(x) = e^x \frac{\mathrm{d}(x^n\, e^{-x})}{\mathrm{d}x^n}, \qquad n = 1, 2, \ldots.$$

We list the first six polynomials:

$$\begin{array}{ll} L_0(x) = 1; & 3!\, L_3(x) = -x^3 + 9x^2 - 18x + 6; \\ L_1(x) = -x + 1; & 4!\, L_4(x) = x^4 - 16x^3 + 72x^2 - 96x + 24; \\ 2!\, L_2(x) = x^2 - 4x + 2; & 5!\, L_5(x) = -x^5 + 25x^4 - 200x^3 + 600x^2 - 600x + 120. \end{array}$$

Laguerre polynomials form an orthogonal system with weight e^{-x} on $[0, \infty)$:

$$\int_0^{\infty} L_m(x) L_n(x)\, e^{-x}\, \mathrm{d}x = \begin{cases} 0, & \text{if } m \ne n, \\ (n!)^2, & \text{if } m = n. \end{cases}$$

Since the orthogonality relation contains multiple $(n!)^2$, it is convenient to consider normalized polynomials

$$l_n(x) = \frac{L_n(x)}{n!} = \frac{e^x}{n!} \frac{\mathrm{d}(x^n\, e^{-x})}{\mathrm{d}x^n}, \qquad l_0 = L_0 = 1.$$

This allows us to find coefficients in the Laguerre expansion:

$$f(x) = \sum_{n \geqslant 0} c_n\, l_n(x), \qquad 0 < x < \infty, \tag{12.4.20}$$

of a smooth function $f(x)$ on the semi-infinite interval $(0, \infty)$:

$$c_n = \int_0^{\infty} f(x)\, l_n(x)\, e^{-x}\, \mathrm{d}x, \qquad n = 0, 1, 2, \ldots. \tag{12.4.21}$$

Mathematica has a default command for the normalized Laguerre polynomials: `LaguerreL[n, x]`; *Maple* uses `L(n,x)` when the package `orthopoly` is invoked. Similarly, MATLAB uses `laguerreL(n, x)` and *Maxima* has `laguerre(n, x)`. SymPy utilizes `laguerre(n,a,x)`, where $a = 0$ corresponds to the ordinary Laguerre polynomials.

and so on $c_2 \approx 0.123969$, $c_3 \approx 0.0497043$, $c_4 \approx 0.0134013$, $c_5 \approx -0.00378451$, $c_6 \approx -0.011112$, ... The above figure shows approximation (in blue) of the function $f(x) = (1+x^2)^{-1}$ (in black) by partial Laguerre sum with $N = 7$ terms.

Problems

1. Using generating functions $g_T(x,t) = \dfrac{1-xt}{1-2xt+t^2}$ and $g_U(x,t) = \dfrac{1}{1-2xt+t^2}$, show that

$$T_n(x) = \frac{1}{2}\left[\left(x + \sqrt{x^2-1}\right)^n + \left(x - \sqrt{x^2-1}\right)^n \right],$$

$$U_n(x) = \frac{1}{2\sqrt{x^2-1}} \left[\left(x + \sqrt{x^2-1}\right)^{n+1} - \left(x - \sqrt{x^2-1}\right)^{n+1} \right].$$

Example 12.4.5: Now we expand the function $f(x) = (1+x^2)^{-1}$ on the interval $(0,\infty)$ using Laguerre polynomials. First, we calculate a few of the first coefficients in the expansion (12.4.20):

$$c_0 = \int_0^\infty \frac{e^{-x}}{1+x^2}\,dx \approx 0.62145,$$

$$c_1 = \int_0^\infty \frac{e^{-x}}{1+x^2}\,(1-x)\,dx \approx 0.278072,$$

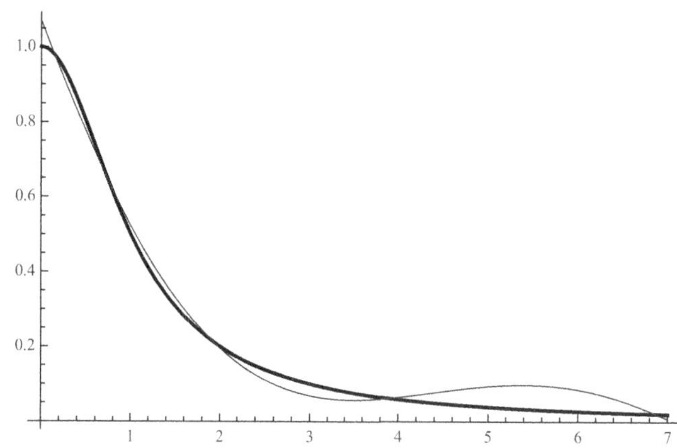

2. Show that for nonnegative integers n and m,

$$T_n(x)\,T_m(x) = \frac{1}{2}\left[T_{n+m}(x) + T_{|m-n|}(x)\right].$$

3. Show that both Chebyshev polynomials, $T_n(x)$ and $U_n(x)$, satisfy the recurrence

$$y_{n+1} = 2x\,y_n - y_{n-1}, \qquad n = 1,2,\ldots.$$

4. Prove the following recurrence formulas:

   (a) $T_{n+1}^2(x) - T_n(x)\,T_{n+2}(x) = 1 - x^2$, **(b)** $(1-x^2)\,T_n'(x) = n\left[T_{n-1}(x) - x\,T_n(x)\right]$.

5. Show that the Chebyshev equation $\left(1-x^2\right)y'' - x\,y' + n^2 y = 0$ can be reduced to the harmonic equation $y'' + n^2 y = 0$ using substitution $x = \cos\theta$.

6. Using the exponential generating function $\hat{g}(x,t) = \exp\left(2xt - t^2\right) = \sum_{n\geqslant 0} H_n(x)\,\frac{t^n}{n!}$, prove the recurrence for Hermite polynomials: $H_{n+1}(x) = 2x\,H_n(x) - 2n\,H_{n-1}(x)$.

7. Using the generating function for the Laguerre polynomials $g(x,t) = \frac{1}{1-t}\exp\left\{-\frac{xt}{1-t}\right\} = \sum_{n\geqslant 0} L_n(x)\,t^n$, prove the recurrence $(n+1)\,L_{n+1}(x) = (2n+1-x)\,L_n(x) - n\,L_{n-1}(x)$.

8. Prove the recurrence relations (12.4.10) – (12.4.11).

9. Using the generating function

$$g(x,t) = \frac{1}{\sqrt{1-2xt+t^2}} = \sum_{n\geqslant 0} P_n(x)\,t^n,$$

   prove the recurrence (12.4.12).

10. Prove the formula $(x^2-1)\,P_n'(x) = nx\,P_n(x) - n\,P_{n-1}(x)$.

11. Prove the formula $(x^2-1)\,P_n'(x) = -(n+1)x\,P_n(x) + (n+1)\,P_{n+1}(x)$.

12. Prove the recurrence for Legendre's polynomials: $x\,P_n'(x) - P_{n-1}'(x) - n\,P_n(x) = 0$.

13. Use the recurrence relation for the Legendre polynomials to prove that for all integers $n \geqslant 0$:

    (a) $P_n(-1) = (-1)^n$;

    (b) $P_n(0) = \begin{cases} 0, & n \text{ is odd}, \\ (-1)^{n/2}\,\frac{1\cdot 3\cdot 5\cdots(n-1)}{2\cdot 4\cdot 6\cdots n}, & n \text{ is even}. \end{cases}$

14. Show that

$$\int_{-1}^{1} P_n(x)\,dx = \begin{cases} 0, & \text{if } n > 0, \\ 2, & \text{if } n = 0. \end{cases}$$

15. Let $w = (x^2-1)^n$, and let $w^{(n)}$ denote the nth derivative of w.

    (a) Use integration by parts to prove that

$$\int_{-1}^{1} w^{(n)} w^{(n)}\,dx = (2n)! \int_{-1}^{1} (1-x^2)^n\,dx.$$

(b) Prove that

$$\int_{-1}^{1} (1-r^2)^n \, dr = \frac{(n!)^2 \, 2^{2n+1}}{(2n)! \, (2n+1)}$$

16. Using the previous result, prove Theorem 12.3.
17. Prove Theorem 12.4.
18. Show that $Q_3(x) = \frac{1}{2} P_3(x) \ln \frac{x+1}{x-1} - \frac{5}{2} x^2 + \frac{2}{3}$.
19. Prove Corollary 12.1 (page 651).
20. Prove Theorem 12.5 (page 651).
21. Obtain the Legendre polynomial from Laplace's integral formula

$$P_n(x) = \frac{1}{\pi} \int_0^{\pi} \left(x + \sqrt{x^2-1} \, \cos\theta \right)^n \, d\theta.$$

22. Find first three coefficients in the expansion (12.4.13) of the function

$$f(x) = \begin{cases} x, & 0 \leqslant x \leqslant 1, \\ 0, & -1 \leqslant x \leqslant 0. \end{cases}$$

23. Prove the Maclaurin expansion for the Legendre polynomial:

$$P_n(x) = \frac{1}{2^n} \sum_{k=0}^{n} \binom{n}{k}^2 (x-1)^{n-k} (x+1)^k = 2^n \sum_{k=0}^{n} x^k \binom{n}{k} \binom{\frac{n-k+1}{2}}{n}.$$

24. Find the first three coefficients in the expansion of the function

$$f(\theta) = \begin{cases} \cos\theta, & 0 \leqslant \theta \leqslant \pi/2, \\ 0, & \pi/2 \leqslant \theta \leqslant \pi \end{cases}$$

in a series of the form

$$f(\theta) = \sum_{n\geqslant 0} a_n \, P_n\left(\cos\theta\right), \qquad 0 \leqslant \theta \leqslant \pi.$$

25. Obtain the Legendre functions of the second kind $Q_0(x)$ and $Q_1(x)$ by means of

$$Q_n(x) = P_n(x) \int \frac{dx}{\left[P_n(x)\right]^2 (1-x^2)}.$$

26. Expand $\arccos x$ into the series with respect to Chebyshev polynomials of the first kind.
27. In each exercise, solve the given initial boundary value problem.

(a) $u_t = u_{xx} + 1 \ (0 < x < 2)$,
$u(0,t) = 1, \quad u(2,t) = 3$,
$u(x,0) = x^2 - x + 1$.

(b) $u_t = u_{xx} + \pi/4 \ (0 < x < \pi/2)$,
$u_x(0,t) = 0, \quad u(\pi/2,t) = 1$,
$u(x,0) = \cos(3x) - \cos 2x$.

(c) $u_t = u_{xx} - x \ (0 < x < 1/2)$,
$u(0,t) = 0, \quad u_x(1/2,t) = 1$,
$u(x,0) = \sin(\pi x)$.

(d) $u_t = u_{xx} + e^{-t} \ (0 < x < \pi)$,
$u_x(0,t) = 1, \quad u_x(\pi,t) = -1$,
$u(x,0) = \sin x$.

28. Consider a vertically hanging string of length ℓ subject to the horizontal force with harmonically density distribution $F(x,t) = A \sin\omega t$ per unit length. Let $u(x,t)$ be the horizontal displacement of the spring from the vertical equilibrium position at the point x and time t. Then $u(x,t)$ is a solution of the following partial differential equation

$$\frac{\partial^2 u}{\partial t^2} = c^2 \frac{\partial}{\partial x}\left(x \, \frac{\partial u}{\partial x} \right) + F(x,t)/\rho, \qquad u(0,t) < \infty, \quad u(\ell,t) = 0,$$

where the density of the string is assumed to be $\rho = 1$. Find $u(x,t)$ assuming that $u(x,0) = \dot{u}(x,0) = 0$.

12.5 Nonhomogeneous Boundary Value Problems

In this section, we introduce nonhomogeneous boundary value problems. This topic is very important in applications and covered thoroughly in many books. Previously in §11.2, we showed how to solve nonhomogeneous equations using eigenvalue expansion (see also Problems 1 and 2 on page 665). We start with the following statement, known as the **Fredholm**[122] **alternative theorem**.

Theorem 12.6: For a given value of μ, either the nonhomogeneous problem

$$L[x,\mathtt{D}]y = \mu\rho(x)\,y(x) + f(x), \qquad B_0[y]=0, \quad B_\ell[y]=0, \tag{12.5.1}$$

where $L[x,\mathtt{D}] = -\mathtt{D}p(x)\mathtt{D} + q(x)$, $\mathtt{D} \stackrel{\text{def}}{=} \mathrm{d}/\mathrm{d}x$, and the boundary operators are $B_0[y] = \alpha_0 y(0) - \alpha_1 y'(0)$, $B_\ell[y] = \beta_0 y(\ell) + \beta_1 y'(\ell)$, has a unique solution for each smooth function $f(x)$ from the domain of $L[x,\mathtt{D}]$ (if μ is not an eigenvalue of the corresponding homogeneous Sturm–Liouville boundary value problem), or else the homogeneous problem has a nontrivial solution. If $\mu = \lambda_m$ is the eigenvalue, the nonhomogeneous boundary value problem (12.5.1) has no solutions unless f is orthogonal to $\phi_m(x)$.

PROOF: We assume that the solution $y = \phi(x)$ of the nonhomogeneous problem (12.5.1) admits eigenfunction expansion:

$$\phi(x) = \sum_{n\geqslant 1} a_n \phi_n(x), \tag{12.5.2}$$

with respect to known set $\{\phi_n(x)\}_{n\geqslant 1}$ of all normalized eigenfunctions $\left(\text{so } \|\phi_n\|^2 = \int_0^\ell \rho(x)\,\phi_n^2(x)\,\mathrm{d}x = 1\right)$ corresponding to distinct eigenvalues $\lambda_1 < \lambda_2 < \cdots < \lambda_n < \cdots$ for the homogeneous Sturm–Liouville boundary value problem: $L[x,\mathtt{D}]y = \lambda\rho(x)\,y(x)$, $B_0[y] = B_\ell[y] = 0$. Substituting the series (12.5.2) into the differential expression $L[x,\mathtt{D}]\phi = L[\phi](x)$ and using the equation $L[\phi_n](x) = \lambda_n \rho(x)\,\phi_n(x)$, we obtain

$$L[\phi] = L\left[\sum a_n\phi_n\right](x) = \sum_{n\geqslant 1} a_n\, L[\phi_n](x) = \rho(x)\sum_{n\geqslant 1} a_n\lambda_n\,\phi_n(x),$$

where the interchange of summation and differentiation is assumed to be justified. We substitute this series into the differential equation (12.5.1):

$$\sum_{n\geqslant 1} a_n\lambda_n\rho(x)\,\phi_n(x) = \mu\rho(x)\sum_{n\geqslant 1} a_n\phi_n(x) + f(x).$$

Expanding $f(x)/\rho(x)$ into the series with respect to eigenfunctions, we get

$$\frac{f(x)}{\rho(x)} = \sum_{n=1}^{\infty} c_n\phi_n(x),$$

where

$$c_n = \int_0^\ell \frac{f(x)}{\rho(x)}\,\rho(x)\,\phi_n(x)\,\mathrm{d}x = \int_0^\ell f(x)\,\phi_n(x)\,\mathrm{d}x, \qquad n = 1,2,\ldots. \tag{12.5.3}$$

After substituting the series for $\phi(x)$, $L[\phi](x)$, and $f(x)$, we find that

$$\sum_{n\geqslant 1} a_n\lambda_n\rho(x)\,\phi_n(x) = \mu\rho(x)\sum_{n\geqslant 1} a_n\phi_n(x) + \rho(x)\sum_{n\geqslant 1} c_n\phi_n(x).$$

Upon collecting terms and canceling the common nonzero factor $\rho(x)$, we obtain

$$\sum_{n=1}^{\infty}\left[(\lambda_n - \mu)\,a_n - c_n\right]\phi_n(x) = 0. \tag{12.5.4}$$

[122]The Swedish mathematician Erik Ivar Fredholm (1866–1927) is best remembered for his work on integral equations and spectral theory.

If Eq. (12.5.4) is to hold for each x in the interval $[0, \ell]$, then the coefficient of $\phi_n(x)$ must be zero for each n:

$$(\lambda_n - \mu)\, a_n = c_n \qquad \Longrightarrow \qquad a_n = \frac{c_n}{\lambda_n - \mu} \ (\mu \ne \lambda_n), \qquad n = 1, 2, 3, \ldots. \tag{12.5.5}$$

Therefore, if $\mu \ne \lambda_n$ for $n = 1, 2, 3, \ldots$, the solution becomes

$$y = \phi(x) = \sum_{n \geqslant 1} \frac{c_n}{\lambda_n - \mu}\, \phi_n(x), \tag{12.5.6}$$

where the coefficients c_n are determined from Eq. (12.5.3). While we did not prove that the series (12.5.6) converges uniformly to a smooth function that possesses two continuous derivatives, it can be done with even less stringent conditions on the forcing term $f(x)$. Thus we obtain a formal solution, and it is reasonable to expect that the series (12.5.6) pointwise converges.

Now suppose that μ is equal to one of the eigenvalues of the corresponding homogeneous problem, say, $\mu = \lambda_m$. In this case, the coefficient of ϕ_m in the expansion (12.5.4) becomes $-c_m \phi_m(x)$, which forces us to assume that $c_m = 0$. Again, we must consider two cases.

In the event that $\mu = \lambda_m$ and $c_m \ne 0$, there is no value of a_m that satisfies Eq. (12.5.5), and therefore the nonhomogeneous problem Eq. (12.5.1) has no solution.

When $\mu = \lambda_m$ and $c_m = \int_0^\ell f(x)\, \phi_m(x)\, \mathrm{d}x = 0$, Eq. (12.5.5) is satisfied regardless of the value of a_m; in other words, a_m remains arbitrary. In this case, the nonhomogeneous two-point boundary value problem (12.5.1) has infinitely many solutions.

Example 12.5.1: Solve the boundary value problem

$$y'' + 4y = -x^2, \qquad y(0) - y'(0) = 0, \quad y(1) = 0.$$

Solution. We seek the solution as a series (12.5.2) with respect to the set of eigenfunctions $\{\phi_n(x)\}$ of the corresponding homogeneous Sturm–Liouville problem:

$$y'' + \lambda y = 0, \qquad y(0) - y'(0) = 0, \quad y(1) = 0.$$

Since the general solution of the equation $y'' + \lambda y = 0$ is $y = c_1 \cos\left(\sqrt{\lambda} x\right) + c_2 \sin\left(\sqrt{\lambda} x\right)$, with arbitrary constants $c_{1,2}$, the given boundary conditions yield

$$c_1 - \sqrt{\lambda}\, c_2 = 0, \qquad c_1 \cos\left(\sqrt{\lambda}\right) + c_2 \sin\left(\sqrt{\lambda}\right) = 0.$$

In order for the Sturm–Liouville boundary value problem, $y'' + \lambda y = 0$ and $y(0) - y'(0) = 0, \quad y(1) = 0$, to have a nontrivial solution, the parameter $\mu = \sqrt{\lambda}$ must be a solution of the following transcendent equation:

$$\mu \cos \mu + \sin \mu = 0. \tag{12.5.7}$$

This equation has infinite many roots, μ_n, $n = 1, 2, \ldots$. The first few eigenvalues can be found numerically,

$$\lambda_1 = (\mu_1)^2 \approx (2.02876)^2 \approx 4.115858365694522, \quad \lambda_2 \approx 24.1393, \quad \lambda_3 \approx 63.66.$$

For large n, their values are approximately

$$\lambda_n \approx \frac{(2n-1)^2 \pi^2}{4} \quad \text{for } n = 4, 5, 6, \ldots.$$

Assuming that the solution is given by Eq. (12.5.2)

$$y = \sum_{n \geqslant 1} a_n \phi_n(x), \qquad \phi_n(x) = \sqrt{\lambda_n}\, \cos\left(\sqrt{\lambda_n} x\right) + \sin\left(\sqrt{\lambda_n} x\right),$$

we find its coefficients from Eq. (12.5.3):

$$a_n = \frac{c_n}{\lambda_n - 4},$$

where the c_n are expansion coefficients of the forcing term $f(x) = x^2$:

$$c_n = \frac{1}{\|\phi_n\|^2} \int_0^1 x^2\, \phi_n(x)\, \mathrm{d}x = \frac{4}{\lambda_n} \cdot \frac{\lambda_n^{3/2} \sin\sqrt{\lambda_n} - 2 + (2 + \lambda_n)\cos\sqrt{\lambda_n}}{2\sqrt{\lambda_n}\,(2 + \lambda_n) - 2\sqrt{\lambda_n}\, \cos 2\sqrt{\lambda_n} + (\lambda_n - 1)\sin 2\sqrt{\lambda_n}}.$$

Example 12.5.2: (Forced vibrations) Consider a mass m attached to a coil spring of length ℓ_0, the upper end of which is securely fastened (see Fig. 12.5). The mass causes an elongation ℓ of the spring in the downward (positive) direction. At that point the mass remains at rest. There are two forces acting at the point where the mass is attached: the gravitational force, or weight of the mass acting downward, and the spring force that acts upward. The gravitational force has magnitude mg, where g is the acceleration due to gravity. The spring force is proportional to the elongation ℓ: $F_s = -\kappa\ell$, which is known as Hooke's law. The constant of proportionality κ is called the spring constant. Since the mass is in equilibrium, we have

$$mg - \kappa\ell = 0.$$

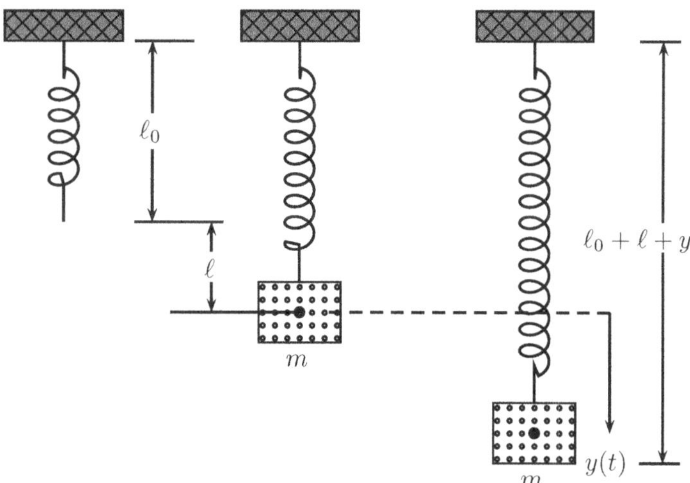

Figure 12.5: A spring-mass system.

We will now assume that the spring-mass system is also subject to an external periodic force $A\,\sin\omega t$. In this case, the mass will undergo forced vibrations. Let $y(t)$ measures positive downward displacement of the mass from its equilibrium position at time t. To describe the future movement of the mass, we assume that it moves along a vertical line through the center of gravity, and its direction is always from the mass toward the point of equilibrium.

Newton's second law of motion states that the force F acting on this particle moving with varying velocity v is equal to the time rate of change of the momentum mv:

$$F = \frac{\mathrm{d}(mv)}{\mathrm{d}t} = m\,\dot{v}, \qquad \text{where} \quad v = \dot{y}$$

is the velocity. Equating the two forces and applying Hooke's law and the external force, we get

$$m\,\frac{\mathrm{d}^2 y}{\mathrm{d}t^2} = -\kappa y + f(t) \qquad \Longleftrightarrow \qquad m\,\ddot{y} + \kappa\,y = A\,\sin\omega t.$$

For simplicity, we ignore friction and air resistance. We again suppose that the mass is at its equilibrium position initially, and when $t = T$ seconds, and seek a solution to the nonhomogeneous equation subject to the Dirichlet boundary conditions:

$$m\,\ddot{y} + \kappa\,y = A\,\sin\omega t, \qquad y(0) = y(T) = 0. \tag{12.5.8}$$

Following the paradigm we have set up for solving nonhomogeneous boundary value problems, we consider the accompanied Sturm–Liouville problem:

$$m\,\phi'' + \kappa\,y = -\lambda\,\phi, \qquad \phi(0) = \phi(T) = 0.$$

Since the differential operator $m\,\mathtt{D}^2 + \kappa$ in the left-hand side is negative, we assign the right-hand side to $-\lambda\,\phi$. The eigenvalues are obtained by setting $\sqrt{(\kappa+\lambda)/m}$ equal to a multiple of π/T; hence

$$\lambda_n = m\left(\frac{n\pi}{T}\right)^2 - \kappa, \qquad n = 1, 2, 3, \ldots.$$

The corresponding eigenfunctions have the form

$$\phi_n(t) = \sin \frac{n\pi t}{T}, \qquad n = 1, 2, 3, \ldots .$$

Expanding the forcing term into the Fourier series, we get

$$A \sin \omega t = \sum_{n \geqslant 1} f_n \frac{n\pi t}{T},$$

where the Fourier coefficients are obtained according to the Euler–Fourier formulas (10.5.2), page 582:

$$f_n = \frac{2A}{T} \int_0^T \sin \omega t \, \sin \frac{n\pi t}{T} \, \mathrm{d}t = A \times \begin{cases} \dfrac{2n\pi(-1)^n \sin(\omega T)}{\omega^2 T^2 - n^2\pi^2}, & \text{if } T\omega \neq n\pi, \\ 1, & \text{if } T\omega = k\pi \text{ for some } k. \end{cases}$$

We seek a solution of the given boundary value problem in the form of infinite series:

$$y(t) = \sum_{n \geqslant 1} c_n \sin \frac{n\pi t}{T}.$$

This series satisfies the homogeneous boundary conditions because every eigenfunction $\phi_n(t)$ does. Now we substitute these series into the differential equation $\left(m\,\mathtt{D}^2 + \kappa\right) y = f$ to obtain

$$-m \sum_{n \geqslant 1} c_n \left(\frac{n\pi}{T}\right)^2 \sin \frac{n\pi t}{T} + \kappa \sum_{n \geqslant 1} c_n \sin \frac{n\pi t}{T} = \sum_{n \geqslant 1} f_n \sin \frac{n\pi t}{T}.$$

Uniting these three series into one, we have

$$\sum_{n \geqslant 1} \left[-c_n\, m \left(\frac{n\pi}{T}\right)^2 + \kappa\, c_n - f_n \right] \sin \frac{n\pi t}{T} = 0.$$

Due to uniqueness of Fourier series, we equate every coefficient to zero and get

$$-c_n\, m \left(\frac{n\pi}{T}\right)^2 + \kappa\, c_n - f_n = 0 \qquad \Longleftrightarrow \qquad c_n = \frac{f_n T^2}{T^2 \kappa - m n^2 \pi^2} = \frac{A T^2 2n\pi(-1)^n \sin(\omega T)}{T^2\kappa - m n^2 \pi^2},$$

if $T\omega \neq n\pi$, for all $n = 1, 2, 3, \ldots$. When $T\omega = k\pi$ for some k, we get the particular solution

$$y(t) = \frac{A\,T^2}{T^2\kappa - k^2\pi^2 m} \sin\left(\frac{k\pi t}{T}\right),$$

provided that $T^2\kappa \neq k^2\pi^2 m$. If $T^2\kappa = k^2\pi^2 m$, the boundary value problem has no solution.

Example 12.5.3: Consider a physical problem that serves as a crude model of a centrifuge. Suppose that we have a horizontally mounted tube of length ℓ that is rotated about a fixed pivot with a constant angular acceleration. A particle having mass m is initially injected with velocity v_0 at the pivot point of the tube. It can slide freely in a rotated tube. The particle migrates radially outward and its motion within the frictionless tube can be described in terms of polar coordinates r (distance from the pivot) and θ (angle).

Since the particle location is $\mathbf{r} = r\mathbf{e}_r$, its velocity vector becomes

$$\frac{\mathrm{d}}{\mathrm{d}t}\mathbf{r} = \frac{\mathrm{d}}{\mathrm{d}t}(r\mathbf{e}_r) = \dot{r}\mathbf{e}_r + r\,\dot{\theta}\,\mathbf{e}_\theta.$$

Using equations (12.5.9), we find the acceleration vector to be

$$\ddot{\mathbf{r}} = \left(\ddot{r} - r\,\dot{\theta}^2\right)\mathbf{e}_r + \left(r\,\ddot{\theta} + 2\dot{r}\dot{\theta}\right)\mathbf{e}_\theta.$$

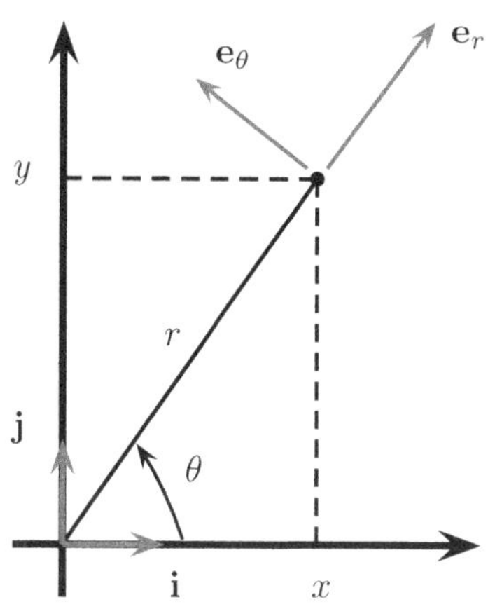

In polar coordinates (r, θ), the rectangular projections are $x = r\cos\theta$ and $y = r\sin\theta$, where both r and θ vary as functions of time t. As shown in the figure at left, the unit vectors in radial and angular directions are

$$\mathbf{e}_r = (\cos\theta)\mathbf{i} + (\sin\theta)\mathbf{j}, \qquad \mathbf{e}_\theta = -(\sin\theta)\mathbf{i} + (\cos\theta)\mathbf{j}. \tag{12.5.9}$$

In contrast to the unit vectors $\mathbf{i}$ and $\mathbf{j}$, along the abscissa and ordinate, respectively, the unit vectors $\mathbf{e}_r(t)$ and $\mathbf{e}_\theta(t)$ vary with time. They remain mutually perpendicular and of unit length, but they change direction. Their differentiation yields

$$\frac{\mathrm{d}}{\mathrm{d}t}\,\mathbf{e}_r = -\dot{\theta}(\sin\theta)\mathbf{i} + \dot{\theta}(\cos\theta)\mathbf{j} = \dot{\theta}\mathbf{e}_\theta, \quad \frac{\mathrm{d}}{\mathrm{d}t}\,\mathbf{e}_\theta = -\dot{\theta}\mathbf{e}_r.$$

Applying Newton's second law of motion in the radial direction, we obtain the following differential equation

$$\ddot{r}(t) - r(t)\left(\dot{\theta}\right)^2 = f(t),$$

where $f(t)$ is an external force acting on the particle (we assume for simplicity that the particle has unit mass). If the tube begins to rotate from rest with a constant angular acceleration, then $\dot{\theta} = \alpha t$, for some constant α. In this case the radial equation of motion becomes

$$\ddot{r}(t) - r(t)\,(\alpha t)^2 = f(t).$$

Suppose we are interested in finding the initial velocity $v_0 = \dot{r}(0)$ of the particle that lead the particle to exit the tube at some prescribed later time $t = T$. This gives us the boundary conditions $r(0) = 0$, $r(T) = \ell$ and finally the boundary value problem:

$$\ddot{r}(t) - r(t)\,(\alpha t)^2 = f(t), \qquad r(0) = 0, \quad r(T) = \ell. \tag{12.5.10}$$

To reduce the problem (12.5.10) to one with homogeneous boundary conditions, we choose a function $v(t)$ that satisfies the given boundary conditions, $v(0) = 0$ and $v(T) = \ell$. While there exist infinite many such functions, a particular choice does not effect the final answer because the boundary value problem (12.5.10) has a unique solution. So we set $v(t) = t\ell/T$. Then, subtracting $v(t)$ from $r(t)$, we get the difference $y(t) = r(t) - v(t)$ that is a solution of a similar problem with homogeneous boundary conditions

$$\ddot{y} - y(t)\,\alpha^2 t^2 = g(t), \qquad y(0) = 0, \quad y(T) = 0, \tag{12.5.11}$$

where $g(t) = f(t) + \alpha^2 \ell t^3/T$ is a known function. The two-point boundary value problem (12.5.11) has the solution presented in quadrature form

$$y(t) = \int_0^T G(t,s)\,g(s)\,\mathrm{d}s,$$

where $G(t,s)$ is the corresponding Green's function. The homogeneous equation $\ddot{r}(t) - r(t)\,(\alpha t)^2 = 0$ has two linearly independent solutions

$$r_1(t) = \sqrt{t}\,I_{1/4}\left(\frac{\alpha t^2}{2}\right), \qquad r_2(t) = \sqrt{t}\,K_{1/4}\left(\frac{\alpha t^2}{2}\right), \tag{12.5.12}$$

where $I_{1/4}(z)$ and $K_{1/4}(z)$ are modified Bessel functions of order $1/4$ (see §4.9.3, page 250). To use the explicit formula (12.1.8) on page 632, we need to determine two linearly independent solutions $\phi(t)$ and $\psi(t)$ of the homogeneous equation $\ddot{y} - y(t)\,\alpha^2 t^2 = 0$ that satisfy the boundary conditions

$$\phi(0) = 0 \qquad \text{and} \qquad \psi(T) = 0,$$

respectively. Since $\phi(t) = r_1(t) = \sqrt{t}\,I_{1/4}\left(\frac{\alpha t^2}{2}\right)$ is already known, we construct $\psi(t)$ as a linear combination of functions (12.5.12):

$$\psi(t) = r_2(t) - c\,r_1(t) = \sqrt{t}\,K_{1/4}\left(\frac{\alpha t^2}{2}\right) - c\sqrt{t}\,I_{1/4}\left(\frac{\alpha t^2}{2}\right).$$

From the boundary condition $\psi(T) = 0$, it follows that

$$c = r_2(T)/r_1(T) = K_{1/4}\left(\frac{\alpha T^2}{2}\right)/I_{1/4}\left(\frac{\alpha T^2}{2}\right), \qquad I_{1/4}\left(\frac{\alpha T^2}{2}\right) \neq 0.$$

Since their Wronskian is $W[\phi(t), \psi(t)] = -2$, we can use formula (12.1.8), page 632, to construct the Green's function explicitly.

Example 12.5.4: Consider a heat conduction problem for a straight bar of uniform cross section and homogeneous material. Let the x-axis be chosen to lie along the axis of the bar and let $x = 0$ and $x = \ell$ denote the end points of the bar. Suppose further that the temperature $u(x,t)$ at the end points $x = 0$ and $x = \ell$ is maintained fixed. Then the temperature distribution within the bar is a solution of the following initial boundary value problem:

$$\frac{\partial u}{\partial t} = c^2 \frac{\partial^2 u}{\partial x^2}, \qquad u(x=0) = T_0, \quad u(x=\ell) = T_\ell, \quad u(t=0) = f(x).$$

Applying the Laplace transform, we get the boundary value problem for the ordinary differential equation

$$\left(\lambda - c^2 \frac{\mathrm{d}^2}{\mathrm{d}x^2}\right) u^L = f(x), \quad 0 < x < \ell, \quad u^L(0) = T_0/\lambda, \quad u^L(\ell) = T_\ell/\lambda,$$

where $u^L(x,\lambda)$ is the Laplace transform of the unknown function $u(x,t)$. Its solution is

$$u^L(x,\lambda) = \frac{T_0}{\lambda}\cosh\left(\frac{x}{c}\sqrt{\lambda}\right) + \left(\frac{T_\ell}{\lambda} - \frac{T_0}{\lambda}\cosh\left(\frac{\ell}{c}\sqrt{\lambda}\right)\right)\left(\sinh\left(\frac{\ell}{c}\sqrt{\lambda}\right)\right)^{-1}\sinh\left(\frac{x}{c}\sqrt{\lambda}\right).$$

Here $\sinh z = \frac{1}{2}e^z - \frac{1}{2}e^{-z}$ and $\cosh z = \frac{1}{2}e^z + \frac{1}{2}e^{-z}$ are hyperbolic functions. Using properties of hyperbolic functions, we rewrite the Laplace transform $u^L(x,\lambda)$ in the more convenient form:

$$u^L(x,\lambda) = \frac{T_\ell}{\lambda}\,\frac{\sinh\left(\frac{x}{c}\sqrt{\lambda}\right)}{\sinh\left(\frac{\ell}{c}\sqrt{\lambda}\right)} - \frac{T_0}{\lambda}\,\frac{\sinh\left(\frac{x-\ell}{c}\sqrt{\lambda}\right)}{\sinh\left(\frac{\ell}{c}\sqrt{\lambda}\right)}. \tag{12.5.13}$$

Therefore, we need to find the inverse Laplace transform of the function:

$$\Phi(t,a,b) = \mathcal{L}^{-1}\left[\frac{\sinh a\sqrt{\lambda}}{\lambda \sinh b\sqrt{\lambda}}\right] = \mathcal{L}^{-1}\left[\frac{e^{(a-b)\sqrt{\lambda}} - e^{-(a+b)\sqrt{\lambda}}}{\lambda\left(1 - e^{-2b\sqrt{\lambda}}\right)}\right] \qquad (a < b),$$

depending on two parameters a, b. Using the sum of geometric series $(1-q)^{-1} = \sum_{k\geqslant 0} q^k$ with $q = e^{-2b\sqrt{\lambda}}$, we represent the required function $\Phi(t,a,b)$ as the series:

$$\Phi(t,a,b) = \mathcal{L}^{-1}\left[\frac{1}{\lambda}e^{-(b-a)\sqrt{\lambda}}\sum_{k\geqslant 0} e^{-2kb\sqrt{\lambda}} - \frac{1}{\lambda}e^{-(a+b)\sqrt{\lambda}}\sum_{k\geqslant 0} e^{-2kb\sqrt{\lambda}}\right] \qquad (a < b).$$

The inverse Laplace transform of the basic element in this series is represented via the error function:

$$\mathcal{L}^{-1}\left[\frac{1}{\lambda}e^{-\alpha\sqrt{\lambda}}\right] = 1 - \operatorname{erf}\left(\frac{\alpha}{2\sqrt{t}}\right) = \operatorname{Erf}\left(\frac{\alpha}{2\sqrt{t}}\right),$$

where $\operatorname{Erf}(x) = \frac{2}{\sqrt{\pi}}\int_x^\infty e^{-x^2}\,\mathrm{d}x = 1 - \operatorname{erf}(x)$ is the complementary error function (also commonly denoted as 'erfc'). This allows us to find its "explicit" expression:

$$\Phi(t,a,b) = \sum_{k\geqslant 0}\operatorname{Erf}\left(\frac{2kb - a + b}{2\sqrt{t}}\right) - \sum_{k\geqslant 0}\operatorname{Erf}\left(\frac{2kb + a + b}{2\sqrt{t}}\right) \qquad (a < b).$$

Then the solution of the given initial boundary value problem becomes

$$u(x,t) = T_\ell\,\Phi\left(t, \frac{x}{c}, \frac{\ell}{c}\right) - T_0\,\Phi\left(t, \frac{x-\ell}{c}, \frac{\ell}{c}\right).$$

Problems

1. Solve the Dirichlet problem for the unit disc $\{(x,y) : x^2+y^2 \leqslant 1\}$

$$\frac{\partial^2 u(r,\theta)}{\partial r^2} + \frac{1}{r}\frac{\partial u(r,\theta)}{\partial r} + \frac{1}{r^2}\frac{\partial^2 u(r,\theta)}{\partial \theta^2} = 0, \qquad u(1,\theta) = f(\theta),$$

where the boundary function $f(\theta)$ is defined by

(a) $f(\theta) = \theta + |\theta|$;
(b) $f(\theta) = \cos(\theta/2)$;
(c) $f(\theta) = \theta^2$;
(d) $f(\theta) = \sin^2\theta$.

2. Show that the Dirichlet problem for the disc $\{(x,y) : x^2+y^2 \leqslant a^2\}$ of radius a:

$$\nabla^2 u(r,\theta) = 0, \qquad u(a,\theta) = f(\theta),$$

where $f(\theta)$ is a given function, has the solution

$$u(r,\theta) = \frac{1}{2\pi}\int_{-\pi}^{\pi} \frac{a^2-r^2}{a^2 - 2ar\cos(\theta-\vartheta) + r^2}\, f(\vartheta)\,\mathrm{d}\vartheta.$$

The above formula is known as the Poisson integral.

3. Solve each of the given problems by means of an eigenfunction expansion.

(a) $y'' + y = x^2,\ \ y(0) = y(2) = 0$;
(b) $y'' + 2y = -x^3,\ \ y(0) = y'(\pi) = 0$;
(c) $y'' + 3y = x(3-x),\ \ y'(0) = y'(\pi) = 0$;
(d) $y'' + 4y = x,\ \ y(0) = y'(1) + y(1) = 0$.

4. In each exercise, determine a formal eigenfunction series expansion for the solution of the given problem. State the value of μ for which the solution exists. Assume that the problem has a unique solution for given function $f(x)$.

(a) $y'' + \mu\, y = -f,\ y'(0) = y\left(\frac{\pi}{2}\right) = 0$;
(b) $y'' + \mu\, y = -f,\ y(0) = y'\left(\frac{\pi}{2}\right) = 0$;
(c) $y'' + \mu\, y = -f,\ y'(0) = y'(\pi) = 0$;
(d) $y'' + \mu\, y = -f,\ y(0) = y'(1) + y(1) = 0$.

5. In each exercise, determine whether there is any value of the constant c for which the problem has a solution. Find the solution for each such case.

(a) $y'' + \pi^2 y = c - x + x^2,\ \ y(0) = y(1) = 0$;
(b) $y'' + 4\pi^2 y = c - x + x^2,\ y(0) = y(1) = 0$;
(c) $y'' + \pi^2 y = c + x,\ \ y(0) = y'\left(\frac{1}{2}\right) = 0$;
(d) $y'' + 4\,y = c,\ \ y'(0) = y'(\pi) = 0$.

6. The steady temperature distribution $u(r,\theta)$ in spherical coordinates not depending on azimuthal angle satisfies the equation $\nabla^2 u = 0$, where

$$\nabla^2 = \frac{1}{r^2}\frac{\partial}{\partial r}\left(r^2\frac{\partial}{\partial r}\right) + \frac{1}{r^2\sin\theta}\frac{\partial}{\partial\theta}\left(\sin\theta\,\frac{\partial}{\partial\theta}\right).$$

Show that its solution, which is regular at $\theta = 0, \pi$, is

$$u(r,\theta) = \sum_{n\geqslant 0}\left(A_n r^n + B_n r^{-n-1}\right) P_n(\cos\theta),$$

for some constants A_n and B_n.

7. Show that the azimuthal symmetric solution of the inner Dirichlet problem for sphere of radius $r = a$

$$\nabla^2 u = 0 \quad (r < a), \qquad u(a,\theta) = f(\theta),$$

where $f(\theta)$ is a given function, is

$$u(r,\theta) = \sum_{n\geqslant 0}\left(n+\frac{1}{2}\right)\left(\frac{r}{a}\right)^n P_n(\cos\theta)\int_0^{\pi} f(\vartheta)\,P_n(\cos\vartheta)\,\sin\vartheta\,\mathrm{d}\vartheta.$$

8. Consider the steady state temperature distribution $u(x, y)$ in a rectangular slab $0 \leqslant x \leqslant 2$, $0 \leqslant y \leqslant 1$. The left and bottom edges are in direct contact a heat sink maintained at zero degrees, but the right edge is partially shielded from the sink through leaky insulation, giving rise to the boundary condition of the third kind: $\frac{\partial u}{\partial x} + \kappa u = 0$, for some given positive constant κ. The temperature of the upper edge is maintained at the generic prescribed temperature $u(x, 1) = f(x)$. Solve the corresponding boundary value problem for the Laplace equation $\nabla^2 u = 0$ inside the slab.

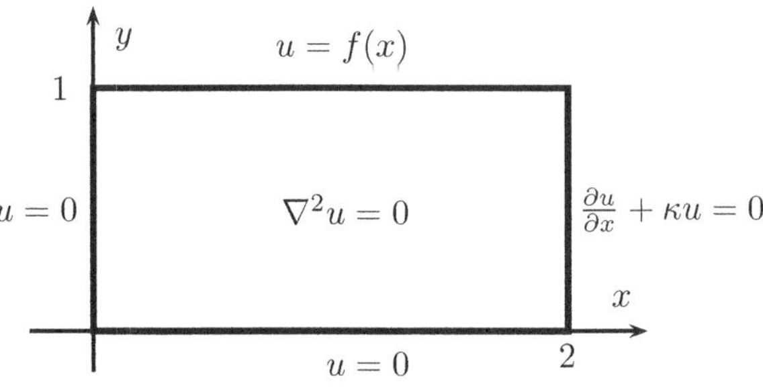

Summary for Chapter 12

1. Consider the linear two-point boundary value problem for self-adjoint (positive if $q(x) > 0$) differential operator $L[x, \texttt{D}] = -\texttt{D}p(x)\texttt{D} + q(x)$, depending on the derivative operator $\texttt{D} = \mathrm{d}/\mathrm{d}x$:

$$-\frac{\mathrm{d}}{\mathrm{d}x}\left[p(x)\,\frac{\mathrm{d}y}{\mathrm{d}x}\right] + q(x)\,y = f(x), \qquad \alpha_0 y(0) - \alpha_1 y'(0) = 0, \quad \beta_0 y(\ell) + \beta_1 y'(\ell) = 0, \tag{12.1.4}$$

on the interval $0 < x < \ell$. The functions $p(x) > 0$, $p'(x)$, $q(x)$, and $f(x)$ are assumed to be continuous on the closed interval $[0, \ell]$, and

$$|\alpha_0| + |\alpha_1| > 0, \qquad |\beta_0| + |\beta_1| > 0.$$

It is convenient to introduce two boundary operators

$$B_0[y] = \alpha_0 y(0) - \alpha_1 y'(0) \qquad \text{and} \qquad B_\ell[y] = \beta_0 y(\ell) + \beta_1 y'(\ell).$$

Then the boundary value problem can be written in compact form:

$$L[x, \texttt{D}]y = f, \qquad B_0[y] = 0, \quad B_\ell[y] = 0. \tag{12.1.5}$$

2. Let $\phi(x)$ and $\psi(x)$ be two linearly independent solutions to the homogeneous equation $L[x, \texttt{D}]y = 0$. If the boundary value problem $L[x, \texttt{D}]y = 0$, $B_0[y] = 0$, $B_\ell[y] = 0$ has only trivial solution, then the nonhomogeneous equation $L[x, \texttt{D}]y = f$ has a solution expressed in quadrature form

$$y(x) = \int_0^\ell G(x, t)\, f(t)\,\mathrm{d}t, \tag{12.1.3}$$

where the kernel $G(x, t)$,

$$G(x, t) = \begin{cases} \dfrac{-\phi(t)\psi(x)}{p(t)\,W[\phi, \psi](t)}, & 0 \leqslant t \leqslant x, \\ \dfrac{-\phi(x)\psi(t)}{p(t)\,W[\phi, \psi](t)}, & x \leqslant t \leqslant \ell, \end{cases} \tag{12.1.8}$$

is called the **Green's function**. The denominator $p(t)\,W[\phi, \psi](t)$ in Eq. (12.1.8) is a constant for the self-adjoint differential operator $L[x, \texttt{D}] = -\texttt{D}p(x)\texttt{D} + q(x)$.

3. The boundary value problem (12.1.4) can be reformulated in vector form:

$$\mathbf{y}'(x) = \mathbf{P}(x)\mathbf{y}(x) + \mathbf{f}(x), \qquad \mathbf{B}^{[0]}\mathbf{y}(0) + \mathbf{B}^{[\ell]}\mathbf{y}(\ell) = \boldsymbol{\alpha}, \tag{12.2.1}$$

where $\mathbf{P}(x)$ is an $(n \times n)$ continuous matrix function, $\mathbf{B}^{[0]}$ and $\mathbf{B}^{[\ell]}$ are given constant $(n \times n)$ matrices, and $\mathbf{f}(x)$ and $\boldsymbol{\alpha}$ are n-column vectors.

4. The Sturm–Liouville boundary value problem

$$L[x, \texttt{D}]y = \lambda\rho(x)\,y \quad a < x < b, \quad \alpha_0 y(a) - \beta_0 y'(a) = 0, \qquad \alpha_1 y(b) + \beta_1 y'(b) = 0,$$

is said to be regular if the functions $p(x)$, $p'(x)$, $q(x)$, $1/p(x)$, and $\rho(x)$ are continuous, and $p(x) > 0$, $\rho(x) > 0$ on the closed interval $[a, b]$.

5. When one or both of the boundary points in a Sturm–Liouville problem goes to $\pm\infty$, or when the coefficient $1/p(x)$ in the differential operator (12.3.1) diverges at an endpoint, the problem becomes singular.

6. The **Fredholm alternative theorem:** For a given value of μ, either the nonhomogeneous problem

$$L[x, \texttt{D}]y = \mu\rho(x)\,y(x) + f(x), \qquad B_0[y] = 0, \quad B_\ell[y] = 0, \tag{12.5.1}$$

where $L[x, \texttt{D}] = -\texttt{D}p(x)\texttt{D} + q(x)$ and the boundary operators are $B_0[y] = \alpha_0 y(0) - \alpha_1 y'(0)$, $B_\ell[y] = \beta_0 y(\ell) + \beta_1 y'(\ell)$, has a unique solution for each continuous function $f(x)$ (if μ is not an eigenvalue of the corresponding homogeneous Sturm–Liouville boundary value problem), or else the homogeneous problem has a nontrivial solution.

Review Questions for Chapter 12

Section 12.1

1. Derive the Green's function for the given two-point boundary value problem.

   (a) $-y'' = f(x), \quad y(0) - y'(0) = 0, \; y(1) + y'(1) = 0.$
   (b) $-y'' - y = f(x), \quad y'(0) = 0, \; y(\pi) + y'(\pi) = 0.$
   (c) $-y'' + y = f(x), \quad y(0) = 0, \; y'(1) = 0.$
   (d) $-y'' - y = f(x), \quad y(0) = y'(0), \; y(\pi) = 0.$
   (e) $\left(x^{-2}y'(x)\right)' + 2x^{-4}y(x) = 0, \quad y(-1) = 0, \; y(1) = f(x).$

2. Find the Green's function for the given differential operator $L[x, \mathtt{D}]$, where $\mathtt{D} = \mathrm{d}/\mathrm{d}x$.

   (a) $\mathtt{D}\, e^{-3x}\mathtt{D} + 2\, e^{-3x}$;
   (b) $\mathtt{D}\, x^3\mathtt{D} + 1/16$;
   (c) $\mathtt{D}\, x\, \mathtt{D} - 4/x$;
   (d) $\mathtt{D}\,(x+1)^2\mathtt{D}$.

3. Find the Green's function for the nonself-adjoint two-point boundary value problem, given two linearly independent solutions for the corresponding homogeneous equation.

   (a) $(x-1)^2\, y'' - 2(x-1)\, y' + 2y = f(x), \quad y'(-1) = 0, \; y(0) = 0$, given $y_1 = x - 1$ and $y_2 = x^2 - 1$.
   (b) $(x \sin x + \cos x)\, y'' - x\, \cos x\, y' + y\, \cos x = f(x), \quad y(0) = 0, \; y(\pi) = 0$, given $y_1 = x$, $y_2 = \cos x$.

4. Show that the eigenvalues of the Sturm–Liouville two-point boundary value problem (see Eq. (12.1.4) on page 631),

$$L[x, \mathtt{D}]y = \lambda \rho(x)\, y(x), \qquad B_0[y] = 0, \quad B_\ell[y] = 0,$$

   are positive, provided that $\rho(x) > 0$, $q(x) \geqslant 0$ and the coefficients in the boundary operators B_0 and B_ℓ are not negative.

5. In each exercise, the Green's function is given for the boundary value problem (12.1.4):

$$-\left(p(x)\, y'\right)' + q(x)\, y = f(x), \qquad \alpha_0 y(0) - \alpha_1 y'(0) = 0, \quad \beta_0 y(1) + \beta_1 y'(1) = 0.$$

   Determine the functions $p(x)$, $q(x)$, and possible values of the constants α_0, α_1, β_0, β_1.

   (a) $G(x,t) = \frac{-1}{6} \times \begin{cases} (x-4)(2+t), & 0 \leqslant t \leqslant x, \\ (t-4)(2+x), & x \leqslant t \leqslant 1; \end{cases}$

   (b) $G(x,t) = \frac{1}{\cos 1} \times \begin{cases} \sin t\, \cos(1-x), & 0 \leqslant t \leqslant x, \\ \sin x\, \cos(1-t), & x \leqslant t \leqslant 1; \end{cases}$

   (c) $G(x,t) = \frac{1}{4}\left(1 + e^{-4}\right) \times \begin{cases} (\cosh 2x - (\tanh 2)\, \sinh 2x)(\cosh 2t + \sinh 2t), & 0 \leqslant t \leqslant x, \\ (\cosh 2x + \sinh 2x)(\cosh 2t - \tanh 2\, \sinh 2t), & x \leqslant t \leqslant 1; \end{cases}$

   (d) $G(x,t) = \frac{1}{\cos 1} \times \begin{cases} \cos(1-x)\, \sin t, & 0 \leqslant t \leqslant x, \\ \sin x\, \cos(1-t), & x \leqslant t \leqslant 1. \end{cases}$

6. Find the Green's function for the nonself-adjoint two-point boundary value problem, given two linearly independent solutions for the corresponding homogeneous equation.

   (a) $\left(x^2+4\right) y'' - 2x\, y' + 2y = f(x), \quad y'(0) = 0, \; y(2) = 0$, given $y_1 = x$ and $y_2 = x^2 - 4$.
   (b) $(x \cos x - \sin x - \cos x)\, y'' + (x-1) \sin x\, y' - y\, \sin x = f(x), \quad y(0) = 0, \; y(1) = 0$, given $y_1 = x - 1$, $y_2 = \sin x$.
   (c) $\left(x^2+4\right) y'' + (4-2x)\, y' + 2y = f(x), \quad y'(0) = 0, \; y(2) = 0$, given $y_1 = x^2$, $y_1 = x - 2$.
   (d) $\left(\cos^2 x\right) y'' + 2\, \sin 2x\, y' + \left(1 + \sin^2 x\right) y = f(x), \quad y(0) = 0, \; y'(\pi) = 0$, given $y_1 = x\, \cos x$, $y_2 = \cos x$.

7. Show that the solution $y(x) = \sin x + c\, \cos x + 2x - 1 - \pi$ of the boundary value problem

$$y'' + y = 2x - 1 - \pi, \qquad y'(0) = 3, \quad y(\pi/2) = 0,$$

   cannot be obtained as the sum $y = y_h(x) + y_p(x)$, where y_h is the solution of the homogeneous equation $y'' + y = 0$ subject to the given boundary conditions and y_p is the solution of the given differential equation subject to the homogeneous boundary conditions, $y'(0) = y(\pi/2) = 0$.

Section 12.2 of Chapter 12 (Review)

1. Express the given boundary value problem as an equivalent boundary value problem for a first order system (12.2.1).

   (a) $\left(x^2y'\right)' - 2y = -f(x) \quad 1 < x < 3, \quad y(1) - y'(1) = 0, \; y(3) + 3\, y'(3) = 0.$

(b) $x^2(2+x)\,y'' + 2x\,y' - 2\,y = -f(x) \quad 1 < x < 2, \quad y(1) - y'(1) = 0, \;\; y(2) = 0.$

(c) $\left(e^{-3x}\,y'\right)' - 10\,e^{-3x}\,y = -f(x) \quad 0 < x < 1, \quad 5\,y(0) - y'(0) = 0, \;\; y'(1) = 0.$

(d) $y'' + x\,y' = -f(x) \quad 0 < x < 1, \quad y'(0) = 0, \;\; y(1) = 0.$

2. In each exercise from the previous problem, determine the Green's matrix for the corresponding first order system.

3. In each exercise, you are given boundary conditions for the two-point boundary value problem (12.2.1), where

$$\mathbf{P} = \begin{bmatrix} 2 & -5 \\ 4 & -2 \end{bmatrix}, \quad \mathbf{y}(x) = \begin{bmatrix} y_1(x) \\ y_2(x) \end{bmatrix}, \quad \boldsymbol{\alpha} = \begin{bmatrix} \alpha_1 \\ \alpha_2 \end{bmatrix}.$$

Note that the fundamental matrix for the corresponding homogeneous equation is given to be $e^{\mathbf{P}\,x}$. Form the matrix $\mathbf{B}$ from Eq. (12.2.3) and determine whether the boundary value problem has a unique solution for every $\mathbf{f}(x)$ and $\boldsymbol{\alpha} = \langle \alpha_1, \alpha_2 \rangle$.

(a) $y_2(0) = \alpha_1, \;\; y_2(\pi) = \alpha_2;$

(b) $y_1(0) = \alpha_1, \;\; y_2(\pi) = -\alpha_1;$

(c) $y_1(0) - y_2(0) = 0, \;\; y_1(\pi) + y_2(\pi) = \alpha_2;$

(d) $y_1(0) + y_2(0) = 0, \;\; y_1(\pi) + y_2(\pi) = \alpha_2.$

Section 12.4 of Chapter 12 (Review)

1. Obtain the first two Legendre coefficients for $f(x) = e^{ax}$:

$$\begin{aligned} a_0 &= \frac{1}{2}\int_{-1}^{1} e^{ax}\,\mathrm{d}x = \frac{1}{2a}\left(e^{a} - e^{-a}\right) = \frac{\sinh a}{a}, \\ a_1 &= \frac{3}{2}\int_{-1}^{1} e^{ax}\,x\,\mathrm{d}x = 3\left(\frac{\cosh a}{a} - \frac{\sinh a}{a^2}\right). \end{aligned}$$

2. Suppose that on an isolated sphere of radius a the electrostatic potential varies as $V(a,\theta) = V_0 e^{\alpha\cos\theta}$ (with some constants V_0 and α). Assuming that the electrostatic potential in charge-free space satisfies the Laplace equation with axial symmetry (no ϕ dependence), derive its series representation

$$V(r,\theta) = \sum_{n\geqslant 0} \frac{b_n}{r^{n+1}}\,P_n(\cos\theta)$$

and find the first two values of the coefficients b_0 and b_1 explicitly.

Section 12.5 of Chapter 12 (Review)

1. Let $G(x,t)$ be the Green's function for the regular Sturm–Liouville boundary value problem

$$L[y] = \mu\,y + f, \qquad B_0[y] = 0, \quad B_\ell[y] = 0,$$

where $L[x,\mathtt{D}] = -\mathtt{D}p(x)\mathtt{D} + q(x)$, depending on the derivative operator $\mathtt{D} = \mathrm{d}/\mathrm{d}x$, and $B_0[y] = \alpha_0 y(0) - \alpha_1 y'(0)$, $B_\ell[y] = \beta_0 y(\ell) + \beta_1 y'(\ell)$. Using the method of eigenfunction expansions, derive the eigenfunction formula for $G(x,t)$:

$$G(x,t) = \sum_{n\geqslant 1} \frac{\phi_n(x)\,\phi_n(t)}{\lambda_n - \mu},$$

where $\{\phi_n\}_{n\geqslant 1}$ is an orthonormal system of eigenfunctions with corresponding eigenvalues $\{\lambda_n\}_{n\geqslant 1}$ for

$$L[y] = \lambda\,y, \qquad B_0[y] = 0, \quad B_\ell[y] = 0.$$

Assume that μ is not an eigenvalue.

2. In each exercise, find a formal eigenfunction expansion for the solution to the given nonhomogeneous boundary value problem, if it exists.

(a) $y'' + 3\,y = 4\,\sin 2x - 23\,\sin 7x, \quad y(0) = 0, \;\; y(\pi) = 0.$

(b) $y'' + 9\,y = \sin 3x - 16\,\sin 5x, \quad y(0) = 0, \;\; y(\pi) = 0.$

(c) $y'' + \pi^2 y = 8\,\sin 3\pi x, \quad y(0) = 0, \;\; y'(1/2) = 0.$

(d) $y'' + 3y = 1 - 4x^2, \quad y'(0) = 0, \;\; y(1/2) = 0.$

3. Find a formal solution to the vibrating string problem governed by the given initial boundary value problem:

$$\ddot{u} = u_{xx} + xt \;(0 < x < \pi), \quad u(0,t) = u(\pi,t) = 0, \quad u(x,0) = \sin x, \;\; u_t(x,0) = 5\,\sin 2x - 3\,\sin 5x.$$

Bibliography

[1] Arino, Ovide and Kimmel, Marek, Stability analysis of models of cell production systems, *Mathematical Modeling*, **7**, 1269–1300, 1986.

[2] Bender, Carl M., and Orszag, Steven A., *Advanced Mathematical Methods for Scientists and Engineers: Asymptotic Methods and Perturbation Theory*, Springer-Verlag, New York, 1999.

[3] Bluman, G. W., and Kumel. S., *Symmetries and Differential Equations*, Springer-Verlag, New York, 1989.

[4] Bocher, Maxime, Certain Cases in which the Vanishing of the Wronskian Is a Sufficient Condition for Linear Dependence, *Transactions of the American Mathematical Society* (Providence, R.I.: American Mathematical Society), **2** (2): 139–149, 1901.

[5] Burden, Richard L., and Faires, Douglas J., *Numerical Analysis*, Cengage Learning, Boston, 9th edition, 2010.

[6] Butcher, J. C., *Numerical Methods for Ordinary Differential Equations*, John Wiley & Sons, Hoboken, NJ, 2003.

[7] Chen, P. J., *Selected Topics in Wave Propagation*, Noordhoff, Leyden, p. 29, 1976.

[8] Churchill, Ruel V., *Modern Operational Mathematics in Engineering*, McGraw-Hill, New York, 1944.

[9] Coddington, Earl E., and Levinson, Norman, *Theory of Ordinary Differential Equations*, McGraw-Hill, New York, 1955.

[10] Curtis, Dan, What's My Domain? *The College Mathematics Journal*, **41**, No. 2, 113–121, 2010.

[11] Davis, P. J., *Interpolation and Approximation*, Dover Publications, New York, 1975.

[12] Debnath, Lokenath, *Integral Transforms and Their Applications*, CRC Press, Boca Raton, FL, 1995.

[13] Dobrushkin, Vladimir, *Methods in Algorithmic Analysis*, CRC Press, Boca Raton, FL, 2010.

[14] Dobrushkin, Vladimir, *Modeling with Differential Equations*, CRC Press, Boca Raton, FL, forthcoming.

[15] Duffy, Dean G. *Green's Functions with Applications*, Chapman and Hall/CRC, 2001.

[16] Eustice, Dan, and Klamkin, M. S., On the Coefficients of a Partial Fraction Decomposition, *American Mathematical Monthly*, **86**, No. 6, 478–480, 1979.

[17] Gill, S., A Process for the Step-by-Step Integration of Differential Equations in an Automatic Digital Computing Machine, *Proc. Cambridge Phil. Soc.*, **47**, 96–108, 1951.

[18] Gottman, J. M., *Why Marriages Succeed or Fail*, Simon and Schuster, New York, 1994.

[19] Halmos, Paul, R., *I Want to Be a Mathematician*, Springer-Verlag, New York, 1987.

[20] Henrici, Peter, *Discrete Variable Methods in Ordinary Differential Equations*, Wiley, New York, 1962.

[21] Hewitt, Edwin, and Hewitt, Robert E., The Gibbs-Wilbraham Phenomenon: An Episode in Fourier Analysis, *Archive for History of Exact Science*, **21**, No. 2, 129–160, 1979.

[22] Higham, Nicholas J., *Functions of Matrices: Theory and Computation*, Cambridge University Press, Cambridge, 2008.

[23] Hoffman, K., and Kunze, R., *Linear Algebra*, second edition, Prentice-Hall, Upper Saddle River, NJ, 1971.

[24] Hubbard, J. H., and West, B. H., *Differential Equations*, Springer-Verlag, New York, 1990.

[25] Isaacson, E., and Keller, H. B., *Analysis of Numerical Methods*, Dover Publications, New York, Reprint edition, 1994.

[26] Kermack, W., and McKendrick, A. G., Contributions to the Mathematical Theory of Epidemics, *Proceedings of the Royal Society*, **A 115**, 700 – 721, 1927; **138**, 55–83, 1932; **141**, 93 – 122, 1933.

[27] Korner, T. W., *Fourier Analysis*, Cambridge University Press, Cambridge, 1988.

[28] Kuznetsov, Yuri, *Elements of Applied Bifurcation Theory*, Springer-Verlag, New York, 2010.

[29] Lebedev, Nikolai Nikolaevich, *Special Functions and Their Applications*, Dover Publications, New York, 1972.

[30] Leonard, I. E., The Matrix Exponential, *SIAM Review*, **38**, No. 3, 507–512, 1996.

[31] Levenson, Norman, and Smith, Oliver K., A General Equation for Relaxation Oscillations, *Duke Mathematical Journal*, **9**, No. 2, 382–403, 1942.

[32] Littlejohn, Lance L. and Krall, Allan M., Orthogonal Polynomials and Singular Sturm–Liouville Systems, *Rocky Mountain Journal of Mathematics*, **16**, No. 3, 435–479, 1986.

[33] Ma, Zhien, and Li, Jia, *Dynamical Modeling and Analysis of Epidemics*, World Scientific, Cleveland, OH, 2009.

[34] Mason, J. C., and Handscomb, D. C., *Chebyshev Polynomials*, Chapman & Hall/CRC, Boca Raton, FL, 2003.

[35] Moler, Cleve, and van Loan, Charles, Nineteen Dubious Ways to Compute the Exponential of a Matrix, Twenty-Five Years Later, *SIAM Review*, **45**, No. 1, 3–49, 2003.

[36] Peano, Giuseppe, Sull' integrabilita della equazioni differenziali di primo ordine, *Atti. Acad. Sci. Torino*, **21**, 677–685, 1886.

[37] Polyanin, Andrei D., and Zaitsev, Valentin F., *Handbook of Exact Solutions for Ordinary Differential Equations*, Second Edition, Chapman & Hall/CRC, Boca Raton, FL, 2002.

[38] Polking, John, Boggess, Albert, and Arnold, David, *Differential Equations*, second edition, Pearson Prentice Hall, Upper Saddle River, NJ, 2005.

[39] Polking, John, Boggess, Albert, and Arnold, David, *Differential Equations with Boundary Value Problems*, second edition, Pearson Prentice Hall, Upper Saddle River, NJ, 2005.

[40] Pipkin, A. C., *A Course on Integral Equations*, Springer-Verlag, New York, 1991.

[41] Press, W. H., Teukolsky, S. A., Vetterling, W. T., and Flannery, B.P., *Numerical Recipies in C*, second edition, Cambridge University Press, Cambridge, 1999.

[42] Reed, Michael, and Simon, Barry, *Methods of Modern Mathematical Physics, Volume 1. Functional Analysis*, Academic Press, Boston, 1981.

[43] Richardson, Lewis F., Generalized Foreign Politics, *British J. Psychol. Monograph Suppl.*, **23**, 1939.

[44] Saaty, T. L., *Mathematical Models of Arms Control and Disarmament*, John Wiley & Sons, Inc., Hoboken, NJ, 1969.

[45] Shampine, Lawrence F., *Numerical Solution of Ordinary Differential Equations*, Chapman & Hall/CRC, Boca Raton, FL, 1994.

[46] Strogatz, Steven H., *Nonlinear Dynamics and Chaos: With Applications to Physics, Biology, Chemistry, and Engineering*, Addison-Wesley Publishing Company, Cambridge, MA, 1994.

[47] Thomson, William Tyrrell, *Laplace Transformation*, Prentice-Hall, Englewood Cliffs, NJ, 1960.

[48] Watkins, David, *Fundamentals of Matrix Computations*, 3 edition, John Wiley & Sons Inc., Hoboken, NJ, 2010.

[49] Zwillinger, Daniel, *Handbook of Differential Equations*, Boston: Academic Press, 1992.

Index

Printed and bound by CPI Group (UK) Ltd, Croydon, CR0 4YY
17/10/2024
01775672-0019